J-holomorphic Curves and Symplectic Topology

Second Edition

American Mathematical Society

Colloquium Publications
Volume 52

J-holomorphic Curves and Symplectic Topology

Second Edition

Dusa McDuff
Dietmar Salamon

AMS AMERICAN MATHEMATICAL SOCIETY

For the first edition of this book, the first author was supported in part by NSF Grants DMS 0072512 and DMS 0305939. For the second edition of the book, the first author was partially supported by NSF Grant DMS0905191.
For the second edition of this book, the second author was partially supported by the Swiss National Science Foundation Grant 200021-127136.

2010 *Mathematics Subject Classification.* Primary 53D05, 53D45, 53D35, 57R17, 37J05, 32Q65; Secondary 53D12, 53D40, 58J05, 14N35.

For additional information and updates on this book, visit
www.ams.org/bookpages/coll-52

Library of Congress Cataloging-in-Publication Data

McDuff, Dusa, 1945–
J-holomorphic curves and symplectic topology / Dusa McDuff, Dietmar Salamon – 2nd ed.
p. cm. — (American Mathematical Society colloquium publications ; v. 52)
Includes bibliographical references and index.
ISBN 978-0-8218-8746-2 (alk. paper) | Softcover ISBN 978-1-4704-7983-1
1. Symplectic and contact topology. 2. Symplectic manifolds. 3. Pseudoholomorphic curves. I. Salamon, D. (Dietmar). II. Title.

QA613.59.M34 2012
516.3′6–dc23

2012016161

∞ The paper used in this book is acid-free and falls within the guidelines established to ensure permanence and durability.
Visit the AMS home page at http://www.ams.org/

11 10 9 8 7 6 5 4 3 2 29 28 27 26 25 24

Contents

Preface to the second edition ix

Preface xi

Chapter 1. Introduction 1
- 1.1. Symplectic manifolds 1
- 1.2. Moduli spaces: regularity and compactness 4
- 1.3. Evaluation maps and pseudocycles 7
- 1.4. The Gromov–Witten invariants 10
- 1.5. Applications and further developments 13

Chapter 2. J-holomorphic Curves 17
- 2.1. Almost complex structures 17
- 2.2. The nonlinear Cauchy-Riemann equations 19
- 2.3. Unique continuation 21
- 2.4. Critical points 26
- 2.5. Somewhere injective curves 30
- 2.6. The adjunction inequality 35

Chapter 3. Moduli Spaces and Transversality 39
- 3.1. Moduli spaces of simple curves 39
- 3.2. Transversality 49
- 3.3. A regularity criterion 56
- 3.4. Curves with pointwise constraints 61
- 3.5. Implicit function theorem 68

Chapter 4. Compactness 75
- 4.1. Energy 76
- 4.2. The bubbling phenomenon 80
- 4.3. The mean value inequality 85
- 4.4. The isoperimetric inequality 91
- 4.5. Removal of singularities 96
- 4.6. Convergence modulo bubbling 99
- 4.7. Bubbles connect 105

Chapter 5. Stable Maps 115
- 5.1. Stable maps 115
- 5.2. Gromov convergence 122
- 5.3. Gromov compactness 126
- 5.4. Uniqueness of the limit 134
- 5.5. Gromov compactness for stable maps 139

5.6. The Gromov topology 147

Chapter 6. Moduli Spaces of Stable Maps 153
6.1. Simple stable maps 155
6.2. Transversality for simple stable maps 158
6.3. Transversality for evaluation maps 165
6.4. Semipositivity 168
6.5. Pseudocycles 171
6.6. Gromov–Witten pseudocycles 177
6.7. The pseudocycle of graphs 182

Chapter 7. Gromov–Witten Invariants 201
7.1. Counting pseudoholomorphic spheres 203
7.2. Variations on the definition 210
7.3. Counting pseudoholomorphic graphs 220
7.4. Rational curves in projective spaces 225
7.5. Axioms for Gromov–Witten invariants 239

Chapter 8. Hamiltonian Perturbations 257
8.1. Trivial bundles 258
8.2. Locally Hamiltonian fibrations 264
8.3. Pseudoholomorphic sections 270
8.4. Pseudoholomorphic spheres in the fiber 277
8.5. The pseudocycle of sections 279
8.6. Counting pseudoholomorphic sections 285

Chapter 9. Applications in Symplectic Topology 295
9.1. Periodic orbits of Hamiltonian systems 296
9.2. Obstructions to Lagrangian embeddings 310
9.3. The nonsqueezing theorem 323
9.4. Symplectic 4-manifolds 329
9.5. The group of symplectomorphisms 344
9.6. Hofer geometry 353
9.7. Distinguishing symplectic structures 359

Chapter 10. Gluing 369
10.1. The gluing theorem 370
10.2. Connected sums of J-holomorphic curves 373
10.3. Weighted norms 376
10.4. Cutoff functions 380
10.5. Construction of the gluing map 382
10.6. The derivative of the gluing map 392
10.7. Surjectivity of the gluing map 400
10.8. Proof of the splitting axiom 406
10.9. The gluing theorem revisited 413

Chapter 11. Quantum Cohomology 417
11.1. The small quantum cohomology ring 418
11.2. The Gromov–Witten potential 436
11.3. Four examples 442

11.4. The Seidel representation 467
11.5. Frobenius manifolds 478

Chapter 12. Floer Homology 487
12.1. Floer's cochain complex 488
12.2. Ring structure 499
12.3. Poincaré duality 503
12.4. Spectral invariants 505
12.5. The Seidel representation 514
12.6. Donaldson's quantum category 519
12.7. The symplectic vortex equations 524

Appendix A. Fredholm Theory 531
A.1. Fredholm theory 531
A.2. Determinant line bundles 533
A.3. The implicit function theorem 538
A.4. Finite dimensional reduction 545
A.5. The Sard–Smale theorem 547

Appendix B. Elliptic Regularity 549
B.1. Sobolev spaces 549
B.2. The Calderon–Zygmund inequality 562
B.3. Regularity for the Laplace operator 568
B.4. Elliptic bootstrapping 571

Appendix C. The Riemann–Roch Theorem 579
C.1. Cauchy–Riemann operators 579
C.2. Elliptic estimates 586
C.3. The boundary Maslov index (by Joel Robbin) 593
C.4. Proof of the Riemann–Roch theorem 598
C.5. The Riemann mapping theorem 604
C.6. Nonsmooth bundles 613
C.7. Almost complex structures 614

Appendix D. Stable Curves of Genus Zero 619
D.1. Möbius transformations and cross ratios 619
D.2. Trees, labels, and splittings 622
D.3. Stable curves 629
D.4. The Grothendieck–Knudsen manifold 631
D.5. The Gromov topology 640
D.6. Cohomology 643
D.7. Examples 648

Appendix E. Singularities and Intersections (written with Laurent Lazzarini) 653
E.1. The main results 654
E.2. Positivity of intersections 658
E.3. Integrability 664
E.4. The Hartman–Wintner theorem 668
E.5. Local behaviour 673

E.6. Contact between branches 678
E.7. Singularities of J-holomorphic curves 686

Bibliography 695

List of Symbols 711

Index 715

Preface to the second edition

This revision has two main purposes: first to correct various errors that crept into the first edition and second to update our discussions of current work in the field. Since the first edition of this book appeared in 2004, symplectic geometry has developed apace. It has found new applications in low dimensional topology, via Heegaard Floer theory [**318**] and the newly understood relations of embedded contact homology to gauge theory [**74, 219**]. Several important books have been published that develop powerful new ideas and techniques: Seidel [**371**] on the Fukaya category, Fukaya–Oh–Ohta–Ono [**128**] on Lagrangian Floer homology, and Cieliebak and Eliashberg [**63**] on the relations between complex and symplectic manifolds. Another exciting development is the introduction of sheaf-theoretic methods for proving fundamental rigidity results in symplectic geometry by Tamarkin [**388**] and Guillermou–Kashiwara–Shapira [**166**]. There has also been great progress on particular problems; for example Taubes [**394**] solved the Weinstein conjecture for 3-dimensional contact manifolds using Seiberg–Witten–Floer theory, Hingston [**176**] and Ginzburg [**143**] solved the Conley conjecture by new advances in Hamiltonian dynamics and Floer theory, and the nearby Lagrangian conjecture has been partially solved (by Fukaya–Seidel–Smith [**131**] and Abouzaid [**1**] among others) using Fukaya categories. A comprehensive exposition of Hamiltonian Floer theory is now available with the book by Audin–Damian [**24**], which presents all the basic analysis needed to set up Hamiltonian Floer theory for manifolds with $c_1 = 0$ as well as in the monotone case. Finally, the long series of papers and books by Hofer–Wysocki–Zehnder [**184, 185, 186, 187, 188, 189**] develops a new functional analytic approach to the theory of J-holomorphic curves. Their work will eventually give solid foundations to Lagrangian Floer theory and the various forms of Symplectic Field Theory.

We do not say much about the details of these developments. However, we have updated the introductions to the chapters where relevant, and also have extended the discussions of various applications of J-holomorphic curves in Chapters 9, 11 and 12, aiming to give a sense of the main new developments and the main new players rather than to be comprehensive.

Many of the corrections are rather minor. However, we have rewritten Section 4.4 on the isoperimetric inequality, the proof of Theorem 7.2.3, the proof of Proposition 7.4.8, and the proof of the sum formula for the Fredholm index in Theorem C.4.2. In Chapter 10 we added Section 10.9 with a new geometric formulation of the gluing theorem for z-independent almost complex structures, in Appendix C we expanded Section C.5 to include a proof of integrability of almost complex structures in dimension two, and in Appendix D we expanded Section D.4 to include the material previously in Sections D.4 and D.5 and added a new Section D.6 on the cohomology of the moduli space of stable curves of genus zero.

We warmly thank everyone who pointed out mistakes in the earlier edition, but particularly Aleksei Zinger who sent us an especially thorough and useful list of comments.

Dusa McDuff and Dietmar Salamon, April 2012

Preface

The theory of J-holomorphic curves has been of great importance to symplectic topologists ever since its inception in Gromov's paper of 1985. Its applications include many key results in symplectic topology, and it was one of the main inspirations for the creation of Floer homology. It has caught the attention of mathematical physicists since it provides a natural context in which to define Gromov–Witten invariants and quantum cohomology, which form the so-called A-side of the mirror symmetry conjecture. Insights from physics have in turn inspired many fascinating developments, for example highlighting as yet little understood connections between the theory of integrable systems and Gromov–Witten invariants.

Several years ago the two authors of this book wrote an expository account of the field that explained the main technical steps in the theory of J-holomorphic curves. The present book started life as a second edition of that book, but the project quickly grew. The field has been developing so rapidly that there has been little time to consolidate its foundations. Since these involve many analytic subtleties, this has proved quite a hindrance. Therefore the main aim of this book is to establish the fundamental theorems in the subject in full and rigorous detail. We also hope that the book will serve as an introduction to current work in symplectic topology. These two aims are, of course, somewhat in conflict, and in different parts of the book different aspects are predominant.

We have chosen to concentrate on setting up the foundations of the theory rather than attempting to cover the many recent developments in detail. Thus, we limit ourselves to genus zero curves (though we do treat discs as well as spheres). A more serious limitation is that we restrict ourselves to the semipositive case, where it is possible to define the Gromov–Witten invariants in terms of pseudocycles. Our main reason for doing this is that an optimal framework for the general case (which would involve constructing a virtual moduli cycle) has not yet been worked out. Rather than cobbling together a definition that would do for many applications but not suffice in broader contexts such as symplectic field theory, we decided to show what can be done with a simpler, more geometric approach. On the other hand, we give a very detailed proof of the basic gluing theorem. This is the analytic foundation for all subsequent work on the virtual moduli cycle and is the essential ingredient in the proof of the associativity of quantum multiplication. There are also five extensive appendices, on topics ranging from standard results such as the implicit function theorem, elliptic regularity and the Riemann–Roch theorem to lesser known subjects such as the structure of the moduli space of genus zero stable curves and positivity of intersections for J-holomorphic curves in dimension four. We have adopted the same approach to the applications, giving complete proofs of the foundational results and illustrating more recent developments by describing some key examples and giving a copious list of references.

The book is written so that the subject develops in logical order. Chapters 2 through 5 establish the foundational Fredholm theory and compactness results for J-holomorphic spheres and discs; Chapter 6 introduces the concepts need to define the Gromov–Witten pseudocycle for semipositive manifolds; Chapter 7 is the pivotal chapter in which the invariants are defined; and the later chapters discuss various applications. Since there is more detail in Chapters 2 through 6 than can possibly be absorbed at a first reading, we have written the introductory Chapter 1 to describe the outlines of the theory and to introduce the main definitions. It serves as a detailed guide to this book, pointing out where the key arguments occur and where to find the background details needed to understand various examples. Each chapter also has an introduction describing its main contents, which should help to orient the more knowledgeable readers. Wherever possible we have written the sections and chapters to be independent of each other. Hence the reader should feel free to skip parts that seem excessively technical.

We hope that Chapter 1 (supplemented by suitable parts of Chapters 2–6) will provide beginning students with enough of the essential background for understanding the main definitions in Chapter 7. Here is a brief outline of the contents of the remaining chapters. After the basic invariants are defined in Section 7.1 (with important supplemental ideas in Section 7.2 and Section 7.3), Section 7.4 discusses the fundamental example of rational curves in projective space. The chapter ends with a discussion of the Kontsevich–Manin axioms for the genus zero Gromov–Witten invariants, and deduces from them Kontsevich's beautiful iterative formula for the number of degree d rational curves in the projective plane.

Chapter 8 sets up the theory of locally Hamiltonian fibrations over Riemann surfaces and shows how to count sections of such fibrations. This allows us to define Gromov–Witten invariants of arbitrary genus (but where the complex structure of the domain is fixed). It also provides the background for some important applications, for example Gromov's result that every Hamiltonian system on a symplectically aspherical manifold has a 1-periodic orbit (see Theorem 9.1.1), and results about the group of Hamiltonian symplectomorphisms: a taste of Hofer geometry in Section 9.6 and a discussion of the Seidel representation in Sections 11.4 and 12.5.

Chapter 9 describes some of the main applications of J-holomorphic curve techniques in symplectic geometry. Besides the examples mentioned above and a discussion of the basic properties of Lagrangian submanifolds, it gives full proofs of McDuff's results on the structure of rational and ruled symplectic 4-manifolds as well as Gromov's results on the symplectomorphism group of the projective plane and the product of 2-spheres.

The other main application, quantum cohomology, requires a further deep analytic technique, that of gluing. The first rigorous gluing arguments are due to Floer (in the somewhat easier context of Floer homology) and Ruan–Tian (in the context relevant to quantum homology). In Chapter 10 we present a different, perhaps easier, method of gluing and derive from it a proof of the splitting axiom for the Gromov–Witten invariants in semipositive manifolds.

With this in hand, Chapter 11 defines quantum cohomology and explains some of the structures arising from it, such as the Gromov–Witten potential and Frobenius manifolds. As is clear from the examples in Section 11.3, this is the place where symplectic topology makes the deepest contact with other areas such as algebraic geometry, conformal field theory, mirror symmetry, and integrable systems. This chapter should be accessible after Chapter 7. Finally, Chapter 12 is a survey that formulates the main outlines of Floer theory, omitting the analytic underpinnings. It explains the relations between Floer theory and quantum cohomology, using a geometric approach, and also indicates the directions of further developments, both analytic (the vortex equations) and geometric (Donaldson's quantum category).

There are five appendices. The first three set up the foundations of the classical theory of linear elliptic operators that is generalized in Chapters 3 and 4: Fredholm theory and the implicit function theorem for Banach manifolds in Appendix A, Sobolev spaces and elliptic regularity in Appendix B, and the Riemann–Roch theorem for Riemann surfaces with boundary in Appendix C. Appendix D provides background for Chapter 5. It explains the structure of the Grothendieck–Knudson moduli space of genus zero stable curves using cross ratios rather than the usual algebro-geometric approach. Appendix E was written jointly with Laurent Lazzarini. It contains a complete proof of positivity of intersections and the adjunction inequality for J-holomorphic curves in four-dimensional manifolds. Lazzarini provided the first draft of this appendix with complete proofs and we then worked together on the exposition. The results of Appendix E provide the basis for the structure theorems for rational and ruled symplectic 4-manifolds.

Those who wish to use this book as the basis for a graduate course must make some firm decisions about what kind of course they want to teach. As we know from experience, it is impossible in one semester to prove all the main analytic techniques as well as to describe interesting examples. One possibility, explained in more detail in Chapter 1, would be to concentrate on Chapter 1 (amplified by small parts of Chapter 2), Chapter 3 through Section 3.3 (together perhaps with some extra analysis from Appendices B and C), the basic compactness result for spheres with minimal energy in Section 4.2, very selected parts of Chapter 6 (the definition of pseudocycle), and then move to Section 7.1. Then either one could go directly to some of the geometric applications in Chapter 9 (for example, prove the nonsqueezing theorem or some of the results about symplectic 4-manifolds in Section 9.4) or one could discuss the Kontsevich–Manin axioms for Gromov–Witten invariants in Section 7.5 and then move to Chapter 11 to set up quantum cohomology. The idea here would be to develop a familiarity with the main analytic setup, prove some of the basic techniques, and then set them in context by discussing one set of applications

The above outline is perhaps still too ambitious, but there are ways to shorten the preliminaries. For example, it is possible to discuss many of the applications in Chapter 9 directly after the foundational material of Chapters 2–4 (and relevant parts of Chapter 8), without any reference to Chapters 5, 6 and 7. For if one considers only the simplest cases of these applications, rather than proving them in their most general form, the relevant moduli spaces are compact and so the results become accessible without any formal definition of the Gromov–Witten invariants. Alternatively, those aiming at quantum cohomology could state the results on Fredholm theory without proof and instead concentrate on explaining some of the compactness (bubbling) results in Chapters 4 and 5. These combine well with a study of the moduli space of stable maps and hence lead naturally to the Kontsevich–Manin axioms.

As indicated above, a first course, unless it moves incredibly fast or contains almost no applications, cannot both cover Fredholm theory and come to grips with the analytic details of the compactness proof, even less go through all the details of gluing. Even though this proof in the main needs the same analytic background as Chapter 3, the proof of the surjectivity of the gluing map hinges on the deepest result from Chapter 4 (the behaviour of long cylinders with small energy) and relies on several technical estimates. We have written the gluing chapter to try to make accessible the outlines of the construction, together with the main analytic ideas. (These are summarized in Section 10.5.) Hence, for a more analytically sophisticated audience, one might base a course on Chapters 3, 4 and 10, with motivation taken from some of the examples in Chapter 9 or 11.

Despite the length of this book, its subject is so rich that it is impossible to treat all its aspects. We have given many references throughout. Here are some books on related areas that the reader might wish to consult both on their own account and for the further references that they contain: Cox–Katz [**76**] on mirror symmetry and algebraic geometry, Donaldson [**87**] on Floer homology and gauge theory, Manin [**286**] on Frobenius manifolds and quantum cohomology, Polterovich [**330**] on the geometry of the symplectomorphism group, and the paper by Eliashberg–Givental–Hofer [**101**] on symplectic field theory.

This book has been long in the making and would not have been possible without help from many colleagues who shared their insights and knowledge with us. In particular, Coates, Givental, Lalonde, Lazzarini, Polterovich, Popescu, Robbin, Ruan, and Seidel all gave crucial help with various parts of this book. We also wish to thank the many students and others who pointed out various typos and inaccuracies, and especially Eduardo Gonzalez, Sam Lisi, Jake Solomon, and Fabian Ziltener for their meticulous attention to detail.

Dusa McDuff and Dietmar Salamon, December 2003

CHAPTER 1

Introduction

The theory of J-holomorphic curves, introduced by Gromov in 1985, has profoundly influenced the study of symplectic geometry, and now permeates almost all its aspects. The methods are also of interest in the study of Kähler manifolds, since often when one studies properties of holomorphic curves in such manifolds it is useful to perturb the complex structure to be generic. The effect of this is to ensure that one is looking at persistent rather than accidental features of these curves. However, the perturbed structure may no longer be integrable, and so again one is led to the study of curves that are holomorphic with respect to some nonintegrable almost complex structure J.

These curves satisfy a nonlinear analogue of the Cauchy–Riemann equations. Before one can do anything useful with them, one must understand the elements of the theory of these equations; for example, know what conditions ensure that the solution spaces are finite dimensional manifolds and know ways of dealing with the fact that these solution spaces are usually noncompact. As explained in the preface, this chapter introduces all the basic concepts and outlines the ingredients needed to establish these results. Readers, specially those unfamiliar with the theory, should start by reading this chapter and then branch out into more detailed study of whichever aspects of the theory interests them.

We assume that the reader is familiar with the elements of symplectic geometry. There are several introductory books. One possible reference is McDuff–Salamon [**277**], but there are now more elementary treatments such as Berndt [**35**] and Cannas da Silva [**59**] as well as classics such as Arnold [**15**].

1.1. Symplectic manifolds

A symplectic structure on a smooth $2n$-dimensional manifold M is a closed 2-form ω which is nondegenerate in the sense that the top-dimensional form ω^n does not vanish anywhere. By Darboux's theorem, all symplectic forms are locally diffeomorphic to the standard symplectic form

$$\omega_0 = dx_1 \wedge dy_1 + \cdots + dx_n \wedge dy_n$$

on Euclidean space $\mathbb{R}^{2n}$ with coordinates $(x_1, y_1, \ldots, x_n, y_n)$. This makes it hard to get a handle on the global structure of symplectic manifolds. Variational techniques have been developed which allow one to investigate some global questions in Euclidean space and in manifolds such as cotangent bundles which have some linear structure: see [**190, 277**] and the references contained therein. But the method which applies to the widest variety of symplectic manifolds is that of J-holomorphic curves.

Here J is an almost complex structure on M which is tamed by ω. An almost complex structure is an automorphism J of the tangent bundle TM of M which satisfies the identity $J^2 = -\mathbb{1}$. Thus J can be thought of as multiplication by i, and it makes TM into a complex vector bundle of dimension n. The form ω is said to **tame** J if

$$\omega(v, Jv) > 0$$

for all nonzero $v \in TM$. Geometrically, this means that ω restricts to a positive form on each complex line $L = \mathrm{span}\{v, Jv\}$ in the tangent space T_xM. Given ω the set $\mathcal{J}_\tau(M, \omega)$ of almost complex structures tamed by ω is always nonempty and contractible. Note that it is very easy to construct and perturb tame almost complex structures, because they are defined by pointwise conditions. Note also that, because $\mathcal{J}_\tau(M, \omega)$ is path connected, different choices of $J \in \mathcal{J}_\tau(M, \omega)$ give rise to isomorphic complex vector bundles (TM, J). Thus the Chern classes of these bundles are independent of the choice of J and will be denoted by $c_i(TM)$.[1]

In what follows we shall only need to use the first Chern class, and what will be relevant is the value $c_1(A) := \langle c_1(TM), A \rangle$ which it takes on a homology class $A \in H_2(M)$. If A is represented by a smooth map $u : \Sigma \to M$, defined on a closed oriented 2-manifold Σ then $c_1(A) = c_1(E)$ is the first Chern number of the pullback tangent bundle $E := u^*TM \to \Sigma$. But every complex bundle E over a 2-manifold Σ decomposes as a sum of complex line bundles $E = L_1 \oplus \cdots \oplus L_n$. Correspondingly

$$c_1(E) = \sum_i c_1(L_i).$$

Since the first Chern number of a complex line bundle is just the same as its Euler number, it is often easy to calculate the $c_1(L_i)$ directly. For example, if A is the class of the sphere $S = \mathrm{pt} \times S^2$ in $M = V \times S^2$ then it is easy to see that

$$TM\big|_S = TS \oplus L_2 \oplus \cdots \oplus L_n,$$

where the line bundles L_k are trivial. It follows that

$$c_1(A) = c_1(TM\big|_S) = c_1(TS) = \chi(S) = 2$$

where $\chi(S)$ is the Euler characteristic of S.

A smooth map $\phi : (M, J) \to (M', J')$ from one almost complex manifold to another is said to be (J, J')**-holomorphic** if and only if its derivative $d\phi(x) : T_xM \to T_{\phi(x)}M'$ is complex linear, that is

$$d\phi(x) \circ J(x) = J'(\phi(x)) \circ d\phi(x).$$

These are the Cauchy–Riemann equations, and, when (M, J) and (M', J') are both subsets of complex n-space $\mathbb{C}^n$, they are satisfied precisely by the holomorphic maps. An almost complex structure J is said to be **integrable** if it arises from an underlying complex structure on M. This is equivalent to saying that one can choose an atlas for M whose coordinate charts are (J, i)-holomorphic where i is the standard complex structure on $\mathbb{C}^n$. In this case the coordinate changes are holomorphic maps (in the usual sense) between open sets in $\mathbb{C}^n$. When M has dimension 2 a fundamental theorem says that all almost complex structures J on M are integrable: for a proof see Section E.4. However this is far from true in higher dimensions.

---

[1]There is another space of almost complex structures naturally associated to (M, ω), namely the set $\mathcal{J}(M, \omega)$ of ω**-compatible structures** defined in Section 3.1. For the present purposes one can use either space. In fact, to make our results as general as possible, we will often work with $\mathcal{J}(M, \omega)$ because this is very slightly harder: $\mathcal{J}_\tau(M, \omega)$ is open in the space of all almost complex structures, while $\mathcal{J}(M, \omega)$ is not.

The basic example of an almost complex symplectic manifold is standard Euclidean space $(\mathbb{R}^{2n}, \omega_0)$ with its standard almost complex structure J_0 obtained from the usual identification with $\mathbb{C}^n$ via

$$z_j = x_j + iy_j$$

for $j = 1, \ldots, n$. Thus J_0 maps the tangent vector $\partial/\partial x_j$ to $\partial/\partial y_j$ and maps $\partial/\partial y_j$ to $-\partial/\partial x_j$ in the standard basis of $\mathbb{R}^{2n} = T_z\mathbb{R}^{2n}$. In other words, the automorphism $J_0 : \mathbb{R}^{2n} \to \mathbb{R}^{2n}$ is represented by the $2n \times 2n$-matrix

$$J_0 := \begin{pmatrix} 0 & -1 & \cdots & 0 & 0 \\ 1 & 0 & \cdots & 0 & 0 \\ \vdots & \vdots & \ddots & \vdots & \vdots \\ 0 & 0 & \cdots & 0 & -1 \\ 0 & 0 & \cdots & 1 & 0 \end{pmatrix}.$$

Every Kähler manifold gives another example (but not every symplectic manifold admits a Kähler structure).

***J*-holomorphic curves.** A J-holomorphic curve is a (j, J)-holomorphic map

$$u : \Sigma \to M$$

from a Riemann surface[2] (Σ, j) to an almost complex manifold (M, J). Usually, we will take (Σ, j) to be the Riemann sphere $S^2 = \mathbb{C} \cup \{\infty\}$. In accordance with the terminology of complex geometry it is often convenient to think of the 2-sphere as the complex projective line $\mathbb{C}P^1$.

If u is an embedding (that is, an injective immersion) then its image C is a 2-dimensional submanifold of M whose tangent spaces T_xC are J-invariant. Thus each tangent space is a complex line in TM. Further, by the taming condition, ω restricts to a positive form on each such line. Therefore C is a symplectic submanifold of M.[3] Conversely, if $C \subset M$ is a 2-dimensional symplectic submanifold then there is an ω-tame almost complex structure J such that TC is J-invariant. (First define J on TC, then extend to the tangent spaces T_xM for $x \in C$, and finally extend the section to the rest of M.) Since the restriction of J to TC is integrable, it follows that C is the image of a J-holomorphic curve. Thus embedded J-holomorphic curves are essentially the same as 2-dimensional symplectic submanifolds of M. This argument uses the 2-dimensionality of C in an essential way; it is not possible to construct a good theory of J-holomorphic maps from higher dimensional manifolds into (M, J). (The symplectic vortex equations mentioned in Section 12.7 do not contradict this statement, since they have extra structure.)

Observe that according to this definition, a curve u is always parametrized. One should contrast this with the situation in complex geometry, where one often defines a curve by requiring it to be the common zero set of a certain number of holomorphic polynomials. Such an approach makes no sense in the nonintegrable, almost complex world, since when J is nonintegrable there usually are no holomorphic functions $(M, J) \to \mathbb{C}$.[4]

---

[2]A Riemann surface is a 1-dimensional complex manifold.

[3]A submanifold X of M is said to be symplectic if ω restricts to a nondegenerate form on X.

[4]However Donaldson has recently developed a powerful theory of *asymptotically holomorphic* functions $(M, J, \omega) \to \mathbb{C}$: see Donaldson [**84, 85**] and Auroux [**26**].

1.2. Moduli spaces: regularity and compactness

The crucial fact about J-holomorphic curves is that when J is generic they occur in finite dimensional families. These families make up finite dimensional manifolds

$$\mathcal{M}^*(A;J)$$

called **moduli spaces**, whose cobordism classes are independent of the particular J chosen in $\mathcal{J}_\tau(M,\omega)$. Here A is a homology class in $H_2(M,\mathbb{Z})$, and $\mathcal{M}^*(A;J)$ consists of essentially all J-holomorphic curves $u:\Sigma\to M$ that represent the class A. Although the manifold $\mathcal{M}^*(A;J)$ is almost never compact, it usually retains enough elements of compactness for one to be able to use it to define invariants.

Chapters 2–5 of this book are taken up with formulating and proving precise versions of the above statements. The results divide naturally into two groups, the first (discussed in Chapters 2 and 3) concerning issues of regularity and transversality and the second (discussed in Chapters 4 and 5) concerning the compactification of $\mathcal{M}^*(A;J)$. In the next paragraphs we pick out the most important concepts and theorems. For those new to the theory, these results together with their proofs are the ones to focus on first. The introductions to each chapter give further guidance on how to read the book.

Local properties. Chapter 2 concerns the local properties of J-holomorphic curves. The key results for future developments are the energy identity, which is crucial for later results on compactness, and Proposition 2.5.1, which gives a characterization of those curves which are not multiply covered. A curve $u:\Sigma\to M$ is said to be **multiply covered** if it is the composite of a holomorphic branched covering map $(\Sigma,j)\to(\Sigma',j')$ of degree greater than 1 with a J-holomorphic map $\Sigma'\to M$. It is called **simple** if it is not multiply covered. The multiply covered curves are often singular points in the moduli space $\mathcal{M}(A;J)$ of all J-holomorphic A-curves, and so one needs a workable criterion which guarantees that u is simple. We will say that a curve u is **somewhere injective** if there is a point $z\in\Sigma$ such that

$$du(z)\neq 0,\qquad u^{-1}(u(z))=\{z\}.$$

A point $z\in\Sigma$ with this property is called an **injective point** for u. Here is the statement of Proposition 2.5.1. It is an important ingredient of the proof of Theorem 3.1.6 which asserts that the moduli space $\mathcal{M}^*(A;J)$ of simple J-holomorphic A-curves is a smooth manifold for a generic ω-tame J.

PROPOSITION. *Every simple J-holomorphic curve u is somewhere injective. Moreover the set of injective points is open and dense in Σ.*

The proof is elementary, except for an appeal to a deep analytic theorem extending the unique continuation principle to J-holomorphic curves.

Fredholm theory. Fix a Riemann surface Σ of genus g and let $\mathcal{M}^*(A;J)$ denote the set of all *simple* J-holomorphic maps $u:\Sigma\to M$ which represent the class A.[5] Here is an informal version of the main result of Chapter 3. Recall that a subset in a complete metric space is said to be residual (in the sense of Baire) if it

[5]In the first edition of this book we denoted this space simply by $\mathcal{M}(A;J)$. However, it is now customary to use the letter $\mathcal{M}$ for full moduli spaces.

contains a countable intersection of open and dense sets. Baire's category theorem asserts that every such set is dense.

THEOREM A. *There is a residual subset* $\mathcal{J}_{\mathrm{reg}}(A) \subset \mathcal{J}_\tau(M,\omega)$ *such that for each* $J \in \mathcal{J}_{\mathrm{reg}}(A)$ *the space* $\mathcal{M}^*(A;J)$ *is a smooth manifold of dimension*

$$\dim \mathcal{M}^*(A;J) = n(2-2g) + 2c_1(A).$$

This manifold $\mathcal{M}^*(A;J)$ *carries a natural orientation.*

Another important theorem specifies the dependence of $\mathcal{M}^*(A;J)$ on the choice of J (Theorem 3.1.8). The basic reason why these theorems are valid is that the Cauchy–Riemann equation

$$du \circ j = J \circ du$$

is elliptic, and hence its linearization at a J-holomorphic curve is a Fredholm operator, denoted by D_u. The set $\mathcal{J}_{\mathrm{reg}}(A)$ in Theorem A is the set of all those almost complex structures $J \in \mathcal{J}_\tau(M,\omega)$ such that the linearized operator D_u is onto for every $u \in \mathcal{M}^*(A;J)$. The elements of $\mathcal{J}_{\mathrm{reg}}(A)$ are sometimes called **regular** almost complex structures. An interesting fact is that the taming condition on J is irrelevant here; the above results hold for *all* almost complex structures J on *any* compact manifold M.

A bounded linear operator $D : X \to Y$ between Banach spaces is said to be **Fredholm** if it has a finite dimensional kernel and a closed image of finite codimension in Y. The **index** of D is defined to be the difference in dimension between the kernel and cokernel of D:

$$\mathrm{index}\, D = \dim\, \ker D - \dim\, \mathrm{coker} D.$$

An important fact is that the set of Fredholm operators is open with respect to the norm topology and the Fredholm index is constant on each component of the set of Fredholm operators. Thus it does not change as D varies continuously, though of course the dimension of the kernel and cokernel can change.

As we show in Appendix A, Fredholm operators are essentially as well behaved as finite dimensional operators and they play an important role in infinite dimensional implicit function theorems. More precisely, assume that $\mathcal{F} : X \to Y$ is a a C^∞-smooth map between separable Banach spaces whose derivative $d\mathcal{F}(x) : X \to Y$ is Fredholm of index k at each point $x \in X$. If $y \in Y$ is a **regular value** of $\mathcal{F}$ in the sense that $d\mathcal{F}(x)$ is surjective for all $x \in \mathcal{F}^{-1}(y)$, then, just as in the finite dimensional case, the inverse image

$$\mathcal{M} := \mathcal{F}^{-1}(y)$$

is a smooth manifold of dimension k. An infinite dimensional version of Sard's theorem says that almost all points of Y are regular values of $\mathcal{F}$. (Technically, the regular points form a residual set.) This theorem remains true if X and Y are separable Banach manifolds rather than Banach spaces. However it does not extend as stated to other kinds of infinite dimensional vector spaces, such as Fréchet spaces. Therefore, although the set $\mathcal{J}_{\mathrm{reg}}$ mentioned in Theorem A above does consist of the regular values of a Fredholm operator with target $\mathcal{J}_\tau(M,\omega)$, there are some additional technicalities in the proof because $\mathcal{J}_\tau(M,\omega)$ is a Fréchet rather than a Banach manifold.

***J*-holomorphic discs.** Although most of the results in this book concern the properties of J-holomorphic spheres, we formulate and prove the foundational results in chapters 3 and 4 as well as appendices B and C for general compact Riemann surfaces with (possibly empty) boundary and thus include the case of J-holomorphic discs; we show in Chapter 3 that with Lagrangian boundary conditions the delbar operator is again Fredholm. Thus moduli spaces of J-holomorphic discs with boundary in a given Lagrangian submanifold again form finite dimensional manifolds for generic J. This fact is an important ingredient of the inductive proof of the Riemann–Roch theorem given in Appendix C. We give several applications of this theory in Section 9.2, notably the proof that there is no closed exact Lagrangian submanifold in $\mathbb{R}^{2n}$. The fact that Lagrangian boundary conditions are elliptic is also very important in Floer theory: see Chapter 12.

This is the first step in extending the theory to maps with more general domains. As a next step one could consider either closed domains of higher genus or noncompact domains of genus zero. Our work here is sufficient to understand closed domains (Σ, j) of higher genus with fixed complex structure j. Varying j does not significantly affect the Fredholm theory but does cause additional problems when discussing compactness. On the other hand, one needs considerably more sophisticated analysis in order to set up Fredholm theory for noncompact domains. (In this case one often studies perturbed equations: cf. the development of Floer theory in [**116**].) Both of these extensions are beyond the scope of the present book. They are also both needed to understand the full structure of Gromov–Witten invariants as displayed for example in Eliashberg, Givental and Hofer's recent paper [**101**] on symplectic field theory.

Compactness. The next task is to develop an understanding of when the moduli spaces $\mathcal{M}^*(A; J)$ are compact. Here the taming condition plays an essential role. The symplectic form ω and an ω-tame almost complex structure J together determine a Riemannian metric

$$\langle v, w\rangle = \tfrac{1}{2}(\omega(v, Jw) + \omega(w, Jv))$$

on M and the **energy** of a J-holomorphic curve $u : \Sigma \to M$ with respect to this metric is given by the formula

$$E(u) = \tfrac{1}{2}\int_\Sigma |du(z)|_J^2 \,\mathrm{dvol}_\Sigma = \int_\Sigma u^*\omega.$$

Thus the L^2-norm of the derivative of a J-holomorphic curve satisfies a uniform bound which depends only on the homology class A represented by u. This in itself does not guarantee compactness because it is a borderline case for Sobolev estimates. (A uniform bound on the L^p-norms of du with $p > 2$ would guarantee compactness.)

Another manifestation of the failure of compactness can be observed in the fact that the energy $E(u)$ is invariant under conformal rescaling of the metric on Σ. This effect is particularly clear in the case where the domain Σ of our curves is the Riemann sphere $\mathbb{C}P^1$, since here there is a large group of global, rather than local, rescalings. Indeed, the noncompact group $\mathrm{G} = \mathrm{PSL}(2, \mathbb{C})$ acts on the Riemann sphere by conformal transformations

$$z \mapsto \frac{az + b}{cz + d}.$$

Thus each element $u \in \mathcal{M}^*(A;J)$ has a noncompact family of reparametrizations $u \circ \phi$, for $\phi \in \mathrm{G}$, and so $\mathcal{M}^*(A;J)$ itself can never be both compact and nonempty (unless A is the zero class, in which case all the maps u are constant). However, the quotient space $\mathcal{M}^*(A;J)/\mathrm{G}$ will sometimes be compact.

Recall that a homology class $B \in H_2(M)$ is called **spherical** if it is in the image of the Hurewicz homomorphism $\pi_2(M) \to H_2(M)$. Here is a statement of the first important theorem in Chapter 4. Observe that the hypothesis on A implies that no A-curve can be multiply covered, that is $\mathcal{M}^*(A;J) = \mathcal{M}(A;J)$.

THEOREM B. *Assume that there is no spherical homology class $B \in H_2(M)$ such that $0 < \omega(B) < \omega(A)$. Then the moduli space $\mathcal{M}^*(A;J)/\mathrm{G}$ is compact.*

To prove this one shows that if u_ν is a sequence in $\mathcal{M}^*(A;J)$ which has no limit point in $\mathcal{M}^*(A;J)$ then there is a point $z \in \mathbb{C}P^1$ at which the derivatives $du_\nu(z)$ are unbounded. After passage to a subsequence one finds a decreasing sequence U_ν of neighbourhoods of z in $\mathbb{C}P^1$ whose images $u_\nu(U_\nu)$ converge in the limit to a nonconstant J-holomorphic sphere. If B is the homology class represented by this sphere, then either $\omega(B) = \omega(A)$, in which case the maps u_ν can be reparametrized so that they do converge in $\mathcal{M}^*(A;J)$, or $\omega(B)$ lies strictly between 0 and $\omega(A)$. This is the process of "bubbling", which occurs in this context in a simple and geometrically clear way: see Section 4.2.

Theorem B implies that if $\omega(A)$ is already the smallest positive value assumed by ω on spheres then the moduli space $\mathcal{M}^*(A;J)/\mathrm{G}$ is compact. As we shall see below, this is enough for some interesting applications. However, to cope with more general classes A we need to understand the complete limit of the sequence u_ν. The analysis required to solve this challenging problem is developed in the second half of Chapter 4. In particular, in order to see that the limiting curve is connected, we study the properties of long cylinders of small energy in Section 4.7.

The full structure of the limit is formulated in Theorem 5.3.1, a result known as Gromov's compactness theorem. We give a more modern formulation than Gromov's original statement in [**160**], using Kontsevich's language of **stable maps**. Although not strictly necessary for the applications in this book, this language allows us to set up the Gromov–Witten invariants in their natural context and is the basis of many recent applications of the theory especially those involving quantum cohomology. Therefore, we develop it in considerable detail in Chapter 5 and Appendix D.

The best strategy for those new to the subject might be to try to understand the definition of Gromov convergence (Definition 5.2.1) and the statement of Theorem 5.3.1, leaving further details for later. Proposition 4.1.5 is also important; it says that for each J and each $c > 0$ there are only finitely many classes A with J-holomorphic representatives and energy $\omega(A)$ bounded by c.

1.3. Evaluation maps and pseudocycles

The Gromov–Witten invariants are built from the evaluation map

$$\mathcal{M}^*(A;J) \times \mathbb{C}P^1 \to M : (u,z) \mapsto u(z).$$

Note that this factors through the action of the reparametrization group G given by

$$\phi \cdot (u,z) = (u \circ \phi^{-1}, \phi(z)).$$

Hence we get a map defined on the quotient

$$\mathrm{ev} = \mathrm{ev}_J : \mathcal{M}^*(A;J) \times_{\mathrm{G}} \mathbb{C}P^1 \to M.$$

EXAMPLE. Let V be a closed symplectic manifold of dimension $2n-2$ and

$$M := \mathbb{C}P^1 \times V$$

with the product symplectic form. Suppose that $\pi_2(V) = 0$. Then $A := [\mathbb{C}P^1 \times \mathrm{pt}]$ generates the group of spherical 2-classes in M, and so $\omega(A)$ is necessarily the smallest value assumed by ω on the spherical classes. Theorems A and B therefore imply that the space $\mathcal{M}^*(A;J)\times_{\mathrm{G}}\mathbb{C}P^1$ is a compact manifold for generic J. Because $c_1(A) = 2$, in this case the dimension of $\mathcal{M}^*(A;J) \times_{\mathrm{G}} \mathbb{C}P^1$ is $2n$ and agrees with the dimension of M. Moreover, we will see in Chapter 3 (see Theorem 3.1.8) that different choices of J give rise to cobordant maps ev_J. Since cobordant maps have the same degree, this means that the degree of ev_J is independent of all choices. Now if $J = i \times J'$ is a product, where i denotes the standard complex structure on $\mathbb{C}P^1$, then it is easy to see that the elements of $\mathcal{M}^*(A;J)$ have the form

$$u(z) = (\phi(z), v_0)$$

where $v_0 \in V$ and $\phi \in \mathrm{G}$. It follows that the map ev_J has degree 1 for this choice of J and hence for every regular J.

Gromov used this fact in [**160**] to prove his celebrated nonsqueezing theorem.

THEOREM. *Let V be a closed symplectic manifold of dimension $2n-2$ such that $\pi_2(V) = 0$. If ψ is a symplectic embedding of the ball $B^{2n}(r)$ of radius r into the cylinder $B^2(\lambda) \times V$ then $r \le \lambda$.*

SKETCH OF PROOF. Embed the disc $B^2(\lambda)$ into a 2-sphere $\mathbb{C}P^1$ of area $\pi\lambda^2+\varepsilon$, and let ω be the product symplectic structure on $\mathbb{C}P^1 \times V$. Let J' be an ω-tame almost complex structure on $\mathbb{C}P^1 \times V$ which, on the image of ψ, equals the push-forward by ψ of the standard structure J_0 of the ball $B^{2n}(r)$. Since the evaluation map $\mathrm{ev}_{J'}$ has degree 1, there is a J'-holomorphic curve through every point of $\mathbb{C}P^1 \times V$. In particular, there is such a curve, C' say, through the image $\psi(0)$ of the center of the ball. This curve pulls back by ψ to a J_0-holomorphic curve C through the center of the ball $B^{2n}(r)$. Since J_0 is standard, this curve C is holomorphic in the usual sense and so is a minimal surface in $B^{2n}(r)$. But it is a standard result in the theory of minimal surfaces that a properly embedded surface through the center of a ball in Euclidean space has area at least that of the flat disc, namely πr^2. Further, because C is holomorphic, it is easy to check that its area is just given by the integral of the standard symplectic form ω_0 over it. Thus

$$\pi r^2 \le \int_C \omega_0 = \int_{\psi^{-1}(C')} \psi^*(\omega) < \int_{C'} \omega = \omega(A) = \pi\lambda^2 + \varepsilon$$

where the middle inequality holds because $\psi(C)$ is only a part of C'. Since this is true for all $\varepsilon > 0$, the result follows. □

More details of the above argument may be found in [**160, 320**]. In Section 9.3 we shall give full details of a slightly different proof that replaces the appeal to the theory of minimal surfaces by using the symplectic blow up. Chapter 9 also contains complete proofs of some of the other foundational results stated in Gromov's paper [**160**]. For example, in Section 9.2 we use the theory of J-holomorphic discs developed in Chapter 3 to show that there is no closed exact Lagrangian submanifold in $\mathbb{R}^{2n}$ when $n \geq 2$.

The Gromov–Witten pseudocycle. Often it is useful to evaluate the map u at more than one point. If one is evaluating at up to three points then, because G acts triply transitively on the sphere, it does not matter which points one chooses. However, if one is evaluating at more than three points (or if the domain is not a sphere) the choice of points does make a difference. One incorporates these points into the moduli space itself, hence making the basic object of study a tuple $(u, z_1, \ldots, z_k)$ consisting of an element $u \in \mathcal{M}(A; J)$ together with k pairwise distinct **marked points** $z_i \in S^2$. The space of all such tuples is denoted $\widetilde{\mathcal{M}}_{0,k}(A; J)$ and its quotient by G is denoted $\mathcal{M}_{0,k}(A; J)$.[6] As before $\mathcal{M}^*_{0,k}(A; J)$ denotes the subset for which u is simple. It is a manifold of dimension

$$\dim \mathcal{M}^*_{0,k}(A; J) = 2n + 2c_1(A) + 2k - 6 =: \mu(A, k)$$

for generic J. Moreover, the obvious evaluation map

$$\widetilde{\mathcal{M}}_{0,k}(A; J) \to M^k : (u, z_1, \ldots, u_k) \mapsto (u(z_1), \ldots, u(z_k))$$

descends to

$$\mathrm{ev} : \mathcal{M}_{0,k}(A; J) \to M^k.$$

If this map represents a cycle one can define the Gromov-Witten invariants as its intersection numbers with cycles in M^k.

Observe that even though $\mathcal{M}_{0,k}(A; J)$ consists of unparametrized maps it is never compact when $k \geq 2$ since the marked points are distinct by definition. Nevertheless, we would like to define the Gromov–Witten invariants using the d-dimensional homology class represented by the evaluation map

$$\mathrm{ev} : \mathcal{M}^*_{0,k}(A; J) \to M^k,$$

where $d = \mu(A, k)$. Therefore, even if we know that $\mathcal{M}_{0,k}(A; J)$ is a manifold, we must work to show that it represents a cycle. The crucial point here is to understand its compactification. Kontsevich realised that the compactification of $\mathcal{M}_{0,k}(A; J)$ with the best structure is not the space of cusp-curves considered by Gromov but rather the space $\overline{\mathcal{M}}_{0,k}(A; J)$ of **stable maps**: see Definition 5.1.1. We formulate and prove Gromov's compactness theorem in Chapter 5 using this language.

Even with a good understanding of the compactification $\overline{\mathcal{M}}_{0,k}(A; J)$ it is very hard to understand the sense in which ev represents a cycle for general manifolds (M, ω), since multiply covered curves in lower strata can cause these to have too high a dimension.[7] However, if (M, ω) is **semipositive** then we show in Chapter 6 that the evaluation map represents a **pseudocycle** of dimension $d := \mu(A, k)$. Intuitively speaking this means that its image can be compactified by adding a set of dimension at most $d - 2$.[8] Thus its boundary has dimension at most $d - 2$ and so

---

[6] In this notation the 0 in the subscript denotes the genus of the domain. In this book we restrict to the genus 0 case, though much of what we say applies almost word for word in the case when (Σ, j) has arbitrary genus and fixed complex structure j.

[7] This problem has now been solved for general manifolds. By using delicate global perturbations one defines a d-dimensional *virtual moduli cycle*, which should be thought of as the fundamental class of $\mathcal{M}^*_{0,k}(A; J)$. Detailed references may be found in the introduction to Chapter 6. This is beyond the scope of the present book.

[8] A subset B of a smooth manifold X is said to be of **dimension at most** m if it is contained in the image of a smooth map $g : W \to X$ defined on a manifold W of dimension m.

is not seen from homological point of view. More formally we make the following definition.

DEFINITION. *A d-dimensional* **pseudocycle** *in a manifold X is a smooth map*

$$f : V \to X$$

defined on an oriented d-dimensional manifold V such that its image $f(V)$ has compact closure and its limit set Ω_f has dimension $\leq d-2$. Here

$$\Omega_f = \bigcap_{\substack{K \subset V \\ K \text{ compact}}} \overline{f(V \setminus K)}$$

is the set of all limit points of sequences $f(x_\nu)$ where x_ν has no convergent subsequence in V.

This definition is explored in Section 6.5. The following theorem is proved in Section 6.6. Recall that $\mathcal{J}$ denotes the space of of all ω-tame J.

THEOREM C. *Let (M, ω) be a closed semipositive symplectic manifold. Then there is a residual set $\mathcal{J}_{\text{reg}}(M, \omega) \subset \mathcal{J}_\tau(M, \omega)$ with the following significance. If $A \in H_2(M; \mathbb{Z})$ satisfies $c_1(A) > 0$ and $J \in \mathcal{J}_{\text{reg}}(M, \omega)$ then the evaluation map*

$$\text{ev} : \mathcal{M}^*_{0,k}(A; J) \to M^k$$

is a pseudocycle of dimension $\mu(A, k)$. Moreover, its homology class is independent of the choice of $J \in \mathcal{J}_{\text{reg}}(M, \omega)$.

The proof is sketched in the introduction to Chapter 6. It involves transversality results for evaluation maps that are carried out in Section 6.1-3. The reader is advised, in particular, to explore the precise definition of the set $\mathcal{J}_{\text{reg}}(M, \omega)$ in section 6.2 in terms of transversality for edge evaluation maps as without this precise formulation Theorem C cannot be fully understood.

There is an important variant of Theorem C that concerns the restriction of ev to the subspace $\mathcal{M}^*_{0,k}(A; \mathbf{w}, J)$ of $\widetilde{\mathcal{M}}^*_{0,k}(A; J)$ consisting of tuples $(u, z_1, \ldots, z_k)$ for which the marked points z_i associated to the indices $i \in I$ are fixed and set equal to w_i. (Here we assume that $\#I \geq 3$ and denote $\mathbf{w} := \{\mathrm{w}_i\}_{i \in I}$.) To establish that the resulting map

$$\text{ev} : \mathcal{M}^*_{0,k}(A; \mathbf{w}, J) \to M^k$$

is a pseudocycle (this time of dimension $\mu(A, k) - 2(\#I - 3)$), one must introduce a wider class of perturbations of the Cauchy–Riemann equation, looking at solutions of a equation in which $J = \{J_z\}$ depends on $z \in S^2$. Geometrically, this is equivalent to looking at the graphs $z \mapsto (z, u(z))$ of suitable maps $u : S^2 \to M$. A similar approach is needed in order to prove Theorem C above in the case when $c_1(A) = 0$. To understand the extension of Theorem C to these cases one must redo the whole theory in a slightly more general setting. The details are explained in Section 6.7.

1.4. The Gromov–Witten invariants

The **Gromov–Witten invariants** are obtained by taking the intersection of the Gromov–Witten pseudocycle

$$\text{ev} : \mathcal{M}^*_{0,k}(A; J) \to M^k$$

with cycles of complementary dimension in M^k. More precisely, let $a_1, \ldots, a_k \in H^*(M)$ be cohomology class of pure degrees such that

$$m := \sum_{j=1}^{k} \deg(a_j) = \mu(A, k).$$

Then we may choose a cycle $\alpha \subset M^k$ that is Poincaré dual to the cohomology class $a := \pi_1^* a_1 \smile \cdots \smile \pi_k^* a_k$ and is **strongly transverse** to the Gromov–Witten pseudocycle $\mathrm{ev} : \mathcal{M}_{0,k}^*(A; J) \to M^k$ in a sense made precise in Section 6.5. We then define the Gromov–Witten invariant as the oriented intersection number

$$\mathrm{GW}_{A,k}^M(a_1, \ldots, a_k) = \mathrm{ev} \cdot \alpha.$$

We may choose the cycle α to be a product $\alpha_1 \times \cdots \times \alpha_k$ where $\alpha_i \subset M$ is Poincaré dual to the class a_i. In this case the Gromov–Witten invariant is the oriented number of J-holomorphic curves $u : S^2 \to M$ in the homology class A which meet each of the cycles $\alpha_1, \ldots, \alpha_k$.

The invariants $\mathrm{GW}_{A,k}^{M,I}$, in which the marked points z_i are fixed for $i \in I$, are defined similarly by taking the intersection of a suitable evaluation map with α. As a first approximation, one can take this evaluation map to be the restriction of ev to the subspace $\mathcal{M}_{0,k}^*(A; \mathbf{w}, J)$ on which $z_i := \mathrm{w}_i$ for $i \in I$. Note that when $k = 3$ the invariants agree:

$$\mathrm{GW}_{A,3}^M(a_1, a_2, a_3) = \mathrm{GW}_{A,3}^{M,\{1,2,3\}}(a_1, a_2, a_3).$$

However, when $\#I > 3$ the two invariants are different: see Section 7.3. In particular, it turns out that the invariants $\mathrm{GW}_{A,k}^{M,I}$ have very interesting formal properties that reflect the operad structure on the homology $H_*(\overline{\mathcal{M}}_{0,k})$ of the genus zero Grothendieck–Knudsen spaces. (More on this below.)

An important point is that these invariants depend only on the deformation class of (M, ω), i.e. on the component of ω in the space of symplectic forms on M. The proof of this statement uses the existence of the virtual moduli cycle. In this book we prove only enough to show that the invariants do not change if ω varies among semipositive forms. Since every symplectic form on a manifold of dimension at most 6 is semipositive, this implies deformation invariance in these dimensions.

Section 7.1 defines the invariants $\mathrm{GW}_{A,k}^M$ and gives the first examples. This section contains many cross references to earlier results in order to make it accessible to readers who want to start there, referring back to the technical details as needed. The invariants $\mathrm{GW}_{A,k}^{M,I}$ with some fixed marked points are defined in Section 7.3. The ideas are illustrated in Section 7.4 that shows how to count rational curves in projective spaces, while Section 7.5 discusses the so-called **full Gromov–Witten invariant**.

To explain this, observe that when the number k of marked points is at least 3 there is a forgetful map

$$\pi : \mathcal{M}_{0,k}(A; J) \to \overline{\mathcal{M}}_{0,k},$$

where $\overline{\mathcal{M}}_{0,k}$ is the Grothendieck–Knudsen moduli space[9] of stable curves of genus 0 and k marked points and π takes a stable map $[\mathbf{u}, \mathbf{z}]$ to the underlying stable curve

[9]This is also known as the Deligne–Mumford space. However, the latter names are more properly attached to the corresponding spaces for higher genus domains.

$[\mathbf{z}]$. Consider the map

$$\mathrm{ev} \times \pi : \mathcal{M}_{0,k}(A; J) \to M^k \times \overline{\mathcal{M}}_{0,k}$$

This is a pseudocycle only under very special conditions (see Exercise 6.7.13). Nevertheless, one can get very useful insight into the expected properties of the Gromov–Witten invariants by looking at them from this point of view. In particular, as explained at the end of Chapter 6, for each homology class β in $\overline{\mathcal{M}}_{0,k}$ one should be able to define an invariant whose value

$$\mathrm{GW}^M_{A,k}(a; \beta)$$

on the class $a \in H^*(M^k)$ is the intersection number of a cycle α representing the Poincaré dual to a with the restriction of ev to to the preimage under π of a cycle in $\overline{\mathcal{M}}_{0,k}$ representing the class β. When β is the fundamental class $[\overline{\mathcal{M}}_{0,k}]$ we recover the invariant $\mathrm{GW}^M_{A,k}$, while if $\beta = [\mathrm{pt}] \in H_0(\overline{\mathcal{M}}_{0,k}, \mathbb{Q})$ we obtain the invariant $\mathrm{GW}^{M,\{1,\dots,k\}}_{A,k}$ in which all marked points are fixed. The power of this approach is immediately evident. If we represent the class $[\mathrm{pt}] \in H_0(\overline{\mathcal{M}}_{0,k})$ by a point $\mathbf{w}$ in the top stratum $\mathcal{M}_{0,k}$ of $\overline{\mathcal{M}}_{0,k}$, then $\mathbf{w}$ is a tuple of k distinct points in S^2 and the invariant $\mathrm{GW}^M_{A,k}(a; \mathrm{pt})$ is obviously just $\mathrm{GW}^{M,\{1,\dots,k\}}_{A,k}$. However, if we take $\mathbf{w}$ to lie in some other stratum then we are counting curves whose domain is modelled on some fixed tree. If the invariant $\mathrm{GW}^M_{A,k}(a; \beta)$ is to be well defined then these two counts must be the same. This implies that the Gromov–Witten invariants should satisfy the so-called *(Splitting)* axiom.

In Section 7.5 we formulate the Kontsevich–Manin axioms for Gromov–Witten invariants, and interpret them in terms of our invariants $\mathrm{GW}^{M,I}_{A,k}$. All the axioms are easy to establish except for the *(Splitting)* axiom. Since this requires the gluing theorem, its proof is deferred to Section 10.8. The most important special case of this axiom is the decomposition rule

$$\mathrm{GW}^{M,\{1,2,3,4\}}_{A,4}(a_1, a_2, a_3, a_4) = \sum_{B \in H_2(M)} \sum_{\nu\mu} \mathrm{GW}^M_{A-B,3}(a_1, a_2, e_\nu)\, g^{\nu\mu}\, \mathrm{GW}^M_{B,3}(e_\mu, a_3, a_4),$$

where e_ν runs over a basis for the rational cohomology $H^*(M)$ and $g^{\nu\mu}$ is the inverse of the pairing matrix $g_{\nu\mu} := \int_M e_\nu \smile e_\mu$. This turns out to be a crucial ingredient in the proof of associativity of the quantum cup product.

Gluing. The last very technical chapter is Chapter 10 which develops a gluing principle for J-holomorphic curves. This should be thought of as the converse of Gromov's compactness theorem. It asserts that if two (or more) J-holomorphic curves intersect and satisfy a suitable transversality condition then they can be approximated in the sense of Gromov convergence by a sequence of J-holomorphic curves $u^\alpha : S^2 \to M$ representing the sum of their homology classes. This theorem was first proved by Ruan–Tian [**345**]. We give a different argument that mimics constructions used in gauge theory and is based on a careful choice of cutoff functions. Our proof is essentially the same as that in the 1994 version of this book. However, it is given in much greater detail. Although it takes considerable work to establish all the properties of the gluing map, the underlying ideas are not too difficult to understand, and we have tried to write the chapter to make these accessible.

The techniques used in this chapter do not occur elsewhere in the book, and so one can postpone reading it without serious consequences. As one can see from Chapter 9, there are many interesting applications of J-holomorphic curves that do not use gluing. These applications are based on the existence of a J-holomorphic curve with suitable properties. However, as soon as one needs to count the number of curves, one almost always needs to use gluing in the guise of the *(Splitting)* axiom. But often one does not need the full force of this axiom. A special case, which is enough to establish quantum cohomology, is proved in Section 10.9.

1.5. Applications and further developments

There are two main chapters devoted to applications, Chapter 9, which proves many of the results first stated by Gromov in his seminal 1985 paper, and Chapter 11 about quantum cohomology. Both chapters contained detailed proofs of the main results, together with extended discussions of related questions, which are listed in the index under the individual topics and also under the general headings "comments" and "examples". The most accessible applications are the proof of the nonsqueezing theorem in Section 9.3, the results on symplectic 4-manifolds and their groups of symplectomorphisms in Sections 9.4 and 9.5, and the results in Section 9.7 which use Gromov–Witten invariants to distinguish between symplectic structures. None of these sections requires any more than a basic knowledge of the Gromov–Witten invariant $\mathrm{GW}^M_{A,k}$, though they do use the soft methods of symplectic geometry. In particular, the proofs use the statements of Theorems A,B,C above rather than the methods and ideas in their proofs.

Section 9.2 on obstructions to Lagrangian embeddings is also fairly straightforward. It involves studying the properties of a perturbed Cauchy–Riemann equation on a disc with Lagrangian boundary values, and hence needs a little more analytic preparation. On the other hand, the first theorem in this section is Gromov's celebrated result that there are nonstandard symplectic structures on Euclidean space. The proof is a simple and direct argument due to Gromov that is based on properties of Lagrangian submanifolds in Euclidean space. In this situation there are no J-holomorphic spheres, which simplifies the discussion of compactness.

The other applications in Chapter 9 rely more heavily on Chapter 8. This is a preparatory chapter about Hamiltonian fibrations that introduces a geometric framework for studying perturbations of the Cauchy–Riemann equation. Solutions to these perturbed equations can be interpreted as sections of a trivial bundle $S^2 \times M \to S^2$ that are holomorphic for an appropriate almost complex structure $\widetilde{J}_H$. (The difference between $\widetilde{J}_H$ and the product almost complex structure corresponds to the perturbation.) Although trivial bundles suffice for some of the applications (such as those in Section 9.1), the analysis is no harder if one considers general locally Hamiltonian fibrations over arbitrary Riemann surfaces. Just as when studying graphs in Section 6.7 one then needs to reprove the basic regularity and compactness results in this more general setup. The arguments are essentially the same, but as always there are a few tricky points where one needs to take care to get a sharp result. With this, one obtains a useful tool (the Seidel representation) for studying the fundamental group of the group of Hamiltonian symplectomorphisms. One also obtains a definition of higher genus Gromov–Witten invariants, in the case when the complex structure on the domain (Σ, j) is fixed.

Section 9.1 describes important applications of these ideas to Hamiltonian dynamics. The first main result is that every Hamiltonian flow on a semipositive symplectic manifold has at least one 1-periodic orbit. To prove this in full detail one does need most of Chapter 8; although one can restrict to the trivial bundle, one still needs to understand how energy is measured and how to deal with bubbling. The second set of results in this section prove the existence of two distinct 1-periodic orbits on symplectically aspherical manifolds. They are somewhat easier, since the hypothesis implies there is no bubbling. Section 9.6 gives a taste of the Hofer geometry on the Hamiltonian group, describing a simple consequence of the Seidel representation due to Polterovich and Seidel. This representation can really only be defined in the context of quantum or Floer cohomology. Therefore, we return to it in both Chapter 11 and Chapter 12.

Quantum cohomology. Chapter 11 on quantum cohomology is completely independent of Chapter 9, and uses Chapter 8 only in the section on the Seidel representation. Hence it can be read immediately after Chapter 7.

The basic idea in the definition of quantum cohomology is very easy to understand. Let us suppose for simplicity that (M, ω) is (spherically) monotone, that is the restriction of the symplectic class $[\omega]$ to the spherical homology classes in $H_2(M)$ is a positive multiple $\tau c_1(TM)$ of the first Chern class. Further, denote by $H^*(M)$ the integral cohomology of M modulo torsion. Thus one can think of $H^*(M)$ as the image of $H^*(M,\mathbb{Z})$ in $H^*(M,\mathbb{R})$. The advantage of neglecting torsion is that the group $H^k(M)$, for example, may be identified with the dual $\mathrm{Hom}(H^{2n-k}(M),\mathbb{Z})$; in other words a k-dimensional cohomology class a may be described by specifying all the values of its pairings

$$\langle a, c\rangle := \int_M a \smile c$$

with the elements $c \in H^{2n-k}(M)$.

We define the quantum cup product $a * b$ of the classes $a \in H^k(M)$ and $b \in H^\ell(M)$ as the formal sum

$$a * b = \sum_{A \in H_2(M)} (a * b)_A \, q^{c_1(A)}, \tag{1.5.1}$$

where q is an auxiliary variable, considered to be of degree 2, and the cohomology class $(a * b)_A \in H^{k+\ell-2c_1(A)}(M)$ is defined in terms of Gromov–Witten invariants by

$$\langle (a * b)_A, c\rangle = \mathrm{GW}^M_{A,3}(a, b, c), \qquad c \in H^*(M).$$

Note that the classes a, b, c satisfy the dimension condition

$$2n + 2c_1(A) = \deg(a) + \deg(b) + \deg(c) \tag{1.5.2}$$

required for $\mathrm{GW}^M_{A,3}$ to be nonzero. Thus $c_1(A) \le 2n$ and, since $0 \le c_1(A)$ by monotonicity, only finitely many powers of q occur in the formula (1.5.1). Moreover, the classes A which contribute to the coefficient of q^d satisfy $\omega(A) = \tau c_1(A) = \tau d$ and hence, by Proposition 4.1.5, only finitely many can be represented by J-holomorphic curves. This shows that the sum on the right hand side of (1.5.1) is finite. Since only nonnegative exponents of q occur in the sum (1.5.1) it follows that $a * b$ is an element of the group

$$\mathrm{QH}^*(M; \mathbb{Z}[q]) = H^*(M) \otimes \mathbb{Z}[q],$$

where $\mathbb{Z}[q]$ is the polynomial ring in the variable q of degree 2. Extending by linearity, we therefore get a multiplication

$$\mathrm{QH}^*(M;\mathbb{Z}[q]) \otimes \mathrm{QH}^*(M;\mathbb{Z}[q]) \to \mathrm{QH}^*(M;\mathbb{Z}[q]).$$

It turns out that this multiplication is distributive over addition, skew-commutative and associative. The first two properties are obvious, but the last is much more subtle and depends on the *(Splitting)* axiom. If (M, ω) is not monotone, then the sum occurring in (1.5.1) may be infinite. To make sense of it, one must choose a suitable coefficient ring Λ. There are several possible choices for this quantum coefficient ring: see Example 11.1.4.

If $A = 0$, all J-holomorphic curves in the class A are constant. It follows that $\mathrm{GW}^M_{0,3}(a,b,c)$ is just the usual triple intersection $\int_M a \smile b \smile c$. Since $\omega(A) > 0$ for all other A with J-holomorphic representatives, the constant term in $a*b$ is just the usual cup product. Thus $a * b$ is a deformation of the usual cup product.

As an example, let M be complex projective n-space $\mathbb{C}P^n$ with its usual Kähler form. If p is the positive generator of $H^2(\mathbb{C}P^n)$, and if $L \in H_2(\mathbb{C}P^n)$ is the class represented by the line $\mathbb{C}P^1$, then the fact that there is a unique line through any two points is reflected in the identity

$$\langle (p * p^n)_L, p^n \rangle = \mathrm{GW}^{\mathbb{C}P^n}_{L,3}(p, p^n, p^n) = 1.$$

By equation (1.5.2), the other classes $(p * p^n)_A$ vanish for reasons of dimension. Thus $p * p^n = q1$, where $1 \in H^0(\mathbb{C}P^n)$ is the unit. Further one can show that $p^i * p^j = p^{i+j}$ for $i + j \le n$. Hence the quantum cohomology of $\mathbb{C}P^n$ is given by

$$\mathrm{QH}^*(\mathbb{C}P^n;\mathbb{Z}[q]) = \frac{\mathbb{Z}[p,q]}{\langle p^{n+1} = q \rangle}.$$

Example 11.1.12 gives a direct proof. We also explain how to deduce this result from Batyrev's formula for the quantum cohomology of toric manifolds (Theorem 11.3.4); cf. Exercise 11.3.11. The occurrence of the letters p, q is no accident here. In Section 11.3.2, we describe some recent work of Givental and Kim in which they interpret the quantum cohomology ring of flag manifolds as the ring of functions on a Lagrangian variety.

One of the aims of Chapter 11 is to demonstrate how much information is carried by the quantum cohomology. For example, in the case of Grassmannians and flag manifolds the quantum cohomology reflects structures such as the Verlinde algebra and the Toda lattice. These connections to other areas of mathematics are not entirely unexpected because quantum cohomology forms the "A-side" of the mirror symmetry conjecture and these other structures are related to the "B-side". We shall not attempt to explain what this means, but we do discuss some of the important related concepts such as Frobenius manifolds and the Dubrovin connection. We also explain how the information contained in the Gromov–Witten invariants can be encoded by a function Φ called the Gromov–Witten potential. The *(Splitting)* axiom turns out to be equivalent to the fact that Φ satisfies a third order partial differential equation called the WDVV-equation.

Floer theory. The last chapter explains the basic definitions and structures in Floer theory without going into the analytic details. (Many of these are written up in the book Audin–Damian [**24**].) We outline a proof that the ring structure on Floer theory agrees with that in quantum cohomology, and indicate some of the directions of current research. In particular, we define the spectral invariants for Hamiltonian symplectomorphisms due to Schwarz and Oh, and explain how they interact with the Seidel representation. Floer theory has many variants, some of which are outlined in Remark 12.5.5 and in Section 12.6. We end with a brief discussion of the symplectic vortex equations, that form a bridge between the Gromov–Witten invariants and gauge theory.

Concluding remarks. This book explains the fundamentals of the theory of J-holomorphic curves. There are several other quite different techniques used in symplectic topology — for example variational methods, generating functions, and Donaldson's technique of almost-complex geometry — as well as many outgrowths of the kind of elliptic techniques discussed here such as Floer theory, symplectic homology, and symplectic field theory. It is natural to wonder where the theory of J-holomorphic spheres fits in this spectrum.

The purely variational methods and the method of generating functions seem to work only when the underlying space is something like a cotangent bundle and so has some linear structure. For example, as in Hofer [**177**], Viterbo [**404**] or Bialy–Polterovich [**37**], one can use these methods to define various versions of the Hofer norm on the group $\mathrm{Ham}(\mathbb{R}^{2n}, \omega_0)$ of Hamiltonian symplectomorphisms of Euclidean space. Conley and Zehnder's proof in [**75**] of the Arnold conjecture for the torus lifted the problem to Euclidean space and then used variational methods. More recently, Tamarkin [**388**] and Guillermou–Kashiwara–Shapira [**166**] have used sheaf-theoretic methods (a significant generalization of the idea of a generating function) to establish some fundamental nondisplaceability results. Another situation where these methods apply is that of a toric manifold, which as we explain in Chapter 11.3 is the quotient of an open subset of Euclidean space by a (complex) torus. For example, Givental used this approach in [**145**]. When considering general symplectic manifolds, it seems that one must use J-holomorphic curves or some version of Floer homology (which one can think of as a combination of variational with elliptic techniques). For example, Lalonde–McDuff's proof in [**223**] that the Hofer norm is nondegenerate on completely arbitrary symplectic manifolds uses J-holomorphic spheres together with symplectic embedding techniques. Symplectic geometers have combined the available methods in many ingenious ways. We mention some of the possibilities at appropriate places in Chapters 9, 11 and 12.

CHAPTER 2

J-holomorphic Curves

In this chapter we establish the basic properties of J-holomorphic curves. These include the energy identity, which provides the foundation for the compactness arguments of Chapter 4, and the description of simple curves in Proposition 2.5.1, an essential ingredient of the approach to transversality developed in Chapter 3. Another important point is the unique continuation theorem, asserting that two curves with the same ∞-jet at a point must coincide; we give a self-contained proof using the Carleman similarity principle.

This chapter contains many foundational details, included for the sake of completeness but which are not essential for the understanding of the subsequent chapters. In particular, even though our ultimate aim is to understand C^∞ curves, our arguments in Chapters 3 and 6 use Banach manifolds whose elements are curves with less smoothness. Therefore many of the results in the present chapter are proved for C^ℓ almost complex structures. (Following an idea of Floer, one can also develop the entire theory in the smooth category, as is briefly explained in Remark 3.2.7.) Readers who are new to the subject are advised to read the statements of the main theorems at the beginning of the sections as well as the easier proofs, but to defer the more technical parts until they are needed.

2.1. Almost complex structures

Let (M, ω) be a symplectic manifold. An almost complex structure $J : TM \to TM$ is called **ω-tame** if

$$v \neq 0 \qquad \Longrightarrow \qquad \omega(v, Jv) > 0$$

for every $v \in TM$. It is called **ω-compatible** if it is ω-tame and

$$\omega(Jv, Jw) = \omega(v, w)$$

for all $x \in M$ and $v, w \in T_xM$. Every ω-tame almost complex structure determines a Riemannian metric

$$g_J(v, w) := \langle v, w \rangle_J := \tfrac{1}{2}\Big(\omega(v, Jw) + \omega(w, Jv)\Big). \tag{2.1.1}$$

In the ω-compatible case this metric can be expressed in the form

$$\langle \cdot, \cdot \rangle_J = \omega(\cdot, J\cdot).$$

In either case the automorphism J is a skew-adjoint isometry of TM. On a Riemann surface $(\Sigma, j_\Sigma, \mathrm{dvol}_\Sigma)$ the volume (or symplectic) form dvol_Σ and the complex structure j_Σ together determine a metric on Σ. Conversely, if Σ is oriented, every metric on Σ determines both a volume form and a complex structure.

Throughout we shall denote by $\mathcal{J}(M, \omega)$ the space of **ω-compatible almost complex structures** and by $\mathcal{J}_\tau(M, \omega)$ the space of **ω-tame almost complex structures** on M. Thus $\mathcal{J}(M, \omega) \subset \mathcal{J}_\tau(M, \omega)$, the space $\mathcal{J}_\tau(M, \omega)$ is open in the space $\mathcal{J}(M)$ of all almost complex structures on M, and both spaces are contractible (c.f. [**277**]). The above definitions apply to any nondegenerate 2-form ω, whether closed or not. However, we shall always assume that ω is closed unless explicit mention is made to the contrary.

In the following we denote by $N = N_J \in \Omega^2(M, TM)$ the Nijenhuis tensor of J. It is given by[1]

$$N(X,Y) = [X,Y] + J[JX,Y] + J[X,JY] - [JX,JY]$$

for $X, Y \in \mathrm{Vect}(M)$ and measures the nonintegrability of J. More precisely, $N_J = 0$ if and only if J is integrable, i.e. there is an atlas of coordinate charts on M whose transition functions are J-holomorphic. If in addition J is ω-compatible the triple (M, J, g_J) is a Kähler manifold. For a further discussion of the general case we refer to [**277**, Chapter 4] and the references therein.[2] Observe in particular that N_J always vanishes when M has dimension two, so that any almost complex structure on a Riemann surface is integrable. We give two proofs of this special case, one as Corollary C.5.3 in Section C.5 and the other in Section E.3.

Suppose (M, J) is an almost complex manifold and ω is a (not necessarily closed) 2-form on M such that $\omega(\cdot, J\cdot)$ is a Riemannian metric. Let ∇ denote the Levi-Civita connection of this metric. Then the condition $\nabla J = 0$ characterizes the Kähler case, i.e. $\nabla J = 0$ if and only if $d\omega = 0$ and $N_J = 0$. In particular, if (M, ω) is symplectic and $J \in \mathcal{J}(M, \omega)$ is ω-compatible then J is not usually parallel under the Levi-Civita connection ∇. On the other hand, the modified connection

$$\widetilde{\nabla}_X Y := \nabla_X Y - \tfrac{1}{2} J(\nabla_X J) Y$$

always preserves J as well as the metric (i.e. it is Hermitian), though it is torsion free only in the integrable case. In fact, a straightforward calculation shows that its torsion $\widetilde{T}(X,Y)$ can be expressed in terms of the Nijenhuis tensor:

$$\widetilde{T}(X,Y) := \widetilde{\nabla}_Y X - \widetilde{\nabla}_X Y - [X,Y] = -\tfrac{1}{4} N(X,Y).$$

In the ω-tame case $\widetilde{\nabla}$ is still a Hermitian connection however its torsion is no longer the Nijenhuis tensor. In this case there is another complex connection $\widehat{\nabla}$ whose torsion is the Nijenhuis tensor but which need not preserve the metric (see Appendix C). Note that in general there need not be a torsion free connection on TM that preserves J.

EXERCISE 2.1.1. Let J be an almost complex structure on M and ∇ be the Levi-Civita connection of any Riemannian metric on M. Prove that the Lie derivative of J in the direction of a vector field X is given by

$$(\mathcal{L}_X J)Y = (\nabla_X J)Y - \nabla_{JY} X + J\nabla_Y X.$$

---

[1] For consistency, we shall follow the sign conventions of McDuff–Salamon [**277**] and define the Lie bracket of two vector fields $X, Y \in \mathrm{Vect}(M)$ by

$$[X,Y] := \nabla_Y X - \nabla_X Y = -\mathcal{L}_X Y.$$

Here ∇ is the Levi-Civita connection of any Riemannian metric on M. For a rationale, see [**277**, Remark 3.3].

[2] We also point out a recent paper by Hill–Taylor[**173**] that extends the Newlander–Nirenberg integrability theorem to almost complex structures with somewhat less than Lipschitz regularity.

2.2. The nonlinear Cauchy-Riemann equations

Let (M, J) be an almost complex manifold and (Σ, j) be a Riemann surface. A smooth map $u : \Sigma \to M$ is called a ***J*-holomorphic curve** (or in short a ***J*-curve**) if the differential du is a complex linear map with respect to j and J:

$$J \circ du = du \circ j.$$

Sometimes it is convenient to write this equation in the form $\bar{\partial}_J(u) = 0$. Here the 1-form

$$\bar{\partial}_J(u) := \tfrac{1}{2}\Big(du + J \circ du \circ j\Big) \in \Omega^{0,1}(\Sigma, u^*TM)$$

is the complex antilinear part of du with respect to the almost complex structure J. It takes values in the pullback tangent bundle

$$u^*TM := \big\{(z, v) \mid z \in \Sigma,\, v \in T_{u(z)}M\big\}.$$

In more abstract terms there is an infinite dimensional vector bundle $\mathcal{E} \to \mathcal{B}$ over the space $\mathcal{B} := C^\infty(\Sigma, M)$ of smooth maps from Σ to M whose fiber over $u \in \mathcal{B}$ is the space $\mathcal{E}_u := \Omega^{0,1}(\Sigma, u^*TM)$. The nonlinear operator

$$\mathcal{B} \to \mathcal{E} : u \mapsto \big(u, \bar{\partial}_J(u)\big)$$

is a section of this bundle and its zero set is the space of J-holomorphic curves. In the case where Σ has genus zero or one it is interesting to note that the projection $\mathcal{E} \to \mathcal{B}$ and the section $\bar{\partial}_J$ are equivariant under the action of the group G of complex automorphisms of Σ on both $\mathcal{B}$ and $\mathcal{E}$. This will be discussed in detail below. In Chapter 3 we shall see, in particular, that $\bar{\partial}_J$ is a Fredholm section (with respect to suitable Sobolev completions of $\mathcal{B}$ and $\mathcal{E}$). The global form of the equation is useful in order to establish properties of the moduli space of all J-holomorphic curves. In the present chapter we shall mostly examine the local properties of J-holomorphic curves and begin by expressing the equations in local coordinates.

Local coordinates. The integrability theorem implies that there is an atlas $\{U_\alpha, \phi_\alpha\}_\alpha$ on Σ of holomorphic coordinate charts $\phi_\alpha : U_\alpha \to \mathbb{C}$. This means that

$$d\phi_\alpha \circ j = i d\phi_\alpha$$

and so the transition maps $\phi_\beta \circ \phi_\alpha^{-1}$ are holomorphic. Such local coordinates are also called **conformal**. Recall that a map ϕ between open subsets of $\mathbb{C}$ is conformal, i.e. preserves angles and orientation, if and only if it is holomorphic. A smooth map $u : \Sigma \to M$ is J-holomorphic if and only if its local coordinate representations

$$u_\alpha = u \circ \phi_\alpha^{-1} : \phi_\alpha(U_\alpha) \to M$$

are J-holomorphic with respect to the standard complex structure i on the open set $\phi_\alpha(U_\alpha) \subset \mathbb{C}$.

In conformal coordinates $z = s + it$ on Σ the 1-form $\bar{\partial}_J(u)$ is given by

$$\bar{\partial}_J(u_\alpha) = \tfrac{1}{2}\Big(\partial_s u_\alpha + J(u_\alpha)\partial_t u_\alpha\Big)\, ds + \tfrac{1}{2}\Big(\partial_t u_\alpha - J(u_\alpha)\partial_s u_\alpha\Big)\, dt.$$

Hence u is a J-holomorphic curve if and only if in conformal coordinates it satisfies the nonlinear first order partial differential equation

$$\partial_s u_\alpha + J(u_\alpha)\partial_t u_\alpha = 0, \tag{2.2.1}$$

where ∂_s denotes the partial derivative $\partial/\partial s$. If we identify $\mathbb{C}^n = \mathbb{R}^{2n}$ then the standard complex structure $J = i = J_0$ is the constant matrix

$$J_0 = \begin{pmatrix} 0 & -\mathbb{1} \\ \mathbb{1} & 0 \end{pmatrix}.$$

In this case equation (2.2.1), for a smooth map $u = f + ig : \mathbb{C} \to \mathbb{C}^n$, is equivalent to the Cauchy-Riemann equations in their familiar form

$$\partial_s f = \partial_t g, \qquad \partial_s g = -\partial_t f.$$

Thus a J_0-holomorphic curve is holomorphic in the usual sense.

The energy identity. Let (M, ω) be an almost symplectic manifold (i.e. ω is nondegenerate but not necessarily closed), $J \in \mathcal{J}_\tau(M)$ be an ω-tame almost complex structure, and $(\Sigma, j_\Sigma, \mathrm{dvol}_\Sigma)$ be a compact Riemann surface. Then M carries a metric g_J determined by ω and J via (2.1.1), and the Riemann surface carries a metric determined by j_Σ and dvol_Σ. Every J-holomorphic curve $u : \Sigma \to M$ is conformal with respect to g_J, i.e. its differential preserves angles, or, equivalently, it preserves inner products up to a common positive factor. The converse holds when M has dimension two.

The **energy** of a smooth map $u : \Sigma \to M$ is defined as the square of the L^2-norm of the 1-form $du \in \Omega^1(\Sigma, u^*TM)$:

$$E(u) := \tfrac{1}{2} \int_\Sigma |du|_J^2 \, \mathrm{dvol}_\Sigma.$$

Here the norm of the (real) linear map $L := du(z) : T_z\Sigma \to T_{u(z)}M$ is defined by

$$|L|_J := |\zeta|^{-1} \sqrt{|L(\zeta)|_J^2 + |L(j_\Sigma \zeta)|_J^2}$$

for $0 \neq \zeta \in T_z\Sigma$. The right hand side is independent of ζ by Exercise 2.2.3 below. For an explicit formula for the integrand in conformal coordinates on Σ see the proof of Lemma 2.2.1 below.

It is important to note that, while the energy density $|du|_J^2$ depends on the metric on Σ, the energy $E(u)$ depends only on the complex structure j_Σ and not on the volume form. To see this, note that

$$|\alpha|^2 \mathrm{dvol}_\Sigma = \langle \alpha \wedge * \alpha \rangle = -\langle \alpha \wedge (\alpha \circ j_\Sigma) \rangle$$

for every Hermitian vector bundle $E \to \Sigma$ and every 1-form $\alpha \in \Omega^1(\Sigma, E)$. For general smooth maps $u : \Sigma \to M$ the energy $E(u)$ also depends on the metric g_J on M. However, we now show that in the case of J-holomorphic curves in symplectic manifolds the energy is a *topological* invariant that depends only on the homology class of the curve modulo its boundary.

LEMMA 2.2.1. *Let ω be a nondegenerate 2-form on a smooth manifold M. If J is ω-tame then every J-holomorphic curve $u : \Sigma \to M$ satisfies the energy identity*

$$E(u) = \int_\Sigma u^*\omega. \tag{2.2.2}$$

If J is ω-compatible then every smooth map $u : \Sigma \to M$ satisfies

$$E(u) = \int_\Sigma |\bar{\partial}_J(u)|_J^2 \, \mathrm{dvol}_\Sigma + \int_\Sigma u^*\omega. \tag{2.2.3}$$

PROOF. By choosing conformal coordinates $z = s + it$ we may assume without loss of generality that Σ is an open subset of $\mathbb{C}$. In this case

$$\begin{aligned}\tfrac{1}{2}\,|du|_J^2\,\mathrm{dvol}_\Sigma &= \tfrac{1}{2}\Big(|\partial_s u|_J^2 + |\partial_t u|_J^2\Big)\,ds\wedge dt\\ &= \tfrac{1}{2}\big|\partial_s u + J\partial_t u\big|_J^2\,ds\wedge dt - \big\langle \partial_s u, J\partial_t u\big\rangle_J\,ds\wedge dt\\ &= \big|\bar\partial_J(u)\big|_J^2\,\mathrm{dvol}_\Sigma + \tfrac{1}{2}\Big(\omega(\partial_s u,\partial_t u) + \omega(J\partial_s u, J\partial_t u)\Big)\,ds\wedge dt.\end{aligned}$$

In the ω-compatible case the last term on the right is equal to $u^*\omega$. If J is ω-tame and $\partial_s u + J\partial_t u = 0$ then the first term on the right vanishes and the last term on the right is also equal to $u^*\omega$. □

Lemma 2.2.1 shows that if ω is closed, J is ω-compatible and Σ is closed, then J-holomorphic curves minimize energy in their homology class, and hence are harmonic maps. This need not be the case if J is only ω-tame. However in both cases the energy of a J-holomorphic curve is a topological invariant provided that ω is closed. The same holds for Riemann surfaces with boundary if we consider J-holomorphic curves $u : \Sigma \to M$ that map the boundary to a given Lagrangian submanifold $L \subset M$. This observation plays a crucial role in Gromov's compactness theorems for J-holomorphic curves (Chapters 4 and 5). It is the key difference between J-holomorphic curves in general almost complex manifolds and in those where the almost complex structure is *tamed* by a symplectic form. All the other results in this chapter about the local behaviour of J-holomorphic curves refer to general almost complex manifolds.

EXERCISE 2.2.2. Show that if J is ω-tame and u is J-holomorphic then, in the notation of Lemma 2.2.1,

$$\tfrac{1}{2}\,|du|_J^2\,\mathrm{dvol}_\Sigma = \omega(\partial_s u, J\partial_s u)\,ds\wedge dt. \tag{2.2.4}$$

EXERCISE 2.2.3. Let $L : \mathbb{C} \to \mathbb{R}^{2n}$ be any real linear map. Show that the expression

$$|L| := |\zeta|^{-1}\sqrt{|L(\zeta)|^2 + |L(i\zeta)|^2}$$

is independent of the choice of $\zeta \in \mathbb{C}\setminus\{0\}$, where $|\cdot|$ denotes the Euclidean norm on $\mathbb{R}^{2n}$. If we identify $\mathbb{R}^{2n}$ with $\mathbb{C}^n$ and L is complex linear then $|L| = \sqrt{2}|L(1)|$; show that $|L(1)|$ is the operator norm of L; show that $|L|$ is the square root of the sum of the squares of the entries of a real matrix representing L in orthonormal bases. If $\phi : \mathbb{C} \to \mathbb{C}$ is a holomorphic function prove that $|d\phi(z)| = \sqrt{2}|\phi'(z)|$, where $\phi'(z) \in \mathbb{C}$ denotes its complex derivative.

2.3. Unique continuation

In this section we establish a unique continuation theorem for J-holomorphic curves. In the integrable case this is an obvious consequence of the fact that every holomorphic function (of one complex variable) admits a power series expansion. The proof in the almost complex case is considerably harder. Our treatment follows the work of Floer–Hofer–Salamon [**119**]. See also Sikorav [**379**].

Since the question is local, we may take the domain of u to be a ball

$$B_\varepsilon := \{z \in \mathbb{C}\mid |z| < \varepsilon\}$$

in $\mathbb{C}$ and M to be $\mathbb{R}^{2n}$. Hence the almost complex structure J is a function on $\mathbb{R}^{2n}$ with values in the space $\mathrm{GL}(2n,\mathbb{R})$ of linear automorphisms of $\mathbb{R}^{2n}$. For later reference it is convenient to prove the unique continuation theorem under the weakest possible smoothness hypotheses on J. Thus we shall consider functions $u : B_\varepsilon \to \mathbb{R}^{2n}$ that satisfy the equation

$$\partial_s u + J(u)\partial_t u = 0, \tag{2.3.1}$$

where $J : \mathbb{R}^{2n} \to \mathrm{GL}(2n,\mathbb{R})$ is a C^1-function such that $J^2 = -1\!\!1$.

REMARK 2.3.1. Denote by Δ the standard Laplacian

$$\Delta = (\partial_s)^2 + (\partial_t)^2.$$

Using the fact that $\partial_t(J^2) = (\partial_t J)J + J(\partial_t J) = 0$ one can prove that every $W^{1,p}$-solution u of (2.3.1) with $p > 2$ also is a weak solution of the second order quasi-linear equation

$$\Delta u = \big(\partial_t J(u)\big)\partial_s u - \big(\partial_s J(u)\big)\partial_t u. \tag{2.3.2}$$

An elliptic bootstrapping argument then shows that if J is of class C^ℓ for some $\ell \geq 1$ then every J-holomorphic curve u of class $W^{1,p}$ with $p > 2$ is necessarily of class $W^{\ell+1,p}$ (see Theorem B.4.1 and Remark B.4.3). In particular, u is C^1 so that the results of Section 2.2 apply to these curves.

An integrable function $w : B_\varepsilon \to \mathbb{C}^n$ is said to **vanish to infinite order** at the point $z = 0$ if

$$\int_{|z|\leq r} |w(z)| = O(r^k) \qquad \text{for every } k \in \mathbb{N}.$$

If w is smooth this means that the ∞-jet of w vanishes at zero. However, the integral form of this condition is meaningful for every integrable function.

THEOREM 2.3.2 (Unique continuation). *Suppose that $u, v \in C^1(B_\varepsilon, \mathbb{R}^{2n})$ satisfy (2.3.1) for some C^1 almost complex structure $J : \mathbb{R}^{2n} \to \mathrm{GL}(2n,\mathbb{R})$, and that $u - v$ vanishes to infinite order at zero. Then $u \equiv v$.*

COROLLARY 2.3.3. *Assume Σ is a connected Riemann surface and J is a C^1 almost complex structure on a smooth manifold M. If two J-holomorphic curves $u, v : \Sigma \to M$ agree to infinite order at a point $z \in \Sigma$ then $u \equiv v$.*

PROOF. By Theorem 2.3.2, the set of all points $z \in \Sigma$ such that u and v agree to infinite order at z is open and closed. □

Aronszajn's theorem. The function $w := v - u$ in Theorem 2.3.2 is of class $W^{2,p}$ for every $p < \infty$ (see Remark 2.3.1) and vanishes to infinite order at zero. In particular, w is continuously differentiable. Because J and its derivatives are bounded it follows from (2.3.2) that w satisfies a differential inequality of the form

$$|\Delta w(z)| \leq c(|w| + |\partial_s w| + |\partial_t w|) \tag{2.3.3}$$

for all $z = s + it \in B_\varepsilon$. Hence the assertion of Theorem 2.3.2 follows from Aronszajn's unique continuation theorem which we now quote.

THEOREM 2.3.4 (Aronszajn). *Let $\Omega \subset \mathbb{C}$ be a connected open set. Suppose that the function $w \in W^{2,2}_{\mathrm{loc}}(\Omega, \mathbb{R}^m)$ satisfies the pointwise estimate (2.3.3) (almost everywhere) in Ω and that w vanishes to infinite order at some point $z_0 \in \Omega$. Then $w \equiv 0$.*

Here $W^{2,2}_{\text{loc}}(\Omega, \mathbb{R}^m)$ is the Sobolev space of maps whose second derivative is L^2 on each precompact open subset of Ω. (A reader who is unfamiliar with Sobolev spaces may suppose that w is twice continuously differentiable. This is automatically satisfied when J is of class C^2.) Aronszajn's theorem can be viewed as a generalization of the unique continuation theorem for analytic functions. It is proved by Aronszajn in [**18**] and by Hartman and Wintner in [**169**]: cf. Section E.4. In the following we shall give an alternative proof of Theorem 2.3.2 which does not rely on Aronszajn's theorem. The rest of this section can be omitted at first reading.

The Carleman similarity principle. Consider the linear first order elliptic system

$$\partial_s u(z) + J(z)\partial_t u(z) + C(z)u(z) = 0. \tag{2.3.4}$$

Here $z = s + it$, ∂_s denotes $\partial/\partial s$, u is a function $B_\varepsilon \to \mathbb{R}^{2n}$, and J and C are functions on B_ε with values in the space $\text{End}(\mathbb{R}^{2n}) = \mathbb{R}^{2n\times 2n}$ of real linear transformations of $\mathbb{R}^{2n}$. We assume also that $J(z)^2 = -\mathbb{1}$ for all $z \in B_\varepsilon$. Since we may take $J(z) = J'(u(z))$ where J' is an almost complex structure on $\mathbb{R}^{2n}$, any solution of (2.3.1) is a solution of (2.3.4) for a suitable function $J(z)$. However, note that even if the initial lmost complex structure J' on $\mathbb{R}^{2n}$, is C^ℓ-smooth the composite $z \mapsto J(z)$ will be no smoother than u. Hence we shall only assume that the function $z \mapsto J(z)$ is $W^{1,p}$-smooth. The linear term involving C does not appear when translating from (2.3.1) but will be needed for the applications.

This procedure of replacing the nonlinear equation (2.3.1) by a corresponding linear equation (2.3.4) is appropriate when one wants to study the analytic properties of a given solution u to (2.3.1). However, it obviously cannot be used to study the properties of the solution space of (2.3.1) as a whole since the linear equation depends on the solution u.

The most useful tool in understanding the regularity properties of J-holomorphic curves for general nonsmooth J is the **Carleman Similarity Principle**. It says that one can transform any solution u to (2.3.4) into a holomorphic function σ by multiplication by a suitable matrix valued function $z \mapsto \Phi(z)$.

THEOREM 2.3.5. *Fix a constant $p > 2$ and let $C \in L^p(B_\varepsilon, \text{End}(\mathbb{R}^{2n}))$ and $J \in W^{1,p}(B_\varepsilon, \text{GL}(2n, \mathbb{R}))$ be such that $J^2 = -\mathbb{1}$. Suppose that $u \in W^{1,p}(B_\varepsilon, \mathbb{R}^{2n})$ is a solution of (2.3.4) such that $u(0) = 0$. Then there is a $\delta \in (0, \varepsilon)$, a map $\Phi \in W^{1,p}(B_\delta, \text{Hom}_\mathbb{R}(\mathbb{C}^n, \mathbb{R}^{2n}))$, and a holomorphic function $\sigma : B_\delta \to \mathbb{C}^n$ such that $\Phi(z)$ is invertible and*

$$u(z) = \Phi(z)\sigma(z), \qquad \sigma(0) = 0, \qquad \Phi(z)^{-1}J(z)\Phi(z) = i.$$

for every $z \in B_\delta$.

Geometrically, one can think of the map $z \mapsto \Phi^{-1}(z)$ as a trivialization of the bundle $(B_\delta \times \mathbb{R}^{2n}, J)$ that converts the varying family of almost complex structures $J(z)$ on the fibers $z \times \mathbb{R}^{2n}$ into the constant family i, and that transforms the section $\tilde{u} : z \mapsto (z, u(z))$ into a holomorphic section $\widetilde{\sigma} : z \mapsto (z, \sigma(z))$. Thus there is a commutative diagram:

$$\begin{array}{ccc} (B_\delta \times \mathbb{R}^{2n}, J) & \stackrel{\Phi^{-1}}{\longrightarrow} & (B_\delta \times \mathbb{C}^n, i) \\ \tilde{u} \uparrow & & \uparrow \widetilde{\sigma} \\ B_\delta & \stackrel{=}{\longrightarrow} & B_\delta. \end{array}$$

Proof of Theorem 2.3.2. Define $w : B_\varepsilon \to \mathbb{R}^{2n}$ by $w(z) := u(z) - v(z)$ for $z = s + it \in B_\varepsilon$. By assumption, w is continuously differentiable and vanishes to infinite order at zero. Moreover, w satisfies the equation

$$\begin{aligned}\partial_s w(z) + J(u(z))\partial_t w(z) &= \big(J(v(z)) - J(u(z))\big)\partial_t v(z) \\ &= \left(\int_0^1 \frac{d}{d\tau} J\Big(u(z) + \tau\big(v(z) - u(z)\big)\Big)\, d\tau\right) \partial_t v(z) \\ &= -\left(\int_0^1 dJ\Big(u(z) - \tau\big(w(z)\big)\Big) w(z)\, d\tau\right) \partial_t v(z) \\ &= -C(z)w(z),\end{aligned}$$

where the real linear map $C(z) \in \mathrm{End}(\mathbb{R}^{2n})$ is defined by

$$C(z)\xi := \left(\int_0^1 dJ\big(u(z) - \tau w(z)\big)\xi\, d\tau\right) \partial_t v(z)$$

for $z \in B_\varepsilon$ and $\xi \in \mathbb{R}^{2n}$. (Note that $dJ(x)\xi \in \mathrm{End}(\mathbb{R}^{2n})$ for $x, \xi \in \mathbb{R}^{2n}$.) Since J, u, and v are continuously differentible, the function $C : B_\varepsilon \to \mathrm{End}(\mathbb{R}^{2n})$ is continuous and the function $z \mapsto J(u(z))$ is continuously differentiable.

Thus we have proved that $w : B_\varepsilon \to \mathbb{R}^{2n}$ satisfies the hypotheses of Theorem 2.3.5 with $J(z)$ replaced by $J(u(z))$. Hence every point $z_0 \in B_\varepsilon$ has a neighbourhood $B_\delta(z_0)$ in which w has the form $w(z) = \Phi(z)\sigma(z)$, where $\sigma : B_\delta(z_0) \to \mathbb{C}^n$ is holomorphic and $\Phi : B_\delta(z_0) \to \mathrm{Hom}_\mathbb{R}(\mathbb{C}^n, \mathbb{R}^{2n}))$ is in $W^{1,p}$ and hence continuous.

Let Z be the set of points $z_0 \in B_\varepsilon$ such that w vanishes to infinite order at z_0. This set is open: if $z_0 \in Z$ and δ, σ, Φ are as above, then σ vanishes to infinite order at z_0; since σ is holomorphic, this implies that σ, and hence also w, vanishes on $B_\delta(z_0)$; hence $B_\delta(z_0) \subset Z$. And Z is closed: if $z_\nu \in Z$ converges to $z_0 \in B_\varepsilon$ and δ, σ, Φ are as above, then σ vanishes on a sequence converging to z_0; hence σ, and also w, vanish to infinite order at z_0; hence $z_0 \in Z$. Thus Z is an open and closed subset of B_ε. Since $0 \in Z$ it follows that $Z = B_\varepsilon$ and so w vanishes on B_ε. □

Proof of Theorem 2.3.5. The proof has three steps.

Step 1. *It suffices to consider the case* $(\mathbb{R}^{2n}, J(z)) = (\mathbb{C}^n, i)$.

By assumption, there is a function $\Psi \in W^{1,p}(B_\varepsilon, \mathrm{Hom}_\mathbb{R}(\mathbb{C}^n, \mathbb{R}^{2n}))$ such that $\Psi(z)$ is invertible and conjugates $J(z)$ to i, that is $\Psi(z)^{-1}J(z)\Psi(z) = i$ for every $z \in B_\varepsilon$. The corresponding function v defined by $u = \Psi v$ satisfies the following equation:

$$\begin{aligned}0 &= \partial_s u + J\partial_t u + Cu \\ &= (\partial_s\Psi)v + \Psi\partial_s v + J(\partial_t\Psi)v + J\Psi(\partial_t v) + C\Psi v \\ &=: \Psi(\partial_s v + i\partial_t v + \widetilde{C}v),\end{aligned}$$

where

$$\widetilde{C} := \Psi^{-1}(\partial_s\Psi + J\partial_t\Psi + C\Psi) \in L^p(B_\varepsilon, \mathrm{End}_\mathbb{R}(\mathbb{C}^n)). \tag{2.3.5}$$

(Here we denote by $\mathrm{End}_\mathbb{R}(\mathbb{C}^n)$ the space of $\mathbb{R}$-linear transformations of $\mathbb{C}^n$.) Therefore, assuming the result for $J = i$, we can conclude that $v = \widetilde{\Phi}(z)\sigma$ where $\widetilde{\Phi}$ is i-linear and σ is holomorphic. Hence $u = \Phi\sigma$ where $\Phi := \Psi\widetilde{\Phi}$ conjugates J to i.

Step 2. *We may assume in addition that* $C \in \mathrm{End}_\mathbb{R}(\mathbb{C}^n))$ *is* i*-linear.*

We can replace the function $\widetilde{C}$ in (2.3.5) by any L^p function A such that

$$A(z)v(z) = \widetilde{C}(z)v(z)$$

for all z. To find a suitable function A, we decompose $\widetilde{C}$ into its complex linear and complex anti-linear parts:

$$\widetilde{C} = \widetilde{C}^+ + \widetilde{C}^-, \qquad \widetilde{C}^\pm := \tfrac{1}{2}(\widetilde{C} \mp i\widetilde{C}i).$$

Then the function

$$A := \widetilde{C}^+ + \widetilde{C}^- D$$

is complex linear if and only if D is complex anti-linear. Thus we must find a (not necessarily continuous) map $D \in L^\infty(B_\varepsilon, \mathrm{End}_{\mathbb{R}}(\mathbb{C}^n))$ such that $D(z)$ is complex anti-linear and $D(z)v(z) = v(z)$ for all z. An example of such a map is given by the formula

$$D(z)\zeta := \begin{cases} |v(z)|^{-2}\, v(z)v(z)^T\bar{\zeta}, & \text{if } v(z) \neq 0, \\ 0, & \text{if } v(z) = 0. \end{cases}$$

STEP 3. *The result holds when $J(z) = i$ and C is i-linear.*

We must find a family of complex linear automorphisms $\Phi(z)$ such that the function $\sigma(z) := \Phi^{-1}(z)u(z)$ is holomorphic. We do this by considering the Cauchy–Riemann operator on the unit disc B_1. Denote

$$V := \mathbb{C}^{n\times n}, \qquad W := \mathbb{R}^{n\times n}$$

and note that W is a Lagrangian subspace of V. For $\delta \in (0, \varepsilon)$ consider the function $C_\delta \in L^p(B_1, V)$ defined by

$$C_\delta(z) := \begin{cases} C(z), & \text{if } z \in B_\delta, \\ 0, & \text{if } z \in B_1 \setminus B_\delta. \end{cases}$$

Denote

$$W^{1,p}(B_1, V; W) := \{\Phi \in W^{1,p}(B_1, V) \,|\, \Phi(\partial B_1) \subset W\}$$

and consider the Cauchy–Riemann operator

$$D_\delta : W^{1,p}(B_1, V; W) \to L^p(B_1, V)$$

given by

$$D_\delta \Phi := \partial_s \Phi + i\partial_t \Phi + C_\delta \Phi.$$

The operator $W^{1,p}(B_1, V; W) \to L^p(B_1, V) : \Phi \mapsto \partial_s\Phi + i\partial_t\Phi$ is surjective and its kernel consists of the constant functions $\Phi : B_1 \to W$ (see Theorem C.4.1). Hence the operator $W^{1,p}(B_1, V; W) \to L^p(B_1, V) \times W : \Phi \mapsto (\partial_s\Phi + i\partial_t\Phi, \Phi(1))$ is bijective. Since

$$\lim_{\delta\to 0} \|C_\delta\|_{L^p} = 0$$

the operator $\Phi \mapsto (D_\delta\Phi, \Phi(1))$ is also surjective for small δ. Hence, for small δ, there is a unique element $\Phi_\delta \in W^{1,p}(B_1, V; W)$ such that

$$D_\delta \Phi_\delta = 0, \qquad \Phi_\delta(1) = \mathbb{1}.$$

By uniqueness, Φ_δ converges to the constant function $\mathbb{1}$ as $\delta \to 0$. Hence Φ_δ takes values in the invertible elements of V for sufficiently small δ. Now define $\sigma : B_\varepsilon \to \mathbb{C}^n$ by the equation $u =: \Phi_\delta\sigma$. Then it follows as in Step 1 that σ is holomorphic in B_δ. □

EXERCISE 2.3.6. Give a proof of Step 3 that uses the Cauchy–Riemann operator on the 2-sphere. Such a proof does not require elliptic regularity on the boundary.

2.4. Critical points

In this section we establish some useful local properties of J-holomorphic curves. We assume throughout that J is of class C^1 so that, by Remark 2.3.1, u is also. A **critical point** of a J-holomorphic curve $u : \Sigma \to M$ is a point $z \in \Sigma$ such that $du(z) = 0$. The image of a critical point under u is called a **critical value**. If we think of the image $C := u(\Sigma) \subset M$ as an unparametrized J-holomorphic curve, then a critical point on C is a critical value of u. Points on C which are not critical are called regular or nonsingular. In the integrable case critical points of nonconstant holomorphic curves are well known to be isolated. The next lemma asserts this for arbitrary almost complex structures. We give two proofs.

LEMMA 2.4.1. *Let Σ be a compact Riemann surface without boundary, J be a C^1 almost complex structure on a smooth manifold M, and $u : \Sigma \to M$ be a nonconstant J-holomorphic curve of class C^1. Then the set*

$$X := u^{-1}\left(\{u(z) \,|\, z \in \Sigma,\, du(z) = 0\}\right)$$

of preimages of critical values is finite. Moreover, $u^{-1}(x)$ is a finite set for every $x \in M$.

PROOF 1. In the first proof we assume that J is smooth. It suffices to prove that critical points are isolated, and so we may work locally. Thus we assume that $\Omega \subset \mathbb{C}$ is an open neighbourhood of zero, that $u : \Omega \to \mathbb{C}^n$ is J-holomorphic for some almost complex structure $J : \mathbb{C}^n \to \mathrm{GL}(2n, \mathbb{R})$, and that

$$u(0) = 0, \qquad du(0) = 0, \qquad u \not\equiv 0, \qquad J(0) = J_0.$$

Write $z = s + it$. Since u is nonconstant it follows from Corollary 2.3.3 that the ∞-jet of $u(z)$ at $z = 0$ must be nonzero. Hence there exists an integer $\ell \geq 2$ such that

$$u(z) = O(|z|^\ell), \qquad u(z) \neq O(|z|^{\ell+1}).$$

This implies

$$J(u(z)) = J_0 + O(|z|^\ell).$$

Now examine the Taylor expansion of

$$\partial_s u + J(u)\partial_t u = 0$$

up to order $\ell - 1$ to obtain

$$\partial_s T_\ell(u) + J_0 \partial_t T_\ell(u) = 0.$$

Here $T_\ell(u) : \mathbb{C} \to \mathbb{C}^n$ denotes the Taylor expansion of u up to order ℓ. It follows that $T_\ell(u)$ is a holomorphic function and there exists a nonzero vector $a \in \mathbb{C}^n$ such that

$$u(z) = az^\ell + O(|z|^{\ell+1}), \qquad \partial_s u(z) = \ell a z^{\ell-1} + O(|z|^\ell).$$

Hence

$$0 < |z| \leq \varepsilon \qquad \Longrightarrow \qquad u(z) \neq 0, \quad du(z) \neq 0$$

with $\varepsilon > 0$ sufficiently small. Hence critical points are isolated and, since Σ is compact, the set of critical points of u is finite. It also follows that $u^{-1}(p)$ is a finite set for every $p \in M$. Hence X is a finite set. □

PROOF 2. Assume $J : \mathbb{C}^n \to \mathrm{GL}(2n, \mathbb{R})$ is an almost complex structure of class C^1. Let $\Omega \subset \mathbb{C}$ be an open set. Suppose u is such a J-holomorphic curve. Then u satisfies (2.3.4), where $J(z) = J(u(z))$ is of class C^1 and $C(z) = 0$. Hence it follows from Theorem 2.3.5 that $u^{-1}(p)$ is a finite set for every $p \in M$. Moreover, differentiating (2.3.4) with $C = 0$, we find that the function $v := \partial_s u \in W^{1,p}(\Omega, \mathbb{C}^n)$ satisfies the equation

$$\partial_s v(z) + J(z)\partial_t v(z) + \big(\partial_s J(z)\big) J(z) v(z) = 0.$$

Applying Theorem 2.3.5 to v we see that its zeros are isolated. Hence the set of critical points of u is finite and so is the set X. □

We now show how to choose nice coordinates near a regular point of a J-holomorphic curve.

LEMMA 2.4.2. *Let $\ell \geq 2$ and J be a C^ℓ almost complex structure on a smooth manifold M. Let $\Omega \subset \mathbb{C}$ be an open neighbourhood of zero and $u : \Omega \to M$ be a local J-holomorphic curve such that $du(0) \neq 0$. Then there exists a $C^{\ell-1}$ coordinate chart $\psi : U \to \mathbb{C}^n$, defined on an open neighbourhood of $u(0)$, such that*

$$\psi \circ u(z) = (z, 0, \ldots, 0), \qquad d\psi(u(z))J(u(z)) = J_0 d\psi(u(z))$$

for $z \in \Omega \cap u^{-1}(U)$.

PROOF. By Theorem B.4.1 and Remark B.4.3, u is of class $W^{\ell+1,p}$ for every $p < \infty$ and hence, by the Sobolev embedding theorem B.1.11, is of class C^ℓ. Write $z = s + it \in \Omega$ and $w = (w_1, w_2, \ldots, w_n) \in \mathbb{C}^n$ where $w_j = x_j + iy_j$. Shrink Ω if necessary and choose a complex $C^{\ell-1}$ frame of the bundle u^*TM such that

$$Z_1(z), \ldots, Z_n(z) \in T_{u(z)}M, \qquad Z_1 = \frac{\partial u}{\partial s}.$$

Define $\phi : \Omega \times \mathbb{C}^{n-1} \to M$ by

$$\phi(w_1, \ldots, w_n) = \exp_{u(w_1)}\left(\sum_{j=2}^{n} x_j Z_j(w_1) + \sum_{j=2}^{n} y_j J(u(w_1)) Z_j(w_1)\right).$$

Then ϕ is a $C^{\ell-1}$ diffeomorphism of a neighbourhood V of zero in $\mathbb{C}^n$ onto a neighbourhood U of $u(0)$ in M. It satisfies $\phi(w_1, 0, \ldots, 0) = u(w_1)$ and

$$\frac{\partial \phi}{\partial x_j} + J(\phi)\frac{\partial \phi}{\partial y_j} = 0, \qquad j = 1, \ldots, n,$$

at all points $w = (w_1, 0, \ldots, 0)$. Hence the inverse $\psi := \phi^{-1} : U \to V$ has the required properties. □

We next start investigating the intersections of two distinct J-holomorphic curves. The most significant results in this connection occur in dimension four, and are discussed in McDuff [**256, 264**] for example. For now, we prove a useful result which is valid in all dimensions asserting that intersection points of distinct J-holomorphic curves $u : \Sigma \to M$ and $u' : \Sigma' \to M$ can only accumulate at points which are critical on both curves $C = u(\Sigma)$ and $C' = u'(\Sigma')$.[3] For local J-holomorphic curves this statement can be reformulated as follows.

[3]In fact, any pair of distinct closed J-holomorphic curves has only a finite number of intersection points (see Sikorav [**379**] and Appendix E by Lazzarini). However, this is harder to prove and we shall not need it for most of our applications. Our present result is taken from McDuff [**254**].

LEMMA 2.4.3. *Let J be a C^2 almost complex structure on a smooth manifold M. Let $\Omega \subset \mathbb{C}$ be an open neighbourhood of zero and $u, v : \Omega \to M$ be J-holomorphic curves such that*

$$u(0) = v(0), \qquad du(0) \neq 0.$$

Moreover, assume that there exist sequences $z_\nu, \zeta_\nu \in \Omega$ such that

$$u(z_\nu) = v(\zeta_\nu), \qquad \lim_{\nu\to\infty} z_\nu = \lim_{\nu\to\infty} \zeta_\nu = 0, \qquad \zeta_\nu \neq 0.$$

Then there exists a holomorphic function $\phi : B_\varepsilon(0) \to \Omega$ defined in some neighbourhood of zero such that $\phi(0) = 0$ and

$$v = u \circ \phi.$$

PROOF 1. In this proof we assume that J is smooth and use Aronszajn's theorem. By Lemma 2.4.2, we may assume without loss of generality that $M = \mathbb{C}^n$, $J : \mathbb{C}^n \to \mathrm{GL}(2n, \mathbb{R})$ is a smooth almost complex structure, and

$$u(z) = (z, 0), \qquad J(w_1, 0) = i,$$

where $w = (w_1, \tilde{w})$ with $\tilde{w} \in \mathbb{C}^{n-1}$. Write $v(z) =: (v_1(z), \tilde{v}(z))$.

We show first that the ∞-jet of $\tilde{v}$ at $z = 0$ must vanish. Otherwise there would exist an integer $\ell \geq 0$ such that $\tilde{v}(z) = O(|z|^\ell)$ and $\tilde{v}(z) \neq O(|z|^{\ell+1})$. The assumption of the lemma implies $\ell \geq 1$ and hence $J(v(z)) = J_0 + O(|z|^\ell)$. As in the proof of Lemma 2.4.1, consider the Taylor expansion up to order $\ell - 1$ on the left hand side of the equation $\partial_s v + J\partial_t v = 0$ to obtain that $T_\ell(v)$ is holomorphic. Hence

$$v_1(z) = p(z) + O(|z|^{\ell+1}), \qquad \tilde{v}(z) = \tilde{a}z^\ell + O(|z|^{\ell+1}),$$

where $p(z)$ is a polynomial of order ℓ and $\tilde{a} \in \mathbb{C}^{n-1}$ is nonzero. This implies that $\tilde{v}(\zeta) \neq 0$ in some neighbourhood of zero and hence $v(\zeta) \notin \operatorname{im} u$ for every nonzero element ζ in this neighbourhood, in contradiction to the assumption of the lemma. Thus we have proved that the ∞-jet of $\tilde{v}$ at $z = 0$ vanishes.

We prove that $\tilde{v}(z) \equiv 0$. To see this note that, because $J = J_0$ along the axis $\tilde{w} = 0$, we have

$$\frac{\partial J(w_1, 0)}{\partial x_1} = \frac{\partial J(w_1, 0)}{\partial y_1} = 0,$$

for all w_1. Hence

$$\left|\frac{\partial J(w)}{\partial x_1}\right| + \left|\frac{\partial J(w)}{\partial y_1}\right| \leq c\,|\tilde{w}|\,.$$

Using (2.3.2) we obtain an inequality

$$|\Delta\tilde{v}| \leq c(|\tilde{v}| + |\partial_s\tilde{v}| + |\partial_t\tilde{v}|).$$

Hence it follows from Aronszajn's theorem that $\tilde{v} \equiv 0$. Thus $v(z) = (v_1(z), 0)$ which implies that $\phi := v_1$ is holomorphic and $v(z) = u(\phi(z))$ for $z \in B_\delta$ as required. □

PROOF 2. In this proof we consider the general case where J is of class C^ℓ, for some constant $\ell \geq 2$, and use the Carleman similarity principle. By Lemma 2.4.2, we may assume that $J : \mathbb{C}^n \to \mathrm{GL}(2n, \mathbb{R})$, $u(z) = (z, 0)$, and $v = (v_1, \tilde{v})$ are as in the first proof, except that now J and v are only continuously differentiable. We claim that there exists a continuous function $\widetilde{C} : \Omega \to \mathrm{End}_\mathbb{R}(\mathbb{C}^{n-1})$ such that

$$\partial_s\tilde{v} + i\partial_t\tilde{v} + \widetilde{C}\tilde{v} = 0.$$

This equation can be written in the form

$$\widetilde{C}\tilde{v} = \tilde{\pi}((J(v) - i)\partial_t v),$$

where $\tilde{\pi} : \mathbb{C}^n \to \mathbb{C}^{n-1}$ denotes the projection $\tilde{\pi}(w_1, \tilde{w}) := \tilde{w}$. Since $J(v_1, 0) = i$, the function $\widetilde{C}$ is given by

$$\widetilde{C}(z)\tilde{\xi} := \tilde{\pi}\left(\left(\int_0^1 \frac{d}{ds}\bigg|_{s=0} J(v_1(z), \tau\tilde{v}(z) + s\tilde{\xi})\, d\tau\right) \partial_t v(z)\right)$$

for $z \in \Omega$ and $\tilde{\xi} \in \mathbb{C}^{n-1}$. Note that $\widetilde{C}$ is continuous.

It now follows from Theorem 2.3.5 that there is a constant $\delta > 0$, a function $\widetilde{\Phi} \in W^{1,p}(B_\delta, \mathrm{GL}(n-1, \mathbb{C}))$, and a holomorphic function $\tilde{\sigma} : B_\delta \to \mathbb{C}^{n-1}$ such that

$$\tilde{v}(z) = \widetilde{\Phi}(z)\tilde{\sigma}(z), \qquad \tilde{\sigma}(0) = 0.$$

By assumption, there is a sequence $0 \neq \zeta_\nu \to 0$ such that $\tilde{v}(\zeta_\nu) = 0$. Hence $\tilde{\sigma}(\zeta_\nu) = 0$ for ν sufficiently large and this implies $\tilde{v} \equiv 0$. □

We now prove the global version of the preceding lemma, which is a simple exercise in point set topology. It will be needed in Section 3.4 when we discuss the transversality of evaluation maps.

PROPOSITION 2.4.4. *Let J be a C^2 almost complex structure on M and Σ_0, Σ_1 be compact connected Riemann surfaces without boundary. Let $u_0 : \Sigma_0 \to M$ and $u_1 : \Sigma_1 \to M$ be J-holomorphic curves such that $u_0(\Sigma_0) \neq u_1(\Sigma_1)$ and u_0 is nonconstant. Then the set $u_0^{-1}(u_1(\Sigma_1))$ is at most countable and can accumulate only at the critical points of u_0.*

PROOF. If u_1 is constant then, by Lemma 2.4.1, $u_0^{-1}(u_1(\Sigma_1))$ is a finite set and so there is nothing to prove. Hence assume u_1 is nonconstant. For $i = 0, 1$ let $X_i \subset \Sigma_i$ denote the set of critical points of u_i. By Lemma 2.4.1, these sets are finite. We prove that, for every $z_0 \in \Sigma_0 \setminus X_0$, the following are equivalent.

(i) There exists a neighbourhood $U_0 \subset \Sigma_0$ of z_0 such that $u_0(U_0) \subset u_1(\Sigma_1)$.

(ii) There exists a sequence $z_\nu \in u_0^{-1}(u_1(\Sigma_1)) \setminus \{z_0\}$ that converges to z_0.

That (i) implies (ii) is obvious. Hence assume (ii) and choose a sequence $\zeta_\nu \in \Sigma_1$ such that $u_1(\zeta_\nu) = u_0(z_\nu)$ for every ν. Passing to a subsequence, if necessary, we may assume that ζ_ν converges to ζ_0 and so $u_1(\zeta_0) = u_0(z_0)$. Since $z_\nu \neq z_0$ for all ν it follows that $u_0(z_\nu) \neq u_0(z_0)$ and hence $\zeta_\nu \neq \zeta_0$ for large ν. Since z_0 is not a critical point of u_0 there exists, by Lemma 2.4.3, a neighbourhood $U_1 \subset \Sigma_1$ of ζ_0 and a holomorphic map $\phi : U_1 \to \Sigma_0$ such that $\phi(\zeta_0) = z_0$ and $u_1|_{U_1} = u_0 \circ \phi$. Since u_1 is nonconstant, so is ϕ. Hence $U_0 := \phi(U_1)$ is a neighbourhood of z_0 and $u_0(U_0) \subset u_1(\Sigma_1)$. This proves (i).

Now let $W_0 \subset \Sigma_0 \setminus X_0$ be the set of all points z_0 that satisfy (i) and hence also (ii). By (i), this set is open and, by (ii), it is relatively closed in $\Sigma_0 \setminus X_0$. We must prove that $W_0 = \emptyset$. Suppose otherwise that $W_0 \neq \emptyset$. Then $W_0 = \Sigma_0 \setminus X_0$ and hence $u_0(\Sigma_0) \subset u_1(\Sigma_1)$. This implies that the set $W_1 \subset \Sigma_1 \setminus X_1$, defined as above with the roles of Σ_0 and Σ_1 reversed, is also nonempty. (Prove this!) Hence the same argument as above shows that $u_1(\Sigma_1) \subset u_0(\Sigma_0)$ and so $u_0(\Sigma_0) = u_1(\Sigma_1)$. This contradicts our assumption and hence $W_0 = \emptyset$ as claimed. The assertion now follows from the definition of W_0. □

2.5. Somewhere injective curves

Let (Σ, j) be a compact Riemann surface and (M, J) be an almost complex manifold. A J-holomorphic curve $u : \Sigma \to M$ is said to be **multiply covered** if there is a compact Riemann surface (Σ', j'), a J-holomorphic curve $u' : \Sigma' \to M$, and a holomorphic branched covering $\phi : \Sigma \to \Sigma'$ such that

$$u = u' \circ \phi, \qquad \deg(\phi) > 1.$$

The curve u is called **simple** if it is not multiply covered. We shall see in the next chapter that the simple J-holomorphic curves in a given homology class form a smooth finite dimensional manifold for generic J. In other words, the multiply covered curves are the exceptional case and they may be singular points in the moduli space of J-holomorphic curves. The proof of this result is based on the observation that every simple J-holomorphic curve is **somewhere injective** in the sense that

$$du(z) \neq 0, \qquad u^{-1}(u(z)) = \{z\}$$

for some $z \in \Sigma$. A point $z \in \Sigma$ with this property is called an **injective point** of u. Let us denote by

$$Z(u) := \left\{z \in \Sigma \,|\, du(z) = 0 \quad \text{or} \quad \#u^{-1}(u(z)) > 1\right\}$$

the complement of the set of injective points. The next result shows that the set of injective points is open and dense for every simple J-holomorphic curve.

PROPOSITION 2.5.1. *Let J be a C^2 almost complex structure on M, Σ be a compact Riemann surface without boundary, and $u : \Sigma \to M$ be a simple J-holomorphic curve. Then u is somewhere injective. Moreover, the set $Z(u)$ of noninjective points is at most countable and can only accumulate at the critical points of u.*

PROOF. We give two proofs. Both show that any curve $u : \Sigma \to M$ may be expressed as a composition $u' \circ \phi : \Sigma \to \Sigma' \to M$ where $u' : \Sigma' \to M$ is somewhere injective. If u is simple, ϕ must have degree 1 and it follows that u is also somewhere injective. The first proof constructs the domain Σ' from the image curve $u(\Sigma)$ in M, while the second constructs Σ' by means of an equivalence relation on Σ.

Let $X \subset \Sigma$ denote the set of preimages of critical values of u and

$$X' := u(X)$$

the set of critical values. By Lemma 2.4.1, these sets are finite. Let Q be the set of points in

$$Y := u(\Sigma) \setminus X'$$

where distinct branches of $u(\Sigma)$ meet. This means $x \in Q$ if and only if there exist two points $z_1, z_2 \in \Sigma \setminus X$ such that

$$u(z_1) = u(z_2) = x, \qquad z_1 \neq z_2, \qquad u(U_1) \neq u(U_2)$$

for any two sufficiently small neighbourhoods U_j of z_j. By Lemma 2.4.3, Q is a discrete subset of Y (i.e. it has no accumulation points in Y). Thus the set

$$S := Y \setminus Q$$

is an embedded submanifold in M. Let $\iota : S \to M$ denote this embedding. Since only finitely many branches of Y can meet at each point of Q, each such point gives rise to a finite number of ends of S each diffeomorphic to a punctured disc $D \setminus \{\text{pt}\}$. Therefore, we may add a point to each of these ends and extend ι smoothly over the resulting surface S' to ι'. Because ι' is an immersion, there is a unique complex structure on S' with respect to which ι' is J-holomorphic.

The manifold S' still has ends corresponding to the points in X'. In fact, there may be several branches of $\mathrm{Im}\, u$ through each $x \in X'$, and there is an end of S' corresponding to each such branch. It follows that each end of S' is the conformal image of a punctured disc, and so must have the conformal structure of the punctured disc. Therefore we may form a closed Riemann surface Σ' by adding a point to each end of S'. Further, because u extends over the whole of Σ, the map ι' must extend to a J-holomorphic map $u' : \Sigma' \to M$. This map u' is somewhere injective and u factors as $u' \circ \phi$ where ϕ is a holomorphic map $\Sigma \to \Sigma'$. Thus ϕ is a branched cover and has degree one if and only if u is somewhere injective.

Here is a more detailed argument that constructs Σ' from Σ. Consider the set

$$\Gamma_0 \subset (\Sigma \setminus X) \times (\Sigma \setminus X)$$

of all pairs (z, ζ) such that there exist sequences $z_\nu \to z$ and $\zeta_\nu \to \zeta$ that satisfy $u(z_\nu) = u(\zeta_\nu)$, $z_\nu \neq z$, and $\zeta_\nu \neq \zeta$:

$$\Gamma_0 := \{(z,\zeta) \,|\, z, \zeta \in \Sigma \setminus X,\, \exists z_\nu \to z\, \exists \zeta_\nu \to \zeta \ni u(z_\nu) = u(\zeta_\nu),\, z_\nu \neq z,\, \zeta_\nu \neq \zeta\}.$$

In other words, Γ_0 is the set of accumulation points of multiple points of u. Isolated self-intersection points are excluded. We prove that the set Γ_0 has the following properties.

(i) Γ_0 is an equivalence relation on $\Sigma \setminus X$.

(ii) The projection $\pi : \Gamma_0 \to \Sigma \setminus X$ onto the first factor is a local diffeomorphism and each local inverse of π is holomorphic.

(iii) The covering π is proper.

(iv) $\pi^{-1}(z)$ is a finite set for every $z \in \Sigma \setminus X$ and the number

$$m := \#\{\zeta \in \Sigma \setminus X \,|\, (z, \zeta) \in \Gamma_0\}$$

is independent of $z \in \Sigma \setminus X$.

Assertion (i) follows from Lemma 2.4.3. Namely, if $(z_0, z_1) \in \Gamma_0$ and $(z_1, z_2) \in \Gamma_0$ then, by Lemma 2.4.3, there exist open neighbourhoods $U_i \subset \Sigma$ of z_i for $i = 0, 1, 2$ and holomorphic maps $\phi_{10} : U_0 \to U_1$ and $\phi_{21} : U_1 \to U_2$ such that

$$\phi_{10}(z_0) = z_1, \qquad \phi_{21}(z_1) = z_2, \qquad u|_{U_0} = u \circ \phi_{10}, \qquad u|_{U_1} = u \circ \phi_{21}.$$

The maps ϕ_{10} and ϕ_{21} are evidently nonconstant. Since

$$u|_{U_0} = u \circ \phi_{21} \circ \phi_{10}, \qquad \phi_{21} \circ \phi_{10}(z_0) = z_2,$$

it follows that $(z_0, z_2) \in \Gamma_0$. Thus we have proved (i). Assertion (ii) follows from the same argument and the fact that, since $z_0 \notin X$, the holomorphic map ϕ_{10} satisfies $d\phi_{10}(z_0) \neq 0$. To prove (iii) let $z_\nu \in \Sigma \setminus X$ be a sequence converging to $z \in \Sigma \setminus X$ and let $(z_\nu, \zeta_\nu) \in \Gamma_0$. We must prove that ζ_ν has a subsequence converging to a point in $\Sigma \setminus X$. Passing to a subsequence, we may assume without loss of generality that ζ_ν converges to $\zeta \in \Sigma$. Then $u(z) = u(\zeta)$ and hence $\zeta \in u^{-1}(u(\Sigma \setminus X)) = \Sigma \setminus X$. This shows that π is proper. Assertion (iv) follows from (ii) and (iii) and the fact that $\Sigma \setminus X$ is connected.

The strategy now is to prove that if $m \geq 2$ then u is not simple, and if $m = 1$ then u satisfies the requirements of the proposition. Suppose first that $m = 1$. We must prove that the set $Z(u)$ can only accumulate at the set of critical points of u. Suppose otherwise that there exists a sequence $z_\nu \in Z(u)$ converging to a point

$z_0 \in \Sigma$ such that $du(z_0) \neq 0$. Choose an open neighbourhood $U_0 \subset \Sigma$ of z_0 such that the restriction of u to U_0 is an embedding. Suppose, without loss of generality, that $z_\nu \in U_0$ for every ν. Then $du(z_\nu) \neq 0$ for every ν and hence there exists a sequence $\zeta_\nu \in \Sigma \setminus \{z_\nu\}$ such that $u(z_\nu) = u(\zeta_\nu)$. Since $u|_{U_0}$ is an embedding it follows that $\zeta_\nu \notin U_0$. Assume without loss of generality that ζ_ν converges to a point $\zeta_0 \in \Sigma \setminus U_0$. Then, by Lemma 2.4.3, there exist an open neighbourhood $V_0 \subset \Sigma$ of ζ_0 and a holomorphic map $\phi : V_0 \to U_0$ such that $\phi(\zeta_0) = z_0$ and $u|_{V_0} = u \circ \phi$. Shrinking V_0, if necessary, we may assume that $V_0 \setminus \{\zeta_0\} \subset \Sigma \setminus X$. It follows that $(\zeta, \phi(\zeta)) \in \Gamma_0$ for every $\zeta \in V_0 \setminus \{\zeta_0\}$ and hence $m \geq 2$, contradicting our assumption that $m = 1$. This shows that, in the case $m = 1$, the set $Z(u)$ can only accumulate at the critical set of u.

Now suppose that $m \geq 2$ and extend Γ_0 to its closure

$$\Gamma := \mathrm{cl}(\Gamma_0) \subset \Sigma \times \Sigma.$$

This set is an equivalence relation on Σ and we denote

$$z \sim \zeta \qquad \Longleftrightarrow \qquad (z, \zeta) \in \Gamma.$$

The set X is invariant under this equivalence relation. (If $z \in X$ and $z \sim \zeta$ then $\zeta \in X$.) Hence each point $z \in \Sigma$ carries a natural multiplicity $m(z) \geq 1$ defined as follows. If $z \in \Sigma \setminus X$ define $m(z) = 1$. If $z \in X$ and $w \in \Sigma \setminus X$ is sufficiently close to z then all m points in the equivalence class of w are close to X. Define $m(z)$ as the number of points equivalent to w which are close to z. By continuity this number is independent of the choice of w. With this definition we have

$$\sum_{\zeta \sim z} m(\zeta) = m$$

for every $z \in \Sigma$.

We define the topological space Σ' as the quotient

$$\Sigma' := \Sigma / \sim.$$

It remains to find an atlas on Σ' with holomorphic transition maps such that the obvious projection $\phi : \Sigma \to \Sigma'$ is holomorphic. For $z_0 \in \Sigma \setminus X$ the required coordinate chart near z_0 is simply the restriction of a holomorphic coordinate chart on Σ to a sufficiently small neighbourhood of z_0. Now let $z_0 \in X$ and denote $m_0 := m(z_0)$. Then, for each z near z_0, there are m_0 distinct points near z_0 that are equivalent to z (including the point z itself). All the other points in the equivalence class of z are close to other points of X. Choose a neighbourhood $U_0 \subset \Sigma$ of z_0 that is invariant under this local part of the equivalence relation. Then, for every $z \in U_0 \setminus \{z_0\}$, the intersection of U_0 with the equivalence class of z consists of precisely m_0 points. Shrinking U_0, if necessary, we may assume that there exists a holomorphic coordinate chart $w : U_0 \to \mathbb{C}$ such that

$$w(z_0) = 0.$$

Then the required coordinate chart on Σ' is the map $w' : U_0/\sim =: U_0' \to \mathbb{C}$ given by

$$w'([z]) := \prod_{\zeta \in U_0,\, \zeta \sim z} w(\zeta).$$

More precisely, by a standard lifting argument and the removable singularity theorem for holomorphic maps of one complex variable, there exist m_0 holomorphic diffeomorphisms $\psi_j : U_0 \to U_0$, $j = 1, \dots, m_0$, such that $\psi_1 = \mathrm{id}$, $\psi_j(z_0) = z_0$ for every j, and

$$z \sim \zeta \quad \iff \quad \zeta \in \{\psi_1(z), \dots, \psi_{m_0}(z)\}$$

for $z, \zeta \in U_0$. With this notation the map $w' : U_0' \to \mathbb{C}$ is given by

$$w'([z]) = \prod_{j=1}^{m_0} w \circ \psi_j(z).$$

Thus the composition of the projection $U_0 \to U_0'$ with $w' : U_0' \to \mathbb{C}$ is a holomorphic map with a zero of order m_0 at z_0. Hence a nonzero element in the image of w' has precisely m_0 preimages under this map, namely the elements of one equivalence class under the local equivalence relation on U_0. Hence $w' : U_0' \to \mathbb{C}$ is a homeomorphism onto its image. That the transition maps are holomorphic follows from the property (ii) above of the equivalence relation Γ_0.

By construction, $\phi : \Sigma \to \Sigma'$ is a holomorphic map of degree m such that

$$z \sim \zeta \quad \iff \quad \phi(z) = \phi(\zeta). \tag{2.5.1}$$

Hence there exists a unique map $u' : \Sigma' \to M$ such that $u = u' \circ \phi$. The map u' is obviously continuous on Σ' and J-holomorphic on the complement of the finite set $\phi(X)$. Its energy near every point in $\phi(X)$ is finite and hence it follows from the removable singularity theorem for J-holomorphic curves (see Theorem 4.1.2 below) that u' is smooth. Since $\deg(\phi) = m \geq 2$ it follows that u is not simple. □

REMARK 2.5.2. Assume $\Sigma = \mathbb{C}P^1$ in the proof of Proposition 2.5.1. Then Σ' is isomorphic to $\mathbb{C}P^1$ and the holomorphic function $\phi : \Sigma \to \Sigma'$ corresponds to a rational function, still denoted by $\phi : \mathbb{C}P^1 \to \mathbb{C}P^1$, under this isomorphism. Here is a direct argument for the construction of the rational function from the holomorphic equivalence relation Γ, so that (2.5.1) holds. Identify $\mathbb{C}P^1$ with the Riemann sphere $\mathbb{C} \cup \{\infty\}$ in such a way that $0 \not\sim \infty$ and denote

$$P_\infty := \{z \in \mathbb{C} \cup \{\infty\} \,|\, z \sim \infty\}, \qquad P_0 := \{z \in \mathbb{C} \cup \{\infty\} \,|\, z \sim 0\}.$$

Thus $\infty \in P_\infty$, $0 \in P_0$, and $0 \notin P_\infty$, $\infty \notin P_0$. Define $\phi : \mathbb{C} \setminus P_\infty \to \mathbb{C}$ by

$$\phi(z) := \prod_{\zeta \sim z} \zeta^{m(\zeta)}, \qquad z \in \mathbb{C} \setminus P_\infty. \tag{2.5.2}$$

Then ϕ is holomorphic and extends to a continuous function from $\mathbb{C} \cup \{\infty\}$ to itself. (Prove this!) Hence ϕ is a rational function. It has a zero of order $m(z)$ at each point $z \in P_0$ and no other zeros. Hence ϕ has degree m. By definition, it satisfies $z \sim \zeta \implies \phi(z) = \phi(\zeta)$. The converse follows from the fact that ϕ has degree m. Hence ϕ satisfies (2.5.1).

EXERCISE 2.5.3. Let $u : \mathbb{C} \cup \{\infty\} \to \mathbb{C} \cup \{\infty\}$ be a nonconstant rational function such that $u(0) \neq u(\infty)$. Denote by $\sim$ the equivalence relation on the Riemann sphere defined by

$$z \sim \zeta \quad \iff \quad u(z) = u(\zeta).$$

Denote by $m : \mathbb{C} \cup \{\infty\} \to \mathbb{N}$ the function that assigns to each pole z of u the order of z as a pole and to each other point z the order of z as a zero of $u - u(z)$. Let ϕ be given by (2.5.2). Show that u is the composition of ϕ with a Möbius transformation.

COROLLARY 2.5.4. *Let J be a C^2 almost complex structure on M, Σ_0, Σ_1 be compact connected Riemann surfaces without boundary, and $u_j : \Sigma_j \to M$ be simple J-holomorphic curves such that*

$$u_0(\Sigma_0) = u_1(\Sigma_1).$$

Then there exists a holomorphic diffeomorphism $\phi : \Sigma_1 \to \Sigma_0$ such that

$$u_1 = u_0 \circ \phi.$$

PROOF. For $i = 0, 1$ let $Z_i \subset \Sigma_i$ be the set of noninjective points of u_i. Thus Z_i consists of the critical points of u_i together with points $z \in \Sigma_i$ such that $u_i^{-1}(u_i(z))$ contains points other than z. In particular, every preimage of a critical value of u_i that is not itself a critical point satisfies the second condition and so belongs to Z_i. Hence

$$Z_0 = u_0^{-1}(u_0(Z_0)), \qquad Z_1 = u_1^{-1}(u_1(Z_1)).$$

By Proposition 2.5.1 and Lemma 2.4.1, Z_i is at most countable and can accumulate only at its finite subset consisting of critical points of u_i. Since $u_0(\Sigma_0) = u_1(\Sigma_1)$, there exists a unique bijection

$$\phi : \Sigma_1 \setminus u_1^{-1}\big(u_0(Z_0) \cup u_1(Z_1)\big) \to \Sigma_0 \setminus u_0^{-1}\big(u_0(Z_0) \cup u_1(Z_1)\big)$$

such that $u_1 = u_0 \circ \phi$. Since $du_0(\phi(z_1))$ and $du_1(z_1)$ are nonzero and complex linear for every z_1 in the domain of ϕ, it follows from the chain rule that ϕ is a holomorphic diffeomorphism. Moreover, by Lemma 2.4.1, the set $u_1^{-1}(u_0(Z_0) \cup u_1(Z_1))$ is at most countable and can accumulate only at finitely many points. Since ϕ is bounded when expressed in suitable local coordinates, the removable singularity theorem for holomorphic functions of one variable implies that ϕ extends to a holomorphic diffeomorphism from Σ_1 to Σ_0. This proves Corollary 2.5.4. □

COROLLARY 2.5.5. *Let J be a C^2 almost complex structure on M, $\Sigma_0, \ldots, \Sigma_N$ be compact connected Riemann surfaces without boundary, and $u_j : \Sigma_j \to M$ be J-holomorphic curves for $j = 0, \ldots, N$ such that u_0 is simple and*

$$u_0(\Sigma_0) \neq u_j(\Sigma_j) \qquad \textit{for } j > 0.$$

Then, for every $z_0 \in \Sigma_0$ and every open neighbourhood $U_0 \subset \Sigma_0$ of z_0, there exists an annulus $A_0 \subset U_0$ centered at z_0 such that $u_0 : A_0 \to M$ is an embedding and

$$u_0^{-1}(u_0(A_0)) = A_0, \qquad u_0(A_0) \cap u_j(\Sigma_j) = \emptyset \qquad \textit{for } j > 0.$$

PROOF. Let $Z_j := u_0^{-1}(u_j(\Sigma_j))$ for $j > 0$ and $Z_0 \subset \Sigma_0$ denote the set of noninjective points of u_0. By Proposition 2.5.1, the set Z_0 can only accumulate at the critical points of u_0 and, by Proposition 2.4.4, the same holds for the sets $Z_1, \ldots, Z_N$. Choose $A_0 \subset \Sigma_0$ to be any annulus in the complement of of the union $Z_0 \cup \cdots \cup Z_N$. Then A_0 satisfies the requirements of Corollary 2.5.5. □

REMARK 2.5.6. In many situations it is useful to consider J-holomorphic discs

$$u : (\mathbb{D}, \partial\mathbb{D}) \to (M, L)$$

with boundary on a Lagrangian submanifold L. For these, one needs a much more precise notion of a multiply covered object since a disc can wrap partially around itself: think of a disc in S^2 with boundary on the equator that wraps one and a half times around the sphere. The situation here was finally clarified by Lazzarini [**234, 235**]. See also Kwon–Oh [**220**] and Zehmisch [**425**].

2.6. The adjunction inequality

We close this chapter with a discussion of the adjunction inequality for J-holomorphic curves in almost complex four-manifolds. The proof will be given in Appendix E. As a warmup the following exercise examines intersections of J-holomorphic curves with complex submanifolds of real codimension two.

EXERCISE 2.6.1. Suppose that Q is a compact codimension two submanifold of (M, J) that is J-holomorphic in the sense that $JTQ = TQ$, and let $u : B_1 \to (M, J)$ be a J-holomorphic curve such that $u(0) \in Q$.

(i) Show that this intersection point is isolated except in the case $u(B_1) \subset Q$. *Hint:* Find suitable local coordinates as in Lemma 2.4.2 and then imitate the proof of Lemma 2.4.3.

(ii) Suppose $u(B_1) \not\subset Q$. Shrinking B_1, if necessary, we may assume that $u(0)$ is the unique point where $u(B_1)$ meets Q. Define the local intersection number $u \cdot Q$ of u with Q to be the number of points of intersection of v with Q (counted with multiplicities), where $v : B_1 \to Q$ is a generic perturbation of u through smooth maps that equal u on ∂B_1. Show that $u \cdot Q \geq 1$ with equality if and only if u is transverse to Q at zero.

The above exercise is the easiest case of *positivity of intersections*, a phenomenon first noticed by Gromov in [**160**]. It is especially important in the 4-dimensional case since one can get information about the singularities of a curve by looking at its intersections both with other curves and with itself. Exercise 2.6.1 shows that an intersection point of two J-holomorphic curves in dimension four contributes positively to the intersection number provided that it is nonsingular on one of the curves. A rather nontrivial refinement of this argument shows that every intersection point contributes positively and that every singular point on a curve C contributes positively to the (local) self-intersection number $C \cdot C$. The latter result leads to the adjunction formula in Theorem 2.6.4 below. The former result is stated in Theorem 2.6.3. Both theorems are proved with the same methods, but neither implies the other. The proofs will be given in Appendix E.

DEFINITION 2.6.2. *Let (M, J) be an almost complex 4-manifold and for $i = 0, 1$ let (Σ_i, j_i) be a closed Riemann surface and $u_i : \Sigma_i \to M$ a J-holomorphic curve. A* **transverse intersection of** u_0 **and** u_1 *is a pair $(z_0, z_1) \in \Sigma_0 \times \Sigma_1$ such that*

$$u_0(z_0) = u_1(z_1) =: x, \qquad T_xM = \operatorname{im} du_0(z_0) \oplus \operatorname{im} du_1(z_1). \tag{2.6.1}$$

In the case $\Sigma_0 = \Sigma_1 = \Sigma$ and $u_0 = u_1 = u$ such a pair (z_0, z_1) is called a **transverse self-intersection of** u.

The number of all intersection points will be denoted by

$$\delta(u_0, u_1) := \# \{(z_0, z_1) \in \Sigma_0 \times \Sigma_1 \,|\, u_0(z_0) = u_1(z_1)\}$$

and the number of all self-intersections of a curve u by

$$\delta(u) := \tfrac{1}{2} \# \{(z_0, z_1) \in \Sigma \times \Sigma \,|\, z_0 \neq z_1,\, u(z_0) = u(z_1)\}.$$

This number is evidently infinite whenever u is not simple and likewise $\delta(u_0, u_1)$ is infinite when the union of the curves is not simple. That these numbers are finite in the simple case is not obvious from the definition. In that case Lemma 2.4.3 shows that intersection points in $\Sigma_0 \times \Sigma_1$ can only accumulate at pairs of singular points of the two curves. However, Theorems 2.6.3 and 2.6.4 below give topological upper bounds for the numbers $\delta(u_0, u_1)$ and $\delta(u)$ in the simple case. We denote by $c_1(A) = \langle c_1(TM, J), A \rangle$ the first Chern number of a two-dimensional homology class A, by $A_0 \cdot A_1$ the intersection number of two classes A_0 and A_1, and by $\chi(\Sigma)$ the Euler characteristic of a closed Riemann surface Σ. The Riemann surfaces in the next two theorems are closed but not necessarily connected.

THEOREM 2.6.3 (Positivity of intersections). *Let (M, J) be an almost complex 4-manifold and $A_0, A_1 \in H_2(M; \mathbb{Z})$ be homology classes that are represented by simple J-holomorphic curves $u_0 : \Sigma_0 \to M$ and $u_1 : \Sigma_1 \to M$, respectively. Suppose that the union of the two curves is simple (i.e. $u_0(U_0) \neq u_1(U_1)$ for any two nonempty open subsets $U_0 \subset \Sigma_0$ and $U_1 \subset \Sigma_1$). Then*

$$\delta(u_0, u_1) \leq A_0 \cdot A_1$$

with equality if and only if all intersections are transverse.

PROOF. See Sections E.1 and E.2. □

THEOREM 2.6.4 (Adjunction inequality). *Let (M, J) be an almost complex 4-manifold and $A \in H_2(M; \mathbb{Z})$ be a homology class that is represented by a simple J-holomorphic curve $u : \Sigma \to M$. Then*

$$2\delta(u) - \chi(\Sigma) \leq A \cdot A - c_1(A). \tag{2.6.2}$$

with equality if and only if u is an immersion with only transverse self-intersections.

PROOF. See Sections E.1 and E.2. □

An immediate consequence is that that it is impossible for a sequence of embedded J-holomorphic curves $u_\nu : \Sigma \to M^4$ to converge to a simple curve $u : \Sigma \to M^4$ that has singular points or self-intersections.

EXAMPLE 2.6.5. Let J_0 be the standard complex structure on $\mathbb{C}P^2$ and Σ be a connected Riemann surface of genus g. Then a holomorphic curve $u : \Sigma \to \mathbb{C}P^2$ of degree d represents a class A such that $A \cdot A = d^2$ and $c_1(A) = 3d$. Hence the adjunction inequality has the form

$$\delta + g \leq \frac{(d-1)(d-2)}{2}.$$

A disconnected example is a union of d distinct lines in $\mathbb{C}P^2$. In this case Σ has d components, each of genus zero, and so $\chi(\Sigma) = 2d$. Since $2\delta = d(d-1)$, the two sides are equal. If two of the lines agree then $2\delta = d(d-1) - 2$ and so the inequality (2.6.2) still holds (and is strict) even though u is not simple.

EXERCISE 2.6.6. Let (M, J) be an almost complex 4-manifold such that every J-holomorphic curve in M has positive self-intersection number. Show that the inequality $c_1(A) - A \cdot A \leq \chi(\Sigma)$ continues to hold for classes represented by nonsimple curves and is strict unless each J-holomorphic curve $u : \Sigma \to M$ representing the class A is embedded. Find an example of a nonsimple curve with $A \cdot A < 0$ which violates this inequality.

EXERCISE 2.6.7. Show that the assertion of Theorem 2.6.3 follows from Theorem 2.6.4 whenever u_0 and u_1 are mutually transverse immersions with only transverse self-intersections.

The adjunction inequality for J-holomorphic curves was first stated by McDuff in [**256**]. Micallef pointed out that a mistaken choice of coordinates in one lemma affected many of the subsequent details. A corrected and simplified version appears in McDuff [**264**]. The argument shows that if z is a critical point of u then any local J-holomorphic perturbation of u that is immersed must have at least one double point. Alternatively one can analyse the topological type of the singularities of the image curve $u(\Sigma)$. (Here there is no need to restrict to curves in 4-dimensional spaces.) This approach was attempted in McDuff [**261**], but unfortunately that paper contains some serious errors and the proof of Proposition 2.8 is not valid. Even if it were corrected, the method is basically topological and would only classify the germs up to C^0-equivalence. In [**287**] Micallef and White succeeded in showing that the germ of any singularity on a J-curve is C^1-equivalent to a standard holomorphic germ, as a special case of a much more general result on singularities of minimal surfaces. Sikorav gives a more direct treatment of these questions in [**379**], using the weakest possible smoothness assumptions on J. In Appendix E Lazzarini describes a different version of the Micallef–White argument that is valid for smooth J-holomorphic curves.

A related question that has attracted attention is whether any symplectically embedded closed 2-manifold in $\mathbb{C}P^2$ is symplectically isotopic to a complex curve. See Shevchishin [**375**] for some progress on this problem and a list of references.

CHAPTER 3

Moduli Spaces and Transversality

In this chapter we examine the moduli space of J-holomorphic curves in a compact symplectic manifold (M, ω) and prove, under suitable hypotheses, that this space is a smooth finite dimensional manifold for a generic almost complex structure J. We shall consider holomorphic curves defined on a compact Riemann surface Σ with a fixed complex structure j_Σ.

It is important to distinguish two cases. To achieve transversality for all (nonconstant) J-holomorphic curves it is necessary either to allow for families of almost complex structures that are parametrized by Σ or to allow for suitable Hamiltonian perturbations. If the almost complex structure is independent of the base point in Σ, then transversality can only be achieved for simple J-holomorphic curves. In this Chapter we restrict to the case of simple curves; the more general case is discussed in Section 6.7.

In Section 3.1 we introduce the moduli spaces of simple curves and discuss the main results for this case. Section 3.2 deals with Thom-Smale transversality and establishes the relevant transversality theorems for simple curves and generic almost complex structures. It follows that the moduli space is a smooth manifold whenever the linearized operator is surjective for every J-holomorphic curve. We also prove that the moduli space carries a natural orientation. Section 3.3 describes some easy examples in dimension four, establishing surjectivity of the linearized operator directly, and for all almost complex structures. In this special case the argument relies on the Riemann–Roch theorem (see Appendix C); the Sard–Smale theorem is not required.

The last two sections can be omitted at first reading. Section 3.4 concerns moduli spaces of curves that satisfy pointwise constraints. This will be relevant to the discussion in Chapter 6 of transversality for moduli spaces of stable maps. In Section 3.5 we prove a quantitative version of the implicit function theorem for the present context which establishes the existence of J-holomorphic curves near approximate ones under suitable hypotheses. The result is formulated in a form that will be needed in the gluing argument, with uniform estimates for a suitable class of volume forms on Σ.

Although we do not explicitly develop the properties of curves satisfying Lagrangian boundary conditions, we do lay many of the foundations needed to understand this case. Relevant discussions and applications may be found in Section 9.2; cf. in particular Remark 9.2.3.

3.1. Moduli spaces of simple curves

In this section we formulate the main results in the case of simple J-holomorphic spheres, with proofs given in Section 3.2. These two sections show the structure of the main regularity argument. The properties of simple curves are discussed in Section 2.5.

Let (M,ω) be a compact symplectic $2n$-manifold. Fix a compact Riemann surface $(\Sigma, j_\Sigma, \mathrm{dvol}_\Sigma)$ and an ω-tame almost complex structure $J \in \mathcal{J}_\tau(M,\omega)$. We consider the equation

$$\bar{\partial}_J(u) = 0, \tag{3.1.1}$$

where

$$\bar{\partial}_J(u) := \tfrac{1}{2}(du + J \circ du \circ j_\Sigma).$$

Given a homology class $A \in H_2(M;\mathbb{Z})$, the moduli space of solutions of (3.1.1) that represent the class A will be denoted by

$$\mathcal{M}(A,\Sigma;J) := \{u \in C^\infty(\Sigma, M) \mid J \circ du = du \circ j_\Sigma,\ [u] = A\}.$$

For the subspace of simple solutions we write

$$\mathcal{M}^*(A,\Sigma;J) := \{u \in \mathcal{M}(A,\Sigma;J) \mid u \text{ is simple}\}.$$

In the case $\Sigma = \mathbb{C}P^1$ we abbreviate

$$\mathcal{M}(A;J) := \mathcal{M}(A,\mathbb{C}P^1;J), \qquad \mathcal{M}^*(A;J) := \mathcal{M}^*(A,\mathbb{C}P^1;J).$$

The moduli space $\mathcal{M}(A,\Sigma;J)$ can be interpreted as the zero set of a section of an infinite dimensional vector bundle as follows. Let $\mathcal{B} \subset C^\infty(\Sigma, M)$ denote the space of all smooth maps $u : \Sigma \to M$ that represent the homology class $A \in H_2(M;\mathbb{Z})$. This space can be thought of as an infinite dimensional (Fréchet) manifold whose tangent space at $u \in \mathcal{B}$ is the space

$$T_u\mathcal{B} = \Omega^0(\Sigma, u^*TM)$$

of all smooth vector fields $\xi(z) \in T_{u(z)}M$ along u. Consider the infinite dimensional vector bundle $\mathcal{E} \to \mathcal{B}$ whose fiber at u is the space

$$\mathcal{E}_u = \Omega^{0,1}(\Sigma, u^*TM)$$

of smooth J-antilinear 1-forms on Σ with values in u^*TM. Then the complex antilinear part of du defines a section $\mathcal{S} : \mathcal{B} \to \mathcal{E}$ of this vector bundle:

$$\mathcal{S}(u) := (u, \bar{\partial}_J(u)).$$

The moduli space $\mathcal{M}(A,\Sigma;J)$ is the zero set of this section. The subset $\mathcal{M}^*(A,\Sigma;J)$ is the intersection of $\mathcal{M}(A,\Sigma;J)$ with the open subset $\mathcal{B}^* \subset \mathcal{B}$ of all smooth maps $u : \Sigma \to M$ that represent the class A and are somewhere injective.

The operator D_u. Our goal in this book is to explain the construction of the Gromov–Witten invariants for a suitable class of symplectic manifolds. The first step in this program is to prove that, for a generic almost complex structure J, the moduli spaces $\mathcal{M}^*(A,\Sigma;J)$ are smooth finite dimensional manifolds of the appropriate dimensions. To establish this we must show that $\mathcal{S}$ is transverse to the zero section. This means that the image of the differential $d\mathcal{S}(u) : T_u\mathcal{B} \to T_{(u,0)}\mathcal{E}$ is complementary to the tangent space $T_u\mathcal{B}$ of the zero section for every $u \in \mathcal{M}^*(A,\Sigma;J)$. Given $u \in \mathcal{M}^*(A,\Sigma;J)$ let us denote by

$$D_u := D_{J,u} := D\mathcal{S}(u) : \Omega^0(\Sigma, u^*TM) \to \Omega^{0,1}(\Sigma, u^*TM)$$

the composition of the differential $d\mathcal{S}(u) : T_u\mathcal{B} \to T_{(u,0)}\mathcal{E}$ with the projection

$$\pi_u : T_{(u,0)}\mathcal{E} = T_u\mathcal{B} \oplus \mathcal{E}_u \to \mathcal{E}_u.$$

The operator D_u is called the **vertical differential** of the section $\mathcal{S}$ at u. Transversality can now be expressed in the form that the linearized operator D_u is surjective for every $u \in \mathcal{M}^*(A, \Sigma; J)$.

In local coordinates on Σ and M, a J-holomorphic curve is a solution $u : \mathbb{C} \to \mathbb{R}^{2n}$ of the equation

$$\partial_s u + J(u)\partial_t u = 0$$

and a vector field along u is a map $\xi : \mathbb{C} \to \mathbb{R}^{2n}$. A local formula for $D_u\xi$ is obtained by differentiating the left hand side of this equation in the direction ξ. Thus

$$D_u\xi = \eta\, ds - J(u)\eta\, dt, \qquad \eta := \tfrac{1}{2}\big(\partial_s\xi + J(u)\partial_t\xi + \partial_\xi J(u)\partial_t u\big).$$

Using the fact that u is J-holomorphic we find that

$$D_u\xi = \bar{\partial}_J\xi - \tfrac{1}{2}\big(J\partial_\xi J\big)(u)\partial_J(u) \tag{3.1.2}$$

Note that the second term on the right hand side is an antiholomorphic 1-form because

$$\partial_J(u) := \tfrac{1}{2}(du - J \circ du \circ j)$$

is a holomorphic 1-form and J anticommutes with $J(\partial_\xi J)$. The above discussion shows that the operator defined by formula (3.1.2) is independent of the choice of local coordinates. The direct verification of this is left to the reader.

It follows from formula (3.1.2) that D_u is a real linear Cauchy–Riemann operator in the sense of Appendix C and so is Fredholm. This information about D_u is enough to establish most of the results of this Chapter. However, the implicit function theorem in Section 3.5 and the gluing theorem of Chapter 10 rely on sophisticated estimates which are based on a thorough understanding of the operator D_u. Therefore it is worth while to spend some effort in developing concise formulas for D_u.

The definition of D_u can be extended to general smooth maps $u : \Sigma \to M$, J-holomorphic or not, but it now depends on a choice of splitting of the tangent space $T_{(u,\bar{\partial}_J(u))}\mathcal{E}$ into horizontal and vertical subspaces. Such a splitting depends on a connection on TM. We must assume that this connection preserves the almost complex structure J so that the fibers of $\mathcal{E}$ are invariant under pointwise parallel transport.[1] We shall work with the complex linear connection

$$\widetilde{\nabla}_v X := \nabla_v X - \tfrac{1}{2}J(\nabla_v J)X$$

induced by the Levi-Civita connection ∇ of the metric (2.1.1).

Given $\xi \in \Omega^0(\Sigma, u^*TM)$, let

$$\Phi_u(\xi) : u^*TM \to \exp_u(\xi)^*TM$$

denote the complex bundle isomorphism, given by parallel transport with respect to $\widetilde{\nabla}$ along the geodesics $s \mapsto \exp_{u(z)}(s\xi(z))$. (Here, as always, we take geodesics with respect to ∇.) Then define the map

$$\mathcal{F}_u : \Omega^0(\Sigma, u^*TM) \to \Omega^{0,1}(\Sigma, u^*TM)$$

by

$$\mathcal{F}_u(\xi) := \Phi_u(\xi)^{-1}\bar{\partial}_J(\exp_u(\xi)). \tag{3.1.3}$$

---

[1] A section $\alpha_\lambda \in \Omega^{0,1}(\Sigma, u_\lambda^*TM)$ of $\mathcal{E}$ along a smooth curve $\mathbb{R} \to \mathcal{B} : \lambda \mapsto u_\lambda$ is called **parallel** with respect to a connection $\widetilde{\nabla}$ on TM if the vector field $\lambda \mapsto \alpha_\lambda(z;\zeta) \in T_{u_\lambda(z)}M$ along the curve $\lambda \mapsto u_\lambda(z)$ is parallel for every $\zeta \in T_z\Sigma$. If $\xi(\lambda) \in T_{x(\lambda)}M$ is any vector field along a curve in M, we write $\widetilde{\nabla}_\lambda\xi$ for the corresponding covariant derivative.

This map is precisely the vertical part of the section $\mathcal{S} : \mathcal{B} \to \mathcal{E}$ with respect to the trivialization determined by $\widetilde{\nabla}$. It follows that, for every J-holomorphic curve $u : \Sigma \to M$, the differential of $\mathcal{F}_u$ at zero is the operator D_u defined above. For general maps u we shall use this derivative as the definition of D_u. The next proposition gives an explicit formula.

PROPOSITION 3.1.1. *For any smooth map $u : \Sigma \to M$ define the operator*

$$D_u : \Omega^0(\Sigma, u^*TM) \to \Omega^{0,1}(\Sigma, u^*TM)$$

by

$$D_u\xi := d\mathcal{F}_u(0)\xi.$$

Then

$$D_u\xi = \tfrac{1}{2}\Big(\nabla\xi + J(u)\nabla\xi \circ j_\Sigma\Big) - \tfrac{1}{2}J(u)(\nabla_\xi J)(u)\partial_J(u). \tag{3.1.4}$$

for every $\xi \in \Omega^0(\Sigma, u^*TM)$.

PROOF. Consider the path $\mathbb{R} \to C^\infty(\Sigma, M) : \lambda \mapsto u_\lambda := \exp_u(\lambda\xi)$. Then, by definition, $\Phi_u(\lambda\xi)\mathcal{F}_u(\lambda\xi) = \bar\partial_J(u_\lambda)$. Since $\Phi_u(\lambda\xi)$ is given by parallel transport along the geodesic $\lambda \mapsto u_\lambda(z)$, it follows that

$$\begin{aligned}
D_u\xi &= \frac{d}{d\lambda}\mathcal{F}_u(\lambda\xi)\Big|_{\lambda=0} \\
&= \widetilde{\nabla}_\lambda \bar\partial_J(u_\lambda)\Big|_{\lambda=0} \\
&= \tfrac{1}{2}\Big(\widetilde{\nabla}_\lambda du_\lambda + J(u_\lambda)\widetilde{\nabla}_\lambda du_\lambda \circ j_\Sigma\Big)\Big|_{\lambda=0} \\
&= \tfrac{1}{2}\Big(\nabla_\lambda du_\lambda + J(u_\lambda)\nabla_\lambda du_\lambda \circ j_\Sigma\Big)\Big|_{\lambda=0} \\
&\quad - \tfrac{1}{4}\Big(J(u_\lambda)(\nabla_{\partial_\lambda u_\lambda}J)(u_\lambda)du_\lambda - (\nabla_{\partial_\lambda u_\lambda}J)(u_\lambda)du_\lambda \circ j_\Sigma\Big)\Big|_{\lambda=0} \\
&= \tfrac{1}{2}\Big(\nabla\xi + J(u)\nabla\xi \circ j_\Sigma\Big) - \tfrac{1}{2}J(u)\left(\nabla_\xi J\right)(u)\partial_J(u).
\end{aligned}$$

To obtain the last equality we used the fact that ∇ is torsion free. This implies that $\nabla_\lambda\partial_t\gamma = \nabla_t\partial_\lambda\gamma$ for every smooth map $\mathbb{R}^2 \to M : (\lambda, t) \mapsto \gamma(\lambda, t)$, and in particular for the map $\gamma(\lambda, t) = u_\lambda(z(t))$. □

REMARK 3.1.2 (NATURALITY). It is sometimes convenient to indicate the dependence of the operator D_u on the complex structure $j = j_\Sigma$ in the notation and write $D_{j,u} := D_u$. Then the map $(j, u, \xi) \mapsto D_{j,u}\xi$ is equivariant under the action of the diffeomorphism group of Σ, i.e.

$$D_{\phi^*j,\phi^*u}\phi^*\xi = \phi^*(D_{j,u}\xi) \tag{3.1.5}$$

for every smooth map $u : \Sigma \to M$, every vector field $\xi \in \Omega^0(\Sigma, u^*TM)$ along u, every (almost) complex structure j on Σ, and every diffeomorphism $\phi : \Sigma \to \Sigma$. Here we denote $\phi^*u := u\circ\phi$, $\phi^*\xi := \xi\circ\phi$ and $\phi^*\alpha = \alpha\circ d\phi$ for $\alpha \in \Omega^{0,1}(\Sigma, u^*TM)$. In particular, if the complex structure j is fixed and ϕ is a holomorphic diffeomorphism of Σ, we have

$$D_{\phi^*u}\phi^*\xi = \phi^*(D_u\xi).$$

EXERCISE: Prove that $\mathcal{F}_{\phi^*j,\phi^*u}(\phi^*\xi) = \phi^*\mathcal{F}_{j,u}(\xi)$ and deduce the formula (3.1.5). Prove naturality under the action of the diffeomorphism group of M.

REMARK 3.1.3. If J is ω-compatible then the operator D_u can also be expressed in the form

$$D_u\xi = (\widetilde{\nabla}\xi)^{0,1} + \tfrac{1}{4}N_J(\xi, \partial_J(u)), \tag{3.1.6}$$

where N_J denotes the Nijenhuis tensor of J and $(\widetilde{\nabla}\xi)^{0,1}$ the $(0,1)$-part of the 1-form $\widetilde{\nabla}\xi \in \Omega^1(\Sigma, u^*TM)$. Thus

$$(\widetilde{\nabla}\xi)^{0,1} := \tfrac{1}{2}\left(\widetilde{\nabla}\xi + J(u)\widetilde{\nabla}\xi \circ j_\Sigma\right).$$

In the ω-tame case the complex linear part of D_u is determined by the connection

$$\widehat{\nabla}_v X := \widetilde{\nabla}_v X - \tfrac{1}{4}(\nabla_{JX}J)v - \tfrac{1}{4}J(\nabla_X J)v.$$

This connection preserves J, but not necessarily the metric, and its torsion is equal to minus a quarter of the Nijenhuis tensor. The operator D_u is now given by

$$D_u\xi = (\widehat{\nabla}\xi)^{0,1} + \tfrac{1}{4}N_J(\xi, \partial_J(u)) + \tfrac{1}{4}\left(J(\nabla_{\bar\partial_J(u)}J) + \nabla_{J\bar\partial_J(u)}J\right)\xi. \tag{3.1.7}$$

Note that the last three terms in (3.1.7) are complex antilinear while the first term is complex linear. Thus D_u splits as a sum of a complex linear Cauchy–Riemann operator and a zeroth order complex antilinear operator. This continues to hold for general real linear Cauchy–Riemann operators: see Exercise C.1.7. It is shown in Appendix C that every ω-compatible J satisfies

$$\nabla_{Jv}J = -J\nabla_v J$$

and so $\widehat{\nabla} = \widetilde{\nabla}$ and the last two terms in (3.1.7) vanish in this case. For a more detailed discussion see Lemmas C.7.2 and C.7.3.

EXERCISE 3.1.4. Prove that

$$D_u(X \circ u) = \tfrac{1}{2}(\mathcal{L}_X J)(u)du \circ j_\Sigma = -\tfrac{1}{2}J(u)(\mathcal{L}_X J)(u)\partial_J(u)$$

for every J-holomorphic curve $u : \Sigma \to M$ and every vector field $X \in \mathrm{Vect}(M)$. *Hint:* Let $\mathbb{R} \to \mathrm{Diff}(M) : t \mapsto \phi_t$ be the flow of X and denote

$$u_t := \phi_t^{-1} \circ u, \qquad J_t := \phi_t^* J.$$

Then u_t is a J_t-holomorphic curve for every t. Now differentiate the identity $\bar\partial_{J_t}(u_t) = 0$ covariantly with respect to the connection $\widetilde{\nabla}$ at $t = 0$. Alternatively, prove this by direct computation using the formula of Exercise 2.1.1.

So far we have assumed that u is smooth. However, the definition of the map $\mathcal{F}_u$, and hence of the operator D_u, extends to the case where u is of class $W^{k,p}$ for some integer $k \geq 1$ and some real number $p > 2$. In this case $\mathcal{F}_u$ is a map between the appropriate Sobolev completions:

$$\mathcal{F}_u : W^{k,p}(\Sigma, u^*TM) \to W^{k-1,p}(\Sigma, \Lambda^{0,1} \otimes_J u^*TM).$$

This map is as smooth as J and its differential D_u at zero is a bounded linear operator between the same two spaces. Here we denote by $W^{k,p}(M, E)$ the completion of the space $\Omega^0(M, E)$ of smooth sections of the bundle $E \to M$ with respect to the Sobolev $W^{k,p}$-norm. Moreover, $\Lambda^{0,1} := \Lambda^{0,1}T^*\Sigma$ denotes the bundle of 1-forms on Σ of type $(0,1)$. For more details about Sobolev spaces see Appendix B.

Outline of the main argument. The formula (3.1.4) shows that D_u is a real linear Cauchy–Riemann operator as explained in Appendix C. Hence, by Theorem C.1.10, D_u is a *Fredholm* operator. This means that it has a closed image and finite dimensional kernel and cokernel. The Fredholm index of such an operator is defined as the dimension of the kernel minus the dimension of the cokernel. It is stable under compact (lower order) perturbations by Theorem A.1.5 in Appendix A. In the case at hand the Riemann–Roch theorem C.1.10 asserts that the Fredholm index is

$$\operatorname{index} D_u = n(2-2g) + 2c_1(u^*TM),$$

where g is the genus of Σ. From an algebro-geometric point of view this formula can be understood as follows. In equation (3.1.7) we have seen that the complex antilinear part of D_u is a lower order (compact) perturbation and so can be removed without changing the Fredholm index. The leading term in D_u is the complex linear Cauchy–Riemann operator $\xi \mapsto (\widehat{\nabla}\xi)^{0,1}$ and so defines a holomorphic structure on the bundle u^*TM. The index is the Euler characteristic of the Dolbeault cohomology of this bundle and is given by the Riemann–Roch formula.

We next outline an argument which proves that $\mathcal{M}^*(A,\Sigma;J)$ is a manifold of the "correct dimension"

$$\dim \mathcal{M}^*(A,\Sigma;J) = n(2-2g) + 2c_1(u^*TM)$$

for a generic almost complex structure J. For this to hold, J must be allowed to vary in some space $\mathcal{J}$ of almost complex structures on M which is sufficiently large for the transversality argument in Proposition 3.2.1 to work. For example, $\mathcal{J}$ could be any subset of the space of all smooth structures which is open in the C^∞ topology; there is no need for M to be symplectic. The most important cases in the context of this book are those where M has a symplectic form ω and $\mathcal{J}$ is either the space of all ω-compatible or the space of all ω-tame almost complex structures (see Section 2.1). In all cases $\mathcal{J}$ carries the usual C^∞-topology. We write $\mathcal{J}^\ell$ for the corresponding space of C^ℓ almost complex structures.

The basic observation is that the **universal moduli space** of simple curves

$$\mathcal{M}^*(A,\Sigma;\mathcal{J}^\ell) := \big\{(u,J)\,|\,J\in\mathcal{J}^\ell,\ u\in\mathcal{M}^*(A,\Sigma;J)\big\}$$

is a separable Banach manifold when ℓ is sufficiently large (see Proposition 3.2.1). Moreover, the projection

$$\pi : \mathcal{M}^*(A,\Sigma;\mathcal{J}^\ell) \to \mathcal{J}^\ell$$

is a Fredholm map because its differential at a point (u,J) is essentially the operator D_u. In particular, π has the same index as D_u and is surjective at (u,J) precisely when D_u is onto. Because we are in a Banach manifold setting, we may apply the implicit function theorem (Theorem A.3.3) to conclude that $\mathcal{M}^*(A,\Sigma;J)$ is indeed a finite dimensional manifold whose tangent space at u is the kernel of D_u whenever J is a regular value of π. Moreover, by the Sard–Smale theorem (Theorem A.5.1), the set of regular values of π is residual (in the sense of Baire) in $\mathcal{J}^\ell$ for ℓ sufficiently large. Hence $\mathcal{M}^*(A,\Sigma;J)$ is a manifold of the right dimension for a generic element of $\mathcal{J}^\ell$.

An argument of Taubes allows us to show that this statement remains true for generic elements of the space $\mathcal{J}$ of smooth almost complex structures. It is based on the observation that the space of all simple J-holomorphic curves is a countable union of compact sets and for each of these compact sets the corresponding set of regular J is open. Hence the set of all regular J is a countable intersection of open sets $\mathcal{J}_{\mathrm{reg,K}}$ and the substance of the matter is to show that these sets are also dense. The argument outlined above does this by considering Banach manifolds of C^ℓ almost complex structures. In Remark 3.2.7 we shall explain an alternative approach due to Floer that remains within the C^∞ category.

The above statements are proved in Section 3.2. Here we sum up the main conclusions. First a definition.

DEFINITION 3.1.5. *Fix a compact Riemann surface* $(\Sigma, j_\Sigma, \mathrm{dvol}_\Sigma)$ *and a homology class* $A \in H_2(M;\mathbb{Z})$. *An almost complex structure* J *on* M *is called* **regular (for** A **and** Σ**)** *if* D_u *is onto for every* $u \in \mathcal{M}^*(A,\Sigma;J)$. *Given* $\mathcal{J}$ *as above, we denote by*

$$\mathcal{J}_{\mathrm{reg}}(A,\Sigma)$$

the set of all $J \in \mathcal{J}$ *that are regular for* A *and* Σ. *In the case* $\Sigma = S^2$ *we abbreviate*

$$\mathcal{J}_{\mathrm{reg}}(A) := \mathcal{J}_{\mathrm{reg}}(A, S^2).$$

THEOREM 3.1.6. *Assume* $\mathcal{J} = \mathcal{J}(M,\omega)$ *or* $\mathcal{J} = \mathcal{J}_\tau(M,\omega)$. *Fix a compact Riemann surface* $(\Sigma, j_\Sigma, \mathrm{dvol}_\Sigma)$ *and a homology class* $A \in H_2(M;\mathbb{Z})$.

(i) *If* $J \in \mathcal{J}_{\mathrm{reg}}(A,\Sigma)$ *then the space* $\mathcal{M}^*(A,\Sigma;J)$ *is a smooth manifold of dimension*

$$\dim \mathcal{M}^*(A,\Sigma;J) = n(2-2g) + 2c_1(A).$$

It carries a natural orientation.

(ii) *The set* $\mathcal{J}_{\mathrm{reg}}(A,\Sigma)$ *is residual in* $\mathcal{J}$. *This means that it contains an intersection of countably many open and dense subsets of* $\mathcal{J}$.

The next task is to discuss the dependence of the manifolds $\mathcal{M}^*(A,\Sigma;J)$ on the choice of $J \in \mathcal{J}_{\mathrm{reg}}(A,\Sigma)$. A **(smooth) homotopy** of almost complex structures is a smooth map $[0,1] \to \mathcal{J} : \lambda \to J_\lambda$. For any such homotopy define

$$\mathcal{W}^*(A,\Sigma;\{J_\lambda\}_\lambda) = \{(\lambda,u) \,|\, 0 \le \lambda \le 1,\, u \in \mathcal{M}^*(A,\Sigma;J_\lambda)\}.$$

Given $J_0, J_1 \in \mathcal{J}_{\mathrm{reg}}(A,\Sigma)$ denote by $\mathcal{J}(J_0,J_1)$ the space of all smooth homotopies of almost complex structures connecting J_0 to J_1. In general, even though $\mathcal{J}$ is path-connected, there does not exist a homotopy such that $J_\lambda \in \mathcal{J}_{\mathrm{reg}}(A,\Sigma)$ for every λ. In other words the space $\mathcal{M}^*(A,\Sigma;J_\lambda)$ may fail to be a manifold for some values of λ. However, we shall see that there is always a smooth homotopy such that the space $\mathcal{W}^*(A,\Sigma;\{J_\lambda\}_\lambda)$ is a manifold.

DEFINITION 3.1.7. *Fix a compact Riemann surface* $(\Sigma, j_\Sigma, \mathrm{dvol}_\Sigma)$ *and a homology class* $A \in H_2(M;\mathbb{Z})$. *Let* $J_0, J_1 \in \mathcal{J}_{\mathrm{reg}}(A,\Sigma)$. *A homotopy* $[0,1] \to \mathcal{J} : \lambda \to J_\lambda$ *from* J_0 *to* J_1 *is called* **regular (for** A **and** Σ**)** *if*

$$\Omega^{0,1}(\Sigma, u^*TM) = \mathrm{im}\, D_{J_\lambda,u} + \mathbb{R}\, v_\lambda$$

for every $(\lambda,u) \in \mathcal{W}^*(A,\Sigma;\{J_\lambda\}_\lambda)$ *where*

$$v_\lambda := (\partial_\lambda J_\lambda) du \circ j_\Sigma$$

is the image in $\Omega^{0,1}(\Sigma, u^*TM)$ *of the tangent vector to the path* $\lambda \mapsto J_\lambda$. *The space of regular homotopies will be denoted by* $\mathcal{J}_{\mathrm{reg}}(A,\Sigma;J_0,J_1)$.

We will see in Section 3.2 that a homotopy $\lambda \mapsto J_\lambda$ is regular precisely when it is transverse to the projection π. Intuitively speaking, one can think of the space $\mathcal{J}_{\mathrm{reg}}(A,\Sigma)$ of regular almost complex structures as the complement of a subvariety

$\mathcal{S}$ of codimension one in the space $\mathcal{J}$. A smooth homotopy $\lambda \mapsto J_\lambda$ is regular if it is transversal to $\mathcal{S}$.

THEOREM 3.1.8. *Assume $\mathcal{J} = \mathcal{J}(M,\omega)$ or $\mathcal{J} = \mathcal{J}_\tau(M,\omega)$. Fix a compact Riemann surface $(\Sigma, j_\Sigma, \mathrm{dvol}_\Sigma)$ and a homology class $A \in H_2(M;\mathbb{Z})$. Let $J_0, J_1 \in \mathcal{J}_{\mathrm{reg}}(A,\Sigma)$.*

(i) *If $\{J_\lambda\}_\lambda \in \mathcal{J}_{\mathrm{reg}}(A,\Sigma;J_0,J_1)$ then $\mathcal{W}^*(A,\Sigma;\{J_\lambda\}_\lambda)$ is a smooth oriented manifold with boundary*

$$\partial\mathcal{W}^*(A,\Sigma;\{J_\lambda\}_\lambda) = \mathcal{M}^*(A,\Sigma;J_0) \cup \mathcal{M}^*(A,\Sigma;J_1).$$

The boundary orientation agrees with the orientation of $\mathcal{M}^*(A,\Sigma;J_1)$ and is opposite to the orientation of $\mathcal{M}^*(A,\Sigma;J_0)$.

(ii) *The set $\mathcal{J}_{\mathrm{reg}}(A,\Sigma;J_0,J_1)$ is residual in the space of all smooth homotopies $[0,1] \to \mathcal{J} : \lambda \mapsto J_\lambda$ from J_0 to J_1.*

Theorem 3.1.8 shows that the moduli spaces $\mathcal{M}^*(A,\Sigma;J_0)$ and $\mathcal{M}^*(A,\Sigma;J_1)$ are oriented cobordant. However, until we establish some version of compactness this does not mean very much. This question is addressed in Chapter 4. The arguments that we use to establish both transversality and compactness rely on the basic observations about elliptic regularity explained below, but then they diverge and become essentially independent of each other. Nevertheless, in order to understand our notation and to have some perspective on Sobolev spaces, we recommend that the reader who is unfamiliar with these topics continue until the end of the next section before moving on to consider the question of compactness.

REMARK 3.1.9. Perhaps this is a good place to put in a word of warning. In our discussions the word "regular" is used in two rather different ways. In the first (as in *elliptic regularity*), one is talking about the degree of smoothness of a map, for example exactly which $W^{k,p}$ space of functions contains a given map $u : \Sigma \to M$. In the other (as in *regular* J), we are concerned with the question of whether J is a regular, rather than critical, value of the projection map π. People often talk about generic instead of regular J. The difficulty with this terminology is that the word generic has other connotations; for example it seems contradictory to talk about the product almost complex structure on $S^2 \times S^2$ as being generic, since it is so special. On the other hand it is regular for all genus zero curves.

Elliptic regularity. In preparation for the proofs of Theorems 3.1.6 and 3.1.8 we shall first introduce the appropriate Sobolev space framework and discuss the relevant elliptic regularity theorems. The proofs are given in the Appendices B and C.

Later on it will be important to consider almost complex structures of class C^ℓ, rather than smooth ones, in order to obtain a parameter space with a Banach manifold structure. Hence we shall consider the space

$$\mathcal{J}^\ell = \mathcal{J}^\ell_\tau(M,\omega)$$

of all ω-tame almost complex structures of class C^ℓ. We shall assume below that $\ell \geq 2$ so that the metric (2.1.1) determined by ω and J is at least of class C^2 and the results of Section 2.5 apply. However, the regularity results discussed in the present section also hold for $\ell = 1$.

For a real number $p > 2$ and an integer $k \geq 1$ we shall denote by

$$\mathcal{B}^{k,p} := \{u \in W^{k,p}(\Sigma, M) \,|\, [u] = A\}$$

the space of continuous maps $\Sigma \to M$ whose k-th derivatives are of class L^p and which represent the class $A \in H_2(M;\mathbb{Z})$. As explained in more detail in Appendix B, the space $W^{k,p}(\Sigma, M)$ is the completion of $C^\infty(\Sigma, M)$ with respect to a distance function based on the Sobolev $W^{k,p}$-norm. This norm is defined as the sum of the L^p-norms of all derivatives up to order k. The corresponding metric on $C^\infty(\Sigma, M)$ can be defined by embedding M into some Euclidean space $\mathbb{R}^N$ and then using the Sobolev norm on the ambient space $W^{k,p}(\Sigma, \mathbb{R}^N)$. Alternatively, one can use a Riemannian metric on M and covariant derivatives to define the $W^{k,p}$-norm on the tangent space $T_uC^\infty(\Sigma, M) = \Omega^0(\Sigma, u^*TM)$ and then minimize the length of paths in $C^\infty(\Sigma, M)$ with fixed endpoints. Since Σ and M are compact any two such $W^{k,p}$ distance functions are compatible, and so the resulting completion does not depend on the choices. The completion $W^{k,p}(\Sigma, M)$ is a smooth separable Banach manifold with tangent space

$$T_uW^{k,p}(\Sigma, M) = W^{k,p}(\Sigma, u^*TM).$$

Local coordinate charts can be defined by using the exponential map along smooth maps $u : \Sigma \to M$: see the proof of Proposition 3.2.1 below. The space $\mathcal{B}^{k,p}$ is a component of $W^{k,p}(\Sigma, M)$ and hence is also a smooth separable Banach manifold.

In order for the space $\mathcal{B}^{k,p}$ to be well defined we must assume that

$$kp > 2.$$

There are various reasons for this. Firstly, the very definition of $\mathcal{B}^{k,p}$, in terms of local coordinate representations of class $W^{k,p}$, requires this assumption. The point is this: the $W^{k,p}$-norm is well-defined for maps between open sets in Euclidean space, but for a general manifold one needs a space which is invariant under composition with coordinate charts. Now the composition of a $W^{k,p}$-map $u : \mathbb{R}^2 \to \mathbb{R}^{2n}$ with a C^k-map in the target is again of class $W^{k,p}$ precisely when $kp > 2$. Secondly, we will often use the Sobolev embedding theorem which says that under this condition the space $\mathcal{B}^{k,p}$ embeds into the space of continuous maps from $\Sigma \to M$. Moreover, by Rellich's theorem this embedding is compact. Thirdly, the condition $kp > 2$ is required to show that the product of two $W^{k,p}$-functions is again of this class. In other words, the condition $kp > 2$ is needed to deal with the nonlinearities.

The first key observation is that when J is smooth every J-holomorphic curve of class $W^{1,p}$ with $p > 2$ is necessarily smooth. More precisely, we have the following regularity theorem which can be proved by elliptic bootstrapping methods. The details are carried out in Appendix B (see Theorem B.4.1 and Remark B.4.3).

PROPOSITION 3.1.10 (Elliptic regularity). *Assume J is an almost complex structure of class C^ℓ with $\ell \geq 1$. If $u : \Sigma \to M$ is a J-holomorphic curve of class $W^{1,p}$ with $p > 2$, then u is of class $W^{\ell+1,p}$. In particular, u is of class C^ℓ, and if J is smooth (C^∞) then so is u.*

Proposition 3.1.10 shows that, for $J \in \mathcal{J}^\ell$, the moduli space of J-holomorphic curves of class $W^{k,p}$ is independent of the choice of k so long as $k \leq \ell + 1$. When the equations are linearized we lose a derivative of J and so the condition $k \leq \ell$ is needed for the operator D_u to be well defined on the appropriate Sobolev spaces. The next result is a linear version of Proposition 3.1.10. It relates the cokernel of the operator

$$D_u : W^{k,p}(\Sigma, u^*TM) \to W^{k-1,p}(\Sigma, \Lambda^{0,1} \otimes_J u^*TM) \tag{3.1.8}$$

to the kernel of the formal adjoint operator

$$D_u^* : W^{k,p}(\Sigma, \Lambda^{0,1} \otimes_J u^*TM) \to W^{k-1,p}(\Sigma, u^*TM) \tag{3.1.9}$$

and vice versa. The operator D_u^* is defined by the identity

$$\int_\Sigma \langle \eta, D_u\xi \rangle \, \mathrm{dvol}_\Sigma = \int_\Sigma \langle D_u^*\eta, \xi \rangle \, \mathrm{dvol}_\Sigma.$$

for $\xi \in W^{k,p}(\Sigma, u^*TM)$ and $\eta \in W^{k,p}(\Sigma, \Lambda^{0,1} \otimes_J u^*TM)$. It is a first order differential operator and can also be interpreted as a real linear Cauchy–Riemann operator (see Appendix C).

PROPOSITION 3.1.11. *Fix a positive integer ℓ and a constant $p > 2$. Let J be an ω-tame C^ℓ-almost complex structure on M and $u \in W^{\ell,p}(\Sigma, M)$. Let $k \in \{1, \dots, \ell\}$ and $q > 1$ such that $1/p + 1/q = 1$. Then the following holds.*

(i) *The operators (3.1.8) and (3.1.9) are Fredholm with indices*

$$\operatorname{index} D_u = -\operatorname{index} D_u^* = n(2 - 2g) + 2c_1(u^*TM),$$

where g is the genus of Σ.

(ii) *If $\eta \in L^q(\Sigma, \Lambda^{0,1} \otimes_J u^*TM)$ satisfies*

$$\int_\Sigma \langle \eta, D_u\xi \rangle \, \mathrm{dvol}_\Sigma = 0$$

for all $\xi \in W^{k,p}(\Sigma, u^*TM)$, then $\eta \in W^{\ell,p}(\Sigma, \Lambda^{0,1} \otimes_J u^*TM)$ and $D_u^*\eta = 0$.

(iii) *If $\xi \in L^q(\Sigma, u^*TM)$ satisfies*

$$\int_\Sigma \langle \xi, D_u^*\eta \rangle \, \mathrm{dvol}_\Sigma = 0$$

for all $\eta \in W^{k,p}(\Sigma, \Lambda^{0,1} \otimes_J u^*TM)$, then $\xi \in W^{\ell,p}(\Sigma, u^*TM)$ and $D_u\xi = 0$.

PROOF. Assume first that u is smooth and consider the connection

$$\Omega^0(\Sigma, u^*TM) \to \Omega^1(\Sigma, u^*TM) : \xi \mapsto \nabla\xi - \tfrac{1}{2}J(u)(\nabla_\xi J)(u)du$$

on u^*TM. In general, this connection does not preserve the complex structure or the metric. The operator D_u is the composition of this map with the projection $\Omega^1(\Sigma, u^*TM) \to \Omega^{0,1}(\Sigma, u^*TM)$, extended to the appropriate Sobolev completions. Hence D_u is a real linear Cauchy–Riemann operator. If u is of class $W^{\ell,p}$ then the same argument shows that D_u is a real linear Cauchy–Riemann operator of class $W^{\ell-1,p}$ as in Definition C.1.5. Hence the result follows from Theorem C.2.3 with $q = p$ and $1/r + 1/p = 1$. □

Proposition 3.1.11 implies that the kernel and cokernel of D_u do not depend on the precise choice of the space on which the operator is defined. To see this, assume that J is of class C^ℓ and that u is a J-holomorphic curve. Then, by Proposition 3.1.10, u is of class C^ℓ. Now consider the operator (3.1.8) with $k \le \ell$. By Proposition 3.1.11, every element in the kernel of D_u is necessarily of class $W^{\ell,p}$ for any p and so the kernel of D_u does not depend on the choice of k and p as long as $k \le \ell$. The same holds for the cokernel. In particular, the operator D_u is onto for some choice of k and p if and only if it is onto for all such choices.

3.2. Transversality

The proofs of Theorems 3.1.6 and 3.1.8 are based on an infinite dimensional version of Sard's theorem which is due to Smale [**381**] (see Theorem A.5.1). Roughly speaking, this says that if X and Y are separable Banach manifolds and

$$\mathcal{F} : X \to Y$$

is a Fredholm map of index k then the set Y_{reg} of regular values of $\mathcal{F}$ is residual, provided that $\mathcal{F}$ is sufficently differentiable (it should be at least C^{k+1}). As in the finite dimensional case a point $y \in Y$ is called a regular value if the linearized operator $d\mathcal{F}(x) : T_xX \to T_yY$ is surjective whenever $\mathcal{F}(x) = y$. It then follows from the implicit function theorem that $\mathcal{F}^{-1}(y)$ is a k-dimensional manifold for every $y \in Y_{\mathrm{reg}}$ (Theorem A.3.3).

In order to get a result in the smooth category, our strategy is to prove that the set of regular almost complex structures of class C^ℓ is generic (i.e. residual) with respect to the C^ℓ-topology and then to take the intersection of these sets over all ℓ. This approach is due to Taubes. Throughout we shall denote by $\mathcal{J}^\ell$ either the space $\mathcal{J}^\ell(M,\omega)$ of all ω-compatible almost complex structures of class C^ℓ or the space $\mathcal{J}^\ell_\tau(M,\omega)$ of all ω-tame almost complex structures of class C^ℓ. Proposition 3.2.1 below holds for both cases.

The universal moduli space. The universal moduli space

$$\mathcal{M}^*(A,\Sigma;\mathcal{J}^\ell) := \{(u,J) \,|\, J \in \mathcal{J}^\ell,\, u \in \mathcal{M}^*(A,\Sigma;J)\}$$

consists of all simple J-holomorphic curves $u : \Sigma \to M$ representing the class A, where J varies over the space $\mathcal{J}^\ell$. Recall from Proposition 3.1.10 that every J-holomorphic curve is of class C^ℓ whenever J is of class C^ℓ. Hence we may view $\mathcal{M}^*(A,\Sigma;\mathcal{J}^\ell)$ as a subset of $\mathcal{B}^{k,p} \times \mathcal{J}^\ell$ for any $p > 2$ and any $k \in \{1,\ldots,\ell\}$, where $\mathcal{B}^{k,p}$ denotes the space of $W^{k,p}$-maps $u : \Sigma \to M$ representing the class A, as above. Care must be taken in the case $k = 1$, because a $W^{1,p}$-function is not necessarily continuously differentiable. A continuous function $u : \Sigma \to M$ is called **somewhere injective** if there is a point $z \in \Sigma$ and a constant $\delta > 0$ such that, for every $\zeta \in \Sigma$, we have

$$d_M(u(z), u(\zeta)) \geq \delta d_\Sigma(z,\zeta).$$

This condition is independent of the choice of the metrics on M and Σ used to express it. Note also that, if u is continuously differentiable, then this condition agrees with the notion of somewhere injectivity introduced in Section 2.5.

The parameter space $\mathcal{J}^\ell$ is a smooth separable Banach manifold. Its tangent space $T_J\mathcal{J}^\ell$ at J (in the ω-compatible case) consists of C^ℓ-sections Y of the bundle $\mathrm{End}(TM,J,\omega)$ whose fiber at $x \in M$ is the space of linear maps $Y : T_xM \to T_xM$ such that

$$YJ + JY = 0, \qquad \omega(Yv,w) + \omega(v,Yw) = 0.$$

The first equation is the derivative of the identity $J^2 = -\mathbb{1}$, while the second comes from the compatibility condition. Equivalently one could express these conditions as

$$Y = Y^* = JYJ$$

where Y^* denotes the adjoint operator with respect to the metric g_J. The space of C^ℓ-sections of the bundle $\mathrm{End}(TM,J,\omega) \to M$ is a Banach space and gives rise to a local model for the space $\mathcal{J}^\ell$ via $Y \mapsto J\exp(-JY)$. The corresponding space of C^∞-sections is not a Banach space but only a Fréchet space. Our convention is that spaces with no superscripts consist of elements which are C^∞-smooth.

PROPOSITION 3.2.1. *Fix a homology class $A \in H_2(M, \mathbb{Z})$, an integer $\ell \geq 2$, a real number $p > 2$, and an integer $k \in \{1, \ldots, \ell\}$. Then the universal moduli space $\mathcal{M}^*(A, \Sigma; \mathcal{J}^\ell)$ is a separable $C^{\ell-k}$-Banach submanifold of $\mathcal{B}^{k,p} \times \mathcal{J}^\ell$.*

As we shall see this proposition holds because the tangent space $T_J\mathcal{J}$ is sufficiently large. In particular, the next lemma shows that it acts transitively on T_xM at each point $x \in M$.

LEMMA 3.2.2. *Let $\xi, \eta \in \mathbb{R}^{2n}$ be two nonzero vectors. Then there exists a matrix $Y \in \mathbb{R}^{2n\times 2n}$ such that*

$$Y = Y^T = J_0 Y J_0, \qquad Y\xi = \eta.$$

PROOF. An explicit formula for Y is

$$\begin{aligned} Y \quad := \quad & \frac{1}{|\xi|^2}\Big(\eta\xi^T + \xi\eta^T + J_0(\eta\xi^T + \xi\eta^T)J_0\Big) \\ & - \frac{1}{|\xi|^4}\Big(\langle\eta,\xi\rangle(\xi\xi^T + J_0\xi\xi^T J_0) + \langle\eta, J_0\xi\rangle(J_0\xi\xi^T - \xi\xi^T J_0)\Big). \end{aligned}$$

This formula is constructed in stages. The first term is a matrix that takes ξ to η, the second makes it symmetric while the third and fourth lead to the equation $Y = J_0 Y J_0$. But the resulting matrix no longer takes ξ to η; the last four terms adjust for this. □

PROOF OF PROPOSITION 3.2.1. Consider the $C^{\ell-k}$-Banach space bundle

$$\mathcal{E}^{k-1,p} \to \mathcal{B}^{k,p} \times \mathcal{J}^\ell,$$

whose fiber over (u, J) is the space

$$\mathcal{E}^{k-1,p}_{(u,J)} := W^{k-1,p}(\Sigma, \Lambda^{0,1} \otimes_J u^*TM)$$

of J-antilinear 1-forms on Σ of class $W^{k-1,p}$ with values in the bundle u^*TM. The $C^{\ell-k}$-Banach manifold structure of $\mathcal{E}^{k-1,p}$ can be constructed as follows. Fix any smooth metric on M to identify a neighbourhood $\mathcal{N}(u)$ of a smooth element $u : \Sigma \to M$ in $\mathcal{B}^{k,p}$ with a neighbourhood of zero in the Banach space $W^{k,p}(\Sigma, u^*TM)$ via $\xi \mapsto \exp_u(\xi)$. This gives coordinate charts on $\mathcal{B}^{k,p}$ with smooth transition maps. Now trivialize the bundle $\mathcal{E}^{k-1,p}$ over such a coordinate chart by using the Hermitian connection $\widetilde{\nabla}$. Since J is of class C^ℓ, the connection $\widetilde{\nabla}$ and its parallel transport maps are of class $C^{\ell-1}$. Hence, if we differentiate the transition maps arising from these trivializations $\ell - k$ times, the result is still well defined in $W^{k-1,p}$. This constructs local trivializations over open sets $\mathcal{N}(u)$ in the slice $\mathcal{B}^{k,p} \times \{J\}$. To extend this over neighbourhoods of the form $\mathcal{N}(u) \times \mathcal{N}(J)$ first trivialize over a suitable neighbourhood $\{u\} \times \mathcal{N}(J)$ via the isomorphism

$$\Lambda^{0,1} \otimes_J u^*TM \to \Lambda^{0,1} \otimes_{J'} u^*TM : \quad \alpha \mapsto \tfrac{1}{2}(\alpha + J' \circ \alpha \circ j),$$

and then extend this trivialization over each slice $\mathcal{N}(u) \times \{J'\}$ using parallel translation as before. This shows that $\mathcal{E}^{k-1,p}$ is a Banach space bundle of class $C^{\ell-k}$.

The map $(u, J) \mapsto \bar{\partial}_J(u)$ defines a $\mathcal{C}^{\ell-k}$-section of the bundle $\mathcal{E}^{k-1,p} \to \mathcal{B}^{k,p} \times \mathcal{J}^\ell$. Denote this section by

$$\mathcal{F} : \mathcal{B}^{k,p} \times \mathcal{J}^\ell \to \mathcal{E}^{k-1,p}, \qquad \mathcal{F}(u, J) = \bar{\partial}_J(u).$$

We must prove that the vertical differential

$$D\mathcal{F}(u,J): W^{k,p}(\Sigma, u^*TM) \times C^\ell(M, \mathrm{End}(TM, J, \omega)) \\ \to W^{k-1,p}(\Sigma, \Lambda^{0,1} \otimes_J u^*TM)$$

is surjective whenever u is simple and $\mathcal{F}(u,J) = 0$. By Proposition 3.1.1, this differential is given by the formula

$$D\mathcal{F}(u,J)(\xi, Y) = D_u\xi + \tfrac{1}{2}Y(u) \circ du \circ j.$$

Since D_u is Fredholm the operator $D\mathcal{F}(u,J)$ has a closed image and it suffices to prove that its image is dense whenever $\bar{\partial}_J(u) = 0$.

We prove this first in the case $k = 1$. If the image is not dense then, by the Hahn–Banach theorem, there exists a nonzero section $\eta \in L^q(\Lambda^{0,1} \otimes_J u^*TM)$, with $1/p + 1/q = 1$, which annihilates the image of $D\mathcal{F}(u,J)$. This means that

$$\int_\Sigma \langle \eta, D_u\xi \rangle \,\mathrm{dvol}_\Sigma = 0 \tag{3.2.1}$$

for every $\xi \in W^{1,p}(\Sigma, u^*TM)$ and

$$\int_\Sigma \langle \eta, Y(u) \circ du \circ j \rangle \,\mathrm{dvol}_\Sigma = 0 \tag{3.2.2}$$

for every $Y \in C^\ell(M, \mathrm{End}(TM, J, \omega))$, where $j := j_\Sigma$. It follows from (3.2.1) and Proposition 3.1.11 that η is of class $W^{1,p}$ and $D_u^*\eta = 0$.

Since u is simple, Proposition 2.5.1 asserts that the set of injective points of u is open and dense in Σ. Let $z_0 \in \Sigma$ be such an injective point, i.e.

$$du(z_0) \neq 0, \qquad u^{-1}(u(z_0)) = \{z_0\}.$$

We shall prove that η vanishes at z_0. Assume, by contradiction, that $\eta(z_0) \neq 0$. Then, since $du(z_0) \neq 0$, it follows from Lemma 3.2.2 that there exists an endomorphism $Y_0 \in \mathrm{End}(T_{u(z_0)}M, J_{u(z_0)}, \omega_{u(z_0)})$ such that

$$\langle \eta(z_0), Y_0 \circ du(z_0) \circ j(z_0) \rangle > 0.$$

Now choose any section $Y \in C^\ell(M, \mathrm{End}(TM, J, \omega))$ such that $Y(u(z_0)) = Y_0$. Then the scalar function $\langle \eta, Y(u) \circ du \circ j \rangle$ on Σ is positive in some open neighbourhood $V_0 \subset \Sigma$ of z_0. Since z_0 is an injective point of u, the compact set $u(\Sigma \setminus V_0)$ does not contain the point $u(z_0)$. Hence there exists an open neighbourhood $U_0 \subset M$ of $u(z_0)$ such that $u(\Sigma \setminus V_0) \cap U_0 = \emptyset$ and hence

$$u^{-1}(U_0) \subset V_0.$$

Now choose a smooth cutoff function $\beta : M \to [0,1]$ supported in U_0 such that $\beta(u(z_0)) = 1$. Then the function $\beta(u)\langle \eta, Y(u) \circ du \circ j \rangle$ on Σ is supported in V_0, is nonnegative, and is positive somewhere. Hence the integral on the left hand side of (3.2.2), with Y replaced by βY, does not vanish. This contradiction shows that $\eta(z_0) = 0$. Since this holds for every injective point z_0 of u, and the set of injective points is dense in Σ (Proposition 2.5.1), it follows that η vanishes almost everywhere. Since η is continuous we obtain $\eta \equiv 0$. Thus we have proved that $D\mathcal{F}(u,J)$ has a dense image and is therefore onto in the case $k = 1$.

To prove surjectivity for general k let $\eta \in W^{k-1,p}(\Sigma, \Lambda^{0,1} \otimes_J u^*TM)$ be given. Then, by surjectivity for $k = 1$, there exists a pair

$$(\xi, Y) \in W^{1,p}(\Sigma, u^*TM) \times C^\ell(M, \mathrm{End}(TM, J, \omega))$$

such that $D\mathcal{F}(u,J)(\xi,Y)=\eta$. Now the above formula for $D\mathcal{F}(u,J)$ shows that

$$D_u\xi = \eta - \tfrac{1}{2}Y(u)\circ du\circ j \in W^{k-1,p}$$

and, by elliptic regularity, $\xi\in W^{k,p}(\Sigma,u^*TM)$ (Theorem C.2.3). Hence $D\mathcal{F}(u,J)$ is onto for every pair $(u,J)\in\mathcal{M}^*(A,\Sigma;\mathcal{J}^\ell)$.

Because D_u is a Fredholm operator it follows from Lemma A.3.6 that the operator $D\mathcal{F}(u,J)$ has a right inverse. Hence it follows from the infinite dimensional implicit function theorem (Theorem A.3.3) that the space $\mathcal{M}^*(A,\Sigma;\mathcal{J}^\ell)$ is a $\mathcal{C}^{\ell-k}$ Banach submanifold of $\mathcal{B}^{k,p}\times\mathcal{J}^\ell$. Since $\mathcal{B}^{k,p}\times\mathcal{J}^\ell$ is separable so is $\mathcal{M}^*(A,\Sigma;\mathcal{J}^\ell)$. This proves Proposition 3.2.1 □

REMARK 3.2.3. The assertion of Proposition 3.2.1 continues to hold for more general sets $\mathcal{J}$ of almost complex structures. For example, it suffices to consider the space of ω-compatible almost complex structures that agree with a given almost complex structure J_0 on the complement of an open set $U\subset M$, provided that every J-holomorphic curve $u:\Sigma\to M$ representing the class A passes through the set U (for every J). However in this case the proof requires Aronszajn's theorem 2.3.4. Namely, if $D_u^*\eta=0$ then

$$0 = D_uD_u^*\eta = \Delta\eta + \text{lower order terms},$$

and it follows from Aronszajn's theorem that, if η vanishes on some open set, then $\eta\equiv 0$. Now it suffices to choose infinitesimal almost complex structures Y with support in the set U to guarantee that the annihilator η in the proof of Proposition 3.2.1 vanishes on some open set. We will encounter an application of this idea later in Lemmas 3.4.3 and 3.4.6 where we consider moduli spaces of curves whose values are constrained at a finite set of points in the domain.

Proofs of the main theorems. With this preparation we are now ready to prove the main theorems of this chapter.

PROOF OF THEOREM 3.1.6 (I). Suppose given an almost complex structure $J\in\mathcal{J}_{\mathrm{reg}}(A,\Sigma)$ and a J-holomorphic curve $u\in\mathcal{M}^*(A,\Sigma;J)$. By Proposition 3.1.10, u is smooth. Given an integer $k\geq 1$ and a real number $p>2$, consider the map

$$\mathcal{F}_u : W^{k,p}(\Sigma,u^*TM)\to W^{k-1,p}(\Sigma,\Lambda^{0,1}\otimes_J u^*TM)$$

defined by (3.1.3). Then a $W^{k,p}$-neighbourhood of zero in $\mathcal{F}_u^{-1}(0)$ is diffeomorphic to a $W^{k,p}$-neighbourhood of u in $\mathcal{M}^*(A,\Sigma;J)$ via the map $\xi\mapsto\exp_u(\xi)$. By Proposition 3.1.1 and the definition of $\mathcal{J}_{\mathrm{reg}}(A,\Sigma)$, the differential $d\mathcal{F}_u(0)=D_u$ is surjective. Since $\mathcal{F}_u$ is a smooth map between Banach spaces, it follows from Theorem A.3.3 that $\mathcal{F}_u^{-1}(0)$ intersects a sufficiently small neighbourhood of zero in a smooth submanifold of dimension $n(2-2g)+2c_1(A)$. The image of this submanifold under the map $\xi\mapsto\exp_u(\xi)$ is a smooth submanifold of $W^{k,p}(\Sigma,M)$ and agrees with a neighbourhood of u in $\mathcal{M}^*(A,\Sigma;J)$. Hence $\mathcal{M}^*(A,\Sigma;J)$ is a smooth submanifold of $W^{k,p}(\Sigma,M)$. We emphasize that the coordinate charts on $\mathcal{M}^*(A,\Sigma;J)$ obtained in this way are independent of k and p, because all elements in $\mathcal{F}_u^{-1}(0)$ near zero are necessarily smooth (Proposition 3.1.10).

To complete the proof of Theorem 3.1.6(i) we need to show that the moduli space has a canonical orientation. To understand why, observe first that the tangent space $T_u\mathcal{M}^*(A,\Sigma;J)$ is the kernel of D_u. Now, according to the formula (3.1.7), the operator D_u is the sum of the operator $\xi\mapsto(\widehat{\nabla}\xi)^{0,1}$ and an operator of the form $\xi\mapsto T(\xi,\partial_J(u))$. The first of these has order one and commutes with J, while the second has order zero and anticommutes with J (see Remark 3.1.3). Hence the kernel of D_u will in general not be invariant under J and so J might not determine an almost complex structure on $T_u\mathcal{M}^*(A,\Sigma;J)$. However, by multiplying the second part of D_u by a constant which tends to zero, one can homotop D_u through Fredholm operators D_u^t to a Fredholm operator which does commute with J, namely the complex linear Cauchy–Riemann operator $\xi \mapsto (\widehat{\nabla}\xi)^{0,1}$.

Now the determinant

$$\det(D) = \Lambda^{\max}(\ker D) \otimes \Lambda^{\max}(\ker D^*)$$

of a Fredholm operator $D : X \to Y$ between complex Banach spaces carries a natural orientation whenever the operator D is complex linear (see Section A.2). In our case the complex antilinear part of the operator D_u is compact and hence the determinant line $\det(D_u) = \Lambda^{\max}(\ker D_u)$ inherits a natural orientation from the complex linear part of D_u. These orientations of $\det(D_u)$ for $u \in \mathcal{M}^*(A,\Sigma;J)$ determine the orientation of $\mathcal{M}^*(A,\Sigma;J)$. This proves Theorem 3.1.6 (i). □

REMARK 3.2.4. Similar arguments were used by Donaldson in [**81**] for the orientation of Yang-Mills moduli spaces and also by Ruan [**342**] in the present context. A slightly different line of argument was used by McDuff in [**254**] where she established the existence of a canonical stable almost complex structure[2] on compact subsets of the moduli space $\mathcal{M}^*(A,\Sigma;J)$.

REMARK 3.2.5. If the index is zero, then $\mathcal{M}^*(A,\Sigma;J)$ is a discrete set and the orientation of each point is a sign $\nu(u) = \pm 1$. This sign can be interpreted in terms of **crossing numbers** as in Proposition A.2.4. More precisely, choose a smooth family of zeroth order (compact) operators $P^t : \Omega^0(\Sigma, u^*TM) \to \Omega^{0,1}(\Sigma, u^*TM)$ such that $P^0 = 0$, the operator $D_u^1 := D_u + P^1$ is complex linear and bijective, and the operator family $t \mapsto D_u^t := D_u + P^t$ has only simple crossings. Then each crossing is a value of t where D_u^t has a nonzero (one-dimensional) kernel and $\nu(u) = (-1)^m$, where m is the number of crossings. Thus the orientation is determined by the mod-2 spectral flow of the operators family $t \mapsto D_u^t$. Note that this definition of the sign only applies to simple J-holomorphic curves in $\mathcal{M}^*(A,\Sigma;J)$ in the zero dimensional case. In certain situations (for example when counting isolated embedded tori in symplectic 4-manifolds), one has to take account of multiply covered curves and it is a more subtle matter to determine their contributions to the invariant (see Taubes [**389**] and Ionel–Parker [**198**]).

REMARK 3.2.6. Note that if J is integrable, then D_u commutes with J, and so J induces an (integrable) almost complex structure on $\mathcal{M}^*(A,\Sigma;J)$. By definition, this complex structure is compatible with the orientation described in the above proof of Theorem 3.1.6 (i). In particular if the index is zero and J is regular and integrable every element of the moduli space $\mathcal{M}^*(A,\Sigma;J)$ gives the contribution $+1$ to the count of of J-holomorphic A curves. Hence in situations when this number is negative, one cannot find regular integrable J (see Example 3.3.7).

PROOF OF THEOREM 3.1.6 (II). We now must show that the set $\mathcal{J}_{\mathrm{reg}}(A,\Sigma)$ of smooth regular almost complex structures J is residual in $\mathcal{J}$. The argument is

[2] A manifold X is said to have a stable almost complex structure if there is a number k such that the Whitney sum $TX \oplus \mathbb{R}^k$ of the tangent bundle TX with the trivial k-dimensional real bundle carries an almost complex structure.

based on the properties of the projection

$$\pi : \mathcal{M}^*(A, \Sigma; \mathcal{J}^\ell) \to \mathcal{J}^\ell.$$

By Proposition 3.2.1 with $k = 1$, this is a $C^{\ell-1}$-map between separable $C^{\ell-1}$-Banach manifolds. The tangent space

$$T_{(u,J)}\mathcal{M}^*(A, \Sigma; \mathcal{J}^\ell) \subset W^{1,p}(\Sigma, u^*TM) \times C^\ell(M, \mathrm{End}(TM, J, \omega))$$

consists of all pairs (ξ, Y) such that

$$D_u\xi + \tfrac{1}{2}Y(u) \circ du \circ j_\Sigma = 0.$$

Moreover, the derivative

$$d\pi(u, J) : T_{(u,J)}\mathcal{M}^*(A, \Sigma; \mathcal{J}^\ell) \to T_J\mathcal{J}^\ell$$

is the projection $(\xi, Y) \mapsto Y$. Hence the kernel of $d\pi(u, J)$ is isomorphic to the kernel of D_u. By Lemma A.3.6, its cokernel also is isomorphic to the cokernel of D_u. It follows that $d\pi(u, J)$ is a Fredholm operator with the same index as D_u Moreover the operator $d\pi(u, J)$ is onto precisely when D_u is onto. Hence a regular value J of π is an almost complex structure with the property that D_u is onto for every simple J-holomorphic curve $u \in \mathcal{M}^*(A, \Sigma; J) = \pi^{-1}(J)$. In other words the set

$$\mathcal{J}^\ell_{\mathrm{reg}}(A, \Sigma) := \left\{J \in \mathcal{J}^\ell \,|\, D_u \text{ is onto for all } u \in \mathcal{M}^*(A, \Sigma; J)\right\}$$

of regular almost complex structures is precisely the set of regular values of π. By the Sard–Smale theorem (Theorem A.5.1), this set is residual in the sense of Baire (a countable intersection of open and dense sets). Here we use the fact that the manifold $\mathcal{M}^*(A, \Sigma; \mathcal{J}^\ell)$ and the projection π are of class $C^{\ell-1}$. Hence we can apply the Sard–Smale theorem whenever

$$\ell - 2 \geq \mathrm{index}\,\pi = \mathrm{index}\,D_u = n(2 - 2g) + 2c_1(A).$$

Thus we have proved that the set $\mathcal{J}^\ell_{\mathrm{reg}}$ is dense in $\mathcal{J}^\ell$ with respect to the C^ℓ-topology for ℓ sufficiently large. We shall now use an argument due to Taubes, to deduce that $\mathcal{J}_{\mathrm{reg}}(A, \Sigma)$ is residual in $\mathcal{J}$ with respect to the C^∞-topology.

Consider the set

$$\mathcal{J}_{\mathrm{reg,K}}(A, \Sigma) \subset \mathcal{J}$$

of all smooth almost complex structures $J \in \mathcal{J}$ such that the operator D_u is onto for every J-holomorphic curve $u : \Sigma \to M$ representing the class A which satisfies

$$\|du\|_{L^\infty} \leq K \tag{3.2.3}$$

and for which there exists a point $z \in \Sigma$ such that

$$\inf_{\zeta \neq z} \frac{d(u(z), u(\zeta))}{d(z, \zeta)} \geq \frac{1}{K}. \tag{3.2.4}$$

Condition (3.2.4) guarantees that u is simple. Moreover, every simple J-holomorphic curve $u : \Sigma \to M$ satisfies both conditions for some value of $K > 0$. Hence

$$\mathcal{J}_{\mathrm{reg}}(A, \Sigma) = \bigcap_{K>0} \mathcal{J}_{\mathrm{reg,K}}(A, \Sigma).$$

We shall prove that each set $\mathcal{J}_{\mathrm{reg,K}}(A, \Sigma)$ is open and dense in $\mathcal{J}$ with respect to the C^∞-topology. We first prove that $\mathcal{J}_{\mathrm{reg,K}}(A, \Sigma)$ is open or, equivalently, that its complement is closed. Hence assume that the sequence $J_\nu \notin \mathcal{J}_{\mathrm{reg,K}}(A, \Sigma)$ converges to $J \in \mathcal{J}$ in the C^∞-topology. Then there exist, for every ν, a point $z_\nu \in \Sigma$

and a J_ν-holomorphic curve $u_\nu \in \mathcal{M}^*(A,\Sigma;J_\nu)$ which satisfies (3.2.3) and (3.2.4), with z replaced by z_ν, such that the operator D_{u_ν} is not surjective. It follows from standard elliptic bootstrapping arguments that there exists a subsequence u_{ν_i} which converges, uniformly with all derivatives, to a smooth J-holomorphic curve $u:\Sigma\to M$ (see Theorem B.4.2). Choose the subsequence such that z_{ν_i} converges to z. Then the limit curve u satisfies the conditions (3.2.3) and (3.2.4) for this point z and, moreover, since the operators $D_{u_{\nu_i}}$ are not surjective, it follows that D_u cannot be surjective either. This shows that $J\notin\mathcal{J}_{\mathrm{reg,K}}(A,\Sigma)$ and thus we have proved that the complement of $\mathcal{J}_{\mathrm{reg,K}}(A,\Sigma)$ is closed in the C^∞-topology.

Next we prove that the set $\mathcal{J}_{\mathrm{reg,K}}(A,\Sigma)$ is dense in $\mathcal{J}$ with respect to the C^∞-topology. To see this note first that

$$\mathcal{J}_{\mathrm{reg,K}}(A,\Sigma)=\mathcal{J}^\ell_{\mathrm{reg,K}}(A,\Sigma)\cap\mathcal{J},$$

where $\mathcal{J}^\ell_{\mathrm{reg,K}}(A,\Sigma)\subset\mathcal{J}^\ell$ is the set of all $J\in\mathcal{J}^\ell(M,\omega)$ such that the operator D_u is onto for every J-holomorphic curve $u:\Sigma\to M$ of class C^ℓ that represents the class A and satisfies (3.2.3) and (3.2.4) for some $z\in\Sigma$. The same argument as above shows that $\mathcal{J}^\ell_{\mathrm{reg,K}}(A,\Sigma)$ is open in $\mathcal{J}^\ell$ with respect to the C^ℓ-topology.

Now let $J\in\mathcal{J}$. Since $\mathcal{J}^\ell_{\mathrm{reg}}(A,\Sigma)$ is dense in $\mathcal{J}^\ell$ for large ℓ, there exists a sequence $J_\ell\in\mathcal{J}^\ell_{\mathrm{reg}}(A,\Sigma)$, $\ell\ge\ell_0$, such that

$$\|J-J_\ell\|_{C^\ell}\le 2^{-\ell}.$$

Since $J_\ell\in\mathcal{J}^\ell_{\mathrm{reg,K}}(A,\Sigma)$ and $\mathcal{J}^\ell_{\mathrm{reg,K}}(A,\Sigma)$ is open in the C^ℓ-topology, there exists an $\varepsilon_\ell>0$ such that for every $J'\in\mathcal{J}^\ell$,

$$\|J'-J_\ell\|_{C^\ell}<\varepsilon_\ell \qquad\Longrightarrow\qquad J'\in\mathcal{J}^\ell_{\mathrm{reg,K}}(A,\Sigma).$$

Choose $J'_\ell\in\mathcal{J}$ to be any smooth element such that

$$\|J'_\ell-J_\ell\|_{C^\ell}<\min\left\{\varepsilon_\ell,2^{-\ell}\right\}.$$

Then

$$J'_\ell\in\mathcal{J}^\ell_{\mathrm{reg,K}}(A,\Sigma)\cap\mathcal{J}=\mathcal{J}_{\mathrm{reg,K}}(A,\Sigma)$$

and the sequence J'_ℓ converges to J in the C^∞-topology. This shows that the set $\mathcal{J}_{\mathrm{reg,K}}(A,\Sigma)$ is dense in $\mathcal{J}$ as claimed. Thus $\mathcal{J}_{\mathrm{reg}}(A,\Sigma)$ is the intersection of the countable number of open dense sets $\mathcal{J}_{\mathrm{reg,K}}(A,\Sigma)$, $K\in\mathbb{N}$, and so is residual as required. This proves Theorem 3.1.6 (ii). □

Proof of Theorem 3.1.8. The proof is almost word by word the same as that of Theorem 3.1.6. Let $\mathcal{J}^\ell(J_0,J_1)$ denote the space of C^ℓ-homotopies

$$[0,1]\to\mathcal{J}^\ell(M,\omega):\lambda\mapsto J_\lambda$$

from J_0 to J_1. One then considers the universal moduli space

$$\mathcal{W}^*(A,\Sigma;\mathcal{J}^\ell)\subset[0,1]\times\mathcal{B}^{1,p}\times\mathcal{J}^\ell(J_0,J_1)$$

of all triples (λ,u,J), where $\lambda\in[0,1]$, $J\in\mathcal{J}^\ell(J_0,J_1)$, and $(\lambda,u)\in\mathcal{W}^*(A,\Sigma;J)$. As in the proof of Proposition 3.2.1 and Theorem 3.1.6 one can show that $\mathcal{W}^*(A,\Sigma;\mathcal{J}^\ell)$ is a $C^{\ell-1}$-Banach submanifold of $[0,1]\times\mathcal{B}^{1,p}\times\mathcal{J}^\ell(J_0,J_1)$. Its boundary is the intersection of $\mathcal{W}^*(A,\Sigma;\mathcal{J}^\ell)$ with $\{0,1\}\times\mathcal{B}^{1,p}\times\mathcal{J}^\ell(J_0,J_1)$. The projection

$$\mathcal{W}^*(A,\Sigma;\mathcal{J}^\ell)\to\mathcal{J}^\ell(J_0,J_1):(\lambda,u,J)\mapsto J \tag{3.2.5}$$

is again a $C^{\ell-1}$-Fredholm map and the regular values form the set $\mathcal{J}^\ell_{\mathrm{reg}}(A,\Sigma;J_0,J_1)$ of regular homotopies of class C^ℓ. By the Sard–Smale theorem, the set of regular values is dense in $\mathcal{J}^\ell(J_0, J_1)$. The reduction of the smooth case to the C^ℓ-case is as in the proof of Theorem 3.1.6. It follows that $\mathcal{J}_{\text{reg}}(A, \Sigma; J_0, J_1)$ is a residual subset of $\mathcal{J}(J_0, J_1)$. If $\{J_\lambda\}_\lambda \in \mathcal{J}_{\text{reg}}(A, \Sigma; J_0, J_1)$ is a smooth regular homotopy, then the smooth moduli space $\mathcal{W}^*(A, \Sigma; \{J_\lambda\}_\lambda)$ is the preimage of $\{J_\lambda\}_\lambda$ under the projection (3.2.5) for every ℓ, by elliptic regularity (Proposition 3.1.10). Hence, for every integer ℓ, it is a $C^{\ell-1}$ submanifold of $[0, 1] \times \mathcal{B}^{1,p}$. Hence it is a smooth submanifold of $[0, 1] \times \mathcal{B}^{1,p}$ whose boundary is the intersection with $\{0, 1\} \times \mathcal{B}^{1,p}$. This proves Theorem 3.1.8. □

REMARK 3.2.7. The key point in the proof of Theorem 3.1.6 (ii) is to show that the set $\mathcal{J}_{\text{reg,K}}(A, \Sigma)$ is dense in $\mathcal{J}$ for every K. Our proof uses the Sard–Smale theorem for C^ℓ Banach manifolds to establish this. One can instead use an argument of Floer that remains within the C^∞-category. More precisely, one can consider a suitable Banach space $C^\infty_\varepsilon(M, \text{End}(TM, J, \omega)) \subset C^\infty(M, \text{End}(TM, J, \omega))$ of all sections $Y \in C^\infty(M, \text{End}(TM, J, \omega))$ such that the norm

$$\|Y\|_\varepsilon := \sum_{\ell=1}^{\infty} \varepsilon_\ell \|Y\|_{C^\ell}$$

is finite. It is easy to see that if the sequence ε_ℓ converges to zero sufficiently rapidly, then there exist sections $Y \in C^\infty_\varepsilon(M, \text{End}(TM, J, \omega))$ which are supported in arbitrarily small neighbourhoods of any point $x_0 \in M$ and take arbitrary values at x_0. (For details, see Floer [**114**].) Hence the proof of Proposition 3.2.1 goes through if one replaces $\mathcal{J}^\ell$ by the space $\mathcal{J}_\varepsilon$ of all almost complex structures of the form $J = J_0 \exp(-J_0 Y)$, where $Y \in C^\infty_\varepsilon(M, \text{End}(TM, J_0, \omega))$. Since $\mathcal{J}_\varepsilon$ is a smooth separable Banach manifold, the Sard–Smale theorem applies. It then follows as in the proof of Theorem 3.1.6 (ii) that every $J_0 \in \mathcal{J}$ can be approximated in the C^∞-topology by a sequence of regular almost complex structures.

REMARK 3.2.8. The proof of Theorem 3.1.6 shows that a point (u, J) in the (smooth) universal moduli space $\mathcal{M}^*(A, \Sigma; \mathcal{J})$ is regular in the sense of Definition 3.1.5 if and only if the projection $d\pi(u, J) : T_{(u,J)}\mathcal{M}^*(A, \Sigma; \mathcal{J}) \to T_j\mathcal{J}$ is surjective. By the implicit function theorem, this implies that any smooth path $[0, 1] \to \mathcal{J} : t \mapsto J_t$ which starts at $J_0 = J$ can be lifted, on some interval $[0, \varepsilon)$, to a path $[0, \varepsilon) \to \mathcal{M}^*(A, \Sigma; \mathcal{J}) : t \mapsto (u_t, J_t)$ in the universal moduli space which starts at $u_0 = u$. Thus any regular curve (u, J) persists when J is perturbed. In contrast, if (u, J) is not regular there might be no nearby curve when J is perturbed (see Example 3.3.6).

3.3. A regularity criterion

Before proceeding further with the general theory, we give some examples to illustrate the concept of regularity. This becomes particularly transparent in the integrable case and when $\Sigma = S^2 = \mathbb{C}P^1$. Grothendieck [**162**] proved that any holomorphic vector bundle E over $\mathbb{C}P^1$ is holomorphically equivalent to a direct sum of holomorphic line bundles. Moreover, this splitting is unique up to the order of the summands. Hence

$$E = L_1 \oplus \cdots \oplus L_n$$

is completely characterized by the set of Chern numbers $c_1(L_1), \ldots, c_1(L_n)$. Here and throughout we identify the Chern class $c_1(L)$ with the corresponding Chern number $\langle c_1(L), [S^2] \rangle$. In particular, Grothendieck's theorem applies to the vector bundle $E = u^*TM$ whenever $u : S^2 \to M$ is a J-holomorphic sphere for an (integrable) complex structure J. Note that the sum

$$c_1(E) = \sum_i c_1(L_i)$$

is a topological invariant, but that the numbers $c_1(L_1), \ldots, c_1(L_n)$ in the decomposition of $E = u^*TM$ may vary as $u : S^2 \to M$ varies in a connected component of the space of J-holomorphic spheres.

The next result follows immediately from part (iii) of Theorem C.1.10, which asserts that a real linear Cauchy–Riemann operator on a complex line bundle over $\mathbb{C}P^1$ is either injective or surjective or both. Although this is proved in full detail in Appendix C, we outline an argument here that is based on observations from algebraic geometry to illustrate how our analytic framework based on elliptic PDEs fits in with classical results about holomorphic vector bundles.

LEMMA 3.3.1. *Assume J is integrable and let $u : \mathbb{C}P^1 \to M$ be a J-holomorphic sphere. If every summand of u^*TM has Chern number $c_1 \geq -1$, then D_u is onto.*

PROOF. Since J is integrable, the operator D_u is complex linear and is exactly the Dolbeault derivative $\bar\partial$ determined by the holomorphic coordinate charts of M. Hence it respects the splitting of u^*TM into holomorphic line bundles, and we can consider each line subbundle L of (TM, J) separately. Further, the cokernel of $D_u = \bar\partial : \Omega^0(\mathbb{C}P^1, L) \to \Omega^{0,1}(\mathbb{C}P^1, L)$ is precisely the Dolbeault cohomology group $H^{0,1}_{\bar\partial}(\mathbb{C}P^1, L)$. Therefore it suffices to show that this group vanishes whenever $c_1(L) \geq -1$. Now, for any holomorphic line bundle $L \to \mathbb{C}P^1$,

$$H^{0,1}_{\bar\partial}(\mathbb{C}P^1, L) \cong \left(H^{1,0}_{\bar\partial}(\mathbb{C}P^1, L^*)\right)^*.$$

Here $H^{1,0}_{\bar\partial}(\mathbb{C}P^1, L^*)$ is the space of holomorphic 1-forms with values in the dual bundle L^* and so is isomorphic to the space $H^0(\mathbb{C}P^1, L^* \otimes K)$ of holomorphic sections of the bundle $L^* \otimes K$ where $K := T^*\mathbb{C}P^1$ is the canonical bundle. This is an easy special case of Kodaira–Serre duality (cf. [**159**, Ch 1§2]) which can be checked directly by considering the transition maps. (Note that the fiber of the bundle $L^* \otimes K$ at z consists of the homomorphisms $T_zS^2 \to L_z^*$, i.e. the L_z^*-valued 1-forms at z.) Now a line bundle L' over $\mathbb{C}P^1$ has nonzero holomorphic sections if and only if $c_1(L') \geq 0$. This is a special case of the Kodaira vanishing theorem, and holds by positivity of intersections and the interpretation of $c_1(L')$ as the self-intersection number of the zero section. Thus D_u is surjective if and only if $c_1(L^* \otimes K) < 0$ for every holomorphic summand L of E. Since $c_1(L^* \otimes K) = -c_1(L) - 2$, this is equivalent to $c_1(L) \geq -1$. This proves Lemma 3.3.1. □

The previous argument generalizes to almost complex manifolds as follows.

LEMMA 3.3.2. *Let $E \to S^2$ be a complex vector bundle of rank n and*

$$D : \Omega^0(S^2, E) \to \Omega^{0,1}(S^2, E)$$

be a real linear Cauchy–Riemann operator. Suppose that there exists a filtration $E_1 \subset E_2 \subset \cdots \subset E_n = E$ such that each E_k is a D-invariant complex subbundle of rank k. Then D is surjective if and only if $c_1(E_k/E_{k-1}) \geq -1$ for every k.

PROOF. The proof is by induction on n. For $n = 1$ the assertion follows from Theorem C.1.10 (iii). Let $n \geq 2$ and assume it holds for $n - 1$. Then the restriction

$$D' : \Omega^0(S^2, E_{n-1}) \to \Omega^{0,1}(S^2, E_{n-1})$$

of the operator D to sections of E_{n-1} is surjective if and only if $c_1(E_k/E_{k-1}) \geq -1$ for $k = 1, \ldots, n-1$. Moreover, by Theorem C.1.10 (iii), the quotient operator

$$D'' : \Omega^0(S^2, E_n/E_{n-1}) \to \Omega^{0,1}(S^2, E_n/E_{n-1}),$$

determined by D, is surjective if and only if $c_1(E_n/E_{n-1}) \geq -1$. Hence the induction step follows from the fact that D is surjective if and only if D' and D'' are surjective. This proves Lemma 3.3.2. □

LEMMA 3.3.3. *Let J be an almost complex structure on a 4-manifold M and $u : \mathbb{C}P^1 \to M$ be an immersed J-holomorphic sphere. Then D_u is onto if and only if $c_1(u^*TM) \geq 1$.*

PROOF. It follows from the definition of D_u in (3.1.4) that $D_u(du \circ \zeta) = du \circ \bar{\partial}_j \zeta$ for every vector field $\zeta \in \mathrm{Vect}(\Sigma)$ and every J-holomorphic curve u. If u is an immersion it follows that the complex subbundle $L := \mathrm{im}\, du \subset u^*TM$ is invariant under D_u. When $\Sigma = \mathbb{C}P^1$ we have $c_1(L) = 2$ and $c_1(u^*TM/L) = c_1(u^*TM) - 2$. Both Chern numbers are greater than or equal to -1 if and only if $c_1(u^*TM) \geq 1$. Hence Lemma 3.3.3 follows from Lemma 3.3.2. □

COROLLARY 3.3.4. *Let J be an almost complex structure on a 4-manifold M and C be an embedded J-holomorphic sphere with self-intersection number $C \cdot C = p$. Then J is regular for the class $A = [C]$ if and only if $p \geq -1$.*

PROOF. By Theorem 2.6.4, all the curves in the moduli space $\mathcal{M}(A; J)$ are embedded. Hence the result follows from Lemma 3.3.3. □

COROLLARY 3.3.5. *Let $\widetilde{M}$ be the product of S^2 with a symplectic manifold (M, ω) and $\widetilde{A} \in H_2(\widetilde{M}; \mathbb{Z})$ be the homology class represented by the spheres $S^2 \times \{\mathrm{pt}\}$. Then, for every $J \in \mathcal{J}(M, \omega)$, the product almost complex structure $\widetilde{J} := i \times J$ is regular for $\widetilde{A}$.*

PROOF. Every $\widetilde{J}$-holomorphic $\widetilde{A}$-sphere has the form $\tilde{u}(z) = (\phi(z), x_0)$, where ϕ is a fractional linear transformation and $x_0 \in M$. Hence $D_{\tilde{u}}$ is complex linear and $\tilde{u}^*T\widetilde{M}$ decomposes into a direct sum of $D_{\tilde{u}}$-invariant line bundles one of which has Chern number two and the others have Chern number zero. Hence, by Lemma 3.3.2, $D_{\tilde{u}}$ is surjective. □

EXAMPLE 3.3.6. Consider the product manifold

$$M = S^2 \times S^2$$

with its product almost complex structure $J_0 = j \times j$. Then the projection maps π_i to the two factors are holomorphic so that any holomorphic map $u : S^2 \to S^2 \times S^2$ has the form $z \mapsto (v_1(z), v_2(z))$, where $v_i : S^2 \to S^2$ is holomorphic and has nonnegative degree d_i. It follows that u^*TM splits holomorphically as a sum $L_1 \oplus L_2$ where $L_i := v_i^*TS^2$ has degree $2d_i$. Thus all these curves are regular. However, $S^2 \times S^2$ supports other tamed complex structures that admit nonregular curves.

The best way to see this is to consider Hirzebruch's complex structures J_{2k} obtained by identifying $S^2 \times S^2$ with the projectivization $\mathbf{P}(L \oplus \mathbb{C})$, where L is a holomorphic line bundle of degree $2k > 0$. The subbundle L corresponds to a section C_L of $\mathbf{P}(L \oplus \mathbb{C})$ with normal bundle L^*. Thus $c_1(L^*) < -1$ when $k > 0$, so that C_L is not regular. Now, one can identify $\mathbf{P}(L \oplus \mathbb{C})$ with $S^2 \times S^2$ in such a way that C_L is taken to the antidiagonal $(z, A(z))$, where A is the antipodal map. (This can be done explicitly by looking at local trivializations.) Moreover, there are Kähler forms on $\mathbf{P}(L \oplus \mathbb{C})$ that are taken under this identification to the symplectic forms $\omega^\lambda := \lambda\pi_1^*(\sigma) \oplus \pi_2^*(\sigma)$, where σ is an area form on S^2 and $\lambda > 1$. The upshot is that when $\lambda > 1$ the symplectic manifold $(S^2 \times S^2, \omega^\lambda)$ has an ω_λ-compatible J for which there is a nonregular curve, namely the antidiagonal. (See [**5**] for example for a construction of these manifolds as reduced spaces; an extensive discussion of ruled surfaces as symplectic manifolds appears in [**262**].)

Another way to construct nonregular curves is by blowing up. Take $\mathbb{C}P^2$ and blow up twice, choosing the second blowup point to lie in the exceptional divisor C of the first blowup X_1. Then C is regular in the one point blowup X_1, since it has normal bundle of Chern class -1. However, its proper transform C' in the two point blowup X_2 has normal bundle of Chern class -2 and so is not regular. In fact, we can also obtain this example by blowing up the Hirzebruch surface $(S^2 \times S^2, J_2)$ at a point off the anti-diagonal. However in this second picture it is easy to see how nonregular curves can disappear when J varies (cf. Remark 3.2.8). For if we move the second blowup point off the exceptional divisor C, then there is no longer any curve in X_2 with normal bundle of Chern class -2.

This section is the first place in our discussion where it is important to restrict to spheres. Our other results apply to J-holomorphic curves with fixed domain (Σ, j), but these are almost never regular because, even when J is integrable, the condition $H^{0,1}_{\bar\partial}(\Sigma, u^*TM) = 0$ is usually not satisfied. Indeed when u is embedded and the tangent bundle TC of its image $C := u(\Sigma)$ is a holomorphic summand of TM, then $H^{0,1}_{\bar\partial}(\Sigma, u^*TM)$ contains the nontrivial space $H^{0,1}_{\bar\partial}(\Sigma, T\Sigma)$, i.e. the tangent space of the Teichmüller space of Σ at j. Thus when J varies on M one usually must vary j on Σ in order to find a corresponding deformation of the curve. To deal with this, one has to set up the Fredholm theory in such a way that j is allowed to vary in Teichmüller space. This presents no problem at this stage, but it does make the discussion of compactness more complicated, since Teichmüller space, even when quotiented out by the mapping class group, is not compact.

Much of the above discussion applies to embedded curves of higher genus. For example, if one knows that the holomorphic bundle u^*TM is a sum of line bundles, then the arguments in Lemma 3.3.2 still work. However, one must be somewhat careful because the question of whether a line bundle over a curve of genus g has holomorphic sections is no longer purely topological depending on the Chern number alone, unless this number is bigger than $2g - 2$ (see Theorem C.1.10). In the 4-dimensional embedded case, this condition for the line bundle u^*TM/TC is equivalent to the requirement that $c_1(A) > 0$.

In dimension four the genus of embedded J-holomorphic curves representing a class A is determined by the adjunction formula

$$2 - 2g + A \cdot A = c_1(A).$$

For any genus one can prove that this moduli space is regular, provided that $c_1(A) > 0$. The analytic argument was first suggested by Gromov and carried out by Hofer–Lizan–Sikorav [**179**]. One uses positivity of intersections to show that the moduli space consists entirely of embedded curves and then argues as in the proof of Lemma 3.3.3. Here, as mentioned above, one must reformulate the theory to allow the complex structure j on Σ to vary freely in Teichmüller space. Then the argument reduces to showing that a real linear Cauchy–Riemann operator on a complex line bundle of negative degree over a Riemann surface is always injective (see Theorem C.1.10 (iii)). In the case of immersed spheres, one can obtain similar results by using geometric comparison arguments: see McDuff [**260**]. Ivashkovich–Shevchishin [**202**] extend the discussion in [**179**] to the interesting case of curves with singularities.

The next remark illustrates some of the new features that arise when one considers embedded curves of higher genus.

EXAMPLE 3.3.7 (Elliptic surfaces). Consider the elliptic surfaces $E(n)$ for $n \geq 1$. $E(1)$ is simply a nine point blow up of $\mathbb{C}P^2$, where the nine points are chosen so that $E(1)$ admits the structure of a Lefschetz fibration over the projective line whose generic fiber is is an elliptic curve. For this it suffices to choose the nine points as the intersection points of two distinct degree 3 curves $f_1 = 0$ and $f_2 = 0$ in $\mathbb{C}P^2$, since then these points lie on the family of curves

$$\Gamma_\lambda := \{\lambda_1 f_1 + \lambda_2 f_2 = 0\}, \qquad [\lambda_1 : \lambda_2] \in \mathbb{C}P^1,$$

(see for example [**277**, Example 7.6]). When these nine points are blown up, the curves Γ_λ become distinct and so form the fibers of a map to S^2 that is generically a smooth fiber bundle with fiber a torus, but that has a finite number of special fibers over points λ for which Γ_λ is singular. In general, $E(n)$ is defined to be n-fold branched cover of $E(1)$, which in the symplectic context is the same as the n-fold fiber sum $E(1)\#_F \dots \#_F E(1)$. In particular, $E(2)$ is the $K3$-surface: see [**277**, Exercise 7.31].

Consider the fiber class F and the question of counting the number of embedded tori that represent this class. If the moduli problem is set up so that the complex structure j_Σ on the domain is allowed to vary, then its index is zero. This is the index $2n(1-g) + 2c_1(A) + 2$ of the relevant Cauchy–Riemann operator (where the summand 2 is the dimension of Teichmüller space) minus the dimension 2 of the group of holomorphic automorphisms (translations) of the torus. Moreover, the proof of Lemma 3.3.1 can be adapted to show that an embedded torus in a complex surface is regular precisely when the dual of its normal bundle $L = u^*TM/T\Sigma$ has no sections. When $c_1(L) = 0$ this is equivalent to saying that L is nontrivial. Now Kodaira's classification of complex surfaces implies that when $n > 2$ every complex structure on $E(n)$ admits a fibration over S^2 whose generic fiber is a torus. (For details see Beauville [**33**, Ch IX].) Thus for integrable J there is always a 2-(real)-parameter family of embedded tori in class F which are not regular. In contrast, although the product structure on the manifold $T^2 \times S^2$ is not regular for the class $F = [T^2 \times pt]$, this manifold does admit complex structures that are regular for F since it can be identified with the ruled surface $\mathbf{P}(L \oplus \mathbb{C})$ where $L \to T^2$ is a nontrivial bundle with $c_1(L) = 0$. (This ruled surface has precisely two regular holomorphic tori in the class F, namely $\mathbf{P}(L \oplus \{0\})$ and $\mathbf{P}(\{0\} \oplus \mathbb{C})$.)

One important difference between these cases is that, as is explained in McDuff [**265**], the (signed) number of tori in $E(n)$ of class F is $2-n$, and so is negative when $n > 2$; the corresponding number for $T^2 \times S^2$ is positive, namely 2. It follows from Remark 3.2.6 that no complex structure J on $E(n)$, $n > 2$, can be regular for

F. On the other hand, a generic (nonintegrable) perturbation of J admits a finite number of embedded tori whose signs add up to $2 - n$. This can also be explained by using obstruction bundle techniques.

3.4. Curves with pointwise constraints

This section is of preparatory nature and can be omitted at first reading; the results proved below are used in Chapter 6. The main technical result is Lemma 3.4.3 which extends the argument of Proposition 3.2.1 to cover the case when the vector fields $\xi \in \Omega^0(\Sigma, u^*TM)$ vanish at one or more points. The proof of Lemma 3.4.3 is based on new analytic ideas developed in Lemma 3.4.4. The final result (Lemma 3.4.6) presents a more constructive and geometric approach to these questions. Because these geometric ideas are applicable in many contexts it is worth explaining them, even though they work only in the smooth case.

Throughout we fix a class $A \in H_2(M;\mathbb{Z})$ and a closed Riemann surface Σ. We shall not assume that Σ is connected, unless otherwise mentioned, and denote by $\mathcal{M}^*(A,\Sigma;J)$ the moduli space of J-holomorphic curves $u:\Sigma\to M$ in the class A such that the restriction of u to each component of Σ is simple and no two components have the same image. Thus $\mathcal{M}^*(A,\Sigma;J)$ is the moduli space of simple maps with smooth but disconnected domains, a space that will be important in Chapter 6.

There are many situations in which one wants to consider curves $u:\Sigma\to M$ that satisfy some geometric constraints. Here we shall fix a finite sequence of pairwise distinct points

$$\mathbf{w} := (\mathrm{w}_1,\dots,\mathrm{w}_m) \in \Sigma^m$$

and consider the evaluation map $\mathrm{ev}_{\mathbf{w}}:\mathcal{M}^*(A,\Sigma;J)\to M^m$ defined by

$$\mathrm{ev}_{\mathbf{w}}(u) := (u(\mathrm{w}_1),\dots,u(\mathrm{w}_m)).$$

Given a smooth submanifold $X\subset M^m$ we examine the moduli space

$$\mathcal{M}^*(A,\Sigma;\mathbf{w},X;J) := \{u\in\mathcal{M}^*(A,\Sigma;J)|\mathrm{ev}_{\mathbf{w}}(u)\in X\}$$

of simple curves that take the tuple $(\mathrm{w}_1,\dots,\mathrm{w}_m)$ to X. As above we denote by $\mathcal{J}$ either the space of ω-tame or the space of ω-compatible almost complex structures. An almost complex structure $J\in\mathcal{J}$ is called **regular for the tuple** $(A,\Sigma,\mathbf{w},X)$ if $J\in\mathcal{J}_{\mathrm{reg}}(A,\Sigma)$ and the evaluation map $\mathrm{ev}_{\mathbf{w}}:\mathcal{M}^*(A,\Sigma;J)\to M^m$ is transverse to X. The set of such regular almost complex structures will be denoted by $\mathcal{J}_{\mathrm{reg}}(A,\Sigma;\mathbf{w},X)$.

THEOREM 3.4.1. *The set $\mathcal{J}_{\mathrm{reg}}(A,\Sigma;\mathbf{w},X)$ is residual in $\mathcal{J}$. Moreover, if $J\in\mathcal{J}_{\mathrm{reg}}(A,\Sigma;\mathbf{w},X)$ then the moduli space $\mathcal{M}^*(A,\Sigma;\mathbf{w},X;J)$ is a finite-dimensional smooth manifold of dimension $d=\dim\mathcal{M}^*(A,\Sigma;J)-\mathrm{codim}\,X$.*

Theorem 3.4.1 extends in a straightforward fashion to other spaces $\mathcal{J}$ of almost complex structures. For example, sometimes one wants to restrict $\mathcal{J}$ so that its elements satisfy some geometric constraints. This question is discussed in Remark 3.4.5.

The proof of Theorem 3.4.1 relies on the observation that the evaluation map on the universal moduli space has a surjective differential at every point. For this statement to make sense, we first need to check that the universal moduli space $\mathcal{M}^*(A,\Sigma;\mathcal{J}^\ell)$ is a manifold when Σ is disconnected. This is a very slight generalization of Proposition 3.2.1, and the details of its proof are at this stage left to the reader as an exercise. They are carried out later in the proof of Proposition 6.2.7. Now consider the universal evaluation map

$$\mathrm{ev}_{\mathbf{w},\mathcal{J}} : \mathcal{M}^*(A,\Sigma;\mathcal{J}^\ell) \to M^m$$

defined by

$$\mathrm{ev}_{\mathbf{w},\mathcal{J}}(u,J) := (u(\mathrm{w}_1),\dots,u(\mathrm{w}_m)).$$

PROPOSITION 3.4.2. *Every point in M^m is a regular value of* $\mathrm{ev}_{\mathbf{w},\mathcal{J}}$.

The proof of this proposition is quite complex and occupies the remainder of this section. We first show that it implies Theorem 3.4.1.

PROPOSITION 3.4.2 IMPLIES THEOREM 3.4.1. Proposition 3.4.2 implies that the universal evaluation map $\mathrm{ev}_{\mathbf{w},\mathcal{J}} : \mathcal{M}^*(A,\Sigma;\mathcal{J}^\ell) \to M^m$ is transverse to every submanifold of M^m. Hence the universal moduli space

$$\mathcal{M}^*(A,\Sigma;\mathbf{w},X;\mathcal{J}^\ell) := \left\{(u,J) \in \mathcal{M}^*(A,\Sigma;\mathcal{J}^\ell) \,|\, \mathrm{ev}_{\mathbf{w},\mathcal{J}}(u,J) \in X\right\}$$

is a $C^{\ell-1}$ Banach manifold for every $\ell \geq 2$. The rest of the argument follows the proof of Theorem 3.1.6. Namely, one shows that the projection

$$\pi : \mathcal{M}^*(A,\Sigma;\mathbf{w},X;\mathcal{J}^\ell) \to \mathcal{J}^\ell$$

is a Fredholm map of the expected index and then uses the Sard–Smale theorem and Taubes' argument. The details are left to the reader. Full details of the special case when X is the multidiagonal Δ^E corresponding to a tree T are given in the proof of Theorem 6.2.6 in Section 6.2. This proves Theorem 3.4.1. □

Next we give an elementary proof that every m-tuple $\mathbf{x} = (x_1,\dots,x_m) \in M^m$ of *pairwise distinct* points in M is a regular value of $\mathrm{ev}_{\mathbf{w},\mathcal{J}}$. The argument is based on the observation that the identity component of the symplectomorphism group of M acts transitively on each component of M. It acts on $\mathcal{M}^*(A,\Sigma;\mathcal{J}^\ell)$ via

$$\phi^*(u,J) := (\phi^{-1}\circ u, \phi^*J)$$

and the evaluation map $\mathrm{ev}_{\mathbf{w},\mathcal{J}}$ is equivariant under this action.

Let $(u,J) \in \mathcal{M}^*(A,\Sigma;\mathcal{J}^\ell)$ and suppose that the points $u(\mathrm{w}_1),\dots,u(\mathrm{w}_m)$ are pairwise distinct. Then, for any m-tuple of tangent vectors $v_i \in T_{u(\mathrm{w}_i)}M$, there exists a Hamiltonian function $H : M \to \mathbb{R}$ such that

$$X_H(u(\mathrm{w}_i)) + v_i = 0, \qquad i = 1,\dots,m.$$

Let ϕ_t denote the Hamiltonian flow of H. Then, by Exercise 3.1.4, the pair $(-X_H(u), \mathcal{L}_{X_H}J)$ is tangent to $\mathcal{M}^*(A,\Sigma;\mathcal{J}^\ell)$ and

$$d\mathrm{ev}_{\mathbf{w},\mathcal{J}}(u,J)\big(-X_H(u),\mathcal{L}_{X_H}J\big) = \left.\frac{d}{dt}\right|_{t=0} \mathrm{ev}_{\mathbf{w},\mathcal{J}}(\phi_t^{-1}\circ u, \phi_t^*J) = (v_1,\dots,v_m).$$

Hence $d\mathrm{ev}_{\mathbf{w},\mathcal{J}}(u,J)$ is surjective whenever $u(\mathrm{w}_1),\dots,u(\mathrm{w}_m)$ are pairwise distinct.

This suffices to prove Theorem 3.4.1 whenever X is disjoint from the **fat diagonal** $\Delta^m \subset M^m$ of all tuples $\mathbf{x} = (x_1,\dots,x_m) \in M^m$ that are not pairwise distinct. However, there are cases in which it is essential to consider submanifolds that intersect the fat diagonal. For example, in Chapter 6 we need to know that evaluation maps on distinct moduli spaces intersect transversally. The simplest case is where $\Sigma = \Sigma_1 \cup \Sigma_2$ has two components of genus zero and $\mathbf{w} = (\mathrm{w}_1,\mathrm{w}_2)$ with $\mathrm{w}_i \in \Sigma_i$. In this case Theorem 3.4.1 asserts that, for a generic almost complex structure J, the evaluation map $\mathrm{ev}_{\mathbf{w}} : \mathcal{M}^*(A,\Sigma;J) \to M \times M$ is transverse to the diagonal $X = \Delta \subset M \times M$. It follows that the moduli space of intersecting pairs of distinct simple J-holomorphic spheres is a smooth manifold of the predicted dimension for a generic almost complex structure J.[3] This observation also plays an important role in the gluing theorem for J-holomorphic spheres. To sum up, there are important examples in which $X \cap \Delta^m \neq \emptyset$ and in this situation the proof of Theorem 3.4.1 requires Proposition 3.4.2. For some purposes one could get by with the easier special case in which the points w_i lie on different components of Σ, but we will prove this result in full generality. The proof is based on the next lemma.

LEMMA 3.4.3. *Let $p > 2$ and $\ell \geq 2$ be an integer or $\ell = \infty$. Let $J \in \mathcal{J}^\ell$ and $u : \Sigma \to M$ be a simple J-holomorphic curve defined on a connected Riemann surface Σ. Suppose that $\mathrm{w}_0, \mathrm{w}_1, \ldots, \mathrm{w}_m \in \Sigma$ are pairwise distinct. Then, for every $\varepsilon > 0$ and every collection of tangent vectors*

$$v_i \in T_{u(\mathrm{w}_i)}M, \qquad i = 1, \ldots, m,$$

there is a vector field $\xi \in W^{\ell,p}(\Sigma, u^*TM)$ and a section $Y \in C^\ell(M, \mathrm{End}(TM, J, \omega))$ such that

$$\xi(\mathrm{w}_i) = v_i, \qquad D_u\xi + \tfrac{1}{2}Y(u)du \circ j = 0, \qquad \mathrm{supp}\, Y \subset B_\varepsilon(u(\mathrm{w}_0)), \tag{3.4.1}$$

for $i = 1, \ldots, m$.

LEMMA 3.4.3 IMPLIES PROPOSITION 3.4.2. Recall that Σ need not be connected and denote its connected components by Σ_j. The tangent space of the universal moduli space $\mathcal{M}^*(A, \Sigma; \mathcal{J}^\ell)$ at (u, J) consists of all pairs

$$(\xi, Y) \in W^{\ell,p}(\Sigma, u^*TM) \times C^\ell(M, \mathrm{End}(TM, J, \omega))$$

that satisfy the second equation in (3.4.1), and

$$\mathrm{dev}_{\mathbf{w},\mathcal{J}}(u, J)(\xi, Y) = (\xi(\mathrm{w}_1), \ldots, \xi(\mathrm{w}_m)).$$

Choose a constant $\varepsilon > 0$ and a collection of points $\mathrm{w}_{0j} \in \Sigma_j$, one for each connected component of Σ, such that the balls $B_\varepsilon(u(\mathrm{w}_{0j}))$ are pairwise disjoint and do not intersect $u(\Sigma_i)$ unless $i = j$. Here we use the fact that u is simple, i.e. its restriction to each component Σ_j is simple and the images of distinct components are distinct. Given any collection of vectors $v_i \in T_{u(\mathrm{w}_i)}M$, apply Lemma 3.4.3 to each connected component Σ_j of Σ to obtain elements (ξ_j, Y_j) tangent to $\mathcal{M}^*(A, \Sigma_j; \mathcal{J}^\ell)$ such that

$$\mathrm{supp} Y_j \subset B_\varepsilon(u(\mathrm{w}_{0j}))$$

and $\xi_j(\mathrm{w}_i) = v_i$ whenever w_i lies on Σ_j. Then, if ξ denotes the element of $W^{\ell,p}(\Sigma, u^*TM)$ whose restriction to Σ_j is ξ_j for each j, the pair $(\xi, \sum_j Y_j)$ is tangent to $\mathcal{M}^*(A, \Sigma; \mathcal{J}^\ell)$. Hence $\mathrm{dev}_{\mathbf{w},\mathcal{J}}(u, J)$ is surjective for every $(u, J) \in \mathcal{M}^*(A, \Sigma; \mathcal{J}^\ell)$. This proves Proposition 3.4.2 □

The above proof shows the importance of the precise control of the support of Y given by Lemma 3.4.3. To obtain this precision we need an improved regularity result that applies when the operator D_u is restricted to sections $\xi \in \Omega^0(\Sigma, u^*TM)$ that vanish at finitely many points. Recall from Proposition 3.1.10 that if J is of class C^ℓ then every J-holomorphic curve u is of class $W^{\ell+1,p}$ and the linearized operator D_u is well defined from $W^{\ell,p}$ to $W^{\ell-1,p}$.

---

[3] The elements of such a moduli space are examples of *stable maps*. Some references use the term *cusp curves*.

LEMMA 3.4.4. *Let $p > 2$ and $\ell \geq 2$ be an integer or $\ell = \infty$. Let $J \in \mathcal{J}^\ell$, $u : \Sigma \to M$ be a simple J-holomorphic curve, defined on a connected Riemann surface Σ, and $Z \subset \Sigma$ be a finite set. Let $q > 1$ such that $1/p + 1/q = 1$ and suppose that $\eta \in L^q(\Sigma, \Lambda^{0,1} \otimes_J u^*TM)$ satisfies*

$$\xi(Z) = 0 \quad \Longrightarrow \quad \int_\Sigma \langle \eta, D_u\xi \rangle \,\mathrm{dvol}_\Sigma = 0 \tag{3.4.2}$$

for every $\xi \in W^{1,p}(\Sigma, u^*TM)$. Then $\eta \in W^{\ell,p}_{\mathrm{loc}}(\Sigma \setminus Z, \Lambda^{0,1} \otimes_J u^*TM)$ and $D_u^*\eta = 0$ on $\Sigma \setminus Z$. Moreover, if η vanishes on some nonempty open set then $\eta \equiv 0$.

LEMMA 3.4.4 IMPLIES LEMMA 3.4.3. We begin by examining the image of the operator $(\xi, Y) \mapsto D_u\xi + \frac{1}{2}Y(u)du \circ j_\Sigma$, restricted to the set of all pairs $(\xi, Y) \in W^{k,p}(\Sigma, u^*TM) \times T_J\mathcal{J}^\ell$, where ξ vanishes on the finite set

$$Z := \{\mathrm{w}_1, \ldots, \mathrm{w}_m\}$$

and Y is supported near w_0. Thus we define the subspace

$$\mathcal{Z}^k := \left\{ D_u\xi + \tfrac{1}{2}Y(u)du \circ j_\Sigma \;\middle|\; \begin{array}{l} \xi \in W^{k,p},\ Y \in C^\ell,\ \xi(Z) = 0, \\ \operatorname{supp} Y \subset B_\varepsilon(u(\mathrm{w}_0)) \end{array} \right\}.$$

We claim that

$$\mathcal{Z}^k = W^{k-1,p}(\Sigma, \Lambda^{0,1} \otimes_J u^*TM) \tag{3.4.3}$$

for $k = 1, \ldots, \ell$. To see this, assume first that $k = 1$. Since

$$D_u : W^{1,p}(\Sigma, u^*TM) \to L^p(\Sigma, \Lambda^{0,1} \otimes_J u^*TM)$$

is a Fredholm operator, the image of the subspace $\{\xi \in W^{1,p}(\Sigma, u^*TM) \,|\, \xi(Z) = 0\}$ under D_u is closed and has finite codimension. Hence the subspace $\mathcal{Z}^1$ is closed in $L^p(\Sigma, \Lambda^{0,1} \otimes_J u^*TM)$. It remains to prove that it is dense.

To this end assume that $\eta \in L^q(\Sigma, \Lambda^{0,1} \otimes_J u^*TM)$ annihilates $\mathcal{Z}^1$, where $1/p + 1/q = 1$. Then η satisfies (3.4.2). Hence it follows from Lemma 3.4.4 that

$$\eta \in W^{\ell,p}_{\mathrm{loc}}(\Sigma \setminus Z, \Lambda^{0,1} \otimes_J u^*TM), \qquad D_u^*\eta = 0 \text{ on } \Sigma \setminus Z.$$

Moreover,

$$\int_\Sigma \langle \eta, Y(u)du \circ j_\Sigma \rangle \,\mathrm{dvol}_\Sigma = 0$$

for every $Y \in C^\ell(M, \mathrm{End}(TM, J, \omega))$ with support in $B_\varepsilon(u(\mathrm{w}_0))$. Since u is simple, there exists an injective point $z \in \Sigma$ such that $d(u(z), u(\mathrm{w}_0)) < \varepsilon$. If $\eta(z) \neq 0$, one argues as in the proof of Proposition 3.2.1 that there exists an infinitesimal almost complex structure $Y \in C^\ell(M, \mathrm{End}(TM, J, \omega))$ with support in $B_\varepsilon(u(\mathrm{w}_0))$ such that the function $\langle \eta, Y(u)du \circ j_\Sigma \rangle$ is nonnegative everywhere and is positive at the point z. Hence the integral above does not vanish, a contradiction. This shows that η vanishes at every injective point near w_0, hence η vanishes on some open set. Hence, by Lemma 3.4.4, $\eta = 0$. Thus the annihilator of $\mathcal{Z}^1$ is zero and so, by the Hahn–Banach theorem, $\mathcal{Z}^1$ is dense in $L^p(\Sigma, \Lambda^{0,1} \otimes_J u^*TM)$ as claimed.

Thus we have proved (3.4.3) in the case $k = 1$. To prove it for $k > 1$ let $\eta \in W^{k-1,p}(\Sigma, \Lambda^{0,1} \otimes_J u^*TM)$ be given. Then, by what we just proved, there exists a pair $(\xi, Y) \in W^{1,p}(\Sigma, u^*TM) \times T_J\mathcal{J}^\ell$ satisfying

$$\xi(Z) = 0, \qquad D_u\xi + \tfrac{1}{2}Y(u)du \circ j_\Sigma = \eta, \qquad \operatorname{supp} Y \subset B_\varepsilon(u(\mathrm{w}_0)). \tag{3.4.4}$$

It follows from (3.4.4) that $D_u\xi \in W^{k-1,p}$. Since $k \leq \ell$ it follows from Theorem C.2.3 that $\xi \in W^{k,p}$. This proves (3.4.3) for general k.

To prove the lemma, choose any vector field $\xi' \in W^{\ell,p}(\Sigma, u^*TM)$ along u such that $\xi'(\mathrm{w}_i) = v_i$ for $i = 1, \dots, m$. By (3.4.3), there exists a pair $(\xi'', Y) \in W^{\ell,p}(\Sigma, u^*TM) \times T_J\mathcal{J}^\ell$ such that (3.4.4) holds with $\eta := -D_u\xi'$ and ξ replaced by ξ''. Hence the pair $(\xi' + \xi'', Y)$ satisfies (3.4.1). This proves Lemma 3.4.3. □

PROOF OF LEMMA 3.4.4. To prove the regularity statement, choose conformal coordinates $s + it$ in an open set $\Omega \subset \mathbb{C}$ on $\Sigma \setminus Z$ so that $\mathrm{dvol}_\Sigma = \lambda^2\, ds \wedge dt$. Write.

$$\eta =: \xi'\, ds - J(u)\xi'\, dt, \qquad \xi'(s,t) \in T_{u(s,t)}M.$$

Then, if $\eta \in W^{1,p}(\Sigma, u^*TM)$ has support in Ω, equation (3.4.2) has the form

$$\begin{aligned} 0 &= \int_\Omega \langle \eta, D_u\xi \rangle \lambda^2\, dsdt \\ &= \int_\Omega \langle \xi', \nabla_s\xi + J(u)\nabla_t\xi + (\nabla_\xi J)(u)\partial_t u \rangle \lambda^2\, dsdt \\ &= \int_\Omega \langle \xi', \nabla_s\xi + J(u)\nabla_t\xi \rangle \lambda^2\, dsdt + \int_\Omega \langle B_{s,t}(\xi', \partial_t u), \xi \rangle \lambda^2\, dsdt. \end{aligned}$$

Here $B_{s,t} : T_{u(s,t)}M \times T_{u(s,t)}M \to T_{u(s,t)}M$ is the bilinear map defined by

$$\langle B_{s,t}(v_1, v_2), v_3 \rangle := \langle v_1, (\nabla_{v_3} J)(u)v_2 \rangle.$$

Note that, since J and u are of class C^ℓ, the section $B_{s,t}(\xi', \partial_t u)$ belongs to the same Sobolev space as ξ' so long as the number of derivatives is at most $\ell - 1$. To begin with, ξ' belongs to L^q. Hence, by Lemma B.4.6 with $r = q$, we have $\xi' \in W^{1,q}_{\mathrm{loc}} \subset L^{q'}_{\mathrm{loc}}$, where $q' := 2q/(2-q) > 2$ and $W^{1,q}_{\mathrm{loc}} := W^{1,q}_{\mathrm{loc}}(\Omega, u^*TM)$. Applying Lemma B.4.6 again, with r and q both replaced by q', we obtain $\xi' \in W^{1,q'}_{\mathrm{loc}} \subset L^p_{\mathrm{loc}}$. Now apply Proposition B.4.9 inductively to obtain $\xi' \in W^{k,p}_{\mathrm{loc}}$ for $k = 1, \dots, \ell$. This proves regularity. The equation $D_u^*\eta = 0$ now follows by integration by parts.

To prove the last assertion note that the equation $D_u^*\eta = 0$ has the form

$$-\nabla_s\xi' + J(u)\nabla_t\xi' + C\xi' = 0$$

in local coordinates, where $C : \Omega \to \mathbb{R}^{2n\times 2n}$ is a matrix valued function of class $C^{\ell-1}$. Hence it follows from the Carleman similarity principle (Theorem 2.3.5) that the set of points $z \in \Sigma \setminus Z$ at which ξ' vanishes to infinite order is open and closed. By assumption, this set is nonempty and so η vanishes on $\Sigma \setminus Z$. This proves Lemma 3.4.4. □

REMARK 3.4.5. Sometimes it is useful to prove regularity results for a restricted set of almost complex structures. For example, when constructing the relative Gromov–Witten invariants of a pair (M, V), where V is a symplectic submanifold of M of codimension 2, it is useful to restrict attention to the space $\mathcal{J}_V$ of ω-tame (or compatible) almost complex structures for which V is J-holomorphic, i.e. $J(TV) \subset TV$. (One may need to impose more restrictions: see for example Ionel–Parker [**200**].) Whatever the codimension of V, it is easy to check that Lemma 3.4.3 still holds in this case, provided that one adds the condition that $u(\mathrm{w}_0) \notin V$. Hence the moduli space

$$\mathcal{M}^*(A, \Sigma; V, \mathbf{w}, X; J) := \{u \in \mathcal{M}^*(A, \Sigma; J) | \mathrm{ev}_{\mathbf{w}}(u) \in X,\ u(\Sigma) \not\subset V\}$$

of simple curves that take the tuple $(\mathrm{w}_1, \dots, \mathrm{w}_m)$ to X and do not lie entirely in V is a smooth manifold of the correct dimension for generic $J \in \mathcal{J}_V$.

We end this section by proving a smooth version of Lemma 3.4.3. It yields an infinitesimal almost complex structure Y that is supported in a neighbourhood of the image under u of a union of small annuli centered at the points w_i rather than in some unrelated set $u(B_\varepsilon(\mathrm{w}_0))$. The idea is to construct a local solution ξ of $D_u\xi = 0$ near the points w_i and then modify ξ by a cutoff function to make it vanish outside a small neighbourhood of these points. We then have to compensate for the effect of the cutoff function by introducing an infinitesimal almost complex structure Y which is supported in a neighbourhood of the image under u of small annuli centered at the w_i. The argument can be viewed as a linearized version of McDuff [**254**, Proposition 4.1] in which the holomorphic curve u itself is perturbed near the points w_i.

LEMMA 3.4.6. *Let $u : \Sigma \to M$ be a simple J-holomorphic curve for some smooth almost complex structure J, and $\mathrm{w}_1, \dots, \mathrm{w}_m$ be pairwise distinct points in Σ. Then there is a constant $\delta > 0$ such that the following holds. For every set of tangent vectors $v_i \in T_{u(\mathrm{w}_i)}M$, $1 \le i \le m$, and every pair of real numbers $0 < \rho < r < \delta$ there is a smooth vector field $\xi \in \Omega^0(\Sigma, u^*TM)$ along u and an infinitesimal almost complex structure $Y \in \Omega^0(M, \mathrm{End}(TM, J, \omega))$ such that*

$$\xi(\mathrm{w}_i) = v_i, \qquad D_u\xi + \tfrac{1}{2}Y(u)du \circ j_\Sigma = 0, \qquad \mathrm{supp}\,\xi \subset \bigcup_{i=1}^m B_r(\mathrm{w}_i),$$

for $i = 1, \dots m$. Moreover Y is supported in an arbitrarily small neighbourhood of the union $\bigcup_{i=1}^m u(B_r(\mathrm{w}_i) \setminus B_\rho(\mathrm{w}_i))$.

PROOF. First consider the case $m = 1$. Choose conformal coordinates $s + it$ on Σ near w_1 so that the point w_1 has coordinates $s = t = 0$. Then

$$D_u\xi + \tfrac{1}{2}Y(u)du \circ j_\Sigma = \tfrac{1}{2}\Big(\eta\, ds - J(u)\eta\, dt\Big),$$

where

$$\eta := \nabla_s\xi + J(u)\nabla_t\xi + \nabla_\xi J(u)\partial_t u + Y(u)\partial_t u.$$

The first step in the argument is to transform this equation into one where $J(u)$ is replaced by the constant matrix

$$J_0 := \begin{pmatrix} 0 & -\mathbb{1} \\ \mathbb{1} & 0 \end{pmatrix}.$$

To this end choose a (smooth) unitary trivialization $\Phi(s,t) : \mathbb{R}^{2n} \to T_{u(s,t)}M$ such that

$$J(u(s,t))\Phi(s,t) = \Phi(s,t)J_0, \quad \omega(\Phi v, \Phi w) = (J_0 v)^T w,$$

where the superscript T denotes the transpose, and write

$$\xi(s,t) = \Phi(s,t)\xi_0(s,t), \qquad v_1 = \Phi(0,0)v_0, \qquad \partial_t u(s,t) = -\Phi(s,t)\zeta_0(s,t).$$

Then we have

$$D_u\xi + \tfrac{1}{2}Y(u)du \circ j_\Sigma = \Phi(\eta_0\, ds - J_0\eta_0\, dt)$$

where

$$\eta_0 = \partial_s\xi_0 + J_0\partial_t\xi_0 + A_0\xi_0 - Y_0\zeta_0$$

and

$$Y_0 := \Phi^{-1}Y(u)\Phi, \qquad A_0 := \Phi^{-1}\Big(\nabla_s\Phi + J(u)\nabla_t\Phi + (\nabla_\Phi J)(u)\partial_t u\Big).$$

The compatibility condition $Y(x) \in \mathrm{End}(T_xM, J_x, \omega_x)$ for $x \in M$ translates into

$$Y_0 = Y_0^T = J_0 Y_0 J_0.$$

In this language, our problem is now to find such a function $Y_0 : B_1 \to \mathbb{R}^{2n\times 2n}$ together with a vector field $\xi_0 : B_1 \to \mathbb{R}^{2n}$ such that

$$\partial_s \xi_0 + J_0 \partial_t \xi_0 + A_0 \xi_0 = Y_0 \zeta_0, \qquad \xi_0(0) = v_0,$$

where the vector field ζ_0 and vector v_0 are specified. Note that Y_0 must be the pullback via Φ of a function defined on M. Thus it must be supported on a subset of B_1 on which u is injective.

First consider the equation

$$\partial_s \xi_0 + J_0 \partial_t \xi_0 + A_0 \xi_0 = 0, \qquad \xi_0(0) = v_0. \tag{3.4.5}$$

If $A_0 = 0$ and we impose an appropriate Lagrangian boundary condition on the boundary of the unit disc then equation (3.4.5) has a unique solution in B_1 satisfying the boundary condition. For example, in the case $n = 1$ consider the boundary condition $\xi_0(e^{i\theta}) \in e^{i\theta/2}\mathbb{R}$. Then every holomorphic map $\xi_0 : B_1 \to \mathbb{C}$ satisfying this condition has the form $\xi_0(z) = a + \bar{a}z$ for some $a \in \mathbb{C}$: see Step 2 in the proof of Theorem C.4.1. For general n we can take the direct sum of these boundary loops in $\mathbb{R}^{2n} = \mathbb{C}^n$. It follows that, for every value of v_0, equation (3.4.5) with $A_0 = 0$ has a unique solution in B_1 satisfying the boundary condition. This continues to hold for A_0 sufficiently small. By a rescaling argument this can be extended to arbitrary A_0 but with the domain of the solution ξ_0 restricted to B_δ where δ depends on A_0.

Now let ρ and r be two real numbers such that $0 < \rho < r < \delta$. Shrinking the annulus $B_r \setminus B_\rho$, if necessary, we may assume, by Corollary 2.5.5, that the restriction of u to $B_r \setminus B_\rho$ is an embedding and $u^{-1}(u(B_r \setminus B_\rho)) = B_r \setminus B_\rho$. Here we use the fact that u is simple. Now choose a cutoff function $\beta : \mathbb{C} \to [0, 1]$ such that $\beta(z) = 1$ for $|z| \le \rho$ and $\beta(z) = 0$ for $|z| \ge r$. Then

$$\partial_s(\beta\xi_0) + J_0\partial_t(\beta\xi_0) + A_0(\beta\xi_0) = (\partial_s\beta)\xi_0 + (\partial_t\beta)J_0\xi_0 =: \eta_0.$$

So we need a function $Y_0 : B_1 \to \mathbb{R}^{2n\times 2n}$ with support in $B_r \setminus B_\rho$ such that

$$Y_0\zeta_0 = \eta_0, \qquad Y_0 = Y_0^T = J_0 Y_0 J_0, \tag{3.4.6}$$

where $\zeta_0 = -\Phi^{-1}\partial_t u$. Notice that η_0 is supported in the annulus $B_r \setminus B_\rho$ and that $\zeta_0 \ne 0$ on the closure of $B_r \setminus B_\rho$. Hence Lemma 3.2.2 provides an explicit formula for a function Y_0 that satisfies (3.4.6). (Note that this formula depends smoothly on $\eta(z)$ provided that $\xi(z) \ne 0$.)

The final step is to check that Y_0 is the pullback

$$Y_0 = \Phi^{-1} Y(u) \Phi$$

of a section $Y \in C^\infty(M, \mathrm{End}(TM, J, \omega))$ over M. This holds because the restriction of u to $B_r \setminus B_\rho$ is an embedding. Further, since $B_r \setminus B_\rho = u^{-1}(u(B_r \setminus B_\rho))$ it is possible to choose the support of Y so close to $u(B_r \setminus B_\rho)$ that $Y(u)$ vanishes on the complement of $B_r \setminus B_\rho$. It follows that Y and $\xi := \beta\Phi\xi_0$ satisfy the requirements of the lemma.

This completes the proof when $m = 1$. The only difficulty in extending this to arbitrary m lies in ensuring that the different elements $Y_i \in C^\infty(M, \mathrm{End}(TM, J, \omega))$ needed to correct for the cutoff functions at each w_i do not interfere with one another. More precisely, each section $Y_i(u)$ should be supported in a suitable annulus around w_i. This is possible by Corollary 2.5.5. □

EXERCISE 3.4.7. Prove Proposition 3.4.2 from Lemma 3.4.6 and Corollary 2.5.5. Then prove Theorem 3.4.1 using Floer's ideas as outlined in Remark 3.2.7.

EXERCISE 3.4.8. The argument in the proof of Lemma 3.4.6 loses derivatives. Suppose that the almost complex structure J in Lemma 3.4.6 is of class C^ℓ. How smooth can you make Y_0 and ξ_0? Which steps in the construction lose smoothness?

3.5. Implicit function theorem

In this section we shall establish a refined version of the implicit function theorem. Roughly speaking, it asserts that if u is an *approximate J-holomorphic curve*, in the sense that $\bar{\partial}_J(u)$ is sufficiently small in the L^p-norm, and if the operator D_u is surjective with a uniformly bounded right inverse then there exists an actual J-holomorphic curve near u. This theorem will play an important role in the gluing construction in Chapter 10 and hence we shall explain it carefully. This section may be skipped at a first reading. We shall assume throughout that J is a fixed smooth almost complex structure.

In view of Proposition 3.1.10 it suffices to work with the Sobolev space $\mathcal{B}^{1,p}$ of $W^{1,p}$-maps $u : \Sigma \to M$ for some fixed number $p > 2$. Consider the infinite dimensional Banach space bundle $\mathcal{E}^p \to \mathcal{B}^{1,p}$ whose fiber at $u \in \mathcal{B}^{1,p}$ is the space

$$\mathcal{E}^p_u = L^p(\Sigma, \Lambda^{0,1} \otimes_J u^*TM)$$

of complex antilinear 1-forms on Σ of class L^p which take values in the pullback tangent bundle u^*TM. The nonlinear Cauchy-Riemann equations determine a section $\bar{\partial}_J : \mathcal{B}^{1,p} \to \mathcal{E}^p$ of this bundle whose vertical differential at a J-holomorphic curve u is the operator

$$D_u : W^{1,p}(\Sigma, u^*TM) \to L^p(\Sigma, \Lambda^{0,1} \otimes_J u^*TM)$$

introduced in Section 3.1 (see equation (3.1.4)).

We shall now turn to the question of surjectivity of the operator D_u. By Proposition 3.1.11, this is equivalent to the injectivity of the formal adjoint operator D_u^*. For the implicit function theorem it is important to have a quantitative expression of surjectivity. Roughly speaking, this means that the norm of a suitable right inverse does not get too large. One possibility for constructing such a right inverse is to take the operator

$$Q_u = D_u^*(D_u D_u^*)^{-1} : L^p(\Sigma, \Lambda^{0,1} \otimes_J u^*TM) \to W^{1,p}(\Sigma, u^*TM).$$

One can prove that an inequality of the form

$$\|\eta\|_{W^{1,q}} \le c \|D_u^*\eta\|_{L^q}$$

implies a similar inequality for the operator D_u restricted to the image of D_u^*, a complement of the kernel of D_u. (See for example Lemma 4.5 in [**92**].) An alternative technique for constructing a right inverse is to reduce the domain $W^{1,p}(\Sigma, u^*TM)$ of D_u, by imposing pointwise conditions on ξ, so that the resulting operator is bijective, and then taking Q_u to be the inverse of this restricted operator.

The next theorem is a refined version of the implicit function theorem discussed in the beginning of this section. We prove that if u is an *approximate* J-holomorphic curve with *sufficiently surjective* operator D_u, then there are J-holomorphic curves near u and they can be modelled on a neighbourhood of zero in the kernel of D_u. More explicitly, for every sufficiently small section $\xi \in \ker D_u$ we can find a unique J-holomorphic curve of the form $v_\xi = \exp_u(\xi + Q_u\eta)$.

For our applications in the gluing argument it is crucial that the volume forms on Σ are allowed to vary. The purpose of the following remark is to introduce a class of volume forms for which we can obtain uniform estimates in the implicit function theorem.

REMARK 3.5.1. Let (Σ, j_Σ) be a compact complex 2-manifold. Given a constant $p > 2$ and a positive volume form dvol_Σ, we denote by $c_p(\mathrm{dvol}_\Sigma)$ the norm of the Sobolev embedding $W^{1,p}(\Sigma) \hookrightarrow C^0(\Sigma)$ (see Theorem B.1.11). Thus

$$c_p(\mathrm{dvol}_\Sigma) := \sup_{0 \neq f \in C^\infty(\Sigma)} \frac{\|f\|_{L^\infty}}{\|f\|_{W^{1,p}}}.$$

Now let $E \to \Sigma$ be a Riemannian vector bundle and ∇ be any Riemannian connection on E. Then, for every $W^{1,p}$-section $\xi \in W^{1,p}(\Sigma, E)$, the scalar function $z \mapsto |\xi(z)|$ belongs to $W^{1,p}(\Sigma)$ and

$$\int_\Sigma |d\,|\xi||^p \,\mathrm{dvol}_\Sigma \le \int_\Sigma |\nabla \xi|^p \,\mathrm{dvol}_\Sigma$$

for any metric on Σ. To see this, take the limit of the functions $z \mapsto \sqrt{|\xi(z)|^2 + \varepsilon^2}$ as $\varepsilon \to 0$ and use the fact that

$$|d|\xi|| = \frac{|\langle \nabla \xi, \xi \rangle|}{|\xi|} \le |\nabla \xi|$$

when ξ is nonzero. It follows that

$$\sup_\Sigma |\xi| \le c_p(\mathrm{dvol}_\Sigma) \left(\int_\Sigma (|\xi|^p + |\nabla \xi|^p) \,\mathrm{dvol}_\Sigma \right)^{1/p}$$

for every $W^{1,p}$-section ξ of every Riemannian vector bundle $E \to \Sigma$ and every Riemannian connection ∇.

In the next theorem we shall fix a compact complex 2-manifold (Σ, j_Σ) and consider all volume forms dvol_Σ for which the constant $c_p(\mathrm{dvol}_\Sigma)$ satisfies a uniform upper bound. The volume form dvol_Σ and the complex structure j_Σ together determine a metric on Σ with respect to which all relevant norms are to be understood. The Sobolev norms for sections of the bundle $u^*TM \to \Sigma$ are understood with respect to the Levi-Civita connection ∇ of the metric determined by ω and J.

THEOREM 3.5.2. *Fix a constant $p > 2$ and a closed Riemann surface (Σ, j_Σ). Then, for every constant $c_0 > 0$, there exists a constant $\delta > 0$ such that the following holds for every volume form dvol_Σ on Σ such that $c_p(\mathrm{dvol}_\Sigma) \le c_0$. Suppose $u \in W^{1,p}(\Sigma, M)$ and $\xi_0 \in W^{1,p}(\Sigma, u^*TM)$ satisfy*

$$\|du\|_{L^p} \le c_0, \qquad \|\xi_0\|_{W^{1,p}} < \frac{\delta}{8}, \qquad \left\|\bar\partial_J(\exp_u(\xi_0))\right\|_{L^p} < \frac{\delta}{4c_0}.$$

Moreover, suppose that $Q_u : L^p(\Sigma, \Lambda^{0,1} \otimes_J u^*TM) \to W^{1,p}(\Sigma, u^*TM)$ is a right inverse of D_u such that

$$D_u Q_u = \mathbb{1}, \qquad \|Q_u\| \le c_0.$$

Then there exists a unique section $\xi = Q_u \eta \in W^{1,p}(\Sigma, u^*TM)$ such that

$$\bar\partial_J(\exp_u(\xi_0 + \xi)) = 0, \qquad \|\xi_0 + \xi\|_{W^{1,p}} < \delta.$$

Moreover, $\|\xi\|_{W^{1,p}} \le 2c_0 \left\|\bar\partial_J(\exp_u(\xi_0))\right\|_{L^p}$.

The proof is an application of the implicit function theorem. Consider the spaces

$$\mathcal{X}_u := W^{1,p}(\Sigma, u^*TM), \qquad \mathcal{Y}_u := L^p(\Sigma, \Lambda^{0,1} \otimes_J u^*TM)$$

and the map $\mathcal{F}_u : \mathcal{X}_u \to \mathcal{Y}_u$ defined by (3.1.3), i.e.

$$\mathcal{F}_u(\xi) := \Phi_u(\xi)^{-1}\big(\bar{\partial}_J(\exp_u(\xi))\big)$$

for $\xi \in \mathcal{X}_u$, where $\Phi_u(\xi) : u^*TM \to \exp_u(\xi)^*TM$ is defined by parallel transport with respect to the Hermitian connection $\widetilde{\nabla}$ along the geodesics $s \mapsto \exp_{u(z)}(s\xi(z))$. Then, by Proposition 3.1.1, the differential of $\mathcal{F}_u$ at zero is given by

$$d\mathcal{F}_u(0) = D_u.$$

The next result can be interpreted as a quadratic estimate for $\mathcal{F}_u$ (Remark 3.5.5). We phrase it in terms of an estimate for the differential $d\mathcal{F}_u(\xi) : \mathcal{X}_u \to \mathcal{Y}_u$ of the map $\mathcal{F}_u(\xi) : \mathcal{X}_u \to \mathcal{Y}_u$. Note that the volume form on Σ does not enter into the definition of the map $\mathcal{F}_u$.

PROPOSITION 3.5.3. *Fix a constant $p > 2$ and a closed Riemann surface (Σ, j_Σ). Then, for every constant $c_0 > 0$, there exists a constant $c > 0$ such that the following holds for every volume form* dvol_Σ *on Σ such that $c_p(\mathrm{dvol}_\Sigma) \le c_0$. If $u \in W^{1,p}(\Sigma, M)$ and $\xi \in W^{1,p}(\Sigma, u^*TM)$ satisfy*

$$\|du\|_{L^p} \le c_0, \qquad \|\xi\|_{L^\infty} \le c_0, \tag{3.5.1}$$

then

$$\|d\mathcal{F}_u(\xi) - D_u\| \le c\,\|\xi\|_{W^{1,p}}. \tag{3.5.2}$$

Here $\|\cdot\|$ denotes the operator norm on $\mathcal{L}(\mathcal{X}_u, \mathcal{Y}_u)$.

PROOF. The proof is by direct calculation. Given $x \in M$ and $\xi \in T_xM$, we define the (bi)linear maps

$$E_x(\xi) : T_xM \to T_{\exp_x(\xi)}M, \qquad \Psi_x(\xi) : T_xM \times T_xM \to T_{\exp_x(\xi)}M$$

by

$$E_x(\xi)\xi' := \frac{d}{dt}\exp_x(\xi + t\xi')\Big|_{t=0}, \qquad \Psi_x(\xi;\xi',\eta) := \widetilde{\nabla}_t\left(\Phi_u(\xi + t\xi')\eta\right)\Big|_{t=0}.$$

Note that $\Psi_x(0;\xi',\eta) = 0$ since when $\xi = 0$ we are taking the covariant derivative along the same geodesic used to define $\Phi_u(t\xi')$.

Now differentiate the identity

$$\Phi_u(\xi + t\xi')\mathcal{F}_u(\xi + t\xi') = \bar{\partial}_J(\exp_u(\xi + t\xi'))$$

covariantly (with respect to $\widetilde{\nabla}$) at $t = 0$ to obtain

$$\Phi_u(\xi)d\mathcal{F}_u(\xi)\xi' + \Psi_u(\xi;\xi',\mathcal{F}_u(\xi)) = D_{\exp_u(\xi)}\left(E_u(\xi)\xi'\right).$$

We write this identity in the form

$$d\mathcal{F}_u(\xi)\xi' = \Phi_u(\xi)^{-1}D_{\exp_u(\xi)}\left(E_u(\xi)\xi'\right) - \Phi_u(\xi)^{-1}\Psi_u(\xi;\xi',\mathcal{F}_u(\xi)). \tag{3.5.3}$$

Choose a constant $c_1 > 0$ such that the inequalities

$$|E_x(\xi)| \le c_1, \qquad |\Psi_x(\xi;\xi',\eta)| \le c_1\,|\xi|\,|\xi'|\,|\eta|$$

hold for every $x \in M$, every $\xi \in T_xM$ such that $|\xi| \le c_0$, and every $\eta \in T_xM$. Here $|\cdot|$ denotes the operator norm on $\mathcal{L}(T_xM, T_{\exp_x(\xi)}M)$ whenever appropriate, and the second inequality holds because $\Psi_x(0;\xi',\eta)=0$. The second term in (3.5.3) then satisfies the pointwise estimate

$$\left|\Phi_u(\xi)^{-1}\Psi_u(\xi;\xi',\mathcal{F}_u(\xi))\right| \le c_1\,|d\exp_u(\xi)|\,|\xi|\,|\xi'|,$$

where $d\exp_u(\xi)$ is the derivative of the map $\exp_u(\xi):S^2\to M$. There is a constant c_2, depending only c_0 and the metric on M, such that

$$|d\exp_u(\xi)| \le c_2\left(|du|+|\nabla\xi|\right)$$

whenever u and ξ satisfy (3.5.1). Hence

$$\left\|\Phi_u(\xi)^{-1}\Psi_u(\xi;\xi',\mathcal{F}_u(\xi))\right\|_{L^p} \le c_1c_2\left(\|du\|_{L^p}+\|\nabla\xi\|_{L^p}\right)\|\xi\|_{L^\infty}\|\xi'\|_{L^\infty}.$$

Since $\|du\|_{L^p}\le c_0$, $\|\xi\|_{L^\infty}\le c_0$, and $\|\xi\|_{L^\infty}\le c_0\|\xi\|_{W^{1,p}}$ it follows that

$$\left\|\Phi_u(\xi)^{-1}\Psi_u(\xi;\xi',\mathcal{F}_u(\xi))\right\|_{L^p} \le c_3\,\|\xi\|_{W^{1,p}}\|\xi'\|_{W^{1,p}}, \tag{3.5.4}$$

where $c_3:=(c_0+1)c_0^2c_1c_2$.

We show below that the first term in (3.5.3) may be estimated as follows: there is a constant $c_4>0$, depending only on c_0 (and J, ω, and j_Σ), such that

$$\left\|\Phi_u(\xi)^{-1}D_{\exp_u(\xi)}\left(E_u(\xi)\xi'\right)-D_u\xi'\right\|_{L^p} \le c_4\,\|\xi\|_{W^{1,p}}\|\xi'\|_{W^{1,p}} \tag{3.5.5}$$

whenever u and ξ satisfy (3.5.1). These two estimates (3.5.4) and (3.5.5) immediately imply the desired conclusion (3.5.2).

Hence it remains to prove (3.5.5). To this end, abbreviate $u_\xi:=\exp_u(\xi)$ and consider the identity

$$\begin{aligned} D_{u_\xi}\left(E_u(\xi)\xi'\right)-\Phi_u(\xi)D_u\xi' &= \Big(\nabla(E_u(\xi)\xi')-\Phi_u(\xi)\nabla\xi'\Big)^{0,1} \\ &\quad -\tfrac{1}{2}J(u_\xi)\Big((\nabla_{E_u(\xi)\xi'}J)(u_\xi)du_\xi-\Phi_u(\xi)(\nabla_{\xi'}J)(u)du\Big)^{0,1}. \end{aligned}$$

We consider the two summands on the right separately. Since J is an isometry of TM, the terms in the second summand can be estimated pointwise by

$$\begin{aligned} &\left|(\nabla_{E_u(\xi)\xi'}J)(u_\xi)du_\xi-\Phi_u(\xi)(\nabla_{\xi'}J)(u)du\right| \\ &\quad\le \left|(\nabla_{E_u(\xi)\xi'}J)(u_\xi)du_\xi-(\nabla_{E_u(\xi)\xi'}J)(u_\xi)\Phi_u(\xi)du\right| \\ &\qquad+\left|(\nabla_{E_u(\xi)\xi'}J)(u_\xi)\Phi_u(\xi)du-\Phi_u(\xi)(\nabla_{\xi'}J)(u)du\right| \\ &\quad\le c_1\,\|\nabla J\|_{L^\infty}|\xi'|\,|du_\xi-\Phi_u(\xi)du|+c_5\,|du|\,|\xi|\,|\xi'| \\ &\quad\le c_6\left(|du|\,|\xi|+|\nabla\xi|\right)|\xi'|. \end{aligned}$$

Here the constant c_6 depends only on c_0 and the metric on M. The pointwise estimate for the first summand takes the form

$$\begin{aligned} |\nabla(E_u(\xi)\xi')-\Phi_u(\xi)\nabla\xi'| &\le |(E_u(\xi)-\Phi_u(\xi))\nabla\xi'|+|\nabla(E_u(\xi)\xi')-E_u(\xi)\nabla\xi'| \\ &\le c_7\left(|\xi|\,|\nabla\xi'|+|du|\,|\xi|\,|\xi'|+|\nabla\xi|\,|\xi'|\right). \end{aligned}$$

Putting things together we obtain the pointwise inequality

$$\left|D_{u_\xi}\left(E_u(\xi)\xi'\right)-\Phi_u(\xi)D_u\xi'\right| \le (c_6+c_7)\left(|\xi|\,|\nabla\xi'|+|du|\,|\xi|\,|\xi'|+|\nabla\xi|\,|\xi'|\right).$$

Now take the pth power and integrate, using the L^p-norm on the factors du, $\nabla\xi$, and $\nabla\xi'$, respectively, and the L^∞ norm on the other factors. Since

$$\|\xi\|_{L^\infty}\le c_0\,\|\xi\|_{W^{1,p}}$$

this proves (3.5.5) and Proposition 3.5.3. □

PROOF OF THEOREM 3.5.2. By assumption, the operator $D_u = d\mathcal{F}_u(0)$ is onto and has a right inverse $Q_u : \mathcal{Y}_u \to \mathcal{X}_u$ such that

$$\|Q_u\| \leq c_0.$$

Now let c be the constant of Proposition 3.5.3 and choose $\delta \in (0, 1)$ such that

$$c\delta \leq 1/2c_0.$$

Then, by (3.5.2),

$$\|\xi\|_{W^{1,p}} < \delta \qquad \Longrightarrow \qquad \|d\mathcal{F}_u(\xi) - D_u\| \leq \frac{1}{2c_0}.$$

Hence the hypotheses of Proposition A.3.4 are satisfied with $X = \mathcal{X}_u$, $Y = \mathcal{Y}_u$, $f = \mathcal{F}_u$, $x_0 = 0$, $c = c_0$, and this constant δ. Hence, if $\xi_0 \in \mathcal{X}_u$ satisfies

$$\|\xi_0\|_{W^{1,p}} < \frac{\delta}{8}, \qquad \|\mathcal{F}_u(\xi_0)\|_{L^p} = \left\|\bar{\partial}_J(\exp_u(\xi_0))\right\|_{L^p} < \frac{\delta}{4c_0},$$

then, by Proposition A.3.4, there exists a unique $\xi \in \operatorname{im} Q_u$ such that

$$\mathcal{F}_u(\xi_0 + \xi) = 0, \qquad \|\xi_0 + \xi\|_{W^{1,p}} < \delta.$$

Moreover, this vector ξ satisfies

$$\|\xi\|_{W^{1,p}} \leq 2c_0 \|\mathcal{F}_u(\xi_0)\|_{L^p} = 2c_0 \left\|\bar{\partial}_J(\exp_u(\xi_0))\right\|_{L^p}.$$

This proves Theorem 3.5.2. □

REMARK 3.5.4. We have stated Theorem 3.5.2 for fixed almost complex structures $J \in \mathcal{J}_\tau(M, \omega)$. In Chapter 10 we shall need a version of Theorem 3.5.2 for almost complex structures that depend on the base point $z \in \Sigma$. (See Equation (6.7.4) and Remark 6.7.3 in Section 6.7). Let us fix a smooth map $\Lambda \to \mathcal{J}_\tau(M, \omega) : \lambda \mapsto J_\lambda$, defined on a compact manifold Λ. We wish to modify Theorem 3.5.2 to allow for smooth families of almost complex structures of the form $J_z := J_{\lambda(z)}$, where $\lambda : \Sigma \to \Lambda$ is a smooth map. The important point is that the derivatives of the map $\lambda : \Sigma \to \Lambda$ do not enter in the quadratic estimates of Proposition 3.5.3. Hence the assertion Theorem 3.5.2 continues to hold, with uniform constants c_0 and δ, for every smooth map $\lambda : \Sigma \to \Lambda$ and every J_λ-holomorphic curve $u : \Sigma \to M$ for which the relevant right inverse Q_u exists.

REMARK 3.5.5. The proof of Proposition 3.5.3 shows that there is an estimate

$$\|d\mathcal{F}_u(\xi)\xi' - D_u\xi'\|_{L^p} \leq c\left(\|\xi\|_{L^\infty} \|\xi'\|_{W^{1,p}} + \|\xi\|_{W^{1,p}} \|\xi'\|_{L^\infty}\right)$$

whenever u and ξ satisfy (3.5.1). A slightly more complicated argument shows that

$$\|d\mathcal{F}_u(\xi_1)\xi - d\mathcal{F}_u(\xi_0)\xi\|_{L^p} \leq c\left(\|\xi_1 - \xi_0\|_{L^\infty} \|\xi\|_{W^{1,p}} + \|\xi_1 - \xi_0\|_{W^{1,p}} \|\xi\|_{L^\infty}\right)$$

whenever u, ξ_0, and ξ_1 satisfy (3.5.1). Integrating this gives the quadratic estimate

$$\|\mathcal{F}_u(\xi_0 + \xi) - \mathcal{F}_u(\xi_0) - d\mathcal{F}_u(\xi_0)\xi\|_{L^p} \leq c\|\xi\|_{L^\infty} \|\xi\|_{W^{1,p}}$$

whenever u, ξ_0, and ξ satisfy (3.5.1). All these estimates hold uniformly for all volume forms on Σ that satisfy $c_p(\mathrm{dvol}_\Sigma) \leq c_0$.

Using the estimates of Remark 3.5.5 we can refine the uniqueness statement of Theorem 3.5.2 as follows. This refined version will not be used in this book.

COROLLARY 3.5.6. *Let (M, ω, J), (Σ, j_Σ), and $p > 2$ be as in Theorem 3.5.2. Then, for every $c_0 > 0$, there exists a $\delta > 0$ such that the following holds for every volume form* dvol_Σ *such that $c_p(\mathrm{dvol}_\Sigma) \le c_0$. If $u \in W^{1,p}(\Sigma, M)$, $\xi_0, \xi_1 \in W^{1,p}(\Sigma, u^*TM)$, and D_u has a right inverse Q_u such that*

$$\bar{\partial}_J(\exp_u(\xi_0)) = 0, \qquad \bar{\partial}_J(\exp_u(\xi_1)) = 0, \qquad \xi_1 - \xi_0 \in \mathrm{im}\, Q_u,$$
$$\|du\|_{L^p} \le c_0, \qquad \|Q_u\| \le c_0, \qquad \|\xi_0\|_{W^{1,p}} \le \delta, \qquad \|\xi_1\|_{L^\infty} \le \delta,$$

then $\xi_0 = \xi_1$.

PROOF. Denote $\xi := \xi_1 - \xi_0$ so that $\xi = Q_u D_u \xi$ and $\mathcal{F}_u(\xi_0) = \mathcal{F}_u(\xi_0 + \xi) = 0$. Moreover, $\|\xi\|_{L^\infty} \le \|\xi_0\|_{L^\infty} + \|\xi_1\|_{L^\infty} \le (c_0 + 1)\delta$. Choose $c > 0$ such that the estimates of Proposition 3.5.3 and Remark 3.5.5 hold. Then

$$\begin{aligned}
\|\xi\|_{W^{1,p}} &\le c_0 \|D_u \xi\|_{L^p} \\
&\le c_0 \|\mathcal{F}_u(\xi_0 + \xi) - \mathcal{F}_u(\xi_0) - d\mathcal{F}_u(\xi_0)\xi\|_{L^p} + c_0 \|(d\mathcal{F}_u(\xi_0) - D_u)\xi\|_{L^p} \\
&\le c_0 c \left(\|\xi\|_{L^\infty} + \|\xi_0\|_{W^{1,p}}\right) \|\xi\|_{W^{1,p}} \\
&\le (c_0 + 2) c_0 c \delta \|\xi\|_{W^{1,p}}.
\end{aligned}$$

If $(c_0 + 2)c_0 c\delta < 1$ it follows that $\xi = 0$. This proves Corollary 3.5.6 □

CHAPTER 4

Compactness

Because any manifold V is cobordant to the empty manifold via the noncompact cobordism $V \times [0, 1)$, Theorem 3.1.8 is useless unless one can establish some kind of compactness. Now in the case $\Sigma = S^2$ the manifold $\mathcal{M}(A; J)$ itself cannot be compact (unless it consists of constant maps) since the noncompact group $\mathrm{G} = \mathrm{PSL}(2, \mathbb{C})$ of biholomorphic maps of S^2 acts on this space by reparametrization. However, in some cases the moduli space $\mathcal{M}(A; J)/\mathrm{G}$ of unparametrized J-holomorphic spheres is compact.

If J is tamed by a symplectic form ω, then it follows from the energy identity (Lemma 2.2.1) that there is a uniform bound on the energy, and hence on the $W^{1,2}$-norm, of all J-holomorphic curves in a given homology class. This is the Sobolev borderline case and, as a result, the space of such curves will in general not be compact. In fact, the standard elliptic bootstrapping argument asserts that any sequence u^ν in $\mathcal{M}(A; J)$ which is bounded in the $W^{1,p}$-norm for some $p > 2$ has a subsequence which converges uniformly with all derivatives. On the other hand, if the first derivatives of u^ν are only bounded in L^2, then the conformal invariance of the energy in two dimensions leads to the phenomenon of bubbling, which was first discovered by Sacks and Uhlenbeck in the context of minimal surfaces [**349**]. Indeed a simple geometric argument using conformal rescaling allows one to construct a J-holomorphic map $v : \mathbb{C} \to M$ with finite area which, by "removal of singularities", can be extended to $S^2 = \mathbb{C} \cup \{\infty\}$. This is the phenomenon of "bubbling off of spheres."

The arguments used below are very close to those of Floer. Gromov's original approach was somewhat different. He argued geometrically, using isoperimetric inequalities and the Schwartz Lemma for conformal maps. More details of his proofs have been written up by Pansu [**320**]. The flavour of his arguments may be sampled in the proof of Theorem 4.1.2 (removal of singularities) given in Sections 4.1 and 4.5 below. The mean value inequality in Section 4.3 first appeared in Salamon [**351**] and our proofs in Sections 4.6 and 4.7 follow the exposition of Gromov compactness in Hofer–Salamon [**180**]. Other accounts of this subject can be found in Parker–Wolfson [**321**], Ye [**423**], and Hummel [**193**]. In recent years, the whole subject has developed signficantly as more complicated holomorphic objects are studied. For a comprehensive treatment see the paper [**46**] by Bourgeous, Eliashberg, Hofer, Wysocki and Zehnder on compactness results in symplectic field theory.

In Section 4.1 we discuss the important notion of the energy of a J-holomorphic curve and state the basic compactness and removable singularity theorems. Section 4.2 gives a first introduction to the bubbling phenomenon and discusses several examples. It also contains a proof of the fact that moduli spaces of unparametrized minimal energy J-holomorphic spheres are compact. In Section 4.3 we prove a pointwise estimate for the first derivatives of J-holomorphic curves with small energy. Granted this, the proof of the removable singularity theorem in Section 4.5 is fairly easy. However, it also uses the isoperimetric inequality, and so we devote Section 4.4 to this topic. In Section 4.6 we return to the bubbling phenomenon and show that every sequence of J-holomorphic curves with uniformly bounded energy has a subsequence which "converges modulo bubbling". In Section 4.7 we explain a refined conformal rescaling argument which gives rise to a bubble that is connected to the limit of the original sequence. This result plays a crucial role in the proof of Gromov compactness for stable maps in Chapter 5. For the sake of completeness, we have in most places included a discussion of the behaviour of discs.

4.1. Energy

We begin with a discussion of the energy identity. It distinguishes the pseudo-holomorphic curves in symplectic manifolds from those in general almost complex manifolds. In the former case the energy is a topological invariant, while in the latter case the energy does not satisfy a universal bound and hence no general compactness results are available. Nevertheless, as we shall point out some results do hold in this more general setting.

Let (M, ω) be a symplectic manifold equipped with an ω-tame almost complex structure J (see Section 2.1). Recall that ω and J determine a Riemannian metric (2.1.1) on M and that the energy of a J-holomorphic curve $u : \Sigma \to M$, defined on a compact Riemann surface $(\Sigma, j_\Sigma, \mathrm{dvol}_\Sigma)$, is given by

$$E(u) := \tfrac{1}{2}\int_\Sigma |du(z)|_J^2 \,\mathrm{dvol}_\Sigma = \int_\Sigma u^*\omega$$

(see Lemma 2.2.1). Hence the energy of a J-holomorphic curve defined on a closed Riemann surfaces is a topological invariant. The same holds for compact Riemann surfaces with boundary if the boundary is mapped to a Lagrangian submanifold $L \subset M$. Thus we obtain a uniform L^2-bound on the first derivatives for all J-holomorphic curves that represent the same homology class. We shall see below that such L^2-bounds do not suffice to establish compactness. However, if we have a uniform L^p-bound on the first derivatives of the elements of a sequence u_ν of J-holomorphic curves then standard "elliptic bootstrapping" techniques can be used to establish the existence of a convergent subsequence. This is formulated more precisely in the following theorem. Here the domain Σ may be noncompact. Recall that a submanifold L of an almost complex manifold (M, J) is said to be **totally real** if $TL \cap J(TL) = \{0\}$.

THEOREM 4.1.1. *Let (M, J) be a compact almost complex manifold, $L \subset M$ be a compact totally real submanifold, and J^ν be a sequence of almost complex structures on M that converges in the C^∞-topology to J. Moreover, let $(\Sigma, j_\Sigma, \mathrm{dvol}_\Sigma)$ be a Riemann surface, $\Omega^\nu \subset \Sigma$ be an increasing sequence of open sets that exhaust Σ, and $u^\nu : (\Omega^\nu, \Omega^\nu \cap \partial\Sigma) \to (M, L)$ be a sequence of J^ν-holomorphic curves such that*

$$\sup_\nu \|du^\nu\|_{L^\infty(K)} < \infty$$

for every compact subset $K \subset \Sigma$. Then u^ν has a subsequence which converges uniformly with all derivatives on compact subsets of Σ to a J-holomorphic curve $u : (\Sigma, \partial\Sigma) \to (M, L)$.

The proof of Theorem 4.1.1 will be carried out in Appendix B. The conclusion continues to hold when the maps u^ν are uniformly bounded in the $W^{1,p}$-norm for some $p > 2$ (see Theorem B.4.2). However, the energy identity of Lemma 2.2.1 only guarantees a uniform $W^{1,p}$-bound for $p = 2$. Although this $W^{1,2}$-bound does not imply that the derivatives are uniformly bounded, it does give some control over what can happen when the derivatives blow up. As we shall see in Section 4.2, the conformal invariance of the energy leads in this case to the formation of *bubbles*.

We next formulate the removable singularity theorem, which is an important analytical tool for understanding the bubbling phenomenon. In the integrable case this is a familiar result from complex analysis. Its proof for J-holomorphic curves in symplectic manifolds will be carried out in Section 4.5. We give here a proof of continuity in the ω-compatible case using the monotonicity property of minimal surfaces.

Throughout we denote by $\mathbb{D} := \{z \in \mathbb{C} \,|\, |z| \le 1\}$ the closed unit disc and by $\mathbb{H} := \{z \in \mathbb{C} \,|\, \operatorname{Im} z \ge 0\}$ the closed upper half plane.

THEOREM 4.1.2 (Removal of singularities). *Let (M, ω) be a compact symplectic manifold, $L \subset M$ be a compact Lagrangian submanifold, and J be an ω-tame almost complex structure on M with associated metric g_J.*

(i) *If $u : \mathbb{D} \setminus \{0\} \to M$ is a J-holomorphic curve with finite energy $E(u) < \infty$, then u extends to a smooth map $\mathbb{D} \to M$.*

(ii) *If $u : (\mathbb{D} \cap \mathbb{H} \setminus \{0\}, \mathbb{D} \cap \mathbb{R} \setminus \{0\}) \to (M, L)$ is a J-holomorphic curve with finite energy $E(u) < \infty$, then u extends to a smooth map $\mathbb{D} \cap \mathbb{H} \to M$.*

PROOF OF CONTINUITY IN (I). Here we follow essentially the line of argument in Gromov's original work. We assume that J is compatible with ω so that the J-holomorphic curves minimize the energy. As pointed out by Pansu in [**320**, §37], one can prove the removable singularity theorem in this case by using the monotonicity theorem for minimal surfaces. This states that there are constants $c > 0$ and $\varepsilon_0 > 0$ (which depend on M and the metric g_J) such that, for every nonconstant compact connected minimal surface S in (M, g_J) with nonempty boundary, every $x \in M$, and every constant $0 < \varepsilon < \varepsilon_0$, the following holds (see [**232**, 3.15]):

$$x \in S, \quad \partial S \cap B(x, \varepsilon) = \emptyset \quad \Longrightarrow \quad \operatorname{area}_{g_J}(S) \ge c\epsilon^2. \tag{4.1.1}$$

Now let $u : \mathbb{D} \to M$ be a J-holomorphic curve. For $0 < r < 1$ define the loop $\gamma_r : \mathbb{R}/2\pi\mathbb{Z} \to M$ by $\gamma_r(\theta) := u(re^{i\theta})$ and denote by $\ell(\gamma_r)$ the length of γ_r. We claim that there is a decreasing sequence $0 < r_\nu \le 1$ such that

$$\lim_{\nu\to\infty} \ell(\gamma_{r_\nu}) = 0, \qquad \lim_{\nu\to\infty} r_\nu = 0. \tag{4.1.2}$$

To see this, note that $|du(re^{i\theta})| = \sqrt{2}|\dot\gamma_r(\theta)|/r$ by conformality of u. Hence

$$\begin{aligned} E(u) &= \int_0^1 \int_0^{2\pi} \frac{|\dot\gamma_r(\theta)|^2}{r^2} r\,dr \wedge d\theta \\ &\ge \int_0^1 \left(\int_0^{2\pi} |\dot\gamma_r(\theta)|\,d\theta \right)^2 \frac{1}{2\pi r} dr \\ &= \int_0^1 \frac{\ell(\gamma_r)^2}{2\pi r} dr. \end{aligned}$$

Since u has finite energy, this proves the existence of a decreasing sequence $r_\nu \in (0, 1]$ that satisfies (4.1.2).

Since M is compact there exists a subsequence, still denoted by r_ν, and a point $p \in M$ such that

$$\lim_{\nu\to\infty} \max_\theta d(\gamma_{r_\nu}(\theta), p) = 0$$

We claim that $u(z)$ converges to p as z tends to zero. Otherwise there exists a sequence $z_k \to 0$ such that $u(z_k)$ converges to a point $q \neq p$. Choose a constant $\delta > 0$ so small that $\delta < \varepsilon_0$ and $\delta < d(p,q)$. Choose ν_0 so large that $\max_\theta d(\gamma_{r_\nu}(\theta), p) < (d(p,q) - \delta)/2$ for $\nu \geq \nu_0$. Given $\nu \geq \nu_0$ choose k such that $|z_k| < r_\nu$ and $d(u(z_k), q) < (d(p,q) - \delta)/2$. Given z_k choose $\nu' \geq \nu_0$ such that $r_{\nu'} < |z_k|$. Denote

$$A(r_{\nu'}, r_\nu) := \{z \in \mathbb{C} \,|\, r_{\nu'} \leq |z| \leq r_\nu\}$$

Then, for every $z \in \partial A(r_{\nu'}, r_\nu)$, we have

$$d(u(z), u(z_k)) \geq d(p,q) - d(p, u(z)) - d(u(z_k), q) > \delta.$$

Hence it follows from (4.1.1) with $S := u(A(r_{\nu'}, r_\nu))$ and $x := u(z_k)$ that

$$\int_{A(r_{\nu'}, r_\nu)} |du|^2 \geq c\delta^2$$

for every $\nu \geq \nu_0$. Now choose an increasing subsequence $\nu_k, k \geq 0$, with $\nu_k' < \nu_{k+1}$ for all k. Then the annuli $A(r_{\nu_k'}, r_{\nu_k}), k \geq 1$, are disjoint. Since u has energy at least $c\delta^2$ on each of them, this contradicts the fact that u has finite energy. Thus we have proved that u extends to a continuous function on $\mathbb{D}$, assuming the monotonicity theorem for minimal surfaces. □

REMARK 4.1.3. The above argument works only when ω is closed since otherwise J-holomorphic curves are not minimal: see Lemma 2.2.1. Our later argument is more analytic, but still uses the closedness of ω in the proof of the isoperimetric inequality of Section 4.4. In [**423**] Ye proves an analogue of the local isoperimetric inequality in Lemma 4.5.1 for arbitrary ω that allows him to establish removal of singularities for J-holomorphic curves in a compact almost complex manifold with Hermitian metric.

In Sections 4.2 and 4.6 we shall explain how Theorems 4.1.1 and 4.1.2 together with the energy identity can be used to understand the bubbling phenomenon for pseudoholomorphic curves. We shall see that bubbling, for a suitably chosen subsequence, can only occur near finitely many points. This follows from the fact that the energy of a nonconstant J-holomorphic sphere (or a J-holomorphic disc with Lagrangian boundary condition) cannot be arbitrarily small. We close this section with a proof of this observation which is based on the apriori estimate in Lemma 4.3.1 below. This apriori estimate establishes a pointwise estimate of the first derivatives of pseudoholomorphic curves with sufficiently small energy.

PROPOSITION 4.1.4. *Let (M, J) be a compact almost complex manifold and L be a compact totally real submanifold. Suppose that M is equipped with any Riemannian metric. Then there exists a constant $\hbar > 0$ such that*

$$E(u) \geq \hbar$$

for every nonconstant J-holomorphic sphere $u : S^2 \to M$ and every nonconstant J-holomorphic disc $u(\mathbb{D}, \partial\mathbb{D}) \to (M, L)$.

PROOF. The 2-sphere is conformally diffeomorphic (via stereographic projection) to the Riemann sphere $\mathbb{C} \cup \{\infty\}$ and the 2-disc is conformally diffeomorphic to $\mathbb{H} \cup \{\infty\}$. Hence it suffices to prove the assertions for finite energy holomorphic maps $u : \mathbb{C} \to M$ and $u : (\mathbb{H}, \mathbb{R}) \to (M, L)$. The apriori estimate of Lemma 4.3.1 below asserts that there are constants $c, \delta > 0$ such that

$$\int_{B_r(z)} |du|^2 < \delta \qquad \Longrightarrow \qquad |du(z)|^2 \leq \frac{c}{r^2} \int_{B_r(z)} |du|^2$$

for every J-holomorphic curve $u : \mathbb{C} \to M$ and every $z \in \mathbb{C}$ and

$$\int_{B_{2r}(z)\cap\mathbb{H}} |du|^2 < \delta \qquad \Longrightarrow \qquad \sup_{B_r(z)\cap\mathbb{H}} |du|^2 \leq \frac{2c}{r^2} \int_{B_{2r}(z)\cap\mathbb{H}} |du|^2$$

for every J-holomorphic curve $u : (\mathbb{H}, \mathbb{R}) \to (M, L)$ and every $z \in \mathbb{H}$. Here we denote by $B_r(z) \subset \mathbb{C}$ the ball of radius r centered at $z \in \mathbb{C}$. If $E(u) < \delta$, then these estimates hold for every $r > 0$, so u is constant. This proves Proposition 4.1.4. □

If J is ω-tame one can also prove Proposition 4.1.4 using removal of singularities and the bubbling argument explained in the next section. Namely, if $u_\nu : S^2 \to M$ is a sequence of J-holomorphic spheres with bounded energy then a composition of u_ν with a suitable sequence of holomorphic maps $\phi_\nu : \mathbb{C} \to S^2$ has a subsequence which converges to a nonconstant finite energy J-holomorphic plane $u : \mathbb{C} \to M$. By the Removable Singularity Theorem 4.1.2, u extends to a J-holomorphic sphere and hence must have positive energy. Hence the sequence $E(u_\nu)$ cannot converge to zero. A similar argument works for J-holomorphic discs. However, the proof of Theorem 4.1.2 in Section 4.5 uses the same mean value inequality as the above proof of Proposition 4.1.4, so that this argument is in the end no easier than before. Still in the ω-tame case, Proposition 4.1.4 is also a corollary of the next result, which asserts that only finitely many homology classes can be represented by J-holomorphic spheres with a fixed upper bound on the energy. One can derive Proposition 4.1.5 from Proposition 4.1.4 and Gromov compactness. However, it has an alternative, much more elementary proof due to Joel Fish, that we now explain.

PROPOSITION 4.1.5. *Let (M, ω) be a compact symplectic manifold and J be an ω-tame almost complex structure on M. Fix a constant $c > 0$. Then the set of all homology classes $A \in H_2(M; \mathbb{Z})$ that can be represented by a J-holomorphic sphere and satisfy $\omega(A) \leq c$ is finite.*

PROOF. Denote by $g_J := \frac{1}{2}(\omega(\cdot, J\cdot) - \omega(J\cdot, \cdot))$ the Riemannian metric determined by ω and J. Let $\beta \in \Omega^2(M)$ be any 2-form. Then, for every $x \in M$ and every tangent vector $\xi \in T_xM$, we have

$$|\beta_x(\xi, J\xi)| \leq |\beta_x|_{g_J} |\xi|_{g_J}^2 = |\beta_x|_{g_J} \omega(\xi, J\xi).$$

This implies the pointwise inequalty $|u^*\beta| \leq \|\beta\|_{L^\infty} u^*\omega$ and hence

$$\left| \int_{S^2} u^*\beta \right| \leq \|\beta\|_{L^\infty} \int_{S^2} u^*\omega \tag{4.1.3}$$

for every J-holomorphic curve $u : S^2 \to M$. Now choose finitely many closed 2-forms $\beta_1, \ldots, \beta_\ell \in \Omega^2(M)$ whose cohomology classes form a basis of $H^2(M; \mathbb{R})$. Let

$A \in H_2(M;\mathbb{Z})$ be a homology class that can be represented by a J-holomorphic sphere and satisfies $\omega(A) \le c$. Then it follows from (4.1.3) that

$$|\beta_j(A)| \le \|\beta_j\|_{L^\infty} c, \qquad j = 1,\dots,\ell. \tag{4.1.4}$$

Since the function $H_2(M;\mathbb{R}) \to \mathbb{R} : A \mapsto \max_j |\beta_j(A)|$ is a norm, there are only finitely many integral homology classes satisfying (4.1.4). This proves Proposition 4.1.5. □

In the following we shall denote by $\hbar = \hbar(M,\omega,L,J)$ the largest constant which satisfies the requirements of Proposition 4.1.4:

$$\hbar(M,\omega,L,J) := \min\left\{ \inf_{\substack{u:S^2\to M \\ \bar\partial_J(u)=0,\, E(u)>0}} E(u), \inf_{\substack{u:(B,\partial B)\to(M,L) \\ \bar\partial_J(u)=0,\, E(u)>0}} E(u) \right\}.$$

If $L = \emptyset$, then

$$\hbar(M,\omega,J) := \hbar(M,\omega,\emptyset,J) = \inf_{\substack{u:S^2\to M \\ \bar\partial_J(u)=0,\, E(u)>0}} E(u).$$

4.2. The bubbling phenomenon

Let (M,ω) be a compact symplectic manifold and $J \in \mathcal{J}_\tau(M,\omega)$ be an ω-tame almost complex structure on M. Let $(\Sigma, j_\Sigma, \mathrm{dvol}_\Sigma)$ be a closed Riemann surface. In Section 4.1 we have seen that every sequence of J-holomorphic curves $u^\nu : \Sigma \to M$ with uniformly bounded first derivatives has a C^∞-convergent subsequence (Theorem 4.1.1). Hence compactness can only fail if the sequence du_ν is unbounded:

$$\sup_\nu \|du^\nu\|_{L^\infty} = \infty.$$

If, in addition, the energy is uniformly bounded, i.e.

$$\sup_\nu E(u^\nu) < \infty,$$

then a conformal rescaling argument shows that a holomorphic sphere *bubbles off*. Here is how this works. Let $z^\nu \in \Sigma$ be a point at which the real valued function $|du^\nu|$ attains its maximum:

$$|du^\nu(z^\nu)| = \|du^\nu\|_{L^\infty(\Sigma)} =: c^\nu.$$

Passing to a subsequence, if necessary, we may assume that z^ν converges to a point $z_0 \in \Sigma$ and that c^ν diverges to infinity. Now choose a holomorphic coordinate chart $\phi : \Omega \to \Sigma$ defined on an open neighbourhood $\Omega \subset \mathbb{C}$ of zero such that $\phi(0) = z_0$. Then the pullback volume form is

$$\phi^*\mathrm{dvol}_\Sigma = \lambda^2 ds \wedge dt$$

for some function $\lambda : \Omega \to (0,\infty)$. We may assume without loss of generality that $z^\nu \in \Omega$ for all ν and

$$\lambda(0) = 1, \qquad \tfrac{1}{2} \le \lambda(z) \le 2,$$

for all $z \in \Omega$. Consider the sequences

$$u^\nu_{\mathrm{loc}} := u^\nu \circ \phi : \Omega \to M, \qquad z^\nu_{\mathrm{loc}} := \phi^{-1}(z^\nu).$$

In the following we shall drop the subscript loc. Then

$$c^\nu = \frac{|du^\nu(z^\nu)|}{\lambda(z^\nu)} = \sup_\Omega \frac{|du^\nu|}{\lambda}, \qquad \lim_{\nu\to\infty} z^\nu = 0.$$

Now choose $\varepsilon > 0$ such that $B_\varepsilon(z^\nu) \subset \Omega$ for every ν and consider the reparametrized sequence $v^\nu : B_{\varepsilon c^\nu} \to M$ defined by

$$v^\nu(z) := u^\nu(z^\nu + z/c^\nu).$$

This sequence satisfies

$$|dv^\nu(0)| \ge 1/2, \qquad \|dv^\nu\|_{L^\infty(B_{\varepsilon c^\nu})} \le 2$$

and

$$E(v^\nu; B_{\varepsilon c^\nu}) = E\big(u^\nu; B_\varepsilon(z^\nu)\big) \le E(u^\nu),$$

where $E(u; B)$ denotes the energy of the restriction of u to B. The last identity follows from the conformal invariance of the energy. By Theorem 4.1.1, there exists a subsequence, still denoted by v^ν, that converges uniformly with all derivatives on compact sets. The limit function $v : \mathbb{C} \to M$ is again a J-holomorphic curve such that

$$|dv(0)| \ge 1/2, \qquad 0 < E(v) = \int_{\mathbb{C}} v^*\omega \le \sup_\nu E(u^\nu).$$

The conformal invariance of the energy implies that the map

$$\mathbb{C} \setminus \{0\} \to M : z \mapsto v(1/z)$$

has finite energy and so, by the removable singularity theorem 4.1.2, it extends smoothly over 0. Hence v extends to a nonconstant J-holomorphic map from the Riemann sphere $S^2 \cong \mathbb{C} \cup \{\infty\}$ to M. A J-holomorphic sphere v constructed in this way is called a *bubble*. The energy $E(v)$ of this bubble is positive and the energy of u^ν in an arbitrarily small neighbourhood of the point z_0 is at least $E(v)$ in the limit $\nu \to \infty$, i.e.

$$\liminf_{\nu\to\infty} E\big(u^\nu; B_\varepsilon(z_0)\big) \ge E(v) \tag{4.2.1}$$

for every $\varepsilon > 0$. To see this note that

$$\begin{aligned} E(v; B_R) &= \lim_{\nu\to\infty} E(v^\nu; B_R) \\ &= \lim_{\nu\to\infty} E\big(u^\nu; B_{R/c^\nu}(z^\nu)\big) \\ &\le \lim_{\nu\to\infty} E\big(u^\nu; B_\varepsilon(z_0)\big) \end{aligned}$$

for every $R > 0$ and every $\varepsilon > 0$. The inequality (4.2.1) follows by taking the limit $R \to \infty$.

The group of Möbius transformations. The above rescaling argument hinges on the fact that the energy is conformally invariant. A first example where the bubbling phenomenon can be observed explicitly is in the group $\mathrm{G} := \mathrm{PSL}(2, \mathbb{C})$ of conformal automorphisms of the Riemann sphere. This is also known as the group of **Möbius transformations**. Its elements are fractional linear transformations

$$\phi(z) = \frac{az + b}{cz + d}, \qquad ad - bc = 1, \tag{4.2.2}$$

where $a, b, c, d \in \mathbb{C}$. These are precisely the diffeomorphisms of the 2-sphere that preserve the complex structure and hence they form the moduli space of holomorphic maps from S^2 to $M = S^2$ of degree one as described in Chapter 3.

The group G is noncompact; a maximal compact subgroup of G is the group $\mathrm{SO}(3) = \mathrm{SU}(2)/\{\pm 1\!\!1\}$ of isometries of S^2. The noncompactness of G is related to the conformal invariance of the energy and can be interpreted in terms of bubbling.

Examples and exercises.

EXERCISE 4.2.1. Identify the unit sphere $S^2 \subset \mathbb{R}^3$ with the Riemann sphere $\mathbb{C} \cup \{\infty\}$ by stereographic projection from the north pole:

$$S^2 \to \mathbb{C} \cup \{\infty\} : (x_1, x_2, x_3) \mapsto \frac{x_1 + ix_2}{1 - x_3}.$$

Prove that the pullback of the Fubini–Study form $(1 + x^2 + y^2)^{-2} dx \wedge dy$ (where $z = x + iy$) is equal to minus one fourth of the standard area form $x_1 dx_2 dx_3 + x_2 dx_3 dx_1 + x_3 dx_1 dx_2$ on S^2. (The $-$ sign is needed because the orientation of S^2 as the boundary of a ball in $\mathbb{R}^3$ is different from its orientation as $\mathbb{C} \cup \{\infty\}$.)

EXERCISE 4.2.2. Let ϕ be a fractional linear transformation of the form (4.2.2). Prove that ϕ preserves the upper half plane

$$\mathbb{H} := \{z \in \mathbb{C} \,|\, \mathrm{Im}\, z \geq 0\}$$

if and only if $a, b, c, d \in \mathbb{R}$. Prove that the fractional linear transformation $\phi(z) := (i - z)/(i + z)$ identifies the upper half plane with $B \setminus \{-1\}$, where

$$B := \{z \in \mathbb{C} \,|\, |z| \leq 1\}.$$

The inverse map is $B \setminus \{-1\} \to \mathbb{H} : z \mapsto i(1 - z)(1 + z)^{-1}$.

EXERCISE 4.2.3. The derivative of a Möbius transformation ϕ of the form (4.2.2) is given by

$$\phi'(z) = \frac{1}{(cz + d)^2}$$

and the norm of the derivative with respect to the Fubini–Study metric on $S^2 = \mathbb{C} \cup \{\infty\}$ is

$$\frac{1}{\sqrt{2}} |d\phi(z)|_{\mathrm{FS}} = |\phi'(z)| \frac{1 + |z|^2}{1 + |\phi(z)|^2} = \frac{1 + |z|^2}{|az + b|^2 + |cz + d|^2}$$

(Here the Fubini–Study metric is used on both the domain and target. Also see Exercise 2.2.3.) Prove directly, without using Theorem 4.1.1, that every sequence $\phi^\nu \in \mathrm{G}$ which satisfies

$$\sup_\nu \sup_z |d\phi^\nu(z)|_{\mathrm{FS}} < \infty$$

has a subsequence which converges uniformly with all derivatives on all of S^2.

EXERCISE 4.2.4. Use the bubbling argument outlined above to give an alternative proof of Lemma D.1.2: *If $\phi^\nu \in \mathrm{G}$ is a sequence of Möbius transformations which does not have a uniformly convergent subsequence, then there exist points $x, y \in S^2$ and a subsequence ϕ^{ν_i} which converges to y uniformly in compact subsets of $S^2 \setminus \{x\}$.* *Hint:* If you get stuck, consult Theorem 4.6.1 and use the fact that $E(\phi) = \pi = \hbar(S^2, \omega_{\mathrm{FS}}, J_{\mathrm{FS}})$ for $\phi \in \mathrm{G}$.

EXERCISE 4.2.5. The biholomorphic maps $\phi : B \to B$ are the fractional linear transformations of the form

$$\phi(z) = e^{i\theta} \frac{z - z_0}{1 - \bar{z}_0 z}$$

for some $\theta \in \mathbb{R}$ and $z_0 \in \mathrm{int}(B)$. Show that

$$|\phi'(z)| = \frac{1 - |\phi(z)|^2}{1 - |z|^2}.$$

Hence deduce that if one uses the hyperbolic metric $(dx^2 + dy^2)/(1 - x^2 - y^2)^2$ in both domain and target then ϕ is an isometry. Next consider the sequence

$$\phi^\nu(z) := \frac{z - x^\nu}{1 - x^\nu z},$$

where $x^\nu \in (0, 1)$ is a sequence converging to 1. Show that $(\phi^\nu)'(1)$ diverges to infinity, whereas $(\phi^\nu)'(z)$ converges to zero for every $z \neq 1$. Interpret this limit behaviour in terms of bubbling on the boundary for J-holomorphic discs with Lagrangian boundary conditions.

EXERCISE 4.2.6. Consider the case $M = \Sigma = S^2 = \mathbb{C} \cup \{\infty\}$. A holomorphic curve is a rational function

$$u(z) = \frac{p(z)}{q(z)}.$$

If p and q are relatively prime, then the degree of u is the maximum of the degrees of the numerator and the denominator. As in Exercise 4.2.3, the norm of its derivative with respect to the Fubini–Study metric is given by

$$\frac{1}{\sqrt{2}} |du(z)|_{\mathrm{FS}} = |u'(z)| \frac{1 + |z|^2}{1 + |u(z)|^2}.$$

Now consider a sequence of rational maps

$$u_\nu(z) := \frac{zp(z)}{(z - a_\nu)q(z)}$$

of degree $k + 1$. Assume that p and q are relatively prime and $p(0) = q(0) = 1$. Assume also that a_ν converges to 0. Prove that a holomorphic sphere of degree 1 bubbles off at $z = 0$ and u_ν converges uniformly with all derivatives on compact subsets of $S^2 \setminus \{0\}$ to $u = p/q$.

EXERCISE 4.2.7. Consider the sequence of holomorphic maps $u^\nu : \mathbb{C}P^1 \to \mathbb{C}P^2$ given by

$$u^\nu([x : y]) := [x^2 : y^2 : \nu xy].$$

The image is the family of quadrics $XY = Z^2/\nu^2$ that degenerates to the pair of lines $X = 0$ and $Y = 0$ as $\nu \to \infty$. Exhibit each of these limiting curves as a bubble that is formed by a suitable reparametrization of the given sequence u^ν.

EXERCISE 4.2.8. Consider the same question as in Exercise 4.2.7 for the sequence of maps $u^\nu : S^2 \to S^2 \times S^2$ given by

$$u^\nu(z) := (z, \phi^\nu(z)),$$

where ϕ^ν is as in Exercise 4.2.4.

EXAMPLE 4.2.9. Consider the case $\Sigma = \mathbb{T}^2 = \mathbb{C}/\Lambda$ where $\Lambda = \mathbb{Z} + i\mathbb{Z}$. The *Weierstrass p-function* is the meromorphic function $p : \mathbb{T}^2 \to S^2$ defined by

$$p(z) := \sum_{\lambda \in \Lambda} \frac{1}{(z-\lambda)^2}.$$

This map has degree 2 since it is onto and $p(z_1) = p(z_2)$ if and only if $z_1 \pm z_2 \in \Lambda$. Now $c_1(p^*TS^2) = 4$ and so the moduli space $\mathcal{M}_2$ of degree 2 meromorphic maps $T^2 \to S^2$ has real dimension 8. Every $u \in \mathcal{M}_2$ can be represented in the form

$$u(z) = \phi(p(z+z_0)),$$

where $\phi \in \mathrm{PSL}(2,\mathbb{C})$ and $z_0 \in \mathbb{C}$. Consider the sequence

$$u^\nu(z) := \frac{p(z) - p(z_0)}{p(z) - a_\nu p(z_0)},$$

where $p(z_0)$ is a nonzero regular value of p. If a^ν converges to 1 then the derivative of u^ν blows up near $\pm z_0$:

$$(u^\nu)'(\pm z_0) = \frac{p'(\pm z_0)}{p(z_0)(1-a^\nu)}.$$

Thus, holomorphic spheres of degree 1 bubble off at $\pm z_0$ and u^ν converges to the constant 1 in the C^∞-topology on every compact subset of $\mathbb{T}^2 \setminus \{-z_0, z_0\}$.

REMARK 4.2.10. When the domain Σ has positive genus further degenerations are possible if one allows the complex structure j on Σ to vary. A first example in which this phenomenon can be observed explicitly is in the case of cubics in $\mathbb{C}P^2$. Generically, cubics are embedded tori and they can degenerate, for example, into a rational curve with one self-intersection or one singular point, or a triply covered line. An example of a reducible limit is a union of three lines. For a careful study of this limit behaviour see Lizan [**252**].

Spheres with minimal energy. A homology class $A \in H_2(M)$ is called **spherical** if it is in the image of the Hurewicz homomorphism $\pi_2(M) \to H_2(M)$. The set of spherical classes is denoted $H_2^S(M)$. Fix a spherical homology class $A \in H_2(M;\mathbb{Z})$ such that

$$\langle [\omega], A \rangle = \hbar,$$

where $\hbar = \hbar(M,\omega,J)$ is the constant of Proposition 4.1.4. Let $\mathcal{M}(A;J)$ denote the moduli space of J-holomorphic spheres $u : S^2 \to M$ representing the class A. By the removable singularity theorem, we may think of a holomorphic sphere as a finite energy holomorphic curve $u : \mathbb{C} \to M$. Then the function $\mathbb{C}\setminus\{0\} \to M : z \mapsto u(1/z)$ extends smoothly over zero and so u extends to a J-holomorphic map from the Riemann sphere $\mathbb{C}\cup\{\infty\}$ to M. The group $\mathrm{G} = \mathrm{PSL}(2,\mathbb{C})$ of Möbius transformation acts on $\mathcal{M}(A;J)$ by reparametrization. We say that a sequence of J-holomorphic curves $u^\nu : \mathbb{C} \to M$ converges in the C^∞ topology on S^2 if both $u^\nu(z)$ and $u^\nu(1/z)$ converge uniformly with all derivatives on compact subsets of $\mathbb{C}$.

THEOREM 4.2.11. *Let (M,ω) be a compact symplectic manifold and fix $J \in \mathcal{J}_\tau(M,\omega)$. If $A \in H_2(M;\mathbb{Z})$ is a spherical homology class such that $\langle [\omega], A \rangle = \hbar$ then the moduli space $\mathcal{M}(A;J)/\mathrm{G}$ is compact, i.e. for every sequence $u^\nu : S^2 \to M$ of J-holomorphic spheres representing the class A there exists a sequence $\phi^\nu \in \mathrm{G}$ such that the sequence $u^\nu \circ \phi^\nu$ has a C^∞-convergent subsequence.*

PROOF. By composing with a sequence of Möbius transformations in SO(3) we may assume without loss of generality that

$$|du^\nu(0)|_{\mathrm{FS}} = \sup_{z\in\mathbb{C}} |du^\nu(z)|_{\mathrm{FS}}$$

for every ν. Since $|du^\nu(z)|_{\mathrm{FS}} = |du^\nu(z)|(1+|z|^2)$, this implies

$$|du^\nu(0)| = \sup_{z\in\mathbb{C}} |du^\nu(z)|.$$

By conformal rescaling centered at zero, we may further assume that

$$|du^\nu(0)| = 1.$$

By Theorem 4.1.1, u^ν has a subsequence, still denoted by u^ν, which converges uniformly with all derivatives on compact subsets of $\mathbb{C}$ to a nonconstant J-holomorphic curve $u : \mathbb{C} \to M$. Now consider the sequence $z \mapsto u^\nu(1/z)$. This sequence converges uniformly with all derivatives on compact subsets of $\mathbb{C} \setminus \{0\}$ to $u(1/z)$. We claim that the sequence $u^\nu(1/z)$ also converges uniformly near zero. Otherwise, bubbling would occur, and by (4.2.1), this would imply

$$\liminf_{\nu\to\infty} E(u^\nu; \mathbb{C} \setminus B_R) \geq \hbar$$

for every $R > 0$. It would then follow that $E(u; B_R) = 0$, in contradiction to the fact that u is nonconstant. Hence the sequence u^ν converges to u in the C^∞ topology on all of S^2. □

The results of this section suffice for several remarkable applications of J-holomorphic curves in symplectic topology that are due to Gromov [**160**] and do not require the refined compactness results of Chapter 5 (see Chapter 9). These applications give a glimpse of the power of the theory of J-holomorphic curves in symplectic topology.

4.3. The mean value inequality

In this section we shall establish the following mean value inequality for the first derivative of a J-holomorphic curve with small energy. We denote by

$$B_r(z) := \{\zeta \in \mathbb{C} \,|\, |\zeta - z| \leq r\}$$

the closed disc of radius r centered at z. We abbreviate $B_r := B_r(0)$.

LEMMA 4.3.1. *Let (M, J) be a compact almost complex manifold and $L \subset M$ be a compact totally real submanifold. Choose a Riemannian metric on M. Then there are constants $c, \delta > 0$ such that the following holds.*

(i) *If $r > 0$ and $u : B_r \to M$ is a J-holomorphic curve, then*

$$\int_{B_r} |du|^2 < \delta \qquad \Longrightarrow \qquad |du(0)|^2 \leq \frac{c}{r^2}\int_{B_r} |du|^2.$$

(ii) *If $r > 0$ and $u : (B_{2r}\cap\mathbb{H}, B_{2r}\cap\mathbb{R}) \to (M, L)$ is a J-holomorphic curve, then*

$$\int_{B_{2r}\cap\mathbb{H}} |du|^2 < \delta \qquad \Longrightarrow \qquad \sup_{B_r\cap\mathbb{H}} |du|^2 \leq \frac{2c}{r^2}\int_{B_{2r}\cap\mathbb{H}} |du|^2.$$

REMARK 4.3.2. Denote the Riemannian metric in Lemma 4.3.1 by g. The proof of Lemma 4.3.1 shows that the constant δ depends on the choice of an auxiliary Riemannian metric on M with respect to which J is skew-adjoint, JTL is orthogonal to TL, and L is totally real. Call it the **Frauenfelder metric of J and L** and denot it by $g_{J,L}$ (see Lemma 4.3.4 below). Denote the constants of Lemma 4.3.1 for $g = g_{J,L}$ by $\delta_{J,L}$ and $c_{J,L}$. The proof shows that $c_{J,L} = 8/\pi$ while $\delta_{J,L}$ depends continuously on J and $g_{J,L}$ with respect to the C^2 topology. The estimate involves the Levi-Civita connection of $g_{J,L}$, the Riemann curvature tensor of $g_{J,L}$, and the first and second covariant derivatives of J with respect to $g_{J,L}$. The constants c and δ in Lemma 4.3.1 depend only on $c_{J,L}$ and $\delta_{J,L}$ as well as the comparison constant for the metrics $g_{J,L}$ and g. The upshot is that, when $L = \emptyset$, the constant δ in Lemma 4.3.1 depends continuously on J with respect to the C^2 topology and c depends continuously on J with respect to the C^0 topology.

The proof of Lemma 4.3.1 relies on a result about the partial differential inequality $\Delta w \geq -aw^2$ for the energy density $w := \frac{1}{2}|du|^2$, where

$$\Delta := \partial_s{}^2 + \partial_t{}^2$$

is the standard Laplacian on $\mathbb{R}^2$. There is a similar result in every dimension (see e.g. [**336**]), but here we only consider functions of two variables.

LEMMA 4.3.3. *Let $r > 0$ and $a \geq 0$. If $w : B_r \to \mathbb{R}$ is a C^2 function that satisfies the inequalities*

$$\Delta w \geq -aw^2, \qquad w \geq 0, \qquad \int_{B_r} w < \frac{\pi}{8a} \tag{4.3.1}$$

then

$$w(0) \leq \frac{8}{\pi r^2} \int_{B_r} w. \tag{4.3.2}$$

PROOF. The proof consists of five steps.

STEP 1. *If $w : B_r \to \mathbb{R}$ is a C^2-function that satisfies the inequalities*

$$\Delta w \geq -b, \qquad w \geq 0,$$

for some constant $b \geq 0$, then

$$w(0) \leq \frac{br^2}{8} + \frac{1}{\pi r^2} \int_{B_r} w.$$

This is the mean value inequality for the subharmonic function

$$v(s,t) := w(s,t) + \frac{b}{4}(s^2 + t^2).$$

We include a proof for the sake of completeness. By the divergence theorem, we have

$$0 \leq \frac{1}{\rho} \int_{B_\rho} \Delta v = \frac{1}{\rho} \int_{\partial B_\rho} \frac{\partial v}{\partial \nu} = \int_0^{2\pi} \frac{d}{d\rho} v(\rho e^{i\theta}) d\theta = \frac{d}{d\rho} \left(\frac{1}{\rho} \int_{\partial B_\rho} v \right),$$

where $\frac{\partial v}{\partial \nu}$ denotes the normal derivative. Hence

$$\frac{1}{2\pi\sigma} \int_{\partial B_\sigma} v \leq \frac{1}{2\pi\rho} \int_{\partial B_\rho} v$$

for $0 < \sigma \leq \rho$. The term on the left converges to $v(0)$ as σ tends to zero. Hence

$$2\pi\rho\, v(0) \leq \int_{\partial B_\rho} v.$$

Integrating this inequality from 0 to r gives the mean value inequality for v. Hence

$$w(0) = v(0) \leq \frac{1}{\pi r^2}\int_{B_r} v = \frac{br^2}{8} + \frac{1}{\pi r^2}\int_{B_r} w.$$

This proves Step 1.

STEP 2. *It suffices to prove the lemma for $r = 1$.*

Suppose that $w : B_r \to \mathbb{R}$ satisfies (4.3.1) and define $\tilde{w} : B_1 \to \mathbb{R}$ and $\tilde{a} \in \mathbb{R}$ by

$$\tilde{w}(s,t) := w(rs, rt), \qquad \tilde{a} := ar^2.$$

Then

$$\Delta\tilde{w} \geq -\tilde{a}\tilde{w}^2, \qquad \int_{B_1} \tilde{w} = \frac{1}{r^2}\int_{B_r} w \leq \frac{\pi}{8ar^2} = \frac{\pi}{8\tilde{a}}.$$

Hence, assuming the lemma for $r = 1$, we obtain

$$w(0) = \tilde{w}(0) \leq \frac{8}{\pi}\int_{B_1} \tilde{w} = \frac{8}{\pi r^2}\int_{B_r} w.$$

This proves Step 2.

STEP 3. *It suffices to prove the lemma for $a = 1$.*

If $a = 0$ the result follows from Step 1. Hence assume $a > 0$ and suppose that $w : B_r \to \mathbb{R}$ satisfies (4.3.1). Then $\tilde{w} := aw$ satisfies (4.3.1) with a replaced by 1. If the result holds for $a = 1$, then $\tilde{w}$ satisfies (4.3.2) and hence so does w.

STEP 4 (The Heinz trick). *Assume $a = r = 1$ and define $f : [0,1] \to \mathbb{R}$ by*

$$f(\rho) := (1-\rho)^2 \sup_{B_\rho} w$$

for $0 \leq \rho \leq 1$. Since $f(1) = 0$ and f is nonnegative, there exist $\rho^* \in [0,1)$ and $z^* \in B_{\rho^*}$ such that

$$f(\rho^*) = \max_{0\leq\rho\leq 1} f(\rho), \qquad c := w(z^*) = \sup_{B_{\rho^*}} w.$$

Denote

$$\varepsilon := \frac{1-\rho^*}{2}.$$

Then, for $0 \leq \rho \leq \varepsilon$,

$$c \leq 2c^2\rho^2 + \frac{1}{\pi\rho^2}\int_{B_1} w. \tag{4.3.3}$$

To see this, note first that

$$\sup_{B_\varepsilon(z^*)} w \leq \sup_{B_{\rho^*+\varepsilon}} w = \frac{f(\rho^*+\varepsilon)}{(1-\rho^*-\varepsilon)^2} = \frac{4f(\rho^*+\varepsilon)}{(1-\rho^*)^2} \leq \frac{4f(\rho^*)}{(1-\rho^*)^2} = 4c.$$

Hence $\Delta w \geq -w^2 \geq -16c^2$ in $B_\varepsilon(z^*)$ and so (4.3.3) follows from Step 1 with $b := 16c^2$ and $r := \rho \leq \varepsilon$. This proves Step 4.

STEP 5. *The lemma holds for $r = 1$ and $a = 1$.*

Assume $a = r = 1$ and let ρ^*, z^*, c, and ε be as in Step 4. If $4c\varepsilon^2 \geq 1$, then we can choose $\rho := \sqrt{1/4c} \leq \varepsilon$ in (4.3.3) and obtain

$$c = w(z^*) \leq \frac{c}{2} + \frac{4c}{\pi}\int_{B_1} w$$

and hence $\int_{B_1} w \geq \pi/8$, a contradiction. Hence $4c\varepsilon^2 < 1$, and hence it follows from (4.3.3) with $\rho = \varepsilon$ that

$$c \leq 2c^2\varepsilon^2 + \frac{1}{\pi\varepsilon^2}\int_{B_1} w \leq \frac{c}{2} + \frac{1}{\pi\varepsilon^2}\int_{B_1} w.$$

Hence

$$w(0) = f(0) \leq f(\rho^*) = (1-\rho^*)^2 c = 4c\varepsilon^2 \leq \frac{8}{\pi}\int_{B_1} w.$$

This proves Step 5 and Lemma 4.3.3. □

The next lemma was proved by Urs Frauenfelder [**120**] in his diploma thesis. It asserts the existence of a metric which renders a totally real submanifold L of (M, J) totally geodesic.

LEMMA 4.3.4 (Frauenfelder). *Let (M, J) be an almost complex manifold and $L \subset M$ be a totally real submanifold with $2\dim L = \dim M$. Then there exists a Riemannian metric $g = \langle\cdot,\cdot\rangle$ on M such that*

(i) $\langle J(p)v, J(p)w\rangle = \langle v, w\rangle$ *for $p \in M$ and $v, w \in T_pM$,*
(ii) *$J(p)T_pL$ is the orthogonal complement of T_pL for every $p \in L$,*
(iii) *L is totally geodesic with respect to g.*

PROOF. Choose coordinates $x_1, \ldots, x_n$ on L and extend these to coordinates

$$x_1, \ldots, x_n, y_1, \ldots, y_n$$

on M such that

$$J\frac{\partial}{\partial x_i} = \frac{\partial}{\partial y_i}, \qquad i = 1, \ldots, n,$$

on L. Write a metric and the almost complex structure in these coordinates in the form

$$g(x,y) = \begin{pmatrix} a(x,y) & b(x,y)^T \\ b(x,y) & c(x,y) \end{pmatrix}, \qquad J(x,y) = \begin{pmatrix} A(x,y) & B(x,y) \\ C(x,y) & D(x,y) \end{pmatrix},$$

where

$$a(x,y) = a(x,y)^T, \qquad b(x,y), \qquad c(x,y) = c(x,y)^T,$$

and

$$A(x,y), \qquad B(x,y), \qquad C(x,y) \qquad , D(x,y)$$

are real $n \times n$-matrices. Then

$$B(x,0) = -\mathbb{1}, \qquad C(x,0) = \mathbb{1}, \qquad A(x,0) = D(x,0) = 0.$$

The metric g satisfies (i), (ii), and (iii) if and only if

$$a(x,0) = c(x,0), \qquad b(x,0) = 0, \qquad \partial_{n+i}a(x,0) = 0, \qquad J^T gJ = g. \tag{4.3.4}$$

To prove the local existence of such a metric, we first choose a metric g that satisfies the first three identities in (4.3.4) and

$$\partial_{n+i}c(x,0) + \partial_{n+i}C(x,0)^T c(x,0) + c(x,0)\partial_{n+i}C(x,0) = 0,$$

and then replace g by $J^TgJ + g$. Now the set of metrics that satisfy (4.3.4) is invariant under convex combinations and under multiplication by cutoff functions $\beta = \beta(x, y)$ that satisfy

$$\partial_{n+i}\beta(x, 0) = 0.$$

This condition on the cutoff function is intrinsic. It asserts that

$$q \in L,\ v \in T_qL \quad \Longrightarrow \quad d\beta(q)J(q)v = 0. \tag{4.3.5}$$

Hence the result follows by choosing local metrics that satisfy (4.3.4) and patching with a partition of unity consisting of finitely many cutoff functions that satisfy (4.3.5). This proves Lemma 4.3.4. □

Proof of Lemma 4.3.1. Assume that the metric on M is as in Lemma 4.3.4. Under this assumption we prove the lemma with the constant $c = 8/\pi$. For a general metric the result then follows with a suitable constant c, because M is compact and hence any two metrics are comparable.

Let ∇ denote the Levi-Civita connection of the metric in Lemma 4.3.4 and $R \in \Omega^2(M, \mathrm{End}(TM))$ be the curvature tensor of ∇. Abbreviate

$$\xi := \partial_s u, \qquad \eta := \partial_t u, \qquad w := \tfrac{1}{2}|du|^2 : B_r \to \mathbb{R}.$$

Then

$$\xi + J\eta = 0, \qquad w = |\xi|^2 = |\eta|^2,$$

and

$$\Delta w = 2\,|\nabla_s\xi|^2 + 2\,|\nabla_t\xi|^2 + 2\langle \xi, \nabla_s\nabla_s\xi + \nabla_t\nabla_t\xi\rangle. \tag{4.3.6}$$

We must show that Δw satisfies the hypotheses of Lemma 4.3.3. Since

$$\xi = -J\eta, \qquad \nabla_s\eta = \nabla_t\xi$$

we have

$$\nabla_s\xi + \nabla_t\eta = \nabla_t(J\xi) - \nabla_s(J\eta) = (\nabla_\eta J)\xi - (\nabla_\xi J)\eta. \tag{4.3.7}$$

Hence

$$\begin{aligned} \nabla_s\nabla_s\xi + \nabla_t\nabla_t\xi &= \nabla_s(\nabla_s\xi + \nabla_t\eta) + \nabla_t\nabla_s\eta - \nabla_s\nabla_t\eta \\ &= \nabla_s((\nabla_\eta J)\xi - (\nabla_\xi J)\eta) - R(\xi, \eta)\eta, \end{aligned}$$

where the last equality uses the identity

$$R(\xi, \eta)\eta = (\nabla_s\nabla_t - \nabla_t\nabla_s)\eta.$$

Therefore

$$\tfrac{1}{2}\Delta w = |\nabla_s\xi|^2 + |\nabla_t\xi|^2 - \langle R(\xi, \eta)\eta, \xi\rangle + \kappa, \tag{4.3.8}$$

where the error term κ is

$$\begin{aligned} \kappa &= \langle \xi, \nabla_s((\nabla_\eta J)\xi - (\nabla_\xi J)\eta)\rangle \\ &= \langle \xi, (\nabla_\eta J)\nabla_s\xi - (\nabla_\xi J)\nabla_s\eta\rangle + \langle \xi, (\nabla_s(\nabla_\eta J))\xi - (\nabla_s(\nabla_\xi J))\eta\rangle. \end{aligned}$$

Now there is a constant $c = c(M, J, g) > 0$ such that

$$|\nabla_s(\nabla_\eta J)| \le c(|\xi|^2 + |\nabla_t\xi|), \quad |\nabla_s(\nabla_\xi J)| \le c(|\xi|^2 + |\nabla_s\xi|).$$

In the first of these estimates the coefficient of $|\xi|^2 = |\xi|\,|\eta|$ involves estimates for the second derivatives of J while that for $|\nabla_t\xi| = |\nabla_s\eta|$ involves the first derivatives of J. Hence there is a constant $c' = c'(M, J, g) > 0$ such that

$$\begin{aligned}\kappa &\geq -c'|\xi|^4 - c'|\xi|^2(|\nabla_s\xi| + |\nabla_t\xi|)\\ &\geq -\tfrac{1}{2}|\nabla_s\xi|^2 - \tfrac{1}{2}|\nabla_t\xi|^2 - c'(1+c')|\xi|^4.\end{aligned}$$

By (4.3.8), there is a constant $c'' = c''(M, J, g) > 0$ such that

$$\Delta w \geq -c''|\xi|^4 = -c''w^2.$$

But $w = \frac{1}{2}|du|^2$. Hence assertion (i) with $\delta = \pi/4c''$ follows from Lemma 4.3.3.

To prove (ii) we extend the function $w : B_{2r} \cap \mathbb{H} \to \mathbb{R}$ to B_{2r} by reflection, i.e.

$$w(s, -t) := w(s, t)$$

for $t > 0$. We prove that the normal derivative vanishes on the boundary:

$$\begin{aligned}\tfrac{1}{2}\partial_t w(s,0) &= \langle\nabla_t\xi, \xi\rangle(s,0)\\ &= \langle\nabla_s\eta, \xi\rangle(s,0)\\ &= \langle\nabla_s(J\xi), \xi\rangle(s,0)\\ &= \langle J\nabla_s\xi, \xi\rangle(s,0)\\ &= 0.\end{aligned}$$

Here the penultimate equality follows from the fact that the endomorphism $\nabla_s J$ is skew-adjoint. The last equality follows from the fact that L is totally geodesic and $\xi(s,0) \in T_{u(s,0)}L$, and so $\nabla_s\xi(s,0) \in T_{u(s,0)}L$. Since $\partial_t w(s,0) = 0$ the extended function w is of class C^2 and hence (ii) follows from the inequality $\Delta w \geq -c''w^2$ and Lemma 4.3.3. The additional factor two arises because the integral is over the half disc. This proves Lemma 4.3.1. □

EXERCISE 4.3.5. Let S^2 be the Riemann sphere with a symplectic form of area π. Find the largest value of δ such that the mean value estimate in Lemma 4.3.1 holds for maps into S^2.

REMARK 4.3.6. Suppose J is compatible with ω. Then every J-holomorphic curve $u : \mathbb{C} \to M$ is a harmonic map, i.e.

$$\nabla_s\partial_s u + \nabla_t\partial_t u = (\nabla_{\partial_t u}J)\partial_s u - (\nabla_{\partial_s u}J)\partial_t u = 0. \tag{4.3.9}$$

Here the first equation follows from (4.3.7). To prove the second equation in (4.3.9), use the formula $\partial_t u = J\partial_s u$ and the identity $(\nabla_{J\partial_s u}J)\partial_s u = (\nabla_{\partial_s u}J)J\partial_s u$ for ω-compatible J in (C.7.5) in Lemma C.7.1. It follows that the function κ in (4.3.8) vanishes and hence

$$\tfrac{1}{2}\Delta|\partial_s u|^2 = |\nabla_s\partial_s u|^2 + |\nabla_t\partial_s u|^2 - \langle R(\partial_s u, \partial_t u)\partial_t u, \partial_s u\rangle. \tag{4.3.10}$$

If (M, ω, J) has nonpositive sectional curvature, the right hand side is nonnegative, and so the function $|\partial_s u|^2$ is subharmonic. Hence, in this case, the maximum principle (or the mean value inequality) for subharmonic functions shows that every finite energy J-holomorphic curve $u : \mathbb{C} \to M$ is constant. In other words, when J is compatible with ω and the sectional curvature of the metric $\omega(\cdot, J\cdot)$ is nonpositive, the small energy assumption in Lemma 4.3.1 can be dropped and hence there are no nonconstant J-holomorphic spheres. In particular, this holds for Kähler manifolds with nonpositive sectional curvature.

4.4. The isoperimetric inequality

The most basic isoperimetric inequality concerns simple closed curves in the Euclidean plane and is often expressed by saying that among all such curves of a given length the circle encloses the greatest area. Another way of saying this is as follows. If γ is a simple closed curve in $\mathbb{R}^2$ denote by $L(\gamma)$ its length with respect to the usual metric and by $\mathcal{A}(\gamma)$ the enclosed area. Then

$$\mathcal{A}(\gamma) \leq \frac{1}{4\pi} L(\gamma)^2,$$

with equality if and only if γ is a circle. We show in this section that this form of the isoperimetric inequality has a natural extension to symplectic manifolds in which $\mathcal{A}(\gamma)$ is interpreted as the symplectic area of a disc with boundary γ. In general (i.e. if ω does not vanish on $\pi_2(M)$) γ bounds many discs with different symplectic areas, and so we must specify the disc we consider. This is possible only if γ is sufficiently short.

In Theorem 4.4.1 below we establish the isoperimetric inequality in the ω-compatible case with a sharp constant. This can be used to obtain a sharp constant in the exponential estimate of Lemma 4.7.3 for ω-compatible almost complex structures. The general isoperimetric inequality in the ω-tame case (without control of the constant) can be proved with easier methods, as outlined in Remark 4.4.3. This is all that is needed to establish removal of singularities.

Let (M, ω) be a compact symplectic manifold and $J \in \mathcal{J}_\tau(M, \omega)$ be an ω-tame almost complex structure. We assume throughout that M is equipped with the Riemannian metric $g_J = \langle \cdot, \cdot \rangle_J$ determined by ω and J via (2.1.1). Consider a smooth loop $\gamma : \mathbb{R}/2\pi\mathbb{Z} \to M$ whose length

$$\ell(\gamma) := \int_0^{2\pi} |\dot{\gamma}(\theta)|_J \, d\theta$$

is smaller than the injectivity radius of M. Then its image is contained in some geodesic ball $U \subset M$, whose radius is at most half the injectivity radius. Since this ball is contractible, γ admits a smooth local extension $u_\gamma : \mathbb{D} \to U$ such that

$$u_\gamma(e^{i\theta}) = \gamma(\theta)$$

for every $\theta \in \mathbb{R}$. We define the **local symplectic action** of γ as minus the symplectic area[1] of u_γ and denote it by

$$a(\gamma) := -\int_{\mathbb{D}} u_\gamma^* \omega.$$

Since any two such extensions (with images in geodesic balls of radii at most half the injectivity radius) are homotopic relative to their boundary, they have the same symplectic area. Thus the local symplectic action $a(\gamma)$ is well defined provided that γ is sufficiently short. Note also that the restriction of ω to U is exact, so that we may also define the local symplectic action by

$$a(\gamma) := \int \gamma^* \lambda, \qquad \lambda \in \Omega^1(U), \qquad d\lambda = -\omega. \tag{4.4.1}$$

That this agrees with the previous definition follows from Stokes' theorem.

[1]The minus sign seems a little awkward here. However, it is consistent with standard sign conventions for the symplectic action.

Now let $L \subset M$ be a compact Lagrangian submanifold. Then the notion of the local symplectic action extends as follows to paths with boundary points in L. Suppose $\gamma : [0, \pi] \to M$ is a smooth path such that $\gamma(0), \gamma(\pi) \in L$. If γ is sufficiently short then its image is contained in a contractible open subset $U \subset M$ such that $L \cap U$ is also contractible. It follows that there is a half disc $u_\gamma : \mathbb{D} \cap \mathbb{H} \to M$ such that

$$u_\gamma(\mathbb{D} \cap \mathbb{R}) \subset L, \qquad u_\gamma(e^{i\theta}) = \gamma(\theta)$$

for $0 \le \theta \le \pi$. The local symplectic action of γ is defined by

$$a(\gamma) := -\int_{\mathbb{D}\cap\mathbb{H}} u_\gamma^*\omega.$$

Alternatively, $a(\gamma)$ can be defined by (4.4.1), provided that λ vanishes on L. That these two definitions agree follows again from Stokes' theorem. Hence the local symplectic action is independent of the map u_γ and the 1-form λ used to define it.

THEOREM 4.4.1 (Isoperimetric inequality). *Let (M, ω) be a compact symplectic manifold, let $L \subset M$ be a compact Lagrangian submanifold, and let J be an almost complex structure on M. Then the following holds.*

(i) *If J is ω-compatible then, for every constant $c > 1/4\pi$, there exists a constant $\delta > 0$ such that*

$$\ell(\gamma) < \delta \qquad \Longrightarrow \qquad |a(\gamma)| \le c\,\ell(\gamma)^2 \tag{4.4.2}$$

for every smooth loop $\gamma : S^1 \to M$.

(ii) *If J is ω-compatible then, for every constant $c > 1/2\pi$, there exists a constant $\delta > 0$ such that (4.4.2) holds for every smooth path $\gamma : [0, \pi] \to M$ with endpoints in L.*

(iii) *If J is ω-tame then there exist positive constants c, δ such that (4.4.2) holds for every smooth loop $\gamma : S^1 \to M$ and every smooth path $\gamma : [0, \pi] \to M$ with endpoints in L.*

REMARK 4.4.2. Fix an ω-compatible almost complex structure J_0 on M. Denote the length of a loop γ with respect to the metric g_{J_0} by $\ell_0(\gamma)$ and let δ_0 be half the injectivitiy radius of g_{J_0}. Then the local symplectic action $a(\gamma)$ is well defined for every loop of length $\ell_0(\gamma) < \delta_0$. Given $J \in \mathcal{J}_\tau(M, \omega)$ and a loop γ, denote by $\ell(\gamma)$ the length with respect to g_J. Let $c_J \ge 1$ the smallest constant such that

$$\frac{1}{c_J}\ell_0(\gamma) \le \ell(\gamma) \le c_J \ell_0(\gamma)$$

for every loop γ. Theorem 4.4.1 shows that the constant δ depends only on the constant c_J. Hence the constant $\delta = \delta(J) > 0$ in Theorem 4.4.1 can be chosen continuous with respect to the C^0 topology on $\mathcal{J}_\tau(M, \omega)$.

REMARK 4.4.3. It is easy to see that the isoperimetric inequality holds with *some* constant c. Define the local extension $u_\gamma : \mathbb{D} \to M$ of γ by

$$u_\gamma(re^{i\theta}) := \exp_{\gamma(0)}(r\xi(\theta)), \tag{4.4.3}$$

where $\xi(\theta) \in T_{\gamma(0)}M$ is determined by the condition

$$\exp_{\gamma(0)}(\xi(\theta)) = \gamma(\theta).$$

To prove (4.4.2), note that

$$|\partial_r u_\gamma| = |\xi(\theta)| = d\big(\gamma(0), \gamma(\theta)\big) \le \ell(\gamma)$$

and

$$|\partial_\theta u_\gamma| \le c_1|\dot\xi(\theta)| \le c_2|\dot\gamma(\theta)|$$

for all r and θ and some constants c_1 and c_2 depending only on the metric. Hence

$$|a(u_\gamma)| = \left|\int_0^{2\pi}\int_0^1 \omega(\partial_r u_\gamma, \partial_\theta u_\gamma)\, drd\theta\right| \le c_3\ell(\gamma)^2,$$

whenever γ is shorter than half the injectivity radius.

The idea of the proof of Theorem 4.4.1 is to show by direct calculation that it holds in symplectic vector spaces, and then to deduce it for general symplectic manifolds by using Darboux charts. This approach works because the charts do not distort the action and distort the lengths of short curves only a little.

Here are the details. Let (V,ω) be a symplectic vector space and $J \in \mathrm{Aut}(V)$ be an ω-compatible complex structure. Then V is equipped with the inner product

$$\langle v, w\rangle := \omega(v, Jw)$$

as in (2.1.1). Throughout, we identify the unit circle $S^1 \subset \mathbb{C}$ with $\mathbb{R}/2\pi\mathbb{Z}$. Then the action, length, and energy of a loop $\gamma : S^1 \to V$ are given by

$$\begin{aligned} A(\gamma) &:= \tfrac{1}{2}\int_0^{2\pi} \omega\big(\dot\gamma(\theta), \gamma(\theta)\big)\, d\theta, \\ L(\gamma) &:= \int_0^{2\pi} |\dot\gamma(\theta)|\, d\theta, \\ E(\gamma) &:= \tfrac{1}{2}\int_0^{2\pi} |\dot\gamma(\theta)|^2\, d\theta. \end{aligned}$$

The same formulas, with 2π replaced by π, define the action, length, and energy of a smooth path $\gamma : [0,\pi] \to V$. The next lemma establishes the isoperimetric inequality for symplectic vector spaces with a sharp constant.

LEMMA 4.4.4. *Let (V,ω) be a symplectic vector space, $\Lambda \subset V$ be a Lagrangian subspace, and $J \in \mathrm{Aut}(V)$ be an ω-compatible complex structure on V. Then*

$$|A(\gamma)| \le \frac{1}{4\pi} L(\gamma)^2 \tag{4.4.4}$$

for every smooth loop $\gamma : S^1 \to V$ and

$$|A(\gamma)| \le \frac{1}{2\pi} L(\gamma)^2. \tag{4.4.5}$$

for every smooth path $\gamma : [0,\pi] \to V$ that satisfies $\gamma(0), \gamma(\pi) \in \Lambda$.

PROOF. We first prove the estimate

$$|A(\gamma)| \le E(\gamma) \tag{4.4.6}$$

for every loop $\gamma : S^1 \to V$. The proof relies on the identity

$$\int_0^{2\pi} \omega(e^{k\theta J}v, e^{\ell\theta J}Jw)\, d\theta = \begin{cases} 2\pi\langle v, w\rangle, & \text{if } k = \ell, \\ 0, & \text{if } k \ne \ell. \end{cases} \tag{4.4.7}$$

This follows by integrating the formula

$$\begin{aligned} \omega(e^{k\theta J}v, e^{\ell\theta J}Jw) &= \cos(k\theta)\cos(\ell\theta)\,\omega(v, Jw) + \sin(k\theta)\sin(\ell\theta)\,\omega(w, Jv) \\ &\quad + \sin(k\theta)\cos(\ell\theta)\,\omega(Jv, Jw) + \cos(k\theta)\sin(\ell\theta)\,\omega(w, v) \\ &= \cos((k-\ell)\theta)\langle v, w\rangle + \sin((k-\ell)\theta)\omega(v, w). \end{aligned} \tag{4.4.8}$$

over the interval $0 \le \theta \le 2\pi$. (Here the last equation uses the fact that J is ω-compatible. If J is only ω-tame, equation (4.4.7) continues to hold for $k+\ell \neq 0$ but is wrong for $k+\ell = 0$.) Now write γ as a Fourier series

$$\gamma(\theta) = \sum_{k=-\infty}^{\infty} e^{k\theta J} v_k. \tag{4.4.9}$$

Then, by (4.4.7), we have

$$\begin{aligned} A(\gamma) &= -\tfrac{1}{2}\int_0^{2\pi} \omega\big(\gamma(\theta), \dot\gamma(\theta)\big)\, d\theta \\ &= -\tfrac{1}{2}\sum_{k,\ell} \ell \int_0^{2\pi} \omega(e^{k\theta J} v_k, e^{\ell\theta J} J v_\ell)\, d\theta \\ &= -\pi \sum_k k\, |v_k|^2 \end{aligned}$$

and

$$\begin{aligned} E(\gamma) &= \tfrac{1}{2}\int_0^{2\pi} \omega\big(\dot\gamma(\theta), J\dot\gamma(\theta)\big)\, d\theta \\ &= \tfrac{1}{2}\sum_{k,\ell} k\ell \int_0^{2\pi} \omega(e^{k\theta J} v_k, e^{\ell\theta J} J v_\ell)\, d\theta \\ &= \pi \sum_k k^2\, |v_k|^2 . \end{aligned}$$

This implies (4.4.6).

To deduce (4.4.4) from (4.4.6) we first consider the case where γ is immersed and reparametrize it by arc length. Thus assume $\dot\gamma(\theta) \neq 0$ for every θ, and choose the reparametrization $\alpha : \mathbb{R} \to \mathbb{R}$ so that $\alpha(\theta + 2\pi) = \alpha(\theta)$ and $|\partial_\theta(\gamma \circ \alpha)| \equiv L(\gamma)/2\pi$. Then, by (4.4.6), we have

$$|A(\gamma)| = |A(\gamma \circ \alpha)| \le E(\gamma \circ \alpha) = \frac{1}{4\pi} L(\gamma)^2.$$

This proves (4.4.4) for immersed curves. To prove (4.4.4) in general choose a sequence $a_\nu \in V$ that converges to 0 and is such that $a_\nu \neq e^{-\theta J} J\dot\gamma(\theta)$ for every θ and every ν. Then $\gamma_\nu(\theta) := \gamma(\theta) + e^{\theta J} a_\nu$ is a sequence of immersions converging to γ in the C^∞ topology. Hence $|A(\gamma_\nu)| \le L(\gamma_\nu)^2/4\pi$ for every ν and so the result for γ follows by taking the limit $\nu \to \infty$. This proves (4.4.4).

To prove (4.4.5), we observe that

$$\int_0^{\pi} \omega(e^{k\theta J} v, e^{\ell\theta J} J w)\, d\theta = \begin{cases} \pi\langle v, w\rangle, & \text{if } k = \ell, \\ 0, & \text{if } k \neq \ell, \end{cases} \qquad v, w \in \Lambda, \tag{4.4.10}$$

by (4.4.8). Now let $\gamma : [0, \pi] \to V$ be a smooth path with endpoints $\gamma(0), \gamma(\pi) \in \Lambda$ and write γ as a Fourier series (4.4.9) with $v_k \in \Lambda$. Then, by (4.4.10), we have

$$A(\gamma) = -\frac{\pi}{2}\sum_k k\, |v_k|^2, \qquad E(\gamma) = \frac{\pi}{2}\sum_k k^2\, |v_k|^2 .$$

Hence the inequality (4.4.6) continues to hold in the Lagrangian case on the time interval $[0, \pi]$. For immersed curves the estimate (4.4.5) follows from (4.4.6) by reparametrization. To establish this inequality in general use the above approximation argument with $a_\nu \in \Lambda$. This proves Lemma 4.4.4. □

The next example is due to Pavel Giterman. It shows that the isoperimetric inequality for ω-tame complex structures cannot hold with a constant that is independent of J when $\dim V \geq 4$. (The estimates (4.4.4) and (4.4.5) were claimed for ω-tame complex structures J in the 2004 edition of this book. The mistake was noted by Joseph Bernstein.)

EXAMPLE 4.4.5. Write the elements of $V = \mathbb{R}^{2n}$ as $z = (x, y)$ with $x, y \in \mathbb{R}^n$. Denote by $\omega := \sum_i dx_i \wedge dy_i$ the standard symplectic form and let $Q \in \mathrm{GL}(n, \mathbb{R})$. Then the complex structure

$$J_Q := \begin{pmatrix} 0 & -Q^{-1} \\ Q & 0 \end{pmatrix}$$

is ω-tame. The associated norm on $\mathbb{R}^{2n}$ is given by

$$|z|_Q^2 := \omega(z, J_Q z) = \langle x, Qx \rangle + \langle y, Q^{-1} y \rangle.$$

Take the case $n = 2$ and choose

$$Q := \begin{pmatrix} \varepsilon & -\sqrt{1-\varepsilon^2} \\ \sqrt{1-\varepsilon^2} & \varepsilon \end{pmatrix} = (Q^{-1})^T.$$

Then $|z|_Q^2 = \varepsilon |z|^2$ and hence $c_Q := 1/4\pi\varepsilon$ is a sharp constant for the isoperimetric inequality (4.4.4) with the norm on V replaced by $|\cdot|_Q$.

PROOF OF THEOREM 4.4.1. Assume first that J is ω-compatible. Cover M by finitely many Darboux charts $\phi : U \to \mathbb{R}^{2n}$ such that each chart extends smoothly to an open neighbourhood of the closure of U. Now choose $\delta > 0$ such that, for every $x_0 \in M$, there is a Darboux chart ϕ from our finite collection such that:

(i) the ball $B_\delta(x_0)$ is contained in U, and
(ii) for every $x \in B_\delta(x_0)$

$$\|d\phi(x)\|^2 \, \|\phi_* J(\phi(x)) - \phi_* J(\phi(x_0))\| \leq \varepsilon := 4\pi c - 1.$$

Here we denote by $\|d\phi(x)\|$ the operator norm of the linear map $d\phi(x) : T_x M \to \mathbb{R}^{2n}$ with respect to the metric (2.1.1) on $T_x M$ and the Euclidean norm on $\mathbb{R}^{2n}$. The second term $\|\phi_* J(\phi(x)) - \phi_* J(\phi(x_0))\|$ refers to the Euclidean matrix norm on $\mathbb{R}^{2n \times 2n}$.

Let $\gamma : S^1 \to M$ be a loop of length $\ell(\gamma) < \delta$ and $\phi : U \to \mathbb{R}^{2n}$ be a Darboux chart as above that contains the point $x_0 := \gamma(0)$. Then $\gamma(S^1) \subset B_\delta(x_0) \subset U$. Denote by

$$J_0 := \phi_* J(\phi(x_0)) = d\phi(x_0) J(x_0) d\phi(x_0)^{-1}$$

the ω_0-compatible complex structure on $\mathbb{R}^{2n}$ induced by $J(x_0)$. Then, for every $x \in B_\delta(x_0)$ and every $\xi \in T_x M$, we have

$$\begin{aligned} |d\phi(x)\xi|_{J_0}^2 &= |\xi|_J^2 + \omega_0 \left(d\phi(x)\xi, \left(J_0 - d\phi(x) J(x) d\phi(x)^{-1}\right) d\phi(x)\xi \right) \\ &\leq \left(1 + \|d\phi(x)\|^2 \, \|\phi_* J(\phi(x_0)) - \phi_* J(\phi(x))\|\right) |\xi|_J^2 \\ &\leq (1 + \varepsilon) \, |\xi|_J^2 . \end{aligned}$$

Hence, by Lemma 4.4.4,

$$|a(\gamma)| = |A(\phi \circ \gamma)| \leq \frac{1}{4\pi} L(\phi \circ \gamma)^2 \leq \frac{1+\varepsilon}{4\pi} \ell(\gamma)^2 = c\,\ell(\gamma)^2.$$

Here $L(\phi\circ\gamma)$ is understood with respect to the metric on $\mathbb{R}^{2n}$ induced by ω_0 and J_0. This proves (i). The proof of (ii) is exactly the same, except that our Darboux charts must now be chosen such that $\phi(U \cap L) = \phi(U) \cap \mathbb{R}^n$ and $\varepsilon := 2\pi c - 1$. To prove (iii), choose an ω-compatible almost complex structure J_0. Then (i) and (ii) hold for J_0. Since M is compact the lengths of a path γ with respect to J_0 and J are related by an inequalty $\ell_0(\gamma)/c_0 \le \ell(\gamma) \le c_0 \ell_0(\gamma)$ with a constant c_0 that is independent of γ. This proves Theorem 4.4.1. □

EXERCISE 4.4.6. Show that if (M, J) is a compact almost complex manifold with Riemannian metric g then there are positive constants δ and c such that every closed curve γ with length less than δ bounds a disc $u : \mathbb{D} \to M$ of energy $E(u) < c\ell(\gamma)^2$.

4.5. Removal of singularities

The removable singularity theorem asserts that every J-holomorphic curve on the punctured disc with values in a compact symplectic manifold extends smoothly to the whole disc provided that it has finite energy. This also applies to the case of punctures on the boundary provided that u maps the boundary to a compact Lagrangian submanifold $L \subset M$. We now give a proof of this result that is independent of the argument in Section 4.1 and instead relies on the mean value estimate of Lemma 4.3.1. The first step is to extend the isoperimetric inequality of Section 4.4 to the case when the extension u_γ is defined over a punctured disc. This is related to the monotonicity property of minimal surfaces used in Section 4.1. We denote by $\ell(\gamma)$ the length of a path or loop $\gamma : I \to M$.

LEMMA 4.5.1. *Let (M, ω) be a compact symplectic manifold, $L \subset M$ be a compact Lagrangian submanifold, and $J \in \mathcal{J}_\tau(M, \omega)$ be an ω-tame almost complex structure. Fix a constant $R > 0$. Then the following holds.*

(i) *Let $u : B_R \setminus \{0\} \to M$ be a J-holomorphic curve on a punctured disc such that $E(u) < \infty$. Then there are constants $r_0 \in (0, R]$ and $c > 0$ such that*

$$0 < r < r_0 \qquad \Longrightarrow \qquad \tfrac{1}{2}\int_{B_r} |du|^2 \le c\,\ell(\gamma_r)^2,$$

where $\gamma_r : \mathbb{R}/2\pi\mathbb{Z} \to M$ denotes the loop $\gamma_r(\theta) := u(re^{i\theta})$.

(ii) *Let $u : (B_R \cap \mathbb{H} \setminus \{0\}, B_R \cap \mathbb{R} \setminus \{0\}) \to (M, L)$ be a J-holomorphic curve on the punctured half disc such that $E(u) < \infty$. Then there are constants $r_0 \in (0, R]$ and $c > 0$ such that*

$$0 < r < r_0 \qquad \Longrightarrow \qquad \tfrac{1}{2}\int_{B_r \cap \mathbb{H}} |du|^2 \le c\,\ell(\gamma_r)^2,$$

where $\gamma_r : [0, \pi] \to M$ denotes the path $\gamma_r(\theta) := u(re^{i\theta})$.

PROOF. We prove (i). Choose constants $c > 0$ and $\delta > 0$ such that the assertions of Lemma 4.3.1 (i) and Theorem 4.4.1 (iii) hold with these constants δ and c. Choose a constant $0 < r_0 \le R/2$ such that

$$4\pi^2 c\, E(u; B_{2r_0}) < \delta^2.$$

Define the smooth function $\varepsilon : (0, 2r_0] \to \mathbb{R}$ by

$$\varepsilon(r) := E(u; B_r) = \tfrac{1}{2}\int_{B_r} |du|^2$$

for $0 < r \le 2r_0$. Then ε extends to a continuous function on the closed interval $[0, 2r_0]$ such that $\varepsilon(0) = 0$. Moreover, by Lemma 4.3.1 (i), we have

$$\tfrac{1}{2}|du(re^{i\theta})|^2 \le \frac{c}{r^2}\varepsilon(2r)$$

for $r \le r_0$. Now the derivative of γ_r has the norm

$$|\dot{\gamma}_r(\theta)| = \frac{r}{\sqrt{2}}\left|du(re^{i\theta})\right| \le \sqrt{c\varepsilon(2r)},$$

and so the length of the loop γ_r satisfies the inequality

$$\ell(\gamma_r) = \int_0^{2\pi} |\dot{\gamma}_r(\theta)|\, d\theta \le 2\pi\sqrt{c\varepsilon(2r)} < \delta \tag{4.5.1}$$

for $r \le r_0$. Hence $\ell(\gamma_r)$ converges to zero as r tends to zero.

For $0 < \rho < r \le r_0$ let $u_r := u_{\gamma_r} : \mathbb{D} \to M$ be the extension of Remark 4.4.3 and consider the sphere $v_{\rho r} : S^2 \to M$ that is obtained from the restriction of u to the annulus $B_r \setminus B_\rho$ by filling in the boundary circles γ_ρ and γ_r with the discs u_ρ and u_r. This sphere is contractible because it bounds a 3-ball consisting of the union of the 2-discs $u_s : \mathbb{D} \to M$ for $\rho \le s \le r$. Hence

$$E(u; B_r \setminus B_\rho) + \int_{\mathbb{D}} u_\rho^*\omega = \int_{\mathbb{D}} u_r^*\omega$$

for $0 < \rho < r \le r_0$. Take the limit $\rho \to 0$ and use the isoperimetric inequality of Theorem 4.4.1 to obtain the inequality

$$E(u; B_r) = \int_{\mathbb{D}} u_r^*\omega \le c\,\ell(\gamma_r)^2.$$

This proves assertion (i) in Lemma 4.5.1. □

EXERCISE 4.5.2. Prove assertion (ii) of Lemma 4.5.1. The proof is almost word by word the same as that of assertion (i). *Hint:* Replace the local symplectic action of a short loop $\gamma : S^1 \to M$ by the local symplectic action of a sufficiently short arc $\gamma : [0, \pi] \to M$ that maps the boundary points to L. See also the diploma thesis of Urs Frauenfelder [**120**].

PROOF OF THEOREM 4.1.2. We prove (i), continuing to use the notation in the proof of Lemma 4.5.1. In particular, c and δ are positive constants such that the assertions of Lemma 4.3.1 (i) and Theorem 4.4.1 (iii) hold with these constants. Inspecting the proof of Theorem 4.4.1 one finds that $c > 1/4\pi$. Moreover,

$$\varepsilon(r) := E(u; B_r) = \tfrac{1}{2}\int_{B_r} |du|^2 = \tfrac{1}{2}\int_0^r \rho \int_0^{2\pi} \left|du(\rho e^{i\theta})\right|^2 d\theta d\rho.$$

It follows from the isoperimetric inequality of Lemma 4.5.1 that, for r sufficiently small, we have

$$\begin{aligned}
\varepsilon(r) &\le c\ell(\gamma_r)^2 \\
&= \frac{cr^2}{2}\left(\int_0^{2\pi} \left|du(re^{i\theta})\right| d\theta\right)^2 \\
&\le \pi c r^2 \int_0^{2\pi} \left|du(re^{i\theta})\right|^2 d\theta \\
&= 2\pi c r \dot{\varepsilon}(r).
\end{aligned}$$

With $\mu := 1/4\pi c < 1$ this can be rewritten as

$$\frac{2\mu}{r} \leq \frac{\dot{\varepsilon}(r)}{\varepsilon(r)}.$$

Integrating this inequality from r to r_1 we obtain

$$\left(\frac{r_1}{r}\right)^{2\mu} \leq \frac{\varepsilon(r_1)}{\varepsilon(r)}.$$

Hence

$$\varepsilon(r) \leq c_1 r^{2\mu}, \qquad c_1 := r_1^{-2\mu}\varepsilon(r_1),$$

and, by Lemma 4.3.1 (i), we have

$$|du(\rho e^{i\theta})|^2 \leq \frac{c}{\rho^2}\varepsilon(2\rho) \leq \frac{c_2}{\rho^{2-2\mu}}, \qquad c_2 := cc_1 2^{2\mu}, \tag{4.5.2}$$

for ρ sufficiently small. With

$$2 < p < \frac{2}{1-\mu}, \qquad c_3 := 2\pi c_2^{p/2},$$

this implies

$$\begin{aligned} \int_{B_r} |du|^p &= \int_0^r \int_0^{2\pi} \rho |du(\rho e^{i\theta})|^p \, d\theta d\rho \\ &\leq c_3 \int_0^r \rho^{1-p(1-\mu)} \, d\rho \end{aligned}$$

for r sufficiently small. Since $p < 2/(1-\mu)$, we have

$$1 - p(1-\mu) > -1$$

and hence the integral is finite. Moreover, it follows from (4.5.2) that u is uniformly Hölder continuous on the punctured disc. Since M is compact (and therefore complete) this implies that u extends continuously over zero. Now Exercise 4.5.4 shows that the weak first derivatives of u exist on $\mathbb{D}$ and agree with the strong derivatives on $\mathbb{D} \setminus \{0\}$. Hence u belongs to the Sobolev space of $W^{1,p}$-maps from $\mathbb{D}$ to M and so we can apply Theorem B.4.1 to u in a local coordinate chart around $u(0)$ to obtain that u is smooth. This proves (i). For the proof of (ii) see Exercise 4.5.5 below. This proves Theorem 4.1.2. □

Exercise 4.5.3. Prove from (4.5.2) that u is Hölder continuous with the exponent μ.

Exercise 4.5.4. Suppose $u : \mathbb{D} \to \mathbb{R}$ is a continuous function whose restriction to the punctured disc $\mathbb{D} \setminus \{0\}$ is smooth. Moreover, assume that the L^p-norms of $\partial_s u$ and $\partial_t u$ are bounded. Prove that u belongs to the Sobolev space $W^{1,p}(\mathbb{D})$. *Hint:* Show that the L^p-function $\partial_s u : \mathbb{D} \to \mathbb{R}$ agrees with the weak derivative of u, i.e.

$$\int_{\mathbb{D}} (u\partial_s\phi + (\partial_s u)\phi) = 0$$

for every $\phi \in C_0^\infty(\mathbb{D})$. Use Stokes' theorem to express the integral over $\{|z| \geq \delta\}$ as an integral over $\{|z| = \delta\}$ and show that this integral tends to zero as $\delta \to 0$.

EXERCISE 4.5.5. Prove assertion (ii) of Theorem 4.1.2. The proof is almost word by word the same as that of assertion (i). *Hint:* Define

$$\varepsilon(r) := E(u; B_r \cap \mathbb{H})$$

and use assertion (ii) of Lemma 4.5.1 (instead of assertion (i)). See also [**302, 120**].

EXERCISE 4.5.6. Examine holomorphic maps $f : \mathbb{D} \setminus \{0\} \to \mathbb{C}$ that have a singularity at 0. For which value of p can such a map belong to $L^p(\mathbb{D})$ or to $W^{1,p}(\mathbb{D})$? Consider the same question for holomorphic maps into S^2. Think of these as composites of holomorphic maps $\mathbb{D} \setminus \{0\} \to \mathbb{C}$ with the inclusion of $\mathbb{C}$ into the Riemann sphere $\mathbb{C} \cup \{\infty\}$.

4.6. Convergence modulo bubbling

In Section 4.2 we have seen how bubbling occurs for sequences of finite energy J-holomorphic curves whenever the derivatives diverge to infinity. In this section we shall also allow for Lagrangian boundary conditions and show how bubbling near the boundary may lead to J-holomorphic discs with boundary values in the Lagrangian submanifold. Since the J-holomorphic discs or spheres cannot have arbitrarily small energy (Proposition 4.1.4) it follows that, after passing to a suitable subsequence, bubbling can only occur near finitely many points. The next theorem formulates this limiting behaviour more precisely. We assume that $(\Sigma, j_\Sigma, \mathrm{dvol}_\Sigma)$ is a (not necessarily compact) Riemann surface with a fixed complex structure, that (M, ω) is a compact symplectic manifold, that $L \subset M$ is a compact Lagrangian submanifold, that $J \in \mathcal{J}_\tau(M, \omega)$ is an ω-tame almost complex structure, and that $\hbar > 0$ is the constant of Proposition 4.1.4.

THEOREM 4.6.1. *Let $J^\nu \in \mathcal{J}_\tau(M, \omega)$ be a sequence of ω-tame almost complex structures that converges to J in the C^∞-topology, Ω^ν be an increasing sequence of open sets that exhaust Σ, and $u^\nu : (\Omega^\nu, \Omega^\nu \cap \partial\Sigma) \to (M, L)$ be a sequence of J^ν-holomorphic curves such that*

$$\sup_\nu E(u^\nu) < \infty.$$

Then there exists a subsequence (still denoted by u^ν), a J-holomorphic curve $u : (\Sigma, \partial\Sigma) \to (M, L)$, and a finite set $Z = \{z_1, \dots, z_\ell\} \subset \Sigma$ such that the following holds.

(i) *u^ν converges to u uniformly with all derivatives on compact subsets of $\Sigma \setminus Z$.*

(ii) *For every j and every $\varepsilon > 0$ such that $B_\varepsilon(z_j) \cap Z = \{z_j\}$ the limit*

$$m_\varepsilon(z_j) := \lim_{\nu \to \infty} E\big(u^\nu; B_\varepsilon(z_j)\big)$$

exists and is a continuous function of ε, and

$$m(z_j) := \lim_{\varepsilon \to 0} m_\varepsilon(z_j) \geq \hbar.$$

(iii) *For every compact subset $K \subset \Sigma$ with $Z \subset \mathrm{int}(K)$,*

$$E(u; K) + \sum_{j=1}^{\ell} m(z_j) = \lim_{\nu \to \infty} E(u^\nu; K).$$

REMARK 4.6.2. The convergent subsequence u^ν of Theorem 4.6.1 satisfies

$$\lim_{\nu\to\infty} E\big(u^\nu; B_\varepsilon(z_j^\nu)\big) = m_\varepsilon(z_j)$$

for *every* sequence $z_j^\nu \to z_j$. To see this, fix an arbitrarily small number $\delta > 0$ and note that

$$B_{\varepsilon-\delta}(z_j) \subset B_\varepsilon(z_j^\nu) \subset B_{\varepsilon+\delta}(z_j)$$

for ν sufficiently large. Hence the limit lies in the interval $[m_{\varepsilon-\delta}(z_j), m_{\varepsilon+\delta}(z_j)]$ for all $\delta > 0$ and so the identity follows from the continuity of the function $\varepsilon \mapsto m_\varepsilon(z_j)$.

REMARK 4.6.3. In Theorem 4.6.1 bubbling occurs at every point of Z. This means that for every j there exists either a nonconstant J-holomorphic sphere $v : S^2 \to M$ or a nonconstant J-holomorphic disc $v : (B, \partial B) \to (M, L)$ whose image can be approximated by points in $u^\nu(B_\varepsilon(z_j))$ for every $\varepsilon > 0$.

The proof of Theorem 4.6.1 relies on Lemma 4.6.5 below which asserts that there is a concentration of energy greater than or equal to $\hbar$ near every point at which the energy density diverges to infinity, and that at every such point either a J-holomorphic sphere or a J-holomorphic disc bubbles off. The proof of this lemma uses a conformal rescaling argument as in Section 4.2. There we rescaled around local maxima of the energy density. However, in the current situation the domain may be noncompact and there may be no local maxima. We will see that it suffices to rescale around "approximate local maxima". The next lemma, which is due to Hofer, shows how to find them.

LEMMA 4.6.4. *Let (X, d) be a metric space, $f : X \to \mathbb{R}$ be a nonnegative continuous function, and fix $x \in X$ and $\delta > 0$. Suppose that the closed ball $B_{2\delta}(x)$ is complete. Then there exists a $\zeta \in X$ and a positive number $\varepsilon \le \delta$ such that*

$$\text{(i)}\ \ d(x,\zeta) < 2\delta, \qquad \text{(ii)}\ \ \sup_{B_\varepsilon(\zeta)} f \le 2f(\zeta), \qquad \text{(iii)}\ \ \varepsilon f(\zeta) \ge \delta f(x).$$

PROOF. Suppose not. Then we claim that there exists a sequence $x_k \in X$ such that

$$x_0 = x, \quad d(x_k, x_{k+1}) \le \delta/2^k, \qquad f(x_{k+1}) > 2f(x_k)$$

for $k \ge 0$ and this contradicts completeness. We construct the sequence by induction. First choose $x_0 = x$. Then (i) and (iii) hold when $\zeta = x_0$ and $\varepsilon = \delta$. If (ii) is not satisfied then

$$\sup_{B_\delta(x_0)} f > 2f(x_0)$$

and hence there exists a point $x_1 \in X$ such that

$$d(x_0, x_1) \le \delta, \qquad f(x_1) > 2f(x_0).$$

Now repeat the argument, taking $\zeta = x_1$ and $\varepsilon = \delta/2$. More generally, suppose that suitable $x_0, x_1, \ldots, x_k$ have been constructed for some $k \ge 1$. Then

$$d(x, x_k) < 2\delta, \qquad f(x_k) > 2^k f(x_0).$$

Thus (i) and (iii) hold when $\zeta = x_k$ and $\varepsilon = \delta/2^k$. If (ii) is not satisfied, then

$$\sup_{B_{\delta/2^k}(x_k)} f > 2f(x_k)$$

and hence there exists a point $x_{k+1} \in X$ such that $d(x_k, x_{k+1}) \le \delta/2^k$ and $f(x_{k+1}) > 2f(x_k)$. This completes the induction and the proof of Lemma 4.6.4. □

LEMMA 4.6.5. *Let M, ω, L, J^ν and J be as in Theorem 4.6.1 and let $\hbar = \hbar(M,\omega,L,J)$ be the constant of Proposition 4.1.4. Let $\Omega \subset \mathbb{H}$ be an open set and $u^\nu : (\Omega, \Omega\cap\mathbb{R}) \to (M,L)$ be a sequence of J^ν-holomorphic curves such that*

$$\sup_\nu E(u^\nu;\Omega) < \infty.$$

If $|du^\nu(z^\nu)| \to \infty$ for some sequence $z^\nu \to z_0 \in \Omega$, then

$$\liminf_{\nu\to\infty} E\big(u^\nu; B_\varepsilon(z_0)\cap\mathbb{H}\big) \ge \hbar \tag{4.6.1}$$

for every $\varepsilon > 0$ and there exists either a nonconstant J-holomorphic sphere $v : S^2 \to M$ or a nonconstant J-holomorphic disc $u : (B,\partial B) \to (M,L)$.

PROOF. Let $\varepsilon > 0$ be such that $B_\varepsilon(z_0)\cap\mathbb{H} \subset \Omega$. Apply Lemma 4.6.4 to the function $B_\varepsilon(z_0)\cap\mathbb{H} \to \mathbb{R} : z \mapsto |du^\nu(z)|$, the point $x := z^\nu$, and the constant $\delta := \delta^\nu := |du^\nu(z^\nu)|^{-1/2}$ to obtain sequences $\zeta^\nu \in \mathbb{C}$ and $\varepsilon^\nu > 0$ such that

$$\zeta^\nu \to z_0, \qquad \sup_{B_{\varepsilon^\nu}(\zeta^\nu)} |du^\nu| \le 2c^\nu, \qquad \varepsilon^\nu \to 0, \qquad \varepsilon^\nu c^\nu \to \infty,$$

where

$$c^\nu := |du^\nu(\zeta^\nu)|.$$

We consider the following two cases. (When $\Omega\cap\mathbb{R} = \emptyset$ we are always in Case 1.)

CASE 1: *The sequence $c^\nu \mathrm{Im}\,\zeta^\nu$ is unbounded; a J-holomorphic sphere bubbles off.*

CASE 2: *The sequence $c^\nu \mathrm{Im}\,\zeta^\nu$ is bounded; a J-holomorphic disc bubbles off.*

In Case 1 we may assume without loss of generality that $c^\nu\,\mathrm{Im}\,\zeta^\nu \to \infty$. The limit point $z_0 = \lim_{\nu\to\infty}\zeta^\nu$ may either lie in the interior of Ω or in $\Omega\cap\mathbb{R}$. However, in the latter case ζ^ν converges to z_0 with a slower rate than the derivatives diverge to infinity. Because $c^\nu\,\mathrm{Im}\,\zeta^\nu \to \infty$, we can shrink ε^ν if necessary so that, as well as $\varepsilon^\nu c^\nu \to \infty$, we have

$$\varepsilon^\nu \le \mathrm{Im}\,\zeta^\nu, \qquad \varepsilon^\nu + |\zeta^\nu - z_0| \le \varepsilon.$$

Then

$$B_{\varepsilon^\nu}(\zeta^\nu) \subset B_\varepsilon(z_0)\cap\mathbb{H} \subset \Omega$$

for every ν. Consider the renormalized sequence

$$v^\nu(z) = u^\nu\left(\zeta^\nu + z/c^\nu\right).$$

The function v^ν is defined on $B_{\varepsilon^\nu c^\nu}$ for every ν and satisfies

$$\sup_{B_{\varepsilon^\nu c^\nu}} |dv^\nu| \le 2, \qquad |dv^\nu(0)| = 1,$$

and

$$E(v^\nu; B_R) = E\big(u^\nu; B_{R/c^\nu}(\zeta^\nu)\big)$$

for $R \le \varepsilon^\nu c^\nu$. Hence v^ν has uniformly bounded derivatives on arbitrarily large domains (because $\varepsilon^\nu c^\nu \to \infty$) and it follows from Theorem 4.1.1 that there exists a subsequence (still denoted by v^ν) which converges, uniformly with all derivatives on compact subsets of $\mathbb{C}$, to a J-holomorphic curve $v : \mathbb{C} \to M$. The limit function $v : \mathbb{C} \to M$ has finite energy and is nonconstant since $|dv(0)| = 1$. Hence the map $\mathbb{C}\setminus\{0\} \to M : z \mapsto v(1/z)$ has finite energy and so, by Theorem 4.1.2 (i), it extends smoothly over 0. Hence v extends to a smooth J-holomorphic map on the Riemann sphere $S^2 = \mathbb{C} \cup \{\infty\}$ which will still be denoted by v. By Proposition 4.1.4,

$$E(v) \geq \hbar.$$

Hence, for every real number $\alpha \in (0,1)$ there exists a constant $R > 0$ such that $E(v; B_R) > \alpha\hbar$, and hence

$$\begin{aligned} \alpha\hbar &< E(v; B_R) \\ &= \lim_{\nu\to\infty} E(v^\nu; B_R) \\ &= \lim_{\nu\to\infty} E\big(u^\nu; B_{R/c^\nu}(\zeta^\nu)\big) \\ &\leq \liminf_{\nu\to\infty} E\big(u^\nu; B_\varepsilon(z_0)\big). \end{aligned}$$

This holds for every $\alpha < 1$ and this implies (4.6.1).

Now assume Case 2. Then, in particular, $\mathrm{Im}\,\zeta^\nu$ converges to zero and hence $z_0 \in \Omega \cap \mathbb{R}$. Passing to a subsequence we may assume without loss of generality that

$$\varepsilon^\nu + |\mathrm{Re}\,\zeta^\nu - z_0| \leq \varepsilon$$

and the limit

$$t_0 := \lim_{\nu\to\infty} c^\nu \mathrm{Im}\,\zeta^\nu \geq 0$$

exists. Consider the renormalized sequence

$$v^\nu(z) = u^\nu\left(\mathrm{Re}\,\zeta^\nu + z/c^\nu\right).$$

Since $B_{\varepsilon^\nu}(\mathrm{Re}\,\zeta^\nu) \cap \mathbb{H} \subset \Omega$, the function v^ν is defined on $B_{\varepsilon^\nu c^\nu} \cap \mathbb{H}$. It satisfies

$$\sup_{B_{\varepsilon^\nu c^\nu} \cap \mathbb{H}} |dv^\nu| \leq 2, \qquad |dv^\nu(ic^\nu \mathrm{Im}\,\zeta^\nu)| = 1,$$

and

$$E(v^\nu; B_R \cap \mathbb{H}) \leq E\big(u^\nu; B_{R/c^\nu}(\mathrm{Re}\,\zeta^\nu) \cap \mathbb{H}\big)$$

for $R \leq \varepsilon^\nu c^\nu$. Hence v^ν has uniformly bounded derivatives on arbitrarily large domains and it follows again from Theorem 4.1.1 that there exists a subsequence (still denoted by v^ν) which converges, uniformly with all derivatives on compact subsets of $\mathbb{H}$, to a J-holomorphic curve $v : (\mathbb{H}, \mathbb{R}) \to (M, L)$. The limit function $v : \mathbb{H} \to M$ has finite energy and is nonconstant since $|dv(it_0)| = 1$. Hence the map $\mathbb{H} \setminus \{0\} \to M : z \mapsto v(-1/z)$ has finite energy and so, by the Theorem 4.1.2 (ii), it extends smoothly over 0. Hence the function $w : \mathbb{D} \setminus \{-1\} \to M$ defined by

$$w(\zeta) := v\left(\frac{i(1-\zeta)}{1+\zeta}\right)$$

extends to a J-holomorphic disc $w : (\mathbb{D}, \partial\mathbb{D}) \to (M, L)$. By Proposition 4.1.4, $E(v) = E(w) \geq \hbar$. Hence, for every real number $\alpha \in (0,1)$ there exists a constant $R > 0$ such that $E(v; B_R) > \alpha\hbar$, and hence

$$\begin{aligned} \alpha\hbar &< E(v; B_R \cap \mathbb{H}) \\ &= \lim_{\nu\to\infty} E(v^\nu; B_R \cap \mathbb{H}) \\ &\leq \lim_{\nu\to\infty} E\big(u^\nu; B_{R/c^\nu}(\mathrm{Re}\,\zeta^\nu) \cap \mathbb{H}\big) \\ &\leq \liminf_{\nu\to\infty} E\big(u^\nu; B_\varepsilon(z_0) \cap \mathbb{H}\big). \end{aligned}$$

This holds for every $\alpha < 1$ and this implies (4.6.1). Thus Lemma 4.6.5 is proved. □

PROOF OF THEOREM 4.6.1. Call a point $z \in \Sigma$ **singular** for the sequence u^ν if there exists a sequence $z^\nu \to z$ such that $|du^\nu(z^\nu)| \to \infty$. By Lemma 4.6.5, every sequence of J^ν-holomorphic curves $u^\nu : \Sigma \to M$ with $\sup_\nu E(u^\nu) \le c$ has at most $c/\hbar$ singular points.

CLAIM. *There exists a subsequence, still denoted by u^ν, which has a finite set $Z = \{z_1, \dots, z_\ell\}$ of singular points and satisfies*

$$\sup_\nu \|du^\nu\|_{L^\infty(K)} < \infty$$

for every compact set $K \subset \Sigma - Z$.

Suppose, by induction that we have constructed a subsequence, still denoted by u^ν, whose singular set contains the set

$$Z_k = \{z_1, \dots, z_k\}.$$

(We start the induction with $k = 0$ and $Z_k = \emptyset$.) If the sequence $\|du^\nu\|_{L^\infty(K)}$ is bounded for every compact set $K \subset \Sigma \setminus Z_k$ then the claim is proved with $\ell = k$. Otherwise, choose a compact set $K \subset \Sigma \setminus Z_k$ such that the sequence $\|du^\nu\|_{L^\infty(K)}$ is unbounded and choose $z_{k+1}^\nu \in K$ such that

$$\|du^\nu\|_{L^\infty(K)} = |du^\nu(z_{k+1}^\nu)|.$$

By passing to a subsequence, if necessary, we may assume that z_{k+1}^ν converges to a point $z_{k+1} \in \Sigma \setminus Z_k$ and $|du^\nu(z_{k+1}^\nu)| \to \infty$ and hence the singular set of this subsequence contains the set

$$Z_{k+1} := Z_k \cup \{z_{k+1}\}.$$

The induction must terminate after finitely many steps because the number of possible singular points is bounded above. This proves the claim.

With the claim established, it follows from Theorem 4.1.1 that there exists a further subsequence, still denoted by u^ν, which converges, uniformly with all derivatives on compact subsets of $\Sigma \setminus Z$, to a J-holomorphic curve $u : \Sigma \setminus Z \to M$ such that $u(\partial\Sigma \setminus Z) \subset L$. The removable singularity theorem 4.1.2 asserts that u extends to a J-holomorphic curve $u : (\Sigma, \partial\Sigma) \to (M, L)$.

Now we fix numbers $\varepsilon_j > 0$ for $j = 1, \dots, \ell$ such that the balls $B_{\varepsilon_j}(z_j)$ are disjoint, and choose a further subsequence such that the limits

$$m_{\varepsilon_j}(z_j) := \lim_{\nu \to \infty} E\big(u^\nu; B_{\varepsilon_j}(z_j)\big)$$

exist for $j = 1, \dots, \ell$. This subsequence satisfies all the requirements of the theorem. In particular, the limits in (ii) exist and are continuous for $0 < \varepsilon \le \varepsilon_j$ since u^ν converges in the annulus $B_{\varepsilon_j}(z_j) - B_\varepsilon(z_j)$. By Proposition 4.1.4, we have

$$m(z_j) := \lim_{\varepsilon \to 0} m_\varepsilon(z_j) \ge \hbar.$$

To prove (iii), fix a number $\varepsilon \le \min_j \varepsilon_j$ and note that

$$\begin{aligned} E\Big(u; K - \bigcup_{j=1}^{\ell} B_\varepsilon(z_j)\Big) &= \lim_{\nu \to \infty} E(u^\nu; K) - \sum_{j=1}^{\ell} \lim_{\nu \to \infty} E\big(u^\nu; B_\varepsilon(z_j)\big) \\ &= \lim_{\nu \to \infty} E(u^\nu; K) - \sum_{j=1}^{\ell} m_\varepsilon(z_j). \end{aligned}$$

Take the limit $\varepsilon \to 0$ to obtain

$$E(u;K) = \lim_{\nu\to\infty} E(u^\nu;K) - \sum_{j=1}^{\ell} m(z_j)$$

as claimed. This proves Theorem 4.6.1. □

The following lemma establishes a small technical point that will be useful in Chapter 5. It shows that if a curve u is the uniform limit of a sequence u^ν of J-holomorphic curves then all derivatives of u^ν must also converge uniformly. Note that there is no need to pass to a subsequence here.

LEMMA 4.6.6. *Let J^ν and J be as in Theorem 4.6.1. Let $\Omega \subset \mathbb{C}$ be an open set and $u^\nu : \Omega \to M$ be a sequence of J^ν-holomorphic curves. Moreover, suppose that $u : \Omega \to M$ is a continuous function to which u^ν converges uniformly. Then u is a J-holomorphic curve and u^ν converges to u uniformly with all derivatives on every compact subset of Ω.*

PROOF. Suppose first that

$$\sup_\nu \|du^\nu\|_{L^\infty(K)} < \infty \tag{4.6.2}$$

for every compact subset $K \subset \Omega$. Then it follows from the standard elliptic bootstrapping techniques that every subsequence of u^ν has a further subsequence which converges, uniformly with all derivatives on compact subsets of Ω, to some J-holomorphic curve. By assumption, the limit curve always is equal to u. Hence, in this case, the result follows by the usual indirect reasoning. More precisely, if the result does not hold, then there exists a subsequence ν_j and an integer ℓ such that

$$d_{C^\ell}(u^{\nu_j}, u) \geq \delta$$

for all j and some $\delta > 0$. But passing to a further subsequence, we obtain that u^{ν_j} converges to u in the C^ℓ-topology, a contradiction. Thus we have proved the result under the assumption (4.6.2).

Now suppose that (4.6.2) does not hold. Then the proof of Lemma 4.6.5 shows that there are sequences $z^\nu \to z_0 \in \Omega$ and $\varepsilon^\nu \to 0$ such that the rescaling

$$v^\nu(z) = u^\nu(z^\nu + \varepsilon^\nu z)$$

has a subsequence that converges uniformly with all derivatives to a nonconstant J-holomorphic curve $v : B_1 \to M$. On the other hand it follows from the uniform convergence of u^ν to u that, for any $\varepsilon > 0$, there exists a ν_0 such that

$$u^\nu(B_{\varepsilon^\nu}(z^\nu)) \subset B_\varepsilon(u(z_0))$$

for $\nu \geq \nu_0$. Hence the image of the limit curve $v : B_1 \to M$ is contained in the ball $B_\varepsilon(u(z_0))$ for all $\varepsilon > 0$ and hence $v \equiv u(z_0)$, in contradiction to the fact that v is nonconstant. This proves Lemma 4.6.6. □

Theorem 4.6.1 is rather crude in that it only asserts the existence of bubbles and does not give complete information about all the holomorphic spheres or discs that bubble off or about the way in which the sequence u^ν converges to this collection of bubbles. To obtain a better understanding we must refine the above rescaling argument. The next section carries out this discussion in the case of J-holomorphic spheres.

4.7. Bubbles connect

Throughout this section we assume that (M,ω) is a compact symplectic manifold and $J \in \mathcal{J}_\tau(M,\omega)$ is an ω-tame almost complex structure on M. We restrict the discussion to the interior bubbling off of J-holomorphic spheres. The main result, stated in Propositions 4.7.1 and 4.7.2 below, is that with a careful choice of the rescaling sequence ψ^ν the bubble v connects to the limit u of the original sequence of maps u^ν. The key point in the proof is to choose the rescaling sequence so that there is no loss of energy: see identity (iii) below. The importance of the stability condition (iv) will emerge in the next chapter.

In the following proposition we identify S^2 with the Riemann sphere $\mathbb{C} \cup \{\infty\}$ and denote by $\mathrm{G} = \mathrm{PSL}(2,\mathbb{C})$ the group of Möbius transofmrations.

PROPOSITION 4.7.1. *Let $J^\nu \in \mathcal{J}_\tau(M,\omega)$ be a sequence of ω-tame almost complex structures on M that converges to J in the C^∞-topology. Fix a point $z_0 \in \mathbb{C}$ and a number $r > 0$. Suppose that $u^\nu : B_r(z_0) \to M$ is a sequence of J^ν-holomorphic curves and $u : B_r(z_0) \to M$ is a J-holomorphic curve such that the following holds.*

(a) *u^ν C^∞-converges to u on every compact subset of $B_r(z_0) \setminus \{z_0\}$.*

(b) *The limit*

$$m_0 := \lim_{\varepsilon \to 0} \lim_{\nu \to \infty} E(u^\nu; B_\varepsilon(z_0))$$

exists and is positive.

Then there is a subsequence, still denoted by u^ν, a sequence of Möbius transformations $\psi^\nu \in \mathrm{G}$, a J-holomorphic sphere $v : S^2 \to M$, and finitely many distinct points $z_1, \dots, z_\ell, z_\infty \in S^2$, such that the following holds.

(i) *ψ^ν C^∞-converges to z_0 on every compact subset of $S^2 \setminus \{z_\infty\}$.*

(ii) *The sequence $v^\nu := u^\nu \circ \psi^\nu$ converges to v in the C^∞-topology on every compact subset of $S^2 \setminus \{z_1, \dots, z_\ell, z_\infty\}$, and the limits*

$$m_j := \lim_{\varepsilon \to 0} \lim_{\nu \to \infty} E\big(v^\nu; B_\varepsilon(z_j)\big)$$

exist and are positive for $j = 1, \dots, \ell$.

(iii) *$E(v) + \sum_{j=1}^{\ell} m_j = m_0$.*

(iv) *If v is constant then $\ell \geq 2$.*

The next proposition states that the image of u connects to that of v at the point $u(z_0) = v(z_\infty)$. This happens because, by condition (iii) above, there is no energy left to form a new bubble in between.

PROPOSITION 4.7.2. *Let $J^\nu \in \mathcal{J}_\tau(M,\omega)$, $z_0 \in \mathbb{C}$, and $u, u^\nu : B_r(z_0) \to M$ be as in the hypotheses of Proposition 4.7.1 and suppose that $\psi^\nu \in \mathrm{G}$, $v : S^2 \to M$, and $z_1, \dots, z_\ell, z_\infty \in S^2$, satisfy the assertions (i-iii) of Proposition 4.7.1. Then*

$$u(z_0) = v(z_\infty).$$

Moreover, for every $\varepsilon > 0$, there exist constants $\delta > 0$ and $\nu_0 \in \mathbb{N}$ such that

$$d(z, z_0) + d\big((\psi^\nu)^{-1}(z), z_\infty\big) < \delta \qquad \Longrightarrow \qquad d\big(u^\nu(z), u(z_0)\big) < \varepsilon$$

for every integer $\nu \geq \nu_0$ and every $z \in S^2$.

The proofs of both Propositions 4.7.1 and 4.7.2 are based on the following result about the energy of a J-holomorphic curve on an arbitrarily long cylinder. It asserts that if the energy is sufficiently small then it cannot be spread out uniformly but must be concentrated near the ends. We phrase the result in terms of closed annuli

$$A(r,R) := B_R \setminus \mathrm{int}(B_r)$$

for $r < R$. The relative size of the radii r, R will be all important in what follows. One should think of the ratio R/r as being very large, and much bigger than e^T. In (4.7.2) below we denote by d the distance function on M induced by the Riemannian metric g_J given by ω and J.

LEMMA 4.7.3. *Let (M,ω) be a compact symplectic manifold and $J \in \mathcal{J}_\tau(M,\omega)$. Then there exist constants $0 < \mu < 1$, $\delta > 0$, and $c > 0$ such that the following holds. If $0 < r < R$ and $u : A(r,R) \to M$ is a J-holomorphic curve satisfying the energy bound*

$$E(u) = E\big(u; A(r,R)\big) < \delta,$$

then

$$E\big(u; A(e^T r, e^{-T}R)\big) \le \frac{4^\mu}{e^{2\mu T}} E(u) \tag{4.7.1}$$

and

$$\sup_{z,z' \in A(e^T r, e^{-T}R)} d\big(u(z), u(z')\big) \le \frac{c}{e^{\mu T}} \sqrt{E(u)} \tag{4.7.2}$$

for $\log 2 \le T \le \log\sqrt{R/r}$.

REMARK 4.7.4. The proof of Lemma 4.7.3 is deferred to the end of this section. It shows that the constants μ, δ, and c depend explicitly on the constants in Lemma 4.3.1 (i) and in Theorem 4.4.1 (iii). Hence it follows from Remark 4.3.2 and Remark 4.4.2 that the constants μ, δ, and c can be chosen to depend continuously on J with respect to the C^2 topology.

As a consequence of Lemma 4.7.3 we obtain the following result for sequences of J^ν-holomorphic annuli that converge to pairs of J-holomorphic discs without loss of energy. This result is the main step in the proof of Proposition 4.7.2. As usual we identify S^2 with the Riemann sphere $\mathbb{C} \cup \{\infty\}$.

LEMMA 4.7.5. *Let $J^\nu \in \mathcal{J}_\tau(M,\omega)$ be a sequence of ω-tame almost complex structures on M that converges to J in the C^∞ topology. Fix a positive real number r and a positive sequence $\delta^\nu \to 0$. Suppose that $u^\nu : A(\delta^\nu/r, r) \to M$ is a sequence of J^ν-holomorphic curves satisfying the following conditions.*

- *u^ν converges to a J-holomorphic curve $u : B_r \to M$ uniformly on compact subsets of $B_r \setminus \{0\}$.*
- *$u^\nu(\delta^\nu \cdot)$ converges to a J-holomorphic curve $v : S^2 \setminus \mathrm{int}\, B_{1/r} \to M$ uniformly on compact subsets of $\mathbb{C} \setminus \mathrm{int}\, B_{1/r}$.*
- *$\lim_{\rho\to 0} \lim_{\nu\to\infty} E\big(u^\nu, A(\delta^\nu/\rho, \rho)\big) = 0$.*

Then

$$u(0) = v(\infty), \qquad \lim_{\rho\to 0}\lim_{\nu\to\infty} \sup_{\delta^\nu/\rho \le |z| \le \rho} d(u^\nu(z), u(0)) = 0.$$

PROOF. Abbreviate

$$E^\nu(\rho) := E\big(u^\nu, A(\delta^\nu/\rho, \rho)\big), \qquad E(\rho) := \lim_{\nu\to\infty} E^\nu(\rho).$$

Then $E(\rho)$ tends to zero as $\rho \to 0$. Hence we can apply Lemma 4.7.3 to the J^ν-holomorphic curve $u^\nu : A(\delta^\nu/\rho, \rho) \to M$ provided that $\rho > 0$ is sufficiently small and then ν is sufficiently large. If $2\rho \le r$ then $A(\delta^\nu/2\rho, 2\rho) \subset A(\delta^\nu/r, r)$. Applying (4.7.2) to the annulus $A(\delta^\nu/2\rho, 2\rho)$ with $T = \log 2$ we obtain positive constants δ and c such that

$$E^\nu(2\rho) < \delta \qquad \Longrightarrow \qquad \sup_{z,z'\in A(\delta^\nu/\rho,\rho)} d\big(u^\nu(z), u^\nu(z')\big) < c\sqrt{E^\nu(2\rho)}. \tag{4.7.3}$$

Take the limit $\nu \to \infty$ to obtain

$$E(2\rho) < \delta \qquad \Longrightarrow \qquad d(u(\rho), v(1/\rho)) = \lim_{\nu\to\infty} d(u^\nu(\rho), u^\nu(\delta^\nu/\rho)) \le c\sqrt{E(2\rho)}.$$

Now take the limit $\rho \to 0$ to obtain $u(0) = v(\infty) =: x$.

To prove the second assertion we fix a constant $\varepsilon > 0$. We must prove that there is a constant $\rho > 0$ and an integer $\nu_0 > 0$ such that

$$\nu \ge \nu_0, \quad \frac{\delta^\nu}{\rho} \le |z| \le \rho \qquad \Longrightarrow \qquad d\big(u^\nu(z), u(0)\big) < \varepsilon.$$

First choose positive constants δ and c such that (4.7.3) holds for $0 < \rho \le r/2$. Second choose $\rho > 0$ so small that $2\rho \le r$ and

$$E(2\rho) < \delta, \qquad c\sqrt{E(2\rho)} < \varepsilon/3, \qquad d\big(u(\rho), u(0)\big) < \varepsilon/3.$$

Third choose ν_0 so large that, for $\nu \ge \nu_0$, we have

$$E^\nu(2\rho) < \delta, \qquad c\sqrt{E^\nu(2\rho)} < \varepsilon/3, \qquad d\big(u^\nu(\rho), u(\rho)\big) < \varepsilon/3.$$

Now suppose that $\nu \ge \nu_0$ and $\delta^\nu/\rho \le |z| \le \rho$. Then, by the triangle inequality and (4.7.3), we have

$$\begin{aligned} d\big(u^\nu(z), u(0)\big) &\le d\big(u^\nu(z), u^\nu(\rho)\big) + d\big(u^\nu(\rho), u(\rho)\big) + d\big(u(\rho), u(0)\big) \\ &\le c\sqrt{E^\nu(2\rho)} + d\big(u^\nu(\rho), u(\rho)\big) + d\big(u(\rho), u(0)\big) \\ &< \varepsilon/3 + \varepsilon/3 + \varepsilon/3 \\ &= \varepsilon. \end{aligned}$$

This proves Lemma 4.7.5. □

PROOF OF PROPOSITION 4.7.2. In this proposition the rescaling sequence ψ^ν is arbitrary. We shall deduce the result from Lemma 4.7.5. For this we must modify the rescaling sequence ψ^ν so that it has the form $\psi^\nu(z) = \delta^\nu z$ with $\delta^\nu \to 0$ and then verify the hypotheses of Lemma 4.7.5. This will take three steps.

STEP 1. *We may assume without loss of generality that* $z_0 = 0$ *and* $z_\infty = \infty$.

To see this, choose isometries $\phi_0, \phi_\infty \in \mathrm{G}$ such that $\phi_0(z_0) = 0$ and $\phi_\infty(z_\infty) = \infty$. Then replace u^ν, v^ν, ψ^ν, and u, v by

$$\tilde u^\nu = u^\nu \circ \phi_0^{-1}, \qquad \tilde v^\nu = v^\nu \circ \phi_\infty^{-1}, \qquad \tilde\psi^\nu = \phi_0 \circ \psi^\nu \circ \phi_\infty^{-1},$$

$$\tilde u = u \circ \phi_0^{-1}, \qquad \tilde v = v \circ \phi_\infty^{-1}.$$

These functions satisfy the assumptions of the lemma, and if they satisfy the conclusion, then so do the original functions. This proves Step 1.

STEP 2. *We may assume without loss of generality that* $\psi^\nu(z) = \delta^\nu z$, *where* $\delta^\nu > 0$ *and* δ_ν *converges to* 0.

Since $\psi^\nu(0) \to 0$ and $(\psi^\nu)^{-1}(\infty) \to \infty$, there are sequences $\rho_0^\nu, \rho_\infty^\nu \in \mathrm{G}$ that each converge uniformly to multiplication by an element in $\mathbb{C}^*$ and satisfy

$$\rho_0^\nu(\psi^\nu(0)) = 0, \qquad \rho_0^\nu(\infty) = \infty, \qquad \rho_\infty^\nu((\psi^\nu)^{-1}(\infty)) = \infty, \qquad \rho_\infty^\nu(0) = 0,$$

for ν sufficiently large. By passing to a subsequence, we may also assume that

$$\rho_\infty^\nu(1) = 1, \qquad \rho_0^\nu(\psi^\nu(1)) \in \mathbb{R}^+.$$

Now replace u^ν, v^ν, and ψ^ν by

$$\widetilde{u}^\nu := u^\nu \circ (\rho_0^\nu)^{-1}, \qquad \widetilde{v}^\nu := v^\nu \circ (\rho_\infty^\nu)^{-1}, \qquad \tilde{\psi}^\nu := \rho_0^\nu \circ \psi^\nu \circ (\rho_\infty^\nu)^{-1}.$$

These functions satisfy the requirements of Step 2.

STEP 3. $E(\rho) := \lim_{\nu\to\infty} E(u^\nu; B_\rho \setminus B_{\delta^\nu/\rho})$ *converges to zero as* $\rho \to 0$.

It follows from the conformal invariance of the energy that

$$\begin{aligned} E(\rho) &= \lim_{\nu\to\infty} E(u^\nu; B_\rho) - \lim_{\nu\to\infty} E(v^\nu; B_{1/\rho}) \\ &= m_0 + E(u; B_\rho) - E(v; B_{1/\rho}) - \sum_{j=1}^{\ell} m_j \\ &= E(u; B_\rho) + E(v; \mathbb{C} \setminus B_{1/\rho}). \end{aligned}$$

Here we have used the fact that, by assumption, u^ν and v^ν satisfy conditions (a), (b) and (i-iii) in Proposition 4.7.1. The second equality above uses (b) and (ii) and the last equality uses (iii). This proves Step 3.

By Step 3, the sequence u^ν satisfies the hypotheses of Lemma 4.7.5. Hence $u(0) = v(\infty)$. Moreover, since $\psi^\nu(z) = \delta^\nu z$, the condition $\delta^\nu/\rho < |z| < \rho$ translates into $|z| < \rho$ and $|(\psi^\nu)^{-1}(z)| > 1/\rho$, and hence means that z is close to $z_0 = 0$ and $(\psi^\nu)^{-1}(z)$ is close to $z_\infty = \infty$. Hence the second assertion of Proposition 4.7.2 also follows from Lemma 4.7.5. This completes the proof of Proposition 4.7.2. □

PROOF OF PROPOSITION 4.7.1. The proof has four steps.

STEP 1. *We may assume without loss of generality that* $z_0 = 0$ *and that each function* $z \mapsto |du^\nu(z)|$ *assumes its maximum at* $z = 0$.

By (a), the sequence $|du^\nu|$ is uniformly bounded in the annulus $B_r(z_0) \setminus B_\varepsilon(z_0)$ for every $\varepsilon > 0$ and, by (b),

$$\lim_{\nu\to\infty} \sup_{B_\varepsilon(z_0)} |du^\nu| = \infty$$

for every $\varepsilon > 0$. Hence every sequence $z^\nu \in B_r(z_0)$ with

$$|du^\nu(z^\nu)| = \sup_{B_r(z_0)} |du^\nu|$$

converges to z_0. Therefore, we may assume that $z^\nu \in B_{r/2}(z_0)$ for all ν. Consider the maps $\tilde{u}^\nu : B_{r/2} \to M$, defined by $\tilde{u}^\nu(z) := u^\nu(z + z^\nu)$. These satisfy hypotheses (a) and (b) of the proposition with r replaced by $r/2$ and z_0 replaced by 0. Further if they satisfy the conclusion for some sequence ψ^ν then so do the original functions u^ν with ψ^ν replaced by $z \mapsto \psi^\nu(z) + z^\nu$. Therefore replacing u^ν by $\tilde{u}^\nu$ and r by $r/2$ we achieve Step 1.

STEP 2. *Assume Step 1. Let $\hbar$ be the constant of Proposition 4.1.4 and choose a positive constant $0 < \delta \le \hbar$ such that the assertion of Lemma 4.7.3 holds with this constant δ, $\mu = 1/2$, and J replaced by J^ν. Then, for ν sufficiently large, there is a unique constant $\delta^\nu > 0$ such that*

$$E(u^\nu; B_{\delta^\nu}) = m_0 - \frac{\delta}{2}. \tag{4.7.4}$$

Define $\psi^\nu \in \mathrm{G}$ by $\psi^\nu(z) := \delta^\nu z$. Then $\delta^\nu \to 0$, so ψ^ν satisfies (i) with $z_\infty = \infty$. Moreover, there is a subsequence, still denoted by u^ν and ψ^ν, that satisfies (ii).

By Theorem 4.6.1, we have $m_0 \ge \hbar$. Hence it follows from the definition of m_0 that, for large ν, there is a unique $\delta^\nu > 0$ satisfying (4.7.4); moreover δ^ν converges to zero. We have chosen δ^ν so that almost all of the "bubble" energy m_0 that is concentrating at zero lies inside the ball B_{δ^ν}. Hence, for $\varepsilon > 0$ sufficiently small, there is not enough energy in the annulus $B_\varepsilon \setminus B_{\delta^\nu}$ to allow for another bubble.

Since δ^ν converges to zero, the sequence ψ^ν satisfies condition (i) in Proposition 4.7.1 with $z_\infty := \infty$. Now define $v^\nu : B_{r/\delta^\nu} \to M$ by

$$v^\nu(z) := u^\nu(\delta^\nu z).$$

By Step 1, we have

$$|dv^\nu(0)| = \sup_{B_\rho} |dv^\nu| \tag{4.7.5}$$

whenever $\rho < r/\delta^\nu$. Moreover, the energy of v^ν in B_ρ is bounded by the energy of u^ν in B_r (for any fixed ρ and ν sufficiently large). Hence, by Theorem 4.6.1, there exists a subsequence of v^ν which satisfies condition (ii) for some finite set $Z = \{z_1, \dots, z_\ell\} \subset \mathbb{C}$. Condition (4.7.5) guarantees that if $Z \ne \emptyset$ then $0 \in Z$. This proves Step 2.

We now prove that with these choices for the sequences ψ^ν and v^ν conditions (iii) and (iv) hold. In particular, there is no need to pass to a further subsequence. The next step contains the key identity needed to show that no energy gets lost in the limit.

STEP 3. *Let δ and δ^ν be as in Step 2. Then*

$$\lim_{R\to\infty} \lim_{\nu\to\infty} E(u^\nu; B_{R\delta^\nu}) = m_0. \tag{4.7.6}$$

By assumption (b) and Theorem 4.6.1, we have

$$\lim_{\varepsilon\to 0} \lim_{\nu\to\infty} E(u^\nu; B_\varepsilon) = m_0 \ge \hbar > \delta > 0. \tag{4.7.7}$$

Next we construct a nonincreasing sequence of real numbers $\varepsilon^\nu \in (0, r]$ such that

$$\lim_{\nu\to\infty} E(u^\nu; B_{\varepsilon^\nu}) = m_0, \qquad \lim_{\nu\to\infty} \varepsilon^\nu = 0. \tag{4.7.8}$$

Here are the details. It follows from (4.7.7) that, for every $\ell \in \mathbb{N}$, there is a number $\varepsilon_\ell \in (0, r]$ such that

$$\lim_{\nu\to\infty} |E(u^\nu; B_{\varepsilon_\ell}(z_0)) - m_0| = \left|\lim_{\nu\to\infty} E(u^\nu; B_{\varepsilon_\ell}(z_0)) - m_0\right| < \frac{1}{\ell}. \tag{4.7.9}$$

Shrinking ε_ℓ inductively, if necessary, we may also assume that $\varepsilon_{\ell+1} < \varepsilon_\ell < 1/\ell$ for every ℓ. By (4.7.9) there is a sequence of integers $\nu_\ell \in \mathbb{N}$ such that

$$\sup_{\nu \ge \nu_\ell} |E(u^\nu; B_{\varepsilon_\ell}) - m_0| < 1/\ell, \qquad \nu_{\ell+1} > \nu_\ell.$$

Define $\varepsilon^\nu := r$ for $\nu < \nu_1$ and

$$\varepsilon^\nu := \varepsilon_\ell \qquad \text{for} \quad \nu_\ell \le \nu < \nu_{\ell+1}.$$

This sequence satisfies (4.7.8). (Note that we may be unable to choose ε^ν so that $E(u^\nu; B_{\varepsilon^\nu})$ is equal to m_0, because the sequence $u^\nu : B_r \to M$ may consist of maps with energy less than m_0 that converge on $B_r \setminus \{0\}$ to a constant.) We also have

$$\lim_{\nu\to\infty} E(u^\nu; B_{R\varepsilon^\nu}) = m_0 \qquad \text{for every } R \ge 1. \tag{4.7.10}$$

That the limit inferior is greater than or equal to m_0 follows from (4.7.8) and the fact that $R \ge 1$, and that the limit superior is less than or equal to m_0 follows from (4.7.7) and the fact that $R\varepsilon^\nu \to 0$.

Now suppose (4.7.6) is wrong. Define

$$f(R) := \limsup_{\nu\to\infty} E(u^\nu; B_{R\delta^\nu})$$

for $R > 0$. This function is nondecreasing and $f(R) \le m_0$ for all R, by (4.7.7). Moreover, by Step 2, the sequence of functions $z \mapsto u^\nu(\delta^\nu z)$ converges modulo bubbling on B_R; in particular, the sequence $E(u^\nu; B_{R\delta^\nu})$ converges to $f(R)$ for R sufficiently large. Since (4.7.6) is wrong, it follows that $\lim_{R\to\infty} f(R) < m_0$. Hence there is a constant $\rho > 0$ such that

$$\limsup_{\nu\to\infty} E(u^\nu; B_{R\delta^\nu}) \le m_0 - \rho \qquad \text{for every } R > 0. \tag{4.7.11}$$

Given $R > 0$, it follows from (4.7.8) and (4.7.11) that $E(u^\nu, B_{R\delta^\nu}) < E(u^\nu, B_{\varepsilon^\nu})$ and hence $R\delta^\nu < \varepsilon_\nu$ for ν sufficiently large. Thus $\delta^\nu/\varepsilon^\nu$ converges to zero. Moreover, by (4.7.4) and (4.7.10), we have

$$\lim_{\nu\to\infty} E(u^\nu; A(\delta^\nu, R\varepsilon^\nu)) = \delta/2 \qquad \text{for every } R \ge 1.$$

Hence it follows from our choice of δ in Step 2 that the restriction of u^ν to the annulus $A(\delta^\nu, R\varepsilon^\nu)$ satisfies the assumptions of Lemma 4.7.3 with $\mu = 1/2$ for ν sufficiently large. With $T := \log R$, this gives

$$\limsup_{\nu\to\infty} E\big(u^\nu; A(R\delta^\nu, \varepsilon^\nu)\big) \le \frac{2}{R} \lim_{\nu\to\infty} E\big(u^\nu; A(\delta^\nu, R\varepsilon^\nu)\big) = \frac{\delta}{R} \qquad \text{for every } R \ge 2.$$

Hence, by (4.7.11), we obtain

$$\lim_{\nu\to\infty} E(u^\nu, B_{\varepsilon^\nu}) \le m_0 - \rho + \delta/R \qquad \text{for every } R \ge 2.$$

For $R > \delta/\rho$ this contradicts (4.7.8). Thus we have proved Step 3.

STEP 4. *Completion of the proof.*

We prove (iii). By (4.7.6) and the definition of δ^ν and v^ν in Step 2, we have

$$\lim_{R\to\infty}\lim_{\nu\to\infty} E(v^\nu; B_R) = \lim_{R\to\infty}\lim_{\nu\to\infty} E(u^\nu; B_{R\delta^\nu}) = m_0$$

and, for every ν,

$$E(v^\nu; B_1) = E(u^\nu; B_{\delta^\nu}) = m_0 - \frac{\delta}{2} \ge m_0 - \frac{\hbar}{2}.$$

Hence all the bubble points $z_1, \dots, z_\ell$ of the sequence v^ν lie in the closed unit disc B_1. Now fix a number $s > 1$. Then, using (4.7.6) again, we have

$$\begin{aligned} m_0 &= \lim_{R\to\infty}\lim_{\nu\to\infty} E(v^\nu; B_R) \\ &= \lim_{R\to\infty}\lim_{\nu\to\infty} E(v^\nu; B_R \setminus B_s) + \lim_{\nu\to\infty} E(v^\nu; B_s) \\ &= \lim_{R\to\infty} E(v; B_R \setminus B_s) + \lim_{\nu\to\infty} E(v^\nu; B_s) \\ &= E(v; \mathbb{C} \setminus B_s) + \lim_{\varepsilon\to 0}\lim_{\nu\to\infty} E\Big(v^\nu; B_s \setminus \bigcup_{j=1}^{\ell} B_\varepsilon(z_j)\Big) + \sum_{j=1}^{\ell} m_j \\ &= E(v) + \sum_{j=1}^{\ell} m_j. \end{aligned}$$

This proves (iii).

We prove (iv). We have already seen that all the bubble points of v^ν lie inside the closed unit disc. Therefore, if v is constant and $1 < R_1 < R_2$, then

$$\lim_{\nu\to\infty} E\big(v^\nu; A(R_1, R_2)\big) = E\big(v; A(R_1, R_2)\big) = 0.$$

This shows that the limit of $E(v^\nu; B_R) = E(u^\nu; B_{R\delta^\nu})$ as $\nu \to \infty$ is independent of $R > 1$. Hence, by (4.7.6),

$$\lim_{\nu\to\infty} E(v^\nu; B_R) = m_0$$

for every $R > 1$. Since

$$E(v^\nu; B_1) = m_0 - \delta/2,$$

this is only possible if bubbling occurs on the boundary of the unit disc. This means that the singular set Z of v^ν contains a point z_j with $|z_j| = 1$. But we saw in Step 2 that, if $Z \neq \emptyset$, then $0 \in Z$. Hence $\ell = \#Z \geq 2$ as required. This proves Proposition 4.7.1. □

To complete the proofs in this section, it remains to prove Lemma 4.7.3.

Proof of Lemma 4.7.3. Choose positive constants δ and C such that the assertion of Lemma 4.3.1 (i) holds with c replaced by $2C$. Thus, for every $r > 0$ and every J-holomorphic curve $v : B_r \to M$, we have

$$E(v; B_r) < \delta \qquad \Longrightarrow \qquad \tfrac{1}{2}|dv(0)|^2 \leq \frac{C}{r^2} E(v; B_r) \tag{4.7.12}$$

Shrinking δ if necessary and choosing a suitable constant $0 < \mu < 1$, we may also assume that the assertion of Theorem 4.4.1 (iii) holds with δ replaced by $3\pi\sqrt{C\delta}$ and c replaced by $1/4\pi\mu$. Thus

$$\ell(\gamma) < 3\pi\sqrt{C\delta} \qquad \Longrightarrow \qquad |a(\gamma)| \leq \frac{\ell(\gamma)^2}{4\pi\mu} \tag{4.7.13}$$

for every smooth loop $\gamma : S^1 \to M$. Define

$$c := 16\sqrt{C}\left(2\pi + \frac{1}{\mu}\right). \tag{4.7.14}$$

We will show that Lemma 4.7.3 holds with these constant δ and c.

First, we may assume without loss of generality that $r = 1/R$. Namely, define

$$\lambda := \sqrt{rR}$$

and replace u, r, and R, by

$$\widetilde{u}(z) := u(\lambda z), \qquad \widetilde{r} := r/\lambda, \qquad \widetilde{R} := R/\lambda.$$

Then $\widetilde{r}\widetilde{R} = 1$ and, if the lemma has been established for the triple $(\widetilde{u}, \widetilde{r}, \widetilde{R})$, it also holds for the triple (u, r, R).

Let $R > 0$ and $u : A(1/R, R) \to M$ be a a J-holomorphic curve with energy

$$E(u) < \delta.$$

Define the J-holomorphic curve $v : [-\log R, \log R] \times \mathbb{R}/2\pi\mathbb{Z} \to M$ by

$$v(s,t) := u(e^{s+it}), \qquad -\log R < s < \log R, \qquad t \in \mathbb{R}/2\pi\mathbb{Z}.$$

For $s \in [-\log R, \log R]$ define $\gamma_s : \mathbb{R}/2\pi\mathbb{Z} \to M$ by

$$\gamma_s(t) := u(e^{s+it}) = v(s,t).$$

Since $E(v) = E(u) < \delta$, the apriori estimate (4.7.12) with $r = \log 2$ shows that

$$|\dot{\gamma}_s(t)|^2 = |\partial_s v(s,t)|^2 \le \frac{C}{(\log 2)^2} E(u), \qquad |s| \le \log(R/2). \tag{4.7.15}$$

This implies

$$\begin{aligned} \ell(\gamma_s)^2 &= \left(\int_0^{2\pi} |\dot{\gamma}_s(t)| \, dt \right)^2 \\ &\le 2\pi \int_0^{2\pi} |\dot{\gamma}_s(t)|^2 \, dt \\ &\le \frac{4\pi^2 C}{(\log 2)^2} E(u) \\ &\le 9\pi^2 C E(u). \end{aligned}$$

Here we have used the inequality $4/(\log 2)^2 \le 9$.

We prove (4.7.1). Recall from Remark 4.4.3 that the local symplectic action of a sufficiently short loop $\gamma : S^1 \to M$ is defined by

$$a(\gamma) := -\int_{B_1} u_\gamma^* \omega,$$

where $u_\gamma : B_1 \to M$ is a local extension of γ. Since

$$\ell(\gamma_s) \le 3\pi\sqrt{CE(u)} \le 3\pi\sqrt{C\delta}$$

for $|s| \le \log(R/2)$, it follows from (4.7.13) that

$$|a(\gamma_s)| \le \frac{\ell(\gamma_s)^2}{4\pi\mu}, \qquad |s| \le \log(R/2). \tag{4.7.16}$$

Moreover,

$$E(v; [s_0, s_1] \times \mathbb{R}/2\pi\mathbb{Z}) + a(\gamma_{s_1}) - a(\gamma_{s_0}) = 0 \tag{4.7.17}$$

for $-\log(R/2) \le s_0 \le s_1 \le \log(R/2)$. To see this, cap off the ends of the cylinder $v : [s_0, s_1] \times \mathbb{R}/2\pi\mathbb{Z} \to M$ with small discs to obtain a contractible sphere, and then integrate the symplectic form over this sphere.

Now consider the function

$$\varepsilon(T) := E\big(u; A(e^T/R, e^{-T}R)\big), \qquad \log 2 \le T \le \log R.$$

Abbreviate $\rho := \log R$. Then

$$\begin{aligned}\varepsilon(T) &= E(v; [-\rho+T, \rho-T] \times \mathbb{R}/2\pi\mathbb{Z}) \\ &= \int_{-\rho+T}^{\rho-T} \int_0^{2\pi} |\partial_s v(s,t)|^2 \, dt\, ds.\end{aligned}$$

Hence, by (4.7.16) and (4.7.17), we have

$$\begin{aligned}\varepsilon(T) &= a(\gamma_{-\rho+T}) - a(\gamma_{\rho-T}) \\ &\le \frac{1}{4\pi\mu}\left(\ell(\gamma_{-\rho+T})^2 + \ell(\gamma_{\rho-T})^2\right) \\ &\le \frac{1}{2\mu}\int_0^{2\pi} |\partial_s v(-\rho+T,t)|^2 \, dt + \frac{1}{2\mu}\int_0^{2\pi} |\partial_s v(\rho-T,t)|^2 \, dt \\ &= -\frac{1}{2\mu}\dot{\varepsilon}(T).\end{aligned}$$

This implies

$$\dot{\varepsilon}(T) \le -2\mu\varepsilon(T) < 0$$

and hence

$$\varepsilon(T) \le e^{-2\mu(T-\log 2)}\varepsilon(\log 2) \le 4^\mu e^{-2\mu T} E(u) \tag{4.7.18}$$

for $\log 2 \le T \le \log R$. Thus we have proved (4.7.1).

We prove (4.7.2). The first step is to establish the estimate

$$|\partial_s v(s,t)|^2 \le \frac{C 4^\mu e^{2\mu}}{(\log 2)^2} e^{-2\mu(\rho-|s|)} E(v), \qquad |s| \le \rho - \log 2. \tag{4.7.19}$$

To see this, assume first that

$$\rho - \log 2 - 1 \le |s| \le \rho - \log 2.$$

Then, by (4.7.15), we have

$$|\partial_s v(s,t)|^2 \le \frac{C}{(\log 2)^2} E(v) \le \frac{C 4^\mu e^{2\mu}}{(\log 2)^2} e^{-2\mu(\rho-|s|)} E(v).$$

The last inequality uses the fact that $\rho - |s| \le \log 2 + 1$ and hence $e^{2\mu(\rho-|s|)} \le 4^\mu e^{2\mu}$. Now suppose that

$$|s| \le \rho - \log 2 - 1$$

and denote $T := \rho - |s|$. Then $T - 1 \ge \log 2$. Hence

$$\begin{aligned}|\partial_s v(s,t)|^2 &\le CE(v; [-\rho+T-1, \rho-T+1] \times \mathbb{R}/2\pi\mathbb{Z}) \\ &= C\varepsilon(T-1) \\ &\le C 4^\mu e^{2\mu} e^{-2\mu T} E(v).\end{aligned}$$

Here the first inequality follows from (4.7.12) with $r = 1$ and the last inequality follows from (4.7.18) with T replaced by $T - 1$. Since $T = \rho - |s|$ and $(\log 2)^2 < 1$ this implies (4.7.19).

For $s \geq 0$ the estimate (4.7.19) implies

$$0 \leq s \leq \rho - \log 2 \qquad \Longrightarrow \qquad |\partial_s v(s,t)| \leq 8\sqrt{C}e^{-\mu(\rho-s)}\sqrt{E(v)}.$$

Here we have used the inequality $2^\mu e^\mu/\log 2 \leq 2e/\log 2 < 8$. Now fix a point $(s_0, t_0) \in \mathbb{R}^2$ such that $0 \leq s_0 \leq \rho - T$ and $0 \leq t_0 \leq 2\pi$. Integrate the last inequality to obtain

$$\begin{aligned}
d\left(v(0,0), v(s_0,t_0)\right) &\leq \int_0^{s_0} |\partial_s v(s,0)|\, ds + \int_0^{t_0} |\partial_t v(s_0,t)|\, dt \\
&\leq 8\sqrt{C}\left(\int_0^{s_0} e^{-\mu(\rho-s)}\, ds + \int_0^{t_0} e^{-\mu(\rho-s_0)}\, dt\right)\sqrt{E(v)} \\
&\leq 8\sqrt{C}\left(\int_{-\infty}^{-T} e^{\mu s}\, ds + 2\pi e^{-\mu T}\right)\sqrt{E(v)} \\
&= 8\sqrt{C}\left(\frac{1}{\mu} + 2\pi\right) e^{-\mu T}\sqrt{E(v)}.
\end{aligned}$$

By symmetry about $s = 0$, we find

$$d(v(s_0,t_0), v(s_1,t_1)) \leq 16\sqrt{C}\left(\frac{1}{\mu} + 2\pi\right) e^{-\mu T}\sqrt{E(v)}$$

for $\log 2 \leq T \leq \rho$ and $s_0, s_1 \in [-\rho+T, \rho-T]$ and $t_0, t_1 \in [0, 2\pi]$. Thus (4.7.2) holds with c given by (4.7.14). This proves Lemma 4.7.3. □

EXERCISE 4.7.6. Extend the results of this section to J-holomorphic curves with Lagrangian boundary conditions (cf. [**120**]).

CHAPTER 5

Stable Maps

This chapter deals with the limiting behaviour of sequences of J-holomorphic spheres in compact symplectic manifolds. The most important result is that every sequence of J-holomorphic spheres has a subsequence that converges to a *stable map* in the sense of Kontsevich [**215**]. In Section 5.1 we discuss in detail the notion of stable maps. In his original work [**160**] Gromov used the term *cusp-curves*, which corresponds to the image of a stable map, or to a reducible curve in the terminology of algebraic geometry. In Section 5.2 we discuss Gromov convergence and, in particular, prove that the homotopy class is preserved under these limits. We then prove in Section 5.3 that every sequence of J-holomorphic curves with uniformly bounded energy has a Gromov convergent subsequence.

The remainder of the chapter deals with the structure of the moduli space $\overline{\mathcal{M}}_{0,n}(M,A;J)$ of equivalence classes of stable maps and is not used elsewhere in this book. Our aim is to give this space a compact metrizable topology arising from the Gromov convergent sequences. In Section 5.4 we establish the uniqueness of Gromov limits and extend the notion of convergence to sequences of stable maps. Finally in Section 5.6 we introduce the Gromov topology on $\overline{\mathcal{M}}_{0,n}(M,A;J)$ and show that this space is compact and metrizable. Throughout we shall assume for simplicity that the Riemann surface Σ is a sphere. In fact, the arguments go through without essential change to the general case provided that one fixes the complex structure j_Σ on Σ. In all cases, the bubbles that appear are spherical.

5.1. Stable maps

We begin by recalling some definitions from Appendix D. A tree is a connected graph without cycles. We think of it as a finite set T equipped with a relation $E \subset T \times T$ such that two vertices $\alpha, \beta \in T$ are related by E (i.e. $\alpha E \beta$) if and only if they are connected by an edge. An n-labelling Λ of T is a function

$$I \to T : i \mapsto \alpha_i,$$

where $I := \{1, \dots, n\}$. Such an n-labelling can also be described as a partition of the index set I into the disjoint subsets

$$\Lambda_\alpha := \{i \in I \,|\, \alpha_i = \alpha\}.$$

It will sometimes be convenient to allow for more general index sets I (see Section 5.5). For $\alpha, \beta \in T$ the interval $[\alpha, \beta] \subset T$ denotes the set of vertices along the chain of edges connecting α to β. When $\alpha E \beta$ we denote by $T_{\alpha\beta}$ the subtree containing β after removing the edge connecting α and β, i.e. $T_{\alpha\beta} := \{\gamma \in T \,|\, \beta \in [\alpha, \gamma]\}$. We often denote an n-labelled tree by (T, Λ) or simply T. The number of vertices is $\#T$ and the number of unoriented edges is $e(T) := \#E/2 = \#T - 1$.

DEFINITION 5.1.1 (Stable maps). *Let (M,ω) be a compact symplectic manifold, $J \in \mathcal{J}_\tau(M,\omega)$ be an ω-tame almost complex structure, and $n \geq 0$ be a nonnegative integer. Let (T,E,Λ) be an n-labelled tree. A* **stable J-holomorphic map of genus zero in M with n marked points (modelled over (T,E,Λ))** *is a tuple*

$$(\mathbf{u},\mathbf{z}) = \big(\{u_\alpha\}_{\alpha\in T}, \{z_{\alpha\beta}\}_{\alpha E\beta}, \{\alpha_i, z_i\}_{1\leq i\leq n}\big),$$

consisting of a collection of J-holomorphic spheres $u_\alpha : S^2 \to M$ labelled by the vertices $\alpha \in T$, a collection of **nodal points** *$z_{\alpha\beta} \in S^2$ labelled by the oriented edges $\alpha E\beta$, and a sequence of n* **marked points** *$z_1,\dots,z_n \in S^2$, such that the following conditions are satisfied.*

(NODAL POINTS) *If $\alpha,\beta \in T$ with $\alpha E\beta$ then $u_\alpha(z_{\alpha\beta}) = u_\beta(z_{\beta\alpha})$. The set of nodal points on the α-sphere is denoted by*

$$Z_\alpha := \{z_{\alpha\beta}\,|\,\alpha E\beta\}.$$

(SPECIAL POINTS) *For every $\alpha \in T$ the nodal points $z_{\alpha\beta}$ (for $\alpha E\beta$) and the marked points z_i (for $i \in \Lambda_\alpha$) are pairwise distinct. These are called the* **special points**. *The set of special points on the α-sphere is denoted by*

$$Y_\alpha := Z_\alpha \cup \{z_i\,|\,\alpha_i = \alpha\}.$$

(STABILITY) *If u_α is a constant function then $\#Y_\alpha \geq 3$.*

We shall see that the stability condition in Definition 5.1.1 implies that the automorphism group of a stable map is finite. Notice, however, that the underlying labelled tree (T,E,Λ) need not be stable in the sense of Appendix D since the stability condition above only refers to the components on which u_α is constant (see equation (D.2.1) in Section D.2).

The domain of each map u_α is the Riemann sphere $S^2 \cong \mathbb{C}\cup\{\infty\}$. It may sometimes be helpful to think of these domains as abstract complex 2-manifolds Σ_α that are diffeomorphic to the 2-sphere so that Y_α and Z_α are finite subsets of Σ_α. This is important when one considers stable maps of higher genus. However, we shall throughout work with the standard 2-sphere as domains of our J-holomorphic maps and describe the elements of Σ_α as pairs $(\alpha,z) \in T\times S^2$. The domain of a stable map $(\mathbf{u},\mathbf{z})$ can therefore be represented as the quotient

$$\Sigma(\mathbf{z}) := T\times S^2/\sim,$$

where the equivalence relation on $T\times S^2$ is given by $(\alpha,z)\sim(\beta,w)$ if and only if either $\alpha=\beta$ and $z=w$ or $\alpha E\beta$, $z=z_{\alpha\beta}$, and $w=z_{\beta\alpha}$. We denote by $[\alpha,z]\in\Sigma(\mathbf{z})$ the equivalence class of a pair $(\alpha,z)\in T\times S^2$. Then the collection $\{u_\alpha\}_{\alpha\in T}$ gives rise to a map

$$\Sigma(\mathbf{z}) \to M : [\alpha,z] \mapsto u_\alpha(z).$$

The subset $[\alpha\times S^2]\subset\Sigma(\mathbf{z})$ is called the α-component of Σ. If u_α is constant it is called a **ghost bubble** or **ghost component**.

If $(\mathbf{u},\mathbf{z})$ is a stable map we denote

$$E(\mathbf{u}) := \sum_{\alpha\in T} E(u_\alpha), \qquad m_{\alpha\beta}(\mathbf{u}) := \sum_{\gamma\in T_{\alpha\beta}} E(u_\gamma), \tag{5.1.1}$$

for $\alpha,\beta\in T$ with $\alpha E\beta$ (see Figure 1). Thus the energy $E(\mathbf{u})$ of $\mathbf{u}$ vanishes if and only if all the maps u_α are constant. In this case the stable map $(\mathbf{u},\mathbf{z})$ can be considered as a pair consisting of a stable curve in the sense of Definition D.3.1

together with a point in M. For general $\mathbf{u}$, the domain $\Sigma(\mathbf{z})$ together with the marked points $[\alpha_i, z_i]$ is not stable. In this case the nodal curve $\Sigma(\mathbf{z})$ or the tuple $\mathbf{z} := (\{z_{\alpha\beta}\}_{\alpha E \beta}, \{\alpha_i, z_i\}_{1 \le i \le n})$ is called a **prestable curve**. Note that a prestable curve can be stabilized by adding new marked points.

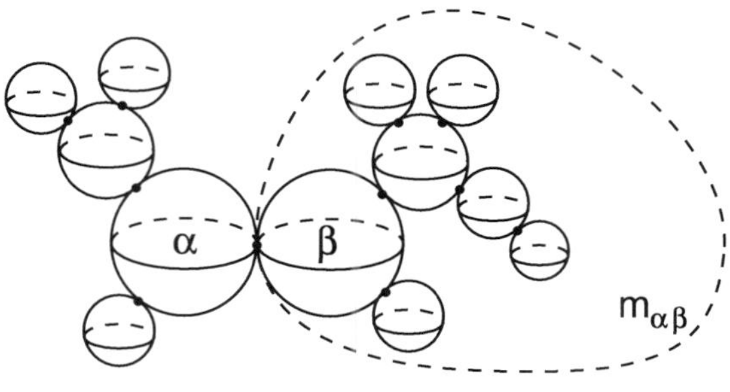

FIGURE 1. Stable maps

EXERCISE 5.1.2. A **weighted tree** is a labelled tree (T, E, Λ) equipped with a weight function $\mathrm{w} : T \to [0, \infty)$. The **total weight** of a weighted tree is the number $W(T) := \sum_{\alpha \in T} \mathrm{w}(\alpha)$. A weighted tree $(T, E, \Lambda, \mathrm{w})$ is called **stable** if every vertex α with $\mathrm{w}(\alpha) = 0$ satisfies the stability condition (D.2.1). Use Lemma D.2.3 to prove that for every stable weighted n-labelled tree T with at least one edge the number of edges is bounded above by

$$e(T) \le \frac{2W(T)}{\hbar} + n - 3,$$

where $\hbar$ is the minimum of its nonzero weights. Show by example that this estimate is sharp when $W(T) = 0$. In particular, if $\hbar$ is the constant of Proposition 4.1.4 and $(\mathbf{u}, \mathbf{z})$ is a stable map with n marked points modelled over T, then

$$e(T) \le \frac{2E(\mathbf{u})}{\hbar} + n - 3. \tag{5.1.2}$$

EXERCISE 5.1.3. (i) Show that the vertices $\alpha_1, \ldots, \alpha_N$ of any tree T can be ordered in such a way that the subset $T_i := \{\alpha_1, \ldots, \alpha_i\}$ is a tree for every i. Show that any vertex $\alpha_1 \in T$ can be chosen as the starting point of such an ordering.

(ii) Given an ordering as in (i) show that for every $i \ge 2$ there is a unique index $j_i < i$ such that $\alpha_{j_i} E \alpha_i$. Hence a tree with N vertices can be identified with an integer vector $\mathbf{j} = (j_2, \ldots, j_N)$ such that $1 \le j_i < i$ for every i.

(iii) Let $V_k := \{i \,|\, j_i = k\}$ and note that $k < i$ for all $i \in V_k$. If $k < i$ show that $\alpha_k E \alpha_i$ if and only if $j_i = k$.

The reparametrization group.

DEFINITION 5.1.4. *Two stable maps* $(\mathbf{u}, \mathbf{z})$ *and* $(\tilde{\mathbf{u}}, \tilde{\mathbf{z}})$*, modelled over the labelled trees* (T, Λ) *and* $(\widetilde{T}, \widetilde{\Lambda})$*, respectively, are called* **equivalent** *if there exists a tree isomorphism* $f : T \to \widetilde{T}$ *and a function* $T \to \mathrm{G} = \mathrm{PSL}(2, \mathbb{C}) : \alpha \mapsto \phi_\alpha$ *(which assigns to each vertex a fractional linear transformation) such that*

$$\tilde{u}_{f(\alpha)} \circ \phi_\alpha = u_\alpha, \qquad \tilde{z}_{f(\alpha) f(\beta)} = \phi_\alpha(z_{\alpha\beta}), \qquad \tilde{z}_i = \phi_{\alpha_i}(z_i). \tag{5.1.3}$$

It follows that $\Lambda_\alpha = \widetilde{\Lambda}_{f(\alpha)}$ *for every* $\alpha \in T$.

For every labelled tree T there is an associated group G_T that acts on the set of stable maps modelled on T and whose orbit space is the corresponding set of equivalence classes. The elements of G_T are tuples

$$g = (f, \{\phi_\alpha\}_{\alpha \in T}),$$

where f is an automorphism of T such that

$$\Lambda_\alpha = \Lambda_{f(\alpha)}$$

and $\phi_\alpha \in \mathrm{G}$ for $\alpha \in T$. Thus, in the case $T = \widetilde{T}$, the map f can only move vertices α at which there are no marked points. The group operation is given by

$$g' \cdot g = \left(f' \circ f, \left\{ \phi'_{f(\alpha)} \circ \phi_\alpha \right\}_{\alpha \in T} \right).$$

It is useful to think of f as the map which assigns to each fractional linear transformation ϕ_α its target sphere, and to think of ϕ_α as a map from the 2-sphere Σ_α to $\Sigma_{f(\alpha)}$. The group G_T acts on the space of stable maps modelled over T by

$$g \cdot (\mathbf{u}, \mathbf{z}) := (\tilde{\mathbf{u}}, \tilde{\mathbf{z}})$$

where $(\tilde{\mathbf{u}}, \tilde{\mathbf{z}})$ is given by (5.1.3). Clearly its orbits precisely define the equivalence classes of stable maps that are modelled on T.

The stability condition in Definition 5.1.1 guarantees that the isotropy subgroup of every stable map is finite. Usually this isotropy group is trivial. However, it is nontrivial if, for example, T has two vertices α, β and no labels and $u_\alpha = u_\beta$.

Exercise 5.1.5. The group G_T acts on the set of domains $\Sigma(\mathbf{z})$ of the stable maps modelled on T. Describe the orbits of this action when T is the tree with five vertices $\alpha_1, \dots, \alpha_5$, where $\alpha_1 E \alpha_i$ for $i > 1$ and there are no labels. Show that G_T acts transitively on the set of such domains $\Sigma(\mathbf{z})$ if and only if $\# Z_\alpha \le 3$ for each α.

Exercise 5.1.6. Let T be an n-labelled tree and $(\mathbf{u}, \mathbf{z})$ be a stable map modelled over T. Show that each component of G_T has complex dimension $3e(T) + 3$. Show that the subgroup $\mathrm{G}_{T,\mathbf{z}} \subset \mathrm{G}_T$ that preserves the special points (but not the maps) has complex dimension $e(T) + 3 - n$ when $n + e(T) \le 3$. Show that $e(T) + 3 - n \le \dim^{\mathbb{C}} \mathrm{G}_{T,\mathbf{z}} \le 2e(T)$ when $n + e(T) \ge 3$. Find examples to show that all nonnegative possibilities occur.

Moduli spaces. Stable maps to M form the objects of a category $\mathcal{SC}_{0,n}(M; J)$. (The letters $\mathcal{SC}$ can be taken to mean "stable category".) A morphism from an object $(\mathbf{u}, \mathbf{z}) \in \mathcal{SC}_{0,n}(M; J)$ to $(\tilde{\mathbf{u}}, \tilde{\mathbf{z}}) \in \mathcal{SC}_{0,n}(M; J)$ is a tuple $(f, \{\phi_\alpha\}_{\alpha \in T})$ that satisfies equation (5.1.3) in Definition 5.1.4. Thus, every morphism is an isomorphism and two stable maps $(\mathbf{u}, \mathbf{z})$ and $(\tilde{\mathbf{u}}, \tilde{\mathbf{z}})$ are equivalent if and only if there exists an isomorphism between them in the category $\mathcal{SC}_{0,n}(M; J)$. We denote the equivalence class of a stable map $(\mathbf{u}, \mathbf{z})$ under the equivalence relation of Definition 5.1.4 by $[\mathbf{u}, \mathbf{z}]$.

Now fix a spherical homology class $A \in H_2(M; \mathbb{Z})$. A stable map $(\mathbf{u}, \mathbf{z}) \in \mathcal{SC}_{0,n}(M; J)$ is said to **represent the class** A if

$$A = \sum_{\alpha \in T} u_{\alpha *}[S^2]. \tag{5.1.4}$$

The subcategory of stable maps that represent the class A will be denoted by

$$\mathcal{SC}_{0,n}(M, A; J) := \{ (\mathbf{u}, \mathbf{z}) \in \mathcal{SC}_{0,n}(M; J) \,|\, \mathbf{u} \text{ satisfies } (5.1.4) \}.$$

By a slight abuse of standard conventions in logic we shall pretend that the equivalence classes form a set and we denote this set by $\overline{\mathcal{M}}_{0,n}(M,A;J)$ or simply by $\overline{\mathcal{M}}_{0,n}(A;J)$. (As explained in Section D.2 this terminology would be perfectly rigorous if we only considered trees whose vertices form a subset of some given set.) The moduli space $\overline{\mathcal{M}}_{0,n}(M,A;J)$ carries a natural topology, called the Gromov topology. The goal of the present chapter is to define this topology and to prove that it renders $\overline{\mathcal{M}}_{0,n}(M,A;J)$ into a compact metrizable space.

Strata. For each labelled tree $T=(T,E,\Lambda)$ and homology class $A\in H_2(M;\mathbb{Z})$ we will denote the space of all stable maps that are modelled on T and represent the class A by $\widetilde{\mathcal{M}}_{0,T}(M,A;J)$ or simply by $\widetilde{\mathcal{M}}_{0,T}(A;J)$. Note that the space $\widetilde{\mathcal{M}}_{0,T}(A;J)$ is a disjoint union of the subsets that correspond to the decompositions of A into sums

$$A=\sum_{\alpha\in T}A_\alpha$$

of integral homology classes. By Proposition 4.1.5, only finitely many of these subsets are nonempty. The quotient by the action of the reparametrization group G_T will be denoted by

$$\mathcal{M}_{0,T}(M,A;J):=\mathcal{M}_{0,T}(A;J):=\widetilde{\mathcal{M}}_{0,T}(A;J)/\mathrm{G}_T. \tag{5.1.5}$$

Thus

$$\overline{\mathcal{M}}_{0,n}(A;J)=\bigcup_T\mathcal{M}_{0,T}(A;J),$$

where the union runs over all isomorphism classes of n-labelled trees. By Exercise 5.1.2 only finitely many sets in this union are nonempty. With slight abuse of terminology we shall sometimes call the sets $\mathcal{M}_{0,T}(A;J)$ the **strata** of $\overline{\mathcal{M}}_{0,n}(A;J)$. This does not imply that $\overline{\mathcal{M}}_{0,n}(A;J)$ is a stratified space in a rigorous meaning of the word. However, we shall prove in Chapter 6 that the subset $\mathcal{M}^*_{0,T}(A;J)\subset\mathcal{M}_{0,T}(A;J)$ of *simple stable elements* is, for a generic ω-compatible almost complex structure J, a manifold that has codimension $2e(T)$ relative to the top stratum.

The stratum corresponding to the n-labelled tree T_n with one vertex is called the **top stratum** and will be denoted by

$$\mathcal{M}_{0,n}(A;J):=\mathcal{M}_{0,T_n}(A;J)=\widetilde{\mathcal{M}}_{0,n}(A;J)/\mathrm{G},\qquad \widetilde{\mathcal{M}}_{0,n}(A;J):=\widetilde{\mathcal{M}}_{0,T_n}(A;J).$$

When $A\neq 0$ and $k=0,1$ these spaces have the following description in the terminology of Chapter 3:

$$\widetilde{\mathcal{M}}_{0,0}(A;J)=\mathcal{M}(A;J),\qquad \widetilde{\mathcal{M}}_{0,1}(A;J)=\mathcal{M}(A;J)\times S^2$$

and

$$\mathcal{M}_{0,0}(A;J)=\mathcal{M}(A;J)/\mathrm{G},\qquad \mathcal{M}_{0,1}(A;J)=\mathcal{M}(A;J)\times_{\mathrm{G}}S^2.$$

These identities no longer hold when $A=0$ because of the stability condition imposed on the elements of $\widetilde{\mathcal{M}}_{0,k}(A;J)$.

In **[160]** Gromov introduced the word **cusp-curve** which corresponds to the image of an element in $\overline{\mathcal{M}}_{0,0}(A;J)$ with at least two components. These may be parametrized by maps whose domain is an arbitrary nodal Riemann surface. This notion is less precise than that of a stable map since the structure of the domain is lost. While it is adequate if all one wants to do is establish the existence of the Gromov–Witten invariants of semipositive manifolds, one cannot appreciate their full structure without the use of stable maps (see Section 7.5).

In algebraic geometry one describes curves by words such as reduced and irreducible. Here the emphasis is again on the image of the curve, or rather on the curve as the solution set to some equations. Thus each curve is a finite union of components, each with some multiplicity. A curve is said to be **irreducible** if it consists of a single component of multiplicity one; otherwise it is **reducible**. It is **reduced** if all its components have multiplicity one. In our language, a curve is irreducible if it is the image of a **simple** map $u : S^2 \to M$. The concept of a reduced curve will also be important to us. Therefore we introduce the idea of a **simple stable map** $(\mathbf{u}, \mathbf{z})$. These are stable maps in which each nonconstant component u_α is simple and also no two images $\operatorname{im} u_\alpha, \operatorname{im} u_\beta$ of nonconstant maps with $\alpha \neq \beta$ coincide. The simple stable maps modelled on T form an open subset $\mathcal{M}^*_{0,T}(A;J)$ of the stratum $\mathcal{M}_{0,T}(A;J)$ whose properties are discussed in Section 6.2. Clearly, a curve is reduced in the sense of algebraic geometry if and only if it is the image of a simple stable map. Finally, we note that algebraic geometers often use the words **rational curves** to mean curves of genus zero.

REMARK 5.1.7 (Notation). Our use of the letter $\mathcal{M}$ to denote spaces of maps in one context and equivalence classes of maps in another is somewhat unsatisfactory. However, both usages are well established. Many papers (as well as the first edition of this book) use notations similar to $\mathcal{M}(A;J)$ to denote spaces of maps, and the notations $\mathcal{M}_{0,n}(A;J)$ and $\overline{\mathcal{M}}_{0,n}(A;J)$ for moduli spaces of equivalence classes of stable maps are also very common. To help readers distinguish between the two notations we will never use subscripts in a case like $\mathcal{M}(A;J)$ where we mean a space of maps. Also, in situations where we need to talk about both kinds of spaces at once we will use the notation $\widetilde{\mathcal{M}}_{0,0}(A;J)$ rather than $\mathcal{M}(A;J)$ for the space of J-holomorphic spheres. We warn readers that many other authors use the word stable map to refer to the elements $[\mathbf{u}, \mathbf{z}]$ of $\overline{\mathcal{M}}_{0,n}(A;J)$ rather than explicitly saying that these elements are equivalence classes.

Stable curves. If the target space M is a point then $\overline{\mathcal{M}}_{0,n} := \overline{\mathcal{M}}_{0,n}(\{\mathrm{pt}\}, 0; J)$ is the Grothendieck–Knudsen moduli space of stable curves of genus zero with n marked points. This moduli space is discussed in detail in Appendix D. It is a smooth algebraic manifold of complex dimension $n-3$ and can be expressed as a disjoint union of strata

$$\overline{\mathcal{M}}_{0,n} = \bigcup_T \mathcal{M}_{0,T},$$

where T runs over the set of all isomorphism classes of stable n-labelled trees. The open stratum $\mathcal{M}_{0,n}$ consists of all configurations of n distinct marked points in S^2 modulo the action of $\mathrm{G} = \mathrm{PSL}(2,\mathbb{C})$. The closure $\overline{\mathcal{M}}_{0,T}$ of each stratum $\mathcal{M}_{0,T}$ is a smooth submanifold of $\overline{\mathcal{M}}_{0,n}$ of complex codimension $e(T)$ and any two such submanifolds intersect cleanly (the intersection is a submanifold and the tangent space of the intersection is equal to the intersection of the tangent spaces). Moreover, the submanifolds $\overline{\mathcal{M}}_{0,T}$ generate the cohomology of $\overline{\mathcal{M}}_{0,n}$ additively [**208**]. The moduli space $\overline{\mathcal{M}}_{0,3}$ is a point, $\overline{\mathcal{M}}_{0,4}$ is diffeomorphic to the 2-sphere, and $\overline{\mathcal{M}}_{0,5}$ is diffeomorphic to the product $S^2 \times S^2$ with three points on the diagonal blown up: see Section D.7.

In Appendix D we show how to embed the Grothendieck–Knudsen moduli space $\overline{\mathcal{M}}_{0,n}$ into a product of 2-spheres. On the open stratum $\mathcal{M}_{0,n}$ the embedding is given by the cross ratios

$$w(z_i, z_j, z_k, z_\ell) = \frac{(z_j - z_k)(z_\ell - z_i)}{(z_i - z_j)(z_k - z_\ell)} \in \mathbb{C} \cup \{\infty\}.$$

This definition continues to make sense when no three of the points z_i, z_j, z_k, z_ℓ are equal and gives rise to a map that is invariant under the action of G. For example, since $w(0, 1, \infty, z) = z$ the cross ratio defines a natural diffeomorphism $\overline{\mathcal{M}}_{0,4} \to S^2$. In Section D.4 the cross ratios are extended to functions

$$w_{ijk\ell} : \overline{\mathcal{M}}_{0,n} \to S^2$$

as follows. For every vertex $\alpha \in T$ and every index $i \in \{1, \ldots, n\}$ we define $z_{\alpha i} \in S^2$ to be the element of the α-sphere which points in the direction of the marked point z_i, i.e.

$$z_{\alpha i} := \begin{cases} z_i, & \text{if } \alpha_i = \alpha, \\ z_{\alpha\beta}, & \text{if } \alpha_i \in T_{\alpha\beta}. \end{cases} \tag{5.1.6}$$

Now for any four indices i, j, k, ℓ there exists a vertex α such that no three of the points $z_{\alpha i}, z_{\alpha j}, z_{\alpha k}, z_{\alpha \ell}$ are equal and, moreover, the resulting cross ratio is independent of the choice of such a vertex α. One should think of the resulting function $w_{ijk\ell}$ as the forgetful map $\overline{\mathcal{M}}_{0,n} \to S^2 := \overline{\mathcal{M}}_{0,4}$ that forgets all the marked points except for those indexed by i, j, k, ℓ. These functions define the required embedding of $\overline{\mathcal{M}}_{0,n}$ into a product of 2-spheres. They can also be used to define a topology on $\overline{\mathcal{M}}_{0,n}$ (see Section D.5). Moreover, the closure $\overline{\mathcal{M}}_{0,T}$ of each stratum $\mathcal{M}_{0,T}$ maps onto a complex submanifold of codimension $e(T)$, and the stratum $\mathcal{M}_{0,n}$ is open and dense in $\overline{\mathcal{M}}_{0,n}$. Thus $\overline{\mathcal{M}}_{0,n}$ is indeed a natural compactification of $\mathcal{M}_{0,n}$.

The forgetful map. For $n \geq 3$ there is a forgetful map

$$\pi : \overline{\mathcal{M}}_{0,n}(A; J) \to \overline{\mathcal{M}}_{0,n}.$$

Its definition requires some care. Deleting the maps u_α one obtains a tuple $\mathbf{z} := (\{z_{\alpha\beta}\}_{\alpha E \beta}, \{\alpha_i, z_i\}_{1 \leq i \leq n})$, which can be thought of as a curve of genus zero with nodal points and n marked points, modelled over the labelled tree T. In general, this curve will no longer be stable, however, it can be reduced uniquely to a stable curve by collapsing every vertex with one or two marked points and making the obvious adjustments to the relation E on T. The resulting stable curve will be denoted by $\mathbf{z}^s$ and we define

$$\pi([\mathbf{u}, \mathbf{z}]) := [\mathbf{z}^s].$$

An alternative definition of the forgetful map can be given in terms of cross ratios. Namely, the definitions of both (5.1.6) and $w_{ijk\ell}$ make sense for stable maps and not just stable curves. Hence the cross ratios define a map from the moduli space $\overline{\mathcal{M}}_{0,n}(A; J)$ into a product of 2-spheres which is equal to the composition of π with the embedding of $\overline{\mathcal{M}}_{0,n}$ into the same product. In Section 5.6 we show that the forgetful map π is continuous with respect to the Gromov topology. Note that π is not defined when $n < 3$ since $\overline{\mathcal{M}}_{0,n}$ is empty in this case.

EXERCISE 5.1.8. Show that there is a collapsing map $[T, \Lambda] \mapsto [T^s, \Lambda^s]$ which assigns to every isomorphism class of prestable trees $[T, \Lambda]$ an isomorphism class of stable trees $[T^s, \Lambda^s]$. Construct the map by deleting unstable vertices. If $[\mathbf{u}, \mathbf{z}]$ is modelled over (T, Λ) prove that $\pi([\mathbf{u}, \mathbf{z}])$ is modelled over (T^s, Λ^s).

EXERCISE 5.1.9. Define a map

$$\overline{\mathcal{M}}_{0,n}(A;J) \to \overline{\mathcal{M}}_{0,n-1}(A;J)$$

that forgets the last marked point, explaining what happens on the tree level (see the discussion of the forgetful map (D.6.1) in Section (D.6)). Compare this map with the forgetful map $\overline{\mathcal{M}}_{0,n}(A;J) \to \overline{\mathcal{M}}_{0,n}$, finding examples of components that are contracted by one map and not the other. Are there ever components contracted by both maps? Prove that the obvious diagram commutes.

EXERCISE 5.1.10. Describe the moduli space $\overline{\mathcal{M}}_{0,n}(\mathbb{C}P^1, 2L; J_0)$ for $0 \le n \le 3$, where $L = [\mathbb{C}P^1]$ and J_0 is the standard complex structure (cf. [**268**]).

EXERCISE 5.1.11. For every $i \in \{1, \dots, n\}$ there is an **evaluation map**

$$\mathrm{ev}_i : \overline{\mathcal{M}}_{0,n}(M, A; J) \to M$$

defined by

$$\mathrm{ev}_i([\mathbf{u}, \mathbf{z}]) = u_{\alpha_i}(z_i) \tag{5.1.7}$$

for $[\mathbf{u}, \mathbf{z}] \in \overline{\mathcal{M}}_{0,n}(M, A; J)$. Prove that this map is well defined, i.e. if $(\mathbf{u}, \mathbf{z})$ is equivalent to $(\tilde{\mathbf{u}}, \tilde{\mathbf{z}})$ then $u_{\alpha_i}(z_i) = \tilde{u}_{\tilde{\alpha}_i}(\tilde{z}_i)$. Use the definition and properties of the Gromov topology in Section 5.6 to prove that ev_i is continuous.

5.2. Gromov convergence

In the remainder of this chapter we develop the notion of a Gromov convergent sequence of stable maps. This is an elaboration of the corresponding notion for stable curves that is treated in Section D.5. That context is considerably simpler since there is no analysis.

Here is a formal definition of what it means for a sequence of J^ν-holomorphic curves $u^\nu : S^2 \to M$ to converge to a stable map $(\mathbf{u}, \mathbf{z}) \in \mathcal{SC}_{0,n}(M; J)$. Recall from Section D.1 that a sequence of maps $v^\nu : S^2 \to M$ is said to **converge u.c.s.** on $S^2 \setminus X$ to a map $v : S^2 \to M$ if it converges to v in the C^∞-topology on every compact subset of $S^2 \setminus X$. Recall further the notation

$$Z_\alpha := \{z_{\alpha\beta} \,|\, \alpha E \beta\}, \qquad Y_\alpha := Z_\alpha \cup \{z_i \,|\, \alpha_i = \alpha\}$$

from Definition 5.1.1.

DEFINITION 5.2.1 (Gromov convergence). *Let J^ν be a sequence of ω-tame almost complex structures on M that converges in the C^∞ topology to $J \in \mathcal{J}_\tau(M, \omega)$, let*

$$(\mathbf{u}, \mathbf{z}) = \big(\{u_\alpha\}_{\alpha \in T}, \{z_{\alpha\beta}\}_{\alpha E \beta}, \{\alpha_i, z_i\}_{1 \le i \le n}\big) \in \mathcal{SC}_{0,n}(M; J)$$

be a stable map and let $u^\nu : S^2 \to M$ be a sequence of J^ν-holomorphic maps with n distinct marked points $z_1^\nu, \dots, z_n^\nu \in S^2$. The sequence $(u^\nu, \mathbf{z}^\nu) := (u^\nu, z_1^\nu, \dots, z_n^\nu)$ is said to **Gromov converge** *to $(\mathbf{u}, \mathbf{z})$ if there exists a collection of Möbius transformations $\{\phi_\alpha^\nu\}_{\alpha \in T}^{\nu \in \mathbb{N}}$ such that the following holds.*

(MAP) *For every $\alpha \in T$ the sequence $u_\alpha^\nu := u^\nu \circ \phi_\alpha^\nu : S^2 \to M$ converges to u_α u.c.s. on $S^2 \setminus Z_\alpha$.*

(ENERGY) *If $\alpha E \beta$ and $m_{\alpha\beta}(\mathbf{u})$ is defined by (5.1.1) then*

$$m_{\alpha\beta}(\mathbf{u}) = \lim_{\varepsilon \to 0} \lim_{\nu \to \infty} E\big(u_\alpha^\nu; B_\varepsilon(z_{\alpha\beta})\big). \tag{5.2.1}$$

(RESCALING) *If $\alpha E\beta$ then the sequence $\phi^\nu_{\alpha\beta} := (\phi^\nu_\alpha)^{-1} \circ \phi^\nu_\beta$ converges to $z_{\alpha\beta}$ u.c.s. on $S^2 \setminus \{z_{\beta\alpha}\}$.*

(MARKED POINT) $z_i = \lim_{\nu\to\infty}(\phi^\nu_{\alpha_i})^{-1}(z_i^\nu)$ *for $i = 1, \dots, n$.*

Since this is such a complicated and important definition we shall take some time trying to explain it. One difficulty in understanding it is that the sequence u^ν itself does not appear explicitly except in the compositions

$$u^\nu_\alpha := u^\nu \circ \phi^\nu_\alpha$$

since we are only interested in its limiting behaviour *modulo* the action of G. Therefore it is best to look at the convergence from the vantage point of some vertex α. Then the *(Map)* axiom says that u^ν_α converges on all compact subsets $S^2 \setminus B_\varepsilon(Z_\alpha)$ of $S^2 \setminus Z_\alpha$. Moreover for any adjacent vertex β the corresponding maps u^ν_β are rescalings of the u^ν_α in the following sense.

First, let us define a **rescaling of S^2 with source x and target y** to be a sequence of fractional linear transformations $\phi^\nu \in$ G that converges to y u.c.s. on $S^2 \setminus \{x\}$. If ϕ^ν_i is a sequence of rescalings with source x_i and target y_i for $i = 0, 1$ and if $y_0 \neq x_1$, then the composite $\phi^\nu_1 \circ \phi^\nu_0$ is a sequence of rescalings with source x_0 and target y_1. Moreover, by Exercise D.1.3, the inverse of a sequence of rescalings of S^2 with source x and target y is a sequence of rescalings with source y and target x. With this terminology, the *(Rescaling)* axiom says that $\phi^\nu_{\alpha\beta}$ is a sequence of rescalings with source $z_{\beta\alpha}$ and target $z_{\alpha\beta}$. Since $z_{\beta\alpha} \neq z_{\beta\gamma}$ for $\alpha \neq \gamma$, the composite

$$\phi^\nu_{\alpha\gamma} = \phi^\nu_{\alpha\beta} \circ \phi^\nu_{\beta\gamma}$$

is also a rescaling (with source $z_{\gamma\beta}$ and target $z_{\alpha\beta}$). The rescaled sequence

$$u^\nu_\beta = u^\nu_\alpha \circ \phi^\nu_{\alpha\beta}$$

converges on the compact subsets $S^2 \setminus B_\varepsilon(Z_\beta)$. It picks out and "magnifies" the restriction of u^ν_α to the small neighbourhood $\phi^\nu_{\alpha\beta}(S^2 \setminus B_\varepsilon(z_{\beta\alpha}))$ of the bubbling point $z_{\alpha\beta}$. Similarly, looking at the α-bubble from the point of view of the β-bubble, we see that u^ν_α is the magnification of the restriction of u^ν_β to the small neighbourhood $\phi^\nu_{\beta\alpha}(S^2 \setminus B_\varepsilon(z_{\alpha\beta}))$ of $z_{\beta\alpha}$.

The *(Marked point)* axiom says that the marked points converge, while the *(Energy)* axiom states that no energy is lost. To explain its importance, let us define the **bubble energy** of a sequence $v^\nu : S^2 \to M$ at the point z as the limit

$$E(\{v^\nu\}; z) := \lim_{\varepsilon\to 0} \lim_{\nu\to\infty} E\big(v^\nu; B_\varepsilon(z)\big).$$

Whenever we use the notion of bubble energy we shall assume that the limit on the right exists and that the sequence v^ν converges u.c.s. on $B_\varepsilon(z) \setminus \{z\}$ for some $\varepsilon > 0$. In this case the sequence v^ν is said to **converge with bubbling at z**. With this terminology the *(Energy)* axiom asserts that the bubble energy of the sequence u^ν_α at the point $z_{\alpha\beta}$ is equal to the energy $m_{\alpha\beta}(\mathbf{u})$ of the limit stable map $(\mathbf{u}, \mathbf{z})$ on the part of the tree separated by the edge from α to β. This means that the rescalings have been chosen so as to lose no energy. In particular, if $m_{\alpha\beta}(\mathbf{u}) > 0$, the sequence

$$\psi^\nu := \phi^\nu_{\alpha\beta}$$

satisfies the requirements of Proposition 4.7.2 with $u^\nu = u^\nu_\alpha$, $v^\nu = u^\nu_\beta$, and

$$z_0 := z_{\alpha\beta}, \qquad z_\infty := z_{\beta\alpha}.$$

The *(Energy)* axiom in Definition 5.2.1 implies that Hypothesis (b) in Proposition 4.7.1 holds. Hence it contains the information that the bubbles corresponding to adjacent vertices α, β are joined to each other. Thus one can prove that a Gromov limit is connected by establishing an energy estimate of this form.

We shall prove in later sections that the limit is unique (up to equivalence) and that every sequence u^ν of J^ν-holomorphic spheres with uniformly bounded energy has a convergent subsequence in the sense of Definition 5.2.1. To begin with we derive some consequences of Gromov compactness that are not quite immediate, e.g. that every limit of a sequence $x^\nu \in u^\nu(S^2)$ lies in the image of the limit curve, and that the homotopy class is preserved under Gromov convergence. Note that in (iii) below the points $z_{\alpha i}$ are defined by (5.1.6). Moreover, assertion (iv) about the cross ratios $w_{ijk\ell}$ shows that the image of a Gromov convergent sequence under the forgetful map converges in the moduli space of stable curves (see Appendix D).

THEOREM 5.2.2. *Suppose that $(u^\nu, \mathbf{z}^\nu)$ Gromov converges to the stable map $(\mathbf{u}, \mathbf{z})$ as in Definition 5.2.1. Then the following holds.*

(i) *If $x^\nu \in u^\nu(S^2)$ converges to $x \in M$ then $x \in u_\alpha(S^2)$ for some $\alpha \in T$.*

(ii) *For sufficiently large ν the map $u^\nu : S^2 \to M$ is homotopic to the connected sum $\#_{\alpha\in T} u_\alpha$.*

(iii) *$z_{\alpha i} = \lim_{\nu\to\infty} (\phi_\alpha^\nu)^{-1}(z_i^\nu)$ for every $\alpha \in T$ and every i.*

(iv) *$w_{ijk\ell}(\mathbf{z}) = \lim_{\nu\to\infty} w_{ijk\ell}(\mathbf{z}^\nu)$ for all i, j, k, ℓ.*

PROOF. We prove (i). Suppose that $x^\nu = u^\nu(z^\nu)$ converges to x. Passing to a subsequence, if necessary, we may assume without loss of generality that $(\phi_\alpha^\nu)^{-1}(z^\nu)$ converges to some point $z_\alpha \in S^2$ for every α. If there exists an $\alpha \in T$ such that $z_\alpha \neq z_{\alpha\beta}$ for all $\beta \in T$ with $\alpha E \beta$, then

$$x = \lim_{\nu\to\infty} u^\nu(z^\nu) = \lim_{\nu\to\infty} u_\alpha^\nu((\phi_\alpha^\nu)^{-1}(z^\nu)) = u_\alpha(z_\alpha)$$

and we are done. If there is no such α then we claim that there is an edge $\alpha E \beta$ in T such that

$$z_\alpha = z_{\alpha\beta}, \qquad z_\beta = z_{\beta\alpha}. \tag{5.2.2}$$

To see this, let us begin with any vertex α_0 and note that there is a unique sequence of vertices $\alpha_0, \alpha_1, \alpha_2, \ldots$ such that $\alpha_i E \alpha_{i+1}$ and $z_{\alpha_i} = z_{\alpha_i \alpha_{i+1}}$. There must be some j with $\alpha_j = \alpha_{j+2}$, since otherwise the α_i would form an infinite sequence of pairwise distinct vertices. Hence the vertices $\alpha = \alpha_j$ and $\beta = \alpha_{j+1}$ satisfy (5.2.2). If $m_{\alpha\beta}(\mathbf{u}) = 0$ then, by (5.2.1) and Theorem 4.6.1, the sequence u_α^ν converges uniformly to u_α on a neighbourhood of $z_{\alpha\beta}$. Hence $x = \lim_{\nu\to\infty} u^\nu(z^\nu) = u_\alpha(z_{\alpha\beta})$ as before. If $m_{\alpha\beta}(\mathbf{u}) \neq 0$, then the assumptions of Proposition 4.7.2 are satisfied for the sequence u_α^ν with

$$u := u_\alpha, \qquad v := u_\beta, \qquad z_0 := z_{\alpha\beta}, \qquad z_\infty := z_{\beta\alpha}, \qquad \psi^\nu := \phi_{\alpha\beta}^\nu.$$

By (5.2.2) we have $z_0 = z_\alpha$ and $z_\infty = z_\beta$ and hence

$$z_0 = \lim_{\nu\to\infty} \zeta_\nu, \qquad z_\infty = \lim_{\nu\to\infty} (\psi^\nu)^{-1}(\zeta^\nu), \qquad \zeta^\nu := (\phi_\alpha^\nu)^{-1}(z^\nu).$$

By Proposition 4.7.2, this implies that the sequence $u^\nu(z^\nu) = u_\alpha^\nu(\zeta^\nu)$ converges to $u(z_0) = v(\infty)$. Thus

$$x = \lim_{\nu\to\infty} u^\nu(z^\nu) = u_\alpha(z_{\alpha\beta}) = u_\beta(z_{\beta\alpha}),$$

which proves (i).

We prove (ii). Fix an edge $\alpha E\beta$. For $r>0$ smaller than the diameter of the 2-sphere and $\nu\in\mathbb{N}$ define the set

$$\begin{aligned} A_r^\nu(\alpha,\beta) &:= \phi_\alpha^\nu\big(B_r(z_{\alpha\beta})\big)\cap\phi_\beta^\nu\big(B_r(z_{\beta\alpha})\big) \\ &= \big\{z\in S^2\,|\,d\big((\phi_\alpha^\nu)^{-1}(z),z_{\alpha\beta}\big)<r,\,d\big((\phi_\beta^\nu)^{-1}(z),z_{\beta\alpha}\big)<r\big\}. \end{aligned}$$

Since $\phi_{\beta\alpha}^\nu$ converges to $z_{\beta\alpha}$ u.c.s. on $S^2\setminus\{z_{\alpha\beta}\}$, the image $\phi_{\beta\alpha}^\nu\big(S^2\setminus B_r(z_{\alpha\beta})\big)$ is a small subdisc of $B_r(z_{\beta\alpha})$ whenever $\nu\ge\nu_0(r)$ is sufficiently large. Hence the set $A_r^\nu(\alpha,\beta)$, which is diffeomorphic to $\phi_{\beta\alpha}^\nu\big(B_r(z_{\alpha\beta})\big)\cap B_r(z_{\beta\alpha})$, is an annulus for $\nu\ge\nu_0(r)$. We prove that, for every $\varepsilon>0$, there exist constants $r>0$ and $\nu_0\in\mathbb{N}$ such that

$$\nu\ge\nu_0\qquad\Longrightarrow\qquad\sup_{z\in A_r^\nu(\alpha,\beta)}d\big(u^\nu(z),u_\alpha(z_{\alpha\beta})\big)<\varepsilon. \tag{5.2.3}$$

First assume $m_{\alpha\beta}(\mathbf{u})=0$. Then there is no bubbling at $z_{\alpha\beta}$, so that u_α^ν converges uniformly to u_α on a neighbourhood of $z_{\alpha\beta}$. Given $\varepsilon>0$, choose $r>0$ so small that $d\big(u_\alpha(\zeta),u_\alpha(z_{\alpha\beta})\big)<\varepsilon/2$ for $\zeta\in B_r(z_{\alpha\beta})$ and u_α^ν converges uniformly to u_α on $B_r(z_{\alpha\beta})$. Choose ν_0 such that $d\big(u_\alpha^\nu(\zeta),u_\alpha(\zeta)\big)<\varepsilon/2$ for all $\zeta\in B_r(z_{\alpha\beta})$ and $\nu\ge\nu_0$. Then (5.2.3) holds by the triangle inequality. Next assume $m_{\alpha\beta}(\mathbf{u})>0$. Then Proposition 4.7.2 applies to the sequences $u^\nu:=u_\alpha^\nu$, $\psi^\nu:=\phi_{\beta\alpha}^\nu$, the map $u:=u^\alpha$, and the points $z_0:=z_{\alpha\beta}$, $z_\infty:=z_{\beta\alpha}$. Given $\varepsilon>0$, choose $\delta>0$ and ν_0 as in Proposition 4.7.2 and let $r:=\delta/2$. If $\nu\ge\nu_0$, $z\in A_r^\nu(\alpha,\beta)$, and $\zeta:=(\phi_\alpha^\nu)^{-1}(z)$, then we have $d(\zeta,z_0)+d((\psi^\nu)^{-1}(\zeta),z_\infty)<2r=\delta$ and hence $d(u^\nu(z),u_\alpha(u_{\alpha\beta}))=d(u_\alpha^\nu(\zeta),u_\alpha(z_0))<\varepsilon$. This proves (5.2.3).

Now choose $\varepsilon>0$ smaller than half the injectivity radius of M and fix constants $r>0$ and $\nu_0\in\mathbb{N}$ such that (5.2.3) holds. Shrinking r, if necessary, we may assume that, for each α, the closed discs $B_r(z_{\alpha\beta})$ for $\alpha E\beta$ are pairwise disjoint. Define

$$S_\alpha:=S^2\setminus\bigcup_{\alpha E\beta}B_r(z_{\alpha\beta})$$

and construct a surface Σ_r by connecting the boundary circles $\partial B_r(z_{\alpha\beta})\subset S_\alpha$ and $\partial B_r(z_{\beta\alpha})\subset S_\beta$ by a standard annulus $A_{\alpha\beta}\cong[0,1]\times S^1$ for each edge $\alpha E\beta$. Thus each arc $[0,1]\times e^{i\theta}\subset A_{\alpha\beta}$ connects a point in $\partial B_r(z_{\alpha\beta})\subset S_\alpha$ to a point in $\partial B_r(z_{\beta\alpha})\subset S_\beta$. Then Σ_r is homeomorphic to S^2. Define $u_r:\Sigma_r\to M$ by

$$u_r|_{S_\alpha}:=u_\alpha$$

and, on the annuli, define u_r by the unique minimal geodesics connecting the given values on opposite boundary points. For $\nu\ge\nu_0$ define $u_r^\nu:\Sigma_r\to M$ by

$$u_r^\nu|_{S_\alpha}:=u_\alpha^\nu=u^\nu\circ\phi_\alpha^\nu$$

and, on the annuli, again connect the values on opposite boundary points by geodesics in M. Then u_r is a connected sum of the u_α and u_r^ν converges uniformly to u_r. We claim that, for ν sufficiently large, there is a homeomorphism

$$\psi_r^\nu:\Sigma_r\to S^2$$

such that u_r^ν is homotopic to $u^\nu\circ\psi_r^\nu$. The restriction of ψ_r^ν to S_α is given by ϕ_α^ν. Its image is the set

$$S_\alpha^\nu:=\psi_r^\nu(S_\alpha)=\phi_\alpha^\nu(S_\alpha)=S^2\setminus\bigcup_{\alpha E\beta}\phi_\alpha^\nu(B_r(z_{\alpha\beta})).$$

The complement $S^2 \setminus \bigcup_\alpha S^\nu_\alpha$ is precisely the union of the annuli $A^\nu_r(\alpha,\beta)$. Thus the map extends over the annuli to an orientation preserving homeomorphism ψ^ν_r from Σ_r to S^2. The composition $u^\nu \circ \psi^\nu_r$ agrees with u^ν_r on S_α and is C^0 close to u^ν_r on each annulus, by (5.2.3). Hence $u^\nu \circ \psi^\nu_r$ is homotopic to u^ν_r. This proves (ii).

We prove (iii). Let $i \in \{1,\dots,n\}$. We must show that $z_{\alpha i} = \lim_{\nu\to\infty}(\phi^\nu_\alpha)^{-1}(z^\nu_i)$ for each vertex α. The proof is by induction on the number k of edges from α_i to α. If $k=0$ then $\alpha = \alpha_i$ and $z_{\alpha i} = z_{\alpha_i i} = z_i$ by (5.1.6), and hence the assertion follows from the *(Marked point)* axiom in Definition 5.2.1. Now let $k>0$ and suppose, by induction, that the assertion has been established for every vertex that can be reached from α_i through a chain of at most $k-1$ edges. Let α be any vertex that can be reached from α_i through a chain $\gamma_0,\dots,\gamma_k$ of k edges with $\gamma_0 = \alpha_i$ and $\gamma_k = \alpha$ (see Section D.2). Then the induction hypothesis asserts that

$$z_{\gamma_{k-1} i} = \lim_{\nu\to\infty} (\phi^\nu_{\gamma_{k-1}})^{-1}(z^\nu_i) \tag{5.2.4}$$

Moreover, by the *(Rescaling)* axiom in Definition 5.2.1, the sequence

$$\phi^\nu_{\gamma_k\gamma_{k-1}} = (\phi^\nu_{\gamma_k})^{-1} \circ \phi^\nu_{\gamma_{k-1}}$$

converges to $z_{\gamma_k\gamma_{k-1}} = z_{\gamma_k i}$ u.c.s. on $S^2 \setminus \{z_{\gamma_{k-1}\gamma_k}\}$. Since $z_{\gamma_{k-1} i} \neq z_{\gamma_{k-1}\gamma_k}$, this implies

$$\begin{aligned} z_{\gamma_k i} &= \lim_{\nu\to\infty} \phi^\nu_{\gamma_k\gamma_{k-1}}(z_{\gamma_{k-1} i}) \\ &= \lim_{\nu\to\infty} \phi^\nu_{\gamma_k\gamma_{k-1}}((\phi^\nu_{\gamma_{k-1}})^{-1}(z^\nu_i)) \\ &= \lim_{\nu\to\infty} (\phi^\nu_{\gamma_k})^{-1}(z^\nu_i). \end{aligned}$$

Here the second equation follows from (5.2.4). Thus we have proved (iii).

Assertion (iv) follows from (iii) and the invariance of the cross ratio under Möbius transformation: see Appendix D. This proves Theorem 5.2.2. □

REMARK 5.2.3. In complex geometry there is a duality between holomorphic sections of k-dimensional vector bundles on manifolds of complex dimension n and the $(n-k)$-dimensional subvarieties formed by their zero sets. When $n-k=1$ these subvarieties are curves, with an alternate description as images of holomorphic maps $u : \Sigma \to M$. In many situations there is quite a close correspondence between the (parametrized) map description and (unparametrized) subvariety description of holomorphic curves; cf. Exercise 4.2.7 for $M = \mathbb{C}P^2$. In other cases (e.g. when $M = \mathbb{C}P^1$) most information is lost when one simply identifies a holomorphic curve with its image. In fact, when M is itself a Riemann surface, algebraic geometers also study holomorphic curves as maps and encode the geometric information in branch points and monodromy data. The compactness results in the present chapter are concerned with the limiting behavior of the maps, explaining this in terms of their domains as well as their images. Gromov, in his original paper [**160**], understood this convergence in rather different terms, analyzing it in terms of the induced metrics on the image subvarieties.

5.3. Gromov compactness

In this section we show that every sequence of J-holomorphic spheres with uniformly bounded energy has a Gromov convergent subsequence.

THEOREM 5.3.1 (Gromov compactness). *Let (M, ω) be a compact symplectic manifold and $J^\nu \in \mathcal{J}_\tau(M, \omega)$ be a sequence of ω-tame almost complex structures on M that converges to $J \in \mathcal{J}_\tau(M, \omega)$ in the C^∞-topology. Let $u^\nu : S^2 \to M$ be a sequence of J^ν-holomorphic spheres such that $\sup_\nu E(u^\nu) < \infty$ and $\mathbf{z}^\nu = (z_1^\nu, \ldots, z_n^\nu)$ be a sequence of n-tuples of pairwise distinct points in S^2. Then $(u^\nu, \mathbf{z}^\nu)$ has a Gromov convergent subsequence.*

It follows from Theorems 5.3.1 and and 5.2.2 that, for every ω-compatible almost complex structure J, only finitely many homotopy classes with a fixed upper bound on the symplectic area can be represented by J-holomorphic spheres. (Proposition 4.1.5 gives an easier proof.) Hence, by Exercise 5.1.2 only finitely many such classes can be represented by stable maps of genus zero. The next corollary extends this observation to families of almost complex structures.

COROLLARY 5.3.2. *Let Λ be a compact metric space and $\Lambda \to \mathcal{J}(M, \omega) : \lambda \mapsto J_\lambda$ be a continuous function with respect to the C^∞-topology on $\mathcal{J}(M, \omega)$. Then, for every $c > 0$, there exist only finitely many homotopy classes $A \in \pi_2(M)$ with $\langle [\omega], A \rangle \leq c$ that can be represented by J_λ-holomorphic stable maps (of genus zero) for some $\lambda \in \Lambda$.*

PROOF. By Proposition 4.1.4 it suffices to prove that only finitely many homotopy classes $A \in \pi_2(M)$ with $\langle [\omega], A \rangle \leq c$ can be represented by J_λ-holomorphic spheres for some λ. Suppose, by contradiction, that this is wrong. Then there is a sequence $\lambda^\nu \in \Lambda$ and a sequence of J_{λ^ν}-holomorphic spheres $u^\nu : S^2 \to M$ in which each element represents a different homotopy class A^ν such that $\langle [\omega], A^\nu \rangle \leq c$. By Theorem 5.3.1, the sequence u^ν has a Gromov convergent subsequence u^{ν_j}. By Theorem 5.2.2, the elements of of the subsequence represent the same homotopy class for sufficiently large j. This contradicts our assumption. □

EXERCISE 5.3.3. Use the method of Proposition 4.1.5 to give an alternative, and much simpler proof of Corollary 5.3.2, even for functions $\Lambda \to \mathcal{J}(M, \omega) : \lambda \mapsto J_\lambda$ that are continuous only with respect to the C^0 topology on $\mathcal{J}(M, \omega)$.

We first prove the compactness theorem when there are no marked points. The argument relies heavily on the results of Section 4.7.

PROOF OF THEOREM 5.3.1 FOR $n = 0$. Let $u^\nu : S^2 \to M$ be a sequence of J^ν-holomorphic spheres. Suppose, without loss of generality, that $E(u^\nu)$ converges and denote the limit by

$$E = \lim_{\nu \to \infty} E(u^\nu)$$

To carry out the induction, it is useful to describe a tree T with N vertices as a vector $\mathbf{j} = (j_2, \ldots, j_N)$ with $j_i < i$ for all i, as in Exercise 5.1.3. By induction, we shall construct

(a) a $(3N-2)$-tuple $\mathbf{u} = (u_1, \ldots, u_N; j_2, \ldots, j_N; z_2, \ldots, z_N)$ consisting of J-holomorphic spheres $u_i : S^2 \to M$, positive integers $j_i < i$ for $i \geq 2$, and complex numbers $z_i \in \mathbb{C}$ with $|z_i| \leq 1$, where u_1 is nonconstant, $z_k \neq z_{k'}$ whenever $j_k = j_{k'}$, and $u_{j_i}(z_i) = u_i(\infty)$ for $i = 2, \ldots, N$,

(b) finite subsets $Z_i \subset B_1$ for $i = 1, \ldots, N$, such that $Z_1 \subset \{0\}$ and $\#Z_i \geq 2$ whenever u_i is constant, and

(c) sequences of Möbius transformations $\{\phi_i^\nu\}_{\nu \in \mathbb{N}}$ for $i = 1, \ldots, N$,

such that a suitable subsequence satisfies the following conditions.

(i) $u^\nu \circ \phi_1^\nu$ converges to u_1 u.c.s. on $S^2 \setminus Z_1$. For $i = 2, \ldots, N$, $u^\nu \circ \phi_i^\nu$ converges to u_i u.c.s. on $\mathbb{C} \setminus Z_i$.

(ii) The limit

$$m_i(z) = \lim_{\varepsilon \to 0} \lim_{\nu \to \infty} E\big(u^\nu \circ \phi_i^\nu; B_\varepsilon(z)\big)$$

exists and is positive for all $z \in Z_i$. If $Z_1 = \{0\}$ then $E = E(u_1) + m_1(0)$. If $i \geq 2$ then $z_i \in Z_{j_i}$ and

$$m_{j_i}(z_i) = E(u_i) + \sum_{z \in Z_i} m_i(z).$$

(iii) For $i = 2, \ldots, N$, $(\phi_{j_i}^\nu)^{-1} \circ \phi_i^\nu$ converges to z_i u.c.s. on $\mathbb{C} = S^2 \setminus \{\infty\}$.

(iv) $Z_N = \emptyset$ and

$$Z_i = \{z_k \,|\, i < k \leq N,\, j_k = i\}, \qquad i = 1, \ldots, N-1.$$

To explain these conditions, note that (i) together with the first part of (ii) says that each sequence $u^\nu \circ \phi_i^\nu$ bubbles precisely at the points of Z_i. Thus the first bubble u_1 has at most one bubble point Z_1 and so is an endpoint of the tree T, while by (a) when $i > 1$ the ith bubble u_i is attached to the j_ith at the point $u_i(\infty) = u_{j_i}(z_i)$. Moreover, the second part of (ii) says that all the energy of u_{j_i} at z_i is captured by the union of u_i with its bubbles at Z_i. Since $j_i = j$ only when $i > j$, the tree T has no cycles. We will start the proof by constructing u_1 to have at most one bubble point and then will proceed by induction, constructing the u_i and Z_i so as to satisfy (i)–(iii). When the induction is complete, (iv) will also be satisfied. In other words for each i all the bubbling of u_i will be captured by the components u_k with $k > i$ and $j_k = i$. Note that (b) contains the stability condition: we construct u_1 to be nonconstant, and require that $\#Z_i \geq 2$ (rather than 3) when u_i is constant because ∞ is an additional nodal point on the ith sphere for $i \geq 1$.

Throughout this proof we shall think of u^ν as a function from $\mathbb{C} \to M$ and denote by $|du^\nu(z)|$ the norm of the differential $du^\nu(z) : \mathbb{C} \to T_{u^\nu(z)}M$ with respect to the usual metric on $\mathbb{C}$. Then, for each ν, the norm of $du^\nu(z)$ with respect to the Fubini–Study metric is bounded and is given by $|du^\nu(z)|_{\mathrm{FS}} = |du^\nu(z)|(1+|z|^2)$ (see Exercise 4.2.3). Thus each function $|du^\nu(z)|$ must converge to 0 as $|z| \to \infty$. Hence there is a sequence $z^\nu \in \mathbb{C}$ such that

$$|du^\nu(z^\nu)| = \sup_{z \in \mathbb{C}} |du^\nu(z)| =: c^\nu.$$

Now observe that the sequence

$$v^\nu(z) := u^\nu(z^\nu + z/c^\nu)$$

satisfies $\sup|dv^\nu| = 1 = |dv^\nu(0)|$, and therefore has a subsequence which converges u.c.s. on $\mathbb{C} = S^2 \setminus \{\infty\}$ to a nonconstant J-holomorphic curve v. Passing to a further subsequence, we may assume that $E(v^\nu; \mathbb{C} \setminus B_1)$ converges. Then the functions

$$u_1(z) = v(1/z), \qquad \phi_1^\nu(z) = z^\nu + \frac{1}{c^\nu z}$$

satisfy (i) and (ii) with $Z_1 = \emptyset$ or $Z_1 = \{0\}$. Condition (iii) is void in the case $N = 1$. If $Z_1 = \emptyset$ then $u^\nu \circ \phi_1^\nu$ converges to u_1 uniformly with all derivatives on all of S^2 and so the theorem is proved. Hence assume $Z_1 = \{0\}$.

Let $\ell \geq 1$ and suppose, by induction, that u_i, j_i, z_i, Z_i, and $\{\phi_i^\nu\}_{\nu \in \mathbb{N}}$ have been constructed for $i \leq \ell$ so as to satisfy (i) - (iii) but not (iv), with N replaced by ℓ. Then there exists a $j \leq \ell$ such that $Z_j \neq \{z_i \,|\, j < i \leq \ell,\, j_i = j\}$. Denote

$$Z_{j;\ell} := Z_j \setminus \{z_i \,|\, j < i \leq \ell,\, j_i = j\}, \tag{5.3.1}$$

choose any element $z_{\ell+1} \in Z_{j;\ell}$, and apply Proposition 4.7.1 to the sequence $u^\nu \circ \phi_j^\nu$ and the point $z_0 = z_{\ell+1}$. Indexing the resulting subsequence by ν in the usual way, we obtain elements $\psi^\nu \in \mathrm{G}$ that satisfy the requirements of Proposition 4.7.1. In particular, $u^\nu \circ \phi_j^\nu \circ \psi^\nu$ converges to a J-holomorphic sphere $u_{\ell+1} : S^2 \to M$, u.c.s. on $S^2 \setminus Z$, where $Z \subset B_1$ is a finite set. Proposition 4.7.1 also shows that ψ^ν converges to $z_{\ell+1}$, u.c.s. on $\mathbb{C} = S^2 \setminus \{\infty\}$. Moreover, by Proposition 4.7.2,

$$u_{\ell+1}(\infty) = u_j(z_{\ell+1}).$$

This shows that conditions (i)-(iii) with $N = \ell + 1$ are satisfied for $i \leq \ell + 1$ with $j_{\ell+1} := j$, $Z_{\ell+1} := Z$, and $\phi_{\ell+1}^\nu := \phi_j^\nu \circ \psi^\nu$. This completes the induction.

We claim that the induction terminates after finitely many steps. To see this, at the ℓth stage consider the tree $T_\ell = (j_2, \ldots, j_\ell)$ with weights

$$m(j;\ell) = E(u_j) + \sum_{z \in Z_{j;\ell}} m_j(z)$$

for $j = 1, \ldots, \ell$, where $Z_{j;\ell}$ is given by (5.3.1). (Weighted trees are defined in Exercise 5.1.2.) Think of this quantity as the energy of u_j together with the energy of those of its bubble points that have not yet been blown up into a bubble u_i with $j < i \leq \ell$. Note that $m(j;\ell) = 0$ only when u_j is constant and all its bubble points have been resolved. By (b), this implies that j is a stable component of the weighted tree T_ℓ. Moreover $\hbar$ is the minimum of the nonzero weights $m(j;\ell)$. We next claim that

$$\sum_{j=1}^{\ell} m(j;\ell) = E. \tag{5.3.2}$$

This is obvious for $\ell = 1$, since $E = E(u_1) + m_1(0) = m(1;1)$. Now suppose (5.3.2) has been proved for some integer $\ell \geq 1$. Note that $m(j;\ell+1) = m(j;\ell)$ whenever $j \neq j_{\ell+1}$ and $j \leq \ell$, while

$$m(j_{\ell+1};\ell+1) = m(j_{\ell+1};\ell) - m_{j_{\ell+1}}(z_{\ell+1}),$$

and, by (ii) with $i = \ell + 1$,

$$m(\ell+1;\ell+1) = m_{j_{\ell+1}}(z_{\ell+1}).$$

Hence (5.3.2) holds with ℓ replaced by $\ell + 1$. Thus we have proved (5.3.2) for every integer $\ell \geq 1$. By Exercise 5.1.2, the number of edges of T_ℓ is bounded above by $2E/\hbar$. Hence the induction terminates.

Thus we have proved the existence of u_i, j_i, z_i, and $\{\phi_i^\nu\}_{\nu \in \mathbb{N}}$ for $i = 1, \ldots, N$ such that a suitable subsequence satisfies (i) - (iv). Under these assumptions we claim that the subsequence u^ν Gromov converges to the stable map $\mathbf{u}$. Let us first check that $\mathbf{u}$ is stable. To see this, note that the vertices j and i with $i > j$ are connected by an edge if and only if $j = j_i$. Further, in the notation of Definition 5.1.1, $z_{j_i i} = z_i$ and $z_{i j_i} = \infty$. Hence the singular set of the vertex $j \in \{2, \ldots, N\}$ is $Z_j \cup \{\infty\}$ with Z_j as in (iv), while the singular set of the vertex 1 is Z_1. Now, if u_i is constant then $i \geq 2$ and $\#Z_i \geq 2$. Hence $\mathbf{u}$ is stable as claimed.

We must verify the four conditions of Definition 5.2.1. The *(Marked point)* axiom is void in the case $n = 0$ and the *(Map)* and *(Rescaling)* axioms follow from (i) and (a). Thus it remains to verify the *(Energy)* axiom.

Denote by T_i the set of vertices i' which can be reached from j_i by a chain of edges, starting with the one from j_i to i. Thus $T_i = T_{j_i i}$ in the notation of Section 5.1. These sets can be defined recursively for $i = N, N-1, \ldots, 1$ by

$$T_i := \{i\} \cup \bigcup_{j_k = i} T_k.$$

In particular, $T_i = \{i\}$ whenever there is no $k > i$ such that $j_k = i$. We first prove, by induction over the number of elements in T_i, that

$$m_{j_i}(z_i) = \sum_{i' \in T_i} E(u_{i'}). \tag{5.3.3}$$

This follows immediately from (ii) whenever $Z_i = \emptyset$ and so $T_i = \{i\}$. Hence assume that this formula has been proved for $\#T_i \le m$. Suppose $\#T_i = m + 1$ and let

$$\{i_1, \ldots, i_s\} := \{k > i \,|\, j_k = i\}.$$

Then the set $\{i_1, \ldots, i_s\} \cup \{j_i\}$ is the set of vertices adjacent to i and $\#T_{i_r} \le m$ for $r = 1, \ldots, s$. Hence, by the induction hypothesis,

$$m_i(z_{i_r}) = \sum_{i' \in T_{i_r}} E(u_{i'}), \qquad r = 1, \ldots, s.$$

Moreover, it follows from (ii) that

$$m_{j_i}(z_i) = E(u_i) + \sum_{r=1}^{s} m_i(z_{i_r}).$$

Inserting the previous formulas and using $T_i = \{i\} \cup \bigcup_r T_{i_r}$ we obtain (5.3.3). This is equivalent to (5.2.1) with $\alpha = j_i$ and $\beta = i$.

Now consider the case $\alpha = i$ and $\beta = j_i$. Then $T_{\alpha\beta} = \{1, \ldots, N\} \setminus T_i$ and $z_{\alpha\beta} = \infty$, so that (5.2.1) takes the form

$$m_i(\infty) := \lim_{R \to \infty} \lim_{\nu \to \infty} E(u^\nu \circ \phi_i^\nu; \mathbb{C} \setminus B_R) = \sum_{i' \notin T_i} E(u_{i'}) \tag{5.3.4}$$

To see this note that

$$\begin{aligned} E - m_i(\infty) &= \lim_{R \to \infty} \lim_{\nu \to \infty} E(u^\nu \circ \phi_i^\nu; B_R) \\ &= E(u_i) + \sum_{z \in Z_i} m_i(z) \\ &= m_{j_i}(z_i) \\ &= \sum_{i' \in T_i} E(u_{i'}). \end{aligned}$$

The first equality follows from the definitions of $E := \lim_{\nu \to \infty} E(u^\nu)$ and $m_i(\infty)$, the second follows from the definition of $m_i(z)$, the third from (ii), and the fourth from (5.3.3). Moreover, by (5.3.2) with $\ell = N$, we have $E = \sum_{i=1}^N E(u_i)$, and this, together with the previous equation, implies (5.3.4). This proves Theorem 5.3.1 in the case $n = 0$. □

It remains to consider what happens when there are marked points. We assume, by induction, that the sequence $(u^\nu, z_1^\nu, \ldots, z_k^\nu)$ Gromov converges to a stable map

$$(\mathbf{u}, \mathbf{z}) = \big(\{u_\alpha\}_{\alpha\in T}, \{z_{\alpha\beta}\}_{\alpha E\beta}, \{z_i\}_{1\le i\le k}\big) \in \mathcal{SC}_{0,k}(M;J)$$

for some $k \geq 0$, with corresponding sequences $\phi_\alpha^\nu \in \mathrm{G}$ for $\alpha \in T$. We will consider what happens when we add another sequence of marked points $\zeta^\nu \in S^2 \setminus \{z_i^\nu, \ldots, z_k^\nu\}$. By passing to a subsequence we may suppose that the limits

$$\zeta_\alpha = \lim_{\nu\to\infty} (\phi_\alpha^\nu)^{-1}(\zeta^\nu) \tag{5.3.5}$$

exist for all $\alpha \in T$. Recall that $Z_\alpha = \{z_{\alpha\beta} \,|\, \beta \in T,\, \alpha E\beta\}$ is the set of points of the α-sphere where the adjacent bubbles attach.

LEMMA 5.3.4. *Suppose that $(u^\nu, z_1^\nu, \ldots, z_k^\nu)$ Gromov converges to a stable map $(\mathbf{u}, \mathbf{z}) \in \mathcal{SC}_{0,k}(M;J)$ via $\phi_\alpha^\nu \in \mathrm{G}$ and that the sequence $\zeta^\nu \in S^2 \setminus \{z_i^\nu, \ldots, z_k^\nu\}$ satisfies (5.3.5). Then precisely one of the following conditions holds.*

(I) *There exists a (unique) vertex $\alpha \in T$ such that*

$$\zeta_\alpha \notin Y_\alpha(\mathbf{u}, \mathbf{z}) = Z_\alpha \cup \{z_i \,|\, \alpha_i = \alpha\}.$$

(II) *There exists a (unique) index $i \in \{1, \ldots, k\}$ such that $\zeta_{\alpha_i} = z_i$.*

(III) *There exists a (unique) edge $\alpha E\beta$ in T such that $\zeta_\alpha = z_{\alpha\beta}$ and $\zeta_\beta = z_{\beta\alpha}$.*

PROOF. First note that the collection $\{\zeta_\alpha\}_{\alpha\in T}$ satisfies

$$\alpha E\gamma,\ \zeta_\alpha \neq z_{\alpha\gamma} \qquad \Longrightarrow \qquad \zeta_\gamma = z_{\gamma\alpha}. \tag{5.3.6}$$

This follows from (5.3.5) and the fact that the sequence $\phi_{\gamma\alpha}^\nu = (\phi_\gamma^\nu)^{-1}\circ\phi_\alpha^\nu$ converges to $z_{\gamma\alpha}$ u.c.s. on $S^2\setminus\{z_{\alpha\gamma}\}$. Now suppose that there is some $\alpha \in T$ such that $\zeta_\alpha \neq z_{\alpha\gamma}$ whenever $\alpha E\gamma$. Then (5.3.6) shows that $\zeta_\beta \in Z_\beta$ for every $\beta \neq \alpha$. Namely, given $\beta \neq \alpha$ choose a chain of edges $\gamma_0, \ldots, \gamma_m \in T$ running from $\gamma_0 = \alpha$ to $\gamma_m = \beta$. Then, by induction using (5.3.6) and $\zeta_{\gamma_0} \neq z_{\gamma_0\gamma_1}$, we see that $\zeta_{\gamma_i} = z_{\gamma_i\gamma_{i-1}}$ for $i = 1, \ldots, m$. This proves uniqueness in both (I) and (II).

We prove uniqueness in (III). Suppose that there are two pairs $\alpha E\beta$ and $\alpha' E\beta'$ with

$$\zeta_\alpha = z_{\alpha\beta}, \qquad \zeta_\beta = z_{\beta\alpha}, \qquad \zeta_{\alpha'} = z_{\alpha'\beta'}, \qquad \zeta_{\beta'} = z_{\beta'\alpha'}.$$

Let $\gamma_0, \ldots, \gamma_m \in T$ be the chain of edges running from $\gamma_0 = \alpha$ to $\gamma_m = \beta'$. By interchanging α and β and adding a term to the chain, if necessary, we may assume that $\gamma_1 = \beta$. Likewise, we may assume that $\gamma_{m-1} = \alpha'$. Suppose that $m \geq 2$. Then we have $\zeta_{\gamma_1} = z_{\gamma_1\gamma_0} \neq z_{\gamma_1\gamma_2}$ and it follows from (5.3.6) that $\zeta_{\gamma_2} = z_{\gamma_2\gamma_1}$. By induction, $\zeta_{\gamma_i} = z_{\gamma_i\gamma_{i-1}}$ for $i \geq 1$ and hence $\zeta_{\alpha'} = z_{\gamma_{m-1}\gamma_{m-2}} \neq z_{\alpha'\beta'}$. This contradiction proves that $m = 1$ and hence $\{\alpha, \beta\} = \{\alpha', \beta'\}$.

Now assume that (I) and (II) do not hold. Then, for each $\alpha \in T$, there exists a $\beta \in T$ such that $\alpha E\beta$ and $\zeta_\alpha = z_{\alpha\beta}$. We prove the existence of a pair $\alpha E\beta$ with $\zeta_\alpha = z_{\alpha\beta}$ and $\zeta_\beta = z_{\beta\alpha}$. Simply start with any pair $\alpha_0 E\alpha_1$ such that $\zeta_{\alpha_0} = z_{\alpha_0\alpha_1}$. If $\zeta_{\alpha_1} \neq z_{\alpha_1\alpha_0}$ then $\zeta_{\alpha_1} = z_{\alpha_1\alpha_2}$ for some $\alpha_2 \neq \alpha_0$ and one can construct a sequence $\alpha_i \in T$ with $\alpha_i E\alpha_{i+1}$ and $\zeta_{\alpha_i} = z_{\alpha_i\alpha_{i+1}}$. This sequence can only terminate if $\zeta_{i+1} = z_{\alpha_{i+1}\alpha_i}$ for some i, and it must terminate since T is a finite set. Thus we have proved that one of the conditions (I), (II) and (III) is satisfied. Note that these conditions are mutually exclusive. □

It remains to show how to accommodate the extra marked point in each of the above three cases. Case (I) is straightforward; however, in the other two cases we must add a new vertex to the tree.

PROOF OF THEOREM 5.3.1 FOR $n > 0$. Let $(u^\nu, z_1^\nu, \dots, z_n^\nu)$ be a sequence of J^ν-holomorphic spheres, each with n distinct marked points. We prove, by induction over k, that a subsequence of $(u^\nu, z_1^\nu, \dots, z_k^\nu)$ Gromov converges to a stable map

$$(\mathbf{u}, \mathbf{z}) = \left(\{u_\alpha\}_{\alpha \in T}, \{z_{\alpha\beta}\}_{\alpha E \beta}, \{z_i\}_{1 \le i \le k}\right) \in \mathcal{SC}_{0,k}(M; J)$$

with corresponding Möbius transformations ϕ_α^ν. We have already proved that this holds for $k = 0$. Let $k \ge 1$ and assume, by induction, that this has been established with k replaced by $k - 1$. Passing to a further subsequence, if necessary, we may assume that the limits

$$z_{\alpha k} := \lim_{\nu \to \infty} (\phi_\alpha^\nu)^{-1}(z_k^\nu) \tag{5.3.7}$$

exist for all α and hence may apply Lemma 5.3.4. Thus precisely one of the conditions (I-III) holds.

If (I) holds then there exists a unique element $\alpha_k \in T$ such that $z_{\alpha_k k} \notin Y_{\alpha_k}(\mathbf{u}, \mathbf{z})$. In this case we introduce a new marked point

$$z_k^{\mathrm{new}} := z_{\alpha_k k}$$

on the α_k-sphere.

If (II) holds then there exists a unique index $i \le k - 1$ such that $z_{\alpha_i k} = z_i$. In this case we choose a sequence of Möbius transformations $\psi^\nu \in \mathrm{G}$ such that

$$\psi^\nu(0) = z_i^\nu, \qquad \psi^\nu(1) = z_k^\nu, \qquad \psi^\nu(\infty) = \phi_{\alpha_i}^\nu(w),$$

where $w \in S^2 \setminus \{z_i\}$ is chosen such that $\phi_{\alpha_i}^\nu(w) \notin \{z_i^\nu, z_k^\nu\}$ for all ν. Then the sequences $(\phi_{\alpha_i}^\nu)^{-1} \circ \psi^\nu(0)$ and $(\phi_{\alpha_i}^\nu)^{-1} \circ \psi^\nu(1)$ both converge to z_i and the sequence $(\psi^\nu)^{-1} \circ \phi_{\alpha_i}^\nu(w)$ is constant and converges to ∞. Hence, by Lemma D.1.4, the sequence $(\phi_{\alpha_i}^\nu)^{-1} \circ \psi^\nu$ converges to z_i u.c.s. on $\mathbb{C} = S^2 \setminus \{\infty\}$.

We now extend the tree T by introducing an additional vertex γ which is connected only to α_i and carries the marked points z_i and z_k. The corresponding sequence ϕ_γ^ν is set equal to ψ^ν. Thus

$$(\phi_\gamma^\nu)^{-1}(z_i^\nu) = 0, \qquad (\phi_\gamma^\nu)^{-1}(z_k^\nu) = 1,$$

and $(\phi_{\alpha_i}^\nu)^{-1} \circ \phi_\gamma^\nu$ converges to z_i u.c.s. on $S^2 \setminus \{\infty\}$. Since $z_i \notin Z_{\alpha_i}$, the sequence $u^\nu \circ \phi_\gamma^\nu$ converges to the constant function $u_\gamma(z) \equiv u_{\alpha_i}(z_i)$ u.c.s. on $\mathbb{C} = S^2 \setminus \{\infty\}$. The new stable map is given by $T^{\mathrm{new}} := T \cup \{\gamma\}$ and

$$z_{\gamma\alpha_i}^{\mathrm{new}} := \infty, \quad z_{\alpha_i\gamma}^{\mathrm{new}} := z_i, \quad \alpha_k^{\mathrm{new}} := \gamma, \quad z_k^{\mathrm{new}} := 1, \quad \alpha_i^{\mathrm{new}} := \gamma, \quad z_i^{\mathrm{new}} := 0.$$

The new bubble is a "ghost", i.e. the corresponding map u_γ is constant. To verify the *(Energy)* axiom of Definition 5.2.1 for the new vertex γ we observe that the γ-sphere contains only one nodal point $z_{\gamma\alpha_i}^{\mathrm{new}} = \infty$. Since the limit u_γ is constant, the sequence $E(u^\nu \circ \phi_\gamma^\nu; B_R)$ converges to zero for every $R > 0$ and hence

$$\lim_{\nu \to \infty} E(u^\nu \circ \phi_\gamma^\nu; \mathbb{C} \setminus B_R) = \lim_{\nu \to \infty} E(u^\nu) = E(\mathbf{u})$$

for every $R > 0$. This proves the *(Energy)* axiom.

If (III) holds then there exists a unique edge $\alpha E\beta$ such that $z_{\alpha k} = z_{\alpha\beta}$ and $z_{\beta k} = z_{\beta\alpha}$. We prove that there exists a sequence $\psi^\nu \in \mathrm{G}$ that satisfies the following conditions.

(i) $(\psi^\nu)^{-1}(z_k^\nu) = 1$ for all ν.

(ii) $(\phi_\alpha^\nu)^{-1} \circ \psi^\nu$ converges to $z_{\alpha\beta}$ u.c.s. on $S^2 \setminus \{\infty\}$ and $(\phi_\beta^\nu)^{-1} \circ \psi^\nu$ converges to $z_{\beta\alpha}$ u.c.s. on $S^2 \setminus \{0\}$.

(iii) $u^\nu \circ \psi^\nu$ converges to $u_\alpha(z_{\alpha\beta}) = u_\beta(z_{\beta\alpha})$ u.c.s. on $S^2 \setminus \{0, \infty\}$.

(iv) For every $r > 0$,

$$\lim_{\nu\to\infty} E(u^\nu \circ \psi^\nu; B_r) = m_{\alpha\beta}(\mathbf{u}), \qquad \lim_{\nu\to\infty} E(u^\nu \circ \psi^\nu; \mathbb{C} \setminus B_r) = m_{\beta\alpha}(\mathbf{u}).$$

To see this, consider the sequences

$$\phi_{\alpha\beta}^\nu := (\phi_\alpha^\nu)^{-1} \circ \phi_\beta^\nu, \qquad \xi^\nu := (\phi_\beta^\nu)^{-1}(z_k^\nu)$$

so that $\xi^\nu \to z_{\beta\alpha}$ and $\phi_{\alpha\beta}^\nu(\xi^\nu) \to z_{\alpha\beta}$. We may assume without loss of generality that

$$z_{\alpha\beta} = 0, \qquad z_{\beta\alpha} = \infty.$$

Otherwise, choose Möbius transformations ψ_α and ψ_β such that $\psi_\alpha(0) = z_{\alpha\beta}$ and $\psi_\beta(\infty) = z_{\beta\alpha}$ and replace ϕ_α^ν by $\phi_\alpha^\nu \circ \psi_\alpha$ and ϕ_β^ν by $\phi_\beta^\nu \circ \psi_\beta$. Moreover, we may assume without loss of generality that $\phi_{\alpha\beta}^\nu(\infty) = \infty$. Otherwise, since $\phi_{\beta\alpha}^\nu(\infty) \to \infty$, we may choose a sequence of isometries $\rho^\nu \in \mathrm{SO}(3)$ converging uniformly to the identity such that $\rho^\nu \circ \phi_{\beta\alpha}^\nu(\infty) = \infty$ and replace ϕ_β^ν by $\phi_\beta^\nu \circ (\rho^\nu)^{-1}$. After these adjustments the sequence $\phi_{\alpha\beta}^\nu$ has the form

$$\phi_{\alpha\beta}^\nu(z) = z^\nu + \varepsilon^\nu z,$$

where the sequences $z^\nu, \varepsilon^\nu \in \mathbb{C}$ converge to zero and $\varepsilon^\nu \neq 0$ for all ν. Now recall that

$$\lim_{\nu\to\infty} \xi^\nu = \infty, \qquad \lim_{\nu\to\infty} \varepsilon^\nu \xi^\nu = 0.$$

Hence the sequence $\rho^\nu \in \mathrm{G}$, defined by

$$\rho^\nu(w) := \frac{w - z^\nu}{\varepsilon^\nu \xi^\nu}$$

for large ν, satisfies the following conditions.

(a) ρ^ν converges to ∞ u.c.s. on $S^2 \setminus \{z_{\alpha\beta} = 0\}$.

(b) $\rho^\nu \circ \phi_{\alpha\beta}^\nu$ converges to 0 u.c.s. on $S^2 \setminus \{z_{\beta\alpha} = \infty\}$.

(c) $\rho^\nu \circ (\phi_\alpha^\nu)^{-1}(z_k^\nu) = 1$ for all ν.

Now let

$$\psi^\nu := \phi_\alpha^\nu \circ (\rho^\nu)^{-1}.$$

Then $(\psi^\nu)^{-1}(z_k^\nu) = \rho^\nu \circ (\phi_\alpha^\nu)^{-1}(z_k^\nu) = 1$ and, by (a) and (b), $(\phi_\alpha^\nu)^{-1} \circ \psi^\nu$ converges to $z_{\alpha\beta}$ u.c.s. on $S^2 \setminus \{\infty\}$ and $(\phi_\beta^\nu)^{-1} \circ \psi^\nu$ converges to $z_{\beta\alpha}$ u.c.s. on $S^2 \setminus \{0\}$. Hence, by Lemma 5.4.2, the sequence $u^\nu \circ \psi^\nu$ converges to the constant $u_\alpha(z_{\alpha\beta}) = u_\beta(z_{\beta\alpha})$ u.c.s. on $S^2 \setminus \{0, \infty\}$ and

$$\lim_{\nu\to\infty} E(u^\nu \circ \psi^\nu; B_r) = m_{\alpha\beta}(\mathbf{u}), \qquad \lim_{\nu\to\infty} E(u^\nu \circ \psi^\nu; \mathbb{C} \setminus B_r) = m_{\beta\alpha}(\mathbf{u})$$

for every $r > 0$. Hence the sequence ψ^ν satisfies (i-iv) as claimed.

We now extend the tree T by introducing an additional vertex γ corresponding to the rescaling $\phi_\gamma^\nu = \psi^\nu$. The corresponding limit map u_γ is constant:

$$u_\gamma(z) \equiv u_\alpha(z_{\alpha\beta}) = u_\beta(z_{\beta\alpha}).$$

This bubble carries the marked point $z_k = 1$. Moreover, for this new sequence ϕ_γ^ν the limits (5.3.7) still exist. In the new stable map α and β are no longer adjacent, but are separated by γ, and we have

$$z_{\alpha\gamma}^{\mathrm{new}} = z_{\alpha\beta}, \quad z_{\beta\gamma}^{\mathrm{new}} = z_{\beta\alpha}, \quad z_{\gamma\alpha}^{\mathrm{new}} = \infty, \quad z_{\gamma\beta}^{\mathrm{new}} = 0, \quad \alpha_k^{\mathrm{new}} = \gamma, \quad z_k^{\mathrm{new}} = 1.$$

It follows from (i-iv) that all the conditions in Definition 5.2.1 hold. This completes the induction and the proof of Theorem 5.3.1. □

5.4. Uniqueness of the limit

This section is devoted to the proof of the following uniqueness statement.

THEOREM 5.4.1. *Let $J^\nu \in \mathcal{J}_\tau(M,\omega)$ be a sequence of ω-tame almost complex structures that converges to $J \in \mathcal{J}_\tau(M,\omega)$ in the C^∞-topology. Let $(u^\nu, z_1^\nu, \dots, z_n^\nu)$ be a sequence of J^ν-holomorphic spheres with n distinct marked points that Gromov converges to two stable maps $(\mathbf{u}, \mathbf{z})$ and $(\tilde{\mathbf{u}}, \tilde{\mathbf{z}})$. Then $(\mathbf{u}, \mathbf{z})$ and $(\tilde{\mathbf{u}}, \tilde{\mathbf{z}})$ are equivalent.*

To prove this we have to compare the behaviour of the two sets of rescaling maps ϕ_α^ν and $\tilde{\phi}_{\tilde{\alpha}}^\nu$ that appear in the definition of Gromov convergence. The first lemma shows that if α, β are adjacent vertices in the tree T of $(\mathbf{u}, \mathbf{z})$, then no rescaling ψ^ν can produce a nontrivial bubble lying between them because there is not enough energy to form such a bubble.

LEMMA 5.4.2. *Let J^ν be as in Theorem 5.4.1 and suppose that the sequence $(u^\nu, z_1^\nu, \dots, z_n^\nu)$ of J^ν-holomorphic spheres with marked points Gromov converges to the stable map $(\mathbf{u}, \mathbf{z}) \in \mathcal{SC}_{0,n}(M;J)$ via the reparametrization sequences $\phi_\alpha^\nu \in \mathrm{G}$. Moreover, let $\alpha E \beta$ be an edge and $\psi^\nu \in \mathrm{G}$ be a sequence such that*

(a) *$(\phi_\alpha^\nu)^{-1} \circ \psi^\nu$ converges to $z_{\alpha\beta}$ u.c.s. on $S^2 \setminus \{w_0\}$.*

(b) *$(\phi_\beta^\nu)^{-1} \circ \psi^\nu$ converges to $z_{\beta\alpha}$ u.c.s. on $S^2 \setminus \{w_1\}$.*

Then $u^\nu \circ \psi^\nu$ converges to $u_\alpha(z_{\alpha\beta}) = u_\beta(z_{\beta\alpha})$ u.c.s. on $S^2 \setminus \{w_0, w_1\}$ and

$$\begin{aligned} \lim_{\nu\to\infty} E(u^\nu \circ \psi^\nu; B_r(w_0)) &= m_{\beta\alpha}(\mathbf{u}), \\ \lim_{\nu\to\infty} E(u^\nu \circ \psi^\nu; B_r(w_1)) &= m_{\alpha\beta}(\mathbf{u}), \end{aligned} \tag{5.4.1}$$

whenever $r < d_{S^2}(w_0, w_1)$. Moreover, if $\alpha_i \in T_{\alpha\beta}$ then $(\psi^\nu)^{-1}(z_i^\nu)$ converges to w_1, and if $\alpha_i \in T_{\beta\alpha}$ then $(\psi^\nu)^{-1}(z_i^\nu)$ converges to w_0.

PROOF. Fix a constant $\varepsilon > 0$. Then, for ν sufficiently large, we have

$$\phi_{\beta\alpha}^\nu(S^2 \setminus B_\varepsilon(z_{\alpha\beta})) \subset B_\varepsilon(z_{\beta\alpha}).$$

Applying ϕ_β^ν to both sides we obtain

$$\phi_\alpha^\nu(S^2 \setminus B_\varepsilon(z_{\alpha\beta})) \cap \phi_\beta^\nu(S^2 \setminus B_\varepsilon(z_{\beta\alpha})) = \emptyset.$$

On the other hand, it follows from (a) and (b) that

$$\phi_\alpha^\nu(S^2 \setminus B_\varepsilon(z_{\alpha\beta})) \subset \psi^\nu(B_\varepsilon(w_0)), \qquad \phi_\beta^\nu(S^2 \setminus B_\varepsilon(z_{\beta\alpha})) \subset \psi^\nu(B_\varepsilon(w_1))$$

for ν sufficiently large. Integrating the form $(u^\nu)^*\omega$ over these two sets we obtain

$$\begin{aligned} E(u^\nu \circ \psi^\nu; B_\varepsilon(w_0)) &\geq E(u^\nu_\alpha; S^2 \setminus B_\varepsilon(z_{\alpha\beta})) = E(\mathbf{u}) - E(u^\nu_\alpha; B_\varepsilon(z_{\alpha\beta})), \\ E(u^\nu \circ \psi^\nu; B_\varepsilon(w_1)) &\geq E(u^\nu_\beta; S^2 \setminus B_\varepsilon(z_{\beta\alpha})) = E(\mathbf{u}) - E(u^\nu_\beta; B_\varepsilon(z_{\beta\alpha})). \end{aligned}$$

Denote $m_\varepsilon(w_i) := \liminf_{\nu\to\infty} E(u^\nu \circ \psi^\nu; B_\varepsilon(w_i))$ for $i = 0, 1$. Then the function $\varepsilon \mapsto m_\varepsilon(w_i)$ is nondecreasing and, by the *(Energy)* axiom in Definition 5.2.1, we have

$$\begin{aligned} m(w_0) := \lim_{\varepsilon\to 0} m_\varepsilon(w_0) &\geq E(\mathbf{u}) - m_{\alpha\beta}(\mathbf{u}) = m_{\beta\alpha}(\mathbf{u}), \\ m(w_1) := \lim_{\varepsilon\to 0} m_\varepsilon(w_1) &\geq E(\mathbf{u}) - m_{\beta\alpha}(\mathbf{u}) = m_{\alpha\beta}(\mathbf{u}). \end{aligned}$$

If $B_\varepsilon(w_0) \cap B_\varepsilon(w_1) = \emptyset$ then

$$m_{\beta\alpha}(\mathbf{u}) \leq m_\varepsilon(w_0) \leq E(\mathbf{u}) - m_\varepsilon(w_1) \leq E(\mathbf{u}) - m_{\alpha\beta}(\mathbf{u}) = m_{\beta\alpha}(\mathbf{u}).$$

This implies (5.4.1) and that $u^\nu \circ \psi^\nu$ converges to a constant u.c.s. on $S^2 \setminus \{w_0, w_1\}$. Next we prove that the constant limit is equal to $u_\alpha(z_{\alpha\beta}) = u_\beta(z_{\beta\alpha})$. To see this, choose any $z \in S^2 \setminus \{w_0, w_1\}$ and consider the sequence

$$\zeta^\nu := (\phi^\nu_\alpha)^{-1} \circ \psi^\nu(z).$$

Then ζ^ν converges to $z_{\alpha\beta}$ and $(\phi^\nu_{\alpha\beta})^{-1}(\zeta^\nu)$ converges to $z_{\beta\alpha}$. Hence the assumptions of Proposition 4.7.2 are satisfied with u^ν replaced by u^ν_α, v^ν replaced by u^ν_β, ψ^ν replaced by $\phi^\nu_{\alpha\beta}$, $z_0 = z_{\alpha\beta}$, $z_\infty = z_{\beta\alpha}$, $u = u_\alpha$, and $v = u_\beta$. Hence the sequence

$$u^\nu \circ \psi^\nu(z) = u^\nu_\alpha(\zeta^\nu)$$

converges to $u_\alpha(z_{\alpha\beta})$.

Now suppose that $\alpha_i \in T_{\alpha\beta}$. Then $z_{\alpha i} = z_{\alpha\beta}$ and, by Theorem 5.2.2 (iii),

$$z_{\beta i} = \lim_{\nu\to\infty} (\phi^\nu_\beta)^{-1}(z^\nu_i) \neq z_{\beta\alpha}.$$

This implies that $(\psi^\nu)^{-1}(z^\nu_i)$ converges to w_1 since otherwise we would obtain the limit $z_{\beta\alpha}$ after applying the Möbius transformations $(\phi^\nu_\beta)^{-1} \circ \psi^\nu$ to this sequence and passing to a suitable subsequence. A similar argument shows that $(\psi^\nu)^{-1}(z^\nu_i)$ converges to w_0 whenever $\alpha_i \in T_{\beta\alpha}$. This proves Lemma 5.4.2. □

The next lemma shows that if ψ^ν is a rescaling such that $u^\nu \circ \psi^\nu$ converges off some finite set, then ψ^ν must be commensurate with one of the ϕ^ν_α in the sense that the composite sequence $(\phi^\nu_\alpha)^{-1} \circ \psi^\nu$ has a convergent subsequence. It is here that the stability condition is crucial.

LEMMA 5.4.3. *Let J^ν, J, $(u^\nu, z^\nu_1, \ldots, z^\nu_n)$, $(\mathbf{u}, \mathbf{z})$, and ϕ^ν_α be as in Lemma 5.4.2. Suppose that ψ^ν is a sequence in G, $Z \subset S^2$ is a finite set, and $v : S^2 \to M$ a J-holomorphic sphere, such that the following holds.*

(a) *$v^\nu := u^\nu \circ \psi^\nu$ converges to v u.c.s. on $S^2 \setminus Z$ with bubbling at Z.*

(b) *For each i the limit $\zeta_i = \lim_{\nu\to\infty}(\psi^\nu)^{-1}(z^\nu_i)$ exists.*

(c) *If v is constant then $\#Y \geq 3$, where $Y := Z \cup \{\zeta_i \,|\, 1 \leq i \leq n\}$.*

Then there exists a vertex $\alpha \in T$ such that the sequence $(\phi^\nu_\alpha)^{-1} \circ \psi^\nu$ has a convergent subsequence.

PROOF. Suppose not and fix any element $\alpha_0 \in T$. Then the sequence

$$\chi^\nu := (\phi^\nu_{\alpha_0})^{-1} \circ \psi^\nu \in \mathrm{G}$$

has no convergent subsequence. Hence, by Lemma D.1.2, there exist $w_0, z_0 \in S^2$ such that a subsequence of χ^ν, still denoted by χ^ν, converges to z_0 u.c.s. on $S^2 \setminus \{w_0\}$. We claim that

$$z_0 \in Z_{\alpha_0}, \qquad w_0 \in Z. \tag{5.4.2}$$

Suppose, by contradiction, that $z_0 \notin Z_{\alpha_0}$. Then $u^\nu_{\alpha_0}$ converges to u_{α_0}, uniformly in a neighbourhood of z_0. Hence $v^\nu = u^\nu_{\alpha_0} \circ \chi^\nu$ converges to the constant $u_{\alpha_0}(z_0)$ u.c.s. on $S^2 \setminus \{w_0\}$. Hence v is constant and $Z \subset \{w_0\}$. Moreover, since $z_0 \notin Z_{\alpha_0}$ there is at most one index i such that $z_{\alpha_0 i} = z_0$. If such an index i exists then, for $j \neq i$, the sequence $(\phi^\nu_{\alpha_0})^{-1}(z^\nu_j)$ converges to $z_{\alpha_0 j} \neq z_0$ and hence

$$w_0 = \lim_{\nu\to\infty} (\chi^\nu)^{-1}\big((\phi^\nu_{\alpha_0})^{-1}(z^\nu_j)\big) = \lim_{\nu\to\infty} (\psi^\nu)^{-1}(z^\nu_j).$$

Hence $Y \subset \{w_0, \zeta_i\}$. If no such i exists then $Y = Z \subset \{w_0\}$. In either case this contradicts (c). Hence $z_0 \in Z_{\alpha_0}$ and a similar argument shows that $w_0 \in Z$.

By (5.4.2), there exists a point $w_0 \in Z$ and a vertex $\alpha_1 \in T$ with $\alpha_0 E \alpha_1$ such that, after passing to a subsequence if necessary,

$$(\phi^\nu_{\alpha_0})^{-1} \circ \psi^\nu \text{ converges to } z_{\alpha_0\alpha_1} \text{ u.c.s. on } S^2 \setminus \{w_0\}.$$

Again, $(\phi^\nu_{\alpha_1})^{-1} \circ \psi^\nu$ has no convergent subsequence. Hence, by the same argument, we may assume, after passing to a further subsequence, that there exist points $z_{\alpha_1\alpha_2} \in Z_{\alpha_1}$ and $w_1 \in Z$ such that

$$(\phi^\nu_{\alpha_1})^{-1} \circ \psi^\nu \text{ converges to } z_{\alpha_1\alpha_2} \text{ u.c.s. on } S^2 \setminus \{w_1\}.$$

We claim that $\alpha_0 \neq \alpha_2$. Otherwise we could apply Lemma 5.4.2 to the vertices $\alpha = \alpha_0$ and $\beta = \alpha_1$ with $\alpha E \beta$. It would then follow that $u^\nu \circ \psi^\nu$ converges to the constant $u_{\alpha_0}(z_{\alpha_0\alpha_1})$, u.c.s. on $S^2 \setminus \{w_0, w_1\}$, and that $(\psi^\nu)^{-1}(z^\nu_i)$ converges to either w_0 or w_1 for all i. But this would imply that v is constant and $Z \subset \{w_0, w_1\}$, in contradiction to (c). Hence $\alpha_0 \neq \alpha_2$, as claimed. By induction, we obtain an infinite sequence $\alpha_0, \alpha_1, \alpha_2, \ldots$ in T, such that $\alpha_i E \alpha_{i+1}$ and $\alpha_i \neq \alpha_{i+2}$ for all i. Any such sequence consists of distinct elements, contradicting the finiteness of T. This proves Lemma 5.4.3. □

PROOF OF THEOREM 5.4.1. Let $\{\phi^\nu_\alpha\}_{\alpha\in T}$ and $\{\tilde\phi^\nu_\alpha\}_{\alpha\in\tilde T}$ be the sequences in the definition of Gromov convergence. We prove in five steps, that $(\mathbf{u}, \mathbf{z})$ is equivalent to $(\tilde{\mathbf{u}}, \tilde{\mathbf{z}})$. We first extend the *(Rescaling)* condition in the definition of Gromov convergence to pairs of vertices α, β that are not joined by an edge. For this we extend the notation $z_{\alpha\beta}$ to arbitrary pairs of distinct vertices $\alpha, \beta \in T$. Namely, if $\alpha \neq \beta$ then there is a unique vertex $\gamma \in T$, adjacent to α, such that β lies in the subset $T_{\alpha\gamma}$ that can be reached from α via γ. We denote

$$z_{\alpha\beta} := z_{\alpha\gamma}, \qquad \beta \in T_{\alpha\gamma}.$$

We also use the notation

$$\phi^\nu_{\alpha\beta} := (\phi^\nu_\alpha)^{-1} \circ \phi^\nu_\beta$$

for any pair $\alpha, \beta \in T$.

STEP 1. *If $\alpha, \beta \in T$ and $\alpha \neq \beta$, then $\phi^\nu_{\alpha\beta}$ converges to $z_{\alpha\beta}$ u.c.s. on $S^2 \setminus \{z_{\beta\alpha}\}$.*

Let $\gamma_0, \ldots, \gamma_m$ denote the chain of edges in T running from $\gamma_0 = \alpha$ to $\gamma_m = \beta$. Then, by assumption, $\phi^\nu_{\gamma_{i-1}\gamma_i}$ converges to $z_{\gamma_{i-1}\gamma_i}$ u.c.s. on $S^2 \setminus \{z_{\gamma_i\gamma_{i-1}}\}$. Since $\gamma_{i-2} \neq \gamma_i$ we have $z_{\gamma_{i-1}\gamma_i} \neq z_{\gamma_{i-1}\gamma_{i-2}}$. Hence $\phi^\nu_{\gamma_{i-2}\gamma_i}$ converges to $z_{\gamma_{i-2}\gamma_i}$ u.c.s. on $S^2 \setminus \{z_{\gamma_i\gamma_{i-2}}\}$. By induction, the sequence $\phi^\nu_{\gamma_0\gamma_i}$ converges to $z_{\gamma_0\gamma_i}$ u.c.s. on $S^2 \setminus \{z_{\gamma_i\gamma_0}\}$. With $i = m$ this is the assertion of Step 1.

STEP 2. *There is a unique bijection $f : T \to \widetilde{T}$ and a subsequence (still denoted by $(u^\nu, z_i^\nu, \phi^\nu_\alpha)$) such that the limit*

$$\psi_\alpha = \lim_{\nu\to\infty} (\phi^\nu_\alpha)^{-1} \circ \tilde{\phi}^\nu_{f(\alpha)} \tag{5.4.3}$$

exists for every $\alpha \in T$. Here the convergence is uniform on all of S^2.

Applying Lemma 5.4.3 to the sequence $\psi^\nu := \tilde{\phi}^\nu_{\tilde{\alpha}}$, we find a vertex $\alpha \in T$ such that the sequence $(\phi^\nu_\alpha)^{-1} \circ \tilde{\phi}^\nu_{\tilde{\alpha}}$ has a uniformly convergent subsequence. Moreover α is unique by Step 1. Now we can apply Lemma 5.4.3 again to this subsequence and any other element $\tilde{\beta} \in \widetilde{T}$, and proceed by induction. This gives rise to a map $g : \widetilde{T} \to T$ and a collection of Möbius transformations $\{\chi_{\tilde{\alpha}}\}_{\tilde{\alpha}\in\widetilde{T}}$ such that

$$\chi_{\tilde{\alpha}} = \lim_{\nu\to\infty} (\phi^\nu_{g(\tilde{\alpha})})^{-1} \circ \tilde{\phi}^\nu_{\tilde{\alpha}}$$

for every $\tilde{\alpha} \in \widetilde{T}$. Reversing the roles of $\mathbf{u}$ and $\tilde{\mathbf{u}}$ we find that g is bijective. The function $f := g^{-1} : T \to \widetilde{T}$ and the Möbius transformations $\psi_\alpha := \chi_{f(\alpha)}$ satisfy the requirements of Step 2.

STEP 3. *Let f and ψ_α be as in Step 2. Then*

$$\tilde{u}_{f(\alpha)} = u_\alpha \circ \psi_\alpha, \qquad \tilde{z}_{f(\alpha)i} = \psi_\alpha^{-1}(z_{\alpha i}), \qquad \tilde{z}_{f(\alpha)f(\beta)} = \psi_\alpha^{-1}(z_{\alpha\beta}), \tag{5.4.4}$$

for $\alpha, \beta \in T$ with $\alpha \neq \beta$ and $i = 1, \ldots, k$. Moreover, for $\alpha E \beta$, the bubble energy of the sequence $u^\nu \circ \tilde{\phi}^\nu_{f(\alpha)}$ at the point $\tilde{z}_{f(\alpha)f(\beta)}$ is given by

$$E(\{u^\nu \circ \tilde{\phi}^\nu_{f(\alpha)}\}; \tilde{z}_{f(\alpha)f(\beta)}) = m_{\alpha\beta}(\mathbf{u}). \tag{5.4.5}$$

The first equation in (5.4.4) follows from (5.4.3), namely

$$u_\alpha \circ \psi_\alpha = \lim_{\nu\to\infty} u^\nu \circ \phi^\nu_\alpha \circ \psi_\alpha = \lim_{\nu\to\infty} u^\nu \circ \tilde{\phi}^\nu_{f(\alpha)} = \tilde{u}_{f(\alpha)}.$$

Here the limit is to be understood as convergence modulo bubbling. The second equation in (5.4.4) also uses (5.4.3):

$$\psi_\alpha^{-1}(z_{\alpha i}) = \psi_\alpha^{-1}\left(\lim_{\nu\to\infty} (\phi^\nu_\alpha)^{-1}(z_i^\nu)\right) = \lim_{\nu\to\infty} (\tilde{\phi}^\nu_{f(\alpha)})^{-1}(z_i^\nu) = \tilde{z}_{f(\alpha)i}.$$

To prove the third equation in (5.4.4), take $\alpha, \beta \in T$ with $\alpha \neq \beta$. Then, by Step 1, $\phi^\nu_{\alpha\beta}$ converges to $z_{\alpha\beta}$ u.c.s. on $S^2 \setminus \{z_{\beta\alpha}\}$. Hence, for $z \neq z_{\beta\alpha}$ we obtain, by (5.4.3),

$$\begin{aligned} \psi_\alpha^{-1}(z_{\alpha\beta}) &= \psi_\alpha^{-1}\left(\lim_{\nu\to\infty} \phi^\nu_{\alpha\beta}(z)\right) \\ &= \lim_{\nu\to\infty} (\tilde{\phi}^\nu_{f(\alpha)})^{-1} \circ \phi^\nu_\beta(z) \\ &= \lim_{\nu\to\infty} \tilde{\phi}^\nu_{f(\alpha)f(\beta)}(\psi_\beta^{-1}(z)). \end{aligned}$$

Thus $\tilde{\phi}^\nu_{f(\alpha)f(\beta)}$ converges to $\psi_\alpha^{-1}(z_{\alpha\beta})$ u.c.s. on $S^2 \setminus \{\psi_\beta^{-1}(z_{\beta\alpha})\}$. This proves (5.4.4).

To prove (5.4.5), note that the limit as $\varepsilon \to 0$ of the function

$$\varepsilon \mapsto \lim_{\nu\to\infty} E\big(u^\nu \circ \tilde{\phi}^\nu_{f(\alpha)}; B_\varepsilon(\tilde{z}_{f(\alpha)f(\beta)})\big)$$

remains the same if the ball $B_\varepsilon(\tilde{z}_{f(\alpha)f(\beta)}) = B_\varepsilon(\psi_\alpha^{-1}(z_{\alpha\beta}))$ is replaced by the set ${\psi_\alpha}^{-1}(B_\varepsilon(z_{\alpha\beta}))$. Now

$$\begin{aligned}
&\lim_{\nu\to\infty} E\left(u^\nu \circ \tilde{\phi}^\nu_{f(\alpha)}; \psi_\alpha^{-1}(B_\varepsilon(z_{\alpha\beta}))\right) \\
&\quad = \lim_{\nu\to\infty} E\left(u^\nu \circ \tilde{\phi}^\nu_{f(\alpha)} \circ \psi_\alpha^{-1}; B_\varepsilon(z_{\alpha\beta})\right) \\
&\quad = \lim_{\nu\to\infty} E\left((u^\nu \circ \phi^\nu_\alpha) \circ ((\phi^\nu_\alpha)^{-1} \circ \tilde{\phi}^\nu_{f(\alpha)} \circ \psi_\alpha^{-1}); B_\varepsilon(z_{\alpha\beta})\right) \\
&\quad = \lim_{\nu\to\infty} E\left(u^\nu \circ \phi^\nu_\alpha; B_\varepsilon(z_{\alpha\beta})\right).
\end{aligned}$$

The last equality follows from the fact that

$$\mathrm{id} = \lim_{\nu\to\infty} (\phi^\nu_\alpha)^{-1} \circ \tilde{\phi}^\nu_{f(\alpha)} \circ {\psi_\alpha}^{-1}.$$

Now (5.4.5) follows by taking the limit as $\varepsilon \to 0$. This proves Step 3.

STEP 4. *Let $\alpha, \beta \in T$. Then $\alpha E \beta$ if and only if*

$$E(\{u^\nu_\alpha\}; z_{\alpha\beta}) + E(\{u^\nu_\beta\}; z_{\beta\alpha}) = E(\mathbf{u}) \tag{5.4.6}$$

and there is no $i \in \{1, \ldots, k\}$ that satisfies both $z_{\alpha i} = z_{\alpha\beta}$ and $z_{\beta i} = z_{\beta\alpha}$.

If $\alpha E \beta$ then it follows from the definition of Gromov convergence that these conditions are satisfied. Conversely, suppose that $\alpha, \beta \in T$ satisfy the two conditions (5.4.6) and that for each i either $z_{\alpha i} \neq z_{\alpha\beta}$ or $z_{\beta i} \neq z_{\beta\alpha}$. Choose a chain of edges $\gamma_0, \ldots, \gamma_m \in T$ running from $\gamma_0 = \alpha$ to $\gamma_m = \beta$. Then (5.4.6) is equivalent to $m_{\gamma_0\gamma_1}(\mathbf{u}) + m_{\gamma_m\gamma_{m-1}}(\mathbf{u}) = E(\mathbf{u})$. This implies

$$m_{\gamma_0\gamma_1}(\mathbf{u}) = m_{\gamma_{m-1}\gamma_m}(\mathbf{u}).$$

If $m \neq 1$ then it follows that $E(u_\gamma) = 0$ for every $\gamma \in T_{\gamma_0\gamma_1} \setminus T_{\gamma_{m-1}\gamma_m}$ and this set is nonempty. Hence there must be a sphere in $T_{\gamma_0\gamma_1} \setminus T_{\gamma_{m-1}\gamma_m}$ which carries a marked point z_i. For this i we obtain $z_{\alpha i} = z_{\alpha\beta}$ and $z_{\beta i} = z_{\beta\alpha}$, in contradiction to our assumption. Hence $m = 1$ and hence $\alpha E \beta$. This proves Step 4.

STEP 5. *f is a tree isomorphism.*

Step 3 shows that if $\alpha E \beta$ then

$$E\big(\{u^\nu \circ \tilde{\phi}^\nu_{f(\alpha)}\}; \tilde{z}_{f(\alpha)f(\beta)}\big) + E\big(\{u^\nu \circ \tilde{\phi}^\nu_{f(\beta)}\}; \tilde{z}_{f(\beta)f(\alpha)}\big) = E(\tilde{\mathbf{u}}).$$

Further, for each i, we have either $\tilde{z}_{f(\alpha)i} \neq \tilde{z}_{f(\alpha)f(\beta)}$ or $\tilde{z}_{f(\beta)i} \neq \tilde{z}_{f(\beta)f(\alpha)}$. Hence Step 4 shows that $f(\alpha)\tilde{E}f(\beta)$. Replacing f by f^{-1}, we find that $f(\alpha)\tilde{E}f(\beta)$ implies $\alpha E \beta$. Hence f is a tree isomorphism. By Step 5 and Step 3, $(\mathbf{u}, \mathbf{z})$ is equivalent to $(\tilde{\mathbf{u}}, \tilde{\mathbf{z}})$. This proves Theorem 5.4.1. □

5.5. Gromov compactness for stable maps

In this section we first define Gromov convergence for sequences of stable maps (of genus zero) and prove that every sequence of stable maps has a Gromov convergent subsequence. We next prove that limits are unique up to the equivalence relation of Definition 5.1.4. We shall then explore a way of measuring the distance between two stable maps $\mathbf{x} = (\mathbf{u}, \mathbf{z})$ and $\mathbf{x}' = (\mathbf{u}', \mathbf{z}')$ in terms of suitable numbers $\rho_\varepsilon(\mathbf{x}, \mathbf{x}') > 0$. These do not define a metric but have the property that $\mathbf{x}^\nu$ converges to $\mathbf{x}$ if and only if $\rho_\varepsilon(\mathbf{x}, \mathbf{x}^\nu)$ converges to zero (as ν tends to infinity) for ε sufficiently small. In Section 5.6 we shall see that these results lead to a natural topology on the moduli space of stable maps.

Gromov convergence. For a stable map $(\mathbf{u}, \mathbf{z}) \in \mathcal{SC}_{0,n}(M; J)$ which is modelled over a tree T, a vertex $\alpha \in T$, and an open set $U_\alpha \subset S^2$ we denote

$$E_\alpha(u; U_\alpha) := E(u_\alpha; U_\alpha) + \sum_{\substack{\beta \in T \\ \alpha E \beta,\ z_{\alpha\beta} \in U_\alpha}} m_{\alpha\beta}(\mathbf{u}).$$

For $z \in S^2$ and $\varepsilon > 0$ we denote by $B_\varepsilon(z) \subset S^2$ the open ball of radius ε with respect to the Fubini–Study metric.

DEFINITION 5.5.1 (Gromov convergence). *Let $J^\nu \in \mathcal{J}_\tau(M, \omega)$ be a sequence of ω-tame almost complex structures on M that converges to $J \in \mathcal{J}_\tau(M, \omega)$. A sequence of stable maps*

$$(\mathbf{u}^\nu, \mathbf{z}^\nu) = \left(\{u_\alpha^\nu\}_{\alpha \in T^\nu}, \{z_{\alpha\beta}^\nu\}_{\alpha E^\nu \beta}, \{\alpha_i^\nu, z_i^\nu\}_{1 \le i \le n}\right) \in \mathcal{SC}_{0,n}(M; J^\nu)$$

is said to **Gromov converge** *to a stable map*

$$(\mathbf{u}, \mathbf{z}) = \left(\{u_\alpha\}_{\alpha \in T}, \{z_{\alpha\beta}\}_{\alpha E \beta}, \{\alpha_i, z_i\}_{1 \le i \le n}\right) \in \mathcal{SC}_{0,n}(M; J)$$

if, for every sufficiently large ν, there exists a surjective tree homomorphism

$$f^\nu : T \to T^\nu$$

and a collection of Möbius transformations $\{\phi_\alpha^\nu\}_{\alpha \in T}$ such that the following holds.

(MAP) *For every $\alpha \in T$ the sequence $u_{f^\nu(\alpha)}^\nu \circ \phi_\alpha^\nu : S^2 \to M$ converges to u_α u.c.s. on $S^2 \setminus Z_\alpha$.*

(ENERGY) *If $\alpha E \beta$ then*

$$m_{\alpha\beta}(\mathbf{u}) = \lim_{\varepsilon \to 0} \lim_{\nu \to \infty} E_{f^\nu(\alpha)}\Big(\mathbf{u}^\nu; \phi_\alpha^\nu(B_\varepsilon(z_{\alpha\beta}))\Big). \tag{5.5.1}$$

(RESCALING) *If $\alpha, \beta \in T$ such that $\alpha E \beta$ and ν_j is a subsequence such that $f^{\nu_j}(\alpha) = f^{\nu_j}(\beta)$ then $\phi_{\alpha\beta}^{\nu_j} := (\phi_\alpha^{\nu_j})^{-1} \circ \phi_\beta^{\nu_j}$ converges to $z_{\alpha\beta}$ u.c.s. on $S^2 \setminus \{z_{\beta\alpha}\}$.*

(NODAL POINT) *If $\alpha, \beta \in T$ such that $\alpha E \beta$ and ν_j is a subsequence such that $f^{\nu_j}(\alpha) \neq f^{\nu_j}(\beta)$, then*

$$z_{\alpha\beta} = \lim_{j \to \infty} (\phi_\alpha^{\nu_j})^{-1}(z_{f^{\nu_j}(\alpha) f^{\nu_j}(\beta)}^{\nu_j}).$$

(MARKED POINT) *$\alpha_i^\nu = f^\nu(\alpha_i)$ and*

$$z_i = \lim_{\nu \to \infty} (\phi_{\alpha_i}^\nu)^{-1}(z_i^\nu)$$

for all $i = 1, \ldots, n$.

The *(Rescaling)*, *(Nodal point)*, and *(Marked point)* axioms deal only with the limiting behaviour of the underlying prestable curves $\mathbf{z}^\nu$, and are identical to the corresponding parts of Definition D.5.1 of GK-convergence for stable curves. The *(Map)* and *(Energy)* axioms are the natural extensions of the corresponding axioms in Definition 5.2.1. Note that because there are only finitely many isomorphism types of trees T' onto which T can surject there are (modulo isomorphisms) only finitely many possibilities for T^ν and f^ν. Hence a Gromov convergent sequence is a finite union of subsequences on each of which the T^ν and f^ν can be taken to be constant. The important results about sequences of stable maps can then be reduced to the results already proved about sequences of holomorphic spheres. The key observation is to relate Gromov convergence for stable maps as in Definition 5.5.1 to Gromov convergence for sequences of individual J-holomorphic spheres as in Definition 5.2.1.

Let $(\mathbf{u}, \mathbf{z}) \in \mathcal{SC}_{0,n}(M; J)$ be a stable map modelled over a labelled tree (T, E, Λ) and let $T_0 \subset T$ be a subtree. The **restriction of** $(\mathbf{u}, \mathbf{z})$ **to** T_0 is a stable map $(\mathbf{u}_0, \mathbf{z}_0)$ modelled over the labelled tree (T_0, E_0, Λ_0), where $E_0 := E \cap (T_0 \times T_0)$ and

$$\Lambda_{0\alpha} := \Lambda_\alpha \cup \{\beta \in T \setminus T_0 \,|\, \alpha E \beta\}$$

for $\alpha \in T_0$. It is defined by $u_{0\alpha} := u_\alpha$ and $z_{0\alpha\beta} := z_{\alpha\beta}$ for $\alpha, \beta \in T_0$ such that $\alpha E \beta$. The marked points indexed by the set

$$I_0 := \big\{ i \in I \,|\, \alpha_i \in T_0 \big\} \cup \big\{ \beta \in T \setminus T_0 \,|\, \exists\, \alpha \in T_0 \text{ such that } \alpha E \beta \big\}$$

and are given by $(\alpha_{0i}, z_{0i}) := (\alpha_i, z_i)$ for $i \in I$ with $\alpha_i \in T_0$, and by $(\alpha_{0\beta}, z_{0\beta}) := (\alpha, z_{\alpha\beta})$ for $\alpha \in T_0$ and $\beta \in T \setminus T_0$ such that $\alpha E \beta$. The following proposition will be proved later on in this section.

PROPOSITION 5.5.2. *Let $(\mathbf{u}^\nu, \mathbf{z}^\nu) \in \mathcal{SC}_{0,n}(M; J^\nu)$ be a sequence of stable maps, modelled over the labelled tree (T', E', Λ'), which Gromov converges to the stable map $(\mathbf{u}, \mathbf{z}) \in \mathcal{SC}_{0,n}(M; J)$, modelled over (T, E, Λ), via the surjective tree homomorphism $f : T \to T'$. Then for each $\alpha' \in T'$ the sequence $(u^\nu_{\alpha'}, Y^\nu_{\alpha'})$ of marked J^ν-holomorphic spheres Gromov converges in the sense of Definition 5.2.1 to the restriction of the stable map $(\mathbf{u}, \mathbf{z})$ to the subtree $f^{-1}(\alpha') \subset T$.*

THEOREM 5.5.3 (Uniqueness of the limit). *Let $J^\nu \in \mathcal{J}_\tau(M, \omega)$ be a sequence of ω-tame almost complex structures on M that converges to $J \in \mathcal{J}_\tau(M, \omega)$ in the C^∞-topology. Let $(\mathbf{u}^\nu, \mathbf{z}^\nu) \in \mathcal{SC}_{0,n}(M; J^\nu)$ be a sequence of stable maps that Gromov converges to two stable maps $(\mathbf{u}, \mathbf{z})$ and $(\tilde{\mathbf{u}}, \tilde{\mathbf{z}})$. Then $(\mathbf{u}, \mathbf{z})$ and $(\tilde{\mathbf{u}}, \tilde{\mathbf{z}})$ are equivalent.*

PROOF. By passing to a subsequence, we may assume by the preceding remarks that all the stable maps $(\mathbf{u}^\nu, \mathbf{z}^\nu)$ are modelled over the same tree T', and that all the surjective tree homomorphisms $f^\nu : T \to T'$, respectively $\tilde{f}^\nu : \widetilde{T} \to T'$, in the definition of Gromov convergence agree with a fixed map $f : T \to T'$, respectively $\tilde{f} : \widetilde{T} \to T'$. Under this assumption each vertex $\alpha' \in T'$ determines a sequence of J^ν-holomorphic curves $u^\nu_{\alpha'}$ with marked points $Y^\nu_{\alpha'}$, consisting of the original marked points together with the nodal points $z^\nu_{\alpha'\beta'}$. By Proposition 5.5.2, the sequence $(u^\nu_{\alpha'}, Y^\nu_{\alpha'})$ Gromov converges in the sense of Definition 5.2.1 both to the restriction of $(\mathbf{u}, \mathbf{z})$ to the subtree $f^{-1}(\alpha')$ and to the restriction of $(\tilde{\mathbf{u}}, \tilde{\mathbf{z}})$ to the subtree $\tilde{f}^{-1}(\alpha')$. Hence, by Theorem 5.4.1, the two reduced curves modelled over $f^{-1}(\alpha')$ and $\tilde{f}^{-1}(\alpha')$ are equivalent. It now follows from Definition 5.1.4 that $(\mathbf{u}, \mathbf{z})$ and $(\tilde{\mathbf{u}}, \tilde{\mathbf{z}})$ are equivalent. This proves Theorem 5.5.3. $\square$

THEOREM 5.5.4 (Convergence of the homotopy class). *Let $J^\nu \in \mathcal{J}_\tau(M,\omega)$ be a sequence of ω-tame almost complex structures on M that converges to $J \in \mathcal{J}_\tau(M,\omega)$ in the C^∞-topology. Let $(\mathbf{u}^\nu, \mathbf{z}^\nu) \in \mathcal{SC}_{0,n}(M;J^\nu)$ be a sequence of stable maps that Gromov converges to a stable map $(\mathbf{u}, \mathbf{z}) \in \mathcal{SC}_{0,n}(M;J^\nu)$. Then the following holds.*

(i) *If $x^\nu \in \bigcup_{\alpha\in T^\nu} u_\alpha^\nu(S^2)$ converges to $x \in M$ then $x \in \bigcup_{\alpha\in T} u_\alpha(S^2)$.*

(ii) *For large ν the connected sum $\#_{\alpha\in T^\nu} u_\alpha^\nu$ is homotopic to $\#_{\alpha\in T} u_\alpha$.*

PROOF. As in the proof of Theorem 5.5.3, it suffices to consider the case of a sequence of stable maps modelled over the same tree T' and equipped with the same surjective tree homomorphism from T to T'. For such sequences the result follows from Proposition 5.5.2 and Theorem 5.2.2, applied to the vertices $\alpha' \in T'$ separately. This proves Theorem 5.5.4. □

THEOREM 5.5.5 (Gromov compactness). *Let $J^\nu \in \mathcal{J}_\tau(M,\omega)$ be a sequence of ω-tame almost complex structures on M that converges to $J \in \mathcal{J}_\tau(M,\omega)$ in the C^∞-topology. Let $(\mathbf{u}^\nu, \mathbf{z}^\nu) \in \mathcal{SC}_{0,n}(M;J^\nu)$ be a sequence of stable maps such that $\sup_\nu E(\mathbf{u}^\nu) < \infty$. Then $(\mathbf{u}^\nu, \mathbf{z}^\nu)$ has a Gromov convergent subsequence.*

PROOF. Let $c := \sup_\nu E(\mathbf{u}^\nu)$. Since the constant $\hbar = \hbar(M,\omega;J)$ of Proposition 4.1.4 depends continuously on J, we have $\inf_\nu \hbar(M,\omega;J^\nu) > 0$. Hence, by Exercise 5.1.2, there are only finitely many trees, up to isomorphism, which correspond to stable J^ν-holomorphic curves with energy bounded by c. Passing to a subsequence, we may assume that the trees T^ν are all isomorphic. We choose an isomorphism $T' := T^1 \to T^\nu$ for each ν.

Now, by Theorem 5.3.1, there exists a subsequence, still denoted by $(\mathbf{u}^\nu, \mathbf{z}^\nu)$, such that, for each $\alpha' \in T'$, the sequence $(u_{\alpha'}^\nu, Y_{\alpha'}^\nu)$ of marked J^ν-holomorphic spheres Gromov converges to a marked stable J-holomorphic map. Then connect the limit curves to a tree by introducing an additional edge for each pair $\alpha' E' \beta'$. This proves Theorem 5.5.5. □

The next lemma will be needed in the proof of Proposition 5.5.2 and also the next section where we give a numerical measure of the distance between two stable maps. It shows that the hypothesis $\alpha E \beta$ can be dropped in the *(Rescaling)* and *(Nodal point)* axioms and also gives a natural extension of the *(Marked point)* axiom. For simplicity, we state it only in the case when the tree homomorphisms $f^\nu : T \to T^\nu$ are fixed and equal to $f : T \to T'$. It will sometimes be convenient to use the notation $\alpha' := f(\alpha)$ for $\alpha \in T$. Because f is a homomorphism of labelled trees, it must take α_i to the component α_i' that carries the ith marked point in T'. Hence our notation is consistent and α_i' has an unambiguous meaning. Moreover, we use the notation $z_{\alpha\gamma} := z_{\alpha\beta}$ for $\gamma \in T_{\alpha\beta}$.

LEMMA 5.5.6. *Let $T = (T, E, \Lambda)$ and $T' = (T', E', \Lambda')$ be n-labelled trees and $f : T \to T'$ be a surjective tree homomorphism. Let $\{\phi_\alpha^\nu\}_{\alpha\in T}^{\nu\in\mathbb{N}}$ be a collection of Möbius transformations that satisfy the (Rescaling), (Nodal point), and (Marked point) axioms in Definition 5.5.1. Then the following holds.*

(i) *If $\alpha, \beta \in T$ such that $\alpha \neq \beta$ and $f(\alpha) = f(\beta)$ then the sequence $\phi_{\alpha\beta}^\nu := (\phi_\alpha^\nu)^{-1} \circ \phi_\beta^\nu$ converges to $z_{\alpha\beta}$ u.c.s. on $S^2 \setminus \{z_{\beta\alpha}\}$.*

(ii) *If $\alpha, \beta \in T$ such that $f(\alpha) \neq f(\beta)$ then $z_{\alpha\beta} = \lim_{\nu\to\infty} (\phi_\alpha^\nu)^{-1}(z_{f(\alpha)f(\beta)}^\nu)$.*

(iii) *For each $\alpha \in T$ and each $i \in \{1, \ldots, n\}$, $z_{\alpha i} = \lim_{\nu\to\infty} (\phi_\alpha^\nu)^{-1}(z_{f(\alpha)i}^\nu)$.*

PROOF. Assertion (i) holds by Step 1 of the proof of Theorem 5.4.1.

We prove (ii). Here we use the notation $\alpha' := f(\alpha)$ for $\alpha \in T$. Let $\gamma_0, \ldots, \gamma_m$ be the chain of edges from α to β. Choose $\ell < m$ such that $\alpha' = \gamma_i'$ for $i \leq \ell$ and $\alpha' \neq \gamma_i'$ for $i > \ell$. Then

$$z_{\alpha\beta} = z_{\gamma_0\gamma_1} = z_{\gamma_0\gamma_\ell}, \qquad z^\nu_{\alpha'\beta'} = z^\nu_{\gamma_\ell'\gamma_{\ell+1}'}.$$

By the *(Nodal point)* axiom in Definition 5.5.1, the sequence $(\phi^\nu_{\gamma_\ell})^{-1}(z^\nu_{\gamma_\ell'\gamma_{\ell+1}'})$ converges to $z_{\gamma_\ell\gamma_{\ell+1}} \neq z_{\gamma_\ell\gamma_0}$. Hence it follows from (i) that

$$z_{\alpha\beta} = z_{\gamma_0\gamma_\ell} = \lim_{\nu\to\infty} (\phi^\nu_{\gamma_0})^{-1}(z^\nu_{\gamma_\ell'\gamma_{\ell+1}'}) = \lim_{\nu\to\infty} (\phi^\nu_\alpha)^{-1}(z^\nu_{\alpha'\beta'}).$$

This proves (ii).

To prove (iii) note that, if $\alpha = \alpha_i$, then $z_{\alpha i} = z_i$ and $z^\nu_{\alpha' i} = z^\nu_i$. Hence in this case the assertion is equivalent to the *(Marked point)* axiom in Definition 5.5.1. If $\alpha \neq \alpha_i$ but $\alpha' = \alpha_i'$ then it follows from (i) that

$$z_{\alpha i} = z_{\alpha\alpha_i} = \lim_{\nu\to\infty} (\phi^\nu_\alpha)^{-1} \circ \phi^\nu_{\alpha_i}(B_\delta(z_i)) = \lim_{\nu\to\infty} (\phi^\nu_\alpha)^{-1}(z^\nu_i).$$

The last equality follows again from the *(Marked point)* axiom in Definition 5.5.1. Finally, if $\alpha \neq \alpha_i$ and $\alpha' \neq \alpha_i'$ then $z_{\alpha i} = z_{\alpha\alpha_i}$ and $z^\nu_{\alpha' i} = z^\nu_{\alpha'\alpha_i'}$ Hence it follows from (ii) that $(\phi^\nu_\alpha)^{-1}(z^\nu_{\alpha' i})$ converges to $z_{\alpha i}$. This proves Lemma 5.5.6. □

PROOF OF PROPOSITION 5.5.2. Let $T_0 := f^{-1}(\alpha')$ and denote by $(\mathbf{u}_0, \mathbf{z}_0)$ the restriction of $(\mathbf{u}, \mathbf{z})$ to T_0. We prove that the sequences $(u^\nu_{\alpha'}, Y^\nu_{\alpha'})$ and $\{\phi^\nu_\alpha\}^{\nu\in\mathbb{N}}_{\alpha\in T_0}$ satisfy the requirements of Definition 5.2.1. The *(Rescaling)* and *(Marked point)* axioms follow directly from the *(Rescaling)*, *(Nodal point)*, and *(Marked point)* axioms for the sequence $(\mathbf{u}^\nu, \mathbf{z}^\nu)$ in Definition 5.5.1.

To establish the *(Map)* and *(Energy)* axioms we observe that, by Lemma 5.5.6, we have

$$\gamma \in T_{\alpha\beta} \qquad \Longleftrightarrow \qquad z_{\alpha\beta} = \lim_{\nu\to\infty} (\phi^\nu_\alpha)^{-1}(z^\nu_{\alpha'\gamma'}). \tag{5.5.2}$$

for $\alpha, \beta, \gamma \in T$ such that $\alpha E \beta$ and $\alpha' = f(\alpha) \neq \gamma' := f(\gamma)$. Now consider the set

$$Z_{0\alpha} := \{z_{\alpha\beta} \,|\, \beta \in T_0,\, \alpha E \beta\}.$$

This is a proper subset of Z_α for some $\alpha \in T_0$, unless $T = T_0$. We must prove that $u^\nu \circ \phi^\nu_\alpha$ converges to u_α u.c.s. on $S^2 \setminus Z_{0\alpha}$ for every $\alpha \in T_0$.

Let $\alpha \in T_0$ and $\beta \in T \setminus T_0$ such that $\alpha E \beta$. Then it follows from (5.5.2) and the *(Energy)* axiom in Definition 5.5.1 that

$$m_{\alpha\beta}(\mathbf{u}) = \lim_{\varepsilon\to 0} \lim_{\nu\to\infty} \Big(E\big(u^\nu_{\alpha'} \circ \phi^\nu_\alpha; B_\varepsilon(z_{\alpha\beta})\big) + m_{\alpha'\beta'}(\mathbf{u}^\nu) \Big)$$

and

$$m_{\beta\alpha}(\mathbf{u}) = \lim_{\varepsilon\to 0} \lim_{\nu\to\infty} \Big(E\big(u^\nu_{\beta'} \circ \phi^\nu_\beta; B_\varepsilon(z_{\beta\alpha})\big) + m_{\beta'\alpha'}(\mathbf{u}^\nu) \Big)$$

where $\beta' := f(\beta)$. Moreover, it follows from the *(Energy)* and *(Map)* axioms in Definition 5.5.1 that

$$\begin{aligned} E(\mathbf{u}) &= E(u_\alpha) + \sum_{\gamma \in T, \alpha E \gamma} m_{\alpha\gamma}(\mathbf{u}) \\ &= \lim_{\varepsilon \to 0} \lim_{\nu \to \infty} \Big(E\big(u^\nu_{\alpha'} \circ \phi^\nu_\alpha; S^2 \setminus \bigcup_{\substack{\gamma \in T \\ \alpha E \gamma}} B_\varepsilon(z_{\alpha\gamma})\big) + \sum_{\substack{\gamma \in T \\ \alpha E \gamma}} E_{\alpha'}\big(\mathbf{u}^\nu; \phi^\nu_\alpha(B_\varepsilon(z_{\alpha\gamma}))\big) \Big) \\ &= \lim_{\nu \to \infty} \Big(E(u^\nu_{\alpha'} \circ \phi^\nu_\alpha) + \sum_{\gamma' \in T', \alpha' E' \gamma'} m_{\alpha'\gamma'}(\mathbf{u}^\nu) \Big) \;=\; \lim_{\nu \to \infty} E(\mathbf{u}^\nu). \end{aligned}$$

Since

$$E(\mathbf{u}) = m_{\alpha\beta}(\mathbf{u}) + m_{\beta\alpha}(\mathbf{u}), \qquad E(\mathbf{u}^\nu) = m_{\alpha'\beta'}(\mathbf{u}^\nu) + m_{\beta'\alpha'}(\mathbf{u}^\nu),$$

it follows that

$$m_{\alpha\beta}(\mathbf{u}) = \lim_{\nu \to \infty} m_{\alpha'\beta'}(\mathbf{u}^\nu), \qquad \lim_{\varepsilon \to 0} \lim_{\nu \to \infty} E\big(u^\nu_{\alpha'} \circ \phi^\nu_\alpha; B_\varepsilon(z_{\alpha\beta})\big) = 0. \tag{5.5.3}$$

This shows that the sequence $u^\nu_{\alpha'} \circ \phi^\nu_\alpha$ exhibits no bubbling near the point $z_{\alpha\beta}$ for every $\beta \in T \setminus T_0$ such that $\alpha E \beta$. Hence $u^\nu_{\alpha'} \circ \phi^\nu_\alpha$ converges to u_α u.c.s. on $S^2 \setminus Z_{0\alpha}$.

To prove the *(Energy)* axiom in Definition 5.2.1 let $\alpha, \beta \in T_0$ such that $\alpha E \beta$. Then, by (5.5.2) and (5.5.3),

$$\begin{aligned} m_{\alpha\beta}(\mathbf{u}_0) &= m_{\alpha\beta}(\mathbf{u}) - \sum_{\substack{\gamma_0, \gamma_1 \in T_{\alpha\beta} \\ \gamma_0 \in T_0,\, \gamma_1 \notin T_0,\, \gamma_0 E \gamma_1}} m_{\gamma_0\gamma_1}(\mathbf{u}) \\ &= \lim_{\varepsilon \to 0} \lim_{\nu \to \infty} \Big(E_{\alpha'}\big(\mathbf{u}^\nu; \phi^\nu_\alpha(B_\varepsilon(z_{\alpha\beta}))\big) - \sum_{\gamma \in T_{\alpha\beta}, \alpha' E' \gamma'} m_{\alpha'\gamma'}(\mathbf{u}^\nu) \Big) \\ &= \lim_{\varepsilon \to 0} \lim_{\nu \to \infty} E\big(u^\nu_{\alpha'} \circ \phi^\nu_\alpha; B_\varepsilon(z_{\alpha\beta})\big). \end{aligned}$$

This proves Proposition 5.5.2. □

Measuring the distance between stable maps. We now define the "distance" $\rho_\varepsilon(\mathbf{x}, \mathbf{x}')$ from one stable map $\mathbf{x}$ to another $\mathbf{x}'$. Despite its name, this is *not* a symmetric function of $\mathbf{x}, \mathbf{x}'$ and it does not satisfy the triangle inequality. Let

$$\mathbf{x} = (\mathbf{u}, \mathbf{z}) = \big(\{u_\alpha\}_{\alpha \in T}, \{z_{\alpha\beta}\}_{\alpha E \beta}, \{\alpha_i, z_i\}_{1 \le i \le n}\big) \in \mathcal{SC}_{0,n}(M, A; J)$$

be a stable J-holomorphic curve of genus zero with n marked points and fix a sufficiently small constant $\varepsilon > 0$. For any other stable map

$$\mathbf{x}' = (\mathbf{u}', \mathbf{z}') = \big(\{u'_{\alpha'}\}_{\alpha' \in T'}, \{z'_{\alpha'\beta'}\}_{\alpha' E' \beta'}, \{\alpha'_i, z'_i\}_{1 \le i \le n}\big) \in \mathcal{SC}_{0,n}(M, A; J)$$

define the real number

$$\rho_\varepsilon(\mathbf{x}, \mathbf{x}') := \inf_{f: T \to T'} \inf_{\{\phi_\alpha\}} \rho_\varepsilon\big(\mathbf{x}, \mathbf{x}'; f, \{\phi_\alpha\}\big),$$

where

$$\begin{aligned}\rho_\varepsilon(\mathbf{x}, \mathbf{x}'; f, \{\phi_\alpha\}) \quad := \quad & \sup_{\alpha E\beta} \left| E_\alpha(\mathbf{u}; B_\varepsilon(z_{\alpha\beta})) - E_{f(\alpha)}(\mathbf{u}'; \phi_\alpha(B_\varepsilon(z_{\alpha\beta}))) \right| \\ & + \sup_{\alpha\in T} \sup_{S^2\setminus B_\varepsilon(Z_\alpha)} d(u'_{f(\alpha)} \circ \phi_\alpha, u_\alpha) \\ & + \sup_{\substack{\alpha\neq\beta \\ f(\alpha)=f(\beta)}} \sup_{S^2\setminus B_\varepsilon(z_{\alpha\beta})} d(\phi_\beta^{-1} \circ \phi_\alpha, z_{\beta\alpha}) \\ & + \sup_{f(\alpha)\neq f(\beta)} d(\phi_\beta^{-1}(z'_{f(\beta)f(\alpha)}), z_{\beta\alpha}) \\ & + \sup_{\substack{\alpha\in T \\ 1\le i\le n}} d(\phi_\alpha^{-1}(z'_{f(\alpha)i}), z_{\alpha i}).\end{aligned}$$

In the definition of ρ_ε the infimum is taken over all tuples $\{\phi_\alpha\}_{\alpha\in T} \in \mathrm{G}^T$ and all surjective tree homomorphisms $f : T \to T'$ which satisfy $f(\alpha_i) = \alpha'_i$ for all i. If no such surjective tree homomorphism exists we set $\rho_\varepsilon(\mathbf{x}, \mathbf{x}') := \infty$. The above formula looks quite formidable. However, each term in the sum is a natural measure of some way in which the two stable maps could differ.

REMARK 5.5.7. By definition, the functions $\mathbf{x}' \mapsto \rho_\varepsilon(\mathbf{x}, \mathbf{x}')$ descend to the moduli space $\overline{\mathcal{M}}_{0,n}(A; J)$ of equivalence classes of stable maps. Thus,

$$\mathbf{x}' \equiv \mathbf{y}' \qquad \Longrightarrow \qquad \rho_\varepsilon(\mathbf{x}, \mathbf{x}') = \rho_\varepsilon(\mathbf{x}, \mathbf{y}')$$

and, moreover,

$$\mathbf{x} \equiv \mathbf{x}' \qquad \Longrightarrow \qquad \rho_\varepsilon(\mathbf{x}, \mathbf{x}') = 0$$

for $\mathbf{x}, \mathbf{x}', \mathbf{y}' \in \mathcal{SC}_{0,n}(M, A; J)$. Here $\equiv$ denotes the equivalence relation on the category $\mathcal{SC}_{0,n}(M, A; J)$.

The functions $\rho_\varepsilon(\mathbf{x}, \mathbf{x}')$ characterize Gromov convergence in the following sense.

LEMMA 5.5.8. *Fix a stable J-holomorphic curve $\mathbf{x} = (\mathbf{u}, \mathbf{z})$. Then there exists a constant $\varepsilon_0 > 0$ such that the following holds for $0 < \varepsilon < \varepsilon_0$. A sequence $\mathbf{x}^\nu = (\mathbf{u}^\nu, \mathbf{z}^\nu)$ Gromov converges to $\mathbf{x}$ if and only if the sequence of real numbers $\rho_\varepsilon(\mathbf{x}, \mathbf{x}^\nu)$ converges to zero.*

PROOF. Choose $\varepsilon_0 > 0$ so small that $E(u_\alpha; B_{\varepsilon_0}(Z_\alpha)) < \hbar/2$ for all $\alpha \in T$ and

$$\alpha E\beta,\ \alpha E\gamma,\ \beta \neq \gamma \quad \Longrightarrow \quad B_{\varepsilon_0}(z_{\alpha\beta}) \cap B_{\varepsilon_0}(z_{\alpha\gamma}) = \emptyset.$$

Then

$$E_\alpha(\mathbf{u}; B_{\varepsilon_0}(z_{\alpha\beta})) = E(u_\alpha; B_{\varepsilon_0}(z_{\alpha\beta})) + m_{\alpha\beta}(\mathbf{u})$$

for all $\alpha, \beta \in T$ such that $\alpha E\beta$. We claim that the assertion of the lemma is satisfied for this choice of ε_0.

If $\mathbf{x}^\nu$ Gromov converges to $\mathbf{x}$ then it follows by the definitions and Lemma 5.5.6 that $\rho_\varepsilon(\mathbf{x}, \mathbf{x}^\nu)$ converges to zero. Conversely, suppose that

$$\lim_{\nu\to\infty} \rho_\varepsilon(\mathbf{x}, \mathbf{x}^\nu) = 0.$$

Then, for $\nu \in \mathbb{N}$ sufficiently large, there exists a surjective tree homomorphism $f^\nu : T \to T^\nu$ and a tuple $\{\phi_\alpha^\nu\}_{\alpha\in T} \in \mathrm{G}^T$ such that $f^\nu(\alpha_i) = \alpha_i^\nu$ and

$$\rho^\nu := \rho_\varepsilon(\mathbf{x}, \mathbf{x}^\nu; f^\nu, \{\phi_\alpha^\nu\}) \le \rho_\varepsilon(\mathbf{x}, \mathbf{x}^\nu) + 2^{-\nu}.$$

We prove that this sequence satisfies all five conditions of Definition 5.5.1. Since there are only finitely many possible tree surjections with domain T it suffices to consider the case when $T^\nu = T^1 =: T'$ and $f^\nu = f^1 =: f : T \to T'$ for all ν. We shall use the notation $\alpha' := f(\alpha)$ for $\alpha \in T$ whenever convenient.

The *(Marked point)* and *(Nodal point)* axioms are obviously satisfied. Moreover, if $\alpha E \beta$ and $\alpha' = \beta'$ then the condition $\rho^\nu \to 0$ implies that $(\phi_\alpha^\nu)^{-1} \circ \phi_\beta^\nu$ converges to $z_{\alpha\beta}$ on $S^2 \setminus B_\varepsilon(z_{\beta\alpha})$, and $(\phi_\beta^\nu)^{-1} \circ \phi_\alpha^\nu$ converges to $z_{\beta\alpha}$ on $S^2 \setminus B_\varepsilon(z_{\alpha\beta})$. By Lemma D.1.4, this implies that $(\phi_\alpha^\nu)^{-1} \circ \phi_\beta^\nu$ converges to $z_{\alpha\beta}$ u.c.s on $S^2 \setminus \{z_{\beta\alpha}\}$. This proves the *(Rescaling)* axiom.

To prove the *(Map)* and *(Energy)* axioms, note that, since ρ^ν converges to zero, the sequence $u_\alpha^\nu := u_{\alpha'}^\nu \circ \phi_\alpha^\nu$ converges to u_α uniformly on $S^2 \setminus \bigcup_{\alpha E \beta} B_\varepsilon(z_{\alpha\beta})$ and that

$$m_{\alpha\beta}(\mathbf{u}) + E\big(u_\alpha; B_\varepsilon(z_{\alpha\beta})\big) = \lim_{\nu\to\infty} E_{\alpha'}\big(\mathbf{u}^\nu, \phi_\alpha^\nu(B_\varepsilon(z_{\alpha\beta}))\big) \tag{5.5.4}$$

whenever $\alpha E \beta$. Hence, by Lemma 4.6.6, u_α^ν converges uniformly *with all derivatives* on compact subsets of $S^2 \setminus \bigcup_{\alpha E \beta} B_\varepsilon(z_{\alpha\beta})$.

It remains to show that the sequence u_α^ν has no bubbles in $B_\varepsilon(z_{\alpha\beta}) \setminus B_\delta(z_{\alpha\beta})$ for any $\delta > 0$. To see this we observe that, by Lemma 5.5.6 (ii),

$$z_{\alpha\gamma} = \lim_{\nu\to\infty} (\phi_\alpha^\nu)^{-1}(z_{\alpha'\gamma'}^\nu)$$

for every $\gamma \in T$ such that $\alpha' \neq \gamma'$. This implies that, for every $\delta > 0$, there exists a constant $\nu_0 = \nu_0(\delta)$ such that

$$E_{\alpha'}\Big(\mathbf{u}^\nu; \phi_\alpha^\nu\big(B_\varepsilon(z_{\alpha\beta}) \setminus B_\delta(z_{\alpha\beta})\big)\Big) = E\Big(u_\alpha^\nu; B_\varepsilon(z_{\alpha\beta}) \setminus B_\delta(z_{\alpha\beta})\Big)$$

for $\nu \geq \nu_0$. Hence it follows from (5.5.4) that

$$E\Big(u_\alpha; B_\varepsilon(z_{\alpha\beta}) \setminus B_\delta(z_{\alpha\beta})\Big) = \lim_{\nu\to\infty} E\Big(u_\alpha^\nu; B_\varepsilon(z_{\alpha\beta}) \setminus B_\delta(z_{\alpha\beta})\Big)$$

for $0 < \delta < \varepsilon$. By choice of ε, this limit is bounded by $\hbar/2$. Hence no bubbling can occur in the domain $B_\varepsilon(z_{\alpha\beta}) \setminus B_\delta(z_{\alpha\beta})$. This implies that every subsequence of u_α^ν has a further subsequence which converges u.c.s. on $B_\varepsilon(z_{\alpha\beta}) \setminus \{z_{\alpha\beta}\}$ to some J-holomorphic curve. By the unique continuation theorem, the limit curve is always equal to u_α, and hence the sequence u_α^ν itself converges to u_α u.c.s. on $B_\varepsilon(z_{\alpha\beta}) \setminus \{z_{\alpha\beta}\}$. This proves the *(Map)* axiom. The *(Energy)* axiom then follows from (5.5.4). Thus we have proved that the sequence $\mathbf{x}^\nu = (\mathbf{u}^\nu, \mathbf{z}^\nu)$ Gromov converges to $\mathbf{x} = (\mathbf{u}, \mathbf{z})$. This proves Lemma 5.5.8. □

The functions ρ_ε satisfy the following substitute for the triangle inequality.

LEMMA 5.5.9. *Let $\mathbf{x} \in \mathcal{SC}_{0,n}(M, A; J)$ and $0 < \varepsilon < \varepsilon_0$, where ε_0 is chosen as in Lemma 5.5.8. If $\mathbf{x}' \in \mathcal{SC}_{0,n}(M, A; J)$ satisfies $\rho_\varepsilon(\mathbf{x}, \mathbf{x}') < \varepsilon$ and the sequence $\mathbf{x}^\nu \in \mathcal{SC}_{0,n}(M, A; J)$ Gromov converges to $\mathbf{x}'$, then*

$$\limsup_{\nu\to\infty} \rho_\varepsilon(\mathbf{x}, \mathbf{x}^\nu) \leq \rho_\varepsilon(\mathbf{x}, \mathbf{x}').$$

PROOF. By assumption, there is a surjective tree homomorphism $g : T \to T'$ and a collection of Möbius transformations $\{\psi_\alpha\}_{\alpha\in T}$ such that $g(\alpha_i) = \alpha_i'$ and

$$\rho_\varepsilon(\mathbf{x}, \mathbf{x}'; g, \{\psi_\alpha\}) < \varepsilon. \tag{5.5.5}$$

Whenever convenient we shall use the notation $\alpha' := g(\alpha)$ for $\alpha \in T$. It follows from (5.5.5) and the definition of ρ_ε that, for all $\alpha, \beta \in T$ with $\alpha' \neq \beta'$ and all $i \in \{1, \dots, n\}$, we have

$$d\big(\psi_\alpha^{-1}(z_{\alpha'\beta'}'), z_{\alpha\beta}\big) < \varepsilon, \qquad d\big(\psi_\alpha^{-1}(z_{\alpha' i}'), z_{\alpha i}\big) < \varepsilon.$$

Hence

$$Z_{\alpha'}(\mathbf{x}') \subset \psi_\alpha\big(B_\varepsilon(Z_\alpha(\mathbf{x}))\big) \tag{5.5.6}$$

for every $\alpha \in T$. Now suppose that $\mathbf{x}^\nu$ Gromov converges to $\mathbf{x}'$ with corresponding surjective tree homomorphisms $f^\nu : T' \to T^\nu$ and Möbius transformations $\phi^\nu_{\alpha'} \in \mathrm{G}$. Then we claim that

$$\rho_\varepsilon\big(\mathbf{x}, \mathbf{x}'; g, \{\psi_\alpha\}\big) = \lim_{\nu\to\infty} \rho_\varepsilon\big(\mathbf{x}, \mathbf{x}^\nu; f^\nu \circ g, \{\phi^\nu_{g(\alpha)} \circ \psi_\alpha\}\big), \tag{5.5.7}$$

where g and $\{\psi_\alpha\}$ satisfy (5.5.5) as above. To prove this it suffices as usual to consider the case where $T^\nu = T^1 =: T''$ and $f^\nu = f^1 =: f : T' \to T''$ for all ν. For $\alpha \in T$ we shall abbreviate

$$\alpha'' := f \circ g(\alpha), \qquad \chi^\nu_\alpha := \phi^\nu_{\alpha'} \circ \psi_\alpha \in \mathrm{G}, \qquad u^\nu_\alpha := u^\nu_{\alpha''} \circ \chi^\nu_\alpha, \qquad Z_\alpha := Z_\alpha(\mathbf{x}).$$

By (5.5.6), the sequence $u^\nu_{\alpha''} \circ \phi^\nu_{\alpha'}$ converges to $u'_{\alpha'}$ uniformly on $S^2 \setminus \psi_\alpha(B_\varepsilon(Z_\alpha))$. Hence u^ν_α converges to $u'_{\alpha'} \circ \psi_\alpha$ uniformly on $S^2 \setminus B_\varepsilon(Z_\alpha)$, so that

$$\sup_{S^2\setminus B_\varepsilon(Z_\alpha)} d(u'_{\alpha'} \circ \psi_\alpha, u_\alpha) = \lim_{\nu\to\infty} \sup_{S^2\setminus B_\varepsilon(Z_\alpha)} d(u^\nu_\alpha, u_\alpha).$$

If $\alpha, \beta \in T$ such that $\alpha \neq \beta$ and $\alpha' = \beta'$, then $\psi_\beta^{-1} \circ \psi_\alpha = (\chi^\nu_\beta)^{-1} \circ \chi^\nu_\alpha$ and so

$$\sup_{S^2\setminus B_\varepsilon(z_{\alpha\beta})} d\big(\psi_\beta^{-1} \circ \psi_\alpha, z_{\beta\alpha}\big) = \sup_{S^2\setminus B_\varepsilon(z_{\alpha\beta})} d\left((\chi^\nu_\beta)^{-1} \circ \chi^\nu_\alpha, z_{\beta\alpha}\right).$$

If $\alpha, \beta \in T$ such that $\alpha' \neq \beta'$ and $\alpha'' = \beta''$ then, by Lemma 5.5.6 (i) and (5.5.6), $(\phi^\nu_{\beta'})^{-1} \circ \phi^\nu_{\alpha'}$ converges to $z'_{\beta'\alpha'}$ uniformly on $S^2 \setminus \psi_\alpha(B_\varepsilon(z_{\alpha\beta}))$. Hence $(\chi^\nu_\beta)^{-1} \circ \chi^\nu_\alpha$ converges to $\psi_\beta^{-1}(z'_{\beta'\alpha'})$ uniformly on $S^2 \setminus B_\varepsilon(z_{\alpha\beta})$, and hence

$$d\big(\psi_\beta^{-1}(z'_{\beta'\alpha'}), z_{\beta\alpha}\big) = \lim_{\nu\to\infty} \sup_{S^2\setminus B_\varepsilon(z_{\alpha\beta})} d\left((\chi^\nu_\beta)^{-1} \circ \chi^\nu_\alpha, z_{\beta\alpha}\right).$$

If $\alpha, \beta \in T$ such that $\alpha'' \neq \beta''$ then $(\phi^\nu_{\beta'})^{-1}(z^\nu_{\beta''\alpha''})$ converges to $z'_{\beta'\alpha'}$ and hence

$$d\big(\psi_\beta^{-1}(z'_{\beta'\alpha'}), z_{\beta\alpha}\big) = \lim_{\nu\to\infty} d\left((\chi^\nu_\beta)^{-1}(z^\nu_{\beta''\alpha''}), z_{\beta\alpha}\right).$$

By Lemma 5.5.6 (iii), the sequence $(\phi^\nu_{\alpha'})^{-1}(z^\nu_{\alpha''i})$ converges to $z'_{\alpha'i}$ for every $\alpha' \in T'$ and every $i \in \{1, \dots, n\}$. Hence

$$d\big(\psi_\alpha^{-1}(z'_{\alpha'i}), z_{\alpha i}\big) = \lim_{\nu\to\infty} d\left((\chi^\nu_\alpha)^{-1}(z^\nu_{\alpha''i}), z_{\alpha i}\right)$$

for every $\alpha \in T$ and every $i \in \{1, \dots, n\}$. Finally, if $\alpha, \beta \in T$ such that $\alpha E \beta$ then, by Definition 5.5.1,

$$E_{\alpha'}\big(\mathbf{u}', \psi_\alpha(B_\varepsilon(z_{\alpha\beta}))\big) = \lim_{\nu\to\infty} E_{\alpha''}\big(\mathbf{u}^\nu, \chi^\nu_\alpha(B_\varepsilon(z_{\alpha\beta}))\big).$$

These equations put together imply (5.5.7). Thus we have proved Lemma 5.5.9. □

5.6. The Gromov topology

We wish to use Definition 5.5.1 to define a topology on $\overline{\mathcal{M}}_{0,n}(M,A;J)$, called the **Gromov topology**. A sequence of equivalence classes $[\mathbf{u}^\nu,\mathbf{z}^\nu]$ should converge to $[\mathbf{u},\mathbf{z}]$ in this topology if $(\mathbf{u}^\nu,\mathbf{z}^\nu)$ Gromov converges to $(\mathbf{u},\mathbf{z})$. A subset $F\subset\overline{\mathcal{M}}_{0,n}(M,A;J)$ is called **Gromov closed** if the limit of every Gromov convergent sequence in F lies again in F. A subset $U\subset\overline{\mathcal{M}}_{0,n}(M,A;J)$ is called **Gromov open** if its complement is Gromov closed. That this defines a topology on $\overline{\mathcal{M}}_{0,n}(M,A;J)$ is obvious. However, the proof that convergence with respect to this topology is equivalent to Gromov convergence requires some preparation.

Topology and sequences. Let $(X,\mathcal{U})$ be a Hausdorff topological space which is first countable (i.e. every point in X has a countable neighbourhood basis). Under these hypotheses limits are unique and, moreover, the topology $\mathcal{U}$ is fully described by the set of convergent sequences. This means that the closure $\mathrm{cl}(A)$ of a subset $A\subset X$ is precisely the set of limit points of convergent sequences in A. To elaborate this point let us denote by

$$\mathcal{C}=\mathcal{C}(\mathcal{U})\subset X\times X^{\mathbb{N}}$$

the set of all pairs $(x_0,(x_n)_n)$ of elements $x_0\in X$ and sequences $x_n\in X$ such that x_n converges to x_0. The collection $\mathcal{C}$ of convergent sequences has the following properties.

(Constant) If $x_n=x_0$ for all $n\in\mathbb{N}$ then $(x_0,(x_n)_n)\in\mathcal{C}$.

(Subsequence) If $(x_0,(x_n)_n)\in\mathcal{C}$ and $g:\mathbb{N}\to\mathbb{N}$ is strictly increasing, then $(x_0,(x_{g(n)})_n)\in\mathcal{C}$.

(Subsubsequence) If, for every strictly increasing function $g:\mathbb{N}\to\mathbb{N}$, there is a strictly increasing function $f:\mathbb{N}\to\mathbb{N}$ such that $(x_0,(x_{g\circ f(n)})_n)\in\mathcal{C}$, then $(x_0,(x_n)_n)\in\mathcal{C}$.

(Diagonal) If $(x_0,(x_k)_k)\in\mathcal{C}$ and $(x_k,(x_{k,n})_n)\in\mathcal{C}$ for every k then there exist sequences $k_i,n_i\in\mathbb{N}$ such that $(x_0,(x_{k_i,n_i})_i)\in\mathcal{C}$.

(Uniqueness) If $(x_0,(x_n)_n)\in\mathcal{C}$ and $(y_0,(x_n)_n)\in\mathcal{C}$ then $x_0=y_0$.

Now let us turn the problem around and try to define a topology on X starting from an arbitrary collection of sequences $\mathcal{C}\subset X\times X^{\mathbb{N}}$. Given $\mathcal{C}$ we define

$$\mathcal{U}(\mathcal{C})\subset 2^X$$

to be the set of all subsets $U\subset X$ that satisfy

$$(x_0,(x_n)_n)\in\mathcal{C}\cap(U\times X^{\mathbb{N}})\qquad\Longrightarrow\qquad\exists N\in\mathbb{N}\ \forall n\in\mathbb{N}\ \ (n\ge N\Longrightarrow x_n\in U).$$

Exercise 5.6.1 below shows that $\mathcal{U}(\mathcal{C})$ is always a topology. However, we would like $\mathcal{U}(\mathcal{C})$ to have the property that the set of sequences $\mathcal{C}(\mathcal{U}(\mathcal{C}))$ that converge with respect to $\mathcal{U}(\mathcal{C})$ is precisely the original collection $\mathcal{C}$. We shall see that the *(Subsequence)* axiom guarantees that, for every subset $F\subset X$,

$$X\setminus F\in\mathcal{U}\qquad\Longleftrightarrow\qquad\Big((x_0,(x_n)_n)\in\mathcal{C}\cap(X\times F^{\mathbb{N}})\Longrightarrow x_0\in F\Big).$$

The *(Constant)* and *(Diagonal)* axioms are needed to show that the closure of a subset $A\subset X$ consists of all elements $x_0\in X$ for which there exists a sequence $x_n\in A$ such that $(x_0,(x_n)_n)\in\mathcal{C}$. All five axioms are needed to prove that $\mathcal{C}(\mathcal{U}(\mathcal{C}))=\mathcal{C}$.

EXERCISE 5.6.1. Let $\mathcal{C}$ be any subset of $X \times X^{\mathbb{N}}$. Prove that $\mathcal{U}(\mathcal{C})$ is a topology. Prove that $\mathcal{C} \subset \mathcal{C}(\mathcal{U}(\mathcal{C}))$ and $\mathcal{U}(\mathcal{C}) = \mathcal{U}(\mathcal{C}(\mathcal{U}(\mathcal{C})))$.

EXERCISE 5.6.2. Suppose $\mathcal{C}$ satisfies the *(Subsequence)* axiom and let $\mathcal{F} \subset 2^X$ be the collection of all subsets $F \subset X$ which satisfy

$$(x_0, (x_n)_n) \in \mathcal{C} \cap (X \times F^{\mathbb{N}}) \qquad \Longrightarrow \qquad x_0 \in F$$

Prove that

$$F \in \mathcal{F} \qquad \Longleftrightarrow \qquad X \setminus F \in \mathcal{U}(\mathcal{C}).$$

Hint: The *(Subsequence)* axiom is only needed for $F \in \mathcal{F} \Longrightarrow X \setminus F \in \mathcal{U}$.

LEMMA 5.6.3. *Suppose that $\mathcal{C}$ satisfies the (Constant), (Subsequence), and (Diagonal) axioms and let $A \subset X$ be any subset. Then the closure of A with respect to $\mathcal{U}(\mathcal{C})$ is given by*

$$\mathrm{cl}(A) = \left\{x_0 \in X \,|\, \exists (x_n)_n \in A^{\mathbb{N}} \text{ such that } (x_0, (x_n)_n) \in \mathcal{C}\right\}.$$

PROOF. Let F be given by the right hand side of the equation. The *(Constant)* axiom shows that $A \subset F$ and it follows from the definitions that every closed set containing A also contains F. By Exercise 5.6.2 and the *(Diagonal)* axiom, F is closed. This proves Lemma 5.6.3. □

LEMMA 5.6.4. *Suppose that $\mathcal{C}$ is a collection of sequences in X satisfying the (Constant), (Subsequence), (Subsubsequence), (Diagonal), and (Uniqueness) axioms. Then X has a unique topology $\mathcal{U}$ with $\mathcal{C}$ as its set of convergent sequences.*

PROOF. The preceding discussion shows that it suffices to check that $\mathcal{C} = \mathcal{C}(\mathcal{U}(\mathcal{C}))$. By Exercise 5.6.1, $\mathcal{C} \subset \mathcal{C}(\mathcal{U}(\mathcal{C}))$, so we must prove that $\mathcal{C}(\mathcal{U}(\mathcal{C})) \subset \mathcal{C}$. We argue indirectly and assume that $(x_0, (x_n)_n) \notin \mathcal{C}$. By the *(Subsubsequence)* axiom there exists a strictly increasing function $g : \mathbb{N} \to \mathbb{N}$ such that, for every strictly increasing function $f : \mathbb{N} \to \mathbb{N}$,

$$(x_0, (x_{g \circ f(n)})_n) \notin \mathcal{C}. \tag{5.6.1}$$

By the *(Constant)* axiom, this implies that there exists an $N \in \mathbb{N}$ such that, for all $n \in \mathbb{N}$,

$$n \geq N \qquad \Longrightarrow \qquad x_{g(n)} \neq x_0. \tag{5.6.2}$$

Define

$$F := \mathrm{cl}(\{x_{g(n)} \,|\, n \geq N\}).$$

We claim that $x_0 \notin F$. Suppose otherwise that $x_0 \in F$. Then, by Lemma 5.6.3, there exists a sequence of integers $n_i \geq N$ such that $(x_0, (x_{g(n_i)})_i) \in \mathcal{C}$. If the sequence n_i were bounded then, by the *(Subsequence)* axiom, we could assume that $n_i = n \geq N$ for all i. By the *(Constant)* axiom, this would imply that $(x_{g(n)}, (x_{g(n_i)})_i) \in \mathcal{C}$ and hence, by the *(Uniqueness)* axiom, $x_0 = x_{g(n)}$, contradicting (5.6.2). Thus we have proved that the sequence n_i is unbounded. By the *(Subsequence)* axiom, we may therefore assume that $n_{i+1} > n_i$ for all i. Thus we have found a strictly increasing function $f : \mathbb{N} \to \mathbb{N}$ with $(x_0, (x_{g \circ f(i)})_i) \in \mathcal{C}$, in contradiction to (5.6.1). This proves that $x_0 \notin F$, as claimed. Thus we have found an open neighbourhood $U := X \setminus F$ of x_0 and a strictly increasing function $g : \mathbb{N} \to \mathbb{N}$ such that $x_{g(n)} \notin U$ for all $n \geq N$. Hence x_n does not converge to x_0 with respect to the topology $\mathcal{U}(\mathcal{C})$. This proves Lemma 5.6.4. □

To gain a better understanding of the topology $\mathcal{U}(\mathcal{C})$, for example to figure out when it is Hausdorff, we need to know that it is first countable, i.e. that every point in X has a countable neighbourhood basis. This requires further assumptions such as those in Lemma 5.6.5 below.

The Gromov topology. Consider the space $X = \overline{\mathcal{M}}_{0,n}(A; J)$ of equivalence classes of stable maps. The definition of Gromov convergence for a sequence of stable maps is phrased in such a way that it descends to $\overline{\mathcal{M}}_{0,n}(A; J)$. In other words, a given sequence $\mathbf{x}^\nu$ Gromov converges to $\mathbf{x}$ if and only if all equivalent sequences Gromov converge to $\mathbf{y}$ where $\mathbf{y}$ is any stable map that is equivalent to $\mathbf{x}$. Therefore it makes sense to talk about Gromov convergent sequences $[\mathbf{x}^\nu]$ in $\overline{\mathcal{M}}_{0,n}(A; J)$. Moreover, by Remark 5.5.7 and Lemmas 5.5.8 and 5.5.9, this convergence may be described in terms of a family of functions $\rho_\varepsilon(\mathbf{x}, \mathbf{x}')$ satisfying the properties assumed in the next lemma. Therefore, we may apply the results of the next lemma to the case when $X = \overline{\mathcal{M}}_{0,n}(A; J)$ and $\mathcal{C}$ is the set of all Gromov convergent sequences $([\mathbf{x}], [\mathbf{x}^\nu]_\nu)$.

LEMMA 5.6.5. *Let X be a set and $\mathcal{C} \subset X \times X^{\mathbb{N}}$ be a collection of sequences in X that satisfies the (Uniqueness) axiom. Suppose that for every $x \in X$ there exists a constant $\varepsilon_0(x) > 0$ and a collection of functions $X \to [0, \infty] : x' \mapsto \rho_\varepsilon(x, x')$ for $0 < \varepsilon < \varepsilon_0(x)$ satisfying the following conditions.*

(a) *If $x \in X$ and $0 < \varepsilon < \varepsilon_0(x)$ then $\rho_\varepsilon(x, x) = 0$.*

(b) *If $x \in X$, $0 < \varepsilon < \varepsilon_0(x)$, and $(x_n)_n \in X^{\mathbb{N}}$ then*
$$(x, (x_n)_n) \in \mathcal{C} \qquad \iff \qquad \lim_{n\to\infty} \rho_\varepsilon(x, x_n) = 0.$$

(c) *If $x \in X$, $0 < \varepsilon < \varepsilon_0(x)$, and $(x', (x_n)_n) \in \mathcal{C}$, then*
$$\rho_\varepsilon(x, x') < \varepsilon \qquad \Longrightarrow \qquad \limsup_{n\to\infty} \rho_\varepsilon(x, x_n) \le \rho_\varepsilon(x, x').$$

Then $\mathcal{C} = \mathcal{C}(\mathcal{U}(\mathcal{C}))$. Moreover, the topology $\mathcal{U}(\mathcal{C})$ is first countable and Hausdorff.

PROOF. It follows from (a) and (b) that $\mathcal{C}$ satisfies the *(Constant)*, *(Subsequence)*, and *(Subsubsequence)* axioms. To prove the *(Diagonal)* axiom, assume that $(x, (x_k)_k) \in \mathcal{C}$ and $(x_k, (x_{k,n})_n) \in \mathcal{C}$ for $k \ge 1$. Choose a real number ε such that $0 < \varepsilon < \varepsilon_0(x)$. Then, by (b), there is a sequence of integers k_i such that $\rho_\varepsilon(x, x_{k_i}) \le 2^{-i}\varepsilon$. Hence, by (c), there is a sequence of integers n_i such that $\rho_\varepsilon(x, x_{k_i,n_i}) \le 2^{1-i}\varepsilon$. Hence, by (b), $(x, (x_{k_i,n_i})_i) \in \mathcal{C}$. This shows that $\mathcal{C}$ satisfies all five axioms and it follows from Lemma 5.6.4 that $\mathcal{C} = \mathcal{C}(\mathcal{U}(\mathcal{C}))$.

We prove that every point in X has a countable neighbourhood basis. Given x, choose a real number ε such that $0 < \varepsilon < \varepsilon_0(x)$ and consider the sets
$$U_n(x) = \{x' \in X \,|\, \rho_\varepsilon(x, x') < \varepsilon/n\}, \qquad n \in \mathbb{N}.$$
We claim that these form a basis of open neighbourhhoods for x. To see this note first that, if $y_0 \in U_n(x)$ and $(y_0, (y_k)_k) \in \mathcal{C}$, then it follows from (c) that $\limsup_{k\to\infty} \rho_\varepsilon(x, y_k) \le \rho_\varepsilon(x, y_0) < \varepsilon/n$. Hence $y_k \in U_n(x)$ for k sufficiently large. By definition of $\mathcal{U}(\mathcal{C})$, this shows that $U_n(x)$ is open. In particular, $U_n(x)$ is a neighbourhood of x for every $x \in X$ and every $n \in \mathbb{N}$. To see that the sets $U_n(x)$ form a neighbourhood basis we argue by contradiction. Suppose to the contrary that there is an open neighbourhood U of x that contains none of the sets $U_n(x)$. Then, for each n, there is a point $x_n \in U_n(x) \setminus U$. It follows from (b) that

$(x,(x_n)_n)\in\mathcal{C}$. Since U is an open set containing x we must have $x_n\in U$ for large n, which contradicts our assumption.

Thus we have proved that $\mathcal{U}(\mathcal{C})$ is first countable. To see that it is Hausdorff we must show that any two distinct points x,y have disjoint open neighbourhoods. We prove that if $x\neq y$ then there is an integer $n=n(x,y)$ such that $U_n(x)\cap U_n(y)=\emptyset$. If not, there exists a sequence $x_n\in X$ such that $x_n\in U_n(x)\cap U_n(y)$ for all n. Then, by (b), the sequence x_n converges to both x and y which contradicts our assumption that limits are unique. This proves Lemma 5.6.5. □

The **Gromov topology** on the moduli space $\overline{\mathcal{M}}_{0,n}(M,A;J)$ is now defined to be the collection $\mathcal{U}(\mathcal{C})$, where $\mathcal{C}$ is the set of all Gromov convergent sequences. The following theorem shows that this topology has all the expected properties.

THEOREM 5.6.6. (i) *A sequence in $\overline{\mathcal{M}}_{0,n}(M,A;J)$ converges with respect to the Gromov topology if and only if it Gromov converges.*

(ii) *$\overline{\mathcal{M}}_{0,n}(M,A;J)$ is a compact metrizable space. In particular, it is Hausdorff and has a countable dense subset.*

(iii) *The evaluation maps*

$$\mathrm{ev}_i:\overline{\mathcal{M}}_{0,n}(M,A;J)\to M,$$

the projection

$$\pi:\overline{\mathcal{M}}_{0,n}(M,A;J)\to\overline{\mathcal{M}}_{0,n-1}(M,A;J),$$

and the forgetful map

$$\pi:\overline{\mathcal{M}}_{0,n}(M,A;J)\to\overline{\mathcal{M}}_{0,n}$$

of Section 5.1 are continuous with respect to the Gromov topology.

PROOF. By Theorem 5.5.3, Gromov convergent sequences have unique limits. Hence it follows from Remark 5.5.7 and Lemmas 5.5.8 and 5.5.9 that Gromov convergent sequences satisfy all the hypotheses of Lemma 5.6.5. This proves (i) and shows that the Gromov topology is first countable and Hausdorff.

We prove that the Gromov topology on $\overline{\mathcal{M}}_{0,n}(M,A;J)$ has a countable basis. To see this, fix an n-labelled tree $T=(T,E,\Lambda)$. Then the set $\widetilde{\mathcal{M}}_{0,T}(M,A;J)$ of all $\mathbf{x}\in\mathcal{SC}_{0,n}(M,A;J)$ that are modelled on the tree T is a countable union of compact subsets of a separable Banach manifold, and hence contains a countable dense subset. Therefore each space $\widetilde{\mathcal{M}}_{0,T}(M,A;J)$ has a countable dense subset and hence so does the quotient stratum $\mathcal{M}_{0,T}(M,A;J)=\widetilde{\mathcal{M}}_{0,T}(M,A;J)/\mathrm{G}_T$. Since there are only finitely many strata, the same holds for $\overline{\mathcal{M}}_{0,n}(M,A;J)$. Hence the Gromov topology on $\overline{\mathcal{M}}_{0,n}(M,A;J)$ has a countable dense subset. The union of the countable neighbourhood bases of the elements of this countable dense subset is a countable basis for the Gromov topology.

Next observe that, by Theorem 5.5.5, every sequence in $\overline{\mathcal{M}}_{0,n}(M,A;J)$ has a Gromov convergent subsequence. Hence the Gromov topology is sequentially compact. Since every sequentially compact topological space with a countable basis is compact ([**338**, Chapter 9, Section 2]), we conclude that $\overline{\mathcal{M}}_{0,n}(M,A;J)$ is compact.[1] Now every compact Hausdorff space is normal (disjoint closed subsets

[1]Here is a direct argument. Let X be a sequentially compact topological space with a countable basis. Then every open cover has a countable refinement $\{U_n\}_{n\in\mathbb{N}}$, consisting of all those elements from the basis which are contained in some element of the cover. Suppose that this refinement does not have a finite subcover. Passing to a subsequence that still covers X we

can be separated by open sets). Urysohn's metrization theorem [**338**, Chapter 8, Section 3] asserts that every normal topological space with a countable basis is metrizable. This proves (ii).

The continuity of the evaluation map follows directly from the definitions. The continuity of the forgetful map $\overline{\mathcal{M}}_{0,n}(M, A; J) \to \overline{\mathcal{M}}_{0,n}$ follows from Theorem 5.2.2 (iv) and Proposition 5.5.2. The proof of continuity of the projection $\overline{\mathcal{M}}_{0,n}(M, A; J) \to \overline{\mathcal{M}}_{0,n-1}(M, A; J)$ is left as an exercise. This proves Theorem 5.6.6. □

---

may assume that $U_n \not\subset U_1 \cup \cdots \cup U_{n-1}$ for every n. Hence there is a sequence $x_n \in U_n$ such that $x_n \notin U_m$ for $m < n$. Choose a convergent subsequence $x_{n_i} \to x$. Then $x \in U_m$ for some m. Hence there exists an $i_0 \in \mathbb{N}$ such that $x_{n_i} \in U_m$ for $i \geq i_0$. Thus $x_{n_i} \in U_m$ for some $n_i > m$, a contradiction. Hence the refinement has a finite subcover, and so does the original open cover.

CHAPTER 6

Moduli Spaces of Stable Maps

Let (M, ω) be a closed symplectic manifold and $J \in \mathcal{J}_\tau(M, \omega)$ be an ω-tame almost complex structure. Fix a spherical homology class $A \in H_2(M)$ and a non-negative integer k. For $A \neq 0$ denote by

$$\mathcal{M}^*_{0,k}(A; J) := \frac{\mathcal{M}^*(A; J) \times (S^2)^k \setminus \Delta_k}{\mathrm{G}}$$

the moduli space of equivalence classes of tuples $(u, z_1, \dots, z_k)$, where $u : S^2 \to M$ is a simple J-holomorphic curve representing the class A and the z_i are pairwise distinct points on S^2. The reader is referred to Section 3.1 for the notation $\mathcal{M}^*(A; J)$, and $\Delta^k \subset (S^2)^k$ is the fat diagonal consisting of all ordered k-tuples of points in S^2 such that at least two of them agree. The equivalence relation is given by the diagonal action of the reparametrization group $\mathrm{G} = \mathrm{PSL}(2, \mathbb{C})$. When $k \geq 3$ we extend the notation to the class $A = 0$ by declaring the constant maps to be simple. When $k < 3$ we define $\mathcal{M}^*_{0,k}(0; J)$ as the empty set. Slightly abusing language we call $\mathcal{M}^*_{0,k}(A; J)$ the **moduli space of simple k-pointed J-holomorphic spheres in the class** A, even when $A = 0$. In the notation of Section 5.1, this moduli space is an open subset of the top stratum $\mathcal{M}_{0,k}(A; J)$ of the moduli space $\overline{\mathcal{M}}_{0,k}(A; J)$ of stable maps of genus zero with k marked points representing the class A. It follows from the results of Chapter 3 that, for a generic almost complex structure J, the moduli space $\mathcal{M}^*_{0,k}(A; J)$ is a (noncompact) smooth manifold of dimension

$$\dim \mathcal{M}^*_{0,k}(A; J) = \mu(A, k) := 2n + 2c_1(A) + 2k - 6.$$

The goal of this chapter is to establish conditions under which the evaluation map

$$\mathrm{ev} : \mathcal{M}^*_{0,k}(A; J) \to M^k, \qquad \mathrm{ev}([u, z_1, \dots, z_k]) := (u(z_1), \dots u(z_k)),$$

defines a homology class $\mathrm{ev}_*([\mathcal{M}_{0,k}(A; J)])$ in M^k. This class will be used in Chapter 7 to define the Gromov–Witten invariants.

Intuitively speaking, a map $f : V \to X$ whose domain is a noncompact d-manifold represents a d-dimensional homology class in X if its limit set $f(\overline{V} \setminus V)$ has dimension at most $d - 2$. Here $\overline{V}$ is to be understood as a suitable compactification of V to which f extends continuously and the dimension condition means that the image of the "boundary" $\partial V := \overline{V} \setminus V$ under f is contained in the image of a smooth map, defined on a manifold whose components have dimensions less than or equal to $d-2$. We call such a map $f : V \to X$ a d-dimensional pseudocycle. The precise definition is given in Section 6.5. To establish the pseudocycle condition for the evaluation map $\mathrm{ev} : \mathcal{M}^*_{0,k}(A; J) \to M^k$, we must examine the image of the boundary $\overline{\mathcal{M}}_{0,k}(A; J) \setminus \mathcal{M}^*_{0,k}(A; J)$ under the extended evaluation map. This boundary contains elements of two kinds: the nonsimple elements in $\mathcal{M}_{0,k}(A; J)$ and the stable maps belonging to strata $\mathcal{M}_{0,T}(A; J)$ with $e(T) > 0$ (where $e(T)$

is the number of edges in the tree T). As we shall see, the *virtual dimension* of each stratum $\mathcal{M}_{0,T}(A;J)$ is $\mu(A,k)-2e(T)$. Therefore, if we could ignore the non-simple elements and regularize each stratum by choosing a generic J, the required dimensional condition would be satisfied.

Of course, in general this is not possible. Just as in Chapter 3, the relevant transversality results for generic almost complex structures J can only be established if one restricts attention to the subspaces $\mathcal{M}^*_{0,T}(\{B_\alpha\};J)$ of **simple** stable maps. Proposition 6.1.2 asserts that one can indeed compactify $\mathcal{M}^*_{0,k}(A;J)$ using only simple strata $\mathcal{M}^*_{0,T}(\{B_\alpha\};J)$ provided that one considers all collections $\{B_\alpha\}_{\alpha\in T}$ such that

$$A=\sum_{\alpha\in T} m_\alpha B_\alpha$$

for some integers $m_\alpha>0$. The problem now is that there is no guarantee in general that the dimension $\mu(B,k)-2e(T)$ of the stratum $\mathcal{M}^*_{0,T}(\{B_\alpha\};J)$ is small enough, i.e. is at most $\mu(A,k)-2$. For this to hold for all classes A we must assume that (M,ω) is **semipositive**, which means that there are no holomorphic spheres with negative Chern number for generic J. This is a symplectic version of the NEF condition in algebraic geometry. In the case $e(T)=0$ care must also be taken with holomorphic spheres of Chern number zero.

The end result is that, when (M,ω) is semipositive, the evaluation map

$$\mathrm{ev}:\mathcal{M}^*_{0,k}(A;J)\to M^k$$

is a pseudocycle, called the **Gromov–Witten pseudocycle**. We prove this in Section 6.6. The main ingredient in the proof is Theorem 6.2.6 which states that, for a generic J, the moduli space $\mathcal{M}^*_{0,T}(\{B_\alpha\};J)$ is a finite dimensional manifold of dimension

$$\dim \mathcal{M}^*_{0,T}(\{B_\alpha\};J)=\mu(B,k)-2e(T),\qquad B:=\sum_{\alpha\in T} B_\alpha.$$

If (M,ω) is semipositive then this number is less than or equal to $\mu(A,k)-2$ for every collection of homology classes $\{B_\alpha\}_\alpha$ that satisfy $\sum_{\alpha\in T} m_\alpha B_\alpha = A$ with $m_\alpha>0$ and every tree T such that $e(T)>0$. In the case $e(T)=0$ one only obtains a lower dimensional moduli space if $c_1(A)>0$. If $A=mB\in H_2(M)$, $m>1$, is a multiple class with Chern number zero, one needs a wider class of perturbations to control the boundary of $\mathcal{M}^*_{0,k}(A;J)$. We show in Section 6.7 that it is possible to define a pseudocycle $\mathrm{ev}:\mathcal{M}^*_{0,k}(A;\{J_z\})\to M^k$ in this case, but now the almost complex structure J_z in the defining equation depends on the base point $z\in S^2$. This device also allows us to define a pseudocycle in which some of the marked points are fixed.

The chapter ends with a brief discussion of the product of the evaluation map with the forgetful map π:

$$\mathrm{ev}\times\pi:\overline{\mathcal{M}}_{0,k}(A;J)\to M^k\times\overline{\mathcal{M}}_{0,k}.$$

All the pseudocycles mentioned above, both with fixed and with varying marked points, can be derived as restrictions of this product map followed by the projection onto the first factor. However, to prove that $\mathrm{ev}\times\pi$ itself is a pseudocycle for arbitrary semipositive manifolds would require an even wider range of perturbations, and so we treat only a very special case: see Exercise 6.7.13.

In order to construct the Gromov–Witten invariants for general compact symplectic manifolds one has to regularize the whole of the compactification $\overline{\mathcal{M}}_{0,k}(A; J)$ including the multiply covered curves. This leads to the construction of the *virtual fundamental cycle* in Gromov–Witten theory, which has been the subject of extensive research since the mid nineties (see Fukaya-Ono [**127**], Li-Tian [**239**], Ruan [**343**], Liu–Tian [**249**], Siebert [**376**], McDuff [**267**]).

In a recent series of papers, Hofer–Wysocki–Zehnder have recast the whole theory in terms of scale calculus and polyfolds; see [**184, 185, 186, 187, 188, 189**] A more traditional treatment based on the analysis discussed in the present book may be found in McDuff and Wehrheim [**281**]. We do not discuss the virtual fundamental cycle here, but instead restrict attention to the semipositive case, where the Gromov–Witten invariants can be defined directly by *counting J-holomorphic curves.*

As in previous chapters we will restrict most of our arguments to the genus zero case. However, some of them continue to hold with minor changes in the case when the domain (Σ, j_Σ) of u is a closed Riemann surface with a fixed complex structure.

6.1. Simple stable maps

This section introduces the notion of a simple stable map and constructs for every stable map a simple one with the same image. This is the key property that allows us to form a useful compactification of the moduli space $\mathcal{M}^*_{0,k}(A; J)$ that involves only simple strata.

DEFINITION 6.1.1. *A stable map*

$$(\mathbf{u}, \mathbf{z}) \in \mathcal{SC}_{0,k}(M, A, J),$$

modelled over a k-labelled tree $T = (T, E, \Lambda)$, is called **simple** *if every nonconstant map u_α is simple and*

$$u_\alpha(S^2) \neq u_\beta(S^2)$$

for any two vertices $\alpha \neq \beta$ such that u_α and u_β are nonconstant.

Let $T = (T, E, \Lambda)$ be a k-labelled tree and

$$T \to H_2(M; \mathbb{Z}) : \alpha \mapsto A_\alpha$$

be a function such that

$$A_\alpha = 0 \quad \Longrightarrow \quad \#\big(\Lambda_\alpha \cup \{\beta \in T \,|\, \alpha E \beta\}\big) \geq 3. \tag{6.1.1}$$

We denote by $\widetilde{\mathcal{M}}^*_{0,T}(\{A_\alpha\}; J)$ the set of simple stable maps $(\mathbf{u}, \mathbf{z})$ modelled over T such that

$$[u_\alpha] = A_\alpha$$

for every $\alpha \in T$. It follows easily from the definitions that the reparametrization group G_T acts freely and properly on $\widetilde{\mathcal{M}}^*_{0,T}(\{A_\alpha\}; J)$ and we denote the quotient by

$$\mathcal{M}^*_{0,T}(\{A_\alpha\}; J) := \widetilde{\mathcal{M}}^*_{0,T}(\{A_\alpha\}; J)/\mathrm{G}_T.$$

When T has no edges, this quotient is just the space $\mathcal{M}^*_{0,k}(A; J)$ of simple k-pointed J-holomorphic spheres mentioned in the introduction to this chapter. Because of the stability condition this space is empty when $A = 0$ and $k < 3$.

An arbitrary stable map may fail to be simple for one of two reasons: it might contain some multiply covered components and/or it might contain several components with the same image. The rest of this section is devoted to showing that every stable map covers a simple one with the same image. Intuitively, this is clear; one simply deletes components with the same image and replaces each multiple covering with an underlying simple map. However, we need to show that the resulting stable map is modelled over a tree, and also must keep track of the marked points. The next proposition spells out the details. Recall that ev_i denotes evaluation at the ith marked point, as in Exercise 5.1.11.

PROPOSITION 6.1.2. *Let $(\mathbf{u},\mathbf{z}) \in \widetilde{\mathcal{M}}^*_{0,T}(\{A_\alpha\};J)$ be a stable map modelled over a k-labelled tree (T,E,Λ) and representing the classes $A_\alpha \in H_2(M;\mathbb{Z})$ for $\alpha \in T$. Then there is a k-labelled tree $T' = (T',E',\Lambda')$, two functions*

$$T' \to H_2(M;\mathbb{Z}) : \alpha' \mapsto A'_{\alpha'}, \qquad T' \to \mathbb{N} : \alpha' \mapsto m_{\alpha'}$$

and a simple stable map $(\mathbf{u}',\mathbf{z}') \in \widetilde{\mathcal{M}}^*_{0,T'}(\{A'_{\alpha'}\};J)$ such that

$$\bigcup_{\alpha'\in T'} u'_{\alpha'}(S^2) = \bigcup_{\alpha\in T} u_\alpha(S^2), \qquad A = \sum_{\alpha'\in T'} m_{\alpha'}A'_{\alpha'},$$

and

$$\mathrm{ev}_i(\mathbf{u}',\mathbf{z}') = \mathrm{ev}_i(\mathbf{u},\mathbf{z})$$

for $i = 1,\dots,k$.

PROOF. We prove the result in the case $k=0$, i.e. there are no marked points, leaving the general case as an exercise. Let us define a **weighted prestable map** (without marked points) as a tuple

$$\big(T,f,m,\mathbf{u},\mathbf{z}\big) = \big(T,f,\{m_\alpha\}_{\alpha\in T},\{u_\alpha\}_{\alpha\in T},\{z_\alpha\}_{\alpha\in T\setminus\{0\}}\big)$$

consisting of a finite totally ordered set $(T,\preceq)$ with minimal element 0, a function $f : T\setminus\{0\} \to T$ (that plays the role of the vector $\mathbf{j}$ in Exercise 5.1.3), positive integer weights $m_\alpha \in \mathbb{N}$ for $\alpha \in T$, J-holomorphic spheres $u_\alpha : S^2 = \mathbb{C}\cup\{\infty\} \to M$ for $\alpha \in T$, and complex numbers $z_\alpha \in \mathbb{C}$ for $\alpha \in T$, such that

$$f(\alpha) \prec \alpha, \qquad u_{f(\alpha)}(z_\alpha) = u_\alpha(\infty)$$

for every $\alpha \in T\setminus\{0\}$. We assume further that the set

$$T_\alpha := \{\beta \in T \,|\, \alpha \prec \beta,\ f(\beta) = \alpha\}$$

consists of at least two points whenever u_α is constant. Think of this as an inductive description of a tree of J-holomorphic spheres. Each new sphere u_α is connected to one of the previous spheres, namely $u_{f(\alpha)}$, at the point $u_\alpha(\infty) = u_{f(\alpha)}(z_\alpha)$. Thus the points z_α should be thought of as belonging to the sphere labelled by $f(\alpha)$. The number m_α is understood as the *multiplicity* of u_α and the **homology class represented by the weighted prestable map** $\big(T,f,m,\mathbf{u},\mathbf{z}\big)$ is defined by

$$A := \sum_{\alpha\in T} m_\alpha A_\alpha, \qquad A_\alpha := [u_\alpha].$$

A prestable map need not be stable, because the nodal points on the same sphere are not required to be pairwise distinct.

Call a weighted prestable map $(T, f, m, \mathbf{u}, \mathbf{z})$ **stable** if

$$\alpha \neq \beta, \;\; f(\alpha) = f(\beta) \qquad \Longrightarrow \qquad z_\alpha \neq z_\beta$$

for all $\alpha, \beta \in T \setminus \{0\}$. Call it **simple** if u_α is simple whenever u_α is nonconstant, and if $u_\alpha(S^2) \neq u_\beta(S^2)$ whenever $\alpha \neq \beta$ and u_α and u_β are nonconstant. We prove the proposition by showing, in three steps, that for every weighted prestable map $(T, f, m, \mathbf{u}, \mathbf{z})$ there is a simple weighted stable map $(T', f', m', \mathbf{u}', \mathbf{z}')$ with the same image and representing the same homology class.

STEP 1. *Let $(T, f, m, \mathbf{u}, \mathbf{z})$ be a weighted prestable map. Then there is a weighted stable map $(T', f', m', \mathbf{u}', \mathbf{z}')$ and an injection $\iota : T \to T'$ such that*

$$u_\alpha = u'_{\iota(\alpha)}, \qquad m_\alpha = m'_{\iota(\alpha)}$$

for every $\alpha \in T$ and $u'_{\alpha'}$ is constant for every $\alpha' \in T' \setminus \iota(T)$. In particular, both maps have the same image and represent the same homology class.

Suppose that there exist vertices $\alpha_0, \alpha_1 \in T \setminus \{0\}$ such that

$$\alpha_0 \neq \alpha_1, \quad f(\alpha_0) = f(\alpha_1) =: \gamma, \quad z_{\alpha_0} = z_{\alpha_1}.$$

We then define a new tree $T' := T \cup \{\gamma'\}$ by inserting a new element γ' in T as the successor to γ. Thus the new ordering is

$$\alpha \prec' \beta \qquad \Longleftrightarrow \qquad \begin{cases} \alpha \prec \beta, & \text{if } \alpha, \beta \in T, \\ \alpha \preceq \gamma, & \text{if } \beta = \gamma' \text{ and } \alpha \in T, \\ \gamma \prec \beta, & \text{if } \alpha = \gamma' \text{ and } \beta \in T. \end{cases}$$

The new weights and holomorphic curves are

$$m'_\alpha := \begin{cases} m_\alpha, & \text{if } \alpha \in T, \\ 1, & \text{if } \alpha = \gamma' \end{cases} \qquad u'_\alpha := \begin{cases} u_\alpha, & \text{if } \alpha \in T, \\ u_\gamma(z_{\alpha_0}), & \text{if } \alpha = \gamma'. \end{cases}$$

The function $f' : T' \setminus \{0\} \to T'$ and the nodal points are defined by

$$f'(\alpha) := \begin{cases} f(\alpha), & \text{if } \alpha \in T \setminus \{\alpha_0, \alpha_1\}, \\ \gamma', & \text{if } \alpha \in \{\alpha_0, \alpha_1\}, \\ \gamma, & \text{if } \alpha = \gamma', \end{cases} \qquad z'_\alpha := \begin{cases} z_\alpha, & \text{if } \alpha \in T \setminus \{\alpha_0, \alpha_1\}, \\ 0, & \text{if } \alpha = \alpha_0, \\ 1, & \text{if } \alpha = \alpha_1, \\ z_{\alpha_0}, & \text{if } \alpha = \gamma'. \end{cases}$$

Continue by induction to construct a stable weighted map with the same image as the original prestable one. This proves Step 1.

STEP 2. *Let $(T, f, m, \mathbf{u}, \mathbf{z})$ be a weighted prestable map. Then there is a weighted stable map $(T', f', m', \mathbf{u}', \mathbf{z}')$ with the same image and representing the same homology class such that $u'_{\alpha'}$ is simple whenever $u'_{\alpha'}$ is nonconstant.*

Suppose that u_γ is nonconstant and multiply covered for some $\gamma \in T$. Then, by Proposition 2.5.1, there is a simple J-holomorphic sphere $v : S^2 \to M$ and a rational map $\phi : S^2 \to S^2$ of degree $d := \deg(\phi) > 1$ such that $u_\gamma = v \circ \phi$ and $\phi(\infty) = \infty$. This gives rise to a weighted prestable map $(T, f, m', \mathbf{u}', \mathbf{z}')$, defined by

$$m'_\alpha := \begin{cases} m_\alpha, & \text{if } \alpha \neq \gamma, \\ dm_\gamma, & \text{if } \alpha = \gamma, \end{cases} \qquad u'_\alpha := \begin{cases} u_\alpha, & \text{if } \alpha \neq \gamma, \\ v, & \text{if } \alpha = \gamma, \end{cases} \qquad \alpha \in T,$$

$$z'_\alpha := \begin{cases} z_\alpha, & \text{if } f(\alpha) \neq \gamma, \\ \phi(z_\alpha), & \text{if } f(\alpha) = \gamma, \end{cases} \qquad \alpha \in T \setminus \{0\}.$$

Apply Step 1 to make the map stable and continue by induction. This proves Step 2.

STEP 3. *The proposition holds when $k = 0$.*

Let $(T, f, m, \mathbf{u}, \mathbf{z})$ be a weighted stable map such that u_α is simple for every α. Suppose that there exist vertices $\gamma_0, \gamma_1 \in T$ such that u_{γ_0} and u_{γ_1} are nonconstant and

$$\gamma_0 \prec \gamma_1, \qquad u_{\gamma_0}(S^2) = u_{\gamma_1}(S^2).$$

Then, by Corollary 2.5.4, there exists a Möbius transformation

$$\phi : S^2 \to S^2$$

such that

$$u_{\gamma_1} = u_{\gamma_0} \circ \phi.$$

Define a new prestable map by

$$T' := T \setminus \{\gamma_1\}$$

with the ordering induced by $(T, \prec)$, $u'_\alpha := u_\alpha$ for $\alpha \in T'$,

$$f'(\alpha) := \begin{cases} f(\alpha), & \text{if } f(\alpha) \neq \gamma_1, \\ \gamma_0, & \text{if } f(\alpha) = \gamma_1, \end{cases} \qquad z'_\alpha := \begin{cases} z_\alpha, & \text{if } f(\alpha) \neq \gamma_1, \\ \phi(z_\alpha), & \text{if } f(\alpha) = \gamma_1, \end{cases}$$

for $\alpha \in T' \setminus \{0\}$, and

$$m'_\alpha := \begin{cases} m_\alpha, & \text{if } \alpha \neq \gamma_0, \\ m_{\gamma_0} + m_{\gamma_1}, & \text{if } \alpha = \gamma_0, \end{cases}$$

for $\alpha \in T'$. Then use Step 1 to make the map stable and continue by induction. This proves Proposition 6.1.2 in the case $k = 0$. □

EXERCISE 6.1.3. Prove Proposition 6.1.2 in the case $k > 0$. *Hint:* It remains to choose marked points on the simple stable map whose images agree with those of the marked points on the original curve. This may require the introduction of additional (ghost) vertices.

EXERCISE 6.1.4. Show that the reparametrization group G_T acts freely and properly on the space $\widetilde{\mathcal{M}}^*_{0,T}(\{A_\alpha\}, J)$.

6.2. Transversality for simple stable maps

Let us fix a k-labelled tree $T = (T, E, \Lambda)$ and a function

$$T \to H_2(M; \mathbb{Z}) : \alpha \mapsto A_\alpha$$

that satisfies the stability condition (6.1.1). Denote

$$A := \sum_{\alpha \in T} A_\alpha.$$

We wish to prove that, for a generic almost complex structure $J \in \mathcal{J}_\tau(M, \omega)$, the moduli space $\mathcal{M}^*_{0,T}(\{A_\alpha\}; J)$ is a smooth oriented finite dimensional manifold of the "correct" dimension. As a first step we formulate the conditions that J must satisfy in order for this to hold. Clearly, J should be regular in the sense of Definition 3.1.5 for each class A_α so that each moduli space $\mathcal{M}^*(A_\alpha; J)$ is a manifold. However this is not sufficient: we also need the different spaces $\mathcal{M}^*(A_\alpha; J)$ to intersect transversally.

To understand what is required here, let $\{A_\alpha\}_{\alpha\in T}$ be any finite collection of spherical homology classes in $H_2(M;\mathbb{Z})$ and consider the space

$$\mathcal{M}^*(\{A_\alpha\};J)\subset\prod_{\alpha\in T}\mathcal{M}^*(A_\alpha;J)$$

of all tuples $\mathbf{u}=\{u_\alpha\}_{\alpha\in T}$ of J-holomorphic spheres such that $[u_\alpha]=A_\alpha$, each nonconstant u_α is simple, and $u_\alpha(S^2)\neq u_\beta(S^2)$ whenever $\alpha\neq\beta$ and u_α and u_β are nonconstant. Thus $\mathcal{M}^*(\{A_\alpha\};J)$ is a subset of the space of simple curves with disconnected domain $\Sigma:=T\times S^2$ that is denoted by $\mathcal{M}^*(A,\Sigma;J)$ in Section 3.4. Then consider the open set

$$Z(T)\subset(S^2)^E\times(S^2)^k$$

of all tuples

$$\mathbf{z}=(\{z_{\alpha\beta}\}_{\alpha E\beta},\{z_i\}_{1\leq i\leq k})$$

such that, for every α, the points $z_{\alpha\beta}$ for $\alpha E\beta$ and z_i for $\alpha_i=\alpha$ are pairwise distinct. Here we can think of the elements of $E\subset T\times T$ as representing oriented edges. However, as explained in Section D.2, our general convention is that edges are unoriented, unless explicit mention is made to the contrary. There is an evaluation map

$$\mathrm{ev}^E:\mathcal{M}^*(\{A_\alpha\};J)\times Z(T)\to M^E$$

defined by

$$\mathrm{ev}^E(\mathbf{u},\mathbf{z}):=\{u_\alpha(z_{\alpha\beta})\}_{\alpha E\beta}\,. \tag{6.2.1}$$

Let us denote by $\Delta^E\subset M^E$ the diagonal determined by the edge relation:

$$\Delta^E:=\Big\{\{x_{\alpha\beta}\}_{\alpha E\beta}\in M^E\,|\,x_{\alpha\beta}=x_{\beta\alpha}\Big\}.$$

DEFINITION 6.2.1. *An ω-tame almost complex structure $J\in\mathcal{J}_\tau(M;\omega)$ is called* **regular for T and $\{A_\alpha\}$** *if the following holds.*

(i) *D_{u_α} is surjective for every $\mathbf{u}=\{u_\alpha\}\in\mathcal{M}^*(\{A_\alpha\};J)$ and every $\alpha\in T$.*

(ii) *The evaluation map ev^E is transverse to Δ^E.*

Note that condition (ii) makes sense because condition (i) implies that the domain $\mathcal{M}^*(\{A_\alpha\};J)\times Z(T)$ of ev^E is a manifold. Here is a more explicit version.

LEMMA 6.2.2. *The edge evaluation map ev^E is transverse to Δ^E if and only if, for every simple stable map*

$$(\mathbf{u},\mathbf{z})=\big(\{u_\alpha\}_{\alpha\in T},\{z_{\alpha\beta}\}_{\alpha E\beta},\{z_i\}_{1\leq i\leq k}\big)\in\widetilde{\mathcal{M}}^*_{0,T}(\{A_\alpha\};J)$$

and every collection of tangent vectors $v_{\alpha\beta}\in T_{u_\alpha(z_{\alpha\beta})}M$ satisfying

$$v_{\alpha\beta}+v_{\beta\alpha}=0,$$

there exist vector fields $\xi_\alpha\in\ker D_{u_\alpha}$ and tangent vectors $\zeta_{\alpha\beta}\in T_{z_{\alpha\beta}}S^2$ such that

$$v_{\alpha\beta}=\xi_\alpha(z_{\alpha\beta})+du_\alpha(z_{\alpha\beta})\zeta_{\alpha\beta}-\xi_\beta(z_{\beta\alpha})-du_\beta(z_{\beta\alpha})\zeta_{\beta\alpha} \tag{6.2.2}$$

for all $\alpha,\beta\in T$ such that $\alpha E\beta$.

PROOF. Write the edge evaluation map in the form

$$\mathrm{ev}^E = \{\mathrm{ev}_{\alpha\beta}\}_{\alpha E\beta},$$

where the map

$$\mathrm{ev}_{\alpha\beta} : \mathcal{M}^*(\{A_\alpha\}; J) \times Z(T) \to M$$

is defined by

$$\mathrm{ev}_{\alpha\beta}(\mathbf{u}, \mathbf{z}) := u_\alpha(z_{\alpha\beta}).$$

Then

$$d\mathrm{ev}_{\alpha\beta}(\mathbf{u}, \mathbf{z})(\xi, \zeta) = \xi_\alpha(z_{\alpha\beta}) + du_\alpha(z_{\alpha\beta})\zeta_{\alpha\beta}$$

for $\xi = \{\xi_\alpha\}_{\alpha\in T}$ and $\zeta = \{\zeta_{\alpha\beta}\}_{\alpha E\beta}$. Hence equation (6.2.2) is equivalent to

$$v_{\alpha\beta} = d\,\mathrm{ev}_{\alpha\beta}(\mathbf{u}, \mathbf{z})(\xi, \zeta) - d\,\mathrm{ev}_{\beta\alpha}(\mathbf{u}, \mathbf{z})(\xi, \zeta)$$

for $\alpha E\beta$. With this understood the result follows from Exercise 6.2.3. □

EXERCISE 6.2.3. Let V be a vector space, (T, E) be a tree, denote by $\Delta^E \subset V^E$ the edge diagonal, and let $W \subset V^E$ be a subspace. Prove that

$$V^E = \Delta^E + W$$

if and only if for every tuple $\{v_{\alpha\beta}\}_{\alpha E\beta} \in V^E$ satisfying $v_{\alpha\beta} + v_{\beta\alpha} = 0$ there exists a tuple $\{w_{\alpha\beta}\}_{\alpha E\beta} \in W$ such that $v_{\alpha\beta} = w_{\alpha\beta} - w_{\beta\alpha}$ for $\alpha E\beta$.

EXERCISE 6.2.4. Suppose $\mathcal{M}^*(\{A_\alpha\}; J) \neq \emptyset$. Show that condition (i) in Definition 6.2.1 holds if and only if J is regular in the sense of Definition 3.1.5 for all A_α. Note that $\mathcal{M}^*(\{A_\alpha\}; J) = \emptyset$ whenever there exists a pair of vertices $\alpha, \beta \in T$ such that $\alpha \neq \beta$, $A_\alpha = A_\beta$, and there is only one J-holomorphic sphere in class A_α (up to reparametrization). In what other situations is the moduli space of simple tuples empty?

The next exercise shows that adding marked points does not effect the regularity of the almost complex structure.

EXERCISE 6.2.5. Let $T = (T, E, \Lambda)$ be a k-labelled tree and suppose that the map $T \to H_2(M; \mathbb{Z}) : \alpha \mapsto A_\alpha$ satisfies the stability condition (6.1.1). Let (T', E') be the reduced tree obtained by forgetting the marked points and deleting the vertices with $A_\alpha = 0$ which carry less than three nodal points. Show that J is regular for T and $\{A_\alpha\}_{\alpha\in T}$ if and only if it is regular for T' and $\{A_\alpha\}_{\alpha\in T'}$. *Hint:* Show that it suffices to consider the case when $T = T' \cup \{\gamma\}$ and there are at most two edges with endpoint γ. Use the fact that when u_α is constant, $\ker D_{u_\alpha}$ consists of the constant sections of the trivial bundle $u_\alpha^*TM = S^2 \times T_{u_\alpha(z)}M$. Related arguments are used in the proof given in Section 6.3 of Theorem 6.3.1.

In the following we denote by $\mathcal{J}_{\mathrm{reg}}(T, \{A_\alpha\})$ the set of ω-tame almost complex structures that are regular for T and $\{A_\alpha\}$ and by

$$\mathcal{J}_{\mathrm{reg}}(M, \omega) := \bigcap_{T}\bigcap_{\{A_\alpha\}} \mathcal{J}_{\mathrm{reg}}(T, \{A_\alpha\})$$

the set of almost complex structures that are regular for every T and every $\{A_\alpha\}$. Note that this is a countable intersection. An almost complex structure $J \in \mathcal{J}_\tau(M, \omega)$ belongs to the set $\mathcal{J}_{\mathrm{reg}}(M, \omega)$ if and only if the operator D_u is surjective for every simple J-holomorphic sphere u and the evaluation map (6.2.1) is transverse to Δ^E for every T and every $\{A_\alpha\}$. After this preparation, we are now ready to state the main theorem in this section.

THEOREM 6.2.6. *Let T be a k-labelled tree and fix a map*

$$T \to H_2(M;\mathbb{Z}) : \alpha \mapsto A_\alpha$$

satisfying the stability condition (6.1.1).

(i) *If $J \in \mathcal{J}_{\mathrm{reg}}(T,\{A_\alpha\})$ then the moduli space $\mathcal{M}^*_{0,T}(\{A_\alpha\};J)$ is a smooth oriented manifold of dimension*

$$\dim \mathcal{M}^*_{0,T}(\{A_\alpha\},J) = \mu(A,T) := 2n + 2c_1(A) + 2k(T) - 6 - 2e(T),$$

where $k(T)$ is the number of marked points and $e(T)$ is the number of edges.

(ii) *The set $\mathcal{J}_{\mathrm{reg}}(T,\{A_\alpha\})$ is residual in $\mathcal{J}_\tau(M,\omega)$.*

PROOF OF THEOREM 6.2.6 (I). Fix $J \in \mathcal{J}_{\mathrm{reg}}(T,\{A_\alpha\})$. Then, by (i) in Definition 6.2.1 and Theorem 3.1.6, the moduli space $\mathcal{M}^*(\{A_\alpha\};J)$ is a smooth oriented manifold of dimension

$$\dim \mathcal{M}^*(\{A_\alpha\};J) = 2n(e(T)+1) + 2c_1(A).$$

By (ii) in Definition 6.2.1, the evaluation map

$$\mathrm{ev}^E : \mathcal{M}^*(\{A_\alpha\};J) \times Z(T) \to M^E$$

is transverse to Δ^E and hence the preimage

$$\widetilde{\mathcal{M}}^*_{0,T}(\{A_\alpha\};J) = (\pi^E)^{-1}(\Delta^E)$$

is a smooth oriented manifold. Since $Z(T)$ has dimension $4e(T) + 2k(T)$ and Δ^E has codimension $2ne(T)$, it follows that

$$\dim \widetilde{\mathcal{M}}^*_{0,T}(\{A_\alpha\};J) = 2n + 2c_1(A) + 2k(T) + 4e(T)$$

Since G_T has dimension $6e(T)+6$ and acts freely and properly by orientation preserving diffeomorphisms on $\widetilde{\mathcal{M}}^*(\{A_\alpha\};J)$, this proves (i) in Theorem 6.2.6. □

The proof of (ii) is based on the following generalization of Proposition 3.2.1. Let T be any finite set and $\{A_\alpha\}_{\alpha\in T}$ be any collection of spherical homology classes. Let $\mathcal{J}^\ell = \mathcal{J}^\ell_\tau(M,\omega)$ denote the Banach manifold of ω-tame almost complex structures of class C^ℓ. Associated to these data is the universal moduli space

$$\mathcal{M}^*(\{A_\alpha\};\mathcal{J}^\ell) := \left\{(\mathbf{u},J) \mid J \in \mathcal{J}^\ell,\ \mathbf{u} \in \mathcal{M}^*(\{A_\alpha\};J)\right\}.$$

As in Section 3.4 we may think of this space as a subset of the Banach manifold $\mathcal{B}^{k,p} \times \mathcal{J}^\ell$, where

$$\mathcal{B}^{k,p} := \left\{\{u_\alpha\}_{\alpha\in T} \mid u_\alpha \in W^{k,p}(S^2,M),\ [u_\alpha] = A_\alpha\right\}.$$

PROPOSITION 6.2.7. *Let $\ell \geq 2$ be an integer, $p > 2$ a real number, $\{A_\alpha\}_{\alpha\in T}$ be a finite collection of spherical homology classes in $H_2(M;\mathbb{Z})$, and $k \in \{1,\ldots,\ell\}$. Then $\mathcal{M}^*(\{A_\alpha\}_{\alpha\in T};\mathcal{J}^\ell)$ is a $C^{\ell-k}$ Banach submanifold of $\mathcal{B}^{k,p} \times \mathcal{J}^\ell$.*

PROOF. The proof is almost word by word the same as that of Proposition 3.2.1. The only difference is that in the proof of transversality we have to deal simultaneously with a collection of $(0,1)$-forms $\eta_\alpha \in \Omega^{0,1}(S^2,u_\alpha^*TM)$ (of class L^q), and must show that they all vanish under the assumptions that $D^*_{u_\alpha}\eta_\alpha = 0$ for every $\alpha \in T$ and that, for every $Y \in T_J\mathcal{J}^\ell$,

$$\sum_{\alpha\in T}\int_{S^2} \langle \eta_\alpha, Y(u_\alpha)du_\alpha \circ i\rangle \mathrm{dvol}_{S^2} = 0. \tag{6.2.3}$$

The proof then proceeds as in Proposition 3.2.1 except that more care must be taken in the choice of Y. Let $Z(u_\alpha) \subset S^2$ denote the set of noninjective points of u_α. Then, by Proposition 2.5.1, the set $Z(u_\alpha)$ is at most countable and can only accumulate at the critical points of u_α. By Proposition 2.4.4, the same holds for the set $u_\alpha^{-1}(u_\beta(S^2))$ whenever $\beta \neq \alpha$. Now it follows from (6.2.3) as in the proof of Proposition 3.2.1 that η_α must vanish at every point in the complement of $Z(u_\alpha) \cup \bigcup_{\beta\neq\alpha} u_\alpha^{-1}(u_\beta(S^2))$. Since this complement is open and dense in S^2 it follows that $\eta_\alpha \equiv 0$ for every α. With this understood, the argument proceeds as in the proof of Proposition 3.2.1. This proves Proposition 6.2.7. □

Now let us fix a k-labelled tree $T = (T, E, \Lambda)$ and consider the universal edge evaluation map

$$\mathrm{ev}^E : \mathcal{M}^*(\{A_\alpha\}; \mathcal{J}^\ell) \times Z(T) \to M^E$$

defined by

$$\mathrm{ev}^E(\mathbf{u}, \mathbf{z}, J) := \{u_\alpha(z_{\alpha\beta})\}_{\alpha E \beta}.$$

PROPOSITION 6.2.8. *For every $\ell \geq 2$, every k-labelled tree $T = (T, E, \Lambda)$, and every collection $\{A_\alpha\}_{\alpha\in T}$ of spherical homology classes in $H_2(M;\mathbb{Z})$, the universal edge evaluation map ev^E is transverse to Δ^E.*

PROOF. See Section 6.3. □

When T has just one edge, Δ^E is a submanifold of M^2 and so this proposition follows from Proposition 3.4.2. In general, Δ^E is a union of cleanly intersecting submanifolds of various dimensions and has the structure of a stratified space. For example, if there are two edges Δ^E is the union of the two submanifolds of M^4 given by the equations $x_1 = x_2$ and $x_3 = x_4$. The extension to this case is not difficult — we will do it by induction over the number of edges — but will be deferred to the next section in order not to interrupt the flow of the current argument.

PROOF OF THEOREM 6.2.6 (II). By Proposition 6.2.7, $\mathcal{M}^*(\{A_\alpha\}; \mathcal{J}^\ell)$ is a separable $C^{\ell-1}$-Banach manifold and, by Proposition 6.2.8, the universal evaluation map ev^E is transverse to Δ^E. Hence it follows from the implicit function theorem A.3.3 that the preimage of Δ^E under ev^E is a separable $C^{\ell-1}$-Banach manifold. This preimage is the universal moduli space

$$\widetilde{\mathcal{M}}^*_{0,T}(\{A_\alpha\}; \mathcal{J}^\ell) := \left\{(\mathbf{u}, \mathbf{z}, J) \,|\, J \in \mathcal{J}^\ell,\ (\mathbf{u}, \mathbf{z}) \in \widetilde{\mathcal{M}}^*_{0,T}(\{A_\alpha\}; J)\right\} \tag{6.2.4}$$

of simple stable maps (for some $J \in \mathcal{J}^\ell$) modelled over the tree T and representing the homology classes A_α. Now consider the projections

$$p^\ell : \mathcal{M}^*(\{A_\alpha\}; \mathcal{J}^\ell) \to \mathcal{J}^\ell, \qquad \pi^\ell : \widetilde{\mathcal{M}}^*_{0,T}(\{A_\alpha\}; \mathcal{J}^\ell) \to \mathcal{J}^\ell.$$

These are Fredholm maps of indices

$$\mathrm{index}(p^\ell) = 2n(1 + e(T)) + 2c_1(A), \qquad \mathrm{index}(\pi^\ell) = \mu(A, T) + \dim \mathrm{G}_T.$$

Hence, by the Sard–Smale theorem A.5.1, the set $\mathcal{J}^\ell_{\mathrm{reg}}(T, \{A_\alpha\})$ of common regular values of p^ℓ and π^ℓ is residual in $\mathcal{J}^\ell$ for ℓ sufficiently large. Moreover, an almost complex structure $J \in \mathcal{J}^\ell$ is a common regular value of p^ℓ and π^ℓ if and only if it satisfies the conditions of Definition 6.2.1.

Now, for every $K > 0$, consider the subset $\mathcal{M}^*_K(\{A_\alpha\};\mathcal{J}^\ell) \subset \mathcal{M}^*(\{A_\alpha\};\mathcal{J}^\ell)$ of all tuples $(\mathbf{u}, J) \in \mathcal{M}^*(\{A_\alpha\};\mathcal{J}^\ell)$ that satisfy

$$\|du_\alpha\|_{L^\infty} \le K \tag{6.2.5}$$

and

$$\inf_{\zeta\in S^2\setminus\{z_\alpha\}} \frac{d(u_\alpha(z_\alpha), u_\alpha(\zeta))}{d(z_\alpha,\zeta)} \ge \frac{1}{K}, \qquad \inf_{\zeta\in S^2} d(u_\alpha(z_\alpha), u_\beta(\zeta)) \ge \frac{1}{K} \tag{6.2.6}$$

for every $\alpha \in T$, $\beta \in T\setminus\{\alpha\}$, and some collection of points $\{z_\alpha\}_{\alpha\in T}$ in S^2. Likewise, let $Z_K(T) \subset Z(T)$ be the set of all tuples $\mathbf{z} \in Z(T)$ that satisfy

$$d(z_{\alpha\beta}, z_{\alpha\gamma}) \ge \frac{1}{K}, \qquad d(z_i, z_j) \ge \frac{1}{K}, \qquad d(z_{\alpha\beta}, z_i) \ge \frac{1}{K}$$

for all α, $\beta \ne \gamma$, $i \ne j$ with $\alpha E\beta$, $\alpha E\gamma$, and $\alpha_i = \alpha_j = \alpha$, and denote

$$\widetilde{\mathcal{M}}^*_{0,T;K}(\{A_\alpha\};J) := \widetilde{\mathcal{M}}^*_{0,T}(\{A_\alpha\};J) \cap \Big(\mathcal{M}^*_K(\{A_\alpha\};\mathcal{J}^\ell) \times Z_K(T)\Big).$$

Then the projections

$$p^\ell_K : \mathcal{M}^*_K(\{A_\alpha\};\mathcal{J}^\ell) \to \mathcal{J}^\ell, \qquad \pi^\ell_K : \widetilde{\mathcal{M}}^*_{0,T;K}(\{A_\alpha\};\mathcal{J}^\ell) \to \mathcal{J}^\ell$$

are proper Fredholm maps and so the set $\mathcal{J}^\ell_{\mathrm{reg,K}}(T,\{A_\alpha\})$ of common regular values of p^ℓ_K and π^ℓ_K is open and dense in $\mathcal{J}^\ell$ for ℓ sufficiently large. By the same reasoning the set

$$\mathcal{J}_{\mathrm{reg,K}}(T,\{A_\alpha\}) := \mathcal{J}^\ell_{\mathrm{reg,K}}(T,\{A_\alpha\}) \cap \mathcal{J}_\tau(M,\omega)$$

is open in $\mathcal{J}_\tau(M,\omega)$. Moreover, since $\mathcal{J}^\ell_{\mathrm{reg,K}}(T,\{A_\alpha\})$ is dense in $\mathcal{J}^\ell$ for ℓ sufficiently large it follows as in the proof of Theorem 3.1.6 that $\mathcal{J}_{\mathrm{reg,K}}(T,\{A_\alpha\})$ is dense in $\mathcal{J}_\tau(M,\omega)$. Hence the set

$$\mathcal{J}_{\mathrm{reg}}(T,\{A_\alpha\}) = \bigcap_{K>0} \mathcal{J}_{\mathrm{reg,K}}(T,\{A_\alpha\})$$

is a countable intersection of open and dense sets and is therefore a residual subset of $\mathcal{J}_\tau(M,\omega)$. This proves (ii) in Theorem 6.2.6. □

EXERCISE 6.2.9. Show that the projection $\pi^\ell : \mathcal{M}^*_{0,T}(\{A_\alpha\};\mathcal{J}^\ell) \to \mathcal{J}^\ell$ is a Fredholm map of index index $\mu(A,T) + \dim \mathrm{G}_T = 2n + 2c_1(A) + 2k + 4e(T)$.

As in Chapter 3, we shall introduce an appropriate class of regular homotopies of almost complex structures in order to obtain smooth cobordisms between moduli spaces of simple stable maps. Suppose $J_0, J_1 \in \mathcal{J}_{\mathrm{reg}}(T,\{A_\alpha\})$ and let $[0,1] \to \mathcal{J}_\tau(M,\omega) : \lambda \mapsto J_\lambda$ be a smooth homotopy of ω-tame almost complex structures on M. Associated to any such homotopy is the moduli space

$$\widetilde{\mathcal{W}}^*_{0,T}(\{A_\alpha\};\{J_\lambda\}) := \Big\{(\lambda,\mathbf{u},\mathbf{z})\,|\,0 \le \lambda \le 1,\ (\mathbf{u},\mathbf{z}) \in \widetilde{\mathcal{M}}^*_{0,T}(\{A_\alpha\};J_\lambda)\Big\}$$

of simple stable maps and its quotient by the reparametrization group:

$$\mathcal{W}^*_{0,T}(\{A_\alpha\};\{J_\lambda\}) := \widetilde{\mathcal{W}}^*_{0,T}(\{A_\alpha\};\{J_\lambda\})/\mathrm{G}_T.$$

We denote by $\mathcal{J}_\tau(J_0,J_1)$ the set of all smooth homotopies of ω-tame almost complex structures on M from J_0 to J_1.

DEFINITION 6.2.10. *A homotopy* $\{J_\lambda\} \in \mathcal{J}_\tau(J_0, J_1)$ *is called* **regular (for** T **and** $\{A_\alpha\}$**)** *if*

$$J_0, J_1 \in \mathcal{J}_{\mathrm{reg}}(T, \{A_\alpha\})$$

and the following holds.

(i) *For every* $\lambda \in [0,1]$, *every tuple*

$$\mathbf{u} \in \mathcal{M}^*(\{A_\alpha\}; J_\lambda),$$

and every $\alpha \in T$,

$$\Omega^{0,1}(S^2, u_\alpha^*TM) = \operatorname{im} D_{u_\alpha} + \mathbb{R}\, v_\lambda,$$

where $v_\lambda := (\partial_\lambda J_\lambda) du_\alpha \circ j$ *is the image of the tangent vector to the path* $\lambda \mapsto J_\lambda$.

(ii) *The evaluation map*

$$\mathcal{W}^*(\{A_\alpha\}; \{J_\lambda\}) \times Z(T) \to M^E : (\lambda, \mathbf{u}, \mathbf{z}) \mapsto \{u_\alpha(z_{\alpha\beta})\}_{\alpha E \beta}$$

is transverse to Δ^E.

As in the case of Definition 6.2.1, condition (ii) makes sense because (i) implies that the domain of the evaluation map is a manifold. Explicitly, this condition reads as follows: for every

$$(\lambda, \mathbf{u}, \mathbf{z}) \in \widetilde{\mathcal{W}}_{0,T}^*(\{A_\alpha\}; \{J_\lambda\})$$

and every collection of tangent vectors $v_{\alpha\beta} \in T_{u_\alpha(z_{\alpha\beta})}M$ satisfying

$$v_{\alpha\beta} + v_{\beta\alpha} = 0,$$

there exist vector fields $\xi_\alpha \in \Omega^0(S^2, u_\alpha^*TM)$ and tangent vectors $\zeta_{\alpha\beta} \in T_{z_{\alpha\beta}}S^2$ such that $D_{u_\alpha}\xi_\alpha \in \mathbb{R}\, v_\lambda$ and

$$v_{\alpha\beta} = \xi_\alpha(z_{\alpha\beta}) + du_\alpha(z_{\alpha\beta})\zeta_{\alpha\beta} - \xi_\beta(z_{\beta\alpha}) - du_\beta(z_{\beta\alpha})\zeta_{\beta\alpha}$$

for all $\alpha, \beta \in T$ such that $\alpha E \beta$.

Let $\mathcal{J}_{\mathrm{reg}}(T, \{A_\alpha\}; J_0, J_1)$ denote the set of regular homotopies for T and $\{A_\alpha\}$ and $\mathcal{J}_{\mathrm{reg}}(M, \omega; J_0, J_1)$ the intersection of these sets over all trees T and all $\{A_\alpha\}$. The following result is a straightforward extension of Theorem 6.2.6. It is proved by combining the arguments in the proof of Theorem 6.2.6 with those in the proof of Theorem 3.1.8. The details are left as an exercise to the reader.

THEOREM 6.2.11. *Let* T *be a* k*-labelled tree, fix a map*

$$T \to H_2(M; \mathbb{Z}) : \alpha \mapsto A_\alpha$$

satisfying the stability condition (6.1.1), *and suppose that* $J_0, J_1 \in \mathcal{J}_{\mathrm{reg}}(T, \{A_\alpha\})$.

(i) *If* $\{J_\lambda\} \in \mathcal{J}_{\mathrm{reg}}(T, \{A_\alpha\}; J_0, J_1)$ *then the moduli space*

$$\mathcal{W}_{0,T}^*(\{A_\alpha\}, \{J_\lambda\})$$

is a smooth oriented manifold with boundary

$$\partial \mathcal{W}_{0,T}^*(\{A_\alpha\}; \{J_\lambda\}) = \mathcal{M}_{0,T}^*(\{A_\alpha\}; J_1) \cup \big(-\mathcal{M}_{0,T}^*(\{A_\alpha\}; J_0)\big).$$

(ii) *The set* $\mathcal{J}_{\mathrm{reg}}(T, \{A_\alpha\}; J_0, J_1)$ *is residual in* $\mathcal{J}_\tau(J_0, J_1)$.

6.3. Transversality for evaluation maps

This section is devoted to the proof of Proposition 6.2.8 stating that for any tree T the evaluation map

$$\mathrm{ev}^E : \mathcal{M}^*(\{A_\alpha\}; \mathcal{J}^\ell) \times Z(T) \to M^E$$

is transverse to the diagonal Δ^E. It turns out to be convenient to prove this result for objects more general than trees, where the hypothesis that any two vertices can be connected by a chain of edges has been dropped. A finite set T with a relation $E \subset T \times T$ is called a **forest** if it satisfies the *(Symmetry)*, *(Antireflexivity)*, and *(No cycles)* conditions in Section D.2. The *(Connectedness)* axiom is not required. In this case we shall denote $[\alpha, \beta] := \emptyset$ if there is no chain of edges connecting α to β. All the definitions in this chapter carry over to forests in the obvious fashion. For example, the universal moduli space space of simple stable maps is the preimage

$$\widetilde{\mathcal{M}}^*_{0,T}(\{A_\alpha\}; \mathcal{J}^\ell) = (\pi^E)^{-1}(\Delta^E) \subset \widetilde{\mathcal{M}}^*(\{A_\alpha\}; \mathcal{J}^\ell) \times Z(T)$$

of the diagonal under the universal evaluation map as in (6.2.4), and the reparametrization group G_T is defined as before. The space $\widetilde{\mathcal{M}}^*_{0,T}(\{A_\alpha\}; \mathcal{J}^\ell)$ is a $C^{\ell-1}$ Banach manifold whenever ev^E is transverse to the edge diagonal Δ^E of the labelled forst (T, E, Λ). In this case the group G_T acts freely and properly by $C^{\ell-1}$ diffeomorphisms on $\widetilde{\mathcal{M}}^*_{0,T}(\{A_\alpha\}; \mathcal{J}^\ell)$, because each u_α is of class $W^{\ell+1,p}$ for every $(\mathbf{u}, \mathbf{z}, J) \in \widetilde{\mathcal{M}}^*_{0,T}(\{A_\alpha\}; \mathcal{J}^\ell)$. Hence the quotient

$$\mathcal{M}^*_{0,T}(\{A_\alpha\}; \mathcal{J}^\ell) := \widetilde{\mathcal{M}}^*_{0,T}(\{A_\alpha\}; \mathcal{J}^\ell)/\mathrm{G}_T.$$

is a $C^{\ell-1}$ Banach manifold whenever ev^E is transverse to Δ^E.

The proof of Proposition 6.2.8 is based on the following analogue of Proposition 3.4.2. As in Section 3.4, the weaker statement that the complement of the diagonal consists entirely of regular values of the evaluation map ev_{ij} has an elementary proof, which uses the action of $\mathrm{Diff}(M, \omega)$ on the universal moduli space $\mathcal{M}^*_{0,T}(\{A_\alpha\}; \mathcal{J}^\ell)$. Details of this argument are left to the reader. We explain here the much more sophisticated argument needed to show that every point on the diagonal is also a regular value.

THEOREM 6.3.1. *Let $T = (T, E, \Lambda)$ be a k-labelled forest and*

$$T \to H_2(M; \mathbb{Z}) : \alpha \mapsto A_\alpha$$

be a map that satisfies the stability condition (6.1.1). Suppose that the edge evaluation map

$$\mathrm{ev}^E : \mathcal{M}^*(\{A_\alpha\}; \mathcal{J}^\ell) \times Z(T) \to M^E$$

is transverse to Δ^E. Let $i, j \in \{1, \dots, k\}$ such that the vertices α_i, α_j belong to different components of T, i.e. $[\alpha_i, \alpha_j] = \emptyset$. Then every point in $M \times M$ is a regular value of the evaluation map

$$\mathrm{ev}_{ij} : \mathcal{M}^*_{0,T}(\{A_\alpha\}; \mathcal{J}^\ell) \to M \times M,$$

defined by

$$\mathrm{ev}_{ij}([\mathbf{u}, \mathbf{z}, J]) := (u_{\alpha_i}(z_i), u_{\alpha_j}(z_j)).$$

PROOF. By Proposition 6.2.7, $\mathcal{M}^*(\{A_\alpha\};\mathcal{J}^\ell)$ is a Banach manifold and, by assumption, the universal evaluation map ev^E is transverse to Δ^E. Hence it follows from the implicit function theorem that the universal moduli space

$$\widetilde{\mathcal{M}}^*_{0,T}(A,\mathcal{J}^\ell) = (\mathrm{ev}^E)^{-1}(\Delta^E)$$

of simple stable maps is a Banach manifold. Its tangent space at a point $(\mathbf{u},\mathbf{z},J)$ is the set of all tuples

$$(\xi,\zeta,Y) = (\{\xi_\alpha\}_{\alpha\in T},\{\zeta_{\alpha\beta}\}_{\alpha E\beta},\{\zeta_i\}_{1\le i\le k},Y),$$

where $\xi_\alpha \in W^{\ell,p}(S^2,u_\alpha^*TM)$, $\zeta_{\alpha\beta}\in T_{z_{\alpha\beta}}S^2$, $\zeta_i\in T_{z_i}S^2$, and $Y\in T_J\mathcal{J}^\ell$ satisfy

$$D_{u_\alpha}\xi_\alpha + \tfrac{1}{2}Y(u_\alpha)\circ du_\alpha\circ j = 0, \tag{6.3.1}$$

$$\xi_\alpha(z_{\alpha\beta}) + du_\alpha(z_{\alpha\beta})\zeta_{\alpha\beta} = \xi_\beta(z_{\beta\alpha}) + du_\beta(z_{\beta\alpha})\zeta_{\beta\alpha} \tag{6.3.2}$$

for all $\alpha,\beta\in T$ such that $\alpha E\beta$. Note that (6.3.2) is simply the derivative of the condition $u_\alpha(z_{\alpha\beta}) = u_\beta(z_{\beta\alpha})$. The differential of the evaluation map ev_{ij} at $(\mathbf{u},\mathbf{z},J)$ is given by

$$d\,\mathrm{ev}_{ij}(\mathbf{u},\mathbf{z},J)(\xi,\zeta,Y) = \left(\xi_{\alpha_i}(z_i) + du_{\alpha_i}(z_i)\zeta_i, \xi_{\alpha_j}(z_j) + du_{\alpha_j}(z_j)\zeta_j\right). \tag{6.3.3}$$

We prove that this differential is onto for every $(\mathbf{u},\mathbf{z},J)\in\widetilde{\mathcal{M}}^*_{0,T}(A,\mathcal{J}^\ell)$. By symmetry in i and j it suffices to prove that the subspace $T_{u_{\alpha_i}(z_i)}M\times\{0\}$ is contained in the image of $d\mathrm{ev}_{ij}(\mathbf{u},\mathbf{z},J)$. Hence let us fix a vector $v_i\in T_{u_{\alpha_i}(z_i)}M$. We prove that the pair $(v_i,0)$ is contained in the image of $d\mathrm{ev}_{ij}(\mathbf{u},\mathbf{z},J)$.

Assume first that u_{α_i} is nonconstant. Then, by Propositions 2.5.1 and 2.4.4, there exists an injective point $\mathrm{w}_i\in S^2$ for u_{α_i} such that $u_{\alpha_i}(\mathrm{w}_i)\notin u_\beta(S^2)$ for $\beta\ne\alpha_i$. Choose $\varepsilon>0$ such that

$$\beta\ne\alpha_i \qquad\Longrightarrow\qquad B_\varepsilon(u_{\alpha_i}(\mathrm{w}_i))\cap u_\beta(S^2)=\emptyset$$

for every $\beta\in T$. Then, by Lemma 3.4.3, there exists a pair (ξ_{α_i},Y) satisfying (6.3.1) for $\alpha=\alpha_i$ and

$$\xi_{\alpha_i}(z_i)=v_i,\qquad \xi_{\alpha_i\beta}(z_{\alpha_i\beta})=0,\qquad \operatorname{supp}Y\subset B_\varepsilon(u_{\alpha_i}(\mathrm{w}_i)),$$

for every $\beta\in T$ such that $\alpha_iE\beta$. It follows that $Y\circ u_\beta\equiv 0$ for $\beta\ne\alpha_i$. Hence the tuple (ξ,ζ,Y), defined by $\zeta:=0$ and $\xi_\beta:=0$ for $\beta\ne\alpha_i$, satisfies (6.3.1) and (6.3.2). Moreover, $\alpha_j\ne\alpha_i$ and hence $d\,\mathrm{ev}_{ij}(\mathbf{u},\mathbf{z},J)(\xi,\zeta,Y)=(v_i,0)$.

Now suppose that u_{α_i} is constant and consider the subtree

$$T_i := \{\alpha\in T\,|\,[\alpha_i,\alpha]\ne\emptyset,\ A_\beta=0\ \forall\beta\in[\alpha_i,\alpha]\}.$$

It consists of all vertices that can be connected to α_i by a chain of ghost components. Note that $u_\alpha\equiv u_{\alpha_i}(z_i)$ for every $\alpha\in T_i$. Let us denote by

$$V_i := \{\beta\in T\setminus T_i\,|\,\exists\alpha\in T_i \text{ such that } \alpha E\beta\}$$

the set of vertices that are adjacent to T_i. Then $A_\beta\ne 0$ for every $\beta\in V_i$ and there is a unique map $f:V_i\to T_i$ such that $f(\beta)E\beta$ for every $\beta\in V_i$. Moreover, it follows from Propositions 2.5.1 and 2.4.4 that, for every $\beta\in V_i$, there exists an injective point $\mathrm{w}_\beta\in S^2$ for u_β such that $u_\beta(\mathrm{w}_\beta)\notin u_\gamma(S^2)$ for every $\gamma\in T\setminus\{\beta\}$. Choose $\varepsilon>0$ such that, for all $\beta,\gamma\in T$, we have

$$\beta\in V_i,\quad \beta\ne\gamma \qquad\Longrightarrow\qquad B_\varepsilon(u_\beta(\mathrm{w}_\beta))\cap u_\gamma(S^2)=\emptyset.$$

Now let $\beta \in V_i$. Then, by Lemma 3.4.3, there exists a pair (ξ_β, Y_β) that satisfies (6.3.1) and

$$\xi_\beta(z_{\beta f(\beta)}) = v_i, \qquad \xi_\beta(z_{\beta\gamma}) = 0, \qquad \operatorname{supp} Y_\beta \subset B_\varepsilon(u_\beta(\mathrm{w}_\beta)),$$

for every $\gamma \in T \setminus \{f(\beta)\}$ such that $\beta E \gamma$. It follows that $Y_\beta \circ u_\gamma \equiv 0$ for $\beta \in V_i$ and $\gamma \neq \beta$. Define the tuple (ξ, ζ, Y) by

$$\zeta := 0, \qquad \xi_\alpha := \begin{cases} v_i, & \text{for } \alpha \in T_i, \\ 0, & \text{for } \alpha \in T \setminus (T_i \cup V_i), \end{cases} \qquad Y := \sum_{\beta \in V_i} Y_\beta.$$

Then $\xi_\alpha \in \ker D_{u_\alpha}$ and $Y \circ u_\alpha = 0$ for every $\alpha \in T \setminus V_i$. Moreover, $Y \circ u_\beta = Y_\beta \circ u_\beta$ for every $\beta \in V_i$. Hence the tuple (ξ, ζ, Y) satisfies (6.3.1) and (6.3.2). Moreover, $\alpha_j \notin T_i \cup V_i$ and hence $d\mathrm{ev}_{ij}(\mathbf{u}, \mathbf{z}, J)(\xi, \zeta, Y) = (v_i, 0)$. This proves Theorem 6.3.1. □

PROOF OF PROPOSITION 6.2.8. We prove a stronger result, namely that ev^E is transverse to Δ^E for every forest (T, E, Λ). The assertion is vacuous when the forest has no edges. Let N be a nonnegative integer and suppose, by induction, that the assertion has been proved for all forests with at most N edges. Let T be a k-labelled forest with $N+1$ edges. Choose any edge $\alpha E \beta$ and consider the labelled forest $T' = (T', E', \Lambda')$ obtained by cutting the edge and adding the two points $z_{\alpha\beta}, z_{\beta\alpha}$ to the collection of marked points. It is defined by

$$T' := T, \quad E' := E \setminus \{(\alpha, \beta), (\beta, \alpha)\}, \quad \Lambda'_\gamma := \begin{cases} \Lambda_\gamma, & \text{if } \gamma \notin \{\alpha, \beta\}, \\ \{\alpha\beta\} \cup \Lambda_\gamma, & \text{if } \gamma = \alpha, \\ \{\beta\alpha\} \cup \Lambda_\gamma, & \text{if } \gamma = \beta. \end{cases}$$

The labels are indexed by the set

$$I' := \{\alpha\beta, \beta\alpha\} \cup \{1, \ldots, k\}.$$

Note that $Z(T') = Z(T)$. The induction hypothesis asserts that the universal evaluation map

$$\mathrm{ev}^{E'} : \mathcal{M}^*(\{A_\gamma\}; \mathcal{J}^\ell) \times Z(T) \to M^{E'}$$

is transverse to $\Delta^{E'}$. So $\widetilde{\mathcal{M}}^*_{0,T'}(\{A_\alpha\}; \mathcal{J}^\ell) = (\mathrm{ev}^{E'})^{-1}(\Delta^{E'})$ is a $C^{\ell-1}$-Banach manifold. Consider the evaluation map

$$\mathrm{ev}_{\alpha\beta} : \widetilde{\mathcal{M}}^*_{0,T'}(\{A_\alpha\}; \mathcal{J}^\ell) \to M \times M,$$

defined by

$$\mathrm{ev}_{\alpha\beta}(\mathbf{u}, \mathbf{z}, J) := (u_\alpha(z_{\alpha\beta}), u_\beta(z_{\beta\alpha})).$$

By Theorem 6.3.1, every point in $M \times M$ is a regular value of $\mathrm{ev}_{\alpha\beta}$. Hence $\mathrm{ev}_{\alpha\beta}$ is transverse to the diagonal $\Delta \subset M \times M$. By Exercise 6.3.2 below, it follows that $\mathrm{ev}^E = \mathrm{ev}^{E'} \times \mathrm{ev}_{\alpha\beta}$ is transverse to $\Delta^E = \Delta^{E'} \times \Delta$. This completes the induction and the proof of Proposition 6.2.8. □

EXERCISE 6.3.2. Let $f_0 : X \to Y_0$ and $f_1 : X \to Y_1$ be C^ℓ-maps between C^ℓ (Banach) manifolds and $Z_0 \subset Y_0$ and $Z_1 \subset Y_1$ be C^ℓ submanifolds. Suppose that f_0 is transverse to Z_0 and that the restriction of f_1 to $f_0^{-1}(Z_0)$ is transverse to Z_1. Prove that $f_0 \times f_1$ is transverse to $Z_0 \times Z_1$.

Consider a tree with one vertex and one marked point. Let $\mathrm{ev} : \mathcal{M}^*_{0,1}(A; \mathcal{J}) \to M$ be the corresponding (smooth) universal evaluation map. Although this evaluation map is always a submersion, it does not follows that there is an almost complex structure $J \in \mathcal{J}$ such that the evaluation map

$$\mathrm{ev}_J : \mathcal{M}^*_{0,1}(A; J) \to M$$

is a submersion. For example, the moduli space $\mathcal{M}^*_{0,1}(A; J)$ may simply have too small a dimension. Other examples may be constructed in the situation of the next exercise. This exercise also illustrates the need to restrict to the simple stable maps in Theorem 6.2.6. It is based on the fact that when E is the class of the exceptional divisor in $\mathbb{C}P^2\#\overline{\mathbb{C}P}^2$ the moduli space $\mathcal{M}_{0,0}(E, J)$ contains precisely one element for generic compatible J (see Example 7.1.15). Hence, by positivity of intersections, the class $2E$ has no simple representatives.

EXERCISE 6.3.3. Show that $\mathcal{M}_{0,0}(\mathbb{C}P^2\#\overline{\mathbb{C}P}^2, 2E, J)$ is a manifold of dimension 4 for generic J, while its virtual dimension is 2. Calculate the actual dimension of the stratum $\mathcal{M}_{0,T}(\mathbb{C}P^2\#\overline{\mathbb{C}P}^2, 2E, J)$ when T has two vertices and no marked points and compare it with the virtual dimension.

6.4. Semipositivity

In this section we examine hypotheses that allow us to control the dimensions of the higher strata of $\overline{\mathcal{M}}_{0,k}(A; J)$. The key to these conditions are the values of the first Chern numbers $c_1(A) = \langle c_1(TM, J), A\rangle$ for homology classes $A \in H_2(M; \mathbb{Z})$ that can be represented by J-holomorphic spheres. Indeed, as explained in the introduction to this chapter, we need conditions that ensure that for generic J there are no J-holomorphic spheres with negative Chern number.

A symplectic manifold (M, ω) is called **(spherically) monotone** if there is a number $\tau > 0$ such that

$$c_1(A) = \tau\omega(A)$$

for every $A \in \pi_2(M)$; it is called **semimonotone** if the above equation holds for $\tau \geq 0$. This is a natural but very restrictive condition. As examples, consider the **Fano varieties**. These are complex manifolds (M, J) with ample anticanonical bundle $K^* = \Lambda^{n,0}TM$. In the simply connected case, this is equivalent to the existence of a Kähler metric such that the Kähler form ω is spherically monotone.[1] More relevant to us is the following ad hoc, but useful, definition.

DEFINITION 6.4.1. *A symplectic manifold (M, ω) is called* **semipositive** *if, for every $A \in \pi_2(M)$,*

$$\omega(A) > 0, \quad c_1(A) \geq 3 - n \qquad \Longrightarrow \qquad c_1(A) \geq 0.$$

Thus semimonotone symplectic manifolds are semipositive[2] as are smooth projective varieties that are NEF (i.e. *numerically effective*). Conversely many examples of semipositive manifolds (M, ω) are semimonotone for some symplectic form ω' but this may well be different from ω. In order to characterize the semipositive manifolds we make the following definition.

[1] In general, a Kähler manifold (M, J) has an ample canonical bundle if and only if there is a Kähler form in the class c_1. Thus $[\omega] = c_1$ for all classes A, not just the spherical ones.

[2] In the first edition of this book, we used the term **weakly monotone** instead of semipositive. However, the latter term seems to be more common now in the literature. Note also that because this definition is adopted out of convenience, it is sometimes modified to suit the circumstances.

DEFINITION 6.4.2. *Let* (M,ω) *be a compact symplectic manifold. Then the* **minimal Chern number** *of* (M,ω) *is the integer*

$$N = \inf\left\{k > 0 \,\middle|\, \exists\, A \in \pi_2(M),\, c_1(A) = k\right\}.$$

If $c_1(A) = 0$ *for every* $A \in \pi_2(M)$ *we call* $N = \infty$ *the minimal Chern number. If* $N \neq \infty$ *then*

$$\langle c_1, \pi_2(M)\rangle = N\mathbb{Z}.$$

EXERCISE 6.4.3. Prove that a symplectic manifold (M,ω) is semipositive if and only if one of the following conditions is satisfied.

(i) (M,ω) is semimonotone.

(ii) $c_1(A) = 0$ for every $A \in \pi_2(M)$.

(iii) The minimal Chern number N satisfies $N \geq n-2$.

These conditions are not mutually exclusive. We have emphasized the case when $c_1 = 0$ on $\pi_2(M)$ since this includes the important case of Calabi–Yau manifolds. But this is just the case $N = \infty$ of (iii). Note also that (iii) implies that every symplectic manifold of dimension less than or equal to six is semipositive. It also includes some cases when the restriction of c_1 to $\pi_2(M)$ is a negative multiple of $[\omega]$.

The following lemma highlights the most important consequence of semipositivity.

LEMMA 6.4.4. *Let* (M,ω) *be a semipositive symplectic manifold,* J *be an* ω*-tame almost complex structure, and* $u: S^2 \to M$ *be a* J*-holomorphic sphere. If the cokernel of* D_u *has dimension zero or one then* $c_1(A) \geq 0$*, where* $A \in \pi_2(M)$ *is the homotopy class of* u.

PROOF. If u is constant then $c_1(A) = 0$. Hence assume that u is nonconstant. Then $\omega(A) = E(u) > 0$. Moreover, if $\zeta \in \mathrm{Vect}(S^2)$ is an infinitesimal Möbius transformation then $du \circ \zeta \in \ker D_u$. Hence $\dim \ker D_u \geq 6$. Since $\dim\mathrm{coker}\, D_u \leq 1$ this implies $\mathrm{index}\, D_u = 2n + 2c_1(A) \geq 5$. Hence it follows from the semipositivity hypothesis that $c_1(A) \geq 0$. This proves Lemma 6.4.4. □

Although the condition that (M,ω) be semipositive depends only on the cohomology classes $[\omega]$ and $c_1 = c_1(TM,J)$, what we shall need in practice is the conclusion of the previous lemma that refers to properties of specific almost complex structures J. Therefore it is useful to make the following definition.

DEFINITION 6.4.5. *Let* (M,ω) *be a symplectic manifold and let* $\kappa > 0$*. An* ω*-tame almost complex structure* $J \in \mathcal{J}_\tau(M,\omega)$ *is called* κ-**semipositive** *if every* J*-holomorphic sphere* $u: \mathbb{C}P^1 \to M$ *with energy* $E(u) \leq \kappa$ *has nonnegative Chern number. It is called* **semipositive** *if it is* κ*-semipositive for every* κ*. This means that every* J*-holomorphic sphere has nonnegative Chern number.*

Denote by $\mathcal{J}_+(M,\omega;\kappa)$ the set of all κ-semipositive almost complex structures on M that are ω-tame and by

$$\mathcal{J}_+(M,\omega) := \bigcap_{\kappa>0} \mathcal{J}_+(M,\omega;\kappa)$$

the set of semipositive ω-tame almost complex structures. (Note that these sets may be empty.)

REMARK 6.4.6. We are working here with the space $\mathcal{J}_\tau(M,\omega)$ of all ω-tame almost complex structures, and so have used these J in the above definitions. However, all the results of this section remain valid if we consider instead the space $\mathcal{J}(M,\omega)$ of all ω-compatible almost complex structures, replacing the word ω-tame by ω-compatible wherever it occurs. The advantage of working with ω-tame almost complex structures is that one can vary ω without changing J.

LEMMA 6.4.7. *Let (M,ω) be a compact symplectic $2n$-manifold. Then, for every $\kappa > 0$, the set $\mathcal{J}_+(M,\omega;\kappa)$ is open in $\mathcal{J}_\tau(M,\omega)$ with respect to the C^∞-topology.*

PROOF. Let $J_\nu \in \mathcal{J}_\tau(M,\omega)$ be a sequence of almost complex structures on M that are not κ-semipositive. Assume that J_ν converges to $J \in \mathcal{J}_\tau(M,\omega)$ in the C^∞-topology. Then there exists a sequence $u_\nu : \mathbb{C}P^1 \to M$ of J_ν-holomorphic spheres with $c_1(u_\nu) < 0$ and $E(u_\nu) \le \kappa$. By Theorem 5.3.1, u_ν has a subsequence that Gromov converges to a stable map $\mathbf{u}$. By Theorem 5.2.2, one of its components u_α must have negative Chern number and they all have energy $E(u_\alpha) \le \kappa$. Hence J is not κ-semipositive. This proves Lemma 6.4.7. □

If J is semipositive then the size of a neighbourhood which consists of κ-semipositive structures may depend on κ. Hence the set of semipositive almost complex structures on a compact symplectic manifold need not be open. However, it is residual. Moreover, though it need not be path connected itself, the approximating sets $\mathcal{J}_+(M,\omega;\kappa)$ are, which turns out to be sufficient for our purposes: see the proof of Theorem 6.6.1.

LEMMA 6.4.8. *Let (M,ω) be a compact semipositive symplectic manifold. Then $\mathcal{J}_+(M,\omega)$ is dense in $\mathcal{J}_\tau(M,\omega)$ and, for every $\kappa > 0$, the set $\mathcal{J}_+(M,\omega;\kappa)$ is open and path connected.*

PROOF. Consider the set $\mathcal{J}_r$ of all ω-tame almost complex structures J on M such that for every J-holomorphic sphere u the cokernel of the operator D_u has dimension at most one. By Lemma 6.4.4, this set is contained in $\mathcal{J}_+(M,\omega)$ and, by Theorem 3.1.8, it is residual. Hence $\mathcal{J}_+(M,\omega)$ is dense in $\mathcal{J}_\tau(M,\omega)$.

Now let $\kappa > 0$ and $J_0, J_1 \in \mathcal{J}_+(M,\omega;\kappa)$. Since $\mathcal{J}_+(M,\omega;\kappa)$ is open in $\mathcal{J}_\tau(M,\omega)$ we may assume without loss of generality that $J_0, J_1 \in \mathcal{J}_{\mathrm{reg}}(M,\omega)$. Now Theorem 3.1.8 asserts that J_0 and J_1 can be connected by a regular homotopy $\{J_\lambda\}$ in the sense of Definition 3.1.7. It follows from the definition of a regular homotopy that $\dim\operatorname{coker} D_u \le 1$ for every λ and every J_λ-holomorphic sphere $u : S^2 \to M$. Hence $J_\lambda \in \mathcal{J}_r \subset \mathcal{J}_+(M,\omega)$ for every $\lambda \in [0,1]$. This proves Lemma 6.4.8. □

As a final remark, we observe that there are interesting examples of symplectic manifolds that are not semipositive but where the conclusion of Lemma 6.4.4 still holds for some (but not all) regular ω-tame almost complex structures. In such cases the techniques developed in this book allow the definition of invariants of the manifold (M,ω) that *a priori* might depend on the choice of J: see Remark 6.6.2. If we used the fact that Gromov–Witten invariants are defined for all symplectic manifolds, we could conclude that the invariants are independent of J. But even within the framework of the present book we still get interesting invariants of the triple (M,ω,J). This is particularly relevant in the context of Kähler geometry, where one is often more interested in the properties of the underlying complex manifold (M,J) than in a particular choice of the Kähler form ω.

As an example, consider a product of projective spaces,

$$M = \mathbb{C}P^{n_1} \times \cdots \times \mathbb{C}P^{n_k},$$

with the symplectic form

$$\omega_\lambda = \sum_{i=1}^{k} \lambda_i \omega_i,$$

where ω_i is (the pullback to M of) the usual Kähler form on $\mathbb{C}P^{n_i}$, normalized so that it integrates to π over each complex line. Then M is monotone if and only if the ratios $\lambda_i/(1+n_i)$ are all equal. In most other cases the manifold (M, ω_λ) is not even semipositive.

EXERCISE 6.4.9. Find necessary and sufficient conditions for the product manifold $(\mathbb{C}P^{n_1} \times \cdots \times \mathbb{C}P^{n_k}, \omega_\lambda)$ to be semipositive. Show on the other hand that if $J_{split} = J_1 \times \cdots \times J_k$ is a product of almost complex structures J_i on the factors, then all nonconstant J_{split}-holomorphic spheres have positive Chern numbers.

EXERCISE 6.4.10. Consider the manifold

$$M = S^2 \times S^2$$

with the symplectic form

$$\omega_\lambda := \lambda \pi_1^* \sigma \oplus \pi_2^* \sigma.$$

(See Example 3.3.6.) The standard complex structure on this manifold is semipositive. When $\lambda = 1$ then $(S^2 \times S^2, \omega_\lambda)$ is monotone so that every ω_λ-compatible J is positive in the sense that every J-holomorphic curve has positive Chern number.

Now identify S^2 with $\mathbb{C} \cup \{\infty\}$ so that

$$\sigma = \frac{dx \wedge dy}{(1 + x^2 + y^2)^2}$$

(see Exercise 4.2.1). For $m > 0$ define the embedding $u_m : S^2 \to S^2 \times S^2$ by

$$u_m(z) := (z, \bar{z}^m).$$

Show that u_m is a symplectic embedding with respect to ω_λ if and only if $\lambda > m^2$. Deduce that for every $\lambda > m^2$ there exists an ω_λ-compatible J which admits a J-holomorphic sphere with Chern number $2 - m$. (In fact, as is explained in Exercise 3.3.6, the previous statement holds whenever $\lambda > m$, but this is hard to see using our present very elementary methods.)

6.5. Pseudocycles

The next step towards defining the Gromov–Witten invariants is to develop a suitable notion of a cycle. Let X be a smooth m-dimensional manifold. An arbitrary subset $B \subset X$ is said to be of **dimension at most** d if it is contained in the image of a map $g : W \to X$ which is defined on a manifold W whose components have dimension less than or equal to d.[3] In this case we write $\dim B \leq d$.

[3] All our manifolds are σ-compact. This means that they can be covered by countably many compact sets.

DEFINITION 6.5.1. *A d-dimensional* **pseudocycle** *in X is a smooth map*

$$f : V \to X$$

defined on an oriented d-dimensional manifold V such that $f(V)$ has a compact closure and

$$\dim \Omega_f \leq \dim V - 2, \qquad \Omega_f := \bigcap_{\substack{K \subset V \\ K \text{ compact}}} \overline{f(V \setminus K)}.$$

The set Ω_f is called the **(omega) limit set** *of f. Two d-dimensional pseudocycles $f_0 : V_0 \to X$ and $f_1 : V_1 \to X$ are called* **bordant** *if there is a $(d+1)$-dimensional oriented manifold W with boundary $\partial W = V_1 \cup (-V_0)$ and a smooth map $F : W \to X$ such that $\overline{F(W)}$ is compact and*

$$F|_{V_0} = f_0, \qquad F|_{V_1} = f_1, \qquad \dim \Omega_F \leq d - 1.$$

The omega limit set can be interpreted, roughly speaking, as the image of the *boundary of* V under f. The next exercise shows that this holds when V is the interior of a manifold $\overline{V}$ with boundary and f extends to a continuous function on $\overline{V}$. However, in general it would be too restrictive to impose such a condition.

EXERCISE 6.5.2. (i) Show that $x \in \Omega_f$ if and only if x is the limit point of a sequence $f(v_\nu)$ where $v_\nu \in V$ has no convergent subsequence. (This agrees with the notion of the omega limit set in dynamical systems.) Show that Ω_f is always compact.

(ii) If V is the interior of a compact manifold $\overline{V}$ with boundary $\partial\overline{V}$ and f extends to a continuous map $f : \overline{V} \to X$ then $\Omega_f = f(\partial\overline{V})$. Give an example to show that Ω_f may be strictly larger than the set $\overline{f(V)} \setminus f(V)$ even when $\dim V \leq \dim X$. (When $d \geq \dim X + 2$ then *every* map $f : V \to X$ is a pseudocycle.)

Bordism classes of pseudocycles in X form an abelian group with addition given by disjoint union. The neutral element is the empty map defined on the empty manifold $V = \emptyset$. The inverse of $f : V \to X$ is given by reversing the orientation of V. One could define the pseudohomology of X as the group of bordism classes of pseudocycles. It has now been shown by Schwarz [**359**] and Zinger [**429**] that the resulting homology groups agree with singular homology. However we shall not use this.

REMARK 6.5.3. In order to represent a d-dimensional singular homology class β by a pseudocycle $f : V \to X$, represent it first by a map $f : P \to X$ defined on a d-dimensional finite oriented simplicial complex P without boundary. This condition means that the oriented faces of its top dimensional simplices cancel each other out in pairs. Thus P carries a fundamental homology class $[P]$ of dimension d and β is by definition the class $\beta = f_*[P]$. Now approximate f by a map f' that is smooth on each simplex. Then consider the union of the d and $(d-1)$-dimensional faces of P as a smooth d-dimensional manifold V and approximate f' by a map which is smooth across the $(d-1)$-dimensional simplices. Note that we may construct this smoothing f'' so that it extends continuously over P. Hence Ω_f is simply the image $f''(P - V)$ of the faces of P of codimension at least 2. By slight abuse of language we will call such pseudocycles $f'' : V \to M$ **smooth cycles**. Note that any two smooth cycles that represent β are bordant.

Of course, things become easier if we work with rational homology $H_*(X;\mathbb{Q})$. A theorem of Thom [**395**] asserts that, for every integral singular homology class $\alpha \in H_*(X;\mathbb{Z})$, there exists an integer k such that $k\alpha$ can be represented by a compact oriented submanifold of X (without boundary). Hence there is a basis of $H_*(X;\mathbb{Q})$ consisting of elements that are represented by smooth closed submanifolds of X.

EXERCISE 6.5.4. Let X be a smooth manifold of dimension $\dim X \geq 3$. Prove that every 2-dimensional integral homology class $A \in H_2(X;\mathbb{Z})$ can be represented by an oriented embedded surface. *Hint:* By definition of singular homology, A can be represented by a continuous map defined on a compact 2-dimensional simplicial complex without boundary. Every such complex can be given the structure of a smooth 2-dimensional compact manifold without boundary (which in the case of integer coefficients is orientable). Hence A is represented by a continuous map $f : \Sigma \to X$ defined on a closed 2-manifold Σ. Approximate f by a smooth map and use a general position argument to make f an immersion (with finitely many transverse self-intersections in the case $\dim X = 4$). In the cases $\dim X = 3$ or $\dim X = 4$ remove the self-intersections by using a local surgery argument that appropriately changes Σ. In the case $\dim X > 4$ use a general position argument to obtain an embedding.

Two pseudocycles $e : U \to X$ and $f : V \to X$ are called **strongly transverse** if

$$\Omega_e \cap \overline{f(V)} = \emptyset, \qquad \overline{e(U)} \cap \Omega_f = \emptyset,$$

and

$$e(u) = f(v) = x \qquad \Longrightarrow \qquad T_xX = \operatorname{im} de(u) + \operatorname{im} df(v).$$

If e and f are strongly transverse then the set $\{(u,v) \in U \times V \,|\, e(u) = f(v)\}$ is a compact manifold of dimension $\dim U + \dim V - \dim X$. In particular, this set is finite if U and V are of complementary dimension.

LEMMA 6.5.5. *Let $e : U \to X$ and $f : V \to X$ be pseudocycles of complementary dimension.*

(i) *There exists a residual set $\mathrm{Diff}_{\mathrm{reg}}(X,e,f) \subset \mathrm{Diff}(X)$ such that e is strongly transverse to $\phi \circ f$ for every $\phi \in \mathrm{Diff}_{\mathrm{reg}}(X,e,f)$.*

(ii) *If e is strongly transverse to f then the set $\{(u,v) \in U \times V \,|\, e(u) = f(v)\}$ is finite. In this case define* **the intersection number of** e **and** f *by*

$$e \cdot f = \sum_{\substack{u \in U,\, v \in V \\ e(u) = f(v)}} \nu(u,v),$$

where $\nu(u,v)$ is the intersection number of $e(U)$ and $f(V)$ at $e(u) = f(v)$.

(iii) *The intersection number $e \cdot f$ depends only on the bordism classes of e, f.*

PROOF. This is proved by standard arguments in differential topology as in Milnor [**289**] and Guillemin–Pollack [**165**]. Here are the main points. The first assertion can be proved by the same techniques as the results of Section 6.2 with parameter space $\mathrm{Diff}(X)$ instead of $\mathcal{J}$. One first shows that the map

$$U \times V \times \mathrm{Diff}(X) \to X \times X : (u,v,\phi) \mapsto \big(e(u), \phi(f(v))\big)$$

is transverse to the diagonal and hence the *universal space*

$$\mathcal{M} = \{(u,v,\phi) \,|\, e(u) = \phi(f(v))\} \subset U \times V \times \mathrm{Diff}(X)$$

is a manifold. Now consider the regular values of the projection $\mathcal{M} \to \mathrm{Diff}(X)$ to obtain that e and $\phi\circ f$ are transverse for generic ϕ. As in Section 6.2, one has to use a completion to a Banach manifold of C^ℓ-diffeomorphisms to make this precise. Now choose smooth maps $e_0 : U_0 \to X$ and $f_0 : V_0 \to X$ such that $\dim U_0 = \dim U - 2$ and $\Omega_e \subset e_0(U_0)$ and similarly for f. Apply the same argument as above to the pairs (e_0, f), (e, f_0), (e_0, f_0) to conclude that

$$\overline{e(U)} \cap \phi(\Omega_f) = \emptyset, \qquad \Omega_e \cap \overline{\phi(f(V))} = \emptyset$$

for a generic ϕ. This proves (i).

Statement (ii) follows directly from the definition of strongly transverse. To prove (iii) assume that the pseudocycle $f : V \to X$ is bordant to the empty set with corresponding bordism $F : W \to X$. As in the proof of (i) one can show that this bordism can be chosen such that it is strongly transverse to e and, moreover,

$$\Omega_e \cap \overline{F(W)} = \emptyset, \qquad \overline{e(U)} \cap \Omega_F = \emptyset.$$

Hence the set

$$Y := \{(u, v) \in U \times W, | \, e(u) = F(v)\}$$

is a compact oriented 1-manifold with boundary

$$\partial Y = \{(u, v) \in U \times V \,|\, e(u) = f(v)\}.$$

Thus $e \cdot f = 0$ and this proves Lemma 6.5.5. □

Every d-dimensional pseudocycle $e : U \to X$ determines a homomorphism

$$\Phi_e : H_{m-d}(X; \mathbb{Z}) \to \mathbb{Z}$$

as follows. Represent the class $\beta \in H_{m-d}(X; \mathbb{Z})$ by a smooth cycle $f : V \to X$ as in Remark 6.5.3. Any two such representations are bordant and hence, by Lemma 6.5.5 (iii), the intersection number

$$\Phi_e(\beta) = e \cdot f \tag{6.5.1}$$

is independent of the choice of the cycle f representing β. Likewise, if e_0 and e_1 are bordant, then $e_0 \cdot f = e_1 \cdot f$ for every cycle f. This proves the following.

LEMMA 6.5.6. *The homomorphism Φ_e depends only on the bordism class of e.*

Thus we have proved that every d-dimensional pseudocycle $e : U \to X$ determines a homomorphism $\Phi_e : H_{m-d}(X; \mathbb{Z}) \to \mathbb{Z}$ and hence an element of the space

$$H^{m-d}(X) := H^{m-d}(X; \mathbb{Z})/\text{torsion} = \mathrm{Hom}(H_{m-d}(X; \mathbb{Z}), \mathbb{Z}).$$

We denote it by

$$a_e := [\Phi_e] \in H^{m-d}(X).$$

This defines a homomorphism from bordism classes of pseudocycles to $H^{m-d}(X)$. If e is actually a smooth cycle (as in Remark 6.5.3) then the homology class of e is the Poincaré dual of $a_e = \mathrm{PD}([e])$. In general, e is only a weak representative of the homology class $\alpha_e := \mathrm{PD}(a_e) \in H_d(X)$. Here we call a d-dimensional pseudocycle $e : U \to M$ a **weak representative** of the homology class $\alpha \in H_d(X)$ if $e \cdot f = \alpha \cdot \beta$ for every homology class $\beta \in H_{m-d}(X, \mathbb{Z})$ and every smooth cycle f representing β. That this is a meaningful definition requires the proof that the formula (6.5.1) continues to hold when $f : V \to X$ is an arbitrary pseudocycle that represents the class β in the weak sense just defined. This assertion is not obvious because two weak representatives of a homology class $\beta \in H_d(X)$ may not be bordant.[4]

LEMMA 6.5.7. *Let $e : U \to X$ be a d-dimensional pseudocycle. If the $(m-d)$-dimensional pseudocycle $f : V \to X$ is a weak representative of the homology class $\beta \in H_{m-d}(X)$ then $\Phi_e(\beta) = e \cdot f$.*

PROOF. It suffices to prove the assertion in the case $\beta = 0$. Hence assume that f has intersection number zero with every d-dimensional smooth cycle. We must prove that f has intersection number zero with every d-dimensional pseudocycle. To see this assume, by Lemma 6.5.5, that the d-dimensional pseudocycle $e : U \to X$ is in general position, i.e. is strongly transverse to f. Then, $\overline{e(U)}$ does not intersect the (compact) limit set Ω_f for dimensional reasons. Now choose a sufficiently small open neighbourhood $W \subset X$ of Ω_f with smooth boundary, transverse to f. Then $V_0 := f^{-1}(X \setminus W)$ is a compact manifold with boundary and the restriction of f to V_0 is a smooth map $f_0 : (V_0, \partial V_0) \to (X \setminus W, \partial W)$. This map has intersection number zero with every smooth d-dimensional cycle in $X \setminus W$. Hence there is an integer $\ell > 0$ such that the ℓ-fold multiple of f_0 is a boundary in $H_{m-d}(X \setminus W, \partial W; \mathbb{Z})$. This implies that the pseudocycle $e : U \to X \setminus W$ has intersection number zero with f_0 and hence with f. This proves Lemma 6.5.7. □

REMARK 6.5.8 (Mod 2 coefficients). Although we shall usually use integral or rational coefficients, sometimes it is convenient to work modulo 2. We saw above that every d-dimensional pseudocycle $e : U \to X$ determines a homomorphism $\Phi_e : H_{m-d}(X; \mathbb{Z}) \to \mathbb{Z}$ given by the intersection product: if $f : V \to X$ represents the class β then $\Phi_e(\beta) := e \cdot f$. Reducing modulo 2 gives a homomorphism

$$\Phi_e : H_{m-d}(X; \mathbb{Z}) \to \mathbb{Z}/2\mathbb{Z}$$

such that $\Phi_e(\beta) = e \cdot f \pmod 2$. Hence e determines a unique element $a_e \in H^{m-d}(X; \mathbb{Z}/2\mathbb{Z})$, and so, by Poincaré duality, a unique element in $H_d(X; \mathbb{Z}/2\mathbb{Z})$. The reader may check that Lemma 6.5.7 continues to hold if we interpret $H_*(X)$ and $H^*(X)$ as (co)homology with $\mathbb{Z}/2\mathbb{Z}$ coefficients.

Products and transversality. Later we shall be interested in the relation of pseudocycles $\pi \times e : U \to X \times Y$ in product manifolds to pseudocycles in the factors X and Y. In the remainder of this section we discuss some relevant functorial properties.

LEMMA 6.5.9. *Let $f : Y \to Z$ be a smooth map between smooth manifolds. If $e : U \to Y$ is a pseudocycle in Y then $f \circ e : U \to Z$ is a pseudocycle in Z.*

PROOF. It follows from the definition that $\Omega_{f \circ e} = f(\Omega_e)$. Note that we need to use the fact that $e(U)$ has compact closure in Y. □

Lemma 6.5.9 shows that if Y is a submanifold of Z then a map $e : U \to Y$ is a pseudocycle in Y if and only if its composite with the inclusion is also a pseudocycle in Z. Moreover, if $\pi \times e : U \to X \times Y$ is a pseudocycle then so are $\pi : U \to X$ and $e : U \to Y$. However, the converse need not be true. For example if U has dimension $d = 4$ and X and Y both have dimension 2 then *any* smooth maps $\pi : U \to X$ and $e : U \to Y$ are pseudocycles, but the product $\pi \times e : U \to X \times Y$

[4] In fact, by the above mentioned results of Schwarz [**359**] and Zinger [**429**] they are bordant only if $H_d(X)$ has no torsion.

may not be a pseudocycle. To deal with such products we introduce the following notion of transversality.

DEFINITION 6.5.10. *Let X be a manifold and $X_T \subset X$ be a closed submanifold of codimension ℓ. A d-dimensional pseudocycle $f : U \to X$ is called* **weakly transverse to** *X_T if there exists a finite collection of smooth maps $g_i : W_i \to X$, $i = 0, \dots, d-2$, satisfying the following conditions.*

(i) *W_i is a manifold of dimension i for every i and*

$$\Omega_f \subset \bigcup_{i=0}^{d-2} g_i(W_i).$$

(ii) *$f : U \to X$ is transverse to X_T.*

(iii) *$g_i : W_i \to X$ is transverse to X_T for $d-2-\ell < i \le d-2$.*

Here the transversality conditions in (ii) and (iii) are understood in the usual sense. In particular, we do not require that $\Omega_f \cap X_T = \emptyset$. So the requirements of Definition 6.5.10 are weaker than the strong transversality defined before Lemma 6.5.5 but they are stronger than the standard notion of transversality expressed in (ii).

EXAMPLE 6.5.11. Let $U := \mathbb{C}$ and $X := S^2$. Let $f : U \to S^2$ be the stereographic enbedding of $\mathbb{C}$ into $S^2 \setminus \{x_\infty\}$. Then $\Omega_f = \{x_\infty\}$ and so f is a pseudocycle. It is weakly transverse to $X_T := \{x_0\}$ if and only if $x_0 \ne x_\infty$. It is weakly transverse to an embedded circle $X_T \subset S^2$ if and only if $x_\infty \notin X_T$.

EXAMPLE 6.5.12. If $d = \dim U \ge \dim X + 2$ then every smooth map $f : U \to X$ is a pseudocycle. In this case f is weakly transverse to a closed submanifold $X_T \subset X$ if and only if it is transverse to X_T in the usual sense. To see this, choose $W_i = \emptyset$ for $i \ne \dim X$, and $W_i := X$ and $g_i := \mathrm{id}_X$ for $i = \dim X \le d-2$.

EXERCISE 6.5.13. Show that a pseudocycle $f : U \to X \times Y$ is weakly transverse to $\{x\} \times Y$ for a generic point $x \in X$. Similarly, any submanifold W of X can be perturbed so that f is weakly transverse to $W \times Y$.

LEMMA 6.5.14. *Let $X_T \subset X$ be a closed submanifold of codimension ℓ and $f : U \to X$ be a d-dimensional pseudocycle that is weakly transverse to X_T. Then the restriction $f_T : U_T \to X$ of f to the submanifold $U_T := f^{-1}(X_T) \subset U$ is a $(d-\ell)$-dimensional pseudocycle in X_T and hence also in X.*

PROOF. Denote $W_{Ti} := W_i$ for $i \le d-2-\ell$ and $W_{Ti} := g_i^{-1}(X_T)$ for $i > d-2-\ell$. Then $\dim W_{Ti} \le d-\ell-2$ for every i and $\Omega_{f_T} \subset \bigcup_i g_i(W_{Ti})$. □

LEMMA 6.5.15. *Let X and Y be manifolds, $X_T \subset X$ be a closed submanifold, and $\pi \times e : U \to X \times Y$ be a pseudocycle. If $\pi \times e$ is weakly transverse to $X_T \times Y$ then $\pi : U \to X$ is weakly transverse to X_T.*

PROOF. If $g_i \times h_i : W_i \to X \times Y$ is transverse to $X_T \times Y$ then $g_i : W_i \to X$ is transverse to X_T. □

EXAMPLE 6.5.16. The converse of Lemma 6.5.15 need not be true. For example, let $U := \mathbb{C}^2$, $X = Y := S^2$, $f : \mathbb{C} \to S^2$ be the pseudocycle of Example 6.5.11, and define $\pi : U \to S^2$ and $e : U \to S^2$ by $\pi(z_1, z_2) := f(z_1)$ and $e(z_1, z_2) := f(z_2)$. Then $\pi \times e$ and π are pseudocycles with limit sets

$$\Omega_{\pi \times e} = \left(\{x_\infty\} \times S^2\right) \cup \left(S^2 \times \{x_\infty\}\right), \qquad \Omega_\pi = S^2.$$

So $\pi \times e$ is not weakly transverse to $\{x_\infty\} \times S^2$, however, π is weakly transverse to $X_T := \{x_\infty\}$ since x_∞ does not belong to the image of π (see Example 6.5.12).

PROPOSITION 6.5.17. *Let X and Y be smooth closed manifolds and $X_T \subset X$ be a closed submanifold of codimension ℓ. Suppose that*

$$\pi \times e : U \to X \times Y$$

is a d-dimensional pseudocycle that is weakly transverse to $X_T \times Y$. Then the restriction

$$e_T : U_T := \pi^{-1}(X_T) \to Y$$

is a $(d-\ell)$-dimensional pseudocycle. Moreover, if $f : V \to Y$ is a pseudocycle of dimension $\dim V = \dim Y + \ell - d$, then

$$e_T \cdot_Y f = (\pi \times e) \cdot_{X \times Y} ([X_T] \times f). \tag{6.5.2}$$

PROOF. By Lemma 6.5.14, the restriction $\pi_T \times e_T : U_T \to X \times Y$ of $\pi \times e$ to U_T is a $(d-\ell)$-dimensional pseudocycle. Hence, by Lemma 6.5.9, e_T is a pseudocycle. To prove the assertion about intersection numbers isotop f_Y so that it is strongly transverse to the pseudocycle $e_T : U_T \to Y$. Then the result follows directly from the definitions. □

6.6. Gromov–Witten pseudocycles

We now explain the main result of this chapter, namely that under suitable conditions the evaluation map

$$\mathrm{ev} : \mathcal{M}^*_{0,k}(A; J) \to M^k$$

is a pseudocycle. The argument presented here does not work when A is a multiple mB, $m > 1$, of a class with $c_1(B) = 0$. This case is best treated by replacing the curves $u : S^2 \to M$ by their graphs, and is considered in the next section.

Recall from Section 6.2 the definition of the set

$$\mathcal{J}_{\mathrm{reg}}(M, \omega) := \bigcap_T \bigcap_{\{A_\alpha\}} \mathcal{J}_{\mathrm{reg}}(T, \{A_\alpha\}).$$

It consists of all ω-tame almost complex structures J on M such that D_u is surjective for every simple J-holomorphic sphere $u : \mathbb{C}P^1 \to M$ and the evaluation map $\mathrm{ev}^E : \mathcal{M}^*(\{A_\alpha\}; J) \times Z(T) \to M^E$ is transverse to Δ^E for every labelled tree T and every collection $\{A_\alpha\}$ of spherical homology classes. By Theorem 6.2.6, the set $\mathcal{J}_{\mathrm{reg}}(M, \omega)$ is residual in $\mathcal{J}_\tau(M, \omega)$.

THEOREM 6.6.1. *Let (M, ω) be a closed semipositive symplectic $2n$-manifold and let $J \in \mathcal{J}_{\mathrm{reg}}(M, \omega)$. Let $A \in H_2(M; \mathbb{Z})$ such that*

$$A = mB, \quad c_1(B) = 0 \quad \Longrightarrow \quad m = 1 \tag{6.6.1}$$

for every positive integer m and every spherical homology class $B \in H_2(M; \mathbb{Z})$. Then the evaluation map

$$\mathrm{ev} : \mathcal{M}^*_{0,k}(A; J) \to M^k$$

is a pseudocycle of dimension $\mu(A, k) = 2n + 2c_1(A) + 2k - 6$. Its bordism class is independent of J.

PROOF. Gromov's Compactness Theorem (Theorem 5.3.1) implies that the limit set

$$\Omega_{\mathrm{ev}} := \bigcap_{\substack{K\subset\mathcal{M}^*_{0,k}(A;J)\\ K\ compact}} \overline{\mathrm{ev}(\mathcal{M}^*_{0,k}(A;J)\setminus K)}$$

of the evaluation map ev is covered by the union of the images of the strata $\mathcal{M}_{0,T}(A;J)$ under the evaluation maps

$$\mathrm{ev}_T : \mathcal{M}_{0,T}(A;J) \to M^k, \qquad \mathrm{ev}_T(\mathbf{u},\mathbf{z}) := (u_{\alpha_1}(z_1),\dots,u_{\alpha_k}(z_k)),$$

where T ranges over all k-labelled trees with at least two edges, together with the image under ev of the space $\mathcal{M}_{0,k}(A;J)\setminus\mathcal{M}^*_{0,k}(A;J)$ of multiply covered curves. Next observe that, by Proposition 6.1.2, we can cover Ω_{ev} by the images of simple stable maps provided that we extend the range of homology classes they represent. In fact,

$$\Omega_{\mathrm{ev}} \subset \bigcup_T \bigcup_{\{B_\alpha\}} \mathrm{ev}_T(\mathcal{M}^*_{0,T}(\{B_\alpha\};J)), \tag{6.6.2}$$

where the union is over all k-labelled trees T and all collections of spherical homology classes $\{B_\alpha\}_{\alpha\in T}$ that satisfy the conditions

$$A = \sum_{\alpha\in T} m_\alpha B_\alpha \tag{6.6.3}$$

and

$$e(T) = 0 \qquad \Longrightarrow \qquad B \neq A \tag{6.6.4}$$

for some collection of positive integers $\{m_\alpha\}_{\alpha\in T}$.

This formula holds for every compact symplectic manifold (M,ω) and every almost complex structure $J\in\mathcal{J}(M,\omega)$. Since $J\in\mathcal{J}_{\mathrm{reg}}(M,\omega)$, it follows from Theorem 6.2.6 that the moduli space $\mathcal{M}^*_{0,T}(\{B_\alpha\};J)$ is a smooth manifold of dimension

$$\dim \mathcal{M}^*_{0,T}(\{B_\alpha\};J) = \mu(B,k) - 2e(T), \qquad B := \sum_{\alpha\in T} B_\alpha.$$

By Lemma 6.4.4, the semipositivity hypothesis guarantees that the moduli space $\mathcal{M}^*_{0,T}(\{B_\alpha\};J)$ is empty unless

$$c_1(B_\alpha) \geq 0$$

for every $\alpha\in T$. By (6.6.3), this implies $\mu(B,k)\leq\mu(A,k)$ for every nonempty stratum. Hence the dimensional condition

$$\mu(B,k) - 2e(T) \leq \mu(A,k) - 2$$

is satisfied for every tree with $e(T) > 0$. If $e(T) = 0$ then, by (6.6.3) and (6.6.4), we have then $A = mB$ for some integer $m\geq 2$. By (6.6.1), this implies $c_1(B)\neq 0$. Hence, by semipositivity, $c_1(B) > 0$, and so $\mu(B,k)\leq\mu(A,k)-2$. This shows that the evaluation map $\mathrm{ev}:\mathcal{M}^*_{0,k}(A;J)\to M^k$ is a pseudocycle.

Now let $J_0, J_1\in\mathcal{J}_{\mathrm{reg}}(M,\omega)$ and consider the corresponding evaluation maps

$$\mathrm{ev}_0 : \mathcal{M}^*_{0,k}(A;J_0) \to M^k, \qquad \mathrm{ev}_1 : \mathcal{M}^*_{0,k}(A;J_1) \to M^k.$$

We prove that these maps are bordant as pseudocycles in the sense of Definition 6.5.1. Recall that $\mathcal{J}_{\mathrm{reg}}(M,\omega;J_0,J_1)$ denotes the set of all smooth homotopies $\{J_\lambda\}_{0\leq\lambda\leq 1}$ of ω-tame almost complex structures that are regular in the sense of Definition 6.2.10 for every tree T and every collection $\{A_\alpha\}_{\alpha\in T}$ of homology classes in $H_2(M;\mathbb{Z})$. By Theorem 6.2.11, the set $\mathcal{J}_{\mathrm{reg}}(M,\omega;J_0,J_1)$ is nonempty. Fix a regular homotopy

$$\{J_\lambda\} \in \mathcal{J}_{\mathrm{reg}}(M,\omega;J_0,J_1)$$

and consider the evaluation map

$$\mathrm{ev} : \mathcal{W}^*_{0,k}(A;\{J_\lambda\}) \to M^k.$$

By Theorem 6.2.11, the space $\mathcal{W}^*_{0,k}(A;\{J_\lambda\})$ is an oriented smooth manifold with boundary

$$\partial\mathcal{W}^*_{0,k}(A;\{J_\lambda\}) = \mathcal{M}^*_{0,k}(A;J_1) \cup \big(-\mathcal{M}^*_{0,k}(A;J_0)\big).$$

It has dimension

$$\dim \mathcal{W}^*_{0,k}(A;\{J_\lambda\}) = \mu(A,k) + 1$$

and the restrictions of the evaluation map ev to the two parts of the boundary at $\lambda = 0$ and $\lambda = 1$ are obviously given by ev_0 and ev_1, respectively.

Now it follows again from Theorem 5.3.1 and Proposition 6.1.2 that

$$\bigcap_{\substack{K\subset\mathcal{W}^*_{0,k}(A;\{J_\lambda\})\\ K\ \textit{compact}}} \overline{\mathrm{ev}(\mathcal{W}^*_{0,k}(A;\{J_\lambda\})\setminus K)} \subset \bigcup_T\bigcup_{B_\alpha} \mathrm{ev}_T(\mathcal{W}^*_{0,T}(\{B_\alpha\};\{J_\lambda\})),$$

where the union is over all k-labelled trees T and all collections of spherical homology classes $\{B_\alpha\}_{\alpha\in T}$ that satisfy (6.6.3) and (6.6.4). By Theorem 6.2.11, the moduli spaces in this union have dimensions

$$\dim \mathcal{W}^*_{0,T}(\{B_\alpha\};\{J_\lambda\}) = \mu(B,k) + 1 - 2e(T) \le \mu(A,k) - 1.$$

The last inequality follows from the same arguments as in the first part of the proof. Note that here we must apply Lemma 6.4.4 to J_λ that need not be regular but are such that $\operatorname{coker} D_u$ has dimension at most one. Hence the evaluation map $\mathrm{ev} : \mathcal{W}^*_{0,k}(A;\{J_\lambda\}) \to M^k$ is the required bordism from ev_0 to ev_1. This proves Theorem 6.6.1. □

REMARK 6.6.2. The assertions of Theorem 6.6.1 remain valid in the nonsemipositive case provided that one considers only almost complex structures that belong to the open set $\mathcal{J}_+(M,\omega;\kappa) \subset \mathcal{J}_\tau(M,\omega)$, where $\kappa = \omega(A)$ (see Lemma 6.4.7). In this case the evaluation map is a pseudocycle for every

$$J \in \mathcal{J}_{\mathrm{reg}}(M,\omega) \cap \mathcal{J}_+(M,\omega;\kappa).$$

However, in general $\mathcal{J}_+(M,\omega;\kappa)$ could be empty, and if it is nonempty it might be disconnected. The assertion about the independence of J holds only for those $J \in \mathcal{J}_{\mathrm{reg}}(M,\omega)\cap\mathcal{J}_+(M,\omega;\kappa)$ that belong to the same path component of $\mathcal{J}_+(M,\omega;\kappa)$.

EXERCISE 6.6.3. Let (M,ω) be a compact semipositive symplectic manifold and $J \in \mathcal{J}_{\mathrm{reg}}(M,\omega)$. Let T be a k-labelled tree and $\{A_\alpha\}_{\alpha\in T}$ be a collection of spherical homology classes satisfying the stability condition (6.1.1). Suppose each class A_α satisfies (6.6.1). Show that the evaluation map

$$\mathrm{ev}_T : \mathcal{M}^*_{0,T}(\{A_\alpha\};J) \to M^k$$

is a pseudocycle of dimension $\mu(A,k) - 2e(T)$.

Fixing the marked points. A finite collection $\mathbf{w} := (\mathrm{w}_1, \dots, \mathrm{w}_k)$ of distinct points on S^2 determines an evaluation map

$$\mathrm{ev}_{\mathbf{w}} : \mathcal{M}^*(A; J) \to M^k.$$

One might expect this to be a pseudocycle (of dimension $2n + 2c_1(A)$) for a generic almost complex structure. However, this does not seem to be the case unless one imposes rather restrictive conditions on (M, ω) and on the number of marked points. As the following example shows, the problem is that one loses control of the positions of the marked points on strata containing multiply covered curves.

EXAMPLE 6.6.4. Let (M, ω) be a manifold of dimension 6 that contains a class A with $c_1(A) = 2$. For instance, we could take $M = S^2 \times \mathbb{C}P^2$ and $A = [S^2 \times pt]$. Consider the evaluation map

$$\mathrm{ev}_{\mathbf{w}} : \mathcal{M}^*(2A; J) \to M^5.$$

It certainly may happen that for a regular J a sequence of simple J-holomorphic curves in class $2A$ converges to a doubly covered A curve: see below. Therefore the limit set of $\mathrm{ev}_{\mathbf{w}}$ contains doubly covered A-curves of the form $u \circ \phi$ where $u : S^2 \to M$ is simple and $\phi : S^2 \to S^2$ has degree 2. Now, the space of holomorphic self maps of S^2 of degree 2 has complex dimension 5 — they may be written in homogeneous coordinates as $[z : w] \mapsto [az^2 + bzw + cw^2 : dz^2 + ezw + fw^2]$ — and it is easy to check that there is such a map that takes any set $\mathbf{w}$ of 5 distinct points to any other set of 5 distinct points. Hence, if we attempt to follow the proof of Theorem 6.6.1 and replace the map $u \circ \phi$ by its underlying simple map u, we must allow the evaluation point $\mathbf{w}$ to vary freely over S^2 in order to cover this boundary stratum. Hence this stratum is covered by the evaluation map with domain $\mathcal{M}_{0,5}(A; J)$, which has the same dimension as $\mathcal{M}^*(2A; J)$.

To see that this problem is not just an artefact of our method of proof, consider the example where L_{-1} is the line bundle over $\mathbb{C}P^2$ whose Chern class evaluates to -1 over the line ℓ and $M := \mathbb{P}(L_{-1} \oplus \mathbb{C})$. Then M contains a copy Z of $\mathbb{C}P^2$ with normal bundle L_{-1}. If A is the class of the line ℓ in Z, then $c_1(A) = 2$. Moreover, the nearby curves in the class $2A$ are the conics in $Z = \mathbb{C}P^2$ and one can check that for any set of 5 points $x_1, \dots, x_5$ on ℓ and any $\mathbf{w}$ there is a sequence of embedded conics $u^\nu : S^2 \to Z$ such that $u^\nu(\mathrm{w}_i) \to x_i$, $i = 1, \dots, 5$, as $\nu \to \infty$. To see this, choose points y_i^ν converging to x_i that for each ν lie on a pair of lines that are close to ℓ, and then slightly perturb this pair of lines to a nondegenerate conic.

Note in the above example that, although the dimension of the moduli space of A-curves is smaller than that of the $2A$-curves (because $c_1(A) > 0$), this decrease in dimension is not enough to compensate for the lack of control of the marked points. However, if one assumes that the number k of marked points is sufficiently small with respect to the minimal Chern number N, one can show that $\mathrm{ev}_{\mathbf{w}}$ is a pseudocycle. A sample result is given in Exercise 6.6.5 below. This solution to the problem is obviously unsatisfactory. Although, as in the first edition of this book, one can get sharper results by a more careful discussion, a much better idea is to extend the range of allowed perturbations of J so as to avoid the troublesome multiply covered curves altogether. This is the approach taken in the next section.

The next example imposes very stringent restriction on the manifold M. In exchange, it applies to a yet more general evaluation map in which only some of the marked points are fixed in the domain.

EXERCISE 6.6.5. Let (M, ω) be a monotone symplectic manifold with minimal Chern number N and let $J \in \mathcal{J}_{\mathrm{reg}}(M, \omega)$. Fix a homology class $A \in H_2(M; \mathbb{Z})$, an integer $k \geq 3$, an index set $I \subset \{1, \ldots, k\}$ such that

$$3 \leq \#I \leq N + 2,$$

and a finite collection

$$\mathbf{w} := \{\mathrm{w}_i\}_{i \in I}$$

of distinct points on S^2. Consider the moduli space $\mathcal{M}^*_{0,k}(A; \mathbf{w}, J)$ of all tuples $(u, z_1, \ldots, z_k)$ where $u : S^2 \to M$ is a simple J-holomorphic sphere in the class A and $z_1, \ldots, z_k$ are pairwise distinct points on S^2 such that $z_i = \mathrm{w}_i$ for $i \in I$. Prove that the evaluation map

$$\mathrm{ev}_{\mathbf{w}} : \mathcal{M}^*(A; \mathbf{w}, J) \to M^k,$$

defined by

$$\mathrm{ev}_{\mathbf{w}}(u, z_1, \ldots, z_k) := (u(z_1), \ldots, u(z_k)),$$

is a pseudocycle of dimension $2n + 2c_1(A) + 2(k - \#I)$.

Hint: Let $(u^\nu, z_1^\nu, \ldots, z_k^\nu)$ be a sequence in $\mathcal{M}^*_{0,k}(A; \mathbf{w}, J)$ that has no convergent subsequence with limit in $\mathcal{M}^*_{0,k}(A; \mathbf{w}, J)$. By Theorem 5.3.1 it does have a subsequence which Gromov converges to some stable map $(\mathbf{u}, \mathbf{z})$ modelled over a k-labelled tree (T, E, Λ). Show that there is a unique special vertex $0 \in T$, characterized by the condition that the points z_{0i}, $i \in I$, are pairwise distinct. Show also that the J-holomorphic sphere $u_0 : S^2 \to M$ associated to the special vertex can be parametrized such that $z_{0i} = \mathrm{w}_i$ for every $i \in I$. Next consider the case where the limit stable map is simple and the number of edges is

$$1 \leq e(T) \leq \#I - 3.$$

Show that under this assumption the special vertex carries at least

$$\#I - e(T) \geq 3$$

marked points from the index set I. Show that, after removing $\#I - e(T) - 2$ of these points and allowing the remaining ones (at least two of them) to vary freely, the resulting map remains stable and belongs to a moduli space of dimension $2n + 2c_1(A) + 2(k - \#I) - 2$. In all other cases show that the reduced simple stable map (with arbitrary marked points) belongs to a moduli space of dimension strictly less than $2n + 2c_1(A) + 2(k - \#I)$.

EXERCISE 6.6.6. Let $M = \mathbb{C}P^2$, $L = [\mathbb{C}P^1]$ and J_0 be the standard complex structure on M.

(i) Show that the nonsimple elements $\mathcal{M}_{0,5}(2L; J_0) \setminus \mathcal{M}^*_{0,5}(2L; J_0)$ form a manifold and calculate its dimension. Compare this with the dimension of the underlying simple stratum $\mathcal{M}_{0,5}(L; J_0)$.

(ii) Show that the evaluation map $\mathrm{ev} : \mathcal{M}^*_{0,k}(2L; J_0) \to M^k$ is a pseudocycle for every k.

(iii) Consider the case $k = |I| = 4$ as in Exercise 6.6.5 and let $\mathbf{w} := (0, 1, \infty, -1)$. Show that the evaluation map $\mathrm{ev}_{\mathbf{w}} : \mathcal{M}^*_{0,4}(2L; \mathbf{w}, J_0) \to M^4$ is a pseudocycle. Note that ev_{w} is a map between manifolds of the same dimension.

6.7. The pseudocycle of graphs

The main reason why the results of the previous section do not work for all classes A in semipositive manifolds is that there may be nonsimple elements in the primary moduli space $\mathcal{M}_{0,k}(A;J)$. This happens only if the class A is itself a multiple class. There are several possible ways of dealing with this problem. The idea is to change the basic equation $\bar{\partial}_J(u) = 0$ so that it no longer has any multiply covered solutions, either by adding an inhomogeneous perturbation term or by allowing J to vary with the points z in the domain. In this section we adopt the latter approach. This is a special case of a more general construction described in Chapter 8 that also involves some geometrically interesting perturbations.

We shall carry out the arguments as much as possible for a domain (Σ, j_Σ) of arbitrary genus. However, the main result in this section (Theorem 6.7.1) only holds for genus zero. The reason is that constant curves of higher genus cannot be made transverse by a generic (z-dependent) perturbation of the almost complex structure; as explained in Chapter 8 one needs more general perturbations. This phenomenon is clearly illustrated by the example of degree one curves of genus one in the 2-sphere: there are no such curves but the Gromov–Witten invariant, defined in terms of Hamiltonian perturbations, is nonzero (Example 8.6.12).

The pseudocycle of graphs. The fundamental object of study in this section is the space of solutions $u : \Sigma \to M$ of the equation

$$du(z) + J_z(u(z)) \circ du(z) \circ j_\Sigma(z) = 0 \tag{6.7.1}$$

for $z \in \Sigma$, where $j = j_\Sigma$ is a fixed complex structure on Σ and

$$J = \{J_z\}_{z\in\Sigma}$$

is a smooth family of ω-tame (or ω-compatible) almost complex structures on M, parametrized by the points of Σ. Throughout we shall denote by

$$\mathcal{J}_\tau(\Sigma; M, \omega) := C^\infty(\Sigma, \mathcal{J}_\tau(M, \omega))$$

the space of smooth maps

$$\Sigma \to \mathcal{J}_\tau(M, \omega) : z \mapsto J_z.$$

Since $\mathcal{J}_\tau(M,\omega)$ is contractible, so is the space $\mathcal{J}_\tau(\Sigma; M, \omega)$. This continues to hold in the ω-compatible case. In fact all the results of this section hold both for the ω-compatible and the ω-tame case. We formulate them for ω-tame almost complex structures because that is more convenient for our applications.

Recall the notation $\mathcal{J}_+(M,\omega;\kappa)$ for the set of κ-semipositive and ω-tame almost complex structures on M. Thus an almost complex structure $J \in \mathcal{J}_\tau(M,\omega)$ belongs to $\mathcal{J}_+(M,\omega;\kappa)$ precisely when every J-holomorphic sphere $v : S^2 \to M$ with ω-energy $E(v) \le \kappa$ has nonnegative Chern number. If (M,ω) is semipositive then, by Lemma 6.4.8, the set $\mathcal{J}_+(M,\omega;\kappa)$ is nonempty, open and path connected for every $\kappa > 0$. Hence the set of all $J = \{J_z\} \in \mathcal{J}(\Sigma; M, \omega)$ such that $J_z \in \mathcal{J}_+(M,\omega;\kappa)$ for every $z \in \Sigma$ is also nonempty and open. We denote by

$$\mathcal{J}_+(\Sigma; M, \omega; \kappa) := \big\{\{J_z\} \in \mathcal{J}_\tau(\Sigma; M, \omega) \,|\, J_z \in \mathcal{J}_+(M,\omega;\kappa)\, \forall z \in \Sigma,\ \{J_z\} \sim \mathrm{const}\big\}$$

the component containing the constant maps.

Now assume $\Sigma = S^2$ is the 2-sphere with its standard complex structure. Given an almost complex structure $\{J_z\} \in \mathcal{J}_+(S^2; M, \omega; \kappa)$, a homology class $A \in H_2(M;\mathbb{Z})$ such that $\omega(A) \le \kappa$, a subset $I \subset \{1,\dots,k\}$, and a tuple $\mathbf{w} = \{\mathrm{w}_i\}_{i\in I}$ of pairwise distinct points on S^2 consider the moduli space

$$\mathcal{M}^*_{0,k}(A; \mathbf{w}, \{J_z\})$$

of all tuples $(u, \mathbf{z})$ consisting of a solution $u : S^2 \to M$ of (6.7.1) in the class A and a tuple $\mathbf{z} = (z_1,\dots,z_k)$ of pairwise distinct points on S^2 such that $z_i = \mathrm{w}_i$ for $i \in I$. This moduli space carries an evaluation map

$$\mathrm{ev} : \mathcal{M}^*_{0,k}(A; \mathbf{w}, \{J_z\}) \to M^k, \qquad \mathrm{ev}(u, \mathbf{z}) := (u(z_1),\dots,u(z_k)). \tag{6.7.2}$$

THEOREM 6.7.1. *Let (M,ω) be a compact semipositive symplectic manifold, $A \in H_2(M;\mathbb{Z})$, $k \ge 0$ be an integer, and $I \subset \{1,\dots,k\}$. Let $\mathbf{w} = \{\mathrm{w}_i\}_{i\in I}$ be a tuple of pairwise distinct points on S^2 and $\kappa \ge \omega(A)$. Then there is a residual subset $\mathcal{J}_{\mathrm{reg}}(S^2; M, \omega; \mathbf{w})$ of $\mathcal{J}_\tau(S^2; M, \omega)$ such that the following holds for every*

$$J = \{J_z\}_{z\in S^2} \in \mathcal{J}_{\mathrm{reg}}(S^2; M, \omega; \mathbf{w}) \cap \mathcal{J}_+(S^2; M, \omega; \kappa).$$

(i) *The evaluation map (6.7.2) is a pseudocycle of dimension*

$$\dim \mathcal{M}^*_{0,k}(A; \mathbf{w}, \{J_z\}) = 2n + 2c_1(A) + 2(k - 2\#I).$$

(ii) *The bordism class of the pseudocycle in (i) is independent of $\{J_z\}$ and $\mathbf{w}$.*

(iii) *If A satisfies the hypotheses of Theorem 6.6.1 and $\#I = 3$ then the pseudocycle in (i) is bordant to the Gromov–Witten pseudocycle* $\mathrm{ev} : \mathcal{M}^*_{0,k}(A; J) \to M^k$.

The precise conditions satisfied by the elements of $\mathcal{J}_{\mathrm{reg}}(S^2; M, \omega; \mathbf{w})$ are spelled out in Definition 6.7.10 below. The proof is given at the end of this section, after we have developed suitable language to describe it.

The above theorem refers only to curves in M and one could prove it simply by extending all the arguments of the previous chapters to the case when J depends on z. However, this task is made much easier if one associates to each element of the moduli space $\mathcal{M}(A, \Sigma; \{J_z\})$ of solutions of (6.7.1) in the class A not the corresponding curve in M but its graph in the product manifold

$$\widetilde{M} := \Sigma \times M.$$

Indeed, solutions of (6.7.1) can be thought of as pseudoholomorphic curves in $\widetilde{M}$ as follows. Each family $J = \{J_z\} \in \mathcal{J}(\Sigma; M, \omega)$ gives rise to a unique almost complex structure $\widetilde{J}$ on $\widetilde{M}$ via the formula

$$\widetilde{J}(z, x) := j(z) \oplus J_z(x) \in \mathrm{Aut}(T_z\Sigma \oplus T_xM). \tag{6.7.3}$$

Evidently a smooth function $u : \Sigma \to M$ is a solution of (6.7.1) if and only if its graph $\widetilde{u} : \Sigma \to \widetilde{M}$, given by

$$\widetilde{u}(z) := (z, u(z)),$$

is a $\widetilde{J}$-holomorphic curve. This shows that the compactness arguments of Chapters 4 and 5 carry over verbatim in the genus zero case and with minor modifications in the higher genus case, as long as the complex structure on Σ is fixed. However, the transversality arguments of Chapter 3 must be adapted to the present situation for two reasons. First, we consider only $\widetilde{J}$-holomorphic sections and it is therefore natural to examine the linearized operator on the space of vertical vector fields.

Second, we shall only consider almost complex structures $\widetilde{J}$ on $\widetilde{M}$ that are compatible with the fibration $\pi : \widetilde{M} \to \Sigma$ in the sense that π is holomorphic and the fibers of π are almost complex submanifolds of $\widetilde{M}$. The almost complex structures of the form (6.7.3) satisfy these conditions.

REMARK 6.7.2. Given a homology class $A \in H_2(M;\mathbb{Z})$ define $\widetilde{A} \in H_2(\widetilde{M};\mathbb{Z})$ by

$$\widetilde{A} := \iota_* A + [\Sigma \times \mathrm{pt}],$$

where $\iota : M \to \widetilde{M}$ denotes the inclusion of the fiber. It is instructive to compare the moduli space $\mathcal{M}(A,\Sigma;\{J_z\})$ with the moduli space $\mathcal{M}(\widetilde{A},\Sigma;\widetilde{J})/\mathrm{G}_\Sigma$ of all $\widetilde{J}$-holomorphic representatives of the class $\widetilde{A}$ divided by the complex automorphisms group G_Σ of (Σ, j_Σ). If $\widetilde{u} \in \mathcal{M}(\widetilde{A},\Sigma;\widetilde{J})$ then the composition $\phi := \pi\circ\widetilde{u}$ is a complex automorphism of Σ and so $\widetilde{u}\circ\phi^{-1}$ is a $\widetilde{J}$-holomorphic section of $\pi : \widetilde{M} \to \Sigma$. In other words, there is a $u \in \mathcal{M}(A,\Sigma;\{J_z\})$ and a unique automorphism $\phi \in \mathrm{G}_\Sigma$ such that

$$\widetilde{u}(z) := (\phi(z), u\circ\phi(z)).$$

Thus there is a natural bijection from the moduli space $\mathcal{M}(A,\Sigma;\{J_z\})$ to the quotient $\mathcal{M}(\widetilde{A},\Sigma;\widetilde{J})/\mathrm{G}_\Sigma$. When $g = 0$ this quotient is our space $\mathcal{M}_{0,0}(\widetilde{A};\widetilde{J})$. In the case $g > 0$ the latter notation is usually reserved for the space where the complex structure of Σ varies freely in Teichmüller space, while our methods treat only the case when j_Σ is fixed. This variation in Teichmüller space is the geometric meaning of the cokernel of the operator $D_{\widetilde{u}}$ noted in Remark 6.7.5 below. Observe also that if (Σ, j_Σ) is a generic Riemann surface of genus $g > 1$ it has no automorphisms, and that the automorphism group of $(\mathbb{T}^2, j)$ has two components for generic j.

The moduli space of graphs. Given a compact Riemann surface (Σ, j_Σ), an almost complex structure $\{J_z\} \in \mathcal{J}_\tau(\Sigma; M,\omega)$, and a homology class $A \in H_2(M;\mathbb{Z})$, we shall denote the moduli space of $\{J_z\}$-holomorphic curves $u : \Sigma \to M$ that represent the class A by

$$\mathcal{M}(A,\Sigma;\{J_z\}) := \big\{u : \Sigma \to M \,|\, u \text{ satisfies (6.7.1)}, [u] = A\big\}.$$

In the case $\Sigma = S^2$ we shall abbreviate

$$\mathcal{M}(A;\{J_z\}) := \mathcal{M}(A, S^2;\{J_z\}).$$

To understand this moduli space we must examine the linearized operator

$$D_u = D_{u,J} : \Omega^0(\Sigma, u^*TM) \to \Omega^{0,1}(\Sigma, u^*TM).$$

As in Section 3.1, this operator is given by the formula

$$D_u\xi := \tfrac{1}{2}\Big(\nabla\xi + J_z(u)\nabla\xi\circ j_\Sigma\Big) - \tfrac{1}{2}J_z(u)(\nabla_\xi J_z)\partial_J(u), \tag{6.7.4}$$

but now $\nabla = \nabla^z$ denotes the Levi-Civita connection of the Riemannian metric

$$\langle\cdot,\cdot\rangle_z := \tfrac{1}{2}\big(\omega(\cdot, J_z\cdot) - \omega(J_z\cdot,\cdot)\big)$$

associated to ω and J_z via (2.1.1). Thus for each $z \in \Sigma$ we evaluate $(D_u\xi)(z)$ using the connection ∇^z and the almost complex structure J_z. As before, this is a real linear Cauchy–Riemann operator and hence, by Theorem C.1.10, is Fredholm and has index

$$\mathrm{index}\, D_u = n(2-2g) + 2c_1(A),$$

where g is the genus of Σ.

REMARK 6.7.3. The motivation for introducing this operator is as in Section 3.1. For every $u \in C^\infty(\Sigma, M)$ there is a map

$$\mathcal{F}_u : \Omega^0(\Sigma, u^*TM) \to \Omega^{0,1}(\Sigma, u^*TM),$$

given by

$$\mathcal{F}_u(\xi) := \Phi_u(\xi)^{-1}\bar{\partial}_J(\exp_u(\xi)),$$

where $\exp_{u(z)} : T_{u(z)}M \to M$ denotes the exponential map of the metric $\langle\cdot,\cdot\rangle_z$ and

$$\Phi_{u(z)}(\xi(z)) : T_{u(z)}M \to T_{\exp_{u(z)}(\xi(z))}M$$

denotes parallel transport with respect to the Hermitian connection $\widetilde{\nabla}^z$ associated to ∇^z as in Section 2.1. With these conventions we have

$$D_u = d\mathcal{F}_u(0).$$

The proof is word by word the same as that of Proposition 3.1.1 and will be omitted. This setting extends to suitable Sobolev completions as in Section 3.1.

REMARK 6.7.4. If J_z is ω-compatible for every z then the complex linear part of D_u assigns to every $\xi \in \Omega^0(\Sigma, u^*TM)$ the complex linear part of the 1-form $\widetilde{\nabla}\xi \in \Omega^1(\Sigma, u^*TM)$ defined by

$$\widetilde{\nabla}\xi := \nabla\xi - \tfrac{1}{2}J(\nabla_{du}J)\xi - \tfrac{1}{2}J\dot{J}\xi.$$

Here the linear map

$$T\Sigma \to \Omega^0(M, \mathrm{End}(TM)) : \zeta \mapsto \dot{J}_\zeta$$

denotes the differential of the map $z \mapsto J_z$. The resulting decomposition of D_u into the complex linear and the complex anti-linear part (as in (3.1.6) in Remark 3.1.3) has the form

$$D_u\xi = \big(\widetilde{\nabla}\xi\big)^{0,1} + \frac{1}{4}N_J(\xi, \partial_J(u)) + \tfrac{1}{2}(J\dot{J}\xi)^{0,1}.$$

Here N_J denotes, for each $z \in \Sigma$, the Nijenhuis tensor of J_z. The proof of this formula is a straightforward extension of the argument in the proof of Lemma C.7.3 and will be omitted.

REMARK 6.7.5. If $u : \Sigma \to M$ is any smooth map and $\widetilde{u} : \Sigma \to \widetilde{M}$ is defined by $\widetilde{u}(z) := (z, u(z))$ then there is an operator

$$D_{\widetilde{u}} : \Omega^0(\Sigma, \widetilde{u}^*T\widetilde{M}) \to \Omega^{0,1}(\Sigma, \widetilde{u}^*T\widetilde{M})$$

associated to $\widetilde{u}(z) = (z, u(z))$ as in Section 3.1. This operator has index

$$\mathrm{index}\, D_{\widetilde{u}} = (n+1)(2-2g) + 2c_1(\widetilde{A}) = \mathrm{index}\, D_u + 6 - 6g.$$

It is related to D_u as follows. There is a natural isomorphism

$$\widetilde{u}^*T\widetilde{M} \cong T\Sigma \oplus u^*TM$$

of complex vector bundles. In general, neither the vertical, nor the horizontal tangent bundle will be invariant under $D_{\widetilde{u}}$. However, if u is a solution of (6.7.1) and $\widetilde{J}$ is given by (6.7.3), then examining the operator in local coordinates we find

$$D_{\widetilde{u}}(\zeta, \xi) = \Big(\bar{\partial}\zeta, D_u\xi - \tfrac{1}{2}J_z(u)\dot{J}_\zeta(u)\partial_J(u)\Big) \tag{6.7.5}$$

for $\zeta \in \Omega^0(\Sigma, T\Sigma)$ and $\xi \in \Omega^0(\Sigma, u^*TM)$. The formula (6.7.5) shows that the vertical subbundle $u^*TM \subset \widetilde{u}^*T\widetilde{M}$ is invariant under the operator $D_{\widetilde{u}}$ and that D_u is the restriction of $D_{\widetilde{u}}$ to this subbundle. Moreover, if $\Sigma = S^2$ and $u : S^2 \to M$

is a solution of (6.7.1) then D_u is surjective if and only if $D_{\widetilde{u}}$ is. To see this note that the operator $\bar{\partial} : \Omega^0(S^2, TS^2) \to \Omega^{0,1}(S^2, TS^2)$ is surjective, that its kernel consists of holomorphic vector fields, and that

$$\tfrac{1}{2} J_z(u) \dot{J}_\zeta(u) \partial_J(u) = D_u(du \circ \zeta)$$

whenever $u : S^2 \to M$ is a solution of (6.7.1) and $\zeta \in \mathrm{Vect}(S^2)$ is a holomorphic vector field. Hence, in the case $\bar{\partial}\zeta = 0$, the term $J_z(u)\dot{J}_\zeta(u)\partial_J(u)$ contributes nothing that is not already contained in the image of D_u. On the other hand the operator $D_{\widetilde{u}}$ is never surjective in the case $g > 0$ since $\bar{\partial}$ has nonzero cokernel. In the genus zero case the higher dimensional kernel of $D_{\widetilde{u}}$ accounts precisely for the tangent space of the orbit of $\widetilde{u}$ under the reparametrization group $\mathrm{G} = \mathrm{PSL}(2, \mathbb{C})$.

An important observation is that the transversality arguments in Chapter 3 go through for all solutions u of (6.7.1), whether or not they are somewhere injective. Intuitively, this holds because graphs are always embedded. As a warmup we prove that the constant solutions of (6.7.1) are regular in the case $\Sigma = S^2$.

LEMMA 6.7.6. *Let (Σ, j_Σ) be a closed connected Riemann surface, $u : \Sigma \to M$ be a constant map with value $x \in M$, and $J = \{J_z\}$ be a smooth family of ω-tame almost complex structures on M. Let $D_{x,J}$ be the operator defined by (6.7.4). Then the kernel of $D_{x,J}$ consists of the constant maps $\xi : \Sigma \to T_xM$. Moreover, $D_{x,J}$ is onto if and only if Σ has genus zero.*

PROOF. The operator $D_{x,J} : \Omega^0(\Sigma, T_xM) \to \Omega^{0,1}(\Sigma, T_xM)$ is given by

$$(D_{x,J}\xi)(z) = \tfrac{1}{2}\Big(\nabla\xi(z) + J(z)\nabla\xi(z) \circ j_\Sigma(z)\Big)$$

for $\xi : \Sigma \to T_xM$ and $z \in \Sigma$. If J were independent of z the first statement would be obvious since it would be equivalent to saying that any holomorphic function on Σ is constant. In the general case one argues as follows. Since $u(z) \equiv x$, the covariant derivative of ξ is just the ordinary derivative. Denote by $|\nabla\xi(z)|$ the norm of the linear map $\nabla\xi(z) : T_z\Sigma \to T_xM$ with respect to the inner product on T_xM determined by ω and J_z via (2.1.1) (see Exercise 2.2.3). Then

$$D_x\xi = 0 \qquad \Longrightarrow \qquad \tfrac{1}{2}\left|\nabla\xi\right|^2 \mathrm{dvol}_{S^2} = \xi^*\omega_x. \tag{6.7.6}$$

To see this, choose conformal coordinates $s + it$ on Σ and suppose that the volume form is $\mathrm{dvol}_\Sigma = \lambda^2(ds^2 + dt^2)$ in these coordinates. If $D_x\xi = 0$ then $\partial_s\xi + J_{s,t}(x)\partial_t\xi = 0$ and hence

$$\begin{aligned}
\lambda^2 \left|\nabla\xi\right|^2 &= \left|\partial_s\xi\right|^2 + \left|\partial_t\xi\right|^2 \\
&= \omega(\partial_s\xi, J_{s,t}(x)\partial_s\xi) + \omega(\partial_t\xi, J_{s,t}(x)\partial_t\xi) \\
&= 2\omega(\partial_s\xi, \partial_t\xi).
\end{aligned}$$

This proves (6.7.6). Integrating over Σ we deduce that every $\xi \in \ker D_{x,J}$ satisfies $\nabla\xi \equiv 0$ and hence is constant. This proves the first assertion. To prove the second assertion note that, by Theorem C.1.10, the operator $D_{x,J}$ has Fredholm index $n(2 - 2g)$, where g is the genus of Σ. By what we just proved, its kernel has dimension $2n$. Hence $D_{x,J}$ is onto if and only if $g = 0$. This proves Lemma 6.7.6. □

The relevant set of **regular families** of almost complex structures is denoted by[5]

$$\mathcal{J}_{\mathrm{reg}}(\Sigma; A) := \Big\{\{J_z\} \in \mathcal{J}_\tau(\Sigma; M, \omega) \,|\, u \in \mathcal{M}(A, \Sigma; \{J_z\}) \Longrightarrow D_u \text{ is onto}\Big\}.$$

Given two families of almost complex structures $J^0 = \{J_z^0\}$ and $J^1 = \{J_z^1\}$ in $\mathcal{J}_\tau(\Sigma; M, \omega)$ we denote by $\mathcal{J}_\tau(\Sigma; M, \omega; J^0, J^1)$ the space of smooth homotopies $[0,1] \times \Sigma \to \mathcal{J}_\tau(M, \omega) : (\lambda, z) \mapsto J_z^\lambda$ from J^0 to J^1. Given such a homotopy consider the moduli space

$$\mathcal{W}(A, \Sigma; \{J_z^\lambda\}) := \Big\{(\lambda, u) \,|\, 0 \le \lambda \le 1,\, u \in \mathcal{M}(A, \Sigma; J^\lambda)\Big\}.$$

A homotopy $\{J_z^\lambda\}$ is called **regular (for A)** if $\{J_z^0\}, \{J_z^1\} \in \mathcal{J}_{\mathrm{reg}}(\Sigma; A)$ and

$$\Omega^{0,1}(\Sigma, u^*TM) = \operatorname{im} D_u + \mathbb{R}(\partial_\lambda J_z^\lambda(u)) du \circ j_\Sigma$$

for every $(\lambda, u) \in \mathcal{W}(A, \Sigma; \{J_z^\lambda\})$. The set of regular homotopies from J^0 to J^1 will be denoted by $\mathcal{J}_{\mathrm{reg}}(\Sigma; A; J^0, J^1)$.

PROPOSITION 6.7.7. *Assume $A \ne 0$ or $\Sigma = S^2$.*

(i) *The set $\mathcal{J}_{\mathrm{reg}}(\Sigma; A)$ is residual in $\mathcal{J}_\tau(\Sigma; M, \omega)$.*

(ii) *If $\{J_z\} \in \mathcal{J}_{\mathrm{reg}}(\Sigma; A)$ then $\mathcal{M}(A, \Sigma; \{J_z\})$ is a smooth oriented manifold of dimension*

$$\dim \mathcal{M}(A, \Sigma; \{J_z\}) = n(2 - 2g) + 2c_1(A).$$

(iii) *Let $\{J_z^0\}, \{J_z^1\} \in \mathcal{J}_{\mathrm{reg}}(\Sigma; A)$. Then the set $\mathcal{J}_{\mathrm{reg}}(\Sigma; A; J^0, J^1)$ is residual in $\mathcal{J}_\tau(\Sigma; M, \omega; J^0, J^1)$. Moreover, if $\{J_z^\lambda\} \in \mathcal{J}_{\mathrm{reg}}(\Sigma; A; J^0, J^1)$ then the moduli space $\mathcal{W}(A, \Sigma; \{J_z^\lambda\})$ is a smooth oriented manifold with boundary*

$$\partial \mathcal{W}(A, \Sigma; \{J_z^\lambda\}) = \mathcal{M}(A, \Sigma; \{J_z^1\}) \cup \big(-\mathcal{M}(A, \Sigma; \{J_z^0\})\big).$$

PROOF. For $\Sigma = S^2$ and $A = 0$ the result is an obvious consequence of Lemma 6.7.6. For $A \ne 0$ the proof follows the same pattern as the proofs of Theorems 3.1.6 and 3.1.8. One first shows that the universal moduli space is a manifold (the analogue of Proposition 3.2.1) and then proves proves (i) and (ii) by considering the projection from this universal space to the parameter space $\mathcal{J}_\tau(\Sigma; M, \omega)$ of almost complex structures.

More precisely, denote by $\mathcal{J}^\ell = \mathcal{J}_\tau^\ell(\Sigma; M, \omega)$ the space of families of ω-tame almost complex structures on M of class C^ℓ; this means that the induced complex structure $\widetilde{J}$ on $\widetilde{M}$ is of class C^ℓ. Now consider the universal moduli space

$$\mathcal{M}(A, \Sigma; \mathcal{J}^\ell) := \Big\{(u, \{J_z\}) \,|\, \{J_z\} \in \mathcal{J}^\ell,\, u \in \mathcal{M}(A, \Sigma; \{J_z\})\Big\}.$$

We prove that this space is a $C^{\ell-1}$ Banach submanifold of $W^{1,p}(\Sigma, M) \times \mathcal{J}^\ell$. Following the argument in the proof of Proposition 3.2.1 we must show that the linearized operator

$$W^{1,p}(\Sigma, u^*TM) \times T_{\{J_z\}}\mathcal{J}^\ell : (\xi, \{Y_z\}) \mapsto D_u\xi + \tfrac{1}{2} Y_z(u) du \circ j_\Sigma$$

[5] We emphasize that $\mathcal{J}_{\mathrm{reg}}(\Sigma; A)$ is not equal to the set $\mathcal{J}_{\mathrm{reg}}(A, \Sigma)$ considered in Section 3.1, hat consists of all z-independent almost complex structures J for which the simple J-holomorphic curves $u : \Sigma \to M$ in the class A are regular. Since the entire discussion in the present section is for z-dependent almost complex structures, this notational similarity should not lead to any confusion.

is surjective for every pair $(u, \{J_z\}) \in \mathcal{M}(A, \Sigma; \mathcal{J}^\ell)$. To see this let $q > 1$ such that $1/p + 1/q = 1$ and suppose that $\eta \in L^q(\Sigma, T^*\Sigma \otimes_{\{J_z\}} u^*TM)$ annihilates the image of this operator. Then $\eta \in W^{\ell,p}$,

$$D_u^*\eta = 0,$$

and, for every $\{Y_z\} \in T_{\{J_z\}}\mathcal{J}^\ell$,

$$\int_\Sigma \langle \eta, Y_z(u)du \circ j_\Sigma \rangle \, \mathrm{dvol}_\Sigma = 0. \tag{6.7.7}$$

Since $A \neq 0$ the function u cannot be constant. The critical point analysis of Lemma 2.4.1 carries over and shows that du can only vanish at finitely many points. We show that $\eta(z) = 0$ for every $z \in \Sigma$ such that $du(z) \neq 0$. Suppose otherwise that $du(z_0) \neq 0$ and $\eta(z_0) \neq 0$ for some $z_0 \in \Sigma$. Then, by Lemma 3.2.2, there exists an infinitesimal almost complex structure $\{Y_z\} \in T_{\{J_z\}}\mathcal{J}^\ell$ such that the map $z \mapsto \langle \eta(z), Y_z(u(z))du(z) \circ j_\Sigma \rangle$ is positive at z_0 and hence in some neighbourhood $U \subset \Sigma$ of z_0. Choose a smooth cutoff function $\beta : \Sigma \to [0,1]$ with support in U such that $\beta(z_0) = 1$. Then the infinitesimal almost complex structure $z \mapsto \beta(z)Y_z$ violates the condition (6.7.7). This contradiction shows that η vanishes almost everywhere and hence $\eta \equiv 0$. Thus we have proved that the linearized operator is surjective and so the universal moduli space $\mathcal{M}(A, \Sigma; \mathcal{J}^\ell)$ is a separable $C^{\ell-1}$ Banach manifold. The remainder of the proof is word by word the same as that of Theorems 3.1.6 and 3.1.8 and will be omitted. This proves Proposition 6.7.7. □

***J*-holomorphic spheres in the fiber.** The Gromov–Kontsevich compactification of the moduli space $\mathcal{M}(A, \Sigma; \{J_z\})$ involves J_z-holomorphic spheres in M. This is apparent from the rescaling property in the definition of Gromov convergence. If $u_\nu : \Sigma \to M$ is a sequence of solutions of (6.7.1) whose first derivatives diverge near a point $z \in \Sigma$ and $\phi_\nu : B_{R_\nu} \to \Sigma$ is a sequence of holomorphic embeddings that converges to z on every compact set, then the limit of the sequence $u_\nu \circ \phi_\nu$ (if it exists) will be a J_z-holomorphic sphere. Therefore we must examine the moduli space of pairs (z, v), where $v : S^2 \to M$ is a J_z-holomorphic sphere. One can think of these as **vertical $\widetilde{J}$-holomorphic spheres** in $\widetilde{M} = \Sigma \times M$.

Given a spherical homology class $A \in H_2(M; \mathbb{Z})$ and an almost complex structure $J = \{J_z\} \in \mathcal{J}(\Sigma; M, \omega)$, consider the moduli space

$$\mathcal{M}^{\mathrm{Vert},*}(A; \{J_z\}) := \{(z, v) \,|\, z \in \Sigma,\, v \in \mathcal{M}^*(A; J_z)\}$$

of simple vertical J_z-holomorphic spheres in M representing the class A. The corresponding linearized operator

$$D_{z,v} : T_z\Sigma \oplus \Omega^0(S^2, v^*TM) \to \Omega^{0,1}(S^2, v^*TM) \tag{6.7.8}$$

is given by

$$D_{z,v}(\zeta, \xi) = D_{v,J_z}\xi - \tfrac{1}{2}J_z(v)\dot{J}_\zeta(v)\partial_{J_z}(v),$$

where the linear map $T\Sigma \to \Omega^0(M, \mathrm{End}(TM)) : \zeta \mapsto \dot{J}_\zeta$ denotes the differential of the map $z \mapsto J_z$, as in Remark 6.7.4. To justify this formula, let us first fix two paths $\mathbb{R} \to \Sigma : \lambda \mapsto z(\lambda)$ and $\mathbb{R} \to M : \lambda \mapsto x(\lambda)$. Then the path $\lambda \mapsto z(\lambda)$ defines a complex structure on the pullback tangent bundle x^*TM given by $J_{z(\lambda)}(x(\lambda))$ on $T_{x(\lambda)}M$. The formula

$$\widetilde{\nabla}_\lambda\xi := \nabla_\lambda\xi - \tfrac{1}{2}J(\nabla_{\dot{x}(\lambda)}J)\xi - \tfrac{1}{2}J\dot{J}_{\dot{z}(\lambda)}\xi$$

for a vector field $\xi(\lambda) \in T_{x(\lambda)}M$ along the path x defines a connection on x^*TM which preserves this complex structure. Here $\nabla_\lambda \xi$ denotes the covariant derivative of ξ in the λ direction with respect to the Levi-Civita connection associated to the metric g_{J^λ}. Given $\zeta \in T_z\Sigma$ and $\xi \in T_xM$, denote by $\Phi_{z,x}(\zeta,\xi) : T_xM \to T_{\exp_x(\xi)}M$ parallel transport along the path $\lambda \mapsto \exp_x(\lambda\xi)$ with respect to the connection $\widetilde{\nabla}$ associated to the path $\lambda \mapsto \exp_z(\lambda\zeta)$. Then, as in Remark 6.7.3, there is a map

$$\mathcal{F}_{z,v} : T_z\Sigma \oplus \Omega^0(S^2, v^*TM) \to \Omega^{0,1}(S^2, v^*TM)$$

which assigns to a pair (ζ, ξ) the $(0,1)$-form

$$\mathcal{F}_{z,v}(\zeta,\xi) := \Phi_{z,v}(\zeta,\xi)^{-1}\bar{\partial}_{J_{\exp_z(\zeta)}}(\exp_v(\xi)).$$

With this notation one can prove as in Proposition 3.1.1 that $D_{z,v} = d\mathcal{F}_{z,v}(0)$.

REMARK 6.7.8. Given a smooth map $v : S^2 \to M$ and a point $z \in \Sigma$, define a map

$$\widetilde{v} : S^2 \to \widetilde{M}, \qquad \widetilde{v}(z') := (z, v(z')).$$

Then v is a J_z-holomorphic sphere in M if and only if $\widetilde{v}$ is a $\widetilde{J}$-holomorphic sphere in $\widetilde{M}$. There is a corresponding linearized operator

$$D_{\widetilde{v}} : \Omega^0(S^2, \widetilde{v}^*T\widetilde{M}) \to \Omega^{0,1}(S^2, \widetilde{v}^*TM).$$

Both operators $D_{\widetilde{v}}$ and $D_{z,v}$ are Fredholm and have the same index

$$\operatorname{index} D_{z,v} = \operatorname{index} D_{\widetilde{v}} = 2n + 2 + 2c_1(A).$$

Moreover, the operators are related as follows. Let us denote by $\pi : \widetilde{M} \to \Sigma$ the obvious projection. Then

$$d\pi(\widetilde{v}) \circ D_{\widetilde{v}} = \bar{\partial} \circ d\pi(\widetilde{v}),$$

where $\bar{\partial} : \Omega^0(S^2, T_z\Sigma) \to \Omega^{0,1}(S^2, T_z\Sigma)$ is the Cauchy–Riemann operator on the trivial bundle $S^2 \times T_z\Sigma \to S^2$. This identity shows that the operator $D_{\widetilde{v}}$ maps the subspace $\{\tilde{\xi} \in \Omega^0(S^2, \widetilde{v}^*T\widetilde{M}) \,|\, d\pi(\widetilde{v})\tilde{\xi} = \text{const}\}$ to the kernel of $d\pi(\widetilde{v})$. The restriction of the operator $D_{\widetilde{v}}$ to this subspace is precisely the operator $D_{z,v}$. Since $\bar{\partial}$ is surjective and its kernel is the space of constant functions $S^2 \to T_z\Sigma$, it follows that $D_{\widetilde{v}}$ and $D_{z,v}$ have the same kernel and have isomorphic cokernels. In particular, $D_{\widetilde{v}}$ is onto if and only if $D_{z,v}$ is onto. Thus the vertical sphere $\widetilde{v}$ is regular as a curve in the product $\widetilde{M}$ if and only if it is parametric regular in the sense defined by Buse [**54**]. See Example 8.4.3 for a related example.

The next proposition spells out the relevant transversality result for vertical J_z-holomorphic spheres. We denote the space of regular almost complex structures for vertical J_z-holomorphic spheres in the class A by

$$\mathcal{J}^{\mathrm{Vert}}_{\mathrm{reg}}(\Sigma; A) := \Big\{\{J_z\} \in \mathcal{J}_\tau(\Sigma; M, \omega) \,|\, z \in \Sigma,\ v \in \mathcal{M}^*(A; J_z) \Longrightarrow D_{z,v} \text{ is onto}\Big\}.$$

A homotopy $\{J_z^\lambda\} \in \mathcal{J}_\tau(\Sigma; M, \omega; J^0, J^1)$ determines a moduli space

$$\mathcal{W}^{\mathrm{Vert},*}(A; \{J_z^\lambda\}) := \Big\{(\lambda, z, v) \,|\, 0 \le \lambda \le 1,\ z \in \Sigma,\ v \in \mathcal{M}^*(A; J_z^\lambda)\Big\}.$$

The homotopy is called **vertically regular (for** A**)** if $J^0, J^1 \in \mathcal{J}^{\mathrm{Vert}}_{\mathrm{reg}}(\Sigma; A)$ and

$$\Omega^{0,1}(\Sigma, v^*TM) = \operatorname{im} D_{z,v} + \mathbb{R}(\partial_\lambda J_z^\lambda(v))dv \circ j_{S^2} \tag{6.7.9}$$

for every $(\lambda, z, v) \in \mathcal{W}^{\mathrm{Vert},*}(A; \{J_z^\lambda\})$. The set of vertically regular homotopies from J^0 to J^1 will be denoted by $\mathcal{J}_{\mathrm{reg}}^{\mathrm{Vert}}(\Sigma; A; J^0, J^1)$. The next result can be interpreted as a parametrized version of Theorems 3.1.6 and 3.1.8. The proof is a straightforward adaptation of the arguments in Section 3.2 and will be omitted.

PROPOSITION 6.7.9. *Let $A \in H_2(M; \mathbb{Z})$ be a spherical homology class.*

(i) *The set $\mathcal{J}_{\mathrm{reg}}^{\mathrm{Vert}}(\Sigma; A)$ is residual in $\mathcal{J}_\tau(\Sigma; M, \omega)$.*

(ii) *If $\{J_z\} \in \mathcal{J}_{\mathrm{reg}}^{\mathrm{Vert}}(\Sigma; A)$ then $\mathcal{M}^{\mathrm{Vert},*}(A; \{J_z\})$ is a smooth oriented manifold of dimension*

$$\dim \mathcal{M}^{\mathrm{Vert},*}(A; \{J_z\}) = 2n + 2 + 2c_1(A).$$

(iii) *Let $J^0, J^1 \in \mathcal{J}_{\mathrm{reg}}^{\mathrm{Vert}}(\Sigma; A)$. Then the set $\mathcal{J}_{\mathrm{reg}}^{\mathrm{Vert}}(\Sigma; A; J^0, J^1)$ is residual in $\mathcal{J}_\tau(\Sigma; M, \omega; J^0, J^1)$. Moreover, if $\{J_z^\lambda\} \in \mathcal{J}_{\mathrm{reg}}^{\mathrm{Vert}}(\Sigma; A; J^0, J^1)$ then the moduli space $\mathcal{W}^{\mathrm{Vert},*}(A; \{J_z^\lambda\})$ is a smooth oriented manifold with boundary*

$$\partial \mathcal{W}^{\mathrm{Vert},*}(A; \{J_z^\lambda\}) = \mathcal{M}^{\mathrm{Vert},*}(A; \{J_z^1\}) \cup \left(-\mathcal{M}^{\mathrm{Vert},*}(A; \{J_z^0\})\right).$$

Stable maps. At this point it is convenient for the exposition to restrict attention to the case $\Sigma = S^2$ so that the notation and results of Chapter 5 can be applied directly without any modification. The adaptation of these results to the higher genus case will be discussed extensively in Chapter 8 in the context of Hamiltonian perturbations.

Let us fix a smooth family $J = \{J_z\}_{z \in S^2} \in \mathcal{J}_\tau(S^2; M, \omega)$ of ω-tame almost complex structures on M and denote by $\widetilde{J}$ the corresponding almost complex structure on the product manifold $\widetilde{M} := S^2 \times M$. Then the solutions of (6.7.1) as well as the J_z-holomorphic spheres in M can be expressed as $\widetilde{J}$-holomorphic spheres in $\widetilde{M}$. Let $T = (T, E, \Lambda)$ be a k-labelled tree with a **special vertex** or **root** $0 \in T$ and $\{A_\alpha\}_{\alpha \in T}$ be a collection of homology classes in $H_2(M; \mathbb{Z})$. Let $\iota : M \to \widetilde{M}$ be the inclusion of the fiber and denote

$$\widetilde{A}_\alpha := \begin{cases} \iota_* A_0 + [S^2 \times \mathrm{pt}], & \text{if } \alpha = 0, \\ \iota_* A_\alpha, & \text{if } \alpha \neq 0. \end{cases}$$

Then

$$\pi_* \widetilde{A}_\alpha = \begin{cases} [S^2], & \text{if } \alpha = 0, \\ 0, & \text{if } \alpha \neq 0, \end{cases}$$

where $\pi : \widetilde{M} \to S^2$ denotes the projection. For each $\alpha \in T \setminus \{0\}$, there is a one-to-one correspondence between $\widetilde{J}$-holomorphic spheres $\widetilde{u}_\alpha$ in the class $\widetilde{A}_\alpha$ and pairs (z_α, u_α), where u_α is a J_{z_α}-holomorphic sphere in the class A_α. Likewise, there is a one-to-one correspondence between $\widetilde{J}$-holomorphic spheres $\widetilde{u}_0$ in the class $\widetilde{A}_0$ such that $\pi \circ \widetilde{u}_0 = \mathrm{id}$ and solutions u_0 of (6.7.1) in the class A_0.

Now consider the moduli space of simple stable maps

$$\mathcal{M}_{0,T}^*(\widetilde{M}, \{\widetilde{A}_\alpha\}; \widetilde{J}) := \widetilde{\mathcal{M}}_{0,T}^*(\widetilde{M}, \{\widetilde{A}_\alpha\}; \widetilde{J}) / \mathrm{G}_T$$

as defined in Chapter 5. In the present section we shall depart from the notation of Chapter 5 in one respect: we shall always assume that the component $\widetilde{u}_0 : S^2 \to \widetilde{M}$ of a stable map $(\tilde{\mathbf{u}}, \mathbf{z}) \in \widetilde{\mathcal{M}}_{0,T}^*(\widetilde{M}, \{\widetilde{A}_\alpha\}; \widetilde{J})$ satisfies

$$\pi \circ \widetilde{u}_0 = \mathrm{id}$$

and so is a $\widetilde{J}$-holomorphic section of $\widetilde{M}$. Thus the reparametrization group G_T consists of tuples $(f, \{\phi_\alpha\}_{\alpha \in T})$, where $f : T \to T$ is a tree automorphism, such that

$f(0) = 0$, $\phi_0 = \mathrm{id}$, and $\phi_\alpha \in \mathrm{G} = \mathrm{PSL}(2, \mathbb{C})$ for $\alpha \neq 0$. It acts on the moduli space of stable maps as in Definition 5.1.4.

We shall now consider a subset of the moduli space where some of the marked points are fixed. Fix any index set $I \subset \{1, \ldots, k\}$ and a tuple

$$\mathbf{w} = \{\mathrm{w}_i\}_{i \in I}$$

of pairwise distinct points on S^2 indexed by I. Any such tuple determines a subset

$$\mathcal{M}^*_{0,T}(\widetilde{M}, \{\widetilde{A}_\alpha\}; \mathbf{w}, \widetilde{J}) := \widetilde{\mathcal{M}}^*_{0,T}(\widetilde{M}, \{\widetilde{A}_\alpha\}; \mathbf{w}, \widetilde{J})/\mathrm{G}_T$$

consisting of all equivalence classes $[\tilde{\mathbf{u}}, \mathbf{z}] \in \mathcal{M}^*_{0,T}(\widetilde{M}, \{\widetilde{A}_\alpha\}; \widetilde{J})$ of stable maps that satisfy

$$i \in I \quad \Longrightarrow \quad \pi(\widetilde{u}_{\alpha_i}(z_i)) = \mathrm{w}_i. \tag{6.7.10}$$

This means that $z_i = \mathrm{w}_i$ whenever $\alpha_i = 0$, i.e. whenever the point z_i lies on the special component, and that otherwise the image of the corresponding bubble $\widetilde{u}_{\alpha_i}$ is contained in the fiber $\{\mathrm{w}_i\} \times M$.

This moduli space can be described as the preimage of an evaluation map in the following way. As in Section 6.2, there is a moduli space $\mathcal{M}^*(\widetilde{M}, \{\widetilde{A}_\alpha\}; \widetilde{J})$ of simple tuples $\tilde{\mathbf{u}} = \{\widetilde{u}_\alpha\}_{\alpha \in T}$ of $\widetilde{J}$-holomorphic spheres representing the classes $\widetilde{A}_\alpha$. This space is an oriented finite dimensional manifold whenever $J \in \mathcal{J}_{\mathrm{reg}}(\Sigma; A_0)$ and $J \in \mathcal{J}^{\mathrm{Vert}}_{\mathrm{reg}}(\Sigma; A_\alpha)$ for every $\alpha \in T \setminus \{0\}$. As in Section 6.2 define

$$Z(T) \subset (S^2)^E \times (S^2)^k$$

as the set of all tuples $\mathbf{z} = (\{z_{\alpha\beta}\}_{\alpha E \beta}, \{z_i\}_{1 \le i \le k})$ such that the points $z_{\alpha\beta} \in S^2$ for $\alpha E \beta$ and $z_i \in S^2$ for $\alpha_i = \alpha$ are pairwise distinct for every $\alpha \in T$. Then there is an evaluation map

$$\widetilde{\mathrm{ev}}^E \times \pi^I : \mathcal{M}^*(\widetilde{M}, \{\widetilde{A}_\alpha\}; \widetilde{J}) \times Z(T) \to \widetilde{M}^E \times (S^2)^I, \tag{6.7.11}$$

defined by

$$\widetilde{\mathrm{ev}}^E(\tilde{\mathbf{u}}, \mathbf{z}) := \{\widetilde{u}_\alpha(z_{\alpha\beta})\}_{\alpha E \beta}, \qquad \pi^I(\mathbf{u}, \mathbf{z}) := \{\pi(\widetilde{u}_{\alpha_i}(z_i))\}_{i \in I}.$$

The moduli space of simple stable maps satisfying (6.7.10) is the preimage of the submanifold $\widetilde{\Delta}^E \times \{\mathbf{w}\}$ under $\widetilde{\mathrm{ev}}^E \times \pi^I$. This leads to the following definition.

DEFINITION 6.7.10. *A smooth family $J = \{J_z\} \in \mathcal{J}_\tau(S^2; M, \omega)$ of ω-tame almost complex structures on M is called* **regular for w** *if it satisfies the following conditions.*

(H) *D_u is onto for every solution $u : S^2 \to M$ of (6.7.1).*

(V) *$D_{z,v}$ is onto for each $z \in S^2$ and each J_z-holomorphic sphere $v : S^2 \to M$.*

(E) *The evaluation map (6.7.11) is transverse to $\widetilde{\Delta}^E \times \{\mathbf{w}\}$ for every k-labelled tree T with a special vertex 0 and every collection of homology classes $\{A_\alpha\}_{\alpha \in T}$ in $H_2(M; \mathbb{Z})$.*

The set of families of almost complex structures that are regular for $\mathbf{w}$ *will be denoted by $\mathcal{J}_{\mathrm{reg}}(S^2; M, \omega; \mathbf{w})$.*

With this preparation, we can state the by now standard regularity theorem. The proof is not quite routine because the almost complex structures $\widetilde{J}$ are required to preserve the fibered structure of $\widetilde{M}$: see Step 2.

THEOREM 6.7.11. *Let* $\mathbf{w} = \{\mathrm{w}_i\}_{i\in I}$ *be a tuple of pairwise distinct points on* S^2.

(i) *If* $J \in \mathcal{J}_{\mathrm{reg}}(S^2; M, \omega; \mathbf{w})$ *then, for every tree* T *with a special vertex* 0 *and every collection* $\{A_\alpha\}_{\alpha\in T}$ *of spherical homology classes in* $H_2(M;\mathbb{Z})$, *the moduli space* $\mathcal{M}^*_{0,T}(\widetilde{M}, \{\widetilde{A}_\alpha\}; \mathbf{w}, \widetilde{J})$ *is a smooth oriented manifold of dimension*

$$\dim \mathcal{M}^*_{0,T}(\widetilde{M}, \{\widetilde{A}_\alpha\}; \mathbf{w}, \widetilde{J}) = 2n + 2c_1(A) + 2k - 2|I| - 2e(T),$$

where $A := \sum_{\alpha\in T} A_\alpha$.

(ii) *The set* $\mathcal{J}_{\mathrm{reg}}(S^2; M, \omega; \mathbf{w})$ *is residual in* $\mathcal{J}_\tau(S^2; M, \omega)$.

PROOF. By Definition 6.7.10, the map $\widetilde{\mathrm{ev}}^E \times \pi^I$ is transverse to $\widetilde{\Delta}^E \times \{\mathbf{w}\}$ and

$$\widetilde{\mathcal{M}}^*_{0,T}(\widetilde{M}, \{\widetilde{A}_\alpha\}; \mathbf{w}, \widetilde{J}) = \left(\widetilde{\mathrm{ev}}^E \times \pi^I\right)^{-1}\left(\widetilde{\Delta}^E \times \{\mathbf{w}\}\right).$$

By Propositions 6.7.7 and 6.7.9, the moduli space $\mathcal{M}^*(\widetilde{M}, \{\widetilde{A}_\alpha\}; \widetilde{J})$ has dimension

$$\dim \mathcal{M}^*(\widetilde{M}, \{\widetilde{A}_\alpha\}; \widetilde{J}) = 2n + (2n+2)e(T) + 2c_1(A).$$

Since $\dim Z(T) = 4e(T) + 2k$ and $\mathrm{codim}\,(\widetilde{\Delta}^E \times \{\mathbf{w}\}) = (2n+2)e(T) + 2|I|$, we find

$$\dim \widetilde{\mathcal{M}}^*_{0,T}(\widetilde{M}, \{\widetilde{A}_\alpha\}; \mathbf{w}, \widetilde{J}) = 2n + 2c_1(A) + 2k - 2|I| + 4e(T).$$

Hence (i) follows from the fact that G_T has dimension $6e(T)$ and acts on the moduli space $\widetilde{\mathcal{M}}^*_{0,T}(\widetilde{M}, \{\widetilde{A}_\alpha\}; \mathbf{w}, \widetilde{J})$ by orientation preserving diffeomorphisms.

The proof of (ii) proceeds in four steps, using an inductive argument over forests as in the proof of Proposition 6.2.8 given in Section 6.3. Step 2 is the most interesting since that is where we show that our setup has enough flexibility to provide the needed transversality.

STEP 1. *The universal moduli space* $\mathcal{M}^*(\widetilde{M}, \{\widetilde{A}_\alpha\}; \mathcal{J}^\ell)$ *is a Banach manifold, where* $\mathcal{J}^\ell = \mathcal{J}^\ell_\tau(S^2; M, \omega)$ *denotes the space of all families of* ω*-tame almost complex structures on* M *of class* C^ℓ.

This is an adaptation of the arguments in the proofs of Propositions 3.2.1 and 6.2.7.

STEP 2. *Let* $T = (T, E, \Lambda)$ *be a* k*-labelled forest, suppose that the edge evaluation map* $\widetilde{\mathrm{ev}}^{E,\mathcal{J}} : \mathcal{M}^*(\widetilde{M}, \{\widetilde{A}_\alpha\}; \mathcal{J}^\ell) \times Z(T) \to \widetilde{M}^E$ *is transverse to* $\widetilde{\Delta}^E$, *and consider the universal moduli space*

$$\widetilde{\mathcal{M}}^*_{0,T}(\widetilde{M}, \{\widetilde{A}_\alpha\}; \mathcal{J}^\ell) := (\widetilde{\mathrm{ev}}^{E,\mathcal{J}})^{-1}(\widetilde{\Delta}^E).$$

Then each tuple $\mathbf{w} = \{\mathrm{w}_i\}_{i\in I}$ *of pairwise distinct points on* S^2 *is a regular value of the universal projection* $\pi^{I,\mathcal{J}} : \widetilde{\mathcal{M}}^*_{0,T}(\widetilde{M}, \{\widetilde{A}_\alpha\}; \mathcal{J}^\ell) \to (S^2)^I$.

The space $\widetilde{\mathcal{M}}^*_{0,T}(\widetilde{M}, \{\widetilde{A}_\alpha\}; \mathcal{J}^\ell)$ consists of tuples $(u_0, \{z_\alpha, u_\alpha\}, \{z_{\alpha\beta}\}, \{z_i\}, \{J_z\})$ where z_α labels the fiber in which the bubble u_α lies. These tuples satisfy the conditions

$$\begin{array}{lcl} \bar\partial_{\{J_z\}}(u_0) = 0, & & \bar\partial_{J_{z_\alpha}}(u_\alpha) = 0 \quad (\alpha \in T \setminus \{0\}), \\ 0E\alpha & \Longrightarrow & u_0(z_{0\alpha}) = u_\alpha(z_{\alpha 0}), \quad z_\alpha = z_{0\alpha}, \\ \alpha E\beta,\ \alpha \neq 0,\ \beta \neq 0 & \Longrightarrow & u_\alpha(z_{\alpha\beta}) = u_\beta(z_{\beta\alpha}), \quad z_\alpha = z_\beta. \end{array}$$

In order to understand the tangent space observe that if $S \subset T \setminus \{0\}$ is a subtree, i.e. any two vertices in S are connected by a chain of edges in S, then all the bubbles in S must lie in the same fiber. Hence the vectors ζ_α that describe the infinitesimal movement of the fiber must be the same for all $\alpha \in S$. Moreover, if one of the vertices $\beta \in S$ is connected to zero by an edge, then ζ_β must be equal to the vector $\zeta_{0\beta}$ that describes the infinitesimal movement of the attaching point $z_{0\beta}$ on the root component. Hence the tangent space consists of tuples $(\xi_0, \{\zeta_\alpha, \xi_\alpha\}, \{\zeta_{\alpha\beta}\}, \{\zeta_i\}, \{Y_z\})$ satisfying

$$D_{u_0}\xi_0 + \tfrac{1}{2}Y_z(u_0)du_0 \circ j = 0, \tag{6.7.12}$$

$$D_{u_\alpha}\xi_\alpha + \tfrac{1}{2}Y_{z_\alpha}(u_\alpha)du_\alpha \circ j = \tfrac{1}{2}J_{\zeta_\alpha}(u_\alpha)du_\alpha \circ j \quad (\alpha \neq 0), \tag{6.7.13}$$

$$\xi_0(z_{0\alpha}) + du_0(z_{0\alpha})\zeta_{0\alpha} = \xi_\alpha(z_{\alpha 0}) + du_\alpha(z_{\alpha 0})\zeta_{\alpha 0}, \qquad \zeta_\alpha = \zeta_{0\alpha}, \tag{6.7.14}$$

$$\xi_\alpha(z_{\alpha\beta}) + du_\alpha(z_{\alpha\beta})\zeta_{\alpha\beta} = \xi_\beta(z_{\beta\alpha}) + du_\beta(z_{\beta\alpha})\zeta_{\beta\alpha}, \qquad \zeta_\alpha = \zeta_\beta. \tag{6.7.15}$$

Equation (6.7.14) holds for $0E\alpha$ and (6.7.15) holds for $\alpha, \beta \in T\backslash\{0\}$ such that $\alpha E\beta$. The term on the right hand side of (6.7.13) comes from the horizontal movement of the αth bubble in the direction ζ_α.

The differential of π^I is given by

$$(\xi_0, \{\zeta_\alpha, \xi_\alpha\}, \{\zeta_{\alpha\beta}\}, \{\zeta_i\}, \{Y_z\}) \mapsto \left\{ \begin{array}{ll} \zeta_i, & \text{if } \alpha_i = 0 \\ \zeta_{\alpha_i}, & \text{if } \alpha_i \neq 0 \end{array} \right\}_{i \in I},$$

where, as usual, α_i labels the component containing the marked point z_i. Suppose $\pi^I(\xi_0, \{\zeta_\alpha, \xi_\alpha\}, \{\zeta_{\alpha\beta}\}, \{\zeta_i\}, \{Y_z\})) = \mathbf{w}$ so that

$$\mathrm{w}_i = \left\{ \begin{array}{ll} z_i, & \text{if } \alpha_i = 0 \\ z_{\alpha_i}, & \text{if } \alpha_i \neq 0 \end{array} \right., \qquad i \in I.$$

Given any tuple of tangent vectors $\widehat{\mathrm{w}}_i \in T_{\mathrm{w}_i}S^2$, $i \in I$, we must show that there exists a tuple $(\xi_0, \{\zeta_\alpha, \xi_\alpha\}, \{\zeta_{\alpha\beta}\}, \{\zeta_i\}, \{Y_z\})$ satisfying (6.7.12-6.7.15) such that $\zeta_i = \widehat{\mathrm{w}}_i$ whenever $i \in I$ and $\alpha_i = 0$ and $\zeta_{\alpha_i} = \widehat{\mathrm{w}}_i$ whenever $i \in I$ and $\alpha_i \neq 0$. For $\alpha_i = 0$ this is obvious: simply choose $\zeta_i := \widehat{\mathrm{w}}_i$ and define all the other components of the tuple $(\xi_0, \{\zeta_\alpha, \xi_\alpha\}, \{\zeta_{\alpha\beta}\}, \{\zeta_i\}, \{Y_z\})$ to be zero. For $\alpha_i \neq 0$ we argue as follows.

Let $i \in I$ such that $\alpha_i \neq 0$ and consider the corresponding subtree

$$T_i := \{\alpha \in T \setminus \{0\} \,|\, z_\alpha = \mathrm{w}_i,\ [\alpha, \alpha_i] \neq \emptyset\}.$$

Note that there is at most one edge connecting the vertex 0 to T_i. (But there may be none because in this step we assume that T is a forest rather than a tree.) Choose

$$\zeta_{\alpha\beta} := \left\{ \begin{array}{ll} \widehat{\mathrm{w}}_i, & \text{if } \alpha = 0,\ \alpha E\beta,\ \beta \in T_i, \\ 0, & \text{otherwise,} \end{array} \right. \qquad \zeta_\alpha := \left\{ \begin{array}{ll} \widehat{\mathrm{w}}_i, & \text{if } \alpha \in T_i, \\ 0, & \text{if } \alpha \in T \setminus (T_i \cup \{0\}), \end{array} \right.$$

and

$$\xi_\alpha := 0, \qquad \alpha \in T \setminus T_i.$$

Then (6.7.15) holds for $\alpha, \beta \in T \setminus (T_i \cup \{0\})$ and (6.7.14) holds for $\alpha \in T \setminus T_i$ such that $0E\alpha$. Hence (6.7.14) and (6.7.15) are equivalent to

$$\xi_\beta(z_{\beta 0}) = du_0(\mathrm{w}_i)\widehat{\mathrm{w}}_i \quad (\beta \in T_i,\ 0E\beta), \qquad \xi_\alpha(z_{\alpha\beta}) = \xi_\beta(z_{\beta\alpha}) \quad (\alpha, \beta \in T_i). \tag{6.7.16}$$

Thus it remains to choose Y_z and $\{\xi_\alpha\}_{\alpha \in T_i}$ such that (6.7.16) holds and

$$Y_z(u_0(z)) \equiv 0, \qquad Y_{z_\alpha}(u_\alpha) \equiv 0 \quad (\alpha \in T \setminus (T_i \cup \{0\})), \tag{6.7.17}$$

$$D_{u_\alpha}\xi_\alpha + \tfrac{1}{2}Y_{\mathrm{w}_i}(u_\alpha)du_\alpha \circ j = \tfrac{1}{2}J_{\widehat{\mathrm{w}}_i}(u_\alpha)du_\alpha \circ j \quad (\alpha \in T_i). \tag{6.7.18}$$

We now show that these exist by using equation (3.4.3) in the proof of Lemma 3.4.3. Recall the notation $Z_\alpha := \{z_{\alpha\beta} \,|\, \alpha E\beta\}$.

Assume first that there is no edge from 0 to T_i. Then we may proceed as follows. We shall define Y_z as

$$Y_z := \rho(z) \sum_{\alpha \in T_i} Y_\alpha. \tag{6.7.19}$$

where $\rho : S^2 \to [0,1]$ is a smooth cutoff function with support near w_i and such that $\rho(\mathrm{w}_i) = 1$, and the Y_α are chosen as follows. In the case when u_α is a ghost, i.e. $du_\alpha \equiv 0$, define $\xi_\alpha := 0$ and $Y_\alpha := 0$. In the case $du_\alpha \not\equiv 0$, use (3.4.3) to find a pair (ξ_α, Y_α) that satisfies (6.7.18), $\xi_\alpha(Z_\alpha) = 0$, $Y_\alpha(u_\beta) \equiv 0$ for $\beta \in T \setminus \{0, \alpha\}$ with $z_\beta = \mathrm{w}_i$, and such that Y_α vanishes near $u_0(\mathrm{w}_i)$. Then the tuple $(\{Y_z\}, \{\xi_\alpha\}_{\alpha \in T_i})$ satisfies (6.7.16), (6.7.17), and (6.7.18).

Now suppose that there is an edge from 0 to T_i and denote by $\beta_i \in T_i$ the unique vertex such that $0E\beta_i$. If β_i is a ghost, denote by $S_i \subset T_i$ the subtree of all ghost vertices in T_i that can be connected to β_i by a chain of ghost vertices. Let $V_i \subset T_i$ be the set of all vertices that are connected to S_i by an edge and denote by $f : V_i \to S_i$ the unique map such that $\alpha E f(\alpha)$ for $\alpha \in V_i$. Note that u_α is nonconstant for every $\alpha \in V_i$. If $du_{\beta_i} \not\equiv 0$, define $S_i := \emptyset$ and $V_i := \{\beta_i\}$, and write $f(\beta_i) := 0$. Observe that $u_\alpha \equiv u_0(\mathrm{w}_i)$ for every $\alpha \in S_i$. We define

$$\xi_\alpha := du_0(\mathrm{w}_i)\widehat{\mathrm{w}}_i, \qquad Y_\alpha := 0 \qquad (\alpha \in S_i).$$

Thus (6.7.18) holds for every $\alpha \in S_i$. For $\alpha \in V_i$ choose any $\xi'_\alpha \in W^{\ell,p}$ such that

$$\xi'_\alpha(z_{\alpha f(\alpha)}) = du_0(\mathrm{w}_i)\widehat{\mathrm{w}}_i, \qquad \xi'_\alpha(Z_\alpha \setminus \{z_{\alpha f(\alpha)}\}) = 0.$$

Then, by (3.4.3), there exists a $\xi''_\alpha \in W^{\ell,p}$ and a $Y_\alpha \in C^\ell$ such that

$$\xi''_\alpha(Z_\alpha) = 0, \qquad D_{u_\alpha}(\xi'_\alpha + \xi''_\alpha) + \tfrac{1}{2} Y_\alpha(u_\alpha) du_\alpha \circ j = \tfrac{1}{2} \dot{J}_{\widehat{\mathrm{w}}_i}(u_\alpha) du_\alpha \circ j,$$

Y_α vanishes near $u_0(\mathrm{w}_i)$, and $Y_\alpha(u_\beta) \equiv 0$ for every $\beta \in T \setminus \{0, \alpha\}$ with $z_\beta = \mathrm{w}_i$. We define

$$\xi_\alpha := \xi'_\alpha + \xi''_\alpha.$$

Now consider the remaining vertices $\alpha \in T_i \setminus (V_i \cup S_i)$. They are not connected to the set $S_i \cup \{0\}$ by an edge and so we may proceed as in the first case: if $du_\alpha \equiv 0$ define $\xi_\alpha := 0$ and $Y_\alpha := 0$; if $du_\alpha \not\equiv 0$, use (3.4.3) to find a pair (ξ_α, Y_α) that satisfies (6.7.18), $\xi_\alpha(Z_\alpha) = 0$, $Y_\alpha(u_\beta) \equiv 0$ for $\beta \in T_i \setminus \{\alpha\}$, and such that Y_α vanishes near $u_0(\mathrm{w}_i)$. As before define Y_z by (6.7.19) for a suitable cutoff function $\rho : S^2 \to [0,1]$ with support near w_i. Then $(\{Y_z\}, \{\xi_\alpha\}_{\alpha \in T_i})$ satisfies (6.7.16), (6.7.17), and (6.7.18) as required.

STEP 3. *For each k-labelled forest $(T = (T, E, \Lambda)$ the edge evaluation map $\widetilde{\mathrm{ev}}^{E,\mathcal{J}} : \mathcal{M}^*(\widetilde{M}, \{\widetilde{A}_\alpha\}; \mathcal{J}^\ell) \times Z(T) \to \widetilde{M}^E$ is transverse to $\widetilde{\Delta}^E$.*

As in Proposition 6.2.8, one proves this by induction over forests, at each stage converting one pair $(z_{\alpha\beta}, z_{\beta\alpha})$ of marked points in T' to the endpoints of an edge in the new forest T. The inductive step consists of showing that the corresponding evaluation map

$$\mathrm{ev}_{\alpha\beta} : \mathcal{M}^*_{0,T'}(\widetilde{M}, \{\widetilde{A}_\alpha\}; \mathcal{J}^\ell) \to \widetilde{M} \times \widetilde{M}$$

is transverse to the diagonal. By Step 2, we know that the composition of $\pi_{\alpha\beta}$ with the projection onto $S^2 \times S^2$ is transverse to the diagonal $\Delta_S \subset S^2 \times S^2$. Hence, by Exercise 6.3.2, it suffices now to show that the map

$$\pi_M \circ \mathrm{ev}_{\alpha\beta} : \mathcal{M}^*_{0,T'}(\widetilde{M}, \{\widetilde{A}_\alpha\}; \Delta_S, \mathcal{J}^\ell) := \pi_{\alpha\beta}^{-1}(\Delta_S) \to M \times M$$

is transverse to the diagonal $\Delta_M \subset M \times M$ where $\pi_M : \widetilde{M} \times \widetilde{M} \to M \times M$ is the obvious projection. This follows from the argument in the proof of Theorem 6.3.1.

STEP 4. *We prove (ii).*

By Propositions 6.7.7 and 6.7.9, the set of all almost complex structures $J \in \mathcal{J}_\tau(S^2; M, \omega)$ that satisfy conditions (H) and (V) in Definition 6.7.10 is residual in $\mathcal{J}_\tau(S^2; M, \omega)$. Now let T and $\{A_\alpha\}_{\alpha\in T}$ be as in the statement of the theorem. By Steps 2 and 3, there is yet another universal moduli space

$$\widetilde{\mathcal{M}}^*_{0,T}(\widetilde{M}, \{\widetilde{A}_\alpha\}; \mathbf{w}, \mathcal{J}^\ell) := (\pi^{I,\mathcal{J}})^{-1}(\mathbf{w}).$$

Consider the projection

$$\pi_{T,\{A_\alpha\}} : \widetilde{\mathcal{M}}^*_{0,T}(\widetilde{M}, \{\widetilde{A}_\alpha\}; \mathbf{w}, \mathcal{J}^\ell) \to \mathcal{J}^\ell.$$

This is a Fredholm map and an almost complex structure $J \in \mathcal{J}^\ell$ is a regular value of $\pi_{T,\{A_\alpha\}}$ if and only if it satisfies condition (E) in Definition 6.7.10 for T and $\{A_\alpha\}$. The Sard–Smale theorem asserts that this set is residual in $\mathcal{J}^\ell$, provided that ℓ is sufficiently large. The reduction of the C^∞ case to the C^ℓ case then uses Taubes' argument again as in the proof of Theorem 3.1.6. Hence the set of all smooth regular values of the projection $\pi_{T,\{A_\alpha\}}$ is residual in $\mathcal{J}_\tau(S^2; M, \omega)$. Taking the intersection over all T and $\{A_\alpha\}$ we deduce that $\mathcal{J}_{\mathrm{reg}}(S^2; M, \omega; \mathbf{w})$ is residual in $\mathcal{J}_\tau(S^2; M, \omega)$. The details of all this are a repetition of arguments carried out elsewhere in the book and are left to the reader to verify in the present case. This proves Theorem 6.7.11. □

PROOF OF THEOREM 6.7.1. The first assertion follows immediately from Theorem 6.7.11 (i) and Proposition 6.1.2. It just remains to observe that every stable map in the compactification of the moduli space $\mathcal{M}^*_{0,k}(A; \mathbf{w}, J)$ is modelled over a tree with at least one edge, and that every vertical J_z-holomorphic sphere has nonnegative Chern number.

To prove (ii) suppose that $J^i = \{J^i_z\} \in \mathcal{J}_{\mathrm{reg}}(S^2; M, \omega; \mathbf{w}^i) \cap \mathcal{J}_+(S^2; M, \omega; \kappa)$ for $i = 0, 1$. A homotopy $\{\mathbf{w}^\lambda, J^\lambda_z\}$ from $(\mathbf{w}^0, J^0)$ to $(\mathbf{w}^1, J^1)$ is called **regular** if it satisfies the following conditions.

- **(HV)** $\{J^\lambda_z\} \in \mathcal{J}_{\mathrm{reg}}(S^2; A; J^0, J^1) \cap \mathcal{J}^{\mathrm{Vert}}_{\mathrm{reg}}(S^2; A; J^0, J^1)$ for every $A \in H_2(M; \mathbb{Z})$.
- **(E)** The evaluation map

  $$\widetilde{\mathrm{ev}}^E \times \pi^I : \mathcal{W}^*(\widetilde{M}, \{\widetilde{A}_\alpha\}; \{\widetilde{J}^\lambda\}) \times Z(T) \to \widetilde{M}^E \times [0, 1] \times (S^2)^I,$$

  is transverse to $\widetilde{\Delta}^E \times \{(\lambda, \mathbf{w}^\lambda)\}_\lambda$ for every k-labelled tree T with a special vertex 0 and every collection of homology classes $\{A_\alpha\}_{\alpha\in T}$ in $H_2(M; \mathbb{Z})$. Here $\mathcal{W}^*(\widetilde{M}, \{\widetilde{A}_\alpha\}; \{\widetilde{J}^\lambda\})$ denotes the moduli space of pairs $(\lambda, \tilde{\mathbf{u}})$, where $0 \le \lambda \le 1$ and $\tilde{\mathbf{u}} \in \mathcal{M}^*(\widetilde{M}, \{\widetilde{A}_\alpha\}; \widetilde{J}^\lambda)$.
- **(P)** $\{J^\lambda_z\}_{z\in S^2} \in \mathcal{J}_+(S^2; M, \omega; \kappa)$ for every $\lambda \in [0, 1]$.

Note that condition (E) is meaningful because condition (HV) implies that the moduli space $\mathcal{W}^*(\widetilde{M}, \{\widetilde{A}_\alpha\}; \{\widetilde{J}^\lambda\})$ is a smooth finite dimensional manifold. That there is a regular homotopy $\{\mathbf{w}^\lambda, J^\lambda\}$ from $(\mathbf{w}^0, J^0)$ to $(\mathbf{w}^1, J^1)$ follows from similar arguments as in the proof of Theorem 6.7.11. The argument also uses the fact that the space $\mathcal{J}_+(S^2; M, \omega; \kappa)$ is open and path connected. Associated to a regular homotopy $\{\mathbf{w}^\lambda, J^\lambda_z\}$ is a moduli space $\mathcal{W}^*_{0,k}(A; \{\mathbf{w}^\lambda, J^\lambda_z\})$ of tuples $(\lambda, u, \mathbf{z})$, consisting of a number $\lambda \in [0, 1]$, a solution $u \in \mathcal{M}(A, S^2; J^\lambda)$ of (6.7.1) with $J_z = J^\lambda_z$ such that $[u] = A$, and a tuple $\mathbf{z} = (z_1, \ldots, z_k)$ of pairwise distinct points on S^2 such that $z_i = \mathrm{w}_i^\lambda$ for $i \in I$. It follows already from condition (HV) and Proposition 6.7.7 that this moduli space is a smooth oriented manifold with boundary of dimension

$$\dim \mathcal{W}^*_{0,k}(A; \{\mathbf{w}^\lambda, J_z^\lambda\}) = 2n + 2c_1(A) + 2k - 2|I| + 1$$

and that its boundary is

$$\partial \mathcal{W}^*_{0,k}(A; \{\mathbf{w}^\lambda, J_z^\lambda\}) = \mathcal{M}^*_{0,k}(A; \mathbf{w}^1, \{J_z^1\}) \cup \left(-\mathcal{M}^*_{0,k}(A; \mathbf{w}^0, \{J_z^0\})\right).$$

All the conditions (HV), (E), and (P) together imply that the obvious evaluation map

$$\mathrm{ev} : \mathcal{W}^*_{0,k}(A; \{\mathbf{w}^\lambda, J_z^\lambda\}) \to M^k \tag{6.7.20}$$

is the required cobordism of pseudocycles. This proves (ii).

Assertion (iii) follows from a similar argument. The only difference is that now J^0 is independent of z and the moduli space for $\lambda = 0$ consists only of the simple J^0-holomorphic spheres. The three marked points indexed by $i \in I$ are used to remove the action of the reparametrization group $\mathrm{G} = \mathrm{PSL}(2, \mathbb{C})$ for $\lambda = 0$. With this modification understood it follows as above the evaluation map (6.7.20) is a bordism of pseudocycles for a suitable homotopy $\{J_z^\lambda\}$. This proves Theorem 6.7.1. □

REMARK 6.7.12. Here is a sketch of an alternative proof of Theorem 6.7.1 (iii), based on the implicit function theorem. The proof only shows that both pseudocycles (weakly) represent the same homology class. Fix a triple

$$\mathbf{w} = (\mathrm{w}_1, \mathrm{w}_2, \mathrm{w}_3)$$

of pairwise distinct points on S^2 and an almost complex structure $J \in \mathcal{J}_{\mathrm{reg}}(M, \omega)$. Let $X \subset M^k$ be a compact oriented submanifold of M^k of codimension

$$\mathrm{codim}\, X = 2n + 2c_1(A) + 2k - 6.$$

The submanifold can be chosen such that the evaluation map

$$\mathrm{ev}_{J,T} : \mathcal{M}^*_{0,T}(B; J) \to M^k$$

is transverse to X for every T and every B. Then the evaluation map

$$\mathrm{ev}_J : \mathcal{M}^*_{0,k}(A; \mathbf{w}, J) \to M^k$$

is strongly transverse to X in the sense of pseudocycles. Now choose a sequence $\{J_z^\nu\} \in \mathcal{J}_{\mathrm{reg}}(S^2; M, \omega; \mathbf{w})$ converging to J. Then, by Gromov compactness (Theorem 5.5.5), the evaluation maps on all the strata in the compactification of the moduli space $\mathcal{M}^*_{0,k}(A; \mathbf{w}, J^\nu)$ miss X for large ν. Moreover, by the implicit function theorem, any given compact subset of $\mathcal{M}^*_{0,k}(A; \mathbf{w}, J)$ embeds into $\mathcal{M}^*_{0,k}(A; \mathbf{w}, J^\nu)$ for large ν. Using Theorem 5.3.1, one can then show that there is a one-to-one correspondence between intersection points of $\mathrm{ev}_J : \mathcal{M}^*_{0,k}(A; \mathbf{w}, J) \to M^k$ and those of $\mathrm{ev}_{J^\nu} : \mathcal{M}^*_{0,k}(A; \mathbf{w}, J^\nu) \to M^k$ with X. Thus both pseudocycles have the same intersection number with every submanifold. Hence they are weak representatives of the same homology class.

The full Gromov–Witten pseudocycle. We close this section with some remarks about more general Gromov–Witten invariants and explain how the results of this book fit into that context. The basic idea is to consider the product

$$\mathrm{ev} \times \pi : \overline{\mathcal{M}}_{0,k}(A;J) \to M^k \times \overline{\mathcal{M}}_{0,k}$$

of the evaluation map with the forgetful map. The target space of this map is a smooth compact manifold (Theorem D.4.2) and the source is a compact metrizable space (Theorem 5.6.6). If we pretend for a moment that the source is also a manifold, then it carries a fundamental class $[\overline{\mathcal{M}}_{0,k}(A;J)]$ and the Gromov–Witten invariants can be defined by pulling back cohomology classes from $M^k \times \overline{\mathcal{M}}_{0,k}$ and evaluating them on this fundamental class. Via Poincaré duality this can be interpreted as follows. Choose a submanifold $W \subset \overline{\mathcal{M}}_{0,k}$ that is dual to the class that one wants to pull back from $\overline{\mathcal{M}}_{0,k}$ and is transverse to π. Then consider its preimage

$$\overline{\mathcal{M}}_{0,k}(A;W,J) := \pi^{-1}(W) = \left\{(\mathbf{u},\mathbf{z}) \in \overline{\mathcal{M}}_{0,k}(A;J) \,|\, \pi(\mathbf{u},\mathbf{z}) \in W\right\}$$

and integrate the pullbacks of cohomology classes on M^k under the restricted evaluation map

$$\mathrm{ev}_W : \overline{\mathcal{M}}_{0,k}(A;W,J) \to M^k.$$

The constructions in Exercise 6.6.5 and Theorem 6.7.1 can be interpreted in this light. Here one has to choose $W \subset \overline{\mathcal{M}}_{0,k}$ as the submanifold obtained by fixing the marked points $\mathbf{w} = \{\mathrm{w}_i\}_{i\in I}$. In the language of Appendix D this corresponds to fixing all the cross ratios with indices from the set I. The Gromov–Witten pseudocycle of Theorem 6.6.1 corresponds to the case $W = \overline{\mathcal{M}}_{0,k}$ and the evaluation map with fixed marked points would correspond to the case $W = \{\mathrm{pt}\}$.

All this remains a heuristic discussion until one can prove that the moduli space $\overline{\mathcal{M}}_{0,k}(A;J)$ indeed carries a natural fundamental cycle $[\overline{\mathcal{M}}_{0,k}(A;J)]^{\mathrm{virt}}$ and that the homology class of this cycle is independent of J (i.e. the moduli space associated to any path $\{J_t\}_{0\le t\le 1}$ in $\mathcal{J}_\tau(M,\omega)$ carries a chain whose boundary is the difference of the fundamental cycles associated to J_0 and J_1). If this were true, then for every submanifold W of $\overline{\mathcal{M}}_{0,k}$ one could define the homology class

$$\mathrm{ev}_*\big([\overline{\mathcal{M}}_{0,k}(A,[W];J)]^{\mathrm{virt}}\big) \in H_*(M^k)$$

by setting

$$\mathrm{ev}_*\big([\overline{\mathcal{M}}_{0,k}(A,[W];J)]^{\mathrm{virt}}\big) \cdot X := (\mathrm{ev}\times\pi)_*[\overline{\mathcal{M}}_{0,k}(A;J)]^{\mathrm{virt}} \cdot [X \times W]$$

for every submanifold X of M^k. This would give a well defined rational homology class in M^k that depends only on the homology class of W. The problem then would be to find appropriate interpretations of the class $[\overline{\mathcal{M}}_{0,k}(A,[W];J)]^{\mathrm{virt}}$ for particular choices of W. This is precisely the purpose of much of the recent work on the Gromov–Witten invariants for general compact symplectic manifolds: see [**127, 239, 249, 343, 376, 189, 281**]. For some work that approaches this question in the spirit of the above discussion, see Cieliebak–Mohnke [**68**] and Ionel [**197**].

As we show in Proposition 7.4.3, there are special examples (Kähler manifolds with a transitive action of a compact Lie group) where every genus zero stable map is regular. Using integrability of the complex structure one can prove in this case that the moduli space $\overline{\mathcal{M}}_{0,k}(A;J)$ admits naturally the structure of a complex orbifold. (See Fulton–Pandharipande [**133**], Behrend–Manin [**34**], Robbin–Ruan–Salamon [**333**].) Hence in this case $\overline{\mathcal{M}}_{0,k}(A;J)$ does carry a natural rational fundamental cycle. However, it is far from obvious that the integrals over this fundamental cycle are actually deformation invariants of the symplectic structure.

In the current book, we do not try to make sense of the homology of the compactification $\overline{\mathcal{M}}_{0,k}(A;J)$. Instead we have proved in Theorem 6.6.1 that in the semipositive case the restriction of the evaluation map to the top stratum is a pseudocycle for a generic almost complex structure. Similarly, by Exercise 6.6.3, its restriction to each stratum $\mathcal{M}^*_{0,T}(\{A_\alpha\};J)$ is a pseudocycle. Further, by extending the class of almost complex structures to families $\{J_z\}$ that depend on the points $z \in S^2$, we showed in Theorem 6.7.1 that the restriction of the evaluation map to the subset $\mathcal{M}^*_{0,k}(A;\mathbf{w},\{J_z\})$ of the top stratum on which the points $\mathrm{w}_i, i \in I$, are fixed is also a pseudocycle.

Consider what this means in the simple case when $k=4$. Then the top stratum in $\overline{\mathcal{M}}_{0,4}$ can be identified with $S^2 \setminus \{0,1,\infty\}$, and there are three nontrivial strata each consisting of a point. Therefore, taking $[W]=[\mathrm{pt}]$, we have two different ways of defining the cycle that should correspond to $[\overline{\mathcal{M}}_{0,4}(A,[\mathrm{pt}];J)]^{\mathrm{virt}}$. The first is

$$\mathrm{ev}_T : \mathcal{M}^*_{0,T}(\{A_\alpha\};J) \to M^4,$$

where T is a nontrivial stratum, and the second is

$$\mathrm{ev} : \mathcal{M}^*_{0,4}(A;\mathbf{w},\{J_z\}) \to M^4,$$

where $I=\{1,2,3,4\}$ and $\mathbf{w}$ is a fixed point in the top stratum of $\overline{\mathcal{M}}_{0,4}$. It is not at all clear from these definitions that these two cycles represent the same homology class: the first is represented by intersecting pairs of curves with two marked points on each curve, while the second is represented by single curves that carry 4 marked points with fixed cross ratio. That these two cycles are homologous is a deep result, a special case of the gluing theorem. We shall formulate this statement in more generality in Theorem 7.5.10 and prove it in Chapter 10. A simple example is worked out in Section 7.2: see Remark 7.2.10.

It is natural to wonder whether the product

$$\mathrm{ev} \times \pi : \mathcal{M}^*_{0,k}(A;J) \to M^k \times \overline{\mathcal{M}}_{0,k}$$

is also a pseudocycle. By Exercise 6.5.13 and Lemma 6.5.14 this would mean that

$$\mathrm{ev}_{\mathbf{w}} : \mathcal{M}^*(A;J) \to M^k$$

is a pseudocycle for a generic element $\mathbf{w} \in \mathcal{M}_{0,k}$. We discussed how this might be proved in Example 6.6.4. There we restricted to a fixed z-independent almost complex structure J and encountered the difficulty that the reduction to an underlying simple stable map in Proposition 6.1.2 cannot in general be chosen so as to respect the cross ratios of the marked points. As we saw in Theorem 6.7.1 one can get around this problem by allowing J to depend on $z \in S^2$. This is enough to show that each $\mathrm{ev}_{\mathbf{w}}$ is a pseudocycle. However, the statement that $\mathrm{ev} \times \pi$ is a pseudocycle is even stronger, and involves controlling limit points of $\mathrm{im}(\mathrm{ev} \times \pi)$ that lie over all strata in $\overline{\mathcal{M}}_{0,k}$. To do this for general semipositive (M,ω) one would need to consider yet more general perturbations of the equation.

We shall not pursue this further here, because this general construction would not produce new invariants; the homology of the moduli space $\overline{\mathcal{M}}_{0,k}$ of stable curves is generated by the submanifolds associated to labelled trees (Theorem D.6.4) and we know already that these give rise to pseudocycles by Exercise 6.6.3. Nevertheless, it is worth pointing out that if one makes rather restrictive assumptions on

(M, ω) then arguments similar to those in Exercise 6.6.5 do show that the corresponding map to $M^k \times \overline{\mathcal{M}}_{0,k}$ is pseudocycle. Moreover, in this situation there is a natural interpretation of the associated pseudocycles $\mathrm{ev}_{\mathbf{w}}$ via the construction in Proposition 6.5.17. The next exercise invites the reader to check the details.

EXERCISE 6.7.13. Let (M, ω) be a monotone symplectic manifold with minimal Chern number N and let $J \in \mathcal{J}_{\mathrm{reg}}(M, \omega)$. Fix a homology class $A \in H_2(M; \mathbb{Z})$, an integer $k \geq 3$, and an index set $I \subset \{1, \dots, k\}$ such that

$$3 \leq \#I \leq N + 2.$$

Denote by $\overline{\mathcal{M}}_{0,I}$ the (smooth) moduli space of stable curves of genus zero with index set I for the marked points (see Appendix D).

(i) Prove that the product map

$$\mathrm{ev} \times \pi_I : \mathcal{M}^*_{0,k}(A; J) \to M^k \times \overline{\mathcal{M}}_{0,I}$$

is a pseudocycle, where π_I is the forgetful map that deletes the maps $\mathbf{u}$ and the marked points z_i, $i \neq I$. *Hint:* First consider the case $I = \{1, \dots, k\}$. The limit set of $\mathrm{ev} \times \pi$ consists of points $(\mathbf{x}, \mathbf{w}) \in M^k \times \overline{\mathcal{M}}_{0,k}$ that satisfy

$$\mathbf{w} = \lim_{\nu \to \infty} \pi([\mathbf{z}^\nu]), \qquad x_i = \lim_{\nu \to \infty} u^\nu(z_i^\nu),$$

for $1 \leq i \leq k$, where the sequence $(u^\nu, \mathbf{z}^\nu) \in \mathcal{M}^*_{0,k}(A; J)$ has no convergent subsequence with limit in $\mathcal{M}^*_{0,k}(A; J)$. Now analyse what happens to the marked points in the limit. The argument is much like that in Exercise 6.6.5 except that the structure of the limiting stable curve $\mathbf{w}$ is more complicated than before.

(ii) Let $\mathbf{w} \in \mathcal{M}_{0,I} \subset \overline{\mathcal{M}}_{0,I}$ be a point in the open stratum. Prove that $\mathrm{ev} \times \pi_I$ is weakly transverse to $M^k \times \{\mathbf{w}\}$. Interpret the pseudocycle $\mathrm{ev}_{\mathbf{w}}$ of Exercise 6.6.5 as the restriction of the evaluation map to the submanifold $\mathcal{M}^*_{0,k}(A; \mathbf{w}, J) = \pi_I^{-1}(\mathbf{w})$ of $\mathcal{M}^*_{0,k}(A; J)$ (see Proposition 6.5.17).

CHAPTER 7

Gromov–Witten Invariants

The Gromov–Witten invariants can be interpreted as counting isolated J-holomorphic curves in a symplectic manifold. Here the notion of an *isolated curve* means that it belongs to a zero dimensional moduli space. In the notation of Kontsevich–Manin [**216**] the geometric picture is the following.

Let (M, ω) be a compact symplectic manifold and fix a homology class $A \in H_2(M; \mathbb{Z})$ and two nonnegative integers k and g. Given an ω-tame almost complex structure $J \in \mathcal{J}_\tau(M, \omega)$, we are interested in the compactified moduli space $\overline{\mathcal{M}}_{g,k}(A; J)$ of J-holomorphic curves of genus g with k marked points in M representing the homology class A. In this general setting one considers all J-holomorphic curves of genus g without fixing the complex structure on a particular Riemann surface Σ and the compactification is to be understood in the Gromov–Kontsevich sense as stable maps of genus g with k marked points (see Chapter 5 for $g = 0$). This moduli space carries a natural evaluation map to M^k and a natural projection, the forgetful map, to the Deligne–Mumford space $\overline{\mathcal{M}}_{g,k}$ of stable curves of genus g with k marked points. The latter admits the structure of a compact orbifold and, in genus zero, the structure of a smooth compact manifold. (An exposition for $g = 0$ is given in Appendix D and for all g in [**337**].) Consider the diagram

$$\begin{array}{ccc} \overline{\mathcal{M}}_{g,k}(A; J) & \xrightarrow{\ \mathrm{ev}\ } & M^k \\ {\scriptstyle \pi}\big\downarrow & & \\ \overline{\mathcal{M}}_{g,k} & & \end{array}$$

pretending for the moment that all three spaces are compact smooth manifolds. Then the Gromov–Witten invariants can be interpreted as a homomorphism

$$\mathrm{GW}^M_{g,k,A} : H^*(M; \mathbb{Q})^{\otimes k} \otimes H_*(\overline{\mathcal{M}}_{g,k}; \mathbb{Q}) \to \mathbb{Q}$$

given by

$$\mathrm{GW}^M_{g,k,A}(a_1, \ldots, a_k; \beta) := \int_{\overline{\mathcal{M}}_{g,k}(A;J)} \mathrm{ev}_1^* a_1 \smile \cdots \smile \mathrm{ev}_k^* a_k \smile \pi^*\mathrm{PD}(\beta).$$

Geometrically, one can think of this invariant as counting the number of J-holomorphic curves of genus g with k marked points that pass, at the ith marked point, through a cycle $X_i \subset M$ Poincaré dual to a_i, and such that the image of the curve under the projection π is restricted to a cycle $Y \subset \overline{\mathcal{M}}_{g,k}$ representing the class β. This construction should in principle result in a zero dimensional reduced moduli space whenever the classes a_i and β satisfy the dimensional condition

$$\sum_{i=1}^{k} \deg(a_i) = n(2 - 2g) + 2c_1(A) + \deg(\beta).$$

The resulting invariants are rational; both moduli spaces $\overline{\mathcal{M}}_{g,k}$ and $\overline{\mathcal{M}}_{g,k}(A;J)$ have orbifold singularities, in the former case because Riemann surfaces can have nontrivial automorphisms and in the latter case because of presence of nontrivial isotropy subgroups at multiply covered curves.

The precise definition of these invariants in full generality requires the construction of a rational fundamental cycle on the compactified moduli space $\overline{\mathcal{M}}_{g,k}(A;J)$. This is a long story which goes much beyond the scope of the present book (cf. the discussion at the very end of Chapter 6.) Our goal in this chapter is to explain only a small part of this picture. Firstly, we restrict attention to the genus zero case, secondly we consider only semipositive symplectic manifolds (M,ω), and thirdly we consider only special homology classes β in the moduli space $\overline{\mathcal{M}}_{0,k}$, namely those classes that arise from fixing the marked points associated to a subset $I \subset \{1,\dots,k\}$ of the index set. The latter restriction is not really essential and can easily be removed, because the homology of $\overline{\mathcal{M}}_{0,k}$ is well understood. However, the other two restrictions (genus zero and semipositivity) are essential within the analytic framework developed here. We emphasize that the genus zero invariants in the semipositive case take integer values on integral (co)homology classes.

The genus zero invariants for semipositive manifolds were first introduced into symplectic geometry by Ruan [**342, 340**]. Special cases were previously used by Gromov in [**160**] and by McDuff in [**259**] in order to get information on the structure of symplectic manifolds. The invariants with fixed marked points arise in the context of sigma models and were considered by Witten in [**421**]. The genus zero invariants associated to the homology class $\beta_{k,I} \in H_{2(k-\#I)}(\overline{\mathcal{M}}_{0,k};\mathbb{Z})$, whose Poincaré dual is the pull back of the generator of the top dimensional group $H^{\#I-3}(\overline{\mathcal{M}}_{0,I};\mathbb{Z})$ under the projection $\pi_{k,I} : \overline{\mathcal{M}}_{0,k} \to \overline{\mathcal{M}}_{0,I}$, correspond to the so-called *mixed invariants* of Ruan–Tian [**345**] in which one fixes only the marked points of the index set I. In [**216**] Kontsevich and Manin express the Gromov–Witten invariants as homomorphisms

$$\mathrm{GW}^M_{g,k,A} : H^*(M;\mathbb{Q})^{\otimes k} \to H^*(\overline{\mathcal{M}}_{g,k};\mathbb{Q}).$$

In their terminology the invariants with varying marked points ($\beta = [\overline{\mathcal{M}}_{0,k}]$) are the codimension zero classes and those with fixed marked points ($\beta = [\mathrm{pt}]$) are the classes of highest codimension while the mixed invariants constitute an intermediate case ($\beta_{k,I}$ corresponds to codimension $2\#I$). We shall drop the argument β and the reference to the genus, and denote the genus zero invariants associated to the class $\beta_{k,I}$ by $\mathrm{GW}^{M,I}_{A,k}$. When $\#I = 3$ we abbreviate $\mathrm{GW}^M_{A,k} := \mathrm{GW}^{M,I}_{A,k}$. When the manifold M is understood from the context we drop the superscript M.

The chapter begins with a careful definition of the counting invariant GW and the first examples. For the convenience of the reader it includes a discussion of some relevant results from earlier chapters. Section 7.2 develops some variations on the main construction. For example we define the invariants for curves modelled on a fixed tree or with a fixed subset of marked points. We also discuss the obstruction bundle, which gives a way of calculating the invariants by using nonregular (but sufficiently nice) almost complex structures. Section 7.3 introduces the invariants GW^I associated to pseudoholomorphic graphs and relates them to GW. Section 7.4 illustrates these ideas with the example of curves in projective spaces. The last section describes the Kontsevich–Manin axioms for the Gromov–Witten invariants and interprets them in terms of the GW^I. To demonstrate their power we describe

Kontsevich's calculation of the numbers N_d of degree d rational curves in $\mathbb{C}P^2$ through $3d-1$ generic points. Throughout we restrict attention to the genus zero case. Some higher genus examples are considered in Chapter 8; see Example 8.6.12.

7.1. Counting pseudoholomorphic spheres

In this section we define the Gromov–Witten invariants $\mathrm{GW}_{A,k}$ and calculate them in some easy cases. Before giving the formal definition we begin with some remarks about Poincaré duality and pseudocycles.

As in Section 6.5 we shall use the notation

$$H^*(M) := H^*(M;\mathbb{Z})/\mathrm{Tor}$$

for the free part of $H^*(M;\mathbb{Z})$ and similarly for the homology groups. By definition, a singular cohomology class $a \in H^*(M)$ is uniquely determined by its values on the homology classes in $H_*(M)$ and can be identified with the corresponding homomorphism $H_*(M) \to \mathbb{Z}$. A $(2n-k)$-dimensional pseudocycle $f : U \to M$ is called **(Poincaré) dual** to a if

$$\int_X a = f \cdot X$$

for every closed oriented k-dimensional submanifold $X \subset M$. Then, if $f : U \to M$ is dual to a and $g : V \to M$ is dual to b, we have

$$\int_M a \wedge b = f \cdot g$$

(see Lemma 6.5.7). In the terminology of Section 6.5 a pseudocycle dual to a cohomology class a is a *weak representative* of the homology class $\mathrm{PD}(a)$.

Now suppose that (M,ω) is a closed semipositive symplectic $2n$-manifold, let $A \in H_2(M;\mathbb{Z})$ be a spherical homology class satisfying (6.6.1), and let k be a nonnegative integer. Then Theorem 6.6.1 asserts that the evaluation map

$$\mathrm{ev}_J : \mathcal{M}^*_{0,k}(A;J) \to M^k$$

is a pseudocycle of dimension

$$\mu(A,k) = 2n + 2c_1(A) + 2k - 6$$

for every $J \in \mathcal{J}_{\mathrm{reg}}(M,\omega)$ (see Definition 6.2.1).

THEOREM 7.1.1. *Let (M,ω) be a closed semipositive symplectic $2n$-manifold, k be a nonnegative integer, and $A \in H_2(M;\mathbb{Z})$ be a spherical homology class that is not a nontrivial integer multiple of a spherical homology class B with Chern number $c_1(B) = 0$. Then the homomorphism*

$$\mathrm{GW}^M_{A,k} : H^*(M)^{\otimes k} \to \mathbb{Z}$$

defined by

$$\mathrm{GW}^M_{A,k}(a_1,\dots,a_k) := f \cdot \mathrm{ev}_J, \tag{7.1.1}$$

is independent of the almost complex structure J and the pseudocycle f used to define it. Here $J \in \mathcal{J}_{\mathrm{reg}}(M,\omega)$, $f : U \to M^k$ is a pseudocycle Poincaré dual to $\pi_1^*a_1 \smile \cdots \smile \pi_k^*a_k$, and $\pi_i : M^k \to M$ denotes the projection onto the ith factor.

PROOF. By Theorem 6.6.1, the bordism class of ev_J is independent of J. Hence, by Lemma 6.5.5, the intersection number $f \cdot \mathrm{ev}_J$ is independent of J. By Lemma 6.5.7, it depends only on the homology class represented by the pseudocycle f. This proves Theorem 7.1.1. □

The above homomorphism $\mathrm{GW}^M_{A,k} : H^*(M)^{\otimes k} \to \mathbb{Z}$ is called the **k-pointed genus zero Gromov–Witten invariant of (M, ω) in the homology class A.**

EXERCISE 7.1.2. Show that $\mathrm{GW}^M_{A,k}(a_1, \ldots, a_k)$ can be defined as the intersection number of ev with a product cycle $f_1 \times \cdots \times f_k$ where the $f_j : U_j \to M$ are pseudocycles dual to the cohomology classes a_j, respectively. In other words, transversality can be achieved within the class of product cycles. *Hint:* This is a simple adaptation of Lemma 6.5.5 (i) to the case where f is a product cycle and ϕ is a product diffeomorphism.

Recall from Chapter 6 that the moduli space $\mathcal{M}^*_{0,k}(A; J)$ (in the case $A \neq 0$) is the set of equivalence classes of tuples $(u, z_1, \ldots, z_k)$, where $u : S^2 \to M$ is a simple J-holomorphic sphere representing the class A and the z_i are pairwise distinct points on S^2. The equivalence relation is given by the obvious action of the reparametrization group $\mathrm{G} = \mathrm{PSL}(2, \mathbb{C})$. By Theorem 3.1.6, this space is a smooth oriented manifold of dimension $\mu(A, k)$ for every $J \in \mathcal{J}_{\mathrm{reg}}(A)$. The stronger assertion of Theorem 6.6.1, that it is a pseudocycle for every $J \in \mathcal{J}_{\mathrm{reg}}(M, \omega)$, means that all the strata appearing in the compactification have dimensions at most $\mu(A, k) - 2$. Now suppose that each class $a_i \in H^*(M)$ is dual to an oriented submanifold $X_i \subset M$ in general position. Then the Gromov–Witten invariant $\mathrm{GW}^M_{A,k}(a_1, \ldots, a_k)$ is the number of J-holomorphic spheres in the class A passing through the submanifolds X_i and counted with appropriate signs:

$$\mathrm{GW}^M_{A,k}(a_1, \ldots, a_k) = \# \left\{ [u, z_1, \ldots, z_k] \in \mathcal{M}^*_{0,k}(A; J) \,|\, u(z_i) \in X_i \right\}.$$

This formula is rigorous as a mod-2 invariant whenever the degrees of the classes a_i sum up to the dimension of the moduli space:

$$\sum_{i=1}^{k} \deg(a_i) = \mu(A, k) = 2n + 2c_1(A) + 2k - 6. \tag{7.1.2}$$

In all other cases the invariant is zero by definition. Note that the general position hypothesis means not only that the evaluation map is transverse to $X_1 \times \cdots \times X_k$, but also that this product does not intersect the images of the evaluation maps associated to the lower dimensional strata in the compactification.

EXAMPLE 7.1.3. The case $A = 0$ deserves a special discussion. Since every constant component must carry at least three special points, it follows that $\mathcal{M}^*_{0,k}(0; J)$ is empty for $k < 3$. When $k \geq 3$ it consists of tuples $(x, z_1, \ldots, z_k)$, where $x \in M$ (thought of as the constant J-holomorphic sphere $u(z) \equiv x$) and the z_i are pairwise distinct points on S^2. In the case $k > 3$, the formula (7.1.2) says that the cycles X_i dual to the classes a_i have total codimension bigger than $2n$ and hence do not intersect when in general position. Hence the Gromov–Witten invariants for $A = 0$ vanish for $k \neq 3$. For $k = 3$ the number $\mathrm{GW}^M_{0,3}(a_1, a_2, a_3)$ counts triple intersections of the Poincaré duals. This shows that

$$\mathrm{GW}^M_{0,3}(a_1, a_2, a_3) = \int_M a_1 \smile a_2 \smile a_3.$$

EXAMPLE 7.1.4. Another interesting special case is $k = 0$. When $A = 0$ as well, there is nothing to count and we define the invariant to be zero. However, when $A \neq 0$ we can define a consistent and interesting invariant by considering the manifold M^k to be a point. Then the evaluation map $\mathrm{ev}_J : \mathcal{M}^*_{0,0}(A; J) \to \{\mathrm{pt}\}$ is automatically a pseudocycle whenever $J \in \mathcal{J}_{\mathrm{reg}}(A)$ and the dimension $\mu(A, 0) = 2n + 2c_1(A) - 6$ of the moduli space $\mathcal{M}^*_{0,0}(A; J)$ is greater than zero. However in this case its homology class is zero for dimensional reasons. Thus, the invariant can only be nonzero if

$$2n + 2c_1(A) = 6$$

so that the moduli space $\mathcal{M}^*_{0,0}(A; J)$ is zero dimensional. If $J \in \mathcal{J}_{\mathrm{reg}}(M, \omega)$ then all the boundary strata in the compactification have negative virtual dimensions and are therefore empty. It then follows that $\mathcal{M}^*_{0,0}(A; J)$ is a finite set and the invariant

$$\mathrm{GW}(A) := \mathrm{GW}_{A,0}(1) \in \mathbb{Z}$$

is the number of elements of $\mathcal{M}^*_{0,0}(A; J)$ counted with appropriate signs. In the case $M = S^2$ and $c_1(A) = 2$, this number is one. If $\dim M = 4$, $c_1(A) = 1$, and $A \cdot A = -1$, the invariant counts the number of exceptional divisors representing the class A (see Example 7.1.15 below). By positivity of intersections this number is either zero or one. However, if $A \cdot A \geq 0$ then the invariant may have other values than zero or one (see Exercise 7.1.17 below). The most important case is that of Calabi–Yau manifolds ($\dim M = 6$ and $c_1(A) = 0$). Here it is essential to get rid of the hypothesis that A has to be primitive: see Remark 7.3.8.

We end this discussion with two exercises that illustrate some basic geometric properties of the invariants: cf. the *(Divisor)* and *(Fundamental class)* axioms listed in Section 7.5.

EXERCISE 7.1.5. Let (M, ω) be a compact symplectic 4-manifold and A be a spherical homology class such that $c_1(A) = 1$. Prove that if $a_k \in H^2(M)$ and $k \geq 1$ then

$$\mathrm{GW}_{A,k}(a_1, \dots, a_k) = \mathrm{GW}_{A,k-1}(a_1, \dots, a_{k-1}) \cdot \int_A a_k.$$

The geometric idea here is that every A-curve must meet the Poincaré dual cycle $\mathrm{PD}(a_k)$ a total of $\int_A a_k$ times. Therefore there is no need to keep track of these intersections.

EXERCISE 7.1.6. Let (M, ω) be a compact semipositive symplectic manifold and $A \neq 0$. Prove that

$$\mathrm{GW}_{A,k}(a_1, \dots, a_{k-1}, 1) = 0$$

for $k \geq 1$, where $1 := \mathrm{PD}([M]) \in H^0(M)$. The idea here is that the invariant vanishes because the last marked point is unconstrained, so that the relevant intersection points are never isolated.

Comments on the definition.

Semipositivity. A compact symplectic manifold (M, ω) is called semipositive if there are no spherical homology classes $A \in H_2(M; \mathbb{Z})$ such that

$$\omega(A) > 0, \qquad 2 - n < c_1(A) < 0.$$

In particular, this holds for every symplectic manifold of dimension $2n \leq 6$. As explained in Section 6.4, this condition guarantees that, for generic J, there are no J-holomorphic spheres with negative first Chern numbers. To understand the significance of this, recall first that a stable map is called simple if each of its nonconstant components is simple (i.e. not multiply covered) and if no two of the nonconstant components have the same images. Secondly, the moduli space $\mathcal{M}^*_{0,T}(B;J)$ of simple stable maps modelled over a k-labelled tree T is a smooth manifold of dimension $\mu(B,k) - 2e(T)$ whenever $J \in \mathcal{J}_{\mathrm{reg}}(M,\omega)$ (Theorem 6.2.6). Thirdly, for every nonsimple stable map in the class A there exists an underlying simple stable map in some class B with the same image as the original stable map and the same values at the marked points (Proposition 6.1.2). Now, if every J-holomorphic sphere has a nonnegative Chern number, then $c_1(B) \le c_1(A)$ and so all the strata in the compactification have dimensions less than $\mu(A,T)$. This is the essential ingredient in the proof that the evaluation map is a pseudocycle for $J \in \mathcal{J}_{\mathrm{reg}}(M,\omega)$ (Theorem 6.6.1). The semipositivity condition can be relaxed if one restricts attention to those almost complex structures for which all relevant J-holomorphic spheres have nonnegative Chern numbers (Remark 6.6.2).

Determining signs. Another important ingredient in the computation of the Gromov–Witten invariants is the determination of the correct signs. If J is integrable, we saw in Remark 3.2.6 that the moduli spaces have complex structures that are compatible with their orientations. Hence in this case it is usually easy to figure out the signs of intersection points (see Example 7.1.15 below). However, if J is not integrable, the question is in general much more delicate.

Recognizing regular almost complex structures. In order to calculate the invariants, one has to be able to recognize when a given almost complex structure J belongs to the set $\mathcal{J}_{\mathrm{reg}}(M,\omega)$. This means that the linearized operators D_u are surjective for all simple J-holomorphic spheres and that the edge evaluation maps are transverse to the edge diagonals for all trees (see Definition 6.2.1). These conditions imply that all the moduli spaces of simple stable maps have the predicted dimensions and so $\mathrm{ev}_J : \mathcal{M}^*_{0,k}(A;J) \to M^k$ is a pseudo-cycle for every A and k. For the computation of the invariant $\mathrm{GW}_{A,k}$, it suffices to verify these conditions for those trees T and homology classes $\{B_\alpha\}$ that actually appear in the compactification of the moduli space $\mathcal{M}^*_{0,k}(A;J)$. These classes must satisfy $\sum_{\alpha\in T} m_\alpha B_\alpha = A$ for some collection of positive integers m_α. A particularly simple case is when A does not admit any such nontrivial decomposition.

DEFINITION 7.1.7. *A class A is said to be J-**indecomposable** if it cannot be written as a sum $A = A_1 + \cdots + A_k$, where $k \ge 2$ and each A_i has a nonconstant spherical J-holomorphic representative.*

LEMMA 7.1.8. *Let $A \in H_2(M;\mathbb{Z})$ and $J \in \mathcal{J}_{\mathrm{reg}}(A)$ (i.e. D_u is onto for every simple J-holomorphic sphere u representing the class A). If A is J-indecomposable then, for every integer $k \ge 0$, the evaluation map $\mathrm{ev}_J : \mathcal{M}^*_{0,k}(A;J) \to M^k$ is a pseudocycle and $\mathrm{GW}_{A,k}$ can be calculated using ev_J.*

PROOF. Since $J \in \mathcal{J}_{\mathrm{reg}}(A)$ the operator D_u is surjective for every simple J-holomorphic sphere $u : S^2 \to M$ representing the class A. Hence the moduli space $\mathcal{M}^*_{0,0}(A;J) = \mathcal{M}^*(A;J)/\mathrm{G}$ of simple unmarked J-curves in the class A is a manifold of the predicted dimension $2n + 2c_1(A)$. We claim that it is compact. To see this, note first that because A is J-indecomposable every J-curve representing the class A is simple. Moreover, by Theorem 5.3.1, every sequence in $\mathcal{M}_{0,0}(A;J)$

has a Gromov convergent subsequence. Since there are no marked points, its limit has only one component and so must belong to $\mathcal{M}_{0,0}(A;J) = \mathcal{M}^*_{0,0}(A;J)$. (Note that if A has minimal energy among all classes with J-holomorphic representatives then the same result follows from the much easier Theorem 4.2.11.)

When $k \geq 2$ the corresponding space $\mathcal{M}^*_{0,k}(A;J)$ will not be compact, since marked points must always be distinct. However, the stable maps that lie in the boundary strata consist of one main component that represents the class A together with some ghost components to accommodate the marked points. Such strata are always regular and have dimensions $\mu - 2e(T)$ (see Exercise 7.1.9). Hence all strata in the compactification have the predicted dimensions and Lemma 7.1.8 follows. □

EXERCISE 7.1.9. Suppose that A is J-indecomposable and $J \in \mathcal{J}_{\mathrm{reg}}(A, \mathbb{C}P^1)$. Then every stable map in the class A contains one main component representing the class A together possibly with some ghost components containing marked points. Verify that $J \in \mathcal{J}_{\mathrm{reg}}(A, T)$ for every tree T. (Cf. Exercise 6.2.5.)

EXERCISE 7.1.10. Let $A \in H_2(M;\mathbb{Z})$. Show that the set of all ω-tame almost complex structures J for which A is J-indecomposable is open in $\mathcal{J}_\tau(M,\omega)$.

REMARK 7.1.11. A **deformation** of a symplectic form ω is a smooth 1-parameter family $[0,1] \to \Omega^2(M) : t \mapsto \omega_t$ of symplectic forms starting at $\omega_0 = \omega$. In distinction to the notion of isotopy, we do not require that the cohomology class remain constant. Because the taming condition is open, an almost complex structure J that is tamed by ω is also tamed by all symplectic forms sufficiently close to ω. Hence the Gromov–Witten invariants $\mathrm{GW}^M_{A,k}$ do not change under a deformation ω_t of ω, provided that[1] (M, ω_t) is semipositive for each $t \in [0,1]$. This condition on ω_t is quite restrictive unless it happens that dim $M \leq 6$ or c_1 vanishes on π_2. In these cases, (M, ω_t) is semipositive for one form ω_t if and only if it is semipositive for all forms ω_t in the deformation. Hence in these cases the Gromov–Witten invariants depend only on the deformation type of (M,ω). In dimension 4, more is true: Taubes showed that the Gromov invariants (of embedded curves of possibly higher genus) coincide with the Seiberg–Witten invariants and hence these Gromov invariants depend only on the smooth structure of M. However this is not true in dimension 6: see the discussion of Ruan's examples in Chapter 9.

REMARK 7.1.12 (Invariants for Homology). So far we have defined the Gromov–Witten invariants in terms of the cohomology of M. However, in some situations it is convenient to think of them as determined by homology classes. Thus given $\alpha_1, \ldots, \alpha_k \in H_*(M)$ we define

$$\mathrm{GW}^M_{A,k}(\alpha_1, \ldots, \alpha_k) := \mathrm{GW}^M_{A,k}(\mathrm{PD}(\alpha_1), \ldots, \mathrm{PD}(\alpha_k)).$$

This counts the intersection number of $\mathrm{ev} : \mathcal{M}^*_{0,k}(A;J) \to M^k$ with the product class $\alpha_1 \times \cdots \times \alpha_k \in H^*(M^k)$.

Examples.

EXAMPLE 7.1.13. Consider the product manifold

$$\widetilde{M} := S^2 \times M$$

[1]Of course, once one has constructed the virtual moduli cycle this proviso is unnecessary. Therefore, it is now known that the invariants depend only on the deformation type of (M, ω).

with a product symplectic form and assume that $\widetilde{M}$ is semipositive. For example, this holds when dim $M \le 4$ or $\pi_2(M) = 0$. Consider the spherical homology class

$$\widetilde{A} := [S^2 \times \mathrm{pt}] \in H_2(S^2 \times M; \mathbb{Z})$$

and suppose $\widetilde{J}$ is an $\widetilde{\omega}$-tame product almost complex structure. Then, by Corollary 3.3.5, $\widetilde{J} \in \mathcal{J}_{\mathrm{reg}}(\widetilde{A})$ and the class $\widetilde{A}$ is $\widetilde{J}$-indecomposable. Moreover, there is precisely one $\widetilde{J}$-holomorphic $\tilde{A}$-sphere through each point in $\widetilde{M}$. Hence it follows from Lemma 7.1.8 that the evaluation map ev : $\mathcal{M}_{0,1}(\tilde{A}; \widetilde{J}) \to \widetilde{M}$ is a diffeomorphism. Thus

$$\mathrm{GW}^{\widetilde{M}}_{\widetilde{A},1}(\widetilde{a}) = \int_{\widetilde{M}} \widetilde{a}.$$

EXAMPLE 7.1.14 (Lines in Projective space). Any two points in $\mathbb{C}P^n$ lie on a unique line. This should translate into the formula

$$\mathrm{GW}^{\mathbb{C}P^n}_{L,2}(c^n, c^n) = 1,$$

where

$$L := [\mathbb{C}P^1] \in H_2(\mathbb{C}P^n; \mathbb{Z}),$$

c is the positive generator of $H^2(\mathbb{C}P^n; \mathbb{Z})$, and $c^n = \mathrm{PD}([\mathrm{pt}]) \in H^{2n}(\mathbb{C}P^n; \mathbb{Z})$. To verify this, note first that $c_1(L) = n + 1$ and $k = 2$, and hence the dimensional condition (7.1.2) is satisfied with $\mu(L, 2) = 4n$. Moreover, the standard complex structure J_0 satisfies the hypotheses of Lemma 3.3.2. In fact, if $u : \mathbb{C}P^1 \to \mathbb{C}P^n$ is a line then the pullback tangent bundle $u^*T\mathbb{C}P^n \to \mathbb{C}P^1$ splits as a direct sum of holomorphic line bundles, one of which has Chern number two and the others have Chern number one. The line bundles with Chern number one can be realized as the normal bundles of $\mathbb{C}P^{k-1}$ in $\mathbb{C}P^k$ for $k = 2, \ldots, n$. This shows that $J_0 \in \mathcal{J}_{\mathrm{reg}}(L)$. Since L is a J_0-indecomposable class, Lemma 7.1.8 asserts that we can use J_0 to calculate $\mathrm{GW}_{L,2}$. Hence $\mathrm{GW}_{L,2}(c^n, c^n)$ is exactly the number of lines through two points and so is equal to one.

Another elementary Gromov–Witten invariant is $\mathrm{GW}^{\mathbb{C}P^3}_{L,3}(c^2, c^2, c^3)$. It counts the number of lines in 3-space that go through a point and two generic lines. The dimensional condition (7.1.2) is satisfied, so that this number could be nonzero. Moreover there is a unique line satisfying these incidence conditions, namely the line of intersection of the two planes that go through the point and one of the lines. To conclude that

$$\mathrm{GW}^{\mathbb{C}P^3}_{L,3}(c^2, c^2, c^3) = 1$$

one needs to confirm that the evaluation map ev : $\mathcal{M}(L, J_0) \to (\mathbb{C}P^3)^3$ is transverse to a generic representative of the class $[pt] \times L \times L$. This may be done by hand, but also follows from Proposition 7.4.5 below.

EXAMPLE 7.1.15 (Exceptional divisors). Let $\overline{\mathbb{C}P}^2$ denote the complex projective plane with the reversed orientation. Note that this is not a complex manifold. If $E \in H_2(\overline{\mathbb{C}P}^2; \mathbb{Z})$ denotes the homology class of the submanifold $\mathbb{C}P^1$ (with its standard orientation) then $E \cdot E = -1$.

Let (M, ω) be a complex Kähler surface, and let $(\widetilde{M}, \widetilde{J})$ denote its blowup at the point x. This means that x is replaced by the set of all lines through x. This set of lines is a copy of $\mathbb{C}P^1$ and is called the exceptional divisor Σ. The normal bundle to Σ is the tautological line bundle over $\mathbb{C}P^1$ which has Euler class -1, and it follows that $\widetilde{M}$ is diffeomorphic to the connected sum $M \# \overline{\mathbb{C}P}^2$ by a diffeomorphism that identifies Σ with a complex line in $\overline{\mathbb{C}P}^2$. This construction also works in the symplectic category (see Section 9.3) and, in particular, the complex blowup $(\widetilde{M}, \widetilde{J})$ is Kähler. For more details see Section 9.3.

Let $E = [\Sigma]$ denote the homology class of the exceptional divisor. Then, the moduli space $\mathcal{M}(E; \widetilde{J})$ consists of a single $\widetilde{J}$-holomorphic curve up to reparametrization, namely the exceptional divisor itself. To see this, note that any other $\widetilde{J}$-holomorphic curve in this class would have to intersect Σ with intersection number -1, but this is impossible because the intersection number of any two distinct $\widetilde{J}$-holomorphic curves is nonnegative. Moreover the curve Σ is regular by Lemma 3.3.3, which is consistent with the dimension formula $\dim \mathcal{M}^*(E; \widetilde{J}) = 4 + 2c_1(E) = 6$. Hence it follows from Lemma 7.1.8 and Exercise 7.1.5 that

$$\mathrm{GW}^{M\#\overline{\mathbb{C}P}^2}_{E,1}(\mathrm{PD}(E)) = E \cdot E = -1.$$

It follows that the class E must have a J-holomorphic representative for every $\widetilde{\omega}$-tame J on $\widetilde{M}$. It is possible for this representative to be reducible. For example, if $M = \mathbb{C}P^2$ then $M\#\overline{\mathbb{C}P}^2$ is diffeomorphic to the projectivized bundle $\mathbb{P}(\mathcal{L}_3 \oplus \mathbb{C})$ where the line bundle $\mathcal{L}_3 \to \mathbb{C}P^1$ has Chern number three. The subbundle $\mathcal{L}_3$ gives rise to a section of $\mathbb{P}(\mathcal{L}_3 \oplus \mathbb{C})$ that has self-intersection -3 and represents the class $2E - L \in H_2(\mathbb{C}P^2\#\overline{\mathbb{C}P}^2; \mathbb{Z})$. Thus, for a suitable almost complex structure J, the class E is represented by the union of this (-3)-sphere with the fiber $L - E$ of the projection $\mathbb{P}(\mathcal{L}_3 \oplus \mathbb{C}) \to \mathbb{C}P^1$. However, because the $(2E - L)$-curve cannot be regular, this does not happen for a generic J.

EXERCISE 7.1.16. (i) Show that there is a unique degree 2 homogeneous polynomial f through any five points in $\mathbb{C}P^2$, no three of which lie on a line. Show by direct calculation that its zero set C_f is nonsingular and is the image of a degree 2 map $u : \mathbb{C} \cup \{\infty\} \to \mathbb{C}P^2$. Deduce that $\mathrm{GW}^{\mathbb{C}P^2}_{2L,5}(c^2, c^2, c^2, c^2, c^2) = 1$. *Hint:* Some details of the relevant calculations are worked out in Example 7.2.9.

(ii) Show that $\mathrm{GW}^{\mathbb{C}P^3}_{L,4}(c^2, c^2, c^2, c^2) = 2$. This is the number of lines in 3-space that meet 4 generic lines. *Hint:* Recall from Example 7.1.14 that there is a unique line through a point and two generic lines. Thus, if we fix three generic lines, there is a family of lines that meet all three (let the point move along one of the lines). Show that the set of points on this family form a quadric surface, i.e. is the zero set of a degree 2 homogeneous polynomial. This Gromov–Witten invariant is denoted $N_1(4, 0)$ and may also be calculated using the WDVV equation: see Exercise 7.5.17.

EXERCISE 7.1.17. Let $M = \mathbb{C}P^2$ and $L = [\mathbb{C}P^1]$ as in Example 7.1.14. Show by a dimension count that

$$k \neq 8 \qquad \Longrightarrow \qquad \mathrm{GW}_{3L,k}(c^2, \ldots, c^2) = 0.$$

It turns out that

$$\mathrm{GW}^{\mathbb{C}P^2}_{3L,8}(c^2, \ldots, c^2) = 12$$

(see Proposition 7.5.11). By blowing up $\mathbb{C}P^2$ at eight points, construct an example of a symplectic 4-manifold M and a class A such that $c_1(A) = 1$ and $\mathrm{GW}(A) := \mathrm{GW}^M_{A,0}(1) > 1$. (For notation, see Example 7.1.4.)

7.2. Variations on the definition

In this section we first discuss ways to calculate the Gromov–Witten invariants in cases when the manifold M has a natural (for example, integrable) but nonregular almost complex structure. We then define some related invariants, for stable maps modelled on a fixed tree or with fixed marked points.

In practice, one does not need to assume that an almost complex structure J_0 is regular, that is in $\mathcal{J}_{\text{reg}}(M,\omega)$, in order to be able to use it to calculate $\text{GW}_{A,k}$. The important point is that the evaluation map $\text{ev}_{J_0} : \mathcal{M}^*_{0,k}(A;J_0) \to M^k$ should be a pseudocycle that represents the same homology class as in the case of regular J. For this to hold, J_0 just needs to satisfy the following two conditions.

(a) J_0 is regular for A-curves in the sense that the linearized operator D_u is surjective for every J_0-holomorphic sphere representing the class A. It then follows that the moduli space $\mathcal{M}^*_{0,k}(A;J_0)$ has the correct dimension $d := 2n + 2c_1(A) + 2k - 6$ for any k.

(b) The dimension (in the sense of Section 6.5) of the image under ev_{J_0} of the "boundary" $\overline{\mathcal{M}}_{0,k}(A;J_0) \setminus \mathcal{M}^*_{0,k}(A;J)$ is at most $d-2$.

These conditions imply that ev_{J_0} is a pseudocycle of dimension d (see Definition 6.5.1).

LEMMA 7.2.1. *Suppose that J_0 is an ω-tame almost complex structure on a closed semipositive symplectic manifold (M,ω) that satisfies conditions (a) and (b) above. Then the evaluation map $\text{ev}_{J_0} : \mathcal{M}^*_{0,k}(A;J_0) \to M^k$ is a weak representative for the Gromov–Witten pseudocycle given by ev_J for $\mathcal{J}_{\text{reg}}(M,\omega)$.*

PROOF. By Lemma 6.5.7, we need to see that the intersection number of ev_{J_0} with every smooth map $f: X \to M^k$ defined on a compact oriented smooth manifold X of dimension $2nk - d$ is the same as that of ev_J, $J \in \mathcal{J}_{\text{reg}}(M,\omega)$. By assumption, we may choose $f: X \to M^k$ so that

$$\text{ev}_{J_0}\Big(\overline{\mathcal{M}}_{0,k}(A;J_0) \setminus \mathcal{M}^*_{0,k}(A;J_0)\Big) \cap f(X) = \emptyset.$$

Suppose that $J^\nu \in \mathcal{J}_{\text{reg}}(M,\omega)$ converges to J_0. Then, by the implicit function theorem applied to ev_{J_0}, there are for each large ν at least as many intersections of ev_{J^ν} with f as there are of ev_{J_0} with f. Suppose there are more for all large ν. Then these must converge to a J_0-holomorphic stable map that intersects $\text{im}\, f$ but (by the uniqueness part of the implicit function theorem) does not lie in $\text{ev}_{J_0}(\mathcal{M}^*_{0,k}(A;J_0))$. But this contradicts the fact that f misses the image of $\overline{\mathcal{M}}_{0,k}(A;J_0) \setminus \mathcal{M}^*_{0,k}(A;J)$. Hence there is a bijective correspondence between the intersections of ev_{J^ν} and ev_{J_0} with f that preserves orientations by the implicit function theorem. Thus $\text{ev}_{J^\nu} \cdot f = \text{ev}_{J_0} \cdot f$ for large ν. This proves Lemma 7.2.1. □

EXAMPLE 7.2.2. Let $M = S^2 \times S^2$, denote $A_1 = [S^2 \times \{\text{pt}\}]$, $A_2 = [\{\text{pt}\} \times S^2]$, take A to be the diagonal class $A_1 + A_2$, and choose J_0 so that there is an embedded J_0-holomorphic representative C of the antidiagonal class $A_1 - A_2$. Then J_0 is regular for the class A by Lemma 3.3.3, so that $\mathcal{M}^*_{0,3}(A;J_0)$ has dimension 12. Since the antidiagonal class has no regular representatives (by Lemma 3.3.2), the almost complex structure J_0 does not belong to the set $\mathcal{J}_{\text{reg}}(M,\omega)$. On the other hand, positivity of intersections implies that $\mathcal{M}_{0,0}(A_1;J_0)$ is empty and that C is unique. Therefore the only elements in $\overline{\mathcal{M}}_{0,3}(A;J_0) \setminus \mathcal{M}^*_{0,3}(A;J_0)$ are unions of the antidiagonal curve C with some representative of the class $2A_2$, either two distinct spheres or a 2-fold cover of a single sphere. Therefore, its image in M^3 has dimension at most 10. Hence J_0 satisfies conditions (a) and (b) above and we may use it to calculate $\mathrm{GW}_{A,3}$.

Obstruction bundles. Even if J_0 does not satisfy the regularity conditions of Theorem 7.1.1, one can sometimes compute the Gromov–Witten invariants of A by studying the nonregular moduli space $\mathcal{M}(A; J_0)$. An analogous situation in differential topology is a computation of the Euler class of a bundle via its functorial properties, rather than by counting the zeros of a transverse section. We shall confine our discussion to the very simple situation in which the moduli space $\widehat{\mathcal{M}}_{J_0} := \mathcal{M}(A; J_0)/\mathrm{G}$ of (unparametrized) A-curves is a compact smooth manifold and the tangent space of $\mathcal{M}(A; J_0)$ at each curve u is equal to the kernel of D_u. In this case, the cokernels have constant rank and fit together to form a bundle over $\widehat{\mathcal{M}}_{J_0}$ that is called the obstruction bundle. If in addition the rank of this bundle equals the dimension of $\widehat{\mathcal{M}}_{J_0}$ then the index of the problem is zero, i.e. one would expect there to be a finite number of A-curves. The next result shows that the resulting invariant is precisely the Euler number of the obstruction bundle. More precisely, we consider the following situation.

(H) *Let (M, ω) be a compact semipositive symplectic $2n$-manifold, J_0 be an ω-tame almost complex structure on M, and $A \in H_2(M; \mathbb{Z})$ be a homology class satisfying*

$$2n + 2c_1(A) - 6 = 0.$$

Suppose that every J_0-holomorphic stable map representing the class A is a simple J_0-holomorphic sphere and that $\mathcal{M}(A; J_0)$ is an orientable smooth submanifold of $C^\infty(S^2, M)$, that the quotient $\widehat{\mathcal{M}}_{J_0} := \mathcal{M}(A; J_0)/\mathrm{G}$ is compact, and that

$$T_u\mathcal{M}(A; J_0) = \ker D_u$$

for every $u \in \mathcal{M}(A; J_0)$.

Under these assumptions the group $\mathrm{G} = \mathrm{PSL}(2, \mathbb{C})$ acts freely on $\mathcal{M}(A; J_0)$ and so $\widehat{\mathcal{M}}_{J_0}$ is a compact smooth manifold. Denote by $E_{J_0} \to \mathcal{M}(A; J_0)$ the orientable real vector bundle with fibers

$$E_{u,J_0} := (\operatorname{im} D_u)^\perp \cong \operatorname{coker} D_u = \Omega^{0,1}(S^2, u^*TM)/\operatorname{im} D_u. \tag{7.2.1}$$

Here the orthogonal complement is understood with respect to the L^2-inner product on $\Omega^{0,1}(S^2, u^*TM)$, determined by the Riemannian metric on M associated to ω and J_0. Since the inner product on the space of 1-forms depends only on the complex structure on S^2, and not on the volume form, the bundle E_{J_0} is invariant under the action of $\mathrm{G} = \mathrm{PSL}(2, \mathbb{C})$ (see Remark 3.1.2). Since the Fredholm index of D_u is $2n + 2c_1(A) = 6$, the quotient bundle $\widehat{E}_{J_0} := E_{J_0}/\mathrm{G} \to \widehat{\mathcal{M}}_{J_0}$ has rank

$$\operatorname{rank} \widehat{E}_{J_0} = \dim \operatorname{coker} D_u = \dim \ker D_u - 6 = \dim \widehat{\mathcal{M}}_{J_0} =: d.$$

The bundle $\widehat{E}_{J_0} \to \widehat{\mathcal{M}}_{J_0}$ is called the **obstruction bundle**.

THEOREM 7.2.3 (OBSTRUCTION BUNDLE). *Assume (H). Then the Gromov–Witten invariant is equal to the Euler number of the obstruction bundle:*

$$\mathrm{GW}^M_{A,0} = \int_{\widehat{\mathcal{M}}_{J_0}} e(\widehat{E}_{J_0}). \tag{7.2.2}$$

To understand this formula, note that the tangent space of the total space of the vector bundle $E_{J_0} \to \mathcal{M}(A; J_0)$ at an element of the zero section is the direct sum of the tangent space of the base and the fiber of E_{J_0}, i.e. $T_{(u,0)}E_{J_0} = \ker D_u \oplus \operatorname{coker} D_u$ for every $u \in \mathcal{M}(A; J_0)$. Thus the orientation of the determinant line of D_u determines an orientation of the total space $\widehat{E}_{J_0}$ of our obstruction bundle. While the moduli space $\mathcal{M}(A; J_0)$ need not have a natural orientation, we may fix any orientation of $\mathcal{M}(A; J_0)$ and hence of $\widehat{\mathcal{M}}_{J_0}$. Reversing the orientation of $\widehat{\mathcal{M}}_{J_0}$ then also reverses the orientation of the fibers of $\widehat{E}_{J_0}$. So the right hand side of (7.2.2) remains unchanged.

The identity (7.2.2) is closely related to a result in Cieliebak–Mundet–Salamon [**66**] about the (equivariant) Euler class of Fredholm sections of Hilbert space bundles with compact zero sets. However, their arguments do not apply immediately to the present situation since they assume that the reparametrization group G is compact and acts smoothly on the relevant Hilbert space bundles. To circumvent this problem we use *balanced* maps $u : S^2 \to M$.

EXERCISE 7.2.4. Let $x_1, x_2, x_3 : S^2 \to \mathbb{R}$ denote the standard coordinate functions on $S^2 \subset \mathbb{R}^3$. A 2-form $\sigma \in \Omega^2(S^2)$ is called **balanced** if

$$\int_{S^2} x_i \sigma = 0, \qquad i = 1, 2, 3. \tag{7.2.3}$$

This is equivalent to saying that the center of gravity of (S^2, σ) when considered as a subset of $\mathbb{R}^3$ is at the origin. A smooth map $u : S^2 \to M$ is called **balanced** if $u^*\omega$ is a balanced 2-form on S^2. Let $\sigma \in \Omega^2(S^2)$ with $\int_{S^2} \sigma \neq 0$. Show that there is an element $\phi \in \mathrm{PSL}(2, \mathbb{C})$, unique up to right multiplication by an element of SO(3), such that $\phi^*\sigma$ is balanced. (See Cieliebak–Gaio–Salamon [**64**, Proposition 3.3]).

PROOF OF THEOREM 7.2.3. The proof has six steps.

STEP 1 (THE BALANCING CONDITION). *Let $\mathcal{B} \subset C^\infty(S^2, M)$ be the space of smooth maps $u : S^2 \to M$ representing the homology class A and*

$$\mathcal{B}_0 := \left\{ u \in C^\infty(S^2, M) \,\middle|\, [u] = A \text{ and } u^*\omega \text{ satisfies (7.2.3)} \right\}$$

be the space of balanced maps representing the class A. Then $\mathcal{B}_0$ is a smooth codimension three Fréchet submanifold of $\mathcal{B}$ whose tangent space at u consists of all vector fields $\xi \in \Omega^0(S^2, u^*TM)$ along u that satisfy the condition

$$\int_{S^2} dx_i \wedge \omega(\xi, du(\cdot)) = 0, \qquad i = 1, 2, 3. \tag{7.2.4}$$

The inclusion of $\mathcal{B}_0$ into $\mathcal{B}$ induces a bijection $\mathcal{B}_0/\mathrm{SO}(3) \cong \mathcal{B}/\mathrm{PSL}(2, \mathbb{C})$.

The Fréchet manifold $\mathcal{B}$ carries a symplectic form given by $(\xi, \eta) \mapsto \int_{S^2} \omega(\xi, \eta) \mathrm{dvol}_{S^2}$ for $\xi, \eta \in \Omega^0(S^2, u^*TM) = T_u\mathcal{B}$. The action of SO(3) on $\mathcal{B}$ is a Hamiltonian group action generated by the equivariant moment map

$$\mathcal{B} \to \mathbb{R}^3 : u \mapsto \left(\int_{S^2} x_1 u^*\omega, \int_{S^2} x_2 u^*\omega, \int_{S^2} x_3 u^*\omega \right).$$

Since $\omega(A) > 0$, it follows from Exercise 7.2.4 that, for every $u \in \mathcal{B}$, there is a Möbius transformation $\phi \in \mathrm{G}$, unique up to right composition with an orientation preserving isometry, such that $u \circ \phi$ is balanced. Moreover, since SO(3) acts on $\mathcal{B}$ with finite isotropy, zero is a regular value of the moment map and hence $\mathcal{B}_0$ is a codimension three submanifold of $\mathcal{B}$. This proves Step 1.

STEP 2. *The space $\mathcal{M}_{J_0} := \mathcal{M}(A; J_0) \cap \mathcal{B}_0$ of balanced J_0-holomorphic curves representing the class A is a compact smooth submanifold of $\mathcal{B}_0$ of dimension $d+3$. The group* $\mathrm{SO}(3)$ *acts freely on $\mathcal{M}_{J_0}$ and $\widehat{\mathcal{M}}_{J_0} = \mathcal{M}_{J_0}/\mathrm{SO}(3)$. The restriction of E_{J_0} to $\mathcal{M}_{J_0}$ (still denoted by E_{J_0}) is an* $\mathrm{SO}(3)$*-equivariant vector bundle and the obstruction bundle $\widehat{E}_{J_0} \to \widehat{\mathcal{M}}_{J_0}$ is the quotient bundle $E_{J_0}/\mathrm{SO}(3)$.*

Step 2 follows immediately from hypothesis (H) and Step 1.

STEP 3 (EXTENDING THE OBSTRUCTION BUNDLE). *Fix a constant $p > 2$, abbreviate $\mathcal{J} := \mathcal{J}_\tau(M, \omega)$, and denote by $\mathcal{E} \to \mathcal{B} \times \mathcal{J}$ the vector bundle with fibers*

$$\mathcal{E}_{u,J} := \Omega_J^{0,1}(S^2, u^*TM).$$

Then there is an $\mathrm{SO}(3)$*-invariant neighbourhood $\mathcal{U}_0 \subset \mathcal{B}_0$ of $\mathcal{M}_{J_0}$, open in the $W^{1,p}$ topology on $\mathcal{B}_0$, and a neighbourhood $\mathcal{V}_0 \subset \mathcal{J}$ of J_0, open in the C^0 topology on $\mathcal{J}$, such that the vector bundle $E_{J_0} \to \mathcal{M}_{J_0}$ extends to an* $\mathrm{SO}(3)$*-equivariant real rank-d subbundle of $\mathcal{E}$ over $\mathcal{U}_0 \times \mathcal{V}_0$. This extended subbundle will be denoted $E \to \mathcal{U}_0 \times \mathcal{V}_0$.*

For $u \in \mathcal{B}$ and $J \in \mathcal{J}$ denote by $\pi_{u,J} : \Omega^1(S^2, u^*TM) \to \Omega_J^{0,1}(S^2, u^*TM)$ the obvious projection. Then the map $(u, \eta) \mapsto \pi_{u,J}\eta$ is G-equivariant. Let M be equipped with the Riemannian metric determined by ω and J_0 and denote by $\mathcal{N}$ the normal bundle of $\mathcal{M}_{J_0}$ in $\mathcal{B}$. An element of $\mathcal{N}$ is a pair (u, ξ), where $u \in \mathcal{M}_{J_0}$ and $\xi \in \Omega^0(S^2, u^*TM) = T_u\mathcal{B}$ is L^2 orthogonal to the intersection of the kernel of the operator D_u with $T_u\mathcal{B}_0$, i.e.

$$D_u\xi_0 = 0, \quad \xi_0 \text{ satisfies (7.2.4)} \qquad \Longrightarrow \qquad \int_{S^2} \langle \xi, \xi_0 \rangle \mathrm{dvol}_{S^2} = 0.$$

The normal bundle $\mathcal{N}$ is invariant under the action of SO(3) and the map

$$\mathcal{N} \to \mathcal{B} : (u, \xi) \mapsto \exp_u(\xi) \tag{7.2.5}$$

is SO(3)-equivariant. We emphasize that the $W^{1,p}$-norm of ξ, defined in terms of the covariant derivative, is invariant under the SO(3)-action on $\mathcal{N}$. Now the implicit function theorem asserts that, for $\delta > 0$ sufficiently small, the restriction of the exponential map (7.2.5) to the set

$$\mathcal{N}_\delta := \left\{ (u, \xi) \in \mathcal{N} \,\middle|\, \|\xi\|_{L^p} + \|\nabla\xi\|_{L^p} < \delta \right\}$$

is a diffeomorphism from $\mathcal{N}_\delta$ onto an SO(3)-invariant neighbourhood $\mathcal{U}_\delta \subset \mathcal{B}$ of $\mathcal{M}_{J_0}$, open with respect to the $W^{1,p}$ topology. For $(u, \xi) \in \mathcal{N}_\delta$ denote by

$$\Phi(u, \xi) : \Omega^1(S^2, u^*TM) \to \Omega^1(S^2, \exp_u(\xi)^*TM)$$

the isomorphism given by parallel transport along the geodesics $t \mapsto \exp_{u(z)}(t\xi(z))$. If $\delta > 0$ is sufficiently small and $J \in \mathcal{J}$ is sufficiently close to J_0 in the C^0 topology, then, for every pair $(u, \xi) \in \mathcal{N}_\delta$, the restriction of the linear map

$$\pi_{\exp_u(\xi),J}\Phi(u, \xi) : \Omega^1(M, u^*TM) \to \Omega_J^{0,1}(M, \exp_u(\xi)^*TM)$$

to the subspace E_{u,J_0} is injective. Thus there is a neighbourhood $\mathcal{V}_0 \subset \mathcal{J}$ of J_0, open in the C^0 topology, and a $\delta > 0$ such that the formula

$$E_{\exp_u(\xi),J} := \pi_{\exp_u(\xi),J}\Phi(u, \xi)E_{u,J_0}, \qquad (u, \xi) \in \mathcal{N}_\delta, \qquad J \in \mathcal{V}_0, \tag{7.2.6}$$

defines an SO(3)-equivariant smooth vector bundle E of rank d over $\mathcal{U}_\delta \times \mathcal{V}_0$ whose restriction to $\mathcal{M}_{J_0} \times \{J_0\}$ agrees with E_{J_0}. Define $\mathcal{U}_0 := \mathcal{U}_\delta \cap \mathcal{B}_0$ and restrict E to the product $\mathcal{U}_0 \times \mathcal{V}_0$. This vector bundle satisfies the requirements of Step 3.

STEP 4 (THE KEY ESTIMATE). *Fix a constant $p > 2$ and let $E \to \mathcal{U}_0 \times \mathcal{V}_0$ be the subbundle (7.2.6) of $\mathcal{E}$ constructed in Step 3. Then, for every integer $k \geq 0$ and every pair $(u, J) \in \mathcal{U}_0 \times \mathcal{V}_0$, there exists a constant $c_k = c_k(u, J) > 0$, such that*

$$\|\eta\|_{W^{k+1,p}} \leq c_k(u, J) \|\eta\|_{W^{k,p}} \qquad \text{for all } \eta \in E_{u,J}. \tag{7.2.7}$$

The constant c_k depends only on the $W^{k+1,p}$ norm of u and the C^{k+1} norm of J.

This is obvious for $J = J_0$ and $u \in \mathcal{M}_{J_0}$, because E_{u,J_0} is a finite dimensional vector space consisting of smooth $(0, 1)$-forms for every $u \in \mathcal{M}_{J_0}$, and $\mathcal{M}_{J_0}$ is compact. To prove it in general, choose an element $(\exp_u(\xi), J) \in \mathcal{U}_0 \times \mathcal{V}_0$ and consider the isomorphism $\pi_{\exp_u(\xi),J}\Phi(u, \xi) : E_{u,J_0} \to E_{\exp_u(\xi),J}$. The operator norm of this isomorphism and its inverse with respect to the $W^{k+1,p}$ norms on both spaces is controlled by the $W^{k+1,p}$ norm of ξ and the C^{k+1} norm of J. This implies Step 4.

STEP 5 (CHOOSING NEIGHBOURHOODS). *Let $E \to \mathcal{U}_0 \times \mathcal{V}_0$ be the rank-d subbundle of $\mathcal{E}$ constructed in the proof of Step 3. There exist neighbourhoods $\mathcal{U} \subset \mathcal{U}_0$ of $\mathcal{M}_{J_0}$ and $\mathcal{V} \subset \mathcal{V}_0$ of J_0, open with respect to the C^∞ topology, such that the following holds. Denote by $\overline{\mathcal{U}}$ and $\overline{\mathcal{V}}$ the C^∞ closures of $\mathcal{U}$ and $\mathcal{V}$, respectively.*

(a) *The bundle E extends over $\overline{\mathcal{U}} \times \overline{\mathcal{V}}$, $\mathcal{V}$ is connected, and* SO(3) *acts freely on $\mathcal{U}$.*

(b) *For $J \in \mathcal{V}$ define*

$$\mathcal{M}_{J,E} := \left\{ u \in \overline{\mathcal{U}} \,|\, \bar{\partial}_J(u) \in E_{u,J} \right\} \subset \mathcal{B}_0. \tag{7.2.8}$$

Then $\mathcal{M}_{J_0,E} = \mathcal{M}_{J_0}$ and $\mathcal{M}_{J,E} \subset \mathcal{U}$ for every $J \in \mathcal{V}$.

(c) *The set $\bigcup_{J \in \mathcal{K}} \mathcal{M}_{J,E} \subset \mathcal{U}$ is compact for every compact set $\mathcal{K} \subset \mathcal{V}$.*

(d) *For every $J \in \mathcal{V}$ denote by $\mathcal{E}_J \to \mathcal{U}$ the infinite dimensional vector bundle with fibers $\mathcal{E}_{u,J} := \Omega^{0,1}_J(S^2, u^*TM)$ and by $E_J \to \mathcal{U}$ the rank-d subbundle with fibers $E_{u,J}$ over $u \in \mathcal{U}$. Then the section $\mathcal{U} \to \mathcal{E}_J : u \mapsto \bar{\partial}_J(u)$ is transverse to E_J.*

(e) *If $J \in \mathcal{V}$, then every J-holomorphic stable map representing the class A is a simple J-holomorphic sphere contained in $\mathcal{U}$.*

Assertion (a) is obvious. We prove that $\mathcal{U}$ can be chosen such that (b) and (d) hold for $J = J_0$. Fix two open sets $\mathcal{U} \subset \mathcal{U}_0$ and $\mathcal{V} \subset \mathcal{V}_0$ such that (a) holds. By (7.2.1), the subbundle $E_{J_0} \subset \mathcal{E}_{J_0}$ is transverse to the section

$$\mathcal{U} \to \mathcal{E}_{J_0} : u \mapsto \bar{\partial}_{J_0}(u)$$

along the submanifold $\mathcal{M}_{J_0} \subset \mathcal{M}_{J_0,E}$. By the implicit function theorem in the appropriate Banach manifold setting, this implies that $\mathcal{M}_{J_0}$ is an isolated subset of $\mathcal{M}_{J_0,E}$, i.e. there is a neighbourhood of $\mathcal{M}_{J_0}$ in $\mathcal{U}$, open with respect to the $W^{1,p}$ topology, which contains no other elements of $\mathcal{M}_{J_0,E}$. Shrinking $\mathcal{U}$, if necessary, we obtain $\mathcal{M}_{J_0,E} = \mathcal{M}_{J_0}$. Hence (b) and (d) hold for $J = J_0$.

We prove that $\mathcal{V}$ can be chosen such that (b) holds for every $J \in \mathcal{V}$. Assume, by contradiction, that (b) does not hold for any C^∞ neighbourhood of J_0. Then there are sequences $J^\nu \in \mathcal{J}$ and $u^\nu \in \overline{\mathcal{U}} \setminus \mathcal{U}$ such that J^ν converges to J_0 in the C^∞ topology and $\bar{\partial}_{J^\nu}(u^\nu) \in E_{u^\nu,J^\nu}$. The sequence u^ν is uniformly bounded in $W^{1,p}$ by assumption. Hence it follows from the estimate (7.2.7) in Step 4 and elliptic bootstrapping that the sequence u^ν is uniformly bounded in $W^{k,p}$ for every $k \geq 0$. This implies that u^ν has a C^∞ convergent subsequence whose limit $u_0 \in \overline{\mathcal{U}} \setminus \mathcal{U}$ satisfies $\bar{\partial}_{J_0}(u_0) \in E_{u_0,J_0}$, in contradiction to the fact that $\mathcal{M}_{J_0,E} = \mathcal{M}_{J_0}$. Thus $\mathcal{V}$ can be chosen such that (b) holds for every $J \in \mathcal{V}$.

We prove that $\mathcal{U}$ and $\mathcal{V}$ satisfy (c). Let $\mathcal{K} \subset \mathcal{V}$ be a compact set and (u^ν, J^ν) be a sequence with $J^\nu \in \mathcal{K}$ and $u^\nu \in \mathcal{M}_{J^\nu,E}$. Passing to a subsequence we may assume that J^ν converges in the C^∞ topology to $J \in \mathcal{K}$. Hence it follows again from (7.2.7) and elliptic bootstrapping that the sequence u^ν is uniformly bounded in $W^{k,p}$ for every integer $k \geq 0$. Hence, passing to a further subsequence, we may assume without loss of generality that u^ν converges in the C^∞ topology to some element $u \in \overline{\mathcal{U}}$. The limit satisfies $\bar{\partial}_J(u) \in E_{u,J}$ and hence belongs to the set $\mathcal{M}_{J,E} \subset \mathcal{U}$. Thus we have proved that $\mathcal{U}$ and $\mathcal{V}$ satisfy (c).

Shrinking $\mathcal{V}$ further, if necessary, we obtain (d), because transversality is an open condition, and we obtain (e) by Gromov compactness and hypothesis (H). This proves Step 5.

STEP 6. *We prove the theorem.*

Let $E \to \mathcal{U}_0 \times \mathcal{V}_0$ be the rank-d subbundle (7.2.6) of $\mathcal{E}$ constructed in the proof of Step 3. Let $\mathcal{U} \subset \mathcal{U}_0$ and $\mathcal{V} \subset \mathcal{V}_0$ be as in Step 5. For $J \in \mathcal{V}$ let $\mathcal{M}_{J,E} \subset \mathcal{B}_0$ be defined by (7.2.8). By (b), (c), and (d), $\mathcal{M}_{J,E}$ is a compact manifold without boundary of dimension $d+3$ for every $J \in \mathcal{V}$. Moreover, $\mathcal{M}_{J,E}$ carries a smooth free action of SO(3). The rank d bundle $E_J \to \mathcal{M}_{J,E}$ is SO(3)-equivarient and descends to a quotient bundle

$$\widehat{E}_J := E_J/\mathrm{SO}(3) \to \widehat{\mathcal{M}}_{J,E} := \mathcal{M}_{J,E}/\mathrm{SO}(3).$$

By (a) and (c), the Euler number of the bundle $\widehat{E}_J \to \widehat{\mathcal{M}}_{J,E}$ is independent of J. Moreover, for each $J \in \mathcal{V}$, there is a smooth SO(3)-equivariant section

$$S_J : \mathcal{M}_{J,E} \to E_J, \qquad S_J(u) := \bar{\partial}_J(u) \in E_{u,J}$$

which vanishes identically when $J = J_0$. Now let $J \in \mathcal{V}$ be a regular almost complex structure in the sense that $D_{u,J}$ is onto for every $u \in \mathcal{M}(A;J)$. (This exists by Theorem 3.1.6.) Then the induced section

$$\widehat{S}_J : \widehat{\mathcal{M}}_{J,E} \to \widehat{E}_J$$

is transverse to the zero section and, by (e), the number of zeros, counted with signs, is the required Gromov–Witten invariant. Hence

$$\int_{\widehat{\mathcal{M}}_{J_0}} e(\widehat{E}_{J_0}) = \int_{\widehat{\mathcal{M}}_{J_0,E}} e(\widehat{E}_{J_0}) = \int_{\widehat{\mathcal{M}}_{J,E}} e(\widehat{E}_J) = \mathrm{GW}^M_{A,0}.$$

This proves Theorem 7.2.3. □

The following example illustrates the use of the formula (7.2.2).

EXAMPLE 7.2.5. Let $L \to S^2 \times S^2$ be a line bundle of bidegree $(1,-2)$. In other words, setting $B_1 := [S^2 \times pt]$ and $B_2 := [pt \times S^2]$, we have $\langle c_1(L), B_1\rangle = 1$ and $\langle c_1(L), B_2\rangle = -2$. Then define (M, J_0) to be the projectivization

$$(M, J_0) := (\mathbb{P}(L \oplus \mathbb{C}), J_0).$$

Explicitly, M is a reduced space formed from the action of $\mathbb{T}^3 = S^1 \times S^1 \times S^1$ on $\mathbb{C}^6$ by

$$(\lambda_1, \lambda_2, \lambda_3) \cdot (z_0, z_1, w_0, w_1, v_0, v_1) := \big(\lambda_1 z_0, \lambda_1 z_1, \lambda_2 w_0, \lambda_2 w_1, \lambda_1\lambda_2^{-2}\lambda_3 v_0, \lambda_3 v_1\big).$$

This action has moment map

$$\mu : (z, w, v) \mapsto \big(|z|^2 + |v_0|^2,\ |w|^2 - 2|v_0|^2,\ |v|^2\big)$$

where $|z|^2 := |z_0|^2 + |z_1|^2$ and so on. We define M to be the symplectic quotient $\mu^{-1}(3,1,1)/\mathbb{T}^3$. Thus M is a toric symplectic manifold. Note that its symplectic form ω is integral; $H_2(M;\mathbb{Z})$ is generated by the spheres

$$\begin{array}{lcl} [z_0 : z_1] & \mapsto & [z_0 : z_1 : 0 : 1 : 0 : 1], \\ [w_0 : w_1] & \mapsto & [3 : 0 : w_0 : w_1 : 0, 1], \\ [v_0 : v_1] & \mapsto & [0 : \sqrt{3 - |v_0|^2} : 0 : \sqrt{1 + 2|v_0|^2} : v_0 : v_1], \end{array}$$

with symplectic areas $3, 1, 1$ respectively.

The B_2-spheres $w \mapsto (z, w)$ in $S^2 \times S^2$ lift to a moduli space of spheres u_z in M by

$$u_z(w) = [z_0 : z_1 : w_0 : w_1 : 0 : 1].$$

These spheres all lie in the section $V_- = \{v_0 = 0\}$ of $M \to S^2 \times S^2$ that has normal bundle L. Thus if A denotes the class represented by the lifted spheres we have $A \cdot [V_-] = -2$. Moreover $\omega(A) = 1$. Since the section V_- is holomorphic, positivity of intersections (see Exercise 2.6.1) implies that the only J_0-holomorphic A-spheres in M are those described above that lie entirely in V_-. Hence, the moduli space $\widehat{\mathcal{M}} := \mathcal{M}(A; J_0)/\mathrm{G}$ of these spheres is compact and can be identified with S^2. On the other hand because the restriction of L to each A-sphere has Chern number -2, $c_1(A) = 0$ and we are in the situation of Example 7.1.4 in which $\mu(A, 0) = 0$. Thus our aim is to calculate the number $\mathrm{GW}^M_{A,0}$ of J-holomorphic representatives of the class A.

The natural complex structure J_0 on M is not regular. However, for each $u \in \mathcal{M}(A; J_0)$, the normal bundle to the curve $C_z = \operatorname{im} u_z$ decomposes as $L_0 \oplus L|_C \cong L_0 \oplus L_{-2}$, where we denote by L_k the holomorphic line bundle over S^2 with Chern number k. Therefore the kernel of D_u consists of the space of holomorphic sections of the bundle $L_2 \oplus L_0$, where we think of L_2 as the tangent space to C_z. Thus $\ker D_u$ can be identified with $T_u\mathcal{M}(A; J_0)$ as required by Theorem 7.2.3, and we may calculate $\mathrm{GW}^M_{A,0}$ using the obstruction bundle E.

By definition, the fiber of E at u is just $\operatorname{coker} D_u$. By Lemma 3.3.1, if $u = u_z$ we may identify the dual $(\operatorname{coker} D_u)^*$ with the space

$$H^0(S^2, L^*|_{C_z} \otimes K_{C_z})$$

of holomorphic sections of the trivial bundle $L^*|_{C_z} \otimes K_{C_z}$, where $K_C = T^*C$ denotes the canonical bundle of C. Hence for each u the vector space $(\operatorname{coker} D_u)^*$ has complex dimension 1. Since the line bundles K_{C_z} fit together to form a holomorphic line bundle $K \to V_- = S^2 \times S^2$ of bidegree $(0, -2)$, the corresponding spaces $(\operatorname{coker} D_u)^*$ also form the fibers of a holomorphic line bundle $\widehat{E}^* \to \widehat{\mathcal{M}}$. Moreover, because $L^* \otimes K \to V^-$ is isomorphic to the pullback of $L_{-1} \to S^2$ by projection onto the first factor, and because the elements of $(\operatorname{coker} D_u)^*$ are constant sections, the fiber of E^* at u_z can be identified with the fiber L_z^{-1} of L_{-1} at z. Thus the obstruction bundle $\widehat{E} \to \widehat{\mathcal{M}}$ may be identified with the bundle $L_1 \to S^2$ with Euler (or Chern) number 1. Hence, by (7.2.2), the Gromov–Witten invariant is $\mathrm{GW}^M_{A,0} = 1$.

EXERCISE 7.2.6. Extend the result in Theorem 7.2.3 to the case when $2n + 2c_1(A) - 6 = \mu > 0$, the moduli space $\mathcal{M}(A; J_0)$ is a smooth manifold of dimension $\mu + 6 + d > \mu + 6$ with tangent bundle formed by the kernels $\ker D_u$, and the quotient $\mathcal{M}(A; J_0)/\mathrm{G}$ is compact.

The result of the previous exercise is still very limited. In many cases of interest the space $\mathcal{M}(A; J_0)/\mathrm{G}$, though a smooth manifold with excess dimension $d > 0$, is not compact. In this case it is very important to understand the structure of the extension of the obstruction bundle to the compactification, since its Euler class (and hence the relevant Gromov–Witten invariants) can be affected by twisting over the boundary. Algebraic geometers have developed many sophisticated techniques to deal with this problem; it would take us too far afield to discuss them here. Interested readers might start by consulting Okounkov–Pandharipande [**311, 312**] and Zinger [**426, 427, 428**] and the references contained therein. Note also that the enumerative invariants of interest to algebraic geometers do not always coincide with the Gromov–Witten invariants: cf. Example 8.6.12.

We end this section by discussing some slightly different invariants, that count curves or stable maps with special properties.

Invariants for trees. Suppose (M, ω) is a compact semipositive symplectic manifold. Fix a k-labelled tree $T = (T, E, \Lambda)$ and a collection $\{A_\gamma\}_{\gamma\in T}$ of spherical homology classes in $H_2(M; \mathbb{Z})$ satisfying the stability condition (6.1.1) of Section 6.1 and (6.6.1) in Theorem 6.6.1. (For example, if (M, ω) were monotone, (6.6.1) would be satisfied for all nonzero A_γ.) In Exercise 6.6.3 we have seen that the evaluation map

$$\mathrm{ev}_{T,J,\{A_\gamma\}} : \mathcal{M}^*_{0,T}(\{A_\gamma\}; J) \to M^k$$

is a pseudocycle of dimension $\mu(A, k) - 2e(T)$ for every $J \in \mathcal{J}_{\mathrm{reg}}(M, \omega)$, where $A := \sum\{A_\gamma\}$. Thus we can define an invariant

$$\mathrm{GW}^M_{\{A_\gamma\},T} : H^*(M)^{\otimes k} \to \mathbb{Z}$$

by

$$\mathrm{GW}^M_{\{A_\gamma\},T}(a_1, \ldots, a_k) := f \cdot \mathrm{ev}_{T,J,\{A_\gamma\}},$$

where $f : U \to M^k$ is a pseudocycle dual to $\pi_1^* a_1 \smile \cdots \smile \pi_k^* a_k$. It follows as in Theorem 7.1.1 that this number is independent of J. It can only be nonzero if the classes a_i satisfy the dimensional condition

$$\sum_{i=1}^{k} \deg(a_i) = 2n + 2c_1(A) + 2k - 6 - 2e(T). \tag{7.2.9}$$

Further, we define

$$\mathrm{GW}^M_{A,T}(a_1, \ldots, a_k) := \sum_{\{A_\gamma\}:\sum\{A_\gamma\}=A} \mathrm{GW}^M_{\{A_\gamma\},T}(a_1, \ldots, a_k).$$

Geometrically, the invariant $\mathrm{GW}^M_{A,T}(a_1, \ldots, a_k)$ counts all stable maps $(\mathbf{u}, \mathbf{z})$ representing the class A and modelled over T, such that the ith marked point passes through a cycle Y_i dual to the class a_i.

EXAMPLE 7.2.7. Let $M = \mathbb{C}P^2$, T be a tree with two vertices 0 and 1 and labels $\Lambda_0 = \{1, 2\}$, $\Lambda_1 = \{3, 4\}$, and let $A_0 = A_1 = L$. Given four distinct points $x_1, x_2, x_3, x_4 \in \mathbb{C}P^2$ such that no three of them lie on the same line, there is a unique pair of distinct lines $u_0, u_1 : \mathbb{C}P^1 \to \mathbb{C}P^2$ such that u_0 passes through x_1 and x_2 and u_1 passes through x_3 and x_4. These lines form a stable map, modelled over T, since their intersection point is not equal to any of the x_i. It follows that in the notation of Example 7.1.14

$$\mathrm{GW}^{\mathbb{C}P^2}_{\{L,L\},T}(c^2,c^2,c^2,c^2) = 1.$$

Note that the dimension of the relevant moduli space is 16 and so the dimensional condition (7.2.9) is satisfied in this case. Further, in this case the other splittings of $2L$ (into $\{0,2L\}$ and $\{2L,0\}$) do not contribute so that $\mathrm{GW}^{\mathbb{C}P^2}_{2L,T}(c^2,c^2,c^2,c^2)$ is also equal to 1.

EXERCISE 7.2.8. Choose a basis $\{e_0,\dots,e_N\}$ of the cohomology $H^*(M)$, and denote by $g^{\nu\mu}$ the matrix inverse to

$$g_{\nu\mu} := \int_M e_\nu \smile e_\mu.$$

Then the cohomology class $\sum_{\nu\mu} e_\nu g^{\nu\mu} e_\mu$ is Poincaré dual to the diagonal of $M\times M$. If T is as in the previous example, show that

$$\mathrm{GW}^M_{A,T}(a_1,\dots,a_4) = \sum_{A_0+A_1=A}\sum_{\nu\mu} \mathrm{GW}^M_{A_0,3}(a_1,a_2,e_\nu)\, g^{\nu\mu}\, \mathrm{GW}^M_{A_1,3}(e_\mu,a_3,a_4).$$

Fixed marked points. In the next section we shall define the Gromov–Witten invariants $\mathrm{GW}^{M,I}_{A,k}$ where the marked points associated to the index set I are fixed. These are the so-called *mixed invariants* of Ruan–Tian [**345**]. The definition requires modifying the Cauchy–Riemann equations using smooth families of almost complex structures parametrized by the base point $z\in S^2$ as in Section 6.7. As a warmup we discuss the construction of such invariants with z-independent almost complex structures. To do this we must impose rather restrictive hypotheses on the symplectic manifold and the cardinality of the index set I.

Assume (M,ω) is monotone with minimal Chern number N. Fix a homology class $A\in H_2(M;\mathbb{Z})$ that satisfies (6.6.1), an integer $k\ge 3$, a subset $I\subset\{1,\dots,k\}$ such that

$$3 \le \#I \le N+2,$$

and a tuple $\mathbf{w}=\{\mathrm{w}_i\}_{i\in I}$ of distinct points on S^2. Recall from Exercise 6.6.5 that the evaluation map

$$\mathrm{ev}_{\mathbf{w},J} : \mathcal{M}^*_{0,k}(A,\mathbf{w};J)\to M^k$$

is a pseudocycle of dimension $2n+2c_1(A)+2(k-\#I)$ for $J\in\mathcal{J}_{\mathrm{reg}}(M,\omega)$. Thus we can define an invariant $\mathrm{GW}^{M,I}_{A,k} : H^*(M)^{\otimes k}\to\mathbb{Z}$ by

$$\mathrm{GW}^{M,I}_{A,k}(a_1,\dots,a_k) := f\cdot \mathrm{ev}_{\mathbf{w},J}. \tag{7.2.10}$$

Here $f:U\to M^k$ is a pseudocycle dual to $\pi_1^*a_1\smile\cdots\smile\pi_k^*a_k$ and $\pi_i : M^k\to M$ is the projection onto the ith factor. It follows as in Theorem 7.1.1 that this number is independent of J and $\mathbf{w}$. It can only be nonzero if the classes a_i satisfy the dimensional condition

$$\sum_{i=1}^k \deg(a_i) = 2n+2c_1(A)+2k-2\#I. \tag{7.2.11}$$

Geometrically, the invariant $\mathrm{GW}^{M,I}_{A,k}(a_1,\dots,a_k)$ counts (with appropriate signs) the tuples $(u,\mathbf{z}) = (u,\{z_i\}_{i\notin I})$, where $u:S^2\to M$ is a J-holomorphic sphere representing the class A and the z_i for $i\in\{1,\dots,k\}\setminus I$ are distinct points on

$S^2 \setminus \{\mathrm{w}_i \,|\, i \in I\}$, such that $u(\mathrm{w}_i) \in X_i$ for $i \in I$ and $u(z_i) \in X_i$ for $i \notin I$. Thus the marked points associated to the indices in I are fixed on S^2 while the remaining marked points are allowed to vary freely. Note that there is a strong restriction $\#I \le N + 2$ on the number of fixed marked points in the above definition. In the case $N \ge 2$ this still allows us to deal with the important case $\#I = 4$ for monotone symplectic manifolds.

EXAMPLE 7.2.9. As an illustration of the difference between GW and GW^I consider conics in the complex projective plane $M = \mathbb{C}P^2$. Note that when $A = 2L$ and all the a_i are dual to points, the dimensional condition (7.2.11) for GW^I with $I = \{1, \ldots, k\}$ is satisfied when $k = 4$ while the condition (7.1.2) for GW is satisfied when $k = 5$. Classically, a conic is thought of as the zero set C_f of a homogeneous polynomial f of degree 2. The well known fact that there is a unique conic through five points in general position (i.e. no three collinear) implies that

$$\mathrm{GW}^{\mathbb{C}P^2}_{2L,5}(c^2, c^2, c^2, c^2, c^2) = 1.$$

(For more details, see Exercise 7.1.16 as well as the proof of Theorem 7.4.1.) It follows that the family of all conics which go through four fixed points $x_0, \ldots, x_3$ in $\mathbb{C}P^2$ in general position has real dimension two. Now one can pick out a unique member of this family either by requiring the curve to go through a further point in $\mathbb{C}P^2$ (which leads to the invariant GW) or by fixing the cross ratio of the preimages $\mathrm{w}_0, \ldots, \mathrm{w}_3$ of the points $x_0, \ldots, x_3$. This second approach leads to a calculation of the invariant $\mathrm{GW}^{\mathbb{C}P^2,\{1,\ldots,4\}}_{2L,4}(c^2, \ldots, c^2)$.

Here are the analytic details. Given $\mathrm{w} \in \mathbb{C}\setminus\{0,1\}$ there is a unique holomorphic curve $u_\mathrm{w} : \mathbb{C} \cup \{\infty\} \to \mathbb{C}P^2$ of degree two such that

$$u_\mathrm{w}(0) = [1:0:0] =: x_0, \qquad u_\mathrm{w}(1) = [0:1:0] =: x_1,$$

$$u_\mathrm{w}(\infty) = [0:0:1] =: x_2, \qquad u_\mathrm{w}(\mathrm{w}) = [1:1:1] =: x_3,$$

namely

$$u_\mathrm{w}(z) = [\mathrm{w}(z-1) : z(\mathrm{w}-1) : z(z-1)].$$

Further, the quadruple (x_0, x_1, x_2, x_3) is a regular value of the evaluation map $\mathcal{M}^*(\mathbb{C}P^2, 2L; J_0) \to (\mathbb{C}P^2)^4 : u \mapsto (u(0), u(1), u(\infty), u(\mathrm{w}))$. (This can be proved either by direct calculation or by considering the family of maps $g \circ u_\mathrm{w}$, where $g \in \mathrm{PSL}(3, \mathbb{C})$ as in Proposition 7.4.5.) This shows that

$$\mathrm{GW}^{\mathbb{C}P^2,\{1,\ldots,4\}}_{2L,4}(c^2, \ldots, c^2) = 1.$$

Now, for every point $x_4 = [1 : z_1 : z_2] \in \mathbb{C}P^2$ such that $z_1 \ne 1$ and $z_2 \ne 0$, there exists a unique pair (w, z) such that $u_\mathrm{w}(z) = x_4$, namely

$$\mathrm{w} = \frac{z_1 - z_2}{z_2(z_1 - 1)}, \qquad z = \frac{z_1 - z_2}{z_1 - 1}.$$

Moreover, the differential of the map $(\mathrm{w}, z) \mapsto u_\mathrm{w}(z)$ is surjective at every point (w, z) such that $z \notin \{0, 1, \infty, \mathrm{w}\}$. Hence the point $x := (x_0, x_1, x_2, x_3, x_4)$ is a regular value of the evaluation map $\mathrm{ev} : \mathcal{M}^*_{0,5}(\mathbb{C}P^2, 2L; J_0) \to (\mathbb{C}P^2)^5$ and has a unique preimage whenever $x_4 = [1 : z_1 : z_2]$ such that $z_1 \ne 1$ and $z_2 \ne 0$. in the case $n = 2$. Thus $\mathrm{GW}^{\mathbb{C}P^2}_{2L,5}(c^2, c^2, c^2, c^2, c^2) = 1$.

REMARK 7.2.10. The above argument shows that

$$\mathrm{GW}^{\mathbb{C}P^2,\{1,\ldots,4\}}_{2L,4}(c^2,\ldots,c^2) = 1.$$

The fact that this equals the corresponding invariant $\mathrm{GW}^{\mathbb{C}P^2}_{2L,T}(c^2,c^2,c^2,c^2)$ for the tree T of Example 7.2.7 is a consequence of the gluing theorem as explained in the discussion of the full Gromov–Witten pseudocycle at the very end of Section 6.7.

7.3. Counting pseudoholomorphic graphs

In this section we examine the Gromov–Witten invariants of a compact semipositive symplectic manifold (M,ω) associated to a smooth family of almost complex structures

$$S^2 \to \mathcal{J}_\tau(M,\omega) : z \mapsto J_z$$

as in Section 6.7. Fix a constant $\kappa > 0$ and recall that $\mathcal{J}_+(M,\omega;\kappa) \subset \mathcal{J}_\tau(M,\omega)$ denotes the space of ω-tame almost complex structures J on M such that every J-holomorphic sphere $u : S^2 \to M$ with energy $E(u) = \int_{S^2} u^*\omega \le \kappa$ has nonnegative Chern number. By Lemma 6.4.8, this is a nonempty and path-connected open subset of $\mathcal{J}_\tau(M,\omega)$. Recall further that

$$\mathcal{J}(S^2;M,\omega;\kappa) \subset C^\infty(S^2,\mathcal{J}_+(M,\omega;\kappa))$$

denotes the set of all contractible smooth maps $S^2 \to \mathcal{J}_+(M,\omega;\kappa) : z \mapsto J_z$. This is again a nonempty and path-connected open subset of $C^\infty(S^2,\mathcal{J}_\tau(M,\omega))$. Geometrically one can think of $\mathcal{J}(S^2;M,\omega;\kappa)$ as the set of all smooth families of almost complex structures that are sufficiently close to constant ones. If one wants to work instead with arbitrary families of ω-tame almost complex structures on M one has to impose a stronger condition than semipositivity (namely that $\omega(A) > 0$ and $c_1(A) \ge 2-n$ imply $c_1(A) \ge 0$ for every $A \in H_2(M;\mathbb{Z})$).

Now fix a nonnegative integer k, an index set $I \subset \{1,\ldots,k\}$, and a tuple $\mathbf{w} = \{\mathrm{w}_i\}_{i\in I}$ of distinct points on S^2. Let

$$J = \{J_z\}_{z\in S^2} \in \mathcal{J}_{\mathrm{reg}}(S^2;M,\omega;\mathbf{w},\kappa) := \mathcal{J}_{\mathrm{reg}}(S^2;M,\omega;\mathbf{w}) \cap \mathcal{J}_+(S^2;M,\omega;\kappa)$$

be a regular family of almost complex structures as defined in Section 6.7 (see Definition 6.7.10). Let $A \in H_2(M;\mathbb{Z})$ be a homology class such that $\omega(A) \le \kappa$. Then the moduli space

$$\mathcal{M}(A;J) := \{u : S^2 \to M \,|\, du(z)\circ j = J_z(u(z))\circ du(z),\ [u] = A\}$$

of $\{J_z\}$-holomorphic spheres $u : S^2 \to M$ in the class A is a smooth manifold of dimension $2n + 2c_1(A)$. The corresponding moduli space

$$\mathcal{M}_{0,k}(A;\mathbf{w},J) := \{(u,\mathbf{z}) \in \mathcal{M}(A;J)\times(S^2)^k \,|\, z_i \ne z_j \text{ for } i \ne j,\ z_i = \mathrm{w}_i\ \forall\, i \in I\}$$

of marked spheres has dimension $\mu(A,k,I) := 2n + 2c_1(A) + 2(k - \#I)$. In Theorem 6.7.1 it is shown that the evaluation map

$$\mathrm{ev}_{\mathbf{w},J} : \mathcal{M}_{0,k}(A;\mathbf{w},J) \to M^k$$

is a pseudocycle whenever $J = \{J_z\} \in \mathcal{J}_{\mathrm{reg}}(S^2;M,\omega;\mathbf{w},\kappa)$ and $\omega(A) \le \kappa$. This gives rise to a Gromov–Witten invariant of pseudoholomorphic graphs as follows. We denote by $\pi_i : M^k \to M$ the projection onto the ith factor.

THEOREM 7.3.1. *Let (M, ω) be a closed semipositive symplectic $2n$-manifold, $\kappa > 0$, and $A \in H_2(M; \mathbb{Z})$ be a spherical homology class such that $\omega(A) \leq \kappa$. Let k be a nonnegative integer, $I \subset \{1, \dots, k\}$ be an index set, and $\mathbf{w} = \{\mathrm{w}_i\}_{i \in I}$ be a tuple of pairwise distinct points in S^2. Let $J \in \mathcal{J}_{\mathrm{reg}}(S^2, M, \omega; \mathbf{w}, \kappa)$. Then the homomorphism*

$$\widetilde{\mathrm{GW}}^{M,I}_{A,k} : H^*(M)^{\otimes k} \to \mathbb{Z}$$

defined by

$$\widetilde{\mathrm{GW}}^{M,I}_{A,k}(a_1, \dots, a_k) := f \cdot \mathrm{ev}_{\mathbf{w},J} \tag{7.3.1}$$

for $a_i \in H^*(M)$, where $f : U \to M^k$ is a pseudocycle dual to $\pi_1^* a_1 \smile \cdots \smile \pi_k^* a_k$, is independent of the tuple $\mathbf{w}$ of fixed marked points, the regular almost complex structure $J \in \mathcal{J}_{\mathrm{reg}}(S^2, M, \omega; \mathbf{w}, \kappa)$, and the pseudocycle f used to define it.

PROOF. By Theorem 6.7.1, the bordism class of $\mathrm{ev}_{\mathbf{w},J}$ is independent of $\mathbf{w}$ and J. Hence, by Lemma 6.5.5, the intersection number $f \cdot \mathrm{ev}_{\mathbf{w},J}$ is independent of $\mathbf{w}$ and J. By Lemma 6.5.7, it depends only on the cohomology class represented by the pseudocycle f. This proves Theorem 7.3.1. □

The invariant defined in (7.3.1) is called the **Gromov–Witten invariant of k-pointed graphs in class A with fixed marked points indexed by I**. Geometrically, it can be interpreted as follows. Choose cycles X_i in general position that are Poincaré dual to the classes a_i. Moreover, fix a tuple $\mathbf{w} = \{\mathrm{w}_i\}_{i \in I}$ of pairwise distinct points on S^2. Then the invariant (7.3.1) counts with signs the number of tuples $(u, \mathbf{z}) = (u, \{z_i\}_{i \notin I})$, where $u : S^2 \to M$ is a $\{J_z\}$-holomorphic sphere representing the class A and the z_i for $i \in \{1, \dots, k\} \setminus I$ are pairwise distinct points in $S^2 \setminus \{\mathrm{w}_i \mid i \in I\}$ such that the marked points are mapped to the cycles X_i. That this is well defined as an intersection number of pseudocycles is precisely the assertion of Theorem 7.3.1. We emphasize the following features of this definition.

- To define the invariant one can assume that the pseudocycle f is actually an oriented submanifold $X = X_1 \times \cdots \times X_k \subset M^k$. The price one has to pay is to replace the cohomology classes a_i by suitable integer multiples. To obtain integer invariants one may need more general representatives of the a_i.

- The invariant $\widetilde{\mathrm{GW}}^{M,I}_{A,k}(a_1, \dots, a_k)$ can only be nonzero if the classes a_i satisfy the dimensional condition

$$\sum_{i=1}^{k} \deg(a_i) = 2n + 2c_1(A) + 2k - 2\#I. \tag{7.3.2}$$

- The invariant $\widetilde{\mathrm{GW}}^{M,I}_{A,k}$ vanishes for $\#I < 3$ unless $A = 0$ and $k = \#I$ (see Exercise 7.3.4). This is due to the symmetry breaking nature of the perturbation. If we reparametrize a $\{J_z\}$-holomorphic sphere by an element of $\mathrm{G} = \mathrm{PSL}(2, \mathbb{C})$ it will no longer satisfy the same equation. Thus we cannot divide by the action of G to reduce the moduli space to its "correct" dimension. This is reflected in our set up: the graphs of u and of $u \circ \phi$ are different unless u is constant (and hence $A = 0$).

- If $\#I = 3$ and A satisfies (6.6.1) then the invariant $\widetilde{\mathrm{GW}}^{M,I}_{A,k}$ agrees with $\mathrm{GW}^M_{A,k}$ (see Proposition 7.3.2).

PROPOSITION 7.3.2. *Let (M,ω) be a compact semipositive symplectic manifold, $k \geq 3$ be an integer, and suppose that $A \in H_2(M;\mathbb{Z})$ satisfies (6.6.1). Then*

$$\widetilde{\mathrm{GW}}_{A,k}^{M,\{1,2,3\}} = \mathrm{GW}_{A,k}^{M} : H^*(M)^{\otimes k} \to \mathbb{Z}.$$

PROOF. This follows from Theorem 6.7.1 (iii). □

EXERCISE 7.3.3. Suppose (M,ω) is monotone with minimal Chern number N, let k be a positive integer, and suppose that $I \subset \{1,\dots,k\}$ is an index set such that $3 \leq \#I \leq N+2$. Fix a homology class $A \in H_2(M;\mathbb{Z})$ and let $a_1,\dots,a_k \in H^*(M)$ be a collection of cohomology classes satisfying (7.3.2). Prove that

$$\widetilde{\mathrm{GW}}_{A,k}^{M,I}(a_1,\dots,a_k) = \mathrm{GW}_{A,k}^{M,I}(a_1,\dots,a_k).$$

EXERCISE 7.3.4. Assume $\#I < 3$. Prove that $\widetilde{\mathrm{GW}}_{A,k}^{M,I} = 0$ unless $A = 0$ and $k = \#I$. Prove also that

$$\widetilde{\mathrm{GW}}_{0,1}^{M,\{1\}}(a) = \int_M a, \quad \widetilde{\mathrm{GW}}_{0,2}^{M,\{1,2\}}(a_1,a_2) = \int_M a_1 \smile a_2.$$

Hint: To prove the first statement choose a sequence of almost complex structures $\{J_z^\nu\}$ converging to a z-independent almost complex structure $J \in \mathcal{J}_{\mathrm{reg}}(M,\omega)$. Show that, for dimensional reasons, there is no stable map for J that represents the class A and passes through generic cycles dual to the a_i. To prove the second show by considering the energy that the only solutions of (6.7.1) in class $A = 0$ are constant. Constant solutions are regular by Lemma 6.7.6.

EXAMPLE 7.3.5 (Cup product). In the case $A = 0$ the invariant $\widetilde{\mathrm{GW}}$ is given by the cup-product, namely

$$\widetilde{\mathrm{GW}}_{0,k}^{M,\{1,\dots,k\}}(a_1,\dots,a_k) = \int_M a_1 \smile \cdots \smile a_k,$$

whenever the classes $a_i \in H^*(M)$ satisfy $\sum_i \deg(a_i) = 2n$. To see this, note that every $\{J_z\}$-holomorphic sphere representing the class $A = 0$ is constant and so the moduli space $\mathcal{M}(0,\mathbf{w};J)$ is diffeomorphic to M. By Exercise 7.3.4, this formula holds for all $k \geq 1$. Compare also with Example 7.1.3.

The next proposition compares the invariant $\widetilde{\mathrm{GW}}$ of Theorem 7.3.1 with the Gromov–Witten invariants of the product manifold $\widetilde{M} := S^2 \times M$ in the case where $\widetilde{M}$ is itself semipositive.

PROPOSITION 7.3.6. *Assume that the product manifold $\widetilde{M} := S^2 \times M$ is semipositive, let $A \in H_2(M;\mathbb{Z})$ be a spherical homology class, and denote*

$$\widetilde{A} := [S^2 \times \mathrm{pt}] + \iota_* A \in H_2(\widetilde{M};\mathbb{Z}),$$

where $\iota : M \to S^2 \times M$ is the inclusion of the fiber. Then

$$\widetilde{\mathrm{GW}}_{A,k}^{M,I}(a_1,\dots,a_k) = \mathrm{GW}_{\widetilde{A},k}^{\widetilde{M}}(\widetilde{a}_1,\dots,\widetilde{a}_k)$$

for $a_i \in H^*(M)$, where

$$\widetilde{a}_i := \begin{cases} \mathrm{pr}^* a_i \smile \pi^*\sigma, & \text{if } i \in I, \\ \mathrm{pr}^* a_i, & \text{if } i \notin I. \end{cases}$$

Here $\mathrm{pr} : \widetilde{M} \to M$ is the projection onto the second factor, $\pi : \widetilde{M} \to S^2$ is the projection onto the first factor, and σ is the positive integral generator of $H^2(S^2)$.

PROOF. Suppose that the classes a_i satisfy (7.3.2). Then the classes $\widetilde{a}_i$ satisfy (7.1.2) with M and A replaced by $\widetilde{M}$ and $\widetilde{A}$, respectively:

$$\sum_{i=1}^{k} \deg(\widetilde{a}_i) = \dim\, M + 2c_1(A) + 2k = \dim\, \widetilde{M} + 2c_1(\widetilde{A}) + 2k - 6.$$

The condition $\pi_*\widetilde{A} = [S^2]$ guarantees that $\widetilde{A}$ is not a nontrivial multiple of any other homology class and so satisfies the condition (6.6.1) for the definition of the genus zero Gromov–Witten invariants in Theorem 7.1.1.

Fix a tuple $\mathbf{w} = \{\mathrm{w}_i\}_{i\in I}$ of distinct points on S^2 and a constant $\kappa \geq \omega(A)$, and let $J = \{J_z\}_{z\in S^2} \in \mathcal{J}_{\mathrm{reg}}(S^2, M, \omega; \mathbf{w}, \kappa)$. Then the product almost complex structure $\widetilde{J} \in \mathcal{J}(\widetilde{M})$ determined by J and the standard complex structure j on S^2 is compatible with the symplectic form $\widetilde{\omega} := \mathrm{pr}^*\omega + \pi^*\mathrm{dvol}_{S^2} \in \Omega^2(\widetilde{M})$. Moreover, it follows from the definition of regular families $J = \{J_z\}_{z\in S^2}$ that $\widetilde{J}$ satisfies the requirements of Definition 6.2.1 for every tree T with a special vertex 0 and every collection $\{\widetilde{A}_\alpha\}_{\alpha\in T}$ of homology classes in $H_2(\widetilde{M};\mathbb{Z})$ that satisfy $\pi_*\widetilde{A}_0 = [S^2]$ and $\pi_*\widetilde{A}_\alpha = 0$ for every $\alpha \in T \setminus \{0\}$. While this does not necessarily mean that $\widetilde{J} \in \mathcal{J}_{\mathrm{reg}}(\widetilde{M}, \widetilde{\omega})$, it suffices to construct the invariant of Theorem 7.1.1 in the class $\widetilde{A}$ with the almost complex structure $\widetilde{J}$.

Now suppose that a_i is Poincaré dual to an oriented submanifold $X_i \subset M$ for $i = 1, \ldots, k$. Then the class $\widetilde{a}_i$ is dual to the submanifold

$$\widetilde{X}_i := \begin{cases} \{\mathrm{w}_i\} \times X_i, & \text{if } i \in I, \\ S^2 \times X_i, & \text{if } i \notin I. \end{cases}$$

Moreover, every $\widetilde{J}$-holomorphic curve $\widetilde{u}_0 : S^2 \to \widetilde{M}$, which represents a homology class $\widetilde{A}_0$ such that $\pi_*\widetilde{A}_0 = [S^2]$, can be uniquely reparametrized to a $\widetilde{J}$-holomorphic section of $\widetilde{M}$. Thus the invariant $\mathrm{GW}^{\widetilde{M}}_{\widetilde{A},k}(\tilde{a}_1, \ldots, \tilde{a}_k)$ counts tuples $(u, \{z_i\}_{i\notin I})$ such that $u : S^2 \to M$ is a $\{J_z\}$-holomorphic sphere representing the class A, $u(\mathrm{w}_i) \in X_i$ for $i \in I$, and $u(z_i) \in X_i$ for $i \notin I$. That the two invariants are equal follows from the observation that each tuple $(u, \{z_i\}_{i\notin I})$ with these properties contributes the same sign to both invariants. This proves Proposition 7.3.6. □

Proposition 7.3.2 and Exercise 7.3.3 suggest the following definition for the Gromov–Witten invariants of a semipositive symplectic manifold (M, ω) for any homology class $A \in H_2(M;\mathbb{Z})$, whether or not it satisfies the condition (6.6.1). Moreover, this definition allows for any collection of fixed marked points $I \subset \{1, \ldots, k\}$, and thus removes the very restrictive hypotheses in the definition of $\mathrm{GW}^{M,I}_{A,k}$ in (7.2.10).

DEFINITION 7.3.7. *Let (M, ω) be a closed semipositive symplectic manifold, $k \geq 3$ be an integer, $I \subset \{1, \ldots, k\}$ such that $\#I \geq 3$, and $A \in H_2(M;\mathbb{Z})$. The* **genus zero Gromov–Witten invariant of M in the class A with k marked points** *is defined as the homomorphism*

$$\mathrm{GW}^M_{A,k} := \widetilde{\mathrm{GW}}^{M,\{1,2,3\}}_{A,k} : H^*(M)^{\otimes k} \to \mathbb{Z},$$

and the invariant **with fixed marked points indexed by I** *is defined by*

$$\mathrm{GW}^{M,I}_{A,k} := \widetilde{\mathrm{GW}}^{M,I}_{A,k} : H^*(M)^{\otimes k} \to \mathbb{Z}.$$

Here $\widetilde{\mathrm{GW}}^{M,I}_{A,k}$ is as in Theorem 7.3.1.

REMARK 7.3.8. In the case $k < 3$ and $A \neq 0$ the invariant GW of Section 7.1 cannot be directly recovered from $\widetilde{\mathrm{GW}}$: cf. Exercise 7.3.4. On the other hand, whenever it is defined the invariant $\mathrm{GW}^M_{A,k}$ with $k < 3$ and $A \neq 0$ can be recovered from the 3-point invariants via the divisor axiom (see Exercise 7.1.5). Hence when $A \neq 0$ we can give a consistent definition of the zero-, one-, and two-point invariants as follows. Given nonzero $A \in H_2(M;\mathbb{Z})$ choose $h \in H^2(M;\mathbb{Z})$ such that $h(A) = \int_A h \neq 0$ and define

$$\begin{aligned}\mathrm{GW}^M_{A,0} &:= \frac{1}{h(A)^3}\widetilde{\mathrm{GW}}^{M,\{1,2,3\}}_{A,3}(h,h,h),\\ \mathrm{GW}^M_{A,1}(a) &:= \frac{1}{h(A)^2}\widetilde{\mathrm{GW}}^{M,\{1,2,3\}}_{A,3}(a,h,h),\\ \mathrm{GW}^M_{A,2}(a,b) &:= \frac{1}{h(A)}\widetilde{\mathrm{GW}}^{M,\{1,2,3\}}_{A,3}(a,b,h),\end{aligned}$$

In particular, this gives a way to count the number of all curves in a given homology class (including the multiply covered ones) when M is a 6-dimensional manifold with $c_1 = 0$. (See Section 11.3.4.) When $A = 0$ the invariants $\mathrm{GW}^M_{0,k}$ and $\mathrm{GW}^{M,I}_{0,k}$ for $0 \leq k < 3$ are defined in Examples 7.1.3 and 7.3.5 and Exercise 7.3.4. They all vanish except when $\#I = k \neq 0$ in which case they are given by the cup product.

REMARK 7.3.9. The invariants GW^M and $\mathrm{GW}^{M,I}$ can also be defined, within the analytic context of this book, for some nonsemipositive symplectic manifolds (M,ω). We can assume instead that there exists an almost complex structure $J_0 \in \mathcal{J}_\tau(M,\omega)$ which is κ-semipositive in the sense that every J_0-holomorphic sphere $u : S^2 \to M$ with $\int u^*\omega \leq \kappa$ has nonnegative Chern number and then define invariants as before using the pseudocycle ev_{J_0}. As noted in Remark 6.6.2 the resulting invariants might now depend on the almost complex structure J_0, because the space of ω-tame κ-semipositive almost complex structures might be disconnected. Therefore we must change our perspective, thinking of the almost complex structure J_0 as given data rather than the symplectic form ω. Thus this extension is of particular interest in the Kähler case, when the complex structure rather than the symplectic form is the primary data.[2] Here is a precise formulation of the result. It follows immediately from Remark 6.6.2.

Suppose that (M,ω) is a compact symplectic manifold and $J_0 \in \mathcal{J}_\tau(M,\omega)$ is κ-semipositive with respect to ω. Let $A \in H_2(M;\mathbb{Z})$ such that $\omega(A) \leq \kappa$. Then there is a neighbourhood $\mathcal{N}(J_0)$ of J_0 in the space of all almost complex structures on M such that the homomorphism

$$\mathrm{GW}^M_{A,k} = \mathrm{GW}^{M;J}_{A,k} : H^*(M^k) \to \mathbb{Z},$$

given by

$$\mathrm{GW}^{M;J}_{A,k}(a) = f \cdot \mathrm{ev}_J, \qquad \mathrm{ev}_J : \mathcal{M}^*_{0,k}(A;J) \to M^k,$$

for every pseudocycle $f : U \to M^k$ dual to a, is well defined and independent of the choice of the almost complex structure $J \in \mathcal{J}_{\mathrm{reg}}(M,\omega) \cap \mathcal{N}(J_0)$ used in its definition. It is also independent of the taming form ω provided that $\kappa \geq \omega(A)$. It depends on J_0 up to deformations through κ-semipositive (almost) complex structures.

[2]As in Remark 7.1.11, our scruples are in fact unnecessary; the existence of the virtual moduli cycle shows that these invariants do not in fact depend on the choice of J_0.

REMARK 7.3.10. Although we shall work almost exclusively with the coefficients $\mathbb{Z}$ or $\mathbb{Q}$, it is sometimes convenient in applications to forget orientations and to count curves modulo 2. In this case, as in Remark 6.5.8, we should interpret the notation $H_*(M), H^*(M)$ to mean the homology groups $H_*(M;\mathbb{Z}/2\mathbb{Z}), H^*(M;\mathbb{Z}/2\mathbb{Z})$. Almost all the results in the previous two sections have obvious adaptations to this setting. For example, the Gromov–Witten invariants defined in Definition 7.3.7 above are now homomorphisms $H^*(M;\mathbb{Z}/2\mathbb{Z})^{\otimes k} \to \mathbb{Z}/2\mathbb{Z}$. However, there are some situations in which it is essential to use rational coefficients, for example when counting multiply covered curves as in Remark 7.3.8.

7.4. Rational curves in projective spaces

Curves in projective space have intricate and fascinating properties, and their study has a long history. However, as we shall see most of their complicated behaviour is irrelevant if all one wants to do is calculate the genus zero Gromov–Witten invariants in $\mathbb{C}P^n$.

The case $n = 1$ is uninteresting because, when $M = \mathbb{C}P^1$, the dimension of the moduli space of holomorphic spheres (i.e. of rational maps) of degree $d > 1$ with k marked points is $\mu(dL, k) = 2(k + 2d - 2) > 2k$, and so the Gromov–Witten pseudocycle $\mathrm{ev} : \mathcal{M}^*_{0,k}(dL; J) \to (\mathbb{C}P^1)^k$ is trivial unless $d = 1$. However, one does get interesting invariants by counting curves of higher genus, These are called the Riemann–Hurwitz numbers and have deep connections with the Toda lattice: see Okounkov–Pandharipande [**311, 312**].

In the following we assume $n \geq 2$. Then the virtual dimension of the moduli space of holomorphic spheres of degree d with k marked points is

$$\dim \mathcal{M}^*_{0,k}(dL; J_0) = 2m, \qquad m := d(n+1) + n + k - 3.$$

We shall see below that this is the actual dimension and that J_0 is regular in the sense of Definition 6.2.1. To obtain nonzero invariants we must assume $m \leq nk$ or, equivalently,

$$(n-1)k \geq d(n+1) + n - 3.$$

As shown by Example 7.4.11 it is possible for the invariant to vanish even though the dimensional condition is satisfied. However we will see that it is always nonzero when $k = 3d - 1$. Other cases in which it is nonzero are discussed at the end of the section. First, we prove the following theorem.

THEOREM 7.4.1. *Let $n \geq 2$ and $d, k \geq 1$. Then the evaluation map*

$$\mathrm{ev}_{J_0} : \mathcal{M}^*_{0,k}(dL; J_0) \to (\mathbb{C}P^n)^k$$

is a pseudocycle of real dimension $2d(n+1) + 2n + 2k - 6$. Moreover, if $k = 3d - 1$ it represents a nonzero homology class.

The two parts of this theorem have distinct proofs. The first assertion is elementary and holds because the complex manifold $(\mathbb{C}P^n, J_0)$ has a transitive group $\mathrm{PSL}(n+1)$ of complex automorphisms. The second assertion is somewhat harder to establish. It is easy to deduce from the case $n = 2$, which follows from Kontsevich's iterative formula for the number of degree d spheres in $\mathbb{C}P^2$ (Proposition 7.5.11). But to establish this, one must either use the gluing theorem of Chapter 10 or some quite sophisticated methods from algebraic geometry. The proof of Theorem 7.4.1 given here is complete, though it does require the adjunction formula of Theorem 2.6.4 in the following form.

REMARK 7.4.2. Let $u : \Sigma \to M^4$ be a simple J-holomorphic curve. Recall that an unordered pair of distinct points $z_1, z_2 \in \Sigma$ is called a transverse self-intersection of u if $u(z_1) = u(z_2) =: x$ and $T_xM = \operatorname{im} du(z_1) \oplus \operatorname{im} du(z_2)$. If $u : \Sigma \to \mathbb{C}P^2$ is a simple J_0-holomorphic curve of genus g and degree d with δ transverse self-intersections then, by Theorem 2.6.4,

$$g + \delta \leq \frac{(d-1)(d-2)}{2}$$

with equality if and only if u is an immersion and has only transverse self-intersections. Here multiple intersection points are allowed, that is, there might be k points $z_1, \ldots, z_k$ such that each unordered pair $z_i, z_j, i \neq j$, satisfies the above conditions.

Regularity of the standard complex structure. We shall prove that J_0 belongs to the set $\mathcal{J}_{\mathrm{reg}}(\mathbb{C}P^n, \omega_0)$, where ω_0 denotes the symplectic form of the Fubini–Study metric. The proof carries over to any Kähler manifold (M, J_0, ω_0) whose automorphism group G acts transitively. Examples include Grassmannians and symmetric spaces G/T. In the case $M = \mathbb{C}P^n$ the relevant group is $\mathrm{G} = \mathrm{PSL}(n+1)$. Recall that $J_0 \in \mathcal{J}_{\mathrm{reg}}(M, \omega_0)$ if and only if D_u is surjective for every simple J-holomorphic sphere $u : S^2 \to M$ and the edge evaluation map $\mathrm{ev}^E : \mathcal{M}^*(\{A_\alpha\}; J_0) \times Z(T) \to M^E$ is transverse to Δ^E for every labelled tree T and every collection $\{A_\alpha\}_{\alpha \in T}$ of spherical homology classes satisfying the stability condition (6.1.1). The proof below shows that under our assumptions even the multiply covered J_0-holomorphic spheres are regular.

PROPOSITION 7.4.3. *Let (M, J_0, ω_0) be a compact Kähler manifold and* G *be a Lie group that acts transitively on M by holomorphic diffeomorphisms. Then*

$$J_0 \in \mathcal{J}_{\mathrm{reg}}(M, \omega_0)$$

and every nonconstant J_0-holomorphic sphere in M has Chern number at least two.

PROOF. The proof has four steps.

STEP 1. *The infinitesimal action $\mathfrak{g} \to T_xM$ is surjective for every $x \in M$.*

This holds because the subbundle spanned by $\mathfrak{g}$ in TM is a distribution whose integral submanifolds are the orbits of G. By assumption, each orbit is the whole manifold M.

STEP 2. *Every J_0-holomorphic sphere $u : S^2 \to M$ is regular.*

By Step 1, the tangent bundle TM is spanned at each point by the global holomorphic vector fields

$$X_\xi(x) := \left.\frac{d}{dt}\right|_{t=0} \exp(t\xi)x$$

for $\xi \in \mathfrak{g}$. Now let $u : S^2 \to M$ be a J_0-holomorphic sphere. By Grothendieck's theorem, the pullback tangent bundle splits as a direct sum

$$u^*TM = L_1 \oplus \cdots \oplus L_n$$

of holomorphic line bundles. Let $\pi_i : u^*TM \to L_i$ denote the projection. Then every holomorphic vector field $X \in \mathrm{Vect}(M)$ descends to a holomorphic section $s_i := \pi_i X(u)$ of L_i. Hence, by Step 1, each subbundle L_i admits a nonzero holomorphic section and so must have nonnegative degree. Hence, by Lemma 3.3.2, D_u is surjective.

STEP 3. *Every nonconstant J_0-holomorphic sphere has Chern number at least two.*

Let $A \in H_2(M;\mathbb{Z})$ be a homology class that can be represented by a nonconstant J_0-holomorphic sphere. Then the moduli space $\mathcal{M}_{0,1}(A;J_0)$ (with one marked point) is nonempty and, by Step 2, it is regular and has dimension $2n+2c_1(A)-4$. Since the action of G is transitive, it follows that the evaluation map $\mathrm{ev}:\mathcal{M}^*_{0,1}(A;J_0)\to M$ is surjective. Hence the dimension of this moduli space is at least $2n$, and hence $c_1(A)\geq 2$.

STEP 4. *For every labelled tree T and every collection $\{A_\alpha\}_{\alpha\in T}$ of spherical homology classes in $H_2(M;\mathbb{Z})$ the edge evaluation map*

$$\mathrm{ev}^E:\mathcal{M}(\{A_\alpha\};J_0)\times Z(T)\to M^E$$

is transverse to Δ^E.

This is an immediate consequence of Step 1. The following argument spells out the details. Given

$$\mathbf{u}=\{u_\alpha\}_{\alpha\in T}\in\mathcal{M}(\{A_\alpha\};J_0),\qquad \mathbf{z}=(\{z_{\alpha\beta}\}_{\alpha E\beta},\{z_i\}_{1\leq i\leq k})\in Z(T)$$

denote

$$\mathbf{x}:=\mathrm{ev}^E(\mathbf{u},\mathbf{z})\in\Delta^E.$$

This means that $\mathbf{x}=\{x_{\alpha\beta}\}_{\alpha E\beta}$, where $x_{\alpha\beta}:=u_\alpha(z_{\alpha\beta})$ and $x_{\alpha\beta}=x_{\beta\alpha}$ for $\alpha E\beta$. Denote

$$V_{\mathbf{x}}:=\left\{\{X_{\xi_\alpha}(x_{\alpha\beta})\}_{\alpha E\beta}\,\middle|\,\xi_\alpha\in\mathfrak{g}\right\}\subset T_{\mathbf{x}}M^E.$$

We shall prove that

$$V_{\mathbf{x}}+T_{\mathbf{x}}\Delta^E=T_{\mathbf{x}}M^E. \tag{7.4.1}$$

Since the group G^T acts on the moduli space $\mathcal{M}(\{A_\alpha\};J_0)$ the subspace $V_{\mathbf{x}}$ is contained in the image of the differential $d\mathrm{ev}^E(\mathbf{u},\mathbf{z})$. Hence equation (7.4.1) shows that ev^E is transverse to Δ^E as claimed.

A complement to $T_{\mathbf{x}}\Delta^E$ is spanned by vectors $\mathbf{v}=\{v_{\alpha\beta}\}_{\alpha E\beta}\in T_{\mathbf{x}}M^E$ satisfying

$$\alpha E\beta\qquad\Longrightarrow\qquad v_{\alpha\beta}+v_{\beta\alpha}=0.$$

Therefore to prove (7.4.1) it suffices to show that for each such $\mathbf{v}$ there is a tuple $\{\xi_\alpha\}_{\alpha\in T}\in\mathfrak{g}^T$ such that

$$v_{\alpha\beta}=X_{\xi_\alpha}(x_{\alpha\beta})-X_{\xi_\beta}(x_{\beta\alpha}) \tag{7.4.2}$$

for $\alpha E\beta$. To prove this we argue by induction over the number of edges. If there are no edges then there is nothing to prove. Hence assume $e(T)>0$ and suppose that (7.4.2) has been established for every subtree $T_0\subset T$ with at most $e(T)-1$ edges. Let $\alpha_0\in T$ be an endpoint of the tree and $\beta_0\in T_0:=T\setminus\{\alpha_0\}$ be the unique vertex such that $\alpha_0E\beta_0$. Then the induction hypothesis asserts that there exists a tuple $\{\xi_\alpha\}_{\alpha\in T_0}\in\mathfrak{g}^{T_0}$ such that (7.4.2) holds for all $\alpha,\beta\in T_0$ such that $\alpha E\beta$. By Step 1, there exists a $\xi_{\alpha_0}\in\mathfrak{g}$ such that (7.4.2) holds for $\alpha=\alpha_0$ and $\beta=\beta_0$. Since there is only one edge adjacent to α_0, equation (7.4.2) then holds for all $\alpha E\beta$. This proves (7.4.1). □

REMARK 7.4.4. The holomorphic vector fields X_ξ in the proof of Proposition 7.4.3 have the following explicit form in the case $M = \mathbb{C}P^n$ and $\mathrm{G} = \mathrm{PSL}(n+1)$. Think of an element $\ell \in \mathbb{C}P^n$ as a line in $\mathbb{C}^{n+1}$ and identify the tangent space with

$$T_\ell \mathbb{C}P^n = \mathrm{Hom}(\ell, \mathbb{C}^{n+1}/\ell).$$

Then the infinitesimal action of the Lie algebra $\mathfrak{g} := \mathrm{Lie}(\mathrm{G}) = \mathfrak{sl}(n+1)$ is given by

$$X_\xi(\ell) = \pi_\ell \circ \xi \circ \iota_\ell$$

for $\xi \in \mathfrak{g}$, where $\iota_\ell : \ell \to \mathbb{C}^{n+1}$ is the inclusion and $\pi_\ell : \mathbb{C}^{n+1} \to \mathbb{C}^{n+1}/\ell$ is the projection. These vector fields are holomorphic sections of the tangent bundle and they obviously span the tangent space $T_\ell \mathbb{C}P^n$ for each ℓ.

The next result gives a sufficient condition for the Gromov–Witten invariant to be nonzero. The idea is that when the cohomology classes a_i are Poincaré dual to complex submanifolds X_i whose product is transverse to all the evaluation maps $\mathrm{ev}_{T,J_0} : \mathcal{M}^*_{0,T}(A; J_0) \to M^k$ then each J_0-holomorphic sphere passing through the X_i contributes positively to the Gromov–Witten invariants. Hence it suffices to find a single solution.

PROPOSITION 7.4.5. *Let (M, J_0, ω_0) and G be as in Proposition 7.4.3 and suppose that $A \in H_2(M;\mathbb{Z})$ and $a_1, \dots, a_k \in H^*(M)$ satisfy the dimension condition (7.1.2). Suppose further that there exist compact complex submanifolds $X_1, \dots, X_k \subset M$, a J_0-holomorphic sphere $u : \mathbb{C}P^1 \to M$, and a tuple $\mathbf{z} = (z_1, \dots, z_k)$ of pairwise distinct points in $\mathbb{C}P^1$ such that*

$$a_i = \mathrm{PD}([X_i]), \qquad u(z_i) \in X_i$$

for $i = 1, \dots, k$, and the evaluation map $\mathrm{ev}_{J_0} : \mathcal{M}^*_{0,k}(A; J_0) \to M^k$ is transverse to the product $X_1 \times \cdots \times X_k$ at $[u, \mathbf{z}]$. Then $\mathrm{GW}^M_{A,k}(a_1, \dots, a_k) > 0$.

PROOF. Suppose first that the product $\mathbf{X} := X_1 \times \cdots \times X_k$ is transverse to the evaluation map $\mathrm{ev}_{T,J_0} : \mathcal{M}^*_{0,T}(B; J_0) \to \mathbf{M} := M^k$ for every k-labelled tree T and every spherical homology class $B \in H_2(M;\mathbb{Z})$. Then the set

$$\mathcal{M}^*_{0,k}(A, \mathbf{X}; J_0) := \mathrm{ev}_{J_0}^{-1}(\mathbf{X}) = \left\{[u, \mathbf{z}] \in \mathcal{M}^*_{0,k}(A; J_0) \,|\, u(z_i) \in X_i\right\}$$

is finite and, by definition, the number of its elements counted with appropriate signs equals $\mathrm{GW}^M_{A,k}(a_1, \dots, a_k)$. Now by assumption, $\mathcal{M}^*_{0,k}(A, \mathbf{X}; J_0)$ is nonempty. Moreover, since J_0 is integrable and $\mathbf{X}$ is a complex submanifold of $\mathbf{M}$, each space in the direct sum decomposition

$$T_{\mathbf{x}}\mathbf{M} = T_{\mathbf{x}}\mathbf{X} \oplus \mathrm{im}\, d\mathrm{ev}_{J_0}([u, \mathbf{z}])$$

for $[u, \mathbf{z}] \in \mathcal{M}^*_{0,k}(A, \mathbf{X}; J_0)$ and $\mathbf{x} := (u(z_1), \dots, u(z_k)) \in \mathbf{X}$ is a complex subspace of $T_{\mathbf{x}}\mathbf{M}$. Hence each solution $[u, \mathbf{z}] \in \mathcal{M}^*_{0,k}(A, X; J_0)$ contributes 1 to the invariant and so

$$\mathrm{GW}^M_{A,k}(a_1, \dots, a_k) = \#\mathcal{M}^*_{0,k}(A, \mathbf{X}; J_0) > 0.$$

Now consider an arbitrary submanifold $\mathbf{X}$ and let $\mathbf{G}_{\mathrm{reg}} \subset \mathbf{G} := \mathrm{G}^k$ denote the set of all tuples $\mathbf{g} = (g_1, \dots, g_k) \in \mathbf{G}$ such that $\mathbf{g}\mathbf{X}$ is transverse to $\mathrm{ev}_{T,J_0} : \mathcal{M}^*_{0,T}(B; J_0) \to \mathbf{M}$ for every T and every B. We prove that $\mathbf{G}_{\mathrm{reg}}$ is residual in $\mathbf{G}$. More generally, consider instead of ev_{T,J_0} any smooth map $f : U \to \mathbf{M}$ defined on a smooth manifold U and form the space

$$Y_f := \{(\mathbf{g}, \mathbf{x}, u) \in \mathbf{G} \times \mathbf{X} \times U \,|\, f(u) = \mathbf{g}\mathbf{x}\}.$$

This space is a smooth submanifold of $\mathbf{G} \times \mathbf{X} \times U$. To see this note that, since $\mathbf{G}$ acts transitively on $\mathbf{M}$, the map

$$\mathbf{G} \times \mathbf{X} \times U \to \mathbf{M} \times \mathbf{M} : (\mathbf{g}, \mathbf{x}, u) \mapsto (\mathbf{gx}, f(u))$$

is transverse to the diagonal. Now an element $\mathbf{g} \in \mathbf{G}$ is a regular value of the projection $Y_f \to \mathbf{G}$ if and only if $\mathbf{gX}$ is transverse to f. Hence, by Sard's theorem, the set $\mathbf{G}_{\mathrm{reg}}(f)$ of all $\mathbf{g} \in \mathbf{G}$ such that $\mathbf{gX}$ is transverse to f is residual for every smooth map $f : U \to \mathbf{M}$. Hence the above set $\mathbf{G}_{\mathrm{reg}}$ is residual in $\mathbf{G}$ as claimed. Now it follows from our assumptions that, if $\mathbf{g} \in \mathbf{G}_{\mathrm{reg}}$ is sufficiently close to $\mathbb{1}$, then $\mathcal{M}^*_{0,k}(A, \mathbf{gX}; J_0) \neq \emptyset$. Hence it follows from the first part of the proof that the invariant is positive. This proves Proposition 7.4.5. □

REMARK 7.4.6. One can get some useful insight about the rational curves in $\mathbb{C}P^n$ by considering the action of the general linear group $\mathrm{G} = \mathrm{PSL}(n+1, \mathbb{C})$. Each element u in the moduli space $\widetilde{\mathcal{M}}_{0,0}(\mathbb{C}P^n, dL; J_0)$ has the form

$$u(z) = [u_0(z) : \cdots : u_n(z)], \quad z \in \mathbb{C} \cup \{\infty\}$$

where the u_i are polynomials of degree $\leq d$. Thus, using the coefficients of the u_i as homogeneous coordinates, we may identify $\widetilde{\mathcal{M}}_{0,0}(\mathbb{C}P^n, dL; J_0)$ with an open dense subset of the projective space $\mathbb{C}P^N$ where $N = (d+1)(n+1) - 1$. We do not get the whole of $\mathbb{C}P^N$ since some elements (such as $[1 : z^k : 0 : \cdots : 0]$ when $k < d$) are degenerate and do not represent the correct homology class.

Given such u let $F(u)$ be the $(n+1) \times (d+1)$ matrix whose ith row is given by the coefficients of the polynomials u_i. This is well defined up to multiplication by a scalar. Let $r(u)$ denote the rank of $F(u)$. Then $r(u) - 1$ is the minimum dimension of a projective subspace containing the curve $\mathrm{im}(u)$. Since $r(u) - 1 \leq d$, every conic (i.e. curve of degree 2) lies in a 2-plane, every cubic lies in a 3-plane and so on.

In terms of the matrix $F(u)$, the elements of G act by matrix multiplication:

$$F(g \circ u) = gF(u), \qquad g \in \mathrm{G}.$$

When $d \leq n$ this action is transitive on the set of all u with $r(u) = d + 1$. In particular, if $d = n$ this orbit is spanned by the *rational normal curve* C_d given by

$$z \mapsto [1 : z : z^2 : \cdots : z^d].$$

(When $d = 3$ this curve is also known as the *twisted cubic.*)

EXERCISE 7.4.7. Consider a rational curve in $\mathbb{C}P^n$ with $r(u) - 1 < n$. Show that it lies in an $(r(u) - 1)$-dimensional linear subspace H of $\mathbb{C}P^n$. Show further that it is the image of a curve with maximal rank $r(u) = \max(d + 1, n + 1)$ under a suitable projection

$$\pi_Q : \mathbb{C}P^n \setminus Q \to H \subset \mathbb{C}P^n,$$

where Q is an $(n - r(u))$-dimensional plane in $\mathbb{C}P^n \setminus H$ and $\pi_Q(q)$ denotes the unique point of intersection of H with the linear subspace of $\mathbb{C}P^n$ spanned by Q and the point $q \neq Q$. *Hint:* First consider the case when $d = n$.

Counting curves in $\mathbb{C}P^2$ through $3d - 1$ points. In this section it is important to emphasize the ambient manifold and so we shall use the notation

$$\mathcal{M}_{0,k}(M, A; J) := \mathcal{M}_{0,k}(A; J)$$

for the moduli space of marked J-holomorphic spheres in M representing the class A. Consider the evaluation map

$$\mathrm{ev}_{J_0} : \mathcal{M}^*_{0,k}(\mathbb{C}P^n, dL; J_0) \to (\mathbb{C}P^n)^k.$$

Because J_0 is integrable, the moduli space $\mathcal{M}^*_{0,k}(\mathbb{C}P^n, dL; J_0)$ has a natural complex structure (see Remark 3.2.6) and, moreover, the evaluation map ev_{J_0} is holomorphic. In the case $n = 2$ and $k = 3d - 1$ the moduli space has complex dimension

$$\dim^{\mathbb{C}} \mathcal{M}^*_{0,3d-1}(\mathbb{C}P^2, dL; J_0) = 6d - 2.$$

Note that this is also the dimension of $(\mathbb{C}P^2)^{3d-1}$. The corresponding Gromov–Witten invariant

$$N_d := \mathrm{GW}^{\mathbb{C}P^2}_{dL,3d-1}(c^2, \ldots, c^2)$$

counts the number of degree d curves in $\mathbb{C}P^2$ through $3d-1$ generic points. (Recall from Exercise 7.1.14 that c denotes the positive generator of $H^2(\mathbb{C}P^2; \mathbb{Z})$.) This number can also be interpreted as the *degree* of the evaluation map ev_{J_0} which is well-defined because, by Proposition 7.4.3 and Theorem 6.6.1, this map is a pseudocycle. Since holomorphic maps preserve orientations at regular points, it follows that $N_d \geq 0$ with equality if and only if ev_{J_0} is nowhere of maximal rank.

PROPOSITION 7.4.8. *$N_d > 0$ for every integer $d \geq 0$.*

We shall explain a proof of this result by Harris–Morrison [**168**]. The arguments may serve as an illustration of the difference in viewpoints between the symplectic approach to holomorphic curves explained in this book and the algebro-geometric approach. Since we do not assume that the reader has any knowledge of algebraic geometry, we explain many elementary details.

Let us assume that (M, J_0, ω_0) is a Kähler surface. Then an **effective divisor** in M is a formal sum

$$C = \sum_i m_i C_i,$$

where the m_i are positive integers and each C_i is an irreducible complex 1-dimensional subvariety of M. In the language of this book this means that

$$C_i = u_i(\Sigma_i)$$

is the image of a simple J_0-holomorphic curve $u_i : \Sigma_i \to M$, defined on a compact connected Riemann surface Σ_i without boundary. In the above expression for C we assume that $C_i \neq C_j$ for $i \neq j$. Every effective divisor C can be described as the zero set of a nonzero holomorphic section s of a holomorphic line bundle $L \to M$ and, moreover, the pair (L, s) is uniquely determined by C up to complex bundle isomorphisms. Thus each divisor determines a cohomology class $c_1(L) \in H^2(M; \mathbb{Z})$ which, in our terminology, is the Poincaré dual of the homology class $A = \sum_i m_i A_i$ represented by the formal sum of the J_0-holomorphic curves. This duality between holomorphic curves and holomorphic sections of line bundles is a powerful tool for the study of holomorphic curves in Kähler surfaces and is not available to the same extent in the almost complex or symplectic category.

As an example consider the case $M = \mathbb{C}P^2$ and let $L \to \mathbb{C}P^2$ denote the dth tensor power of the tautological line bundle. Explicitly L is the quotient of

$\mathbb{C}^3 \times \mathbb{C}$ by $\mathbb{C}^*$, where $\lambda \in \mathbb{C}^*$ acts by $\lambda \cdot (z_0, z_1, z_2, \zeta) := (\lambda z_0, \lambda z_1, \lambda z_2, \lambda^d \zeta)$. Thus a holomorphic section of L is a homogeneous polynomial

$$f(z_0, z_1, z_2) = \sum_{i+j+k=d} c_{ijk} {z_0}^i {z_1}^j {z_2}^k$$

of degree d with complex coefficients. The space of such polynomials will be denoted by $\mathcal{F}_d$. To describe the relation between holomorphic sections and divisors one needs the following three basic facts from algebraic geometry. An excellent reference is Atiyah–MacDonald [**21**].

(Unique factorization) The polynomial ring $\mathbb{C}[z_0, \ldots, z_n]$ is a unique factorization domain. This means the following. A polynomial f is called **irreducible** if $f = f_0 f_1$ implies that either f_0 or f_1 is constant. So, by definition, every polynomial factors into irreducibles. The unique factorization property asserts that this factorization is unique up to reordering and multiplication by nonzero constants.

(Noetherian) The polynomial ring $\mathbb{C}[z_0, \ldots, z_n]$ is Noetherian. This means that every ideal is finitely generated.

(Nullstellensatz) For every ideal $\mathcal{I} \subset \mathbb{C}[z_0, \ldots, z_n]$ let $\mathcal{V}(\mathcal{I}) \subset \mathbb{C}^{n+1}$ denote the common zero set of the elements of $\mathcal{I}$ and let $\mathcal{I}(\mathcal{V}(\mathcal{I}))$ denote the ideal of all polynomials in $\mathbb{C}[z_0, \ldots, z_n]$ that vanish on $\mathcal{V}(\mathcal{I})$. Hilbert's Nullstellensatz asserts that

$$\mathcal{I}(\mathcal{V}(\mathcal{I})) = \sqrt{\mathcal{I}},$$

where the **radical** $\sqrt{\mathcal{I}}$ is the set of all polynomials $f \in \mathbb{C}[z_0, \ldots, z_n]$ such that $f^k \in \mathcal{I}$ for some positive integer k.

Let us now return to the case $n = 2$. Then the zero set of a homogeneous polynomial is invariant under scalar multiplication and can therefore be identified with a subset of $\mathbb{C}P^2$. Secondly, the irreducible factors of a homogeneous polynomial are also homogeneous. Thirdly, one can prove that the zero set

$$C_f := \left\{ [z_0 : z_1 : z_2] \in \mathbb{C}P^2 \,|\, f(z_0, z_1, z_2) = 0 \right\}$$

of a nonzero irreducible homogeneous polynomial $f \in \mathcal{F}_d$ is the image of a J_0-holomorphic curve $u : \Sigma \to \mathbb{C}P^2$, defined on a closed connected Riemann surface Σ (see Exercise 7.1.16 for a simple example).

Hilbert's Nullstellensatz, applied to the principle ideal generated by f, asserts that when f is irreducible every polynomial on $\mathbb{C}^3$ that vanishes on C_f is divisible by f. Hence an irreducible homogeneous polynomial f of degree d is uniquely determined up to a constant factor by its zero set C_f. In the reducible case we shall denote by C_f the formal sum of the zero sets of the irreducible factors of f. This is an effective divisor in $\mathbb{C}P^2$ and, by the unique factorization property and the Nullstellensatz, each nonzero polynomial $f \in \mathcal{F}_d$ is uniquely determined by its divisor C_f up to a constant factor. Hence the set of effective divisors of degree d in $\mathbb{C}P^2$ can be naturally identified with the projective space

$$\mathcal{P}_d := \mathbb{P}\mathcal{F}_d.$$

Note that $\mathcal{P}_d$ has complex dimension

$$\dim_{\mathbb{C}} \mathcal{P}_d = m := \tfrac{1}{2} d(d+3).$$

The tangent space of $\mathcal{P}_d$ at an element $[f]$ can be identified with the quotient space $\mathcal{F}_d / \mathbb{C}f$. However, the identification is not canonical but depends on f: replacing f by λf results in replacing a tangent vector $\phi \in \mathcal{F}_d / \mathbb{C}f$ by $\lambda \phi$.

If one wants to recover the moduli space of J_0-holomorphic spheres from the algebro-geometric description of effective divisors one is led to consider those nonzero polynomials $f \in \mathcal{F}_d$ whose zero sets C_f are the images of simple J_0-holomorphic spheres in $\mathbb{C}P^2$ representing the class dL. This is precisely the so-called **Severi variety**

$$\mathcal{V}_d := \{[f] \in \mathcal{P}_d \mid f \text{ is irreducible, } C_f \text{ has genus zero}\}$$

(see Harris–Morrison [**168**, Ch 1.F]). The set $\mathcal{V}_d$ is a rather complicated object and it is therefore interesting to focus on the open subset $\mathcal{U}_d \subset \mathcal{V}_d$ of all irreducible effective degree d divisors of genus zero all of whose singularities are simple nodes. The zero set C_f of a representative $f \in \mathcal{F}_d$ of an element in $\mathcal{U}_d$ is an immersed sphere with transverse self-intersections and at most two branches through each point. Remark 7.4.2 asserts that in this case the number of transverse self-intersections is

$$\delta_d := \tfrac{1}{2}(d-1)(d-2), \tag{7.4.3}$$

and that, conversely, every irreducible J_0-holomorphic curve of degree d with δ_d transverse self-intersections must be an immersed sphere. Hence $\mathcal{U}_d$ has the following elementary description. Given a nonzero homogeneous polynomial $f \in \mathcal{F}_d$ denote by $S_f \subset C_f$ the set of all **simple nodes** of C_f, i.e.

$$S_f := \left\{ [z_0 : z_1 : z_2] \in \mathbb{C}P^2 \,\middle|\, f(z) = 0,\ \frac{\partial f}{\partial z_i}(z) = 0,\ \operatorname{rank}\left(\frac{\partial^2 f}{\partial z_i \partial z_j}(z)\right) = 2 \right\}.$$

(For example, $C_{z_0 z_1}$ has a simple node at $[0:0:1]$ while $C_{z_0^2 - z_1^3}$ does not.) Then

$$\mathcal{U}_d = \{[f] \in \mathcal{P}_d \mid f \text{ is irreducible, } \#S_f = \delta_d\}. \tag{7.4.4}$$

Now let $\mathcal{U}_d^* \subset \mathcal{U}_d$ denote the subset of all equivalence classes $[f] \in \mathcal{U}_d$ of irreducible polynomials whose nodal set S_f is in general position, i.e.

$$\mathcal{U}_d^* := \{[f] \in \mathcal{U}_d \mid S_f \text{ is regular}\}. \tag{7.4.5}$$

Here a finite set $S \subset \mathbb{C}P^2$ is called **regular** if $\dim \mathcal{F}_d/\mathcal{H}_S = \#S$, where $\mathcal{H}_S \subset \mathcal{F}_d$ denotes the linear subspace

$$\mathcal{H}_S := \{\phi \in \mathcal{F}_d \mid \phi(p) = 0 \ \forall\, p \in S\}. \tag{7.4.6}$$

Given a homogeneous function ϕ on $\mathbb{C}^3$ and a point $p \in \mathbb{C}P^2$, the equation $\phi(p) = 0$ means that ϕ vanishes on some, and hence every, lift of p to $\mathbb{C}^3$. If we choose a lift $z_p \in \mathbb{C}^3$ of each $p \in S$, then evaluation at the points z_p for $p \in S$ gives a linear map $\mathcal{F}_d \to \mathbb{C}^S$ with kernel $\mathcal{H}_S$. Hence the subset S is regular exactly if this map is surjective. For example, a set of four distinct collinear points in $\mathbb{C}P^2$ is regular for $d \geq 3$, but not for $d \leq 2$. This is because the space of polynomials in one variable with four prescribed zeros has codimension four in the space of all polynomials of degree d precisely when $d \geq 3$.

Our proof of Proposition 7.4.8 has five steps. It begins with the observation (in Steps 1 and 2) that $\mathcal{U}_d^*$ is a smooth submanifold of $\mathcal{P}_d$ of complex dimension $3d-1$. To see that $\mathcal{U}_d^*$ is nonempty, we show in Steps 3 and 4 how to construct elements of $\mathcal{U}_d^*$ from unions of d lines by resolving $d-1$ nodes. This is an algebraic version of the symplectic gluing construction of Chapter 10. Finally we show that the evaluation map from $\mathcal{U}_d^*$ has positive degree by a transversality argument that uses Hilbert's Nullstellensatz.

PROOF OF PROPOSITION 7.4.8. Consider the set

$$\mathcal{S}_d := \left\{([f],p) \in \mathcal{P}_d \times \mathbb{C}P^2 \,|\, f(p) = 0,\ df(p) = 0\right\}.$$

It consists of all pairs $([f],p) \in \mathcal{P}_d \times \mathbb{C}P^2$, where p is a singular point of C_f.

STEP 1. *$\mathcal{S}_d$ is a smooth closed codimension 3 submanifold of $\mathcal{P}_d \times \mathbb{C}P^2$. Moreover, the projection $\pi : \mathcal{S}_d \to \mathcal{P}_d$ is an immersion near $([f],p) \in \mathcal{S}_d$ if and only if p is a simple node of C_f. In this case the image of the differential of π at $([f],p)$ is given by*

$$\operatorname{im} d\pi([f],p) = \mathcal{H}_p/\mathbb{C}f, \qquad \mathcal{H}_p := \{\phi \in \mathcal{F}_d \,|\, \phi(p) = 0\}.$$

An **affine coordinate chart** on $\mathbb{C}P^2$ is an affine map $\mathbb{C}^2 \to \mathbb{C}^3$ whose image misses the origin. In such coordinates $(x,y) \in \mathbb{C}^2$ a polynomial in $\mathcal{F}_d$ has the form

$$f_a(x,y) = \sum_{i+j \le d} a_{ij} x^i y^j.$$

There are $m+1$ coefficients a_{ij}, where $m := (d^2+3d)/2$. If $([f_a],x,y) \in \mathcal{S}_d$ then f_a cannot be linear. Hence we can normalize f_a by picking two indices $i_0, j_0 \ge 0$ such that $1 < i_0 + j_0 \le d$ and imposing the condition $a_{i_0 j_0} = 1$. This gives rise to local coordinates $(a,x,y) \in \mathbb{C}^m \times \mathbb{C}^2$ on $\mathcal{P}_d \times \mathbb{C}P^2$. In these coordinates a point $([f_a],x,y)$ belongs to $\mathcal{S}_d$ if and only if

$$f_a(x,y) = 0, \qquad \partial_x f_a(x,y) = 0, \qquad \partial_y f_a(x,y) = 0.$$

We emphasize that, given any point $p \in \mathbb{C}P^2$, we may choose the affine coordinate chart so that p has coordinates $x = 0$ and $y = 0$.

Now consider the map $F : \mathbb{C}^m \times \mathbb{C}^2 \to \mathbb{C}^3$, defined by

$$F(a,x,y) := (f_a(x,y), \partial_x f_a(x,y), \partial_y f_a(x,y)).$$

The zero set of this map is $\mathcal{S}_d$. Suppose without loss of generality that $F(a,0,0) = 0$. Then $a_{00} = a_{10} = a_{01} = 0$ and the differential of F at $(a,0,0)$ is given by

$$dF(a,0,0) = \begin{pmatrix} 1 & 0 & 0 & 0 & \cdots & 0 & 0 & 0 \\ 0 & 1 & 0 & 0 & \cdots & 0 & 2a_{20} & a_{11} \\ 0 & 0 & 1 & 0 & \cdots & 0 & a_{11} & 2a_{02} \end{pmatrix},$$

where the first three columns are the derivatives with respect to a_{00}, a_{10}, and a_{01} and the last two are the derivatives with respect to x and y. Hence $dF(a,0,0)$ is surjective and so $\mathcal{S}_d$ is a codimension 3 submanifold of $\mathcal{P}_d \times \mathbb{C}P^2$.

Still assuming $F(a,0,0) = 0$, one finds that the singularity of $C := f_a^{-1}(0)$ at the origin is a simple node if and only if

$$a_{11}^2 - 4a_{20}a_{02} \ne 0.$$

The formula for $dF(a,0,0)$ shows that this is the case if and only if the projection

$$\ker\, dF(a,0,0) \to \mathbb{C}^m : (\alpha, \xi, \eta) \mapsto \alpha$$

is injective. Moreover, the image of this projection is the set of vectors $\alpha \in \mathbb{C}^m$ such that $\alpha_{00} = 0$. In intrinsic terms this means that $([f],p) \in \mathcal{S}_d$ has a simple node at p if and only if the differential of the projection $\pi : \mathcal{S}_d \to \mathcal{P}_d$ is injective at the point $([f],p)$ and that, if this holds, the image of $d\pi([f],p)$ is tangent to the submanifold of all polynomials in $\mathcal{P}_d$ that vanish at the point p. This completes the proof of Step 1.

STEP 2. *$\mathcal{U}_d^*$ is a complex submanifold of $\mathcal{P}_d$ of codimension $\delta := \delta_d$ and*

$$T_{[f]}\mathcal{U}_d^* = \mathcal{H}_{S_f}/\mathbb{C}f$$

for every $[f] \in \mathcal{U}_d^*$.

Let $[f] \in \mathcal{U}_d^*$ and denote by $\{p_1, \dots, p_\delta\} := S_f \subset C_f$ the set of simple nodes of C_f. Then $([f], p_j) \in \mathcal{S}_d$ for $j = 1, \dots, \delta$ and, by Step 1, the projection $\pi : \mathcal{S}_d \to \mathcal{P}_d$ has an injective derivative at $([f], p_j)$ for every k. Choose disjoint open neighbourhoods $\mathcal{N}_j \subset \mathcal{S}_d$ of $([f], p_j)$ such that the restriction of π to $\mathcal{N}_j$ is an embedding for every j. Since S_f is regular, these restrictions intersect transversally, provided that the neighbourhoods are chosen sufficiently small. Hence the set

$$\mathcal{W} := \bigcap_{j=1}^{\delta} \pi(\mathcal{N}_j)$$

is a submanifold of $\mathcal{P}_d$ of complex codimension δ. Now recall from Remark 7.4.2 that C_f has no singularities other than the p_k. Hence, if $f' \in \mathcal{F}_d$ is close to f and also has δ simple nodes, then these simple nodes must be close to the points p_j and hence $[f']$ must belong to $\mathcal{W}$. Conversely, $\mathcal{W}$ is, by definition, contained in $\mathcal{U}_d$. Hence $\mathcal{W}$ is a neighbourhood of $[f]$ in $\mathcal{U}_d^*$. This proves Step 2.

We next show how to make an irreducible curve out of the (reducible) union $C_{f_0} \cup \dots \cup C_{f_k}$ of $k+1$ curves by resolving (i.e. smoothing) k intersection points, one in each of the sets $C_{f_i} \cap C_{f_0}$, $i = 1, \dots, k$.

STEP 3. *Let $f_i \in \mathcal{F}_{d_i}$ be irreducible for $i = 0, \dots, k$. Define $f := f_0 f_1 \cdots f_k \in \mathcal{P}_d$ where $d := d_0 + d_1 + \dots + d_k$, and suppose that the only singularities of C_f are simple nodes. Choose a simple node $p_i \in C_{f_0} \cap C_{f_i}$ of f for each $i = 1, \dots, k$. Then there exist neighbourhoods $\mathcal{N}_i \subset \mathcal{S}_d$ of $([f], p_i)$ and $\mathcal{W} \subset \mathcal{P}_d$ of $[f]$ such that*

$$[f'] \in \mathcal{W} \setminus \bigcup_{i=1}^{k} \pi(\mathcal{N}_i) \qquad \Longrightarrow \qquad f' \text{ is irreducible.}$$

By Step 1, we can choose $\mathcal{N}_i$ so small that the projection $\pi : \mathcal{N}_i \to \mathcal{P}_d$ is an embedding. For each $i = 1, \dots, k$ let $U_i \subset \mathbb{C}P^2$ be an open neighbourhood of p_i and $\mathcal{W} \subset \mathcal{P}_d$ be an open neighbourhood of f such that $(\mathcal{W} \times U_i) \cap \mathcal{S}_d \subset \mathcal{N}_i$. This means that, if $[f'] \in \mathcal{W} \setminus \bigcup_i \pi(\mathcal{N}_i)$ then $([f'], p') \notin \mathcal{S}_d$ for every $p' \in U_i$, and so $C_{f'}$ has no singular point in U_i. Since the only singular points of C_f are simple nodes, by shrinking $\mathcal{W}$ we may assume that the only singularities of $C_{f'}, [f'] \in \mathcal{W}$, are simple nodes, that moreover lie near the unresolved simple nodes of C_f. Now, for each i, fix two regular points $q_i \in C_{f_i}$ and $q_{i0} \in C_{f_0}$ on the boundary of U_i. Then, if f' is sufficiently close to f and $[f'] \notin \pi(\mathcal{N}_i)$, there are nearby points $q_i', q_{i0}' \in C_{f'}$ that can be connected by a regular path in $C_{f'}$. It follows easily that any two regular points in $C_{f'}$ can be connected by a regular path in $C_{f'}$ and hence f' is irreducible. Hence we can choose $\mathcal{W}$ so that the conclusion of Step 3 holds.

STEP 4. *$\mathcal{U}_d^*$ is nonempty.*

Choose d lines $L_1, \dots, L_d \subset \mathbb{C}P^2$ in general position (i.e. pairwise distinct and without triple intersections). For $i = 1, \dots, d$ choose a nonzero linear polynomial $f_i \in \mathcal{F}_1$ such that $C_{f_i} = L_i$. Then the polynomial

$$f := f_1 \cdots f_d \in \mathcal{F}_d$$

has $\delta + d - 1 = \frac{1}{2}d(d-1)$ simple nodes and no other singularities. (Recall that $\delta = \delta_d = \frac{1}{2}(d-1)(d-2)$ by equation (7.4.3).) Denote the set of simple nodes of f by

$$S_f =: \{p_{ij} \,|\, 1 \le i < j \le d\}, \qquad p_{ij} = L_i \cap L_j.$$

Then, for $i < j$, the function $g_{ij} := \prod_{k \ne i,j} f_k$ has the zero set $\bigcup_{k \ne i,j} L_k$ and hence vanishes at all the points of S_f except for p_{ij}. Choose quadratic polynomials $q_{ij} \in \mathcal{F}_2$ such that q_{ij} does not vanish on p_{ij}. Then the polynomials $q_{ij} g_{ij} \in \mathcal{F}_d$ descend to $\#S_f$ linearly independent elements of the quotient space $\mathcal{F}_d/\mathcal{H}_{S_f}$. Hence the set S_f is regular.

Now choose small neighbourhoods $\mathcal{N}_{ij} \subset \mathcal{S}_d$ of $([f], p_{ij})$. Then, since S_f is regular, the sets

$$\mathcal{W}' := \bigcap_{2 \le i < j \le d} \pi(\mathcal{N}_{ij}), \qquad \mathcal{W}_j'' := \mathcal{W}' \cap \pi(\mathcal{N}_{1j}), \qquad 2 \le j \le d,$$

are submanifolds of $\mathcal{P}_d$ such that $\operatorname{codim}^{\mathbb{C}} \mathcal{W}' = \delta = \delta_d$ and $\operatorname{codim}^{\mathbb{C}} \mathcal{W}_j'' = \delta + 1$ for each j. Hence $\mathcal{W}'' := \mathcal{W}_2'' \cup \cdots \cup \mathcal{W}_d''$ is a proper subset of $\mathcal{W}'$. By Step 3, every element $f' \in \mathcal{F}_d$ that is sufficiently close to f and satisfies $[f'] \in \mathcal{W}' \setminus \mathcal{W}''$ is irreducible. Moreover, for any such f', the curve $C_{f'}$ has precisely δ simple nodes and the set of simple nodes of f' is regular. Hence it follows from Remark 7.4.2 that $C_{f'}$ is an immersed sphere and hence $[f'] \in \mathcal{U}_d^*$. This proves Step 4.

STEP 5. *We prove the proposition.*

Let $[f] \in \mathcal{U}_d^*$. Then, by the Nullstellensatz, every polynomial in $\mathcal{F}_d$ that vanishes on C_f must be a scalar multiple of f. This can be expressed in the form

$$\bigcap_{p \in C_f} \mathcal{H}_p = \mathbb{C}f.$$

By definition of $\mathcal{U}_d^*$, the set S_f is regular and so the intersection of the hyperplanes $\mathcal{H}_p$ over all $p \in S_f$ is $\mathcal{H}_{S_f}$. It has dimension

$$\dim \mathcal{F}_d - \#S_f = \tfrac{1}{2}d(d+3) + 1 - \delta = \tfrac{1}{2}d(d+3) + 1 - \tfrac{1}{2}(d-1)(d-2) = 3d.$$

Hence it follows by induction that there exists a finite set $X \subset C_f \setminus S_f$ such that

$$\mathcal{H}_{S_f} \cap \mathcal{H}_X = \mathbb{C}f, \qquad \#X = 3d - 1.$$

But by Step 2, $\mathcal{U}_d^*$ is a smooth manifold of dimension $3d-1$ with tangent space $T_f\mathcal{U}_d^* = \mathcal{H}_{S_f}/\mathbb{C}f$. It follows that $\mathcal{U}_d^*$ is transverse to the projective space $\mathbb{P}\mathcal{H}_X \subset \mathcal{P}_d$ at f. Hence $\mathcal{U}_d \cap \mathbb{P}\mathcal{H}_Y \ne \emptyset$ for every finite set $Y \subset \mathbb{C}P^2$ such that $\#Y = 3d-1$ and Y is sufficiently close to X. Hence the image of the evaluation map

$$\mathrm{ev}_{J_0} : \mathcal{M}^*_{0,3d-1}(\mathbb{C}P^2, dL; J_0) \to (\mathbb{C}P^2)^{3d-1}$$

contains an open subset of $(\mathbb{C}P^2)^{3d-1}$. Hence there exists a regular value in the image of ev_{J_0} and so it follows from Proposition 7.4.5 that $N_d > 0$. This proves Proposition 7.4.8. □

In [**215**] Kontsevich found a beautiful recursion formula that allows one to calculate the numbers N_d. We describe his idea in Proposition 7.5.11 below.

Higher dimensions. We must show that the Gromov–Witten pseudocycle

$$\mathrm{ev}_{J_0} : \mathcal{M}_{0,3d-1}(\mathbb{C}P^n, dL; J_0) \to (\mathbb{C}P^n)^{3d-1}$$

(of complex dimension $n(d+1)+4(d-1)$) represents a nonzero homology class for every d and every $n \geq 3$. As a warmup we begin with the case $d = 2$. We give two proofs, the first geometric and the second with more details.

PROPOSITION 7.4.9.

$$\mathrm{GW}^{\mathbb{C}P^n}_{2L,5}(c^2, c^2, c^n, c^n, c^n) = 1. \tag{7.4.7}$$

FIRST PROOF. We first claim that there is a unique conic (sphere of degree 2) through two generic planes X_1, X_2 of complex codimension 2 and 3 generic points p_3, p_4, p_5. To see this, note that the three points determine a unique 2-plane P. By Example 7.2.9 there is a unique conic in P that goes through these three points as well as the points $X_1 \cap P, X_2 \cap P$ where the lines meet this plane. Since every conic lies in a plane by Remark 7.4.6, there is precisely one conic u that satisfies the given conditions. It remains to check that the evaluation map

$$\mathrm{ev} : \mathcal{M}(\mathbb{C}P^n, 2L; J_0) \to (\mathbb{C}P^n)^5$$

meets the constraints transversally at u. But at the intersection points $X_i \cap P$ we just need to cover directions tangent to P, which is possible by the arguments in Example 7.2.9, while transversality at the three points holds simply by moving these points and using the uniqueness of the conic. The reader is invited to make this argument more explicit by using the techniques of the second proof. □

SECOND PROOF. The second proof is by induction on n. The case $n = 2$ follows from Example 7.2.9. Therefore let $n \geq 3$ and assume, by induction, that the result has been established for $n - 1$. Choose five projective subspaces

$$X_1, X_2, X_3, X_4, X_5 \subset \mathbb{C}P^{n-1}$$

such that X_i has dimension $n-3$ for $i = 1, 2$ and has dimension zero for $i = 3, 4, 5$. The proof of Proposition 7.4.5 shows that the X_i can be chosen such that their product

$$X := X_1 \times \cdots \times X_5 \subset (\mathbb{C}P^{n-1})^5$$

is transverse to the evaluation map $\mathrm{ev}_{T,J_0} : \mathcal{M}^*_{0,T}(\mathbb{C}P^{n-1}, B; J_0) \to (\mathbb{C}P^{n-1})^5$ for every 5-labelled tree T and every homology class $B \in H_2(\mathbb{C}P^{n-1}; \mathbb{Z})$. Since the only decomposition of the class $2L$ is into a pair of intersecting lines, this means that each line through one of the three points X_3, X_4, X_5 and through the two $(n-3)$-planes X_1 and X_2 misses the line through the other two points and that each line through two of the points misses the two $(n-3)$-planes. By the induction hypothesis, there is precisely one conic in $\mathbb{C}P^{n-1}$ passing through $X_1, \ldots, X_5$.

Now define the projective spaces $Y_1, \ldots, Y_5 \subset \mathbb{C}P^n$ by $Y_i := X_i$ for $i = 3, 4, 5$ and, for $i = 1, 2$, let Y_i be the unique $(n-2)$-plane containing X_i and the point $p := [0 :, \cdots : 0 : 1] \in \mathbb{C}P^n \setminus \mathbb{C}P^{n-1}$. We claim that every stable map of degree two in $\mathbb{C}P^n$ passing through the Y_i must be a conic in $\mathbb{C}P^{n-1}$. Firstly, if it is a conic it is the image of a holomorphic map $u = [u_0 : \cdots : u_n]$ of degree two that meets the hyperplane $u_n = 0$ at three points and so must be identically zero. This leaves the case of a pair of intersecting lines. One of these lines must contain two of the three points, say Y_4 and Y_5. The other line must pass through Y_1, Y_2, Y_3. By assumption on the cycles X_i this line cannot be contained in $\mathbb{C}P^{n-1}$. Hence it intersects $\mathbb{C}P^{n-1}$

only in the point Y_3. But this means that the two lines don't intersect and so do not form a stable map. Thus we have proved that there is precisely one stable map of degree two in $\mathbb{C}P^n$ passing through the Y_i and this is a conic in $\mathbb{C}P^{n-1}$.

It remains to prove that the evaluation map

$$\mathrm{ev} : \mathcal{M}(\mathbb{C}P^n, 2L; J_0) \to (\mathbb{C}P^n)^5$$

is transverse to $Y := Y_1 \times \cdots \times Y_5$ at the unique point $[u, \mathbf{z}] \in \mathcal{M}(\mathbb{C}P^n, 2L; J_0)$ such that $u(z_i) \in Y_i$. To see this, note that every degree 2 curve in $\mathbb{C}P^n$ near u has the form $v = [v_0 : \cdots : v_n]$ where $[v_0 : \cdots : v_{n-1}] : \mathbb{C}P^1 \to \mathbb{C}P^{n-1}$ is a conic in $\mathbb{C}P^{n-1}$ close to u and v_n is a quadratic polynomial. This polynomial is uniquely determined by its values at the three marked points z_3, z_4, z_5. This gives precisely the additional three parameters needed to prove transversality to Y in $\mathbb{C}P^n$. It follows that the invariant is one. This proves Proposition 7.4.9. □

Proof of Theorem 7.4.1. We prove, by induction on n, that

$$\mathrm{GW}^{\mathbb{C}P^n}_{dL,3d-1}(c^2, \ldots, c^2, c^n, \ldots, c^n) \geq N_d, \tag{7.4.8}$$

where c^2 occurs $2d - 2$ times and c^n occurs $d + 1$ times. For $n = 2$ this holds by definition of N_d. When $n \geq 3$, one may argue as follows. Since the invariant counts degree d curves through $d + 1$ generic points $p_1, \ldots, p_{d+1}$, each relevant curve lies in the d-plane spanned by $p_1, \ldots, p_{d+1}$. The remaining constraints meet this plane in $2d - 2$ points. By the 2-dimensional result, there are N_d such curves. Hence let $n \geq 3$ and assume, by induction, that the result has been established for $n - 1$. Choose $3d - 1$ projective subspaces

$$X_1, \ldots, X_{3d-1} \subset \mathbb{C}P^{n-1}$$

such that X_i has dimension $n - 3$ for $i = 1, \ldots, 2d - 2$ and has dimension zero for $i = 2d - 1, \ldots, 3d - 1$. The proof of Proposition 7.4.5 shows that the X_i can be chosen such that their product

$$X := X_1 \times \cdots \times X_{3d-1} \subset (\mathbb{C}P^{n-1})^{3d-1}$$

is transverse to the evaluation map $\mathrm{ev}_{T,J_0} : \mathcal{M}^*_{0,T}(\mathbb{C}P^{n-1}, B; J_0) \to (\mathbb{C}P^{n-1})^{3d-1}$ for every tree T and every homology class $B \in H_2(\mathbb{C}P^{n-1}; \mathbb{Z})$. Now the induction hypothesis asserts that there exists a holomorphic sphere $u : \mathbb{C}P^1 \to \mathbb{C}P^{n-1}$ of degree d and a collection of $3d - 1$ distinct points $z_1, \ldots, z_{3d-1} \in \mathbb{C}P^1$ such that

$$u(z_i) \in X_i$$

for every i. In fact, the induction hypothesis asserts that there exist at least N_d such tuples $(u, z_1, \ldots, z_{3d-1})$. Now let $p := [0 : \cdots : 0 : 1] \in \mathbb{C}P^n \setminus \mathbb{C}P^{n-1}$ and let $Y_i \subset \mathbb{C}P^n$ be the projective subspaces defined by

$$Y_i \cap \mathbb{C}P^{n-1} = X_i, \qquad p \in Y_i, \qquad \dim^{\mathbb{C}} Y_i = n - 2, \qquad i = 1, \ldots, 2d - 2,$$

and

$$Y_i := X_i, \qquad i = 2d - 1, \ldots, 3d - 1.$$

We claim that the evaluation map $\mathrm{ev}_{J_0} : \mathcal{M}^*_{0,k}(\mathbb{C}P^n, dL; J_0) \to (\mathbb{C}P^n)^{3d-1}$ is transverse to the product $Y := Y_1 \times \cdots \times Y_{3d-1}$ at the point $[u, \mathbf{z}]$, where here we think of u as a map into $\mathbb{C}P^n$ by identifying $\mathbb{C}P^{n-1}$ with the hyperplane $y_n = 0$ in $\mathbb{C}P^n$. To see this write u in the form

$$u([s : t]) = [u_0(s, t) : \cdots : u_{n-1}(s, t) : 0],$$

where each u_i is a homogeneous polynomial of degree d and the u_i have no common zeros in $\mathbb{C}^2 \setminus \{0\}$. Then every holomorphic curve of degree d near u has the form $v(z) = [v_0(z) : \cdots : v_n(z)]$ where v_i is close to u_i for $i \le n-1$ and v_n is close to zero. Since v_n has degree d it is uniquely determined by its values at the $d+1$ points $z_{2d-1}, \ldots, z_{3d-1}$. Combining this observation with the fact that X is transverse to $\mathrm{ev}_{J_0} : \mathcal{M}^*_{0,k}(\mathbb{C}P^{n-1}, dL; J_0) \to (\mathbb{C}P^{n-1})^{3d-1}$ we find that Y is transverse to the evaluation map $\mathrm{ev}_{J_0} : \mathcal{M}^*_{0,k}(\mathbb{C}P^n, dL; J_0) \to (\mathbb{C}P^n)^{3d-1}$. Hence Theorem 7.4.1 follows from Proposition 7.4.5. □

REMARK 7.4.10. (i) The above proof can be considerably shortened if one is willing to use the fact that any intersection of two holomorphic subvarieties of complementary dimension in $\mathbb{C}P^n$ contributes positively to the intersection number. For then one need not check that the curves intersect the constraints transversally. Hence one can specialize to the case when the $d+1$ point constraints lie in a 2-plane Y. Then, by positivity of intersections, the degree d curve must also lie in Y. Since the other constraints intersect Y in $2d-2$ points, the 2-dimensional result implies that there are N_d curves through these constraints.

(ii) The inductive argument used to prove Theorem 7.4.1 relates the Gromov–Witten invariants of $\mathbb{C}P^n$ to those of $\mathbb{C}P^{n-1}$ (and eventually of $\mathbb{C}P^2$) by projection. Suppose we know that the number of degree d curves in $\mathbb{C}P^{n-1}$ through the constraints

$$p_1, \ldots, p_\ell, Z_1, \ldots, Z_m$$

is N, where the p_i are points and the Z_i are complex planes of suitable dimensions. If $p_0 \in \mathbb{C}P^n$ is any point not on $\mathbb{C}P^{n-1}$, then the obvious projection

$$\pi : \mathbb{C}P^n \setminus \{p\} \to \mathbb{C}P^{n-1}$$

identifies $\mathbb{C}P^n \setminus \{p\}$ with the total space of a line bundle over $\mathbb{C}P^{n-1}$. Its pullback via a degree d map $u : \mathbb{C}P^1 \to \mathbb{C}P^{n-1}$ has degree d. Therefore the space of degree d maps $v : \mathbb{C}P^1 \to \mathbb{C}P^n$ that lift a given u can be identified with the space of sections of a degree d bundle over $\mathbb{C}P^1$. There is one such section through $d+1$ points. Therefore, provided that $\ell > d+1$, there is a bijective correspondence between curves u in $\mathbb{C}P^{n-1}$ going through the constraints $p_1, \ldots, p_\ell, Z_1, \ldots, Z_m$ and curves v in $\mathbb{C}P^n \setminus \{p\}$ that go through $d+1$ points $q_1, \ldots, q_{d+1}$ where $\pi(q_i) = p_i$ as well as meeting $\pi^{-1}(p_j), d+1 < j \le \ell$, and the sets $\pi^{-1}(Z_j), 1 \le j \le m$. Therefore the corresponding Gromov–Witten invariant in $\mathbb{C}P^n$ is at least as large as that in $\mathbb{C}P^{n-1}$.

It may be strictly larger because curves that go through p_0 have been ignored. For example, if one chooses 4 generic points $q_1, \ldots, q_4$ and a set of 4 lines $\ell_1, \ldots, \ell_4$ that all pass through the point $p_0 \in \mathbb{C}P^3 \setminus \mathbb{C}P^2$ then there are cubics through $q_1, \ldots, q_4, p_0$ that might well contribute to the invariant

$$N_3(4,4) := \mathrm{GW}^{\mathbb{C}P^3}_{3L,8}(c^2, c^2, c^2, c^2, c^3, c^3, c^3, c^3),$$

but do not correspond to curves in $\mathbb{C}P^2$. Indeed, we show in Exercise 7.5.17 that $N_3(4,4) = 30 > 12$.

EXERCISE 7.4.11. Prove that $\mathrm{GW}^{\mathbb{C}P^3}_{2L,4}(c^3, c^3, c^3, c^3) = 0$, even though the dimensional condition (7.1.2) is satisfied. *Hint:* Use the fact that every conic lies in a 2-plane by Remark 7.4.6.

There are nonzero Gromov–Witten invariants with $k \neq 3d - 1$. For instance, when $d = 1$ there is a nonzero number of lines through two points and any additional number of hyperplanes. To get a less trivial example, consider the moduli space of cubics in $\mathbb{C}P^3$ that meet five generic points and two lines has dimension zero. To decide if there are any such curves one can consider the special case where four of the points are coplanar. Then there is a 1-parameter family of conics that lie in this plane and go through the four points, and there is a line through the other point and the two skew lines. Therefore there is a reducible cubic curve consisting of the connected union of a line and a conic that go through this set of seven cycles. To show that this curve persists when the cycles are perturbed to general positions is a nontrivial task. One can do this using the technique of gluing that we discuss in Chapter 10. However, one can also use elementary methods as sketched in the next exercise.

EXERCISE 7.4.12. Show that

$$\mathrm{GW}^{\mathbb{C}P^3}_{3L,7}(c^2, c^2, c^3, c^3, c^3, c^3, c^3) > 0.$$

Hint: Fix five distinct points $x_1, \ldots, x_5 \in \mathbb{C}P^3$ such that no three of them lie on a line and no four of them lie in a plane. Prove that for every tuple $\mathbf{z} = (z_1, \ldots, z_5)$ of five distinct points in $\mathbb{C}P^1$ there is a unique holomorphic curve $u_{\mathbf{z}} : \mathbb{C}P^1 \to \mathbb{C}P^3$ of degree three such that

$$u_{\mathbf{z}}(z_i) = x_i, \qquad i = 1, \ldots, 5.$$

Fix the points z_1, z_2, z_3 and show that the differential of the map $(z_4, z_5, z_6, z_7) \mapsto (u_{\mathbf{z}}(z_6), u_{\mathbf{z}}(z_7))$ has rank 4 at some quadruple (z_4, z_5, z_6, z_7). Deduce that this map is transverse to a product $X_6 \times X_7$ of two generic lines such that $u_{\mathbf{z}}(z_6) \in X_6$ and $u_{\mathbf{z}}(z_7) \in X_7$. Hence prove that the submanifold

$$\{(x_1, \ldots, x_5)\} \times X_6 \times X_7 \subset (\mathbb{C}P^3)^7$$

is transverse to the evaluation map $\mathrm{ev}_{J_0} : \mathcal{M}^*_{0,7}(\mathbb{C}P^3, 3L; J_0) \to (\mathbb{C}P^3)^7$. Then use Proposition 7.4.5. (In Exercise 7.5.17 we shall see that the value of the invariant is five.)

7.5. Axioms for Gromov–Witten invariants

Let us return to the heuristic discussion in the beginning of this chapter where the Gromov–Witten invariant for a homology class $A \in H_2(M; \mathbb{Z})$ and an integer $k \geq 3$ is described as the homomorphism

$$\mathrm{GW}^M_{A,k} : H^*(M)^{\otimes k} \otimes H_*(\overline{\mathcal{M}}_{0,k}) \to \mathbb{Q},$$

defined formally by

$$\mathrm{GW}^M_{A,k}(a_1, \ldots, a_k; \beta) := \int_{\overline{\mathcal{M}}_{0,k}(A;J)} \mathrm{ev}_1^* a_1 \smile \cdots \smile \mathrm{ev}_k^* a_k \smile \pi^* \mathrm{PD}(\beta).$$

Recall that $\mathrm{ev}_i : \overline{\mathcal{M}}_{0,k}(A; J) \to M$ is the evaluation map at the ith marked point and

$$\pi : \overline{\mathcal{M}}_{0,k}(A; J) \to \overline{\mathcal{M}}_{0,k}$$

is the forgetful map which assigns to every stable map the underlying reduced stable curve. Integration over $\overline{\mathcal{M}}_{0,k}(A; J)$ is to be understood as evaluation of a cohomology class on the *virtual fundamental cycle* of the moduli space. In the generality discussed here it is important to allow for rational values of the Gromov–Witten invariants, even when evaluating them on integral (co)homology classes. However, if (M, ω) is semipositive then the genus zero invariants take integral values on integral homology classes. Rational values appear in the higher genus case, because the moduli space $\overline{\mathcal{M}}_{g,k}$ of curves is only an orbifold for $g > 0$, and in the nonsemipositive case because then multiply covered curves are relevant for the counting invariants defined in terms of the virtual moduli cycle While we do not set up the theory in this form, it is interesting to examine the formal properties of the Gromov–Witten invariants in this formulation. We prove in Appendix D that the Grothendieck–Knudsen moduli space $\overline{\mathcal{M}}_{0,k}$ of stable genus zero curves is a smooth manifold. For each stable labelled tree $T = (T, E, \Lambda)$ the closure of the set of stable curves modelled on T forms a smooth submanifold $\overline{\mathcal{M}}_{0,T}$ and, by a result of Keel [**208**], the collection of all such submanifolds generates the homology of $\overline{\mathcal{M}}_{0,k}$. (See Theorem D.4.7 and Theorem D.6.4. The submanifolds V_S mentioned there have the form $\overline{\mathcal{M}}_{0,T}$ by part (iv) of Theorem D.4.7.)

As pointed out in the introduction to this chapter, the invariant $\mathrm{GW}^M_{A,k}$ should be understood geometrically as counting the algebraic number of stable maps in M with k marked points that represent the class A, pass at the marked points through cycles $X_i \subset M$ Poincaré dual to the classes a_i, and such that the reduced stable curve belongs to a cycle $Y \subset \overline{\mathcal{M}}_{0,k}$ representing the class β. In Sections 7.1 and 7.3 we have given rigorous definitions of these invariants under the assumption that (M, ω) is semipositive and that the class β is represented either by the submanifold $\overline{\mathcal{M}}_{0,T} \subset \overline{\mathcal{M}}_{0,k}$ associated to a k-labelled tree T or by the submanifold $Y_{k,I}$ obtained by fixing the marked points $\mathbf{w} := \{\mathrm{w}_i\}_{i \in I}$.

In [**216**] Kontsevich and Manin listed the following axioms for the Gromov–Witten invariants. They considered invariants for all genera. Here we specialize to the genus zero case. We shall also suppose that $k \geq 3$. (See Remark 7.5.2 for the case $0 \leq k \leq 2$.) As in Theorem D.4.2, we identify $\overline{\mathcal{M}}_{0,k}$ with its image $\overline{M}_{0,k}$ in $(S^2)^N$ using the cross ratio functions $\mathbf{z} \mapsto w_{ijk\ell}(\mathbf{z})$, and hence will denote its elements by tuples $\{w_{ijk\ell}\}$.

(Effective) If $\omega(A) < 0$ then $\mathrm{GW}^M_{A,k} = 0$.

(Symmetry) For each permutation $\sigma \in S_k$

$$\mathrm{GW}^M_{A,k}(a_{\sigma(1)}, \ldots, a_{\sigma(k)}; \sigma_*\beta) = \varepsilon(\sigma; a)\, \mathrm{GW}^M_{A,k}(a_1, \ldots, a_k; \beta).$$

Here

$$\varepsilon(\sigma; a) := (-1)^{\#\{i<j \,|\, \sigma(i)>\sigma(j),\, \deg(a_i)\deg(a_j) \in 2\mathbb{Z}+1\}}$$

denotes the sign of the induced permutation on the classes of odd degree and $\sigma_*\beta$ denotes the pushforward of a homology class $\beta \in H_*(\overline{\mathcal{M}}_{0,k})$ under the diffeomorphism of $\overline{\mathcal{M}}_{0,k}$ given by

$$\{w_{i_0 i_1 i_2 i_3}\} \mapsto \sigma_*\{w_{i_0 i_1 i_2 i_3}\} := \{w_{\sigma(i_0)\sigma(i_1)\sigma(i_2)\sigma(i_3)}\}.$$

(Grading) If $\mathrm{GW}^M_{A,k}(a_1, \ldots, a_k; \beta) \neq 0$ then

$$\sum_{i=1}^{k} \deg(a_i) - \deg(\beta) = 2n + 2c_1(A).$$

(HOMOLOGY) For every $A \in H_2(M;\mathbb{Z})$ and every integer $k \geq 3$ there exists a homology class

$$\sigma_{A,k} \in H_{2n+2c_1(A)+2k-6}(M^k \times \overline{\mathcal{M}}_{0,k})$$

such that

$$\mathrm{GW}^M_{A,k}(a_1,\dots,a_k;\beta) = \langle \pi_1^* a_1 \smile \cdots \smile \pi_k^* a_k \smile \pi_0^* \mathrm{PD}(\beta), \sigma_{A,k}\rangle,$$

where $\pi_i : M^k \times \overline{\mathcal{M}}_{0,k} \to M$ denotes the projection onto the ith factor and the map $\pi_0 : M^k \times \overline{\mathcal{M}}_{0,k} \to \overline{\mathcal{M}}_{0,k}$ denotes the projection onto the last factor.

(FUNDAMENTAL CLASS) Let

$$\pi_{0,k} : \overline{\mathcal{M}}_{0,k} \to \overline{\mathcal{M}}_{0,k-1}$$

denote the map which forgets the last marked point and let $1 := \mathrm{PD}([M]) \in H^0(M)$. If $(A,k) \neq (0,3)$ then

$$\mathrm{GW}^M_{A,k}(a_1,\dots,a_{k-1},1;\beta) = \mathrm{GW}^M_{A,k-1}(a_1,\dots,a_{k-1};(\pi_{0,k})_*\beta),$$

In particular, this invariant vanishes if $\beta = [\overline{\mathcal{M}}_{0,k}]$.

(DIVISOR) If $(A,k) \neq (0,3)$ and $\deg(a_k) = 2$ then

$$\begin{aligned}\mathrm{GW}^M_{A,k}&(a_1,\dots,a_k;\mathrm{PD}(\pi_{0,k}^*\mathrm{PD}(\beta))) \\ &= \mathrm{GW}^M_{A,k-1}(a_1,\dots,a_{k-1};\beta)\int_A a_k.\end{aligned}$$

(ZERO) If $A = 0$ then $\mathrm{GW}^M_{0,k}(a_1,\dots,a_k;\beta) = 0$ whenever $\deg(\beta) > 0$, and

$$\mathrm{GW}^M_{0,k}(a_1,\dots,a_k;[\mathrm{pt}]) = \int_M a_1 \smile \cdots \smile a_k.$$

(SPLITTING) Fix a basis $e_0,\dots,e_N$ of the cohomology $H^*(M)$, let

$$g_{\nu\mu} := \int_M e_\nu \smile e_\mu,$$

and denote by $g^{\nu\mu}$ the inverse matrix. Moreover, fix a partition of the index set $\{1,\dots,k\} = S_0 \cup S_1$ such that $k_i := \#S_i \geq 2$ for $i = 0,1$ and denote by

$$\phi_S : \overline{\mathcal{M}}_{0,k_0+1} \times \overline{\mathcal{M}}_{0,k_1+1} \to \overline{\mathcal{M}}_{0,k}$$

the canonical map which identifies the last marked point of a stable curve in $\overline{\mathcal{M}}_{0,k_0+1}$ with the first marked point of a stable curve in $\overline{\mathcal{M}}_{0,k_1+1}$. The remaining indices have the unique ordering such that the relative order is preserved, the first k_0 points in $\mathcal{M}_{0,k_0+1}$ are mapped to the points indexed by S_0, and the last k_1 points in $\mathcal{M}_{0,k_1+1}$ are mapped to the points indexed by S_1. Fix two homology classes $\beta_0 \in H_*(\overline{\mathcal{M}}_{0,k_0+1})$ and $\beta_1 \in H_*(\overline{\mathcal{M}}_{0,k_1+1})$. Then

$$\begin{aligned}\mathrm{GW}^M_{A,k}(a_1,\dots,a_k;\phi_{S*}(\beta_0\otimes\beta_1)) &= \varepsilon(S,a)\sum_{A_0+A_1=A} \\ \mathrm{GW}^M_{A_0,k_0+1}(\{a_i\}_{i\in S_0},e_\nu;\beta_0)\,&g^{\nu\mu}\,\mathrm{GW}^M_{A_1,k_1+1}(e_\mu,\{a_j\}_{j\in S_1};\beta_1),\end{aligned}$$

where we use the double summation convention and define

$$\varepsilon(S,a) := (-1)^{\#\{j<i\,|\,i\in S_0,j\in S_1,\,\deg(a_i)\deg(a_j)\in 2\mathbb{Z}+1\}}.$$

REMARK 7.5.1. (i) The *(Effective)* axiom follows from the fact that the invariant is based on the moduli space $\overline{\mathcal{M}}_{0,k}(A;J)$ of J-holomorphic stable maps, representing the class A. If $\omega(A) < 0$ then the energy identity shows that this moduli space is empty.

(ii) The *(Symmetry)* axiom follows from the fact that the symmetric group S_k acts in the obvious way on the moduli space $\overline{\mathcal{M}}_{0,k}(A;J)$ by permuting the marked points and that the evaluation map $\mathrm{ev} : \overline{\mathcal{M}}_{0,k}(A;J) \to M^k$ and the projection $\pi : \overline{\mathcal{M}}_{0,k}(A;J) \to \overline{\mathcal{M}}_{0,k}$ are equivariant under this action.

(iii) The *(Grading)* axiom follows from the dimension formula for the moduli space $\overline{\mathcal{M}}_{0,k}(A;J)$. It also follows from the *(Homology)* axiom.

(iv) The homology class $\sigma_{A,k} \in H_*(M^k \times \overline{\mathcal{M}}_{0,k})$ in the *(Homology)* axiom is to be understood as the pushforward of the *virtual fundamental class*

$$[\overline{\mathcal{M}}_{0,k}(A;J)]^{\mathrm{virt}} \in H_*(\overline{\mathcal{M}}_{0,k}(A;J))$$

under the product map

$$\mathrm{ev}_J \times \pi : \overline{\mathcal{M}}_{0,k}(A;J) \to M^k \times \overline{\mathcal{M}}_{0,k}.$$

In the context of this book this axiom remains a heuristic statement because $\mathrm{ev}_J \times \pi$ is a pseudocycle only under very restrictive hypotheses: cf. the discussion preceding Exercise 6.7.13. As mentioned there, to establish the axiom in general one needs analytic techniques that go beyond the scope of this book.

However, if (M,ω) is semipositive, then the pushforward of the class $\sigma_{A,k}$ under the projection $M^k \times \overline{\mathcal{M}}_{0,k} \to M^k$ has a rigorous meaning in the present context, namely, as the homology class

$$\tau_{A,k} \in H_{2n+2c_1(A)+2k-6}(M^k)$$

represented by the Gromov–Witten pseudocycle

$$\mathrm{ev}_J : \mathcal{M}^*_{0,k}(A;J) \to M^k$$

of Theorem 7.1.1. Further, if T is a k-labelled tree then the pairing (or slant product) of $\sigma_{A,k}$ with the cohomology class $\mathrm{PD}([\overline{\mathcal{M}}_{0,T}]) \in H^{2e(T)}(\overline{\mathcal{M}}_{0,k})$ can be interpreted as the homology class

$$\tau_{A,T} \in H_{2n+2c_1(A)+2k-6-2e(T)}(M^k)$$

represented by the pseudocycle

$$\mathrm{ev}_{T,J} : \mathcal{M}^*_{0,T}(A;J) \to M^k$$

discussed in Exercise 6.6.3 and Section 7.2. Similarly, when $\#I \geq 3$ the class

$$[Y_{k,I}] \in H_{2(k-\#I)}(\overline{\mathcal{M}}_{0,k})$$

represented by a submanifold on which the marked points $\mathbf{w} = \{\mathrm{w}_i\}_{i\in I}$ are fixed has Poincaré dual equal to

$$\mathrm{PD}([Y_{k,I}]) = \pi^*_{k,I}\mathrm{PD}([\mathrm{pt}]) \in H^{2\#I-6}(\overline{\mathcal{M}}_{0,k}),$$

where $\pi_{k,I} : \overline{\mathcal{M}}_{0,k} \to \overline{\mathcal{M}}_{0,I}$ is the forgetful map. Hence, if there were such a class as $\sigma_{A,k}$, its pairing with the cohomology class $\pi^*_{k,I}\mathrm{PD}([\mathrm{pt}]) \in H^{2\#I-6}(\overline{\mathcal{M}}_{0,k})$ would push forward to the class

$$\tau_{A,I} \in H_{2n+2c_1(A)+2(k-\#I)}(M^k)$$

represented by the pseudocycle $\mathrm{ev}_{\mathbf{w},J} : \mathcal{M}_{0,k}(A,\mathbf{w};J) \to M^k$ of Theorem 7.3.1.

Note that we have shown that this pseudocycle is independent of the choice of $\mathbf{w}$ only when this point lies in the top stratum of $\overline{\mathcal{M}}_{0,I}$. In other words, because we cannot assume the existence of the class $\sigma_{A,k}$, our definition of $\tau_{A,I}$ might in principle depend on the choice of representative chosen for the class $[Y_{k,I}]$. As we shall see, the *(Splitting)* axiom implies that we can allow not completely arbitrary representatives of $[Y_{k,I}]$ but at least an arbitrary choice of point $\mathbf{w} \in \overline{\mathcal{M}}_{0,I}$.

(v) In the semipositive case the *(Fundamental class)*, *(Divisor)*, and *(Zero)* axioms are easy consequences of the definition of the Gromov–Witten invariants in Sections 7.1 and 7.3 (see Examples 7.1.3 and 7.3.5 and Exercises 7.1.5 and 7.1.6). We reformulate these axioms for the invariants GW and GW^I below.

(vi) The most important property of the Gromov–Witten invariants is the *(Splitting)* axiom. Its name arises from the observation that it can be interpreted as a decomposition formula for the homology class $\sigma_{A,k}$ associated to a splitting of the index set $\{1, \dots, k\}$. That the sum in this axiom is finite follows from the *(Effective)* and *(Zero)* axioms. Its proof uses the gluing theorem for J-holomorphic curves; we shall discuss it in more detail below and in Chapters 10 and 11. For now, note that it gives a formula not for the "usual" Gromov–Witten invariant in which one counts all A-curves but for the restricted invariant in which the domain lies in a boundary stratum of the form $\beta = \phi_{S*}(\beta_0 \otimes \beta_1)$ in $\overline{\mathcal{M}}_{0,k}$. The simplest case is when $k = 4$ and β_0, β_1 both equal $[\mathrm{pt}] = [\overline{\mathcal{M}}_{0,3}]$. Then $\phi_{S*}(\beta_0 \otimes \beta_1) = [\mathrm{pt}] \in H_0(\overline{\mathcal{M}}_{0,4})$ so that

$$\mathrm{GW}^M_{A,4}(a_1, \dots, a_4; \phi_{S*}(\beta_0 \otimes \beta_1)) = \mathrm{GW}^{M,\{1,2,3,4\}}_{A,4}(a_1, \dots, a_4),$$

the 4-point invariant with fixed cross ratio. The *(Splitting)* axiom shows that this reduces to a product of 3-point invariants. Moreover, as explained in Exercise 7.2.8 this product is precisely the invariant $\mathrm{GW}^M_{A,T}(a_1, \dots, a_4)$ corresponding to the stratum T given by the splitting $\{1, 2, 3, 4\} = S_0 \cup S_1$. In this simple case, the stratum T is a single point and so the product can also be thought of as defined by the pseudocycle $\mathrm{ev}_{\mathbf{w}}$ where $\mathbf{w}$ is not in the top stratum of $\overline{\mathcal{M}}_{0,4}$. It is in this sense that the *(Splitting)* axiom is relevant to the discussion in (iv) above about the *(Homology)* axiom. (Also, compare the remarks just before Exercise 6.7.13.)

REMARK 7.5.2. Later on it will sometimes be useful to consider the k-pointed Gromov–Witten invariants $\mathrm{GW}^M_{A,k}(a_1, \dots, a_k; \beta)$ with $0 \le k \le 2$. Since there is no stable curve of genus zero with less than three marked points, the space $\overline{\mathcal{M}}_{0,k}$ is empty in this case, and so these invariants do not fit into the above framework. However, if we interpret $\overline{\mathcal{M}}_{0,k}$ formally as a *space of negative dimension* $2k-6$ with fundamental cycle $\beta_k \in H_{2k-6}(\overline{\mathcal{M}}_{0,k})$, and define $\mathrm{GW}^M_{A,k}(a_1, \dots, a_k; \beta_k)$ to be the invariant $\mathrm{GW}^M_{A,k}(a_1, \dots, a_k)$ of Remark 7.3.8, then the *(Grading)* axiom holds by definition. Moreover, for $k \ge 4$ we have $\beta_k = \mathrm{PD}(\pi^*_{0,k}\mathrm{PD}(\beta_{k-1}))$. If we formally extend this identity to $k \le 3$, then the *(Divisor)* axiom reduces to the statement in Exercise 7.1.5 and so holds, again by definition. Now consider the *(Fundamental class)* axiom. It is easy to check directly that $\mathrm{GW}^M_{A,k}(a_1, \dots, a_{k-1}, 1; \beta_k) = 0$ if $A \ne 0$ and $k \le 3$. We claim further that

$$\mathrm{GW}^M_{0,k}(a_1, \dots, a_k; \beta_k) = 0, \qquad k = 0, 1, 2.$$

This holds for dimensional reasons. The invariant should count the number of constant spheres meeting cycles Poincaré dual to the classes $a_1, \dots, a_k$. The *(Grading)* axiom implies that

$$\sum_i \deg(a_i) = \dim M + 2k - 6 < \dim M.$$

Hence these cycles do not intersect, and the invariant vanishes. (Compare Example 7.1.3 where we reached a similar conclusion from another point of view.) This extends the *(Fundamental class)* as well as the *(Zero)* axiom. The *(Effective)* and *(Symmetry)* axioms still hold, while the *(Homology)* and *(Splitting)* axioms do not apply. The next exercise shows that some of these additional invariants are nonzero.

EXERCISE 7.5.3. Denote by L the class in $\mathbb{C}P^n$ represented by a line, and by c the element of $H^2(\mathbb{C}P^n)$ such that $c(L) = 1$. Check that

$$\mathrm{GW}^{\mathbb{C}P^1}_{L,0}(\beta_0) = \mathrm{GW}^{\mathbb{C}P^1}_{L,1}(c;\beta_1) = 1, \qquad \mathrm{GW}^{\mathbb{C}P^n}_{L,2}(c^n, c^n;\beta_2) = 1$$

for every $n \geq 1$, and that $\mathrm{GW}^{\mathbb{C}P^n}_{L,k} = 0$ for $k = 0, 1$ and $n \geq 2$.

Gromov–Witten invariants as integrals. The formulation of the Gromov–Witten invariants as integrals can be made rigorous in the semipositive case provided that we represent the classes a_i by differential forms that are supported in sufficiently small neighbourhoods of dual cycles that are transverse to the evaluation map on the moduli space $\mathcal{M}^*_{0,k}(A;J)$. Then the pullback of the product $a_1 \wedge \cdots \wedge a_k$ under the evaluation map has compact support, so that the integral has meaning. To prove that the integral is independent of the choice of the form one would have to show that the pullback of any exact form integrates to zero, i.e.

$$\int_{\mathcal{M}^*_{0,k}(A;J)} \mathrm{ev}^* d\sigma = 0.$$

Intuitively, this should be the case because the boundary of $\mathcal{M}^*_{0,k}(A;J)$ has codimension two and so the integral of σ over it should vanish. However, to make this precise is a nontrivial problem. In the case of intersection theory this problem can be avoided by using Sard's theorem.

The integral formulation of the invariants highlights the similarities between the Gromov–Witten invariants and other gauge theoretic invariants (see for example Ruan [**342**], Givental [**151**], and Manin [**286**]).

Algebraic reformulations. In [**216**] Kontsevich and Manin describe the Gromov–Witten invariants as a homomorphism

$$I_{A,k} : H^*(M)^{\otimes k} \to H^*(\overline{\mathcal{M}}_{0,k})$$

for $k \geq 2$. The invariant $\mathrm{GW}_{A,k}$ is then obtained by pairing the homomorphism $I_{A,k}$ with homology classes in $H_*(\overline{\mathcal{M}}_{0,k})$:

$$\mathrm{GW}_{A,k}(a_1, \ldots, a_k;\beta) = \langle I_{A,k}(a_1, \ldots, a_k), \beta \rangle.$$

The gluing rules can then be thought of as giving $H^*(M)$ the structure of a *Cohomological Field Theory* (see [**286**, Definition 0.3.1]). A second reformulation emphasizes the action of the homology of $\overline{\mathcal{M}}_{0,k+1}$ on the cohomology of M and is obtained by partially dualizing the homomorphism $I_{A,k}$. We get multilinear maps

$$m_{A,k} : H_*(\overline{\mathcal{M}}_{0,k+1}) \otimes H^*(M)^{\otimes k} \to H^*(M)$$

defined by

$$\langle m_{A,k}(\beta; a_1, \ldots, a_k), \mathrm{PD}(a_{k+1}) \rangle := \langle \beta, I_{A,k+1}(a_1, \ldots, a_{k+1}) \rangle.$$

In particular, if $k = 2$ then $\overline{\mathcal{M}}_{0,3}$ is a point, and taking $\beta = [\overline{\mathcal{M}}_{0,3}]$ gives a bilinear map

$$m_A : H^*(M) \otimes H^*(M) \to H^*(M).$$

One consequence of the *(Splitting)* axiom is that the maps m_A, when summed over A, define an associative multiplication on $H^*(M)$ which is often called the *small quantum multiplication.*[3] The associativity of this multiplication was conjectured by physicists; in fact it is only the first of the set of equations that they thought the maps $m_{A,k}$ should satisfy. The proof of the easiest case (with $\beta = [\overline{\mathcal{M}}_{0,3}]$ and semipositive (M, ω)) was first given by Ruan–Tian in [**346**]. The case of other classes β and higher genus was dealt with in [**348**]. However, the case of general (M, ω) had to wait until the virtual moduli cycle was understood. We will give a much more detailed description of quantum multiplication in Chapter 11 and will prove its associativity (in the semipositive case) in Chapter 10.

REMARK 7.5.4. There are natural cohomology classes on $\overline{\mathcal{M}}_{0,k}$, which interact with the Gromov-Witten invariants in interesting ways. These are the first Chern classes of the canonical line bundles

$$\mathcal{L}_i \to \overline{\mathcal{M}}_{0,k}, \qquad i = 1, \ldots, k.$$

The fiber of $\mathcal{L}_i$ at $\mathbf{w}$ is the cotangent space of the stable curve associated to $\mathbf{w}$ at the ith marked point. Equivalently, $\mathcal{L}_i$ is the conormal bundle of $\overline{\mathcal{M}}_{0,k}$ as a submanifold of $\overline{\mathcal{M}}_{0,k+1}$, where the embedding is determined by the ith marked point. The first Chern classes of the $\mathcal{L}_i$ give rise to the so-called *gravitational ancestors.* The *gravitational descendants* arise from similar line bundles on the moduli space of stable maps. For further information, see Coates [**72**], Givental [**151**] and Manin [**286**].

The axioms for GW **and** GW^I**.** We shall now take a closer look at the axioms, reformulating them for the invariants GW and GW^I defined in Sections 7.1 and 7.3. If one could show that Gromov–Witten invariants did exist with all the properties discussed above then these invariants GW and GW^I would have the following interpretations:

$$\mathrm{GW}^M_{A,k}(a_1, \ldots, a_k) =: \mathrm{GW}^M_{A,k}(a_1, \ldots, a_k; [\overline{\mathcal{M}}_{0,k}]),$$

while, for index sets $I \subset \{1, \ldots, k\}$ such that $\#I \geq 3$,

$$\mathrm{GW}^{M,I}_{A,k}(a_1, \ldots, a_k) =: \mathrm{GW}^M_{A,k}(a_1, \ldots, a_k; \beta_{k,I}).$$

Here the homology class $\beta_{k,I} \in H_*(\overline{\mathcal{M}}_{0,k})$ is defined by

$$\beta_{k,I} = \mathrm{PD}(\pi^*_{k,I}\mathrm{PD}([\mathrm{pt}]) \in H_{2(k-\#I)}(\overline{\mathcal{M}}_{0,k}),$$

where

$$\pi_{k,I} : \overline{\mathcal{M}}_{0,k} \to \overline{\mathcal{M}}_{0,I}$$

is the map which forgets the marked points not belonging to the index set I. This means that $\beta_{k,I}$ can be represented by a cycle in $\overline{\mathcal{M}}_{0,k}$ obtained by fixing the marked points $\mathbf{w} := \{\mathrm{w}_i\}_{i \in I}$. Indeed, we show in Lemma 7.5.5 below that it is represented by the inverse image $\pi^{-1}_{k,I}(\mathbf{w})$ of any regular value $\mathbf{w} \in \overline{\mathcal{M}}_{0,I}$ of $\pi_{k,I}$. Note that $\beta_{k,I}$ is the fundamental cycle of $\overline{\mathcal{M}}_{0,k}$ whenever $\#I = 3$.

[3] To define this properly one needs to extend the coefficients of $H = H^*(M)$ by the Novikov ring to keep track of A (see Chapter 11).

We now show that GW and GW^I do have the expected properties. The proof is based on the properties of the homology classes $\beta_{k,I}$. Often, as in Section 7.3 one chooses $\mathbf{w}$ to belong to the open stratum $\mathcal{M}_{0,I}$ of $\overline{\mathcal{M}}_{0,I}$, i.e. one takes the $\mathrm{w}_i, i \in I$, to be pairwise distinct points on S^2. However, it will be convenient for later arguments to make a more canonical choice in which the w_i are as separated as possible, in other words we choose $\mathbf{w}$ to be modelled on a tree with the maximum possible number of components. To this end, write

$$I =: \{i_0, \dots, i_\ell\}$$

where $\ell := \#I - 1$ and let T be the tree with $\ell - 1$ vertices $\alpha_1, \dots, \alpha_{\ell-1}$ such that $\alpha_j E \alpha_{j+1}$ for $j = 1, \dots, \ell-2$. We define $\mathbf{w}_I \in \overline{\mathcal{M}}_{0,I}$ to be the unique element which corresponds to the I-labelled tree (T, Λ, E) where

$$i_0, i_1 \in \Lambda_{\alpha_1}, \qquad i_j \in \Lambda_{\alpha_j}, \qquad i_{\ell-1}, i_\ell \in \Lambda_{\alpha_{\ell-1}}. \tag{7.5.1}$$

Thus $\mathbf{w}_I$ has components $w_{i_{j_0} i_{j_1} j_{j_2} i_{j_3}} = \infty$ for $0 \le j_0 < j_1 < j_2 < j_3 \le \ell$.

LEMMA 7.5.5. *Let $I \subset \{1, \dots, k\}$ be such that $\#I \ge 3$.*

(i) *The class $\beta_{k,I}$ may be represented by the cycle $Y_{k,I} := \pi_{k,I}^{-1}(\mathbf{w}_I)$.*

(ii) *Denote by $\pi_{0,k} : \overline{\mathcal{M}}_{0,k} \to \overline{\mathcal{M}}_{0,k-1}$ the projection which forgets the last marked point. Then $\beta_{k,I} = \mathrm{PD}(\pi_{0,k}^* \mathrm{PD}(\beta_{k-1,I}))$ when $k \notin I$. Moreover*

$$(\pi_{0,k})_* \beta_{k,I} = \begin{cases} \beta_{k-1, I \setminus \{k\}} & k \in I, \ \#I \ge 4, \\ 0 & k \notin I. \end{cases}$$

PROOF. Let $f : X \to Y$ be a smooth map between closed oriented manifolds. Then one can use Poincaré duality to define a homology pullback (or "upper shriek")

$$f^! : H_*(Y) \to H_*(X), \qquad f^!(a) := \mathrm{PD}_X f^* \mathrm{PD}_Y(a).$$

Geometrically, if a homology class in $H_*(Y)$ is represented by an oriented submanifold $Z \subset Y$ that is transverse to f, then its image under $f^!$ is represented by the oriented submanifold $f^{-1}(Z) \subset X$. The homology pullback is functorial in the sense that $(f \circ g)^! = g^! \circ f^!$. In this language $\beta_{k,I} = \pi_{k,I}^!([\mathrm{pt}])$.

To prove (i), we need to check that $\mathbf{w}_I$ is a regular value of $\pi_{k,I}$. For this it suffices to show that for each point $\mathbf{z} \in Y_{k,I}$ there are local coordinates near $\mathbf{w}_I$ in $\overline{\mathcal{M}}_{0,I}$ that extend to a system of local coordinates near $\mathbf{z}$. But each such point $\mathbf{z}$ lies in a submanifold $\overline{\mathcal{M}}_{0,T,\Lambda}$ where Λ is a k-labelling of T that satisfies (7.5.1). Hence the assertion follows from the proof of Theorem D.4.2. Near $\mathbf{w}_I$ the coordinates on $\overline{\mathcal{M}}_{0,I}$ correspond to the edges in the quotient tree for the index set I. They can be extended to a coordinate system on $\overline{\mathcal{M}}_{0,k}$ near $\mathbf{z}$ by choosing coordinates on $\overline{\mathcal{M}}_{0,T,\Lambda}$ according to the recipe in the proof of Theorem D.4.2. This proves (i).

To prove (ii), assume first that $k \notin I$. Then $\pi_{k,I} = \pi_{k-1,I} \circ \pi_{0,k}$ and so the identity $\beta_{k,I} = \mathrm{PD}(\pi_{0,k}^* \mathrm{PD}(\beta_{k-1,I}))$ follows from the functoriality of homology pullback. Now consider the pushforward $(\pi_{0,k})_* \beta_{k,I}$. Clearly, $\pi_{0,k}(Y_{k,I}) = Y_{k-1, I \setminus \{k\}}$, where $Y_{k,I} := \pi_{k,I}^{-1}(\mathbf{w}_I)$. If $k \in I$ and $\#I \ge 4$ then both cycles have the same dimension $2(k - 2\#I)$ and $\pi_{0,k}$ restricts to a holomorphic diffeomorphism from $Y_{k,I}$ to $Y_{k-1, I \setminus \{k\}}$. This proves the second assertion. The last assertion follows from the fact that, if $k \notin I$ then $\pi_{0,k}$ maps $Y_{k,I}$ onto a manifold of lower dimension. This proves Lemma 7.5.5. □

Here is a restatement of the *(Divisor)*, *(Fundamental class)*, and *(Zero)* axioms for the invariant GW^M.

PROPOSITION 7.5.6. *Let (M,ω) be a compact semipositive symplectic manifold, $A \in H_2(M;\mathbb{Z})$, $k \geq 1$, and $a_1,\dots,a_k \in H^*(M)$. Then the following holds.*

(DIVISOR) *If $(A,k) \neq (0,3)$ and $\deg(a_k) = 2$ then*

$$\mathrm{GW}^M_{A,k}(a_1,\dots,a_k) = \mathrm{GW}^M_{A,k-1}(a_1,\dots,a_{k-1}) \cdot \int_A a_k.$$

(FUNDAMENTAL CLASS) *If $(A,k) \neq (0,3)$ then*

$$\mathrm{GW}^M_{A,k}(a_1,\dots,a_{k-1},1) = 0.$$

(ZERO) *If $k \neq 3$ then $\mathrm{GW}^M_{0,k} = 0$. If $k = 3$ then*

$$\mathrm{GW}^M_{0,3}(a_1,a_2,a_3) = \int_M a_1 \smile a_2 \smile a_3.$$

PROOF. Exercise. For the *(Zero)* axiom a detailed discussion is contained in Examples 7.1.3 and 7.3.5. The geometric intuition behind the other axioms is explained in Exercises 7.1.6 and 7.1.5. □

Here are the corresponding results for $\mathrm{GW}^{M,I}$. Note that the fundamental class axiom is now more subtle, reflecting the structure of the axiom for the full set of invariants $\mathrm{GW}^M_{A,k}(a_1,\dots,a_k;\beta)$.

PROPOSITION 7.5.7. *Let (M,ω) be a compact semipositive symplectic manifold, $A \in H_2(M;\mathbb{Z})$, $k \geq 3$, $I \subset \{1,\dots,k\}$ such that $\#I \geq 3$, and $a_1,\dots,a_k \in H^*(M)$. Then the following holds.*

(DIVISOR) *If $k \notin I$ and $\deg(a_k) = 2$ then*

$$\mathrm{GW}^{M,I}_{A,k}(a_1,\dots,a_k) = \mathrm{GW}^{M,I}_{A,k-1}(a_1,\dots,a_{k-1}) \cdot \int_A a_k.$$

(FUNDAMENTAL CLASS) *If $k \in I$ and $\#I \geq 4$ then*

$$\mathrm{GW}^{M,I}_{A,k}(a_1,\dots,a_{k-1},1) = \mathrm{GW}^{M,I\setminus\{k\}}_{A,k-1}(a_1,\dots,a_{k-1}).$$

If $k \notin I$ then $\mathrm{GW}^{M,I}_{A,k}(a_1,\dots,a_{k-1},1) = 0$.

(ZERO) *If $\#I < k$ then $\mathrm{GW}^{M,I}_{0,k} = 0$. If $\#I = k$ then*

$$\mathrm{GW}^{M,\{1,\dots,k\}}_{0,k}(a_1,\dots,a_k) = \int_M a_1 \smile \cdots \smile a_k.$$

PROOF. Exercise. □

EXERCISE 7.5.8. Formulate and prove the *(Symmetry)* axiom for the invariants GW and GW^I. Because the permutation group S_k acts on $\mathcal{M}_{0,k}(A;J)$ by diffeomorphisms that are isotopic to the identity, the sign in the formula appears from the action of S_k on the cohomology of the product M^k.

Next we examine the product map

$$\phi_S : \overline{\mathcal{M}}_{0,k_0+1} \times \overline{\mathcal{M}}_{0,k_1+1} \to \overline{\mathcal{M}}_{0,k}$$

which appears in the *(Splitting)* axiom. Intuitively this joins a stable curve in $\overline{\mathcal{M}}_{0,k_0+1}$ to one in $\overline{\mathcal{M}}_{0,k_1+1}$ by identifying the last marked point in the first curve with the first one in the second. The resulting curve has a nodal point at the join, and hence only $k_0 + k_1$ marked points. We now show that in the situation where some of the marked points are fixed we must assume that the two points that will be glued are also fixed: otherwise the relationship between the fixed points on the first curve and those on the second could vary. This is the content of the definition given below of the subsets $I_{Si}, i = 1, 2$.

Let $I_0, I_1 \subset \{1, \dots, k\}$ be disjoint sets such that

$$\#I_0 \geq 2, \qquad \#I_1 \geq 2$$

and denote their union by

$$I := I_0 \cup I_1.$$

Let $\mathcal{S}(I_0, I_1)$ denote the set of all splittings $S = (S_0, S_1)$ of the index set $\{1, \dots, k\}$ such that $I_0 \subset S_0$ and $I_1 \subset S_1$. Given a splitting

$$S = (S_0, S_1) \in \mathcal{S}(I_0, I_1)$$

let $k_i := \#S_i$ and let

$$\sigma_0 : \{1, \dots, k_0\} \to \{1, \dots, k\}, \qquad \sigma_1 : \{2, \dots, k_1 + 1\} \to \{1, \dots, k\}$$

be the unique order preserving injections such that $\operatorname{im} \sigma_0 = S_0$ and $\operatorname{im} \sigma_1 = S_1$. Define the subsets $I_{Si} \subset \{1, \dots, k_i + 1\}$ by

$$I_{S0} := \sigma_0^{-1}(I) \cup \{k_0 + 1\}, \qquad I_{S1} := \{1\} \cup \sigma_1^{-1}(I).$$

LEMMA 7.5.9 (Keel).

$$\beta_{k,I} = \sum_{S \in \mathcal{S}(I_0, I_1)} \phi_{S*}\left(\beta_{k_0+1, I_{S0}} \otimes \beta_{k_1+1, I_{S1}}\right).$$

PROOF. We begin with an explicit description of the map ϕ_S. It is given by

$$\phi_S(\mathbf{w}^0, \mathbf{w}^1) := \mathbf{w}^S,$$

where $\mathbf{w}^S$ is defined by the following conditions: $w^S_{ii'jj'} := \infty$ whenever $i, i' \in S_0$ and $j, j' \in S_1$;

$$w^S_{\sigma_0(i)\sigma_0(i')\sigma_0(i'')\sigma_1(j)} := w^0_{ii'i''k_0+1}; \qquad w^S_{\sigma_0(i)\sigma_1(j)\sigma_1(j')\sigma_1(j'')} := w^1_{1jj'j''}$$

for $i, i', i'' \in \{1, \dots, k_0\}$ and $j, j', j'' \in \{2, \dots, k_1 + 1\}$; and

$$w^S_{i_0 i_1 i_2 i_3} := w^i_{\sigma_i^{-1}(i_0)\sigma_i^{-1}(i_1)\sigma_i^{-1}(i_2)\sigma_i^{-1}(i_3)}, \qquad i_0, i_1, i_2, i_3 \in S_i, \qquad i = 0, 1.$$

Now consider the submanifold $Y_{k,I} \subset \overline{\mathcal{M}}_{0,k}$ defined in the proof of Lemma 7.5.5. We claim that

$$Y_{k,I} = \bigcup_{S \in \mathcal{S}(I_0, I_1)} \phi_S(Y_{k_0+1, I_{S0}} \times Y_{k_1+1, I_{S1}}).$$

This is obvious from the geometric description of the gluing map ϕ_S given earlier. To see this using the explicit coordinatewise description given above, it suffices to note that

$$Y_{k,I} := \left\{\mathbf{w} \in \overline{\mathcal{M}}_{0,k} \,|\, i_0, i_1, i_2, i_3 \in I, i_0 < i_1 < i_2 < i_3 \Longrightarrow w_{i_0 i_1 i_2 i_3} = \infty\right\}.$$

Now observe that the submanifolds $Y_{k,I}$ and $Y_{k_0+1, I_{S0}} \times Y_{k_1+1, I_{S1}}$ have the same dimension $2(k - \#I)$. Moreover the restriction of ϕ_S to $Y_{k_0+1, I_{S0}} \times Y_{k_1+1, I_{S1}}$ is a holomorphic diffeomorphism onto a union of components of of $Y_{k,I}$. This proves Lemma 7.5.9. □

Combining Lemma 7.5.9 with the (*Splitting*) axiom we obtain the following gluing theorem for the Gromov–Witten invariants $\mathrm{GW}^{M,I}$.

THEOREM 7.5.10 (Gluing). *Suppose* (M,ω) *is semipositive. Then*

$$\begin{aligned}\mathrm{GW}^{M,I}_{A,k}(a_1,\dots,a_k) &= \sum_{S\in\mathcal{S}(I_0,I_1)} \varepsilon(S,a) \sum_{A_0+A_1=A}\sum_{\nu,\mu} \\ &\qquad \mathrm{GW}^{M,I_{S0}}_{A_0,k_0+1}(\{a_i\}_{i\in S_0},e_\nu)\, g^{\nu\mu}\, \mathrm{GW}^{M,I_{S1}}_{A_1,k_1+1}(e_\mu,\{a_j\}_{j\in S_1}).\end{aligned}$$

In particular,

$$\begin{aligned}\mathrm{GW}^{M,\{1,2,3,4\}}_{A,k}(a_1,\dots,a_k) &= \sum_{S\in\mathcal{S}(\{1,2\},\{3,4\})} \varepsilon(S,a) \sum_{A_0+A_1=A}\sum_{\nu,\mu} \\ &\qquad \mathrm{GW}^{M}_{A_0,k_0+1}(\{a_i\}_{i\in S_0},e_\nu)\, g^{\nu\mu}\, \mathrm{GW}^{M}_{A_1,k_1+1}(e_\mu,\{a_j\}_{j\in S_1}),\end{aligned}$$

and, for $k_0+k_1=k$ *and* $k_0,k_1\ge 2$*,*

$$\begin{aligned}\mathrm{GW}^{M,\{1,\dots,k\}}_{A,k}(a_1,\dots,a_k) &= \sum_{A_0+A_1=A}\sum_{\nu,\mu} \\ &\quad \mathrm{GW}^{M,\{1,\dots,k_0+1\}}_{A_0,k_0+1}(a_1,\dots,a_{k_0},e_\nu)\, g^{\nu\mu}\, \mathrm{GW}^{M,\{1,\dots,k_1+1\}}_{A_1,k_1+1}(e_\mu,a_{k_0+1},\dots,a_k).\end{aligned}$$

To explain what these rules mean, let us consider the case

$$k = \#I = 4.$$

Then $\overline{\mathcal{M}}_{0,4}$ is a 2-sphere with three special points $\mathbf{w}_0$, $\mathbf{w}_1$, $\mathbf{w}_\infty$ corresponding to the three trees with two vertices. If $\mathbf{w}$ is not equal to one of the three special points then the invariant $\mathrm{GW}^{M,\{1,2,3,4\}}_{A,4}(a_1,\dots,a_4)$ counts A-curves with four fixed marked points w_i going through cycles dual to the a_i. Moving two of the marked points together, or equivalently, considering one of the three special points in $\overline{\mathcal{M}}_{0,4}$, one can show that the same invariant counts stable maps with two components such that each component passes through two of the cycles. To prove that these two descriptions indeed give the same invariant, one must prove a gluing theorem which shows how to form a single A-curve from a pair of components. This result will be proved in Chapter 10.

Kontsevich's recursion formula. We now illustrate the importance of Theorem 7.5.10 by showing how to use it to compute the number

$$N_d := \mathrm{GW}^{\mathbb{C}P^2}_{dL,3d-1}(c^2,\dots,c^2)$$

of rational curves of degree d in $\mathbb{C}P^2$ passing through $3d-1$ generic points.

PROPOSITION 7.5.11 (Kontsevich). *The numbers* N_d *are determined by the initial condition* $N_1 = 1$ *and the recursion formula*

$$N_d = \sum_{k+\ell=d} N_k N_\ell \left(k^2\ell^2 \binom{3d-4}{3k-2} - k^3\ell \binom{3d-4}{3k-1} \right), \qquad d\ge 2. \tag{7.5.2}$$

The first few values are

$$N_2 = 1, \qquad N_3 = 12, \qquad N_4 = 620, \qquad N_5 = 87,304,$$

$$N_6 = 26,312,976, \qquad N_7 = 14,616,808,192.$$

PROOF. Since the cohomology of $\mathbb{C}P^2$ has even degree the signs in Theorem 7.5.10 are all positive. Consider the standard basis

$$e_\nu := c^\nu, \qquad \nu = 0, 1, 2,$$

of $H^*(\mathbb{C}P^2)$. Then

$$g^{20} = g^{02} = g^{11} = 1$$

and $g^{\nu\mu} = 0$ for all other values of ν and μ. Therefore there are three pairs (e_ν, e_μ) that contribute to the gluing formula, namely $(1, c^2)$, (c, c), and $(c^2, 1)$. We will apply the gluing formula with $I = \{1, 2, 3, 4\}$ so that I_0, I_1 both have two elements. Therefore the Gromov–Witten invariants occurring on the right hand side of the gluing formula have the form GW^I where $\#I = 3$. Hence the *(Fundamental class)* axiom implies that any term involving 1 vanishes unless the corresponding class A_i is zero.

The next observation is that, by Proposition 7.5.6,

$$\mathrm{GW}^{\mathbb{C}P^2}_{dL,3d+k}(c, \dots, c, c^2, \dots, c^2) = d^{k+1} N_d, \tag{7.5.3}$$

where the first $k+1$ arguments are equal to c. With a different number of variables c it equals 0 for dimensional reasons.

We now count the number of curves with four fixed marked points through two lines ℓ_1, ℓ_2 and $3d - 2$ points $p_1, \dots, p_{3d-2}$. The gluing rules show how this decomposes when the index set $\{1, 2, 3, 4\}$ is divided into the two pairs $I_0 := \{1, 2\}$ and $I_1 := \{3, 4\}$. There are two essentially different ways in which to divide up the constraints $\mathrm{PD}(a_i)$: one can either require that the marked points $\mathrm{w}_1, \mathrm{w}_2 \in I_0$ lie on the lines or require that one lie on a line and one at a point. As we now see, this gives the iteration formula.

Let us apply Theorem 7.5.10 using the index sets

$$I_0 := \{1, 2\}, \qquad I_1 := \{3, 4\},$$

and with the constraints ordered so that $\mathrm{w}_1, \mathrm{w}_2$ lie on lines. This gives:

$$\begin{aligned}
&\mathrm{GW}^{\mathbb{C}P^2, \{1,2,3,4\}}_{dL,3d}(c, c, c^2, \dots, c^2) \\
&= \sum_{d_0+d_1=d} \binom{3d-4}{3d_0} \mathrm{GW}^{\mathbb{C}P^2}_{d_0L,3d_0+3}(1, c, c, c^2, \dots, c^2)\mathrm{GW}^{\mathbb{C}P^2}_{d_1L,3d_1-1}(c^2, \dots, c^2) \\
&\quad + \sum_{d_0+d_1=d} \binom{3d-4}{3d_0-1} \mathrm{GW}^{\mathbb{C}P^2}_{d_0L,3d_0+2}(c, c, c, c^2, \dots, c^2)\mathrm{GW}^{\mathbb{C}P^2}_{d_1L,3d_1}(c, c^2, \dots, c^2) \\
&= N_d + \sum_{d_0+d_1=d} \binom{3d-4}{3d_0-1} d_0^3 d_1 N_{d_0} N_{d_1}.
\end{aligned}$$

The first step above expands the sum over the pairs $(e_\nu, e_\mu) = (1, c^2), (c, c), (c^2, 1)$ that represent the diagonal in the gluing formula, keeping (7.5.3) in mind. The first binomial coefficient is the number of partitions of the index set $\{1, \dots, 3d\}$ into two subsets S_0 and S_1 such that $1, 2 \in S_0$, $3, 4 \in S_1$ and $\#S_0 = 3d_0 + 2$, $\#S_1 = 3d_1 - 2$. (By (7.5.3) the invariants vanish if these numerical conditions do not hold.) The second binomial coefficient is the corresponding number with $\#S_0 = 3d_0 + 1$ and $\#S_1 = 3d_1 - 1$. The third summand in which $(\nu, \mu) = (c^2, 1)$ vanishes by the *(Fundamental class)* axiom in Proposition 7.5.6. This explains the first identity. The second identity uses the same fact from Proposition 7.5.6: since $\mathrm{GW}^{\mathbb{C}P^2}_{d_0L,3d_0+3}(1, c, c, c^2, \dots, c^2) = 0$ unless $d_0 = 0$, the only contribution from the first sum counts stable maps whose first component is a ghost that contains precisely two marked points w_1, w_2 and maps to the intersection $\ell_1 \cap \ell_2$ of the two lines, and whose second component is a degree $d_1 = d$ curve through the $3d-1$ points $\ell_1 \cap \ell_2, p_1 \dots, p_{3d-2}$. The evaluation of the second sum uses the *(Divisor)* axiom.

As we remarked above, the invariant remains unchanged if we permute the first four of the a_i. This has the effect of dividing up the constraints differently in the splitting, since w_1 and w_3 now lie on the two lines. We find by reasoning as above that:

$$\begin{aligned}
&N_d + \sum_{d_0+d_1=d} \binom{3d-4}{3d_0-1} d_0^3 d_1 N_{d_0} N_{d_1} \\
&= \mathrm{GW}^{\mathbb{C}P^2,\{1,2,3,4\}}_{dL,3d}(c, c^2, c, c^2, \dots, c^2) \\
&= \sum_{d_0+d_1=d} \binom{3d-4}{3d_0-2} \mathrm{GW}^{\mathbb{C}P^2}_{d_0L,3d_0+1}(c, c, c^2, \dots, c^2) \mathrm{GW}^{\mathbb{C}P^2}_{d_1L,3d_1+1}(c, c, c^2, \dots, c^2) \\
&= \sum_{d_0+d_1=d} \binom{3d-4}{3d_0-2} d_0^2 d_1^2 N_{d_0} N_{d_1}.
\end{aligned}$$

This proves the recursion formula of Proposition 7.5.11. □

EXERCISE 7.5.12. Use the formula (7.5.2) to check that $N_2 = 1, N_3 = 12$. In both cases give explicit geometric descriptions of the stable maps that appear in the calculation. Note that one geometric configuration may contribute several times, both because it may interact in different ways with the constraints and because it has different possible parametrizations. For example, the union of a line with a conic can be parametrized as a stable map of degree 3 in two ways since the unique nodal point in the domain could go to either of the two intersections of the line with the conic.

In **[216]** Kontsevich and Manin also explain how the Gromov–Witten invariants of higher dimensional projective spaces can be computed, in principle, from the gluing theorem. They introduce the numbers

$$N_d(k_2, \dots, k_n) := \mathrm{GW}^{\mathbb{C}P^n}_{dL,k}(c^2, \dots, c^2, c^3 \dots, c^3, \dots, c^n, \dots, c^n),$$

where $k := k_2 + \dots + k_n$ and the cohomology class c^ν occurs k_ν times. For example, $N_d(k_2, k_3)$ is the number of degree d rational curves in $\mathbb{C}P^3$ through k_2 lines and k_3 points. These numbers can only be nonzero when the cohomology classes satisfy the dimensional condition (7.1.2) which in the present case has the form

$$\sum_{\nu=2}^{n} (\nu - 1) k_\nu = n + d(n+1) - 3. \tag{7.5.4}$$

The numbers $N_d(k_2, \dots, k_n)$ appear as the coefficients of the **Gromov–Witten potential**. This is a generating function which combines the information about all the (genus zero) Gromov–Witten invariants into a single expression. For $\mathbb{C}P^n$ it has the following form. Write a cohomology class in $H^*(\mathbb{C}P^n)$ with complex coefficients as a formal linear combination

$$a_t := t_0 + t_1 c + t_2 c^2 + \dots + t_n c^n, \qquad t := (t_0, \dots, t_n).$$

Then the Gromov–Witten potential for $\mathbb{C}P^n$ is the formal power series

$$\Phi^{\mathbb{C}P^n}(s,t_0,\ldots,t_n) := \sum_{d\geq 0}\sum_{k\geq 0}\frac{s^d}{k!}\mathrm{GW}^{\mathbb{C}P^n}_{dL,k}(a_t,\ldots,a_t). \tag{7.5.5}$$

Here we have included terms with $k<3$ (as defined in Remark 7.5.2) since this enhances the formal properties of Φ. By Proposition 7.5.6 and Remark 7.5.2, the class $d=0$ contributes the following cubic polynomial to this power series:

$$\begin{aligned}\phi^{\mathbb{C}P^n}(t) &:= \tfrac{1}{6}\int_M a_t\smile a_t\smile a_t \\ &= \tfrac{1}{6}\sum_{i+j+k=n} t_it_jt_k \\ &= \sum_{i<j<n-i-j} t_it_jt_{n-i-j} + \tfrac{1}{2}\sum_{2i\leq n\neq 3i} t_i^2t_{n-2i} + \tfrac{1}{6}t_{n/3}^3.\end{aligned}$$

Here the last term only appears when n is divisible by three.

LEMMA 7.5.13 (Kontsevich–Manin). *When* $n\geq 2$

$$\Phi^{\mathbb{C}P^n}(s,t) = \phi^{\mathbb{C}P^n}(t) + \sum_{d>0}\sum_{k_2,\ldots,k_n} N_d(k_2,\ldots,k_n)\frac{t_2^{k_2}\cdots t_n^{k_n}}{k_2!\cdots k_n!}e^{dt_1}s^d.$$

PROOF. We drop the superscript $\mathbb{C}P^n$. By the *(Zero)* axiom,

$$\phi(t) = \tfrac{1}{6}\mathrm{GW}_{0,3}(a_t,a_t,a_t) = \sum_{k\geq 3}\frac{1}{k!}\mathrm{GW}_{0,k}(a_t,\ldots,a_t).$$

Moreover, by the *(Fundamental class)* axiom and Exercise 7.5.3, the difference $\Phi_0 := \Phi-\phi$ is independent of t_0 and involves k-point invariants for $k\geq 2$. Hence

$$\begin{aligned}\Phi_0(s,t) &= \sum_{d>0}\sum_{k\geq 2}\frac{s^d}{k!}\mathrm{GW}_{dL,k}(a_t,\ldots,a_t) \\ &= \sum_{d>0}\sum_{k\geq 2}\sum_{\nu_1,\ldots,\nu_k}\frac{s^d}{k!}\mathrm{GW}_{dL,k}(c^{\nu_1},\ldots,c^{\nu_k})t_{\nu_1}\cdots t_{\nu_k} \\ &= \sum_{d>0}\sum_{k\geq 2}\sum_{k_1+\cdots+k_n=k}\frac{s^dt_1^{k_1}\cdots t_n^{k_n}}{k_1!\cdots k_n!}\mathrm{GW}_{dL,k}(c^{1\otimes k_1},c^{2\otimes k_2},\ldots,c^{n\otimes k_n}) \\ &= \sum_{d>0}\sum_{k_1+\cdots+k_n\geq 2} d^{k_1}N_d(k_2,\ldots,k_n)\frac{s^dt_1^{k_1}\cdots t_n^{k_n}}{k_1!\cdots k_n!} \\ &= \sum_{d>0}\sum_{k_2+\cdots+k_n\geq 2} N_d(k_2,\ldots,k_n)\frac{t_2^{k_2}\cdots t_n^{k_n}}{k_2!\cdots k_n!}e^{dt_1}s^d.\end{aligned}$$

The fourth equation follows from the *(Divisor)* axiom. The last equation follows from the fact that $N_d(k_2,\ldots,k_n)=0$ whenever $d>0$ and $k_2+\cdots+k_n<2$. This proves Lemma 7.5.13. □

In terms of the Gromov–Witten potential, the *(Zero)* axiom for the Gromov–Witten invariants can be expressed as the partial differential equation

$$\partial_0\partial_i\partial_j\Phi = g_{ij}. \tag{7.5.6}$$

Here $\partial_i = \partial/\partial t_i$ and the constants g_{ij} are defined as in the *(Splitting)* axiom. In the case of the standard basis of $H^*(\mathbb{C}P^n)$ they are given by

$$g_{ij} = g^{ij} := \begin{cases} 1, & \text{if } i+j=n, \\ 0, & \text{if } i+j \neq n. \end{cases}$$

Moreover, it turns out that the function $\Phi = \Phi^{\mathbb{C}P^n}$ satisfies the following remarkable quadratic third order partial differential equation for all quadruples of integers $i, j, k, \ell \in \{0, \dots, n\}$:

$$\sum_{\nu,\mu=0}^{n} \partial_i \partial_j \partial_\nu \Phi \cdot g^{\nu\mu} \cdot \partial_\mu \partial_k \partial_\ell \Phi = \sum_{\nu,\mu=0}^{n} \partial_j \partial_k \partial_\nu \Phi \cdot g^{\nu\mu} \cdot \partial_\mu \partial_i \partial_\ell \Phi. \tag{7.5.7}$$

This is called the **WDVV-equation** (see Dubrovin [**93**]). It is a highly overdetermined system of partial differential equations and the surprising fact is that it has a solution at all. In [**216**] Kontsevich and Manin explain how the *(Splitting)* axiom implies that the Gromov–Witten potential satisfies the WDVV-equation. That the genus zero Gromov–Witten invariants satisfy the *(Splitting)* axiom in the semipositive case was first proved by Ruan–Tian [**345**] (and a little later by Liu [**248**] and in the 1994 version of this book for the monotone case). Here we show, following the argument of Kontsevich–Manin, that Theorem 7.5.10 implies that $\Phi^{\mathbb{C}P^n}$ satisfies (7.5.7). For a more extensive discussion see Chapter 11.

PROPOSITION 7.5.14. *The Gromov–Witten potential $\Phi^{\mathbb{C}P^n}$ of $\mathbb{C}P^n$ satisfies the WDVV-equation.*

PROOF. We show that the WDVV-equation (for $\mathbb{C}P^n$) follows from the gluing rules in Theorem 7.5.10. To see this note that the third derivatives of Φ can be expressed in the form

$$\partial_i \partial_j \partial_\nu \Phi(s,t) = \sum_{d \geq 0} \sum_{m \geq 0} \frac{s^d}{m!} \mathrm{GW}_{dL,m+3}(a_t, \dots, a_t, c^i, c^j, c^\nu).$$

Hence the left hand side of (7.5.7) has the form

$$\sum_{\nu=0}^{n} \partial_i \partial_j \partial_\nu \Phi(s,t) \cdot \partial_{n-\nu} \partial_k \partial_\ell \Phi(s,t) = \sum_{d \geq 0} \sum_{m \geq 0} \frac{s^d}{m!} \Psi^{d,m}_{ijk\ell}(t),$$

where $\Psi^{d,m}_{ijk\ell}$ is the homogeneous polynomial of degree m in t given by

$$\Psi^{d,m}_{ijk\ell}(t) := \sum_{d_0+d_1=d} \; \sum_{m_0+m_1=m} \; \sum_{\nu=0}^{n} \binom{m}{m_0}$$
$$\mathrm{GW}_{d_0L,m_0+3}(a_t, \dots, a_t, c^i, c^j, c^\nu) \mathrm{GW}_{d_1L,m_1+3}(c^{n-\nu}, c^k, c^\ell, a_t, \dots, a_t).$$

The WDVV-equation asserts that

$$\Psi^{d,m}_{ijk\ell}(t) = \Psi^{d,m}_{jki\ell}(t).$$

Since all cohomology classes have even degrees, this equation follows from the second identity in Theorem 7.5.10. This proves Proposition 7.5.14. □

The WDVV-equation equation can, in principle, be used to compute the coefficients $N_d(k_2, \dots, k_n)$ of the Gromov–Witten potential for $\mathbb{C}P^n$. In [**216**, Theorem 3.1] Kontsevich and Manin show that the coefficients are uniquely determined by this equation and the fact that $N_1(0, \dots, 0, 2) = 1$. In the proof of Theorem 7.4.1

we have seen that $N_d(2d-2,0,\dots,0,d+1) \geq N_d$. In the next three exercises it is convenient to use the abbreviation

$$\Phi_{ijk} := \partial_i\partial_j\partial_k\Phi$$

for $i,j,k \in \{0,1,\dots,n\}$.

EXERCISE 7.5.15. Assume $n=1$ and show that

$$\Phi^{\mathbb{C}P^1}(s,t) = \frac{t_0^2 t_1}{2} + e^{t_1}s.$$

Note that $\Phi_{001} = 1$. Prove that the WDVV-equation for $\mathbb{C}P^1$ is a tautology.

EXERCISE 7.5.16. Assume $n=2$ and show that

$$\Phi^{\mathbb{C}P^2}(s,t) = \frac{t_0^2 t_2 + t_0 t_1^2}{2} + \sum_{d>0} N_d \frac{t_2^{3d-1}}{(3d-1)!} e^{dt_1} s^d.$$

Note that $\Phi_{011} = \Phi_{002} = 1$. Prove that the WDVV equation for $\mathbb{C}P^2$ is equivalent to a single partial differential equation, namely

$$\Phi_{111}\cdot\Phi_{122} + \Phi_{222} = (\Phi_{112})^2.$$

Prove, via comparison of coefficients, that this equation is equivalent to the recursion formula (7.5.2)

EXERCISE 7.5.17. Assume $n=3$. Show that the invariant $N_d(k_2,k_3)$ can only be nonzero if $k_2+2k_3=4d$. Deduce that

$$\Phi^{\mathbb{C}P^3}(s,t) = t_0t_1t_2 + \frac{t_0^2t_3}{2} + \frac{t_1^3}{6} + \sum_{d=1}^{\infty}\sum_{k=0}^{2d} N_d(4d-2k,k)\frac{t_2^{4d-2k}t_3^k}{(4d-2k)!k!}e^{dt_1}s^d.$$

Note that $\Phi_{012} = \Phi_{003} = 1$ and $\Phi_{0ij} = 0$ for all other values of i and j with $i \leq j$. There are six nontrivial WDVV-equations. They are given by (7.5.7) with the constants

$$g^{03} = g^{12} = g^{21} = g^{30} = 1$$

and the indices

$$ijk\ell = 1122,\ 1123,\ 2213,\ 1233,\ 1133,\ 2233.$$

These equations are

$$\begin{aligned}
2\Phi_{123} &= \Phi_{111}\cdot\Phi_{222} - \Phi_{112}\cdot\Phi_{122}, \\
\Phi_{133} &= \Phi_{111}\cdot\Phi_{223} - \Phi_{113}\cdot\Phi_{122}, \\
\Phi_{233} &= \Phi_{222}\cdot\Phi_{113} - \Phi_{112}\cdot\Phi_{223}, \\
\Phi_{333} &= (\Phi_{123})^2 + \Phi_{113}\cdot\Phi_{223} - \Phi_{122}\cdot\Phi_{133} - \Phi_{112}\cdot\Phi_{233}, \\
2\Phi_{123}\cdot\Phi_{113} &= \Phi_{111}\cdot\Phi_{233} + \Phi_{112}\cdot\Phi_{133}, \\
2\Phi_{123}\Phi_{223} &= \Phi_{222}\cdot\Phi_{133} + \Phi_{122}\cdot\Phi_{233}.
\end{aligned} \tag{7.5.8}$$

Equation (7.5.7) is a tautology whenever one of the indices i,j,k,ℓ is equal to zero. Verify that the last two equations in (7.5.8) follow from the first three, so that the WDVV-equations for $\mathbb{C}P^3$ are equivalent to the first four equations in (7.5.8).

Use comparison of coefficients to translate the WDVV-equations into recursion formulas for the numbers

$$N_{d,k} := N_d(4d - 2k, k), \qquad d \geq 1, \qquad k = 0, 1, 2, \ldots, 2d.$$

Thus $N_{d,k}$ is the number of rational degree d curves in $\mathbb{C}P^3$ through k generic points and $4d - 2k$ generic lines. Show that the first equation in (7.5.8) gives the recursion formula

$$\begin{aligned} &2dN_{d,k+1} - N_{d,k} \\ &= \sum_{\substack{d_0+d_1=d \\ k_0+k_1=k}} \binom{k}{k_0} \left(d_0^3 \binom{4d-2k-3}{4d_0 - 2k_0} - d_0^2 d_1 \binom{4d-2k-3}{4d_0-2k_0-1} \right) N_{d_0,k_0} N_{d_1,k_1} \end{aligned}$$

for $0 \leq k \leq 2d - 3/2$. Show that the second equation in (7.5.8) gives

$$\begin{aligned} &dN_{d,k+1} - N_{d,k} \\ &= \sum_{\substack{d_0+d_1=d \\ k_0+k_1=k}} \binom{4d-2k-2}{4d_0 - 2k_0} \left(d_0^3 \binom{k-1}{k_0} - d_0^2 d_1 \binom{k-1}{k_1} \right) N_{d_0,k_0} N_{d_1,k_1} \end{aligned}$$

for $1 \leq k \leq 2d - 1$. Show that the third equation in (7.5.8) gives

$$N_{d,k+1} = \sum_{\substack{d_0+d_1=d \\ k_0+k_1=k}} \binom{k-1}{k_0} \left(d_1^2 \binom{4d-2k-3}{4d_1 - 2k_1} - d_0^2 \binom{4d-2k-3}{4d_0-2k_0-1} \right) N_{d_0,k_0} N_{d_1,k_1}$$

for $1 \leq k \leq 2d-3/2$. These formulas demonstrate again that the WDVV-equations are highly overdetermined. Use them, together with the fact that there is a unique line in $\mathbb{C}P^3$ through any pair of distinct points, to prove that

$$\begin{array}{lllll} N_{1,2} = 1, & N_{1,1} = 1, & N_{1,0} = 2, & & \\ N_{2,4} = 0, & N_{2,3} = 1, & N_{2,2} = 4, & N_{2,1} = 18, & N_{2,0} = 92, \\ N_{3,6} = 1, & N_{3,5} = 5, & N_{3,4} = 30, & N_{3,3} = 190. & \end{array}$$

Examine the geometric meaning of these numbers. Verify the formulas $N_{1,1} = 1$ and $N_{1,0} = 2$ directly. (See Example 7.1.14 and Exercise 7.1.16.) Find the values of $N_{3,2}$, $N_{3,1}$, and $N_{3,0}$.

CHAPTER 8

Hamiltonian Perturbations

It is often useful to consider solutions of a perturbed Cauchy–Riemann equation

$$(*) \qquad \bar{\partial}_J(u) = \nu(u),$$

where $\nu(u) \in \Omega^{0,1}(\Sigma, u^*TM)$. A natural way to generate such perturbations is as follows. Let ν be a section of the bundle $\pi_1^*\Lambda^{0,1}T^*\Sigma \otimes \pi_2^*TM \to \Sigma \times M$ whose fiber at (z, x) is the space of complex anti-linear homomorphisms $T_z\Sigma \to T_xM$. Then define $\nu(u) \in \Omega^{0,1}(\Sigma, u^*TM)$ to be the pullback of ν under the embedding $\widetilde{u} : \Sigma \to \Sigma \times M$ given by

$$\widetilde{u}(z) := (z, u(z)).$$

In this situation Gromov showed in **[160]** how to define an almost complex structure $\widetilde{J}_\nu$ on the product $\widetilde{M} := \Sigma \times M$ such that the $\widetilde{J}_\nu$-holomorphic sections of $\widetilde{M}$ are precisely the graphs $\widetilde{u}$ of solutions of $(*)$. Therefore the solutions to $(*)$ can be analysed using our previous methods.

The virtual moduli cycle is usually constructed so that its elements are solutions to $(*)$ for some general perturbation ν. In this situation one is only interested in counting solutions, so that their precise properties are irrelevant. The focus of the current chapter is somewhat different. We restrict attention to a geometrically natural class of perturbations for which $(*)$ is a coordinate free version of Floer's equation. In this case the solutions have a geometric meaning; we will see some applications in the next chapter.

Observe that the restriction of every perturbing 1-form ν to $\{z\} \times M$ is the complex antilinear part of a linear map from $T_z\Sigma$ into the space of vector fields on M. A specially nice case is when the vector fields in the image of this map are Hamiltonian. In this case ν is called a **Hamiltonian perturbation**. We show that the associated almost complex structure $\widetilde{J}_\nu$ has a natural geometric interpretation in terms of a connection on the fiber bundle $\Sigma \times M \to \Sigma$.

At this point there is no need to restrict to the trivial bundle; one can instead consider any locally Hamiltonian fibration. Thus we are led to consider sections of a general fiber bundle $\widetilde{M} \to \Sigma$ over a Riemann surface, a framework that has many interesting applications (see Gromov **[160]**, Floer **[113, 114, 116]**, Seidel **[362, 363]**, Polterovich **[325, 326, 327]**, Entov **[104]**, Akveld–Salamon **[11]**, Piunikhin–Salamon–Schwarz **[323]**, McDuff **[269]**). However, many of the results in this book do not require the extension of the theory to symplectic fiber bundles and we have written this chapter so that this extension of the theory can be omitted.

The first section introduces the Cauchy–Riemann equations with Hamiltonian perturbations and explains how these are related to symplectic connections on a product bundle. It sets up the notation used in the rest of the chapter. Section 8.2

extends these notions to more general fiber bundles. Section 8.3 discusses the moduli space of holomorphic sections and outlines proofs of the appropriate transversality results. Section 8.4 examines holomorphic spheres in the fibers and Section 8.5 introduces the pseudocycle of pseudoholomorphic sections. In Section 8.6 we define the corresponding Gromov–Witten invariants and explain a few basic examples. The whole chapter can be viewed as an expanded version of Sections 6.7 and 7.3 in a slightly more general setting. Throughout we work with a base manifold Σ of arbitrary genus. Thus we end up defining a Gromov–Witten invariant that counts the number of J-holomorphic sections $u : \Sigma \to \widetilde{M}$, where the base has a fixed complex structure j_Σ. Our last example illustrates some of the new features that appear in this context.

8.1. Trivial bundles

This section discusses the solutions of the perturbed Cauchy–Riemann equation $(*)$ where the perturbation $\nu(u) \in \Omega^{0,1}(\Sigma, u^*TM)$ is Hamiltonian. This means that ν is the complex antilinear part of a 1-form on Σ with values in the space of Hamiltonian vector fields on M. Any such perturbation is uniquely generated by a 1-form $H \in \Omega^1(\Sigma, C_0^\infty(M))$ on Σ with values in the space $C_0^\infty(M)$ of smooth functions on M with mean value zero. We write H in the form

$$T_z\Sigma \to C_0^\infty(M) : \zeta \mapsto H_\zeta.$$

The corresponding 1-form with values in the space of Hamiltonian vector fields is given by $\zeta \mapsto X_{H_\zeta}$. Given a smooth map $u : \Sigma \to M$, we denote by $X_H(u) \in \Omega^1(\Sigma, u^*TM)$ the 1-form

$$T_z\Sigma \to T_{u(z)}M : \zeta \mapsto X_{H_\zeta}(u(z))$$

along u with values in the pullback tangent bundle. Now fix a smooth family of ω-tame almost complex structures $\Sigma \to \mathcal{J}_\tau(M,\omega) : z \mapsto J_z$. Then the perturbed Cauchy–Riemann equation $(*)$ has the form

$$\bar{\partial}_J(u) + X_H(u)^{0,1} = 0, \tag{8.1.1}$$

where $X_H(u)^{0,1}$ denotes the complex anti-linear part of $X_H(u)$:

$$X_H(u)^{0,1}(z) := \tfrac{1}{2}\Big(X_H(u(z)) + J_z \circ X_H(u(z)) \circ j_\Sigma(z)\Big).$$

Note that the first summand in the above expression is a Hamiltonian vector field while the second is a gradient vector field. Taken together with $\bar{\partial}_J(u)$ as in the left hand side of (8.1.1) they form the complex antilinear part of the *covariant derivative*

$$d_H(u) := du + X_H(u)$$

and will be denoted by

$$\bar{\partial}_{J,H}(u) := d_H(u)^{0,1} = \bar{\partial}_J(u) + X_H(u)^{0,1}.$$

Hamiltonian connections. To understand equation (8.1.1) geometrically, it is useful to work on the product manifold $\widetilde{M} := \Sigma \times M$ and replace u by its graph $\widetilde{u} : \Sigma \to \Sigma \times M$, given by

$$\widetilde{u}(z) := (z, u(z)).$$

Thus $\widetilde{u}$ is a section of $\widetilde{M}$ and its covariant derivative $d_H(u)$ can be interpreted as a 1-form on the base with values in the horizontal tangent bundle along $\widetilde{u}$ with respect to a symplectic connection associated to the perturbation H. More precisely, every $H \in \Omega^1(\Sigma, C_0^\infty(M))$ gives rise to a (real valued) 1-form $\widetilde{\sigma}_H \in \Omega^1(\widetilde{M})$, defined by

$$\widetilde{\sigma}_H(\zeta, \xi) := H_\zeta(x)$$

for $\zeta \in T_z\Sigma$ and $\xi \in T_xM$. Observe that this is not the pullback of H to $\widetilde{M}$ (which would have values in $C_0^\infty(M)$ instead of $\mathbb{R}$) but is obtained from the pullback by evaluating the relevant functions at the points of M. We think of $\widetilde{M}$ as fibered over Σ by the obvious projection

$$\pi : \widetilde{M} = \Sigma \times M \quad \longrightarrow \quad \Sigma.$$

The 1-form $\widetilde{\sigma}_H$ is **horizontal** in the sense that it vanishes on the vertical tangent bundle $\ker d\pi$. However, $d\widetilde{\sigma}_H$ does not have this property: though it vanishes on pairs of vertical vectors, it may well be nonzero when only one vector is vertical. Consider the 2-form

$$\widetilde{\omega} = \widetilde{\omega}_H := \mathrm{pr}^*\omega - d\widetilde{\sigma}_H \in \Omega^2(\widetilde{M}), \tag{8.1.2}$$

where $\mathrm{pr} : \widetilde{M} \to M$ denotes the obvious projection. On a coordinate patch U on Σ with local conformal coordinates $z = s + it \in \mathbb{C}$ the 1-form $H \in \Omega^1(\Sigma, C_0^\infty(M))$ may be written as

$$H = F\,ds + G\,dt,$$

where $F, G \in C^\infty(U, C_0^\infty(M))$. Hence

$$\widetilde{\omega}_H = \omega - d'F \wedge ds - d'G \wedge dt + (\partial_t F - \partial_s G) ds \wedge dt, \tag{8.1.3}$$

where d' denotes the exterior derivative with respect to the variables in M.

The 2-form $\widetilde{\omega}_H$ is a closed extension of the fiberwise form ω and hence is a connection form in the terminology of Guillemin–Lerman–Sternberg [**163**] and McDuff–Salamon [**277**, Chapter 6]. The corresponding horizontal distribution in $T\widetilde{M}$ is given at the points $\widetilde{x}$ of $\widetilde{M}$ by

$$\mathrm{Hor}_{H;\widetilde{x}} := \left\{ \widetilde{v} \in T_{\widetilde{x}}\widetilde{M} \,|\, \widetilde{\omega}_H(\widetilde{v}, \widetilde{w}) = 0 \;\forall\, \widetilde{w} \in \ker d\pi(\widetilde{x}) \right\}. \tag{8.1.4}$$

Although $\widetilde{\omega}_H$ is never a symplectic form on $\widetilde{M}$ (see Exercise 8.1.3), its restriction to each fiber is symplectic and, when M is compact, one can make it symplectic by adding a suitably large multiple of the pullback of an area form of the base Σ. Hence the restriction of $\widetilde{\omega}_H$ to every hypersurface in $\widetilde{M}$ of the form $\pi^{-1}(\gamma)$, where γ is an embedded arc in Σ, is nondegenerate. Moreover, by the definition of $\mathrm{Hor}_{H;\widetilde{x}}$, the kernel of the restriction of $\widetilde{\omega}_H$ to this hypersurface is spanned by the horizontal lifts of the tangent vectors to γ. Hence parallel transport along γ is given by a suitable parametrization of the characteristic flow along this hypersurface and so preserves the symplectic form on the fibers. As the following lemma shows, it is in fact given by a family of Hamiltonian symplectomorphisms of M.

LEMMA 8.1.1. *Let $H \in \Omega^1(\Sigma, C_0^\infty(M))$. Then parallel transport along every path in Σ with respect to the connection (8.1.4) determined by $\widetilde{\omega}_H$ is a Hamiltonian symplectomorphism of M.*

PROOF. Let $\gamma : [0,1] \to \Sigma$ be a smooth path and $x : [0,1] \to M$ be a horizontal lift. This means that

$$\begin{aligned} 0 &= \widetilde{\omega}_H\big((\dot\gamma(t), \dot x(t)), (0,\xi)\big) \\ &= \omega(\dot x, \xi) - d\widetilde{\sigma}_H\big((\dot\gamma, \dot x), (0,\xi)\big) \\ &= \omega(\dot x, \xi) + dH_{\dot\gamma}(x)\xi \\ &= \omega\big(\dot x + X_{H_\gamma}(x), \xi\big) \end{aligned}$$

for every $t \in [0,1]$ and every $\xi \in T_{x(t)}M$. Equivalently, x is a solution of the time dependent Hamiltonian differential equation

$$\dot x + X_{H_{\dot\gamma}}(x) = 0.$$

Since parallel transport along γ is the time-1 map $x(0) \mapsto x(1)$, this proves the lemma. □

Curvature. Hamiltonian connections on $\Sigma \times M$ are special cases of connections on a trivial G-bundle $P = \Sigma \times \mathrm{G}$. Here the structure group G is the group of Hamiltonian symplectomorphisms of M, and its Lie algebra $\mathfrak{g} = \mathrm{Lie}(\mathrm{G})$ is the space $C_0^\infty(M)$ of Hamiltonian functions with mean value zero and Lie algebra structure given by the Poisson bracket. The connection potential is the 1-form H on Σ with values in the Lie algebra, and its curvature is a 2-form with values in the Lie algebra. It is convenient to write the curvature form as $R_H \mathrm{dvol}_\Sigma$ so that $R_H : \Sigma \times M \to \mathbb{R}$ is a smooth function. It is given by

$$R_H \mathrm{dvol}_\Sigma = dH + \tfrac{1}{2}\{H \wedge H\} \in \Omega^2(\Sigma, C_0^\infty(M)).$$

Geometrically, the value of the curvature on a pair of vector fields $v, w \in \mathrm{Vect}(\Sigma)$ is the vertical part of the Lie bracket $[v^\sharp, w^\sharp]$ of their horizontal lifts $v^\sharp, w^\sharp \in \mathrm{Vect}(\widetilde{M})$. The restriction of this vertical part to a fiber $\{z\} \times M$ is the Hamiltonian vector field on M associated to the Hamiltonian function $R_H(z,\cdot)\mathrm{dvol}_\Sigma(v(z), w(z)) \in C_0^\infty(M)$.

EXERCISE 8.1.2. With $\widetilde{\omega}_H$ given in local coordinates by (8.1.3), show that the curvature form is given by

$$R_H \lambda = \partial_s G - \partial_t F + \{F, G\},$$

where $\lambda ds \wedge dt$ is the volume form on U and $\{\cdot,\cdot\}$ denotes the Poisson bracket on M. Deduce that

$$R_H \mathrm{dvol}(v, w) = -\widetilde{\omega}_H(v^\sharp, w^\sharp)$$

for any pair of vector fields v, w on Σ. *Hint:* Use the fact that the vector fields

$$\partial_s^\sharp = (\partial_s, -X_F), \qquad \partial_t^\sharp = (\partial_t, -X_G)$$

are horizontal and calculate the vertical part of $[\partial_s^\sharp, \partial_t^\sharp]$.

EXERCISE 8.1.3. Use the results of Exercise 8.1.2 to show prove that the curvature $R_H : \Sigma \times M \to \mathbb{R}$ is determined by the formula

$$\frac{\widetilde{\omega}_H^{n+1}}{(n+1)!} = -R_H\, \pi^* \mathrm{dvol}_\Sigma \wedge \mathrm{pr}^* \frac{\omega^n}{n!},$$

where $\pi : \Sigma \times M \to \Sigma$ and $\mathrm{pr} : \Sigma \times M \to M$ denote the obvious projections. Since R_H has zero mean on each fiber, this formula shows that the 2-form $\widetilde{\omega}_H \in \Omega^2(\widetilde{M})$ cannot be a symplectic form. However, if $\kappa : \Sigma \to \mathbb{R}$ is a smooth function such that

$$\kappa(z) > R_H(z, x) \tag{8.1.5}$$

for all $z \in \Sigma$ and $x \in M$, then the 2-form

$$\widetilde{\omega}_{H,\kappa} := \widetilde{\omega}_H + \pi^*(\kappa\, \mathrm{dvol}_\Sigma)$$

is a symplectic form on $\widetilde{M}$.

Almost complex structures. We now explain a construction of Gromov which shows that the graphs of solutions to the perturbed Cauchy–Riemann equations can be interpreted as holomorphic curves with respect to a suitable almost complex structure on $\widetilde{M}$. The relevant almost complex structure $\widetilde{J}_H$ on $\widetilde{M}$ has the following properties. Its restriction to the fiber $\{z\} \times M$ agrees with J_z, the projection π is holomorphic with respect to $\widetilde{J}_H$ and j_Σ, and $\widetilde{J}_H$ is adapted to the connection form $\widetilde{\omega}_H$ in the sense that it preserves the horizontal distribution Hor_H. Once $\{J_z\}$ and j_Σ are given, these conditions determine the almost complex structure $\widetilde{J}_H$ uniquely. It is given by

$$\widetilde{J}_H(z,x) := \left(\begin{array}{cc} j_\Sigma(z) & 0 \\ J(z,x) \circ X_H(z,x) - X_H(z,x) \circ j_\Sigma(z) & J(z,x) \end{array} \right),$$

where we denote by $X_H(z,x) : T_z\Sigma \to T_xM$ the linear map $\zeta \mapsto X_{H_\zeta}(x)$ and write $J(z,x) := J_z(x)$.

EXERCISE 8.1.4. Let $\kappa : \Sigma \to \mathbb{R}$ be a smooth function that satisfies (8.1.5). Prove that $\widetilde{J}_H$ is tamed by the symplectic form $\widetilde{\omega}_{H,\kappa}$ if and only if J_z is ω-tame for every z. Likewise, $\widetilde{J}_H$ is compatible with $\widetilde{\omega}_{H,\kappa}$ if and only if J_z is ω-compatible for every z. *Hint:* Use the formula

$$\begin{aligned} &\widetilde{\omega}_H\big((\sigma,\tau,\xi),(\sigma',\tau',\xi')\big) = \\ &\qquad \omega\big(\xi + \sigma X_F + \tau X_G, \xi' + \sigma' X_F + \tau' X_G\big) - (\sigma\tau' - \tau\sigma')R_H. \end{aligned}$$

for $\sigma, \tau, \sigma', \tau' \in \mathbb{R}$ and $\xi, \xi' \in T_xM$.

EXERCISE 8.1.5. Prove that a smooth map $u : \Sigma \to M$ is a solution of (8.1.1) if and only if its graph $\widetilde{u} : \Sigma \to \widetilde{M}$, given by $\widetilde{u}(z) := (z, u(z))$, is a $\widetilde{J}_H$-holomorphic section of $\widetilde{M} = \Sigma \times M$.

Local coordinates. Choose conformal coordinates $z = s+it \in U \subset \mathbb{C}$, defined on an open subset of Σ. In such coordinates the 1-form $H \in \Omega^1(\Sigma, C_0^\infty(M))$ may be written $H = F\,ds + G\,dt$, where $F, G \in C^\infty(U, C_0^\infty(M))$. As we saw above,

$$\widetilde{\omega}_H = \omega - d'F \wedge ds - d'G \wedge dt + (\partial_t F - \partial_s G) ds \wedge dt,$$

where d' denotes the exterior derivative with respect to the variables in M, and the local formula for the curvature is

$$R_H\lambda = \partial_s G - \partial_t F + \{F, G\},$$

where $\lambda ds \wedge dt$ denotes the volume form on U. In the frame ∂_s, ∂_t of TU the horizontal subspaces are spanned by the vectors $(1, 0, -X_F)$ and $(0, 1, -X_G)$. Hence $\widetilde{J}_H$ is given by the formula

$$\widetilde{J}_H = \left(\begin{array}{ccc} 0 & -1 & 0 \\ 1 & 0 & 0 \\ JX_F - X_G & JX_G + X_F & J_{s,t} \end{array} \right). \tag{8.1.6}$$

In our local coordinates the covariant derivative $d_H(u)$ has the form

$$d_H(u) = (\partial_s u + X_F(u))\, ds + (\partial_t u + X_G(u))\, dt,$$

where $X_F = X_{F_{s,t}}$ and $X_G = X_{G_{s,t}}$ denote the Hamiltonian vector fields of F and G, respectively. Hence a smooth map $u : \Sigma \to M$ satisfies (8.1.1) if and only if, in local coordinates, it is a solution of the equation

$$\partial_s u + X_F(s,t,u) + J(s,t,u)\big(\partial_t u + X_G(s,t,u)\big) = 0. \tag{8.1.7}$$

The reader may check $u : U \to M$ satisfies this equation if and only if the map $(s,t) \mapsto (s,t,u(s,t))$ is a $\widetilde{J}_H$-holomorphic curve.

The energy identity. Fix a Hamiltonian perturbation $H \in \Omega^1(\Sigma, C_0^\infty(M))$ and a smooth family of ω-tame almost complex structures

$$J = \{J_z\}_{z\in\Sigma} \in C^\infty(\Sigma, \mathcal{J}_\tau(M,\omega)).$$

Then the energy of a smooth map $u : \Sigma \to M$ is defined as half the L^2-norm of the covariant derivative $d_H(u)$:

$$E_H(u) := \tfrac{1}{2}\int_\Sigma |d_H(u)|_J^2 \,\mathrm{dvol}_\Sigma. \tag{8.1.8}$$

When $H = 0$ this restricts to the usual notion of energy. But now the energy of nonconstant maps can vanish. In fact $E_H(u) = 0$ if and only if $\widetilde{u}(z) = (z, u(z))$ is a horizontal section of $\widetilde{M}$. The energy identity involves the curvature R_H and takes the following form.

LEMMA 8.1.6. *Let $u : \Sigma \to M$ be a solution of (8.1.1). Then*

$$E_H(u) = \int_\Sigma u^*\omega + \int_\Sigma R_H(\widetilde{u}) \,\mathrm{dvol}_\Sigma. \tag{8.1.9}$$

PROOF. In local coordinates on U the integrand has the form

$$\begin{aligned}
|\partial_s u + X_F(u)|^2 &:= \omega\left(\partial_s u + X_F(u), \partial_t u + X_G(u)\right) \\
&= \omega(\partial_s u, \partial_t u) + d'F(u)\partial_t u - d'G(u)\partial_s u + \{F,G\}\circ u \\
&= \widetilde{\omega}_H((1,0,\partial_s u),(0,1,\partial_t u)) + (\partial_s G - \partial_t F + \{F,G\})\circ u \\
&= \widetilde{\omega}_H(\partial_s \widetilde{u}, \partial_t \widetilde{u}) + \frac{R_H \circ \widetilde{u}}{\lambda},
\end{aligned}$$

where $\lambda ds\wedge dt$ is the volume form on U. Since $\widetilde{\omega}_H$ differs from $\mathrm{pr}^*\omega$ by an exact form, the integral of $\widetilde{u}^*\widetilde{\omega}_H$ over Σ agrees with the integral of $u^*\omega$. This proves (8.1.9). □

REMARK 8.1.7. The **Hofer norm** of the curvature $R = R_H$ is defined by

$$\|R_H\| := \int_\Sigma \left(\max_{x\in M} R_H(z,x) - \min_{x\in M} R_H(z,x)\right) \mathrm{dvol}_\Sigma(z).$$

Note that this norm is independent of the volume form on Σ. It follows from the energy identity (8.1.9) that

$$E_H(u) \le \int_\Sigma u^*\omega + \|R_H\|$$

for every solution of (8.1.1). This upper bound for the energy is no longer a purely topological quantity but depends also on the curvature of the connection. Its virtue is that it is defined and bounded in situations when the full energy of the graph

$\widetilde{u}(z) = (z, u(z))$ is not: for instance if the domain is noncompact, as in the Floer equations considered in Remark 8.1.9.

EXERCISE 8.1.8. Let u be a solution of (8.1.1) and consider its graph $\widetilde{u} : \Sigma \to \widetilde{M}$. Denote by $E_{H,\kappa}(\widetilde{u})$ its energy with respect to the metric determined by $\widetilde{\omega}_{H,\kappa}$ and $\widetilde{J}_H$, where $\kappa : \Sigma \to \mathbb{R}$ satisfies (8.1.5) (see Exercises 8.1.3, 8.1.4, 8.1.5). Prove that

$$\begin{aligned} E_{H,\kappa}(\widetilde{u}) &= \int_\Sigma \left(\tfrac{1}{2}\left|d_H u\right|_J^2 + \kappa - R_H(\widetilde{u})\right) \mathrm{dvol}_\Sigma \\ &= \int_\Sigma \widetilde{u}^*\widetilde{\omega}_{H,\kappa} = \int_\Sigma u^*\omega + \int_\Sigma \kappa \mathrm{dvol}_\Sigma. \end{aligned}$$

This yields the estimate $E_H(u) \le \int_\Sigma u^*\omega + \int_\Sigma \kappa \mathrm{dvol}_\Sigma$.

REMARK 8.1.9 (The Floer equation). Let us consider the perturbed Cauchy–Riemann equations (8.1.7) on the cylinder $\Sigma := \mathbb{R} \times S^1$ with the standard complex structure. In this case there are global holomorphic coordinates $s + it$ where $s \in \mathbb{R}$ and $t \in S^1 = \mathbb{R}/\mathbb{Z}$. We assume that the almost complex structure and Hamiltonian perturbation have additional symmetries associated to the $\mathbb{R}$-action, namely, that $J_{s,t} = J_t$ is independent of the s-variable and that the Hamiltonian perturbation has the form $F\,ds + G\,dt$, where $F_{s,t} = 0$ and $G_{s,t} = -H_t$ for all s and t. Then equation (8.1.7) reduces to the equation studied by Floer:

$$\partial_s u + J_t(u)(\partial_t u - X_{H_t}(u)) = 0.$$

Thus our perturbed Cauchy–Riemann equations can be interpreted as an s-dependent version of the Floer equations. We will make use of this observation in some of the applications in Chapter 9.

REMARK 8.1.10 (Lagrangian boundary conditions). The above discussion of J-holomorphic sections extends to noncompact fibers (M, ω). It also generalizes to compact Riemann surfaces Σ with boundary $\partial\Sigma$. In this case there is an additional boundary term in the energy identity coming from the integral of $\widetilde{u}^* d\widetilde{\sigma}_H$. The energy identity now reads

$$E_H(u) = \int_\Sigma u^*\omega + \int_\Sigma R_H(u)\,\mathrm{dvol}_\Sigma - \int_{\partial\Sigma} H(u) \tag{8.1.10}$$

for every solution $u : \Sigma \to M$ of (8.1.1), where $H(u) \in \Omega^1(\Sigma)$ denotes the 1-form $T_z\Sigma \to \mathbb{R} : \zeta \mapsto H_\zeta(u(z))$. In the boundary case it is interesting to consider a Lagrangian subbundle $\widetilde{L} \subset \partial\Sigma \times M$ such that the fiber

$$L_z := \left\{x \in M \,\middle|\, (z, x) \in \widetilde{L}\right\}$$

is a Lagrangian submanifold of M for every $z \in \partial\Sigma$. In this situation the form $\widetilde{\omega}_H = \widetilde{\omega} - d\widetilde{\sigma}_H$ vanishes on $T\widetilde{L}$ if and only $\widetilde{L}$ is invariant under parallel transport along $\partial\Sigma$ (see [**11**, Lemma 3.1]). If this holds one can consider solutions of (8.1.1) that satisfy the boundary condition

$$u(z) \in L_z \qquad \text{for} \quad z \in \partial\Sigma. \tag{8.1.11}$$

In Akveld–Salamon [**11**] the solutions of (8.1.1) and (8.1.11) were used to estimate (from below) the Hofer lengths of loops of Lagrangian submanifolds in a fixed Hamiltonian isotopy class.

In applications we often assume that the almost complex structure $J = \{J_z\}$ is fixed and that the perturbations H vary in the set $\mathcal{H}(\kappa) \subset \Omega^1(\Sigma, C_0^\infty(M))$ of all 1-forms whose curvature is bounded by κ. This gives rise to a family of almost complex structures $\widetilde{J}_H$ on $\widetilde{M}$ and we shall see in Section 8.3 that this family is large enough for the usual transversality arguments to go through. In other words, for generic $H \in \mathcal{H}(\kappa)$ the moduli space $\mathcal{M}(A; J, H)$ of solutions of (8.1.1) that represent the homology class $A \in H_2(M; \mathbb{Z})$ is a smooth manifold. Then the arguments in Section 6.7 apply to the current situation and hence permit one to define Gromov–Witten invariants using solutions to the perturbed Cauchy–Riemann equations. The details are worked out later on in this chapter in the more general case of locally Hamiltonian fibrations $\pi : \widetilde{M} \to \Sigma$. This generalization is not needed for many of the applications in Chapter 9 and the reader may wish to go directly to Section 8.3.

8.2. Locally Hamiltonian fibrations

In this section we extend the notions introduced above to the case of a nontrivial fiber bundle

$$\pi : \widetilde{M} \to \Sigma.$$

We begin by discussing the appropriate notions of Hamiltonian fibration, connection, and almost complex structure.

DEFINITION 8.2.1. *Let (M, ω) be a compact symplectic manifold of dimension $2n$ and $(\Sigma, \mathrm{dvol}_\Sigma, j_\Sigma)$ be a compact connected Riemann surface. A* **locally Hamiltonian fibration** *over Σ with fiber (M, ω) is a triple $(\widetilde{M}, \pi, \widetilde{\omega})$ with the following properties.*

(i) *$\widetilde{M}$ is a compact smooth manifold and $\pi : \widetilde{M} \to \Sigma$ is a submersion.*

(ii) *$\widetilde{\omega}$ is a closed 2-form on $\widetilde{M}$ and its restriction ω_z to the fiber $M_z := \pi^{-1}(z)$ is nondegenerate for every $z \in \Sigma$.*

(iii) *For every $z \in \Sigma$ the fiber (M_z, ω_z) is symplectomorphic to (M, ω).*

(iv) *$\pi_* \widetilde{\omega}^{n+1} = 0$, where $\pi_* : \Omega^{2n+2}(\widetilde{M}) \to \Omega^2(\Sigma)$ denotes integration over the fiber.*

Condition (i) asserts that the projection $\pi : \widetilde{M} \to \Sigma$ is a locally trivial fibration. The 2-form $\widetilde{\omega}$ in Definition 8.2.1 is a connection form in the terminology of Guillemin–Lerman–Sternberg [**163**] and McDuff–Salamon [**277**, Chapter 6]. The corresponding horizontal distribution in $T\widetilde{M}$ is given by

$$\mathrm{Hor}_{\widetilde{x}} := \{\widetilde{v} \in T_{\widetilde{x}}\widetilde{M} \,|\, \widetilde{\omega}(\widetilde{v}, \widetilde{w}) = 0,\ \forall\, \widetilde{w} \in \ker d\pi(\widetilde{x})\}. \tag{8.2.1}$$

As we shall see in Proposition 8.2.2, the fact that $\widetilde{\omega}$ is closed implies that the holonomy around every contractible loop in Σ is a Hamiltonian symplectomorphism of the fiber. Hence, in the case $\Sigma = \mathbb{C}P^1$, a locally Hamiltonian fibration is Hamiltonian in the sense of Lalonde–McDuff [**227**], i.e. the holonomy around *every* loop in the base is a Hamiltonian symplectomorphism. However this need not be the case if $\pi_1(\Sigma) \neq 0$. For example, if $f : \Sigma \to \Sigma$ is an area preserving diffeomorphism that is not isotopic to the identity and

$$Y_f := \mathbb{R} \times \Sigma / (t + 1, z) \sim (t, f(z))$$

is its mapping torus, then it is easy to give the 4-manifold

$$X_f := S^1 \times Y_f \to \mathbb{T}^2$$

the structure of a locally Hamiltonian fibration. But its structure group does not reduce to the group $\mathrm{Ham}(\Sigma)$ of Hamiltonian symplectomorphisms since this is connected. Condition (iv) is a normalization condition, which implies that the 2-form $\widetilde{\omega}$ is uniquely determined by the connection that it defines; it is the **coupling form** of Guillemin–Lerman–Sternberg [**163**].

PROPOSITION 8.2.2. *Let $(\widetilde{M}, \pi, \widetilde{\omega})$ be a locally Hamiltonian fibration over a compact Riemann surface $(\Sigma, \mathrm{dvol}_\Sigma, j_\Sigma)$ with fiber (M, ω). Then there is an open cover $\{U_\alpha\}_\alpha$ of Σ and a collection of local trivializations $\widetilde{\psi}_\alpha : \pi^{-1}(U_\alpha) \to U_\alpha \times M$ satisfying the following conditions.*

(i) *The transition maps have the form*

$$\widetilde{\psi}_\beta \circ \widetilde{\psi}_\alpha^{-1}(z, x) = (z, \psi_{\beta\alpha}(z)(x))$$

for $z \in U_\alpha \cap U_\beta$ and $x \in M$, where $\psi_{\beta\alpha} : U_\alpha \cap U_\beta \to \mathrm{Diff}(M, \omega)$ assigns to every $z \in U_\alpha \cap U_\beta$ a symplectomorphism $\psi_{\beta\alpha}(z)$ of (M, ω).

(ii) *For every α the 2-form $\widetilde{\omega}_\alpha \in \Omega^2(U_\alpha \times M)$, defined by*

$$\psi_\alpha^* \widetilde{\omega}_\alpha = \widetilde{\omega},$$

has the form

$$\widetilde{\omega}_\alpha = \omega - d'F_\alpha \wedge ds - d'G_\alpha \wedge dt + (\partial_t F_\alpha - \partial_s G_\alpha) ds \wedge dt, \tag{8.2.2}$$

where $s + it : U_\alpha \to \mathbb{C}$ is a holomorphic coordinate chart on Σ, F_α and G_α are real valued functions on $U_\alpha \times M$ with mean value zero over $\{z\} \times M$ for every $z \in U_\alpha$, and d' denotes the differential with respect to the second factor.

PROOF. Since $\widetilde{\omega}$ is closed, it follows from Cartan's formula that parallel transport along a smooth curve $\gamma : [0,1] \to \Sigma$ preserves the symplectic forms on the fibers. Namely, if $u_t : \Sigma \to \widetilde{M}$ is a horizontal lift of γ then $\partial_t u_t^* \widetilde{\omega} = d\alpha_t$, where $\alpha_t := \widetilde{\omega}(\partial_t u_t, du_t \cdot) = 0$. (See also the discussion before Lemma 8.1.1.) Hence one can construct trivializations satisfying the conditions in (i) by parallel transport along suitable families of paths. Given any such trivialization over a contractible open set $U_\alpha \subset \Sigma$, and conformal coordinates $s + it$ on U_α, it follows by integrating around loops in the base that the horizontal lifts $(\partial_s, -X_{s,t})$ and $(\partial_t, -Y_{s,t})$ of the vector fields ∂_s and ∂_t to $U_\alpha \times M$ are Hamiltonian along M. This means that $X_{s,t}$ and $Y_{s,t}$ are Hamiltonian vector fields on M for all s and t and so are generated by families of functions $F_\alpha(s,t,\cdot)$ and $G_\alpha(s,t,\cdot)$ on M, respectively. (For a detailed proof see [**277**, Theorem 6.21].) These functions can be chosen to have mean value zero on M. It then follows from the definition of the horizontal distribution that

$$\widetilde{\omega}_\alpha = \omega - d'F_\alpha \wedge ds - d'G_\alpha \wedge dt + f ds \wedge dt.$$

Since $\widetilde{\omega}_\alpha$ is closed we obtain $d'f = \partial_t d' F_\alpha - \partial_s d' G_\alpha$ and this determines f up to a function of $s + it$. Finally, one checks that the coupling condition (iv) in Definition 8.2.1 is equivalent to f having zero mean over each fiber $\{s + it\} \times M$. Hence $f = \partial_t F_\alpha - \partial_s G_\alpha$ as in (8.2.2). □

Curvature. The curvature of the connection on $\widetilde{M}$ determined by $\widetilde{\omega}$ is a 2-form $R\mathrm{dvol}_\Sigma$ on Σ with values in the bundle over Σ whose fiber at z is the space $C_0^\infty(M_z)$ of smooth functions on M_z with mean value zero with respect to the volume form ω_z^n. This accords with the usual notion of curvature since $C_0^\infty(M)$ (or, equivalently, the space of Hamiltonian vector fields on M) is the Lie algebra of the structure group of the fiber bundle. Again we write the curvature form as $R\mathrm{dvol}_\Sigma$ so that $R : \widetilde{M} \to \mathbb{R}$ is a smooth function. The value of the curvature on any pair of vector fields $v, w \in \mathrm{Vect}(\Sigma)$ is the vertical part of the Lie bracket $[v^\sharp, w^\sharp]$ of their horizontal lifts $v^\sharp, w^\sharp \in \mathrm{Vect}(\widetilde{M})$ (see [**277**, Remark 6.31]), where the vertical part is defined as the image of $[v^\sharp, w^\sharp]$ under the projection $\Pi : T\widetilde{M} \to T^{\mathrm{Vert}}\widetilde{M}$ along the horizontal distribution. Thus, for each $z \in \Sigma$, the restriction to M_z of $[v^\sharp, w^\sharp]^{vert}$ is the Hamiltonian vector field of the function $R(z)\mathrm{dvol}_\Sigma(v(z), w(z)) \in C_0^\infty(M_z)$. By considering the vector fields $v := \partial_s$ and $w := \partial_t$ one can prove that in the local trivializations of Proposition 8.2.2 the curvature forms are given by

$$R_\alpha \lambda_\alpha = \partial_s G_\alpha - \partial_t F_\alpha + \{F_\alpha, G_\alpha\}.$$

Here $R_\alpha : U_\alpha \times M \to \mathbb{R}$ is defined by $R_\alpha \circ \widetilde{\psi}_\alpha = R$ and $\lambda_\alpha : U_\alpha \to \mathbb{R}$ determines the volume form on Σ via $\mathrm{dvol}_\Sigma = \lambda_\alpha \, ds \wedge dt$. Hence, by (8.2.4) in Exercise 8.2.3 below, we have

$$R\mathrm{dvol}(v, w) = -\widetilde{\omega}(v^\sharp, w^\sharp) \tag{8.2.3}$$

for any pair of vectors v, w on Σ. (See [**277**, Chapter 6] for more details.[1])

EXERCISE 8.2.3. With the notation of Proposition 8.2.2, show that the value of $\widetilde{\omega}_\alpha$ on the horizontal vector fields

$$\partial_s^\sharp = (\partial_s, -X_{F_\alpha}), \qquad \partial_t^\sharp = (\partial_t, -X_{G_\alpha})$$

is given by

$$\widetilde{\omega}_\alpha(\partial_s^\sharp, \partial_t^\sharp) = -\partial_s G_\alpha + \partial_t F_\alpha - \{F_\alpha, G_\alpha\}, \tag{8.2.4}$$

where $\{\cdot, \cdot\}$ denotes the Poisson bracket on M.

EXERCISE 8.2.4. Let $\psi_{\beta\alpha} : U_\alpha \cap U_\beta \to \mathrm{Diff}(M, \omega)$ be as in Proposition 8.2.2. Show that these families of symplectomorphisms are generated by Hamiltonian vector fields. (The symplectomorphisms $\psi_{\beta\alpha}(z)$ may not be Hamiltonian, but they are Hamiltonian isotopic to each other.) Given conformal coordinates $s + it$ on $U_\alpha \cap U_\beta$, define functions $F_{\beta\alpha}, G_{\beta\alpha} : (U_\alpha \cap U_\beta) \times M \to \mathbb{R}$ by

$$\partial_s \psi_{\beta\alpha} + X_{F_{\beta\alpha}} \circ \psi_{\beta\alpha} = 0, \qquad \partial_t \psi_{\beta\alpha} + X_{G_{\beta\alpha}} \circ \psi_{\beta\alpha} = 0,$$

and the condition that $F_{\beta\alpha}$ and $G_{\beta\alpha}$ have mean value zero over $\{z\} \times M$ for every $z \in U_\alpha \cap U_\beta$. Prove that

$$F_\alpha = (F_\beta - F_{\beta\alpha}) \circ \psi_{\beta\alpha}, \qquad G_\alpha = (G_\beta - G_{\beta\alpha}) \circ \psi_{\beta\alpha}$$

over $U_\alpha \cap U_\beta$. Deduce that

$$\partial_s G_\alpha - \partial_t F_\alpha + \{F_\alpha, G_\alpha\} = (\partial_s G_\beta - \partial_t F_\beta + \{F_\beta, G_\beta\}) \circ \psi_{\beta\alpha}.$$

---

[1]Although our overall sign conventions are the same as those in [**277**], the functions called F, G here are equivalent to $-H_i, -H_j$ in [**277**, Exercise 6.30]. Also there is a sign mistake in [**277**, Lemma 6.28]; all the terms in the formula for $d\tau$ should appear with a plus sign.

EXERCISE 8.2.5. Let $\kappa : \Sigma \to \mathbb{R}$ be a smooth function such that $\kappa \circ \pi - R : \widetilde{M} \to \mathbb{R}$ vanishes nowhere. Prove that the 2-form

$$\widetilde{\omega}_\kappa := \widetilde{\omega} + \pi^*(\kappa\, \mathrm{dvol}_\Sigma)$$

is a symplectic form on $\widetilde{M}$. (See Exercise 8.1.3.)

Almost complex structures. We shall be interested in pseudoholomorphic sections of locally Hamiltonian fibrations. For this it is important to consider almost complex structures that are adapted to the fibration in the following sense. We assume that $(\widetilde{M}, \pi, \widetilde{\omega})$ is a locally Hamiltonian fibration over a compact Riemann surface $(\Sigma, \mathrm{dvol}_\Sigma, j_\Sigma)$ with fiber symplectomorphic to (M, ω).

DEFINITION 8.2.6. *An almost complex structure $\widetilde{J}$ on $(\widetilde{M}, \widetilde{\omega})$ is called* **compatible with the fibration** *if it satisfies the following conditions.*

(i) *The projection $\pi : \widetilde{M} \to \Sigma$ is holomorphic, i.e.*

$$d\pi \circ \widetilde{J} = j_\Sigma \circ d\pi.$$

(ii) *For every $z \in \Sigma$ the restriction J_z of $\widetilde{J}$ to $M_z = \pi^{-1}(z)$ is tamed by ω_z.*

(iii) *The horizontal distribution* Hor $\subset T\widetilde{M}$ *is invariant under $\widetilde{J}$. The set of almost complex structures on $\widetilde{M}$ that are compatible with the fibration will be denoted by $\mathcal{J}(\widetilde{M}, \pi, \widetilde{\omega})$.*

Sometimes we shall be interested only in the vertical part of $\widetilde{J}$. Hence we make the following definition. Note the convention that $\widetilde{J}$ denotes an almost complex structure on the whole of $T\widetilde{M}$ while $J = \{J_z\}$ just involves its vertical subbundle.

DEFINITION 8.2.7. *An* **$\widetilde{\omega}$-tame vertical almost complex structure** *on $\widetilde{M}$ is an automorphism J of the vertical tangent bundle*

$$T^{\mathrm{Vert}}\widetilde{M} := \ker\, d\pi \subset TM$$

with square $-\mathbb{1}$ such that $\widetilde{\omega}(\widetilde{v}, J\widetilde{v}) > 0$ for every $\widetilde{v} \in T^{\mathrm{Vert}}\widetilde{M}$. The space of such almost complex structures will be denoted by $\mathcal{J}^{\mathrm{Vert}}(\widetilde{M}, \pi, \widetilde{\omega})$.

Alternatively, a vertical $\widetilde{\omega}$-tame almost complex structure can be described as a smooth family $J = \{J_z\}_{z\in\Sigma}$ of almost complex structures on the fibers M_z such that $J_z \in \mathcal{J}_\tau(M_z, \omega_z)$ for every $z \in \Sigma$. If $J \in \mathcal{J}^{\mathrm{Vert}}(\widetilde{M}, \pi, \widetilde{\omega})$ then the formula (2.1.1) defines an inner product on $T^{\mathrm{Vert}}\widetilde{M}$.

LEMMA 8.2.8. *For every $J \in \mathcal{J}^{\mathrm{Vert}}(\widetilde{M}, \pi, \widetilde{\omega})$ there exists a unique almost complex structure $\widetilde{J} \in \mathcal{J}(\widetilde{M}, \pi, \widetilde{\omega})$ which is compatible with the fibration and restricts to J on the vertical subbundle $T^{\mathrm{Vert}}\widetilde{M}$.*

PROOF. This follows directly from the definitions. □

The above notions are consistent with those introduced in Section 8.1. Hence in the local trivializations of Proposition 8.2.2 the almost complex structure $\widetilde{J}$ is given by formula (8.1.6), and the corresponding $\widetilde{J}$-holomorphic curves by (8.1.7). Now let $\widetilde{J} \in \mathcal{J}(\widetilde{M}, \pi, \widetilde{\omega})$ and $\widetilde{u} : \Sigma \to \widetilde{M}$ be a $\widetilde{J}$-holomorphic section. The **vertical energy** of $\widetilde{u}$ is defined as half the L^2-norm of the vertical part $\Pi d\widetilde{u} \in \Omega^1(\Sigma, \widetilde{u}^*T^{\mathrm{Vert}}\widetilde{M})$ of the differential $d\widetilde{u} : T\Sigma \to \widetilde{u}^*T\widetilde{M}$. Here the projection $\Pi : T\widetilde{M} \to T^{\mathrm{Vert}}\widetilde{M}$ is given by

$$\Pi(\widetilde{x})\widetilde{v} := \widetilde{v} - (d\pi(\widetilde{x})\widetilde{v})^\sharp \tag{8.2.5}$$

for $\widetilde{v} \in T_{\widetilde{x}}\widetilde{M}$, where $\zeta^\sharp \in T_{\widetilde{x}}\widetilde{M}$ denotes the horizontal lift of $\zeta \in T_{\pi(\widetilde{x})}\Sigma$. Thus

$$E^{\mathrm{Vert}}(\widetilde{u}) := \tfrac{1}{2}\int_\Sigma |\Pi d\widetilde{u}|^2_{\widetilde{J}}\,\mathrm{dvol}_\Sigma.$$

The vertical energy vanishes if and only if $\widetilde{u}$ is a horizontal section of $\widetilde{M}$. The vertical energy identity involves the curvature of the connection determined by $\widetilde{\omega}$, and has the same form as before: see (8.1.9).

LEMMA 8.2.9. *Let $(\widetilde{M}, \pi, \widetilde{\omega})$ be a locally Hamiltonian fibration over Σ, let $\widetilde{J} \in \mathcal{J}(\widetilde{M}, \pi, \widetilde{\omega})$, and let $\widetilde{u} : \Sigma \to \widetilde{M}$ be a $\widetilde{J}$-holomorphic section. Then*

$$E^{\mathrm{Vert}}(\widetilde{u}) = \int_\Sigma \widetilde{u}^*\widetilde{\omega} + \int_\Sigma R(\widetilde{u})\mathrm{dvol}_\Sigma.$$

PROOF. In a local chart $U_\alpha \subset \Sigma$ with coordinate $z = s + it$ we have

$$\begin{aligned}
E^{\mathrm{Vert}}(\widetilde{u}; U_\alpha) &= \tfrac{1}{2}\int_\Sigma |\Pi d\widetilde{u}|^2_{\widetilde{J}}\,\mathrm{dvol}_\Sigma \\
&= \int_{U_\alpha} \widetilde{\omega}(\Pi\partial_s\widetilde{u}, \Pi\widetilde{J}\partial_s\widetilde{u})\mathrm{dvol}_\Sigma \\
&= \int_{U_\alpha} \Big(\widetilde{\omega}(\Pi\partial_s\widetilde{u} + \partial_s^\sharp, \Pi\partial_t\widetilde{u} + \partial_t^\sharp) - \widetilde{\omega}(\partial_s^\sharp, \partial_t^\sharp)\Big)\,\mathrm{dvol}_\Sigma \\
&= \int_{U_\alpha} \widetilde{\omega}(\partial_s\widetilde{u}, \partial_t\widetilde{u}) + R(\widetilde{u})\mathrm{dvol}_\Sigma.
\end{aligned}$$

Here the first equation uses the fact that Π commutes with $\widetilde{J}$, the second that the horizontal and vertical parts of a vector are $\widetilde{\omega}$-orthogonal by definition, and the last uses equation (8.2.3). Alternatively one could derive this formula from (8.1.9) using the local trivializations of Proposition 8.2.2. □

REMARK 8.2.10 (Families of connection forms). Given a connection form $\widetilde{\omega}$ and a vertical almost complex structure $J = \{J_z\}$ there is, by Lemma 8.2.8, a unique $\widetilde{\omega}$-compatible almost complex structure $\widetilde{J}$ and hence a uniquely determined Cauchy–Riemann equation. In order to prove regularity results, one needs to consider a family of equations. One way of obtaining such a family is to perturb J. However, another very useful approach is to fix J, instead perturbing the connection form $\widetilde{\omega}$. We are therefore led to consider families of connection forms on $\widetilde{M} \to \Sigma$ that agree with the given form $\widetilde{\omega}$ on the fibers. Every such connection form can be written as an exact perturbation $\widetilde{\omega}_H := \widetilde{\omega} - d\widetilde{\sigma}_H$ of $\widetilde{\omega}$, where the 1-form $\widetilde{\sigma}_H$ vanishes on the fibers. Moreover, by Proposition 8.2.2 (ii) both $\widetilde{\omega}$ and $\widetilde{\omega}_H$ are given in local coordinates by expressions of the form (8.2.2). Therefore the real valued 1-form $\widetilde{\sigma}_H$ must be of the type considered in Section 8.1. In other words, it is obtained from a 1-form H on Σ with values in the bundle whose fiber at $z \in \Sigma$ is $C_0^\infty(M_z)$. If the latter form is denoted by

$$T_z\Sigma \to \mathbb{C}_0^\infty(M_z) : \zeta \mapsto H_\zeta,$$

then the associated 1-form $\widetilde{\sigma}_H$ on $\widetilde{M}$ is given by

$$\widetilde{\sigma}_H(\widetilde{x};\widetilde{v}) = H_\zeta(\widetilde{x}), \qquad \zeta := d\pi(\widetilde{x})\widetilde{v},$$

for $\widetilde{v} \in T_{\widetilde{x}}\widetilde{M}$. With this notation, the horizontal subspace at $\widetilde{x} \in \widetilde{M}$ associated to the connection form $\widetilde{\omega} - d\widetilde{\sigma}_H$ is given by

$$\mathrm{Hor}_{H;\widetilde{x}} = \left\{\widetilde{v} - X_{H_\zeta}(\widetilde{x}) \,|\, \widetilde{v} \in \mathrm{Hor}_{\widetilde{x}},\, \zeta = d\pi(\widetilde{x})\widetilde{v}\right\},$$

where $\mathrm{Hor}_{\widetilde{x}} \subset T_{\widetilde{x}}\widetilde{M}$ denotes the horizontal subspace associated to $\widetilde{\omega}$ (given by equation (8.2.1)), and $X_{H_\zeta} \in \mathrm{Vect}(M_z, \omega_z)$ denotes the Hamiltonian vector field associated to the Hamiltonian function $H_\zeta : M_z \to \mathbb{R}$ for $\zeta \in T_z\Sigma$. It follows that $\widetilde{J}_H$ is given by

$$\widetilde{J}_H\widetilde{v} = \widetilde{J}\widetilde{v} + JX_{H_\zeta}(\widetilde{x}) - X_{H_{j\zeta}}(\widetilde{x}), \qquad \zeta := d\pi(\widetilde{x})\widetilde{v},$$

for $\widetilde{v} \in T_{\widetilde{x}}\widetilde{M}$, where $j := j_\Sigma$.

Loops of Hamiltonian symplectomorphisms. We end this section by constructing explicit connection forms on (locally trivial) Hamiltonian fibrations

$$\pi : \widetilde{M} \to S^2.$$

The most important point is Remark 8.2.11 (iv) which will be used in Section 9.6. By Proposition 8.2.2, we can trivialize the fiber bundle $\pi : \widetilde{M} \to S^2$ over each hemisphere and hence express the total space $\widetilde{M}$ as a union of two trivial bundles over two discs that are glued together over their common boundary. This gluing map is a loop of Hamiltonian symplectomorphisms. Conversely, given such a loop one can construct a fibration over S^2 as follows.

Let (M, ω) be a compact symplectic manifold and

$$\mathbb{R}/\mathbb{Z} \to \mathrm{Diff}(M, \omega) : t \mapsto \psi_t$$

be a loop of Hamiltonian symplectomorphisms. Associated to the loop $\psi = \{\psi_t\}$ is a Hamiltonian fibration

$$M \hookrightarrow \widetilde{M}_\psi \xrightarrow{\pi} S^2$$

defined by gluing together two copies of $\mathbb{C} \times M$ via the loop ψ:

$$\widetilde{M}_\psi := \{\pm 1\} \times \mathbb{C} \times M/\sim \tag{8.2.6}$$

where

$$(1, e^{2\pi(s+it)}, x) \;\sim\; (-1, e^{2\pi(-s-it)}, \psi_t^{-1}(x)).$$

The projection $\pi : \widetilde{M}_\psi \to S^2 = \mathbb{C} \cup \{\infty\}$ is given by $\pi([1, z, x]) := z$. We will often use polar coordinates s, t on $\mathbb{C} \setminus \{0\}$, setting $z = e^{2\pi(s+it)}$. In this notation a section of $\widetilde{M}_\psi$ is a pair $\widetilde{u} = (u^+, u^-)$, where $u^\pm : \mathbb{R} \times \mathbb{R}/\mathbb{Z} \to M$ are smooth functions such that

$$u^+(s,t) = \psi_t(u^-(-s,-t))$$

and the maps $\mathbb{C} \setminus \{0\} \to M : e^{2\pi(s+it)} \mapsto u^\pm(s,t)$ extend smoothly over the origin.

REMARK 8.2.11. **(i)** A connection form $\widetilde{\omega} \in \Omega^2(\widetilde{M}_\psi)$ can be expressed as pair of 2-forms $\omega^\pm \in \Omega^2(\mathbb{C} \times M)$. We write these 2-forms in polar coordinates as

$$\omega^\pm = \omega - d'F^\pm \wedge ds - d'G^\pm \wedge dt + (\partial_t F^\pm - \partial_s G^\pm)\, ds \wedge dt,$$

where d' denotes the differential on M and the functions

$$e^{2\pi(s+it)} \mapsto e^{2\pi(-s-it)}(F^\pm_{s,t} + iG^\pm_{s,t})$$

extend smoothly over the origin. The functions $F^\pm$ and $G^\pm$ are related by

$$F^+_{s,t}\circ\psi_t + F^-_{-s,-t} = 0, \qquad G^+_{s,t}\circ\psi_t + G^-_{-s,-t} = -H_t\circ\psi_t,$$

where the Hamiltonian functions $H_t = H_{t+1} : M \to \mathbb{R}$ generate the Hamiltonian loop ψ via $\partial_t\psi_t = X_{H_t}\circ\psi_t$ and satisfy $\int_M H_t\omega^n = 0$ for each t.

(ii) Let $\sigma \in \Omega^2(S^2)$ be the standard volume form with volume π on $S^2 = \mathbb{C}\cup\{\infty\}$. In polar coordinates $\mathbb{R}\times\mathbb{R}/\mathbb{Z}\to\mathbb{C} : (s,t)\mapsto e^{2\pi(s+it)}$ this volume form is given by

$$\sigma = \frac{4\pi^2 e^{4\pi s}}{(1+e^{4\pi s})^2}\, ds\wedge dt.$$

(iii) The curvature of a connection form $\widetilde{\omega}\in\Omega^2(\widetilde{M}_\psi)$ as in (i) is the 2-form $R_{\widetilde{\omega}}\sigma$ on S^2 with values in the vector bundle of functions on the fibers with mean value zero. Here the function $R = R_{\widetilde{\omega}} : \widetilde{M}_\psi \to \mathbb{R}$ is given by

$$R^\pm(s,t,x) := \frac{(1+e^{4\pi s})^2}{4\pi^2 e^{4\pi s}}\left(\partial_s G^\pm_{s,t}(x) - \partial_t F^\pm_{s,t}(x) + \{F^\pm_{s,t}, G^\pm_{s,t}\}(x)\right)$$

in the two trivializations.

(iv) Let $\kappa\in\Omega^0(S^2)$ and denote $\kappa^\pm(s,t) := \kappa(e^{\pm 2\pi(s+it)})$ and $\omega^\pm_\kappa := \omega^\pm + \kappa^\pm\pi^*\sigma$. The formula

$$\frac{(\omega^\pm_\kappa)^{n+1}}{(n+1)!} = \left(\kappa^\pm - R^\pm_\tau\right)\sigma\wedge\frac{\omega^n}{n!}$$

shows that the function κ can be recovered from the 2-form $\widetilde{\omega}_\kappa := \widetilde{\omega} + \pi^*(\kappa\sigma)$ as the fiber integral

$$\kappa\sigma := \frac{1}{\mathrm{Vol}(M)}\int_{\text{fiber}} \frac{\widetilde{\omega}_\kappa^{n+1}}{(n+1)!}.$$

The formula also shows that $\widetilde{\omega}_\kappa$ is a symplectic form if and only if the function $\kappa\circ\pi - R_{\widetilde{\omega}}$ never vanishes. We write $\widetilde{\omega}_\kappa^{n+1} > 0$ iff $\kappa\circ\pi > R_{\widetilde{\omega}}$ and $\widetilde{\omega}_\kappa^{n+1} < 0$ iff $\kappa\circ\pi < R_{\widetilde{\omega}}$.

8.3. Pseudoholomorphic sections

Our next goal is to examine the moduli space of pseudoholomorphic sections of a locally Hamiltonian fibration. Throughout $\pi : \widetilde{M}\to\Sigma$ is a locally Hamiltonian fibration over a closed Riemann surface $(\Sigma, \mathrm{dvol}_\Sigma, j_\Sigma)$ equipped with a connection form $\widetilde{\omega}\in\Omega^2(\widetilde{M})$ as in Definition 8.2.1. We shall fix a vertical almost complex structure $J = \{J_z\}$ as in Definition 8.2.7 and prove that the moduli space of pseudoholomorphic sections of $\widetilde{M}$ is smooth for a generic exact perturbation of $\widetilde{\omega}$ that preserves the symplectic forms on the fibers. Most of the proofs are very close to those in Section 6.7 and we shall only sketch them. Although we frame the results for a general locally Hamiltonian fibration, the reader can specialize to the case of the trivial bundle $\pi : \widetilde{M} = \Sigma\times M\to\Sigma$ discussed in Section 8.1. In this case, the initial connection form $\widetilde{\omega}$ should be taken to be the pullback $\widetilde{\omega}_0 := \mathrm{pr}^*\omega$ of ω under the obvious projection $\mathrm{pr} : \widetilde{M}\to M$. Further the space $\mathcal{J}^{\mathrm{Vert}}(\widetilde{M},\pi,\widetilde{\omega})$ of vertical almost complex structures can be identified with the space $C^\infty(\Sigma, \mathcal{J}_\tau(M,\omega))$ of families $J := \{J_z\}_{z\in\Sigma}$. Each such family gives rise to an almost complex structure $\widetilde{J}$ on $(\widetilde{M},\widetilde{\omega}_0)$ that equals (j_Σ, J_z) on $T_{z,x}(\Sigma\times M)$.

The moduli space of holomorphic sections. Fix a vertical almost complex structure $J \in \mathcal{J}^{\mathrm{Vert}}(\widetilde{M}, \pi, \widetilde{\omega})$ and let $\widetilde{J} \in \mathcal{J}(\widetilde{M}, \pi, \widetilde{\omega})$ be the corresponding almost complex structure on $\widetilde{M}$. Given a homology class $\widetilde{A} \in H_2(\widetilde{M}; \mathbb{Z})$ consider the moduli space

$$\mathcal{M}(\widetilde{A}; \widetilde{J}) := \left\{ \widetilde{u} : \Sigma \to \widetilde{M} \,|\, \bar{\partial}_{\widetilde{J}}(\widetilde{u}) = 0,\ \pi \circ \widetilde{u} = \mathrm{id}_\Sigma,\ [\widetilde{u}] = \widetilde{A} \right\}.$$

of $\widetilde{J}$-holomorphic sections $\widetilde{u} : \Sigma \to \widetilde{M}$ that represent the class $\widetilde{A}$. Note that this space can only be nonempty when $\pi_* \widetilde{A} = [\Sigma]$. If this holds and $\widetilde{u} : \Sigma \to \widetilde{M}$ is any $\widetilde{J}$-holomorphic curve representing the class $\widetilde{A}$, then $\pi \circ \widetilde{u} : \Sigma \to \Sigma$ is a holomorphic diffeomorphism. Hence $\widetilde{u} \circ (\pi \circ \widetilde{u})^{-1}$ is a $\widetilde{J}$-holomorphic section. Thus the moduli space $\mathcal{M}(\widetilde{A}; \widetilde{J})$ can be identified with the quotient of the space of all $\widetilde{J}$-holomorphic curves in $\widetilde{M}$ representing the class $\widetilde{A}$ by the group of holomorphic automorphisms of Σ. In particular, if the only such automorphism is the identity, then the notation used here agrees with the one in Section 3.1.

Throughout we shall fix the vertical almost complex structure J and the vertical symplectic forms ω_z. We shall then consider exact perturbations

$$\widetilde{\omega}_H = \widetilde{\omega} - d\widetilde{\sigma}_H$$

of the connection form as in equation (8.1.2) and Remark 8.2.10. Here H is a 1-form on Σ with values in the vector bundle of smooth functions on the fibers of $\widetilde{M}$ with mean value zero. We denote by $\mathcal{H}$ the space of such Hamiltonian perturbations. If $\widetilde{M}$ is the product $\Sigma \times M$, then one can set $\widetilde{\omega} = \widetilde{\omega}_0$ so that $\omega_z = \omega$ for every $z \in \Sigma$, and then use the formulas from Section 8.1.

Given $H \in \mathcal{H}$ let $\widetilde{J}_H \in \mathcal{J}(\widetilde{M}, \pi, \widetilde{\omega}_H)$ denote the almost complex structure induced by J. As explained in Exercise 8.1.5 the $\widetilde{J}_H$-holomorphic sections of $\widetilde{M}$ are locally the graphs of solutions of a perturbed Cauchy–Riemann equation. Moreover, Remark 8.2.10 shows that a section $\widetilde{u} : \Sigma \to \widetilde{M}$ is a $\widetilde{J}_H$-holomorphic curve if and only if it satisfies the equation

$$\bar{\partial}_{\widetilde{J}}(\widetilde{u}) + X_H(\widetilde{u})^{0,1} = 0, \tag{8.3.1}$$

where $X_H(\widetilde{u}) \in \Omega^1(\Sigma, \widetilde{u}^* T^{\mathrm{Vert}} \widetilde{M})$ is the 1-form with values in the pullback of the vertical tangent bundle $T^{\mathrm{Vert}} \widetilde{M} \subset T\widetilde{M}$, given by

$$T_z \Sigma \to T^{\mathrm{Vert}}_{\widetilde{u}(z)} \widetilde{M} : \zeta \mapsto X_{H_\zeta}(\widetilde{u}(z)).$$

Let $\mathcal{M}(\widetilde{A}; J, H) := \mathcal{M}(\widetilde{A}; \widetilde{J}_H)$ denote the moduli space of $\widetilde{J}_H$-holomorphic sections of $\widetilde{M}$ that represent the class $\widetilde{A}$. We wish to prove that, given $J \in \mathcal{J}^{\mathrm{Vert}}(\widetilde{M}, \pi, \widetilde{\omega})$, this moduli space is a smooth finite dimensional manifold for a generic Hamiltonian perturbation. The relevant set of regular perturbations will be denoted by

$$\mathcal{H}_{\mathrm{reg}}(\widetilde{A}, J) := \left\{ H \in \mathcal{H} \,|\, \widetilde{u} \in \mathcal{M}(\widetilde{A}; J, H) \Longrightarrow D_{J,H,\widetilde{u}} \text{ is onto} \right\}. \tag{8.3.2}$$

Here we denote by $T^{\mathrm{Vert}} \widetilde{M} \subset T\widetilde{M}$ the vertical tangent bundle and by

$$D_{\widetilde{u}} = D_{J,H,\widetilde{u}} : \Omega^0(\Sigma, \widetilde{u}^* T^{\mathrm{Vert}} \widetilde{M}) \to \Omega^{0,1}(\Sigma, \widetilde{u}^* T^{\mathrm{Vert}} \widetilde{M})$$

the linearized operator associated to the Cauchy–Riemann equation (8.3.1). The precise definition of this operator will be given below. The next theorem is the main result of this section. The proof will be sketched at the end of the section. We write c_1^{Vert} for the first Chern class of the vertical tangent bundle.

THEOREM 8.3.1. *Fix a homology class $\widetilde{A} \in H_2(\widetilde{M};\mathbb{Z})$ and a vertical almost complex structure $J \in \mathcal{J}^{\mathrm{Vert}}(\widetilde{M},\pi,\widetilde{\omega})$.*

(i) *If $H \in \mathcal{H}_{\mathrm{reg}}(\widetilde{A},J)$ then the space $\mathcal{M}(\widetilde{A};J,H)$ is a smooth manifold of dimension*

$$\dim \mathcal{M}(\widetilde{A};J,H) = n(2-2g) + 2c_1^{\mathrm{Vert}}(\widetilde{A}).$$

It carries a natural orientation.

(ii) *The set $\mathcal{H}_{\mathrm{reg}}(\widetilde{A},J)$ is residual in $\mathcal{H}$.*

The proof is almost the same as that of Theorem 3.1.6. There are two differences. One is that, as in Section 6.7, we only consider sections, so that the tangent space $T_{\widetilde{u}}\mathcal{B}$ to the space $\mathcal{B}$ of sections consists of vertical vector fields along $\widetilde{u}$. The second is that we are here considering a restricted set $\mathcal{J}_{\mathcal{H}}$ of almost complex structures rather than the set of all those that are $\widetilde{\omega}$-tame (or $\widetilde{\omega}$-compatible). However, $\mathcal{J}_{\mathcal{H}}$ is a manifold, and it is easy to check from the explicit formulas given in Remark 8.2.10 that at each point $(z,x) \in M_z \subset \widetilde{M}$ the tangent vectors $(X_h(\widetilde{u}))^{0,1}$ to $\mathcal{J}_{\mathcal{H}}$ span the vertical tangent space T_xM_z. This is the analogue of Lemma 3.2.2, and is precisely what is needed to prove that the appropriate universal moduli space is a manifold: see Remark 3.2.3 and the proof of Theorem 8.3.1 given below.

REMARK 8.3.2. A smooth section $\widetilde{u}:\Sigma\to\widetilde{M}$ is called **horizontal for $\widetilde{\omega}_H$** if

$$d\pi(\widetilde{u}(z))\widetilde{v} = 0 \qquad \Longrightarrow \qquad \widetilde{\omega}_H(d\widetilde{u}(z)\zeta,\widetilde{v}) = 0$$

for every $\zeta \in T_z\Sigma$ and every $\widetilde{v} \in T_{\widetilde{u}(z)}\widetilde{M}$. Such a section is $\widetilde{J}_H$-holomorphic for every vertical almost complex structure J. In many interesting examples the horizontal sections are the only $\widetilde{J}_H$-holomorphic sections of $\widetilde{M}$ that one can write down explicitly and they can sometimes be used to compute the invariants (see, for example, the work of Seidel [**362, 363**], Polterovich [**325, 326**], and Akveld–Salamon [**11**]). Now one can prove a similar transversality result as Theorem 8.3.1 for generic vertical almost complex structures J with fixed H (and hence fixed horizontal distribution), however, the argument only applies to the nonhorizontal sections. Apart from the technicalities of the proof it is easy to see why transversality for horizontal sections cannot be achieved by perturbing J. If $\widetilde{u}:\Sigma\to\widetilde{M}$ is a horizontal section for $\widetilde{\omega}_H$ and the operator $D_{H,\widetilde{u}}$ has negative index, then $\widetilde{u}$ is $\widetilde{J}_H$-holomorphic for every J, but $D_{H,\widetilde{u}}$ can never be surjective.

Cobordisms. As always, we need to show that the cobordism class of the moduli space is independent of choices. The following discussion is routine and is included for the sake of completeness.

Suppose $H_0 \in \mathcal{H}_{\mathrm{reg}}(\widetilde{A},J_0)$ and $H_1 \in \mathcal{H}_{\mathrm{reg}}(\widetilde{A},J_1)$. Fix any smooth homotopy $[0,1]\to\mathcal{J}^{\mathrm{Vert}}(\widetilde{M},\pi,\widetilde{\omega}):\lambda\to J_\lambda$ of vertical almost complex structures from J_0 to J_1. For a smooth homotopy $[0,1]\to\mathcal{H}:\lambda\to H_\lambda$ from H_0 to H_1 define

$$\mathcal{W}(\widetilde{A};\{J_\lambda,H_\lambda\}_\lambda) = \Big\{(\lambda,\widetilde{u}) \,|\, 0\le\lambda\le 1,\, \widetilde{u}\in\mathcal{M}(\widetilde{A};J_\lambda,H_\lambda)\Big\}.$$

Denote by $\mathcal{H}(H_0,H_1)$ the space of all smooth homotopies of Hamiltonian perturbations connecting H_0 to H_1. A homotopy $\{H_\lambda\}_\lambda \in \mathcal{H}(H_0,H_1)$ is called **regular (for $\widetilde{A}$ and $\{J_\lambda\}_\lambda$)** if $H_0 \in \mathcal{H}_{\mathrm{reg}}(\widetilde{A},J_0)$, $H_1 \in \mathcal{H}_{\mathrm{reg}}(\widetilde{A},J_1)$, and

$$\Omega^{0,1}(\Sigma,\widetilde{u}^*T^{\mathrm{Vert}}\widetilde{M}) = \operatorname{im} D_{J_\lambda,H_\lambda,\widetilde{u}} + \mathbb{R}\left(\tfrac{1}{2}(\partial_\lambda J_\lambda)d_{H_\lambda}(\widetilde{u})j_\Sigma + (X_{\partial_\lambda H_\lambda}(\widetilde{u}))^{0,1}\right)$$

for every $(\lambda, \widetilde{u}) \in \mathcal{W}(\widetilde{A}; \{J_\lambda, H_\lambda\}_\lambda)$. Here

$$d_H(\widetilde{u}) := \Pi_H(\widetilde{u})d\widetilde{u} : T\Sigma \to \widetilde{u}^*T^{\mathrm{Vert}}\widetilde{M}$$

denotes the composition of $d\widetilde{u}$ with the projection Π_H onto the vertical tangent bundle along the horizontal distribution determined by $\widetilde{\omega} - d\widetilde{\sigma}_H$ for $H \in \mathcal{H}$. The space of regular homotopies will be denoted by $\mathcal{H}_{\mathrm{reg}}(\widetilde{A}, \{J_\lambda\}_\lambda; H_0, H_1)$. The proof of the following result is analogous to that of Theorem 8.3.1 (at the end of this section) and will be omitted.

THEOREM 8.3.3. *Let $\widetilde{A} \in H_2(\widetilde{M}; \mathbb{Z})$, $H_0 \in \mathcal{H}_{\mathrm{reg}}(\widetilde{A}, J_0)$, $H_1 \in \mathcal{H}_{\mathrm{reg}}(\widetilde{A}, J_1)$, and suppose that $[0,1] \to \mathcal{J}^{\mathrm{Vert}}(\widetilde{M}, \pi, \widetilde{\omega}) : \lambda \to J_\lambda$ is a smooth homotopy of vertical almost complex structures from J_0 to J_1.*

(i) *If $\{H_\lambda\}_\lambda \in \mathcal{H}_{\mathrm{reg}}(\widetilde{A}, \{J_\lambda\}_\lambda; H_0, H_1)$ then $\mathcal{W}(\widetilde{A}; \{J_\lambda, H_\lambda\}_\lambda)$ is a smooth oriented manifold with boundary*

$$\partial\mathcal{W}(\widetilde{A}; \{J_\lambda, H_\lambda\}_\lambda) = \mathcal{M}(\widetilde{A}; J_1, H_1) - \mathcal{M}(\widetilde{A}; J_0, H_0).$$

The minus sign indicates the reversed orientation.

(ii) *The set $\mathcal{H}_{\mathrm{reg}}(\widetilde{A}, \{J_\lambda\}_\lambda; H_0, H_1)$ is residual in $\mathcal{H}(H_0, H_1)$.*

In preparation of the proof of Theorem 8.3.1 we shall now discuss the analytic setup for pseudoholomorphic sections of locally Hamiltonian fibrations. We shall begin by defining a natural connection on the vertical tangent bundle $T^{\mathrm{Vert}}\widetilde{M} \to \widetilde{M}$, which we will use to derive an explicit formula for the linearization

$$D_{J,H,\widetilde{u}} : \Omega^0(\Sigma, \widetilde{u}^*T^{\mathrm{Vert}}\widetilde{M}) \to \Omega^{0,1}(\Sigma, \widetilde{u}^*T^{\mathrm{Vert}}\widetilde{M}).$$

As we pointed out at the beginning of Section 3.1, one does not in fact need such a formula in order to prove Theorem 8.3.1; if $\widetilde{u}$ is $\widetilde{J}_H$-holomorphic then $D_{J,H,\widetilde{u}}$ does not depend on a connection and can be understood as a vertical differential. All that is needed is that $D_{J,H,\widetilde{u}}$ is a Cauchy–Riemann operator and is therefore Fredholm. Granted this the proof of Theorem 8.3.1 can be understood without the explicit formula given below.

The vertical Levi-Civita connection. The vector bundle $T^{\mathrm{Vert}}\widetilde{M} \to \widetilde{M}$ carries a connection ∇, induced by the vertical almost complex structure J and the horizontal distribution of $\widetilde{\omega}$. It is called the vertical Levi–Civita connection and is characterized by the following axioms.

(VERTICAL) The restriction of ∇ to each fiber (M_z, ω_z) is the Levi-Civita connection of the metric $g_z = \frac{1}{2}(\omega_z(\cdot, J_z\cdot) - \omega_z(J_z\cdot, \cdot))$.

(HORIZONTAL) Let $\gamma : \mathbb{R} \to \Sigma$ be a smooth path and denote by $\psi_t : M_{\gamma(0)} \to M_{\gamma(t)}$ parallel transport with respect to the horizontal distribution of $\widetilde{\omega}$. Suppose that $(\widetilde{x}, \xi) : \mathbb{R} \to T^{\mathrm{Vert}}\widetilde{M}$ is a horizontal lift of γ, i.e.

$$\pi(\widetilde{x}(t)) = \gamma(t), \qquad \widetilde{x}(t) = \psi_t(\widetilde{x}(0)), \qquad \xi(t) = d\psi_t(\widetilde{x}(0))\xi(0).$$

Then the vertical vector field ξ along $\widetilde{x}$ is parallel with respect to ∇.

(NATURAL) If $\psi : \widetilde{M}_0 \to \widetilde{M}_1$ is a fiberwise diffeomorphism such that $\psi^*\widetilde{\omega}_1 = \widetilde{\omega}_0$ and $\psi^*J_1 = J_0$ then $\psi^*\nabla_1 = \nabla_0$.

(HAMILTONIAN) Let $\widetilde{\sigma}_H \in \Omega^1(\widetilde{M})$ be a horizontal 1-form. Then the vertical Levi-Civita connection ∇_H associated to J and $\widetilde{\omega} - d\widetilde{\sigma}_H$ is given by

$$\nabla_{H,\widetilde{v}}\xi(\widetilde{x}) = \nabla_{\widetilde{v}}\xi(\widetilde{x}) + \nabla_\xi X_{H_\zeta}(\widetilde{x}), \qquad \zeta := d\pi(\widetilde{x})\widetilde{v}, \tag{8.3.3}$$

for every vertical vector field $\xi : \widetilde{M} \to T^{\mathrm{Vert}}\widetilde{M}$ and every $\widetilde{v} \in T_{\widetilde{x}}\widetilde{M}$.

The vertical Levi-Civita connection is uniquely determined by the first two axioms. In other words, ∇ is simply a combination of the fiberwise Levi-Civita connections with the symplectic connection in the fiber bundle $T^{\mathrm{Vert}}\widetilde{M} \to \Sigma$. That the first two conditions are invariant under fiberwise diffeomorphisms is obvious and so they are consistent with the *(Natural)* axiom. On a trivial bundle $\widetilde{M} = \Sigma \times M$ the *(Hamiltonian)* axiom can also be viewed as a definition of the vertical Levi-Civita connection and one then has to verify that it satisfies the *(Horizontal)* axiom. This is a simple consequence of the formula for the horizontal distribution of $\widetilde{\omega}_H$ in Remark 8.2.10. With this understood one can use the *(Hamiltonian)* axiom to define the connection in the local trivializations of Proposition 8.2.2, and use the *(Natural)* axiom to show that it is globally well defined.

EXERCISE 8.3.4. Let ∇ be the vertical Levi-Civita connection of $\widetilde{\omega}$. Let $H \in \mathcal{H}$ be a Hamiltonian perturbation and suppose that ∇_H is the connection on $T^{\mathrm{Vert}}\widetilde{M} \to \widetilde{M}$ defined by (8.3.3). Prove that ∇_H satisfies the *(Horizontal)* axiom for $\widetilde{\omega}_H$. *Hint:* Consider the case $\widetilde{M} = \Sigma \times M$ and $\widetilde{\omega} = \mathrm{pr}^*\omega$.

EXERCISE 8.3.5. Let ∇ be the vertical Levi-Civita connection on $T^{\mathrm{Vert}}\widetilde{M}$ induced by $\widetilde{\omega}$ and J. Let $\mathbb{R}^2 \to M : (s,t) \mapsto \widetilde{x}(s,t)$ be a smooth path such that $d\pi(\widetilde{x})\partial_t\widetilde{x} = 0$. Prove that

$$\nabla_s\partial_t\widetilde{x} = \nabla_s(\Pi(\widetilde{x})\partial_t\widetilde{x}),$$

where $\Pi : T\widetilde{M} \to T^{\mathrm{Vert}}\widetilde{M}$ denotes the projection onto the vertical subspace along the horizontal distribution Hor (see equation (8.2.5)).

In general, the vertical Levi-Civita connection does not preserve the vertical inner product along directions that are not themselves vertical. Just as in Section 3.1, the next step is to define a connection $\widetilde{\nabla}$ which does preserve the almost complex structure as well as the inner product. The next lemma shows that this can be defined by the same formula as before.

LEMMA 8.3.6. *Let $(\widetilde{M}, \pi, \widetilde{\omega})$ be a locally Hamiltonian fibration over a Riemann surface $(\Sigma, j_\Sigma, \mathrm{dvol}_\Sigma)$ and $J \in \mathcal{J}^{\mathrm{Vert}}(\widetilde{M}, \pi, \widetilde{\omega})$. Let ∇ be the corresponding vertical Levi-Civita connection on $T^{\mathrm{Vert}}\widetilde{M}$. Then the formula*

$$\widetilde{\nabla}_{\widetilde{v}}X := \nabla_{\widetilde{v}}X - \tfrac{1}{2}J(\nabla_{\widetilde{v}}J)X$$

for $X \in \Omega^0(\widetilde{M}, T^{\mathrm{Vert}}\widetilde{M})$ defines a Hermitian connection on $T^{\mathrm{Vert}}\widetilde{M}$.

PROOF. It suffices to consider the trivial bundle $\mathbb{R}\times M$. Then a vertical almost complex structure is a smooth map $\mathbb{R} \to \mathcal{J}(M,\omega) : t \mapsto J_t$ and

$$\widetilde{\omega}_H = \omega - dH \wedge dt$$

where $H : \mathbb{R}\times M \to \mathbb{R}$. We write $H_t(x) := H(t,x)$ whenever convenient and denote by $X_t := X_{H_t} \in \mathrm{Vect}(M,\omega)$ the Hamiltonian vector field of H_t. Consider a path

$\mathbb{R} \mapsto \mathbb{R} \times M : t \mapsto (t, x(t))$ and a vertical vector field $\xi(t) \in T_{x(t)}M$ along this path. Then

$$\nabla_{H,t}\xi = \nabla_t\xi + \nabla_\xi X_t(x),$$

where $\nabla = \nabla^t$ denotes the Levi-Civita connection of the metric $\omega(\cdot, J_t\cdot)$ on TM. An easy calculation shows that

$$\nabla_{H,t}J(x) = \nabla_{\dot{x}+X_t(x)}J_t(x) + (\partial_t J_t)(x) - \mathcal{L}_{X_t}J_t(x),$$

where $(\mathcal{L}_X J)\xi = (\nabla_X J)\xi - \nabla_{J\xi}X + J\nabla_\xi X$ denotes the Lie derivative of J in the direction of X. Hence

$$\begin{aligned}\widetilde{\nabla}_{H,t}\xi &= \nabla_{H,t}\xi - \tfrac{1}{2}J(\nabla_{H,t}J)\xi \\ &= \nabla_t\xi - \tfrac{1}{2}J(\partial_t J)\xi - \tfrac{1}{2}J(\nabla_{\dot{x}+X}J)\xi + \nabla_\xi X + \tfrac{1}{2}J(\mathcal{L}_X J)\xi.\end{aligned}$$

Here we have dropped the subscript t and the argument x. The formula

$$\frac{d}{dt}\langle\xi_1, \xi_2\rangle = \langle\nabla_t\xi_1, \xi_2\rangle + \langle\xi_1, \nabla_t\xi_2\rangle - \langle\xi_1, J(\partial_t J)\xi_2\rangle$$

shows that $\nabla_t - \frac{1}{2}J(\partial_t J)$ is a Riemannian connection. Moreover, the endomorphism $J(\nabla_{\dot{x}+X}J)$ of T_xM is skew-adjoint. The formula

$$\langle\xi_1, \nabla_{\xi_2}X\rangle + \langle\nabla_{\xi_1}X, \xi_2\rangle + \langle\xi_1, J(\mathcal{L}_X J)\xi_2\rangle = 0$$

(for Hamiltonian vector fields) shows that the endomorphism

$$\xi \mapsto \nabla_\xi X + \tfrac{1}{2}J(\mathcal{L}_X J)\xi$$

is skew adjoint as well. Hence $\widetilde{\nabla}_H$ preserves the Riemannian metric. That it preserves the almost complex structure follows directly from the definition. □

The linearized operator. We are now ready to introduce the linearized operator for $\widetilde{J}$-holomorphic sections. Fix a compact Riemann surface $(\Sigma, j_\Sigma, \mathrm{dvol}_\Sigma)$ without boundary, a locally Hamiltonian fibration $\pi : \widetilde{M} \to \Sigma$ with connection form $\widetilde{\omega}$, and a vertical almost complex structure $J \in \mathcal{J}^{\mathrm{Vert}}(\widetilde{M}, \pi, \widetilde{\omega})$. For any smooth section $\widetilde{u} : \Sigma \to \widetilde{M}$ define the linear operator

$$D_{\widetilde{u}} = D_{J,\widetilde{u}} : \Omega^0(\Sigma, \widetilde{u}^*T^{\mathrm{Vert}}\widetilde{M}) \to \Omega^{0,1}(\Sigma, \widetilde{u}^*T^{\mathrm{Vert}}\widetilde{M})$$

by

$$D_{\widetilde{u}}\xi := (\nabla\xi)^{0,1} - \tfrac{1}{2}J(\nabla_\xi J)\partial_{\widetilde{J}}(\widetilde{u}),$$

where ∇ denotes the vertical Levi-Civita connection and $\widetilde{J}$ the almost complex structure on $\widetilde{M}$ induced by $\widetilde{\omega}$ and J. As in Section 3.1, this is a real linear Cauchy–Riemann operator. If J and $\widetilde{\omega}$ are of class C^ℓ and u is of class $W^{\ell,p}$ for some integer ℓ and some constant $p > 2$ then $D_{\widetilde{u}}$ is well defined as an operator from $W^{k,p}$-sections of $\widetilde{u}^*T^{\mathrm{Vert}}\widetilde{M}$ to $(0,1)$-forms on Σ of class $W^{k-1,p}$ with values in $\widetilde{u}^*T^{\mathrm{Vert}}\widetilde{M}$ for every $k \in \{1, \ldots, \ell\}$. It is a real linear Cauchy–Riemann operator of class $W^{\ell-1,p}$ and so Theorem C.2.3 applies. Thus Proposition 3.1.11 extends to the operator

$$D_{\widetilde{u}} : W^{k,p}(\Sigma, \widetilde{u}^*T^{\mathrm{Vert}}\widetilde{M}) \to W^{k-1,p}(\Sigma, \Lambda^{0,1}T^*\Sigma \otimes_J \widetilde{u}^*T^{\mathrm{Vert}}\widetilde{M}).$$

In particular, the operator is Fredholm and its index is

$$\mathrm{index}\, D_{\widetilde{u}} = n(2 - 2g) + 2\langle c_1(T^{\mathrm{Vert}}\widetilde{M}), [\widetilde{u}]\rangle.$$

REMARK 8.3.7. The geometric motivation for introducing this operator arises from an abstract setting analogous to that of Section 3.1. For any section $\widetilde{u} : \Sigma \to \widetilde{M}$, $\widetilde{J}$-holomorphic or not, there is a map

$$\mathcal{F}_{\widetilde{u}} : \Omega^0(\Sigma, \widetilde{u}^*T^{\mathrm{Vert}}\widetilde{M}) \to \Omega^{0,1}(\Sigma, \widetilde{u}^*T^{\mathrm{Vert}}\widetilde{M}), \qquad \mathcal{F}_{\widetilde{u}} := \Phi_{\widetilde{u}}(\xi)^{-1}\bar{\partial}_{\widetilde{J}}(\exp_{\widetilde{u}}(\xi)).$$

Here $\exp_{\widetilde{x}} : T_{\widetilde{x}}M_z \to M_z$ denotes the vertical exponential map associated to the metric g_z determined by ω_z and J_z and $\Phi_{\widetilde{x}}(\xi) : T_{\widetilde{x}}M_z \to T_{\exp_{\widetilde{x}}(\xi)}M_z$ denotes parallel transport along vertical geodesics with respect to $\widetilde{\nabla}$. As in Proposition 3.1.1, the differential of $\mathcal{F}_{\widetilde{u}}$ at zero is $d\mathcal{F}_{\widetilde{u}}(0) = D_{\widetilde{u}}$.

REMARK 8.3.8. Let $\widetilde{\sigma}_H \in \Omega^1(\widetilde{M})$ be a horizontal 1-form as in Remark 8.2.10. Denote the corresponding nonlinear Cauchy–Riemann operator (the left hand side of equation (8.3.1)) by

$$\bar{\partial}_{J,H}(\widetilde{u}) := \bar{\partial}_{\widetilde{J}}(\widetilde{u}) + (X_H(\widetilde{u}))^{0,1}.$$

Then the linearized operator

$$D_{H,\widetilde{u}} = D_{J,H,\widetilde{u}} : \Omega^0(\Sigma, \widetilde{u}^*T^{\mathrm{Vert}}\widetilde{M}) \to \Omega^{0,1}(\Sigma, \widetilde{u}^*T^{\mathrm{Vert}}\widetilde{M})$$

associated to J and $\widetilde{\omega} - d\widetilde{\sigma}_H$ is given by

$$\begin{aligned} D_{H,\widetilde{u}}\xi &= (\nabla_H \xi)^{0,1} - \tfrac{1}{2}J(\nabla_\xi J)\partial_{J,H}(\widetilde{u}) \\ &= D_{\widetilde{u}}\xi + \left(\nabla_\xi X_H(\widetilde{u}) - \tfrac{1}{2}J(\nabla_\xi J)X_H(\widetilde{u})\right)^{0,1}. \end{aligned}$$

Here ∇_H denotes the vertical Levi-Civita connection on $T^{\mathrm{Vert}}\widetilde{M}$ associated to the connection form $\widetilde{\omega} - d\widetilde{\sigma}_H$ and the vertical almost complex structure J.

PROOF OF THEOREM 8.3.1. That $\mathcal{M}(\widetilde{A}; J, H)$ is a manifold of the appropriate dimension for every $H \in \mathcal{H}_{\mathrm{reg}}(\widetilde{A}, J)$ follows immediately from the definitions and the implicit function theorem A.3.3. That it carries a natural orientation follows as in the proof of Theorem 3.1.6 (i). To prove (ii) we fix an integer $\ell \geq 2$, denote by $\mathcal{H}^\ell$ the Banach space of Hamiltonian perturbations of class C^ℓ, and consider the universal moduli space

$$\mathcal{M}(\widetilde{A}; J, \mathcal{H}^\ell) := \left\{(\widetilde{u}, H) \,|\, H \in \mathcal{H}^\ell, \widetilde{u} \in \mathcal{M}(\widetilde{A}; J, H)\right\}.$$

This space is the zero set of a $\mathbb{C}^{\ell-1}$ section of a $C^{\ell-1}$ Banach space bundle over the Banach manifold $\mathcal{B}^{1,p} \times \mathcal{H}^\ell$, where $\mathcal{B}^{1,p}$ denotes the space of $W^{1,p}$-sections $\widetilde{u} : \Sigma \to \widetilde{M}$ that represent the class $\widetilde{A}$. In proving that this universal moduli space is itself a $C^{\ell-1}$ Banach manifold we encounter the linearized operator

$$W^{1,p}(\Sigma, \widetilde{u}^*T^{\mathrm{Vert}}\widetilde{M}) \times \mathcal{H}^\ell \to L^p(\Sigma, \Lambda^{0,1}T^*\Sigma \otimes_J \widetilde{u}^*T^{\mathrm{Vert}}\widetilde{M}),$$

given by

$$(\xi, h) \mapsto D_{H,\widetilde{u}}\xi + (X_h(\widetilde{u}))^{0,1}.$$

We must prove that this operator is surjective for every $H \in \mathcal{H}^\ell$ and every $\widetilde{u} \in \mathcal{M}(\widetilde{A}; J, H)$. To see this let $q > 1$ such that $1/p + 1/q = 1$ and suppose that the section $\eta \in L^q(\Sigma, \Lambda^{0,1}T^*\Sigma \otimes_J \widetilde{u}^*T^{\mathrm{Vert}}\widetilde{M})$ annihilates the image of this operator. Since $\widetilde{J}_H$ is an almost complex structure of class $C^{\ell-1}$ it follows from Proposition 3.1.10 that $\widetilde{u}$ is of class $W^{\ell,p}$ and hence $D_{H,\widetilde{u}}$ is a real linear Cauchy–Riemann operator of class $W^{\ell-1,p}$. Hence it follows from Theorem C.2.3 that $\eta \in W^{\ell,p}$,

$$D_{H,\widetilde{u}}^*\eta = 0,$$

and

$$\int_\Sigma \langle \eta, X_h(\widetilde{u}) \rangle \, \mathrm{dvol}_\Sigma = 0 \tag{8.3.4}$$

for every $h \in \mathcal{H}^\ell$. We prove that η must vanish. Suppose otherwise that $\eta(z_0) \neq 0$ for some point $z_0 \in \Sigma$. Then there is a linear map $T_{z_0}\Sigma \to C_0^\infty(M_{z_0}) : \zeta \mapsto h_{0,\zeta}$ such that

$$\langle \eta(z_0), X_{h_0}(\widetilde{u}(z_0)) \rangle > 0.$$

Let $h \in \mathcal{H}^\ell$ be any Hamiltonian perturbation which agrees with h_0 at $z = z_0$. Then the scalar function $\langle \eta, X_h(\widetilde{u}) \rangle$ on Σ is positive in some neighbourhood $U_0 \subset \Sigma$ of z_0. Choose a smooth cutoff function $\beta : \Sigma \to [0,1]$ with support in U_0 such that $\beta(z_0) = 1$. Then the integral on the left hand side of (8.3.4) with h replaced by βh is positive, a contradiction. Thus we have proved that η must vanish, and hence the linearized operator is surjective for every pair $(\widetilde{u}, H) \in \mathcal{M}(\widetilde{A}; J, \mathcal{H}^\ell)$. It follows from the implicit function theorem A.3.3 that $\mathcal{M}(\widetilde{A}; J, \mathcal{H}^\ell)$ is a $C^{\ell-1}$-Banach manifold. As a submanifold of a separable Banach manifold it is itself separable, i.e. it contains a countable dense subset. The proof is now based on the Sard–Smale theorem for the projection $\pi : \mathcal{M}(\widetilde{A}; J, \mathcal{H}^\ell) \to \mathcal{H}^\ell$. The argument is exactly the same as in Theorem 3.1.6 and will be omitted. This proves Theorem 8.3.1. □

8.4. Pseudoholomorphic spheres in the fiber

Our goal is to prove that a suitable evaluation map on the moduli space of pseudoholomorphic sections is a pseudocycle. As in the case of graphs, discussed in Section 6.7, the stable maps in the boundary of the space of sections consist of a section together with spheres that lie in the fibers. The analysis of these spheres for general fibrations $\widetilde{M} \to \Sigma$ is almost identical to the case when $\widetilde{M}$ is a product, and so we will content ourselves with stating the results.

Let $J \in \mathcal{J}^{\mathrm{Vert}} := \mathcal{J}^{\mathrm{Vert}}(\widetilde{M}, \pi, \widetilde{\omega})$ be a vertical almost complex structure, denote by $\widetilde{J}$ the almost complex structure on $\widetilde{M}$ determined by J and $\widetilde{\omega}$, and let $\widetilde{A} \in H_2(\widetilde{M}; \mathbb{Z})$ be a spherical homology class such that

$$\pi_* \widetilde{A} = 0.$$

Then there exists a unique collection of spherical homology classes $A_z \in H_2(M_z; \mathbb{Z})$, indexed by $z \in \Sigma$, such that $\iota_{z*} A_z = \widetilde{A}$, where $\iota_z : M_z \to \widetilde{M}$ denotes the inclusion of the fiber. The condition $\pi_* \widetilde{A} = 0$ implies that, for every $\widetilde{J}$-holomorphic sphere $\widetilde{v} : S^2 \to \widetilde{M}$, the composition $\pi \circ \widetilde{v}$ is constant. Moreover,

$$c_1^{\mathrm{Vert}}(\widetilde{A}) = \langle c_1(TM_z), A_z \rangle = \langle c_1(T\widetilde{M}), \widetilde{A} \rangle = c_1(\widetilde{A}).$$

In the following we shall consider the moduli space

$$\mathcal{M}^*(\widetilde{A}; \widetilde{J}) := \{(z, v) \mid z \in \Sigma,\, v \in \mathcal{M}^*(A_z; J_z)\}$$

of simple J-holomorphic spheres in the fibers representing the class $\widetilde{A}$. Associated to a $\widetilde{J}$-holomorphic sphere $\widetilde{v} : S^2 \to \widetilde{M}$ is a Cauchy–Riemann operator

$$D_{\widetilde{v}} = D_{\widetilde{J}, \widetilde{v}} : \Omega^0(S^2, \widetilde{v}^* T\widetilde{M}) \to \Omega^{0,1}(S^2, \widetilde{v}^* T\widetilde{M})$$

as introduced in Section 3.1. We emphasize that the almost complex structure $\widetilde{J}$ and the operator $D_{\widetilde{v}}$ depend on the connection form $\widetilde{\omega}$, even though the moduli space $\mathcal{M}^*(\widetilde{A}; \widetilde{J})$ only depends on the vertical almost complex structure J. However, as in Remark 6.7.8, the kernel of $D_{\widetilde{v}}$ is independent of this choice for every (vertical) $\widetilde{J}$-holomorphic sphere $\widetilde{v} : S^2 \to \widetilde{M}$. Namely, since every such sphere satisfies $\pi \circ \widetilde{v} \equiv \text{const}$ it is interesting to consider the restricted operator

$$D_{\widetilde{v}}^{\text{Vert}} : \left\{ \tilde{\xi} \in \Omega^0(S^2, \widetilde{v}^* T\widetilde{M}) \,|\, d\pi(\widetilde{v})\tilde{\xi} \equiv \text{const} \right\} \to \Omega^{0,1}(S^2, \widetilde{v}^* T^{\text{Vert}}\widetilde{M}).$$

The formula $d\pi(\widetilde{v}) \circ D_{\widetilde{v}} = \bar{\partial} \circ d\pi(\widetilde{v})$ shows that this operator is well defined, that it has the same kernel as $D_{\widetilde{v}}$, and that both operators have isomorphic cokernels. Thus $D_{\widetilde{v}}$ and $D_{\widetilde{v}}^{\text{Vert}}$ are Fredholm operators of the same index

$$\text{index}\, D_{\widetilde{v}}^{\text{Vert}} = \text{index} D_{\widetilde{v}} = 2n + 2 + 2c_1(\widetilde{A}).$$

and $D_{\widetilde{v}}^{\text{Vert}}$ is surjective if and only if $D_{\widetilde{v}}$ is surjective. We leave it to the reader to check that the operator $D_{\widetilde{v}}^{\text{Vert}}$ is independent of the choice of $\widetilde{\omega}$. Note that when the bundle is trivial, $D_{\widetilde{v}}^{\text{Vert}}$ is just the operator $D_{z,v}$ of equation (6.7.8) and the preceding claims were proved in Remark 6.7.8.

Let $\widetilde{A} \in H_2(\widetilde{M};\mathbb{Z})$ such that $\pi_* \widetilde{A} = 0$. A vertical almost complex structure $J \in \mathcal{J}^{\text{Vert}}$ is called **regular for** $\widetilde{A}$ if the operator $D_{\widetilde{v}}$ is surjective for every simple $\widetilde{J}$-holomorphic sphere $\widetilde{v} : S^2 \to \widetilde{M}$ that represents the class $\widetilde{A}$. The set of regular vertical almost complex structures for $\widetilde{A}$ will be denoted by $\mathcal{J}_{\text{reg}}^{\text{Vert}} = \mathcal{J}_{\text{reg}}^{\text{Vert}}(\widetilde{A})$. The proof of the following theorem is almost word by word the same as that of Theorem 3.1.6 and will be omitted.

THEOREM 8.4.1. *Let $\widetilde{A} \in H_2(\widetilde{M};\mathbb{Z})$ such that $\pi_* \widetilde{A} = 0$.*

(i) *If $J \in \mathcal{J}_{\text{reg}}^{\text{Vert}}(\widetilde{A})$ then $\mathcal{M}^*(\widetilde{A};\widetilde{J})$ is a smooth oriented manifold of dimension*

$$\dim \mathcal{M}^*(\widetilde{A};\widetilde{J}) = 2n + 2 + 2c_1(\widetilde{A}).$$

(ii) *The set $\mathcal{J}_{\text{reg}}^{\text{Vert}}(\widetilde{A})$ is residual in $\mathcal{J}^{\text{Vert}}$.*

Now let $J_0, J_1 \in \mathcal{J}_{\text{reg}}^{\text{Vert}}(\widetilde{A})$ and $[0,1] \to \mathcal{J}^{\text{Vert}} : \lambda \mapsto J_\lambda$ be a smooth homotopy from J_0 to J_1. For any such homotopy consider the moduli space

$$\mathcal{W}^*(\widetilde{A};\{\widetilde{J}_\lambda\}_\lambda) = \left\{ (\lambda, \widetilde{v}) \,|\, 0 \le \lambda \le 1,\, \widetilde{v} \in \mathcal{M}^*(\widetilde{A};\widetilde{J}_\lambda) \right\}.$$

A homotopy $\{J_\lambda\}_\lambda$ is called **regular** if $J_0, J_1 \in \mathcal{J}_{\text{reg}}^{\text{Vert}}(\widetilde{A})$ and

$$\Omega^{0,1}(S^2, \widetilde{v}^* T^{\text{Vert}}\widetilde{M}) = \text{im}\, D_{J_\lambda,\widetilde{v}}^{\text{Vert}} + \mathbb{R}(\partial_\lambda J_\lambda) d\widetilde{v} \circ i$$

for every $\lambda \in [0,1]$ and every simple $\widetilde{J}_\lambda$-holomorphic sphere $\widetilde{v} : S^2 \to \widetilde{M}$ representing the class $\widetilde{A}$. The space of regular homotopies will be denoted by $\mathcal{J}_{\text{reg}}^{\text{Vert}}(\widetilde{A}; J_0, J_1)$ and the space of all smooth homotopies by $\mathcal{J}^{\text{Vert}}(J_0, J_1)$. The proof of the following theorem is almost word by word the same as that of Theorem 3.1.8 and can safely be omitted.

THEOREM 8.4.2. *Let $J_0, J_1 \in \mathcal{J}_{\text{reg}}^{\text{Vert}}(\widetilde{A})$ and $\widetilde{A} \in H_2(\widetilde{M};\mathbb{Z})$ such that $\pi_* \widetilde{A} = 0$.*

(i) *If $\{J_\lambda\}_\lambda \in \mathcal{J}_{\text{reg}}^{\text{Vert}}(\widetilde{A}; J_0, J_1)$ then $\mathcal{W}^*(\widetilde{A};\{\widetilde{J}_\lambda\}_\lambda)$ is a smooth oriented manifold with boundary*

$$\partial\mathcal{W}^*(\widetilde{A};\{\widetilde{J}_\lambda\}_\lambda) = \mathcal{M}^*(\widetilde{A};\widetilde{J}_1) - \mathcal{M}^*(\widetilde{A};\widetilde{J}_0).$$

(ii) *The set $\mathcal{J}_{\text{reg}}^{\text{Vert}}(\widetilde{A}; J_0, J_1)$ is residual in $\mathcal{J}^{\text{Vert}}(J_0, J_1)$.*

EXAMPLE 8.4.3. Let $M = S^2 \times S^2$ with the product symplectic structure $\omega_\lambda := \lambda\pi_1^*\sigma + \pi_2^*\sigma$, where $\lambda > 1$, σ is the standard area form on S^2 and π_i is the projection to the ith factor. Consider the fibration $\pi_1 : M \to S^2$. There is a circle action $\{\psi_t\}_{t\in S^1}$ on M that rotates each fiber, fixing the points on the diagonal $\{(x,x)\,|\,x \in S^2 \subset \mathbb{R}^3\}$ and antidiagonal $\{(x,-x)\,|\,x \in S^2 \subset \mathbb{R}^3\}$. (A formula for this action is given in equation (9.7.1) below.) We show in Proposition 9.7.2 (ii) that when $\lambda > 1$ the loop $\{\psi_t\}$ is isotopic to a loop $\{\psi_t^\lambda\}$ in $\mathrm{Ham}(M,\omega_\lambda)$. Consider the bundle $(M,\omega) \to \widetilde{M} \to S^2$ formed from this loop as in the discussion before Remark 8.2.11. The class $\widetilde{A}$ of the antidiagonal in M has self-intersection -2 in M and Chern number 0, and hence has no J-holomorphic representative for generic ω^λ-tame J. (Cf. Example 7.2.2.) However, it is shown in Le–Ono [**236**] that $c_1(\widetilde{M})(\widetilde{A}) = 1$, and that

$$\mathrm{GW}^{\widetilde{M}}_{\widetilde{A},0} = 1.$$

They interpret this result in terms of parametric Gromov–Witten invariants. (Also see Buse [**54**].) The analytic point here is that if the curve

$$\widetilde{v} = \iota \circ v : \; S^2 \xrightarrow{v} M \overset{\iota}{\hookrightarrow} \widetilde{M}$$

lies in a single fiber, then the image of the linearized operator $D_{\widetilde{v}}^{\mathrm{Vert}}$ described above differs from that of the corresponding operator D_v in M by some terms coming from the derivatives of the family J_z with respect to the base variable $z \in \Sigma$. (The precise formula is given in equation (6.7.9).)

REMARK 8.4.4. One of the applications of the Seidel representation which will be constructed later using the methods we are presently developing (see Example 8.6.8) is to show that the rational homology of the total space of a Hamiltonian fibration $(\widetilde{M},\widetilde{\omega}) \to \Sigma$ is a product:

$$H_*(\widetilde{M};\mathbb{Q}) \cong H_*(\Sigma;\mathbb{Q}) \otimes H_*(M;\mathbb{Q}).$$

In the case of semipositive (M,ω) this is proved in Proposition 11.4.5, following Lalonde–McDuff–Polterovich [**228**]. The proof in the general case may be found in McDuff [**269**]. It follows that the rational class $A_z \in H_2(M_z;\mathbb{Q})$ is uniquely determined by $\widetilde{A}$, and can be identified with a single class $A \in H_2(M;\mathbb{Q})$. (Here we use the fact that each fiber M_z of $\widetilde{M} \to \Sigma$ can be identified with a reference fiber by a symplectomorphism that is well defined up to a Hamiltonian isotopy.)

The above statement about the structure of $H_*(\widetilde{M};\mathbb{Q})$ does not hold for general locally Hamiltonian fibrations over surfaces: consider, for example, a mapping torus $Y_f \to \mathbb{T}^2$ as in Section 8.2 that is built from the map $f : S^2 \times S^2$ that interchanges the two spheres.

The ideas in this section can be developed in a much broader context, since if one is only interested in spheres in the fiber there is no need to restrict to fibrations over surfaces; indeed the base need not be a symplectic manifold. See Seidel [**372**], in particular Example 1.13, for some very interesting applications.

8.5. The pseudocycle of sections

This section is devoted to constructing pseudocycles in $\widetilde{M}^k$ from the moduli space of sections $\mathcal{M}(\widetilde{A}; J, H)$. Many of the needed concepts are very close to those in Section 6.7. However, our current perspective is somewhat different; the setup in Section 6.7 emphasized the curves u in M rather than their graphs $\widetilde{u}$, while the present discussion is entirely in terms of $\widetilde{u}$.

Let us fix a vertical almost complex structure $J \in \mathcal{J}^{\mathrm{Vert}}$, a homology class $\widetilde{A} \in H_2(\widetilde{M};\mathbb{Z})$ such that

$$\pi_*\widetilde{A} = [\Sigma],$$

and a regular Hamiltonian perturbation $H \in \mathcal{H}_{\mathrm{reg}}(\widetilde{A}, J)$. These data determine a smooth moduli space $\mathcal{M}(\widetilde{A}; J, H)$ of $\widetilde{J}_H$-holomorphic sections that represent the class $\widetilde{A}$. We denote by $\mathcal{M}_{\Sigma,k}(\widetilde{A}; J, H)$ the moduli space of tuples

$$\mathcal{M}_{\Sigma,k}(\widetilde{A}; J, H) := \left\{(\widetilde{u}, z_1, \ldots, z_k) \,|\, z_i \neq z_j \; \forall i \neq j\right\} \subset \mathcal{M}(\widetilde{A}; J, H) \times \Sigma^k.$$

It has dimension

$$\dim \mathcal{M}_{\Sigma,k}(\widetilde{A}; J, H) = \mu(\widetilde{A}, g, k) := n(2-2g) + 2c_1^{\mathrm{Vert}}(\widetilde{A}) + 2k,$$

where $c_1^{\mathrm{Vert}} \in H^2(\widetilde{M};\mathbb{Z})$ denotes the first Chern class of the vertical tangent bundle and g the genus of Σ (see Theorem 8.3.1). We aim to find conditions under which the evaluation map

$$\widetilde{\mathrm{ev}} = \widetilde{\mathrm{ev}}_{J,H} : \mathcal{M}_{\Sigma,k}(\widetilde{A}; J, H) \to \widetilde{M}^k,$$

given by $\widetilde{\mathrm{ev}}(\widetilde{u}, z_1, \ldots, z_k) := (\widetilde{u}(z_1), \ldots, \widetilde{u}(z_k))$, is a pseudocycle.

We shall need a condition on the fibers that is slightly stronger than semipositivity. The condition must guarantee that, in the regular case, all J-holomorphic spheres in the fibers have nonnegative Chern numbers.[2] Since we have assumed that the base Σ is connected, all the fibers are symplectomorphic to a single manifold (M,ω) and the relevant condition on (M,ω) is that it is either semimonotone, or its first Chern class vanishes on spheres, or its minimal Chern number is $N \geq n-1$. This condition can also be expressed in the form

$$\omega(A) > 0, \quad c_1(A) \geq 2-n \qquad \Longrightarrow \qquad c_1(A) \geq 0 \tag{8.5.1}$$

for every $A \in \pi_2(M)$, where $c_1 := c_1(TM, J) \in H^2(M;\mathbb{Z})$. It implies that the semipositivity condition in Definition 6.4.1 holds for the classes represented by spheres in the fibers of $(\widetilde{M}, \widetilde{\omega})$. If Σ has positive genus, then it follows that $(\widetilde{M}, \widetilde{\omega}_\kappa)$ is a semipositive symplectic manifold whenever $\kappa : \Sigma \to \mathbb{R}$ satisfies (8.1.5). However, in the case $\Sigma = S^2$, condition (8.5.1) is weaker than the semipositivity of $\widetilde{M}$. Our main application in Section 9.6 concerns Hamiltonian fibrations over the 2-sphere where the fibers are monotone but the total space is not semipositive.

THEOREM 8.5.1. *Let $(\widetilde{M}, \pi, \widetilde{\omega})$ be a locally Hamiltonian fibration over a compact Riemann surface $(\Sigma, j_\Sigma, \mathrm{dvol}_\Sigma)$ of genus g with compact fibers of dimension $2n$ that satisfy (8.5.1). Fix an integer $k \geq 0$ and a homology class $\widetilde{A} \in H_2(\widetilde{M};\mathbb{Z})$ such that $\pi_*\widetilde{A} = [\Sigma]$. Then there exists a residual set $\mathcal{JH}_{\mathrm{reg}}(\widetilde{M}, \pi, \widetilde{\omega}) \subset \mathcal{J}^{\mathrm{Vert}} \times \mathcal{H}$ with the following properties.*

(i) *The evaluation map $\widetilde{\mathrm{ev}}_{J,H} : \mathcal{M}_{\Sigma,k}(\widetilde{A}; J, H) \to \widetilde{M}^k$ is a pseudocycle of dimension $n(2-2g) + 2c_1^{\mathrm{Vert}}(\widetilde{A}) + 2k$ for every pair $(J, H) \in \mathcal{JH}_{\mathrm{reg}}(\widetilde{M}, \pi, \widetilde{\omega})$.*

(ii) *If $(J_0, H_0), (J_1, H_1) \in \mathcal{JH}_{\mathrm{reg}}(\widetilde{M}, \pi, \widetilde{\omega})$ then $\widetilde{\mathrm{ev}}_{J_0,H_0}$ is bordant to $\widetilde{\mathrm{ev}}_{J_1,H_1}$.*

[2]This stronger assumption was not needed in Section 6.7 since there we could assume that the family $\{J_z\}$ was almost constant; cf. the definition of $\mathcal{J}_+(S^2; M, \omega; \kappa)$.

The elements of $\mathcal{JH}_{\mathrm{reg}}(\widetilde{M}, \pi, \widetilde{\omega})$ satisfy analogues of the conditions in Definition 6.7.10 in Section 6.7. The precise formulation is given in Definition 8.5.2 below.

Stable maps. The appropriate notion of stable maps is essentially the same as that in Section 6.7. Let us fix a vertical almost complex structure $J \in \mathcal{J}^{\mathrm{Vert}}$ and a Hamiltonian perturbation $H \in \mathcal{H}$. Let us also fix a k-labelled tree $T = (T, E, \Lambda)$ with a **special vertex** $0 \in T$ and a collection of spherical homology classes $\{\widetilde{A}_\alpha\}_{\alpha \in T}$ that satisfy the stability condition (6.1.1) and

$$\pi_* \widetilde{A}_0 = [\Sigma], \qquad \pi_* \widetilde{A}_\alpha = 0, \tag{8.5.2}$$

for every $\alpha \in T \setminus \{0\}$. Associated to these data is a moduli space of stable maps

$$\mathcal{M}_{\Sigma,T}(\{\widetilde{A}_\alpha\}; J, H) := \widetilde{\mathcal{M}}_{\Sigma,T}(\{\widetilde{A}_\alpha\}; J, H)/\mathrm{G}_T$$

defined as follows. An element of $\widetilde{\mathcal{M}}_{\Sigma,T}(\{\widetilde{A}_\alpha\}; J, H)$ is a stable map

$$(\tilde{\mathbf{u}}, \mathbf{z}) = (\{\widetilde{u}_\alpha\}_{\alpha \in T}, \{z_{\alpha\beta}\}_{\alpha E \beta}, \{z_i\}_{1 \le i \le k})$$

such that $\widetilde{u}_\alpha : \Sigma_\alpha \to \widetilde{M}$ is a $\widetilde{J}_H$-holomorphic curve representing the class $\widetilde{A}_\alpha$ for every $\alpha \in T$, the points $z_{\alpha\beta} \in \Sigma_\alpha$ for $\alpha E \beta$ and $z_i \in \Sigma_\alpha$ for $\alpha = \alpha_i$ are pairwise distinct for every $\alpha \in T$, and $\widetilde{u}_\alpha(z_{\alpha\beta}) = \widetilde{u}_\beta(z_{\beta\alpha})$ for $\alpha E \beta$. Moreover, we impose the conditions $\Sigma_0 = \Sigma$, $\pi \circ \widetilde{u}_0 = \mathrm{id}_\Sigma$, and $\Sigma_\alpha = \mathbb{C}P^1$ for $\alpha \neq 0$. This is the natural extension of the notion of stable maps to pseudoholomorphic sections. The reparametrization group G_T consists of tuples $(f, \{\phi_\alpha\}_{\alpha \in T})$, where $f : T \to T$ is a tree automorphism, such that $f(0) = 0$, $\phi_0 = \mathrm{id}$, and $\phi_\alpha \in \mathrm{G} = \mathrm{PSL}(2, \mathbb{C})$ for $\alpha \neq 0$. It acts on the moduli space of stable maps as in Definition 5.1.4. In the case $\Sigma = \mathbb{C}P^1$, this is just a special case of the definitions in Chapter 5.

Theorem 8.5.1 refers only to the properties of moduli spaces of sections where all the marked points are allowed to vary freely. The pseudocycles so obtained define Gromov–Witten invariants in which there is no restriction on the marked points. We shall see in Proposition 8.6.5 that the invariants for fixed marked points can then be recovered by choosing cohomology classes in $\widetilde{M}$ whose Poincaré duals can be represented by submanifolds of a fiber. However, to compute these invariants and to prove the relevant gluing theorems, it is sometimes useful to suppose that some of the marked points are fixed. This leads us to examine moduli spaces of stable maps with some fixed marked points. (In later applications we shall only consider the extreme cases when either no points are fixed or all points are fixed. The general case is included here for completeness.)

Fix a subset $I \subset \{1, \ldots, k\}$ and a tuple $\mathbf{w} = \{\mathrm{w}_i\}_{i \in I} \in \Sigma^I$ of pairwise distinct points on Σ indexed by I. Any such tuple determines a subset

$$\mathcal{M}_{\Sigma,T}(\{\widetilde{A}_\alpha\}; \mathbf{w}, J, H) := \widetilde{\mathcal{M}}_{\Sigma,T}(\{\widetilde{A}_\alpha\}; \mathbf{w}, J, H)/\mathrm{G}_T$$

consisting of all stable maps $(\tilde{\mathbf{u}}, \mathbf{z})$ that satisfy

$$i \in I \qquad \Longrightarrow \qquad \pi(\widetilde{u}_{\alpha_i}(z_i)) = \mathrm{w}_i. \tag{8.5.3}$$

The corresponding moduli space of simple stable maps will be denoted by

$$\mathcal{M}^*_{\Sigma,T}(\{\widetilde{A}_\alpha\}; \mathbf{w}, J, H) := \widetilde{\mathcal{M}}^*_{\Sigma,T}(\{\widetilde{A}_\alpha\}; \mathbf{w}, J, H)/\mathrm{G}_T.$$

It is defined exactly as in Section 6.1. In particular

$$\mathcal{M}_{\Sigma,k}(\widetilde{A}; \mathbf{w}, J, H) := \mathcal{M}(\widetilde{A}; J, H) \times \Sigma^k(\mathbf{w}), \tag{8.5.4}$$

where $\Sigma^k(\mathbf{w}) \subset \Sigma^k$ is the set of k-tuples $\mathbf{z} = (z_1, \ldots, z_k)$ of pairwise distinct points in Σ such that $z_i = \mathrm{w}_i$ for $i \in I$.

As in Section 6.2, the set of simple stable maps can be described as the preimage of an evaluation map. There is a moduli space $\mathcal{M}^*(\{\widetilde{A}_\alpha\}; J, H)$ of simple tuples $\{\widetilde{u}_\alpha\}$ representing the classes $\widetilde{A}_\alpha$. This space is an oriented finite dimensional manifold whenever $J \in \mathcal{J}^{\mathrm{Vert}}_{\mathrm{reg}}(\widetilde{A}_\alpha)$ for every $\alpha \in T \setminus \{0\}$ and $H \in \mathcal{H}_{\mathrm{reg}}(\widetilde{A}_0, J)$ (see equation (8.3.2)). Moreover, let $Z(T)$ denote the set of tuples

$$\mathbf{z} = (\{z_{\alpha\beta}\}_{\alpha E \beta}, \{z_i\}_{1 \le i \le k})$$

such that the points $z_{\alpha\beta} \in \Sigma_\alpha$ for $\alpha E \beta$ and $z_i \in \Sigma_\alpha$ for $\alpha_i = \alpha$ are pairwise distinct for every $\alpha \in T$. Then there is an evaluation map

$$\widetilde{\mathrm{ev}}^E \times \pi^I : \mathcal{M}^*(\{\widetilde{A}_\alpha\}; J, H) \times Z(T) \to \widetilde{M}^E \times \Sigma^I, \tag{8.5.5}$$

defined by

$$\widetilde{\mathrm{ev}}^E(\tilde{\mathbf{u}}, \mathbf{z}) := \{\widetilde{u}_\alpha(z_{\alpha\beta})\}_{\alpha E \beta}, \qquad \pi^I(\tilde{\mathbf{u}}, \mathbf{z}) := \{\pi(\widetilde{u}_{\alpha_i}(z_i))\}_{i \in I}.$$

The moduli space of simple stable maps satisfying (8.5.3) is the preimage of the submanifold $\widetilde{\Delta}^E \times \{\mathbf{w}\}$ under $\widetilde{\mathrm{ev}}^E \times \pi^I$. This leads to the following definition.

Definition 8.5.2. *Let $\mathbf{w} \in \Sigma_I$. A pair $(J, H) \in \mathcal{J}^{\mathrm{Vert}} \times \mathcal{H}$ is called* **regular for w** *if it satisfies the following conditions.*

(H) *$H \in \mathcal{H}_{\mathrm{reg}}(\widetilde{A}, J)$ for every $\widetilde{A} \in H_2(\widetilde{M}; \mathbb{Z})$ with $\pi_* \widetilde{A} = [\Sigma]$.*

(V) *$J \in \mathcal{J}^{\mathrm{Vert}}_{\mathrm{reg}}(\widetilde{A})$ for every $\widetilde{A} \in H_2(\widetilde{M}; \mathbb{Z})$ with $\pi_* \widetilde{A} = 0$.*

(EW) *The evaluation map (8.5.5) is transverse to $\widetilde{\Delta}^E \times \{\mathbf{w}\}$ for every k-labelled tree $T = (T, E, \Lambda)$ with a special vertex 0 and every collection $\{\widetilde{A}_\alpha\}_{\alpha \in T}$ of spherical homology classes in $H_2(\widetilde{M}; \mathbb{Z})$ that satisfy the stability conditions (6.1.1) and (8.5.2).*

The set of regular pairs (J, H) will be denoted in general by $\mathcal{J}\mathcal{H}_{\mathrm{reg}}(\widetilde{M}, \pi, \widetilde{\omega}; \mathbf{w})$, and by $\mathcal{J}\mathcal{H}_{\mathrm{reg}}(\widetilde{M}, \pi, \widetilde{\omega})$ when $I = \emptyset$.

Proof of the main result.

Theorem 8.5.3. *For every $\mathbf{w} \in \Sigma_I$ the set $\mathcal{J}\mathcal{H}_{\mathrm{reg}}(\widetilde{M}, \pi, \widetilde{\omega}; \mathbf{w})$ is residual in $\mathcal{J}^{\mathrm{Vert}} \times \mathcal{H}$.*

Sketch of a proof. Given a k-labelled forest $T = (T, E, \Lambda)$ and a collection of homology classes $\{\widetilde{A}_\alpha\}_{\alpha \in T}$ satisfying (8.5.2), let us denote by $\mathcal{J}\mathcal{H}^\ell_{\mathrm{reg}}(T, \{\widetilde{A}_\alpha\}; \mathbf{w})$ the set of all pairs $(J, H) \in \mathcal{J}^{\mathrm{Vert}, \ell} \times \mathcal{H}^\ell$ of class C^ℓ that satisfy the requirements of Definition 8.5.2 for T and $\{\widetilde{A}_\alpha\}$. One can prove in four steps that the set $\mathcal{J}\mathcal{H}^\ell_{\mathrm{reg}}(T, \{\widetilde{A}_\alpha\}; \mathbf{w})$ is dense in $\mathcal{J}^{\mathrm{Vert}, \ell} \times \mathcal{H}^\ell$ for all T and $\{\widetilde{A}_\alpha\}$. The first step is to observe that the universal moduli space $\mathcal{M}^*(\{\widetilde{A}_\alpha\}; \mathcal{J}^{\mathrm{Vert}, \ell}, \mathcal{H}^\ell)$ is a Banach manifold. This follows by combining the arguments in the proofs of Proposition 3.2.1, Proposition 6.2.7, and Theorem 8.3.1. The second step is to show that, if the universal evaluation map

$$\widetilde{\mathrm{ev}}^{E, \mathcal{J}, \mathcal{H}} : \mathcal{M}^*(\{\widetilde{A}_\alpha\}; \mathcal{J}^{\mathrm{Vert}, \ell}, \mathcal{H}^\ell) \times Z(T) \to \widetilde{M}^E$$

is transverse to $\widetilde{\Delta}^E$ then each tuple $\mathbf{w} \in \Sigma_I$ is a regular value of the universal projection

$$\pi^{I, \mathcal{J}, \mathcal{H}} : \mathcal{M}^*_{\Sigma, T}(\{\widetilde{A}_\alpha\}; \mathcal{J}^{\mathrm{Vert}, \ell}, \mathcal{H}^\ell) := \left(\widetilde{\mathrm{ev}}^{E, \mathcal{J}, \mathcal{H}}\right)^{-1} (\widetilde{\Delta}^E) \to \Sigma^I.$$

This follows as in Step 2 in the proof of Theorem 6.7.11. The third step is to prove that the evaluation map $\widetilde{\mathrm{ev}}^{E,\mathcal{J},\mathcal{H}}$ is indeed transverse to $\widetilde{\Delta}^E$. This follows by induction over the number of edges in the forest. The induction step combines the second step with the argument in the proof of Theorem 6.3.1 (as in Step 3 in the proof of Theorem 6.7.11). The fourth step is to consider the projection

$$\pi^\ell : \mathcal{M}^*_{\Sigma,T}(\{\widetilde{A}_\alpha\}; \mathbf{w}, \mathcal{J}^{\mathrm{Vert},\ell}, \mathcal{H}^\ell) := \left(\pi^{I,\mathcal{J},\mathcal{H}}\right)^{-1}(\mathbf{w}) \to \mathcal{J}^{\mathrm{Vert},\ell} \times \mathcal{H}^\ell.$$

This is a Fredholm map and a pair $(J, H) \in \mathcal{J}^{\mathrm{Vert},\ell} \times \mathcal{H}^\ell$ is a regular value of π^ℓ if and only if $(J, H) \in \mathcal{JH}^\ell_{\mathrm{reg}}(T, \{\widetilde{A}_\alpha\}; \mathbf{w})$. Hence the Sard–Smale theorem asserts that the set $\mathcal{JH}^\ell_{\mathrm{reg}}(T, \{\widetilde{A}_\alpha\}; \mathbf{w})$ is dense in $\mathcal{J}^{\mathrm{Vert},\ell} \times \mathcal{H}^\ell$.

The reduction of the C^∞ case to the C^ℓ case again uses Taubes' argument as in the proof Theorem 3.1.6. Finally, one takes the countable intersection over all T and $\{\widetilde{A}_\alpha\}$ to conclude that the set $\mathcal{JH}_{\mathrm{reg}}(\widetilde{M}, \pi, \widetilde{\omega}; \mathbf{w})$ is residual in $\mathcal{J}^{\mathrm{Vert}} \times \mathcal{H}$. This proves Theorem 8.5.3. □

PROOF OF THEOREM 8.5.1. The proof is analogous to that of Theorem 6.7.1 and we shall only sketch the main points. Let $(J, H) \in \mathcal{JH}_{\mathrm{reg}}(\widetilde{M}, \pi, \widetilde{\omega})$. For any k-labelled tree T with special vertex $0 \in T$ and any collection of spherical homology classes $\{\widetilde{B}_\alpha\}_{\alpha\in T}$ in $H_2(\widetilde{M}; \mathbb{Z})$ satisfying (6.1.1) and (8.5.2) we consider the evaluation map

$$\widetilde{\mathrm{ev}}_T : \mathcal{M}^*_{\Sigma,T}(\{\widetilde{B}_\alpha\}) \to \widetilde{M}^k,$$

given by

$$\widetilde{\mathrm{ev}}_T(\tilde{\mathbf{u}}, \mathbf{z}) := (\widetilde{u}_{\alpha_1}(z_1), \ldots, \widetilde{u}_{\alpha_k}(z_k)).$$

By a simple dimension counting argument we have

$$\dim \mathcal{M}^*_{\Sigma,T}(\{\widetilde{B}_\alpha\}) = \mu(\widetilde{B}, g, k) - 2e(T), \qquad \widetilde{B} := \sum_{\alpha\in T} \widetilde{B}_\alpha. \tag{8.5.6}$$

Moreover, by (8.5.1), the moduli space is empty unless $c_1(\widetilde{B}_\alpha) \geq 0$ for every $\alpha \in T$. Now it follows from Gromov compactness (Theorem 5.3.1 in the case $\Sigma = \mathbb{C}P^1$ and its obvious extension in the higher genus case) and Proposition 6.1.2 that the limit set of the evaluation map $\widetilde{\mathrm{ev}}$ can be covered by the union of the images of the above evaluation maps $\widetilde{\mathrm{ev}}_T$ over all k-labelled trees T with special vertex 0 and all collections of spherical homology classes $\{\widetilde{B}_\alpha\}_{\alpha\in T}$ that satisfy (6.1.1), (8.5.2), $e(T) > 0$, and

$$\widetilde{A} = \sum_{\alpha\in T} m_\alpha \widetilde{B}_\alpha, \tag{8.5.7}$$

for some collection of positive integers $\{m_\alpha\}_{\alpha\in T}$. The key observation is that, by the rescaling axiom for Gromov convergence, each bubble that appears in the limit stable map takes values in a single fiber of $\widetilde{M}$. Alternatively, one can argue that, for each $\widetilde{J}_H$-holomorphic curve $\widetilde{u}_\alpha : \Sigma_\alpha \to \widetilde{M}$ in the limit stable map, the projection $\pi \circ \widetilde{u}_\alpha : \Sigma_\alpha \to \Sigma$ is holomorphic and so has nonnegative degree. Since $\pi_* \widetilde{A} = [\Sigma]$ it follows that the sum of the degrees is one. Hence one of the maps $\pi \circ \widetilde{u}_\alpha$ has degree one and the others have degree zero. Since the component with degree one is simple, the same holds for the underlying simple stable map. Now the dimension formula (8.5.6) shows that all these moduli spaces have dimensions at most $\mu(\widetilde{A}, g, k) - 2$. This proves (i).

Now let $(J_i, H_i) \in \mathcal{JH}_{\mathrm{reg}}(\widetilde{M}, \pi, \widetilde{\omega})$ for $i = 0, 1$. Consider the corresponding evaluation maps

$$\widetilde{\mathrm{ev}}_0 : \mathcal{M}_{\Sigma,k}(\widetilde{A}; J_0, H_0) \to \widetilde{M}^k, \qquad \widetilde{\mathrm{ev}}_1 : \mathcal{M}_{\Sigma,k}(\widetilde{A}; J_1, H_1) \to \widetilde{M}^k.$$

We prove that $\widetilde{\mathrm{ev}}_1$ is bordant to $\widetilde{\mathrm{ev}}_0$ in the sense of Definition 6.5.1. This requires a homotopy argument which we only sketch. Choose a homotopy

$$[0, 1] \to \mathcal{J}^{\mathrm{Vert}} \times \mathcal{H} : \lambda \mapsto (J_\lambda, H_\lambda)$$

that satisfies the following conditions.

(H) $\{H_\lambda\} \in \mathcal{H}_{\mathrm{reg}}(\widetilde{A}, \{J_\lambda\}; H_0, H_1)$ for every $\widetilde{A} \in H_2(\widetilde{M}; \mathbb{Z})$ with $\pi_*\widetilde{A} = [\Sigma]$.

(V) $\{J_\lambda\} \in \mathcal{J}^{\mathrm{Vert}}_{\mathrm{reg}}(\widetilde{A}; J_0, J_1)$ for every $\widetilde{A} \in H_2(\widetilde{M}; \mathbb{Z})$ with $\pi_*\widetilde{A} = 0$.

(EW) The evaluation map

$$\widetilde{\mathrm{ev}}^E : \mathcal{W}^*(\{\widetilde{A}_\alpha\}; \{J_\lambda, H_\lambda\}) \times Z(T) \to \widetilde{M}^E,$$

is transverse to $\widetilde{\Delta}^E$ for every k-labelled tree $T = (T, E, \Lambda)$ with a special vertex 0 and every collection $\{\widetilde{A}_\alpha\}_{\alpha \in T}$ of spherical homology classes in $H_2(\widetilde{M}; \mathbb{Z})$ that satisfy (6.1.1) and (8.5.2).

That such a regular homotopy exists follows as in the proof of Theorem 8.5.3. Once the regular homotopies have been chosen one considers the moduli space

$$\mathcal{W}_{\Sigma,k}(\widetilde{A}; \{J_\lambda, H_\lambda\}) := \left\{ (\lambda, \widetilde{u}, \mathbf{z}) \,|\, 0 \le \lambda \le 1,\, (\widetilde{u}, \mathbf{z}) \in \mathcal{M}_{\Sigma,k}(\widetilde{A}; J_\lambda, H_\lambda) \right\}.$$

By Theorem 8.3.3, this space is a smooth oriented manifold with boundary

$$\partial \mathcal{W}_{\Sigma,k}(\widetilde{A}; \{J_\lambda, H_\lambda\}) = \mathcal{M}_{\Sigma,k}(\widetilde{A}; J_1, H_1) \cup \left(-\mathcal{M}_{\Sigma,k}(\widetilde{A}; J_0, H_0))\right).$$

It has dimension $\mu(\widetilde{A}, g, k) + 1$ and carries an evaluation map

$$\widetilde{\mathrm{ev}} : \mathcal{W}_{\Sigma,k}(\widetilde{A}; \{J_\lambda, H_\lambda\}) \to \widetilde{M}^k, \tag{8.5.8}$$

given by

$$\widetilde{\mathrm{ev}}(\lambda, \widetilde{u}, z_1, \ldots, z_k) := (\widetilde{u}(z_1), \ldots, \widetilde{u}(z_k)).$$

It follows again from Gromov compactness and Proposition 6.1.2 that the limit set of the evaluation map (8.5.8) can be covered by the union of the images of the evaluation maps

$$\widetilde{\mathrm{ev}}_T : \mathcal{W}^*_{\Sigma,T}(\{\widetilde{B}_\alpha\}; \{J_\lambda, H_\lambda\}) \to \widetilde{M}^k$$

over all k-labelled trees T and all collections of spherical homology classes $\{\widetilde{B}_\alpha\}_{\alpha \in T}$ that satisfy (6.1.1), (8.5.2), $e(T) > 0$, and (8.5.7). The moduli spaces in this union have dimensions $\mu(\widetilde{B}, g, k) + 1 - 2e(T) \le \mu(\widetilde{A}, g, k) - 1$. Hence the evaluation map (8.5.8) is a pseudocycle and is the required bordism from $\widetilde{\mathrm{ev}}_0$ to $\widetilde{\mathrm{ev}}_1$. This completes the sketch of the proof of Theorem 8.5.1. □

The following exercise defines an analogous pseudocycle in the case when some of the marked points are fixed.

EXERCISE 8.5.4. Fix an integer $k \ge 0$ and a subset $I \subset \{1, \ldots, k\}$. Any tuple $\mathbf{w} = \{\mathrm{w}_i\}_{i \in I}$ of pairwise distinct points in Σ determines a submanifold

$$\widetilde{M}^k(\mathbf{w}) := \left\{ \widetilde{\mathbf{x}} = (\widetilde{x}_1, \ldots, \widetilde{x}_k) \in \widetilde{M}^k \,|\, \pi(\widetilde{x}_i) = \mathrm{w}_i \text{ for } i \in I \right\}$$

of the product $\widetilde{M}^k$ and a moduli space

$$\mathcal{M}_{\Sigma,k}(\widetilde{A}; \mathbf{w}, J, H)$$

as in (8.5.4). Show that if $(J,H) \in \mathcal{JH}_{\mathrm{reg}}(\widetilde{M},\pi,\widetilde{\omega};\mathbf{w})$, then the evaluation map

$$\widetilde{\mathrm{ev}}_{\mathbf{w}} = \widetilde{\mathrm{ev}}_{\mathbf{w},J,H} : \mathcal{M}_{\Sigma,k}(\widetilde{A};\mathbf{w},J,H) \to \widetilde{M}^k(\mathbf{w}), \tag{8.5.9}$$

given by

$$\widetilde{\mathrm{ev}}_{\mathbf{w}}(\widetilde{u},z_1,\dots,z_k) := (\widetilde{u}(z_1),\dots,\widetilde{u}(z_k)),$$

is a pseudocycle of dimension $n(2-2g)+2c_1^{\mathrm{Vert}}(\widetilde{A})+2k-2|I|$. Deduce that if $\widetilde{\mathrm{ev}}_I$ denotes the composite of $\widetilde{\mathrm{ev}}_{\mathbf{w}}$ with the inclusion $\widetilde{M}^k(\mathbf{w}) \to \widetilde{M}^k$ the map

$$\widetilde{\mathrm{ev}}_I : \mathcal{M}_{\Sigma,k}(\widetilde{A};\mathbf{w},J,H) \to \widetilde{M}^k, \tag{8.5.10}$$

is a pseudocycle of the same dimension.

8.6. Counting pseudoholomorphic sections

We now examine the Gromov–Witten invariants associated to symplectic fiber bundles. Let $(\widetilde{M},\pi,\widetilde{\omega})$ be a locally Hamiltonian fibration over a compact Riemann surface $(\Sigma, j_\Sigma, \mathrm{dvol}_\Sigma)$ of genus g whose fibers (M_z,ω_z) are symplectomorphic to a compact symplectic manifold (M,ω) of dimension $2n$. We shall assume throughout that the fibers satisfy the strong semipositivity condition (8.5.1) which guarantees that all pseudoholomorphic spheres associated to a generic 2-parameter family of ω-tame almost complex structures on M have nonnegative Chern numbers. Choose $\widetilde{A} \in H_2(\widetilde{M};\mathbb{Z})$ such that $\pi_*\widetilde{A} = [\Sigma]$, let k be a nonnegative integer, and $(J,H) \in \mathcal{JH}_{\mathrm{reg}}(\widetilde{M},\pi,\widetilde{\omega})$ be a regular pair as in Definition 8.5.2. Then, by Theorem 8.5.1, the evaluation map

$$\widetilde{\mathrm{ev}}_{J,H} : \mathcal{M}_{\Sigma,k}(\widetilde{A};J,H) \to \widetilde{M}^k$$

is a pseudocycle of dimension

$$\mu(\widetilde{A},g,k) = n(2-2g)+2c_1^{\mathrm{Vert}}(\widetilde{A})+2k.$$

This gives rise to a Gromov–Witten invariant of pseudoholomorphic sections as follows.

THEOREM 8.6.1. *Let (M,ω) be a closed symplectic manifold of dimension $2n$ that satisfies (8.5.1), and let $(\widetilde{M},\pi,\widetilde{\omega})$ be a locally Hamiltonian fibration over a compact Riemann surface $(\Sigma,j_\Sigma,\mathrm{dvol}_\Sigma)$ with fibers symplectomorphic to (M,ω). Let $\widetilde{A} \in H_2(\widetilde{M};\mathbb{Z})$ be such that $\pi_*\widetilde{A} = [\Sigma]$, k be a nonnegative integer, and $(J,H) \in \mathcal{JH}_{\mathrm{reg}}(\widetilde{M},\pi,\widetilde{\omega})$. Then the homomorphism*

$$\mathrm{GW}^{\widetilde{M}}_{\widetilde{A},k} : H^*(\widetilde{M}^k) \to \mathbb{Z}$$

defined by

$$\mathrm{GW}^{\widetilde{M}}_{\widetilde{A},k}(\widetilde{a}) := \widetilde{f}\cdot\widetilde{\mathrm{ev}}_{J,H}, \tag{8.6.1}$$

is independent of the regular pair $(J,H) \in \mathcal{JH}_{\mathrm{reg}}(\widetilde{M},\pi,\widetilde{\omega})$ and the pseudocycle $\widetilde{f} : U \to \widetilde{M}^k$ (Poincaré dual to $\widetilde{a}$) used to define it.

PROOF. By Theorem 8.5.1, the bordism class of $\widetilde{\mathrm{ev}}_{J,H}$ is independent of J and H. Hence, by Lemma 6.5.5, the intersection number $\widetilde{f}\cdot\widetilde{\mathrm{ev}}_{J,H}$ is independent of J and H. By Lemma 6.5.7, it depends only on the homology class represented by the pseudocycle $\widetilde{f}$. This proves Theorem 8.6.1. □

The invariant defined by (8.6.1) is called the **Gromov–Witten invariant of k-pointed pseudoholomorphic sections of $\widetilde{M}$ in the class $\widetilde{A}$.** For $\widetilde{a}_1, \ldots, \widetilde{a}_k \in H^*(\widetilde{M})$ we abbreviate

$$\mathrm{GW}^{\widetilde{M}}_{\widetilde{A},k}(\widetilde{a}_1, \ldots, \widetilde{a}_k) := \mathrm{GW}^{\widetilde{M}}_{\widetilde{A},k}(\pi_1^*\widetilde{a}_1 \smile \cdots \smile \pi_k^*\widetilde{a}_k),$$

where $\pi_i : \widetilde{M}^k \to \widetilde{M}$ denotes the projection onto the ith factor. Geometrically, this invariant can be interpreted as follows. Fix k cycles $\widetilde{X}_i \subset \widetilde{M}$ in general position that are Poincaré dual to the cohomology classes $\widetilde{a}_i$. Then the invariant counts with signs the number of tuples $(\widetilde{u}, z_1, \ldots, z_k)$, where $\widetilde{u} : \Sigma \to \widetilde{M}$ is a $\widetilde{J}_H$-holomorphic section, the z_i are pairwise distinct points on Σ, and $\widetilde{u}(z_i) \in \widetilde{X}_i$.

As we shall see below, the fact that the invariants $\mathrm{GW}^{\widetilde{M}}_{\widetilde{A},k}$ count curves that are adapted to the fibration gives them extra structure. It is easiest to explain this in terms of homology rather than cohomology. Hence we carry out most of the following discussion in terms of the homology invariants of Remark 7.1.12: given homology classes $\widetilde{\alpha}_1, \ldots, \widetilde{\alpha}_k$ in $\widetilde{M}$ with Poincaré duals $\widetilde{a}_i := \mathrm{PD}(\widetilde{\alpha}_i)$ we define

$$\mathrm{GW}^{\widetilde{M}}_{\widetilde{A},k}(\widetilde{\alpha}_1, \ldots, \widetilde{\alpha}_k) := \mathrm{GW}^{\widetilde{M}}_{\widetilde{A},k}(\widetilde{a}_1, \ldots, \widetilde{a}_k).$$

EXERCISE 8.6.2. Show that the invariant $\mathrm{GW}^{\widetilde{M}}_{\widetilde{A},k}$ is independent of j_Σ. *Hint:* Adapt the proof of Theorem 8.5.1 to varying complex structures on Σ.

REMARK 8.6.3. If $\Sigma = S^2$ and $(\widetilde{M}, \widetilde{\omega})$ is itself semipositive then the Gromov–Witten invariants $\mathrm{GW}^{\widetilde{M}}_{B,k}$ of $\widetilde{M}$ are defined in Theorem 7.1.1 for all spherical classes B in $H_2(\widetilde{M})$. As our notation indicates, when B is a section class $\widetilde{A}$, these previously defined invariants agree with the new ones of Theorem 8.6.1. In this case the content of Theorem 8.6.1 is to show that the invariants $\mathrm{GW}^{\widetilde{M}}_{\widetilde{A},k}$ may be calculated using the restricted class of $\widetilde{\omega}$-tame almost complex structures that are adapted to the fibration, namely those of the form $\widetilde{J}_H$. In other words, this class of almost complex structures is rich enough to ensure transversality. Since the projection $\widetilde{M} \to S^2$ is $(\widetilde{J}_H, j_{S^2})$-holomorphic, every $\widetilde{J}_H$-holomorphic curve $v : S^2 \to \widetilde{M}$ that represents a section class projects to a holomorphic curve in S^2 of degree one. Hence $\operatorname{im} v$ is a section. This fact is built into the new definitions: the moduli space $\mathcal{M}_{\Sigma,k}(\widetilde{A}; J, H)$ consists only of sections with their natural parametrizations.

Another special situation, again with $\Sigma = S^2$, is when the fibration

$$\widetilde{M} := S^2 \times M$$

is trivial and $\widetilde{\omega} = \mathrm{pr}^*\omega_0$, where $\mathrm{pr} : \widetilde{M} \to M$ is the obvious projection. In this case we claim that the invariants of Theorem 8.6.1 contain the same information as the invariants $\widetilde{\mathrm{GW}}^{M,I}_{A,k}$ defined in Theorem 7.3.1. (Note however that our present point of view is somewhat different since we now evaluate the invariants by intersections in $\widetilde{M}^k$ rather than in M^k.) To explain this, let us fix an index set $I \subset \{1, \ldots, k\}$. Then we claim that

$$\widetilde{\mathrm{GW}}^{M,I}_{A,k}(\alpha_1, \ldots, \alpha_k) = \mathrm{GW}^{\widetilde{M}}_{\widetilde{A},k}(\widetilde{\alpha}_1, \ldots, \widetilde{\alpha}_k) \tag{8.6.2}$$

for $\alpha_i \in H_*(M)$, where the homology classes $\widetilde{\alpha}_i \in H_*(\widetilde{M})$ are defined by

$$\widetilde{\alpha}_i := \begin{cases} \iota_*\alpha_i, & \text{if } i \in I, \\ [S^2] \times \alpha_i, & \text{if } i \notin I. \end{cases}$$

Here $\iota : M \to \widetilde{M}$ denotes the inclusion of a fiber. For the Poincaré dual cohomology classes this means that $\widetilde{a}_i = \mathrm{pr}^* a_i \smile \mathrm{PD}(\text{fiber})$ for $i \in I$ and $\widetilde{a}_i = \mathrm{pr}^* a_i$ for $i \notin I$. In other words, fixing the marked points indexed by I in the $\widetilde{\mathrm{GW}}$ invariants for M corresponds to intersecting with cycles $\iota_* \alpha_i$ that lie in a fiber. This claim is a restatement of Proposition 7.3.6 in terms of homology and may be proved by the same argument. Note that the semipositivity hypothesis in Proposition 7.3.6 is no longer needed because the invariants $\mathrm{GW}^{\widetilde{M}}_{\widetilde{A},k}$ are well defined by virtue of Theorem 8.6.1.

EXERCISE 8.6.4. By (7.3.2) the left hand side of (8.6.2) is nonzero only if

$$2nk - \sum \deg(\alpha_i) = 2n + 2c_1(A) + 2(k - \#I).$$

Check that this agrees with the dimensional condition for nontriviality on the right. Give a geometric explanation of why the counts correspond.

We now show that a similar statement holds for general Hamiltonian fibrations $\widetilde{M} \to \Sigma$. Again it is convenient to explain this in terms of the homology invariants

$$\mathrm{GW}^{\widetilde{M}}_{\widetilde{A},k} : H_*(\widetilde{M}^k) \to \mathbb{Z}.$$

Recall from Exercise 8.5.4 the definition of the submanifold $\widetilde{M}^k(\mathbf{w})$ of $\widetilde{M}^k$:

$$\widetilde{M}^k(\mathbf{w}) := \left\{ \tilde{\mathbf{x}} = (\widetilde{x}_1, \ldots, \widetilde{x}_k) \in \widetilde{M}^k \mid \pi(\widetilde{x}_i) = \mathrm{w}_i \text{ for } i \in I \right\}$$

We will use the pseudocycle

$$\widetilde{\mathrm{ev}}_{\mathbf{w}} = \widetilde{\mathrm{ev}}_{\mathbf{w},J,H} : \mathcal{M}_{\Sigma,k}(\widetilde{A}; \mathbf{w}, J, H) \to \widetilde{M}^k(\mathbf{w}),$$

constructed in Exercise 8.5.4 and denote by

$$\iota_{\mathbf{w}} : \widetilde{M}^k(\mathbf{w}) \to \widetilde{M}^k$$

the obvious inclusion.

PROPOSITION 8.6.5. *Let (M,ω), $(\widetilde{M},\pi,\widetilde{\omega})$, Σ, and $\widetilde{A} \in H_2(\widetilde{M};\mathbb{Z})$ be as in Theorem 8.6.1. Let k be a positive integer, $I \subset \{1,\ldots,k\}$, $\mathbf{w} = \{\mathrm{w}_i\}_{i \in I}$ be a tuple of pairwise distinct points in Σ, and $(J,H) \in \mathcal{JH}_{\mathrm{reg}}(\widetilde{M},\pi,\widetilde{\omega};\mathbf{w})$. Then*

$$\mathrm{GW}^{\widetilde{M}}_{\widetilde{A},k}(\iota_{\mathbf{w}*}\widetilde{\alpha}) = \widetilde{\mathrm{ev}}_{\mathbf{w}} \cdot \widetilde{\alpha}$$

for $\widetilde{\alpha} \in H_*(\widetilde{M}^k(\mathbf{w}))$.

PROOF. We saw in Exercise 8.5.4 that $\widetilde{\mathrm{ev}}_{\mathbf{w}}$ is a pseudocycle. Therefore the right hand side is well defined. Suppose, without loss of generality, that $\widetilde{\alpha}$ can be represented by an oriented submanifold $X \subset \widetilde{M}^k(\mathbf{w})$ (see Remark 6.5.3). By Lemma 6.5.5, the submanifold X can be chosen strongly transverse to the pseudocycle $\widetilde{\mathrm{ev}}_{\mathbf{w}}$. Then the submanifold $\iota_{\mathbf{w}}(X) \subset \widetilde{M}^k$ is strongly transverse to the pseudocycle $\widetilde{\mathrm{ev}} : \mathcal{M}_{\Sigma,k}(\widetilde{A}; J, H) \to \widetilde{M}^k$. (This holds because each element in $\mathcal{M}_{\Sigma,k}(\widetilde{A}; J, H)$ is a section; so the extra transversality in directions normal to the fiber at w_i is obtained by varying the ith marked point.) It follows that $\iota_{\mathbf{w}}(X) \cdot \widetilde{\mathrm{ev}} = X \cdot \widetilde{\mathrm{ev}}_{\mathbf{w}}$ and this proves Proposition 8.6.5. □

Before explaining the meaning of this result, it will be convenient to introduce the following notation for Gromov–Witten invariants in M that are given by counting sections in $\widetilde{M}$. Because we allow fibrations in which $\pi_1(\Sigma)$ acts nontrivially on the (co)homology of the fiber, these invariants may depend on the way in which the fibers are identified with M. For simplicity (and because this is all we will use later) we will restrict to the case $I = \{1, \dots, k\}$.

DEFINITION 8.6.6. *Let (M, ω), $(\widetilde{M}, \pi, \widetilde{\omega})$, Σ, and $\widetilde{A} \in H_2(\widetilde{M}; \mathbb{Z})$ be as in Theorem 8.6.1. Let k be a nonnegative integer and $\mathbf{w} = \{\mathrm{w}_i\}_{1 \le i \le k}$ be a k-tuple of pairwise distinct points in Σ. For $i = 1, \dots, k$ let $\iota_i : M \to \widetilde{M}$ be a symplectic embedding with $\iota_i(M) = \pi^{-1}(\mathrm{w}_i)$ and denote $\iota := \iota_1 \times \cdots \times \iota_k : M^k \to \widetilde{M}^k$. Define the homomorphism*

$$\mathrm{GW}^{\widetilde{M}, \mathbf{w}}_{\widetilde{A}, k} : H_*(M) \to \mathbb{Z}$$

by

$$\mathrm{GW}^{\widetilde{M}, \mathbf{w}}_{\widetilde{A}, k}(\alpha_1, \dots, \alpha_k) := \mathrm{GW}^{\widetilde{M}}_{\widetilde{A}, k}(\iota_*(\alpha_1 \times \cdots \times \alpha_k)) \tag{8.6.3}$$

The corresponding invariants in cohomology are defined via Poincaré duality in M:

$$\mathrm{GW}^{\widetilde{M}, \mathbf{w}}_{\widetilde{A}, k}(a_1, \dots, a_k) := \mathrm{GW}^{\widetilde{M}, \mathbf{w}}_{\widetilde{A}, k}(\mathrm{PD}_M(a_1), \dots, \mathrm{PD}_M(a_k))$$

for $a_1, \dots, a_k \in H^*(M)$.

Proposition 8.6.5 shows that the invariant (8.6.3) can be be expressed either as the intersection number of $\widetilde{\mathrm{ev}}$ with a representative of the class $\iota_*(\alpha_1 \times \cdots \times \alpha_k)$ in $\widetilde{M}^k$ or as the intersection number of $\iota^{-1} \circ \widetilde{\mathrm{ev}}_{\mathbf{w}}$ with a representative of the class $\alpha_1 \times \cdots \times \alpha_k$ in M^k. The effect is that, when $(J, H) \in \mathcal{JH}_{\mathrm{reg}}(\widetilde{M}, \pi, \widetilde{\omega}; \mathbf{w})$ and the $\widetilde{X}_i \subset \pi^{-1}(\mathrm{w}_i)$ are submanifolds representing the homology classes $\iota_{i*}\alpha_i$, we can perturb the $\widetilde{X}_i$ within their respective fibers (rather than the ambient manifold $\widetilde{M}$) to make the product $\widetilde{X} := \widetilde{X}_1 \times \cdots \times \widetilde{X}_k$ strongly transverse to the pseudocycle $\widetilde{\mathrm{ev}}$. It follows that these invariants have the same formal properties as the $\mathrm{GW}^{M,I}_{A,k}$; in particular, they obey an analogue of the *(Splitting)* axiom: see Theorem 11.4.6. This is an important ingredient of our discussion of the Seidel representation.

Examples. We now describe some examples. The first two involve Hamiltonian fibrations in an essential way, but the last does not. Instead, taking advantage of the fact that we have now given a precise definition of the higher genus invariants in the case when the complex structure j_Σ is fixed, we illustrate some of the new features of this invariant.

EXAMPLE 8.6.7 (Lefschetz numbers). This example is due to Ionel and Parker [**199**]. It has interesting connections to symplectic Floer homology, to the Seiberg–Witten invariants of 3-manifolds, and to Turaev's version of Reidemeister torsion. Let (M, ω) be a compact symplectic manifold which satisfies the strong semipositivity condition (8.5.1) and

$$f : M \to M$$

be a symplectomorphism. Denote by

$$M_f := \mathbb{R} \times M/(t, f(x)) \sim (t + 1, x)$$

the mapping torus. Consider the locally Hamiltonian fibration

$$\widetilde{M}_f := S^1 \times M_f \to \mathbb{T}^2.$$

Since f is a symplectomorphism, the pullback of ω under the projection $S^1 \times \mathbb{R} \times M \to M$ descends to a connection form $\widetilde{\omega}_f \in \Omega^2(\widetilde{M}_f)$. Now let $J = \{J_t\}_{t\in\mathbb{R}}$ be any family of ω-tame almost complex structures on M such that $J_{t+1} = f^*J_t$. Then J is a vertical almost complex structure on $\widetilde{M}_f$ and hence, together with $\widetilde{\omega}_f$, determines an almost complex structure $\widetilde{J}_f$ on $\widetilde{M}_f$. Now every fixed point $x \in \mathrm{Fix}(f)$ determines a section $\widetilde{u}_x(s,t) := x$ and hence a homology class

$$\widetilde{A}_x := [\widetilde{u}_x] \in H_2(\widetilde{M}_f;\mathbb{Z}).$$

It follows easily from the definitions that the vertical Fredholm operator $D_{\widetilde{u}_x}$ has index zero. It is bijective if and only if x is a nondegenerate fixed point, i.e.

$$\det(\mathbb{1} - df(x)) \neq 0.$$

Moreover, one can show that the section $\widetilde{u}_x$ contributes with the fixed point index $\operatorname{sign}\det(\mathbb{1} - df(x))$ to the invariant. Since the curvature of $\widetilde{\omega}$ vanishes, the vertical energy identity of Lemma 8.2.9 shows that every $\widetilde{J}_f$-holomorphic section of $\widetilde{M}_f$ that represents the class $\widetilde{A}_x$ must be horizontal and hence correspond to a fixed point of f.

Now assume that all the fixed points of f are nondegenerate and let $\widetilde{A} \in H_2(\widetilde{M}_f;\mathbb{Z})$ be a homology class such that

$$\pi_*\widetilde{A} = [\Sigma], \qquad \widetilde{\omega}_f(\widetilde{A}) = 0, \qquad c_1^{\mathrm{Vert}}(\widetilde{A}) = 0.$$

Let us call such a class **horizontal**. The above discussion shows that the unmarked Gromov–Witten invariant of a horizontal class $\widetilde{A}$ agrees with the Lefschetz number:

$$\mathrm{GW}^{\widetilde{M}_f}_{\widetilde{A},0}(1) = L(f;\widetilde{A}) := \sum_{\substack{x\in\mathrm{Fix}(f)\\ \widetilde{A}_x=\widetilde{A}}} \operatorname{sign}\det(\mathbb{1} - df(x)).$$

Taking the sum over all horizontal classes $\widetilde{A}$ gives the full Lefschetz number

$$\sum_{\widetilde{A}\ \textit{is horizontal}} \mathrm{GW}^{\widetilde{M}_f}_{\widetilde{A},0}(1) = L(f) = \mathrm{trace}(f_{\mathrm{ev}}) - \mathrm{trace}(f_{\mathrm{odd}}),$$

where $f_{\mathrm{odd}} : H_{\mathrm{odd}}(M) \to H_{\mathrm{odd}}(M)$ denotes the induced homomorphism on the odd dimensional homology of M and similarly for f_{ev}. Now the **zeta function** of f is related to the Lefschetz numbers of the iterates f^k by

$$\zeta_f(t) = \exp\left(\sum_{k=1}^{\infty} \frac{L(f^k)t^k}{k}\right) = \frac{\det(\mathbb{1} - tf_{\mathrm{odd}})}{\det(\mathbb{1} - tf_{\mathrm{ev}})}.$$

To interpret the zeta function in terms of a generating function for the Gromov–Witten invariants of $\widetilde{M}_f$ one has to consider homology classes $\widetilde{A} \in H_2(\widetilde{M};\mathbb{Z})$ such that

$$\pi_*\widetilde{A} = k[\Sigma]$$

and examine holomorphic curves in $\widetilde{M}_f$ other than sections. This was carried out by Ionel and Parker [**199**]. We shall not discuss this extension in this book.

EXAMPLE 8.6.8 (The Seidel representation). Let (M,ω) be a compact monotone symplectic manifold. In [**363**] Seidel associated a cohomology class $\mathcal{S}(\psi) \in H^*(M)$ called the **Seidel element** to every loop $\psi = \{\psi_t\}_{t\in\mathbb{R}/\mathbb{Z}}$ of Hamiltonian symplectomorphisms of M. Each such loop defines a Hamiltonian fibration $\widetilde{M}_\psi$ as in (8.2.6), and the class $\mathcal{S}(\psi)$ is obtained by intersecting the moduli space of holomorphic sections of $\widetilde{M}_\psi$ with a fiber. More precisely, we define $\mathcal{S}(\psi)$ as a sum:

$$\mathcal{S}(\psi) := \sum_{\widetilde{A}} \mathcal{S}_{\widetilde{A}}(\psi),$$

where the cohomology class $\mathcal{S}_{\widetilde{A}}(\psi) \in H^{2c}(M)$ is Poincaré dual to the homology class in M represented by the pseudocycle

$$\widetilde{\mathrm{ev}}_{\mathbf{w},J,H} : \mathcal{M}(\widetilde{A}; J, H) \to M$$

of sections in class $\widetilde{A}$ with one fixed marked point w. Here $-c := c_1^{\mathrm{Vert}}(\widetilde{A})$ denotes the vertical Chern number of $\widetilde{A}$ and we identify the fiber of $\widetilde{M}_\psi$ over w with M. The sum runs over all homology classes $\widetilde{A} \in H_2(\widetilde{M}_\psi;\mathbb{Z})$ such that $\pi_*\widetilde{A} = [S^2]$. Thus $\widetilde{A}$ contributes only if $-2n \le -c \le 0$; the monotonicity of (M,ω) implies that this sum is finite. Equivalently, one can express the element $\mathcal{S}_{\widetilde{A}}(\psi)$ in terms of the Gromov–Witten invariants

$$\mathrm{GW}^{\widetilde{M}_\psi,\mathrm{w}}_{\widetilde{A},1} : H^{2n-2c}(M) \to \mathbb{Z}$$

of Definition 8.6.6 via the formula

$$\mathcal{S}_{\widetilde{A}}(\psi) := \sum_{\nu,\mu} \mathrm{GW}^{\widetilde{M}_\psi,\mathrm{w}}_{\widetilde{A},1}(e_\nu) g^{\nu\mu} e_\mu \in H^{2c}(M). \tag{8.6.4}$$

Here we identify M with the fiber at w. As usual, $e_0, \ldots, e_m$ denotes a basis of $H^*(M)$, and $g^{\nu\mu}$ is the inverse of the matrix $g_{\nu\mu}$ defined by

$$g_{\nu\mu} := \int_M e_\nu \smile e_\mu.$$

For general manifolds (M,ω) one must separate out the contributions to $\mathcal{S}(\psi)$ from the different classes $\widetilde{A}$. This is most naturally done in terms of quantum cohomology. Indeed, in Section 11.4 we shall see that $\mathcal{S}(\psi)$ can be interpreted as an invertible element of the quantum cohomology ring $\mathrm{QH}^*(M)$ of (M,ω). Moreover the map

$$\pi_1(\mathrm{Ham}(M,\omega)) \to H^*(M) : \psi \mapsto \mathcal{S}(\psi)$$

is a group homomorphism from the fundamental group of the group of Hamiltonian symplectomorphisms of (M,ω) to the multiplicative group of invertible elements of $\mathrm{QH}^*(M)$. Taking $\mathcal{S}(\phi)$ to act on the Λ-module $\mathrm{QH}^*(M)$ by quantum multiplication, one obtains a representation of $\pi_1(\mathrm{Ham}(M,\omega))$ in the group of automorphisms of $\mathrm{QH}^*(M)$ as a left $\mathrm{QH}^*(M)$-module that is known as the **Seidel representation**.

If the above statements are to hold, then $\mathcal{S}(\psi)$ should equal 1 when the loop $\psi = \{\psi_t\}_{t\in\mathbb{R}/\mathbb{Z}}$ is contractible. To check this, note that in this case the Hamiltonian fibration $\widetilde{M}_\psi$ is fiberwise symplectomorphic to the product $S^2 \times M$. Hence, by Exercise 7.3.4, the invariant vanishes for $\widetilde{A} \neq 0$ and, by Exercise 7.3.5,

$$\mathrm{GW}^{\widetilde{M}_\psi,\mathrm{w}}_{0,1}(a) = \int_M a.$$

Hence $\mathcal{S}(\psi) = \mathcal{S}_0(\psi) = 1$. We will return to this example in Section 9.6.

Consider the special case where the cohomology class of ω vanishes on $\pi_2(M)$. Then there are no J-holomorphic spheres for any J and so the ring structure on the quantum cohomology of M is the usual cup product. In this case the multiplicative property of the Seidel representation takes the following form.

THEOREM 8.6.9 (Seidel). *Let (M, ω) be a compact symplectic manifold such that $[\omega]$ vanishes on $\pi_2(M)$. Then*

$$\mathcal{S}(\psi\phi) = \mathcal{S}(\psi) \smile \mathcal{S}(\phi)$$

for any two loops $\psi = \{\psi_t\}_{t\in\mathbb{R}/\mathbb{Z}}$ and $\phi = \{\phi_t\}_{t\in\mathbb{R}/\mathbb{Z}}$ of Hamiltonian symplectomorphisms of M.

PROOF. See Section 11.4. □

Here is an interesting corollary that will be put into context in Section 9.1. In the case where the first Chern class vanishes on $\pi_2(M)$ it was proved by Schwarz in [**360**], using ideas that go back to Seidel. The main step is to show that for these manifolds the leading term $\mathcal{S}(\psi)_0 \in H^0(M)$ of the Seidel element $\mathcal{S}(\psi)$ is ± 1.

COROLLARY 8.6.10 (Schwarz). *Let (M, ω) be a compact symplectic manifold such that $[\omega]$ vanishes on $\pi_2(M)$ and let $\{\psi_t\}_{t\in\mathbb{R}/\mathbb{Z}}$ be a loop of Hamiltonian symplectomorphisms of M. Let $H_t : M \to \mathbb{R}$ be the Hamiltonian with mean value zero generating ψ_t so that*

$$\partial_t\psi_t = X_{H_t} \circ \psi_t, \qquad \iota(X_{H_t})\omega = dH_t, \qquad \int_M H_t\omega^n = 0.$$

Let $\mathbb{D} \subset \mathbb{C}$ be the closed unit disc and $u : \mathbb{D} \to M$ be a smooth map such that $u(e^{2\pi it}) = \psi_t(x_0)$, where $x_0 := u(1)$. Then

$$\int_{\mathbb{D}} u^*\omega + \int_0^1 H_t(\psi_t(x_0))\,dt = 0.$$

PROOF. Let $\pi : \widetilde{M}_\psi \to S^2$ be the Hamiltonian fibration (8.2.6) and $\widetilde{\omega} \in \Omega^2(\widetilde{M}_\psi)$ be the connection form of Remark 8.2.11 with

$$F^\pm = 0, \quad G^+_{s,t} = -\rho(s)H_t, \quad G^-_{s,t} = -\rho(s)H_{-t} \circ \psi_{-t},$$

where $\rho : \mathbb{R} \to [0,1]$ is a smooth cutoff function such that $\rho(s) = 0$ for $s \le -1$, $\rho(s) = 1$ for $s \ge 1$, and $\rho(s) + \rho(-s) = 1$ for all s. Suppose, without loss of generality, that $u(e^{2\pi(s+it)}) = \psi_t(x_0)$ for $s \ge -1$ and define a section $\widetilde{u} : S^2 \to \widetilde{M}_\psi$ by $u^+(s,t) := u(e^{2\pi(s+it)})$ and $u^-(s,t) := \psi_{-t}(u(e^{2\pi(-s-it)}))$. Then

$$a(\psi) := -\int_{S^2} \widetilde{u}^*\widetilde{\omega} = -\int_{\mathbb{D}} u^*\omega - \int_0^1 H_t(\psi_t(x_0))\,dt.$$

Since $[\omega]$ vanishes over $\pi_2(M)$ this integral is independent of the section $\widetilde{u}$ of $\widetilde{M}_\psi$ and hence in particular of the choice of point x_0. We must prove that $a(\psi) = 0$.

Fix a vertical almost complex structure J on $\widetilde{M}_\psi$. Given any horizontal 1-form $\widetilde{\sigma}_K \in H^1(\widetilde{M}_\psi)$ denote by $\widetilde{J}_K$ the almost complex structure on $\widetilde{M}_\psi$, induced by J and $\widetilde{\omega} - d\widetilde{\sigma}_K$, and by $\mathcal{M}(J,K)$ the moduli space of all $\widetilde{J}_K$-holomorphic sections of $\widetilde{M}_\psi$. Since $[\omega]$ vanishes on $\pi_2(M)$, this moduli space is compact. If K is regular in the sense of (8.3.2) then, by Theorem 8.3.1, $\mathcal{M}(J,K)$ is a compact oriented smooth manifold. Its components may have different dimensions: indeed the component corresponding to the homology class $\widetilde{A} \in H_2(\widetilde{M}_\psi;\mathbb{Z})$ has dimension $2n + 2c_1^{\mathrm{Vert}}(\widetilde{A})$. Let us denote by $\mathcal{M}_0(J,K)$ the union of the components with Chern numbers zero. In the regular case the evaluation map $\mathrm{ev}_K : \mathcal{M}(J,K) \to M$ represents the homology class $\mathrm{PD}(\mathcal{S}(\psi))$, while the restricted evaluation map $\mathrm{ev}_{K,0} : \mathcal{M}_0(J,K) \to M$ represents the leading term

$$\mathcal{S}(\psi)_0 = \deg(\mathrm{ev}_{K,0})1 \in H^0(M).$$

By Theorem 8.6.9, we have $\mathcal{S}(\psi) \smile \mathcal{S}(\psi^{-1}) = 1$. This is only possible if the evaluation map $\mathrm{ev}_{K,0} : \mathcal{M}_0(J,K) \to M$ has degree plus or minus one.

Now consider the evaluation map

$$\widetilde{\mathrm{ev}}_{K,0} : \mathcal{M}_0(J,K) \times S^2 \to \widetilde{M}_\psi,$$

and integrate the pullback of $\widetilde{\omega}^{n+1} \in \Omega^{2n+2}(\widetilde{M}_\psi)$ to obtain

$$\begin{aligned} a(\psi)\deg(\mathrm{ev}_{K,0}) \int_M \frac{\omega^n}{n!} &= a(\psi) \int_{\mathcal{M}_0(J,K)} \frac{\mathrm{ev}_{K,0}^*\omega^n}{n!} \\ &= -\int_{\mathcal{M}_0(J,K)\times S^2} \frac{\widetilde{\mathrm{ev}}_{K,0}^*\widetilde{\omega}^{n+1}}{(n+1)!} \\ &= -\deg(\widetilde{\mathrm{ev}}_{K,0}) \int_{\widetilde{M}_\psi} \frac{\widetilde{\omega}^{n+1}}{(n+1)!} \\ &= 0. \end{aligned}$$

The first and third identities follow from the degree theorem, the second identity follows from Exercise 8.6.11 below, and the last from Exercise 8.1.3. Since $\deg(\mathrm{ev}_{K,0}) \neq 0$ we obtain $a(\psi) = 0$ as claimed. This proves Corollary 8.6.10. □

EXERCISE 8.6.11. Let X be a compact oriented $2n$-manifold and $\tau \in \Omega^2(X \times S^2)$ be a closed 2-form. Let $\iota_X : X \to X \times S^2$ and $\iota_{S^2} : S^2 \to X \times S^2$ be embeddings $\iota_X(x) := (x, z_0)$ and $\iota_{S^2}(z) := (x_0, z)$. Prove that

$$\int_{X\times S^2} \frac{\tau^{n+1}}{(n+1)!} = \int_X \frac{\iota_X^*\tau^n}{n!} \int_{S^2} \iota_{S^2}^*\tau.$$

Hint: Since S^2 is simply connected, the Künneth formula asserts that there are closed 2-forms $\tau_X \in \Omega^2(X)$ and $\tau_{S^2} \in \Omega^2(S^2)$ such that $\tau - \pi_X^*\tau_X - \pi_{S^2}^*\tau_{S^2}$ is exact.

EXAMPLE 8.6.12. Consider the case $M = \mathbb{C}P^n$ and let Σ be a Riemann surface of genus g. Let $d \geq 0$ such that

$$m := (n+1)(d+1-g) + g - 1 \geq 0.$$

This number is half the virtual dimension of the moduli space of holomorphic maps $u : \Sigma \to \mathbb{C}P^n$ of degree d. Let $c \in H^2(\mathbb{C}P^n;\mathbb{Z})$ denote the positive generator and choose integers $m_1, \dots, m_k \in \mathbb{N}$ such that $0 \leq m_i \leq n$ and $\sum_{i=1}^k m_i = m$. Then

$$\mathrm{GW}_{\Sigma,dL,k}^{\mathbb{C}P^n,\{1,\dots,k\}}(c^{m_1}, \dots, c^{m_k}) = (n+1)^g. \tag{8.6.5}$$

As our notation suggests, this invariant can be defined just in terms of genus g stable maps to $\mathbb{C}P^n$. However, it is sometimes useful to identify these with sections of the trivial bundle $\Sigma \times \mathbb{C}P^n$. (For a proof of (8.6.5) see Bertram–Daskalopoulos–Wentworth [**36**] and also Cieliebak–Gaio–Mundet–Salamon [**67**].)

It is interesting to examine the special case $n = g = d = 1$ more closely. In this case $m = 2$ and we may choose $k = 2$ and $m_1 = m_2 = 1$. By (8.6.5), the invariant is 2 even though there are no holomorphic maps of degree one from the 2-torus to the 2-sphere. However, there is a nonempty moduli space of stable maps of degree $d = 1$ and genus $g = 1$ with a fixed complex structure on $\mathbb{T}^2$ and $k = 2$ marked points z_1, z_2 whose projections to the torus are fixed at $\mathrm{w}_1, \mathrm{w}_2 \in \mathbb{T}^2$. Every such stable map consists of a constant component of genus one and a nonconstant rational curve of degree one attached to the torus at a point that we shall call the nodal point. If this nodal point is one of $\{\mathrm{w}_1, \mathrm{w}_2\}$ then there might also be a ghost bubble containing the appropriate marked point. But note that in this case the images in $M = S^2$ of the two marked points coincide. In the product $\mathbb{T}^2 \times S^2$ the constant component appears as a constant graph while the rational curve appears as a holomorphic sphere in the fiber over the nodal point. There are five strata, one where the nodal point on the torus does not belong to $\{\mathrm{w}_1, \mathrm{w}_2\}$, and the other four where it is either w_1 or w_2 and the corresponding marked point lies either on the rational curve or a ghost bubble. It follows easily that there are precisely two stable maps in this space sending the marked points to two distinct prescribed points $x_1, x_2 \in S^2$; one where the torus is mapped to x_1, the nodal point is w_2, and the second marked point lies on the sphere and is mapped to x_2; and one where the torus is mapped to x_2, the nodal point is w_1, and the first marked point lies on the sphere and is mapped to x_1. A generic Hamiltonian perturbation will give rise to two regular solutions $u : \mathbb{T}^2 \to S^2$ of the perturbed Cauchy–Riemann equations satisfying $u(\mathrm{w}_1) = x_1$ and $u(\mathrm{w}_2) = x_2$.

A similar situation occurs when one counts holomorphic curves in $\mathbb{C}P^2$. Here there are classical enumerative invariants that compute the number of genus g degree d curves passing through a suitable number of points. These are related to, but (except in the genus zero case) need not be identical with, the corresponding Gromov–Witten invariants that count all stable maps passing through these constraints. One reason for this difference is the effect of ghosts such as the ghost (or constant) torus in the example discussed above. For further discussion, see for example the papers by Ionel [**196**] and Zinger [**426, 427**].

REMARK 8.6.13. The above discussion of Hamiltonian fibrations has mostly focussed on those aspects of the theory that cast light on the properties of the fiber M. However, there is an interesting general theory of Hamiltonian fibrations over an arbitrary smooth base manifold. One open problem here is the c-splitting conjecture formulated in Lalonde–McDuff [**227**]: is it true that the rational cohomology of the (compact) total space of a smooth Hamiltonian fibration is additively isomorphic to the tensor product of the cohomology of the base and fiber? (We mentioned in Remark 8.4.4 that this can be proved when the base is S^2 by using the Seidel representation. It also holds in the Kähler case.) Even more interesting is the study of symplectic Lefschetz fibrations. These are smooth maps $f : (M, \omega) \to S^2$ whose fibers are symplectic away from a finite number of isolated singularities with local models $(z_1, \dots, z_n) \mapsto \sum_{i=1}^n z_i^2$, where $(z_1, \dots, z_n) \in \mathbb{C}^n$ form the coordinates of a local Darboux chart. Thus they are the complex analog of a Morse function and, by Donaldson [**85**], exist on a suitable blowup of *any* closed symplectic manifold provided only that $[\omega] \in H^2(M; \mathbb{Q})$. These have proved most useful in four dimensions since the the fibers are then two-dimensional. For example, Auroux–Muñoz–Presas [**29**] have used them to study isotopy classes of Lagrangian spheres in a symplectic 4-manifold as well as the symplectic mapping class group.

CHAPTER 9

Applications in Symplectic Topology

This chapter describes applications of J-holomorphic spheres to symplectic topology, giving full details of some basic results and indicating the directions of further development. Many of the applications discussed here are due to Gromov [**160**] though we do not always use his proof. It is impossible to be comprehensive since this area has been so active recently.

The first section deals with periodic orbits of Hamiltonian systems. We prove the existence of one periodic orbit for every Hamiltonian system on a semipositive symplectic manifold. In the symplectically aspherical case we recover a theorem of Schwarz which asserts that there are two periodic orbits with different action unless the time-1 map is the identity, and explain how this result can be used to establish nondegeneracy of the Hofer metric. Section 9.2 deals with Lagrangian submanifolds. We prove Gromov's theorems about the existence of a nonconstant holomorphic disc with boundary in a Lagrangian submanifold and about the nonempty intersection of a Lagrangian submanifold with any of its Hamiltonian deformations. The proofs of all these results are based on the same idea: one constructs an interesting geometric object (such as a periodic orbit, a J-holomorphic disc, a Floer connecting orbit, a Lagrangian intersection point) from a limit of solutions to a family of deformations of the Cauchy–Riemann equation. The deformations in question are Hamiltonian perturbations of the type considered in Chapter 8; however the corresponding bundle is trivial and we have tried to make the discussion accessible to readers who are not familiar with all the details of that chapter.

Next we explain a proof of the nonsqueezing theorem that uses the technique of blowing up rather than the original argument via minimal surfaces. Section 9.4 discusses the structure of rational and ruled surfaces following Gromov [**160**] and McDuff [**258**], while Section 9.5 proves Gromov's results on the symplectomorphism groups of $\mathbb{R}^4$, $\mathbb{C}P^2$ and $S^2 \times S^2$. In Section 9.6 we describe some applications of J-holomorphic curve techniques to Hofer geometry. The first work in this direction is Lalonde–McDuff [**223, 224**]. Here we explain an application of the Seidel homomorphism due to Polterovich [**325, 326, 327**] and Seidel [**363, 364**]. Finally, Section 9.7 discusses some examples of Ruan [**340**] and McDuff [**254**] in which J-holomorphic spheres are used to distinguish between symplectic manifolds.

The applications discussed here are restricted to the genus zero case. and remain fairly close in spirit to Gromov's original ideas. We have concentrated on situations where there is no need to count the number of holomorphic curves (or rather to count beyond one), instead getting information from the existence of a single holomorphic disc or sphere, or perhaps from a family of them, one through each point. In such cases there is no need to use the gluing theorem of Chapter 10. In contrast, the information obtained by counting holomorphic spheres can be assembled into quantum cohomology (Chapter 11) and this does require the gluing theorem. Noncompact J-holomorphic curves such as cylinders and strips were studied by Floer. This led to what is now called Floer homology and has many remarkable applications. (See Chapter 12 for an overview.) Symplectic Floer theory pre-dated the development of quantum cohomology, but is closely related to it and requires another type of gluing theorem, which is not covered in this book. In dimension four, and its three-dimensional counterpart in contact geometry, there are very interesting recent developments based on the study of noncompact J-holomorphic curves that fit together to form finite energy foliations. This approach was initiated by Hofer [**178**] and Hofer–Wysocki–Zehnder [**183**], with further interesting applications by Hind [**174**] and Wendl [**418**]. These results will not be covered in this book, but are close in spirit to some of the ideas described in the present chapter.

9.1. Periodic orbits of Hamiltonian systems

The study of periodic solutions of Hamiltonian systems has a long and remarkable history, attracting the attention of mathematicians such as Weierstrass, Poincaré, and Moser. The existence problem in particular has motivated many developments in symplectic topology and there is a vast literature on this subject that we cannot possibly begin to discuss here. Instead we shall focus on an existence theorem for 1-periodic solutions of periodic Hamiltonian systems under the assumption that there are no J-holomorphic spheres with negative Chern numbers. The argument produces just one periodic orbit (or two under an additional assumption). It should be thought of as a prototype that indicates the relevance of elliptic methods to this question. Our proof is similar to some arguments in Gromov [**160**] in that it studies a family of increasingly large perturbations of the Cauchy–Riemann equation. However the geometric context is different. Gromov considered discs with boundary on a Lagrangian submanifold while here we consider perturbed J-holomorphic spheres which in the limit give rise to solutions of the Floer equation on the cylinder. Moreover, Gromov's conclusion is a little weaker: he asserted only the existence of a fixed point for the time-1 map of the Hamiltonian flow rather than a *contractible* 1-periodic orbit. Nevertheless, his paper was the foundation for the ideas presented here. A few years after it appeared, Floer [**113, 114, 116**] developed a new homology theory, known as Floer homology, that provides a way to count the periodic orbits.

Let (M, ω) be a compact symplectic manifold and $\mathbb{R} \to M : (t, x) \mapsto H_t(x)$ be a smooth Hamiltonian function such that $H_t = H_{t+1}$ for every t. Let $X_{H_t} \in \mathrm{Vect}(M)$ denote the Hamiltonian vector field associated to H_t, i.e.

$$\iota(X_{H_t})\omega = dH_t,$$

and consider the differential equation

$$\dot{x}(t) = X_{H_t}(x(t)). \tag{9.1.1}$$

A solution $x : \mathbb{R} \to M$ of (9.1.1) is called **1-periodic** if $x(t) = x(t+1)$ for every t. We denote by $\mathcal{P}(H)$ the set of 1-periodic solutions of (9.1.1) and by $\mathcal{P}_0(H)$ the set of contractible 1-periodic solutions of (9.1.1).

THEOREM 9.1.1. *Let (M,ω) be a closed connected symplectic manifold. Suppose that there is an ω-tame almost complex structure $J \in \mathcal{J}_\tau(M,\omega)$ such that every J-holomorphic sphere has nonnegative Chern number. Then equation (9.1.1) has a contractible 1-periodic solution for every Hamiltonian function $H : \mathbb{R}/\mathbb{Z} \times M \to \mathbb{R}$.*

Since every Hamiltonian symplectomorphism ϕ is generated by a 1-periodic Hamiltonian, the theorem implies that every element ϕ of the Hamiltonian group $\mathrm{Ham}(M,\omega)$ has a fixed point x. (This was the result that Gromov proved in [**160**] for manifolds where ω vanishes on $\pi_2(M)$.) Moreover, if $\{\phi_t\}_{0\le t\le 1}$ is any path in $\mathrm{Ham}(M,\omega)$ from the identity to ϕ we may choose x so that the loop $t \mapsto \phi_t(x)$ is contractible in M. This has the following somewhat unexpected consequence.

COROLLARY 9.1.2. *Suppose (M,ω) satisfies the hypotheses of Theorem 9.1.1 and let $\mathbb{R}/\mathbb{Z} \to \mathrm{Ham}(M,\omega) : t \mapsto \phi_t$ be a loop of Hamiltonian symplectomorphisms. Then the loop $\mathbb{R}/\mathbb{Z} \to M : t \mapsto \phi_t(x)$ is contractible for every $x \in M$. In other words, the evaluation map $\pi_1(\mathrm{Ham}(M,\omega)) \to \pi_1(M,x)$ given by $\{\phi_t\} \mapsto \{\phi_t(x)\}$ is trivial.*

PROOF. By Theorem 9.1.1, one of the loops $\mathbb{R}/\mathbb{Z} \to M : t \mapsto \phi_t(x)$ is contractible. Hence they are all contractible. □

This corollary implies that for these manifolds (M,ω) one can assign to each fixed point x_0 of a Hamiltonian symplectomorphism ϕ a conjugacy class in $\pi_1(M,x)$, namely the free homotopy class of the loop $\{\phi_t(x_0)\}_{t\in[0,1]}$ where ϕ_t is any Hamiltonian path with time-1 map ϕ. In particular, each such ϕ has a well defined set $\mathrm{Fix}_0(\phi)$ of **contractible fixed points** consisting of points for which this loop is contractible.

Here we have a purely topological result proved by analytic methods. Though there are other proofs of this corollary (cf. [**228**] for example), all those of which we are aware use the theory of J-holomorphic curves in one form or another.

The Arnold conjecture. Before proving Theorem 9.1.1, let us put this result into context. Arnold [**16**] conjectured in the 1960s that

$$\#\mathcal{P}_0(H) \ge \mathrm{Crit}(M) \tag{9.1.2}$$

for every Hamiltonian function $H : \mathbb{R}/\mathbb{Z} \times M \to \mathbb{R}$, where $\mathrm{Crit}(M)$ denotes the minimal number of critical points of a function $f : M \to \mathbb{R}$. Here one can either allow all functions H and f or restrict to the case in which both are nondegenerate. (This means that f is Morse and that the time-1 maps of the linearized flows around the periodic orbits of H have eigenvalues different from 1.) In this form the conjecture is still open.

The homological version of this conjecture has proved more accessible. In the nondegenerate case it asserts that

$$\#\mathcal{P}_0(H) \ge \dim H^*(M;\mathbb{Q}), \tag{9.1.3}$$

whenever the contractible 1-periodic solutions are all nondegenerate. This was first established by Eliashberg [**98**] for Riemann surfaces and by Conley–Zehnder [**75**] for the standard $2n$-torus. The breakthrough came with the work of Floer [**112, 113, 114, 116**] who proved the nondegenerate Arnold conjecture for monotone symplectic manifolds. All the subsequent work on this question was based on Floer's ideas. The extension to the semipositive case was found by Hofer–Salamon [**180**] and Ono [**313**]; their proof is sketched in Section 12.1. The nondegenerate Arnold conjecture, in the form (9.1.3), has now been confirmed for all compact symplectic manifolds by Fukaya–Ono [**127**] and Liu–Tian [**249**]. It follows that the hypothesis about Chern numbers of holomorphic spheres can be removed from Theorem 9.1.1.

REMARK 9.1.3. (i) The estimate (9.1.2) follows from Morse theory when the Hamiltonian function H is independent of t since each critical point of H gives an element of $\mathcal{P}_0(H)$. It follows from the Lefschetz fixed point formula whenever the odd dimensional homology of M is zero. Further, if the Euler characteristic of M is nonzero, the Lefschetz fixed point formula implies that the time-1 map of the flow of H has a fixed point, though it does not assert that this fixed point is contractible (i.e. corresponds to a contractible loop.)

(ii) **(The Conley conjecture)** Conley conjectured that every Hamiltonian system on the $2n$-torus must have infinitely many distinct periodic solutions with integer periods. Such a claim should hold more more general manifolds, though not for all; for example, an irrational rotation of S^2 has only two periodic points. In Salamon–Zehnder [**356**] the existence of infinitely many periodic orbits was established under the hypothesis that $[\omega]$ vanishes on $\pi_2(M)$ and that every 1-periodic solution has at least one Floquet multiplier not equal to one. The full conjecture has now been solved, first for the torus by Hingston [**176**] and then for any closed symplectically aspherical manifold by Ginzburg [**143**]. However, the largest class of manifolds for which it holds is not yet clear.

(iii) **(The Weinstein conjecture)** The original form of this conjecture states that the characteristic flow on any closed contact hypersurface in symplectic Euclidean space has a periodic orbit. This was confirmed by Viterbo [**402**] in an important early paper that provided a key link between Conley–Zehnder's work [**75**] on the Arnold conjecture for the torus and Ekeland–Hofer's variational theory of symplectic capacities [**97**]. The ideas in the proof of Theorem 9.1.1 given below were later extended by Hofer–Viterbo [**182**], Liu–Tian [**251**], and Lu [**253**] to prove the existence of periodic orbits under quite general assumptions. Some new results on this subject can be found in the elegant paper [**123**] by Frauenfelder and Schlenk, and in Albers–Hofer [**12**]. In [**394**], Taubes succeeded in proving the conjecture for Reeb flows on 3-dimensional contact manifolds, using arguments that he extended to construct an isomorphism between Seiberg–Witten–Floer theory and the embedded contact homology of Hutchings and Sullivan [**195**]. However, it is not true that every closed hypersurface in Euclidean space admits such an orbit. Interesting counterexamples were discovered by Ginzburg [**141, 142**] and Herman [**172, 424**].

The proof of Theorem 9.1.1 is based on the nonlinear Cauchy–Riemann equations on the 2-sphere with perturbations of the type considered in Section 8.1. We will assume that the reader is familiar with that section. The perturbed equation has the form

$$\bar{\partial}_J(v) + X_K(v)^{0,1} = 0, \tag{9.1.4}$$

where $J \in C^\infty(S^2, \mathcal{J}_\tau(M,\omega))$ and $K \in \Omega^1(S^2, C_0^\infty(M))$. Here $C_0^\infty(M)$ denotes the space of smooth functions with mean value zero. By Lemma 8.1.6 and Remark 8.1.7, the energy of a solution $v : S^2 \to M$ of (9.1.4) is bounded by

$$E_K(v) := \int_{S^2} |dv + X_K(v)|_J^2 \,\mathrm{dvol}_{S^2} \le \int_{S^2} v^*\omega + \|R_K\|, \tag{9.1.5}$$

where $\|R_K\|$ denotes the Hofer norm of the curvature. Recall the definition of the set $\mathcal{J}_+(M,\omega;\kappa)$ of all ω-tame almost complex structures $J\in\mathcal{J}_\tau(M,\omega)$ such that every J-holomorphic sphere $v:S^2\to M$ with energy $E(v)\le\kappa$ has nonnegative Chern number. By Lemma 6.4.7, this is an open (possibly empty) subset of $\mathcal{J}_\tau(M,\omega)$.

PROPOSITION 9.1.4. *Suppose that (M,ω) is a compact symplectic manifold, $J=\{J_z\}_{z\in S^2}$ is a smooth family of ω-tame almost complex structures, and $K\in\Omega^1(S^2,C_0^\infty(M))$. If there exists a real number $\kappa>\|R_K\|$ such that $J_z\in\mathcal{J}_+(M,\omega;\kappa)$ for every z then there exists a solution $v:S^2\to M$ of (9.1.4) with energy*

$$E_K(v)\le\|R_K\|.$$

Before giving the proof of the proposition, we shall use it to prove the main result.

PROOF OF THEOREM 9.1.1. Let J be an ω-tame almost complex structure on M such that every J-holomorphic sphere has nonnegative Chern number. (Such an almost complex structure exists by assumption.) For every $T>0$ let $\beta_T:\mathbb{R}\to[0,1]$ be a cutoff function that equals one on $[-T,T]$, has support in $[-T-1,T+1]$, and is nonincreasing (respectively nondecreasing) on the interval $[T,T+1]$ (respectively $[-T-1,-T]$). Given a smooth Hamiltonian function

$$\mathbb{R}/\mathbb{Z}\to C_0^\infty(M):t\mapsto H_t$$

we consider the perturbed Cauchy–Riemann equation

$$\partial_s u+J(u)(\partial_t u-\beta_T(s)X_{H_t}(u))=0 \tag{9.1.6}$$

for smooth maps $u:\mathbb{R}\times\mathbb{R}/\mathbb{Z}\to M$. Now identify S^2 with the Riemann sphere $\mathbb{C}\cup\{\infty\}$ via stereographic projection and the cylinder $\mathbb{R}\times\mathbb{R}/\mathbb{Z}$ with $S^2\setminus\{0,\infty\}$ via the holomorphic diffeomorphism $\phi:\mathbb{R}\times\mathbb{R}/\mathbb{Z}\to\mathbb{C}\setminus\{0\}\subset S^2$ given by

$$\phi(s,t):=e^{2\pi(s+it)}.$$

By the removable singularity theorem (Theorem 4.1.2), a solution u of (9.1.6) has finite energy

$$E(u):=\int_0^1\int_{-\infty}^{\infty}|\partial_s u|^2\,dsdt<\infty$$

if and only if the function $v_u:=u\circ\phi^{-1}:S^2\setminus\{0,\infty\}\to M$ extends to a smooth map from S^2 to M. Moreover, v_u is a solution of (9.1.4), where the perturbation $K=K_T$ pulls back to the 1-form $\phi^*K_T=-\beta_T(s)H_t\,dt$ on the cylinder. The crucial observation is that the Hofer norm of the curvature of K_T is given by

$$\|R_{K_T}\|=\int_{-\infty}^{\infty}\int_0^1\left|\dot\beta_T(s)\right|\left(\max_M H_t-\min_M H_t\right)\,dtds=2\|H\|.$$

Hence it follows from Proposition 9.1.4 that, for every $T>0$, there exists a solution $u_T:\mathbb{R}\times\mathbb{R}/\mathbb{Z}\to M$ of (9.1.6) with energy $E(u_T)$ bounded by $2\|H\|$. Restricting u_T to the subset $[-T,T]\times\mathbb{R}$, we find that u_T satisfies the inequality

$$\int_0^1\int_{-T}^{T}|\partial_t u_T-X_{H_t}(u_T)|^2\,dsdt\le E(u_T)\le 2\,\|H\|\,.$$

Hence there is a real number $s_T\in[-T,T]$ such that the loop $x_T(t):=u_T(s_T,t)$ satisfies

$$\int_0^1|\dot x_T(t)-X_{H_t}(x_T(t))|^2\,dt\le\frac{\|H\|}{T}.$$

Since M is compact there exists a sequence $T_\nu \to \infty$ such that x_{T_ν} converges (in the $W^{1,2}$-topology) to a 1-periodic solution $x : \mathbb{R}/\mathbb{Z} \to M$ of (9.1.1). Now x_{T_ν} is contractible for every ν since it is the restriction of a map v_{T_ν} with domain equal to the 2-sphere. Hence so is x. This proves the theorem. □

PROOF OF PROPOSITION 9.1.4. We first prove a slightly less general result which suffices for our applications. Fix a constant $\kappa > 0$ so that the assumptions of the proposition are satisfied. Then the set $\mathcal{J}_+(M,\omega;\kappa)$ is nonempty and, by Lemma 6.4.7, it is an open subset of $\mathcal{J}_\tau(M,\omega)$. Let us fix an element $J_0 \in \mathcal{J}_+(M,\omega;\kappa)$ and denote by $\mathcal{J}_0(\kappa)$ the space of smooth maps $S^2 \to \mathcal{J}_+(M,\omega;\kappa) : z \mapsto J_z$ that are homotopic to the constant map $J_z \equiv J_0$. Denote by $\mathcal{H}_0(\kappa)$ the set of all Hamiltonian perturbations $K \in \Omega^1(S^2, C_0^\infty(M))$ such that the pullback of K under the diffeomorphism $\phi : \mathbb{R}\times\mathbb{R}/\mathbb{Z} \to S^2$, $\phi(s,t) := e^{2\pi(s+it)}$, has the form $\phi^*K = G\,dt$ and the Hofer norm of the curvature is bounded by κ:

$$\|R_K\| = \int_{\mathbb{R}\times\mathbb{R}/\mathbb{Z}} \left(\max_M \partial_s G - \min_M \partial_s G\right) dsdt < \kappa.$$

Note that $\mathcal{H}_0(\kappa)$ is contractible: if $K \in \mathcal{H}_0(\kappa)$ then $\lambda K \in \mathcal{H}_0(\kappa)$ for $0 \le \lambda \le 1$.

Associated to $J \in \mathcal{J}_0(\kappa)$ and $K \in \mathcal{H}_0(\kappa)$ is the moduli space of contractible solutions $v : S^2 \to M$ of (9.1.4):

$$\mathcal{M}_0(J,K) := \left\{v : S^2 \to M \,|\, (9.1.4),\ v \text{ is contractible}\right\}.$$

We claim that $\mathcal{M}_0(J,K)$ is nonempty for a dense set of pairs $(J,K) \in \mathcal{J}_0(\kappa)\times\mathcal{H}_0(\kappa)$. The proof rests on the energy identity (9.1.5) which in the present case takes the form

$$E_K(v) \le \|R_K\| < \kappa.$$

Denote by $\mathcal{JH}_{\mathrm{reg}}(\kappa)$ the set of all pairs $(J,K) \in \mathcal{J}_0(\kappa) \times \mathcal{H}_0(\kappa)$ that are regular in the sense of Definition 8.5.2 for $k = |I| = 1$ and the point $\mathbf{w} = 0$. Theorem 8.5.3 asserts that this set is residual in $\mathcal{J}_0(\kappa) \times \mathcal{H}_0(\kappa)$. If $(J,K) \in \mathcal{JH}_{\mathrm{reg}}(\kappa)$ then, by Theorem 8.5.1, the evaluation map

$$\mathrm{ev}_{J,K} : \mathcal{M}_0(J,K) \to M, \qquad \mathrm{ev}_{J,K}(u) := u(0),$$

is a pseudocycle of dimension $2n = \dim M$, and the homology class in $H_{2n}(M)$ represented by $\mathrm{ev}_{J,K}$ is independent of J and K, because the sets $\mathcal{J}_0(\kappa)$ and $\mathcal{H}_0(\kappa)$ are connected. The corresponding Gromov–Witten invariant is

$$\mathrm{GW}^M_{0,1}(a) = \mathrm{ev}_{J,H} \cdot \mathrm{pt} = 1, \qquad a := \mathrm{PD}([\mathrm{pt}]) \in H^{2n}(M).$$

To see this, consider the zero Hamiltonian. By Lemma 6.7.6, each pair (J,K) with $K = 0$ belongs to the regular set $\mathcal{JH}_{\mathrm{reg}}(\kappa)$ and, for every point $x_0 \in M$, there is precisely one contractible solution of equation (9.1.4) with $K = 0$ that satisfies $v(0) = x_0$, namely the constant solution $v(z) \equiv x_0$. It follows that $\mathcal{M}_0(J,K) \neq \emptyset$ for every $(J,K) \in \mathcal{JH}_{\mathrm{reg}}(\kappa)$.

Now let $J \in \mathcal{J}_0(\kappa)$ and $K \in \mathcal{H}_0(\kappa)$ and choose a regular sequence $(J_\nu, K_\nu) \in \mathcal{JH}_{\mathrm{reg}}(\kappa)$ converging to (J,K) in the C^∞ topology. Then, by what we just proved, $\mathcal{M}_0(J_\nu, K_\nu) \neq \emptyset$ for every ν. Fix any sequence $v_\nu \in \mathcal{M}_0(J_\nu, K_\nu)$. Because the curvature of K_ν is uniformly bounded, it follows from Remark 8.1.7 that there is a symplectic form $\omega_{K,\kappa}$ on the product manifold $S^2 \times M$ that tames the almost complex structure $\widetilde{J}_\nu$ induced by J_ν and K_ν for every ν (See Exercise 8.1.4.) Hence we may apply Theorem 4.6.1 to the sequence of graphs of the v_ν in the product manifold $S^2 \times M$ to conclude that there is a finite set $Z \subset S^2$ and a subsequence

that converges uniformly with all derivatives on every compact subset of $S^2 \setminus Z$ to a solution $v : S^2 \to M$ of (9.1.4). Since $E_{K_\nu}(v_\nu) \le \|R_{K_\nu}\|$ and R_{K_ν} converges to R_K, it follows that the limit solution v satisfies the required energy bound $E_K(v) \le \|R_K\|$.

Thus we have proved the proposition for pairs (J, K) in the set $\mathcal{J}_0(\kappa) \times \mathcal{H}_0(\kappa)$ and this is all that is needed for the proof of Theorem 9.1.1. To prove the proposition for general pairs (J, K) one can extend the above argument as follows. Fix a constant $\kappa > 0$, denote by $\mathcal{H}(\kappa)$ the set of Hamiltonian perturbations $K \in \Omega^1(S^2, C_0^\infty(M))$ that satisfy $\|R_K\| < \kappa$, and by $\mathcal{J}(\kappa)$ the space of all smooth maps

$$S^2 \to \mathcal{J}_+(M, \omega; \kappa) : z \mapsto J_z.$$

As before we denote by $\mathcal{JH}_{\mathrm{reg}}(\kappa)$ the set of all pairs $(J, K) \in \mathcal{J}(\kappa) \times \mathcal{H}(\kappa)$ that satisfy the requirements of Definition 8.5.2 for $k = \#I = 1$ and $\mathbf{w} = 0$. This time we do not know if the set $\mathcal{J}(\kappa) \times \mathcal{H}(\kappa)$ is connected and thus obtain a Gromov–Witten invariant

$$\mathrm{GW}_{0,1}^M(\mathrm{PD}([\mathrm{pt}])) = \mathrm{ev}_{J,H} \cdot \mathrm{pt}$$

for each component of $\mathcal{J}(\kappa) \times \mathcal{H}(\kappa)$. We claim that this invariant is one in each component. The previous argument applies to every component that contains a pair of the form $(J, 0)$. Now assume $R_K = 0$. Then there is a Hamiltonian gauge transformation $S^2 \to \mathrm{Ham}(M, \omega) : z \mapsto \psi_z$ such that $\psi^* K = 0$. Hence the invariant is one for every component of $\mathcal{J}(\kappa) \times \mathcal{H}(\kappa)$ that contains a pair (J, K) with $R_K = 0$. By Exercise 9.1.5 below, each component of $\mathcal{J}(\kappa) \times \mathcal{H}(\kappa)$ does contain such a pair and so the invariant is always one. It follows that $\mathcal{M}_0(J, K) \neq \emptyset$ for every regular pair $(J, K) \in \mathcal{JH}_{\mathrm{reg}}(\kappa)$. To complete the proof one uses the same approximation argument as above. Thus we have proved Proposition 9.1.4. □

The argument in the second part of the proof is a bit cumbersome because in this book we do not define the Gromov–Witten invariants for all compact symplectic manifolds. Thus we are restricted to almost complex structures J for which there are no J-holomorphic spheres with negative Chern-numbers. The resulting space $\mathcal{J}(\kappa)$ need not be connected and we need the additional arguments to prove that the invariant is one in each component. With the general definition of the Gromov–Witten invariants, these additional arguments are unnecessary and the assumption $J_z \in \mathcal{J}_+(M, \omega; \kappa)$ in Proposition 9.1.4 can be dropped.

EXERCISE 9.1.5. Let $K \in \Omega^1(S^2, C_0^\infty(M)$. Prove that there exists a smooth path

$$[0, 1] \to \Omega^1(S^2, C_0^\infty(M)) : \lambda \mapsto K_\lambda$$

such that

$$R_{K_0} = 0, \qquad K_1 = K, \qquad \sup_{0 \le \lambda \le 1} \|R_{K_\lambda}\| \le \|R_K\|.$$

Hint: Consider the pullback

$$\phi^* K = F\, ds + G\, dt$$

of K to $\mathbb{R} \times \mathbb{R}/\mathbb{Z}$ under our embedding $\phi : \mathbb{R} \times \mathbb{R}/\mathbb{Z} \to S^2$. If either F or G vanishes the result is easy: one can simply consider the family λK. Homotop K by a family of gauge transformations to achieve $F_{s,t} = 0$ for $s \le 1$. Next homotop to achieve $F = G = 0$ for $s \le 0$. Now repeat the argument for $s \ge 0$. Note that this result holds only because the underlying Hamiltonian fibration is trivial. In Section 9.6 we show that there are topological obstructions to making $\|R\|$ arbitrarily small.

The symplectic action. We next discuss some extensions of these ideas that involve the symplectic action functional. The most significant results apply only in the case when (M, ω) is a compact connected symplectic manifold such that the cohomology class of ω vanishes on $\pi_2(M)$, i.e.

$$\int_{S^2} v^*\omega = 0$$

for every smooth map $v : S^2 \to M$. We call such a manifold **symplectically aspherical**. This assumption has genuine geometric meaning since without it the action functional is only defined on a covering of the loop space. Moreover, it implies that there are no J-holomorphic spheres so that the proofs become technically easier.

Let us denote by L_0M the space of contractible loops $x : \mathbb{R}/\mathbb{Z} \to M$, and by

$$\widetilde{L_0M}$$

the covering whose elements are pairs $(x, [u])$, where $x \in L_0M$ and $[u]$ is an extension of x to the disc. More precisely, $[u]$ denotes the homotopy class relative to the boundary of a smooth map u from the unit disc

$$\mathbb{D} := \{z \in \mathbb{C} \,|\, |z| \le 1\}$$

to M such that $u(e^{2\pi i t}) = x(t)$ for $t \in \mathbb{R}$. A smooth Hamiltonian function $H : \mathbb{R}/\mathbb{Z} \times M \to \mathbb{R}$ determines a **symplectic action functional**

$$\mathcal{A}_H : \widetilde{L_0M} \to \mathbb{R},$$

defined by

$$\mathcal{A}_H(x, [u]) := -\int_{\mathbb{D}} u^*\omega - \int_0^1 H_t(x(t))\, dt. \tag{9.1.7}$$

Here we have chosen the signs so that $\mathcal{A}_H$ agrees with the classical action integral of $p\,dq - H\,dt$ along a loop in Euclidean space. We show in Lemma 9.1.8 below that the critical points of $\mathcal{A}_H$ are the pairs $(x, [u])$, where x is a 1-periodic orbit of the Hamiltonian isotopy $\{\phi_t^H\}_{0 \le t \le 1}$ generated by H and hence is a solution of (9.1.1). Further, by Lemma 9.1.9, the critical value $\mathcal{A}_H(x, [u])$ depends only on the homotopy class of the Hamiltonian path $\{\phi_t^H\}_{0 \le t \le 1}$ relative to its endpoints provided that H_t is normalized to have zero mean for every t. (To make sense of this, one has to vary u continuously with H.) It follows that the set of critical values

$$\mathrm{Spec}(H) := \{\mathcal{A}_H(x, [u]) \,|\, x \in \mathcal{P}_0(H)\}$$

depends only on the element $\widetilde{\phi}$ defined by H in the universal covering $\widetilde{\mathrm{Ham}}(M, \omega)$ of the Hamiltonian group. Here $\widetilde{\phi} := (\phi, [\phi_t])$, where $[\phi_t]$ denotes the homotopy class with fixed endpoints of the Hamiltonian flow $\{\phi_t\}_{0 \le t \le 1}$, generated by H and ϕ is the time-1 map ϕ_1. The set $\mathrm{Spec}(H)$ is called the **action spectrum** $\mathrm{Spec}(\widetilde{\phi})$ of $\widetilde{\phi}$. As shown by Oh [**305**], Schwarz [**360**], and Entov–Polterovich [**105**] it contains much interesting information about the properties of $\widetilde{\phi}$. For further discussion see Section 12.4.

When (M, ω) is symplectically aspherical the number $\mathcal{A}_H(x, [u])$ is independent of the choice of u and so in this case we shall drop u from our notation. For symplectically aspherical manifolds we shall refine Theorem 9.1.1 as follows.

THEOREM 9.1.6 (Schwarz [**360**]). *Let (M,ω) be a compact connected symplectic manifold that is symplectically aspherical. Let $H : \mathbb{R}/\mathbb{Z} \times M \to \mathbb{R}$ be a smooth Hamiltonian function whose time-1 map is not equal to the identity. Then there are two contractible periodic orbits $x, y \in \mathcal{P}_0(H)$ such that*

$$-\int_0^1 \max_M H_t\,dt \le \mathcal{A}_H(y) < \mathcal{A}_H(x) \le -\int_0^1 \min_M H_t\,dt.$$

In [**360**] Schwarz actually proved that the action spectrum has two distinguished (and distinct) elements $\mathcal{A}_H(x_0)$ and $\mathcal{A}_H(y_0)$ whose choice depends only on the time-1 map ϕ of H. This stronger result uses the spectral invariants constructed in Section 12.4 and does not seem to be accessible via the more elementary methods discussed here. He also showed that if $H_t \in C_0^\infty(M)$ the action $\mathcal{A}_H(x_0)$ depends only on x_0 and ϕ, but not on the Hamiltonian isotopy from the identity to ϕ. Again this is a deep result. Exercise 9.1.11 below shows that the relative action of two fixed points is independent of the Hamiltonian isotopy. This is an elementary observation that has an analogue valid for any manifold. By way of contrast, the corresponding assertion for the absolute action follows from Corollary 8.6.10, which in turn relies on the gluing theorem for J-holomorphic curves. Moreover, it holds only in the aspherical case.

In [**329**] Polterovich discovered remarkable applications of Theorem 9.1.6 to finitely generated group actions on symplectic manifolds. For example, he was able to show that every homomorphism from $\mathrm{SL}(3,\mathbb{Z})$ to the identity component of the group of symplectomorphisms of a Riemann surface of positive genus has to be trivial. His theory applies much more generally to symplectically aspherical manifolds of all dimensions.

Before embarking on the proof of Theorem 9.1.6 we explore some further properties of the symplectic action. Here are first some comments on our sign conventions.

REMARK 9.1.7 (Sign Conventions). Unfortunately, there are many mutually incompatible sign conventions in common use. Our choices are governed by two overriding principles: the symplectic structure of $\mathbb{R}^2 = \mathbb{C}$ with coordinate $z = x+iy$ should be $\omega = dx \wedge dy$ and the Hamiltonian equations generated by the function $H : \mathbb{R}^2 \to \mathbb{R}$ should be

$$\dot{x} = \frac{\partial H}{\partial y}, \qquad \dot{y} = -\frac{\partial H}{\partial x}.$$

It follows that the vector field X_H tangent to this flow must satisfy the equation

$$\omega(X_H, \cdot) = dH. \tag{9.1.8}$$

(The Hamiltonian differential equation in local coordinates is standard in the literature. However, many authors alter both the sign of the standard symplectic form on $\mathbb{R}^{2n}$ and the sign in equation (9.1.8).) With the above choices, the family of Hamiltonian symplectomorphisms $\{\phi_t^H\}_{0\le t\le 1}$, generated by H_t is given by

$$\partial_t \phi_t^H(x) = X_{H_t}(\phi_t^H(x)). \tag{9.1.9}$$

We use these sign conventions whenever we are in the realm of Hamiltonian dynamics. However, in the geometric setting of Hamiltonian connections on symplectic fibrations we have chosen to work with the equation $\partial_t\psi + X_{G_t} \circ \psi_t = 0$ for parallel transport, instead of (9.1.9), to make our discussion compatible with standard conventions in gauge theory. (See Sections 8.1, 8.2, and 9.6.)

The next lemmas establish the most important properties of the symplectic action. We first show that the critical points of $\mathcal{A}_H$ are the contractible periodic solutions of (9.1.1). The proof uses the fact that because $\widetilde{L_0M}$ is a covering of L_0M its tangent space at $(x,[u])$ may be identified with the tangent space to L_0M at x.

LEMMA 9.1.8. *Let (M,ω) be a symplectic manifold and fix a Hamiltonian function $H\in C^\infty(\mathbb{R}/\mathbb{Z}\times M)$. Let $(x,[u])\in\widetilde{L_0M}$ and $\xi\in T_x\widetilde{L_0M}=C^\infty(\mathbb{R}/\mathbb{Z},x^*TM)$. Then*

$$d\mathcal{A}_H(x,[u])\xi=\int_0^1\omega(\dot{x}-X_{H_t}(x),\xi)\,dt.$$

PROOF. Choose two smooth maps

$$\mathbb{R}\times\mathbb{R}/Z\to M:(s,t)\mapsto x^s(t)=v(s,t),\qquad \mathbb{R}\times\mathbb{D}\to M:(s,z)\mapsto u^s(z)$$

such that $x^0(t)=x(t)$, $u^0(z)=u(z)$, $u^s(e^{2\pi it})=x^s(t)$, and $\partial_sv(0,t)=\xi(t)$. Then the map $\mathbb{R}\to\widetilde{L_0M}:s\mapsto(x^s,[u^s])$ is the lift of the path $s\mapsto x^s$ in L_0M that goes through $(x^0,[u^0])=(x,[u])$. Moreover,

$$\mathcal{A}_H(x^{s_0},[u^{s_0}])=\mathcal{A}_0(x^0,[u^0])-\int_0^{s_0}\int_0^1\omega(\partial_sv,\partial_tv)\,dtds-\int_0^1H_t(v(s_0,t))\,dt.$$

Hence

$$\frac{d}{ds}\mathcal{A}_H(x^s,[u^s])=-\int_0^1\omega\big(\partial_sv,\partial_tv-X_{H_t}(v)\big)\,dt,$$

and the result follows by taking $s=0$. This proves Lemma 9.1.8. □

We next prove that the action spectrum $\mathrm{Spec}(H)$ depends only on the element $\widetilde{\phi}$ in the universal cover $\widetilde{\mathrm{Ham}}$ that is generated by H. More precisely, assume that (M,ω) is a compact connected symplectic manifold without boundary and

$$[0,1]\times\mathbb{R}/\mathbb{Z}\times M\to\mathbb{R}:(\lambda,t,x)\mapsto H_t^\lambda(x)$$

is a smooth family of normalized 1-periodic functions, so that

$$\int_M H_t^\lambda\omega^n=0 \tag{9.1.10}$$

for all t and λ. For each λ let $\{\phi_t^\lambda\}_{t\in\mathbb{R}}$ be the Hamiltonian isotopy generated by H^λ via

$$\partial_t\phi_t^\lambda=X_t^\lambda\circ\phi_t^\lambda,\qquad \phi_0^\lambda=\mathrm{id},\qquad \iota(X_t^\lambda)\omega=dH_t^\lambda.$$

Suppose that the time-1 map $\phi:=\phi_1^\lambda$ is independent of λ. The next Lemma asserts that, under these assumptions, the action spectrum $\mathrm{Spec}(H^\lambda)$ is independent of λ.

LEMMA 9.1.9. *Let $x_0\in\mathrm{Fix}_0(\phi)$ be a contractible fixed point. For $\lambda\in[0,1]$ define the loop $x^\lambda:\mathbb{R}/\mathbb{Z}\to M$ by $x^\lambda(t):=\phi_t^\lambda(x_0)$ and let*

$$[0,1]\times\mathbb{D}\to M:(\lambda,z)\mapsto u^\lambda(z)$$

be any smooth map such that $u^\lambda(e^{2\pi it})=x^\lambda(t)$ for all λ and t. Then, with H_t^λ normalized as in (9.1.10), the function

$$\lambda\mapsto\mathcal{A}_{H^\lambda}(x^\lambda,[u^\lambda])$$

is constant. Hence each element $\widetilde{\phi}\in\widetilde{\mathrm{Ham}}(M,\omega)$ has a well defined action spectrum.

PROOF. For each λ, define $f^\lambda : M \to \mathbb{R}$ by

$$f^\lambda(x) := -\int_{[0,1]^2} u_x^*\omega + \int_0^1 \left(H_t^0(\phi_t^0(x)) - H_t^\lambda(\phi_t^\lambda(x))\right) dt,$$

where $u_x : [0,1] \times [0,1] \to M$ is given by $u_x(s,t) := \phi_t^{s\lambda}(x)$. Note that for each $x \in M$ the restriction of u_x to the boundary of the square parametrizes the loop ℓ_x given by first going from x to $\phi(x)$ along the path $\phi_t^\lambda(x)$ and then going back to x along $\phi_{1-t}^0(x)$. Since this loop is generated by the juxtaposition K^λ of H_t^λ with $-H_{1-t}^0$, we find that

$$f^\lambda(x) = \mathcal{A}_{K^\lambda}(\ell_x, [u_x]).$$

Therefore Lemma 9.1.8 implies that $f^\lambda(x)$ is a critical value of $\mathcal{A}_{K^\lambda} : \widetilde{L_0M} \to \mathbb{R}$ for all $x \in M$. Since M is connected, f^λ must be a constant function of x.

Now differentiate f^λ with respect to λ and use Lemma 9.1.8 to obtain

$$\frac{d}{d\lambda} f^\lambda(x) = -\int_0^1 (\partial_\lambda H_t^\lambda)(\phi_t^\lambda(x))\, dt.$$

By what we have just proved this expression is independent of x. Moreover, its integral over M is zero since each function H_t^λ has mean value zero. Hence $\partial_\lambda f^\lambda(x) = 0$ for every λ and every x. Since the derivative of the path $\lambda \mapsto f^\lambda(x_0)$ agrees with that of $\lambda \mapsto \mathcal{A}_{H^\lambda}(x^\lambda, [u^\lambda])$, the lemma is proved. □

EXERCISE 9.1.10. Show by direct calculation that the function $f^\lambda : M \to \mathbb{R}$ in the above proof is constant. *Hint:* Let $\mathbb{R} \to M : r \mapsto x(r)$ be any smooth path and abbreviate $u_r := u_{x(r)}$. Use Cartan's formula and Stokes' theorem to prove that

$$\frac{d}{dr}\int_{[0,1]^2} u_r^*\omega = \frac{d}{dr}\int_0^1 \left(H_t^0(\phi_t^0(x(r))) - H_t^\lambda(\phi_t^\lambda(x(r)))\right) dt.$$

EXERCISE 9.1.11. Let $\mathcal{P}_\omega \subset \mathbb{R}$ denote the subgroup of periods of ω, i.e.

$$\mathcal{P}_\omega := \{\omega(A) \mid A \in \pi_2(M)\}.$$

For each $\phi \in \mathrm{Ham}(M,\omega)$ define the **relative symplectic action** of a pair of contractible fixed points $x^0, x^1 \in \mathrm{Fix}_0(\phi)$ by setting

$$\mathcal{A}_\phi(x^0, x^1) := \int_{[0,1]^2} v^*\omega \in \mathbb{R}/\mathcal{P}_\omega$$

where $v : [0,1] \times [0,1] \to M$ is a smooth map satisfying $v(0,t) = x^0$, $v(1,t) = x^1$, and $v(s,0) = \phi(v(s,1))$. Thus $\mathcal{A}_\phi(x^0,x^1)$ is the area (mod $\mathcal{P}_\omega$) between the path $s \mapsto v(s,0)$ from x^0 to x^1 and its image $v(s,1) = \phi^{-1}(v(s,0))$ by ϕ^{-1}. Show that

$$\mathcal{A}_\phi(x^0, x^1) = \mathcal{A}_H(x^0, [u^0]) - \mathcal{A}_H(x^1, [u^1]) \pmod{\mathcal{P}_\omega},$$

where $H \in C^\infty(\mathbb{R}/\mathbb{Z} \times M)$ is a Hamiltonian function whose Hamiltonian isotopy $t \mapsto \phi_t$ has time-1 map $\phi_1 = \phi$, $x^i(t) := \phi_t(x^i)$ for $i = 0, 1$, and $(x^i, [u^i])$ is any lift of x^i to $\widetilde{L_0M}$. It follows that $\mathcal{A}_\phi(x^0,x^1)$ is independent of v and is well defined as an element of $\mathbb{R}/\mathcal{P}_\omega$. *Hint:* To show that v exists, choose any path $[0,1] \to L_0M : s \mapsto x^s$ from x^0 to x^1 and set $v(s,t) := \phi_t^{-1}(x^s(t))$. Now choose a smooth map $[0,1] \times B \to M : (s,z) \mapsto u^s(z)$ such that $u^s(e^{2\pi i t}) = x^s(t)$ for all s and t. Differentiate the curve $s \mapsto \mathcal{A}_H(x^s, [u^s])$ using Lemma 9.1.8 and show that

$$\frac{d}{ds}\mathcal{A}_H(x^s, [u^s]) = -\int_0^1 \omega(\partial_s v(s,t), \partial_t v(s,t))\, dt.$$

(Abbreviate $u(s,t) := u^s(e^{2\pi i t})$ and differentiate the identity $u(s,t) = \phi_t(v(s,t))$ with respect to s and t.) Alternatively, define $w : [0,1]^2 \to M$ by $w(s,t) := \phi_t(v(s,1))$, prove that $u^0 - u^1 + v + w$ is a 2-cycle (representing a class in the image of the Hurewitz homomorphism $\pi_2(M) \to H_2(M)$), and integrate ω over this cycle. (The integral of $w^*\omega$ gives the Hamiltonian terms.)

The Floer equation. The proof of Theorem 9.1.6 expands on the ideas in the proof of Theorem 9.1.1, though it is easier since there is no bubbling in the aspherical case. Therefore we begin by formalizing some of the concepts introduced there.

Let us fix a smooth 1-parameter family of ω-compatible almost complex structures $J_t = J_{t+1} \in \mathcal{J}(M,\omega)$. Any such family determines an L^2-inner product on each tangent space $T_x L_0 M = C^\infty(\mathbb{R}/\mathbb{Z}, x^*TM)$ via

$$\langle \xi, \eta \rangle := \int_0^1 \omega\big(\xi(t), J_t(x)\eta(t)\big)\, dt.$$

By Lemma 9.1.8, the gradient of $\mathcal{A}_H$ with respect to this inner product is given by $\operatorname{grad} \mathcal{A}_H(x) = J_t(x)(\dot{x} - X_{H_t}(x))$. Hence a (negative) gradient flow line of $\mathcal{A}_H$ is a smooth function $u : \mathbb{R} \times \mathbb{R}/\mathbb{Z} \to M$ satisfying the **Floer equation**

$$\partial_s u + J_t(u)(\partial_t u - X_{H_t}(u)) = 0. \tag{9.1.11}$$

Note that equation (9.1.11) is meaningful even in the case where (M,ω) is not symplectically aspherical and where the J_t are only ω-tame. It was introduced by Floer [**114**] in his study of Floer homology. The energy of a solution u of (9.1.11) is defined by

$$E_H(u) := \tfrac{1}{2} \int_{-\infty}^{\infty} \int_0^1 \left(|\partial_s u|_{J_t}^2 + |\partial_t u - X_{H_t}(u)|_{J_t}^2 \right) dt ds.$$

We denote by $\mathcal{M}^{\mathrm{Floer}}(H,J)$ the set of contractible finite energy solutions of (9.1.11). The next lemma shows that if $\mathcal{M}^{\mathrm{Floer}}(H,J) \neq \emptyset$ (for some J) then $\mathcal{P}_0(H) \neq \emptyset$. Although the lemma is formulated for symplectically aspherical manifolds, this conclusion remains valid in general.

LEMMA 9.1.12. *Let (M,ω) be a compact symplectic manifold that is symplectically aspherical. Let $H : \mathbb{R}/\mathbb{Z} \times M \to \mathbb{R}$ be a smooth Hamiltonian function, $J_t = J_{t+1}$ be a smooth family of ω-tame almost complex structures on M, and $u \in \mathcal{M}^{\mathrm{Floer}}(H,J)$. Then the following holds.*

(i) $\lim_{s\to\pm\infty} \sup_{t\in\mathbb{R}} |\partial_s u(s,t)| = 0$.

(ii) *There exist $x, y \in \mathcal{P}_0(H)$ such that $E_H(u) = \mathcal{A}_H(x) - \mathcal{A}_H(y)$ and*

$$\mathcal{A}_H(x) = \sup_{s\in\mathbb{R}} \mathcal{A}_H(u(s,\cdot)), \qquad \mathcal{A}_H(y) = \inf_{s\in\mathbb{R}} \mathcal{A}_H(u(s,\cdot)).$$

(iii) *If $\mathcal{P}_0(H)$ is a finite set then the periodic solutions x, y in (ii) can be chosen such that $\lim_{s\to-\infty} u(s,t) = x(t)$ and $\lim_{s\to\infty} u(s,t) = y(t)$, uniformly in t.*

PROOF. We prove that $\sup_{s,t} |\partial_s u(s,t)| < \infty$. Since $E_H(u) < \infty$ there exists a $T > 0$ such that the energy of the restriction of u to the set $(\mathbb{R} \setminus [-T,T]) \times [0,1]$ is less than the energy of every holomorphic sphere (for every J_t). Hence the bubbling argument in Section 4.2 shows that the function $(s,t) \mapsto |\partial_s u(s,t)|$ is bounded.

We prove (i). Suppose, otherwise, that there are sequences $s_\nu, t_\nu \in \mathbb{R}$ and a constant $\delta > 0$ such that $|s_\nu| \to \infty$, $0 \le t_\nu \le 1$, and $|\partial_s u(s_\nu, t_\nu)| \ge \delta$. Since $|\partial_s u|$ is bounded, it follows from the elliptic bootstrapping analysis of Appendix B (and Gromov's graph construction that interprets graphs of solutions of (9.1.11) as $\widetilde{J}_H$-holomorphic curves: see Remark 8.1.9)) that there exists a subsequence ν_i such that the sequence $(s,t) \mapsto u(s_{\nu_i} + s, t)$ converges, in the topology of C^∞-convergence on compact sets, to a solution $u^\infty(s,t)$ of (9.1.11). Since for each fixed T the energy of u on the domain $[s_{\nu_i} - T, s_{\nu_i} + T] \times [0,1]$ tends to zero, it follows that the energy of the limit function u^∞ must be zero. Hence $\partial_s u(s_{\nu_i}, t_{\nu_i})$ converges to zero, a contradiction. This proves (i).

We prove (ii). By Lemma 9.1.8, we have

$$
\begin{aligned}
0 &\le E_H(u; [a,b] \times [0,1]) \\
&= \int_a^b \int_0^1 \omega\big(\partial_s u, \partial_t u - X_{H_t}(u)\big)\, dtds \qquad (9.1.12)\\
&= \mathcal{A}_H(u(a,\cdot)) - \mathcal{A}_H(u(b,\cdot))
\end{aligned}
$$

for $a < b$. Hence it remains to prove that the limits of $\mathcal{A}_H(u(s,\cdot))$ for $s \to \pm\infty$ are critical levels of $\mathcal{A}_H$. To see this, note first that $|\partial_t u - X_{H_t}(u)|$ converges to zero, uniformly in t, as $|s|$ tends to infinity. This implies that, for every sequence $s_\nu \to \infty$, there is a subsequence ν_i and a $y \in \mathcal{P}_0(H)$ such that $u(s_{\nu_i}, t)$ converges to $y(t)$ uniformly in t. Since $\partial_t u - X_{H_t}(u)$ tends to zero as $|s| \to \infty$, it follows that $u(s_{\nu_i}, \cdot)$ converges to y in the C^1-topology and hence $\mathcal{A}_H(u(s_{\nu_i}, \cdot))$ converges to $\mathcal{A}_H(y)$. The analogous conclusion holds for sequences $s_\nu \to -\infty$. This proves (ii).

We prove (iii). If $\mathcal{P}_0(H)$ is a finite set then the contractible 1-periodic solutions of (9.1.1) are isolated. Hence the limit x, respectively y, is independent of the sequence $s_\nu \to -\infty$, respectively $s_\nu \to +\infty$. This proves assertion (iii) and Lemma 9.1.12. □

By Lemma 9.1.12, the existence theorem for periodic orbits can be reduced to an existence theorem for solutions of the Floer equation. Recall that the **Hofer norm** of a Hamiltonian function $\mathbb{R}/\mathbb{Z} \to C^\infty(M) : t \mapsto H_t$ is defined by

$$\|H\| := \int_0^1 \left(\max_M H_t\, dt - \min_M H_t \right) dt.$$

The proof of the next theorem will be deferred to the end of the section.

THEOREM 9.1.13. *Suppose (M,ω) is a compact symplectic manifold that is symplectically aspherical. Let $H \in C^\infty(\mathbb{R}/\mathbb{Z} \times M)$ and $\{J_t\}_{t\in\mathbb{R}/\mathbb{Z}}$ be a smooth loop in $\mathcal{J}_\tau(M,\omega)$. Then, for every $x_0 \in M$, there exists a $u \in \mathcal{M}^{\mathrm{Floer}}(H,J)$ such that $u(0,0) = x_0$, $E_H(u) \le \|H\|$, and, for every $s \in \mathbb{R}$,*

$$-\int_0^1 \max_M H_t\, dt \le \mathcal{A}_H(u(s,\cdot)) \le -\int_0^1 \min_M H_t\, dt. \qquad (9.1.13)$$

PROOF OF THEOREM 9.1.6. Let $t \mapsto \phi_t$ be the Hamiltonian isotopy generated by H and suppose that $\phi_1 \ne \mathrm{id}$. Choose $x_0 \in M$ such that $\phi_1(x_0) \ne x_0$ and let $\{J_t\}_{t\in\mathbb{R}/\mathbb{Z}}$ be any smooth family of ω-tame almost complex structures on M. Then, by Theorem 9.1.13, there is a $u \in \mathcal{M}^{\mathrm{Floer}}(H,J)$ satisfying $u(0,0) = x_0$ and (9.1.13). Since x_0 is not a fixed point of ϕ_1 we have $E_H(u) > 0$. Hence the periodic solutions $x, y \in \mathcal{P}_0(H)$ of Lemma 9.1.12 (ii) satisfy the requirements of Theorem 9.1.6. □

Another consequence of Theorem 9.1.13 is that the Hofer metric on the group of Hamiltonian symplectomorphisms of a compact symplectically aspherical manifold is nondegenerate. (For more details on the Hofer metric see Section 9.6 below.) We shall denote by $\phi_H : M \to M$ the Hamiltonian symplectomorphism generated by $H \in C^\infty(\mathbb{R}/\mathbb{Z} \times M)$.

COROLLARY 9.1.14 (The Hofer metric). *Suppose (M, ω) is a compact symplectic manifold that is symplectically aspherical and $\phi \neq \mathrm{id}$ is a Hamiltonian symplectomorphism of M. Then*

$$d(\phi, \mathrm{id}) := \inf\{\|H\| \mid H \in C^\infty(\mathbb{R}/\mathbb{Z} \times M),\ \phi_H = \phi\} > 0.$$

PROOF. Choose a smooth family $\mathbb{R} \to \mathcal{J}_\tau(M, \omega) : t \mapsto J_t$ of almost complex structures such that $J_{t+1} = \phi^* J_t$ for every t. Now consider the equation

$$\partial_s v + J_t(v)\partial_t v = 0, \qquad v(s,t) = \phi(v(s,t+1)) \tag{9.1.14}$$

for smooth functions $v : \mathbb{R}^2 \to M$. The energy of v is defined by

$$E(v) := \int_{-\infty}^{\infty} \int_0^1 |\partial_s v(s,t)|_{J_t}^2 \, dtds.$$

Choose a point $x_0 \in M$ such that $\phi(x_0) \neq x_0$. Then, by compactness,

$$\delta := \inf\{E(v) \mid v \text{ satisfies (9.1.14)}, v(0,0) = x_0,\ E(v) < \infty\} > 0. \tag{9.1.15}$$

Now let $H \in C^\infty(\mathbb{R}/\mathbb{Z} \times M)$ such that $\phi_H = \phi$ and denote by $t \mapsto \phi_t$ the Hamiltonian isotopy generated by H. Then v is a solution of (9.1.14) if and only if the function $u(s,t) := \phi_t(v(s,t))$ is a solution of (9.1.11) with J_t replaced by $\phi_{t*}J_t$. Moreover, $E_H(u) = E(v)$ and $u(0,0) = v(0,0)$. Now it follows from Theorem 9.1.13 that there is a finite energy solution u of (9.1.11) (with J_t replaced by $\phi_{t*}J_t$) such that $u(0,0) = x_0$ and $E_H(u) \leq \|H\|$. Hence

$$v(s,t) := \phi_t^{-1}(u(s,t))$$

is a finite energy solution of (9.1.14) such that $v(0,0) = x_0$. Hence

$$\delta \leq E(v) = E_H(u) \leq \|H\|.$$

This proves Corollary 9.1.14. □

EXERCISE 9.1.15. Show that $\delta > 0$ in (9.1.15).

We now turn to the proof of Theorem 9.1.13. As in the case of Theorem 9.1.1, we construct the desired Floer trajectory as a limit of solutions to the perturbed Cauchy–Riemann equations (9.1.4) on the 2-sphere. Given a smooth family of almost complex structures $J : S^2 \to \mathcal{J}_\tau(M, \omega)$ and a Hamiltonian perturbation $K \in \Omega^1(S^2, C_0^\infty(M))$, we denote by $\mathcal{M}_0(J, K)$ the space of contractible solutions of (9.1.4).

PROPOSITION 9.1.16. *Let (M, ω) be a compact symplectic manifold that is symplectically aspherical. Let $J \in C^\infty(S^2, \mathcal{J}_\tau(M, \omega))$ and $K \in \Omega^1(S^2, C_0^\infty(M))$. Then the evaluation map*

$$\mathrm{ev}_{w,J,K} : \mathcal{M}_0(J, K) \to M, \qquad \mathrm{ev}_{w,J,K}(v) := v(w),$$

is surjective for every $w \in S^2$.

PROOF. By Theorem 4.6.1, the moduli space $\mathcal{M}_0(J,K)$ is compact for every $J \in \mathcal{J} := C^\infty(S^2, \mathcal{J}_\tau(M,\omega))$ and every $K \in \mathcal{H} := \Omega^1(S^2, C_0^\infty(M))$. Given $J \in \mathcal{J}$ denote by $\mathcal{H}_{\mathrm{reg}}(J)$ the set of all Hamiltonian perturbations K for which the (vertical) linearized operator $D_{K,v}$ is onto for every solution $v : S^2 \to M$ of (9.1.4) (see Section 8.3). Then, for $K \in \mathcal{H}_{\mathrm{reg}}(J)$, the moduli space is a compact oriented smooth $2n$-manifold (Theorem 8.3.1) and its bordism class is independent of $J \in \mathcal{J}$ and $K \in \mathcal{H}_{\mathrm{reg}}(J)$ (Theorem 8.3.3). It follows as in the proof of Proposition 9.1.4 that evaluation map $\mathrm{ev}_{w,J,K} : \mathcal{M}_0(J,K) \to M$ has degree one when $J = J_0$ and $K = 0$. Hence this continues to hold for every $J \in \mathcal{J}$ and every $K \in \mathcal{H}_{\mathrm{reg}}(J)$. Since $\mathcal{H}_{\mathrm{reg}}(J)$ is dense in $\mathcal{H}$ (Theorem 8.3.1), it follows from compactness that $\mathrm{ev}_{w,J,K}$ is surjective for every $J \in \mathcal{J}$, every $K \in \mathcal{H}$, and every $w \in S^2$. This proves Proposition 9.1.16 □

PROOF OF THEOREM 9.1.13. As in equation (9.1.6) let $K_T \in \Omega^1(S^2, C_0^\infty(M))$ be the 1-form whose pullback to $\mathbb{R} \times \mathbb{R}/\mathbb{Z}$ is

$$\phi^* K_T = -\beta_T(s) H_t \, dt,$$

where $\beta_T : \mathbb{R} \to [0,1]$ is a compactly supported smooth cutoff function that equals one for $|s| \le T$ and that is nonincreasing for $s \ge 0$ and nondecreasing for $s \le 0$. Fix a point $x_0 \in M$ and use Proposition 9.1.16 with $w = 1 \in \mathbb{C} \subset S^2$ to obtain a contractible solution $v_T : S^2 \to M$ of (9.1.4) with $K = K_T$ and $J = J_T$ such that $v_T(1) = x_0$. Then the map

$$u_T(s,t) := v_T(e^{2\pi(s+it)})$$

satisfies (9.1.11) for $|s| \le T$ and $u_T(0,0) = x_0$. Moreover,

$$-\int_0^1 \max_M H_t \, dt \le \mathcal{A}_H(u_T(s,\cdot)) \le -\int_0^1 \min_M H_t \, dt \tag{9.1.16}$$

for $-T \le s \le T$. To see this, note that, for $s \ge -T$, we have

$$\begin{aligned} \frac{d}{ds}\mathcal{A}_{\beta_T(s)H}(u_T(s,\cdot)) &= -\int_0^1 |\partial_s u_T(s,t)|_{J_T}^2 \, dt - \int_0^1 \dot\beta_T(s) H_t(u_T(s,t)) \, dt \\ &\le -\dot\beta_T(s) \int_0^1 \max_M H_t \, dt. \end{aligned}$$

Now observe that $\mathcal{A}_{\beta_T(s)H}(u_T(s,\cdot))$ converges to zero as $s \to \infty$. Hence the first inequality in (9.1.16) follows by integration over the interval $[s,\infty)$. The second inequality is proved similarly.

It follows from (9.1.16) and (9.1.12) that the energy of u_T on the cylinder $[-T,T] \times \mathbb{R}/\mathbb{Z}$ is bounded by $\|H\|$. Since there are no J-holomorphic spheres, we have compactness without bubbling. Hence the functions $(s,t) \mapsto |\partial_s u_T(s,t)|$ are uniformly bounded on every compact set (see Section 4.2). Hence, by Theorem B.4.2 (applied to the graphs), there is a sequence $T_\nu \to \infty$ such that u_{T_ν} converges to a finite energy solution u of (9.1.11) in the C^∞-topology on every compact set. The limit u satisfies $u(0,0) = x_0$, and the inequality (9.1.13) follows from (9.1.16). Moreover, since u_T is contractible for every T, so is u. This proves Theorem 9.1.13. □

9.2. Obstructions to Lagrangian embeddings

In the following we write a vector in $\mathbb{C}^n$ in the form $z = (z_1, \ldots, z_n)$, assume that $\mathbb{C}^n$ is equipped with the standard symplectic form

$$\omega_0 := \sum_{j=1}^{n} dx_j \wedge dy_j, \qquad z_j =: x_j + iy_j,$$

and denote by $\mathbb{D} := \{s + it \in \mathbb{C} \,|\, s^2 + t^2 \leq 1\}$ the closed unit disc in $\mathbb{C}$. In [**160**] Gromov proved the following generalization of the Riemann mapping theorem.

THEOREM 9.2.1 (Gromov). *Let $L \subset \mathbb{C}^n$ be a compact Lagrangian submanifold. Then there exists a nonconstant holomorphic disc $u : \mathbb{D} \to \mathbb{C}^n$ such that $u(\partial\mathbb{D}) \subset L$.*

He also constructed a symplectic form ω on $\mathbb{C}^n$, for each $n > 1$, which admits a closed Lagrangian submanifold L that is **exact** in the sense that the restriction $\lambda|_L$ to L of some primitive λ for ω is exact. (Since the 1-form λ satisfies $d\lambda = \omega$, this restriction is necessarily closed.) Theorem 9.2.1 implies that this symplectic structure ω is **exotic**, in that $(\mathbb{C}^n, \omega)$ does not embed into standard Euclidean space $(\mathbb{C}^n, \omega_0)$. See [**277**] for a further discussion of this topic, and McLean [**283**] for some recent relevant results.

This theorem also implies that any embedded Lagrangian submanifold L in Euclidean space $(\mathbb{C}^n, \omega_0)$ has $H^1(L; \mathbb{R}) \neq 0$. Indeed, the restriction of the 1-form

$$\lambda := \tfrac{1}{2} \sum_{j=1}^{n} (x_j dy_j - y_j dx_j)$$

to L has a nonzero integral over the loop $\mathbb{R}/2\pi\mathbb{Z} \to L : \theta \mapsto u(e^{i\theta})$ and so represents a nonzero cohomology class. In contrast, there are simply connected immersed Lagrangian manifolds. An example is the immersion $\iota : S^n \to \mathbb{C}^n$, given by

$$\iota(x_0, x_1, \ldots, x_n) := (x_1, x_0 x_1, \ldots, x_n, x_0 x_n),$$

which is Lagrangian and maps the two points $(\pm 1, 0, \ldots, 0)$ to the origin.

However, Theorem 9.2.1 does not provide much other information about which manifolds admit Lagrangian embeddings into Euclidean space, and this question is still not fully understood. There are some easy remarks one can make. For example, if $L \subset \mathbb{C}^n$ is any oriented embedded Lagrangian submanifold it must have zero Euler characteristic. To see this note that the normal bundle of L is isomorphic to the tangent bundle via $T_z L \mapsto T_z L^\perp : \zeta \mapsto i\zeta$. Hence the self-intersection number of L is equal to

$$L \cdot L = (-1)^{n(n-1)/2} \chi(L).$$

But every closed orientable submanifold of $\mathbb{C}^n$ has self-intersection number zero. Hence tori are the only orientable surfaces which admit Lagrangian embeddings into $\mathbb{C}^2$. In the nonorientable case the same argument shows that the Euler characteristic must be even. The nonorientable surfaces with even Euler characteristics are connected sums of $g + 1$ Klein bottles:

$$K_g := K \# \cdots \# K \cong K \# \Sigma_g, \qquad \chi(K_g) = -2g.$$

It has been known for some time that K_g admits a Lagrangian embedding into $\mathbb{C}^2$ for every $g > 0$: see Givental [**144**] and also Lalonde–Sikorav [**230**]. The case $g = 0$ has recently been resolved by Nemirovski [**298**] who showed that the Klein bottle does not admit a Lagrangian embedding into $\mathbb{C}^2$.

REMARK 9.2.2. Theorem 9.2.1 has since been refined, in that estimates were found for the Maslov numbers of holomorphic discs with boundaries in L. In **[22]** Audin conjectured that every Lagrangian embedding of a torus $\iota : \mathbb{T}^n \to \mathbb{C}^n$ admits a (not necessarily holomorphic) disc

$$u : (\mathbb{D}, \partial\mathbb{D}) \to (\mathbb{C}^n, L), \qquad L := \iota(\mathbb{T}^n),$$

with Maslov number two. Viterbo **[403]** proved this for $n = 2$. For monotone Lagrangian submanifolds $L \subset \mathbb{C}^n$ Polterovich **[324]** established the existence of a holomorphic disc with Maslov number $\mu \leq n + 1$, Oh **[304]** found a holomorphic disc with Maslov number 2 under the assumption $n \leq 24$ (also in the monotone case), and Seidel, in an unpublished work, extended this result to all dimensions. In **[42]** Biran and Cieliebak established the existence of a holomorphic disc with Maslov number

$$3 - n \leq \mu \leq n + 1$$

for general compact Lagrangian submanifolds of $\mathbb{C}^n$, and indeed of subcritical Stein manifolds. (For the inequality $\mu \leq n + 1$ see also Remark 9.2.5 below.) Audin's conjecture has now been proved for monotone tori in $\mathbb{R}^{2n}$ by Buhovsky **[50]**. More recently, Damian **[77]** extended the Audin conjecture to a more general class of monotone Lagrangian submanifolds of $\mathbb{C}^n$. However, the non-monotone Audin conjecture is still open at the time of writing. A related open question (pointed out to the authors by Paul Biran) is whether every Lagrangian submanifold of $\mathbb{C}^n$, torus or not, has a nonzero Maslov class.

The idea of the proof of Theorem 9.2.1 is to study the following perturbed Cauchy–Riemann equation for pairs (λ, u), where $0 \leq \lambda \leq 1$ and $u : \mathbb{D} \to \mathbb{C}^n$ is a disc with boundary on L:

$$\partial_s u + J_0 \partial_t u = \nabla H^\lambda_{s,t}(u), \qquad u(\partial\mathbb{D}) \subset L, \qquad [u] = A_0. \tag{9.2.1}$$

Here $A_0 \in \pi_2(\mathbb{C}^n, L)$ denotes the homotopy class of the constant maps and ∇ denotes the gradient with respect to the standard metric. The family of functions $H^\lambda, \lambda \in [0, 1]$, given by

$$[0, 1] \times \mathbb{D} \times \mathbb{C}^n \to \mathbb{R} : (\lambda, s, t, z) \mapsto H^\lambda_{s,t}(z)$$

starts at $H^0_{s,t}(z) = 0$, ends at the linear map $H^1_{s,t}(z) = \langle a, z \rangle$ for a suitably chosen vector $a \in \mathbb{C}^n$, and, for some constant $c > 0$, satisfies the following condition for all intermediate values of λ

$$H^\lambda_{s,t}(z) = \lambda \langle a, z \rangle, \qquad |z| \geq c. \tag{9.2.2}$$

We show in Lemma 9.2.4 that with appropriate values of a and c equation (9.2.1) has no solutions for $\lambda = 1$. On the other hand when $\lambda = 0$ there is precisely one solution (the constant map) through each point of L. We will also see that for generic H the space of all solutions $(\lambda, u), \lambda \in [0, 1]$, to (9.2.1) that go through a generic point in L is a 1-dimensional manifold. The crux of the matter is now this. This solution space cannot be compact since if it were it would be a compact 1-dimensional manifold with one boundary point. But we will see that the only way in which noncompactness can occur is for a J-holomorphic disc or sphere to bubble off. Since $\mathbb{C}^n$ contains no J-holomorphic spheres, there must therefore be some J-holomorphic disc with boundary on L.

REMARK 9.2.3. Note that here we are considering J-holomorphic discs that satisfy Lagrangian boundary conditions. Although we did discuss the bubbling off of such discs in Chapter 4, we never explicitly treated the boundary value problem in Chapter 3. However, as indicated below it is not very different from the case of closed curves. In our applications, the main point is to check that the linearized equation is Fredholm, which is established in Theorem C.1.10. In the current situation it is enough to achieve transversality by making Hamiltonian perturbations as in Chapter 8. (Alternatively, one could allow J to depend on the points of the domain as in Section 6.7.) If one is working in a geometric situation in which it is important to find a generic almost complex structure on M, then, just as in Section 3.2, we need to work with somewhere injective discs. But the structure of nowhere injective discs is much more complicated than in the closed case explained in Section 2.5, since discs can wrap around themselves in intricate ways. The papers by Lazzarini [**234, 235**] and Kwon–Oh [**220**] explain the structure of general discs (see also Zehmisch [**425**]), and Biran–Cornea [**43**, §3] use this to find generic J.

Here is the first key lemma in the proof of Theorem 9.2.1.

LEMMA 9.2.4. *Let $a \in \mathbb{C}^n$, fix a constant $c \geq \sup_{z\in L}|z|$, and let $u : \mathbb{D} \to \mathbb{C}^n$ be such that $u(\partial\mathbb{D}) \subset L$.*

(i) *If u satisfies the equation $\partial_s u + J_0\partial_t u = a$ then $|a| \leq 2c$.*

(ii) *If $\mathbb{D} \times \mathbb{C}^n \to \mathbb{R} : (s,t,z) \mapsto H_{s,t}(z)$ is a smooth function such that*

$$|z| \geq c \qquad \Longrightarrow \qquad H_{s,t}(z) = \langle a, z\rangle$$

and u satisfies the equation $\partial_s u + J_0\partial_t u = \nabla H_{s,t}(u)$, then $\sup_{s,t}|u(s,t)| \leq c$.

PROOF. The proof of the second assertion is based on the observation that every solution of the equation $\partial_s u + J_0\partial_t u = a$ is harmonic and hence satisfies

$$\Delta|u|^2 = 2|\partial_s u|^2 + 2|\partial_t u|^2 \geq 0,$$

where $\Delta := \partial_s^2 + \partial_t^2$ denotes the Laplacian. This means that $|u|^2$ is subharmonic on the open set

$$\Omega := \{s + it \in \mathbb{D} \,|\, |u(s,t)| > c\}.$$

Since $c \geq \sup_{z\in L}|z|$, this set is contained in the interior of $\mathbb{D}$. We claim that $\Omega = \emptyset$. Otherwise it would follow from the mean value inequality for subharmonic functions that $|u|$ is constant on each component of Ω. But this is impossible because $|u| = c$ on $\partial\Omega$. This proves (ii). To prove (i) note that, by the divergence theorem,

$$\begin{aligned}
|a| &= \frac{1}{\pi}\left|\int_{\mathbb{D}} (\partial_s u + J_0\partial_t u)\, dsdt\right| \\
&= \frac{1}{\pi}\left|\int_0^{2\pi} (\cos\theta + \sin\theta J_0)\, u(e^{i\theta})\, d\theta\right| \\
&\leq \frac{1}{\pi}\int_0^{2\pi} \left|u(e^{i\theta})\right| d\theta \\
&\leq 2c.
\end{aligned}$$

This proves Lemma 9.2.4. □

PROOF OF THEOREM 9.2.1. Assume, by contradiction, that there is no non-constant holomorphic disc $u : (\mathbb{D}, \partial\mathbb{D}) \to (\mathbb{C}^n, L)$. Let $a \in \mathbb{C}^n$ be any vector such that

$$|a| > c, \qquad c := 2 \sup_{z \in L} |z|.$$

Denote by $\mathcal{H} \subset C^\infty([0,1] \times \mathbb{D} \times \mathbb{C}^n)$ the set of smooth Hamiltonian functions H such that

$$H^0_{s,t}(z) = 0, \quad H^1_{s,t}(z) = \langle a, z \rangle,$$

and equation (9.2.2) holds. For every $H \in \mathcal{H}$, consider the moduli space

$$\mathcal{M}(H) := \{(\lambda, u) \,|\, 0 \le \lambda \le 1, \, u : \mathbb{D} \to \mathbb{C}^n \text{ satisfies (9.2.1)}\}.$$

We prove that $\mathcal{M}(H)$ is compact for every $H \in \mathcal{H}$. Now there are no holomorphic spheres in $\mathbb{C}^n$ and, by assumption, there are also no holomorphic discs in $\mathbb{C}^n$ with boundary in L. Moreover, by Lemma 9.2.4 (ii), every solution $u : \mathbb{D} \to \mathbb{C}^n$ of (9.2.1) takes values in the compact set $\{z \in \mathbb{C}^n \,|\, |z| \le c\}$. The energy identity takes the form

$$\begin{aligned} E(u) &:= \int_{\mathbb{D}} |\partial_s u|^2 \, dsdt \\ &= \int_{\mathbb{D}} \langle \partial_s u, -J_0 \partial_t u + \nabla H^\lambda_{s,t}(u) \rangle \, dsdt \\ &= \int_{\mathbb{D}} \omega_0(\partial_s u, \partial_t u) \, dsdt + \int_{\mathbb{D}} \partial_s(H^\lambda_{s,t} \circ u) \, dsdt - \int_{\mathbb{D}} (\partial_s H^\lambda_{s,t}) \circ u \, dsdt \\ &= \int_0^{2\pi} \cos\theta \, H^\lambda_{e^{i\theta}}(u(e^{i\theta})) \, d\theta - \int_{\mathbb{D}} (\partial_s H^\lambda_{s,t}) \circ u \, dsdt, \end{aligned}$$

where the last equation follows from the fact that $[u] = 0$. It follows that there is a uniform energy bound on the elements of the moduli space $\mathcal{M}(H)$. Now apply Theorem 4.6.1 to the compact symplectic manifold $\widetilde{M} := \mathbb{D} \times \{z \in \mathbb{C}^n \,|\, |z| \le c\}$, the almost complex structure $\widetilde{J}$ on $\widetilde{M}$ induced by J_0 and H via (8.1.6), and the Lagrangian submanifold $\widetilde{L} := S^1 \times L$ to deduce that the moduli space $\mathcal{M}(H)$ is compact. Here we use the fact that, just as in Section 6.7 and 8.5, if there is bubbling at a point z in the domain, its image must lie in the fiber over z and so must be a holomorphic disc or sphere.

Next observe that the proofs of Theorems 8.3.1 and 8.3.3 carry over in a straightforward fashion to the present case and show that, for a generic Hamiltonian perturbation $H \in \mathcal{H}_{\text{reg}}$, the moduli space $\mathcal{M}(H)$ is a smooth manifold of dimension $n+1$. The key point here is that every solution $u : \mathbb{D} \to \mathbb{C}^n$ of (9.2.1) must pass through the region $\{|z| < c\}$ in which we can perturb H, since $u(\partial\mathbb{D}) \subset L$ is contained in this region. With this understood the usual arguments apply: one first shows that the universal moduli space $\mathcal{M}(\mathcal{H}^\ell)$ of all triples (λ, u, H) with $H \in \mathcal{H}^\ell$ and $(\lambda, u) \in \mathcal{M}(H)$ is a $C^{\ell-1}$ Banach manifold and then considers the set $\mathcal{H}^\ell_{\text{reg}}$ of regular values of the projection $\pi : \mathcal{M}(\mathcal{H}^\ell) \to \mathcal{H}^\ell$. The above compactness result shows that this projection is proper and so its set of regular values is open, and the usual argument using the Sard–Smale theorem implies that this set is dense. Further, it follows from the index formula of Theorem C.1.10 that

$$\dim \mathcal{M}(H) = n + 1$$

for $H \in \mathcal{H}_{\mathrm{reg}}$. This holds because that the boundary Maslov index of the bundle pair $(\mathbb{D}\times\mathbb{C}^n, u^*TL)$ is zero for every smooth map $u : (\mathbb{D}, \partial\mathbb{D}) \to (\mathbb{C}^n, L)$ representing the homotopy class $A_0 = 0$ of the constant maps.

Now suppose $H \in \mathcal{H}_{\mathrm{reg}}$ and consider the evaluation map

$$\mathrm{ev} : \mathcal{M}(H) \to L, \qquad \mathrm{ev}(\lambda, u) := u(1, 0).$$

Let $z_0 \in L$ be a regular value of ev. Then the set

$$\mathcal{M}(H; z_0) := \mathrm{ev}^{-1}(z_0) \subset \mathcal{M}(H)$$

is a smooth compact 1-manifold with boundary

$$\partial\mathcal{M}(H; z_0) = \{(\lambda, u) \in \mathcal{M}(H) \,|\, \lambda \in \{0, 1\},\, u(1, 0) = z_0\}\,.$$

By Lemma 9.2.4 (i), the moduli space $\mathcal{M}(H)$ has no element with $\lambda = 1$. Hence the boundary of $\mathcal{M}(H; z_0)$ consists of a single point $(0, u_0)$ where $u_0(s, t) \equiv z_0$. This contradicts the classification of compact 1-manifolds. Hence our assumption that there was no nonconstant holomorphic disc $u : (\mathbb{D}, \partial\mathbb{D}) \to (\mathbb{C}^n, L)$ must have been wrong. This proves Theorem 9.2.1. □

REMARK 9.2.5. (i) A refinement of the above argument shows that for every compact embedded Lagrangian submanifold $L \subset \mathbb{C}^n$ there is a nonconstant holomorphic disc $v : (\mathbb{D}, \partial\mathbb{D}) \to (\mathbb{C}^n, L)$ with Maslov number

$$\mu(v) \le n + 1.$$

To see this one has to extend the compactness theorem 5.3.1 to holomorphic discs. The key point is that the homotopy class is preserved under Gromov convergence (see Frauenfelder [**120**]). It then follows that failure of compactness results in a (connected) nonempty finite collection of holomorphic discs $u_j : (\mathbb{D}, \partial\mathbb{D}) \to (\mathbb{C}^n, L)$, $j = 1, \ldots, N$, together with a solution $u_0 : (\mathbb{D}, \partial\mathbb{D}) \to (\mathbb{C}^n, L)$ of the perturbed equation such that the sum of the Maslov numbers is zero:

$$\sum_{j=0}^{N} \mu(u_j) = 0.$$

Now transversality for generic Hamiltonian perturbations implies that the moduli space of solutions of the perturbed equation is empty whenever $\mu < -n - 1$. Hence

$$\sum_{j=1}^{N} \mu(u_j) = -\mu(u_0) \le n + 1$$

and so one of the Maslov numbers $\mu(u_1), \ldots, \mu(u_N)$ has to be less than or equal to $n + 1$. Thus, if there is no holomorphic disc with Maslov number $\mu \le n + 1$, it follows that the moduli space is compact and this leads to a contradiction as in the above proof of Theorem 9.2.1.

(ii) Theorem 9.2.1 carries over to displaceable closed Lagrangian submanifolds in Stein domains (exact symplectic manifolds with global plurisubharmonic functions; see Definition 9.2.6 below).

Convexity. In preparation for the next theorem about Lagrangian intersections we discuss the notion of convexity. This was introduced by Eliashberg and Gromov [**100**] in order to extend the theory of J-holomorphic curves to noncompact symplectic manifolds. For in this case one needs a hypothesis to ensure that J-holomorphic curves cannot leave a suitable compact subset of M so that Gromov compactness still applies (see Corollary 9.2.11).

DEFINITION 9.2.6. *A noncompact symplectic manifold (M, ω) is called* **convex (at infinity)** *if there exists a pair (f, J), where J is an ω-compatible almost complex structure and $f : M \to [0, \infty)$ is a proper smooth function such that*

$$\omega_f(v, Jv) \geq 0, \qquad \omega_f := -d(df \circ J), \tag{9.2.3}$$

for every x outside of some compact subset of M and every $v \in T_xM$. Such a function f is called **plurisubharmonic**. *If (M, ω) is a compact symplectic manifold with boundary then the boundary of M is called* **convex** *if there exists a pair (f, J), where $J \in \mathcal{J}(M, \omega)$, $f : M \to (-\infty, 0]$ satisfies (9.2.3) near the boundary, and $\partial M = f^{-1}(0)$. Such almost complex structures J are said to be* **adapted to the boundary.**

REMARK 9.2.7. Some authors give stronger definitions of plurisubharmonic functions by imposing the condition $\omega_f = \omega$ globally on M. There are examples of manifolds that satisfy the weak convexity hypothesis but not the strong one: see Exercise 9.2.14. One advantage of Definition 9.2.6 is that the product of two convex manifolds is again convex. For a comprehensive reference that explores the very close connections between the symplectic and complex notions of convexity, see Cieliebak–Eliashberg [**63**].

The next lemma gives a geometric criterion for a manifold to have a convex boundary.

LEMMA 9.2.8. *Let (M, ω) be a compact symplectic manifold with boundary and $X \in \mathrm{Vect}(M)$ be a vector field which satisfies $\mathcal{L}_X\omega = \omega$ near the boundary and points out at the boundary. Then (M, ω) has convex boundary.*

PROOF. Let $\phi_t \in \mathrm{Diff}(M)$, $t \leq 0$, be the flow of X. Choose an almost complex structure $J = \{J(x)\}_{x \in \partial M}$ along the boundary so that

$$J(x)X(x) \in T_x\partial M, \qquad \omega(v, J(x)X(x)) = 0, \qquad \omega(X(x), J(x)X(x)) = 1, \tag{9.2.4}$$

for all $x \in \partial M$ and $v \in T_x\partial M$. Then extend J to a neighbourhood of the boundary by the flow of X. Thus

$$J(\phi_t(x)) := d\phi_t(x)J(x)d\phi_t(x)^{-1}, \qquad -\varepsilon < t \leq 0.$$

The first two conditions in (9.2.4) imply that X is perpendicular to the boundary with respect to the metric $\omega(\cdot, J\cdot)$ and so the subspace $\{v \in T_xM \mid \omega(X(x), v) = 0\}$ is invariant under $J(x)$ for every $x \in T_xM$. Now choose a smooth function $f : M \to (-\infty, 0]$ such that $f(\phi_t(x)) = e^t$ for $x \in \partial M$ and $-\varepsilon < t \leq 0$. Then its gradient with respect to the metric $\omega(\cdot, J\cdot)$ is $\nabla f = X$ near the boundary. Hence $df = -\iota(JX)\omega$, so that $-d(df \circ J) = d\iota(X)\omega = \mathcal{L}_X\omega = \omega$. This proves Lemma 9.2.8. □

EXERCISE 9.2.9. (i) Show that the flow ϕ_t of a vector field X such that $\mathcal{L}_X\omega = \omega$ expands ω by the constant factor e^t: namely $\phi_t^*\omega = e^t\omega$. Such X are called **Liouville vector fields**.

(ii) If X is a Liouville vector field defined near ∂M and transverse to it show that the 1-form $\alpha := \iota(X)\omega$ is a contact form on ∂M. Moreover if X' is another Liouville vector field that is transverse to ∂M and pointing in the same direction, then the two contact forms α, α' are isotopic, that is they may be joined by a family of contact forms on ∂M. Therefore, there is a unique induced contact structure on ∂M. Boundaries ∂M with this property are said to be **of contact type**.

(iii) Show that if Q is any closed hypersurface in M such that the restriction $\omega|_Q$ is exact, then the integral $\int_Q \alpha \wedge \omega^{n-1}$ is independent of the choice of 1-form α satisfying $\omega|_Q = d\alpha$.

The next lemma shows that the notion of convexity given in Definition 9.2.6 captures the important features of the complex case. We denote by $\Delta := \partial_s^2 + \partial_t^2$ the standard Laplacian.

Lemma 9.2.10. *Let J be an ω-compatible almost complex structure on M, $\Omega \subset \mathbb{C}$ be an open set, $u : \Omega \to M$ be a J-holomorphic curve, and $f : M \to \mathbb{R}$ be a smooth function that satisfies (9.2.3) on a neighbourhood of the image of u. Then $f \circ u$ is subharmonic:*

$$\Delta(f \circ u) = \omega_f(\partial_s u, J\partial_s u) \geq 0.$$

Proof. The identity

$$\Delta(f \circ u) = \omega_f(\partial_s u, J\partial_s u) \tag{9.2.5}$$

for J-holomorphic curves can be established in two ways. Abbreviate $g := f \circ u$. Since u is J-holomorphic, we have

$$u^*(df \circ J) = dg \circ i = (\partial_s g\, ds + \partial_t g\, dt) \circ i = \partial_t g\ ds - \partial_s g\ dt$$

and hence

$$u^*\omega_f = -du^*(df \circ J) = -d\big(\partial_t g\, ds + \partial_s g\, dt\big) = (\Delta g)\, ds \wedge dt.$$

This is equivalent to (9.2.5). The second proof of (9.2.5) is based on the identity

$$\langle \nabla_\xi \nabla f, \xi \rangle + \langle \nabla_{J\xi} \nabla f, J\xi \rangle = \omega_f(\xi, J\xi) \tag{9.2.6}$$

Here ∇ is the Levi-Civita connection of the metric determined by ω and J. To prove (9.2.6), apply the identity $d\alpha(X,Y) = \partial_X(\alpha(Y)) - \partial_Y(\alpha(X)) + \alpha([X,Y])$ to the 1-form $\alpha_f := -df \circ J = \langle J\nabla f, \cdot \rangle$. This gives

$$\begin{aligned}
\omega_f(X,Y) &= \partial_X \langle J\nabla f, Y \rangle - \partial_Y \langle J\nabla f, X \rangle + \langle J\nabla f, [X,Y] \rangle \\
&= \langle J\nabla_X \nabla f, Y \rangle - \langle J\nabla_Y \nabla f, X \rangle + \langle (\nabla_X J)\nabla f, Y \rangle + \langle (\nabla_Y J)X, \nabla f \rangle \\
&= \langle J\nabla_X \nabla f, Y \rangle - \langle J\nabla_Y \nabla f, X \rangle - \langle (\nabla_{\nabla f} J)Y, X \rangle
\end{aligned}$$

for $X, Y \in \mathrm{Vect}(M)$, where the last identity uses equation (C.7.2). Now choose $Y = JX$ and use the fact that $J\nabla_{\nabla f} J$ is a skew-adjoint endomorphism (cf. (C.7.4)) to obtain (9.2.6). Next observe that every J-holomorphic curve $u : \Omega \to M$ is a harmonic map, i.e. it satisfies $\nabla_s \partial_s u + \nabla_t \partial_t u = 0$ (see Remark 4.3.6). Hence

$$\Delta(f \circ u) = \langle \nabla_{\partial_s u} \nabla f, \partial_s u \rangle + \langle \nabla_{\partial_t u} \nabla f, \partial_t u \rangle = \omega_f(\partial_s u, J\partial_s u).$$

The last identity follows from (9.2.6) and the fact that $\partial_t u = J\partial_s u$. This completes the second proof of (9.2.5) and Lemma 9.2.10. □

COROLLARY 9.2.11. *Suppose (M,ω) is a compact symplectic manifold with convex boundary and J is any ω-compatible almost complex structure that is adapted to the boundary. Let Σ be a connected Riemann surface without boundary, $U \subset M$ be a neighbourhood of ∂M, and $u : \Sigma \to M$ be a smooth map whose restriction to $u^{-1}(U)$ is J-holomorphic. Then $u(\Sigma) \cap \partial M \neq \emptyset$ if and only if $u(\Sigma) \subset \partial M$.*

PROOF. The set $\Sigma_0 := \{z \in \Sigma \,|\, u(z) \in \partial M\}$ is closed and, by Lemma 9.2.10 and the mean value inequality for subharmonic functions, it is open. Hence it is either empty or equal to Σ. This proves Corollary 9.2.11. □

EXAMPLE 9.2.12. The function $f(z) = |z|^2/4$ on $\mathbb{C}^n$ is plurisubharmonic. Its gradient vector field $X(z) = z/2$ satisfies $\mathcal{L}_X\omega = \omega$.

EXAMPLE 9.2.13. Let L be a compact Riemannian manifold and denote by ω_{can} the canonical symplectic structure on the cotangent bundle $M := T^*L$. Then X, the radial vector field in the fibers, is a global Liouville vector field on T^*L. To see this, write an element of T^*L in the form (x,y) where $x \in L$ and $y \in T_x^*L$. In local coordinates the canonical 1-form $\lambda_{\mathrm{can}} \in \Omega^1(T^*L)$ and the canonical symplectic form $\omega_{\mathrm{can}} = -d\lambda_{\mathrm{can}} \in \Omega^1(T^*L)$ are given by

$$\lambda_{\mathrm{can}} = \sum_i y_i dx^i, \qquad \omega_{\mathrm{can}} = \sum_i dx^i \wedge dy_i,$$

while $X := \sum_i y_i \partial/\partial y_i$. Hence $\iota(X)\omega_{\mathrm{can}} = -\lambda_{\mathrm{can}}$, so that $\mathcal{L}_X\omega_{\mathrm{can}} = \omega_{\mathrm{can}}$. The metric g on L induces a function ρ_g on T^*L whose value at (x,y) is the square of the length of y. Since X is transverse to the level set $\rho_g = 1$, we can define J on this set as in Lemma 9.2.8 and extend it via the flow of X to the complement of the zero section 0_L in such a way that ρ_g i s plurisubharmonic on $T^*L \setminus 0_L$.

In fact, the metric g defines a canonical J so that ρ_g is J-plurisubharmonic everywhere. This can be checked using the following formulas. In local coordinates the metric on T^*L is given by

$$|(\xi,\eta)|^2 = \sum_{i,j} \xi^i g_{ij} \xi^j + \sum_{i,j}\Big(\eta_i - \sum_{\kappa,\lambda}\Gamma_{i\kappa}^{\lambda}(x) y_\lambda \xi^\kappa\Big) g^{ij} \Big(\eta_j - \sum_{k,\ell}\Gamma_{jk}^{\ell}(x) y_\ell \xi^k\Big),$$

where $\Gamma_{jk}^{\ell}(x)$ are the Christoffel symbols. There is a corresponding almost complex structure J that takes the tangent space to the fiber at (x,y) to its orthogonal complement. It is given by $J(\xi,\eta) = (\widehat{\xi},\widehat{\eta})$, where

$$\widehat{\xi^i} := -\sum_j g^{ij}\Big(\eta_j - \sum_{k,\ell}\Gamma_{jk}^{\ell}(x) y_\ell \xi^k\Big), \qquad \widehat{\eta}_j - \sum_{k,\ell}\Gamma_{jk}^{\ell}(x) y_\ell \widehat{\xi}^k := \sum_k g_{jk}\xi^k.$$

Then the function $f : T^*L \to [0,\infty)$, defined by

$$f(x,y) := \tfrac{1}{2}|y|^2,$$

is plurisubharmonic with respect to J because $df \circ J = \lambda_{\mathrm{can}}$.

EXERCISE 9.2.14. Let $\pi : E \to \Sigma$ be a complex Hermitian line bundle of degree k over a closed oriented 2-manifold Σ and denote by $P \subset E$ the unit circle bundle. Then P is a principal circle bundle, so that it supports an action of S^1 generated by the vector field ∂_θ. Choose an area form $\sigma \in \Omega^2(\Sigma)$, with total area one. Then there is an S^1-invariant 1-form α on P such that

$$\alpha(\partial_\theta) = 1/2\pi, \quad d\alpha = -k\pi^*\sigma.$$

(This is called the *global angular form* in Bott–Tu [**45**, Chapter I, §6], and can be constructed by patching local solutions together by a partition of unity.) Let r denote the radial variable on E. When $k \le 0$ prove that $\omega := \pi^*\sigma + d(\pi r^2 \alpha)$ is a symplectic form on E. Use Lemma 9.2.10 to show it is convex at infinity. Show that ω satisfies the stronger convexity condition in Remark 9.2.7 only when $k < 0$. When $k > 0$ show that ω is not symplectic, but that there is a bounded function $\beta : [0,\infty) \to [0,\infty)$ that equals πr^2 for r near 0 and is such that $\omega_\beta := \pi^*\sigma + d(\beta(r)\alpha)$ is symplectic. However, ω_β is never convex at infinity.

Hint 1: Show that when $k > 0$ the level set $Q_c := \{r = c\}$ in (E, ω_β) is of contact type in the sense of Exercise 9.2.9. However the corresponding Liouville field X points *into* the region $r < r_0$, so that the integral $\int_Q \alpha \wedge \omega_\beta$ in part (iii) of this exercise is a decreasing function of c of the form *const* $- \mathrm{Vol}(\{r \le c\})$. Deduce that (E, ω_β) cannot be convex at infinity. The argument when $k = 0$ is similar.

Hint 2: P is isomorphic to the frame bundle of E and $A := 2\pi i \alpha$ is a connection form. Show that $\omega = \pi^*\sigma + d(\pi r^2\alpha) = \widetilde{\omega}$ in the notation of Exercise 9.2.15 below.

EXERCISE 9.2.15. The previous discussion extends to more general fibrations

$$\pi : \widetilde{M} = P \times_{\mathrm{G}} M \to \Sigma$$

over a closed Riemann surface (Σ, j, σ). Here G is a compact Lie group with Lie algebra $\mathfrak{g} = \mathrm{Lie}(\mathrm{G})$, $\pi_P : P \to \Sigma$ is a principal G-bundle, and (M, ω) is a symplectic manifold equipped with a Hamiltonian G-action. Denote the action by $(g, x) \mapsto gx$ and the infinitesimal action by $\mathfrak{g} \to \mathrm{Vect}(M) : \xi \mapsto X_\xi$. Let $\mu : M \to \mathfrak{g}^*$ be an equivariant moment map for the action so that

$$\omega(X_\xi(x), \hat{x}) = \langle d\mu(x)\hat{x}, \xi\rangle, \qquad \omega(X_\xi(x), X_\eta(x)) = \langle \mu(x), [\xi, \eta]\rangle,$$

for $\hat{x} \in T_xM$ and $\xi, \eta \in \mathfrak{g}$. Let $A \in \mathcal{A}(P) \subset \Omega^1(P, \mathfrak{g})$ be a connection 1-form and denote its curvature by $F_A \in \Omega^2(P, \mathfrak{g})$. For $p \in P$ denote by $\iota_p : M \to \widetilde{M}$ the inclusion of the fiber, given by $\iota_p(x) := [p, x]$.

1. Show that the 2-form $\omega_A := \omega - d\langle \mu, A\rangle \in \Omega^2(P \times M)$ descends to a 2-form $\widetilde{\omega}_A \in \Omega^2(\widetilde{M})$. Show that, for $\hat{x}_i \in T_xM$ and $v_i \in T_pP$,

$$\widetilde{\omega}_A\big([v_1, \hat{x}_1], [v_2, \hat{x}_2]\big) = \omega\big(\hat{x}_1 + X_{A_p(v_1)}(x), \hat{x}_2 + X_{A_p(v_2)}(x)\big) - \big\langle \mu(x), F_A(v_1, v_2)\big\rangle.$$

So $(\widetilde{M}, \widetilde{\omega}_A)$ is a Hamiltonian fibration and $\iota_p^*\widetilde{\omega}_A = \omega$ for $p \in P$.

2. Assume that, for $x \in M$ and $v_i \in T_pP$,

$$\pi_P^*\sigma(v_1, v_2) > 0 \quad \Longrightarrow \quad \big\langle \mu(x), F_A(v_1, v_2)\big\rangle \le 0. \tag{9.2.7}$$

Show that $\widetilde{\omega} := \pi^*\sigma + \widetilde{\omega}_A$ is a symplectic form on $\widetilde{M}$.

3. Let J be a G-invariant ω-tame almost complex structure on X. Show that there is a unique almost complex structure $\widetilde{J}$ on $\widetilde{M}$ such that $\pi : \widetilde{M} \to \Sigma$ and $\iota_p : M \to \widetilde{M}$ are holomorphic for all $p \in P$. Show that $\widetilde{J}$ is $\widetilde{\omega}$-tame.

4. Let $K \subset M$ be a G-invariant compact set, and $f : M \to [0, \infty)$ be a G-invariant proper smooth function that is plurisubharmonic on $M \setminus K$ and satisfies

$$df(x)JX_{\mu(x)}(x) \ge 0 \qquad \text{for } x \in M \setminus K. \tag{9.2.8}$$

Show that the function $\widetilde{f} : \widetilde{M} \to [0, \infty)$, defined by $\widetilde{f}([p, x]) := f(x)$, satisfies (9.2.3) on the complement of the compact set $\widetilde{K} := P \times_{\mathrm{G}} K$. (The inequality (9.2.8) is used in [**135**] to obtain convexity for the symplectic vortex equations.)

Lagrangian intersections. The next theorem asserts the existence of intersection points between a Lagrangian submanifold L and any Hamiltonian deformation of L. It is related to a Lagrangian version of the Arnold conjecture.

THEOREM 9.2.16 (Gromov). *Let (M, ω) be a symplectic manifold without boundary and assume that (M, ω) is convex at infinity. Let $L \subset M$ be a compact Lagrangian submanifold such that $[\omega]$ vanishes on $\pi_2(M, L)$. Let $\psi : M \to M$ be a Hamiltonian symplectomorphism. Then $\psi(L) \cap L \neq \emptyset$.*

In [**160**] Gromov established the existence of a single intersection point. In [**113, 114**] Floer proved a stronger theorem, namely, that under the assumption of Theorem 9.2.16 there is an inequality

$$\#L \cap \psi(L) \geq \dim H_*(L; \mathbb{Z}_2)$$

whenever L and $\psi(L)$ intersect transversally. Thus he settled the Lagrangian version of the Arnold conjecture in the case where ω vanishes on $\pi_2(M, L)$. In his papers Floer considered compact symplectic manifolds; however the proof carries over word for word to the noncompact case whenever (M, ω) is convex at infinity. The argument given here is again informed by Floer's ideas. Thus we give a direct proof of Theorem 9.2.16 using holomorphic strips, rather than deducing it from Theorem 9.2.1 via the "figure of eight trick" in Gromov [**160**, 2.3.B_3'].

The hypothesis that $[\omega]$ vanishes on $\pi_2(M, L)$ cannot be removed in Theorem 9.2.16. For example, a contractible embedded circle on a 2-torus can be disjoined from itself by a Hamiltonian isotopy whenever the area enclosed by the circle is strictly smaller than the area of the complement. However, there are cases when $[\omega]$ does not vanish on $\pi_2(M, L)$ and the assertion of Theorem 9.2.16 still holds. The easiest case is the example of a great circle on the 2-sphere. Another example is the case $M = \mathbb{C}P^n$ and $L = \mathbb{R}P^n$ (see Oh [**303**] and Abreu–Macarini [**6**]). The A^∞ version of Lagrangian Floer homology developed by Fukaya–Oh–Ohta–Ono [**128, 129**] gives further insight into this question, as does the Entov–Polterovich [**106**] theory of symplectic rigidity.

Recall the assertion of Theorem 9.2.1 that a compact Lagrangian submanifold of $\mathbb{C}^n$ cannot be exact. This follows also from Theorem 9.2.16. Namely, if $L \subset \mathbb{C}^n$ is a compact Lagrangian submanifold then there exists a Hamiltonian symplectomorphism $\psi : \mathbb{C}^n \to \mathbb{C}^n$ such that $\psi(L) \cap L = \emptyset$. Hence the hypothesis of Theorem 9.2.16, that $[\omega]$ vanishes on $\pi_2(\mathbb{C}^n, L)$, cannot be satisfied. Hence L is not exact. Thus the assertion of Theorem 9.2.16 can also be interpreted as an obstruction to Lagrangian embeddings rather than as an existence theorem for Lagrangian intersections.

In the case where M is compact and ω vanishes on $\pi_2(M)$, Theorem 9.1.1 also follows from Theorem 9.2.16. Namely, we can apply Theorem 9.2.16 to the product manifold $\widetilde{M} := M \times M$ with the symplectic form $\widetilde{\omega} := \pi_2^*\omega - \pi_1^*\omega$, the diagonal $\widetilde{L} := \Delta$, and the graph $\widetilde{\psi}(x, y) := (x, \psi_H(y))$ of the symplectomorphism generated by H. In fact, we shall see that the proof of Theorem 9.2.16 is based on the same idea as that of Theorem 9.1.1.

The **nearby Lagrangian conjecture** asserts that every exact Lagrangian submanifold of the cotangent bundle of a compact manifold is Hamiltonian isotopic to the zero section. Although the conjecture remains an open question, great progress has recently been made using Fukaya categories; see, for example, Fukaya–Seidel–Smith [**131**], Abouzaid [**1**], and Kragh [**217**]. The following corollary of Theorem 9.2.16 shows that every exact Lagrangian submanifold must at least intersect the zero section. This was proved by Gromov [**160**]. The second assertion about intersections with the fibers was proved by Lalonde–Sikorav [**230**].

COROLLARY 9.2.17 (Gromov, Lalonde–Sikorav). *Let L be a compact manifold and $\Lambda \subset T^*L$ be an exact Lagrangian submanifold (i.e. the restriction of λ_{can} to Λ is exact). Then Λ must intersect the zero section of T^*L and each fiber of the projection $\pi : T^*L \to L$.*

PROOF. First note that the cotangent bundle is convex at infinity, by Example 9.2.13. Second, since Λ is exact, $[\omega_{\mathrm{can}}]$ vanishes on $\pi_2(T^*L, \Lambda)$. Hence the pair (T^*L, Λ) satisfies the hypotheses of Theorem 9.2.16 and so $\psi(\Lambda) \cap \Lambda \neq \emptyset$ for every Hamiltonian symplectomorphism $\psi : T^*L \to T^*L$.

Assume, by contradiction, that Λ does not intersect the zero section. Consider the Lagrangian embeddings $\iota_t : \Lambda \to M$ defined by

$$\iota_t(x, y) := (x, ty)$$

for $(x, y) \in \Lambda$. We construct below a Hamiltonian isotopy $\psi_t : T^*L \to T^*L$ that satisfies $\psi_t(\Lambda) = \iota_t(\Lambda)$. But this contradicts Theorem 9.2.16 since $\iota_t(\Lambda) \cap \Lambda = \emptyset$ for sufficiently large t.

To construct ψ_t, note first that because $\iota_1^*\lambda_{\mathrm{can}}$ is exact there exists a smooth function $H_1 : T^*L \to \mathbb{R}$ (with compact support) such that H_1 vanishes in a neighbourhood of the zero section and

$$\iota_1^*\lambda_{\mathrm{can}} + d(H_1 \circ \iota_1) = 0.$$

Define $H_t : T^*L \to \mathbb{R}$ by

$$H_t(x, y) := tH_1(x, y/t).$$

Then $H_t \circ \iota_t = tH_1 \circ \iota_1$, and so the identity $\iota_t^*\lambda_{\mathrm{can}} = t\iota_1^*\lambda_{\mathrm{can}}$ implies that

$$d(H_t \circ \iota_t) = -\iota_t^*\lambda_{\mathrm{can}} = \omega_{\mathrm{can}}(\partial_t\iota_t, d\iota_t\cdot).$$

Hence $X_{H_t}(\iota_t(q)) \in \partial_t\iota_t(q) + \mathrm{im} d\iota_t(q)$ for every $q \in \Lambda$. Therefore the Hamiltonian isotopy $\psi_t : T^*L \to T^*L$ generated by H_t satisfies $\psi_t(\Lambda) = \iota_t(\Lambda)$ as required.

Now suppose that there is a point $x_0 \in L$ such that Λ does not intersect the fiber over x_0. Choose an open neighbourhood $U \subset L$ of x_0 such that $\Lambda \cap \pi^{-1}(U) = \emptyset$ and let $f : L \to \mathbb{R}$ be a smooth function such that all the critical points of f belong to the set U. Then consider the Hamiltonian isotopy $\psi_t(x, y) := (x, y + tdf(x))$. This isotopy again disjoins Λ from itself for t sufficiently large, which is impossible by Theorem 9.2.16. This proves Corollary 9.2.17. □

The proof of Theorem 9.2.16 is pretty much a carbon copy of the proof of Theorem 9.1.1 in the previous section. The relevant intermediate result (see Proposition 9.1.4) concerns the perturbed Cauchy–Riemann equations for a function $v : \mathbb{D} \to M$ on the unit disc $\mathbb{D} := \{s + it \,|\, t^2 + t^2 \leq 1\} \subset \mathbb{C}$. They have the form

$$\partial_s v + X_{F_{s,t}}(v) + J(v)(\partial_t v + X_{G_{s,t}}(v)) = 0, \qquad v(\partial\mathbb{D}) \subset L. \tag{9.2.9}$$

We abbreviate $K := F\,ds + G\,dt \in \Omega^1(\mathbb{D}, C^\infty(M))$ and denote by $\|R_K\|$ the Hofer norm of the curvature (see Remark 8.1.7). We consider only perturbations $K \in$

$\Omega^1(\mathbb{D}, C^\infty(M))$ that that have compact support in $\mathbb{D} \times M$ and satisfy the condition

$$\zeta \in T_z\partial\mathbb{D} \qquad \Longrightarrow \qquad K_{z,\zeta}|_L \equiv 0.$$

This means that $S^1 \times L$ is a Lagrangian submanifold of $\widetilde{M} := \mathbb{D} \times M$ with respect to the symplectic form $\widetilde{\omega}_K$ determined by the Hamiltonian perturbation K. The set of such perturbations will be denoted by $\mathcal{H}(L) \subset \Omega^1(\mathbb{D}, C^\infty(M))$. Since $[\omega]$ vanishes on $\pi_2(M, L)$, it follows from Lemma 8.2.9, that every solution $v : \mathbb{D} \to M$ of (9.2.9) with $K \in \mathcal{H}(L)$ satisfies the energy bound

$$E_K(v) = \int_{\mathbb{D}} |\partial_s v + X_F(v)|_J^2 \, dsdt \le \|R_K\|.$$

Denote by $\mathcal{M}(L; K)$ the space of all solutions of (9.2.9).

PROPOSITION 9.2.18. *Let (M, ω, L) be as in Theorem 9.2.16 and (f, J) as in Definition 9.2.6. Then, for every $K \in \mathcal{H}(L)$ and every $w \in \partial\mathbb{D}$, the evaluation map $\mathcal{M}(L; K) \to L : v \mapsto v(w)$ is surjective.*

PROOF. Fix a constant $c > 0$ such that (9.2.3) holds in the region $f \ge c$ and L is contained in the set $f \le c$. Denote by $\mathcal{H}^c(L)$ the set of all Hamiltonian perturbations $K = F_{s,t}\, ds + G_{s,t}\, dt \in \mathcal{H}(L)$ such that $F_{s,t}$ and $G_{s,t}$ are supported in the compact set $f \le c$. If $K \in \mathcal{H}^c(L)$ then it follows from Corollary 9.2.11 that every solution v of (9.2.9) takes values in the compact set

$$M^c := \{x \in M \mid f(x) \le c\},$$

because (9.2.3) holds in $M \setminus M^c$ and $L \subset M^c$. Next observe that our assumptions guarantee that there are no holomorphic spheres in M and no holomorphic discs with boundary in L. Hence it follows from Theorem 4.6.1 that $\mathcal{M}(L; K)$ is compact for every $K \in \mathcal{H}^c(L)$. Here we apply Theorem 4.6.1 to the compact manifold $\widetilde{M} := \mathbb{D} \times M^c$ and the Lagrangian submanifold $\widetilde{L} := S^1 \times L$. The manifold $\widetilde{M}$ is equipped with the symplectic form $\widetilde{\omega}_K$ and the almost complex structure $\widetilde{J}_K$ determined by the Hamiltonian perturbation K.

Now Theorem 8.3.1 extends in an obvious way to the boundary value problem (9.2.9) and proves the following. Let us denote by $\mathcal{H}^c_{\mathrm{reg}}(L) \subset \mathcal{H}^c(L)$ the set of all Hamiltonian perturbations $K \in \mathcal{H}^c(L)$ for which the linearized operator

$$D_v : \Omega^0_{v|_{\partial\mathbb{D}}{}^*TL}(\mathbb{D}, v^*TM) \to \Omega^{0,1}(\mathbb{D}, v^*TM)$$

is surjective for every $v \in \mathcal{M}(L; K)$. Then the proof of Theorem 8.3.1 shows that the set $\mathcal{H}^c_{\mathrm{reg}}(L)$ is residual in $\mathcal{H}^c(L)$. Moreover, it shows that the moduli space $\mathcal{M}(L; K)$ is a (compact) smooth manifold of dimension $\dim \mathcal{M}(L; K) = n = \dim L$ whenever $K \in \mathcal{H}^c_{\mathrm{reg}}(L)$.

Now consider the evaluation map $\mathrm{ev}_K : \mathcal{M}(L; K) \to L$, given by $\mathrm{ev}_K(v) := v(w)$ for $v \in \mathcal{M}(L; K)$ and $K \in \mathcal{H}^c(L)$. Combining Theorem 4.6.1 and Theorem 8.3.3 we see that the evaluation maps ev_{K_0} and ev_{K_1} are compactly cobordant for any two regular perturbations $K_0, K_1 \in \mathcal{H}^c_{\mathrm{reg}}(L)$. Hence the mod-2 degree of ev_K is independent of the choice of $K \in \mathcal{H}^c_{\mathrm{reg}}(L)$. Considering the zero perturbation we find that the moduli space consists of the constant maps and hence ev_0 is a diffeomorphism. Therefore $\deg_2(\mathrm{ev}_K) = 1$ and so ev_K is onto whenever $K \in \mathcal{H}^c_{\mathrm{reg}}(L)$. Since $\mathcal{H}^c_{\mathrm{reg}}(L)$ is dense in $\mathcal{H}^c(L)$ the obvious compactness argument shows that ev_K is onto for every $K \in \mathcal{H}^c(L)$. Since c can be chosen arbitrarily large, this completes the proof of Proposition 9.2.18. $\square$

PROOF OF THEOREM 9.2.16. Fix a pair (f, J) as in Definition 9.2.6. For every compactly supported Hamiltonian function $\mathbb{R} \times [0,1] \to M : (s,t,x) \mapsto H_{s,t}(x)$ we consider the following equation for functions $u : \mathbb{R} \times [0,1] \to M$:

$$\partial_s u + J(u)(\partial_t u - X_{H_{s,t}}(u)) = 0, \qquad u(s,0), u(s,1) \in L. \tag{9.2.10}$$

We claim that, for every compactly supported Hamiltonian H and every $x_0 \in L$, there exists a solution $u : \mathbb{R} \times [0,1] \to M$ of (9.2.10) that satisfies $u(0,0) = x_0$ and has an energy bound

$$E(u) := \int_0^1 \int_{-\infty}^{\infty} |\partial_s u|_J^2 \, dsdt \le \|\partial_s H\|.$$

To see this consider the holomorphic diffeomorphism $\phi : \mathbb{R} \times [0,1] \to \mathbb{D} \setminus \{\pm 1\}$, given by

$$\phi(s,t) := \frac{e^{\pi(s+it)} - i}{e^{\pi(s+it)} + i}. \tag{9.2.11}$$

Assume without loss of generality that $H_{s,t}$ has mean value zero for all s and t and let $K \in \Omega^1(\mathbb{D}, C^\infty(M))$ be the unique Hamiltonian perturbation whose pullback under ϕ is $\phi^* K = -H\,dt$. Then $K \in \mathcal{H}(L)$ and $\|R_K\| = \|\partial_s H\|$. Moreover, $v : \mathbb{D} \to M$ is a solution of (9.2.9) if and only if $u := v \circ \phi$ is a solution of (9.2.10). Both solutions have the same energy and $v(-i) = u(0,0)$. Hence the claim follows from Proposition 9.2.18.

The rest of the argument is very similar to that in the proof of Theorem 9.1.1. Namely, we consider the Hamiltonian perturbation

$$H_{T;s,t} := \beta_T(s) H_t : \mathbb{R} \times [0,1] \to \mathbb{R},$$

where $\beta_T : \mathbb{R} \to [0,1]$ is a compactly supported smooth cutoff function satisfying $\beta_T(s) = 1$ for $|s| \le T$ and $\pm\dot{\beta}_T(s) \le 0$ for $\pm s \ge 0$. Then, for every T, there is a solution u_T of (9.2.10) with $H = H_T$ that satisfies

$$E(u_T) \le \|\partial_s H_T\| = 2\|H\|,$$

and hence a path $x_T : [0,1] \to M$ of the form $x_T(t) = u_T(s_T, t)$ with $s_T \in [-T, T]$ such that

$$\int_0^1 |\dot{x}_T(t) - X_{H_t}(x_T(t))|^2 \, dt \le \frac{\|H\|}{T}, \qquad x_T(0), x_T(1) \in L.$$

Taking the limit $T \to \infty$ we obtain a solution $x : [0,1] \to M$ of the Hamiltonian differential equation $\dot{x}(t) = X_{H_t}(x(t))$ satisfying the boundary condition $x(0), x(1) \in L$. Hence $x(1) = \psi(x(0))$ and so $x(1) \in L \cap \psi(L)$ is the required intersection point. This proves Theorem 9.2.16. □

EXERCISE 9.2.19. Prove the existence of at least two intersection points of L and $\psi(L)$ under the assumptions of Theorem 9.2.16. *Hint:* Show that there is a well defined symplectic action functional on the space of paths with endpoints in L that are homotopic to constant paths. Show that the solutions of (9.2.10) with $H_{s,t} = H_t$ are the gradient flow lines of this action functional with respect to the L^2 inner product determined by J.

9.3. The nonsqueezing theorem

Gromov's nonsqueezing theorem is a cornerstone of symplectic topology. It says that a symplectomorphism cannot squeeze a ball into a cylinder of smaller radius. We denote by $B^{2n}(r)$ the closed ball with center zero and radius r in $\mathbb{R}^{2n}$.

THEOREM 9.3.1 (Gromov). *If $\iota : B^{2n}(r) \to \mathbb{R}^{2n}$ is a symplectic embedding such that $\iota(B^{2n}(r)) \subset B^2(R) \times \mathbb{R}^{2n-2}$ then $r \le R$.*

The converse, that this property characterizes symplectomorphisms and anti-symplectomorphisms, was proved independently by Eliashberg [**98**] and Ekeland–Hofer [**97**]. This leads to one manifestation of *symplectic rigidity*: the group of symplectomorphisms is closed with respect to the C^0-topology. This phenomenon is discussed fully in [**277**, Chapter 12] and so will not be further explored here. Rather, we concentrate here on giving a complete proof of the following extension of the nonsqueezing theorem.

THEOREM 9.3.2. *Let (V, τ) be a compact symplectic manifold of dimension $2n-2$ such that $\pi_2(V) = 0$. If there is a symplectic embedding of the ball $(B^{2n}(r), \omega_0)$ into $B^2(R) \times V$ then $r \le R$.*

THEOREM 9.3.2 IMPLIES THEOREM 9.3.1. A symplectic embedding

$$B^{2n}(r) \hookrightarrow B^2(R) \times \mathbb{R}^{2n-2}$$

induces a symplectic embedding $B^{2n}(r) \hookrightarrow B^2(R) \times V$, where $V := \mathbb{R}^{2n-2}/\lambda\mathbb{Z}^{2n-2}$ is the $(2n-2)$-torus and $\lambda > 0$ is sufficiently large. Hence, by Theorem 9.3.2, $R \ge r$. This proves Theorem 9.3.1. □

To prove Theorem 9.3.2 one replaces the 2-disc $B^2(R)$ by the 2-sphere. The argument is then based on the observation that the Gromov–Witten invariant of the product $M := S^2 \times V$ in the class $A := [S^2 \times \mathrm{pt}]$ is $\mathrm{GW}^M_{A,1}(\mathrm{PD}(\mathrm{pt})) = 1$. Since the moduli space $\mathcal{M}(A; J)/G$ is compact this implies that, for every almost complex structure on M which is tamed by the product symplectic form and every point $x_0 \in M$, there must be a J-holomorphic A-sphere that passes through x_0. In particular this holds when $\iota^* J = J_0$ and $x_0 = \iota(0)$. In this special case the relevant J-holomorphic curve pulls back to a minimal surface in $B^{2n}(r)$ which passes through the origin. Now the monotonicity theorem asserts that this surface has to have area at least πr^2. Hence the symplectic area of the class A (which is equal to the area of the 2-sphere) must be at least πr^2. We shall describe a modification of this argument which uses the blowup construction to circumvent the monotonicity theorem for minimal surfaces. A similar argument appears in Lalonde–Pinsonnault [**229**]. This paper also contains an interesting discussion of higher order widths; given a nontrivial element α in the kth homotopy group of the symplectomorphism group of M they consider the maximum radius of a symplectic ball whose image is fixed by all the symplectomorphisms in some representative of α.

The blowup construction. The naive idea of blowing up a complex manifold (M, J) at a point x_0 is to replace the point x_0 by the set of all complex lines through x_0. To make this definition more formal, consider the total space L of the tautological line bundle over CP^{n-1}:

$$L := \{(z, [w_1 : \cdots : w_n]) \in \mathbb{C}^n \times \mathbb{C}P^{n-1} \mid z \in \mathbb{C}w\}.$$

The map $(z,[w_1:\cdots:w_n])\mapsto z$ identifies the complement of the zero section in L with $\mathbb{C}^n\setminus\{0\}$. Thus L is a copy of $\mathbb{C}^n$ in which the origin 0 has been replaced by the zero section of L. Since the zero section is the set of all lines in $\mathbb{C}^n$ through 0, we may think of L as a model for the blowup of $\mathbb{C}^n$ at 0.

To blow up a general complex manifold, one proceeds as follows. Choose a complex embedding $\iota: B^{2n}(\delta)\to M$ such that $\iota(0)=x_0$ and consider the r-ball subbundle U_r of L:

$$U_r := \left\{(z,[w_1:\cdots:w_n])\in\mathbb{C}^n\times\mathbb{C}P^{n-1}\,|\,z\in\mathbb{C}w,\ |z|<r\right\}.$$

Now define the blowup $\widetilde{M}$ as the union

$$\widetilde{M} := (M\setminus\{x_0\})\cup U_r/\sim,$$

where the equivalence relation identifies the point $(z,[z_1:\cdots:z_n])\in U_r\setminus\mathbb{C}P^{n-1}$ with the point $\iota(z)\in M$. Then $\widetilde{M}$ is a complex manifold and there is an embedding $\tilde{\iota}:\mathbb{C}P^{n-1}\to\widetilde{M}$ defined as the composition of the embedding of the zero section $\mathbb{C}P^{n-1}\to U_r:[w_1:\cdots:w_n]\mapsto(0,[w_1:\cdots:w_n])$ with the inclusion $U_r\hookrightarrow\widetilde{M}$. The image of this embedding is the **exceptional divisor**

$$E := \tilde{\iota}(\mathbb{C}P^{n-1})\subset\widetilde{M}.$$

Moreover, there is a holomorphic projection $\pi:\widetilde{M}\to M$ such that $E=\pi^{-1}(x_0)$ and π restricts to a diffeomorphism from $\widetilde{M}\setminus E$ to $M\setminus\{x_0\}$. We will call $(\widetilde{M},\widetilde{J})$ the **complex blowup** of (M,J) at x_0. One can show that the biholomorphism class of the resulting manifold is independent of the choice of embedding ι.

We emphasize that every almost complex structure J on M such that $\iota^*J=J_0$ induces a unique almost complex structure $\widetilde{J}$ on $\widetilde{M}$ with respect to which the projection π is holomorphic. Another useful fact about the complex blowup is that any J-holomorphic curve through x_0 lifts to the blowup. Since this is a local statement it suffices to consider the lift of nonconstant J_0-holomorphic maps $u:(B^2,0)\to(\mathbb{C}^n,0)$. Every such map has the form

$$u(z)=z^k\big(h_1(z),\ldots,h_n(z)\big),\quad k\geq 1, \tag{9.3.1}$$

where $h_i(0)\neq 0$ for some i, and it lifts to

$$\tilde{u}(z)=\big(z^kh_1(z),\ldots,z^kh_n(z),[h_1(z):\cdots:h_n(z)]\big)\in L.$$

If (M,J) supports a symplectic form ω it is not hard to define a symplectic form $\widetilde{\omega}$ on the blowup $\widetilde{M}$ whose integral over the line in the exceptional divisor E is small. However, this approach obscures the close connection between symplectic blowing up and embedded balls in (M,ω). It was pointed out in McDuff [**257**] that the most geometric way to think of the symplectic blowup is as follows: cut out the interior of a symplectically embedded ball $B=\iota(B^{2n}(\rho))$ and collapse the bounding sphere $\partial B\cong S^{2n-1}$ via the Hopf map to an exceptional divisor $E\cong\mathbb{C}P^{n-1}$. This gives rise to an alternative description of the blowup as a (suitably smoothed) quotient

$$\widehat{M}_\rho := M\setminus\iota(\mathrm{int}B)/\sim,$$

where $\iota(z)\sim\iota(\lambda z)$ for $z\in\partial B^{2n}(\rho)$ and $\lambda\in S^1$. In this formulation the exceptional divisor is the quotient of $\iota(\partial B)$ by the S^1-action. Since this S^1 action preserves ω, the form ω descends to the quotient; in other words the exceptional divisor is the corresponding symplectic reduced space, which is none other than $\mathbb{C}P^{n-1}$ provided with a suitable multiple of the Fubini–Study form ω_{FS}. Thus the blowup

$\widehat{M}_\rho$ inherits a symplectic form $\widehat{\omega}_\rho$ from M which agrees with ω on $M \setminus B$ and restricts on the exceptional divisor E to the form $\rho^2\omega_{\mathrm{FS}}$, where ω_{FS} is normalized so that it integrates to π on any line. We give precise formulas for $\widehat{\omega}_\rho$ below.

This description of the symplectic blowup shows its connection with embedded balls and hence makes plausible its relevance to the nonsqueezing theorem. It has been formalized in the notion of symplectic cutting: see Lerman [**237**]. Note also that this process is reversible: given a symplectically embedded copy E of $(\mathbb{C}P^{n-1}, \rho^2\omega_{\mathrm{FS}})$ in (M, ω) whose normal line bundle can be identified with the canonical line one can cut out a small neighbourhood of E and glue in a ball of appropriate radius $\rho + \varepsilon$ to obtain a symplectic manifold called the **blowdown** of (M, ω) along E. (Paradoxically, blowing up reduces the volume of M, while blowing down increases it. The terminology obviously comes from the blow up operation in the complex category which does enlarge the point x_0 to a whole submanifold.)

However, because this definition does not make it easy to understand the relation between holomorphic curves in M and its blowup, we now explain another approach to the symplectic blowup that was developed in McDuff–Polterovich [**263**].

PROPOSITION 9.3.3. *Let (M, ω) be a symplectic manifold, $\iota : B^{2n}(r) \to M$ be a symplectic embedding, and $J \in \mathcal{J}(M, \omega)$ be an ω-compatible almost complex structure such that $\iota^*J = J_0$. Then, for every $\rho < r$, there exists a symplectic form $\widetilde{\omega}_\rho$ on the complex blowup $(\widetilde{M}, \widetilde{J})$ of (M, J) at $x_0 := \iota(0)$ with the following properties.*

(i) *The 2-form $\pi^*\omega$ agrees with $\widetilde{\omega}_\rho$ on $\pi^{-1}(M \setminus \iota(B_r)) \subset \widetilde{M}$.*

(ii) *$\widetilde{\iota}^*\widetilde{\omega}_\rho = \rho^2\omega_{\mathrm{FS}}$ is the Fubini–Study form (with integral $\rho^2\pi$ over every line).*

(iii) *$\widetilde{\omega}_\rho$ is compatible with $\widetilde{J}$.*

(iv) *If Σ is a closed oriented 2-manifold and $\widetilde{u} : \Sigma \to \widetilde{M}$ is a smooth map, then*
$$\int_\Sigma (\pi \circ \widetilde{u})^*\omega = \int_\Sigma \widetilde{u}^*\widetilde{\omega}_\rho + \pi\rho^2(\widetilde{u} \cdot E).$$

PROOF. We first show that a neighbourhood of the zero section in tautological line bundle $L \to \mathbb{C}P^{n-1}$ is symplectomorphic to a spherical shell in $\mathbb{R}^{2n}$. We saw above that the δ-ball bundle U_δ has coordinates $(z_1, \dots, z_n, [z_1 : \cdots : z_n])$. Since the Fubini–Study form is, in homogeneous coordinates, given by $(i/2)\partial\bar{\partial}(\log|z^2|)$ it follows that the 2-form
$$\begin{aligned}\omega_\rho &:= \frac{i}{2}\partial\bar{\partial}\left(|z|^2 + \rho^2\log|z|^2\right)\\ &= \frac{i}{2}\left(dz \wedge d\bar{z} + \rho^2\frac{dz \wedge d\bar{z}}{|z|^2} - \rho^2\frac{\bar{z} \cdot dz \wedge z \cdot d\bar{z}}{|z|^4}\right)\end{aligned}$$
extends to a form on U_δ that restricts to $\rho^2\omega_{\mathrm{FS}}$ on the exceptional divisor. Here we denote
$$dz \wedge d\bar{z} := \sum_j dz_j \wedge d\bar{z}_j, \qquad \bar{z} \cdot dz := \sum_j \bar{z}_j \cdot dz_j.$$
Now consider the diffeomorphism
$$h_\rho : B^{2n}(\delta) \setminus \{0\} \to B^{2n}(\sqrt{\rho^2 + \delta^2}) \setminus B^{2n}(\rho)$$
given by
$$h_\rho(z) := \sqrt{|z|^2 + \rho^2}\frac{z}{|z|} = \sqrt{1 + \frac{\rho^2}{|z^2|}}\,z.$$

A somewhat tedious but elementary calculation shows that the pullback of the standard symplectic form under h_ρ is given by

$$h_\rho^*\omega_0 = \frac{i}{2}\left(dz\wedge d\bar{z} + \rho^2\frac{dz\wedge d\bar{z}}{|z|^2} - \rho^2\frac{\bar{z}\cdot dz\wedge z\cdot d\bar{z}}{|z|^4}\right) = \omega_\rho.$$

This formula allow us to give a precise description of the blowup $(\widehat{M}_\rho, \widehat{\omega}_\rho)$. We assume that $\iota : B^{2n}(r) \to M$ is a symplectic embedding and, for any $\rho < r$, we define $\widehat{\omega}_\rho$ to equal ω on $M \setminus \iota(B^{2n}(\rho))$. This form extends smoothly over the exceptional divisor since we may identify the spherical shell $\iota(B^{2n}(r)\setminus B^{2n}(\rho))$ with $U_\varepsilon \setminus \mathbb{C}P^{n-1}$ via the map $h_\rho \circ \iota^{-1}$.

We now construct a diffeomorphism $f : \widetilde{M} \to \widehat{M}_\rho$ such that the pullback form

$$\widetilde{\omega}_\rho := f^*\widehat{\omega}_\rho$$

satisfies the requirements of the proposition. Choose $\delta > 0$ such that

$$\rho^2 + \delta^2 < (r-\delta)^2$$

and let $\beta : [0,r] \to [\rho, r]$ be a smooth function such that

$$\beta(s) = \begin{cases} \sqrt{s^2+\rho^2}, & \text{for } 0 \le s \le \delta, \\ s, & \text{for } r-\delta \le s \le r, \end{cases}$$

and

$$0 < s \le r \qquad \Longrightarrow \qquad 0 < \beta'(s) \le 1.$$

Now define $f : \widetilde{M} \to \widehat{M}_\rho$ by

$$f(\widetilde{x}) := \begin{cases} \pi(\widetilde{x}), & \text{if } \pi(\widetilde{x}) \in M \setminus \iota(B^{2n}(r-\delta)), \\ \iota\left(\frac{\beta(|z|)z}{|z|}\right), & \text{if } \pi(\widetilde{x}) = \iota(z),\ 0 < |z| < r, \\ [\iota(w)], & \text{if } \widetilde{x} = \widetilde{\iota}([w]),\ [w] \in \mathbb{C}P^{n-1}. \end{cases}$$

By construction the restriction of the form $\widehat{\omega}_\rho$ to a deleted neighbourhood of the exceptional divisor can be identified with the standard form ω_0 on the spherical shell $B^{2n}(\sqrt{\rho^2+\delta^2})\setminus B^{2n}(\rho)$. Therefore, the pullback form $\widetilde{\omega}_\rho := f^*\widehat{\omega}_\rho$ satisfies (i) and (ii). Moreover, it follows by direct calculation that the pullback of ω_0 under the map $z \mapsto \beta(|z|)z/|z|$ is compatible with J_0 whenever $\beta'(s) > 0$ for $s > 0$. (See Exercise 9.3.5 below.) This implies that $\widetilde{\omega}_\rho$ is compatible with $\widetilde{J}$.

To prove (iv) assume first $\widetilde{u}\cdot E = 0$. By a general position argument we may also assume that $\widetilde{u}$ is transverse to E. Then a surgery construction along curves in E joining intersection points with opposite intersection numbers shows that we may assume $\widetilde{u}(\Sigma)\cap E = \emptyset$ without changing the homology class of $\widetilde{u}$ (though we may decrease the number of connected components of Σ or increase the genus of a connected component in the process of removing intersection points.) It follows that $\widetilde{u}$ is homologous to a map with values in $M \setminus \iota(B^{2n}(r)) \subset \widetilde{M}$. For any such map assertion (iv) is obvious. Thus we have proved (iv) in the case $\widetilde{u}\cdot E = 0$. Now let $\widetilde{u} : \Sigma \to \widetilde{M}$ be any smooth map and denote $k := \widetilde{u}\cdot E$. Let $v : S^2 \to \mathbb{C}P^{n-1}$ be a smooth map in the homology class $k[\mathbb{C}P^1]$ and denote $\widetilde{v} := \widetilde{\iota}\circ v : S^2 \to \widetilde{M}$. Then

$$\widetilde{v}\cdot E = -k, \qquad \int_{S^2}\widetilde{v}^*\widetilde{\omega}_\rho = \rho^2\pi k, \qquad \int_{S^2}(\pi\circ\widetilde{v})^*\omega = 0.$$

Hence the disjoint union of the maps $\widetilde{u}$ and $\widetilde{v}$ has intersection number zero with E. This proves (iv) and Proposition 9.3.3. □

PROOF OF THEOREM 9.3.2. Fix a constant $\varepsilon > 0$, let $\sigma \in \Omega^2(S^2)$ be an area form with area $\int_{S^2} \sigma = \pi R^2 + \varepsilon$, and choose an area preserving embedding of the ball $B^2(R)$ into S^2. Then the given symplectic embedding of $B^{2n}(r)$ into $B^2(R) \times V$ gives rise to a symplectic embedding

$$\iota : B^{2n}(r) \to M := S^2 \times V.$$

Consider the blowup $(\widetilde{M}, \widetilde{\omega}_\rho)$ of the manifold M, determined as in Proposition 9.3.3 by the symplectic embedding ι and a constant $\rho < r$. Let $A := [S^2 \times \{\mathrm{pt}\}] \in H_2(M; \mathbb{Z})$, and denote its lift to $\widetilde{M}$ by $\widetilde{A} \in H_2(\widetilde{M}; \mathbb{Z})$. Thus $\widetilde{A}$ is represented by the submanifold $\pi^{-1}(S^2 \times \{y\})$ for any $y \in V$ such that $x_0 := \iota(0) \notin S^2 \times \{y\}$.

Since A is J-indecomposable for every $J \in \mathcal{J}(M, \omega)$, Lemma 7.1.8 implies that the moduli space $\mathcal{M}(A; J)/\mathrm{G}$ is compact for every $J \in \mathcal{J}(M, \omega)$. By choosing a product almost complex structure on $M = S^2 \times V$ we see that the evaluation map $\mathrm{ev} : \mathcal{M}_{0,1}(A; J) \to M$ has degree one for some, and hence every, $J \in \mathcal{J}_{\mathrm{reg}}$. Hence $\mathrm{GW}^M_{A,1}(\mathrm{PD}(\mathrm{pt})) = 1$. Since the moduli space is compact this implies that, for every $J \in \mathcal{J}(M, \omega)$ and every $x \in M$, there exists a J-holomorphic sphere in M which represents the class A and passes through x.

Now there exists an ω-compatible almost complex structure on M such that $\iota^* J$ is equal to the standard complex structure J_0 on $B^{2n}(r)$. Let $u_0 : S^2 \to M$ be a J-holomorphic sphere such that $[u_0] = A$ and $x_0 \in u_0(S^2)$. By Lemma 2.4.1, the set $Z_0 := u_0^{-1}(x_0)$ is finite. Consider the map

$$\widetilde{u}_0 := \pi^{-1} \circ u_0 : S^2 \setminus Z_0 \to \widetilde{M}.$$

Because one can always lift holomorphic curves to the complex blowup, $\widetilde{u}_0$ extends to a $\widetilde{J}$-holomorphic sphere in $\widetilde{M}$ which will still be denoted by $\widetilde{u}_0$. By construction, this extended $\widetilde{J}$-holomorphic sphere satisfies $\widetilde{u}_0(Z_0) \subset E$. Hence, by Exercise 2.6.1 (the easy case of positivity of intersections),

$$\widetilde{u}_0 \cdot E > 0.$$

Now it follows from Proposition 9.3.3 (iv) that

$$\pi R^2 + \varepsilon = \mathrm{Vol}(S^2) = \int_{S^2} u_0^* \omega = \int_{S^2} (\pi \circ \widetilde{u}_0)^* \omega = \int_{S^2} \widetilde{u}_0^* \widetilde{\omega}_\rho + \pi \rho^2 (\widetilde{u}_0 \cdot E) \geq \pi \rho^2.$$

Since this holds for every $\varepsilon > 0$ and every $\rho < r$ we deduce that $R \geq r$. □

REMARK 9.3.4. What is needed in the proof of Theorem 9.3.2 is an almost complex structure J on $M = S^2 \times V$ such that $\iota^* J$ is integrable and tamed by $f_\beta^* \omega_0$, where $f_\beta : B^{2n}(r) \setminus \{0\} \to B^{2n}(r) \setminus B^{2n}(\rho)$ is given by

$$f_\beta(z) := \beta(|z|)|z|^{-1} z.$$

In general, the projection $S^2 \times V \to S^2$ will not be holomorphic for any such J. For this reason we have imposed the condition $\pi_2(V) = 0$, instead of just assuming that V is semipositive. (The argument would also work if (M, ω) were semipositive.) To extend the proof to all compact symplectic manifolds one needs techniques that guarantee the existence of a stable J-holomorphic curve in the class $[S^2 \times \mathrm{pt}]$ through every point in $S^2 \times V$ for every almost complex structure on $S^2 \times V$. This can be achieved with the virtual moduli cycle.

EXERCISE 9.3.5. Let $f : (\mathbb{C}^n, 0) \to (\mathbb{C}^n, 0)$ be a diffeomorphism of the form

$$f(z) = \frac{\beta(|z|)}{|z|} z, \qquad \beta'(r) \geq 0.$$

Prove that $\omega := f^*\omega_0$ is compatible with J_0. *Hint:* Differentiate f to obtain

$$df(z)\zeta = \beta'(|z|)\zeta_0 + \frac{\beta(|z|)}{|z|}\zeta_1,$$

where ζ_0 is the orthogonal projection of ζ onto $\mathbb{R}z$ and $\zeta_0 + \zeta_1 = \zeta$.

The discussion so far has not mentioned the relation between the Gromov–Witten invariants of (M, ω) and its blowup $(\widetilde{M}, \widetilde{\omega})$. This problem has been solved in great generality in Hu [**191**] using the symplectic sum formula of Li–Ruan [**238**] and Ionel–Parker [**201**]. It is hard to formulate very general theorems with the tools at hand since the blowup of a semipositive manifold M is almost never semipositive unless $\dim M \leq 6$. We leave the following statement as an exercise for the reader.

EXERCISE 9.3.6. Let $(\widetilde{M}, \widetilde{\omega})$ be the blowup of a symplectic manifold (M, ω) of dimension less than or equal to six. Suppose that $a_1, \ldots, a_k \in H^*(M)$ are classes of positive degrees and denote $\widetilde{a}_i := \pi^* a_i \in H^*(\widetilde{M})$. Further given $A \in H_2(M)$ with representing cycle $Z_A \subset M \setminus \{x_0\}$ denote by $\widetilde{A} \in H_2(\widetilde{M})$ the class in $\widetilde{M}$ represented by $\pi^{-1}(Z_A)$. Then

$$\mathrm{GW}^{\widetilde{M}}_{\widetilde{A},k}(\widetilde{a}_1, \ldots, \widetilde{a}_k) = \mathrm{GW}^{M}_{A,k}(a_1, \ldots, a_k).$$

Hint: Consider almost complex structures $\widetilde{J}, J$ such that J is integrable near x_0 and the blow down map $\pi : (\widetilde{M}, \widetilde{J}) \to (M, J)$ is holomorphic. If C is a J-holomorphic curve through x_0 then its lift to $\widetilde{M}$ represents the class $\widetilde{A} - kL$, where k is as in (9.3.1) and L is the class of the line in the exceptional divisor E. Therefore the exercise reduces to finding an almost complex structure J as above that is regular for A and is such that no J-holomorphic curve that contributes to $\mathrm{GW}^M_{A,k}(a_1, \ldots a_k)$ goes through the point x_0.

EXERCISE 9.3.7. Suppose that Z is a symplectic submanifold of (M, ω) of codimension $2k$. Make explicit a construction for the symplectic blowup of M in directions normal to Z. *Hint:* Perhaps the neatest way to do this is to identify a neighbourhood of Z with a neighbourhood of the zero section in the bundle $P \times_{\mathrm{U}(k)} \mathbb{C}^k$, where $P \to Z$ is a principal $\mathrm{U}(k)$-bundle, and then replace the fiber $\mathbb{C}^k$ by its blowup at the origin: see Lerman [**237**].

Generalizations of the nonsqueezing theorem. The first generalization is to consider embedded balls of radius r in the product manifold

$$(M, \omega) := (S^2 \times V, \pi_1^*\sigma + \pi_2^*\tau),$$

where (V, τ) is an arbitrary symplectic manifold. If V is closed and we assume the existence of Gromov–Witten invariants for arbitrary closed symplectic manifolds then the previous argument goes through without essential change and yields the expected inequality (9.3.2) (see Exercise 9.3.8). Even without this assumption Lalonde–McDuff [**223**] succeeded in proving this inequality by a rather complicated geometric construction. This construction applies to completely arbitrary, even noncompact, symplectic manifolds V.

A second generalization is to symplectic 2-sphere bundles $S^2 \hookrightarrow M \to V$ over compact symplectic $(2n-2)$-manifolds V. In this situation the same techniques apply and prove the inequality (9.3.2) with A equal to the class of the fiber whenever there is a symplectic embedding of the standard ball $(B^{2n}(r), \omega_0)$ into (M, ω). This version of the nonsqueezing theorem plays an important role in the work of Biran [**38, 39**].

Next one could consider more general Hamiltonian bundles over S^2 with fiber (V, τ). Thus we assume that there is a submersion $\pi : M \to S^2$ such that the restriction of ω to each fiber is nondegenerate and symplectomorphic to (V, τ). In this case one can measure the size of (M, ω, π) by the ratio

$$\mathrm{area}(M, \omega) := \frac{\mathrm{Vol}(M, \omega)}{\mathrm{Vol}(V, \tau)} = \frac{\int_M \omega^n}{n \int_V \tau^{n-1}}.$$

The nonsqueezing theorem is said to hold for (M, ω, π) if the radius r of any symplectically embedded ball satisfies the inequality $\pi r^2 \le \mathrm{area}(M, \omega)$. This no longer holds in all cases. However, it does hold if the fiber (V, τ) is symplectically aspherical. For a proof and further results see McDuff [**271**]. Further developments of these ideas are discussed in Section 9.6.

One could also consider more general base manifolds. However, this does not seem a very promising line of inquiry since the nonsqueezing theorem does not even hold for all products $(\Sigma \times V, \pi_1^*\sigma + \pi_2^*\tau)$ when Σ is a Riemann surface of positive genus (see [**277**, Exercise 12.4]).

EXERCISE 9.3.8. Let (V, τ) be a symplectic manifold of dimension $2n-2$, $\sigma \in \Omega^2(S^2)$ be a symplectic form, $\iota : B^{2n}(r) \to S^2 \times V$ be a symplectic embedding, and $J \in \mathcal{J}(S^2 \times V, \pi_1^*\sigma + \pi_2^*\tau)$ be an almost complex structure such that $\iota^*J = J_0$. Suppose that there exists a J-holomorphic stable map representing the homology class $A := [S^2 \times \mathrm{pt}] \in H_2(S^2 \times V; \mathbb{Z})$ and passing through the point $x_0 := \iota(0)$. Prove that

$$\pi r^2 \le \int_{S^2} \sigma. \tag{9.3.2}$$

REMARK 9.3.9 (Symplectic embedding problems). Another direction in which one could generalize the nonsqueezing theorem is to change the domain of the embedding as well as its target. For example, one can consider embedding many equal balls, an arbitrary collection of balls, a symplectic ellipsoid, or a polydisc. Many of these problems have now been solved in dimension four, though what happens in higher dimensions is little understood. Readers interested in such questions might start by looking at the following papers [**65, 167, 175, 278, 231, 56**].

9.4. Symplectic 4-manifolds

In this section we prove some classical uniqueness results for symplectic structures on 4-manifolds. There are many related questions, some discussed in Remark 9.4.3 and some deferred until Section 9.7.

In [**160**] Gromov proved that every noncompact symplectic 4-manifold (M, ω) which is symplectomorphic to $(\mathbb{R}^4, \omega_0)$ at infinity and is such that ω vanishes on $\pi_2(M)$ must be symplectomorphic to $\mathbb{R}^4$. He also proved a uniqueness result for symplectic forms on $S^2 \times S^2$: any symplectic form ω on $S^2 \times S^2$ that is nondegenerate on two spheres

$$C_A := S^2 \times \{pt\}, \qquad C_B := \{pt\} \times S^2$$

is diffeomorphic to a standard product form $\pi_1^*\sigma + \pi_2^*\sigma$ provided that the spheres have the same area. From this he deduced the uniqueness statement in Theorem 9.4.1 (iii) below. McDuff [**258**] built on his ideas to prove the following symplectic classification theorem for rational and ruled surfaces. Recall that a **symplectic ruled surface** is a symplectic 4-manifold (M, ω) that admits the structure of a locally trivial fibration $\pi : M \to \Sigma$ over a Riemann surface Σ whose fibers are 2-spheres on which ω is nondegenerate.

THEOREM 9.4.1 (McDuff). *Let (M, ω) be a closed connected symplectic 4-manifold which contains a symplectically embedded 2-sphere S with nonnegative self-intersection number but no such sphere with self-intersection number -1. Then the following holds.*

(i) *There is a symplectically embedded 2-sphere C with self-intersection number either zero or one.*

(ii) *If there is a symplectically embedded 2-sphere C with self-intersection number $C \cdot C = 0$ then M is symplectomorphic to a ruled surface.*

(iii) *If there is a symplectically embedded 2-sphere C with self-intersection number $C \cdot C = 1$ then the pair (M, C) is symplectomorphic to $(\mathbb{C}P^2, \mathbb{C}P^1)$.*

Gromov's basic idea was very simple. He showed that each of the spheres C_A, C_B belongs to a family of spheres, one through each point of $M = S^2 \times S^2$. Moreover each A-sphere intersects each B-sphere exactly once transversally. It follows that the spheres are embedded, and hence give new coordinates on M. The final step is to show that ω is a product with respect to these new coordinates. This proof is carried out in Theorem 9.4.7 below. The first and last of these steps are fairly straightforward. However, the middle step uses Theorem 2.6.3 on positivity of intersections , which requires much more work to prove. (Unfortunately, the easy case described in Exercise 2.6.1 is not sufficient, since one has to rule out intersections that are singular on both curves.) Theorem 9.4.1 needs yet more preparation. Besides using general properties of J-holomorphic curves, the proof relies on McDuff's adjunction formula (Theorem 2.6.4), as well as on results about symplectic blowing up and down from McDuff [**257**]. The proof given below largely follows the exposition in Lalonde–McDuff [**226**]. However, some of the more geometric arguments (such as [**257**, Lemma 3.5]) are here replaced by analytic arguments that use properties of real linear Cauchy–Riemann operators.

Combining Theorem 9.4.1 with Theorem 0.3.C in Gromov [**160**], we obtain the following result. We call a subset $V \subset \mathbb{R}^m$ **star-shaped** if $x \in V$ implies $tx \in V$ for every $t \in [0, 1]$.

THEOREM 9.4.2 (Gromov–McDuff). *Let (M, ω) be a connected symplectic 4-manifold and $K \subset M$ be a compact subset such that the following holds.*

(i) *There is no symplectically embedded 2-sphere $S \subset M$ with self-intersection number $S \cdot S = -1$.*

(ii) *There exists a symplectomorphism $\psi : \mathbb{R}^4 \setminus V \to M \setminus K$, where $V \subset \mathbb{R}^4$ is a star-shaped compact set.*

Then (M, ω) is symplectomorphic to $(\mathbb{R}^4, \omega_0)$. Moreover, for every open neighbourhood $U \subset M$ of K, the symplectomorphism can be chosen equal to ψ^{-1} on $M \setminus U$.

Before proving Theorems 9.4.1 and 9.4.2 we discuss various related developments and place these results into context.

REMARK 9.4.3. (i) Let $M \to \Sigma$ be a ruled surface. It was proved by McDuff [**258**] for $\Sigma = S^2$, and by Lalonde–McDuff [**225**] in general, that any two cohomologous symplectic forms on M that admit symplectically embedded spheres in the class of the fiber are diffeomorphic. Moreover, the diffeomorphism can be chosen to identify the two symplectic spheres in the class of the fiber. For example, if the bundle $M \to S^2$ is trivial then ω is diffeomorphic to a product form $\pi_1^*\sigma_1 + \pi_2^*\sigma_2$. This work can be interpreted as a strengthening of assertion (ii) in Theorem 9.4.1. It uses a technique called symplectic inflation to control changes of the cohomology class of the symplectic form. (See the proof of Proposition 9.7.2. For further applications of this method, see McDuff [**266**] and Biran [**38, 39**].) In the case of the product $M = S^2 \times S^2$ one can avoid using inflation by assuming the existence of another symplectically embedded sphere. The arguments needed here are described in Theorem 9.4.7 and are essentially due to Gromov (apart from the technicalities caused by possible degenerations of the A, B curves when the two spheres do not have equal size). They form the basis for the proof of Theorem 9.4.1 (iii).

(ii) In [**392, 393**] Taubes proved that for every symplectic form ω on $\mathbb{C}P^2$ and every almost complex structure $J \in \mathcal{J}(\mathbb{C}P^2, \omega)$ there exists a J-holomorphic sphere representing the ω-positive generator of $H_2(\mathbb{C}P^2; \mathbb{Z})$. Combining this result with Theorem 9.4.1 (iii), one finds that every symplectic form on $\mathbb{C}P^2$ with volume $\pi^2/2$ is symplectomorphic to the Fubini–Study form ω_{FS}. We remark that the reversed form $-\omega_{\mathrm{FS}}$ is symplectomorphic to ω_{FS} via the orientation preserving diffeomorphism $f : \mathbb{C}P^2 \to \mathbb{C}P^2$ given by $f([z_0 : z_1 : z_2]) := ([\bar{z}_0 : \bar{z}_1 : \bar{z}_2])$.

Similarly, it was proved by Li–Liu [**240**], using Taubes–Seiberg–Witten theory, that every symplectic form on a ruled surface M admits an embedded symplectic sphere in the class of the fiber. Combining this with the work of Lalonde–McDuff [**258, 225, 226**] discussed above we deduce that any two cohomologous symplectic forms on M are diffeomorphic.

(iii) A symplectically embedded 2-sphere in (M^4, ω) with self-intersection -1 is called an **exceptional sphere**. We say that (M, ω) is **minimal** if it contains no exceptional spheres. McDuff showed in [**258**] that if one blows down a maximal collection of disjoint exceptional spheres in an arbitrary symplectic 4-manifold (M, ω) then the resulting symplectic manifold $(\overline{M}, \overline{\omega})$ is minimal. Moreover, using topological properties of various moduli spaces of curves, she showed in [**260**] that the symplectomorphism type of the blow down $(\overline{M}, \overline{\omega})$ is independent of all choices unless $(\overline{M}, \overline{\omega})$ is rational or ruled. Nowadays, this result has been generalized to the smooth case and is best seen as a consequence of Taubes–Seiberg–Witten theory (cf. Li–Liu [**240**], Li [**241**]).

(iv) Two symplectic manifolds (M, ω) and (M', ω') are called **deformation equivalent**[1] or **symplectomorphic up to deformation** if there exists a diffeomorphism $f : M \to M'$ and a smooth family of symplectic forms $[0, 1] \to \Omega^2(M) : t \mapsto \omega_t$ (not necessarily in the same cohomology class) such that $\omega_0 = \omega$ and $\omega_1 = f^*\omega'$. It is now known that, for any two integers $N \geq 2$ and $k \geq 1$, there is a $4k$-manifold M with symplectic forms $\omega_1, \ldots, \omega_N$ such that no two of the corresponding symplectic manifolds $(M, \omega_1), \ldots, (M, \omega_N)$ are deformation equivalent. McMullen–Taubes [**284**] gave the first examples with $N = 2$ and $k = 1$, while Smith [**382**] proved the result stated above. His manifolds are distinguished by the

[1]Some authors call this relation equivalence, while others call it weak deformation equivalence.

divisibility properties of the first Chern class $c_1(J)$ of the elements in $\mathcal{J}(M,\omega_i)$. By way of contrast, the results mentioned in (i) and (ii) above imply that any two symplectic forms on a ruled surface are symplectomorphic up to deformation.

(v) Let M be a closed oriented $2n$-manifold and $a \in H^2(M;\mathbb{R})$ be a cohomology class that can be represented by a symplectic form. Denote by $\Omega_a \subset \Omega^2(M)$ the space of all symplectic forms representing the class a. A natural question to ask is if this space is connected. A positive answer would imply, by Moser isotopy, that any two symplectic forms representing the class a are diffeomorphic. In dimension two the space Ω_a is always convex and hence connected. In dimension six there is an example, due to McDuff, where Ω_a is disconnected (see Section 9.7). However, in dimension four this question is completely open in either direction. There is not a single example of a smooth 4-manifold M and a cohomology class $a \in H^2(M;\mathbb{R})$ where Ω_a is known to be disconnected, nor is there any example where Ω_a is nonempty and known to be connected. In the case $M = \mathbb{C}P^2$ it follows from the results of Gromov, McDuff, and Taubes discussed above and in Section 9.5, that Ω_a is connected if and only if every diffeomorphism of $\mathbb{C}P^2$ that induces the identity on homology is isotopic to the identity. In the opposite direction one could ask if two symplectic forms in the same cohomology class a can have different Chern classes. This would, of course, imply that Ω_a is disconnected. In the case of the projective plane with any number of points blown up it is known that all symplectic forms in the same cohomology class also have the same first Chern class [**240**]. In the examples of Smith discussed in (iv) it is not known if the symplectic forms with different first Chern classes can be chosen in the same cohomology class.

(vi) It has been a longstanding conjecture in symplectic topology that every symplectic form on the 4-torus can be deformed in the same cohomology class to one with constant coefficients. This would imply a positive answer to the uniqueness problem posed in (v). At the time of writing this is still an open problem. A promising approach to tackling this conjecture, and the equivalent one for the $K3$ surface, via a new geometric flow equation was suggested by Donaldson [**86**]. (See also Tosatti–Weinkove [**397**] and the references therein.) A result of Taubes [**390, 391**], based on Seiberg–Witten theory, shows that every symplectic form on the 4-torus or the $K3$ surface has first Chern class zero. In the case of the $K3$ surface a refinement of this argument due to Donaldson, using cohomological orientation, shows that any two symplectic forms on the $K3$ surface (compatible with the standard orientation) belong to the same homotopy class of nondegenerate 2-forms. Even this weaker question is open for $\mathbb{T}^4$. (Homotopy classes of nondegenerate 2-forms with vanishing first Chern class are classified by an invariant in $H^3(M;\mathbb{Z}) \oplus \mathbb{Z}_2$.)

(vii) It is interesting to examine the relation between symplectic 4-manifolds and Kähler surfaces. There are two natural questions.

(I) Given a closed symplectic 4-manifold (M,ω), does there exist an integrable complex structure J compatible with ω?

(II) Given a closed symplectic 4-manifold (M,ω), does there exist a Kähler form on M, representing the same cohomology class as ω?

By Theorem 9.4.1 and the uniqueness results mentioned in (ii), the answer to question (I) is positive for $\mathbb{C}P^2$ and for ruled surfaces. Kodaira's classification theorem for complex surfaces implies that every Kähler form on the 4-torus is diffeomorphic to a linear form. A slightly weaker problem than the one discussed in (vi) would be to show that every symplectic form on the 4-torus is diffeomorphic to a linear form. By Kodaira classification, that would be equivalent to a positive answer to question (I) for the 4-torus.

The weaker question (II) concerns the relation between the **symplectic cone** of a manifold M (which is the set of cohomology classes with a symplectic representative) and its Kähler cone (the set of classes represented by a form that is Kähler with respect to some complex structure on M). A classical example of Thurston shows that that the Kähler cone of a symplectic manifold can be empty, in which case, of course, the answer to question (I) is negative for every symplectic form ω. There are now many such examples, including simply connected ones. However, if both cones are nonempty, there are very few examples where they are known to differ. Even in the case of a blow up of $\mathbb{C}P^2$ at nine or more points, the answer to this question is unknown (see Biran's work on the Nagata conjecture [**40**]). However, the two cones are different when M is the one point blowup of the 4-torus (see Latschev–McDuff–Schlenk [**231**, Corollary 1.3]) or the one point blowup of the nontrivial S^2 bundle over the 2-torus (see Cascini–Panov [**60**]).

Another related problem, proposed by Tian-Jun Li in [**243**], is to understand symplectic 4-manifolds of Kodaira dimension zero, i.e. with

$$\omega \cdot c_1(TM) = c_1(TM)^2 = 0.$$

The only known manifolds of this type are $K3$ surfaces, Enriques and hyperelliptic surfaces, and $\mathbb{T}^2$-bundles over $\mathbb{T}^2$. (Among these are Thurston's examples with an empty Kähler cone.) Although it is not known whether this list is complete, Bauer [**32**] and Li [**244**] show that every such manifold has the same cohomology ring as one of the members of this list. See also Smith [**383**] for further information about torus bundles, and Li [**245**] for a discussion of other similar problems. The new flow studied by Streets–Tian in [**386**], a symplectic version of the Kähler Ricci flow, might eventually cast light on some of these questions.

(viii) Finally we observe that the methods and results in this section are special to dimension 4: the proofs make heavy use of the adjunction formula which shows that in this dimension homological information gives geometric information. In higher dimensions an argument like that in Proposition 9.4.4 could at best determine the degree of an evaluation map rather than showing it is a diffeomorphism. For some results along these lines see McDuff [**259**]. Among other things, this paper proves a generalization of Theorem 9.4.2 due to Eliashberg–Floer–McDuff asserting that any manifold (M, ω) for which ω vanishes on $\pi_2(M)$ and that is symplectomorphic to Euclidean space outside a compact set must be contractible, though when $n > 2$ it is unknown whether the symplectomorphism to $\mathbb{R}^{2n}$ that is defined near infinity actually extends to a symplectomorphism defined on the whole of M.

After these comments, we now turn to the proofs of the theorems. We shall first prove assertions (i) and (ii) of Theorem 9.4.1, then Theorem 9.4.2, and afterwards assertion (iii) of Theorem 9.4.1. The argument is based on two fundamental results. The first is concerned with moduli spaces of embedded A-spheres with self-intersection p where p is nonnegative and minimal. In this case, Proposition 9.4.4 states that the moduli space $\mathcal{M}(A; J)/\mathrm{G}$ of unmarked spheres is compact and the evaluation map (with an appropriate number of marked points) is a diffeomorphism. It then follows that either $p = 1$ and $M = \mathbb{C}P^2$, or $p = 0$ and M is a ruled surface.

The second basic result is Theorem 9.4.7, a careful analysis of symplectic forms on $S^2 \times S^2$ that are standard near the set $X := S^2 \times \{\mathrm{pt}\} \cup \{\mathrm{pt}\} \times S^2$.

In the next proposition, we denote by $\Delta^k \subset M^k$ the fat diagonal of all k-tuples $\mathbf{x} = (x_1, \dots, x_k)$ in M^k that are not pairwise distinct.

PROPOSITION 9.4.4. *Let (M, ω) be a compact connected symplectic 4-manifold, J be any ω-tame almost complex structure on M, and $A \in H_2(M;\mathbb{Z})$ such that $A \cdot A =: p \geq 0$. Suppose A can be represented by a symplectically embedded 2-sphere and that there is no symplectically embedded 2-sphere S such that $-1 \leq S \cdot S < p$. Assume further that every nonconstant J-holomorphic sphere has positive Chern number. Then every J-holomorphic sphere representing the class A is embedded (and hence regular), the moduli space $\mathcal{M}(A;J)/\mathrm{G}$ is compact, and the evaluation map*

$$\mathrm{ev} : \mathcal{M}_{0,p+1}(A;J) \to M^{p+1} \setminus \Delta^{p+1} \tag{9.4.1}$$

is a diffeomorphism.

Note here that J is an arbitrary tame almost complex structure such that every nonconstant J-holomorphic sphere has positive Chern number: we do not assume it is regular. In dimension 4 this assumption is slightly more general than regularity: indeed the proof of Lemma 6.4.4 shows that if $J \in \mathcal{J}_{\mathrm{reg}}(M, \omega)$ every nonconstant J-holomorphic sphere $u : S^2 \to M$ has positive Chern number. Observe further that the hypothesis on A concerns the properties of embedded J-holomorphic curves. Therefore we shall need to use the adjunction formula to conclude that A is J-indecomposable (which implies the compactness of $\mathcal{M}(A;J)/\mathrm{G}$ as in Lemma 7.1.8). Our proof that the evaluation map is a diffeomorphism uses the Riemann–Roch theorem for real linear Cauchy–Riemann operators as in the proof of the regularity result Lemma 3.3.2.

COROLLARY 9.4.5 (Gromov). *Let $M := \mathbb{C}P^2$ and $\omega = \omega_{\mathrm{FS}}$ be the standard symplectic form. Then, for every $J \in \mathcal{J}_\tau(\mathbb{C}P^2, \omega_{\mathrm{FS}})$ and any two distinct points $x_1, x_2 \in \mathbb{C}P^2$, there exists a J-holomorphic sphere $u : S^2 \to \mathbb{C}P^2$, unique up to reparametrization, such that $[u] = [\mathbb{C}P^1]$ and $x_1, x_2 \in u(S^2)$.*

PROOF. The hypotheses of Proposition 9.4.4 are satisfied with $p = 1$. In particular, every nonconstant J-holomorphic sphere has positive Chern number because $\mathbb{C}P^2$ is monotone. □

The proof of Proposition 9.4.4 is based on the following observation.

LEMMA 9.4.6. *Let (M, ω, J) and A be as in the hypotheses of Proposition 9.4.4. Then A is J-indecomposable.*

PROOF. Suppose that this does not hold. Then there exist homology classes $A_i \in H_2(M;\mathbb{Z})$ and positive integers m_i such that each class A_i is represented by a simple J-holomorphic sphere $u_i : S^2 \to M$ and

$$A = \sum_{i=1}^N m_i A_i, \qquad \sum_{i=1}^N m_i > 1, \qquad A_i \neq A_j \quad (i \neq j).$$

We prove that, for all i and j,

$$A_i \cdot A_j \geq 0, \qquad A_i \cdot A_i \geq c_1(A_i) > 0. \tag{9.4.2}$$

The first inequality follows from positivity of intersections (Theorem 2.6.3.) To prove the second we observe that, since every J-holomorphic sphere in M has positive Chern number and $\sum_i m_i > 1$, we have

$$0 < c_1(A_i) < c_1(A)$$

for every i. Moreover, by the adjunction inequality (Theorem 2.6.4),

$$A_i \cdot A_i \geq c_1(A_i) - 2$$

for every i, with equality if and only u_i is an embedding. If u_i is not embedded then $A_i \cdot A_i > c_1(A_i) - 2$ and, since $A_i \cdot A_i - c_1(A_i)$ is always even, this implies $A_i \cdot A_i \geq c_1(A_i)$. Thus the second inequality in (9.4.2) holds for each i such that u_i is not an embedding. On the other hand, if u_i is an embedding, then $A_i \cdot A_i = c_1(A_i) - 2$. Combining this with the equation $A \cdot A = c_1(A) - 2$ and the inequality $1 \leq c_1(A_i) < c_1(A)$, we obtain $-1 \leq A_i \cdot A_i < A \cdot A$, which is excluded by our assumption. Thus we have proved (9.4.2).

Now consider the identity

$$\begin{aligned} -2 &= A \cdot A - c_1(A) \\ &= \sum_{i=1}^{N} (m_i^2 - m_i) A_i \cdot A_i + 2 \sum_{i<j} m_i m_j A_i \cdot A_j + \sum_{i=1}^{N} m_i (A_i \cdot A_i - c_1(A_i)). \end{aligned}$$

By (9.4.2), each summand on the right is nonnegative, a contradiction. Hence our assumption that A is not J-indecomposable must have been wrong. □

Proof of Proposition 9.4.4. Assume first that the moduli space $\mathcal{M}(A; J)$ is nonempty. Since A can be represented by a symplectically embedded sphere we have

$$c_1(A) = A \cdot A + 2.$$

Hence, by Theorem 2.6.4, every simple J-holomorphic sphere representing the class A is embedded. Moreover, by Lemma 9.4.6, A is J-indecomposable. Hence every J-holomorphic sphere in the class A is simple and therefore embedded. It now follows from Lemma 3.3.3 that the operator D_u is onto for every J-holomorphic sphere $u \in \mathcal{M}(A; J)$. Hence, by Theorem 3.1.6, the moduli space $\mathcal{M}(A; J)$ is a smooth manifold of dimension dim $\mathcal{M}(A; J) = 2p + 8$. Since A is J-indecomposable it follows from Theorem 5.3.1 that the quotient space $\mathcal{M}(A; J)/\mathrm{G}$ is compact.

We prove that the evaluation map (9.4.1) is injective. If $\mathrm{ev}(u, \mathbf{z}) = \mathrm{ev}(u', \mathbf{z}')$ then the curves u and u' intersect at $p + 1$ distinct points. Since $A \cdot A = p$ and the curves are embedded, Exercise 2.6.7 asserts that $u(S^2) = u'(S^2)$. Hence, by Corollary 2.5.4, there exists a unique Möbius transformation $\phi : S^2 \to S^2$ such that $u \circ \phi = u'$. Since u and u' are embeddings this implies $\phi(z_i') = z_i$ for $i = 0, \ldots, p$ and so $[u, \mathbf{z}] = [u', \mathbf{z}']$.

It is less trivial to see that the differential $d\mathrm{ev}$ is everywhere surjective. To this end, note that the tangent space of the moduli space $\widetilde{\mathcal{M}}_{0,p+1}(A; J)$ of pointed maps at $(u, \mathbf{z}) = (u, z_0, \ldots, z_p)$ is given by

$$T_{(u,\mathbf{z})} \widetilde{\mathcal{M}}_{0,p+1}(A; J) = \ker D_u \oplus T_{z_0} S^2 \oplus \cdots \oplus T_{z_p} S^2,$$

and that the differential of the evaluation map has components

$$d\widetilde{\mathrm{ev}}_i(u, \mathbf{z})(\xi, \zeta) = \xi(z_i) + du(z_i)\zeta_i,$$

where $\xi \in \ker D_u$ and $\zeta_i \in T_{z_i}(S^2)$. Now let $L_0 := \operatorname{im} du \subset u^*TM$ be the tangent bundle to the curve $\operatorname{im} u$ and denote its orthogonal complement by $L_1 \subset u^*TM$. Then L_0 has Chern number two and L_1 has Chern number p. Moreover, the operator D_u splits as

$$D_u(\xi_0 + \xi_1) = D_{00}\xi_0 + D_{01}\xi_1 + D_{11}\xi_1,$$

where $D_{jj} : \Omega^0(S^2, L_j) \to \Omega^{0,1}(S^2, L_j)$ is a real linear Cauchy–Riemann operator and $D_{01} : \Omega^0(S^2, L_1) \to \Omega^{0,1}(S^2, L_0)$ is a zeroth order operator.

We first show that the map

$$(\ker D_{11}) \to L_{1z_0} \oplus \cdots \oplus L_{1z_p} : \xi_1 \mapsto (\xi_1(z_0), \ldots, \xi_1(z_p)). \tag{9.4.3}$$

is bijective. To see this note first that (9.4.3) is a (real) linear map between vector spaces of the same dimension $2p+2$. Hence it suffices to show it is injective. But if not there exists a nonzero section $\xi_1 \in \ker D_{11}$ which vanishes at $p+1$ distinct points. However, because D_{11} is a real linear Cauchy–Riemann operator, each zero of ξ_1 has positive intersection index. (As we saw at the end of the proof of Theorem C.1.10 this follows from the fact that ξ_1 is in the kernel of a complex linear Cauchy–Riemann operator and hence is holomorphic for a suitable complex structure on L_1.) Since $L_1 \to S^2$ is a complex line bundle of degree p it follows that ξ_1 has at most p zeros, a contradiction. Hence the map (9.4.3) is bijective.

With this understood it follows immediately that the kernel of $d\widetilde{\mathrm{ev}}(u, \mathbf{z})$ is the tangent space of the orbit of $(u, \mathbf{z})$ under the action of the reparametrization group $\mathrm{PSL}(2, \mathbb{C})$ on $\widetilde{\mathcal{M}}_{0,p+1}(A; J)$. Hence the differential of ev on the quotient space is injective. Since ev is a smooth map between manifolds of the same dimension $4p+4$, its differential is bijective at every point.

The next step is to show that the evaluation map (9.4.1) is proper. To this end, let $[u^\nu, \mathbf{z}^\nu]$ be a sequence in $\mathcal{M}_{0,p+1}(A; J)$ such that

$$\mathbf{x}^\nu := \mathrm{ev}(u^\nu, \mathbf{z}^\nu) \in M^{p+1} \setminus \Delta^{p+1}$$

converges to a point $\mathbf{x} \in M^{p+1} \setminus \Delta^{p+1}$. Since $\mathcal{M}(A; J)/\mathrm{G}$ is compact, we may assume, reparametrizing u^ν if necessary, that $(u^\nu, \mathbf{z}^\nu)$ has a convergent subsequence. Let $(u, \mathbf{z}) \in \mathcal{M}(A; J) \times (S^2)^{p+1}$ be the limit. Since $\mathbf{x} \in M^{p+1}$ is a tuple of pairwise distinct points the tuple $\mathbf{z} \in (S^2)^{p+1}$ also consists of pairwise distinct points. Hence the subsequence converges in $\mathcal{M}_{0,p+1}(A; J)$. Thus we have proved that the evaluation map (9.4.1) is a proper injective map such that the differential is bijective at every point. This shows that (9.4.1) is a diffeomorphism whenever $\mathcal{M}(A; J) \neq \emptyset$.

It remains to show $\mathcal{M}(A; J)$ is nonempty for every ω-tame J such that every nonconstant J-holomorphic sphere has positive Chern number. To see this, observe first that because the ω-compatible almost complex structures on M are sections of a bundle with contractible fibers, we may construct $J \in \mathcal{J}(M, \omega)$ so that C is J-holomorphic. By Lemma 3.3.3, C is a regular J-holomorphic curve and so persists under small perturbations of J. Therefore there is a nearby almost complex structure $J' \in \mathcal{J}_{\mathrm{reg}}(M, \omega)$ that admits a J'-holomorphic sphere in the class A. Since J' is regular, every nonconstant J'-holomorphic sphere has positive Chern number. We showed above that under these circumstances the corresponding evaluation map (9.4.1) is a diffeomorphism. Hence we obtain a nontrivial Gromov–Witten invariant

$$\mathrm{GW}^M_{A,p+1}(a, \ldots, a) = \pm 1, \qquad a := \mathrm{PD}([\mathrm{pt}]) \in H^4(M).$$

This in turn implies that the class A can be represented by a J-holomorphic stable map for *every* $J \in \mathcal{J}_\tau(M,\omega)$. If every nonconstant J-holomorphic sphere has positive Chern number then, by Lemma 9.4.6, the only such stable maps are embedded J-holomorphic spheres and hence, in this case, $\mathcal{M}(A;J) \neq \emptyset$. This proves Proposition 9.4.4. □

PROOF OF THEOREM 9.4.1 (I) AND (II). Let $C \subset M$ be a symplectically embedded 2-sphere with self-intersection number $C\cdot C =: p \geq 0$ and suppose that there is no symplectically embedded 2-sphere with self-intersection number between -1 and $p-1$. Let $A := [C]$ and $J \in \mathcal{J}_{\mathrm{reg}}(M,\omega)$. Then, by Proposition 9.4.4, every A-curve is embedded and the evaluation map $\mathrm{ev} : \mathcal{M}_{0,p+1}(A;J) \to M^{p+1} \setminus \Delta^{p+1}$ is a diffeomorphism.

We prove that $p=0$ or $p=1$ by showing that otherwise there is a nonembedded curve in class A. Thus assume $p \geq 2$. Fix $p-1$ distinct points $x_0, x_3, \ldots, x_p \in M$ and choose $v_1, v_2 \in T_{x_0}M$ such that the vectors v_1, Jv_1, v_2, Jv_2 form a basis of $T_{x_0}M$. Consider the sequences

$$x_1^\nu := \exp_{x_0}(2^{-\nu}v_1), \qquad x_2^\nu := \exp_{x_0}(2^{-\nu}v_2).$$

Since (9.4.1) is a diffeomorphism there exists, for every ν, a unique J-holomorphic sphere $u^\nu : S^2 \to M$ representing the class A such that

$$u^\nu(0) = x_0, \qquad u^\nu(1) = x_1^\nu, \qquad u^\nu(2) = x_2^\nu, \qquad x_i \in u^\nu(S^2) \quad (i \geq 3).$$

Since $\mathcal{M}(A;J)/\mathrm{G}$ is compact there exists a sequence of Möbius transformations $\phi^\nu \in \mathrm{G}$ such that $\phi^\nu(0) = 0$ and $u^\nu \circ \phi^\nu$ has a convergent subsequence. Passing to this subsequence we may assume that the limit $u := \lim_{\nu\to\infty} u^\nu \circ \phi^\nu$ exists in the C^∞-topology. Passing to a further subsequence, we may assume that the limits

$$z_1 := \lim_{\nu\to\infty} (\phi^\nu)^{-1}(1), \qquad z_2 := \lim_{\nu\to\infty} (\phi^\nu)^{-1}(2)$$

exist. Then

$$u(z_1) = \lim_{\nu\to\infty} u^\nu \circ \phi^\nu((\phi^\nu)^{-1}(1)) = \lim_{\nu\to\infty} u^\nu(1) = \lim_{\nu\to\infty} x_1^\nu = x_0$$

and, similarly, $u(z_2) = x_0$. We claim that z_1 and z_2 cannot both be equal to zero. Namely, if $z_1^\nu := (\phi^\nu)^{-1}(1) \to 0$ we may assume, by passing to a further subsequence, that the quotient $z_1^\nu/|z_1^\nu|$ converges in S^1. Examining the difference quotient

$$\frac{u_\nu \circ \phi_\nu(z_1^\nu) - u_\nu \circ \phi_\nu(0)}{|z_1^\nu|} = \frac{x_1^\nu - x_0}{|z_1^\nu|}$$

in local coordinates we find that the vector $v_1 \in T_{x_0}M$ belongs to the image of $du(0)$. The same argument applies to the sequence $z_2^\nu := (\phi^\nu)^{-1}(2)$. But by construction the vectors v_1 and v_2 are complex linearly independent and so cannot both belong to the image of $du(0)$. Hence $z_1 \neq 0$ or $z_2 \neq 0$. In either case u has a self-intersection. But this contradicts the fact that all J-holomorphic spheres representing the class A are embedded. Hence our assumption that $p \geq 2$ must have been wrong. Therefore $p \leq 1$ and this proves (i).

We prove (ii). Let A be the homology class of the symplectically embedded sphere C with $C \cdot C = 0$ and let $J \in \mathcal{J}_{\mathrm{reg}}(M,\omega)$. Then every nonconstant J-holomorphic sphere has positive Chern number and hence, by Proposition 9.4.4 and Theorem 2.6.3, the moduli space $\mathcal{M}(A;J)$ has the following properties.

(a) Every $u \in \mathcal{M}(A;J)$ is an embedding. Moreover, if $u_0, u_1 \in \mathcal{M}(A;J)$ such that $u_0(S^2) \cap u_1(S^2) \neq \emptyset$ then $u_0(S^2) = u_1(S^2)$.

(b) D_u is onto for every $u \in \mathcal{M}(A;J)$.

(c) The quotient $\mathcal{M}(A;J)/\mathrm{G}$ is compact.

(d) The evaluation map $\mathrm{ev} : \mathcal{M}_{0,1}(A;J) \to M$ is a diffeomorphism.

Now $\mathcal{M}_{0,1}(A;J)$ is a bundle over the compact oriented 2-manifold $\mathcal{M}(A;J)/\mathrm{G}$ with fibers diffeomorphic to S^2. Moreover, the pullback form $\mathrm{ev}^*\omega$ is nondegenerate on these fibers since they map to J-holomorphic curves in M. This proves (ii) in Theorem 9.4.1. □

The proof of Theorem 9.4.1 (iii) is based on a structure theorem for symplectic forms on $S^2 \times S^2$ which we now explain.

THEOREM 9.4.7. *Let (M,ω) be a compact connected symplectic 4-manifold that does not contain any symplectically embedded 2-sphere with self-intersection number -1. Let $A, B \in H_2(M;\mathbb{Z})$ be such that*

$$A \cdot A = B \cdot B = 0, \qquad A \cdot B = 1,$$

and suppose that A and B are represented by symplectically embedded 2-spheres. Let $\sigma \in \Omega^2(S^2)$ be an area form such that $\int_{S^2} \sigma = 1$. Then the following holds.

(i) *There is a diffeomorphism $\psi : S^2 \times S^2 \to M$ such that*

$$\psi^*\omega = a\pi_1^*\sigma + b\pi_2^*\sigma, \qquad a := \int_A \omega, \qquad b := \int_B \omega.$$

(ii) *If $U \subset S^2$ is an open disc, $\iota : (S^2 \times U) \cup (U \times S^2) \to M$ is an embedding such that*

$$\iota^*\omega = a\pi_1^*\sigma + b\pi_2^*\sigma, \qquad \iota_*[S^2 \times \{w\}] = A, \qquad \iota_*[\{z\} \times S^2] = B,$$

for all $z, w \in U$, and $D \subset U$ is any compact set, the diffeomorphism ψ in (i) can be chosen to agree with ι on $(S^2 \times D) \cup (D \times S^2)$.

PROOF. Let $J \in \mathcal{J}_{\mathrm{reg}}(M,\omega)$. We have seen in the proof of Theorem 9.4.1 (ii) that the elements of the moduli space $\mathcal{M}(A;J)$ are all embedded spheres whose images are either disjoint or coincide. The same applies to the moduli space $\mathcal{M}(B;J)$. By positivity of intersections (Theorem 2.6.3), every J-holomorphic sphere $u \in \mathcal{M}(A;J)$ has precisely one transverse intersection with every J-holomorphic sphere $v \in \mathcal{M}(B;J)$. Hence there is a well defined map

$$\Psi : \mathcal{M}(A;J)/\mathrm{G} \times \mathcal{M}(B;J)/\mathrm{G} \to M$$

given by

$$\Psi([u],[v]) := \mathrm{im}\,[u] \cap \mathrm{im}\,[v].$$

We prove that this map is a diffeomorphism. By Proposition 9.4.4 with $p = 0$, the evaluation maps $\mathrm{ev}_A : \mathcal{M}_{0,1}(A;J) \to M$ and $\mathrm{ev}_B : \mathcal{M}_{0,1}(B;J) \to M$ are diffeomorphisms. Hence there is precisely one A-curve and one B-curve through every point in M. So Ψ is bijective and its inverse can be expressed as the composition

$$\Psi^{-1} = (\pi_A \times \pi_B) \circ (\mathrm{ev}_A \times \mathrm{ev}_B)^{-1} \circ \iota,$$

where $\pi_A : \mathcal{M}_{0,1}(A;J) \to \mathcal{M}(A;J)/\mathrm{G}$ and $\pi_B : \mathcal{M}_{0,1}(B;J) \to \mathcal{M}(B;J)/\mathrm{G}$ are the obvious projections and $\iota : M \to M \times M$ is the inclusion of the diagonal. Hence Ψ^{-1} is smooth. Moreover, the differential of $\mathrm{ev}_A \times \mathrm{ev}_B$ at a point $([u,z],[v,w])$

with $u(z) = v(w)$ maps the vertical space to the subspace $\operatorname{im} du(z) \times \operatorname{im} dv(w)$, which is transverse to the diagonal. Hence the differential of $(\mathrm{ev}_A \times \mathrm{ev}_B)^{-1}$ maps the tangent space of the diagonal to a horizontal space. So the differential of Ψ^{-1} is bijective at every point and hence Ψ is a diffeomorphism.

Now fix maps $u_* \in \mathcal{M}(A; J), v_* \in \mathcal{M}(B; J)$. Since Ψ is a diffeomorphism, its restriction to the subset $\mathcal{M}(A; J)/\mathrm{G} \times \{[v_*]\} \subset \mathcal{M}(A; J)/\mathrm{G} \times \mathcal{M}(B; J)/\mathrm{G}$ induces a diffeomorphism

$$g : \mathcal{M}(A; J)/\mathrm{G} \to \operatorname{im} v_*$$

onto the image of v_* that takes each (unparametrized) curve $[u]$ to its point of intersection with $\operatorname{im} v_*$. Precomposing its inverse with v_* we get a diffeomorphism

$$g^{-1} \circ v_* : S^2 \to \mathcal{M}(A; J)/\mathrm{G} : w \mapsto [u_w],$$

where $[u_w]$ is the unique element in $\mathcal{M}(A; J)/\mathrm{G}$ through the point $v_*(w)$. Similarly, there is a diffeomorphism $S^2 \to \mathcal{M}(B; J)/\mathrm{G}$ that takes z to the unique element $[v_z] \in \mathcal{M}(B; J)/\mathrm{G}$ that intersects the image of u_* at the point $u_*(z)$. The composition of the diffeomorphism $(z, w) \mapsto ([u_w], [v_z])$ with Ψ is a diffeomorphism

$$\psi = \psi_J : S^2 \times S^2 \to M, \qquad \psi(z, w) := \operatorname{im}[u_w] \cap \operatorname{im}[v_z]. \tag{9.4.4}$$

For every $w \in S^2$ the image of $S^2 \times \{w\}$ under ψ is the embedded J-holomorphic sphere $\operatorname{im}[u_w]$ representing the class A. Hence $S^2 \times \{w\}$ is a ψ^*J-holomorphic sphere in $S^2 \times S^2$ for every $w \in S^2$ and has area a with respect to $\psi^*\omega$. Likewise, for every $z \in S^2$, the submanifold $\{z\} \times S^2$ is ψ^*J-holomorphic with area b. Hence $\psi^*\omega$ is homologous to the form

$$\omega_0 := a\pi_1^*\sigma + b\pi_2^*\sigma.$$

Moreover, the vertical and horizontal subbundles of $T(S^2 \times S^2)$ are symplectic complements of each other with respect to ω_0 and are complex subbundles with respect to ψ^*J. Hence ω_0 tames ψ^*J. Since $\psi^*\omega$ also tames ψ^*J it follows that the 2-form

$$\omega_t := \omega_0 + t(\psi^*\omega - \omega_0)$$

is symplectic for every $t \in [0, 1]$. (This step could also be justified by appealing to Exercise 9.4.8 as we do later.) Now the Moser isotopy argument (see Remark 9.4.9 below) shows that there is a diffeomorphism $f : S^2 \times S^2 \to S^2 \times S^2$ such that

$$f^*\psi^*\omega = \omega_0.$$

Hence the composition $\psi \circ f : S^2 \times S^2 \to M$ is a symplectomorphism. This proves (i).

To prove (ii) we make a more careful choice of J. Shrinking U if necessary, we may assume that the symplectic embedding ι extends smoothly to some open neighbourhood of the closure of $(S^2 \times U) \cup (U \times S^2)$. Let $K \subset M$ denote the image of this closure under ι, and choose the almost complex structure $J \in \mathcal{J}(M, \omega)$ so that the pullback structure ι^*J is the standard product structure J_0 on the subset $(S^2 \times U) \cup (U \times S^2)$. We claim that we may assume in addition that every nonconstant J-holomorphic sphere has positive Chern number. To see this, observe that every J-holomorphic sphere whose image is entirely contained in K pulls back to a holomorphic sphere in $S^2 \times S^2$ (with respect to the standard complex structure) and so has positive Chern number whenever it is nonconstant. This shows that every nonconstant holomorphic sphere with nonpositive Chern number must intersect the complement of K. Hence, because these spheres cannot be regular for dimensional reasons, they can be removed by perturbing J in $M \setminus K$ (cf. Remark 3.2.3).

Now fix $z_*, w_* \in D$ and choose the elements u_*, v_* to equal the restriction of ι to the spheres $S^2 \times \{w_*\}, \{z_*\} \times S^2$, i.e. $u_*(z) := \iota(z, w_*)$ and $v_*(w) := \iota(z_*, w)$. Let $\psi : S^2 \times S^2 \to M$ be the diffeomorphism defined by (9.4.4). Then $\psi(z, w)$ is the unique point of intersection of the A-sphere $[u_w]$ through $\iota(z_*, w)$ with the B-sphere $[v_z]$ through $\iota(z, w_*)$. Hence ψ has the following properties.

(a) If $w \in U$ then $\psi(z, w) \in \iota(S^2 \times \{w\})$, and if $z \in U$ then $\psi(z, w) \in \iota(\{z\} \times S^2)$. Moreover, $\psi(z, w_*) = \iota(z, w_*)$ and $\psi(z_*, w) = \iota(z_*, w)$ for all $z, w \in S^2$.

(b) The submanifolds $S^2 \times \{w\}$ and $\{z\} \times S^2$ are symplectic with respect to $\psi^*\omega$ for all $z, w \in S^2$.

It follows from (a) that ψ agrees with ι on the set $X := S^2 \times \{w_*\} \cup \{z_*\} \times S^2$ and on $U \times U$. The strategy of the proof is now to alter ψ to a diffeomorphism ψ' that agrees with ι on $(S^2 \setminus D) \times D$ and $D \times (S^2 \setminus D)$ and still satisfies (b). Granted this, the rest of the proof goes through as before. Namely, by Exercise 9.4.8 below, we have $\omega_0 \wedge (\psi')^*\omega > 0$ and hence the linear path

$$\omega_t := (1-t)\omega_0 + t(\psi')^*\omega, \qquad 0 \le t \le 1,$$

still consists of symplectic forms. Since $\omega_t = \omega_0$ on $(S^2 \setminus D) \times D \cup D \times (S^2 \setminus D)$ by construction, the Moser argument provides a diffeomorphism f with support in $(S^2 \setminus D) \times (S^2 \setminus D)$ such that $f^*(\psi')^*\omega = \omega_0$ (see Remark 9.4.9). Therefore the map $\psi' \circ f$ satisfies all requirements.

Hence it remains to construct ψ'. We take it to equal ψ outside the two sets $(S^2 \setminus U) \times U$ and $U \times (S^2 \setminus U)$. Because the latter are disjoint, we can deal with them separately. Hence it suffices to construct ψ' on $(S^2 \setminus U) \times U$. Since the image curves $\psi(S^2 \times \{w\})$ and $\iota(S^2 \times \{w\})$ are equal when $w \in U$ we may write

$$\psi(z, w) = \iota(\phi(z, w), w), \qquad z \in S^2, w \in U.$$

Note that $\phi(z, w) = z$ when $z \in U$ or when $w = w_*$.[2] Moreover, for every $z \in S^2$, the image of the sphere $\{z\} \times S^2$ under ψ is the B-curve $\operatorname{im}[v_z]$. Hence

$$\iota^{-1}(\operatorname{im}[v_z]) \cap S^2 \times U = \{(\phi(z, w), w) \mid w \in U\}.$$

Since ι^*J is the standard complex structure on $S^2 \times U$, this set is a complex submanifold of $S^2 \times U$ and so the map $U \to S^2 : w \mapsto \phi(z, w)$ is holomorphic (with respect to the usual complex structures) for every $z \in S^2$. In particular, its Jacobian determinant is everywhere nonnegative, so that it pulls back an area form to a nonnegative form.

Now choose a smooth map $\rho : S^2 \to S^2$ so that

$$\rho(U) \subset U; \quad \rho(D) = w_*; \quad \rho(w) = w, w \notin U; \quad \det(d\rho(w)) \ge 0, w \in S^2.$$

For example, identifying U with the unit disc in $\mathbb{C}$, we may take $\rho(w) = \beta(|w|)w$ where β is a monotone cutoff function. Then we claim that

$$\psi'(z, w) := \iota(\phi(z, \rho(w)), w), \quad z \in S^2, w \in U$$

[2] If $d\psi$ were also equal to the identity along the sphere $w = w_*$ then we could define ψ' by composing ψ with a C^1-small isotopy of the form $(z, w) \mapsto (g_t(z, w), w)$. However, this need not be the case; the correcting isotopy might have to be large with respect to the C^1-norm. A similar problem of straightening out the fibers of a 2-dimensional symplectic bundle has occurred in several contexts. The argument we give here is somewhat special in that it exploits the fact that J is integrable in $(S^2 \setminus U) \times U$. But this is unnecessary: a more general version may be found in McDuff [**262**, Lemma 3.11].

agrees with ψ in $U \times U$ (since $\phi(z,w) = z$ when $z \in U$) and has the required properties. The only point that is not obvious is that ω is nondegenerate on the image of the curves $w \mapsto \psi'(z,w)$ for all z. Since $\iota^*\omega = \omega_0$ in $U \times (S^2 \setminus U)$, it suffices to check that ω_0 is nondegenerate on the image of the curves $w \mapsto (\phi(z,\rho(w)),w)$. But this holds since the first component of this map has nonnegative Jacobian determinant, by construction. This proves Theorem 9.4.7. □

EXERCISE 9.4.8. Let (V,ω_0) be a 4-dimensional symplectic vector space and W be a 2-dimensional symplectic subspace of V. Let $W^\perp$ denote the symplectic complement of W. Let ω be another symplectic form on V whose restrictions to W and $W^\perp$ are positive area forms. Prove that $\omega \wedge \omega_0 > 0$.

REMARK 9.4.9 (Moser isotopy). Let $t \mapsto \omega_t$ be a smooth family of cohomologous symplectic forms on a compact symplectic manifold M. Then there exists a smooth isotopy $t \mapsto f_t$ such that $f_0 = \mathrm{id}$ and $f_t^*\omega_t = \omega_0$ for every t. One first chooses a family β_t of 1-forms such that $d\beta_t = \partial_t\omega_t$. Now the ansatz is to construct f_t as the isotopy associated to a family of vector fields X_t. If these vector fields are chosen such that $\iota(X_t)\omega_t + \beta_t = 0$ then

$$\frac{d}{dt} f_t^*\omega_t = f_t^*\left(\mathcal{L}_{X_t}\omega_t + \frac{d}{dt}\omega_t\right) = f_t^* d\left(\iota(X_t)\omega_t + \beta_t\right) = 0$$

as required. If ω_t is independent of t on a simply connected open subset $U \subset M$, and $K \subset U$ is a compact set, then the 1-forms β_t can be chosen to vanish on K. Hence X_t vanishes on K for every t and hence f_t is the identity on K for every t. Note that the isotopy is determined only after the primitives β_t for the 2-forms $\partial_t\omega_t$ are chosen. One can get rid of this indeterminacy by choosing β_t to be coexact (i.e. β_t belongs to the image of the operator $d^* : \Omega^2(M) \to \Omega^1(M)$) with respect to some metric. The construction then involves no choices at all. For more details see [**277**, Section 3.2].

Instead of completing the proof of Theorem 9.4.1 now, it is more efficient at this point to prove Theorem 9.4.2 on the structure of symplectic forms on Euclidean space. The next lemma is needed to control the support of the symplectomorphisms that we construct.

LEMMA 9.4.10. *Let $V \subset \mathbb{R}^{2n}$ be a star-shaped compact set and assume that $f : \mathbb{R}^{2n} \setminus V \to \mathbb{R}^{2n}$ is a symplectic embedding equal to the identity near infinity. Then, for every open neighbourhood $W \subset \mathbb{R}^{2n}$ of V, there exists a symplectomorphism $g : \mathbb{R}^{2n} \to \mathbb{R}^{2n}$ such that $g|_{\mathbb{R}^{2n}\setminus W} = f$.*

PROOF. Replacing V by its union with a small ball centered at zero we may assume, without loss of generality, that $0 \in \mathrm{int}(V)$. Then there is a constant $T > 1$ such that $f(Tx) = Tx$ for every $x \in \mathbb{R}^{2n} \setminus V$. Define $f_t : \mathbb{R}^{2n} \setminus V \to \mathbb{R}^{2n}$ by

$$f_t(x) := t^{-1}f(tx), \qquad 1 \le t \le T.$$

Then $f_1 = f$, $f_T = \mathrm{id}$, and $f_t^*\omega_0 = \omega_0$ for all $t \ge 1$. Denote

$$V_t := \left\{x \in \mathbb{R}^{2n} \,|\, tx \notin f(\mathbb{R}^{2n} \setminus tV)\right\}, \qquad W_t := \left\{x \in \mathbb{R}^{2n} \,|\, tx \notin f(\mathbb{R}^{2n} \setminus tW)\right\}.$$

These rather tortuous definitions are needed because f is not everywhere defined. In fact, V_t is the compact region in $\mathbb{R}^4$ with boundary $f_t(\partial V) = (1/t)f(t\partial V)$. Therefore V_t is compact, f_t is a diffeomorphism from $\mathbb{R}^{2n} \setminus V$ onto $\mathbb{R}^{2n} \setminus V_t$, W_t is an open neighbourhood of V_t, and f_t maps $\mathbb{R}^{2n} \setminus W$ diffeomorphically onto $\mathbb{R}^{2n} \setminus W_t$.

Define the vector fields $X_t \in \mathrm{Vect}(\mathbb{R}^{2n} \setminus V_t)$ by

$$\frac{d}{dt} f_t = X_t \circ f_t.$$

Since $f_t^*\omega_0 = \omega_0$ the 1-form $\alpha_t := \iota(X_t)\omega_0$ is closed for every t. Moreover, α_t vanishes near infinity. Since $\mathbb{R}^{2n} \setminus V_t$ is a connected open set with compact complement this implies that α_t is exact (even in the case $n = 1$). Hence there exists a smooth family of Hamiltonian functions $F_t \in C^\infty(\mathbb{R}^{2n} \setminus V_t)$ such that

$$\iota(X_t)\omega_0 = dF_t$$

and F_t vanishes near infinity. Now choose a smooth family of cutoff functions $\rho_t : \mathbb{R}^{2n} \to [0, 1]$ such that $\rho_t = 0$ in some neighbourhood of V_t and $\rho_t = 1$ on $\mathbb{R}^{2n} \setminus W_t$. Then the function $G_t := \rho_t F_t : \mathbb{R}^{2n} \to \mathbb{R}$ is well defined on all of $\mathbb{R}^{2n}$; it is equal to zero on V_t and has compact support (uniformly in t). Let g_t be the Hamiltonian isotopy generated by G_t via

$$\frac{d}{dt} g_t = Y_t \circ g_t, \qquad g_T = \mathrm{id}, \qquad \iota(Y_t)\omega_0 = dG_t.$$

Then $g := g_1$ is a symplectomorphism of $\mathbb{R}^{2n}$ and agrees with f on $\mathbb{R}^{2n} \setminus W$. This proves the lemma. □

In the following proof we denote by $D_r := \{z \in \mathbb{C} \mid |z| \le r\}$ the closed disc of radius r in $\mathbb{C}$ and by $\Delta_r := D_r \times D_r$ the closed polydisc of radius r in $\mathbb{C}^2 = \mathbb{R}^4$.

Proof of Theorem 9.4.2. Let R, r, ε be positive real numbers such that $R - \varepsilon > r$ and $V \subset \Delta_r$. Denote

$$M_R := \psi(\mathrm{int}(\Delta_R) \setminus V) \cup K = M \setminus \psi(\mathbb{R}^4 \setminus \mathrm{int}(\Delta_R)).$$

(If ψ were everywhere defined, M_R would just be the image of $\mathrm{int}(\Delta_R)$ under ψ.) Now construct a compact symplectic manifold $\widetilde{M}$ by replacing the interior $\mathrm{int}(\Delta_R)$, understood as a subset of $S^2 \times S^2$, by M_R. More precisely, choose an area form $\sigma \in \Omega^2(S^2)$ such that $\int_{S^2} \sigma = \pi R^2$, let $\iota : \mathrm{int}(D_R) \to S^2 \setminus \{z_*\}$ be an area preserving embedding, and define

$$\widetilde{M} := ((S^2 \times S^2) \setminus (\iota \times \iota)(V)) \cup M_R / \sim.$$

Here the equivalence relation identifies $M_R \setminus K$ with $(\iota \times \iota)(\mathrm{int}(\Delta_R) \setminus V)$ via

$$(\iota(z_1), \iota(z_2)) \sim \psi(z_1, z_2), \qquad (z_1, z_2) \in \mathrm{int}(\Delta_R) \setminus V.$$

Since $\iota \times \iota$ and ψ are symplectomorphisms, there is a unique symplectic form $\widetilde{\omega}$ on $\widetilde{M}$ which agrees with ω on $M_R \setminus K$ and with $\pi_1^*\sigma + \pi_2^*\sigma$ on $(S^2 \times S^2) \setminus (\iota \times \iota)(V)$.

It follows from Theorem 9.4.7 (ii) that there is a diffeomorphism

$$\widetilde{\psi} : S^2 \times S^2 \to \widetilde{M}$$

such that $\widetilde{\psi}$ is equal to the identity on $(S^2 \times S^2) \setminus (\iota \times \iota)(\Delta_{R-\varepsilon})$ and

$$\widetilde{\psi}^*\widetilde{\omega} = \pi_1^*\sigma + \pi_2^*\sigma.$$

Define the map $\psi' : \mathbb{R}^4 \to M$ by

$$\psi'(z_1, z_2) := \begin{cases} \widetilde{\psi}(\iota(z_1), \iota(z_2)), & \text{if } (z_1, z_2) \in \mathrm{int}(\Delta_R), \\ \psi(z_1, z_2), & \text{if } (z_1, z_2) \notin \Delta_{R-\varepsilon}. \end{cases}$$

Then ψ' is a well defined symplectomorphism and agrees with the original map ψ on $\mathbb{R}^4 \setminus \Delta_{R-\varepsilon}$.

Now let U be an open neighbourhood of K and denote $W := V \cup \psi^{-1}(U \setminus K)$ so that

$$\psi(\mathbb{R}^4 \setminus W) = M \setminus U.$$

Consider the symplectic embedding

$$f := (\psi')^{-1} \circ \psi : \mathbb{R}^4 \setminus V \to \mathbb{R}^4.$$

Since V is star-shaped, it follows from Lemma 9.4.10 that there is a symplectomorphism $g : \mathbb{R}^4 \to \mathbb{R}^4$ such that $g|_{\mathbb{R}^4 \setminus W} = f$. Define $\psi'' := \psi' \circ g : \mathbb{R}^4 \to M$. Then ψ'' is a symplectomorphism that agrees with ψ on $\mathbb{R}^4 \setminus W$. It maps $\mathbb{R}^4 \setminus W$ onto $M \setminus U$. This proves the theorem. □

With this in hand, we can now prove the last part of Theorem 9.4.1.

PROOF OF THEOREM 9.4.1 (III). Let $C \subset M$ be a symplectic submanifold that is diffeomorphic to the 2-sphere, has self-intersection number $C \cdot C = 1$, and has area $\int_C \omega = \pi$. Then the symplectic neighbourhood theorem asserts that there is an open neighbourhood $U \subset \mathbb{C}P^2$ of $\mathbb{C}P^1$ and an embedding $\iota : U \to M$ such that

$$\iota^*\omega = \omega_{\mathrm{FS}}, \qquad \iota(\mathbb{C}P^1) = C$$

(cf. [**277**, Theorem 3.30]). Here ω_{FS} denotes the Fubini-Study form on $\mathbb{C}P^2$ with respect to which $\mathbb{C}P^1$ has area π. Now $\mathbb{C}P^2 \setminus \mathbb{C}P^1$ is symplectomorphic to the open unit ball in $\mathbb{R}^4$ (see Exercise 9.4.11 below). Choose any such symplectomorphism

$$\psi : \mathrm{int}(B_1) \to \mathbb{C}P^2 \setminus \mathbb{C}P^1, \qquad \psi^*\omega_{\mathrm{FS}} = \omega_0.$$

Choose a constant $\varepsilon > 0$ such that $\psi(\mathrm{int}(B_1) \setminus B_{1-2\varepsilon}) \subset U$ and consider the symplectic manifold

$$\widetilde{M} := (M \setminus C) \cup (\mathbb{R}^4 \setminus B_{1-2\varepsilon})/\sim,$$

where the equivalence relation is given by

$$x \sim z \quad \Longleftrightarrow \quad x = \iota(\psi(z)), \quad x \in M \setminus C, \quad z \in \mathrm{int}(B_1) \setminus B_{1-2\varepsilon}.$$

By Theorem 9.4.2, there is a symplectomorphism $f : \mathbb{R}^4 \to \widetilde{M}$ which is equal to the identity on $\mathbb{R}^4 \setminus B_{1-\varepsilon}$. Thus f restricts to a symplectomorphism from $\mathrm{int}(B_1)$ to $M \setminus C$ which is equal to $\iota \circ \psi$ on $\mathrm{int}(B_1) \setminus B_{1-\varepsilon}$. Hence the symplectomorphism

$$M \to \mathbb{C}P^2 : x \mapsto \begin{cases} \psi(f^{-1}(x)), & \text{if } x \in M \setminus C, \\ \iota^{-1}(x), & \text{if } x \in M \setminus f(B_{1-\varepsilon}), \end{cases}$$

satisfies the requirements of the theorem. □

EXERCISE 9.4.11. Identify $\mathbb{C}P^1$ with the image of the embedding

$$\mathbb{C}P^1 \to \mathbb{C}P^2 : [z_0 : z_1] \mapsto [0 : z_0 : z_1].$$

Prove that an explicit example of a symplectomorphism $\psi : \mathrm{int}(B_1) \to \mathbb{C}P^2 \setminus \mathbb{C}P^1$ is the composition of the diffeomorphism

$$\mathrm{int}(B_1) \mapsto \mathbb{C}^2 : (\zeta_1, \zeta_2) \mapsto \left(\frac{\zeta_1}{\sqrt{1 - |\zeta_1|^2 - |\zeta_2|^2}}, \frac{\zeta_2}{\sqrt{1 - |\zeta_1|^2 - |\zeta_2|^2}} \right)$$

with the embedding $\mathbb{C}^2 \to \mathbb{C}P^2 : (z_1, z_2) \mapsto [1 : z_1 : z_2]$. *Hint:* Use the formula for ω_{FS} given in the proof of Proposition 9.3.3.

REMARK 9.4.12. In dimensions greater than four the mere existence of a J-holomorphic sphere gives no information because, by Gromov's symplectic embedding theorem, they are both too common and can disappear when J is perturbed; see Gromov[**161**, §3.4(B)] and Eliashberg–Mishachev [**102**]. The existence of a nontrivial Gromov–Witten invariant is more significant, though as yet the geometric and dynamical consequences of this condition are far from being fully understood. So far, most applications have come via properties of the small quantum cohomology ring, which, as we will see in Section 11.1, is obtained from the genus zero 3-point Gromov–Witten invariants.

Another potentially interesting condition is the nonvanishing of a Gromov–Witten invariant of the form

$$\mathrm{GW}^M_{A,k}(\mathrm{PD}(pt), a_2, \dots, a_k) \neq 0.$$

A manifold with this property is said to be **symplectically uniruled**. This condition has various geometric consequences. For example, it follows as in Section 9.3 that the symplectic area $\omega(A)$ of the class of a uniruling curve constraints the size of an embedded symplectic ball, and, as in Hofer–Viterbo [**182**] and Lu [**253**], the uniruling condition helps solve the Weinstein conjecture discussed in Remark 9.1.3 (iii). By Hu–Li–Ruan [**192**], the uniruled property is invariant under blowing up and blowing down, and so, in their language, is a birational symplectic invariant. One interesting open question is whether every monotone symplectic manifold is uniruled, as is the case for Kähler manifolds. (Here, unless we assume that $[\omega] = c_1$ on $H_2(M)$ rather than on $\pi_2(M)$, we should assume also that M is simply connected.) Another problem is to understand what geometric condition might force such a Gromov–Witten invariant to be nonzero. For example, is it enough that for every ω-tame J there is a J-holomorphic sphere through every point? By Theorem 9.4.1 and its extension to blow ups, this holds in dimension 4. In higher dimensions, the only relevant result concerns manifolds with S^1-action: it was shown in McDuff [**273**] that these are uniruled. Kollar and Ruan [**214, 192**] showed that every uniruled smooth projective variety is symplectically uniruled. However, a similar result is not known for the property of being rationally connected, which in the symplectic context means that some invariant $GW^M_{A,k}(\mathrm{PD}(pt), \mathrm{PD}(pt), a_3, \dots, a_k)$ is nonzero. For some very interesting recent work on this kind of question, see Voisin [**410**] and Z. Tian [**396**].

9.5. The group of symplectomorphisms

In this section we shall prove three results of Gromov [**160**] about the symplectomorphism groups of $S^2 \times S^2$ and $\mathbb{C}P^2$ and about the compactly supported symplectomorphisms of $\mathbb{R}^4$. We begin by stating them. Let $\sigma \in \Omega^2(S^2)$ be the standard area form with area one and denote by

$$\omega := \pi_1^*\sigma + \pi_2^*\sigma \in \Omega^2(S^2 \times S^2) \tag{9.5.1}$$

the standard product symplectic form on $S^2 \times S^2$. Since $S^2 \times S^2$ is simply connected, the identity component in the group of symplectomorphisms of $S^2 \times S^2$ is the group of Hamiltonian symplectomorphisms. Throughout we shall identify S^2 with the Riemann sphere $\mathbb{C} \cup \{\infty\}$.

THEOREM 9.5.1 (Gromov). *The group of symplectomorphisms of $S^2 \times S^2$ with the symplectic form (9.5.1) has two connected components. The identity component is homotopy equivalent to* $\mathrm{SO}(3) \times \mathrm{SO}(3)$.

THEOREM 9.5.2 (Gromov). *Let $V \subset \mathbb{R}^4$ be a star-shaped compact set and suppose that $0 \in \mathrm{int}(V)$. Then the group $\mathrm{Symp}_V(\mathbb{R}^4, \omega_0)$ of symplectomorphisms of $\mathbb{R}^4$ with the standard symplectic form ω_0 and with support in V is contractible.*

THEOREM 9.5.3 (Gromov). *The group of symplectomorphisms of $\mathbb{C}P^2$ with the Fubini-Study form ω_{FS} is homotopy equivalent to* $\mathrm{PU}(3)$.

REMARK 9.5.4. (i) Theorems 9.5.1 and 9.5.3 show that, in the cases $M = \mathbb{C}P^2$ and $M = S^2 \times S^2$ with both factors of equal area, a symplectomorphism of M is smoothly isotopic to the identity if and only if it is symplectically isotopic to the identity. By Abreu–McDuff [**5**], this result extends to $S^2 \times S^2$ and to the 1-point blowup of $\mathbb{C}P^2$ with arbitrary symplectic form. However, it is not yet known to hold even for all symplectic forms on the product of S^2 with an arbitrary Riemannn surface, though it does hold in a suitable stable range: cf. McDuff [**270**] and Buse [**55**]. It certainly does not hold for general symplectic 4-manifolds. For example, the group of compactly supported symplectomorphisms of the cotangent bundle T^*S^2 does not have this property by Seidel [**368**]. In [**362**] Seidel produces many examples of closed symplectic 4-manifolds that admit symplectomorphisms that are smoothly, but not symplectically, isotopic to the identity. These examples are squares of generalized Dehn twists τ_L along embedded Lagrangian spheres $L \subset M$. The proof that in most cases $\tau_L \circ \tau_L$ is not symplectically isotopic to the identity is based on Floer homology and goes beyond the scope of this book. (See Remark 12.6.1 for a few more details.) The case of the product $M = S^2 \times S^2$ is a notable exception. Here the antidiagonal $L \subset M$ is a Lagrangian sphere; however Theorem 9.5.1 shows that in this case the symplectomorphism $\tau_L \circ \tau_L$ *is* symplectically isotopic to the identity. The nontrivial elements of $\pi_0(\mathrm{Symp}(M, \omega))$ constructed by Seidel are **fragile** in the sense that they become symplectically isotopic to the identity when the cohomology class of the form ω is slightly perturbed. Seidel explained many other ideas about the symplectic mapping class group $\pi_0(\mathrm{Symp}(M, \omega))$ in [**372**]. Evidence supporting the idea that this group should be generated by generalized Dehn twists may be found in Auroux–Muñoz–Presas [**29**]. See also Li–Wu [**246**], and the references therein. There has been rather little other work on the properties of $\pi_0(\mathrm{Symp}(M, \omega))$ apart from some early results by Ruan [**339**] in the case when M is a del Pezzo surface or Fano 3-fold; see also Li–Li [**242**]. The recent paper by Evans [**108**] contains other results and references.

(ii) Already in [**160**] Gromov pointed out that the homotopy type of the group of symplectomorphisms of $S^2 \times S^2$ changes when one sphere gets larger than the other. He gave an indirect argument to show that in this case there is an element of infinite order in the fundamental group of the symplectomorphism group. It turned out that this element is homotopic to the loop formed by the circle action on the Hirzebruch surface $\mathbf{P}(L_2 \oplus \mathbb{C})$ that is induced by rotating the fibers. (For notation see Example 3.3.6.) As we shall see in Section 9.7, this forms the basis of the construction in [**254**] of nonisotopic symplectic forms on $S^2 \times S^2 \times \mathbb{T}^2$. It also stimulated much work by Abreu [**3**], Abreu–McDuff [**5**], and McDuff [**270**] on the structure of symplectomorphism groups of ruled surfaces.

In the case of ruled surfaces over S^2 the homotopy type changes whenever the ratio of the sizes of the spheres passes an integer. The rational homotopy type is now fully understood; Anjos [**13**] worked out the integral type when this ratio lies in the interval $(1, 2]$. Moreover, in joint work with Granja [**14**] she showed that in this case this group G has the homotopy type of the pushout of the following diagram of groups and group homomorphisms

$$\begin{array}{ccc} \mathrm{SO}(3) & \longrightarrow & S^1 \times \mathrm{SO}(3)\,, \\ {\scriptstyle\iota}\downarrow & & \\ \mathrm{SO}(3) \times \mathrm{SO}(3) & & \end{array}$$

where the vertical homomorphism ι identifies $\mathrm{SO}(3)$ with the diagonal subgroup of $\mathrm{SO}(3) \times \mathrm{SO}(3)$. Thus G is an amalgamated free product (in the homotopy theoretic sense) of $\mathrm{SO}(3) \times \mathrm{SO}(3)$ with $S^1 \times \mathrm{SO}(3)$ over $\mathrm{SO}(3)$. Note that the subgroup $S^1 \times \mathrm{SO}(3)$ appears in G as the automorphisms of the Hirzebruch structure J_2, just as $\mathrm{SO}(3) \times \mathrm{SO}(3)$ is the automorphism group of the product structure J_0. This line of thought culminated in the paper [**4**] by Abreu–Granja–Kitchloo.

One might hope to establish similar results when one blows up a few points of $S^2 \times S^2$. In a study of the group of symplectomorphisms of the one point blowup, Lalonde and Pinsonnault [**229**] found a very interesting interplay between the topology of this group and the ratios of the sizes of the blowup and the two spheres. See Seidel [**372**, Example 1.13] for a different perspective.

We now turn to the proofs of the three theorems stated at the beginning of this section. The logical order of these results is the order in which they are stated in that the proof of each theorem relies on the previous one. Note the parallel to the previous section.

Before proving Theorem 9.5.1 we introduce some notation. Abbreviate

$$M := S^2 \times S^2.$$

Since (M, ω) is monotone every ω-compatible almost complex structure $J \in \mathcal{J} := \mathcal{J}(M, \omega)$ satisfies the hypothesis of Proposition 9.4.4, namely, every nonconstant J-holomorphic sphere has positive Chern number. Denote by

$$A := [S^2 \times \{\mathrm{pt}\}], \qquad B := [\{\mathrm{pt}\} \times S^2]$$

the standard generators of $H_2(M; \mathbb{Z})$. Let

$$\mathrm{Symp}_0 := \mathrm{Symp}_0(M, \omega)$$

denote the group of symplectomorphisms that induce the identity on $H_2(M; \mathbb{Z})$. Since the two spheres can be interchanged, the quotient $\mathrm{Symp}(M, \omega)/\mathrm{Symp}_0(M, \omega)$ has two elements. We must show that the subgroup $\mathrm{Symp}_0(M, \omega)$ is homotopy equivalent to $\mathrm{SO}(3) \times \mathrm{SO}(3)$.

PROOF OF THEOREM 9.5.1. The group Symp_0 acts on $\mathcal{J}$ via

$$\mathrm{Symp}_0 \times \mathcal{J} \to \mathcal{J} : (\psi, J) \mapsto \psi_* J.$$

The stabilizer of the product complex structure $J_0 \in \mathcal{J}$ is $\mathrm{SO}(3) \times \mathrm{SO}(3) \subset \mathrm{Symp}_0$. Hence the action of Symp_0 on J_0 identifies the quotient $\mathrm{Symp}_0/\mathrm{SO}(3) \times \mathrm{SO}(3)$ with the orbit of J_0 via

$$\mathrm{Symp}_0/\mathrm{SO}(3) \times \mathrm{SO}(3) \to \mathcal{J} : \psi \mapsto \psi_* J_0.$$

This map is clearly not surjective; for example every element in its image is integrable. Our aim is to prove that this map has a left inverse.

The basic idea is to assign to each J the two embedded J-holomorphic spheres in M that pass through the point $(0,0) \in M$ and represent the classes A and B. Next we must choose parametrizations of these curves. Given two holomorphic parametrizations we obtain a diffeomorphism $f_J : M \to M$ as in the proof of Theorem 9.4.7 (i) which we can then isotop to a symplectomorphism (in a canonical way) using Moser isotopy. The resulting symplectomorphism depends on the parametrizations of the two J-holomorphic spheres that we have chosen and will not be well defined up to right multiplication by an element of $\mathrm{SO}(3) \times \mathrm{SO}(3)$. We could get around this by choosing balanced parametrizations as in Exercise 7.2.4, but then the coset we obtain from $\psi_* J_0$ may not be equal to the original coset $\psi \circ (\mathrm{SO}(3) \times \mathrm{SO}(3))$ though it will be homotopic to it in a canonical way. Hence this approach constructs a homotopy left inverse.

To construct a strict left inverse we observe that a symplectomorphism $\psi \in \mathrm{Symp}_0$ not only provides us with an almost complex structure $J := \psi_* J_0$ but also with parametrizations of the two relevant J-holomorphic curves, namely $z \mapsto \psi(z, w_*)$ and $w \mapsto \psi(z_*, w)$, where $\psi(z_*, w_*) = (0,0)$. This suggests that we include the parametrizations of the two curves in the parameter space $\mathcal{J}$. In other words, we are led to consider the universal moduli space

$$\mathcal{M}_0 := \{(u, v, J) \mid J \in \mathcal{J}, u \in \mathcal{M}(A; J),\ v \in \mathcal{M}(B; J),\ (0,0) \in \operatorname{im} u \cap \operatorname{im} v\}.$$

This space carries a free right action of the reparametrization group $\mathrm{G} \times \mathrm{G}$. By Proposition 9.4.4 with $p = 0$, the $(\mathrm{G} \times \mathrm{G})$-invariant projection $\mathcal{M}_0 \to \mathcal{J}$ is a submersion and the fibers are the $(\mathrm{G} \times \mathrm{G})$-orbits. In other words, $\mathcal{M}_0$ is a principal bundle with structure group $\mathrm{G} \times \mathrm{G}$ and contractible base $\mathcal{J}$. Hence it is homotopy equivalent to $\mathrm{G} \times \mathrm{G}$ and hence to $\mathrm{SO}(3) \times \mathrm{SO}(3)$.

Consider the commutative diagram

$$\begin{array}{ccccc}
\mathrm{SO}(3) \times \mathrm{SO}(3) & \longrightarrow & \mathrm{Symp}_0 & \longrightarrow & \mathrm{Symp}_0/\mathrm{SO}(3) \times \mathrm{SO}(3)\ , \\
\downarrow & & \mathcal{T}\downarrow & & \downarrow \\
\mathrm{G} \times \mathrm{G} & \longrightarrow & \mathcal{M}_0 & \longrightarrow & \mathcal{M}_0/\mathrm{G} \times \mathrm{G} = \mathcal{J}
\end{array}$$

where the embedding $\mathcal{T} : \mathrm{Symp}_0 \to \mathcal{M}_0$ is given by

$$\mathcal{T}(\psi) := (\psi(\cdot, w_*), \psi(z_*, \cdot), \psi_* J_0), \qquad (z_*, w_*) := \psi^{-1}(0,0).$$

Since $\mathcal{T}$ is equivariant under the right action of $\mathrm{SO}(3) \times \mathrm{SO}(3)$, it induces a continuous map from $\mathrm{Symp}_0/\mathrm{SO}(3) \times \mathrm{SO}(3)$ to the contractible space $\mathcal{M}_0/\mathrm{SO}(3) \times \mathrm{SO}(3)$. Hence the induced map $\mathcal{T} : \mathrm{SO}(3) \times \mathrm{SO}(3) \to \mathcal{M}_0$ is a homotopy equivalence. But $\mathcal{T}$ is injective on $\mathrm{SO}(3) \times \mathrm{SO}(3)$ with image

$$\mathcal{M}_{00} := \{(u, v, J_0) \mid u(z) = (\phi(z), 0),\ v(w) = (0, \phi'(w)),\ \phi, \phi' \in \mathrm{SO}(3)\}.$$

Therefore there is a homotopy $\Psi_t : \mathcal{M}_0 \to \mathcal{M}_0$, $0 \le t \le 1$, such that

$$\Psi_0(\mathcal{M}_0) \subset \mathcal{M}_{00}, \qquad \Psi_1 = \mathrm{id}, \qquad \Psi_t|_{\mathcal{M}_{00}} = \mathrm{id} \quad (0 \le t \le 1).$$

Therefore, if we construct a left inverse $\mathcal{F} : \mathcal{M}_0 \to \mathrm{Symp}_0$ of $\mathcal{T}$, the composition

$$\mathcal{F} \circ \Psi_t \circ \mathcal{T} : \mathrm{Symp}_0 \to \mathrm{Symp}_0$$

will provide the required deformation retraction of Symp_0 onto $\mathrm{SO}(3) \times \mathrm{SO}(3)$.

To construct $\mathcal{F} : \mathcal{M}_0 \to \mathrm{Symp}_0$ we proceed as follows. A triple $(u, v, J) \in \mathcal{M}_0$ determines two smooth maps

$$S^2 \to \mathcal{M}(A; J)/\mathrm{G} : w \mapsto [u_w], \qquad S^2 \to \mathcal{M}(B; J)/\mathrm{G} : z \mapsto [v_z],$$

such that $u(z) \in \mathrm{im}\,[v_z]$ and $v(w) \in \mathrm{im}\,[u_w]$. Define $\psi = \psi_{u,v,J} : M \to M$ by

$$\psi(z, w) := \mathrm{im}\,[u_w] \cap \mathrm{im}\,[v_z].$$

The proof of Theorem 9.4.7 shows that this map is a diffeomorphism and satisfies

$$\psi(z, w_*) = u(z), \qquad \psi(z_*, w) = v(w) \tag{9.5.2}$$

for all $z, w \in S^2$, where $z_*, w_* \in S^2$ are chosen such that $u(z_*) = v(w_*) = (0, 0)$. It also shows that the forms

$$\omega_t := \omega + t(\psi^*\omega - \omega)$$

are symplectic for $0 \le t \le 1$. Since ψ acts as the identity on $H^2(M; \mathbb{R})$, these forms are cohomologous. Hence we can deform ψ to a symplectomorphism by canonical Moser isotopy: by Remark 9.4.9, there is a path $f_t \in \mathrm{Diff}(M)$ depending only on ψ with $f_0 = \mathrm{id}$ and such that

$$f_1^*\psi^*\omega = \omega.$$

Define

$$\mathcal{F}(u, v, J) := \psi_{u,v,J} \circ f_1.$$

Note that $\mathcal{F}(u, v, J) = \psi_{u,v,J}$ whenever $\psi_{u,v,J}$ is a symplectomorphism. Hence it follows from (9.5.2) that $\mathcal{F} \circ \mathcal{T} = \mathrm{id} : \mathrm{Symp}_0 \to \mathrm{Symp}_0$, as required. □

REMARK 9.5.5. In Theorem 9.5.1 it suffices to prove that all the homotopy groups of $\mathrm{Symp}_0/\mathrm{SO}(3) \times \mathrm{SO}(3)$ vanish, because the quotient $\mathrm{Symp}_0/\mathrm{SO}(3) \times \mathrm{SO}(3)$ has the weak homotopy type of a (countable) CW complex.

Since there is no obvious reference for this classical result, we sketch the details of the argument here. By Milnor [**288**, Theorem 1], we need only show that Symp has the homotopy type of an absolute neighbourhood retract (ANR) when M is a closed symplectic manifold. By the Moser isotopy argument, $\mathrm{Symp}(M)$ is a deformation retract of an open subset of $\mathrm{Diff}(M)$. Therefore, it suffices to check that $\mathrm{Diff}(M)$ is an ANR. But, as pointed out by Kriegl–Michor[**218**], $\mathrm{Diff}(M)$ acts freely on the contractible space $\mathrm{Emb}(M, \mathcal{H})$ of smooth embeddings of M into a fixed Hilbert space $\mathcal{H}$. This space of embeddings is an open subset of a Fréchet space, and so is an ANR. Moreover, $\mathrm{Diff}(M)$ is the fiber of the projection from $\mathrm{Emb}(M, \mathcal{H})$ to the smooth Fréchet manifold $\mathcal{M}(M)$ consisting of all embedded submanifolds in $\mathcal{H}$ that are diffeomorphic to M. Hence $\mathrm{Diff}(M)$ in turn is a retract of an open subset in $\mathrm{Emb}(M, \mathcal{H})$, and so is itself an ANR.

In order to clarify the relation between the symplectomorphism groups of $S^2 \times S^2$ and $\mathbb{R}^4$, let us consider the groups $\mathrm{Symp}_{\Delta_r}(\mathbb{R}^{2n}, \omega_0)$ of symplectomorphisms with support in the closed polydisc $\Delta_r := D_r \times D_r$. We can identify this group with

$$\mathcal{G}_\varepsilon := \left\{\psi \in \mathrm{Symp}(S^2 \times S^2, \omega) \,|\, \psi|_{X_\varepsilon} = \mathrm{id}\right\},$$

where $X_\varepsilon := (S^2 \times D_\varepsilon) \cup (D_\varepsilon \times S^2)$ and $D_\varepsilon \subset \mathbb{C} \subset S^2$. The values of ε and r are immaterial.

LEMMA 9.5.6. *The inclusion*

$$\mathrm{Symp}_{\Delta_r}(\mathbb{R}^{2n}, \omega_0) \hookrightarrow \mathrm{Symp}_{\Delta_R}(\mathbb{R}^{2n}, \omega_0)$$

is contractible for $r < R$.

PROOF. We must prove that for $\delta < \varepsilon$ the inclusion $\mathcal{G}_\varepsilon \hookrightarrow \mathcal{G}_\delta$ is homotopic to a constant map. Consider the set

$$\mathcal{J}_\varepsilon := \left\{ J \in \mathcal{J}(S^2 \times S^2, \omega) \mid J|_{X_\varepsilon} = J_0 \right\}.$$

We shall construct a smooth map $\mathcal{F} : \mathcal{J}_\varepsilon \to \mathcal{G}_\delta$ such that $\mathcal{F}(\psi^* J_0) = \psi$ for $\psi \in \mathcal{G}_\varepsilon$. Then the result follows from the fact that $\mathcal{J}_\varepsilon$ is contractible.

To construct $\mathcal{F}$ we use the argument in the proof of Theorem 9.4.7 (ii). Since in the current situation X_ε is a neighbourhood of $(S^2 \times \{0\}) \cup (\{0\} \times S^2)$ we now take $z_* = w_* = 0$ and define u_*, v_* by

$$u_*(z) = (z, 0) \quad v_*(w) = (0, w).$$

Then for each $J \in \mathcal{J}_\varepsilon$ we define $\psi_J \in \operatorname{Diff}(S^2 \times S^2)$ as before. Thus

$$\psi_J(z, w) = \operatorname{im} u_w \cap \operatorname{im} v_z,$$

where $[u_w] \in \mathcal{M}(A; J)$ is defined by $(0, w) \in \operatorname{im} u_w$ and $[v_z] \in \mathcal{M}(B; J)$ is defined by $(z, 0) \in \operatorname{im} v_z$. This map has the form

$$\psi_J(z, w) = \begin{cases} (z, \phi_2(z, w)), & \text{if } z \in D_\varepsilon, \\ (\phi_1(z, w), w), & \text{if } w \in D_\varepsilon, \end{cases}$$

where $\phi_1(z, 0) = z$ and $\phi_2(0, w) = w$ for all $z, w \in S^2$. To make it the identity on X_δ we modify it by choosing once and for all a smooth map $\rho : S^2 \to S^2$ such that $\rho(z) = z$ for $z \notin D_\varepsilon$, $\rho = 0$ in a neighbourhood of D_δ, and $\det(d\rho(z)) \geq 0$ for $z \in S^2$. Then define $\psi'_J : M \to M$ by $\psi'_J := \psi_J$ on $M \setminus X_\varepsilon$ and on X_ε by

$$\psi'_J(z, w) := \begin{cases} (z, \phi_2(\rho(z), w)), & \text{if } z \in D_\varepsilon, \\ (\phi_1(z, \rho(w)), \mathrm{w}), & \text{if } w \in D_\varepsilon. \end{cases}$$

As before, $\psi'_J = \mathrm{id}$ in a neighbourhood of X_δ and the submanifolds $\{z\} \times S^2$ and $S^2 \times \{w\}$ are symplectic with respect to ${\psi'_J}^*\omega$ for all $z, w \in S^2$. Hence

$$\omega \wedge {\psi'_J}^*\omega > 0$$

(Exercise 9.4.8). Therefore we can use Moser isotopy. Namely, the 1-form β_J with

$$d^*\beta_J = 0, \qquad d\beta_J = {\psi'_J}^*\omega - \omega,$$

is exact on a neighbourhood of X_δ and therefore can be canonically modified to vanish on X_δ. As a result we find a diffeomorphism $f_J : M \to M$, depending only on ψ'_J, such that

$$f_J^* {\psi'_J}^* \omega = \omega, \qquad f_J|_{X_\delta} = \mathrm{id}.$$

The map $\mathcal{F} : \mathcal{J}_\varepsilon \to \mathcal{G}_\delta$ is now defined by

$$\mathcal{F}(J) := \psi'_J \circ f_J.$$

It satisfies $\mathcal{F}(\psi_* J_0) = \psi$ for every $\psi \in \mathcal{G}_\varepsilon$; in fact if $J = \psi_* J_0$ then $\psi_J = \psi$ is the identity on X_ε, and hence $\psi'_J = \psi_J$ and $f_J = \mathrm{id}$. This proves Lemma 9.5.6. □

PROOF OF THEOREM 9.5.2. This follows from Lemma 9.5.6. Choose real numbers $r > 0$ and $T > 1$ such that

$$\frac{1}{T} V \subset \Delta_r \subset \operatorname{int}(V),$$

and consider the family of maps $\mathcal{F}_t : \operatorname{Symp}_V(\mathbb{R}^4, \omega_0) \to \operatorname{Symp}_V(\mathbb{R}^4, \omega_0)$, $1 \leq t \leq T$, defined by

$$\mathcal{F}_t(\phi)(x) = \frac{1}{t} \phi(tx).$$

They satisfy $\mathcal{F}_1(\phi) = \phi$ for every $\phi \in \mathrm{Symp}_V(\mathbb{R}^4, \omega_0)$, $\mathcal{F}_t(\mathrm{id}) = \mathrm{id}$ for all t, and

$$\mathcal{F}_T(\mathrm{Symp}_V(\mathbb{R}^4, \omega)) \subset \mathrm{Symp}_{\Delta_r}(\mathbb{R}^4, \omega_0).$$

By Lemma 9.5.6, the inclusion $\mathrm{Symp}_{\Delta_r}(\mathbb{R}^4, \omega_0) \hookrightarrow \mathrm{Symp}_V(\mathbb{R}^4, \omega_0)$ is contractible. Hence we can extend $\mathcal{F}_t : \mathrm{Symp}_V(\mathbb{R}^4, \omega_0) \to \mathrm{Symp}_V(\mathbb{R}^4, \omega_0)$ to a homotopy on the interval $1 \le t \le T+1$ so that $\mathcal{F}_{T+1}(\psi) = \mathrm{id}$ for every $\psi \in \mathrm{Symp}_V(\mathbb{R}^4, \omega_0)$. Therefore $\mathrm{Symp}_V(\mathbb{R}^4, \omega_0)$ is contractible. This proves Theorem 9.5.2. □

Proof of Theorem 9.5.3. To illustrate the main point we prove first that the symplectomorphism group of $\mathbb{C}P^2$ is connected. Denote by $A := [\mathbb{C}P^1] \in H_2(\mathbb{C}P^2; \mathbb{Z})$ the standard generator, by $\omega_{\mathrm{FS}} \in \Omega^2(\mathbb{C}P^2)$ the Fubini–Study form with $\omega_{\mathrm{FS}}(A) = \pi$, and by $J_0 \in \mathcal{J} := \mathcal{J}(\mathbb{C}P^2, \omega_{\mathrm{FS}})$ the standard complex structure.

Let $\psi \in \mathrm{Symp}(\mathbb{C}P^2, \omega_{\mathrm{FS}})$ be given and choose a smooth homotopy of almost complex structures $[0,1] \to \mathcal{J} : t \mapsto J_t$ from J_0 to $J_1 = \psi_* J$. Then, by Corollary 9.4.5, there exists a smooth family of J_t-holomorphic spheres $u_t \in \mathcal{M}(A; J_t)$, unique up to reparametrization, such that

$$u_t(0) = [1:0:0], \quad u_t(\infty) = [0:1:0], \quad u_0(z) = [1:z:0], \quad u_1 = \psi \circ u_0.$$

Here we identify $S^2 \cong \mathbb{C} \cup \{\infty\}$. Since S^2 is simply connected, there is a smooth homotopy of symplectic vector bundle isomorphisms

$$\Psi_t : (u_0^* T\mathbb{C}P^2, J_0, \omega_{\mathrm{FS}}) \to (u_t^* T\mathbb{C}P^2, J_t, \omega_{\mathrm{FS}})$$

from $\Psi_0 := \mathrm{id}$ to $\Psi_1 := d\psi(u_0)$ that is compatible with the maps du_t in the sense that

$$\Psi_0 = \mathrm{id}, \quad \Psi_t du_0 = du_t, \quad \Psi_t J_0 = J_t \Psi_t, \quad \Psi_t^* \omega_{\mathrm{FS}} = \omega_{\mathrm{FS}}. \tag{9.5.3}$$

Call the triple $\{J_t, u_t, \Psi_t\}_{0 \le t \le 1}$ a **framed isotopy** of J-holomorphic curves. We wish to construct a symplectic isotopy $[0,1] \to \mathrm{Symp}(\mathbb{C}P^2, \omega_{\mathrm{FS}}) : t \mapsto \psi_t$, from the identity $\psi_0 = \mathrm{id}$ to $\psi_1 = \psi$, such that

$$\psi_t \circ u_0 = u_t, \qquad d\psi_t(u_0) = \Psi_t. \tag{9.5.4}$$

We shall look for an isotopy that is generated by a family H_t of Hamiltonian functions in the usual way:

$$\frac{d}{dt}\psi_t = X_t \circ \psi_t, \qquad \iota(X_t)\omega_{\mathrm{FS}} = dH_t. \tag{9.5.5}$$

Recall that the symplectic gradient X_t of H_t is related to its gradient with respect to the metric $\omega_{\mathrm{FS}}(\cdot, J_t \cdot)$ by the equation $\mathrm{grad} H_t = J_t X_t$.

The first condition in (9.5.4) translates into the condition

$$X_t(u_t(z)) = \partial_t \psi_t(u_0(z)) = \partial_t u_t(z)$$

for all $z \in S^2$. For the second condition we must control the derivatives of X_t in directions normal to the curve $\mathrm{im}\, u_t$. Indeed for every $v \in T_{u_0(z)}\mathbb{C}P^2$ we need

$$\begin{aligned}
\nabla_t\big(\Psi_t(z)v\big) &= \nabla_t\big(d\psi_t(u_0(z))v\big) \\
&= \nabla_{d\psi_t(u_0(z))v} X_t\big(\psi_t(u_0(z))\big) \\
&= \nabla_{\Psi_t(z)v} X_t\big(u_t(z)\big).
\end{aligned}$$

Here ∇ denotes the Levi-Civita connection of the metric $\omega(\cdot, J_t\cdot)$ and ∇_t denotes covariant differentiation with respect to time. For $\mathrm{grad}H_t = J_tX_t$ these conditions become

$$\begin{aligned} \mathrm{grad}H_t(u_t) &= J_t(u_t)\partial_t u_t, \\ \nabla_{\Psi_t v}(\mathrm{grad}H_t) &= J_t\nabla_t(\Psi_t v) + (\nabla_{\Psi_t(v)}J_t)\partial_t u_t, \end{aligned} \tag{9.5.6}$$

for $v \in u_0^*T\mathbb{C}P^2$. Because $\Psi_t du_0 = du_t$ the second equation reduces to the first when $v \in \mathrm{im}\, du_0(z)$. Thus these conditions are consistent and have a solution H_t.

Now let $\{\psi_t\}_{0\le t\le 1}$ be the Hamiltonian isotopy that that is generated by the Hamiltonian functions H_t via (9.5.5) and satisfies $\psi_1 = \psi$. It follows from (9.5.6) that ψ_t satisfies (9.5.4) for every t. The symplectomorphism ψ_0 need not be equal to the identity. However, it satisfies $\psi_0(x) = x$ and $d\psi_0(x)\xi = \xi$ for every $x \in \mathbb{C}P^1 := \mathrm{im}\, u_0$ and every $\xi \in T_x\mathbb{C}P^2$.

Two further steps are required to isotop ψ_0 to the identity. The first is to make ψ_0 equal to the identity in some neighbourhood of $\mathbb{C}P^1$. This can be done with a Moser isotopy. Namely one first chooses a smooth isotopy from ψ_0 to a diffeomorphism that is equal to the identity near $\mathbb{C}P^1$ and then isotops the pull-back differential forms. This works because ψ_0 already agrees with the identity up to first order along $\mathbb{C}P^1$ so that the isotopies can be taken to be C^1-small. The argument is the same as in the proof of the symplectic neighbourhood theorem (see [**277**, Theorem 3.30]) and we shall not repeat it here. The second step is to isotop the resulting symplectomorphism to the identity, keeping it fixed in a neighbourhood of $\mathbb{C}P^1$. That this is possible follows immediately from Theorem 9.5.2 and Exercise 9.4.11.

Thus we have proved that the symplectomorphism group of $\mathbb{C}P^2$ is connected. The proof that it retracts onto PU(3) is a parametrized version of the same argument. Because $\mathrm{Symp}(\mathbb{C}P^2, \omega_{\mathrm{FS}})$ has the weak homotopy type of a countable CW complex (see Remark 9.5.5 for details) it suffices to show that any smooth family $S \to \mathrm{Symp}(\mathbb{C}P^2, \omega_{\mathrm{FS}}) : s \mapsto \psi_s$ of symplectomorphisms that is parametrized by a compact smooth manifold S is homotopic to a smooth map $S \to \mathrm{PU}(3) : s \mapsto \phi_s$.

As a first step, we must find a suitable candidate for such a map $S \to \mathrm{PU}(3)$. For this we consider the point $x_0 := [1:0:0] = u_0(0) \in \mathbb{C}P^2$ and denote by

$$\mathrm{Sp}_0^+ \subset \mathrm{Sp}\,(T_{x_0}\mathbb{C}P^2, \omega_{\mathrm{FS}})$$

the set of symplectic automorphisms of the tangent space $T_{x_0}\mathbb{C}P^2$ that are symmetric and positive definite with respect to the standard inner product. Thus Sp_0^+ is the set of all automorphisms $\Psi_0 \in \mathrm{Sp}\,(T_{x_0}\mathbb{C}P^2, \omega_{\mathrm{FS}})$ that satisfy

$$\Psi_0^{-1} = -J_0\Psi_0 J_0, \qquad \omega_{\mathrm{FS}}(v, J_0\Psi_0 v) > 0$$

for $0 \ne v \in T_{x_0}\mathbb{C}P^2$. Note that Sp_0^+ is contractible (see [**277**]). Moreover, for every ω_{FS}-compatible complex automorphism J of $T_{x_0}\mathbb{C}P^2$ there is a unique $\Psi_0 \in \mathrm{Sp}_0^+$ such that $\Psi_0 J_0 = J\Psi_0$, namely the square root of $-JJ_0$. Hence there is a unique smooth map $S \to \mathrm{Sp}_0^+ : s \mapsto \Psi_{0s}$ such that

$$\Psi_{0s}J_0 = J_s\Psi_{0s}, \qquad J_s := \psi_{s*}J_0.$$

Because PU(3) is the isometry group of $\mathbb{C}P^2$, there is a unique map $S \to \mathrm{PU}(3) : s \mapsto \phi_s$ such that

$$\phi_s(x_s) = x_0, \qquad d\phi_s(x_s) = (\Psi_{0s})^{-1} d\psi_s(x_s)$$

for every $s \in S$, where $x_s := \psi_s^{-1}(x_0)$.

We now claim that the map

$$S \to \mathrm{Symp}(\mathbb{C}P^2, \omega_{\mathrm{FS}}) : s \mapsto \psi_s \circ \phi_s^{-1}$$

is contractible. To see this, let us denote by $\mathcal{M}_0$ the universal moduli space of all triples (J, u, Ψ), where $J \in \mathcal{J}$, $u : S^2 \to \mathbb{C}P^2$ is a J-holomorphic curve in the class A satisfying $u(0) = x_0$, and $\Psi : u_0^* T\mathbb{C}P^2 \to u^* T\mathbb{C}P^2$ is a framing such that

$$\Psi J_0 = J\Psi, \qquad \Psi^*\omega_{\mathrm{FS}} = \omega_{\mathrm{FS}}, \qquad \Psi du_0 = du, \qquad \Psi(0) \in \mathrm{Sp}_0^+.$$

The space $\mathcal{M}_0$ is contractible (see Exercise 9.5.7). Hence there exists a framed isotopy of J-holomorphic curves

$$S \times [0,1] \to \mathcal{M}_0 : (s,t) \mapsto (J_{s,t}, u_{s,t}, \Psi_{s,t})$$

such that

$$J_{s,0} = J_0, \qquad u_{s,0} = u_0, \qquad \Psi_{s,0} = \mathrm{id},$$
$$J_{s.1} = J_s, \qquad u_{s,1} = \psi_s \circ \phi_s^{-1} \circ u_0, \qquad \Psi_{s,1} = d(\psi_s \circ \phi_s^{-1})(u_0).$$

From this point on the argument proceeds exactly as in the proof of connectedness, except that in each of the three steps the isotopy depends on the additional parameter s. The remaining details are left as an exercise. This proves Theorem 9.5.3. □

We close this section with some exercises about $\mathbb{C}P^2$.

EXERCISE 9.5.7. Prove that the universal moduli space $\mathcal{M}_0$ in the proof of Theorem 9.5.3 is contractible.

EXERCISE 9.5.8. Let $J \in \mathcal{J}_\tau(\mathbb{C}P^2, \omega_{\mathrm{FS}})$ and denote $A := [\mathbb{C}P^1] \in H_2(\mathbb{C}P^2; \mathbb{Z})$. Prove that, for every $x \in \mathbb{C}P^2$ and every complex line $L_x \subset T_x\mathbb{C}P^2$, there exists a unique J-holomorphic curve $[u] \in \mathcal{M}(A; J)/\mathrm{G}$ that passes through x and is tangent to L_x. Deduce that, for every nonzero vector $\xi \in T_x\mathbb{C}P^2$ and every embedded J-holomorphic sphere $C \subset \mathbb{C}P^2$ in the class A such that $x \notin C$, there exists a unique J-holomorphic curve $u_\xi \in \mathcal{M}(A; J)$ such that

$$u_\xi(0) = x, \qquad du_\xi(0)1 = \xi, \qquad u_\xi(\infty) \in C.$$

Show that $u_{\lambda\xi}(z) = u_\xi(\lambda z)$ for $\lambda \in \mathbb{C}$. *Hint:* Use positivity of intersections (Theorem 2.6.3) to prove uniqueness. To prove existence use Corollary 9.4.5 and consider a sequence u^ν whose image passes through x and $\exp_x(2^{-\nu}\xi)$.

EXERCISE 9.5.9. (i) Another way of understanding the above proof (that is reminiscent of the approach taken in the proof of Theorem 9.5.1) is to consider the action of the group $\mathrm{Symp}(\mathbb{C}P^2, \omega_{\mathrm{FS}})$ on the space $\mathcal{CL}$ of **framed complex lines** in $\mathbb{C}P^2$. Here a framed line is a triple (J, u, Ψ), where $J \in \mathcal{J}(\mathbb{C}P^2, \omega_{\mathrm{FS}})$, $u : S^2 \to \mathbb{C}P^2$ is a J-holomorphic sphere in the class $A = [\mathbb{C}P^1]$, and $\Psi : u_0^* T\mathbb{C}P^2 \to u^* T\mathbb{C}P^2$ is a vector bundle isomorphism that satisfies

$$\Psi du_0 = du, \qquad \Psi^*\omega_{\mathrm{FS}} = \omega_{\mathrm{FS}}, \qquad \Psi^* J = J_0$$

(but not necessarily the normalization conditions $u(0) = x_0$ and $\Psi(0) \in \mathrm{Sp}_0^+$). Show that $\mathrm{Symp}(\mathbb{C}P^2, \omega_{\mathrm{FS}})$ acts freely on $\mathcal{CL}$ via

$$\psi_*(J, u, \Psi) := (\psi_* J, \psi \circ u, d\psi(u)\Psi).$$

Show further that the space $\mathcal{CL}$ is homotopy equivalent to $\mathrm{PU}(3)$. Construct a homotopy inverse of the map $\mathrm{Symp}(\mathbb{C}P^2, \omega_{\mathrm{FS}}) \to \mathcal{CL} : \psi \mapsto (\psi_* J_0, \psi \circ u_0, d\psi(u_0))$.

(ii) Define the space $\mathcal{SL}$ of **framed symplectic lines** in $\mathbb{C}P^2$ as the set of pairs (u, Ψ), where $u : S^2 \to \mathbb{C}P^2$ is a symplectic embedding in the class A and $\Psi : u_0^*T\mathbb{C}P^2 \to u^*T\mathbb{C}P^2$ is a vector bundle isomorphism that satisfies $\Psi du_0 = du$ and $\Psi^*\omega_{\mathrm{FS}} = \omega_{\mathrm{FS}}$. Show that $\mathrm{Symp}(\mathbb{C}P^2, \omega_{\mathrm{FS}})$ acts transitively on $\mathcal{SL}$, that the stabilizer of a point is contractible, and that the space $\mathcal{SL}$ is homotopy equivalent to PU(3).

EXERCISE 9.5.10. Show that the moduli space $\mathcal{M}(A; J)/\mathrm{G}$ is diffeomorphic to $\mathbb{C}P^2$ for every $J \in \mathcal{J}_\tau(\mathbb{C}P^2, \omega_{\mathrm{FS}})$ and that the subset $\{[v] \in \mathcal{M}(A; J)/\mathrm{G} \,|\, x \notin \mathrm{im}\, v\}$ is diffeomorphic to $\mathbb{R}^4$ for every $x \in \mathbb{C}P^2$. (Compare McKay [**282**] and Sikorav [**380**].) *Hint 1:* Show that

$$\mathcal{W} := \{(\lambda, [u]) \,|\, \lambda \in [0,1],\, u \in \mathcal{M}(A; J_\lambda)/\mathrm{G})\}$$

is a product cobordism for every smooth path $[0,1] \to \mathcal{J}_\tau(\mathbb{C}P^2, \omega_{\mathrm{FS}}) : \lambda \mapsto J_\lambda$. *Hint 2:* Show that, for every point $x \in \mathbb{C}P^2$, there is a unique dual line L_x which is orthogonal with respect to the Fubini–Study metric to every line passing through x. Show that if $x = [0:0:1]$ then L_x is the image of the embedding $\mathbb{C}P^1 \to \mathbb{C}P^2 : [z_0 : z_1] \mapsto [z_0 : z_1 : 0]$.

9.6. Hofer geometry

In [**177**] Hofer defined a metric on the group of compactly supported symplectomorphisms of Euclidean space, developing its properties by studying periodic orbits of relevant Hamiltonian flows. Lalonde–McDuff [**223, 224**] adopted a more geometric approach, relating the Hofer metric to certain questions concerning Hamiltonian fibrations. In this section we shall explain a beautiful extension of these ideas due to Polterovich. We shall keep the discussion rather narrowly focussed; for general background on this topic the reader can consult Polterovich [**330, 328**] or [**277**, Chapter 12]. We assume throughout that (M, ω) is a compact symplectic manifold. For the most important part of the discussion we shall consider the monotone case. However, many of the results explained below hold in greater generality.

Let $\mathbb{R} \to \mathrm{Symp}(M, \omega) : t \mapsto \psi_t = \psi_{t+1}$ be a loop of Hamiltonian symplectomorphisms. Then there exists a smooth Hamiltonian function $\mathbb{R} \to C^\infty(M) : t \mapsto H_t$ such that

$$\frac{d}{dt}\psi_t = X_{H_t} \circ \psi_t, \qquad H_t = H_{t+1},$$

for every $t \in \mathbb{R}$. For later purposes it will be useful to normalize the Hamiltonian functions such that they all have mean value zero:

$$\int_M H_t \omega^n = 0.$$

The **Hofer length** of the loop $t \mapsto \psi_t$ is defined by

$$\ell(\{\psi_t\}) := \int_0^1 \|H_t\|\, dt, \qquad \|H_t\| := \max_M H_t - \min_M H_t.$$

Note that this length does not depend on the above normalization.

EXAMPLE 9.6.1. Let $\omega_{\mathrm{FS}} \in \Omega^2(\mathbb{C}P^n)$ denote the Fubini–Study form with volume $\pi^n/n!$. Consider the loop

$$\mathbb{R}/\mathbb{Z} \to \mathrm{Symp}(\mathbb{C}P^n, \omega_{\mathrm{FS}}) : t \mapsto \psi_t^k,$$

defined by

$$\psi_t^k([z_0 : \cdots : z_n]) := [z_0 : e^{2\pi i t} z_1 : \cdots : e^{2\pi i t} z_k : z_{k+1} : \cdots : z_n].$$

This loop is generated by the time independent Hamiltonian function

$$H^k([z_0 : \cdots : z_n]) := \pi \left(\frac{k}{n+1} - \frac{|z_1|^2 + \cdots + |z_k|^2}{|z_0|^2 + \cdots + |z_n|^2} \right). \tag{9.6.1}$$

Since $\|H^k\| = \pi$ it follows that

$$\ell(\{\psi_t^k\}) = \pi.$$

Lalonde–McDuff proved in [**224**] that when $n = k = 1$ the loop $t \mapsto \psi_t$ cannot be shortened by a Hamiltonian deformation. Their argument uses embedded balls and a version of the nonsqueezing theorem for Hamiltonian fibrations over S^2. The purpose of the present section is to explain a different proof of this result due to Polterovich that uses the Seidel representation of Example 8.6.8 and works for all $1 \le k \le n$. We also explain some other geometrically defined invariants of a loop that turn out to coincide with its Hofer length.

In [**325, 326, 327**] Polterovich considered three invariants associated to a loop $\psi = \{\psi_t\}_{t \in \mathbb{R}/\mathbb{Z}}$ of Hamiltonian symplectomorphisms on M. The first is the minimal Hofer length of all Hamiltonian loops that are Hamiltonian isotopic to ψ:

$$\lambda(\psi) := \inf_{\psi' \sim \psi} \ell(\psi').$$

The second is the infimum of the Hofer norm of the curvature $R_{\widetilde{\omega}}$ over all connection forms on the Hamiltonian fibration $\widetilde{M}_\psi$, defined by (8.2.6):

$$\kappa(\psi) := \inf_{\widetilde{\omega} \in \mathcal{T}_0(\widetilde{M}_\psi)} \|R_{\widetilde{\omega}}\|.$$

Here we denote by $\mathcal{T}_0(\widetilde{M}_\psi)$ the space of all connection forms $\widetilde{\omega} \in \Omega^2(\widetilde{M}_\psi)$ as in Remark 8.2.11. The third is the **length of the nonsymplectic interval**:

$$\varepsilon(\psi) := \inf_{\widetilde{\omega}_c^{n+1} > 0} \int_{S^2} c\sigma - \sup_{\widetilde{\omega}_c^{n+1} < 0} \int_{S^2} c\sigma.$$

Here the infimum runs over all connection forms $\widetilde{\omega}_c = \widetilde{\omega} + \pi^*(c\sigma) \in \Omega^2(\widetilde{M}_\psi)$ such that $\widetilde{\omega} \in \mathcal{T}_0(\widetilde{M}_\psi)$, $c \in \Omega^0(S^2)$, and $\widetilde{\omega}_c^{n+1} > 0$ (or equivalently $c \circ \pi > R_{\widetilde{\omega}}$). The supremum is understood similarly.

REMARK 9.6.2. This length of the nonsymplectic interval is related to the *area* of a symplectic fiber bundle over S^2 as defined in at the end of Section 9.3. In the case at hand this area is given by

$$\operatorname{area}(\widetilde{M}_\psi, \widetilde{\omega}_c) := \frac{\operatorname{Vol}(\widetilde{M}_\psi, \widetilde{\omega}_c)}{\operatorname{Vol}(M, \omega)} = \int_{S^2} c\sigma.$$

Thus $\varepsilon(\psi) = \varepsilon^+(\psi) + \varepsilon^-(\psi)$, where $\varepsilon^+(\psi)$ is the minimal area of a connection form $\widetilde{\omega}$ on $\widetilde{M}_\psi$ and $\varepsilon^-(\psi) := \varepsilon^+(\psi^{-1})$. The other two invariants defined above also decompose into a sum of two parts since the total variation $\|R\|$ of any normalized function R can be written as $\|R\|_+ + \|R\|_-$, where $\|R\|_+ := \max_M R$ and $\|R\|_- := -\min_M R = \|-R\|_+$. In fact, Polterovich's results refer to these "one sided" invariants. For further details see Polterovich [**325, 326, 327**] and McDuff [**271**].

In [**325, 326, 327**] Polterovich proved that all three invariants are equal. We reproduce his proof of the following weaker statement.

PROPOSITION 9.6.3 (Polterovich). *Let (M,ω) be a compact symplectic manifold and $\psi = \{\psi_t\}_{t\in\mathbb{R}/\mathbb{Z}}$ be a loop of Hamiltonian symplectomorphisms of M. Then*

$$\lambda(\psi) \geq \kappa(\psi) \geq \varepsilon(\psi).$$

PROOF. First, observe that there is a connection form $\widetilde{\omega} \in \mathcal{T}(\widetilde{M}_\psi)$ such that

$$\|R_{\widetilde{\omega}}\| = \ell(\psi).$$

In the notation of Remark 8.2.11, an explicit formula is given by

$$F^{\pm}_{s,t} := 0, \qquad G^+_{s,t} := -\rho(s)H_t, \qquad G^-_{s,t} := -\rho(s)H_{-t}\circ\psi_{-t},$$

where $\rho : \mathbb{R} \to [0,1]$ is a smooth cutoff function which vanishes for $s \leq -1$, is equal to one for $s \geq 1$, and satisfies $\rho(s) + \rho(-s) \equiv 1$.

Second, observe that if two Hamiltonian loops ψ and ψ' are Hamiltonian isotopic then the corresponding Hamiltonian fibrations $\widetilde{M}_\psi$ and $\widetilde{M}_{\psi'}$ are (fiberwise) isomorphic. To see this choose any Hamiltonian isotopy

$$\mathbb{R}\times\mathbb{R}/\mathbb{Z} \to \mathrm{Symp}(M,\omega) : (s,t) \mapsto \psi_{s,t}$$

such that $\psi_{s,t} = \psi_t$ for $s \leq -1$ and $\psi_{s,t} = \psi'_t$ for $s \geq 1$. Now define the diffeomorphism $\Psi : \widetilde{M}_\psi \to \widetilde{M}_{\psi'}$ by

$$\Psi^+(s,t,x^+) := (s,t,\psi_{s,t}\circ\psi_t^{-1}(x^+))$$

on the plus-trivialization and by

$$\Psi^-(s,t,x^-) := (s,t,(\psi'_{-t})^{-1}\circ\psi_{-s,-t}(x^-))$$

on the minus-trivialization. Since $R_{\Psi^*\widetilde{\omega}'} = R_{\widetilde{\omega}'}\circ\Psi$ for every $\widetilde{\omega}' \in \mathcal{T}_0(\widetilde{M}_{\psi'})$ it follows that $\|R_{\Psi^*\widetilde{\omega}'}\| = \|R_{\widetilde{\omega}'}\|$. Since we have already found a connection form such that $\|R_{\widetilde{\omega}'}\| = \ell(\psi')$ this implies that $\kappa(\psi) \leq \lambda(\psi)$.

Third, we prove the inequality $\varepsilon(\psi) \leq \kappa(\psi)$. Fix a constant $\delta > 0$ and choose a connection form $\widetilde{\omega} \in \mathcal{T}_0(\widetilde{M}_\psi)$ such that

$$\|R_{\widetilde{\omega}}\| < \kappa(\psi) + \delta.$$

Then there are smooth functions $c, K : S^2 \to \mathbb{R}$ such that

$$c(z) < \min_{\pi^{-1}(z)} R_{\widetilde{\omega}}(z) \leq \max_{\pi^{-1}(z)} R_{\widetilde{\omega}} < K(z), \qquad \int_{S^2}(K-c)\sigma < \kappa(\psi) + \delta.$$

Denote $\widetilde{\omega}_c := \widetilde{\omega} + \pi^*(c\sigma)$ and $\widetilde{\omega}_K := \widetilde{\omega} + \pi^*(K\sigma)$. Then $\widetilde{\omega}_K^{n+1} > 0$ and $\widetilde{\omega}_c^{n+1} < 0$ and hence

$$\varepsilon(\psi) \leq \int_{S^2}(K-c)\sigma < \kappa(\psi) + \delta.$$

Since this holds for every $\delta > 0$ we deduce that

$$\varepsilon(\psi) \leq \kappa(\psi).$$

This proves Proposition 9.6.3. □

PROPOSITION 9.6.4 (Polterovich–Seidel). *Let $\mathbb{C}P^n \hookrightarrow \widetilde{M}^k \to S^2$ be the Hamiltonian fibration associated to the loop ψ^k of Example 9.6.1. Then, for $k = 1, \dots, n$, there is a homology class $\widetilde{A}^k \in H_2(\widetilde{M}^k; \mathbb{Z})$ such that*

$$\mathcal{S}(\psi^k) = \mathcal{S}_{\widetilde{A}^k}(\psi^k) = c^{n+1-k}, \qquad \int_{\widetilde{A}^k} \widetilde{\omega} = -\frac{\pi(n+1-k)}{n+1}, \tag{9.6.2}$$

for every connection form $\widetilde{\omega} \in \mathcal{T}_0(\widetilde{M}^k)$. Here $c \in H^2(\mathbb{C}P^n)$ denotes the positive generator and $\mathcal{S}(\psi) \in H^*(\mathbb{C}P^n)$ is defined by (8.6.4) in Example 8.6.8.

PROOF. The loop $\psi_t^k \in \mathrm{Symp}(\mathbb{C}P^n, \omega_{\mathrm{FS}})$ preserves the metric and is generated by a time independent Hamiltonian function $H^k : \mathbb{C}P^n \to \mathbb{R}$ (see equation (9.6.1)). Hence the standard complex structure J_0 defines a vertical almost complex structure on $\widetilde{M}^k$. In the notation of Remark 8.2.11, an example of a connection form $\widetilde{\omega} \in \mathcal{T}_0(\widetilde{M}^k)$, is given by the formulas

$$F^{\pm} := 0, \qquad G^{\pm}_{s,t} := -\rho(s)H, \tag{9.6.3}$$

where $\rho : \mathbb{R} \to [0, 1]$ is a smooth cutoff function such that $\rho(s) = 0$ for $s \le -1$, $\rho(s) = 1$ for $s \ge 1$, and $\dot{\rho}(s) \ge 0$ and $\rho(s) + \rho(-s) = 1$ for all s. Note that

$$R_{\widetilde{\omega}}\sigma = -\dot{\rho}(s)H \, ds \wedge dt.$$

Denote by $\widetilde{J}$ the almost complex structure on $\widetilde{M}^k$ determined by J_0 and $\widetilde{\omega}$, and by $\mathcal{M}(\widetilde{A}; J_0, \widetilde{\omega})$ the space of $\widetilde{J}$-holomorphic sections of the Hamiltonian fibration $\widetilde{M}^k \to S^2$.

Now consider the critical manifold

$$Z^k := \{[0 : z_1 : \cdots : z_k : 0 : \cdots : 0] \,|\, (z_1, \dots, z_k) \in \mathbb{C}^k \setminus \{0\}\} \subset \mathbb{C}P^n.$$

The function H^k attains its minimum $\pi(k - n - 1)/(n + 1)$ on this submanifold. Every constant map $u^{\pm}(s, t) \equiv x \in Z^k$ defines a horizontal section $\widetilde{u} : S^2 \to \widetilde{M}^k$ and hence is $\widetilde{J}$-holomorphic. Let $\widetilde{A}^k \in H_2(\widetilde{M}^k; \mathbb{Z})$ be the homology class of these constant sections $\widetilde{u}$. Then, by Lemma 8.2.9,

$$\int_{\widetilde{A}^k} \widetilde{\omega} = E^{\mathrm{Vert}}(\widetilde{u}) - \int_{S^2} (R_{\widetilde{\omega}} \circ \widetilde{u})\sigma = \int_0^1 \int_{-\infty}^{\infty} \dot{\rho}(s)H(x) \, dsdt = \frac{\pi(k - n - 1)}{n + 1}.$$

This shows that the second equation in (9.6.2) holds for some, and hence every, connection form $\widetilde{\omega} \in \mathcal{T}_0(\widetilde{M}^k)$.

Examining the vertical tangent bundle along a horizontal solution with values in Z^k we find that it splits as a direct sum of complex line bundles such that $k - 1$ of them have Chern number zero (these form the normal bundle to the section in directions tangent to Z_k) and $n + 1 - k$ have Chern number -1. So the vertical first Chern number is $c_1^{\mathrm{Vert}}(\widetilde{A}^k) = k - 1 - n$. Moreover, the linearized operator is surjective for each of these constant sections. We prove that there cannot be any other $\widetilde{J}$-holomorphic section in this homology class. Namely, if

$\widetilde{u} = (u^+, u^-) \in \mathcal{M}(\widetilde{A}^k; J_0, \widetilde{\omega})$ then it follows from Lemma 8.2.9 that

$$\begin{aligned} 0 &\le E^{\text{Vert}}(\widetilde{u}) \\ &= \int_{S^2} \widetilde{u}^*\widetilde{\omega} + \int_{S^2} (R_{\widetilde{\omega}} \circ \widetilde{u})\sigma \\ &= \frac{\pi(k-n-1)}{n+1} - \int_0^1 \int_{-\infty}^{\infty} \dot{\rho}(s) H^k(u^{\pm}(s,t))\, dsdt \\ &\le \frac{\pi(k-n-1)}{n+1} - \int_0^1 \int_{-\infty}^{\infty} \dot{\rho}(s) \min_{\mathbb{C}P^n} H^k\, dsdt \\ &= 0. \end{aligned}$$

This implies equality, and so every $\widetilde{J}$-holomorphic curve representing the class $\widetilde{A}^k$ must be horizontal and hence be one of the constant sections already considered. (The other horizontal sections correspond to the minima of H. They have different Chern numbers and hence do not represent the class $\widetilde{A}^k$.) Now the evaluation map $\mathrm{ev}_0 : \mathcal{M}(\widetilde{A}^k; J_0, \widetilde{\omega}) \to \mathbb{C}P^n$ is evidently a diffeomorphism onto Z^k. Since $Z^k \subset \mathbb{C}P^n$ is a projective subspace of complex dimension $k-1$ it follows that

$$\mathrm{GW}^{\widetilde{M}^k,\{1\}}_{\widetilde{A}^k,1}(c^{k-1}) = 1, \qquad \mathcal{S}_{\widetilde{A}^k}(\psi^k) = c^{n+1-k}.$$

To complete the proof of the lemma we must show that

$$\widetilde{A} \neq \widetilde{A}^k \qquad \Longrightarrow \qquad \mathcal{S}_{\widetilde{A}}(\psi^k) = 0$$

To see this note that if $\mathcal{S}_{\widetilde{A}}(\psi^k) \neq 0$ then $\widetilde{A}$ must satisfy $\pi_*\widetilde{A} = [S^2]$ as well as

$$-n \le c_1^{\text{Vert}}(\widetilde{A}) \le 0.$$

The only such class is $\widetilde{A}^k$ whenever $1 \le k \le n$. This proves Proposition 9.6.4. □

THEOREM 9.6.5 (Polterovich). *Let $\psi^k = \{\psi_t^k\}_{t\in\mathbb{R}/\mathbb{Z}}$ be the loop discussed in Example 9.6.1. If $k \in \{1, \dots, n\}$ then*

$$\lambda(\psi^k) = \kappa(\psi^k) = \varepsilon(\psi^k) = \pi.$$

PROOF. By Proposition 9.6.3, we have $\varepsilon(\psi^k) \le \kappa(\psi^k) \le \lambda(\psi^k) \le \pi$. Thus it remains to prove the inequality

$$\varepsilon(\psi^k) \ge \pi.$$

To see this let $\widetilde{\omega} \in \mathcal{T}_0(\widetilde{M}^k)$ and $c \in \Omega^0(S^2)$ such that

$$\widetilde{\omega}_c^{n+1} > 0, \qquad \widetilde{\omega}_c := \widetilde{\omega} + c\sigma.$$

Let $\widetilde{A}^k \in H_2(\widetilde{M}^k; \mathbb{Z})$ be the homology class in Proposition 9.6.4 and fix a point $\mathbf{w} \in S^2$. By Theorem 8.5.3, the set of regular pairs $\mathcal{JH}_{\text{reg}}(\widetilde{M}^k, \pi, \widetilde{\omega}; \mathbf{w})$ is dense in $\mathcal{J}^{\text{Vert}} \times \mathcal{H}$ with respect to the C^∞-topology. Hence, by an arbitrarily small exact perturbation of $\widetilde{\omega}$, we may assume that there is a vertical almost complex structure $J \in \mathcal{J}^{\text{Vert}}(\widetilde{M}^k, \pi, \widetilde{\omega})$ such that the pair $(J, 0)$ is regular for $\mathbf{w}$ (in the sense of Definition 8.5.2) and that the condition $\widetilde{\omega}_c^{n+1} > 0$ is still satisfied. Denote by $\mathcal{M}(\widetilde{A}^k; J, \widetilde{\omega})$ the moduli space of $\widetilde{J}$-holomorphic sections of $\widetilde{M}^k$ representing the class $\widetilde{A}^k$, where $\widetilde{J}$ denotes the almost complex structure on $\widetilde{M}^k$ induced by J and $\widetilde{\omega}$ (see Lemma 8.2.8). Then, by Theorem 8.5.1, the evaluation map

$$\widetilde{\mathrm{ev}}_{\mathrm{w}} : \mathcal{M}(\widetilde{A}; J, \widetilde{\omega}) \to \mathbb{C}P^n, \qquad \widetilde{\mathrm{ev}}_{\mathrm{w}}(\widetilde{u}) := \widetilde{\iota}_{\mathrm{w}}^{-1}(\widetilde{u}(\mathrm{w})),$$

is a pseudocycle and, by Proposition 9.6.4, it represents a nonzero homology class in $\mathbb{C}P^n$. Hence $\mathcal{M}(\widetilde{A}^k; J, \widetilde{\omega}) \neq \emptyset$. Since $\widetilde{\omega}_c$ is a symplectic form on $\widetilde{M}^k$, this implies

$$0 \leq \langle [\widetilde{\omega}_c], \widetilde{A}^k \rangle = \int_{S^2} c\sigma - \frac{\pi(n+1-k)}{n+1}.$$

Here the second equation follows from (9.6.2). Thus we have proved that

$$\inf_{\substack{\widetilde{\omega} \in \mathcal{T}_0(\widetilde{M}^k), c \in \Omega^0(S^2) \\ \widetilde{\omega}_c^{n+1} > 0}} \int_{S^2} c\sigma \geq \frac{\pi(n+1-k)}{n+1}.$$

Since the loop $t \mapsto \psi^k_{-t}$ is Hamiltonian isotopic to $t \mapsto \psi_t^{n+1-k}$, it follows that

$$-\sup_{\substack{\widetilde{\omega} \in \mathcal{T}_0(\widetilde{M}^k), c \in \Omega^0(S^2) \\ \widetilde{\omega}_c^{n+1} < 0}} \int_{S^2} c\sigma = \inf_{\substack{\widetilde{\omega} \in \mathcal{T}_0(\widetilde{M}^{n+1-k}), c \in \Omega^0(S^2) \\ \widetilde{\omega}_c^{n+1} > 0}} \int_{S^2} c\sigma \geq \frac{\pi k}{n+1}.$$

Combining these two inequalities we obtain $\varepsilon(\psi^k) \geq \pi$ as claimed. This proves Theorem 9.6.5. □

REMARK 9.6.6. (i) The Hofer norm gives rise to a bi-invariant Finsler metric on $\mathrm{Ham}(M)$. (Very often, by slight abuse of language, the distance $d(\mathrm{id}, \phi)$ in this metric from the identity element to $\phi \in \mathrm{Ham}(M)$ is called the **Hofer norm** of ϕ.) Building on work by Ostrover–Wagner [**317**], Buhovsky and Ostrover [**51**] recently proved that it is the unique such metric that is C^∞ continuous. However, there are other bi-invariant metrics on $\mathrm{Ham}(M, \omega)$ that are not of Finsler type; cf. Exercise 12.4.8 and Viterbo [**404**].

(ii) In [**61**] Chekanov extended the Hofer metric to spaces of Lagrangian submanifolds; he shows that if (P, ω) is a closed symplectic manifold and $\mathcal{L}(P, \omega, L)$ is the orbit of a given Lagrangian submanifold L under the action of the Hamiltonian group $\mathrm{Ham}(P, \omega)$ then one can take the distance between two Lagrangian submanifolds L_0, L_1 to be the infimum of the lengths of the paths $\{\phi_t\}_{t \in [0,1]}$ in $\mathrm{Ham}(P, \omega)$ such that $\phi_0 = \mathrm{id}$ and $\phi_1(L_0) = L_1$.

(iii) Polterovich's approach to estimating the length of Hamiltonian loops also extends to the Lagrangian case. In [**11**] Akveld and Salamon defined analogous invariants for loops of Lagrangian submanifolds by considering symplectic fibrations over the disc with Lagrangian boundary conditions. In favourable cases the relative Gromov–Witten invariants give rise to lower bounds for the Hofer length in a given Hamiltonian isotopy class of Lagrangian loops. As a result one can obtain explicit length estimates, in the spirit of Theorem 9.6.5, for loops of real projective spaces in $\mathbb{C}P^n$.

(iv) By replacing a symplectomorphism of (M, ω) by its graph one can embed the Hamiltonian group into the space of embedded Lagrangian submanifolds of $(M \times M, \pi_2^*\omega - \pi_1^*\omega)$. A beautiful recent example due to Ostrover [**315**] shows that the inclusion of $\mathrm{Ham}(M, \omega)$ into $\mathcal{L}(M \times M, \pi_2^*\omega - \pi_1^*\omega, \mathrm{diag})$ can severely distort the metric: he constructs an example of a set of infinite diameter in $\mathrm{Ham}(\mathbb{T}^{2n}, \omega_0)$ whose image in the space of Lagrangians is bounded. Nevertheless, the results of Akveld–Salamon [**11**] show that Polterovich's length estimates for loops are sufficiently robust to survive in the Lagrangian setting at least in some cases. Usher [**400**] describes other situations in which subsets in $\mathrm{Ham}(M)$ are not metrically distorted when embedded into the space of Lagrangians.

(v) The argument in the proof of Proposition 9.6.4 has been generalized in McDuff–Tolman [**279**] to many other circle actions; some of the consequences are mentioned in Examples 11.4.8 and 11.4.11. These results show that Hamiltonian circle actions that are semifree in the sense that the stabilizer of each point is either the trivial group or the whole circle minimize Hofer length among the set of all homotopic loops. Similarly, paths $\{\phi_t^H\}_{0\le t\le T}$ in the Hamiltonian group that are generated by time independent Hamiltonians are absolutely length minimizing for sufficiently small $T > 0$: see McDuff [**271**]. These proofs identify the Hofer norm with certain invariants of the action spectrum $\mathrm{Spec}(H)$ of Section 9.1.[3] The properties of $\mathrm{Spec}(H)$ are best investigated in the context of Floer theory as in Section 12.4.

(vi) Questions about the topology, geometry and general properties of the group of Hamiltonian symplectomorphisms $\mathrm{Ham}(M,\omega)$ are a subject of very active research. One interesting open question is whether this group always has infinite diameter with respect to the Hofer metric. For recent results on this question see Usher [**400**] and McDuff [**274**]. Another set of questions concerns the existence and properties of quasimorphisms on $\mathrm{Ham}(M,\omega)$. Some of these (such as those in [**105**]) are constructed from the spectral invariants discussed in Section 12.4, while others come from other differential geometric constructions: see, for example, Py [**332**] and Shelukhin [**374**]. As shown by the recent work of Polterovich and his collaborators [**107, 52, 331**], quasimorphisms have very interesting connections with quantum physics and lead to new symplectic rigidity phenomena. Work by Müller and Oh [**308**] has given rise to new approaches to C^0-symplectic topology, illustrated in Viterbo [**407**] and Buhovsky–Seyfaddini [**53**]. For further questions and references, see Polterovich [**330**] or McDuff [**272**].

9.7. Distinguishing symplectic structures

We end this chapter by discussing situations in which J-holomorphic spheres have been used to distinguish symplectic manifolds. There are fewer such examples than one might expect. One problem is the difficulty of calculating the Gromov–Witten invariants. (This is easier now that the symplectic sum formula is in hand: see Ionel–Parker [**200, 201**] and Li–Ruan [**238**].) A more fundamental problem is that the behaviour of spheres has often not been relevant to the questions under study.

One basic fact is that in four dimensions, the work of Taubes [**390, 391, 392, 393**] and Ionel–Parker [**198**] shows that the most elementary Gromov–Witten invariants (namely those that count the number of embedded J-holomorphic curves of arbitrary genus through an appropriate number of points) contain the same information as the Seiberg–Witten invariants and so depend only on the smooth structure of M. Therefore in dimension four we cannot use Gromov–Witten invariants to distinguish between different symplectic structures on the same smooth manifold. Moreover, although homeomorphic 4-manifolds become diffeomorphic when one takes the product with S^2, we will see in Example 9.7.1 below that differences in Gromov–Witten invariants persist under this operation. This, together with Donaldson's work [**85**] on Lefschetz pencils, is evidence that, in contrast to the smooth case, the topology of symplectic manifolds does not get easier to understand

[3]Although in general the Hofer and spectral norms differ, they coincide in this special case.

as the dimension increases. In fact, Donaldson conjectured that two symplectic 4-manifolds (M, ω) and (M', ω') are diffeomorphic if and only if the 6-manifolds $M \times S^2$ and $M' \times S^2$ with their product symplectic structures are deformation equivalent. (It is interesting to note that Smith in [**382**] has constructed examples showing that the "only if" part of this conjecture does not hold if the 2-sphere is replaced by any symplectic manifold (such as the 2-torus or $\mathbb{C}P^2$) whose first Chern class is divisible by an integer $N > 2$.)

Therefore much recent work on understanding the structure of symplectic manifolds has concentrated on understanding the 4-dimensional case,[4] and is subsumed into the larger problem of understanding smooth ones. Much effort has gone into constructing interesting examples. For early work in this direction, see Gompf [**154**], Fintushel–Stern [**110**], Szabo [**387**] and the references therein. More recently, there have been many attempts to find exotic smooth or symplectic structures on manifolds homeomorphic to k-point blowups of $\mathbb{C}P^2$. The minimal value of k for which such structures exist is still unknown at the time of writing, with $k = 2$ being the current record: see Akhmedov–Park [**10**], Baldridge–Kirk [**30**] and Fintushel–Stern [**111**]. Another important task (the **geography problem**) is to investigate the range of the traditional invariants given by the characteristic numbers of the ω-tame almost complex structures on a symplectic 4-manifold (M, ω) with given fundamental group.

In any case, the genus zero Gromov–Witten invariants that are considered in this book give no new information on such questions because, by Li–Liu [**240**], they are nonzero only for blowups of rational and ruled manifolds (with their standard smooth structures). So far, the only invariant that has been used to detect differences between symplectic forms on a given smooth 4-manifold X is the Chern class of the corresponding tamed almost complex structures: see Remark 9.4.3 (iv). It is possible that developments stemming from Donaldson–Smith's paper [**89**] will make higher genus J-holomorphic curves more useful for understanding symplectic 4-manifolds. One possible way to detect more subtle differences between symplectic 4-manifolds would be to think of them as branched covers of $\mathbb{C}P^2$ as in the work of Auroux–Katzarkov [**28**]: see the survey by Auroux [**27**].

This section describes just two examples, both in dimension six. They can be extended to higher dimensions by taking products with copies of S^2. We first summarize some work by Ruan which surely influenced Donaldson in making his conjecture. Secondly we explain some work by McDuff which constructs two cohomologous symplectic forms that can be joined by a deformation but not by an isotopy. Here the invariant uses the homotopy, rather than homology, information carried by the evaluation map. See Seidel [**373**] for a radically new approach to this question using properties of Fukaya categories.

EXAMPLE 9.7.1 (Nondeformation equivalent 6-manifolds). This example is due to Ruan [**340**] and is just one of many. Let X be $\mathbb{C}P^2$ with 8 points blown up, and Y be Barlow's surface. These manifolds are simply connected and have the same intersection form. Hence they are homeomorphic, but they are not diffeomorphic. By results of Wall, the stabilized manifolds $X' := X \times S^2$ and $Y' := Y \times S^2$ are diffeomorphic. Moreover, each of them is Kähler, and one can choose the

[4]There are beginning to be interesting new constructions in 6 dimensions; see Fine–Panov [**109**].

diffeomorphism

$$\psi : X' \to Y'$$

so that it preserves the Chern classes of the respective complex structures. The latter statement implies that ψ takes the product complex structure J_X on X' to a structure $\psi_* J_X$ which is homotopic to the product complex structure J_Y on Y'. Moreover an easy argument using cup products shows that $\psi_*[X \times \mathrm{pt}] = [Y \times \mathrm{pt}]$. This can be used to show that the homomorphism $\psi_* : H_2(X') \to H_2(Y')$ identifies the subset $H_2(X) \subset H_2(X')$ with $H_2(Y) \subset H_2(Y')$ and that the resulting isomorphism $H_2(X) \to H_2(Y)$ preserves the intersection form.

Now, on any Kähler manifold, the set of Kähler forms is a convex cone, and so any two such forms can be joined by a deformation. Hence the Kähler manifolds X', Y' carry symplectic forms $\omega_{X'}$, $\omega_{Y'}$ that are well defined up to deformation, and it is natural to ask whether the pairs $(X', \omega_{X'})$, $(Y', \omega_{Y'})$ are deformation equivalent. Ruan shows that they are not, by proving that the Gromov–Witten invariant

$$\mathrm{GW}^{X'}_{E',1}(\mathrm{PD}(E')) = -1,$$

where $E' \in H_2(X'; \mathbb{Z})$ is the image of the class $E \in H_2(X; \mathbb{Z})$ of one of the blown up points in X under the homomorphism $H_2(X; \mathbb{Z}) \to H_2(X'; \mathbb{Z})$, while the corresponding invariant for $F' := \psi_* E'$ vanishes:

$$\mathrm{GW}^{Y'}_{F',1}(\mathrm{PD}(F')) = 0$$

Since Gromov–Witten invariants do not change under deformation equivalence, this shows that $(X', \omega_{X'})$ and $(Y', \omega_{Y'})$ are inequivalent.

It remains to calculate the two Gromov–Witten invariants. Since $c_1(E') = 1$ and $\dim X' = 2n = 6$, we have

$$\dim \mathcal{M}_{0,1}(E') = 2n + 2c_1(E') + 2 - 6 = 4 = 2n - \deg(E'),$$

and so the dimension requirement for the definition of $\mathrm{GW}^{X'}_{E',1}(\mathrm{PD}(E'))$ is satisfied. Moreover, there is exactly one holomorphic sphere C in X which represents E. Hence the E'-curves in $X' = X \times S^2$ are precisely the curves $C \times \{z\}$ for $z \in S^2$. Hence the image of the evaluation map $\mathrm{ev} : \mathcal{M}_{0,1}(E'; J_X) \to X'$ is the submanifold $C \times S^2 \subset X'$. Therefore $\mathrm{ev} \cdot E' = -1$, which gives

$$\mathrm{GW}^{X'}_{E',1}(\mathrm{PD}(E')) = -1.$$

However, the Barlow surface Y is minimal and so has no embedded holomorphic spheres with self-intersection number -1. Moreover, by the above discussion, the homology class

$$F' := \psi_* E' \in H_2(Y')$$

is the image of a class $F \in H_2(Y)$, with self-intersection number $F \cdot F = -1$, under the homomorphism $H_2(Y) \to H_2(Y')$. Since J_Y is homotopic to $\psi_* J_X$, it follows that $c_1(F) = 1$. Now let us fix an ω_Y-compatible almost complex structure on Y such that every holomorphic sphere has positive Chern number and denote by J_Y the corresponding product structure on $Y' = Y \times S^2$. We claim that there are no J_Y-holomorphic stable maps of genus zero in Y' representing the class F'. Namely, for every component of such a stable map the composition with the projection $Y' \to S^2$ has nonnegative degree. Since the image of F' under the induced map on homology $H_2(Y') \to H_2(S^2)$ is zero, it follows that every component of our stable map projects to a point in S^2. Hence there would have to be a stable map in Y representing the class F. Since each component of such a stable map has positive Chern number, the stable map can only have one component with Chern number one and self-intersection number minus one. By the adjunction inequality, such a holomorphic sphere would have to be an embedding, in contradiction to the fact that Y is minimal. Thus F' cannot be represented by a J_Y-holomorphic stable map and hence the Gromov–Witten invariant $\mathrm{GW}^{Y'}_{F',1}$ (for the symplectic form $\omega_{Y'}$) vanishes.

Our second example is based on the properties of the following loop of diffeomorphisms of $S^2 \times S^2$: for $t \in \mathbb{R}/\mathbb{Z}$ we define $\psi_t : S^2 \times S^2 \to S^2 \times S^2$ by setting

$$\psi_t(z, w) = (z, \Phi_{z,t}(w)), \tag{9.7.1}$$

where $\Phi_{z,t}$ denotes the rotation through angle $2\pi t$ of the unit sphere $S^2 \subset \mathbb{R}^3$ about the axis through the points $z, -z$. Thus ψ_t fixes all points on the diagonal $\{(z,z)|z \in S^2\}$ of $S^2 \times S^2$ as well as those on the antidiagonal $\{(z,-z)|z \in S^2\}$.

We denote by σ the standard area form on S^2 with total area 4π and consider the family of symplectic forms on $S^2 \times S^2$ given by

$$\omega_\lambda := \lambda \pi_1^*\sigma + \pi_2^*\sigma, \quad \lambda \geq 1.$$

Here we have chosen the first sphere to be the larger one; we think of it as the base of the fiber bundle $\pi_1 : S^2 \times S^2 \to S^2$. Let $\mathcal{H}$ denote the space of continuous maps from $S^2 \times S^2$ to itself that induce the identity on homology.

PROPOSITION 9.7.2. (i) *The loop* $\{\psi_t\}$ *has infinite order in* $\pi_1(\mathcal{H}, \mathbb{1})$.

(ii) *The loop* $\{\psi_t\}$ *is isotopic to a loop in* $\mathrm{Ham}(S^2 \times S^2, \omega_\lambda)$ *if and only if* $\lambda > 1$.

Observe that the "only if" statement in (ii) follows immediately from (i) and the fact that $\mathrm{Ham}(S^2 \times S^2, \omega_1)$ is homotopy equivalent to $\mathrm{SO}(3) \times \mathrm{SO}(3)$ (see Theorem 9.5.1). This means that $\pi_1(\mathrm{Ham}(S^2 \times S^2, \omega_1))$ is finite, so that it cannot contain an element $\{\psi_t\}$ that has infinite order in some larger space. It is also true that $\{\psi_t\}$ cannot be homotoped into $\mathrm{Ham}(S^2 \times S^2, \omega_\lambda)$ for $0 < \lambda < 1$, but this requires some further argument. This is not relevant here, since we are only looking at the case $\lambda \geq 1$. The proof of the "if" statement in (ii) will be deferred to the end of this section. Another proof of (ii), that uses properties of the Seidel representation rather than Theorem 9.4.7, is sketched in Example 11.4.9.

Assertion (i) is an immediate consequence of the following lemma which will also be proved at the end of this section. It is based on the properties of a generalized Hopf invariant for maps $Y \to S^2$ whose domain is a compact oriented smooth 3-manifold Y and that induce the zero map $H_2(Y) \to H_2(S^2)$. For short, we say that these maps are zero on H_2.

LEMMA 9.7.3. *Fix a point* $w_0 \in S^2$ *and, for* $k \in \mathbb{Z}$, *define* $F^k : S^1 \times S^2 \to S^2$ *by*

$$F^k(t, z) := \Phi_{z,kt}(w_0).$$

Then F^k *is bordant to* F^ℓ *by a map that is zero on* H_2 *if and only if* $k = \ell$.

PROOF OF PROPOSITION 9.7.2 (I). Every loop $\mathbb{R}/\mathbb{Z} \to \mathcal{H} : t \mapsto \psi_t$ determines a continuous map $F_\psi : S^1 \times S^2 \to S^2$ via

$$F_\psi(t, z) := \pi_2(\psi_t(z, w_0)),$$

where $\pi_2 : S^2 \times S^2 \to S^2$ denotes the projection onto the second factor. If two loops ψ and ψ' are homotopic then so are the corresponding maps F_ψ and $F_{\psi'}$. Moreover, because F_ψ and $F_{\psi'}$ are themselves zero on H_2 any joining homotopy is also zero on H_2. Now let $\psi^k = \{\psi_t^k\}$ denote the loop $t \mapsto \psi_{kt}$, where ψ_t is defined by (9.7.1). Then $F_{\psi^k} = F^k$. Hence, by Lemma 9.7.3, ψ^k is homotopic to ψ^ℓ if and only if $k = \ell$. This proves (i) in Proposition 9.7.2. □

Consider the following two families of forms on $X := S^2 \times S^2 \times \mathbb{T}^2$. The first is the product form

$$\tau_\lambda^0 := \lambda \pi_1^* \sigma + \pi_2^* \sigma + ds \wedge dt,$$

and the second is $\tau_\lambda^1 := \Psi_* \tau_\lambda^0$, where the diffeomorphism $\Psi : X \to X$ is defined by

$$\Psi(z, w, s, t) = (z, \Phi_{z,t}(w), s, t)$$

and the pushforward Ψ_* on forms is defined by $(\Psi^{-1})^*$.

THEOREM 9.7.4. *The symplectic forms τ_λ^0 and τ_λ^1 on $S^2 \times S^2 \times \mathbb{T}^2$ are isotopic for $\lambda > 1$. For $\lambda = 1$ they are cohomologous and can be joined by a path of symplectic forms but are not isotopic.*

PROOF. That the forms are isotopic for $\lambda > 1$ is a consequence of Proposition 9.7.2. Namely, for $\lambda > 1$, Ψ is isotopic to a diffeomorphism Ψ' of the form $(z, w, s, t) \mapsto (\psi_t'(z, w), s, t)$ where $(\psi_t')^* \omega_\lambda = \omega_\lambda$. Thus $\Psi'_* \tau_\lambda^0$ has the form $\tau_\lambda^0 + dt \wedge \pi^* \alpha_t$, where α_t is a smooth family of exact 1-forms on $S^2 \times S^2$ and $\pi : X \to S^2 \times S^2$ denotes the obvious projection. Hence the linear path $\mu \Psi'_* \tau_\lambda^0 + (1-\mu) \tau_\lambda^0$ consists of cohomologous symplectic forms. Now apply Moser isotopy (Remark 9.4.9).

It follows that the symplectic forms τ_1^0 and τ_1^1 can be joined by a path of symplectic forms. Thus they define the same Chern classes, and they are manifestly cohomologous. Hence the "classical invariants" of this pair of forms are the same.

To see that τ_1^0 and τ_1^1 are not isotopic, we investigate the topology of the moduli space $\mathcal{M}(A; J)$ of spheres in the class

$$A := [S^2 \times \text{pt} \times \text{pt}].$$

Since $\lambda = 1$, A is an indecomposable class. Hence $\mathcal{M}(A; J)/\mathrm{G}$ is compact for every $J \in \mathcal{J}(X, \tau_1^0)$. If J is regular for A and compatible with τ_1^0, then $\mathcal{M}(A; J)/\mathrm{G}$ is a compact smooth 4-manifold. The obvious product structure J^0 is regular for A and compatible with τ_1^0. The moduli space $\mathcal{M}(A; J^0)/\mathrm{G}$ can be identified with $S^2 \times \mathbb{T}^2$ via the diffeomorphism

$$S^2 \times \mathbb{T}^2 \to \mathcal{M}(A; J^0)/\mathrm{G} : (w, s, t) \mapsto [u_{w,s,t}], \qquad u_{w,s,t}(z) := (z, w, s, t).$$

Now consider the τ_1^1-compatible almost complex structure $J^1 := \Psi_* J^0$. Since this is diffeomorphic to J^0, it also is regular for A. The elements of $\mathcal{M}(A; J^1)$ are the equivalence classes $[\Psi \circ u_{w,s,t}]$.

Assume, by contradiction, that the symplectic forms τ_1^0 and τ_1^1 are isotopic via a smooth family of cohomologous symplectic forms τ_1^r, $0 \le r \le 1$. Then there exists a regular family of almost complex structures $J^r \in \mathcal{J}(X, \tau_1^r)$, $0 \le r \le 1$, so that the corresponding moduli space

$$\mathcal{W}(A; \{J^r\}) := \{(r, u) \,|\, 0 \le r \le 1,\, u \in \mathcal{M}(A; J^r)\}$$

gives rise to a compact oriented cobordism $\mathcal{W}(A; \{J^r\})/\mathrm{G}$ from $\mathcal{M}(A; J^0)/\mathrm{G}$ to $\mathcal{M}(A; J^1)/\mathrm{G}$. The corresponding evaluation map

$$\mathrm{ev} : \mathcal{W}(A; \{J^r\}) \times_{\mathrm{G}} S^2 \to [0, 1] \times X, \qquad \mathrm{ev}([r, u, z]) := (r, u(z)),$$

is a smooth map between two compact oriented 7-manifolds with boundary. Consider the embedded circles $\gamma^0, \gamma^1 \subset X$, defined by

$$\gamma^0 := \left\{(z_0, w_0, s_0, t) \,|\, t \in S^1\right\}, \qquad \gamma^1 := \left\{(z_0, \Phi_{z_0,t}(w_0), s_0, t) \,|\, t \in S^1\right\} = \Psi(\gamma^0).$$

These submanifolds are both diffeomorphic to the circle and they are oriented cobordant, i.e. there exists an oriented 2-manifold $\sigma \subset [0,1] \times X$ with boundary

$$\partial\sigma = \sigma \cap (\{0,1\} \times X) = (\{0\} \times \gamma^0) \cup (\{1\} \times \gamma^1).$$

Note that the restriction of ev to the boundary of $\mathcal{W}(A; \{J^r\}) \times_{\mathrm{G}} S^2$ is transverse to $\partial\sigma$. Perturbing σ in the interior of $[0,1] \times X$, if necessary, we may also assume that ev is transverse to σ. It then follows that $\Sigma := \mathrm{ev}^{-1}(\sigma)$ is a compact oriented smooth 2-manifold with boundary. It is given by $\Sigma = \mathcal{W}(A, \sigma; \{J^r\})/\mathrm{G}$, where

$$\mathcal{W}(A, \sigma; \{J^r\}) := \left\{(r, u, z) \,|\, 0 \le r \le 1,\, u \in \mathcal{M}(A; J^r),\, z \in S^2,\, (r, u(z)) \in \sigma\right\},$$

and has boundary $\partial\Sigma = (\{0\} \times \Gamma^0) \cup (\{1\} \times \Gamma^1)$, where $\Gamma^0 \subset \mathcal{M}(A; J^0) \times_{\mathrm{G}} S^2$ and $\Gamma^1 \subset \mathcal{M}(A; J^1) \times_{\mathrm{G}} S^2$ are the embedded circles

$$\Gamma^0 := \left\{[u_{w_0,s_0,t}, z_0] \,|\, t \in S^1\right\}, \qquad \Gamma^1 := \left\{[\Psi \circ u_{w_0,s_0,t}, z_0] \,|\, t \in S^1\right\}.$$

Now consider the moduli space

$$W := \mathcal{W}(A, \sigma; \{J^r\}) \times_{\mathrm{G}} S^2$$

with boundary $\partial W = (\{0\} \times Y^0) \cup (\{1\} \times Y^1)$, where Y^0 and Y^1 are diffeomorphic to $S^1 \times S^2$. The moduli space W carries a projection $\pi : W \to \Sigma$ with fibers S^2 and an evaluation map $F : W \to S^2$, given by $F([r, u, z, z']) := \pi_2(u(z'))$, where $\pi_2 : S^2 \times S^2 \times \mathbb{T}^2 \to S^2$ denotes the projection onto the second factor:

$$\begin{array}{ccc} W & \xrightarrow{\;\;F\;\;} & S^2 . \\ {\scriptstyle\pi}\big\downarrow & & \\ \Sigma & & \end{array}$$

The restrictions of F to the two boundary components, under the identifications with $S^1 \times S^2$, are given by

$$F^0(t, z') = w_0, \qquad F^1(t, z') = \Phi_{z',t}(w_0).$$

Since W is a locally trivial fibration over Σ, the restriction of F to the union of the components of W with nonempty boundaries is zero on H_2 and is a bordism from F^0 to F^1. This contradicts Lemma 9.7.3 and proves Theorem 9.7.4 □

Proof of Lemma 9.7.3. Let Y be a compact oriented smooth 3-manifold. The Hopf invariant for smooth maps from S^3 to S^2 generalizes as follows to smooth maps $F : Y \to S^2$ that are zero on H_2. Let x, y be two regular values of F and denote their preimages by

$$\alpha := F^{-1}(x), \qquad \beta := F^{-1}(y).$$

Since the induced map $F_* : H_2(Y) \to H_2(S^2)$ is zero, the oriented 1-manifolds α and β are both homologous to zero. Choose an oriented smooth cycle $\Sigma \subset Y$ with oriented boundary $\partial\Sigma = \alpha$ and define the **Hopf number** of F by

$$\chi(F) := \Sigma \cdot \beta.$$

Suppose that Σ' is any other such cycle. Then the union $\Sigma \cup (-\Sigma')$ defines a closed cycle in Y. Since β is homologous to zero, its intersection number with this closed cycle is zero. Hence the number $\chi(F)$ is independent of the choice of Σ. That it is also independent of the choice of x and y follows from the usual arguments in differential topology as in Milnor [**289**].

Now assume that $F_0 : Y_0 \to S^2$ and $F_1 : Y_1 \to S^2$ are bordant by a map that is zero on H_2. The there exists a compact oriented smooth 4-manifold W with boundary $\partial W = \bar{Y}_0 \cup Y_1$ (where $\bar{Y}_0$ indicates the reversal of orientation) and a smooth map $F : W \to S^2$ such that

$$F|_{Y_0} = F_0, \qquad F|_{Y_1} = F_1.$$

Let $x, y \in S^2$ be two distinct common regular values of F_0, F_1, and F. Denote

$$\alpha_0 := F_0^{-1}(x), \qquad \alpha_1 := F_1^{-1}(x), \qquad \sigma := F^{-1}(x),$$

$$\beta_0 := F_0^{-1}(y), \qquad \beta_1 := F_1^{-1}(y), \qquad \tau := F^{-1}(y)$$

Then $\sigma, \tau \subset W$ are oriented 2-manifolds with boundaries $\partial\sigma = \bar{\alpha}_0 \cup \alpha_1$ and $\partial\tau = \bar{\beta}_0 \cup \beta_1$. Choose two oriented cycles $\Sigma_0 \subset Y_0$ and $\Sigma_1 \subset Y_1$ such that $\partial\Sigma_0 = \alpha_0$ and $\partial\Sigma_1 = \alpha_1$. Then $\Sigma := \Sigma_0 \cup \sigma \cup \bar{\Sigma}_1$ is a closed cycle in W. Hence

$$\chi(F_0) - \chi(F_1) = \beta_0 \cdot \Sigma_0 + \beta_1 \cdot \bar{\Sigma}_1 = \tau \cdot \Sigma = 0.$$

The second equation follows by perturbing Σ to a generic cycle in the interior of W and using the fact that $\tau \cap \sigma = \emptyset$. The last equation follows from the fact that F is zero on H_2 and so $F^{-1}(y)$ is homologous to zero in $H_2(W, \partial W)$ for every $y \in S^2$.

We claim that $\chi(F^k) = k$ for each $k \in \mathbb{Z}$. Assume first that $k = 1$ and consider the map $F(t, z) := F^1(t, z) = \Phi_{z,t}(w_0)$. Every point $x \neq w_0$ is a regular value of F. The preimage of $x := -w_0$ is the submanifold

$$\alpha := F^{-1}(-w_0) = \{(t, z) \,|\, t = 1/2,\ z \perp w_0\}.$$

Here we identify S^2 with the unit sphere in $\mathbb{R}^3$ and use the standard inner product in $\mathbb{R}^3$ to define $\perp$. Now let $y \in S^2 \setminus \{\pm w_0\}$ be any other regular value. Then the preimage $\beta := F^{-1}(y)$ is the set of pairs (t, z) such that z is perpendicular to $y - w_0$, and the angle from w_0 to y in the oriented plane perpendicular to z is $2\pi t$. Hence β intersects each hemisphere in $\{1/2\} \times S^2 \setminus \alpha$ transversally in precisely one point. The two intersection points have the form $(1/2, z)$, where z is perpendicular to $y - w_0$ and lies on the great circle through w_0 and y. This shows that $|\chi(F)| = 1$. We leave the verification of the sign to the reader. The formula for the invariant $\chi(F^k)$ with $k > 0$ is proved similarly. Namely, $\alpha_k := (F^k)^{-1}(-w_0)$ is the union of k disjoint circles $\{(2i-1)/2k, z) \,|\, z \perp w_0\}$, $i = 1, \dots, k$. As before $\beta_k := (F^k)^{-1}(y)$ meets each hemisphere in $\{(2i-1)/2k\} \times S^2 \setminus \alpha_k$ precisely once. Replacing k by $-k$ is equivalent to reversing the orientation of $S^1 \times S^2$ while leaving F^k unchanged. Thus there are three sign changes, namely the orientations of α_k, β_k, and $S^1 \times S^2$. Hence $\chi(F^{-k}) = -\chi(F^k)$ and this proves Lemma 9.7.3. $\square$

First proof of Proposition 9.7.2 (ii). We proved the "only if" part just after stating the proposition. Therefore it remains to consider the "if" part. We shall give two proofs, the first by direct calculation and the second more conceptual.

Note that the diffeomorphisms ψ_t define a smooth S^1-action on $S^2 \times S^2$. Averaging the symplectic form ω_λ over this action gives a closed 2-form

$$\omega'_\lambda := \int_0^1 \psi_t^* \omega_\lambda \, dt.$$

By construction, this form represents the same cohomology class as ω_λ and is preserved by ψ_t for every t. A straightforward calculation (Exercise 9.7.5) shows that it is given by

$$
\begin{aligned}
\omega'_{\lambda,(z,w)}((\dot z_1,\dot w_1),(\dot z_2,\dot w_2)) \;=\;& \lambda\langle z\times \dot z_1,\dot z_2\rangle+\langle w\times \dot w_1,\dot w_2\rangle\\
&-\langle z\times \dot z_1,\dot w_2\rangle+\langle z\times \dot z_2,\dot w_1\rangle\\
&+2\langle z,w\rangle\langle z\times \dot z_1,\dot z_2\rangle,
\end{aligned}
\tag{9.7.2}
$$

where $\times$ denotes the vector product and $\langle\cdot,\cdot\rangle$ denotes the standard inner product on $\mathbb{R}^3$. To see that ω'_λ is nondegenerate for every $\lambda>1$ we must prove the following for every pair $(z,w)\in S^2\times S^2$ and every pair of tangent vectors $\dot z_1\in T_zS^2$, $\dot w_1\in T_wZ^2$: if the right hand side of (9.7.2) vanishes for all $\dot w_2$ and $\dot z_2$ then $\dot w_1=\dot z_1=0$. Suppose that we are given such a pair of tangent vectors $(\dot z_1,\dot w_1)$. Setting $\dot z_2=0$ and allowing $\dot w_2$ to vary freely, we find that

$$\dot w_1+w\times(z\times\dot z_1)=0. \tag{9.7.3}$$

Hence

$$\langle z\times\dot z_2,\dot w_1\rangle+\langle z,w\rangle\langle z\times\dot z_1,\dot z_2\rangle=0 \tag{9.7.4}$$

for every $\dot z_2\in T_zS^2$. (Use the formula $(z\times\dot z_1)\times(z\times\dot z_2)=\langle z\times\dot z_1,\dot z_2\rangle z$ for $\dot z_j\perp z$.) If we now suppose that $\dot w_2=0$ we find, by (9.7.4), that

$$(\lambda+\langle z,w\rangle)\,\langle z\times\dot z_1,\dot z_2\rangle=0$$

for every $\dot z_2\in T_zS^2$. Since $|\langle z,w\rangle|\le 1<\lambda$, this implies $\dot z_1=0$ and hence, by (9.7.3), $\dot w_1=0$. Thus we have proved that ω'_λ is nondegenerate for $\lambda>1$. A similar argument shows that the 2-form $\omega_{\lambda,s}:=\omega_\lambda+s(\omega'_\lambda-\omega_\lambda)$ is nondegenerate for every $s\in[0,1]$. Hence it follows from Moser isotopy (see Remark 9.4.9) that there is a diffeomorphism ϕ of $S^2\times S^2$ that is isotopic to the identity and satisfies $\phi^*\omega'_\lambda=\omega_\lambda$. Hence the loop $t\mapsto\phi^{-1}\circ\psi_t\circ\phi$ is smoothly isotopic to $t\mapsto\psi_t$ and preserves the symplectic form ω_λ. This completes the first proof of (ii) in Proposition 9.7.2. □

EXERCISE 9.7.5. Derive the formula (9.7.2) in the above proof of Proposition 9.7.2. *Hint:* Identify S^2 with the unit sphere in $\mathbb{R}^3$. Then the standard symplectic form on S^2 (with area 4π) is $\sigma_w(\dot w_1,\dot w_2)=\langle w\times\dot w_1,\dot w_2\rangle$ and the diffeomorphism $\Phi_{z,t}:S^2\to S^2$ is given by

$$\Phi_{z,t}(w)=\cos(2\pi t)\Big(w-\langle z,w\rangle z\Big)+\langle z,w\rangle z+\sin(2\pi t)z\times w.$$

Given a path $\mathbb{R}\to S^2: s\mapsto z(s)$, show that the isotopy $s\mapsto\Phi_{z(s),t}$ of S^2 is generated by the vector fields

$$X_{z,\dot z,t}(w):=\Big(\sin(2\pi t)\dot z+(1-\cos(2\pi t))z\times\dot z\Big)\times w$$

for $z\in S^2$, $\dot z\in T_zS^2$, and $t\in\mathbb{R}$. Prove that $\Phi_{z,t}^*X_{z,\dot z,t}=-X_{z,\dot z,-t}$. Show that the pullback of the vertical symplectic form $\pi_2^*\sigma$ on $S^2\times S^2$ under the diffeomorphism $\psi_t(z,w)=(z,\Phi_{z,t}(w))$ has the form

$$(\psi_t^*\pi_2^*\sigma)_{z,w}\Big((\dot z_1,\dot w_1),(\dot z_2,\dot w_2)\Big)=\sigma_w\Big(\dot w_1-X_{z,\dot z_1,-t}(w),\dot w_2-X_{z,\dot z_2,-t}(w)\Big).$$

Compute and average over t.

Second proof of Proposition 9.7.2 (ii). There are several ways of seeing that $\{\psi_t\}$ can be homotoped into $\mathrm{Ham}(S^2 \times S^2, \omega_\lambda)$ when $\lambda > 1$. The proof in [**254**, Lemma 3.1] uses properties of Hirzebruch surfaces. Perhaps the easiest argument from our current perspective is to identify $S^2 \times S^2$ with the total space of the fibration $(\widetilde{M}_{\phi^2}, \widetilde{\omega}_c) \to S^2$ constructed as in Section 9.6 from the loop $\phi_t^k \in \mathrm{Ham}(S^2, \sigma)$ with $k = 2$ (and $n = 1$). Thus $\widetilde{M}_{\phi^2} = S^3 \times_{S^1} S^2$, where S^1 acts on $S^3 \times S^2$ by

$$t \cdot (z_1, z_2; [w_0 : w_1]) = (e^{2\pi i t} z_1, e^{2\pi i t} z_2; [w_0 : e^{4\pi i t} w_1]).$$

Because the loop ϕ^2 is contractible, there is a commutative diagram

$$\begin{array}{ccc} \widetilde{M}_{\phi^2} & \xrightarrow{\quad f \quad} & S^2 \times S^2\,, \\ & {\scriptstyle\pi}\searrow \quad \swarrow{\scriptstyle\pi_1} & \\ & S^2 & \end{array}$$

where f is a diffeomorphism. As in the proof of Proposition 9.6.4 the fibration $\pi : \widetilde{M}_{\phi^2} \to S^2$ has two natural sections given by the two critical points $[1:0], [0:1]$ of the generating Hamiltonian, and we may define f to take these to the diagonal and antidiagonal sections of the trivial bundle $\pi_1 : S^2 \times S^2 \to S^2$. Therefore in this model $\widetilde{M}_{\phi^2} = S^3 \times_{S^1} S^2$ of $S^2 \times S^2$ the loop $\{\psi_t\}$ rotates the S^2 factor:

$$t \cdot [z_1, z_2; [w_0 : w_1]] = [z_1, z_2; [w_0 : e^{2\pi i t} w_1]].$$

Clearly, this action preserves the symplectic form $\widetilde{\omega}_c$ if we take this to be $\widetilde{\omega} + \kappa\pi^*\sigma$ where $\widetilde{\omega}$ is as in (9.6.3). Since $H : S^2 \to \mathbb{R}$ takes values in $[-1, 1]$ we can choose $\kappa = \lambda$ for any $\lambda > 1$. Then $f_*\widetilde{\omega}_\lambda$ (where we write f_* instead of $(f^{-1})^*$) is cohomologous to ω_λ and it remains to prove that these forms are isotopic. We do this by an explicit application of the process of symplectic inflation. This is a general process that allows one to convert a deformation into an isotopy on all symplectic 4-manifolds that have sufficiently many nonzero Gromov invariants: see McDuff [**266**]. It was first introduced by Lalonde and McDuff [**258, 221, 225**] and was later used by Biran [**38, 39**] in connection with the symplectic packing problem.

First isotop f so that $f_*\widetilde{\omega}_\lambda = \omega_\lambda$ in some neighbourhood U of the diagonal Δ and then consider the linear path $tf_*\widetilde{\omega}_\lambda + (1-t)\omega_\lambda$, $0 \le t \le 1$. This consists of forms that are nondegenerate on the fibers of π_1. Therefore by adding suitable multiples of the pullback $\pi_1^*\sigma$ one obtains a family of (noncohomologous) symplectic forms τ_t joining $\tau_0 = \omega_\lambda$ to $\tau_1 = f_*\widetilde{\omega}_\lambda$ that equal $c_t\pi_1^*\sigma + \pi_2^*\sigma$ in U where $c_t \ge \lambda$ with equality for t near $0, 1$. The idea now is to covert the deformation τ_t into an isotopy

$$\omega_{\lambda,t} = \frac{1}{(1+\kappa_t)}(\tau_t + \kappa_t\rho_t), \quad t \in [0,1],$$

from $\omega_{\lambda,0} = \omega_\lambda$ to $\omega_{\lambda,1} = f_*\widetilde{\omega}_\lambda$, where the form ρ_t has support near Δ and represents the class Poincaré dual to Δ and where the constant $\kappa_t := (c_t - \lambda)/(\lambda - 1)$ is chosen so that $[\omega_t]$ is constant. (It is here that we need $\lambda > 1$.) Therefore, we must find ρ_t such that arbitrarily large multiples can be added to τ_t without affecting nondegeneracy. These forms exist because the τ_t do not vanish on Δ.

Here are the details. Denote by $\pi : L \to S^2$ the complex line bundle with Chern number 2, and let α be the connection 1-form on the corresponding S^1 bundle so that $-d\alpha = 2\pi^*\sigma$. If $r : L \to \mathbb{R}$ is the radial coordinate in the fibers, and $\beta : [0,1] \to [0,1]$ is a nonincreasing cutoff function that equals 1 near $r = 0$, then the compactly supported form $-d(\beta(r)\alpha)$ integrates to 1 over each fiber and to 2 over the zero section L_0. By the symplectic neighbourhood theorem, there is a smooth family of embeddings of $g_t : (U, \Delta) \to (L, L_0)$ such that

$$(g_t)_*(c_t\pi_1^*\sigma + \pi_2^*\sigma) = \frac{c_t}{2}(d(r^2 - 1)\alpha), \qquad t \in [0,1].$$

Now choose a family of cutoff functions β_t with support in the image of g_t for each t and set $\rho_t = -d(\beta_t(r)\alpha)$. Their pullbacks via the g_t have all the required properties. This completes the second proof of (ii) in Proposition 9.7.2. □

A similar argument shows that no two of the forms $\tau_1^k := (\Psi^{\circ k})_*\tau_1^0, \quad k \geq 0$ are isotopic, though they are all cohomologous and deformation equivalent. One can also improve this construction to obtain a family of deformation equivalent, cohomologous but nondiffeomorphic forms on the 8-dimensional manifold $\widetilde{Y}$ obtained by first multiplying $S^2 \times S^2 \times \mathbb{T}^2$ by a further S^2 factor to get Y and then blowing up along the submanifold $Z := S^2 \times \{w_1\} \times \mathbb{T}^2 \times \{x_1\}$. Because Z has codimension 4, Gromov's symplectic embedding theorem (see [**161**] and also Eliashberg—Mishenko [**102**]) implies that it can be symplectically embedded in Y with respect to $\tau_1^j + \pi_3^*\sigma$ for both $j = 0, 1$. Therefore, these forms blow up to give two symplectic forms $\widetilde{\tau}_1^j$ on $\widetilde{Y}$ that are nondegenerate on the blowup $\widetilde{Z}$ of Z. Suppose now that there is a diffeomorphism $f : (\widetilde{Y}, \widetilde{\tau}_1^0) \to (\widetilde{Y}, \widetilde{\tau}_1^1)$. By looking at the moduli space of curves in the class E of the exceptional divisor, one shows (roughly speaking) that the map from $\widetilde{Z}$ to the second S^2 factor in $\widetilde{Y}$ given by $\pi_2 \circ f$ must be nullhomotopic. On the other hand the submanifold $W = S^2 \times S^2 \times \mathbb{T}^2 \times \{x_2\}$ of $(\widetilde{Y}, \widetilde{\tau}_1^j)$ contains 1-parameter families of holomorphic A-curves that twist j-times around the second S^2-factor. Since Z can be homotoped into W one can derive a contradiction by comparing this fact with the information obtained from the E-curves.

There is one analytic point that is worthy of note. Because $\widetilde{\tau}_1^j(E)$ is small, the class $A \in H_2(\widetilde{Y})$ is no longer indecomposable, and the moduli space $\mathcal{M}(A; J)$ has a nonempty boundary. Since for each J we are interested in the properties of a 1-parameter family of A-curves, we must consider 2-parameter families of elements in $\mathcal{M}(A; J_t), t \in [0,1]$. Therefore these families will not in general be compact; we can no longer avoid meeting the boundary by perturbing the path $\{J_t\}$. Nevertheless, one can analyze the limiting stable maps and show that they do not affect the invariant χ. The basic reason is that the projection $\pi_2 : \widetilde{Y} \to S^2$ has zero degree on each component of the limiting stable maps. The argument would fail if an A-curve could decompose as the sum of an $(A - B)$-curve with a B curve, where B denotes the class of the second sphere. This is possible only when $\lambda > 1$; and in this case we know that the argument does fail. For more details see [**254**].

CHAPTER 10

Gluing

In this chapter we give a proof of the splitting axiom for the Gromov–Witten invariants as discussed in Chapter 7 (Theorem 7.5.10). This is based on the observation that two J-holomorphic spheres that intersect transversally in the appropriate sense can be *glued together*, thereby giving rise to a family of J-holomorphic spheres that represent the sum of the homology classes of the original spheres. This is a kind of converse to Gromov compactness. To prove it one first constructs an approximate J-holomorphic sphere by a connected sum construction and then uses the implicit function theorem to establish the existence of an actual J-holomorphic sphere nearby. In the context of pseudoholomorphic curves the first gluing theorem of this kind is due to Floer [**113, 114, 116**], who considered moduli spaces of trajectories joining two periodic orbits. This case is somewhat easier because the periodic orbits are assumed to be nondegenerate. The gluing theorem for J-holomorphic spheres was first proved by Ruan–Tian [**345**] and then Liu [**248**] using methods somewhat different from ours. The argument presented here is a refinement of that in the first version of this book. Subsequently many other versions of this argument have appeared: see for example Fukaya–Ono [**127**], Ionel–Parker [**201**], Zinger [**426**].

In the first section we set up the notation and state the main gluing theorem (Theorem 10.1.2). The next six sections are devoted to its proof. Section 10.8 contains the proof of the Splitting Theorem 7.5.10. Finally in Theorem 10.9.1 we use the main gluing theorem to describe the structure of a neighbourhood of the set of simple stable maps with one node in the compactified moduli space $\overline{\mathcal{M}}_{0,4}(A,J)$. We also show that this simplified gluing theorem is sufficient to establish the associativity of quantum multiplication. These last two sections only require the statement of the gluing theorem, not its proof, and hence can be understood immediately after Section 10.1.

The basic idea of the proof of Theorem 10.1.2 is to use cutoff functions to piece together objects that are very close to those we are looking for, and then to use a Newton type iteration to get a precise solution. The first step is to build a family of approximately J-holomorphic spheres u^R (for large R) from two intersecting J-holomorphic spheres. This step is called pregluing, and is discussed in Section 10.2. The next section introduces the appropriate weighted Sobolev norms for vector fields and 1-forms along the preglued curves u^R. To apply the implicit function theorem of Section 3.5, we need uniform estimates (with respect to these weighted norms) for suitable right inverses of the linearized operators D_{u^R}. Following Donaldson–Kronheimer [**88**], we first construct an approximate right inverse T_{u^R} and then obtain an exact right inverse Q_{u^R} with the same image as T_{u^R}.

For this to work we must use delicate cutoff functions β whose properties are established in Section 10.4. The main construction is described in Section 10.5.

There we define the families of approximate inverses T_{u^R} and of true inverses Q_{u^R}, and then apply the inverse function theorem to define the gluing map $\iota_c^{\delta,R}$. We also establish most of its properties, modulo some details that are deferred to the technical Section 10.6. The remaining part of the proof of Theorem 10.1.2 concerns its surjectivity. To establish this one must understand the sense in which gluing is the converse of compactness, and for this we shall need Lemma 4.7.3 on the properties of long cylinders with small energy. The details appear in Section 10.7. Sections 10.6 and 10.7 are unavoidably technical. To get a good overall picture of the construction, the reader should first concentrate on Sections 10.1, 10.2 and 10.5.

10.1. The gluing theorem

The gluing theorem for J-holomorphic curves can be viewed as a converse of Gromov's compactness theorem. It asserts, roughly speaking, that if two (or more) J-holomorphic curves intersect and satisfy a suitable transversality condition then they can be approximated, in the sense of Gromov convergence (see Definition 5.2.1), by a sequence of J-holomorphic curves representing the sum of their homology classes. Our proof adapts the method used by Donaldson–Kronheimer [**88**, pp 287–295] to establish the gluing theorem for anti-self-dual instantons on 4-manifolds. The precise formulation of the theorem requires some preparation. We shall describe the gluing map in the setting of Section 6.7, where J depends on $z \in S^2$. This makes no essential difference to the analysis because J varies in a compact family, and it gives a result that applies more widely. When J is independent of z there are additional symmetries. This leads to a slightly simpler version of the gluing theorem, which can be stated in the language of stable maps and Gromov convergence. That version will be discussed in Section 10.9, where we also explain how it can be used to prove the easiest version of the splitting principle.

The moduli space of connected pairs. As explained above, we assume that $J^0 = \{J^0_z\}_{z\in S^2}$ and $J^\infty = \{J^\infty_z\}_{z\in S^2}$ are two smooth families of ω-tame almost complex structures on M and $\kappa > 0$ is a constant such that

$$|z| < \kappa \quad \Longrightarrow \quad J^0_z = J^\infty_{1/z} =: J. \tag{10.1.1}$$

Fix two homology classes $A^0, A^\infty \in H_2(M;\mathbb{Z})$. Recall that $\mathcal{M}(A^0;\{J^0_z\})$ denotes the space of $\{J^0_z\}$-holomorphic curves in M representing the class A^0, and similarly for $\mathcal{M}(A^\infty;\{J^\infty_z\})$. Readers may if they wish restrict attention to J that are independent of z, but in this case they should consider only simple maps.

Definition 10.1.1. *A pair* $(J^0, J^\infty) \in \mathcal{J}_\tau(S^2;M,\omega)\times\mathcal{J}_\tau(S^2;M,\omega)$ *is called* **regular for** (A^0, A^∞) *if* $J^0 \in \mathcal{J}_{\mathrm{reg}}(S^2;A^0)$ *and* $J^\infty \in \mathcal{J}_{\mathrm{reg}}(S^2;A^\infty)$ *satisfy (10.1.1) for some* $\kappa > 0$ *and the evaluation map*

$$\begin{aligned} &\mathrm{ev}: \mathcal{M}(A^0;J^0)\times\mathcal{M}(A^\infty;J^\infty) \to M\times M,\\ &\mathrm{ev}(u^0,u^\infty) := (u^0(0), u^\infty(\infty)),\end{aligned} \tag{10.1.2}$$

is transverse to the diagonal. The set of such regular pairs will be denoted by $\mathcal{J}_{\mathrm{reg}}(S^2;A^{0,\infty})$.

By Proposition 6.7.7, the sets $\mathcal{J}_{\mathrm{reg}}(S^2;A^0)$ and $\mathcal{J}_{\mathrm{reg}}(S^2;A^\infty)$ are residual in the space of all ω-tame families $\{J_z\} \in \mathcal{J}_\tau(S^2;M,\omega)$ that satisfy (10.1.1). Moreover, the proof of Theorem 6.3.1 shows that, for a generic pair (J^0, J^∞), the evaluation map (10.1.2) is transverse to the diagonal. Hence $\mathcal{J}_{\mathrm{reg}}(S^2;A^{0,\infty})$ is residual in the space of all pairs $(J^0, J^\infty) \in \mathcal{J}_\tau(S^2; M, \omega) \times \mathcal{J}_\tau(S^2; M, \omega)$ that satisfy (10.1.1). Moreover, if $(J^0, J^\infty) \in \mathcal{J}_{\mathrm{reg}}(S^2; A^{0,\infty})$ then the moduli space

$$\mathcal{M}(A^{0,\infty}; J^{0,\infty}) := \left\{ (u^0, u^\infty) \in \mathcal{M}(A^0; J^0) \times \mathcal{M}(A^\infty; J^\infty) \,|\, u^0(0) = u^\infty(\infty) \right\}$$

is a smooth finite dimensional manifold of dimension $2n + 2c_1(A^0 + A^\infty)$.

Given a constant $c > 0$ we denote by

$$\mathcal{M}(c) := \mathcal{M}(A^{0,\infty}; J^{0,\infty}, c)$$

the compact subset of all pairs $(u^0, u^\infty) \in \mathcal{M}(A^{0,\infty}; J^{0,\infty})$ that satisfy

$$\left\| du^0 \right\|_{L^\infty} \le c, \qquad \left\| du^\infty \right\|_{L^\infty} \le c.$$

In Theorem 10.1.2 we shall consider the evaluation map

$$\mathrm{ev}^0 : \mathcal{M}(c) \times \left(S^2 \setminus \mathrm{int}\, B_{1/c}(0) \right) \to M, \qquad \mathrm{ev}^0(u^0, u^\infty, z) := u^0(z) \tag{10.1.3}$$

on the first sphere. Throughout the following discussion c should be considered a fixed but arbitrarily large constant.

Statement of gluing theorem. In the following we fix two homology classes $A^0, A^\infty \in H_2(M; \mathbb{Z})$ and denote

$$A := A^0 + A^\infty.$$

We also fix a regular pair $(J^0, J^\infty) \in \mathcal{J}_{\mathrm{reg}}(S^2; A^{0,\infty})$. We shall prove that, for every $c > 0$ and every sufficiently large number $R > 0$, the moduli space $\mathcal{M}(A^{0,\infty}; J^{0,\infty}, c)$ can be embedded into the moduli space $\mathcal{M}(A; J^R)$, where

$$J^R_z := \begin{cases} J^0_z, & \text{if } |z| \ge 1/R, \\ J^\infty_{R^2 z}, & \text{if } |z| \le 1/R. \end{cases} \tag{10.1.4}$$

This formula for J^R exhibits how gluing affects the domain: each element in $\mathcal{M}(A; J^R)$ is a map from the Riemann sphere $S^2 = \mathbb{C} \cup \{\infty\}$ to M, where the disc $\{|z| \ge 1/R\}$ should be identified with the corresponding disc in the domain of u^0 while the disc $\{|z| \le 1/R\}$, rescaled by $z \mapsto R^2 z$, should be identified with the disc $\{|z| \le R\}$ in the domain of u^∞. Thus R appears as a rescaling parameter (see Remark 10.2.4).

The gluing map $\mathcal{M}(c) \to \mathcal{M}(A; J^R)$ depends on another parameter δ whose reciprocal corresponds to the "length of the neck". Its role is most clearly illustrated in formula (10.5.4). The three constants c, R and δ are related as follows. Given a large constant $c > 0$ one fixes a small constant $\delta < \delta_0(c)$ and then considers all sufficiently large numbers R. In fact the relevant condition on R has the form $\delta R > 1/\delta_0(c)$ and the parameters δ and R determine an annulus $\delta/R \le |z| \le 1/\delta R$ on which the important analysis takes place. For each $\delta_0 > 0$ we denote the **set of annulus parameters** $\mathcal{A}(\delta_0)$ by

$$\mathcal{A}(\delta_0) := \{(\delta, R) \,|\, 0 < \delta < \delta_0,\ \delta R > \frac{1}{\delta_0}\}. \tag{10.1.5}$$

The next theorem states the main properties of the resulting gluing map

$$\iota_c^{\delta,R} : \mathcal{M}(c) \to \mathcal{M}(A; J^R).$$

For each pair $(\delta, R) \in \mathcal{A}(\delta_0)$, it is a smooth embedding that preserves orientation by (iv). Property (i) states that the gluing map depends smoothly on R. Property (ii) states that the glued curve $\widetilde{u}^R$ converges to its constituent parts (u^0, u^∞) as $R \to \infty$, while (v) shows that its image includes all J^R-holomorphic maps that, when appropriately rescaled, are sufficiently close to $\mathcal{M}(c-1)$. Finally (iii) shows that the evaluation map converges in the C^1-topology. (Observe that C^0-convergence follows from (ii), however, C^1-convergence is needed to control the transversality of ev^R to given cycles in M: see Section 10.8.)

THEOREM 10.1.2 (Gluing). *Fix two homology classes $A^0, A^\infty \in H_2(M;\mathbb{Z})$, a regular pair of almost complex structures $(J^0, J^\infty) \in \mathcal{J}_{\mathrm{reg}}(S^2; A^{0,\infty})$, and a constant $c > 1$. Denote $\mathcal{M}(c) := \mathcal{M}(A^{0,\infty}; J^{0,\infty}, c)$ as above. Then there are constants $\delta_0 > 0$ and $\varepsilon > 0$ and a family of maps*

$$\iota_c^{\delta,R} = \iota_c^R : \mathcal{M}(c) \to \mathcal{M}(A; J^R),$$

one for each pair $(\delta, R) \in \mathcal{A}(\delta_0)$, satisfying the following conditions.

(i) *Fix a constant $0 < \delta < \delta_0$. Then the map*

$$(1/\delta\delta_0, \infty) \times \mathcal{M}(c) \to C^\infty(S^2, M) : (R, u^0, u^\infty) \mapsto \iota_c^R(u^0, u^\infty)$$

is smooth and $\iota_c^R : \mathcal{M}(c) \to \mathcal{M}(A; J^R)$ is an embedding for every $R > 1/\delta\delta_0$.

(ii) *Fix a constant $0 < \delta < \delta_0$. Let (u_ν^0, u_ν^∞) be a sequence in $\mathcal{M}(c)$ converging to $(u^0, u^\infty) \in \mathcal{M}(c)$ in the C^∞ topology and suppose $R_\nu > 1/\delta\delta_0$ diverges to infinity. Denote $\widetilde{u}_\nu := \iota_c^{\delta,R_\nu}(u_\nu^0, u_\nu^\infty)$. Then $\widetilde{u}_\nu(z)$ converges to $u^0(z)$ u.c.s. on $S^2 \setminus \{0\}$ and $\widetilde{u}_\nu(z/R_\nu^2)$ converges to $u^\infty(z)$ u.c.s. on $S^2 \setminus \{\infty\}$ as ν tends to ∞.*

(iii) *Fix a constant $0 < \delta < \delta_0$ and let $\mathrm{ev}^R : \mathcal{M}(A; J^R) \times S^2 \to M$ be the obvious evaluation map for $R > 1/\delta\delta_0$. Then the composition*

$$\mathrm{ev}^R \circ (\iota_c^{\delta,R} \times \mathrm{id}) : \mathcal{M}(c) \times \left(S^2 \setminus \mathrm{int}\, B_{1/c}(0)\right) \to M$$

converges to the evaluation map (10.1.3) in the C^1 topology as R tends to ∞.

(iv) *The embedding $\iota_c^{\delta,R}$ is orientation preserving for every pair $(\delta, R) \in \mathcal{A}(\delta_0)$.*

(v) *If $(\delta, R) \in \mathcal{A}(\delta_0)$, $(u^0, u^\infty) \in \mathcal{M}(c-1)$, and $v \in \mathcal{M}(A; J^R)$ satisfy*

$$\sup_{|z|\geq 1} d(v(z), u^0(z)) < \varepsilon, \qquad \sup_{|z|\leq 1} d(v(z/R^2), u^\infty(z)) < \varepsilon,$$

then v belongs to the image of $\iota_c^{\delta,R}$.

REMARK 10.1.3. (i) The construction in the proof of Theorem 10.1.2 shows that the map $\iota_c^{\delta,R} : \mathcal{M}(c) \to \mathcal{M}(A; J^R)$ is independent of c. In other words, if $0 < c < c'$ and $\delta_0 > 0$ is chosen sufficiently small, then the maps $\iota_c^{\delta,R}$ and $\iota_{c'}^{\delta,R}$ agree on their common domain $\mathcal{M}(c)$ for every pair $(\delta, R) \in \mathcal{A}(\delta_0)$.

(ii) Our construction is symmetric with respect to the two components u^0, u^∞ modulo appropriate rescaling. Hence assertion (iii) of Theorem 10.1.2 implies the following. Consider the evaluation map $\mathrm{ev}^\infty : \mathcal{M}(c) \times B_c(0) \to M$, defined by $\mathrm{ev}^\infty(u^0, u^\infty, z) := u^\infty(z)$. Then the evaluation maps

$$\mathcal{M}(c) \times B_c(0) \to M : (u^0, u^\infty, z) \mapsto u_c^{\delta,R}(z/R^2),$$

with $u_c^{\delta,R} := \iota_c^{\delta,R}(u^0, u^\infty)$, converge to ev^∞ in the C^1 topology as $R \to \infty$.

The surjectivity statement in part (v) of Theorem 10.1.2 is somewhat subtle. For a generic J^∞ that depends on $z \in S^2$, a map v can be J^R-holomorphic for at most one R. However, if this condition does not hold, each v will satisfy the conditions of (v) for some open subset of values of R, and (v) implies that the images of the gluing maps $\iota_c^{\delta,R}$ will overlap as R changes. Note also that $|dv(0)|$ is bounded by a constant times R^2 for $v \in \mathrm{im}\, \iota_c^{\delta,R}$. Therefore the image of $\iota_c^{\delta,R}$ for fixed R does not contain the whole of a Gromov convergent sequence; as the following corollary shows one must allow $R \to \infty$ in order to capture the whole of such a sequence. In Section 10.9 we formulate a version of the gluing theorem for z-independent almost complex structures that involves the gluing maps for all values of R and relates the choice of R to the cross ratio of four marked points.

COROLLARY 10.1.4. *Let $A^0, A^\infty \in H_2(M;\mathbb{Z})$ and $(J^0, J^\infty) \in \mathcal{J}_{\mathrm{reg}}(S^2; A^{0,\infty})$, and fix two constants $c > 1$ and $0 < \delta < \delta_0(c)$. Let $R_\nu \to \infty$ and $u_\nu \in \mathcal{M}(A; J^{R_\nu})$ be a sequence that converges to $u^0(z)$ u.c.s. on $S^2 \setminus \{0\}$ and is such that its rescalings $z \mapsto u_\nu(z/R_\nu^2)$ converge to $u^\infty(z)$ u.c.s. on $S^2 \setminus \{\infty\}$, where $(u^0, u^\infty) \in \mathcal{M}(c-1)$. Then there is a constant ν_0 such that $u_\nu \in \operatorname{im} \iota_c^{\delta, R_\nu}$ for $\nu \geq \nu_0$.*

PROOF. The pair (u^ν, R^ν) satisfies the hypotheses of Theorem 10.1.2 (v) for ν sufficiently large. □

The strategy for the proof of Theorem 10.1.2 is as follows. Given two curves $u^0 \in \mathcal{M}(A^0; J^0)$ and $u^\infty \in \mathcal{M}(A^\infty; J^\infty)$ such that $u^0(0) = u^\infty(\infty)$ we shall consider a family of approximate J^R-holomorphic curves $u^R = u^0 \#_R u^\infty$ (for R sufficiently large) and prove that nearby there is an actual J^R-holomorphic curve $\widetilde{u}^R$. The argument involves a proof that the Fredholm operator D_{u^R} is onto and an estimate for a right inverse that rely on some elementary, but delicate, observations about cutoff functions. It is then fairly easy to show that the resulting maps ι_c^R have properties (i-iii) above. These steps in the argument are carried out in Section 10.5. Establishing (iv) and (v) requires several further estimates, though no essentially new ideas. To make it easier for the reader to understand the main points, this part of the argument is relegated to Sections 10.6 and 10.7.

10.2. Connected sums of J-holomorphic curves

In this section we shall construct a family of maps

$$\mathcal{M}(A^{0,\infty}; J^{0,\infty}) \to C^\infty(S^2, M) : (u^0, u^\infty) \mapsto u^R := u^0 \#^R u^\infty$$

that assigns to each connected pair (u^0, u^∞) of J-holomorphic spheres an approximate J^R-holomorphic sphere u^R. More precisely, the map will only be defined on a compact subset of the moduli space $\mathcal{M}(A^{0,\infty}; J^{0,\infty})$ and, for each compact subset, it is defined for R sufficiently large. The approximate J^R-holomorphic curve $u^R : S^2 = \mathbb{C} \cup \{\infty\} \to M$ agrees with u^0 away from 0 and with the rescaled curve $u^\infty(R^2 z)$ very close to 0. The condition $u^0(0) = u^\infty(\infty)$ guarantees that these maps are approximately equal on the central circle $|z| = \frac{1}{R}$. We shall thicken this circle to an annulus of width approximately $1/\delta R$ and use cutoff functions to construct a glued curve u^R that is equal to the intersection point $u^0(0) = u^\infty(\infty)$ on this annulus. In the limit $R \to \infty$ we obtain the converse of Gromov's compactness: the curves u^R will converge in the sense of Definition 5.2.1 to the pair (u^0, u^∞).

Here are the details. Let us fix constants $c, \kappa > 0$ and suppose that J^0 and J^∞ satisfy (10.1.1) with this constant κ. Let $\varepsilon > 0$ be smaller than the injectivity radius of M with the metrics determined via (2.1.1) by ω and the almost complex structure J in (10.1.1). Fix a smooth cutoff function $\rho : \mathbb{C} \to [0,1]$ such that

$$\rho(z) = \begin{cases} 0, & \text{if } |z| \leq 1, \\ 1, & \text{if } |z| \geq 2. \end{cases}$$

For every pair $(u^0, u^\infty) \in \mathcal{M}(c)$ and any two real numbers δ and R such that $0 < \delta < 1$ and δR is sufficiently large, we construct an *approximate* J^R*-holomorphic curve*

$$u^{\delta,R} := u^R := u^0 \#^R u^\infty : S^2 \to M$$

which satisfies

$$u^R(z) = \begin{cases} u^\infty(R^2 z), & \text{if } |z| \le \frac{\delta}{2R}, \\ u^0(0) = u^\infty(\infty), & \text{if } \frac{\delta}{R} \le |z| \le \frac{1}{\delta R}, \\ u^0(z), & \text{if } \frac{2}{\delta R} \le |z|. \end{cases} \tag{10.2.1}$$

Thus, u^R is a rescaling of u^∞ when z is very close to 0, it is constant on a central annulus around $|z| = 1/R$ and equals u^0 sufficiently far outside this annulus. To define the function u^R on the two remaining annuli we denote the point of intersection of the curves u^0 and u^∞ by

$$x := u^0(0) = u^\infty(\infty)$$

and use the exponential map in a neighbourhood of this point. We claim that

$$|z| < \tan(\varepsilon/c), \quad |z| < \kappa \qquad \Longrightarrow \qquad d_J(u^0(z), x) < \varepsilon,$$

$$|z| > \cot(\varepsilon/c), \quad |z| > 1/\kappa \qquad \Longrightarrow \qquad d_J(u^\infty(z), x) < \varepsilon.$$

To see this, note that $J^0_z = J^\infty_{1/z} = J$ for $|z| < \kappa$, by (10.1.1), that $\|du^0\|_{L^\infty} \le c$ and $\|du^\infty\|_{L^\infty} \le c$ (so u^0 and u^∞ are Lipschitz continuous with constant c), and that the Fubini–Study metric on S^2 is $d_{\mathrm{FS}}(z, 0) = \tan^{-1}(|z|)$.

Define $\zeta^0(z) \in T_x M$ for $|z| < \min\{\kappa, \tan(\varepsilon/c)\}$ and $\zeta^\infty(z) \in T_x M$ for $|z| > \max\{1/\kappa, \cot(\varepsilon/c)\}$ to be the unique tangent vectors of norm less than ε such that

$$u^0(z) = \exp_x(\zeta^0(z)), \qquad u^\infty(z) = \exp_x(\zeta^\infty(z)).$$

Now define $u^R := u^{\delta,R}(z)$ by

$$u^R(z) := \exp_x\left(\rho(\delta R z)\zeta^0(z) + \rho\left(\frac{\delta}{Rz}\right)\zeta^\infty(R^2 z)\right) \tag{10.2.2}$$

for $\frac{\delta}{2R} \le |z| \le \frac{2}{\delta R}$. This is well defined whenever

$$0 < \delta < 1, \qquad \delta R > \frac{2}{\tan(\varepsilon/c)}, \qquad \delta R > \frac{2}{\kappa}. \tag{10.2.3}$$

Moreover, the first term in (10.2.2) is nonzero only when $|z| \ge \frac{1}{\delta R}$, while the second is nonzero only if $|z| \le \frac{\delta}{R}$. The map u^R is not a J^R-holomorphic curve, however we shall see that the error converges to zero as $R \to \infty$. Our goal is to construct a *true* J^R-holomorphic curve $\widetilde{u}^R := \widetilde{u}^{\delta,R}$ near u^R.

REMARK 10.2.1. Fix a constant $0 < \delta < 1$. Then the curves $u^R := u^{\delta,R}$ converge to the pair (u^0, u^∞) in the sense of Definition 5.2.1. More precisely, $u^R(z)$ converges to $u^0(z)$ uniformly with all derivatives on compact subsets of $S^2 \setminus \{0\}$ and $u^R(z/R^2)$ converges to $u^\infty(z)$ uniformly with all derivatives on compact subsets of $S^2 \setminus \{\infty\}$. The J^R-holomorphic curves $\widetilde{u}^R$ which we construct below will converge to the pair (u^0, u^∞) in the same way.

REMARK 10.2.2. We shall find it useful to consider a family of pairs of curves that mediate between the original pair u^0, u^∞ and its pregluing u^R. To this end, set $r := \delta R$ and consider the curves $u^{0,r}, u^{\infty,r} : S^2 \to M$ defined by

$$u^{0,r}(z) := \begin{cases} u^R(z), & \text{if } |z| \geq 1/r, \\ u^0(0), & \text{if } |z| \leq 1/r, \end{cases}$$

$$u^{\infty,r}(z) := \begin{cases} u^R(z/R^2), & \text{if } |z| \leq r, \\ u^\infty(\infty), & \text{if } |z| \geq r. \end{cases}$$

Thus $u^{0,r}$, respectively $u^{\infty,r}$, is a small perturbation of u^0, respectively u^∞, that is flattened out (i.e. made constant) near $z = 0$, respectively $z = \infty$. It follows from Lemma 10.4.3 below that $u^{0,r}$ converges to u^0 and $u^{\infty,r}$ converges to u^∞ in the $W^{1,p}$-norm as $r \to \infty$. The convergence is uniform over all pairs $(u^0, u^\infty) \in \mathcal{M}(c)$ because the derivatives of du^0 and du^∞ are uniformly bounded by c. Even though $u^{0,r}$ and $u^{\infty,r}$ do not converge in the C^1 norm, this $W^{1,p}$ convergence suffices to show that $D_{u^{0,r}}$ and $D_{u^{\infty,r}}$ converge to D_{u^0} and D_{u^∞} in the operator norm as $r \to \infty$ (after identifying domain and target spaces using parallel transport). This is immediate from the formulas given for D_u in Section 3.5.

EXAMPLE 10.2.3. Consider the case $M = S^2 = \mathbb{C} \cup \{\infty\}$ with the standard complex structure. Then the holomorphic curves

$$u^0(z) = 1 + z, \qquad u^\infty(z) = 1 + 1/z$$

satisfy $u(0) = v(\infty) = 1$ as required. The above maps $u^R = u^{\delta,R} : S^2 \to S^2$ satisfy

$$u^R(z) = \begin{cases} 1 + 1/R^2 z, & \text{if } |z| \leq \delta/2R, \\ 1, & \text{if } \delta/R \leq |z| \leq 1/\delta R, \\ 1 + z, & \text{if } 2/\delta R \leq |z|. \end{cases}$$

Nearby J-holomorphic curves are given by

$$\widetilde{u}^R(z) = u^\infty(R^2 z) + u^0(z) - 1 = \frac{z^2 + z + 1/R^2}{z}$$

and these converge to the pair (u^0, u^∞) as in Remark 10.2.1.

REMARK 10.2.4. One should think of gluing as the converse to Gromov convergence. The general procedure is rather complicated because of the difficulty of constructing J-holomorphic curves. However, it still applies in the much simpler context of stable curves where all maps are trivial. For example, consider the moduli space $\overline{\mathcal{M}}_{0,4}$ of genus zero stable curves with four marked points. As explained in Section D.7, this may be identified with S^2 via the cross ratio. The open stratum $S^2 \setminus \{0, 1, \infty\}$ consists of the elements $\tau_z = [0, 1, \infty, z]$ and there are three special elements $\tau_0, \tau_1, \tau_\infty$ whose domains have two components. In this context, Gromov convergence describes the way in which the elements τ_z converge to τ_a as $z \to a$ (for $a = 0, 1, \infty$) while gluing describes how to build a neighbourhood of τ_a in $\overline{\mathcal{M}}_{0,4}$.

Since this neighbourhood has two real dimensions, its elements are described by a complex parameter ε. In terms of the current notation and taking $a = 0$, we may identify τ_0 with the stable curve $[z_1^0, z_2^0, z_3^0, z_4^0]$ whose domain is the disjoint union of two copies S_0^2, S_∞^2 of S^2, where $0 \in S_0^2$ is identified with $\infty \in S_\infty^2$ and with marked points $z_2^0 = 1, z_3^0 = \infty$ in S_0^2 and $z_1^0 = 0, z_4^0 = 1$ in S_∞^2. Then, for a given parameter $|\varepsilon| < 1$, τ_ε is the stable curve $[z_1^\varepsilon, z_2^\varepsilon, z_3^\varepsilon, z_4^\varepsilon]$, where

$$\begin{array}{lll} z_2^\varepsilon = z_2^0 = 1 & z_3^\varepsilon = z_3^0 = \infty & \text{in } S_0^2 \\ z_1^\varepsilon = \varepsilon z_1^0 = 0 & z_4^\varepsilon = \varepsilon z_4^0 = \varepsilon & \text{in } S_\infty^2. \end{array}$$

Thus $\tau_\varepsilon = [0, 1, \infty, \varepsilon]$. Note that when $\varepsilon = |\varepsilon|$ is real, this stable tuple can be naturally identified with the domain of the pregluing u^R, for $\varepsilon = \frac{1}{R^2}$.

We will carry out the main gluing construction with real gluing parameters, but our results extend immediately to the complex parameters $\varepsilon = |\varepsilon|e^{i\theta}$ since multiplication by $e^{i\theta}$ is an isometry that rotates the spheres relative to each other. In fact, if the families $\{J_z^0, J_z^\infty\}$ depend only on $|z|$ then the whole construction is S^1-invariant. (See McDuff [**268**] and the proof of Theorem 10.9.1 in Section 10.9.)

10.3. Weighted norms

In order to apply the implicit function theorem we must specify the norm in which $\bar{\partial}_J(u^R)$ is small. The guiding principle for our choice of norms is the observation that the curves u^0 and u^∞ play equal roles in this gluing argument. However, the map u^∞ appears in rescaled form and is concentrated in a ball of radius $1/R$. So in order to give u^0 and u^∞ equal weight we shall consider a family of R-dependent metrics on the 2-sphere such that the volume of the ball of radius $1/R$ is approximately equal to the volume of S^2 with respect to the standard (Fubini–Study) metric. The rescaled metric is of the form

$$g^R := (\theta^R)^{-2}(ds^2 + dt^2), \tag{10.3.1}$$

where

$$\theta^R(z) := \begin{cases} R^{-2} + R^2|z|^2, & \text{if } |z| \le 1/R, \\ 1 + |z|^2, & \text{if } |z| \ge 1/R. \end{cases}$$

The corresponding volume form is $\mathrm{dvol}^R := (\theta^R)^{-2} ds \wedge dt$ and the area of S^2 with respect to this metric is given by

$$\mathrm{Vol}^R(S^2) = \int_{\mathbb{C}} (\theta^R)^{-2} ds \wedge dt \le 2\pi. \tag{10.3.2}$$

There are two ways of understanding this metric. First note that the involution $z \mapsto 1/(R^2 z)$ is an isometry with respect to g^R that interchanges the two discs $\{|z| \le 1/R\}$ and $\{|z| \ge 1/R\}$. It follows that many calculations need only be done on the outer disc $\{|z| \ge 1/R\}$ where the induced metric is the usual Fubini–Study metric. It is also useful to consider the rescaling $z \mapsto R^2 z$. This induces an isometry from the disc $\{|z| \le 1/R\}$ with metric g^R to the disc $\{|z| \le R\}$ with the Fubini–Study metric g^1. Thus, the effect of this metric on the L^p and $W^{1,p}$ norms of vector fields along u^R is as follows. If we rescale the vector field $\xi(z) \in T_{u^R(z)}M$ with support in $B_{\delta/2R}$ to obtain a vector field $\xi^\infty(z) := \xi(z/R^2) \in T_{u^\infty(z)}M$ along u^∞ then the standard norms of this rescaled vector field ξ^∞ agree with the weighted norms of the original vector field ξ.

More explicitly, recall that the almost complex structure $J^R = \{J_z^R\}_{z \in S^2}$ is defined by (10.1.4) and define the weighted norms

$$\|\xi\|_{0,p,R} = \left(\int_{\mathbb{C}} \theta^R(z)^{-2} |\xi(z)|^p_{J_z^R} \right)^{1/p},$$

$$\|\xi\|_{1,p,R} = \left(\int_{\mathbb{C}} \theta^R(z)^{-2} |\xi(z)|^p_{J_z^R} + \theta^R(z)^{p-2} |\nabla\xi(z)|^p_{J_z^R} \right)^{1/p}.$$

Here $\nabla = \nabla^{R,z}$ denotes the Levi-Civita connection of the metric on M determined by ω and J_z^R via (2.1.1). Moreover, we use the coordinate $z = s + it$ on $\mathbb{C}$ and write

$|\nabla\xi(z)|_J^2 := |\nabla_s\xi(z)|_J^2 + |\nabla_t\xi(z)|_J^2$. For $\eta = \eta_1 ds + \eta_2 dt \in \Omega^1(S^2, u^{R^*}TM)$, write $|\eta|^2 := |\eta_1|^2 + |\eta_2|^2$, $|\nabla\eta|_J^2 := |\nabla_s\eta_1|_J^2 + |\nabla_t\eta_1|_J^2 + |\nabla_s\eta_2|_J^2 + |\nabla_t\eta_2|_J^2$, and define

$$\|\eta\|_{0,p,R} = \left(\int_{\mathbb{C}} \theta^R(z)^{p-2}|\eta(z)|_{J_z^R}^p\right)^{1/p},$$

$$\|\eta\|_{0,\infty,R} = \sup_{z\in\mathbb{C}} \theta^R(z)|\eta(z)|_{J_z^R},$$

$$\|\eta\|_{1,p,R} = \left(\int_{\mathbb{C}} \theta^R(z)^{p-2}|\eta(z)|_{J_z^R}^p + \theta^R(z)^{2p-2}|\nabla\eta(z)|_{J_z^R}^p\right)^{1/p}.$$

In the case $R = 1$ and $J_z^R \equiv J$ these are the usual L^p and $W^{1,p}$ norms with respect to the Fubini–Study metric on $\mathbb{C}$ (as a coordinate patch of $\mathbb{C}P^1$). For general R these norms should be considered in two parts. In the domain $|z| \geq 1/R$ they are still the usual norms and in the domain $|z| \leq 1/R$ they agree with the usual norms of the rescaled vector field

$$\xi^\infty(z) := \xi(z/R^2)$$

or 1-form

$$\eta^\infty := R^{-2}\left(\eta_1(z/R^2)ds + \eta_2(z/R^2)dt\right)$$

along $u^\infty(z)$ in the ball of radius R, again with respect to the Fubini–Study metric on $\mathbb{C}$. Hence, as the next lemma shows, we obtain the usual Sobolev estimates with constants that are independent of R.

LEMMA 10.3.1. *Let $A^0, A^\infty \in H_2(M;\mathbb{Z})$ and $J^0, J^\infty \in \mathcal{J}(S^2; M, \omega)$ be given such that* (10.1.1) *holds. Fix two constants $c > 0$ and $p \geq 1$ with $p \neq 2$. Then there exists a constant $c_0 = c_0(p,c) > 0$ such that the following holds for all δ and R that satisfy* (10.2.3) *and all pairs $u = (u^0, u^\infty) \in \mathcal{M}(c)$, where $u^R = u^{\delta,R} : S^2 \to M$ is given by* (10.2.1) *and* (10.2.2).

(i) *If $p > 2$ then*

$$\|\xi\|_{L^\infty} \leq c_0 \|\xi\|_{1,p,R}, \qquad \|\eta\|_{0,\infty,R} \leq c_0 \|\eta\|_{1,p,R}$$

for every $\xi \in \Omega^0(S^2, (u^R)^*TM)$ and every $\eta \in \Omega^1(S^2, (u^R)^*TM)$.

(ii) *If $1 \leq p < 2$ and $1 \leq q \leq 2p/(2-p)$ then*

$$\|\xi\|_{0,q,R} \leq c_0 \|\xi\|_{1,p,R}, \qquad \|\eta\|_{0,q,R} \leq c_0 \|\eta\|_{1,p,R}$$

for every $\xi \in \Omega^0(S^2, (u^R)^*TM)$ and every $\eta \in \Omega^1(S^2, (u^R)^*TM)$.

PROOF. Both assertions are proved separately for the domains $\{|z| \geq 1/R\}$ and $\{|z| \leq 1/R\}$. We first consider the domain $\{|z| \geq 1/R\}$ and use the fact that the metric in this domain is the Fubini–Study metric on $S^2 \setminus B_{1/R}$ (and so is independent of R). For $p > 2$ the result for functions now follows from the results in Appendix B. Each point $z_0 \in S^2 \setminus B_{1/R}$ is contained in a disc D of radius $\pi/4$ in $S^2 \setminus B_{1/R}$. All such discs are isometric and, because $d_{\mathrm{FS}}(z,0) = \tan^{-1}(|z|)$, can be identified with the unit disc $B_1(0)$ in $(\mathbb{C}, g^1)$. Since this metric is equivalent to the Euclidean metric, the result follows immediately from Lemma B.1.16 and Proposition B.1.9. The corresponding result for vector fields ξ and 1-forms η can be proved as in Remark 3.5.1.

For the case $p < 2$ we examine the proof of Lemma B.1.17. This shows that the estimate in the domain $\{|z| \geq 1/R\}$ with a uniform constant can be obtained for functions that vanish outside the unit ball by integrating along rays parallel to the real and imaginary axes starting from outside the unit ball. Now the uniform estimate on the whole of $S^2 \setminus B_{1/R}$ is obtained by using cutoff functions to separate into a part near $|z| = 1/R$ with support in B_1 and a part with support in $S^2 \setminus B_{1/2}$. The previous argument also works for functions supported in $S^2 \setminus B_{1/2}$ since this can again be identified with a ball centered at the origin in $\mathbb{C}$ whose metric is equivalent to the Euclidean metric.

This proves the estimates in the domain $\{|z| \ge 1/R\}$. In the domain $\{|z| \le 1/R\}$ the results now follow from the involutive symmetry in the formula for the definition of the maps u^R and the weighting functions θ^R (see Exercise 10.5.2 below). This proves Lemma 10.3.1. □

The next lemma plays a key role in our application of the implicit function theorem (Theorem 3.5.2). We formulated the latter so that it applied when one has estimates that are uniform over all metrics on the Riemann surface $\Sigma = \mathbb{C}P^1$ with uniformly bounded Sobolev embedding constants and over all smooth maps $u : \Sigma \to M$ whose first derivatives satisfy a uniform L^p-bound. Lemma 10.3.1 asserts that the metrics $(\theta^R)^{-2}(ds^2 + dt^2)$ satisfy the first hypothesis and Lemma 10.3.2 below asserts that the connected sums u^R satisfy the second. The method of proof is a paradigm for later more elaborate calculations.

LEMMA 10.3.2. *Let A^0, A^∞ and J^0, J^∞ be as in Lemma 10.3.1. Fix two constants $c > 0$ and $p > 2$. Then there exists a constant $c_0 = c_0(c, p) > 0$ such that the following holds. If δ and R satisfy* (10.2.3) *then, for every pair $(u^0, u^\infty) \in \mathcal{M}(c)$, the function u^R defined by* (10.2.1) *and* (10.2.2) *satisfies the estimates*

$$\left\|du^R\right\|_{0,\infty,R} \le c_0, \qquad \left\|du^R\right\|_{0,p,R} \le c_0, \qquad \left\|\bar\partial_{J^R}(u^R)\right\|_{0,p,R} \le c_0\,(\delta R)^{-2/p}.$$

PROOF. The proof of the first two estimates is an easy exercise. It uses the symmetry of the formula (10.2.2) in u^0 and u^∞ (spelled out in Exercise 10.5.2 below) and the fact that, because $\zeta^0(0) = 0$, the product $\delta R\zeta^0(z)$ is uniformly bounded in the domain $|z| \le 2/\delta R$ by a constant that is independent of δ, R, and (u^0, u^∞).

To prove the third estimate, set $r := \delta R$ and recall from Remark 10.2.2 that $u^R(z) = u^{0,r}(z)$ for $|z| \ge 1/R$ and that $u^{0,r}$ converges to u^0 in the $W^{1,p}$ norm as r tends to infinity. Again using symmetry, one concludes that

$$\left\|\bar\partial_{J^R}(u^R)\right\|_{0,p,R} \to 0.$$

To obtain the more precise estimate of the convergence rate observe that

$$u^{0,r}(z) = \exp_x(\rho(rz)\zeta^0(z)), \qquad \exp_x(\zeta^0(z)) = u^0(z),$$

for $|z| \le 2/r$ and $r = \delta R$ sufficiently large (see equation (10.2.2)). Hence the term $\bar\partial_{J^0}(u^{0,r})$ is supported in the annulus $B_{2/r} \setminus B_{1/r}$ and in this region can be expressed as a sum of two terms. One involves the first derivatives of the cutoff function $\rho(rz)$ (which grow like r) multiplied by the function $\zeta^0(z)$, which is bounded by a constant times $1/r$ provided that $|z| \le 2/r$. Thus this product is uniformly bounded. The second term involves the first derivatives of $\zeta^0(z)$ and these are again uniformly bounded. Since the integral for this estimate is over a region of area at most $4\pi r^{-2}$, this proves Lemma 10.3.2. □

EXERCISE 10.3.3. This exercise shows that the operators D_{u^R}, as defined in Proposition 3.1.1, are uniformly bounded. Let $A^0, A^\infty \in H_2(M;\mathbb{Z})$ and choose $(J^0, J^\infty) \in \mathcal{J}_{\mathrm{reg}}(S^2; A^{0,\infty})$. Fix two constants $c > 0$ and $p > 2$. Prove that there is a constant c_0 such that the following holds for all δ and R that satisfy (10.2.3). If $(u^0, u^\infty) \in \mathcal{M}(c)$ and $u^R : S^2 \to M$ is defined by (10.2.2) then, for every $\xi \in \Omega^0(S^2, (u^R)^*TM)$,

$$\|D_{u^R}\xi\|_{0,p,R} \le c_0 \|\xi\|_{1,p,R}.$$

REMARK 10.3.4. Fix a pair $(u^0, u^\infty) \in \mathcal{M}(A^{0,\infty}; J^{0,\infty})$. For every pair of real numbers δ and R that satisfy (10.2.3) define $u^R : S^2 \to M$ by (10.2.2). Then each operator D_{u^R} satisfies estimates of the form

$$\|\xi\|_{k+1,p,R} \le c_{k,p,\delta,R}\left(\|D_{u^R}\xi\|_{k,p,R} + \|\xi\|_{k,p,R}\right) \tag{10.3.3}$$

for $\xi \in \Omega^0(S^2, (u^R)^*TM)$. One might now ask if the constants $c_{k,p,\delta,R}$ can be chosen independent of R. Remarkably, the answer is yes for $k = 0$ and no for $kp > 2$.

The result for $k = 0$ is rather nontrivial. It follows from Proposition 10.5.1 below by considering first the case where u^0 and u^∞ are constant and then reducing the general case to this via a refinement of Lemma 10.4.2 for functions on an annulus that vanish at a point on the inner boundary. We shall not carry out the details because our gluing argument scrupulously avoids the estimate (10.3.3).

To examine the case $kp > 2$ we consider two constant curves $u^0(z) \equiv u^\infty(z) \equiv x$. So the pre-glued curves u^R are also constant and D_{u^R} is the standard Cauchy–Riemann operator for functions from S^2 to T_xM. Thus the question concerns the effect of the weighting functions θ^R on the Calderon–Zygmund inequality. An enlightening example is given by sections of the form

$$\xi^R(z) := \beta^R(z)\left(z + \frac{1}{R^2 z}\right)\xi_0,$$

where $\xi_0 \in T_xM$ is a nonzero vector and multiplication by $i = \sqrt{-1}$ is understood in terms of the almost complex structure $J(x)$ at the point $x \in M$. We assume that the $\beta^R : \mathbb{C} \to [0,1]$ are smooth cutoff functions that are independent of R in the region $|z| \ge 1$ and satisfy

$$\beta^R(z) = \beta^R\left(\frac{1}{R^2 z}\right) = \begin{cases} 1, & \text{if } \frac{1}{R^2} \le |z| \le 1, \\ 0, & \text{if } |z| \ge 2 \text{ or } |z| \le \frac{1}{2R^2}. \end{cases}$$

For symmetry reasons we only need to examine the terms in the estimate in the domain $|z| \ge 1/R$. Since the functions $z \mapsto z + \frac{1}{R^2 z}$ are holomorphic and all their derivatives are uniformly bounded in the domain $1 \le |z| \le 2$ we deduce that

$$\sup_R \left\|\bar\partial\xi^R\right\|_{k,p,R} < \infty$$

for every k and every p. The (k,p,R)-norm of ξ^R is determined by the behaviour near $|z| = 1/R$ and here the term $z \mapsto \frac{1}{R^2 z}$ is the dominating one in the region $1/R \le |z| \le 1$. A simple calculation shows that the (k,p,R)-norms of ξ^R are uniformly bounded whenever $(k-1)p < 2$. The same calculation shows that, for every integer k such that $kp > 2$, there is a constant $c = c(k,p) > 0$ such that

$$c^{-1}R^{k-2/p} \le \left\|\xi^R\right\|_{k+1,p,R} \le cR^{k-2/p}.$$

This implies that the constant $c_{k,p,\delta,R}$ in (10.3.3) cannot be chosen independent of R whenever $k \ge 1$ and $kp > 2$.

10.4. Cutoff functions

So far we used an arbitrary smooth cutoff function ρ to define the pregluing u^R. The rest of our constructions will also be made by patching together objects defined over the regions $\{|z| \ge 1/R\}$ and $\{|z| \le 1/R\}$. However, in order for the required estimates to hold it will be necessary to use a very finely tuned cutoff function that we shall denote by β. Throughout this section all integrals are to be understood with respect to the Lebesgue measure. A reader might well skip this section at first, returning to it after reading in the proof of Proposition 10.5.1 how the various properties of β are used.

LEMMA 10.4.1. *For any two real numbers $0 < \delta < \varepsilon$ there exists a Lipschitz continuous cutoff function $\beta = \beta_{\delta,\varepsilon} : \mathbb{R}^2 \to [0,1]$ such that*

$$\beta(z) = \begin{cases} 1, & \text{if } |z| \le \delta, \\ 0, & \text{if } |z| \ge \varepsilon, \end{cases} \qquad \int_{\mathbb{R}^2} |\nabla\beta|^2 = \frac{2\pi}{\log(\varepsilon/\delta)}.$$

PROOF. Define $\beta : \mathbb{R}^2 \to [0,1]$ by $\beta(z) := 1$ for $|z| \le \delta$, $\beta(z) := 0$ for $|z| \ge \varepsilon$, and

$$\beta(z) := \frac{\log\varepsilon - \log|z|}{\log\varepsilon - \log\delta}, \qquad \delta \le |z| \le \varepsilon. \tag{10.4.1}$$

Then $|\nabla\beta(z)| = 1/|z|\log(\varepsilon/\delta)$ and hence

$$\int_{\mathbb{R}^2} |\nabla\beta(z)|^2 = \int_{\delta\le|z|\le\varepsilon} \frac{1}{|z|^2\log(\varepsilon/\delta)^2} = \int_\delta^\varepsilon \frac{2\pi}{s\log(\varepsilon/\delta)^2}\,ds = \frac{2\pi}{\log(\varepsilon/\delta)}.$$

This proves Lemma 10.4.1. □

We emphasize that the L^2-norm of $\nabla\beta$ in Lemma 10.4.1 depends only the quotient ε/δ. This illustrates the conformal invariance of the L^2-norm of the derivative in dimension two. Moreover, the cutoff function β can evidently be chosen to be smooth if the constant $2\pi/\log(\varepsilon/\delta)$ is replaced by any larger constant. It follows that there is a sequence of compactly supported functions on $\mathbb{R}^2$ which converge to zero in the $W^{1,2}$-norm but not in the L^∞-norm (fix a constant $\varepsilon > 0$ and let δ converge to zero). This is a borderline case for the Sobolev estimates. For $p > 2$ there is an embedding $W^{1,p}(\mathbb{R}^2) \hookrightarrow C(\mathbb{R}^2)$ and hence the assertion of the previous lemma does not hold if we replace the L^2-norm of $\nabla\beta$ by the L^p-norm for $p > 2$. In the following we shall see that the L^2-norm is precisely what is needed for the proof of surjectivity of the first order operator arising in the gluing construction in two dimensions. A similar argument in n dimensions requires the L^n-norm of $\nabla\beta$ to be small and this is again a Sobolev borderline case. The case $n = 4$ is relevant for the gluing of anti-self-dual instantons and this is explained in Donaldson–Kronheimer [**88**, pp 287–295].

Now let β be given by (10.4.1) and $\xi \in W^{1,q}(\mathbb{R}^2)$, where $1 \le q < 2$. Then we claim that, for a suitable constant $r \ge 2$,

$$\|\nabla\beta\cdot\xi\|_{L^q} \le \|\nabla\beta\|_{L^2}\|\xi\|_{L^r} \le c\|\nabla\beta\|_{L^2}\|\xi\|_{W^{1,q}} \le \frac{2\pi c}{\log(\varepsilon/\delta)}\|\xi\|_{W^{1,q}}. \tag{10.4.2}$$

Here the first inequality uses the Hölder inequality

$$\left(\int |uv|^q\right)^{1/q} \le \left(\int |u|^s\right)^{1/s}\left(\int |v|^r\right)^{1/r}, \qquad \frac{1}{s} + \frac{1}{r} = \frac{1}{q},$$

with $s = 2$. Hence we must take $r := 2q/(2-q)$ and so, by Theorem B.1.12, there is a Sobolev embedding $W^{1,q} \hookrightarrow L^r$. This implies the second inequality above. The third follows from Lemma 10.4.1. It follows from (10.4.2) that the multiplication operators $W^{1,q}(\mathbb{R}^2) \to L^q(\mathbb{R}^2) : \xi \mapsto \nabla\beta_{\delta,\varepsilon} \cdot \xi$ converge to zero in the operator norm as δ tends to zero whenever $q < 2$. This assertion, as it stands, no longer holds for $q > 2$. However, in the case $q > 2$ we can consider the restrictions of these operators to the subspace of all functions that vanish at the origin. The next lemma shows that these restricted operators still converge to zero in the operator norm as $\delta \to 0$ when $q > 2$. We denote

$$B_r := \{z \in \mathbb{R}^2 \,|\, |z| \le r\}, \qquad A(r,R) := \left\{z \in \mathbb{R}^2 \,|\, r \le |z| \le R\right\}.$$

LEMMA 10.4.2. *For every $p > 2$ there is a constant $c > 0$ with the following significance. Given $0 < \delta < \varepsilon$ define $\beta : \mathbb{R}^2 \to [0,1]$ by (10.4.1). Then*

$$\|\nabla\beta \cdot \xi\|_{L^p(A(\delta,\varepsilon))} \le \frac{c}{\log(\varepsilon/\delta)^{1-1/p}} \|\xi\|_{W^{1,p}(B_\varepsilon)}$$

for every $\xi \in W^{1,p}(B_\varepsilon)$ such that $\xi(0) = 0$.

PROOF. Recall from the proof of Lemma 10.4.1 that $\nabla\beta$ is supported in the annulus $A(\delta,\varepsilon)$ and that $|\nabla\beta(z)| = 1/|z|\log(\varepsilon/\delta)$ for $\delta \le |z| \le \varepsilon$. Moreover, by Theorem B.1.11, every $\xi \in W^{1,p}(\mathbb{R}^2)$ is Hölder continuous with exponent $\mu = 1 - 2/p$. More precisely, there is a constant $c > 0$, depending only on $p > 2$, such that

$$z_0, z_1 \in B \quad \Longrightarrow \quad |\xi(z_1) - \xi(z_0)| \le c\,\|\xi\|_{W^{1,p}(B)}\, |z_1 - z_0|^{1-2/p} \tag{10.4.3}$$

for every ball $B \subset \mathbb{R}^2$ and every $\xi \in W^{1,p}(B)$ (see the proof of Lemma B.1.16). Hence every function $\xi \in W^{1,p}(B_\varepsilon)$ that vanishes at the point $z = 0$ satisfies the estimate

$$\begin{aligned}
\int_{\delta\le|z|\le\varepsilon} |\nabla\beta(z)|^p|\xi(z)|^p &\le \frac{c^p}{\log(\varepsilon/\delta)^p}\int_{\delta\le|z|\le\varepsilon}\frac{1}{|z|^2}\,\|\xi\|^p_{W^{1,p}(B_\varepsilon)} \\
&= \frac{2\pi c^p}{\log(\varepsilon/\delta)^p}\int_\delta^\varepsilon \frac{1}{s}\,ds\,\|\xi\|^p_{W^{1,p}(B_\varepsilon)} \\
&= \frac{2\pi c^p}{\log(\varepsilon/\delta)^{p-1}}\,\|\xi\|^p_{W^{1,p}(B_\varepsilon)}.
\end{aligned}$$

This proves Lemma 10.4.2. □

We shall also need the following result for an arbitrary smooth cutoff function $\rho : \mathbb{C} \to [0,1]$ that equals zero for $|z| \le 1$ and equals 1 for $|z| \ge 2$.

LEMMA 10.4.3. *Let $p > 2$ and $\xi \in W^{1,p}(B_1)$ be such that $\xi(0) = 0$, and denote $\rho_R(z) := \rho(Rz)$. Then $\lim_{R\to\infty} \|\xi - \rho_R\xi\|_{W^{1,p}(B_1)} = 0$.*

PROOF. We prove that there is a constant $c > 0$ such that

$$\|\xi - \rho_R\xi\|_{W^{1,p}(B_1)} \le c\,\|\xi\|_{W^{1,p}(B_{2/R})} \tag{10.4.4}$$

To see this note first that the estimate holds with $c = 1$ if we replace the $W^{1,p}$-norm by the L^p-norm. Secondly, we have

$$\nabla(\xi - \rho_R\xi) = (1-\rho_R)\nabla\xi - (\nabla\rho_R)\xi$$

The L^p-norm of the first term satisfies the required inequality with $c = 1$. To estimate the second term we observe that $\|\nabla\rho_R\|_{L^\infty} = Rc_0$, where $c_0 := \|\nabla\rho\|_{L^\infty}$. Moreover, by (10.4.3) there is a constant $c_1 > 0$ such that

$$|\xi(z)| \le c_1|z|^{1-2/p}\, \|\xi\|_{W^{1,p}(B_{2/R})}, \qquad |z| \le 2/R.$$

Hence

$$\begin{aligned} \|(\nabla\rho_R)\xi\|_{L^p} &\le c_0c_1R\left(\int_{|z|\le 2/R} |z|^{p-2}\right)^{1/p} \|\xi\|_{W^{1,p}(B_{2/R})} \\ &= c_0c_1\left(\frac{2^{p+1}\pi}{p}\right)^{1/p} \|\xi\|_{W^{1,p}(B_{2/R})}. \end{aligned}$$

This proves (10.4.4) and Lemma 10.4.3. □

10.5. Construction of the gluing map

In this section we define the gluing map ι_c^R and show that it has properties (i), (ii), and (iii) of Theorem 10.1.2. As explained in the introduction, the idea is to find a true holomorphic curve $\widetilde{u}^R := \iota_c^R(u^0, u^\infty)$ near the pregluing $u^R := u^{\delta,R}$ by applying the implicit function theorem. In order to do this, we must prove that the Fredholm operator D_{u^R} is surjective and has a uniformly bounded right inverse provided that $\delta > 0$ is sufficently small and δR is sufficiently large.

An estimate for the inverse. As a first step, we construct an approximate right inverse T_{u^R} by a process used in Donaldson–Kronheimer [**88**]. The next proposition formulates its main properties. Recall from (10.1.5) that $\mathcal{A}(\delta_0)$ denotes the set of pairs of real numbers δ and R such that $0 < \delta < \delta_0$ and $\delta R > 1/\delta_0$.

PROPOSITION 10.5.1. *Let $A^0, A^\infty \in H_2(M;\mathbb{Z})$ and $(J^0, J^\infty) \in \mathcal{J}_{\mathrm{reg}}(S^2; A^{0,\infty})$. Fix two constants $c > 0$ and $p > 2$. Then there are two constants $\delta_0 > 0$ and $c_0 > 0$ and a smooth map which assigns to each element $u = (u^0, u^\infty) \in \mathcal{M}(c)$ and each pair $(\delta, R) \in \mathcal{A}(\delta_0)$ a bounded linear operator*

$$T_{u^R} : L^p(S^2, \Lambda^{0,1} \otimes_{J^R} (u^R)^*TM) \to W^{1,p}(S^2, (u^R)^*TM)$$

along the preglued curve $u^R : S^2 \to M$ defined by (10.2.1) and (10.2.2), such that

$$\|D_{u^R}T_{u^R}\eta - \eta\|_{0,p,R} \le \tfrac{1}{2}\|\eta\|_{0,p,R}, \qquad \|T_{u^R}\eta\|_{1,p,R} \le \frac{c_0}{2}\|\eta\|_{0,p,R} \tag{10.5.1}$$

for every $\eta \in L^p(S^2, \Lambda^{0,1} \otimes_{J^R} (u^R)^*TM)$.

Before embarking on the proof, we set up some simplified notation. To begin let us rephrase the regularity condition for J. Abbreviate

$$W^{1,p}_u := W^{1,p}(S^2, u^*TM), \qquad L^p_{u,J} := L^p(S^2, \Lambda^{0,1} \otimes_J u^*TM)$$

for $u : S^2 \to M$ and $J = \{J_z\}_{z\in S^2}$. Given $u^0, u^\infty : S^2 \to M$ such that $u^0(0) = u^\infty(\infty)$, denote

$$W^{1,p}_{u^{0,\infty}} := \left\{(\xi^0, \xi^\infty) \in W^{1,p}_{u^0} \times W^{1,p}_{u^\infty} \,|\, \xi^0(0) = \xi^\infty(\infty)\right\}$$

Note that the last definition only makes sense for $p > 2$ since it is only in this case that $W^{1,p}$-sections are continuous and can be evaluated at a point. The transversality assumption in Definition 10.1.1 implies that the operator

$$D_{0,\infty} := D_{u^0,u^\infty} : W^{1,p}_{u^{0,\infty}} \to L^p_{u^0,J^0} \times L^p_{u^\infty,J^\infty},$$

defined by

$$D_{0,\infty}(\xi^0, \xi^\infty) := (D_{u^0}\xi^0, D_{u^\infty}\xi^\infty)$$

is onto. To see this, let $\eta^a \in L^p_{u^a,J^a}$ be given for $a = 0, \infty$ and choose $\xi^a \in W^{1,p}_{u^a}$ such that $D_{u^a}\xi^a = \eta^a$. Then choose tangent vectors $\zeta^a \in T_{u^a}\mathcal{M}(A^a; J^a)$ so that

$$\mathrm{dev}(u^0, u^\infty)(\zeta^0, \zeta^\infty) = (\zeta^0(0), \zeta^\infty(\infty)) \in (\xi^0(0), \xi^\infty(\infty)) + T_{(x,x)}\Delta.$$

Then $(\xi^0 - \zeta^0, \xi^\infty - \zeta^\infty)$ is an element of $W^{1,p}_{u^{0,\infty}}$ whose image under $D_{0,\infty}$ is also (η^0, η^∞). Hence $D_{0,\infty}$ has a right inverse that depends smoothly on the pair (u^0, u^∞) and satisfies a uniform bound as (u^0, u^∞) varies in $\mathcal{M}(c) := \mathcal{M}(A^{0,\infty}; J, c)$.

More precisely, let us denote by

$$\mathcal{W}_{u^{0,\infty}} \subset W^{1,p}_{u^{0,\infty}}$$

the L^2-orthogonal complement of the kernel of $D_{0,\infty}$. Then the restriction of $D_{0,\infty}$ to $\mathcal{W}_{u^{0,\infty}}$ is a bijective bounded linear operator. Its inverse is the required operator

$$Q_{0,\infty} := Q_{u^0,u^\infty} := \left(D_{0,\infty}|_{\mathcal{W}_{u^{0,\infty}}}\right)^{-1}.$$

It depends smoothly on the pair (u^0, u^∞). The uniform estimate on its norm follows from the fact that $\mathcal{M}(c)$ is compact. Note that the domains of the operators and their range actually depend on the pair (u^0, u^∞). So, to make this continuous dependence precise, one has to choose an identification of $W^{1,p}_{u^{0,\infty}}$ and $W^{1,p}_{v^{0,\infty}}$ for nearby pairs (u^0, u^∞) and (v^0, v^∞). This can be done by parallel transport along short geodesics as in the proof of Theorem 3.5.2.

Now consider the maps $u^{0,r}, u^{\infty,r}$ defined in Remark 10.2.2, where as usual $r := \delta R$. Since they are $W^{1,p}$-small perturbations of u^0, u^∞, the space $W^{1,p}_{u^{0,\infty}}$ may be interpreted as a limit as $r \to \infty$ of the spaces $W^{1,p}_{u^{0,\infty,r}}$ of vector fields along these flattened curves. Moreover, the operator

$$D_{0,\infty,r} := D_{u^{0,r},u^{\infty,r}}$$

is a $W^{1,p}$-small perturbation of the linearized operator $D_{0,\infty} := D_{u^0,u^\infty}$ and so has a right inverse as well. In the following we shall choose the unique right inverse whose image is the L^2-orthogonal complement of the kernel of $D_{0,\infty,r}$. It will be denoted by

$$Q_{0,\infty,r} := Q_{u^{0,r},u^{\infty,r}} : L^p_{u^{0,r},J^0} \times L^p_{u^{\infty,r},J^\infty} \to W^{1,p}_{u^{0,\infty,r}}.$$

Because the operators $D_{0,\infty,r}$ are small perturbations of $D_{0,\infty}$ they do have uniformly bounded right inverses. However, here we are using a particular family of right inverses $Q_{0,\infty,r}$ whose images are L^2-orthogonal to $\ker D_{0,\infty,r}$. Hence we need to establish uniform estimates for the $Q_{0,\infty,r}$. These are stated and proved in Lemma 10.6.1 below. What we shall need here is the following statement:for every $c > 0$ there are positive constants δ_0, c_0 such that

$$\|Q_{0,\infty,r}\eta\|_{W^{1,p}} \le c_0 \|\eta\|_{L^p}, \tag{10.5.2}$$

for all $u = (u^0, u^\infty) \in \mathcal{M}(c)$, $r > 1/\delta_0$, and

$$\eta = (\eta^0, \eta^\infty) \in L^p_{u^{0,r},J^0} \times L^p_{u^{\infty,r},J^\infty}.$$

Our strategy is now to use the operator $Q_{0,\infty,r}$ to construct, for $r = \delta R$ sufficiently large, approximate right inverses T_{u^R} of the operators D_{u^R} along the preglued curves u^R and to establish a uniform estimate for these approximate right inverses.

PROOF OF PROPOSITION 10.5.1. We begin with the definition of the operator

$$T_{u^R} : L^p_{u^R,J^R} \to W^{1,p}_{u^R}.$$

The construction can be summarized in terms of the following commutative diagram, where the vertical maps are given in terms of cutoff functions:

$$\begin{array}{ccc} W^{1,p}_{u^{0,\infty,r}} & \xleftarrow{\;Q_{0,\infty,r}\;} & L^p_{u^{0,r},J^0} \times L^p_{u^{\infty,r},J^\infty} \\ \big\downarrow & & \big\downarrow \\ W^{1,p}_{u^R} & \xleftarrow{\;T_{u^R}\;} & L^p_{u^R,J^R} \end{array} \tag{10.5.3}$$

Here are the details. Given $\eta \in L^p_{u^R,J^R}$ we first define the pair

$$(\eta^0, \eta^\infty) \in L^p_{u^{0,r},J^0} \times L^p_{u^{\infty,r},J^\infty}$$

by cutting off η along the circle $|z| = 1/R$:

$$\eta^0(z) := \begin{cases} \eta(z), & \text{if } |z| \ge 1/R, \\ 0, & \text{if } |z| \le 1/R, \end{cases} \qquad \eta^\infty(z) := \begin{cases} R^{-2}\eta(z/R^2), & \text{if } |z| \le R, \\ 0, & \text{if } |z| \ge R. \end{cases}$$

The discontinuities in η^0 and η^∞ do not cause problems because only their L^p-norms enter the estimates. Second, define

$$(\xi^0, \xi^\infty) := Q_{0,\infty,r}(\eta^0, \eta^\infty)$$

and note that the vector fields ξ^0, ξ^∞ have the same central value ξ_0:

$$\xi^0(0) = \xi^\infty(\infty) =: \xi_0 \in T_{u^0(0)}M.$$

Third, let $1 - \beta_{\delta,R} : \mathbb{C} \to \mathbb{R}$ denote the cutoff function of Lemma 10.4.1 with δ and ε replaced by δ/R and $1/R$, respectively. Thus, $\beta_{\delta,R}(z) = 0$ for $|z| \le \delta/R$, $\beta_{\delta,R}(z) = 1$ for $|z| \ge 1/R$, and

$$\beta_{\delta,R}(z) := \frac{\log(R|z|/\delta)}{\log(1/\delta)}, \qquad \frac{\delta}{R} \le |z| \le \frac{1}{R}.$$

Fourth, define $T_{u^R}\eta := \xi^R$ by

$$\xi^R(z) := \begin{cases} \xi^0(z), & \text{if } |z| \ge \frac{1}{\delta R}, \\[2mm] \xi^0(z) + \beta_{\delta,R}(\frac{1}{R^2 z})\big(\xi^\infty(R^2 z) - \xi_0\big), & \text{if } \frac{1}{R} \le |z| \le \frac{1}{\delta R}, \\[2mm] \xi^0(z) + \xi^\infty(R^2 z) - \xi_0, & \text{if } |z| = \frac{1}{R}, \\[2mm] \xi^\infty(R^2 z) + \beta_{\delta,R}(z)\big(\xi^0(z) - \xi_0\big), & \text{if } \frac{\delta}{R} \le |z| \le \frac{1}{R}, \\[2mm] \xi^\infty(R^2 z), & \text{if } |z| \le \frac{\delta}{R}. \end{cases} \tag{10.5.4}$$

The easiest way to understand this definition is by considering the case when the central value ξ_0 vanishes and z lies in the annulus $\delta/R \le |z| \le 1/\delta R = 1/r$. In this annulus the maps $u^{0,r}(z)$, $u^{\infty,r}(R^2 z)$, and $u^R(z)$ all take the constant value

$$x := u^0(0) = u^\infty(\infty),$$

and the function $\xi^R : \{\delta/R \le |z| \le 1/\delta R\} \to T_xM$ is simply the superposition of the functions $\beta_{\delta,R}(z)\xi^0(z)$ and $\beta_{\delta,R}(1/R^2z)\xi^\infty(R^2z)$. The important fact is that, in the first term, the cutoff function $\beta_{\delta,R}(z)$ only takes effect in the region $|z| \le 1/R$ where $D_{0,r}\xi^0 := D_{u^{0,r}}\xi^0 = 0$. (See Figure 1.) Similarly for the second term because the construction is completely symmetric in $u^{0,r}$ and $u^{\infty,r}$. In the case $\xi_0 \ne 0$ the formula can be interpreted in the same way, but relative to the "origin" ξ_0.

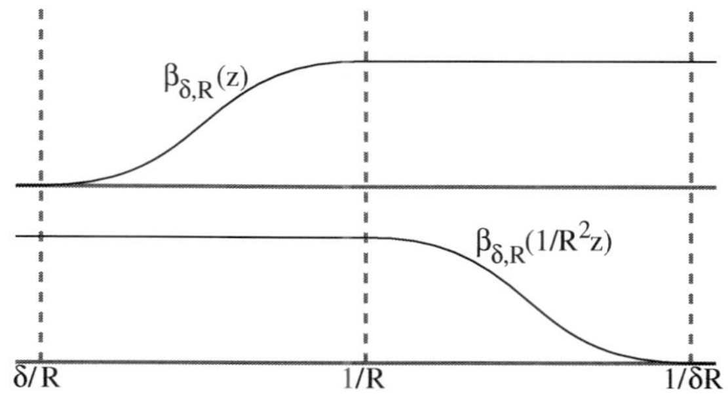

FIGURE 1. The cutoff functions.

Thus for each $\eta \in L^p_{u^R,J^R}$ we have defined an element $T_{u^R}\eta = \xi^R$, and must prove that

$$\left\|D_{u^R}\xi^R - \eta\right\|_{0,p,R} \le \tfrac{1}{2}\left\|\eta\right\|_{0,p,R}. \tag{10.5.5}$$

Since $D_{u^{0,r}}\xi^0 = \eta^0$ and $D_{u^{\infty,r}}\xi^\infty = \eta^\infty$, the term on the left hand side vanishes for $|z| \ge 1/\delta R$ and for $|z| \le \delta/R$. In view of the symmetry of our formulas it suffices to estimate the left hand side in the annulus

$$\delta/R \le |z| \le 1/R.$$

In this region, as remarked above,

$$u^{0,r}(z) = u^{\infty,r}(R^2z) = u^R(z) = x$$

is the constant map. Therefore over this annulus the vector field ξ^R takes values in the fixed vector space T_xM and the corresponding operators $D_{u^{0,r}}$, $D_{u^{\infty,r}}$, and D_{u^R} are all equal to the usual $\bar\partial_J$-operator. In particular, they vanish on the constant function ξ_0. Further, the definition of ξ^∞ implies that $D_{u^R}\xi^\infty(R^2\cdot) = \eta$ in the region $|z| \le 1/R$. Hence, when $|z| \le 1/R$, we find

$$\begin{aligned} D_{u^R}\xi^R - \eta &= D_{u^{0,r}}(\beta_{\delta,R}(\xi^0 - \xi_0)) \\ &= \beta_{\delta,R}D_{u^{0,r}}(\xi^0 - \xi_0) + \bar\partial\beta_{\delta,R} \otimes (\xi^0 - \xi_0) \\ &= \bar\partial\beta_{\delta,R} \otimes (\xi^0 - \xi_0). \end{aligned}$$

Here we have used the crucial fact that by construction $D_{u^{0,r}}\xi^0 = \eta^0 = 0$ in the region $|z| \le R^{-1}$.

Now we must estimate the L^p-norm of this 1-form with respect to the R-dependent metric. The next important point to observe is that the weighting function for 1-forms is $\theta^R(z)^{p-2}$. Since $p > 2$ and $\theta^R(z) \le \theta^1(z) \le 2$ in the region $|z| \le 1/R$, it follows that the $(0,p,R)$-norm of our 1-form is smaller that the ordinary L^p-norm (up to a universal factor less than 2). Hence we obtain the inequality

$$\begin{aligned}
\left\|D_{u^R}\xi^R - \eta\right\|_{0,p,R;B_{1/R}} &\le 2\left\|D_{u^R}\xi^R - \eta\right\|_{L^p(B_{1/R})} \\
&= 2\left\|\bar\partial\beta_{\delta,R}\otimes(\xi^0-\xi_0)\right\|_{L^p(B_{1/R})} \\
&\le \frac{c_2\left\|\xi^0-\xi_0\right\|_{W^{1,p}}}{|\log\delta|^{1-1/p}} \\
&\le \frac{c_3}{|\log\delta|^{1-1/p}}\left(\left\|\eta^0\right\|_{L^p} + \left\|\eta^\infty\right\|_{L^p}\right) \\
&\le \frac{2c_3}{|\log\delta|^{1-1/p}}\left\|\eta\right\|_{0,p,R}.
\end{aligned}$$

Here the third inequality uses the estimate of Lemma 10.4.2 while the fourth follows from the uniform estimate (10.5.2) for the right inverse $Q_{0,\infty,r}$ of $D_{0,\infty,r}$. Here we also need to use the fact that the size of the vector $\xi_0 := \xi^0(0)$ is bounded by the C^0 norm of ξ^0 and hence also by its $W^{1,p}$ norm. The last inequality follows from the definitions of η^0 and η^∞ and of the $(0,p,R)$-norm for 1-forms.

In view of the symmetry, spelled out in Exercise 10.5.2 below, we have a similar estimate in the domain $|z|\ge 1/R$. This proves that the operator T_{u^R} satisfies the first inequality in (10.5.1) provided that δ is sufficiently small and R is sufficiently large. In fact, we can first choose δ so small that $|\log\delta|^{1-1/p}\ge 4c_3$ and then choose R so large that the product δR satisfies (10.2.3). By choosing δ_0 appropriately, these conditions will hold for all pairs $(\delta,R)\in\mathcal{A}(\delta_0)$.

To prove the second inequality in (10.5.1) we examine the L^p-norm of $\nabla\xi$. The two crucial terms are the ones involving the cutoff-function $\beta_{\delta,R}$. By the symmetry of the formula, it suffices to examine one of these terms. We consider the second summand in the fourth term on the right hand side of (10.5.4). Differentiating this term we encounter an expression of the form

$$z\mapsto \nabla\beta_{\delta,R}(z)\otimes(\xi^0(z)-\xi_0)$$

in the domain $\delta/R\le|z|\le 1/R$. Using once again the fact that $\theta^R(z)\le\theta^1(z)\le 2$ in this domain, we obtain from Lemma 10.4.2 that

$$\begin{aligned}
\left\|\nabla\beta_{\delta,R}\otimes(\xi^0-\xi_0)\right\|_{0,p,R} &\le 2^{1-2/p}\left\|\nabla\beta_{\delta,R}\otimes(\xi^0-\xi_0)\right\|_{W^{1,p}} \\
&\le \frac{c_4}{|\log\delta|^{1-1/p}}\left\|\xi^0\right\|_{W^{1,p}} \\
&\le \frac{c_5}{|\log\delta|^{1-1/p}}\left\|\eta\right\|_{0,p,R}.
\end{aligned}$$

A similar inequality holds in the domain $|z|\ge 1/R$. This proves the second estimate in (10.5.1) with $c_0 := 2c_5$ provided that $\delta\le\delta_0(c)<1/e$. □

EXERCISE 10.5.2. Check the symmetry in the formula for $\xi = T_{u^R}\eta$ in the above proof of Proposition 10.5.1 by considering the functions

$$\widetilde{u}^0(z) := u^\infty(1/z),\qquad \widetilde{u}^\infty(z) := u^0(1/z),\qquad \widetilde{u}^R(z) := u^R(1/R^2z)$$

and the vector fields

$$\tilde\xi^0(z) := \xi^\infty(1/z),\qquad \tilde\xi^\infty(z) := \xi^0(1/z),\qquad \tilde\xi(z) := \xi(1/R^2z).$$

Construction of the gluing map. We are now ready to explain the construction of the gluing map $\iota_c^R = \iota_c^{\delta,R}$. Fix two homology classes $A^0, A^\infty \in H_2(M;\mathbb{Z})$, a regular pair of almost complex structures $(J^0, J^\infty) \in \mathcal{J}_{\mathrm{reg}}(S^2; A^{0,\infty})$, and two constants $p > 2$ and $c > 0$. In particular J^0 and J^∞ satisfy (10.1.1) for some constant $\kappa > 0$ and we define $J^R = \{J_z^R\}_{z\in S^2} \in \mathcal{J}_\tau(S^2; M, \omega)$ by (10.1.4) for $R \geq 2/\kappa$. Choose constants $\delta_0 > 0$ and $c_0 \geq 1$ such that the following holds.

(a) Each pair $(\delta, R) \in \mathcal{A}(\delta_0)$ satisfies (10.2.3).

(b) Theorem 3.5.2 holds with c_0 and $\delta = 4c_0^2\delta_0^{2/p}$ and $J = J^R$ for every $R \geq 2/\kappa$. (Although the almost complex structure in Theorem 3.5.2 is z-independent, the proof carries over verbatim to the z-dependent case as explained in Remark 3.5.4.)

(c) Lemmas 10.3.1 and 10.3.2 hold with c_0 for all $(\delta, R) \in \mathcal{A}(\delta_0)$.

(d) Proposition 10.5.1 holds with c_0 and δ_0.

Condition (a) guarantees $R \geq 2/\kappa$ for every pair $(\delta, R) \in \mathcal{A}(\delta_0)$ so that J^R is well defined. Condition (a) also guarantees that that the map u^R in (10.2.2) is well defined for $(\delta, R) \in \mathcal{A}(\delta_0)$. Thus condition (a) is implicitly contained in condition (d), but is listed separately for clarity. Next we construct the map $\iota^{\delta,R} : \mathcal{M}(c) \to \mathcal{M}(A; J^R)$ for $(\delta, R) \in \mathcal{A}(\delta_0)$. As before, we drop the superscript δ.

Fix a pair $(\delta, R) \in \mathcal{A}(\delta_0)$. Define the pregluing map $f^R : \mathcal{M}(c) \to C^\infty(S^2, M)$ by

$$f^R(u^0, u^\infty) := u^R, \tag{10.5.6}$$

where u^R is given by (10.2.1) and (10.2.2). Now fix an element $(u^0, u^\infty) \in \mathcal{M}(c)$ and denote $u^R := f^R(u^0, u^\infty)$. Then, by Lemma 10.3.2, we have

$$\left\|du^R\right\|_{0,p,R} \leq c_0, \qquad \left\|\bar{\partial}_{J^R}(u^R)\right\|_{0,p,R} \leq \frac{c_0}{(\delta R)^{2/p}} < \frac{4c_0^2\delta_0^{2/p}}{4c_0}, \tag{10.5.7}$$

where the last inequality uses $1/\delta R < \delta_0$. By Proposition 10.5.1, the operator

$$Q_{u^R} := T_{u^R}(D_{u^R}T_{u^R})^{-1} = \sum_{k=0}^\infty T_{u^R}(1\!\!1 - D_{u^R}T_{u^R})^k : L^p_{u^R,J^R} \to W^{1,p}_{u^R}$$

satisfies the estimate

$$D_{u^R}Q_{u^R} = 1\!\!1, \qquad \|Q_{u^R}\eta\|_{1,p,R} \leq c_0 \|\eta\|_{0,p,R} \tag{10.5.8}$$

for $\eta \in L^p_{u^R,J^R}$. Note that Q_{u^R} is the unique right inverse of D_{u^R} with the same image as T_{u^R}. Define the map

$$\mathcal{F}_{u^R} : W^{1,p}_{u^R} \to L^p_{u^R,J^R}$$

by

$$\mathcal{F}_{u^R}(\xi) := \Phi^R(u^R, \xi)^{-1}\Big(\bar{\partial}_{J^R}\big(\exp_{u^R}(\xi)\big)\Big),$$

where $\Phi^R(x, \xi) = \Phi^{R,z}(x, \xi) : T_xM \to T_{\exp_x(\xi)}M$ denotes parallel transport with respect to the Hermitian connection

$$\widetilde{\nabla} = \widetilde{\nabla}^{R,z} = \nabla^{R,z} - \tfrac{1}{2}J_z^R(\nabla^{R,z}J_z^R)$$

associated to the metric on M determined by ω and J_z^R as in Section 2.1. Then, by Remark 6.7.3 (see also Proposition 3.1.1), we have

$$\mathcal{F}_{u^R}(0) = \bar{\partial}_{J^R}(u^R), \qquad d(\mathcal{F}_{u^R} \circ Q_{u^R})(0) = 1\!\!1. \tag{10.5.9}$$

With these estimates in place we can apply the Implicit Function Theorem 3.5.2 to the map $\mathcal{F}_{u^R}$. To see this, denote the volume form of the metric (10.3.1) on S^2 by dvol^R. Then it follows from Lemma 10.3.1 that the Sobolev constant $c_p(\mathrm{dvol}^R)$, that appears in Theorem 3.5.2, is bounded above by c_0. Hence it follows from (10.5.7) and (10.5.8), that all the conditions of Theorem 3.5.2 are satisfied when $\Sigma = S^2$, $J = J^R$, $u = u^R$, $Q_u := Q_{u^R}$, $\xi_0 = 0$, c_0 is the above constant, the metric g^R on S^2 is given by (10.3.1), and the constant δ in Theorem 3.5.2 is chosen equal to $4c_0^2\delta_0^{2/p}$. (That the metric g^R on $S^2 = \mathbb{C}\cup\{\infty\}$ is only continuous along the circle $|z| = 1/R$ is immaterial; the proof of Theorem 3.5.2 does not use the derivatives of g^R.) Thus condition (b) asserts that Theorem 3.5.2 is applicable in this situation and hence provides a unique element $\widetilde{\xi}^R \in W^{1,p}_{u^R}$ such that

$$\mathcal{F}_{u^R}(\widetilde{\xi}^R) = 0, \qquad \|\widetilde{\xi}^R\|_{1,p,R} < 4c_0^2\delta_0^{2/p}, \qquad \widetilde{\xi}^R \in \operatorname{im} Q_{u^R}. \tag{10.5.10}$$

Moreover, the theorem implies that $\widetilde{\xi}^R$ is quite a bit smaller than $4c_0^2\delta_0^{2/p}$, namely

$$\|\widetilde{\xi}^R\|_{1,p,R} \le 2c_0 \left\|\bar\partial_{J^R}(u^R)\right\|_{0,p,R} \le \frac{2c_0^2}{(\delta R)^{2/p}}. \tag{10.5.11}$$

We define the map $\iota_c^R = \iota_c^{\delta,R} : \mathcal{M}(c) \to \mathcal{M}(A; J^R)$ by

$$\iota_c^R(u^0, u^\infty) := \widetilde{u}^R := \exp_{u^R}(\widetilde{\xi}^R), \qquad u^R := f^R(u^0, u^\infty), \tag{10.5.12}$$

where $\widetilde{\xi}^R \in \operatorname{im} Q_{u^R}$ is the unique solution of (10.5.10).

REMARK 10.5.3. The constants $p > 2$ and $c > 0$ only enter into the choice of the constants δ_0 and c_0 but not in the definition of the map $\iota_c^{\delta,R} = \iota_{c;p}^{\delta,R}$. To see this, observe that the constants δ_0 and c_0 in Proposition 10.5.1, and Lemmas 10.3.1 and 10.3.2 as well as the constant δ_1 of Theorem 3.5.2 depend continuously on $p > 2$. (Note that $\delta_1 = 1/2c_0c$ where c is as in Proposition 3.5.3.) Now suppose that δ_0 and c_0 satisfy conditions (a), (b), (c), and (d) for every p in some interval $[p_1, p_2]$ with $2 < p_1 < p_2 < \infty$. Then the resulting maps $\iota_{c;p}^{\delta,R}$ agree for $p_1 \le p \le p_2$. This follows directly from the uniqueness statement for the solutions of (10.5.10). That the maps $\iota_c^{\delta,R}$ and $\iota_{c'}^{\delta,R}$ agree on their common domain $\mathcal{M}(c')$ for $0 < c' < c$ is obvious from the definition.

We end this section by showing that the gluing map is an embedding that satisfies conditions (i), (ii), and (iii) in Theorem 10.1.2. The key step in the proof is an estimate for the differential of the gluing map. By construction, the gluing map ι_c^R is the composition of the pregluing map $u \mapsto u^R$ with the perturbation map $u^R \mapsto \widetilde{u}^R := \exp_{u^R}(\widetilde{\xi}^R)$. We must differentiate these maps in directions tangent to the moduli space $\mathcal{M}(c)$ of connected pairs of J-holomorphic curves. We shall see below that the derivative of the pregluing map f^R is fairly easy to understand while that of the perturbation map is much more complicated. The next proposition states the key estimate. Its proof is relegated to the next section.

PROPOSITION 10.5.4. *For every $c > 0$ and every $p > 2$ there are positive constants $\delta_0 > 0$ and $c_0 > 0$ such that the following holds for all $(\delta, R) \in \mathcal{A}(\delta_0)$. If $t \mapsto u_t := (u_t^0, u_t^\infty)$ is a smooth path in $\mathcal{M}(c)$, $u_t^R := f^R(u_t)$, and $\widetilde{\xi}_t^R \in \operatorname{im} Q_{u_t^R}$ such that $\widetilde{u}_t^R := \exp_{u_t^R}(\widetilde{\xi}_t^R) = \iota_c^R(u_t)$, then, for every t, we have*

$$\left\|\nabla_t \widetilde{\xi}_t^R\right\|_{1,p,R} \le c_0(\delta R)^{-2/p}\left(\left\|\partial_t u_t^0\right\|_{L^2} + \left\|\partial_t u_t^\infty\right\|_{L^2}\right). \tag{10.5.13}$$

REMARK 10.5.5. Let M be a compact Riemannian manifold. Then there are two smooth families of endomorphisms $E_i(x,\xi) : T_xM \to T_{\exp_x(\xi)}M$, $i = 1, 2$, that are characterized by the following property. If $\gamma : \mathbb{R} \to M$ is any smooth path in M and $v(t) \in T_{\gamma(t)}M$ is any smooth vector field along this path then the derivative of the path $t \mapsto \exp_{\gamma(t)}(v(t))$ is given by the formula

$$\frac{d}{dt}\exp_\gamma(v) = E_1(\gamma, v)\dot\gamma + E_2(\gamma, v)\nabla_t v.$$

The reader may check that

$$E_1(x,0) = E_2(x,0) = \mathrm{id} : T_xM \to T_xM$$

for every $x \in M$. Hence the $E_i(x,\xi)$ are uniformly invertible for sufficiently small ξ. We shall often apply this when x varies over the points in the image of a smooth map $u : S^2 \to M$ and the exponential map $\exp_{u(z)}$ corresponds to the metric g_{J_z}. In particular, when $u = u^R$ and $J = J^R$ the resulting endomorphisms are denoted $E_i^R = E_i^R(u^R(z), \xi(z))$.

PROOF OF THEOREM 10.1.2 (I-III). Fix two constants $p > 2$ and $c > 0$. Choose positive constants δ_0 and c_0 such that conditions (a), (b), (c), and (d) are satisfied. Fix a constant $0 < \delta < \delta_0$.

We prove Theorem 10.1.2 (i). That the map $\iota_c^R : \mathcal{M}(c) \to C^\infty(S^2, M)$ is smooth for each R is obvious, because the pregluing map $f^R : \mathcal{M}(c) \to C^\infty(S^2, M)$ is smooth by construction, and the map $u^R \mapsto \widetilde{u}^R$ in (10.5.12) is smooth by the implicit function theorem. Smooth dependence on R is slightly more subtle (as the variable R appears in the argument of u^∞ in the definition of f^R). It follows from the fact that all elements of $\mathcal{M}(c)$ are smooth, by elliptic regularity.

Next we prove that ι_c^R is an immersion whenever δR is sufficiently large. Let $t \mapsto u_t := (u_t^0, u_t^\infty)$ be a smooth path in $\mathcal{M}(c)$ and denote

$$\widetilde{u}_t^R := \iota_c^R(u_t) = \exp_{u_t^R}(\widetilde{\xi}_t^R), \qquad u_t^R := f^R(u_t).$$

Then, using the notation of Remark 10.5.5,

$$\partial_t \widetilde{u}_t^R = E_1 \partial_t u_t^R + E_2 \nabla_t \widetilde{\xi}_t^R,$$

where $E_i := E_i^R(u_t^R, \widetilde{\xi}_t^R) : T_{u_t^R(z)}M \to T_{\widetilde{u}_t^R(z)}M$. By (10.5.10) and (10.5.11) the $W^{1,p}$-norm of $\widetilde{\xi}_t^R$ can be made arbitrarily small by choosing $r := \delta R$ sufficiently large. Since the $E_i(x,\xi)$ are uniformly invertible for small ξ, it follows that there are positive constants δ_1 and c_1 such that

$$\big\|\partial_t \widetilde{u}_t^R\big\|_{1,p,R} \ge \delta_1 \big\|\partial_t u_t^R\big\|_{1,p,R} - c_1 \big\|\nabla_t \widetilde{\xi}_t^R\big\|_{1,p,R}. \tag{10.5.14}$$

We must understand each term on the right hand side. Now $\partial_t u_t^R = df^R(u_t)\partial_t u_t$ and, by definition of f^R, we have

$$\big\|df^R(u)(\zeta^0, \zeta^\infty)\big\|_{0,2,R}^2 \ge \big\|\zeta^0\big\|_{L^2(\mathbb{C}\setminus B_{2/\delta R})}^2 + \big\|\zeta^\infty\big\|_{L^2(B_{\delta R/2})}^2,$$

for every $u = (u^0, u^\infty) \in \mathcal{M}(c)$ and every pair $\zeta = (\zeta^0, \zeta^\infty) \in \ker D_u = T_u\mathcal{M}(c)$, where the L^2-norms on the right are understood with respect to the Fubini–Study metric $(1+|z|^2)^{-2}(ds^2+dt^2)$ on the 2-sphere. Since $\zeta^a \in \ker D_{u^a}$ for $a = 0, \infty$, the L^2-norm of ζ^a on the complement of a ball controls the L^2-norm on all of S^2, by unique continuation. Hence there is a constant $c_2 > 0$ such that

$$\big\|\zeta^0\big\|_{L^2} + \big\|\zeta^\infty\big\|_{L^2} \le c_2 \big\|df^R(u)\zeta\big\|_{1,p,R}, \qquad \zeta = (\zeta^0, \zeta^\infty) \in \ker D_u. \tag{10.5.15}$$

The second term on the right hand side of (10.5.14) can be estimated by Proposition 10.5.4:

$$\left\|\nabla_t\widetilde{\xi}_t^R\right\|_{1,p,R} \le c_0(\delta R)^{-2/p}\left(\left\|\partial_t u_t^0\right\|_{L^2}+\left\|\partial_t u_t^\infty\right\|_{L^2}\right).$$

Hence

$$\begin{aligned}\left\|\partial_t\widetilde{u}_t^R\right\|_{1,p,R} &\ge \left(\delta_1/c_2 - c_0c_1(\delta R)^{-2/p}\right)\left(\left\|\partial_t u_t^0\right\|_{L^2}+\left\|\partial_t u_t^\infty\right\|_{L^2}\right)\\ &\ge \delta_2\left(\left\|\partial_t u_t^0\right\|_{W^{1,p}}+\left\|\partial_t u_t^\infty\right\|_{W^{1,p}}\right).\end{aligned}$$

Here the second inequality holds, with a suitable constant $\delta_2 > 0$, whenever $(\delta R)^{2/p} \ge 2c_0c_1c_2/\delta_1$ and uses the fact that the L^2-norm of the tangent vector $(\partial_t u_t^0, \partial_t u_t^\infty) \in T_u\mathcal{M}(c)$ controls its $W^{1,p}$-norm. This estimate shows that the differential of ι_c^R is injective at every point of $\mathcal{M}(c)$ whenever $(\delta R)^{2/p} \ge 2c_0c_1c_2/\delta_1$. Hence ι_c^R is an immersion for δR sufficiently large. Moreover, the uniform estimate for the inverse shows that there is a constant $\varepsilon > 0$ such that the restriction of ι_c^R to every ball of radius ε (in the $W^{1,p}$-topology) is injective for $(\delta R)^{2/p} \ge 2c_0c_1c_2/\delta_1$. At this point we decrease δ_0 so that $\delta_0 \le (\delta_1/2c_0c_1c_2)^{p/2}$.

We next prove that ι_c^R is injective for δR sufficiently large. To see this, fix a pair $u := (u^0, u^\infty) \in \mathcal{M}(c)$ and, for $(\delta, R) \in \mathcal{A}(\delta_0)$, define the maps $u^R, \widetilde{u}^R : S^2 \to M$ by

$$u^R := f^R(u), \qquad \widetilde{u}^R = \exp_{u^R}(\widetilde{\xi}^R) = \iota_c^R(u), \tag{10.5.16}$$

where $\widetilde{\xi}^R \in \mathrm{im}Q_{u^R}$ is the unique solution of (10.5.10). Then, by (10.5.11) and the Sobolev inequality of Lemma 10.3.1, we have

$$\left\|\widetilde{\xi}^R\right\|_{L^\infty} \le c_0\left\|\widetilde{\xi}^R\right\|_{1,p,R} \le \frac{c_3}{(\delta R)^{2/p}},$$

where $c_3 := 2c_0^3$. Hence

$$\sup_{z\in S^2} d_{J_z^R}(u^R(z), \widetilde{u}^R(z)) \le \frac{c_3}{(\delta R)^{2/p}}, \tag{10.5.17}$$

where $d_{J_z^R}$ denotes the distance function associated to the metric $g_{J_z^R}$. Hence it follows from the definition of u^R in (10.2.1) and (10.2.2) that

$$\sup_{|z|\ge 2/\delta R} d_{J_z^0}(u^0(z), \widetilde{u}^R(z)) + \sup_{|z|\le \delta R/2} d_{J_z^\infty}(u^\infty(z), \widetilde{u}^R(z/R^2)) \le \frac{2c_3}{(\delta R)^{2/p}}. \tag{10.5.18}$$

Because the first derivatives of the elements in $\mathcal{M}(c)$ are uniformly bounded, this implies that two elements of $\mathcal{M}(c)$ with the same image under ι_c^R must be C^0-close, and hence $W^{1,p}$-close. We saw above that ι_c^R is injective on each $W^{1,p}$-ball of radius ε. It follows that ι_c^R is globally injective and hence is an embedding for δR sufficiently large.

We prove Theorem 10.1.2 (ii). Fix a constant $0 < \delta < \delta_0$, a sequence $R_\nu > 1/\delta\delta_0$ diverging to infinity, and a sequence $(u_\nu^0, u_\nu^\infty) \in \mathcal{M}(c)$ that converges to a pair $(u^0, u^\infty) \in \mathcal{M}(c)$ in the C^∞ topology. Let

$$u_\nu := f^{R_\nu}(u_\nu^0, u_\nu^\infty), \qquad \widetilde{u}_\nu := \iota_c^{R_\nu}(u_\nu^0, u_\nu^\infty).$$

Then, $\widetilde{u}_\nu$ satisfies (10.5.18) with $R = R_\nu$ and (u^0, u^∞) replaced by (u_ν^0, u_ν^∞). Hence $\widetilde{u}_\nu$ converges to the pair (u^0, u^∞) in the required sense as $\nu \to \infty$. This proves (ii).

We prove Theorem 10.1.2 (iii). That $\mathrm{ev}^R \circ (\iota_c^R \times \mathrm{id})$ converges to ev in the C^0-topology is obvious from the definition. To prove C^1-convergence, fix a smooth path

$$[0,1] \to \mathcal{M}(c) \times S^2 : t \mapsto (u_t^0, u_t^\infty, z_t)$$

such that $z_t \neq 0$ for all t. Now define

$$u_t^R := f^R(u_t^0, u_t^\infty)$$

and choose $\widetilde{\xi}_t^R$ to be the unique small vector field along u_t^R such that

$$\bar{\partial}_{J^R}(\widetilde{u}_t^R) = 0, \qquad \widetilde{u}_t^R := \exp_{u_t^R}(\widetilde{\xi}_t^R), \qquad \widetilde{\xi}_t^R \in \operatorname{im} Q_{u_t^R}.$$

We must prove that the path $t \mapsto \widetilde{u}_t^R(z_t)$ converges to $t \mapsto u_t^0(z_t)$ in the C^1-topology as $R \to \infty$. To see this, we observe that the derivative of this path is given by the formula

$$\frac{d}{dt}\widetilde{u}_t^R(z_t) = E_1^R\big(u_t^R(z_t), \widetilde{\xi}_t^R(z_t)\big)\frac{d}{dt}u_t^R(z_t) + E_2^R\big(u_t^R(z_t), \widetilde{\xi}_t^R(z_t)\big)\nabla_t\widetilde{\xi}_t^R(z_t), \tag{10.5.19}$$

where E_1^R and E_2^R are as in Remark 10.5.5. Choose $\varepsilon > 0$ such that $z_t \notin B_\varepsilon(0)$ for all t and note that u_t^R agrees with u_t^0 on $S^2 \setminus B_\varepsilon(0)$ for R sufficiently large. Hence it follows from (10.5.19) that

$$\frac{d}{dt}\widetilde{u}_t^R(z_t) = E_1^R\big(u_t^0(z_t), \widetilde{\xi}_t^R(z_t)\big)\frac{d}{dt}u_t^0(z_t) + E_2^R\big(u_t^0(z_t), \widetilde{\xi}_t^R(z_t)\big)\nabla_t\widetilde{\xi}_t^R(z_t) \tag{10.5.20}$$

for R sufficiently large. Now observe that, by (10.5.11) and Proposition 10.5.4, the sections $\widetilde{\xi}_t^R$ and $\nabla_t\widetilde{\xi}_t^R$ converge to zero uniformly on $S^2 \setminus B_\varepsilon(0)$ as R tends to infinity. Hence the first term on the right in (10.5.20) converges to

$$\frac{d}{dt}u_t^0(z_t) = (\partial_t u_t^0)(z_t) + du_t^0(z_t)\partial_t z_t$$

and the second term on the right in (10.5.20) converges to zero as R tends to infinity. Thus we have convergence for each path. The C^1 convergence of the evaluation map then follows by invoking uniformity. Namely, the sup norm of $\widetilde{\xi}_t^R$ is bounded by a constant times $(\delta R)^{-2/p}$ and the sup norm of $\nabla_t\widetilde{\xi}_t^R$ is bounded by a constant times $(\delta R)^{-2/p}$ times the sum of the L^2-norms of $\partial_t u_t^0$ and $\partial_t u_t^\infty$, both constants being independent of the chosen path. This proves (iii) in Theorem 10.1.2. □

REMARK 10.5.6. The above techniques can also be used to glue together J-holomorphic maps $u_i : \Sigma_i \to M$ which are defined on Riemann surfaces of higher genus. However, in this case we must allow for the complex structure on the glued Riemann surface

$$\Sigma = \Sigma_1 \# \Sigma_2$$

to vary. The gluing will produce a Riemann surface with a very thin neck. Conversely, it follows from Gromov compactness that a sequence $u_\nu : \Sigma \to M$ of (j_ν, J)-holomorphic curves can only split up into two (or more) surfaces of higher genus if the complex structure j_ν on Σ converges to the boundary of Teichmüller space.

10.6. The derivative of the gluing map

This section begins with some important technical results, that are needed in the proof of Proposition 10.5.4, which was used above to establish parts (i-iii) of the gluing theorem. It ends with a proof of part (iv) concerning the orientation preserving property of the gluing map. The arguments are straightforward, though sometimes long. Throughout we fix two homology classes $A^0, A^\infty \in H_2(M;\mathbb{Z})$ and a regular pair of almost complex structures $(J^0, J^\infty) \in \mathcal{J}_{\mathrm{reg}}(S^2; A^{0,\infty})$. We abbreviate

$$\mathcal{M}(c) := \mathcal{M}(A^{0,\infty}; J^{0,\infty}, c),$$

as before, and shall often denote the elements (u^0, u^∞) of $\mathcal{M}(c)$ simply by u. Further, we denote the product δR by r. Thus the statement that $(\delta, R) \in \mathcal{A}(\delta_0)$ implies that $r > 1/\delta_0$.

The first lemma establishes precise estimates for the right inverse $Q_{0,\infty,r}$ of the linearized Cauchy–Riemann operator on the intermediate family $u^{0,r}, u^{\infty,r}$. In particular we prove the estimate (10.5.2) used above. Care must be taken because we have uniform bounds only on the first derivatives of the $u^{a,r}$ but not on their higher derivatives. Thus we can only work with the L^p and $W^{1,p}$-norms of vector fields along these curves.

LEMMA 10.6.1. *Fix two constants $p > 2$ and $c > 0$. Then there are positive constants δ_0 and c_0 such that, for all $(u^0, u^\infty) \in \mathcal{M}(c)$ and all $(\delta, R) \in \mathcal{A}(\delta_0)$ the following holds for $r := \delta R$.*

(i) *For every $\xi = (\xi^0, \xi^\infty) \in W^{1,p}_{u^{0,\infty,r}}$, we have*

$$\|\xi\|_{W^{1,p}} \le c_0 \left(\|D_{0,\infty,r}\xi\|_{L^p} + \|\xi\|_{L^p} \right).$$

(ii) *For every $\xi = (\xi^0, \xi^\infty) \in W^{1,p}_{u^{0,\infty,r}}$, we have*

$$D_{0,\infty,r}\xi = 0 \qquad \Longrightarrow \qquad \|\xi\|_{W^{1,p}} \le c_0 \|\xi\|_{L^2}.$$

(iii) *For every $\eta = (\eta^0, \eta^\infty) \in L^p_{u^{0,r},J^0} \times L^p_{u^{\infty,r},J^\infty}$ we have*

$$\|Q_{0,\infty,r}\eta\|_{W^{1,p}} \le c_0 \|\eta\|_{L^p},$$

PROOF. For r sufficiently large the operator $D_{0,\infty,r}$ is arbitrarily close in the operator norm to the original operator $D_{0,\infty}$ (after identifying domain and target using parallel transport). Hence the estimate in (i) follows immediately from the corresponding estimate for the operator $D_{0,\infty}$. To prove the remaining estimates we consider the following abstract functional analytic setting. Suppose we have a surjective Fredholm operator

$$D : \mathcal{W} \to \mathcal{L}$$

of index d between two Banach spaces. (Think of the case $\mathcal{W} = W^{1,p}$ and $\mathcal{L} = L^p$ where $D = D_{0,\infty}$ is a first order elliptic operator.) We assume $\mathcal{W}$ is equipped with an inner product $\langle \cdot, \cdot \rangle$ and denote the corresponding norm by

$$|\xi| := \sqrt{\langle \xi, \xi \rangle}.$$

(Think of the L^2 inner product.) We assume further that there are positive constants c and c_D such that

$$|\xi| \le c \|\xi\|_{\mathcal{W}} \tag{10.6.1}$$

for every $\xi \in \mathcal{W}$ and

$$D\xi = 0 \quad \Longrightarrow \quad \|\xi\|_{\mathcal{W}} \le c_D\,|\xi|\,. \tag{10.6.2}$$

This holds in our setting because the kernel of D is finite dimensional. Denote by $Q : \mathcal{L} \to \mathcal{W}$ the right inverse of D whose image is the orthogonal complement of the kernel with respect to the above inner product.

Now suppose that $D' : \mathcal{W} \to \mathcal{L}$ is another bounded linear operator ($D_{0,\infty,r}$ for example) such that

$$\varepsilon := \|D' - D\|\,\|Q\| < 1.$$

Since $DQ = \mathrm{id}$ we have $\|D'Q - \mathrm{id}\| < 1$ and so D' is surjective with right inverse $Q(D'Q)^{-1}$. However, we wish to understand the right inverse $Q' : \mathcal{L} \to \mathcal{W}$ whose image is the orthogonal complement of the kernel of D'.

As a first step we observe that, if $D'\zeta = 0$, then $\|QD\zeta\|_{\mathcal{W}} \le \varepsilon\,\|\zeta\|_{\mathcal{W}}$ and $\|\zeta - QD\zeta\|_{\mathcal{W}} \ge (1-\varepsilon)\,\|\zeta\|_{\mathcal{W}}$. Hence

$$D'\zeta = 0 \quad \Longrightarrow \quad \|QD\zeta\|_{\mathcal{W}} \le \frac{\varepsilon}{1-\varepsilon}\,\|\zeta - QD\zeta\|_{\mathcal{W}}\,. \tag{10.6.3}$$

Since $\zeta - QD\zeta \in \ker D$, we find $\|\zeta - QD\zeta\|_{\mathcal{W}} \le c_D\,|\zeta - QD\zeta| \le c_D\,|\zeta|$. The last inequality holds because $\zeta - QD\zeta$ is the orthogonal projection of ζ onto the kernel of D. But we saw above that $\|(1-\varepsilon)\zeta\|_{\mathcal{W}} \le \|\zeta - QD\zeta\|_{\mathcal{W}}$. Hence

$$D'\zeta = 0 \quad \Longrightarrow \quad \|\zeta\|_{\mathcal{W}} \le \frac{c_D}{1-\varepsilon}\,|\zeta|\,.$$

This proves (ii).

Now assume that $\xi \in \mathcal{W}$ is orthogonal to the kernel of D'. Let $e_1, \dots, e_d$ be an orthonormal basis of $\ker D$ and consider the basis $e_1', \dots, e_d'$ of $\ker D'$ defined by

$$e_i' - QDe_i' = e_i, \qquad i = 1, \dots, d.$$

(The map $\ker D' \to \ker D : \zeta \mapsto \zeta - QD\zeta$ is an isomorphism between the two kernels.) Since $\xi - QD\xi \in \ker D$ we have

$$\xi - QD\xi = \sum_{i=1}^{d} \langle \xi, e_i \rangle e_i.$$

Moreover, $\langle \xi, e_i \rangle = \langle \xi, e_i - e_i' \rangle = \langle \xi, -QDe_i' \rangle$ and hence, by (10.6.3),

$$\left|\langle \xi, e_i \rangle\right| \le c\,\|QDe_i'\|_{\mathcal{W}}\,\|\xi\|_{\mathcal{W}} \le \frac{c\varepsilon}{1-\varepsilon}\,\|e_i\|_{\mathcal{W}}\,\|\xi\|_{\mathcal{W}} \le \frac{cc_D\varepsilon}{1-\varepsilon}\,\|\xi\|_{\mathcal{W}}\,.$$

Here we have also used (10.6.1) and (10.6.2). Combining this with the previous identity we find

$$\|\xi - QD\xi\|_{\mathcal{W}} \le c_D\,|\xi - QD\xi| = c_D \sqrt{\sum_{i=1}^{d} \langle \xi, e_i \rangle^2} \le \frac{\sqrt{d}cc_D^2\varepsilon}{1-\varepsilon}\,\|\xi\|_{\mathcal{W}}\,.$$

Hence

$$\begin{aligned}
\|\xi\|_{\mathcal{W}} &\le \|QD\xi\|_{\mathcal{W}} + \|\xi - QD\xi\|_{\mathcal{W}} \\
&\le \|Q\|\,\|D'\xi\|_{\mathcal{L}} + \|Q\|\,\|D' - D\|\,\|\xi\|_{\mathcal{W}} + \|\xi - QD\xi\|_{\mathcal{W}} \\
&\le \|Q\|\,\|D'\xi\|_{\mathcal{L}} + \varepsilon\left(1 + \frac{\sqrt{d}cc_D^2}{1-\varepsilon}\right)\|\xi\|_{\mathcal{W}}\,.
\end{aligned}$$

If $\varepsilon \le 1/2$ and $\varepsilon(1+2\sqrt{d}cc_D^2) \le 1/2$ we deduce that $\|\xi\|_{\mathcal{W}} \le 2\|Q\|\,\|D'\xi\|_{\mathcal{L}}$ for every $\xi \in \mathcal{W}$ that is orthogonal to the kernel of D'.

Now recall that $\|D'-D\| = \varepsilon/\|Q\|$ and that D' is surjective, so that $D'\xi$ runs over all elements in $\mathcal{L}$. It follows that there is a constant $\delta > 0$ such that

$$\|D'-D\| < \delta \qquad \Longrightarrow \qquad \|Q'\| \le 2\|Q\|.$$

How small δ must be chosen depends only on the operator norm of Q and the constants d, c, c_D. This proves Lemma 10.6.1. □

We next consider the covariant derivatives of the approximate inverse T_{u^R} as u varies along a path $u_t := (u_t^0, u_t^\infty)$ in the moduli space $\mathcal{M}(c)$. To lighten the calculations, we shall use the Levi-Civita connection ∇ on vector fields; but we need the Hermitian connection $\widetilde{\nabla}$ on 1-forms, since it is only the latter that preserves the $(0,1)$-forms. Recall that $\mathcal{A}(\delta_0)$ denotes the set of admissible pairs (δ, R). Thus $(\delta, R) \in \mathcal{A}(\delta_0)$ if and only if $0 < \delta < \delta_0$ and $\delta R > 1/\delta_0$.

LEMMA 10.6.2. *Fix two constants $c > 0$ and $p > 2$. Then there are positive constants δ_0 and $c_0 > 0$ with the following significance. Fix a pair $(\delta, R) \in \mathcal{A}(\delta_0)$, let $\mathbb{R} \to \mathcal{M}(c) : t \mapsto (u_t^0, u_t^\infty)$ be a smooth path and $t \mapsto u_t^R$ be the corresponding path of preglued curves defined by (10.2.2). Let $t \mapsto \eta_t^R$ be a smooth family of $(0,1)$-forms along u_t^R, i.e. $\eta_t^R \in \Omega^{0,1}(S^2, (u_t^R)^*TM)$ for all t. Then*

$$\big\|\nabla_t(T_{u_t^R}\eta_t^R) - T_{u_t^R}\widetilde{\nabla}_t\eta_t^R\big\|_{1,p,R} \le c_0\big\|\partial_t u_t^R\big\|_{1,\infty,R}\big\|\eta_t^R\big\|_{0,p,R}.$$

PROOF. We first prove an analogous estimate in a simpler situation where we have a path $u_t : S^2 \to M$ of spheres and no pregluing map. Thus, let

$$\mathbb{R} \times S^2 \to M : (t,z) \mapsto u_t(z)$$

be a smooth map, denote by D_{u_t} the corresponding family of linearized Cauchy–Riemann operators, and let $\eta_t \in \Omega^{0,1}(S^2, u_t^*TM)$ be a smooth family of $(0,1)$-forms along u_t. We assume throughout that there is a constant c such that $\|du_t\|_{L^\infty} \le c$ for every t. We assume also that the operator $D_t := D_{u_t}$ is onto for every t and denote by $Q_t := Q_{u_t}$ the right inverse whose image is the L^2-orthogonal complement of the kernel of D_t. Define $\xi_t \in \Omega^0(S^2, u_t^*TM)$ by

$$\xi_t := Q_t\eta_t, \qquad \eta_t = D_t\xi_t.$$

Our aim is to obtain the estimate

$$\big\|\nabla_t\xi_t - Q_t\widetilde{\nabla}_t\eta_t\big\|_{W^{1,p}} \le C\big\|\partial_t u_t\big\|_{W^{1,\infty}}\big\|\eta_t\big\|_{L^p}. \tag{10.6.4}$$

We shall make use of the inequalities

$$D_t\xi = 0 \qquad \Longrightarrow \qquad \big\|\xi\big\|_{W^{1,p}} \le c_1\big\|\xi\big\|_{L^2}, \tag{10.6.5}$$

$$\big\|Q_t\eta\big\|_{W^{1,p}} \le c_2\big\|\eta\big\|_{L^p}, \tag{10.6.6}$$

$$\big\|\xi\big\|_{L^2} \le c_3\big\|\xi\big\|_{W^{1,p}}. \tag{10.6.7}$$

We shall also use the commutator estimate

$$\big\|\widetilde{\nabla}_t(D_t\xi_t) - D_t\nabla_t\xi_t\big\|_{L^p} \le c_4\big\|\partial_t u_t\big\|_{W^{1,\infty}}\big\|\xi_t\big\|_{W^{1,p}}, \tag{10.6.8}$$

To prove (10.6.8) differentiate the field of $(0,1)$-forms $t \mapsto D_t\xi_t$ covariantly to obtain

$$\widetilde{\nabla}_t(D_t\xi_t) - D_t\nabla_t\xi_t = B_1(\nabla_t du_t, \xi_t) + B_2(\partial_t u_t, \nabla\xi_t) + B_3(\partial_t u_t, du_t, \xi_t).$$

Here B_1 and B_2 are bilinear forms on TM with values in TM, and B_3 is a trilinear form on TM with values in TM. The multi-linear forms B_i are all uniformly bounded. The inequality (10.6.8) follows by examining the three terms on the right.

We now show that the last four inequalities can be combined to yield (10.6.4). We remind the reader that D_t denotes a family of Cauchy–Riemann operators; differentiation with respect to t is denoted either by ∇_t (when we are differentiating tensors in M) or by ∂_t (when we differentiate a function along a path parametrized by t). Note first that by (10.6.6) and (10.6.8)

$$\begin{aligned}\left\|Q_tD_t\nabla_t\xi_t - Q_t\widetilde{\nabla}_t\eta_t\right\|_{W^{1,p}} &\le c_2\left\|D_t\nabla_t\xi_t - \widetilde{\nabla}_t(D_t\xi_t)\right\|_{L^p}\\ &\le c_2c_4\left\|\partial_tu_t\right\|_{W^{1,\infty}}\left\|\xi_t\right\|_{W^{1,p}}\\ &\le c_2^2c_4\left\|\partial_tu_t\right\|_{W^{1,\infty}}\left\|\eta_t\right\|_{L^p}.\end{aligned}$$

Now suppose that $t\mapsto\zeta_t$ is a path in the kernel of D_t such that $\left\|\zeta_t\right\|_{L^2}=1$. Then, because $D_t\zeta_t=0$, we find

$$\begin{aligned}\left\|Q_tD_t\nabla_t\zeta_t\right\|_{L^p} &\le c_2\left\|D_t\nabla_t\zeta_t - \widetilde{\nabla}_t(D_t\zeta_t)\right\|_{L^p}\\ &\le c_2c_4\left\|\partial_tu_t\right\|_{W^{1,\infty}}\left\|\zeta_t\right\|_{W^{1,p}}\\ &\le c_1c_2c_4\left\|\partial_tu_t\right\|_{W^{1,\infty}}.\end{aligned}$$

Moreover, $0=\partial_t\langle\xi_t,\zeta_t\rangle=\langle\nabla_t\xi_t,\zeta_t\rangle+\langle\xi_t,\nabla_t\zeta_t\rangle$ and hence

$$|\langle\nabla_t\xi_t,\zeta_t\rangle| = |\langle\xi_t,\nabla_t\zeta_t\rangle| = |\langle\xi_t,Q_tD_t\nabla_t\zeta_t\rangle| \le c_1c_2c_3c_4\left\|\partial_tu_t\right\|_{W^{1,\infty}}\left\|\xi_t\right\|_{W^{1,p}}.$$

Now choose an orthonormal frame $\zeta_{1t},\dots,\zeta_{mt}$ of $\ker D_t$ where m denotes the Fredholm index of D. Then

$$\begin{aligned}\left\|\nabla_t\xi_t - Q_tD_t\nabla_t\xi_t\right\|_{W^{1,p}} &\le c_1\left\|\nabla_t\xi_t - Q_tD_t\nabla_t\xi_t\right\|_{L^2}\\ &= c_1\sqrt{\sum_{i=1}^m\langle\nabla_t\xi_t,\zeta_{it}\rangle^2}\\ &\le \sqrt{m}c_1^2c_2c_3c_4\left\|\partial_tu_t\right\|_{W^{1,\infty}}\left\|\xi_t\right\|_{W^{1,p}}\\ &\le \sqrt{m}c_1^2c_2^2c_3c_4\left\|\partial_tu_t\right\|_{W^{1,\infty}}\left\|\eta_t\right\|_{L^p}.\end{aligned}$$

This proves (10.6.4) with $C:=c_2^2c_4(1+\sqrt{m}c_1^2c_3)$.

We must apply the inequality (10.6.4) to the case where u_t is replaced by the connected pair $(u_t^{0,r},u_t^{\infty,r})$ constructed from u_t^0 and u_t^∞ as in Remark 10.2.2 and where we are interested in the operators D,T on the preglued curves u_t^R. Thus let $t\mapsto\eta_t^R$ be the family of $(0,1)$-forms along the pregluing u_t^R in the assumptions of the lemma, denote by

$$\eta_t := (\eta_t^0,\eta_t^\infty)$$

the corresponding pair of $(0,1)$-forms along $u_t^{0,r}$ and $u_t^{\infty,r}$ as defined in the proof of Proposition 10.5.1, abbreviate

$$D_{t,r} := D_{u_t^{0,r},u_t^{\infty,r}},\qquad Q_{t,r} := Q_{u_t^{0,r},u_t^{\infty,r}}$$

and let

$$\xi_t := (\xi_t^0,\xi_t^\infty) := Q_{t,r}(\eta_t^0,\eta_t^\infty),\qquad \xi_t^R := T_{u_t^R}\eta_t^R.$$

Thus ξ_t^R is obtained from the pair ξ_t via the formula (10.5.4) in the proof of Proposition 10.5.1.

By Lemma 10.6.1, the operators $D_{t,r}$ and $Q_{t,r}$ satisfy the inequalities (10.6.5-10.6.8) with suitable constants c_1, c_2, c_3, c_4 that depend only on c and p. Hence, by what we just proved,

$$\left\|\nabla_t \xi_t - Q_{t,r}\widetilde{\nabla}_t \eta_t\right\|_{W^{1,p}} \le C\left\|\partial_t u_t^R\right\|_{1,\infty,R}\left\|\eta_t^R\right\|_{0,p,R}. \tag{10.6.9}$$

Here we have also used the inequalities

$$\left\|\eta_t\right\|_{L^p} \le 2\left\|\eta_t^R\right\|_{0,p,R}, \qquad \left\|\partial_t u_t^{0,r}\right\|_{W^{1,\infty}} + \left\|\partial_t u_t^{\infty,r}\right\|_{W^{1,\infty}} \le 2\left\|\partial_t u_t^R\right\|_{1,\infty,R}.$$

Now the right hand side of (10.5.4) defines a bounded linear operator

$$W^{1,p}_{u^{0,\infty,r}} \to W^{1,p}_{u^R} : \xi = (\xi^0, \xi^\infty) \mapsto \xi^R$$

that commutes with the covariant derivative ∇_t along every path $t \mapsto u_t := (u_t^0, u_t^\infty)$. (This holds because the cutoff function β depends only on the points z in the domain and so is independent of t.) Hence it follows from (10.6.9) that there is a constant $C' > 0$ such that

$$\left\|\nabla_t \xi_t^R - T_{u_t^R}\widetilde{\nabla}_t \eta_t^R\right\|_{1,p,R} \le C'\left\|\partial_t u_t^R\right\|_{1,\infty,R}\left\|\eta_t^R\right\|_{0,p,R}.$$

This proves Lemma 10.6.2. □

The next lemma describes an isomorphism between the tangent spaces of $\mathcal{M}(c)$ and those of the moduli space of preglued curves. We shall use the L^2-norm on elements of the tangent space $T_u\mathcal{M}(c) = \ker D_u$. This is a matter of convenience: all norms on this finite dimensional space are equivalent, with uniform constants as $u = (u^0, u^\infty)$ runs over the compact space $\mathcal{M}(c)$. On the other hand, this norm is relevant because we use the L^2-orthogonal space to $\ker D_u$ to pick out a unique right inverse. To be consistent with the notation in the rest of this section we shall usually write D_u instead of $D_{0,\infty}$. Recall that $Q_{u^R} := T_{u^R}(D_{u^R}T_{u^R})^{-1}$ denotes the right inverse of D_{u^R} whose image is equal to the image of the operator T_{u^R} of Proposition 10.5.1.

LEMMA 10.6.3. *For every $c > 0$ and every $p > 2$ there exist positive constants δ_0 and c_0 with the following significance. If $u = (u^0, u^\infty) \in \mathcal{M}(c)$, and $(\delta, R) \in \mathcal{A}(\delta_0)$ then*

$$\left\|(\xi^0, \xi^\infty)\right\|_{L^2} \le c_0\left\|(\mathbb{1} - Q_{u^R}D_{u^R})df^R(u)(\xi^0, \xi^\infty)\right\|_{1,p,R} \tag{10.6.10}$$

for every $(\xi^0, \xi^\infty) \in T_u\mathcal{M}(c)$. Thus the operator $(\mathbb{1} - Q_{u^R}D_{u^R})df^R(u)$ is an isomorphism from the kernel of $D_u := D_{u^0,u^\infty}$ to the kernel of D_{u^R}.

PROOF. Abbreviate $D^R := D_{u^R}$, $Q^R := Q_{u^R}$, $T^R := T_{u^R}$. We shall prove the estimate

$$\left\|D^R df^R(u)\xi\right\|_{0,p,R} \le c_2(\delta R)^{-2/p}\left\|\xi\right\|_{L^2} \tag{10.6.11}$$

for $u = (u^0, u^\infty) \in \mathcal{M}(c)$ and $\xi = (\xi^0, \xi^\infty) \in \ker D_u = T_u\mathcal{M}(c)$. Combining this with (10.5.8) and the estimate $\left\|\xi\right\|_{L^2} \le c_1\left\|df^R(u)\xi\right\|_{1,p,R}$ in (10.5.15), we obtain

$$\begin{aligned} c_1\left\|(\mathbb{1} - Q^R D^R)df^R(u)\xi\right\|_{1,p,R} &\ge c_1\left\|df^R(u)\xi\right\|_{1,p,R} - c_0c_1\left\|D^R df^R(u)\xi\right\|_{0,p,R} \\ &\ge \left(1 - c_0c_1c_2(\delta R)^{-2/p}\right)\left\|\xi\right\|_{L^2}. \end{aligned}$$

Therefore (10.6.10) follows if δR is sufficiently large.

To prove (10.6.11), abbreviate $\xi^R := df^R(u)\xi$ and note that, by construction, $D_R\xi^R$ vanishes outside the annuli $1/\delta R \le |z| \le 2/\delta R$ and $\delta/2R \le |z| \le \delta/R$.

To estimate $D_R\xi^R$ on the annulus $1/r \le |z| \le 2/r$ (where $r = \delta R$ as usual) we differentiate the defining formula (10.2.2) for u^R along a path in $\mathcal{M}(c)$ to obtain a formula for ξ^R. Denote the path by $t \mapsto u_t := (u_t^0, u_t^\infty)$ and write

$$u_t^0(z) =: \exp_{x_t}(\zeta_t^0(z)), \qquad x_t := u_t^0(0),$$

for $1/r \le |z| \le 2/r$. Then $u_t^R := f^R(u_t^0, u_t^\infty)$ is given by

$$u_t^R(z) = \exp_{x_t}(\rho(rz)\zeta_t^0(z))$$

for $1/r \le |z| \le 2/r$. Differentiating these two identities with respect to t, and using the notation of Remark 10.5.5, we find

$$\partial_t u_t^R = E_1\big(x_t, \rho(rz)\zeta_t^0(z)\big)\,(\partial_t x_t) + \rho(rz)E_2\big(x_t, \rho(rz)\zeta_t^0(z)\big)\,\big(\nabla_t\zeta_t^0(z)\big),$$
$$\partial_t u_t^0 = E_1\big(x_t, \zeta_t^0(z)\big)\,\big(\partial_t u_t^0(0)\big) + E_2\big(x_t, \zeta_t^0(z)\big)\,\big(\nabla_t\zeta_t^0(z)\big).$$

Solve the second equation for $\nabla_t\zeta_t^0(z)$ and insert the result into the first equation. Then set $t = 0$ using $\partial_t u_t^0|_{t=0} = \xi^0$ and $\partial_t u_t^R|_{t=0} = \xi^R$. This gives

$$\xi^R(z) = E_1\big(x, \rho(rz)\zeta^0(z)\big)\xi^0(0) + E_2\big(x, \rho(rz)\zeta^0(z)\big)\widehat{\xi}^0(z), \tag{10.6.12}$$

where $x := u^0(0)$, and $\zeta^0(z), \widehat{\xi}^0(z) \in T_xM$ are defined by

$$u^0(z) =: \exp_x(\zeta^0(z))$$

and

$$\widehat{\xi}^0(z) := \rho(rz)E_2\big(x, \zeta^0(z)\big)^{-1}\Big(\xi^0(z) - E_1\big(x, \zeta^0(z)\big)\xi^0(0)\Big) \tag{10.6.13}$$

for $1/r \le |z| \le 2/r$. This explicit formula shows that the C^1-norm of ξ^R in the annulus $1/r \le |z| \le 2/r$ is bounded by a constant times the C^1-norm of ξ^0. Similarly for the annulus $\delta/2R \le |z| \le \delta/R$. As in the proof of Lemma 10.3.2, we now use the fact that the areas of both annuli in the R-dependent metric are bounded by a constant times $(\delta R)^{-2}$ (where $r = \delta R$). We deduce that

$$\big\|D^R\xi^R\big\|_{0,p,R} \le c_4(\delta R)^{-2/p}\big\|(\xi^0, \xi^\infty)\big\|_{C^1} \le c_5(\delta R)^{-2/p}\big\|(\xi^0, \xi^\infty)\big\|_{L^2}.$$

The last inequality holds because that the L^2-norm bounds the C^1-norm for every element in the kernel of $D_{0,\infty}$. This proves Lemma 10.6.3. □

EXERCISE 10.6.4. Prove that the isomorphism

$$(1 - Q_{u^R}D_{u^R})df^R(u) : \ker D_u \to \ker D_{u^R}$$

of Lemma 10.6.3 is orientation preserving. *Hint:* Assume first that the operators D_u and D_{u^R} are complex linear and estimate the complex anti-linear part of the isomorphism. Reduce the general case to the complex linear case by a homotopy argument.

PROOF OF PROPOSITION 10.5.4. Let $t \mapsto u_t := (u_t^0, u_t^\infty)$ be a smooth path in $\mathcal{M}(c)$, denote $u_t^R := f^R(u_t)$, and choose $\widetilde{\xi}_t^R \in \operatorname{im} Q_{u_t^R}$ such that

$$\widetilde{u}_t^R := \exp_{u_t^R}(\widetilde{\xi}_t^R) = \iota_c^R(u_t).$$

Throughout we abbreviate

$$\widetilde{D}_t^R := D_{\widetilde{u}_t^R}, \qquad D_t^R := D_{u_t^R}, \qquad T_t^R := T_{u_t^R}, \qquad Q_t^R := Q_{u_t^R}.$$

Since $\widetilde{\xi}_t^R \in \operatorname{im} T_t^R = \operatorname{im} Q_t^R$, we have

$$\widetilde{\xi}_t^R = T_t^R\eta_t^R, \qquad \eta_t^R := \big(D_t^RT_t^R\big)^{-1} D_t^R\widetilde{\xi}_t^R.$$

We know from (10.5.11) and (10.5.1) that there is a constant $c_2 > 0$ such that

$$\|\widetilde{\xi}_t^R\|_{1,p,R} + \|\eta_t^R\|_{0,p,R} \le c_2(\delta R)^{-2/p}. \tag{10.6.14}$$

Moreover, it follows from the formulas (10.6.12) and (10.6.13), with ξ^0, ξ^∞, and ξ^R replaced by $\partial_t u_t^0$, $\partial_t u_t^\infty$, and $\partial_t u_t^R$, respectively, that the $(1,\infty,R)$-norm of $\partial_t u_t^R$ can be estimated by the C^1-norm of the pair $(\partial_t u_t^0, \partial_t u_t^\infty)$. Since this pair belongs to the kernel of $D_{u_t^0,u_t^\infty}$, its L^2-norm controls its C^1-norm. Hence there is a constant $c_3 > 0$ such that

$$\|\partial_t u_t^R\|_{1,\infty,R} \le c_3\left(\|\partial_t u_t^0\|_{L^2} + \|\partial_t u_t^\infty\|_{L^2}\right). \tag{10.6.15}$$

It also follows from the definition of the function u_t^R in (10.2.2) that the $(0,1)$-form $D_t^R\partial_t u_t^R$ is supported in the annuli $1/\delta R \le |z| \le 2/\delta R$ and $\delta/2R \le |z| \le \delta/R$ (both of area bounded by $4\pi(\delta R)^{-2}$). Hence there is a constant $c_4 > 0$ such that

$$\|D_t^R\partial_t u_t^R\|_{0,p,R} \le c_4(\delta R)^{-2/p}\|\partial_t u_t^R\|_{1,\infty,R}. \tag{10.6.16}$$

Moreover, by Lemma 10.6.2, there is a constant $c_5 > 0$ such that

$$\|\nabla_t\widetilde{\xi}_t^R - T_t^R\widetilde{\nabla}_t\eta_t^R\|_{1,p,R} \le c_5\|\partial_t u_t^R\|_{1,\infty,R}\|\eta_t^R\|_{0,p,R}. \tag{10.6.17}$$

We now show how the last four inequalities imply the assertion of the lemma. Differentiate the identity

$$\bar{\partial}_{J^R}(\exp_{u_t^R}(\widetilde{\xi}_t^R)) = 0$$

with respect to t to obtain

$$\widetilde{D}_t^R\left(E_1^R\partial_t u_t^R + E_2^R\nabla_t\widetilde{\xi}_t^R\right) = 0, \tag{10.6.18}$$

where the $E_i^R := E_i^R(u_t^R, \widetilde{\xi}_t^R)$ are as in Remark 10.5.5. Now abbreviate

$$\widehat{\xi}_t^R := \nabla_t\widetilde{\xi}_t^R - T_t^R\widetilde{\nabla}_t\eta_t^R. \tag{10.6.19}$$

Then, by (10.6.14), (10.6.15), and (10.6.17), we have

$$\|\widehat{\xi}_t^R\|_{1,p,R} \le c_2c_3c_5(\delta R)^{-2/p}\left(\|\partial_t u_t^0\|_{L^2} + \|\partial_t u_t^\infty\|_{L^2}\right). \tag{10.6.20}$$

Moreover, combining (10.6.18) and (10.6.19), we find

$$\widetilde{D}_t^R E_2^R T_t^R\widetilde{\nabla}_t\eta_t^R = -\widetilde{D}_t^R\left(E_1^R\partial_t u_t^R + E_2^R\widehat{\xi}_t^R\right). \tag{10.6.21}$$

We next claim that the product $\widetilde{D}ET$ occurring in the above formula is a small perturbation of DT and so is invertible. To prove this, denote by

$$\Phi^R := \Phi^R(u_t^R, \widetilde{\xi}_t^R) : T_{u_t^R}M \to T_{\widetilde{u}_t^R}M$$

the complex bundle isomorphism determined by parallel transport along geodesics with respect to $\widetilde{\nabla}$. Then the operator

$$(\Phi^R)^{-1}\widetilde{D}_t^R E_i^R : W^{1,p}_{u_t^R} \to L^p_{u_t^R,J^R}$$

is a small perturbation of D_t^R in the operator norm. More precisely, it follows from Lemma 10.3.1 that, for $i = 1, 2$, there are estimates

$$\|\widetilde{D}_t^R E_i^R(u_t^R, \widetilde{\xi}_t^R)\zeta - \Phi^R(u_t^R, \widetilde{\xi}_t^R)D_t^R\zeta\|_{0,p,R} \le c_6\|\widetilde{\xi}_t^R\|_{1,p,R}\|\zeta\|_{1,p,R} \tag{10.6.22}$$

for $\zeta \in W^{1,p}_{u_t^R}$. Combining this with Proposition 10.5.1 and (10.6.14) we obtain

$$\begin{aligned}\left\|\widetilde{D}_t^R E_2^R T_t^R \eta - \Phi^R \eta\right\|_{0,p,R} &\le \left\|\Phi^R (D_t^R T_t^R \eta - \eta)\right\|_{0,p,R} \\ &\quad + \left\|\widetilde{D}_t^R E_2^R T_t^R \eta - \Phi^R D_t^R T_t^R \eta\right\|_{0,p,R} \\ &\le \tfrac{1}{2}\left\|\eta\right\|_{0,p,R} + c_6 \left\|\widetilde{\xi}_t^R\right\|_{1,p,R}\left\|T_t^R \eta\right\|_{1,p,R} \\ &\le \tfrac{1}{2}\left(1 + c_0 c_2 c_6 (\delta R)^{-2/p}\right)\left\|\eta\right\|_{0,p,R}.\end{aligned}$$

If $c_0 c_2 c_6 (\delta R)^{-2/p} \le 1/2$ it follows that the operator

$$\widetilde{D}_t^R E_2^R T_t^R : L^p_{u_t^R, J^R} \to L^p_{\widetilde{u}_t^R, J^R}$$

is invertible and the norm of its inverse is bounded by 4. Hence, (10.6.21) implies that

$$\begin{aligned}\left\|\widetilde{\nabla}_t \eta_t^R\right\|_{0,p,R} &\le 4\left(\left\|\widetilde{D}_t^R (E_1^R \partial_t u_t^R)\right\|_{0,p,R} + \left\|\widetilde{D}_t^R (E_2^R \widehat{\xi}_t^R)\right\|_{0,p,R}\right) \\ &\le 4\left\|\Phi^R D_t^R \partial_t u_t^R\right\|_{0,p,R} + 4c_6\left\|\widetilde{\xi}_t^R\right\|_{1,p,R}\left\|\partial_t u_t^R\right\|_{1,p,R} + c_7\left\|\widehat{\xi}_t^R\right\|_{1,p,R} \\ &\le 4c_4 (\delta R)^{-2/p}\left\|\partial_t u_t^R\right\|_{1,\infty,R} + 4c_2 c_6 (\delta R)^{-2/p}\left\|\partial_t u_t^R\right\|_{1,p,R} \\ &\quad + c_2 c_3 c_5 c_7 (\delta R)^{-2/p}\left(\left\|\partial_t u_t^0\right\|_{L^2} + \left\|\partial_t u_t^\infty\right\|_{L^2}\right) \\ &\le c_8 (\delta R)^{-2/p}\left(\left\|\partial_t u_t^0\right\|_{L^2} + \left\|\partial_t u_t^\infty\right\|_{L^2}\right).\end{aligned}$$

Here the second inequality uses (10.6.22) and the uniform boundedness of the $\widetilde{D}_t^R$ from Exercise 10.3.3; the third inequality follows from (10.6.14), (10.6.16), and (10.6.20); the last inequality follows from (10.6.15). Taking this together with (10.6.19) we obtain

$$\begin{aligned}\left\|\nabla_t \widetilde{\xi}_t^R\right\|_{1,p,R} &\le \left\|\widehat{\xi}_t^R\right\|_{1,p,R} + \frac{c_0}{2}\left\|\widetilde{\nabla}_t \eta_t^R\right\|_{0,p,R} \\ &\le c_9 (\delta R)^{-2/p}\left(\left\|\partial_t u_t^0\right\|_{L^2} + \left\|\partial_t u_t^\infty\right\|_{L^2}\right).\end{aligned}$$

Here we have also used (10.6.20). This proves Proposition 10.5.4. □

PROOF OF THEOREM 10.1.2 (IV). We sketch the proof that the embedding ι_c^R is orientation preserving. Here we use again the decomposition of ι_c^R into a pregluing map f^R and an approximation map. There is a corresponding decomposition of the derivative

$$d\iota_c^R(u) : \ker D_u \to \ker D_{\widetilde{u}^R}$$

at $u = (u^0, u^\infty) \in \mathcal{M}(c)$ into the isomorphism

$$(1 - Q_{u^R} D_{u^R}) df^R(u) : \ker D_u \to \ker D_{u^R} \tag{10.6.23}$$

of Lemma 10.6.3 with $u^R := f^R(u)$, and an isomorphism

$$\ker D_{u^R} \to \ker D_{\widetilde{u}^R}. \tag{10.6.24}$$

The estimates in the proof of Proposition 10.5.4 show that the isomorphism (10.6.24) is close to the identity in a suitable trivialization of the kernel bundle and so is orientation preserving. Moreover, it follows from Exercise 10.6.4 that the pregluing isomorphism (10.6.23) is orientation preserving. Hence ι_c^R is an orientation preserving embedding whenever δR is sufficiently large. This proves (iv) Theorem 10.1.2. □

10.7. Surjectivity of the gluing map

We now return to the proof of the gluing theorem to complete the final step, the proof of statement (iv) concerning its surjectivity. The problem is that we are considering a J^R-holomorphic curve v that is close to $\mathcal{M}(c-1)$ in a very weak sense: we assume only that there is a pair $(u^0, u^\infty) \in \mathcal{M}(c-1)$ such that the restriction of v to the disc $\{|z| \geq 1\}$ is C^0-close to u^0 and that its restriction to the disc $\{|z| \leq 1/R^2\}$ is C^0-close to the rescaling $u^\infty(R^2 z)$. From this meager information, we must show that v is in the image of ι_c^R. Although the inverse function theorem does include a statement about uniqueness, this statement is very restricted, and applies only if we have already found a preglued element u^R such that v has the form $\exp_{u^R}(\xi)$ where ξ is in the image of the inverse Q_{u^R} and moreover is sufficiently $W^{1,p}$-small. Thus we prove the surjectivity statement by a process of successive approximation, first assuming that v satisfies a rather stringent condition and then considering successively weaker conditions.

Proof of Theorem 10.1.2 (v). The proof consists of four steps.

Step 1. *There is a constant $\varepsilon_1 > 0$ such that, for every $(\delta, R) \in \mathcal{A}(\delta_0)$, the following holds. If $u := (u^0, u^\infty) \in \mathcal{M}(c)$, $u^R := f^R(u)$, and $\xi^R \in \Omega^0(S^2, (u^R)^*TM)$ satisfy*

$$\xi^R \in \operatorname{im} Q_{u^R}, \qquad \bar{\partial}_{J^R}(\exp_{u^R}(\xi^R)) = 0, \qquad \|\xi^R\|_{1,p,R} < \varepsilon_1,$$

then $\exp_{u^R}(\xi^R) = \iota_c^R(u)$.

In view of the discussion preceding the definition of the gluing map ι_c^R in (10.5.12), this follows immediately from the uniqueness statement in Theorem 3.5.2 with $u = u^R$, $\xi_0 = 0$, and $\xi = \xi^R$. (See Remark 3.5.4 for z-dependent almost complex structures.)

Step 2. *For every $\varepsilon_1 > 0$ there are positive constants δ_2 and ε_2 such that, for every $(\delta, R) \in \mathcal{A}(\delta_2)$, the following holds. If $u := (u^0, u^\infty) \in \mathcal{M}(c-1)$, $u^R := f^R(u)$, and $\xi^R \in \Omega^0(S^2, (u^R)^*TM)$ satisfy*

$$v := \exp_{u^R}(\xi^R) \in \mathcal{M}(A; J^R), \qquad \|\xi^R\|_{1,p,R} < \varepsilon_2, \tag{10.7.1}$$

then there is a pair $\bar{u} := (\bar{u}^0, \bar{u}^\infty) \in \mathcal{M}(c)$ and a vector field $\bar{\xi}^R \in \Omega^0(S^2, (\bar{u}^R)^*TM)$ along $\bar{u}^R := f^R(\bar{u})$ such that

$$v = \exp_{\bar{u}^R}(\bar{\xi}^R), \qquad \bar{\xi}^R \in \operatorname{im} Q_{\bar{u}^R}, \qquad \|\bar{\xi}^R\|_{1,p,R} \leq \varepsilon_1. \tag{10.7.2}$$

Fix a pair $(\delta, R) \in \mathcal{A}(\delta_0)$ and an element $v \in \mathcal{M}(A; J^R)$. Assume without loss of generality that $c_0\varepsilon_1$ is smaller than the injectivity radius of M with respect to any of the metrics $g_{J_z^0}$ and $g_{J_z^\infty}$. If necessary, we shall shrink ε_1 further in the course of the proof. Let us denote by

$$\mathcal{U}_1 \subset \mathcal{M}(c)$$

the set of all pairs $u := (u^0, u^\infty) \in \mathcal{M}(c)$ that satisfy

$$\exp_{u^R}(\xi^R) = v, \qquad \|\xi^R\|_{1,p,R} < \varepsilon_1, \tag{10.7.3}$$

for some vector field $\xi^R = \xi^R(u)$. (Note that $\|\xi^R\|_{L^\infty} \leq c_0\|\xi^R\|_{1,p,R} \leq c_0\varepsilon_1$ by Lemma 10.3.1 and hence there is at most one such vector field ξ^R.) We must find a constant $\varepsilon_2 > 0$ with the following significance: if there is an element $u \in \mathcal{U}_1$ such that $\|\xi^R(u)\|_{1,p,R} < \varepsilon_2$ then there is an element $\bar{u} := (\bar{u}^0, \bar{u}^\infty) \in \mathcal{U}_1$ such that the corresponding vector field $\bar{\xi}^R := \xi^R(\bar{u}) \in \Omega^0(S^2, (\bar{u}^R)^*TM)$ belongs to the image of

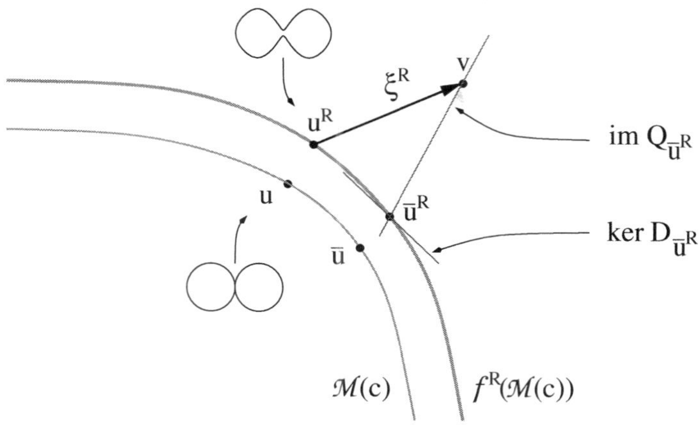

FIGURE 2. The normal bundle of $f^R(\mathcal{M}(c))$.

the operator $Q_{\bar{u}^R}$. To prove this we first observe that $\mathcal{M}(c)$ is a smooth manifold of dimension

$$m := 2n + 2c_1(A).$$

By Lemma 10.6.3, this number is also the dimension of the kernel of D_{u^R} for $u \in \mathcal{M}(c)$ and $u^R := f^R(u)$ with R sufficiently large.

Next we choose a smooth framing

$$u \mapsto \Xi_i(u) \in \ker D_u := D_{u^0,u^\infty}, \qquad i = 1, \dots, m,$$

over $u \in \mathcal{U}_1$. Such a framing exists for ε_1 sufficiently small and δR sufficiently large, because $\mathcal{U}_1$ is then contained in a coordinate chart of $\mathcal{M}(c)$. Define

$$\Xi_i^R(u) := (\mathbb{1} - Q_{u^R} D_{u^R}) df^R(u) \Xi_i(u), \qquad i = 1, \dots, m.$$

We show that this is a framing of ker D_{u^R} for $u \in \mathcal{U}_1$ for δR sufficiently large. In fact, by Lemma 10.6.3 we know that, when δR is suffciently large, the linear map

$$(\mathbb{1} - Q_{u^R} D_{u^R}) df^R(u) : \ker D_u \to \ker D_{u^R}$$

is an isomorphism from the kernel of D_u to the kernel of D_{u^R} for $u \in \mathcal{U}_1$ and $u^R := f^R(u)$, and that there is a uniform bound on the inverse of this isomorphism. Hence there are positive constants c_3 and δ_2 such that the following holds whenever $(\delta, R) \in \mathcal{A}(\delta_2)$. If $u \in \mathcal{U}_1$ and $y = (y_1, \dots, y_m) \in \mathbb{R}^m$ then

$$|y| \le c_3 \Big\| \sum_{i=1}^m y_i \, \Xi_i^R(u) \Big\|_{1,p,R}. \tag{10.7.4}$$

Hence $u \mapsto (\Xi_i^R(u))_{i=1}^m$ is a framing as claimed. Increasing c_3, if necessary, we also have that every smooth path $t \mapsto u_t$ in $\mathcal{U}_1$ satisfies the inequality

$$\big\| \nabla_t \, \Xi_i^R(u_t) \big\|_{1,p,R} \le c_3 \, \|\partial_t u_t\|_{L^2}, \qquad i = 1, \dots, m. \tag{10.7.5}$$

To see this, assume first that $\xi_t = (\xi_t^0, \xi_t^\infty) \in T_{u_t}\mathcal{M}(c)$ is any smooth vector field along u_t. Then the $(1,p,R)$-norm of $\nabla_t(df^R(u_t)\xi_t)$ can be estimated by a constant times $\|\nabla_t \xi_t\|_{W^{1,p}} + \|\xi_t\|_{W^{1,p}} \|\partial_t u\|_{W^{1,p}}$. Now apply this inequality to $\xi_t := \Xi_i(u_t)$ and use Lemma 10.6.2 and (10.6.8) to obtain (10.7.5). (See the discussion for the proof of (10.7.9) below for a similar argument.)

Define the smooth map $h^R = (h_1^R, \dots, h_m^R) : \mathcal{U}_1 \to \mathbb{R}^m$ by

$$\xi^R - Q_{u^R} D_{u^R} \xi^R =: \sum_{i=1}^m h_i^R(u)\, \Xi_i^R(u), \tag{10.7.6}$$

where $u^R := f^R(u)$ and $\xi^R = \xi^R(u)$ is the unique vector field along u^R that satisfies (10.7.3). Thus $h^R(u) \in \mathbb{R}^m$ is the coordinate vector of the projection of $\xi^R(u)$ onto the kernel of D_{u^R} along the image of Q_{u^R}. We need to find a point $\bar{u}$ where $h^R(\bar{u})$ vanishes. By (10.7.4), there is a constant $c_4 > 0$ such that

$$\left|h^R(u)\right| \le c_4 \left\|\xi^R(u)\right\|_{1,p,R}.$$

Hence our assumption implies that there is a point $u \in \mathcal{U}_1$ such that $\left|h^R(u)\right| < c_4\varepsilon_2$, for some small ε_2 that we can choose. To find an actual zero of h we aim to apply Proposition A.3.4. This is a version of the implicit function theorem which asserts that any map h with an approximate zero at a point x_1 has an actual zero very close to x_1 provided that the differential dh is uniformly invertible over a sufficiently large neighbourhood of x_1.

Thus it remains to estimate the inverse of dh^R. The geometric picture is illustrated in Figure 2. The vector $\xi^R = \xi^R(u)$ points from u^R in the direction of v. As u varies in the direction of a tangent vector $\dot{u} \in T_u\mathcal{M}(c)$ the vector $\xi^R \cong v - u$ varies approximately in the direction of $-\dot{u}$. Since $h^R(u)$ represents the projection of $\xi^R(u)$ onto the kernel of D_{u^R}, which is almost parallel to the kernel of D_u, we expect the differential of the map $u \mapsto h^R(u)$ to be approximately equal to minus the identity (after identifying $T_u\mathcal{M}(c)$ with $\mathbb{R}^m$ via our frame $\Xi_1(u), \dots, \Xi_m(u)$). This is made precise in the following discussion.

Let us denote by

$$H(u) := (H_1(u), \dots, H_m(u)) : \ker D_u \to \mathbb{R}^m$$

the isomorphism given by the framing $\Xi_1(u), \dots, \Xi_m(u)$, i.e

$$\sum_{i=1}^m (H_i(u)\xi)\Xi_i(u) = \xi.$$

Then, by definition of $\Xi_i^R(u)$, we have

$$\sum_{i=1}^m (H_i(u)\xi)\Xi_i^R(u) = (\mathbb{1} - Q_{u^R} D_{u^R}) df^R(u)\xi$$

for $\xi \in \ker D_u = T_u\mathcal{M}(c)$. We shall prove that there is a constant $c_5 > 0$ such that

$$\left\|dh^R(u) + H(u)\right\| \le c_5\varepsilon_1 \tag{10.7.7}$$

for every $u := (u^0, u^\infty) \in \mathcal{U}_1$. Here $\|\cdot\|$ denotes the operator norm of a linear map from the kernel of D_u to $\mathbb{R}^m$ with respect to the Euclidean norm on $\mathbb{R}^m$ and the L^2-norm on $\ker D_u$. Since the linear map $H(u)$ is an isomorphism, (10.7.7) implies that $dh^R(u)$ satisfies the conditions of Proposition A.3.4 when ε_1 is sufficiently small. Step 2 is an immediate consequence.

To prove (10.7.7) we differentiate the identity (10.7.6) covariantly along a smooth path $t \mapsto u_t := (u_t^0, u_t^\infty) \in \mathcal{U}_1$. Denote $u_t^R := f^R(u_t)$ and let $\xi_t^R \in \Omega^0(S^2, (u_t^R)^*TM)$ be the unique smooth family of vector fields along u_t^R that satisfy

$$\exp_{u_t^R}(\xi_t^R) = v, \qquad \left\|\xi_t^R\right\|_{1,p,R} \le \varepsilon_1.$$

Then, by (10.7.6), we have

$$\nabla_t(\xi_t^R - QD\xi_t^R) = \sum_{i=1}^m h_i^R(u_t)\nabla_t\Xi_i^R(u_t) + \sum_{i=1}^m \left(\partial_t h_i^R(u_t)\right)\Xi_i^R(u_t),$$

where $Q := Q_{u_t^R}$ and $D := D_{u_t^R}$. Next we differentiate the identity $\exp_{u_t^R}(\xi_t^R) = v$ to obtain

$$E_1\partial_t u_t^R + E_2\nabla_t\xi_t^R = 0,$$

where $E_i := E_i^R(u_t^R, \xi_t^R)$ is as in Remark 10.5.5. Thus

$$\nabla_t\xi_t^R = -E_2^{-1}E_1\partial_t u_t^R = -E_2^{-1}E_1 df^R(u_t)\partial_t u_t,$$

To prove (10.7.7) we need to estimate the functions $\partial_t h_i^R(u_t) + H_i(u_t)\partial_t u_t$. Combining the above identities we find

$$\begin{aligned}&\sum_{i=1}^m \Big(\partial_t h_i^R(u_t) + H_i(u_t)\partial_t u_t\Big)\Xi_i^R(u_t)\\ &\quad = (\mathbb{1} - QD)(\mathbb{1} - E_2^{-1}E_1)df^R(u_t)\partial_t u_t\\ &\qquad + QD\nabla_t\xi_t^R - \nabla_t(QD\xi_t^R) - \sum_{i=1}^m h_i^R(u_t)\nabla_t\Xi_i^R(u_t).\end{aligned} \tag{10.7.8}$$

Since $|h^R(u_t)| \le c_4\left\|\xi^R\right\|_{1,p,R} \le c_4\varepsilon_1$, it follows from (10.7.5) that

$$\Big\|\sum_{i=1}^m h_i^R(u_t)\nabla_t\Xi_i^R(u_t)\Big\|_{1,p,R} \le \sqrt{m}c_3c_4\varepsilon_1\|\partial_t u_t\|_{L^2}.$$

This takes care of the last term on the right in (10.7.8). For the first term on the right in (10.7.8) we can use the pointwise estimate

$$\left|\mathbb{1} - E_2(u_t^R, \xi_t^R)^{-1}E_1(u_t^R, \xi_t^R)\right| \le c_5\left|\xi_t^R\right|$$

There is a similar inequality for the first derivatives of $\mathbb{1} - E_2^{-1}E_1$ in terms of the first derivatives of ξ_t^R (for a fixed t). Hence the first term on the right in (10.7.8) can be estimated by

$$\left\|(\mathbb{1} - E_2^{-1}E_1)df^R(u_t)\partial_t u_t\right\|_{1,p,R} \le c_6\varepsilon_1\left\|\partial_t u_t\right\|_{L^2}.$$

We claim that the remaining term satisfies an estimate of the form

$$\left\|\nabla_t(QD\xi_t^R) - QD\,\nabla_t\xi_t^R\right\|_{1,p,R} \le c_7\left\|\xi_t^R\right\|_{1,p,R}\left\|\partial_t u_t^R\right\|_{1,\infty,R}. \tag{10.7.9}$$

For the operator $T = T_{u_t^R}$ such an estimate was proved in Lemma 10.6.2 and for the operator D we can use (10.6.8) in the proof of Lemma 10.6.2. Combining these estimates we obtain an estimate for the commutator $[DT, \widetilde{\nabla}_t]$ and hence for $[(DT)^{-1}, \widetilde{\nabla}_t]$. This in turn implies the required estimate for the commutator of the operator $QD = T(DT)^{-1}D$ with $\widetilde{\nabla}_t$. Combining (10.7.9) with $\left\|\xi_t^R\right\|_{1,p,R} \le \varepsilon_1$ and with (10.6.15) in the proof of Proposition 10.5.4 we obtain

$$\left\|\nabla_t(QD\xi_t^R) - QD\nabla_t\xi_t^R\right\|_{1,p,R} \le c_8\varepsilon_1\left\|\partial_t u_t\right\|_{L^2}.$$

Thus we have proved that

$$\Big\|\sum_{i=1}^m\Big(\partial_t h_i^R(u_t) + H_i(u_t)\partial_t u_t\Big)\Xi_i^R(u_t)\Big\|_{1,p,R} \le c_9\varepsilon_1\left\|\partial_t u_t\right\|_{L^2}.$$

Hence (10.7.7) follows from (10.7.4). This proves Step 2.

STEP 3. *For every $\varepsilon_2 > 0$ there are positive constants δ_3 and ε_3 such that, for every $(\delta, R) \in \mathcal{A}(\delta_3)$, the following holds. If $(u^0, u^\infty) \in \mathcal{M}^{0,\infty}(c)$ and $v \in \mathcal{M}(A; J^R)$ satisfy*

$$\sup_{|z|\geq 1} d(v(z), u^0(z)) < \varepsilon_3, \qquad \sup_{|z|\leq 1} d(v(z/R^2), u^\infty(z)) < \varepsilon_3, \tag{10.7.10}$$

then there is a vector field $\xi^R \in \Omega^0(S^2, (u^R)^*TM)$ along the curve $u^R := f^R(u^0, u^\infty)$ such that

$$v = \exp_{u^R}(\xi^R), \qquad \|\xi^R\|_{1,p,R} \leq \varepsilon_2. \tag{10.7.11}$$

Suppose, by contradiction, that the assertion is wrong for some constant $\varepsilon_2 > 0$. Then there are sequences

$$\delta_\nu \to 0, \qquad R_\nu \to \infty, \qquad (u^0_\nu, u^\infty_\nu) \in \mathcal{M}^{0,\infty}(c), \qquad v_\nu \in \mathcal{M}^{R_\nu}$$

such that $\delta_\nu R_\nu$ diverges to infinity,

$$\lim_{\nu\to\infty} \sup_{|z|\geq 1} d(v_\nu(z), u^0_\nu(z)) = \lim_{\nu\to\infty} \sup_{|z|\leq 1} d(v_\nu(z/R^2_\nu), u^\infty_\nu(z)) = 0,$$

and, for every ν,

$$\inf\left\{ \|\xi\|_{1,p,R} \mid \xi \in \Omega^0(S^2, (u^{R_\nu}_\nu)^*TM), \, v_\nu = \exp_{u^{R_\nu}_\nu}(\xi) \right\} \geq \varepsilon_2. \tag{10.7.12}$$

By standard convention, we take the infimum over the empty set to be ∞.

Since the moduli space $\mathcal{M}(c)$ is compact we may assume that the sequence (u^0_ν, u^∞_ν) converges in the C^∞ topology to a pair $u := (u^0, u^\infty)$. Moreover, by passing to a subsequence, we may assume that the sequence $v_\nu : S^2 \setminus \{0\} \to M$ converge u.c.s. to some limiting curve v^0 on the complement of a finite set of bubble points (Theorem 4.6.1). (Note that we must work with the restriction of v_ν to $S^2 \setminus \{0\}$ since this is where J^{R_ν} converges.) By assumption, we have $v^0(z) = u^0(z)$ for $|z| \geq 1$ and, by unique continuation (Corollary 2.3.3), this implies that $v^0 = u^0$. The same argument shows that the rescaled sequence $v_\nu(\cdot/R^2_\nu) : \mathbb{C} = S^2 \setminus \{\infty\} \to M$ converges to u^∞ on the complement of a finite set of bubble points. Since v_ν represents the homology class $A = A^0 + A^\infty$ there is no energy left for bubbling. Hence v_ν converges to u^0, uniformly on compact subsets of $S^2 \setminus \{0\}$, and the rescaled sequence $v_\nu(\cdot/R^2_\nu)$ converges to u^∞, uniformly on compact subsets of $S^2 \setminus \{\infty\}$. Moreover, it follows from Lemma 4.7.5 that

$$\lim_{\rho\to 0} \lim_{\nu\to\infty} \sup_{1/\rho R^2_\nu \leq |z| \leq \rho} d(v_\nu(z), x) = 0,$$

where $x := u^0(0) = u^\infty(\infty)$. Since the sequence $u^{R_\nu} := f^{R_\nu}(u^0, u^\infty)$ of pregluings constructed from the limit (u^0, u^∞) satisfies the same conditions, it follows that

$$\lim_{\nu\to\infty} \sup_{z\in S^2} d(v_\nu(z), u^{R_\nu}(z)) = 0.$$

Hence, for every sufficiently large ν, there is a unique small vector field ξ_ν along u^{R_ν} such that

$$v_\nu = \exp_{u^{R_\nu}}(\xi_\nu).$$

We claim that

$$\lim_{\nu\to\infty} \|\xi_\nu\|_{1,p,R_\nu} = 0, \tag{10.7.13}$$

in contradiction to (10.7.12).

To prove (10.7.13), we recall that $J_z^{R_\nu} = J$ for $1/\kappa R_\nu^2 \le |z| \le \kappa$. Hence the formula

$$w_\nu(s,t) := v_\nu(R_\nu^{-1}e^{s+it}), \qquad |s| \le \log\kappa + \log R_\nu,$$

defines a sequence of J-holomorphic curves that converges to $x := u^0(0) = u^\infty(\infty)$, uniformly with all derivatives, on every compact set. Now choose a real number μ such that $1 - 2/p < \mu < 1$. Then, by (4.7.19) in Lemma 4.7.3 there is a constant $c_{10} > 0$ such that

$$|\partial_s w_\nu(s,t)| \le c_{10}e^{-\mu(\log R_\nu - |s|)}, \qquad |s| \le \log R_\nu + \log\kappa - \log 2$$

for ν sufficiently large. Since

$$\sqrt{2}|\partial_s w_\nu(s,t)| = e^s R_\nu^{-1}|dv_\nu(R_\nu^{-1}e^{s+it})|,$$

this translates into the estimate

$$\frac{1}{R_\nu} \le |z| \le \frac{\kappa}{2} \qquad \Longrightarrow \qquad |dv_\nu(z)| \le \frac{\sqrt{2}c_{10}}{|z|^{1-\mu}}.$$

Since $\mu > 1 - 2/p$ we have $p\mu - p + 2 > 0$ and

$$\|dv_\nu\|^p_{L^p\left(A(1/R_\nu,\rho)\right)} \le c_{11}\int_{1/R_\nu \le |z| \le \rho} |z|^{p\mu - p} \le c_{12}\rho^{p\mu-p+2}.$$

This shows that

$$\lim_{\rho\to 0}\lim_{\nu\to\infty} \|dv_\nu\|_{L^p\left(A(1/R_\nu,\rho)\right)} = 0.$$

The function u^{R_ν} satisfies the same condition, by construction, and hence

$$\lim_{\rho\to 0}\lim_{\nu\to\infty} \|\xi_\nu\|_{W^{1,p}\left(A(1/R_\nu,\rho)\right)} = 0.$$

Now the $W^{1,p}$-norm is equivalent to the $(1,p,R_\nu)$-norm in the annulus $A(1/R_\nu,\rho)$ (up to a factor 2). For $|z| \le 1/R_\nu$ we can again use the symmetry of our formulas to obtain a similar statement. This implies

$$\lim_{\rho\to 0}\lim_{\nu\to\infty} \|\xi_\nu\|_{1,p,R_\nu;A(1/\rho R_\nu^2,\rho)} = 0.$$

Combining this with the fact that v_ν converges to u^0 u.c.s. on $S^2\setminus\{0\}$ and $v_\nu(R_\nu^{-2}\cdot)$ converges to u^∞ u.c.s. on $S^2\setminus\{\infty\}$ we obtain (10.7.13). This contradicts (10.7.12) and proves Step 3.

STEP 4. *We prove* (v).

First choose $\varepsilon_1 > 0$ such that the assertion of Step 1 holds. Then choose $\varepsilon_2 > 0$ and $\delta_2 < \delta_0$ such that the assertion of Step 2 holds with this choice of ε_1. Finally, choose ε_3 and $\delta_3 \le \delta_2$ such that the assertion of Step 3 holds with this choice of ε_2. Now assume $0 < \delta < \delta_3$ and $\delta R > 1/\delta_3$. Suppose that $(u^0,u^\infty) \in \mathcal{M}^{0,\infty}(c-1)$ and $v \in \mathcal{M}(A;J^R)$ satisfy (10.7.10). Then, by Step 3, there is a vector field ξ^R along $u^R := f^R(u^0,u^\infty)$ such that $v = \exp_{u^R}(\xi^R)$ and $\|\xi^R\|_{1,p,R} \le \varepsilon_2$. Hence, by Step 2, there is a pair $(\bar{u}^0,\bar{u}^\infty) \in \mathcal{M}^{0,\infty}(c)$ and a vector field $\bar{\xi}^R$ along $\bar{u}^R := f^R(\bar{u}^0,\bar{u}^\infty)$ such that $v = \exp_{\bar{u}^R}(\bar{\xi}^R)$, $\bar{\xi}^R \in \operatorname{im} Q_{\bar{u}^R}$, and $\|\bar{\xi}^R\|_{1,p,R} \le \varepsilon_1$. Hence, by Step 1, $v = \iota_c^R(\bar{u}^0,\bar{u}^\infty)$. This proves (v) in Theorem 10.1.2. $\square$

10.8. Proof of the splitting axiom

We begin by restating the splitting axiom in a slightly different, but equivalent, form. We shall assume throughout that (M, ω) is a compact semipositive symplectic manifold. Let us fix three integers $\ell^0, \ell^\infty \geq 2$ and $k \geq \ell^0 + \ell^\infty$ and denote

$$I^0 := \{0, \dots, \ell^0\}, \qquad I^\infty := \{0, \dots, \ell^\infty\}, \qquad I := \{1, \dots, \ell^0 + \ell^\infty\}.$$

(These sets label the fixed marked points $\mathbf{w}$; the first two correspond in the previous notation to I_{S^0}, I_{S^1}.) Let $\mathcal{S}$ denote the set of all splittings $S = (S^0, S^\infty)$ of the index set $\{1, \dots, k\}$ such that

$$\{1, \dots, \ell^0\} \subset S^0, \qquad \{\ell^0 + 1, \dots, \ell^0 + \ell^\infty\} \subset S^\infty.$$

Given a splitting $S = (S^0, S^\infty) \in \mathcal{S}$, let

$$k^0 := k^0(S) := |S^0| \geq \ell^0, \qquad k^\infty := k^\infty(S) := |S^\infty| \geq \ell^\infty,$$

and denote by

$$\sigma^0 : \{1, \dots, k^0\} \to \{1, \dots, k\}, \qquad \sigma^\infty : \{1, \dots, k^\infty\} \to \{1, \dots, k\}$$

the unique order preserving injections such that $\operatorname{im} \sigma^0 = S^0$ and $\operatorname{im} \sigma^\infty = S^\infty$. Thus $\sigma^0(i) = i$ for $1 \leq i \leq \ell^0$ and $\sigma^\infty(i) = \ell^0 + i$ for $1 \leq i \leq \ell^\infty$. With this notation the splitting axiom for the Gromov–Witten invariants can be expressed in the following form.

THEOREM 10.8.1 (The Splitting Axiom). *For every spherical homology class $A \in H_2(M;\mathbb{Z})$ and every collection of cohomology classes $a_1, \dots, a_k \in H^*(M)$,*

$$\mathrm{GW}^{M,I}_{A,k}(a_1, \dots, a_k) = \sum_{S \in \mathcal{S}} \sum_{A^0 + A^\infty = A} \varepsilon(S, a) \varepsilon(S^0, a) \sum_{\nu,\mu}$$

$$\mathrm{GW}^{M,I^0}_{A^0,k^0+1}(e_\nu, a_{\sigma^0(1)}, \dots, a_{\sigma^0(k^0)})\, g^{\nu\mu}\, \mathrm{GW}^{M,I^\infty}_{A^\infty,k^\infty+1}(e_\mu, a_{\sigma^\infty(1)}, \dots, a_{\sigma^\infty(k^\infty)}),$$

where the e_ν form a basis of $H^*(M)$, $g_{\nu\mu} = \int_M e_\nu \smile e_\mu$, $g^{\nu\mu}$ is the inverse matrix, and

$$\varepsilon(S, a) := (-1)^{\#\{(i,j) \in S^0 \times S^\infty \,|\, j<i,\, \deg(a_i)\deg(a_j) \in 2\mathbb{Z}+1\}},$$

$$\varepsilon(S^0, a) := (-1)^{\#\{i \in S^0 \,|\, \deg(a_i) \in 2\mathbb{Z}+1\}}.$$

REMARK 10.8.2. In Theorem 10.8.1 the invariants GW^{M,I^0} and GW^{M,I^∞} are understood with the index sets $\{0, \dots, k^0\}$ and $\{0, \dots, k^\infty\}$, respectively, for the marked points. The meaning of $\varepsilon(S, a)$ is unchanged. The additional sign $\varepsilon(S^0, a)$ arises because the class e_ν is the first, rather than the last, argument of GW^{M,I^0}. To derive this sign formula we used the fact that the sum of the degrees of the classes on which GW^{M,I^0} is evaluated is always even.

The evaluation map for connected pairs. Let $\mathbf{w}^0 = \{w_i^0\}_{1 \leq i \leq \ell^0}$ be an ℓ^0-tuple of pairwise distinct points in $S^2 \setminus \{0\}$ and $\mathbf{w}^\infty = \{w_j^\infty\}_{1 \leq j \leq \ell^\infty}$ be an ℓ^∞-tuple of pairwise distinct points in $S^2 \setminus \{\infty\}$. Fix an integer $k \geq \ell^0 + \ell^\infty$, let $S = (S^0, S^\infty) \in \mathcal{S}$, denote

$$k^0 := |S^0|, \qquad k^\infty := |S^\infty|,$$

and let σ^0 and σ^∞ be the corresponding reordering maps, as above. Consider the moduli space

$$\mathcal{M}_{0,S}(A^{0,\infty}; \mathbf{w}^{0,\infty}; J^{0,\infty}) := \mathcal{M}_{0,S}(A^0, A^\infty; \mathbf{w}^0, \mathbf{w}^\infty; J^0, J^\infty)$$

of all tuples

$$(u^0, u^\infty, \mathbf{z}^0, \mathbf{z}^\infty) = (u^0, u^\infty, z_1^0, \ldots, z_{k^0}^0, z_1^\infty, \ldots, z_{k^\infty}^\infty),$$

where $(u^0, u^\infty) \in \mathcal{M}(A^{0,\infty}; J^{0,\infty})$, $\mathbf{z}^0 = \{z_i^0\}_{1\le i\le k^0}$ is a k^0-tuple of pairwise distinct points in $S^2 \setminus \{0\}$ with $z_j^0 = \mathrm{w}_j^0$ for $j = 1, \ldots, \ell^\infty$, and similarly for $\mathbf{z}^\infty$. (See Figure 3.) We assume as before that J^0 and J^∞ depends on z so that there is no need to restrict to simple maps.

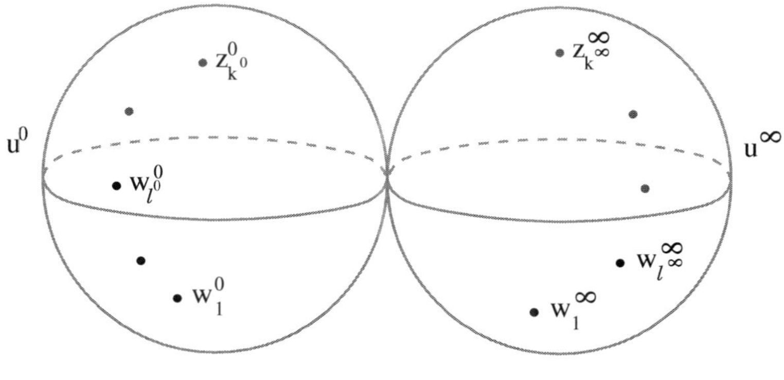

FIGURE 3. A connected pair with marked points.

This space carries an evaluation map

$$\mathrm{ev}_{A^{0,\infty},S} : \mathcal{M}_{0,S}(A^{0,\infty}; \mathbf{w}^{0,\infty}; J^{0,\infty}) \to M^k$$

given by $\mathrm{ev}_S(u^0, u^\infty, \mathbf{z}^0, \mathbf{z}^\infty) := (x_1, \ldots, x_k)$, where

$$x_{\sigma^0(i)} := u^0(z_i^0), \qquad x_{\sigma^\infty(j)} := u^\infty(z_j^\infty)$$

for $i = 1, \ldots, k^0$ and $j = 1, \ldots, k^\infty$. We shall also consider the restriction of this evaluation map to the compact subset

$$\mathcal{M}_{0,S}(A^{0,\infty}; \mathbf{w}^{0,\infty}; J^{0,\infty}, c) \subset \mathcal{M}_{0,S}(A^{0,\infty}; \mathbf{w}^{0,\infty}; J^{0,\infty})$$

of all tuples $(u^0, u^\infty, \mathbf{z}^0, \mathbf{z}^\infty) \in \mathcal{M}_{0,S}(A^{0,\infty}; \mathbf{w}^{0,\infty}; J^{0,\infty})$ that satisfy $\|du^0\|_{L^\infty} \le c$, $\|du^\infty\|_{L^\infty} \le c$, and

$$|z_i^0 - z_{i'}^0| \ge \frac{1}{c}, \qquad |z_i^0| \ge \frac{1}{c}, \qquad |z_j^\infty - z_{j'}^\infty| \ge \frac{1}{c}, \qquad |z_j^\infty| \le c$$

for $i \ne i'$, $j \ne j'$. The next corollary follows immediately from Theorem 10.1.2 (iii).

COROLLARY 10.8.3. *Fix two homology classes $A^0, A^\infty \in H_2(M;\mathbb{Z})$, a regular pair of almost complex structures $(J^0, J^\infty) \in \mathcal{J}_{\mathrm{reg}}(S^2; A^{0,\infty})$, and a constant $c > 0$. Let $A := A^0 + A^\infty$ and let $\delta_0 = \delta_0(c)$ be the constant of Theorem 10.1.2. Suppose that $0 < \delta < \delta_0$ and, for $R > 1/\delta\delta_0$, let*

$$\iota_c^R = \iota_c^{\delta,R} : \mathcal{M}(A^{0,\infty}; J^{0,\infty}, c) \to \mathcal{M}(A; J^R)$$

be the map introduced in Theorem 10.1.2. Let $\mathbf{w}^0 = \{\mathrm{w}_i^0\}_{1\le i\le \ell^0}$ be an ℓ^0-tuple of pairwise distinct points in $S^2 \setminus \{0\}$ and $\mathbf{w}^\infty = \{\mathrm{w}_j^\infty\}_{1\le j\le \ell^\infty}$ be an ℓ^∞-tuple of pairwise distinct points in $S^2 \setminus \{\infty\}$. Fix an integer $k \ge \ell^0 + \ell^\infty$, let $S = (S^0, S^\infty) \in \mathcal{S}$, and define $\mathbf{w}^R$ by $\mathrm{w}_i^R := \mathrm{w}_i^0$ for $i = 1, \ldots, \ell^0$ and $\mathrm{w}_{\ell^0+j}^R := \mathrm{w}_j^\infty / R^2$ for $j = 1, \ldots, \ell^\infty$. Consider the embedding

$$\iota_{c,S}^R : \mathcal{M}_{0,S}(A^{0,\infty}; \mathbf{w}^{0,\infty}; J^{0,\infty}, c) \to \mathcal{M}_{0,k}^*(A; \mathbf{w}^R; J^R),$$

given by $\iota^R_{c,S}(u^0,u^\infty,\mathbf{z}^0,\mathbf{z}^\infty) := (\widetilde{u}^R,\mathbf{z}^R)$, *where*

$$\widetilde{u}^R := \iota^R_c(u^0,u^\infty),\qquad z^R_{\sigma^0(i)} := z^0_i,\qquad z^R_{\sigma^\infty(j)} := z^\infty_j/R^2$$

for $i=1,\dots,k^0$ *and* $j=1,\dots,k^\infty$. *Then the composition of the evaluation map* $\mathrm{ev}^R : \mathcal{M}^*_{0,k}(A;\mathbf{w}^R;J^R)\to M^k$ *with* $\iota^R_{c,S}$ *converges to* $\mathrm{ev}_{A^{0,\infty},S}$ *in the* C^1 *topology as* R *tends to infinity.*

The limit set of the moduli space of pairs. The first step in the proof of Theorem 10.8.1 is to show that the evaluation map

$$\mathrm{ev}_{A^{0,\infty},S} : \mathcal{M}^*_{0,S}(A^{0,\infty};\mathbf{w}^{0,\infty};J^{0,\infty})\to M^k$$

is a pseudocycle. The proof is much the same as that needed to show that the evaluation map for stable maps modelled on a labelled tree T

$$\mathrm{ev}_T : \mathcal{M}^*_{0,T}(\{A_\alpha\};J)\to M^k$$

is a pseudocycle: see Exercise 6.6.3. However, the dependence of J^0 and J^∞ on z gives rise to some extra features that we now explain.

The first step is to understand the limits of Gromov convergent sequences in the domain. Since J^0 and J^∞ varies over the domain S^2, these limits are stable maps with two distinguished components labelled by 0 and ∞ that are $\{J^0_z\}$-holomorphic, respectively $\{J^\infty_z\}$-holomorphic. The other components are bubbles that are holomorphic with respect to the fixed complex structure J^0_z, respectively J^∞_z, associated to the point where they attach to one of the distinguished components. As in Section 6.7 it is useful to think in terms of the convergence of the graphs $\widetilde{u}^0,\widetilde{u}^\infty$ of u^0,u^∞. But now bubbles can form between the two principal components. Therefore there is a third part of the limit that is J-holomorphic, where J is associated to the attaching point of u^0 to u^∞ as in (10.1.1). We will make the above discussion more precise later; for now, it is intended to motivate the following definitions.

Let $\mathbf{w}^0=\{\mathrm{w}^0_i\}_{1\le i\le \ell^0}$ and $\mathbf{w}^\infty=\{\mathrm{w}^\infty_j\}_{1\le j\le\ell^\infty}$ be tuples of pairwise distinct points on S^2 such that $\mathrm{w}^0_i\ne 0$ and $\mathrm{w}^\infty_j\ne\infty$ for all i and j. Let $T=(T,E,\Lambda)$ be a k-labelled tree with two distinct special vertices 0 and ∞. Denote by $T^0\subset T$ and $T^\infty\subset T$ the subtrees defined by

$$T^0 := \{\alpha\in T\,|\,0\in[\alpha,\infty]\},\qquad T^\infty := \{\alpha\in T\,|\,\infty\in[0,\alpha]\}.$$

Thus T^0 is obtained by cutting off the branch pointing from 0 in the direction of ∞, and T^∞ is obtained by cutting off the branch pointing from ∞ in the direction of 0. We assume that the labelling satisfies the condition $\alpha_i\in T^0$ for $i=1,\dots,\ell^0$ and $\alpha_{\ell^0+j}\in T^\infty$ for $j=1,\dots,\ell^\infty$. (See Figure 4.)

Fix a collection of homology classes $B=\{B_\alpha\}_{\alpha\in T}$ in $H_2(M;\mathbb{Z})$ and assume J^0 and J^∞ satisfy (10.1.1). Then the moduli space

$$\mathcal{M}^*_{0,T}(\{B_\alpha\};\mathbf{w}^{0,\infty},J^{0,\infty})$$

consists of tuples

$$\begin{aligned}(\mathbf{u},\mathbf{z}) \;:=\; \big(&u_0,u_\infty,\ \{z_\alpha,u_\alpha\}_{\alpha\in T^0\cup T^\infty\setminus\{0,\infty\}},\\ &\{u_\alpha\}_{\alpha\in T\setminus(T^0\cup T^\infty)},\ \{z_{\alpha\beta}\}_{\alpha E\beta},\ \{\alpha_i,z_i\}_{1\le i\le k}\big)\end{aligned}$$

with the following properties.

(Map) Each $u_\alpha : S^2\to M$ is a smooth map representing the class B_α. Moreover, u_0 is a $\{J^0_z\}$-holomorphic sphere, u_∞ is a $\{J^\infty_z\}$-holomorphic sphere, u_α is

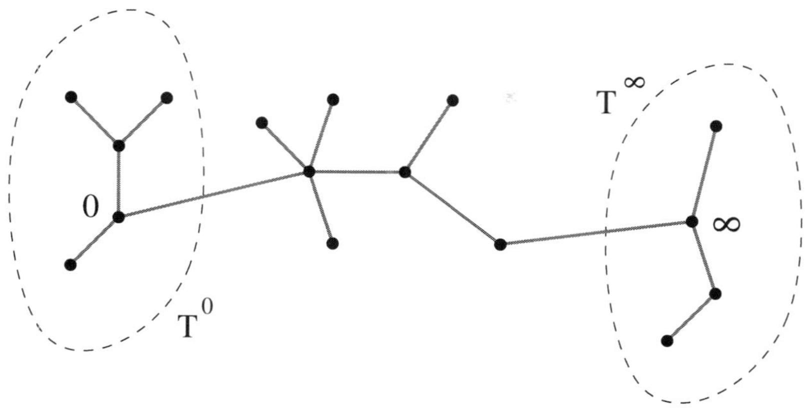

FIGURE 4. A bubble tree with two special vertices.

a $J^0_{z_\alpha}$-holomorphic sphere for $\alpha \in T^0 \setminus \{0\}$, u_α is a $J^\infty_{z_\alpha}$-holomorphic sphere for $\alpha \in T^\infty \setminus \{\infty\}$, and u_α is a J-holomorphic sphere for $\alpha \in T \setminus (T^0 \cup T^\infty)$.

(NODES) $u_\alpha(z_{\alpha\beta}) = u_\beta(z_{\beta\alpha})$ for $\alpha E \beta$. Moreover, $z_\alpha = z_\beta$ whenever $\alpha E \beta$ and $\alpha, \beta \in T^0 \setminus \{0\}$ or $\alpha, \beta \in T^\infty \setminus \{\infty\}$. If $0 E \alpha$ and $\infty E \beta$ then

$$z_{0\alpha} = \begin{cases} z_\alpha, & \text{if } \alpha \in T^0, \\ 0, & \text{if } \alpha \notin T^0, \end{cases} \qquad z_{\infty\beta} = \begin{cases} z_\beta, & \text{if } \beta \in T^\infty, \\ \infty, & \text{if } \beta \notin T^\infty. \end{cases}$$

(STABILITY) The points $z_{\alpha\beta}$ for $\alpha E \beta$ and z_i for $\alpha_i = \alpha$ are pairwise distinct, and there are at least three such points whenever u_α is constant and $\alpha \neq 0, \infty$.

(FIXING) If $i \in \{1, \dots, \ell^0\}$ and $j \in \{1, \dots, \ell^\infty\}$ then

$$\mathrm{w}_i^0 = \begin{cases} z_i, & \text{if } \alpha_i = 0, \\ z_{\alpha_i}, & \text{if } \alpha_i \neq 0, \end{cases} \qquad \mathrm{w}_j^\infty = \begin{cases} z_{\ell^0+j}, & \text{if } \alpha_j = \infty, \\ z_{\alpha_{\ell^0+j}}, & \text{if } \alpha_j \neq \infty. \end{cases}$$

(SIMPLE) The stable maps over the subtrees

$$T_z^0 := \{\alpha \in T^0 \mid z_\alpha = z\}, \qquad T_z^\infty := \{\alpha \in T^\infty \mid z_\alpha = z\}$$

are simple for every $z \in S^2$, and the stable map over the subtree $T \setminus (T^0 \cup T^\infty)$ is simple.

To understand this definition consider the graph $\widetilde{u}_0(z) := (z, u_0(z))$ as a $\widetilde{J}^0$-holomorphic section of $\widetilde{M} := S^2 \times M$. Here $\widetilde{J}^0$ denotes the almost complex structure which is equal to the standard complex structure on each horizontal subspace $T_z S^2 \times \{0\}$ and pulls back to J_z^0 under the embedding $\iota_z : M \to \widetilde{M}$ given by $\iota_z(x) := (z, x)$. The pair (z_α, u_α), for $\alpha \in T^0 \setminus \{0\}$, then gives rise to the vertical $\widetilde{J}^0$-holomorphic sphere $\widetilde{u}_\alpha(z) := (z_\alpha, u_\alpha(z))$. With this notation the *(Map)*, *(Nodes)*, and *(Stability)* conditions assert that the restriction of $(\mathbf{u}, \mathbf{z})$ to the subtree T^0 is a stable map in $(\widetilde{M}, \widetilde{J}^0)$. Similarly, the restriction to T^∞ is a stable map in $(\widetilde{M}, \widetilde{J}^\infty)$. These two stable maps are connected through a third stable map in (M, J), modelled over the subtree $T \setminus (T^0 \cup T^\infty)$. The *(Fixing)* condition asserts that the first ℓ^0 marked points belong to T^0 and project to the points w_i^0, while the next ℓ^∞ marked points belong to T^∞ and project to the w_j^∞.

The last condition asserts the the stable map is simple. As in Chapter 6, this is needed to achieve transversality for generic families of almost complex structures.

Indeed, it follows as in the proof of Theorem 6.7.11 that the almost complex structures J, $\{J_z^0\}_{z\in S^2}$, and $\{J_z^\infty\}_{z\in S^2}$ can be chosen such that (10.1.1) holds and all the relevant edge evaluation maps are transverse to the respective edge diagonals so that each of the moduli spaces $\mathcal{M}^*_{0,T}(\{B_\alpha\};\mathbf{w}^{0,\infty};J^{0,\infty})$ is a smooth manifold of dimension

$$\dim \mathcal{M}^*_{0,T}(\{B_\alpha\};\mathbf{w}^{0,\infty};J^{0,\infty}) = 2n + 2c_1(B) + 2(k-\ell^0-\ell^\infty) - 2(e(T)-1),$$

where $B := \sum_{\alpha\in T} B_\alpha$. This moduli space carries an evaluation map

$$\mathrm{ev}_{\{B_\alpha\},T} : \mathcal{M}^*_{0,T}(\{B_\alpha\};\mathbf{w}^{0,\infty};J^{0,\infty}) \to M^k$$

defined by

$$\mathrm{ev}_{\{B_\alpha\},T}(\mathbf{u},\mathbf{z}) := (u_{\alpha_1}(z_1),\dots,u_{\alpha_k}(z_k)).$$

Note that the evaluation map $\mathrm{ev}_{A^{0,\infty},S}$ corresponds to the special case where the tree T has precisely two vertices 0 and ∞, and the labelling is given by $\alpha_i = 0$ for $i\in S^0$ and $\alpha_i=\infty$ for $i\in S^\infty$. With the above choice of J, J^0, and J^∞ it follows as in the proof of Theorem 6.7.1 that the evaluation map

$$\mathrm{ev}_{A^{0,\infty},S} : \mathcal{M}_{0,S}(A^{0,\infty};\mathbf{w}^{0,\infty};J^{0,\infty}) \to M^k$$

is a pseudocycle for all classes A^0 and A^∞ and all splittings $S\in\mathcal{S}$.

Proof of Theorem 10.8.1. Assume, without loss of generality, that the cohomology classes a_i are Poincaré dual to compact oriented smooth submanifolds $X_i\subset M$:

$$a_i = \mathrm{PD}([X_i]).$$

(This always holds for suitable integer multiples of the classes a_i.) The transversality arguments in the proof of Lemma 6.5.5 show that the submanifolds X_i can be chosen so that their product

$$\mathbf{X} := X_1\times\cdots\times X_k \subset M^k$$

is transverse to each of the evaluation maps $\mathrm{ev}_{\{B_\alpha\},T}$ discussed above. Then $\mathbf{X}$ is strongly transverse, as a pseudocycle, to the evaluation maps $\mathrm{ev}_{A^{0,\infty},S}$ for every splitting $S\in\mathcal{S}$ and every pair of homology classes $A^0, A^\infty\in H_2(M;\mathbb{Z})$.

We claim that

$$\mathbf{X}\cdot \mathrm{ev}_{A^{0,\infty},S} = \varepsilon(S,a)\varepsilon(S^0,a)\sum_{\nu,\mu} \mathrm{GW}^{M,I^0}_{A^0,k^0+1}(e_\nu,a_{\sigma^0(1)},\dots,a_{\sigma^0(k^0)})\,g^{\nu\mu}\,\mathrm{GW}^{M,I^\infty}_{A^\infty,k^\infty+1}(e_\mu,a_{\sigma^\infty(1)},\dots,a_{\sigma^\infty(k^\infty)}). \tag{10.8.1}$$

To see this, consider the moduli space $\widehat{\mathcal{M}}(A^{0,\infty},S)$ of all tuples

$$(\mathbf{u},\mathbf{z}) := (u^0,u^\infty,\mathbf{z}^0,\mathbf{z}^\infty)$$

that satisfy all the conditions for the moduli space $\mathcal{M}_{0,S}(A^{0,\infty};\mathbf{w}^{0,\infty};J^{0,\infty})$, except that $u^0(0)$ is not required to be equal to $u^\infty(\infty)$. This space carries an evaluation map

$$\widehat{\mathrm{ev}}_{A^{0,\infty},S} : \widehat{\mathcal{M}}(A^{0,\infty},S) \to M^k\times M\times M,$$

given by

$$\widehat{\mathrm{ev}}_{A^{0,\infty},S}(\mathbf{u},\mathbf{z}) := \left(\mathrm{ev}_{A^{0,\infty},S}(\mathbf{u},\mathbf{z}),u^0(0),u^\infty(\infty)\right).$$

Our assumptions guarantee that $\widehat{\mathrm{ev}}_{A^{0,\infty},S}$ is a pseudocycle and that the left hand side of (10.8.1) is equal to the intersection number of $\widehat{\mathrm{ev}}_{A^{0,\infty},S}$ with the product $\mathbf{X} \times \Delta$, where $\Delta \subset M \times M$ denotes the diagonal:

$$\mathbf{X} \cdot \mathrm{ev}_{A^{0,\infty},S} = (\mathbf{X} \times \Delta) \cdot \widehat{\mathrm{ev}}_{A^{0,\infty},S}.$$

Now the Poincaré dual of the homology class of the diagonal is

$$\mathrm{PD}([\Delta]) = \sum_{\nu,\mu} g^{\nu\mu}\, \pi_1^* e_\nu \smile \pi_2^* e_\mu, \tag{10.8.2}$$

where $\pi_1, \pi_2 : M \times M \to M$ are the obvious projections. Hence (10.8.1) follows by considering the sign change associated to the relevant permutation in the target space.

By (10.8.1) the assertion of Theorem 10.8.1 can be expressed in the form

$$\mathrm{GW}^{M,I}_{A,k}(a_1, \ldots, a_k) = \sum_{S \in \mathcal{S}} \sum_{A^0 + A^\infty = A} \mathbf{X} \cdot \mathrm{ev}_{A^{0,\infty},S}. \tag{10.8.3}$$

We now prove this using Corollary 10.1.4 and Corollary 10.8.3.

Let us denote by

$$\mathcal{M}^{0,\infty}(\mathbf{X}) := \bigcup_{\substack{A^0 + A^\infty = A \\ S \in \mathcal{S}}} (\mathrm{ev}_{A^{0,\infty},S})^{-1}(\mathbf{X})$$

the set of all tuples

$$(\mathbf{u}, \mathbf{z}) := (u^0, u^\infty, \mathbf{z}^0, \mathbf{z}^\infty) \in \mathcal{M}_{0,S}(A^{0,\infty}; \mathbf{w}^{0,\infty}; J^{0,\infty}),$$

where S varies over all splittings in $\mathcal{S}$ and A^0, A^∞ vary over all homology classes satisfying $A^0 + A^\infty = A$, such that $\mathrm{ev}_{A^{0,\infty},S}(\mathbf{u}, \mathbf{z}) \in \mathbf{X}$. Likewise, denote by

$$\mathcal{M}^R(\mathbf{X}) := (\mathrm{ev}^R)^{-1}(\mathbf{X}) \subset \mathcal{M}_{0,k}(A; \mathbf{w}^R, J^R)$$

the set of all tuples $(v^R, \mathbf{z}^R) \in \mathcal{M}_{0,k}(A; \mathbf{w}^R, J^R)$ that satisfy $\mathrm{ev}^R(v^R, \mathbf{z}^R) \in \mathbf{X}$. By assumption, $\mathcal{M}^{0,\infty}(\mathbf{X})$ is a finite set and so we may choose our constant c so large that each element of the set $\mathcal{M}^{0,\infty}(\mathbf{X})$ belongs to one of the compact subsets $\mathcal{M}_{0,S}(A^{0,\infty}; \mathbf{w}^{0,\infty}; J^{0,\infty}, c-1)$. The transversality assumptions on $\mathbf{X}$ and the C^1-convergence of the evaluation maps in Corollary 10.8.3 show that each of these solutions gives rise to a unique nearby element $(v^R, \mathbf{z}^R) \in \mathcal{M}_{0,k}(A; \mathbf{w}^R, J^R)$ which satisfies $\mathrm{ev}^R(v^R, \mathbf{z}^R) \in \mathbf{X}$ and contributes the same sign to the invariant. Note that the map v^R will not in general equal the gluing $\widetilde{u}^R := \iota^R_c(\mathbf{u})$ since the image of $\widetilde{u}^R$ need not intersect $\mathbf{X}$ at points with the correct cross ratios. However, it will be very close to this gluing. More precisely, for R sufficiently large, there is a unique injective map

$$\mathcal{T}^R : \mathcal{M}^{0,\infty}(\mathbf{X}) \to \mathcal{M}^R(\mathbf{X}), \tag{10.8.4}$$

defined as follows. If

$$(\mathbf{u}, \mathbf{z}) \in \mathcal{M}^{0,\infty}(\mathbf{X}) \cap \mathcal{M}_{0,S}(A^{0,\infty}; \mathbf{w}^{0,\infty}; J^{0,\infty}, c-1)$$

then there is a unique element $(\mathbf{u}', \mathbf{z}') \in \mathcal{M}_{0,S}(A^{0,\infty}; \mathbf{w}^{0,\infty}; J^{0,\infty}, c)$ that is close to $(\mathbf{u}, \mathbf{z})$ and satisfies $\mathrm{ev}^R \circ \iota^R_{c,S}(\mathbf{u}', \mathbf{z}') \in \mathbf{X}$; the element $\mathcal{T}^R(\mathbf{u}, \mathbf{z}) \in \mathcal{M}^R(\mathbf{X})$ is then defined by $\mathcal{T}^R(\mathbf{u}, \mathbf{z}) := \iota^R_{c,S}(\mathbf{u}', \mathbf{z}')$.

It remains to prove that $\mathcal{T}^R$ is surjective for R sufficiently large. Suppose, by contradiction, that this is not the case. Then there exists a sequence $R_\nu \to \infty$ such that $\mathcal{T}^{R_\nu}$ is not surjective. Let $(u_\nu, \mathbf{z}_\nu) \in \mathcal{M}^{R_\nu}(\mathbf{X})$ be an element which does not belong to the image of $\mathcal{T}^{R_\nu}$. We claim that

$$\forall c > 0, \ \exists \nu_c \in \mathbb{N}, \ \forall \nu \geq \nu_c, \ (u_\nu, \mathbf{z}_\nu) \notin \operatorname{im} \iota_{c,S}^{R_\nu}. \tag{10.8.5}$$

Suppose otherwise that (10.8.5) does not hold. Then there is a constant $c > 0$, a sequence $\nu_i \to \infty$, a splitting $S \in \mathcal{S}$, a pair of homology classes $A^0, A^\infty \in H_2(M;\mathbb{Z})$ such that $A^0 + A^\infty = A$, and a sequence

$$(\mathbf{u}_i, \mathbf{z}_i) := (u_i^0, u_i^\infty, \mathbf{z}_i^0, \mathbf{z}_i^\infty) \in \mathcal{M}_{0,S}(A^{0,\infty}; \mathbf{w}^{0,\infty}; J^{0,\infty}, c)$$

such that $(u_{\nu_i}, \mathbf{z}_{\nu_i}) = \iota_{c,S}^{R_{\nu_i}}(\mathbf{u}_i, \mathbf{z}_i)$ for every i. Since $(u_{\nu_i}, \mathbf{z}_{\nu_i}) \in \mathcal{M}^{R_{\nu_i}}(\mathbf{X})$, we have

$$\mathrm{ev}^{R_{\nu_i}} \circ \iota_{c,S}^{R_{\nu_i}}(\mathbf{u}_i, \mathbf{z}_i) \in \mathbf{X}$$

for every i. Moreover, since $\mathcal{M}_{0,S}(A^{0,\infty}; \mathbf{w}^{0,\infty}; J^{0,\infty}, c)$ is compact we may assume, passing to a subsequence if necessary, that the sequence $(\mathbf{u}_i, \mathbf{z}_i)$ converges to an element $(\mathbf{u}, \mathbf{z}) \in \mathcal{M}_{0,S}(A^{0,\infty}; \mathbf{w}^{0,\infty}; J^{0,\infty}, c)$. By Corollary 10.8.3, this limit must belong to $\mathcal{M}^{0,\infty}(\mathbf{X})$. Hence it follows from the definition of $\mathcal{T}^{R_{\nu_i}}$ in (10.8.4) that

$$(u_{\nu_i}, \mathbf{z}_{\nu_i}) = \mathcal{T}^{R_{\nu_i}}(\mathbf{u}, \mathbf{z})$$

for i sufficiently large. But this contradicts our assumption about the original sequence $(u_{\nu_i}, \mathbf{z}_{\nu_i})$. Thus we have proved (10.8.5).

Now it follows from (10.8.5) and Corollary 10.1.4 that $(u_\nu, \mathbf{z}_\nu)$ has a subsequence which Gromov converges to a stable map with more than two components passing through $\mathbf{X}$. More precisely, passing to a subsequence, we may assume that $u_\nu(z)$ converges modulo bubbling to a $\{J_z^0\}$-holomorphic sphere u^0 on $S^2 \setminus \{0\}$ and $u_\nu(z/R_\nu^2)$ converges modulo bubbling to a $\{J_z^\infty\}$-holomorphic sphere u^∞ on $S^2 \setminus \{\infty\}$. (The fact that the second limit exists follows from the definition of J^R.) To understand the limiting stable map we must distinguish three cases in the proof of Theorem 5.3.1. Namely, each component of the limit, except for u^0, is the limit of a sequence of the form $u_\nu(z_\nu + \varepsilon_\nu z)$, where z_ν converges to a point in S^2 and ε_ν converges to zero (see the proof of Proposition 4.7.1, which is the main step in the proof of Theorem 5.3.1). The three cases are:

CASE 1: $\lim_{\nu\to\infty} |z_\nu| > 0$.

CASE 2: $\lim_{\nu\to\infty} |z_\nu| = 0$ and $\lim_{\nu\to\infty} R_\nu^2 |z_\nu| = \infty$.

CASE 3: $\lim_{\nu\to\infty} R_\nu^2 |z_\nu| < \infty$.

In the first case the relevant component of the limit stable map belongs to the subtree T^0, in the third case it belongs to T^∞, and in the intermediate second case to the complement $T \setminus (T^0 \cup T^\infty)$.

We claim that the limiting stable map is modelled over a tree with more than two vertices. It has to have at least two (namely those labelled by 0 and ∞). In fact, if $[u^0] + [u^\infty] \neq A$ then the limit has a nonconstant bubble. If $[u^0] + [u^\infty] = A$ then there is no energy left for bubbling and so the limit has at most two nonconstant components that are associated to the special vertices 0 and ∞. It follows that $u_\nu(z)$ converges to $u^0(z)$, uniformly on the domain $\{|z| \geq 1\}$ and $u^\nu(z/R_\nu^2)$, converges to $u^\infty(z)$, uniformly on the domain $\{|z| \leq 1\}$. Hence, by Corollary 10.1.4, there is a constant $c > 0$ such that $u^\nu \in \operatorname{im} \iota_c^{R_\nu}$ for ν sufficiently large. Now suppose that none of the marked points collide, i.e. either $z_{\nu i} \to z_i^0 \neq 0$ or $R_\nu^2 z_{\nu j} \to z_j^\infty \neq \infty$, and $z_i^0 \neq z_{i'}^0$ and $z_j^\infty \neq z_{j'}^\infty$ for $i \neq i'$ and $j \neq j'$. Then we deduce that $(u^\nu, \mathbf{z}^\nu) \in \operatorname{im} \iota_{c,S}^{R_\nu}$

for some fixed constant $c > 0$, some $S \in \mathcal{S}$, and for ν sufficiently large. But this would contradict (10.8.5). Thus some of the marked points must collide in the case $[u^0] + [u^\infty] = A$, and so there must be a ghost component.

Thus we have proved that the limiting stable map is modelled over a tree with more than two vertices, and hence with more than one edge. The underlying simple stable map still has at least two edges and so belongs to one of the lower dimensional moduli spaces $\mathcal{M}^*_{0,T}(\{B_\alpha\}; \mathbf{w}^{0,\infty}; J^{0,\infty})$ discussed above. However, $\mathbf{X}$ was chosen transverse to all the evaluation maps $\mathrm{ev}_{\{B_\alpha\},T}$ and so these evaluation maps must miss $\mathbf{X}$ for dimensional reasons, a contradiction. This shows that our assumption that $\mathcal{T}^R$ is not surjective for R sufficiently large must have been wrong. Hence $\mathcal{T}^R$ is bijective for large R as claimed. This proves (10.8.3) and Theorem 10.8.1. □

10.9. The gluing theorem revisited

In this section we examine the gluing map in the special case where the almost complex structure $J^0_z = J^\infty_z =: J$ is independent of $z \in S^2$. In this case the methods developed in the present chapter give rise to an embedding of the product of a small 2-disc with a suitable moduli space of nodal curves into $\overline{\mathcal{M}}_{0,4}(A; J)$.

Fix two homology classes $A^0, A^\infty \in H_2(M;\mathbb{Z})$ and write $A := A^0 + A^\infty$. Let (T, E, Λ) denote the 4-labelled tree

$$T := (0, \infty), \qquad \Lambda_0 := (1, 2), \qquad \Lambda_\infty := (3, 4),$$

with a single edge connecting the vertices 0 and ∞. Let J be an ω-tame almost complex structure on M. Recall that $\mathcal{M}^*_{0,T}(A^{0,\infty}; J) \subset \mathcal{M}_{0,T}(A^{0,\infty}; J)$ denotes the moduli space of simple stable maps with four marked points, modelled over T and representing the homology classes A^0 and A^∞. (See Sections 5.1 and 6.1.) Write the elements of $\mathcal{M}_{0,T}(A^{0,\infty}; J)$ as equivalence classes of tuples

$$[\mathbf{u}, \mathbf{z}] = [u^0, z^0, z^0_1, z^0_2; u^\infty, z^\infty, z^\infty_3, z^\infty_4],$$

where $u^0, u^\infty : S^2 \to M$ are J-holomorphic curves representing the homology classes A^0, A^∞, the triples (z^0, z^0_1, z^0_2) and $(z^\infty, z^\infty_3, z^\infty_4)$ both consist of pairwise distinct points on S^2, and

$$u^0(z^0) = u^\infty(z^\infty).$$

The equivalence relation is given by the action of the group $\mathrm{G} \times \mathrm{G}$ of pairs of Möbius transformations $\phi = (\phi^0, \phi^\infty)$ by

$$\phi^*(\mathbf{u}, \mathbf{z}) = (u^0 \circ \phi^0, (\phi^0)^{-1}(z^0_i); u^\infty \circ \phi^\infty, (\phi^\infty)^{-1}(z^\infty_i)).$$

Recall also that $\mathcal{M}_{0,T}(A^{0,\infty}; J)$ is a subset of the moduli space $\overline{\mathcal{M}}_{0,4}(A; J)$ of all stable maps of genus zero with four marked points, representing the class A. Denote by $\iota_T : \mathcal{M}_{0,T}(A^{0,\infty}; J) \to \overline{\mathcal{M}}_{0,4}(A; J)$ the obvious inclusion.

For each tuple $[\mathbf{u}, \mathbf{z}] \in \mathcal{M}_{0,T}(A^{0,\infty}; J)$ there is a unique pair of Möbius transformations $\phi^0, \phi^\infty \in \mathrm{G}$ such that

$$\begin{array}{llll} (10.9.1) & \phi^0(z^0) = 0, & \phi^0(z^0_1) = 1, & \phi^0(z^0_2) = \infty, \\ & \phi^\infty(z^\infty) = \infty, & \phi^\infty(z^\infty_3) = 1, & \phi^\infty(z^\infty_4) = 0. \end{array}$$

For $c > 0$ define

$$\mathcal{M}^*_{0,T}(c) := \left\{ ([\mathbf{u}, \mathbf{z}]) \in \mathcal{M}^*_{0,T}(A^{0,\infty}; J) \,|\, (u^0 \circ \phi^0, u^\infty \circ \phi^\infty) \in \mathcal{M}^*(c) \right\},$$

where ϕ^0, ϕ^∞ are chosen such that (10.9.1) holds and $\mathcal{M}^*(c) \subset \mathcal{M}^*(A^{0,\infty}; J)$ is the exhausting family of compact sets defined in terms of the conditions (6.2.5) and (6.2.6). For $\rho > 0$ let $B_\rho \subset \mathbb{C}$ be the closed disc of radius ρ about the origin.

THEOREM 10.9.1. *Let J be an ω-tame almost complex structure on M that is regular for (A^0, A^∞) in the sense of Definition 6.2.1. For each constant $c > 0$ there is a constant $\rho > 0$ and a continuous embedding*

$$\iota_{T,c} : \mathcal{M}^*_{0,T}(c) \times B_\rho \to \overline{\mathcal{M}}_{0,4}(A; J)$$

with the following properties.

(i) *The restriction of $\iota_{T,c}$ to $\mathcal{M}^*_{0,T}(c) \times \{0\}$ is the obvious inclusion ι_T.*

(ii) *The image of $\iota_{T,c}$ is a neighbourhood of $\iota_T(\mathcal{M}^*_{0,T}(c-1))$ in $\overline{\mathcal{M}}_{0,4}(A; J)$.*

(iii) *The map $\iota_{T,c}$ restricts to a smooth orientation preserving embedding of the manifold $\mathcal{M}^*_{0,T}(c) \times (B_\rho \setminus \{0\})$ into $\mathcal{M}_{0,4}(A; J)$.*

(iv) *For $i = 1, 2, 3, 4$ let $\mathrm{ev}_i : \overline{\mathcal{M}}_{0,4}(A; J) \to M$ denote the evaluation map at the ith marked point. For $\lambda \in B_\rho$ define $\mathrm{ev}_{i,\lambda} : \mathcal{M}^*_{0,T}(c) \to M$ as the composition*

$$\mathrm{ev}_{i,\lambda}([\mathbf{u}, \mathbf{z}]) := \mathrm{ev}_i \circ \iota_{T,c}([\mathbf{u}, \mathbf{z}], \lambda).$$

Then $\mathrm{ev}_{i,\lambda} : \mathcal{M}^*_{0,T}(c) \to M$ converges to $\mathrm{ev}_i \circ \iota_T : \mathcal{M}^*_{0,T}(c) \to M$ in the C^1 topology as λ tends to zero.

REMARK 10.9.2. Since J does not depend on z, regularity can in general only be achieved for *simple* stable maps by a generic choice of J. This is the reason why Theorem 10.9.1 refers to Definition 6.2.1 rather than Definition 10.1.1. This is also the reason why the gluing map of Theorem 10.1.2 will in this case only be defined on the subset $\mathcal{M}^*(c) \subset \mathcal{M}(c)$ rather than on all of $\mathcal{M}(c)$.

PROOF OF THEOREM 10.9.1. Let $0 < \delta_0 < 1$ be as in Theorem 10.1.2 and fix a constant $0 < \delta < \delta_0$. For $R > 1/\delta\delta_0$ denote by $\iota_c^{\delta,R} : \mathcal{M}^*(c) \to \mathcal{M}(A; J)$ the gluing map of Theorem 10.1.2. Choose

$$\rho := (\delta\delta_0)^2$$

and define the map $\iota_{T,c} : \mathcal{M}^*_{0,T}(c) \times B_\rho \to \overline{\mathcal{M}}_{0,4}(A; J)$ as follows. For $\lambda = 0$ define $\iota_{T,c}([\mathbf{u}, \mathbf{z}], 0) := [\mathbf{u}, \mathbf{z}]$. Now fix an element $\lambda \in B_\rho \setminus \{0\}$ and write it in the form

$$\lambda = \frac{e^{i\theta}}{R^2}.$$

The condition $|\lambda| < \rho$ is equivalent to $R > 1/\delta\delta_0$. An equivalence class in $\mathcal{M}^*_{0,T}(c)$ has a unique representative $(\mathbf{u}, \mathbf{z})$ such that

$$z^0 = 0, \quad z_1^0 = 1, \quad z_2^0 = \infty, \quad z^\infty = \infty, \quad z_3^\infty = e^{i\theta}, \quad z_4^\infty = 0. \tag{10.9.2}$$

This representative satisfies $u^0(0) = u^\infty(\infty)$ and $(u^0, u^\infty) \in \mathcal{M}^*(c)$. (This last condition holds because the space $\mathcal{M}^*(c)$ is invariant under the action of $S^1 \times S^1$ by reparametrization.) Define the element $\iota_{T,c}([\mathbf{u}, \mathbf{z}]) = [u_\lambda, \mathbf{z}_\lambda] \in \overline{\mathcal{M}}_{0,4}(A; J)$ by

$$u_\lambda := \iota_c^{\delta,R}(u^0, u^\infty), \qquad \mathbf{z}_\lambda := (1, \infty, \lambda, 0)$$

The map $\mathcal{M}^*_{0,T}(c) \times (B_\rho \setminus \{0\}) \to \overline{\mathcal{M}}_{0,4}(A; J) : ([\mathbf{u}, \mathbf{z}], \lambda) \mapsto [u_\lambda, \mathbf{z}_\lambda]$ is well defined, because each equivalence class has only one representative satisfying (10.9.2). Moreover, it satisfies condition (i) by definition, and it restricts to a smooth map from

$\mathcal{M}^*_{0,T}(c) \times (B_\rho \setminus \{0\})$ to $\mathcal{M}_{0,4}(A;J)$ because the map

$$(1/\delta\delta_0, \infty) \times \mathcal{M}(c) \to \mathcal{M}(A;J) : (R, u^0, u^\infty) \mapsto \iota_c^{\delta,R}(u^0, u^\infty)$$

is smooth, by Theorem 10.1.2 (i).

We prove that the map $\iota_{T,c} : \mathcal{M}^*_{0,T}(c) \times B_\rho \to \overline{\mathcal{M}}_{0,4}(A;J)$ is a continuous embedding. Continuity at points $([\mathbf{u},\mathbf{z}],\lambda)$ with $\lambda \neq 0$ follows from smoothness of the restriction to $\mathcal{M}^*_{0,T}(c) \times (B_\rho \setminus \{0\})$. Continuity at points $([\mathbf{u},\mathbf{z}],0)$ follows from the convergence statement in Theorem 10.1.2 (ii). To show that the map is injective, let $([\mathbf{u},\mathbf{z}],\lambda)$ and $([\mathbf{u}',\mathbf{z}'],\lambda')$ be two elements of $\mathcal{M}_{0,T}(c) \times B_\rho$ such that $\iota_{T,c}([\mathbf{u},\mathbf{z}],\lambda) = \iota_{T,c}([\mathbf{u}',\mathbf{z}'],\lambda')$. Then $\mathbf{z}_\lambda = \mathbf{z}_{\lambda'}$ so that $\lambda = \lambda'$, and it follows from the injectivity of the maps $\iota_c^{\delta,R} : \mathcal{M}^*(c) \to \mathcal{M}(A;J)$ in Theorem 10.1.2 that $[u,\mathbf{z}] = [u',\mathbf{z}']$. Thus $\iota_{T,c}$ is a continuous injection. It is proper, because its domain is compact. Hence $\iota_{T,c}$ is a continuous embedding.

We next show that, for ρ sufficiently small, the image $\iota_{T,c}$ is a neighbourhood of $\iota_T(\mathcal{M}^*_{0,T}(c-1))$. If not, there is a sequence $[u^\nu,\mathbf{z}^\nu] \in \overline{\mathcal{M}}_{0,4}(A;J)$ that is not in the image and Gromov converges to some element $[\mathbf{u},\mathbf{z}] \in \mathcal{M}^*_{0,T}(c-1)$. Then $[u^\nu,\mathbf{z}^\nu]$ cannot belong to the stratum $\mathcal{M}^*_{0,T}(A^{0,\infty};J)$ and hence must be contained in $\mathcal{M}_{0,4}(A;J)$. Assume without loss of generality that the parametrization of $[u^\nu,\mathbf{z}^\nu]$ is chosen such that

$$z_1^\nu = 1, \qquad z_2^\nu = \infty, \qquad z_3^\nu =: \lambda^\nu, \qquad z_4^\nu = 0.$$

Since $[u^\nu,\mathbf{z}^\nu]$ Gromov converges to an element of $\mathcal{M}^*_{0,T}(c-1)$, the sequence λ^ν must converge to 0. Thus $R^\nu := |\lambda^\nu|^{-1/2} \to \infty$ and, passing to a subsequence if necessary, we may assume that $\arg\lambda^\nu =: \theta^\nu$ converges to θ. We may also assume, without loss of generality, that $[\mathbf{u},\mathbf{z}]$ satisfies the normalization condition (10.9.2) with this choice of θ. Then Gromov convergence of the sequence $[u^\nu,\mathbf{z}^\nu]$ to $[\mathbf{u},\mathbf{z}]$ means that u^ν converges to u^0 u.s.c. on $S^2 \setminus \{0\}$ and that there is a sequence ϕ^ν of Möbius transformations such that $u^\nu \circ \phi^\nu$ converges to u^∞ u.s.c. on $S^2 \setminus \{\infty\}$ and

$$\infty = \lim_{\nu\to\infty} (\phi^\nu)^{-1}(\infty), \qquad e^{i\theta} = \lim_{\nu\to\infty} (\phi^\nu)^{-1}(\lambda^\nu), \qquad 0 = \lim_{\nu\to\infty} (\phi^\nu)^{-1}(0).$$

Composing the sequence ϕ^ν with a further sequence of Möbius transformations that converges to the identity, we may assume that

$$\phi^\nu(z) = z/R_\nu^2, \qquad z \in S^2 = \mathbb{C} \cup \{\infty\}.$$

Hence, for large ν, the map u^ν satisfies the conditions of part (v) of Theorem 10.1.2 with $R = R^\nu$ and the pair $(u^0,u^\infty) \in \mathcal{M}^*(c-1)$. This implies that there exists a pair $(\widetilde{u}^0,\widetilde{u}^\infty) \in \mathcal{M}^*(c)$ such that $u^\nu = \iota_c^{\delta,R^\nu}(\widetilde{u}^0,\widetilde{u}^\infty)$. Hence, by construction,

$$[u^\nu,\mathbf{z}^\nu] = \iota_{T,c}\big([\widetilde{u}^0, 0, 1, \infty; \widetilde{u}^\infty, \infty, e^{i\theta_\nu}, 0], \lambda_\nu\big).$$

This contradicts our assumption and proves part (ii) of Theorem 10.9.1. Assertions (iii) and (iv) follow directly from the corresponding properties of the gluing maps $\iota_c^{\delta,R}$ of Theorem 10.1.2. This completes the proof of Theorem 10.9.1. □

The gluing map constructed in the proof of Theorem 10.9.1 has the following additional property. The **gluing parameter**

$$\lambda = w(z_4, z_1, z_2, z_3) = \frac{1}{w(z_1, z_2, z_3, z_4)} \tag{10.9.3}$$

is precisely the cross ratio of the marked points of the element $\iota_{T,c}([\mathbf{u},\mathbf{z}],\lambda)$ in the image of the gluing map. With this additional condition one cannot expect the evaluation maps $\mathrm{ev}_i \circ \iota_{T,c} : \mathcal{M}^*_{0,T}(c) \times B_\rho \to M$ to be continuously differentiable in the almost complex setting. However, continuous differentiability can be achieved by additional scaling.

COROLLARY 10.9.3. *The gluing map $\iota_{T,c} : \mathcal{M}^*_{0,T}(c) \times B_\rho \to \overline{\mathcal{M}}_{0,4}(A;J)$ in Theorem 10.9.1 can be chosen such that the evaluation map*

$$\mathrm{ev}_i \circ \iota_{T,c} : \mathcal{M}^*_{0,T}(c) \times B_\rho \to M$$

is continuously differentiable for $i = 1,2,3,4$.

PROOF. Choose a smooth function $\beta : [0,\infty) \to [0,\infty)$ such that $\dot\beta(0) = 0$ and $\dot\beta(s) > 0$ for $s > 0$. Replace the map $\iota_{T,c}$ in the proof of Theorem 10.9.1 by

$$\iota_{T,c,\beta}([\mathbf{u},\mathbf{z}],\lambda) := \iota_{T,c,\beta}([\mathbf{u},\mathbf{z}],\beta(|\lambda|)\lambda).$$

The derivatives of $\mathrm{ev}_i \circ \iota_{T,c,\beta} : \mathcal{M}^*_{0,T}(c) \times B_\rho \to M$ with respect to the variables in $\mathcal{M}^*_{0,T}(c)$ are continuous over $\mathcal{M}^*_{0,T}(c) \times B_\rho$, by part (iv) of Theorem 10.9.1. The partial derivatives with respect to the variable $\lambda \in B_\rho$ extend continuously over $\mathcal{M}^*_{0,T}(c) \times B_\rho$ and vanish on $\mathcal{M}^*_{0,T}(c) \times \{0\}$. This last step follows from the construction of the gluing map in the proof of Theorem 10.1.2. The details are left as an exercise for the reader. This proves Corollary 10.9.3. □

In this book, the main application of gluing is to prove the splitting axiom for the Gromov–Witten invariants (see Theorem 7.5.10). This was carried out in Section 10.8. The general case discussed there is rather complicated. However, our main application of the splitting axiom is to prove the associativity of quantum multiplication, and for that we only need the splitting axiom when $\#I = 4$; cf. Proposition 11.1.11. In this case, all we need to know is that a count of curves that satisfy four constraints at a set of points with *fixed* cross ratio λ equals a weighted sum of counts of pairs of curves, each of which goes through two of these constraints as well as an appropriate basis class e_ν. The formulation of Theorem 10.9.1 above makes this correspondence transparent. It implies that, when $|\lambda| > 0$ is sufficiently small, the inverse of the gluing map $\iota_{T,c}$ sets up a bijection between the first count and a count of intersecting pairs of curves through the given constraints. To complete the proof it remains to use the standard trick of "splitting the diagonal". In other words, we use the fact that the diagonal in $M \times M$ is Poincaré dual to $\sum_{\nu,\mu} g^{\nu\mu}\pi_1^*(e_\nu) \smile \pi_2^*(e_\mu)$; cf. equation (10.8.2).

EXERCISE 10.9.4. Complete this argument, using Theorem 10.9.1 instead of Theorem 10.1.2, and the relevant details of Theorem 10.8.1 as needed.

CHAPTER 11

Quantum Cohomology

Much of the recent interest in holomorphic spheres and Gromov-Witten invariants has arisen because they may be used to define a new multiplication on the cohomology ring of a compact symplectic manifold. In this chapter, we explain this construction, and point out some of the connections to other subjects such as algebraic geometry, integrable Hamiltonian systems, and mirror symmetry. At the time of writing, this aspect of the subject is developing very rapidly in so many interesting and unexpected directions that it is impossible to mention them all. For more information on quantum cohomology, Frobenius algebras, and mirror symmetry, readers may consult the books by Cox–Katz [**76**], Dubrovin [**94**], Manin [**286**], and Voisin [**409**].

We begin this chapter by discussing the (small) quantum cohomology ring of a semipositive symplectic manifold. Additively, this is the usual cohomology taken with coefficients in a ring Λ, but the multiplication is deformed. One adds to the usual cup product $a \smile b$ of two elements $a, b \in H^*(M)$ various "quantum correction" terms coming from 3-point Gromov–Witten invariants $\mathrm{GW}^M_{A,3}(a, b, \cdot)$; the ring Λ is needed to separate out the contributions from the different homology classes $A \in H_2(M;\mathbb{Z})$. Associativity for the quantum cup product is an easy consequence of the splitting axiom. The quantum cup product uses only a small part of the information provided by the Gromov–Witten invariants. In Section 11.2 we show that this information can all be encoded in a formal power series Φ called the genus zero Gromov–Witten potential. We describe how the splitting axiom translates into a remarkable system of third order quadratic partial differential equations, called the WDVV equations, for Φ.

Section 11.3 discusses four important examples, Fano toric manifolds, Grassmannians, flag manifolds and Calabi–Yau manifolds. Their quantum cohomology rings turn out to be very interesting invariants, with connections to the Toda lattice (in the case of flag manifolds), the Verlinde algebra (in the case of Grassmannians), and the mirror symmetry conjecture (in the Calabi–Yau case).

Next we define the Seidel representation of the fundamental group of the group of Hamiltonian symplectomorphisms of a manifolds M into the group of units of its quantum cohomology ring. As an application, we describe the quantum cohomology of both Fano and nonFano toric manifolds. We shall return to this subject in Chapter 12 in the context of Floer homology.

The last section revisits the discussion in Section 11.2. It uses the Gromov-Witten potential to define the structure of a formal Frobenius manifold on the cohomology of M, a structure that is sometimes known as the big quantum cohomology. It also discusses the Dubrovin connection, showing that its flatness is equivalent to the associativity of the Frobenius products.

11.1. The small quantum cohomology ring

Quantum coefficient rings. The small quantum cohomology ring of a symplectic manifold (M, ω) is, additively, the usual cohomology with coefficients in a suitable ring Λ that is chosen to allow for the definition of a new ring structure. The quantum cup product product $a * b$ of two classes $a, b \in H^*(M)$ is the sum of the usual cup product $a \smile b$ with certain quantum correction terms arising from the 3-point Gromov–Witten invariants $\mathrm{GW}^M_{A,3}(a, b, c)$ with $c \in H^*(M)$. The adjective small indicates that only a small part of the information contained in the Gromov–Witten invariants is used in its definition. The structure of the "big" quantum cohomology incorporates all the genus zero Gromov–Witten invariants in the form of the Gromov–Witten potential discussed in Section 11.2 below, and (depending on the precise definition chosen) may also include all the information from the higher genus invariants (see Manin [**286**] and Usher [**399**, §7]).

There may be infinitely many different homology classes A that contribute to any given product $a * b$. This forces us to work with coefficient rings that allow for infinite sums of cohomology classes labelled by elements $A \in H_2(M)$. There are many possibilities, such as polynomial rings in one or several variables, rings of Laurent series, or Novikov rings determined by the cohomology class of ω. The precise choice of the ring depends on the underlying symplectic manifold and also on how much information on the Gromov–Witten invariants one wants to retain in the ring structure of quantum cohomology. If $c_1(TM) = \lambda[\omega]$ for some real number λ then the three cases $\lambda > 0$ (the monotone case), $\lambda = 0$ (the Calabi–Yau case), and $\lambda < 0$ are rather different. They correspond roughly to the distinction between Fano (positive curvature), elliptic (zero curvature), and general type (negative curvature) in algebraic geometry.

Following Cieliebak–Salamon [**69**] we introduce an axiomatic setup that allows us to deal with all these different choices in a unified framework. The (symplectic) **effective cone** of a symplectic manifold (M, ω) is defined by

$$K^{\mathrm{eff}}(M,\omega) := \Big\{A \in H_2(M) \,\Big|\, \exists\, A_1, \ldots, A_N \in H_2(M) : A = \sum_i A_i,\ \mathrm{GW}^M_{A_i,3} \neq 0\Big\}.$$

This is a cone in $H_2(M) = H_2(M;\mathbb{Z})/\text{torsion}$ lying in the image of $\pi_2(M)$ under the Hurewicz homomorphism. It contains the zero element because $\mathrm{GW}^M_{0,3} \neq 0$. Every homology class in $K^{\mathrm{eff}}(M,\omega)$ can be represented by a finite collection of J-holomorphic spheres for every ω-tame almost complex structure J. In particular ω is positive on every nonzero element in this cone.

The cone $K^{\mathrm{eff}}(M,\omega)$ is a useful conceptual tool, but it can be hard to calculate. Changing point of view from the symplectic to the complex setting, one encounters another natural definition of the effective cone that plays an important role in algebraic geometry and is sometimes easier to calculate.

REMARK 11.1.1. (i) The **effective cone** of an almost complex manifold (M, J) is defined by

$$K^{\mathrm{eff}}(M,J) := \Big\{A \in H_2(M) \,\Big|\, \exists \text{ a } J\text{-holomorphic curve in class } A\Big\}.$$

This is the definition used in algebraic geometry. The holomorphic curves are allowed to be disconnected, of arbitrary genus, and the constant curves are included. If J is tamed by a symplectic form ω, then $K^{\mathrm{eff}}(M,\omega) \subset K^{\mathrm{eff}}(M,J)$ but this inclusion is usually not an equality (even if one replaces $K^{\text{eff}}(M, J)$ by the set of nonnegative linear combinations of classes represented by J-holomorphic spheres). For example, $K^{\text{eff}}(M, \omega) = \{0\}$ for any symplectic form on the $K3$ surface or the 4-torus. For a generic complex structure on a $K3$ surface there are no holomorphic spheres but for some there are. And classical studies show that most complex 4-tori $M_\Lambda = \mathbb{C}^2/\Lambda$ do not have any nonconstant compact holomorphic curves of any genus while those associated to Siegel upper half space admit embeddings into projective space and have a nontrivial effective cone $K^{\text{eff}}(M_\Lambda, i)$, of course only coming from holomorphic curves of genus at least one. (For a recent study of symplectic and complex 4-tori and further references see Latschev–McDuff–Schlenk [**231**].)

(ii) Let (M, J) be a compact complex manifold without boundary of complex dimension n. (M, J) is called a **Fano manifold** if its **anti-canonical line bundle** i.e. the top complex exterior power $L := \Lambda^n_{\mathbb{C}} TM = \Lambda^{n,0} TM$ of the tangent bundle, is **ample** [**170, 233**]. This means that there is an integer $k > 0$ such that the holomorphic sections of $L^{\otimes k}$ define an immersion of M into projective space (i.e. $L^{\otimes k}$ is **very ample**). In this case the pullback of the Fubini-Study form on projective space defines a Kähler form ω on M in the cohomology class $[\omega] = kc_1(TM)$. The **Nakai–Moishezon criterion** asserts (in the present setting) that a holomorphic line bundle $L \to M$ over a closed complex projective manifold is ample if and only if $\langle c_1(L)^i, \alpha \rangle > 0$ for every i and every homology class $\alpha \in H_{2i}(M; \mathbb{Z})$ that can be represented by a complex subvariety of M (see [**171**, Appendix A, Theorem 5.1]). Thus a complex projective surface (M, J) is Fano if and only if $c_1(TM)^2 > 0$ and $c_1(TM)$ takes positive values on all nonzero elements in $K^{\text{eff}}(M, J)$ (see [**171**, Chapter V, Theorem 1.10]).

(iii) A holomorphic line bundle $L \to M$ over a closed complex projective manifold (M, J) is called **NEF** (from the term **numerically effective** in algebraic geometry) if $c_1(L)$ is nonnegative on $K^{\text{eff}}(M, J)$. A theorem of Kleiman asserts that this implies $\langle c_1(L)^i, \alpha \rangle \geq 0$ for every i and every homology class $\alpha \in H_{2i}(M; \mathbb{Z})$ that can be represented by a complex subvariety of M (see [**233**, Theorem 1.4.9]). A complex projective manifold (M, J) is called **NEF** if its anticanonical bundle is NEF, i.e. if $c_1(TM)$ is nonegative on $K^{\text{eff}}(M, J)$.

(iv) A closed Kähler manifold (M, ω, J) is called Fano if the underlying complex manifold is Fano. Then ω may be joined through a path of Kähler forms to one in the class $c_1(TM)$. By Remark 7.3.9, the Gromov–Witten invariants do not change under such a deformation. Hence, for many purposes we may assume that a Fano Kähler manifold is monotone. (See Section 11.3.1 for Fano toric manifolds.)

(v) There are several reasonable ways to generalize these conditions to the symplectic setting. One could define a **symplectic Fano** manifold (M, ω) to be one in which ω may be joined through a path of symplectic forms to a symplectic form ω' in the class $c_1(TM)$. Alternatively, one could require that the class $c_1(TM)$ be positive on the nonzero spherical elements in one of the cones $K^{\text{eff}}(M, \omega)$ or $K^{\text{eff}}(M, J)$. The relation between these various conditions is not at present clear. There are similar generalizations of the NEF condition. In particular, requiring that $c_1(TM)$ be nonnegative on the spherical elements in $K^{\text{eff}}(M, J)$ for some (or every) ω-tame almost complex structure J, is equivalent to saying that J is semipositive in the sense of Definition 6.4.5. For such a manifold, the Gromov–Witten invariants can be defined by our current methods (see Remark 7.3.9).

(vi) For some applications it may be useful to enlarge the cone $K^{\mathrm{eff}}(M,\omega)$ to allow for classes A_i such that $\mathrm{GW}^M_{A_i,k} \neq 0$ for some k, or so that some higher genus invariant of the class A_i is nonzero. Another possibility is to define K^{eff} as a subset of $\pi_2(M)$ (and interpret GW^M_A as the Gromov–Witten invariant of J-holomorphic curves in a fixed homotopy class). However, we shall use the definition given above unless explicit mention is made to the contrary.

The next lemma summarizes the basic properties of the effective cone. The proof is an easy exercise; it uses the important finiteness result Corollary 5.3.2.

LEMMA 11.1.2. *Let (M,ω) be a closed symplectic manifold. Then the effective cone $K^{\mathrm{eff}} := K^{\mathrm{eff}}(M,\omega)$ is an additive semigroup with zero contained in the image of the Hurewicz homomorphism. Moreover, $\omega(A) \geq 0$ for every $A \in K^{\mathrm{eff}}$, with equality if and only if $A = 0$. For every $A \in H_2(M)$ the set $\{B \in K^{\mathrm{eff}} \,|\, A - B \in K^{\mathrm{eff}}\}$ is finite.*

Let R be a commutative ring with unit. In practice R will be one of the standard subrings $\mathbb{Z}, \mathbb{Q}, \mathbb{R}, \mathbb{C}$ of the complex numbers. It should be thought of as the basic (or ground) coefficient ring. One could also work mod 2, but this is most common in the context of Floer rather than quantum cohomology. Consider the $2\mathbb{Z}$-graded commutative ring

$$\Gamma(M,\omega) := \Big\{\lambda : K^{\mathrm{eff}}(M,\omega) \to R \,\Big|\, \sup_{\lambda(A)\neq 0} |c_1(A)| < \infty\Big\}.$$

For $k \in 2\mathbb{Z}$ the degree k part $\Gamma^k \subset \Gamma = \Gamma(M,\omega)$ consists of all maps $\lambda : K^{\mathrm{eff}} \to R$ such that $\lambda(A)$ is nonzero only for classes $A \in K^{\mathrm{eff}}$ with $2c_1(A) = k$. The sum operation on Γ is the obvious one and the product operation is defined by

$$(\lambda\mu)(A) := \sum_{B\in K^{\mathrm{eff}}} \lambda(B)\mu(A-B)$$

for $\lambda, \mu \in \Gamma$. The sum on the right is finite by Lemma 11.1.2. The product respects the grading in that $\lambda\mu \in \Gamma^{k+\ell}$ whenever $\lambda \in \Gamma^k$ and $\mu \in \Gamma^\ell$. The unit is the map $\delta : K^{\mathrm{eff}} \to R$ given by $\delta(0) := 1$ and $\delta(A) := 0$ for $A \neq 0$.

DEFINITION 11.1.3. *Let (M,ω) be a compact symplectic manifold and R be a commutative ring with unit. A* **quantum coefficient ring over** R **for** (M,ω) *is a triple (Λ, ϕ, ι), where Λ is a $2\mathbb{Z}$-graded commutative ring with unit and also an R-module, $\phi : \Gamma(M,\omega) \to \Lambda$ is a ring homomorphism that preserves the R-module structure and the grading, and $\iota : \Lambda \to R$ is an R-module homomorphism that vanishes on elements of nonzero degree and satisfies $\iota(\phi(\lambda)) = \lambda(0)$.*

Given a quantum coefficient ring (Λ, ϕ, ι) as in Definition 11.1.3, we use the notation

$$\phi(\lambda) =: \sum_{A\in K^{\mathrm{eff}}} \lambda(A)e^A$$

for $\lambda \in \Gamma(M,\omega)$. Note that because ϕ is not assumed injective an element in Λ may have several different expressions of this kind. If it is unique, it is often called a formal sum.

An obvious example is, of course, given by $\Lambda := \Gamma(M,\omega)$ and $\phi := \mathrm{id}$. This works for every closed symplectic manifold. However it is often not the best choice: for example we may not know precisely what $K^{\mathrm{eff}}(M,\omega)$ is. Other more interesting examples are listed below. Note that, for every $A \in K^{\mathrm{eff}}$, the element $e^A \in \Lambda$

denotes the image under ϕ of the function $\delta_A : K^{\mathrm{eff}} \to R$ given by $\delta_A(A) := 1$ and $\delta_A(B) := 0$ for $B \neq A$. It has degree $\deg(e^A) = \deg(\delta_A) = 2c_1(A)$. Since ϕ is a ring homomorphism, these elements are subject to the relation

$$e^{A+B} = e^A e^B.$$

The hypotheses in Definition 11.1.3 guarantee that the sum $\lambda := \sum_A \lambda(A) e^A$ is a well defined element of Λ whenever $\lambda \in \Gamma(M,\omega)$. These sums may be infinite, unless (M,ω) is monotone in which case all such sums are finite.

EXAMPLE 11.1.4. (i) If (M,ω) is monotone then

$$\begin{aligned} &A \in K_{\mathrm{eff}}(M,\omega), \quad A \neq 0 && \Longrightarrow && c_1(A) > 0, \\ &\#\big\{A \in K^{\mathrm{eff}}(M,\omega) \,\big|\, c_1(A) \le c\big\} < \infty && && \forall\, c > 0. \end{aligned} \tag{11.1.1}$$

This also holds when there exists an ω-tame complex structure J such that (M,J) is a Fano manifold and, in particular, for every Fano Kähler manifold (M,ω,J) (see Remark 11.1.1). Assuming (11.1.1) we can choose

$$\Lambda := R[q]$$

to be the polynomial ring in one variable q (of degree two). The homomorphism $\phi : \Gamma(M,\omega) \to R[q]$ is given by $e^A := q^{c_1(A)}$ or, equivalently,

$$\phi(\lambda) := \sum_{A \in K^{\mathrm{eff}}} \lambda(A) q^{c_1(A)}.$$

The homomorphism $\iota : R[q] \to R$ assigns to a polynomial the coefficient of $1 = q^0$. A similar example is the ring $\Lambda := R[q, q^{-1}]$ of polynomials in q and q^{-1} or the ring of Laurent series in q or q^{-1}. The ring of Laurent series (in either variable) is a field whenever R is. Note that in the Laurent series example Λ is not the direct sum of the Λ^k.

(ii) Assume (11.1.1). Then we can choose Λ to be the group ring of $H_2(M)$ with coefficients in R. Thus the elements of Λ are finite formal sums of the form

$$\lambda = \sum_{A \in H_2(M)} \lambda(A) e^A \tag{11.1.2}$$

with $\lambda(A) \in R$. By (11.1.1), $\Gamma = \Gamma(M,\omega)$ is a subring of Λ and $\phi : \Gamma \to \Lambda$ is the obvious inclusion. The map $\iota : \Lambda \to R$ is given by $\iota(\lambda) := \lambda(0)$.

An integer basis $A_1, \dots, A_m$ of $H_2(M)$ gives rise to an isomorphism between the group ring of $H_2(M)$ and the polynomial ring

$$\Lambda := R[q_1, \dots, q_m, q_1^{-1}, \dots, q_m^{-1}]$$

in the variables q_i and q_i^{-1} with

$$\deg(q_i) = 2c_1(A_i).$$

In this notation the homomorphism $\Gamma \to \Lambda$ is given by $e^{A_i} := q_i$ and the homomorphism $\iota : \Lambda \to R$ is given by the constant term. If we assume in addition that every element of $K^{\mathrm{eff}}(M,\omega)$ is a sum of the form $A = \sum_i d_i A_i$ with nonnegative coefficients d_i then we may choose instead

$$\Lambda := R[q_1, \dots, q_m]$$

with ϕ and ι defined as above. One can modify this example by either reducing or enlarging the number of variables q_i. Example (i) reappears if we set $q_i := q^{c_1(A_i)}$.

(iii) Now suppose that (M, ω) is any closed symplectic manifold and define

$$\Lambda_\omega := \left\{ \sum_{A \in H_2(M)} \lambda(A) e^A \,\middle|\, \#\left\{ A \,\middle|\, \begin{array}{c} \lambda(A) \neq 0 \\ \omega(A) \leq c \end{array} \right\} < \infty \; \forall\, c > 0 \right\}. \tag{11.1.3}$$

This is a completion of the group ring of $H_2(M)$, called the **Novikov ring of** ω. There are several slightly different definitions of this ring in common use. For example, one can take sums over the spherical homology group $H_2^S(M)$, or over equivalence classes in $H_2(M)$, where $A \sim B$ iff $\omega(A) = \omega(B)$ and $c_1(A) = c_1(B)$. With the above definition $\Gamma(M, \omega)$ is a subring of Λ_ω, ϕ is the obvious inclusion, and $\iota(\lambda) := \lambda(0)$. If the homomorphism $H_2(M) \to \mathbb{R} : A \mapsto \omega(A)$ is injective and R is a field, then Λ_ω is also a field (see Hofer–Salamon [**180**]).

As in example (ii), a basis of $H_2(M)$ gives rise to an identification of Λ_ω with the ring formed by all formal power series $\lambda = \sum_{d \in \mathbb{Z}^m} \lambda_d q^d$ in the variables q_i and q_i^{-1} that satisfy the finiteness condition

$$\#\{d \in \mathbb{Z}^m \,|\, \lambda_d \neq 0,\, \omega(A_d) \leq c\} < \infty, \qquad A_d := \sum_i d_i A_i,$$

for every $c \in \mathbb{R}$. The homomorphisms ϕ and ι are again given by $e^{A_i} := q_i$ and $\iota(\lambda) := \lambda_0$. If $\omega(A_i) > 0$ for all i and every element of K^{eff} is a linear combination of the A_i with nonnegative coefficients, then we can modify this example by choosing Λ to be the ring of formal power series in $q_1, \dots, q_m$. The finiteness condition is then automatically satisfied for every element of Λ.

We emphasize that Λ_ω is not $2\mathbb{Z}$ graded, unless $c_1(TM)$ is a torsion class. Otherwise Λ_ω is not the direct sum of the additive subgroups Λ_ω^k (consisting of all formal sums over classes A that satisfy $2c_1(A) = k$) and hence is not a quantum coefficient ring in the sense of Definition 11.1.3. To restore the $2\mathbb{Z}$ grading one can work with the direct sum of the Λ_ω^k, given by

$$\Lambda_{\omega,c} := \left\{ \lambda \in \Lambda_\omega \,\middle|\, \sup_{\lambda(A) \neq 0} |c_1(A)| < \infty \right\}. \tag{11.1.4}$$

This is a module over its zero graded subring Λ_ω^0 and has a natural $2\mathbb{Z}$ grading. In general, it is not a field. Alternatively, one can consider Λ_ω as an ungraded quantum coefficient ring. This is still useful, but one must bear in mind that then the associated quantum or Floer homology groups will only be $\mathbb{Z}_2$ graded.

(iv) Assume that (M, ω) is a Calabi–Yau manifold so that the first Chern class of the tangent bundle vanishes on $\pi_2(M)$. Then one can choose $\Lambda = \Lambda^0 := \Lambda^{\mathrm{univ}}$ to be the **universal Novikov ring** of all formal power series of the form

$$\lambda = \sum_{\varepsilon \in \mathbb{R}} \lambda_\varepsilon t^\varepsilon, \qquad \#\{\varepsilon \in \mathbb{R} \,|\, \lambda_\varepsilon \neq 0,\, \varepsilon \leq c\} < \infty \text{ for all } c \in \mathbb{R}.$$

The coefficients λ_ε belong to the ground ring R. If R is a field then so is the universal Novikov ring. In this example the homomorphism ϕ is given by

$$\phi(\lambda) := \sum_{A \in K^{\mathrm{eff}}} \lambda(A) t^{\omega(A)} \tag{11.1.5}$$

and $\iota(\lambda) := \lambda_0$ for $\lambda = \sum_\varepsilon \lambda_\varepsilon t^\varepsilon \in \Lambda^{\mathrm{univ}}$.

(v) For general symplectic manifolds (M, ω) one can combine examples (i) and (iv) by choosing $\Lambda := \Lambda^{\mathrm{univ}}[q, q^{-1}]$ as the ring of Laurent polynomials in a variable q of degree two with coefficients in the universal Novikov ring and define

$$\phi(\lambda) := \sum_{A \in K^{\text{eff}}} \lambda(A) t^{\omega(A)} q^{c_1(A)}.$$

Similarly, one can restore the grading in example (iii) by taking $\Lambda = \Lambda_\omega[q, q^{-1}]$ and assigning degree zero to each element e^A.

(vi) If one is prepared to work with mod 2 grading one can use the universal Novikov ring (or Λ_ω as a zero graded ring) even when the first Chern class does not vanish over $\pi_2(M)$. Many authors, for example [**128**], take this approach. Usher in [**399**, §7] defines a "universal quantum coefficient ring" R_M that specializes to many of the rings defined above.

(vii) It would be interesting to work, for example, with real coefficients and choose $\Lambda = \mathbb{R}$. One might be tempted to fix some positive real number t and define ϕ by (11.1.5). However, one then has to restrict Γ to contain only those maps $\lambda : K^{\text{eff}} \to \mathbb{R}$ for which the series (11.1.5) converges absolutely. We shall see below that in our applications the coefficients $\lambda(A)$ will be certain Gromov–Witten invariants of the form $\mathrm{GW}^M_{A,3}(a,b,c)$. This leads to the interesting question as to whether or not the series

$$\sum_{A \in H_2(M)} \mathrm{GW}^M_{A,3}(a,b,c) t^{\omega(A)}$$

converges for a sufficiently small number $t > 0$. This seems to be a hard analytical problem and, at the time of writing, this question is still open. It is of special interest in the context of Calabi–Yau manifolds and mirror symmetry. (See Cox–Katz [**76**, 8.1.3] for a further discussion of ways to deal with this issue.)

REMARK 11.1.5. The notation $e^A := \phi(\delta_A) \in \Lambda$ is meaningful only for $A \in K^{\text{eff}}(M, \omega)$. However, in most examples the homomorphism $\phi : \Gamma(M, \omega) \to \Lambda$ extends naturally to the group ring of $H_2(M)$ and then the notation $e^A := \phi(\delta_A) \in \Lambda$ is meaningful for every $A \in H_2(M)$. (In some cases the restriction $c_1(A) \geq 0$ is required.) In Example 11.1.4 (ii) and (iii) we have

$$\iota(e^A) = \begin{cases} 1, & \text{if } A = 0, \\ 0, & \text{otherwise,} \end{cases}$$

and in Example 11.1.4 (i), (iv), (v) we have

$$\iota(e^A) = \begin{cases} 1, & \text{if } c_1(A) = \omega(A) = 0, \\ 0, & \text{otherwise.} \end{cases}$$

These two formulas agree if we restrict attention to classes $A \in K^{\text{eff}}(M, \omega)$.

REMARK 11.1.6. Novikov [**301**] first introduced a ring of the form Λ_ω as in Example 11.1.4 (iii) in the context of his Morse theory for closed 1-forms. In that case $\pi_2(M)$ (respectively $H_2(M)$ in our formulation) is replaced by the fundamental group $\pi_1(M)$ and the valuation $\pi_1(M) \to \mathbb{R}$ that determines the finiteness condition (11.1.3) is induced by a closed 1-form ω.

As we shall see in Chapter 12 below, the Novikov ring arises in the context of Floer homology from a closed 1-form on the free loop space of M (the symplectic action). This was used by Hofer and Salamon in [**180**] to prove the Arnold conjecture in the semipositive case, and is an indication of the close connection between quantum cohomology and Floer homology.

Quantum cohomology. Let (M,ω) be a closed semipositive symplectic manifold, R be a commutative ring with unit, and (Λ,ϕ,ι) be a quantum coefficient ring over R as in Definition 11.1.3. Recall that $H^*(M)$ denotes the quotient of the integral cohomology group $H^*(M;\mathbb{Z})$ by the torsion submodule. The **quantum cohomology of (M,ω) with coefficients in Λ** is defined by

$$\mathrm{QH}^*(M;\Lambda)=H^*(M)\otimes_{\mathbb{Z}}\Lambda.$$

This is a graded R-module with unit. For $k\in\mathbb{Z}$ the degree-k part of $\mathrm{QH}^*(M;\Lambda)$ is given by

$$\mathrm{QH}^k(M;\Lambda)=\bigoplus_i H^i(M)\otimes_{\mathbb{Z}}\Lambda^{k-i},$$

Each element of $\mathrm{QH}^*(M;\Lambda)$ can be expressed as a finite sum of the form $a=\sum_i a_i\otimes\lambda_i$ with $a_i\in H^*(M)$ and $\lambda_i\in\Lambda$. The quantum cohomology carries an obvious Λ-module structure and the degree-k part $\mathrm{QH}^k(M;\Lambda)$ is a Λ^0-module. In some cases $\mathrm{QH}^*(M;\Lambda)$ is not equal to the direct sum of the Λ^0-modules $\mathrm{QH}^k(M;\Lambda)$ over all $k\in\mathbb{Z}$ (see Example 11.1.4 (i)). It is whenever Λ has the same property.

Note that there is an embedding of the ordinary cohomology $H^*(M)$ into $\mathrm{QH}^*(M;\Lambda)$ given by $a\mapsto a\otimes 1$. Note also that the R-module homomorphism $\iota:\Lambda\to R$ gives rise to an **intersection pairing**

$$\mathrm{QH}^*(M;\Lambda)\otimes\mathrm{QH}^*(M;\Lambda)\to R$$

that restricts to the standard pairing (with values in $\mathbb{Z}\subset R$) on the ordinary cohomology. It is given by

$$\langle a,b\rangle:=\sum_{i,j}\iota(\lambda_i\mu_j)\int_M a_i\smile b_j \tag{11.1.6}$$

for $a=\sum_i a_i\otimes\lambda_i$ and $b=\sum_j b_j\otimes\mu_j$. The pairing (11.1.6) need not be nondegenerate (in the sense that $\langle a,b\rangle=0$ for all b implies $a=0$). It is whenever the pairing $\Lambda\times\Lambda\to R:(\lambda,\mu)\mapsto\iota(\lambda\mu)$ is nondegenerate: cf. Exercise 11.1.21. Moreover, the pairing (11.1.6) is skew commutative in the sense that

$$\langle b,a\rangle=(-1)^{\deg(a)\deg(b)}\langle a,b\rangle,$$

whenever $a,b\in\mathrm{QH}^*(M;\Lambda)$ have pure degree.

REMARK 11.1.7. Let $K^{\mathrm{eff}}\to H^*(M)\otimes_{\mathbb{Z}}R:A\mapsto a_A$ be any function such that $\sup_{a_A\neq 0}|c_1(A)|<\infty$. Then the (possibly infinite) sum

$$a=\sum_A a_A\otimes e^A \tag{11.1.7}$$

has a natural meaning as an element of $\mathrm{QH}^*(M;\Lambda)$. To see this choose an integer basis $e_0,\dots,e_N$ of $H^*(M)$ and write $a_A=:\sum_{\nu=0}^N e_\nu\otimes\lambda_\nu(A)$, where $\lambda_\nu(A)\in R$. By definition, the functions $\lambda_\nu:K^{\mathrm{eff}}\to R$ belong to the ring $\Gamma(M,\omega)$ and we interpret the class a, which is given as an infinite sum over $A\in K^{\mathrm{eff}}$, as the finite sum

$$a=\sum_{A\in K^{\mathrm{eff}}}a_A\otimes e^A:=\sum_{\nu=0}^N e_\nu\otimes\phi(\lambda_\nu).$$

The reader may check that this definition is independent of the choice of the basis. Now assume R is one of the fields $\mathbb{Q},\mathbb{R},\mathbb{C}$ and Λ is either the group ring of $H_2(M)$ or the Novikov ring Λ_ω (see Example 11.1.4 (ii) and (iii)). Then every element of $\mathrm{QH}^*(M;\Lambda)$ is a finite or possibly infinite formal sum of the form (11.1.7) with $a_A \in H^*(M;R)$, where the summation runs over all $A \in H_2(M)$. In the Novikov case the sum satisfies the finiteness condition $\{A \in H_2(M) \,|\, a_A \neq 0,\, \omega(A) \leq c\} < \infty$ for every $c \in \mathbb{R}$.

Quantum cup product. The ring structure on $\mathrm{QH}^*(M;\Lambda)$, also called the quantum cup product, is a bilinear homomorphism of Λ-modules

$$\mathrm{QH}^*(M;\Lambda) \otimes \mathrm{QH}^*(M;\Lambda) \to \mathrm{QH}^*(M;\Lambda) : (a,b) \mapsto a * b.$$

Since $\mathrm{QH}^*(M;\Lambda)$ is generated by the elements of $H^*(M)$ as a Λ-module the quantum cup product is uniquely determined by its values on pairs $a, b \in H^*(M)$. To define the ring structure for such elements let us choose an integer basis $e_0, \ldots, e_n$ of (the $\mathbb{Z}$-module) $H^*(M)$ such that $e_0 = 1 \in H^0(M)$ and each basis element e_ν has pure degree. Define the integer matrix $g_{\nu\mu}$ by

$$g_{\nu\mu} := \int_M e_\nu \smile e_\mu$$

and let $g^{\nu\mu}$ denote the inverse matrix. Then the quantum cup product of two classes $a, b \in H^*(M)$ (of pure degree) is defined by

$$a * b := \sum_{A \in K^{\mathrm{eff}}} \sum_{\nu,\mu} \mathrm{GW}^M_{A,3}(a,b,e_\nu)\, g^{\nu\mu}\, e_\mu \otimes e^A. \tag{11.1.8}$$

Note that the summand on the right for the triple A, ν, μ can only be nonzero if $\deg(e_\nu) + \deg(e_\mu) = \dim M$ and $\dim M + 2c_1(A) = \deg(a) + \deg(b) + \deg(e_\nu)$. Thus we obtain the dimensional condition

$$\deg(a) + \deg(b) - \dim M \leq 2c_1(A) \leq \deg(a) + \deg(b).$$

Using Remark 11.1.7, one finds that the right hand side of (11.1.8) is a well defined element of $\mathrm{QH}^*(M;\Lambda)$. Note that $\deg(a*b) = \deg(a) + \deg(b)$. Another useful way to express the quantum product of a and b is as the sum

$$a * b = \sum_{A \in K^{\mathrm{eff}}} (a*b)_A \otimes e^A, \qquad (a*b)_A := \sum_{\nu,\mu} \mathrm{GW}^M_{A,3}(a,b,e_\nu)\, g^{\nu\mu}\, e_\mu.$$

where the coefficient $(a*b)_A \in H^{\deg(a)+\deg(b)-2c_1(A)}(M)$ of e^A is characterized by the condition

$$\int_M (a*b)_A \smile c := \mathrm{GW}^M_{A,3}(a,b,c) \tag{11.1.9}$$

for $c \in H^*(M)$.

REMARK 11.1.8. In the notation of Remark 11.1.7 the quantum product of two quantum cohomology classes $a = \sum_A a_A \otimes e^A$ and $b = \sum_B b_B \otimes e^B$ is

$$a * b = \sum_{A,B,C,\nu,\mu} \mathrm{GW}^M_{C,3}(a_A, b_B, e_\nu) g^{\nu\mu} e_\mu \otimes e^{A+B+C}$$

and the pairing (11.1.6) is

$$\langle a,b \rangle = \sum_{A,B} \iota(e^{A+B}) \int_M a_A \smile b_B = \alpha(a*b), \qquad \alpha(a) := \sum_A \iota(e^A) \int_M a_A.$$

Here the classes a_A are not required to have pure degree and the integral over M is understood as the integral of the component in degree $2n$. (For the notation $\iota(e^A)$ see Remark 11.1.5.)

REMARK 11.1.9. The cohomology class $(a * b)_A \in H^*(M)$ has the following geometric interpretation. Let $f_0 : X_0 \to M$ and $f_1 : X_1 \to M$ be pseudocycles Poincaré dual to a and b, respectively. Then equation (11.1.9) shows that the A-component $(a * b)_A \in H^{\deg(a)+\deg(b)-2c_1(A)}(M)$ of the quantum product of a and b can be represented by the pseudocycle of all simple J-holomorphic spheres in the class A that pass through the images of f_0 and f_1. Its domain is the set of triples $(x_0, x_1, u) \in X_0 \times X_1 \times \mathcal{M}^*(A; J)$ that satisfy $u(0) = f_0(x_0)$ and $u(1) = f_1(x_1)$, and the pseudocycle is the evaluation map $(x_0, x_1, u) \mapsto u(\infty)$. For a generic triple (f_0, f_1, J) the domain is a manifold of dimension $2n + 2c_1(A) - \deg(a) - \deg(b)$ and the evaluation map is a pseudocycle. By (11.1.9), this pseudocycle is Poincaré dual to the cohomology class $(a * b)_A$.

As an example, consider the projective space $M := \mathbb{C}P^n$ and denote

$$L := [\mathbb{C}P^1] \in H^2(\mathbb{C}P^n), \qquad c := \mathrm{PD}([\mathbb{C}P^{n-1}]) \in H^2(\mathbb{C}P^n).$$

Represent the class $\mathrm{PD}(c^k)$ by an $(n-k)$-plane $X \subset \mathbb{C}P^n$ and $\mathrm{PD}(c^\ell)$ by an $(n-\ell)$-plane $Y \subset \mathbb{C}P^n$. Suppose $k + \ell > n$ and $X \cap Y = \emptyset$. Then the cohomology class $(c^k * c^\ell)_L$ is Poincaré dual to the homology class that is represented by the set of points that lie on lines passing through X and Y. This is a projective plane of complex dimension $2n - k - \ell + 1$ and hence

$$(c^k * c^\ell)_L = c^{k+\ell-n-1}.$$

This is consistent with the discussion in Example 11.1.12 below.

REMARK 11.1.10. The cup product of two classes $a, b \in H^*(M)$ can be expressed as the sum

$$a \smile b = \sum_{\nu,\mu} \left(\int_M a \smile b \smile e_\nu \right) g^{\nu\mu} e_\mu.$$

Hence the triple intersection numbers

$$c_{ijk} := \int_M e_i \smile e_j \smile e_k$$

can be thought of as the structure constants of the cohomology ring. In terms of the structure constants associativity of the cup product can be expressed as the formula

$$\sum_{\nu,\mu} c_{ij\nu} g^{\nu\mu} c_{\mu k\ell} = (-1)^{|e_i|(|e_j|+|e_k|)} \sum_{\nu,\mu} c_{jk\nu} g^{\nu\mu} c_{\mu i\ell}, \tag{11.1.10}$$

where we have used the abbreviation $|e| := \deg(e)$. Similarly, one can think of the Gromov–Witten invariants

$$c_{ijk}(A) := \mathrm{GW}^M_{A,3}(e_i, e_j, e_k)$$

as the structure constants of quantum cohomology, and there is an analogous equation that is equivalent to its associativity. This explains the form of the WDVV equations discussed in Section 11.2.

The next proposition summarizes the main properties of the quantum product. Associativity was first proved by Ruan–Tian [**345, 346**]; subsequent proofs appeared in Liu [**248**] and in the first edition of this book. It is an immediate consequence of the gluing theorem.

PROPOSITION 11.1.11. (i) *The quantum cup product is distributive over addition and skew-commutative in the sense that*

$$b * a = (-1)^{\deg(a)\deg(b)} a * b$$

for elements $a, b \in H^*(M)$ *of pure degree. It is associative and commutes with the action of* Λ.

(ii) *The leading term in the quantum cup product is the standard cup product, i.e.*

$$(a * b)_0 = a \smile b$$

for all $a, b \in \mathrm{QH}^*(M; \Lambda)$. *Moreover, the higher terms vanish whenever* $a \in H^0(M)$ *or* $a \in H^1(M)$, *i.e. in this case* $a * b = a \smile b$ *for all* $b \in H^*(M)$. *Thus the canonical generator* $1 \in H^0(M)$ *is the unit in quantum cohomology.*

(iii) *The pairing* $\mathrm{QH}^*(M;\Lambda) \otimes \mathrm{QH}^*(M;\Lambda) \to R$, *defined by (11.1.6), is compatible with the quantum cup product in the sense that*

$$\langle a * b, c\rangle = \langle a, b * c\rangle$$

for $a, b, c \in \mathrm{QH}^*(M; \Lambda)$.

PROOF. The only nontrivial assertion in (i) is associativity. This follows from the splitting axiom (Theorem 7.5.10) for the Gromov–Witten invariants which was established in Chapter 10. For the invariants with four fixed marked points Theorem 7.5.10 asserts that, with $I := \{1, 2, 3, 4\}$, we have

$$\mathrm{GW}^{M,I}_{A,4}(a_1, a_2, a_3, a_4) = \sum_{A_0 + A_1 = A} \sum_{\nu,\mu} \mathrm{GW}^M_{A_0,3}(a_1, a_2, e_\nu) g^{\nu\mu} \mathrm{GW}^M_{A_1,3}(e_\mu, a_3, a_4).$$

Hence, for any three cohomology classes $a_1, a_2, a_3 \in H^*(M)$ of pure degrees and every $A \in K^{\mathrm{eff}}$, we find

$$\begin{aligned}
((a_1 * a_2) * a_3)_A &= \sum_B ((a_1 * a_2)_B * a_3)_{A-B} \\
&= \sum_B \sum_{\nu,\mu} \mathrm{GW}^M_{B,3}(a_1, a_2, e_\nu)\, g^{\nu\mu}\, (e_\mu * a_3)_{A-B} \\
&= \sum_B \sum_{\nu,\mu} \sum_{i,j} \mathrm{GW}^M_{B,3}(a_1, a_2, e_\nu)\, g^{\nu\mu}\, \mathrm{GW}^M_{A-B,3}(e_\mu, a_3, e_i)\, g^{ij} e_j \\
&= \sum_{i,j} \mathrm{GW}^{M,\{1,2,3,4\}}_{A,4}(a_1, a_2, a_3, e_i)\, g^{ij} e_j \\
&= \varepsilon \sum_{i,j} \mathrm{GW}^{M,\{1,2,3,4\}}_{A,4}(a_2, a_3, a_1, e_i)\, g^{ij} e_j \\
&= \varepsilon((a_2 * a_3) * a_1)_A \\
&= (a_1 * (a_2 * a_3))_A,
\end{aligned}$$

where, using the abbreviation $|a| := \deg(a)$,

$$\varepsilon := (-1)^{|a_1|(|a_2| + |a_3|)}.$$

We have used the fact that, when $a, b, c, d \in H^*(M)$ have pure degree, the invariant $\mathrm{GW}^{M,\{1,2,3,4\}}_{A,4}(a, b, c, d)$ is skew symmetric under permutations of its entries. This proves (i).

The first assertion in (ii) follows from (11.1.9) and the fact that

$$\mathrm{GW}^M_{0,3}(a,b,c) = \int_M a \smile b \smile c.$$

To prove the second assertion we must show that

$$\mathrm{GW}^M_{A,3}(a,b,c) = 0$$

for all $A \in H_2(M) \setminus \{0\}$ and $b,c \in H^*(M)$ whenever $a \in H^i(M)$ for $i = 0,1$. If $a \in H^0(M)$ this is a special case of the *(Fundamental class)* axiom in Proposition 7.5.6. Here is a direct argument. If $\deg(a) \le 1$ and a,b,c satisfy the dimensional condition

$$\deg(a) + \deg(b) + \deg(c) = \dim M + 2c_1(A)$$

then

$$\deg(b) + \deg(c) > \dim M + 2c_1(A) - 2 = \dim \mathcal{M}^*_{0,2}(A;J).$$

Hence, for $J \in \mathcal{J}_{\mathrm{reg}}(M,\omega)$, the evaluation map $\mathrm{ev} : \mathcal{M}^*_{0,2}(A;J) \to M \times M$ misses a generic product $X \times Y$ of two cycles representing the Poincaré duals of b and c. It follows that the invariant is zero.

To prove (iii) note that, when $a,b,c \in H^*(M)$ have pure degrees, $\mathrm{GW}^M_{A,3}(a,b,c)$ is skew symmetric under permutations of the three entries, that is

$$\mathrm{GW}^M_{A,3}(a,b,c) = (-1)^{|a|(|b|+|c|)}\mathrm{GW}^M_{A,3}(b,c,a).$$

By (11.1.9), this means that

$$\langle (a*b)_A, c\rangle = (-1)^{|a|(|b|+|c|)}\langle (b*c)_A, a\rangle = \langle a, (b*c)_A\rangle.$$

This holds for all classes $A \in K^{\mathrm{eff}}$ and all $a,b,c \in H^*(M)$ with

$$|a| + |b| + |c| = \dim M + 2c_1(A).$$

Hence (iii) follows from (11.1.6) and (11.1.9). This poves Proposition 11.1.11. □

Proposition 11.1.11 shows that the small quantum cohomology ring $\mathrm{QH}^*(M;\Lambda)$ is a **Frobenius algebra**. Such algebras have many striking properties, and give rise to flat connections and integrable Hamiltonian systems. We shall discuss this in more detail in Section 11.5.

Examples and exercises.

EXAMPLE 11.1.12 (Complex projective space). Consider the symplectic manifold $M := \mathbb{C}P^n$ with its standard complex structure and with the Kähler form ω associated to the Fubini-Study metric. Let $L \in H_2(\mathbb{C}P^n)$ be the standard generator, represented by the line $\mathbb{C}P^1$. The first Chern class of $\mathbb{C}P^n$ is given by

$$c_1(L) = n+1.$$

Therefore $\mathbb{C}P^n$ is monotone, and we may take Λ to be the polynomial ring $\mathbb{Z}[q]$ where

$$\deg(q) := 2(n+1).$$

The cohomology $H^*(\mathbb{C}P^n)$ is a truncated polynomial ring generated by the class $c \in H^2(\mathbb{C}P^n)$ such that $c(L) = 1$. Therefore the quantum cohomology groups $\mathrm{QH}^k(M) = \mathrm{QH}^k(M;\Lambda)$ vanish when k is odd and have a single generator for each even k, given by $c^i q^\ell$ where $k = 2(i + \ell(n+1))$ and $0 \le i \le n$. This describes the additive structure of $\mathrm{QH}^k(M)$.

To work out the multiplicative structure, it suffices to calculate the products $c^i * c^j$ for $0 \le i, j \le n$. For reasons of dimension the invariant $\mathrm{GW}^{\mathbb{C}P^n}_{dL,3}(a,b,c)$ vanishes unless $d = 0, 1$. By Proposition 11.1.11 (i), the case $d = 0$ corresponds to constant curves and gives the usual cup product. Moreover,

$$\mathrm{GW}^{\mathbb{C}P^n}_{L,3}(c^i, c^j, c^k) = \begin{cases} 1, & \text{if } i+j+k = 2n+1, \\ 0, & \text{otherwise.} \end{cases}$$

To see this, note that the space of lines in $\mathbb{C}P^n$ that meet two generic subspaces of complex dimensions $n-i$ and $n-j$ fills out a subspace of dimension $n-i+n-j+1$. Moreover there is exactly one line through each point of this subspace. Therefore, if the Poincaré dual of c^k has the complementary dimension there is precisely one line that contributes to the Gromov–Witten invariant. It follows that

$$c^i * c^j = \begin{cases} c^{i+j}, & \text{if } i+j \le n \\ c^{i+j-n-1}q, & \text{if } n+1 \le i+j \le 2n. \end{cases}$$

Therefore c^i for $i \le n$ equals the i-fold quantum product $c * \cdots * c =: c^{*i}$ of c, and the relations in the quantum ring are generated by the one new relation:

$$c * c^n = q.$$

This can be expressed in the form

$$\mathrm{QH}^*(\mathbb{C}P^n) = \frac{\mathbb{Z}[p,q]}{\langle p^{n+1} = q \rangle}.$$

Specializing to $q = 0$, we recover the ordinary cohomology ring of $\mathbb{C}P^n$.

The next example, though simple, is an interesting illustration of the consequences of the choice of the coefficient ring Λ for the algebraic structure of the quantum cohomology ring. Although the quantum cup product is determined by the Gromov–Witten invariants which are invariant under deformations of ω (see Remark 7.1.11), the available choices for Λ do depend on the cohomology class of the symplectic form $[\omega]$. Hence one can sometimes use quantum cohomology to get information about $[\omega]$: see Example 11.4.9 below.

EXAMPLE 11.1.13 (The product $S^2 \times S^2$). Consider the manifold $M := S^2 \times S^2$ with its standard symplectic structure $\omega = \pi_1^*\sigma + \pi_2^*\sigma$. Define $A, B \in H_2(M)$ by

$$A := [S^2 \times \mathrm{pt}], \qquad B := [\mathrm{pt} \times S^2],$$

and let a, b be the dual basis of $H^2(M)$. Thus $a = \mathrm{PD}(B)$ and $b = \mathrm{PD}(A)$. Then $H^*(M)$ has an additive basis $1, a, b, ab := a \smile b$. The nontrivial 3-point invariants in M are

$$\mathrm{GW}^M_{A,3}(ab, a, a) = 1, \qquad \mathrm{GW}^M_{B,3}(ab, b, b) = 1, \qquad \mathrm{GW}^M_{A+B,3}(ab, ab, ab) = 1.$$

Correspondingly, we have $a * b = a \smile b$ (so that there is no ambiguity in denoting this element by ab), and the other quantum products are given by

$$\begin{aligned} a * a &= 1 \otimes e^A, & b * b &= 1 \otimes e^B, \\ a * ab &= b \otimes e^A, & b * ab &= a \otimes e^B, \end{aligned}$$
$$ab * ab = 1 \otimes e^{A+B}.$$

The expressions for $a * a$, $a * b$ and $b * b$ determine the other products (such as $a * ab$) by associativity. The reader may check that everything is consistent.

Let us now take a closer look at the coefficient ring Λ. The effective cone is

$$K^{\mathrm{eff}} = \{d_1 A + d_2 B \mid d_1, d_2 \geq 0\}$$

and so $\Gamma = \Gamma(M, \omega)$ is the semigroup ring of K^{eff} and can be identified with the polynomial ring in two variables. We examine the three cases where Λ is the polynomial ring in two variables, the polynomial ring in one variable, and the field of Laurent series in one variable. We shall see that the resulting quantum cohomology rings have rather different algebraic structures.

Consider first the case $\Lambda := \mathbb{C}[q_1, q_2]$, where the isomorphism $\phi : \Gamma \to \Lambda$ is given by

$$e^A := q_1, \qquad e^B := q_2$$

(see Example 11.1.4 (ii)). With this choice we obtain $a * a = q_1$ and $b * b = q_2$, and hence

$$H^*(M; \mathbb{C}) \cong \frac{\mathbb{C}[p_1, p_2]}{\langle p_1^2 = p_2^2 = 0 \rangle}, \qquad \mathrm{QH}^*(M; \Lambda) \cong \frac{\mathbb{C}[p_1, p_2, q_1, q_2]}{\langle p_1^2 = q_1, p_2^2 = q_2 \rangle},$$

where

$$\deg(p_1) = \deg(p_2) = 2, \qquad \deg(q_1) = \deg(q_2) = 4.$$

Hence $\mathrm{QH}^*(M; \Lambda)$ is isomorphic to the polynomial ring $\mathbb{C}[p_1, p_2]$. The latter description obscures the Λ-module structure. However it does make clear that there are no zero divisors.

Since we are in the monotone case, we may also take Λ to be the polynomial ring $\Lambda := \mathbb{C}[q]$ in one variable q of degree four with $e^A := e^B := q$ (Example 11.1.4 (i)). This corresponds to identifying $q_1 = q_2$ in the ring $\mathbb{C}[q_1, q_2]$. With this choice we have $a * a = b * b = q$ and hence

$$\mathrm{QH}^*(M; \Lambda) \cong \frac{\mathbb{C}[p_1, p_2, q]}{\langle p_1^2 = p_2^2 = q \rangle}.$$

This ring does have zero divisors, namely $(p_1 - p_2)(p_1 + p_2) = p_1^2 - p_2^2 = 0$.

Now let Λ be the field of Laurent series in q with complex coefficients. Then the monomial q is invertible in Λ and so

$$e_+ := \frac{q + ab}{2q}, \qquad e_- := \frac{q - ab}{2q}$$

are well defined elements of the quantum cohomology ring. The relations show that these two elements are commuting projections:

$$(e_\pm)^2 = e_\pm, \qquad e_+ e_- = e_- e_+ = 0.$$

It follows that $\mathrm{QH}^*(M; \Lambda)$ is a commutative algebra that is generated as a Λ-vector space by the elements e_+, e_-, ae_+, ae_-. Since $a^2 = q$ it follows that $\mathrm{QH}^*(M; \Lambda)$ is isomorphic as a Λ-algebra to the direct sum

$$e_+ Q \oplus e_- Q,$$

where the field Q is the extension of Λ by the square root of q. This is a special case of a result due to Abrams [**7**] concerning the structure of the quantum cohomology ring $\mathrm{QH}^*(M; \Lambda)$ of certain monotone manifolds with coefficients in a field. These observations were used by Entov–Polterovich [**105**] in their construction of quasimorphisms on the group of Hamiltonian symplectomorphisms.

EXERCISE 11.1.14. Let

$$M := \mathbb{C}P^2 \# \overline{\mathbb{C}P}^2$$

be the one point blowup of $\mathbb{C}P^2$. Show by direct calculation that there are coefficients Λ such that $QH^*(M;\Lambda)$ is a field exactly if M is a *large* blowup, in the sense that $3\omega(E)^2 \geq \omega(L)^2$, where E is the class of the exceptional divisor and L that of the line; cf. Ostrover [**314**, §3.3]. This phenomenon is now fully understood in terms of the displaceability properties of the Lagrangian fibers of the standard toric structure on M: a large blowup has only one nondisplaceable fiber while the others have two. See Fukaya–Oh-Ohta–Ono [**129**, Theorem 1.12] and McDuff [**275**]. For a discussion of some other geometric consequences see Remarks 1.10 and 1.11 of McDuff [**274**].

EXAMPLE 11.1.15 (Moduli space of bundles). Let Σ be a Riemann surface of genus $g \geq 2$ and consider the moduli space M_Σ of stable rank two holomorphic vector bundles over Σ with a fixed determinant line bundle of degree one. This is a simply connected monotone symplectic manifold with $\pi_2 = \mathbb{Z}$. Hence its quantum cohomology ring is well defined. It is worked out in Donaldson [**83**], Munoz [**296**] and Siebert–Tian [**377**]. Since it would take us too far afield to describe this example in detail, we simply point out some of its interesting features. Note first that Donaldson's μ-map provides an isomorphism $H_1(\Sigma;\mathbb{Z}) \to H^3(M_\Sigma;\mathbb{Z})$. Thus there is plenty of odd dimensional cohomology and it turns out the quantum product is deformed on some of these classes. In other words, there are classes $a_1, a_2 \in H^{odd}(M_\Sigma)$, $a_3 \in H^{ev}(M_\Sigma)$ and $A \neq 0 \in H_2(M_\Sigma)$ such that

$$\mathrm{GW}^{M_\Sigma}_{A,3}(a_1, a_2, a_3) \neq 0.$$

Although in principle there is no reason why the Gromov–Witten invariants of odd dimensional classes should vanish, it is not so easy to come up with explicit examples where they do not and this is the only one we give in this book. For precise formulas see Munoz [**296**, §4], for example. This paper also provides an interesting illustration of the difference between giving a presentation for the quantum cohomology ring and describing all the 3-point Gromov–Witten invariants: cf. Remark 11.3.20 below. Finally it is written using a slightly different version of the quantum cohomology ring for a monotone manifold in which one sets $q = 1$ in the polynomial coefficient ring $\mathbb{Z}[q]$. The resulting quantum cohomology ring $\mathrm{QH}^*(M)$ contains precisely the same elements as the usual cohomology ring $H^*(M;\mathbb{Z})$, but has a deformed multiplication and is only graded modulo $2N$ where N is the minimum Chern number of M. Tacitly we were using this version of quantum cohomology in the discussion of the Seidel representation given in Example 8.6.8.

EXAMPLE 11.1.16. In [**83**] Donaldson calculated the small quantum cohomology of the moduli space M_Σ of flat SO(3)-bundles over a closed Riemann surface in the simplest nontrivial case of genus 2. By the Narasimhan–Seshadri theorem, this moduli space is isomorphic to the one discussed in Example 11.1.15. It has real dimension 6 and minimal Chern number $N = 2$. The starting point is the observation that M_Σ is symplectomorphic to a transverse intersection

$$M = X_1 \cap X_2$$

of two quadrics in $\mathbb{C}P^5$. Denote by $h \in H^2(M;\mathbb{Z})$ the pullback of the positive generator of $H^2(\mathbb{C}P^5;\mathbb{Z})$. Then M has Chern classes

$$c_1(TM) = 2h, \qquad c_2(TM) = 3h^2, \qquad c_3(TM) = 0.$$

Hence the Euler characteristic of M is zero and, by the Lefschetz hyperplane theorem, the Betti numbers of M are

$$b_0 = b_2 = b_4 = b_6 = 1, \qquad b_1 = b_5 = 0, \qquad b_3 = 4.$$

Denote the even generators by $h_0 = 1$, $h_2 = h$, h_4, h_6. Then h is dual to a hyperplane section, h_4 is dual to a line in M, and h_6 is dual to a point. Thus

$$h \smile h = 4h_4, \qquad h \smile h_4 = h_6.$$

(The second equation is obvious and the first follows from the identity $h^3 = 4h_6$.) Denote by $(\alpha, \beta) \mapsto \langle \alpha, \beta \rangle$ the intersection pairing on $H^3(M)$ so that

$$\alpha \smile \beta = \langle \alpha, \beta \rangle h_6.$$

Donaldson shows that there are 4 lines in M through a generic point and 2 lines in M through a generic pair of lines. Starting from this geometric information, one can use the splitting axiom to compute the Gromov–Witten invariants for the classes L, $2L$, $3L$, where $L \in H_2(M; \mathbb{Z})$ is the positive generator (represented by the lines in M). This gives the formulas

$$\mathrm{GW}_L(h, h, h_6) = 4, \qquad \mathrm{GW}_L(h, h_4, h_4) = 2, \qquad \mathrm{GW}_L(\alpha, \beta, h_4) = -\langle \alpha, \beta \rangle,$$
$$\mathrm{GW}_{2L}(h, h_6, h_6) = 4, \qquad \mathrm{GW}_{2L}(h_4, h_4, h_6) = 3, \qquad \mathrm{GW}_{3L}(h_6, h_6, h_6) = 4.$$

Hence the quantum products are given by

$$h * h = 4h_4 + 4q, \qquad h * h_4 = h_6 + 2hq, \qquad h * h_6 = 4h_4 q + 4q^2,$$
$$h_4 * h_4 = 2h_4 q + 3q^2, \qquad h_4 * h_6 = 3hq^2, \qquad h_6 * h_6 = 4h_4 q^2 + 4q^3,$$

and

$$h * \alpha = 0, \qquad h_4 * \alpha = -\alpha q, \qquad h_6 * \alpha = 0, \qquad \alpha * \beta = \langle \alpha, \beta \rangle (h_6 - hq)$$

for $\alpha, \beta \in H^3(M)$ where $\deg(q) = 4$. It follows that the quantum cohomology ring of M is generated by an element h (of degree 2), the classes $\alpha \in H^3(M)$ (of degree 3), and an element q (of degree 4) subject to the relations

$$h^4 = 16h^2 q, \qquad h\alpha = 0, \qquad \alpha\beta = \langle \alpha, \beta \rangle \left(\frac{1}{4} h^3 - 4hq \right).$$

The classes h_4 and h_6 are obtained from h and q via

$$h_4 := \frac{1}{4} h^2 - q, \qquad h_6 := \frac{1}{4} h^3 - 3hq.$$

These formulas are independent of the precise choice of the quantum coefficient ring (e.g. polynomials, Laurent polynomials, or Laurent series in q).

Further examples of quantum cohomology are described in Section 11.3.

EXERCISE 11.1.17. Let $a \in H^i(M), b \in H^j(M)$ and write

$$a * b =: \sum_A (a * b)_A \otimes e^A$$

with $(a*b)_A \in H^{i+j-2c_1(A)}(M)$. Show that $(a*b)_A$ is a nonzero element of $H^{2n}(M)$ only if $A = 0$ and $i + j = 2n$. Deduce that

$$\langle a * b, 1 \rangle = \langle a \smile b, 1 \rangle = \int_M a \smile b.$$

This gives another proof of a special case of Proposition 11.1.11 (iii).

We now generalize to iterated products. Let $I := \{1, \ldots, k\}$ and consider the Gromov–Witten invariants with fixed marked points $\mathrm{GW}^{M,I}_{A,k}(a_1, \ldots, a_k)$, $a_i \in H^*(M)$, as defined in Section 7.3. The *(Splitting)* axiom implies that these invariants are determined by the 3-point invariants. The next exercise explains exactly how and shows that they correspond to iterated quantum products. In [**104**], Entov gives a beautiful interpretation of these numbers for Grassmannians in terms of the ABW inequalities for eigenvalues of products of special unitary matrices.

EXERCISE 11.1.18 (Iterated quantum products). Let (M, ω) be a closed semipositive symplectic manifold and $\Lambda = \Lambda_\omega$ be the Novikov ring of Example 11.1.4 (iv) with integer coefficients, i.e. with $R = \mathbb{Z}$. Recall from Remark 11.1.7 that in this case every quantum cohomology class $a \in \mathrm{QH}^*(M; \Lambda)$ has the form $a = \sum_A a_A \otimes e^A$ where the sum runs over all homology classes $A \in H_2(M)$ and $\#\{A \in H_2(M) \,|\, a_A \neq 0,\ \omega(A) \leq c\} < \infty$ for every $c \in \mathbb{R}$. In this case it is natural to extend the definition of the Gromov–Witten invariants with fixed marked points, i.e. for $I_k := \{1, \ldots, k\}$, to a homomorphism

$$\mathrm{GW}^{M,I_k}_{A,k} : \mathrm{QH}^*(M; \Lambda) \times \cdots \times \mathrm{QH}^*(M; \Lambda) \to \mathbb{Z}$$

via the formula

$$\mathrm{GW}^{M,I_k}_{A,k}(a_1, \ldots, a_k) := \sum_{A_1, \ldots, A_k} \mathrm{GW}^{M,I_k}_{A-A_1-\cdots-A_k}(a_{1,A_1}, \ldots, a_{k,A_k}),$$

where $a_i = \sum_B a_{i,B} \otimes e^B \in \mathrm{QH}^*(M; \Lambda)$. By definition, $\mathrm{GW}^{M,I_k}_{A,k}$ descends to $\mathrm{QH}^*(M; \Lambda) \otimes_\Lambda \cdots \otimes_\Lambda \mathrm{QH}^*(M; \Lambda)$. Prove that

$$\mathrm{GW}^{M,I_k}_{A,k}(a_1, \ldots, a_k) = \mathrm{GW}^{M,I_1}_{A,1}(a_1 * \cdots * a_k) = \int_M (a_1 * \cdots * a_k)_A,$$

where $\mathrm{GW}^{M,I_1}_{A,1}$ is the one-point invariant of Exercise 7.3.4 extended to $\mathrm{QH}^*(M; \Lambda)$ as above. It follows that an iterated product can be written in the form

$$a_1 * \cdots * a_k = \sum_A (a_1 * \cdots * a_k)_A \otimes e^A$$

where the cohomology classes $(a_1 * \cdots * a_k)_A \in H^*(M)$ are defined by

$$\langle (a_1 * \cdots * a_k)_A, c \rangle := \mathrm{GW}^{M,I_{k+1}}_{A,k+1}(a_1, \ldots, a_k, c)$$

for $c \in H^*(M)$. *Hint:* Argue by induction and use the gluing formula of Theorem 7.5.10 for the case where all the marked points are fixed.

EXERCISE 11.1.19 (Künneth formula). Suppose that (M_1, ω_1) and (M_2, ω_2) are monotone with the same constant so that $(M_1 \times M_2, \pi_1^*\omega_1 + \pi_2^*\omega_2)$ is also monotone. Let $A_i \in H_2(M_i)$ and $a_i, b_i, c_i \in H^*(M_i)$ for $i = 1, 2$ and denote

$$A := \iota_{1*}A_1 + \iota_{2*}A_2, \quad a := \pi_1^* a_1 + \pi_2^* a_2, \quad b := \pi_1^* b_1 + \pi_2^* b_2, \quad c := \pi_1^* c_1 + \pi_2^* c_2.$$

Show that

$$\mathrm{GW}^{M_1 \times M_2}_{A,3}(a, b, c) = \mathrm{GW}^{M_1}_{A_1,3}(a_1, b_1, c_1) \cdot \mathrm{GW}^{M_2}_{A_2,3}(a_2, b_2, c_2).$$

Deduce that, if we use the one variable polynomial ring $\mathbb{Z}[q]$ as coefficients as in Example 11.1.4 (i), we obtain

$$\mathrm{QH}^*(M_1 \times M_2; \mathbb{Z}[q]) \cong \mathrm{QH}^*(M_1; \mathbb{Z}[q]) \otimes_{\mathbb{Z}[q]} \mathrm{QH}^*(M_2; \mathbb{Z}[q]).$$

Thus the product formula for small quantum cohomology is very simple. It is much more interesting for the large quantum cohomology since this involves k-pointed invariants with $k > 3$. Show that, with the above notation, the following identity holds when $k = 4$:

$$\begin{aligned} \mathrm{GW}^{M_1 \times M_2}_{A,4}(a,b,c,d) &= \mathrm{GW}^{M_1}_{A_1,4}(a_1,b_1,c_1,d_1) \cdot \mathrm{GW}^{M_2,\{1,2,3,4\}}_{A_2,4}(a_2,b_2,c_2,d_2) \\ &\quad + \mathrm{GW}^{M_1,\{1,2,3,4\}}_{A_1,4}(a_1,b_1,c_1,d_1) \cdot \mathrm{GW}^{M_2}_{A_2,4}(a_2,b_2,c_2,d_2). \end{aligned}$$

For a statement of the general Künneth formula, see Kaufmann [**207**].

REMARK 11.1.20 (Quantum homology). In this section we have defined the quantum cup product via Poincaré duality from the Gromov–Witten pseudocycle. Though it is often appropriate to work with cohomology, sometimes it is useful to think in terms of homology because the link with geometry is more direct. The quantum product on the homology of M is a deformation of the intersection product that is defined as the Poincaré dual of the quantum cup product. Thus the quantum product of two homology classes $\alpha, \beta \in H_*(M)$ (of pure degree) has the form

$$\alpha * \beta := \sum_A \sum_{\nu,\mu} \mathrm{GW}^M_{A,3}(\alpha,\beta,\varepsilon_\nu) g^{\nu\mu} \varepsilon_\mu \otimes e^{-A}, \tag{11.1.11}$$

where the ε_ν form a basis of $H_*(M)$ and the $g^{\nu\mu}$ are the entries of the inverse of the intersection matrix $g_{\nu\mu} := \varepsilon_\nu \cdot \varepsilon_\mu$. With the convention $\deg(e^{-A}) := -2c_1(A)$, the quantum product has degree

$$\deg(\alpha * \beta) = \deg(\alpha) + \deg(\beta) - 2n$$

and the leading term (i.e. the coefficient of e^{-A} with $A = 0$) is the intersection product $\alpha \cdot \beta$. We may also write $\alpha * \beta = \sum_A (\alpha * \beta)_A \otimes e^{-A}$ where the homology class $(\alpha * \beta)_A$ is a "fattened" intersection of α with β; as explained more precisely in Remark 11.1.9 $(\alpha * \beta)_A$ is represented by the union of all A-curves that meet α and β.

To make sense of this definition, we need to give a meaning to a (possibly infinite) sum of the form

$$\alpha = \sum_A \alpha_A \otimes e^{-A} \tag{11.1.12}$$

whenever the classes $\alpha_A \in H_{k-2c_1(A)}(M)$ satisfy

$$\#\{A \,|\, \alpha_A \neq 0,\, \omega(-A) \geq c\} < \infty$$

for every $c \in \mathbb{R}$. The quantum homology should have the following features. First, there should be a Poincaré duality isomorphism $\mathrm{PD} : \mathrm{QH}^k \to \mathrm{QH}_{2n-k}$ which restricts to the map

$$a = \sum_A a_A \otimes e^A \mapsto \mathrm{PD}(a) := \sum_A \mathrm{PD}(a_A) \otimes e^{-A}$$

on classes of the form (11.1.7). Second, there should be a (preferably nondegenerate) pairing $\mathrm{QH}^* \times \mathrm{QH}_* \to R$ that satisfies

$$\langle a, \alpha \rangle = \sum_A \int_{\alpha_{-A}} a_A$$

whenever a and α have the form (11.1.7) and (11.1.12), respectively.

To put this on a rigorous footing, we assume that (Λ, ϕ, ι) is a quantum coefficient ring (over a ground ring R) as in Definition 11.1.3. Define the **dual ring** $\check{\Lambda}$ to be equal to Λ as a ring, but with the grading reversed. Thus

$$\check{\Lambda}^k := \Lambda^{-k}.$$

Let $r : \Lambda \to \check{\Lambda}$ denote the obvious (degree reversing) isomorphism and define $\check{\phi} := r \circ \phi : \Gamma(M, \omega) \to \check{\Lambda}$ and $\check{\iota} := \iota \circ r^{-1} : \check{\Lambda} \to R$. The triple $(\check{\Lambda}, \check{\phi}, \check{\iota})$ is called the **dual quantum coefficient ring**. It is convenient to use the notation

$$e^{-A} := r(e^A) = \check{\phi}(\delta_A) \in \check{\Lambda}.$$

Now define

$$\mathrm{QH}_*(M; \check{\Lambda}) := H_*(M) \otimes_{\mathbb{Z}} \check{\Lambda}.$$

As in Remark 11.1.7, the sum (11.1.12) is a well defined element of $\mathrm{QH}_*(M; \check{\Lambda})$ for every function $K^{\mathrm{eff}} \to H_*(M) : A \mapsto \alpha_A$ such that $\sup_{\alpha_A \neq 0} |c_1(A)| < \infty$. It follows that the formula (11.1.11) defines a quantum intersection product on $\mathrm{QH}_*(M; \check{\Lambda})$. Moreover, there is a Poincaré duality isomorphism

$$\mathrm{QH}^k(M; \Lambda) \to \mathrm{QH}_{2n-k}(M; \check{\Lambda}) : a = \sum_i a_i \otimes \lambda_i \mapsto \mathrm{PD}(a) := \sum_i \mathrm{PD}(a_i) \otimes r(\lambda_i)$$

and a pairing $\mathrm{QH}^k(M; \Lambda) \times \mathrm{QH}_k(M; \check{\Lambda}) \to R : (a, \alpha) \mapsto \langle a, \alpha \rangle$ given by

$$\langle a, \alpha \rangle := \sum_{i,j} \langle \lambda_i, \check{\lambda}_j \rangle \int_{\alpha_j} a_i \tag{11.1.13}$$

for $a = \sum_i a_i \otimes \lambda_i$ and $\alpha = \sum_j \alpha_j \otimes \check{\lambda}_j$, where

$$\langle \lambda, \check{\lambda} \rangle := \iota(\lambda r^{-1}(\check{\lambda})).$$

The reader may check that these structures satisfy the above requirements. The pairing between quantum cohomology and quantum homology is isomorphic to the bilinear form (11.1.6) on $\mathrm{QH}^*(M; \Lambda)$ under Poincaré duality. It need not be nondegenerate; it is whenever the pairing

$$\Lambda \times \check{\Lambda} \to R : (\lambda, \check{\lambda}) \mapsto \langle \lambda, \check{\lambda} \rangle := \iota(\lambda r^{-1}(\check{\lambda}))$$

is nondegenerate.

EXERCISE 11.1.21. Prove that the pairing $\Lambda \times \check{\Lambda} \to R$ is nondegenerate for the polynomial ring $\Lambda = R[q, q^{-1}]$ and for the ring of Laurent series in the variable q. Prove that it is degenerate for the polynomial ring $\Lambda = R[q]$ and the formal power series ring in q.

EXERCISE 11.1.22. Show that the pairing (11.1.13) satisfies the identity

$$\langle a, \mathrm{PD}(b) \rangle = \langle a * b, 1 \rangle = \langle a, b \rangle$$

for $a, b \in \mathrm{QH}^*(M)$. Here $\langle a, \mathrm{PD}(b) \rangle$ is defined by (11.1.13), $\langle a, b \rangle$ is defined by (11.1.6), and the signs in the Poincaré duality isomorphism are chosen such that this identity holds on ordinary (co)homology.

11.2. The Gromov–Witten potential

The quantum cup product defined in the previous section uses only a small part of the information provided by the Gromov–Witten invariants. In this section we define a formal power series (called the genus zero Gromov–Witten potential) that encodes all this information. We show how the splitting axiom translates into a system of third order quadratic partial differential equations, called the WDVV equations. The exposition of this section is inspired by the work of Kontsevich and Manin [**216**].

Let (M, ω) be a compact semipositive symplectic manifold and choose as a ground ring R one of the fields $\mathbb{Q}, \mathbb{R}, \mathbb{C}$. The precise choice of R is immaterial, because all the Gromov–Witten invariants (with at least three marked points) are integers in the semipositive case. As in Section 11.1, choose an integer basis $e_0, \dots, e_N$ of $H^*(M)$ such that $e_0 = 1 \in H^0(M)$ and each basis element e_ν has pure degree, define

$$g_{\nu\mu} := \int_M e_\nu \smile e_\mu,$$

and denote by $g^{\nu\mu}$ the inverse matrix. Next we introduce $N+1$ formal variables $t_0, \dots, t_N$ and define the linear polynomial $t \mapsto a_t$ in $t_0, \dots, t_N$ with coefficients in $H^*(M)$ by

$$a_t := t_0 e_0 + t_1 e_1 + \cdots + t_N e_N.$$

The **Gromov–Witten potential** of (M, ω) is a formal power series in the variables $t_0, \dots, t_N$ with coefficients in the Novikov ring Λ_ω. It is defined by

$$\Phi(t) := \sum_A \sum_p \frac{1}{p!} \mathrm{GW}^M_{A,p}(a_t, \dots, a_t)\, e^A, \tag{11.2.1}$$

where the (formal) sum runs over all homology classes $A \in H_2(M)$ and all nonnegative integers p. This is to be understood as follows.

It is important to think of $H^*(M)$ as a **supermanifold** with even and odd coordinates. Call a variable t_ν **odd** if $\deg(e_\nu)$ is odd and **even** otherwise. In supermanifolds one uses the convention that odd variables anti-commute and even variables commute. Thus

$$t_\nu t_\mu = \begin{cases} -t_\mu t_\nu, & \text{if } t_\nu \text{ and } t_\mu \text{ are odd,} \\ t_\mu t_\nu, & \text{otherwise,} \end{cases} \tag{11.2.2}$$

and

$$t_\nu e_\mu = \begin{cases} -e_\mu t_\nu, & \text{if } t_\nu \text{ and } e_\mu \text{ are odd,} \\ e_\mu t_\nu, & \text{otherwise.} \end{cases} \tag{11.2.3}$$

In particular, $(t_\nu)^2 = 0$ for every odd variable. Note that the commutation relations of the t_ν are exactly the same as those of the e_ν as arguments of the Gromov–Witten invariants. Moreover, a formal power series in $t_0, \dots, t_N$ with coefficients in any commutative ring can be written as a sum of the form $\sum_\alpha \lambda_\alpha t^\alpha$ over all integer vectors $\alpha = (\alpha_0, \dots, \alpha_N)$ with $\alpha_\nu \geq 0$ and $\alpha_\nu \in \{0, 1\}$ whenever t_ν is odd.

To understand the formula (11.2.1), let us examine the term $\mathrm{GW}^M_{A,p}(a_t,\dots,a_t)$ on the right more carefully. It can be expressed in the form

$$\begin{aligned}\frac{1}{p!}\mathrm{GW}^M_{A,p}(a_t,\dots,a_t) &= \frac{1}{p!}\sum_{\nu_1,\dots,\nu_p}\mathrm{GW}^M_{A,p}(t_{\nu_1}e_{\nu_1},\dots,t_{\nu_p}e_{\nu_p}) \\ &= \sum_{\nu_1,\dots,\nu_p}\frac{\varepsilon(\nu)}{p!}\mathrm{GW}^M_{A,p}(e_{\nu_1},\dots,e_{\nu_p})t_{\nu_1}\cdots t_{\nu_p} \\ &= \sum_{|\alpha|=p}\frac{\varepsilon(\alpha)}{\alpha!}\mathrm{GW}^M_{A,|\alpha|}(e^\alpha)\,t^\alpha.\end{aligned} \tag{11.2.4}$$

Here the first sum runs over all p-tuples of integers $(\nu_1,\dots,\nu_p)$ with $0\le\nu_i\le N$. The sign $\varepsilon(\nu)=\varepsilon(\alpha)\in\{\pm1\}$ arises from reshuffling the product $t_{\nu_1}e_{\nu_1}\cdots t_{\nu_p}e_{\nu_p}=\varepsilon(\nu)e_{\nu_1}\cdots e_{\nu_p}t_{\nu_1}\cdots t_{\nu_p}$. Thus $\varepsilon(\nu)=r(r+1)/2$ where r is the number of odd variables among $t_{\nu_1},\dots,t_{\nu_p}$. Likewise, we define

$$\varepsilon(\alpha)=(-1)^{r(r+1)/2},\qquad r:=r_\alpha:=\#\{\nu\,|\,\alpha_\nu=1,\ t_\nu\text{ is odd}\}.$$

The sign convention in (11.2.2) is a combinatorial device which guarantees that the sign change caused by a permutation of the e_{ν_i} agrees with that caused by the same permutation of the t_{ν_i}. This gives the last expression in (11.2.4) with $t^\alpha:=(t_0)^{\alpha_0}\cdots(t_N)^{\alpha_N}$, $\alpha!:=\alpha_0!\cdots\alpha_N!$, and

$$\mathrm{GW}^M_{A,p}(e^\alpha):=\mathrm{GW}^M_{A,p}(e_0,\dots,e_0,e_1,\dots,e_1,\dots,e_N,\dots,e_N),$$

where e_ν occurs α_ν times for every ν. The last sum in (11.2.4) runs over all nonnegative integer vectors $\alpha=(\alpha_0,\dots,\alpha_N)$ that satisfy $|\alpha|:=\alpha_0+\cdots+\alpha_N=p$. Inserting (11.2.4) into (11.2.1) we obtain

$$\Phi(t)=\sum_\alpha\sum_A\frac{\varepsilon(\alpha)}{\alpha!}\mathrm{GW}^M_{A,|\alpha|}(e^\alpha)\,e^A\,t^\alpha. \tag{11.2.5}$$

Here the invariants with $|\alpha|<3$ are to be understood as in Remark 7.5.2. The dimension condition

$$\sum_{\nu=0}^N\alpha_\nu\deg(e_\nu)=\dim M+2c_1(A)+2|\alpha|-6 \tag{11.2.6}$$

guarantees that the coefficient of t^α in the series (11.2.5) is an element of Λ_ω of pure degree $d_\alpha:=\sum_{\nu=0}^N\alpha_\nu\,(\deg(e_\nu)-2)+6-\dim M$. To put it differently, if we assign to each t_ν the degree $2-\deg(e_\nu)$, then the power series (11.2.5) is homogeneous of degree $6-\dim M$.

EXAMPLE 11.2.1. Write a cohomology class in $H^*(\mathbb{C}P^n)$ with complex coefficients as a formal linear combination

$$a_t:=t_0+t_1c+t_2c^2+\cdots+t_nc^n,$$

where $t:=(t_0,\dots,t_n)$ and denote the variable corresponding to the generator L of $H_2(\mathbb{C}P^n)$ by q. We saw in Section 7.5 that the Gromov-Witten potential of $\mathbb{C}P^1$ is given by

$$\Phi(t_0,t_1)=\tfrac{1}{2}t_0^2t_1+qe^{t_1},$$

and that the Gromov-Witten potential of $\mathbb{C}P^2$ has the form

$$\Phi^{\mathbb{C}P^2}(t_0, t_1, t_2) = \tfrac{1}{2}\left(t_0^2 t_2 + t_0 t_1^2\right) + \sum_{d>0} N_d \frac{(t_2)^{3d-1}}{(3d-1)!} q^d e^{dt_1},$$

where N_d is the number of degree d genus zero curves through $3d-1$ generic points.

In the following it will be convenient to choose our basis in such a way that $e_1, \ldots, e_m$ form an integer basis of $H^2(M)$. Let us denote by $A_1, \ldots, A_m$ the dual basis of $H_2(M)$ so that

$$\int_{A_j} e_i = \delta_{ij}$$

for $i, j = 1, \ldots, m$. Then we can identify $H_2(M)$ with $\mathbb{Z}^m$ via the isomorphism $\mathbb{Z}^m \mapsto H_2(M) : d \mapsto A_d$ given by

$$A_d := d_1 A_1 + \cdots + d_m A_m$$

for $d = (d_1, \ldots, d_m) \in \mathbb{Z}^m$. Let us also introduce the formal variables $q_i := e^{A_i}$. Then we can write the Gromov–Witten potential as a formal power series in the variables q_i and t_ν, namely

$$\Phi(q, t) := \sum_\alpha \sum_d \frac{\varepsilon(\alpha)}{\alpha!} \mathrm{GW}^M_{A_d, |\alpha|}(e^\alpha)\, q^d t^\alpha. \tag{11.2.7}$$

Equation (11.2.2) guarantees that Φ is affine in each of the odd variables t_ν.

REMARK 11.2.2. Care must be taken when one differentiates a power series with respect to an odd variable. For example, if t_ν and t_μ are distinct odd variables then the partial derivatives of the products $t_\nu t_\mu$ and $t_\mu t_\nu$ with respect to t_ν should have opposite signs and be equal to $\pm t_\mu$. We choose the convention that the partial derivative of a monomial t^α with respect to an odd variable t_ν is given by first moving it to the front and then deleting it. Thus

$$\frac{\partial}{\partial t_\nu} t^\alpha := (-1)^{\varepsilon(\nu,\alpha)} \alpha_\nu t^\alpha / t_\nu$$

whenever $\alpha_\nu > 0$. Here

$$t^\alpha / t_\nu := t^{\alpha'}, \qquad \alpha'_\mu := \begin{cases} \alpha_\mu - 1, & \text{if } \mu = \nu, \\ \alpha_\mu, & \text{if } \mu \neq \nu, \end{cases}$$

and

$$\varepsilon(\nu, \alpha) := (-1)^{\deg(e_\nu) \cdot \#\{\mu < \nu \,|\, t_\mu \text{ is odd}, \alpha_\mu = 1\}} \tag{11.2.8}$$

With this convention the partial derivatives with respect to two different odd variables anticommute while, in all other cases, the partial derivatives commute.

EXERCISE 11.2.3. Given an integer vector α, denote by r_α the number of odd variables for which $\alpha_\nu = 1$, set $\varepsilon(\alpha) = r_\alpha(r_\alpha + 1)/2$ and let α' be the vector that is obtained from α by lowering α_ν by one.

(i) Show that $\varepsilon(\alpha') = (-1)^{r_\alpha |e_\nu|} \varepsilon(\alpha)$, where $|e_\nu|$ denotes the degree of e_ν. Thus $\varepsilon(\alpha) t_\nu t^{\alpha'} = (-1)^{|e_\nu|} \varepsilon(\alpha') t^{\alpha'} t_\nu$.

(ii) Suppose that α, β have no odd variables in common. Show that

$$\varepsilon(\alpha)\varepsilon(\beta) = (-1)^{r_\alpha r_\beta} \varepsilon(\alpha + \beta).$$

The next theorem shows how the axioms for the Gromov–Witten invariants translate into partial differential equations for the Gromov–Witten potential. The *(Fundamental class)* axiom becomes the string equation, the *(Divisor)* axiom the divisor equation, and the *(Splitting)* axiom the WDVV equation (as in Witten-Dijkgraaf-Verlinde-Verlinde). In the present context these equations are very simple. They become much more interesting in the higher genus case: see Manin[**286**].

THEOREM 11.2.4. *Let (M,ω) be a closed semipositive symplectic manifold. Then the Gromov–Witten potential $\Phi = \Phi^M$ satisfies the* **string equation**

$$\partial_{t_0}\Phi(q,t) = \tfrac{1}{2}\int_M a_t \smile a_t, \tag{11.2.9}$$

the **divisor equation**

$$\partial_{t_i}\Phi(q,t) = q_i\partial_{q_i}\Phi(q,t) + \tfrac{1}{2}\int_M e_i \smile a_t \smile a_t \tag{11.2.10}$$

for $i = 1,\ldots,m$, and the **WDVV equation**

$$\sum_{\nu,\mu}\partial_{t_i}\partial_{t_j}\partial_{t_\nu}\Phi\cdot g^{\nu\mu}\cdot\partial_{t_\mu}\partial_{t_k}\partial_{t_\ell}\Phi = \varepsilon_{ijk}\sum_{\nu,\mu}\partial_{t_j}\partial_{t_k}\partial_{t_\nu}\Phi\cdot g^{\nu\mu}\cdot\partial_{t_\mu}\partial_{t_i}\partial_{t_\ell}\Phi, \tag{11.2.11}$$

where $\varepsilon_{ijk} := (-1)^{\deg(e_i)(\deg(e_j)+\deg(e_k))}$.

PROOF. To begin with, write $\Phi(q,t) = \sum_a \Phi_d(t)q^d$, where Φ_d is the formal power series in the variables $t_0,\ldots,t_N$ with rational coefficients defined by

$$\Phi_d(t) := \sum_p \frac{1}{p!}\mathrm{GW}^M_{A_d,p}(a_t,\ldots,a_t) = \sum_\alpha \frac{\varepsilon(\alpha)}{\alpha!}\mathrm{GW}^M_{A_d,|\alpha|}(e^\alpha)\,t^\alpha.$$

With the sign conventions and notation of Remark 11.2.2, the partial derivative of Φ_d with respect to t_ν is given by

$$\begin{aligned}\partial_{t_\nu}\Phi_d(t) &= \sum_{\alpha_\nu\neq 0}\frac{\varepsilon(\alpha)\alpha_\nu}{\alpha!}\mathrm{GW}^M_{A_d,|\alpha|}(e_\nu,e^{\alpha'})\,t^{\alpha'}\\ &= \sum_\alpha(-1)^{|e_\nu|}\frac{\varepsilon(\alpha)}{\alpha!}\mathrm{GW}^M_{A_d,|\alpha|+1}(e^\alpha,e_\nu)\,t^\alpha.\end{aligned}\tag{11.2.12}$$

Here the first equation uses the fact that if we permute the e_ν and t_ν by the same permutation the sign does not change, while the second uses Exercise 11.2.3 (i).

We prove the string equation. By (11.2.12) with $\nu = 0$, we have

$$\begin{aligned}\partial_{t_0}\Phi(q,t) &= \sum_\alpha\sum_d\frac{\varepsilon(\alpha)}{\alpha!}\mathrm{GW}^M_{A_d,|\alpha|+1}(e^\alpha,1)\,q^dt^\alpha\\ &= \sum_{|\alpha|=2}\frac{\varepsilon(\alpha)}{\alpha!}\mathrm{GW}^M_{0,3}(1,e^\alpha)\,t^\alpha\\ &= \tfrac{1}{2}\sum_{i,j}\int_M t_ie_i \smile t_je_j\\ &= \tfrac{1}{2}\int_M a_t \smile a_t.\end{aligned}$$

Here the second equation follows from the *(Fundamental class)* axiom and *(Zero)* axioms.

We prove the divisor equation. By (11.2.12) with $\nu = i \in \{1, \dots, m\}$, we have

$$\begin{aligned}\partial_{t_i}\Phi(q,t) &= \sum_\alpha\sum_d \frac{\varepsilon(\alpha)}{\alpha!}\mathrm{GW}^M_{A_d,|\alpha|+1}(e^\alpha, e_i)\, q^d t^\alpha \\ &= \sum_\alpha\sum_{d\neq 0} \frac{\varepsilon(\alpha)}{\alpha!}\mathrm{GW}^M_{A_d,|\alpha|}(e^\alpha)\left(\int_{A_d} e_i\right) q^d t^\alpha \\ &\qquad + \sum_{|\alpha|=2}\frac{\varepsilon(\alpha)}{\alpha!}\mathrm{GW}^M_{0,3}(e_i, e^\alpha)\, t^\alpha \\ &= q_i\partial_{q_i}\Phi(q,t) + \tfrac{1}{2}\int_M e_i \smile a_t \smile a_t.\end{aligned}$$

Here the second identity follows from the *(Divisor)* and *(Zero)* axioms and the third from the fact that $\int_{A_d} e_i = d_i$ and $d_i q^d = q_i\partial_{q_i} q^d$ for each i.

We prove the WDVV equation. Iterating the identity (11.2.12) we obtain

$$\partial_{t_i}\partial_{t_j}\partial_{t_\nu}\Phi_d(t) = \sum_\alpha (-1)^{r_\alpha}\frac{\varepsilon(\alpha)}{\alpha!}\mathrm{GW}^M_{A,|\alpha|+3}(e^\alpha, e_i, e_j, e_\nu)\, t^\alpha.$$

Here we have used the fact that a Gromov–Witten invariant is nonzero only when evaluated on classes whose total degree is even, and hence r_α has the same parity as $|e_i| + |e_j| + |e_\nu|$. Now denote the left hand side of (11.2.11) by

$$\Phi_{ijk\ell}(t) =: \sum_d \Phi_{ijk\ell,d}(t) q^d.$$

Then, by what we have just observed,

$$\begin{aligned}\Phi_{ijk\ell,d}(t) &= \sum_{d'+d''=d}\sum_{\nu,\mu} \partial_{t_i}\partial_{t_j}\partial_{t_\nu}\Phi_{d'}(t)\cdot g^{\nu\mu}\cdot \partial_{t_\mu}\partial_{t_k}\partial_{t_\ell}\Phi_{d''}(t) \\ &= \sum_{d'+d''=d}\sum_{\nu,\mu}\sum_{\alpha,\beta}(-1)^{r_\alpha+r_\beta}\frac{\varepsilon(\alpha)\varepsilon(\beta)}{\alpha!\beta!} \\ &\qquad \times \mathrm{GW}^M_{A_{d'}}(e^\alpha, e_i, e_j, e_\nu)\, g^{\nu\mu}\,\mathrm{GW}^M_{A_{d''}}(e^\beta, e_\mu, e_k, e_\ell)\, t^\alpha t^\beta.\end{aligned}$$

Here we have dropped the number of marked points from the notation since it is clear from the context (namely $|\alpha|+3$ for $\mathrm{GW}^M_{A_{d'}}$ and $|\beta|+3$ for $\mathrm{GW}^M_{A_{d''}}$). Now define $\varepsilon(\alpha,\beta) \in \{\pm 1\}$ by the identity

$$t^\alpha t^\beta = \varepsilon(\alpha,\beta) t^{\alpha+\beta}.$$

and recall from Exercise 11.2.3 (ii) that

$$\varepsilon(\alpha+\beta) = (-1)^{r_\alpha r_\beta}\varepsilon(\alpha)\varepsilon(\beta).$$

Then we obtain

$$\Phi_{ijk\ell,d}(t) = \sum_\gamma \frac{1}{\gamma!}\Phi_{ijk\ell,d,\gamma} t^\gamma,$$

where

$$\begin{aligned}
\Phi_{ijk\ell,d,\gamma} \quad := \quad & \sum_{\substack{d'+d''=d \\ \alpha+\beta=\gamma}} \sum_{\nu,\mu} (-1)^{r_\alpha + r_\beta + r_\alpha r_\beta} \varepsilon(\gamma)\varepsilon(\alpha,\beta) \frac{\gamma!}{\alpha!\beta!} \times \\
& \qquad \times \mathrm{GW}^M_{A_{d'}}(e^\alpha, e_i, e_j, e_\nu)\, g^{\nu\mu}\, \mathrm{GW}^M_{A_{d''}}(e^\beta, e_\mu, e_k, e_\ell) \\
= \quad & \sum_{\substack{d'+d''=d \\ \alpha+\beta=\gamma}} \sum_{\nu,\mu} (-1)^{r_\alpha + r_\alpha r_\beta} \varepsilon(\gamma)\varepsilon(\alpha,\beta) \frac{\gamma!}{\alpha!\beta!} \times \\
& \qquad \times \mathrm{GW}^M_{A_{d'}}(e^\alpha, e_i, e_j, e_\nu)\, g^{\nu\mu}\, \mathrm{GW}^M_{A_{d''}}(e_\mu, e_k, e_\ell, e^\beta) \\
= \quad & \sum_{\substack{d'+d''=d \\ \alpha+\beta=\gamma}} \sum_{\nu,\mu} (-1)^{r_\alpha r_\gamma} \varepsilon(\gamma)\varepsilon(\alpha,\beta) \frac{\gamma!}{\alpha!\beta!} \times \\
& \qquad \times \mathrm{GW}^M_{A_{d'}}(e^\alpha, e_i, e_j, e_\nu)\, g^{\nu\mu}\, \mathrm{GW}^M_{A_{d''}}(e_\mu, e_k, e_\ell, e^\beta) \\
= \quad & \varepsilon(\gamma) \mathrm{GW}^{M,\{1,2,3,4\}}_{d,|\gamma|+4}(e_i, e_j, e_k, e_\ell, e^\gamma).
\end{aligned}$$

The last equation follows from the *(Splitting)* axiom of Theorem 7.5.10 with four fixed marked points. To understand the sign, note that e^α moves through the six terms $e_i, e_j, e_\nu, e_\mu, e_k, e_\ell$ whose total parity agrees with that of e^γ. This cancels the sign $(-1)^{r_\alpha r_\gamma}$. The sign $\varepsilon(\alpha,\beta)$ gets cancelled when converting the pair e^α, e^β (as arguments of the Gromov–Witten invariant) into e^γ. With this understood, equation (11.2.11) follows from the skew-symmetry of the Gromov–Witten invariants with respect to the first three variables. This proves the theorem. □

EXERCISE 11.2.5. Prove that the string equation can be expressed in the form

$$\partial_{t_0}\partial_{t_i}\partial_{t_j}\Phi = g_{ij}. \tag{11.2.13}$$

Hint: The variables t_i and t_j are either both even or they are both odd.

EXERCISE 11.2.6. Let $\mathcal{E}$ be the formal differential operator

$$\mathcal{E} := \sum_{i=1}^m c^i q_i \partial_{q_i} + \sum_{\nu=0}^N \left(1 - \frac{\deg(e_\nu)}{2}\right) t_\nu \partial_{t_\nu}.$$

where $c_1(TM) = \sum c^i e_i$ is the first Chern class. Show that $\mathcal{E}\Phi = (3-n)\Phi$, where $2n := \dim M$. This holds because of the dimensional condition (11.2.6).

Section 7.5 contains an extensive discussion of the Gromov–Witten potential $\Phi^{\mathbb{C}P^n}$ for $\mathbb{C}P^n$. In particular, we showed in Proposition 7.5.11 that $\Phi^{\mathbb{C}P^2}$ is determined by the WDVV equation and the initial condition $\mathrm{GW}^{\mathbb{C}P^2}_{L,3}(c^2, c^2, c^2) = 1$. Although this result can be generalized (see Theorem 11.5.8 below) it certainly does not always hold; cf. the discussion of Calabi–Yau manifolds in Section 11.3.4. In Section 11.5 we shall explore the geometric meaning of the WDVV equation in more detail and show how it leads to interesting structures on the cohomology of a symplectic manifold such as the Dubrovin connection. In that section we shall restrict attention to the even dimensional cohomology so that the rather tricky sign questions do not interfere with the discussion.

11.3. Four examples

This section discusses the quantum cohomology of Fano toric manifolds, Grassmannians and flag manifolds and explains their connections with the Toda lattice and the Verlinde algebra. All these manifolds are Fano, in the sense that $c_1(A) > 0$ for every class with a holomorphic representative (see Remark 11.1.1.) When this condition is not satisfied new phenomena appear. We illustrate them here in the case of Calabi–Yau 3-folds. Another example occurs in Section 11.4 where we describe the quantum cohomology of nonFano toric manifolds. In the first three cases, we shall explain presentations of the quantum cohomology rings. But, as illustrated in Remark 11.3.20, this does not allow us to derive the formula for even the cubic terms in the Gromov–Witten potential. In the Calabi–Yau example there is a formula for this potential, which was first obtained from mirror symmetry.

11.3.1. Fano Toric manifolds. This subsection describes a formula first stated by Batyrev [**31**] for the small quantum cohomology ring of a Fano toric manifold. The formula plays an important role in the work of Givental [**150**] on mirror symmetry. He sets it into the context of equivariant quantum cohomology. Below we briefly describe a quite different proof by Cieliebak–Salamon [**69**] which is based on the symplectic vortex equations. Another proof by McDuff–Tolman [**279**] is sketched in Section 11.4. We shall begin by describing the structure of toric manifolds. Though this involves some work, one ends up with an explicit and accessible family of nontrivial examples. For further information the reader may consult Audin [**23**], Cox–Katz [**76**], Guillemin–Sternberg [**164**] or Voisin [**409**].

We explain two ways of looking at toric manifolds. One is as a closed symplectic manifold (M, ω) equipped with an effective Hamiltonian torus action, where the torus has half the dimension of M and the effective condition means that there is a point in M on which the torus acts freely. A second description of toric manifolds (or orbifolds) is as a symplectic quotient of a linear torus action on $\mathbb{C}^n$ with proper moment map. To get the full picture one needs to combine both approaches. We shall begin with the second.

Let T be a torus of dimension k, denote by $\mathfrak{t}$ its Lie algebra, by

$$\Lambda := \{\xi \in \mathfrak{t} \mid \exp(\xi) = 1\}$$

the integer lattice, and by

$$\Lambda^* := \{\mathrm{w} \in \mathfrak{t}^* \mid \langle \mathrm{w}, \xi \rangle \in \mathbb{Z} \text{ for } \xi \in \Lambda\}$$

the dual lattice. Suppose T acts on $\mathbb{C}^N$ by a homomorphism

$$\rho = \mathrm{diag}(\rho_1, \dots, \rho_N) : T \to \mathbb{T}^N \subset \mathrm{U}(N)$$

onto the diagonal subgroup $\mathbb{T}^N$ of $\mathrm{U}(N)$. For $\nu = 1, \dots, N$, we write the homomorphism $\rho_\nu : T \to S^1$ in the form

$$\rho_\nu(\exp(\xi)) = e^{-2\pi i \langle \mathrm{w}_\nu, \xi \rangle}, \qquad \xi \in \mathfrak{t}, \tag{11.3.1}$$

where $\mathrm{w}_\nu \in \Lambda^*$. A moment map $\mu : \mathbb{C}^n \to \mathfrak{t}^*$ for this action is given by

$$\mu(x) := \pi \sum_{\nu=1}^{N} |x_\nu|^2 \, \mathrm{w}_\nu. \tag{11.3.2}$$

We assume that the moment map is proper, i.e. there is an open half space in $\mathfrak{t}^*$ which contains all the vectors w_ν. We also assume that it is effective, i.e. the vectors

w_ν span the integer lattice Λ^* in $\mathfrak{t}^*$. The **symplectic quotient** at a regular value $\tau \in \mathfrak{t}^*$ of the moment map is defined by

$$M_\tau := \mathbb{C}^n /\!\!/ T(\tau) := \mu^{-1}(\tau)/T.$$

This quotient is a closed symplectic manifold whenever T acts freely on $\mu^{-1}(\tau)$. Otherwise it is a closed symplectic orbifold. The quotient $\mu^{-1}(\tau)/T$ can also be identified with $U/T_{\mathbb{C}}$, where U is a suitable open subset of $\mathbb{C}^n$ and $T_{\mathbb{C}} \cong (\mathbb{C}^*)^k$ is the complexification of the real torus T. It follows that M_τ is a Kähler manifold, indeed a smooth projective variety, so that we may interpret the word Fano in the projective setting discussed in Remark 11.1.1. For more details see the references cited above or McDuff-Tolman [**280**, §5].

Let us now assume that τ is a regular value of μ and T acts freely on $\mu^{-1}(\tau)$. For every subset $I \subset I_0 := \{1, \ldots, N\}$, consider the subspace

$$E_I := \mathbb{C}^{I_0 \setminus I} = \{x \in \mathbb{C}^N \,|\, x_\nu = 0 \text{ for } \nu \in I\},$$

and denote by the closed positive cone spanned by the vectors $\{\mathrm{w}_\nu\}_{\nu \in I}$ by

$$\mathrm{cone}(I) := \left\{ \sum_{\nu \in I} \eta_\nu \mathrm{w}_\nu \,\middle|\, \eta_\nu \geq 0 \right\} \subset \mathfrak{t}^*.$$

Then, by (11.3.2), we have

$$E_I \cap \mu^{-1}(\tau) \neq \emptyset \quad \iff \quad \tau \in \mathrm{cone}(I_0 \setminus I). \tag{11.3.3}$$

The corresponding submanifolds

$$X_I := \big(E_I \cap \mu^{-1}(\tau)\big)/T = \{[x] \in M_\tau \,|\, x_\nu = 0 \text{ for } \nu \in I\} \tag{11.3.4}$$

of M_τ generate its cohomology in the following sense. If $I = \{\nu\}$ is a singleton then $X_\nu := \{[x] \in M_\tau \,|\, x_\nu = 0\}$ is a smooth divisor (i.e. a complex codimension-1 submanifold) and it determines a cohomology class $\overline{\mathrm{w}}_\nu := \mathrm{PD}([X_\nu]) \in H^2(M_\tau)$. Thus $\overline{\mathrm{w}}_\nu$ is the first Chern class of the line bundle

$$L_\nu := \mu^{-1}(\tau) \times_{\rho_\nu} \mathbb{C} \to M_\tau. \tag{11.3.5}$$

Note that $\overline{\mathrm{w}}_\nu = 0$ if and only if $X_\nu = \emptyset$, which happens when $\tau \notin \mathrm{cone}(I_0 \setminus \{\nu\})$. Moreover, $X_{I \cap I'} = X_I \cap X_{I'}$ and hence

$$X_{\{\nu_1, \ldots, \nu_k\}} = \emptyset \quad \Longrightarrow \quad \overline{\mathrm{w}}_{\nu_1} \smile \cdots \smile \overline{\mathrm{w}}_{\nu_k} = 0$$

By (11.3.5), every relation $\sum_i \eta_\nu \mathrm{w}_\nu = 0$ among the representations ρ_ν gives rise to a corresponding relation $\sum \eta_\nu \overline{\mathrm{w}}_\nu = 0$ among the cohomology classes $\overline{\mathrm{w}}_\nu$. It turns out that these are the only relations among the classes $\overline{\mathrm{w}}_\nu$. A proof of the following theorem can be found in Guillemin–Sternberg [**164**].

THEOREM 11.3.1. *Let $\tau \in \mathfrak{t}^*$ be a regular value of μ and suppose T acts freely on $\mu^{-1}(\tau)$. Then the ring homomorphism $\mathbb{R}[u_1, \ldots, u_N] \to H^*(M_\tau; \mathbb{R}) : u_\nu \mapsto \overline{\mathrm{w}}_\nu$ induces an isomorphism $H^*(M_\tau; \mathbb{R}) \cong \mathbb{R}[u_1, \ldots, u_N]/\mathcal{I}$, where $\mathcal{I} \subset \mathbb{R}[u_1, \ldots, u_N]$ denotes the ideal generated by the relations*

$$\sum_{\nu=1}^N \eta_\nu \mathrm{w}_\nu = 0 \quad \Longrightarrow \quad \sum_{\nu=1}^N \eta_\nu u_\nu = 0, \tag{11.3.6}$$

$$I \subset I_0, \quad \tau \notin \mathrm{cone}\big(I_0 \setminus I\big) \quad \Longrightarrow \quad \prod_{\nu \in I} u_\nu = 0. \tag{11.3.7}$$

REMARK 11.3.2. Let $\tau \in \mathfrak{t}^*$ be a regular value of μ. Then T acts freely on $\mu^{-1}(\tau)$ if and only if the following integrality condition holds: for every subset $J \subset I_0$ of cardinality k such that $\tau \in \mathrm{cone}(J)$ the vectors $\{\mathrm{w}_\nu \,|\, \nu \in J\}$ form a basis for the integral lattice Λ^*. Further, one can show that the first Chern class $c_1(TM_\tau)$ and the cohomology class of the symplectic form $\omega_\tau \in \Omega^2(M_\tau)$ are given by

$$c_1(M_\tau) = \overline{\mathrm{w}}_1 + \cdots + \overline{\mathrm{w}}_N, \qquad [\omega_\tau] = \overline{\tau},$$

where $\overline{\tau}$ is the image of $\tau \in \mathfrak{t}^*$ under the homomorphism

$$\mathfrak{t}^* := \Lambda^* \otimes \mathbb{R} \to H^2(M_\tau; \mathbb{R}) : \mathrm{w} \mapsto \overline{\mathrm{w}}$$

constructed via equations (11.3.1) and (11.3.5) above. Thus M_τ is monotone if and only if $\overline{\tau}$ is a positive multiple of the vector $\overline{\mathrm{w}}_1 + \cdots + \overline{\mathrm{w}}_N$. (See Cieliebak–Salamon [**69**, Appendix C] for more details.)

EXAMPLE 11.3.3. The simplest case of this construction is when $k = 1$. Then every positive $\tau \in \mathfrak{t}^* \equiv \mathbb{R}$ is regular, and $\mu^{-1}(\tau)$ is a sphere S^{2N-1}. The properness condition implies that each weight w_i for $i = 1, \ldots, N$ is a positive multiple $a_i\mathrm{w}$ of a chosen generator w of $\Lambda^* = \mathbb{Z}$. The resulting quotient is called a **weighted projective space**.

Since every nonzero $\tau \in \mathfrak{t}^*$ lies in cone(I) whenever $I \neq \emptyset$, the observations in Remark 11.3.2 imply that S^1 acts freely on $\mu^{-1}(\tau)$ only if $a_i = 1$ for all i. In this case $T = S^1$ acts diagonally on $\mathbb{C}^N$, and the quotient is the standard projective space $\mathbb{C}P^{N-1}$. Theorem 11.3.1 presents its cohomology ring as $\mathbb{R}[u_1, \ldots, u_N]/\mathcal{I}$, where $\mathcal{I}$ is generated by the relations $u_1 = \cdots = u_N$ and $\prod_{i=1}^N u_i = 0$. Note that there is only one multiplicative relation corresponding to the set $I = I_0$. For if $I \neq I_0$ then cone$(I \setminus I_0) = [0, \infty)$ and so contains τ.

Here we have described the structure of M_τ in terms of the elements $\mathrm{w}_1, \ldots, \mathrm{w}_N$ and τ of $\mathfrak{t}^*$. We show below that this description may be translated purely into terms of the moment polytope Δ of the action of the torus $\mathbb{T}^N/\rho(T)$ on the quotient M_τ; cf. Example 11.3.7.

Theorem 11.3.1 implies that $H^2(M_\tau)$ is a quotient of the integer lattice in $\mathfrak{t}^*$. Hence its dual $H_2(M_\tau)$ can be identified with the following sublattice of $\Lambda \subset \mathfrak{t}$:

$$\Lambda(\tau) := \left\{\lambda \in \Lambda \,|\, \tau \notin \mathrm{cone}(I_0 \setminus \{\nu\}) \Longrightarrow \langle \mathrm{w}_\nu, \lambda\rangle = 0\right\}.$$

In many cases $\tau \in \mathrm{cone}(I_0 \setminus \{\nu\})$ for all ν (this is a nondegeneracy condition), so that $\Lambda(\tau)$ is equal to the integral lattice Λ in $\mathfrak{t}$. In general it has dimension $m \leq k$.

Let us denote by $C(\tau) \subset \mathfrak{t}^*$ the connected component of τ in the set of regular values of the moment map μ. This set is called the **chamber of** τ and is an open convex cone in $\mathfrak{t}^*$. It is also known as the **Kähler cone** of M_τ. The dual cone in the integral lattice $\Lambda(\tau)$ is the closed convex cone of all elements $\lambda \in \Lambda(\tau)$ that satisfy $\langle \tau', \lambda\rangle \geq 0$ for every $\tau' \in C(\tau)$; for $\lambda \neq 0$ this condition is equivalent to $\langle \tau', \lambda\rangle > 0$ for every $\tau' \in C(\tau)$, because $C(\tau)$ is open. It is shown in [**76, 69**] that the dual cone is generated by the effective classes, i.e. the classes represented by (possibly constant) holomorphic curves. We denote it by [1]

$$\Lambda^{\mathrm{eff}}(\tau) := \left\{\lambda \in \Lambda(\tau) \,|\, \tau' \in C(\tau) \Longrightarrow \langle \tau', \lambda\rangle \geq 0\right\}.$$

[1]In Section 11.1 we defined two notions of effective cone for a Kähler manifold (M, ω, J), the symplectic cone $K^{\mathrm{eff}}(M, \omega)$ generated by classes with nontrivial Gromov–Witten invariants and the algebraic cone $K^{\mathrm{eff}}(M, J)$ generated by classes represented by holomorphic curves. Thus $\Lambda^{\mathrm{eff}}(\tau) := K^{\mathrm{eff}}(M_\tau, J)$. In the Fano case, it follows from Theorem 11.3.4 that these two cones are

It follows that M_τ is Fano if and only if the vector $c := \sum_{\nu=1}^{N} \mathrm{w}_\nu$ takes positive values on $\Lambda^{\mathrm{eff}}(\tau) \setminus \{0\}$. Equivalently, c belongs to the chamber $C(\tau)$ and so we may as well assume $c = \tau$, i.e. that M_τ is monotone.

In the following we shall assume that the symplectic form τ on the Fano toric manifold is chosen so that M_τ is monotone. We fix an integer basis $v_1, \dots, v_m$ of $\Lambda(\tau)$ so that $\langle \tau, v_i \rangle > 0$ for every i and so that each element of $\Lambda^{\mathrm{eff}}(\tau)$ is a nonnegative linear combination of the v_i. Further, as quantum coefficient ring we choose the polynomial ring $\mathbb{R}[q_1, \dots, q_m]$ as in Example 11.1.4 (iii). Given an integer vector $\ell = (\ell_1, \dots, \ell_N)$ with $\ell_\nu \geq 0$ we denote

$$\overline{\mathrm{w}}^{*\ell} := (\overline{\mathrm{w}}_1 * \cdots * \overline{\mathrm{w}}_1) * \cdots * (\overline{\mathrm{w}}_N * \cdots * \overline{\mathrm{w}}_N) \in \mathrm{QH}^*(M_\tau; \mathbb{R}[q_1, \dots, q_m]),$$

where $\overline{\mathrm{w}}_\nu$ occurs ℓ_ν times. The next theorem gives Batyrev's formula for the quantum cohomology of monotone toric manifolds. Note that the first relation coincides with that for the ordinary cohomology ring described in Theorem 11.3.1.

THEOREM 11.3.4 (Batyrev, Givental, Cieliebak–Salamon). *Let*

$$\tau := \sum_{\nu=1}^{N} \mathrm{w}_\nu$$

be a regular value of μ *and suppose* T *acts freely on* $\mu^{-1}(\tau)$. *Then the ring homomorphism*

$$\mathbb{R}[u_1, \dots, u_N, q_1, \dots, q_m] \to \mathrm{QH}^*(M_\tau; \mathbb{R}[q_1, \dots, q_m]) : u^\ell \mapsto \overline{\mathrm{w}}^{*\ell}$$

induces an isomorphism

$$\mathrm{QH}^*(M_\tau; \mathbb{R}[q_1, \dots, q_m]) \cong \mathbb{R}[u_1, \dots, u_N, q_1, \dots, q_m]/\mathcal{J},$$

where $\mathcal{J} \subset \mathbb{R}[u_1, \dots, u_N, q_1, \dots, q_m]$ *denotes the ideal generated by the relations*

$$\sum_{\nu=1}^{N} \eta_\nu \mathrm{w}_\nu = 0 \quad \Longrightarrow \quad \sum_{\nu=1}^{N} \eta_\nu u_\nu = 0, \tag{11.3.8}$$

$$\tau \notin \mathrm{cone}(I_0 \setminus \{\nu\}) \quad \Longrightarrow \quad u_\nu = 0, \tag{11.3.9}$$

$$\lambda \in \Lambda^{\mathrm{eff}}(\tau), \quad d_\nu^\pm := \max\{\pm\langle \mathrm{w}_\nu, \lambda \rangle, 0\} \quad \Longrightarrow \quad u^{d^+} = q^\lambda u^{d^-}. \tag{11.3.10}$$

Here $q^\lambda := q_1^{\lambda_1} \cdots q_m^{\lambda_m}$ *for* $\lambda = \sum_i \lambda_i v_i \in \Lambda(\tau)$.

The following exercise explains why the last relation (11.3.10) is a deformation of the product relations in the ordinary cohomology.

EXERCISE 11.3.5. Given $\lambda \in \Lambda^{\mathrm{eff}}(\tau)$ let $I := \{\nu \,|\, \langle \mathrm{w}_\nu, \lambda \rangle > 0\}$. Show that $\tau \notin \mathrm{cone}(I_0 \setminus I)$. Deduce that $\overline{\mathrm{w}}^{d^+} = 0$ in $H^*(M_\tau)$.

The proof of this theorem given by Cieliebak and Salamon [**69**, Theorem C] is valid under the additional assumption that M_τ has minimal Chern number at least two. It is based on the properties of the symplectic vortex equations. As we explain in more detail in Section 12.7, solutions to these equations are analogues of J-holomorphic curves for symplectic manifolds with Hamiltonian group actions. In the case at hand the moduli spaces of symplectic vortices are compact and

---

the same. Namely, equation (11.3.10) implies that for each nonzero $\lambda \in \Lambda^{\mathrm{eff}}(\tau)$ some Gromov–Witten invariant $\mathrm{GW}_{\lambda,k} \neq 0$. However, before we have proved that, all we know is that $K^{\mathrm{eff}}(M, \omega)$ is a subset of $K^{\mathrm{eff}}(M, J)$.

it is possible to compute the invariants via a wall crossing formula. A result of Gaio–Salamon [**135**] allows a comparison between the vortex invariants and the Gromov–Witten invariants of the quotient M_τ in the monotone case. This gives rise to relations of the form $\mathrm{GW}_A^{M_\tau}(\overline{\mathrm{w}}^{*d^+}, a) = \mathrm{GW}_{A-A_\lambda}^{M_\tau}(\overline{\mathrm{w}}^{*d^-}, a)$, where λ and $d^\pm$ are related as in (11.3.10) and $A_\lambda \in H_2(M_\tau)$ denotes the homology class associated to $\lambda \in \Lambda(\tau)$ under the isomorphism $\Lambda(\tau) \cong H_2(M_\tau)$. These relations are equivalent to (11.3.10). That the quantum cohomology has the required form then follows from the fact that, if we set $q_i := 0$, the ideal $\mathcal{J}$ in Theorem 11.3.4 is equal to the ideal $\mathcal{I}$ in Theorem 11.3.1. Exercise 11.3.5 shows that $\mathcal{I}$ is contained in the image of $\mathcal{J}$; for a proof of equality see Batyrev [**31**] or [**69**, Theorem C.5]. Another interpretation of these relations in terms of the moment polytope is given in equation (11.3.16) below. A different proof of Theorem 11.3.4, based on properties of the Seidel representation for circle actions, is sketched at the end of Section 11.4.

We now turn to the second description of a toric manifold as a closed symplectic manifold $(\overline{M}, \overline{\omega})$ that supports an effective Hamiltonian group action by a torus of half the dimension of $\overline{M}$. Let us denote this torus by $\overline{T}$ and its Lie algebra by $\bar{\mathfrak{t}}$. The action is generated by a $\overline{T}$-invariant moment map

$$\overline{\mu} : \overline{M} \to \bar{\mathfrak{t}}^*$$

in the sense that the infinitesimal action $X_{\overline{\xi}} \in \mathrm{Vect}(\overline{M})$ of an element $\overline{\xi} \in \bar{\mathfrak{t}}$ is the Hamiltonian vector field of the Hamiltonian function $\langle \overline{\mu}, \overline{\xi} \rangle \colon \overline{M} \to \mathbb{R}$, i.e.

$$\iota(X_{\overline{\xi}})\overline{\omega} = dH_{\overline{\xi}}, \qquad H_{\overline{\xi}} := \langle \overline{\mu}, \overline{\xi} \rangle.$$

A celebrated result of Guillemin–Sternberg and Atiyah asserts that the image

$$\Delta := \overline{\mu}(\overline{M}) \subset \bar{\mathfrak{t}}^*$$

of the moment map is a convex polytope. If $\eta \in \Delta$ belongs to the interior of a k-dimensional face, then its inverse image $\overline{\mu}^{-1}(\eta)$ is a k-dimensional orbit of $\overline{T}$ that is necessarily isotropic. In particular the inverse images of the edges (1-dimensional faces) are 2-spheres. The codimension-1 faces $F_1, \ldots, F_K$ of Δ are called **facets**. Each facet F_ν is given by an equation of the form $\{\eta \in \bar{\mathfrak{t}}^* \mid \langle \eta, \overline{e}_\nu \rangle = \kappa_\nu\}$, where $\kappa_\nu \in \mathbb{R}$ is called a *structure constant* and $\overline{e}_\nu \in \overline{\Lambda} \subset \bar{\mathfrak{t}}$ is a primitive lattice vector called the *normal* to F_ν. We choose the normals $\overline{e}_\nu$ to be inward pointing, so that

$$\Delta = \overline{\mu}(\overline{M}) = \{\eta \in \bar{\mathfrak{t}}^* \mid \langle \eta, \overline{e}_\nu \rangle \geq \kappa_\nu \text{ for } \nu = 1, \ldots, K\}.$$

Delzant [**79**] proved that a convex polytope Δ occurs as the image of the moment map of a toric manifold if and only if the following condition holds for each vertex of Δ: the vertex lies on precisely $n := \dim \overline{T}$ facets and the corresponding set of n normal vectors forms an integral basis for the lattice $\bar{\mathfrak{t}}$. It follows that one can construct the toric manifold $\overline{M}$ as a quotient M_τ as above. We do not prove this here. However a proof may be extracted from the following discussion of the relation between the two constructions. See also Audin [**23**] or [**280**, §5].

In the case $\overline{M} = M_\tau$ the quotient torus

$$\overline{T} := \mathbb{T}^N / \rho(T)$$

acts on $\overline{M}$. Its Lie algebra is the quotient $\bar{\mathfrak{t}} := \mathbb{R}^N / \dot{\rho}(\mathfrak{t})$, where $\dot{\rho} : \mathfrak{t} \to \mathbb{R}^n$ is the induced Lie algebra homomorphism, given by $\dot{\rho}_\nu(\xi) = \langle \mathrm{w}_\nu, \xi \rangle$; cf. equation (11.3.1).

The dual space of $\bar{\mathfrak{t}}$ can be identified with the subspace

$$\bar{\mathfrak{t}}^* := \Big\{\eta \in (\mathbb{R}^N)^* \mid \sum_{\nu=1}^N \eta_\nu \mathrm{w}_\nu = 0\Big\}. \tag{11.3.11}$$

Thus there are exact sequences

$$0 \to \mathfrak{t} \xrightarrow{\dot\rho} \mathbb{R}^N \to \bar{\mathfrak{t}} \to 0, \qquad 0 \to \bar{\mathfrak{t}}^* \to (\mathbb{R}^N)^* \to \mathfrak{t}^* \to 0,$$

where the normal vectors $\bar{e}_\nu$ lie in $\bar{\mathfrak{t}}$ and τ, w_ν lie in $\mathfrak{t}^*$. Note also that that standard basis in $\mathbb{R}^N$ projects to the normals $\bar{e}_\nu$, while that of $(\mathbb{R}^N)^*$ projects to the multiplicity vectors w_ν. Choose a lift $\kappa = (\kappa_1, \dots, \kappa_N) \in (\mathbb{R}^N)^*$ of $-\tau \in \mathfrak{t}^*$, so that

$$\sum_{\nu=1}^N \kappa_\nu \mathrm{w}_\nu = -\tau \in \mathfrak{t}^*.$$

Then the level set $\mu^{-1}(\tau)$ is $\{x \in \mathbb{C}^N : \pi|x_\nu|^2 = -\kappa_\nu \ \forall \nu\}$. Hence, if we denote the T orbit of an element $x \in \mu^{-1}(\tau) \subset \mathbb{C}^N$ by

$$[x] \in \overline{M} = \mu^{-1}(\tau)/T,$$

equation (11.3.2) implies that

$$\bar\mu : \overline{M} \to \bar{\mathfrak{t}}^*, \quad [x] \mapsto (\pi|x_1|^2 + \kappa_1, \dots, \pi|x_N|^2 + \kappa_N)$$

is a moment map. The image of $\bar\mu$ is then the convex polytope

$$\Delta = \bar\mu(\overline{M}) = \Big\{\eta \in (\mathbb{R}^N)^* \Big| \sum_{\nu=1}^N \eta_\nu \mathrm{w}_\nu = 0,\ \eta_\nu \ge \kappa_\nu\Big\} \subset \bar{\mathfrak{t}}^*.$$

Every subset $I \subset I_0 := \{1, \dots, N\}$ determines a (possibly empty) face

$$\Delta_I := \{\eta \in \Delta \mid \eta_\nu = \kappa_\nu \text{ for } \nu \in I\}.$$

Thus $\Delta_\emptyset = \Delta$ and it follows from (11.3.3) that

$$\Delta_I = \emptyset \quad \Longleftrightarrow \quad \tau \notin \mathrm{cone}(I_0 \setminus I). \tag{11.3.12}$$

In particular, the number K of facets of the moment polytope Δ is equal to the number of indices $\nu \in I_0$ such that $\tau \in \mathrm{cone}(I_0 \setminus \{\nu\})$. From now on, we will assume for simplicity that this condition holds[2] for every ν so that $F_\nu = \Delta_\nu \ne \emptyset$ for every ν. This condition also implies that $\Lambda = \Lambda(\tau)$ and $m = k$. Note that in all cases Δ_I is the intersection of the facets Δ_ν for $\nu \in I$. Further, the normal vector $\bar{e}_\nu$ to the facet F_ν is simply the projection of the standard basis vector in $\mathbb{R}^N$ onto the quotient $\bar{\mathfrak{t}} = \mathbb{R}^N/\dot\rho(\mathfrak{t})$.

EXAMPLE 11.3.6. Consider $\mathbb{C}P^2$ as the quotient of $S^5 = \mu^{-1}(1)$ by the diagonal action of S^1 as in Example 11.3.3. Then the weight vectors are $\mathrm{w}_\nu = 1 \in \mathfrak{t} \equiv \mathbb{R}$ for all ν, so that the moment polytope is contained in the hyperplane $\{\sum_{\nu=1}^3 \eta_\nu = 0\} \subset \mathbb{R}^3$ and κ s atisfies $\sum \kappa_\nu = -1$. Suppose that we identify the quotient $\bar{\mathfrak{t}} = \mathbb{R}^3/\mathfrak{t}$ with $\mathbb{R}^2$ by projection onto the first two coordinates. Since the three normal vectors $\bar{e}_\nu$ are the images of the standard basis, we have

$$\bar{e}_1 = (1,0), \quad \bar{e}_2 = (0,1), \quad \bar{e}_3 = (-1,-1).$$

[2] If one starts with the quotient $\overline{M}$ one can always choose the torus T and homomorphism $\rho : T \to \mathbb{T}^N$ so that this is the case. However, it is sometimes useful to express $\overline{M}$ as a quotient M_τ in which some of the F_ν are empty. They are called *ghost facets*.

Further the corresponding inclusion $\mathbb{R}^2 \equiv \bar{\mathfrak{t}}^* \to \{\sum_{\nu=1}^3 \eta_\nu = 0\} \subset \mathbb{R}^3$ is given by $(\xi_1, \xi_2) \mapsto (\xi_1, \xi_2, -\xi_1 - \xi_2)$. Therefore, because $-\kappa_3 = 1 + \kappa_1 + \kappa_2$, the image of the moment polytope in $\mathbb{R}^2$ is the triangle

$$\Delta_\kappa := \big\{(\xi_1, \xi_2) \,|\, \xi_1 \geq \kappa_1,\ \xi_2 \geq \kappa_2,\ \xi_1 \xi_2 \leq 1 + \kappa_1 + \kappa_2\big\}.$$

Changing the lifts κ of $-\tau = -1$ corresponds to translating the polytope, while changing the identification of $\bar{\mathfrak{t}}$ with $\mathbb{R}^2$ alters it by an integral affine transformation. (We always choose this identification so that the integral lattice of $\bar{\mathfrak{t}}$ is $\mathbb{Z}^2 \subset \mathbb{R}^2$.)

The symplectic submanifold X_ν of equation (11.3.4) is now given by

$$X_\nu := \overline{\mu}^{-1}(\Delta_\nu),$$

and we defined the cohomology class $\overline{\mathrm{w}}_\nu$ to be its Poincaré dual. By (11.3.11), the relation (11.3.6) translates into $\sum_\nu \langle \eta, \bar{e}_\nu \rangle \overline{\mathrm{w}}_\nu = 0$ for every $\eta \in \bar{\mathfrak{t}}^*$. On the other hand, the relation (11.3.7) is redundant because $\Delta_\nu \neq \emptyset$ for all ν. In this notation Theorem 11.3.1 asserts that

$$H^*(\overline{M}; \mathbb{R}) \cong \frac{\mathbb{R}[u_1, \ldots, u_N]}{P(\Delta) + SR(\Delta)},$$

where $P(\Delta)$ and $SR(\Delta)$ are the ideals in $\mathbb{R}[u_1, \ldots, u_N]$ defined by

$$P(\Delta) := \Big\langle \sum_\nu \langle \eta, \bar{e}_\nu \rangle u_\nu \,|\, \eta \in \bar{\mathfrak{t}}^* \Big\rangle, \qquad SR(\Delta) = \Big\langle \prod_{\nu \in I} u_\nu \,\Big|\, \Delta_I = \emptyset \Big\rangle.$$

The ideal $SR(\Delta)$ is called the **Stanley–Reisner** ideal. It is generated by the **primitive** subsets $I \subset \{1, \ldots, N\}$ for which $\Delta_I = \emptyset$ but $\Delta_{I'} \neq \emptyset$ for $I' \subsetneq I$. It follows as above that

$$H_2(\overline{M}) \cong \mathcal{D} := \Big\{ d \in \mathbb{Z}^N \,|\, \sum_{\nu=1}^N d_\nu \bar{e}_\nu = 0 \Big\} \cong \Lambda \subset \mathfrak{t}. \tag{11.3.13}$$

EXAMPLE 11.3.7. We saw in Example 11.3.6 that the manifold $\overline{M} = \mathbb{C}P^2$ has a toric structure in which Δ is the triangle in the first quadrant of $\mathbb{R}^2$ cut out by lines normal to the vectors

$$\bar{e}_1 := (1, 0), \qquad \bar{e}_2 := (0, 1), \qquad \bar{e}_3 := (-1, -1).$$

The relations in $P(\Delta)$ imply that $u_1 = u_2 = u_3$, while $SR(\Delta)$ is generated by the product $u_1 u_2 u_3$.

EXERCISE 11.3.8. Fix an index set $J \subset I_0 = \{1, \ldots, N\}$. Show that the vectors $\{\bar{e}_\nu\}_{\nu \in J}$ are linearly independent in $\bar{\mathfrak{t}}$ if and only if the vectors $\{\mathrm{w}_\nu\}_{\nu \in I_0 \setminus J}$ span $\mathfrak{t}^*$. Show that this holds whenever $\Delta_J \neq \emptyset$.

We now explain Batyrev's relations in Theorem 11.3.4 in terms of the moment polytope. As before, we pick an integer basis $v_1, \ldots, v_k$ of Λ and denote by $y_i := \dot{\rho}(v_i)$ the corresponding basis of the subgroup $\mathcal{D} \subset \mathbb{Z}^n$ defined in (11.3.13). Thus y_i has coordinates $\langle \mathrm{w}_\nu, v_i \rangle$ for $\nu = 1, \ldots, N$. Define an isomorphism

$$\Lambda := \mathbb{Z}^k \to \mathcal{D} : \lambda \mapsto d := \sum_{i=1}^k \lambda_i y_i = \dot{\rho}(\lambda) = \begin{pmatrix} \langle \mathrm{w}_1, \lambda \rangle \\ \vdots \\ \langle \mathrm{w}_N, \lambda \rangle \end{pmatrix}. \tag{11.3.14}$$

Note that we now have $k = m$ because of our nondegeneracy assumption.

Here is a sketch of the proof that the ideal $\mathcal{I}$ in Theorem 11.3.1 agrees with $\mathcal{J}$ in Theorem 11.3.4 when we set $q_i = 0$. (For full details see [**69**, Appendices D and E].) We must check that the multiplicative relation (11.3.10) reduces to (11.3.7) when $q = 0$. To see this, consider the fan in $\bar{\mathfrak{t}}$ generated by the vectors $\bar{e}_\nu$, $\nu \in I_0$. For $J \subset I_0$ with $\Delta_J \neq \emptyset$, denote by $\overline{C}_J \subset \bar{\mathfrak{t}}$ the open cone spanned by the vectors $\bar{e}_\nu$, $\nu \in J$. In particular, define $\overline{C}_\emptyset := \{0\}$. It follows from the duality relation between the polytope and the fan that the sets $\{\overline{C}_J \,|\, \Delta_J \neq \emptyset\}$ are disjoint and that their union is $\bar{\mathfrak{t}}$. Fix a primitive index set $I \subset \{1, \ldots, N\}$ so that $\Delta_I = \emptyset$ and $\Delta_{I'} \neq \emptyset$ for $I' \subsetneq I$. Consider the vector $\overline{v}(I) := \sum_{i \in I} \bar{e}_i \in \mathfrak{t}^*$. There is a unique index set $J \subset \{1, \ldots, N\}$ such that $\Delta_J \neq \emptyset$ and $\overline{v}(I) \in \overline{C}_J$. Hence $\overline{v}(I)$ is a positive linear combination of the vectors e_ν, $\nu \in J$. The coefficients are integers, because $\overline{v}(I)$ is integral, and they are uniquely determined by $\overline{v}(I)$ by Exercise 11.3.8. It is convenient to denote the coefficients by $-d_\nu$ for $\nu \in J$ so that $d_\nu < 0$ and $\overline{v}(I) = \sum_{\nu \in J} |d_\nu| \bar{e}_\nu$. It then follows from Exercise 11.3.9 below that $I \cap J = \emptyset$. Define $d_\nu := 1$ for $\nu \in I$ and $d_\nu := 0$ for $\nu \notin I \cup J$. Then the vector $d = d(I) \in \mathbb{Z}^N$ with coordinates d_ν that satisfies the following conditions and is uniquely determined by them:

$$\sum_\nu d_\nu \bar{e}_\nu = 0, \quad d_\nu \begin{cases} = 1, & \text{for } \nu \in I \\ \leq 0, & \text{for } \nu \notin I \end{cases}, \quad \Delta_{\{d_\nu < 0\}} \neq \emptyset \tag{11.3.15}$$

The condition $\sum_\nu d_\nu \bar{e}_\nu = 0$ asserts that $d(I)$ is an element of the subgroup $\mathcal{D}$ defined in equation (11.3.13). Denote by $\lambda(I)$ the corresponding element in Λ via the isomorphism (11.3.14) so that $\langle \mathrm{w}_\nu, \lambda(I) \rangle = d_\nu$. Then the relation (11.3.10) corresponding to the vector $\lambda(I)$ takes the form

$$\prod_{\nu \in I} u_\nu = q^\lambda \prod_{\nu \notin I} u_\nu^{-d_\nu}, \qquad \sum_{i=1}^k \lambda_i y_i = d(I). \tag{11.3.16}$$

By the uniqueness of $d(I)$ when I is primitive, there is precisely one relation of this kind with left hand side equal to $\prod_{\nu \in I} u_\nu$. These relations for primitive I generate the deformation $SR_Q(\Delta) \subset \mathbb{R}[u_1, \ldots, u_N, q_1, \ldots, q_k]$ of the Stanley–Reisner ideal, and Theorem 11.3.4 asserts that

$$\mathrm{QH}^*(\overline{M}; \mathbb{R}[q_1, \ldots, q_k]) \cong \frac{\mathbb{R}[u_1, \ldots, u_N, q_1, \ldots, q_k]}{P(\Delta) + SR_Q(\Delta)} \tag{11.3.17}$$

whenever $\overline{M}$ is Fano. A geometric interpretation of these relations is given at the end of Section 11.4.

Exercise 11.3.9. Suppose that I is primitive and that $\sum_{\nu \in I} \bar{e}_\nu = \sum_{\nu \in J} a_\nu \bar{e}_\nu$, where $a_\nu > 0$ for all $\nu \in J$. Show that $I \cap J = \emptyset$. *Hint:* If $I \cap J \neq \emptyset$ show that there is a relation of this form in which I, J are replaced by $I' \subsetneq I$ and $J' \subset J$. Then $\Delta_{I'} \neq \emptyset$, so that $C_{I'}$ and $C_{J'}$ are disjoint.

Remark 11.3.10. Since $\tau = -\sum_\nu \kappa_\nu \mathrm{w}_\nu$ and $c := c_1 = \sum \mathrm{w}_\nu$, the first Chern number and area of the homology class $A_d \in H_2(M)$ associated to d are given by

$$c_1(A_d) = \sum_{\nu=1}^N d_\nu, \qquad \overline{\omega}(A_d) = -\sum_{\nu=1}^N \kappa_\nu d_\nu. \tag{11.3.18}$$

More generally, one can show that the kth Chern class is given by the kth elementary symmetric polynomial in the u_i. Thus its Poincaré dual is represented by the inverse image under the moment map of the sum of the codimension k faces of Δ; cf. Davis–Januskiewicz [**78**]. The quotient $\overline{M} = M_\tau$ is monotone if and only if there is $\alpha > 0$ such that $\tau = \alpha c$ on all classes A_d, which happens if and only if the relation $\sum_\nu(1+\alpha\kappa_\nu)d_\nu = 0$ is a consequence of the fact that $\sum_\nu d_\nu \overline{e}_\nu = 0$. Hence the quotient is monotone precisely when there is an element $\eta \in \overline{\mathfrak{t}}^*$ such that $\langle \eta, \overline{e}_\nu \rangle = 1 + \alpha\kappa_\nu$ for every ν. In particular, if $\kappa_\nu = -1$ for all ν and $\alpha = 1$ we may take $\eta = 0$. Thus we have expressed monotonicity entirely in terms of the moment polytope $\overline{\Delta}$ (i.e. the vectors $\overline{e}_\nu$ and the constants κ_ν). The Fano condition can also be expressed in terms of $\overline{\Delta}$. Namely, the induced complex structure on M_τ is $\overline{T}$-invariant and the effective cone

$$\Lambda^{\mathrm{eff}}(\tau) \cong K^{\mathrm{eff}}(M_\tau, J)$$

is generated by the spheres represented by the edges of Δ (see Remark 11.1.1). Hence M_τ is Fano when c_1 is positive on all these spheres. Therefore to see that a given toric manifold M_τ can be deformed into a monotone manifold it suffices to check that there is a 1-parameter family of constants $\kappa_\nu^t, t \in [0,1]$, joining $\kappa_\nu^0 = \kappa_\nu$ to $\kappa_\nu^1 = -1$ such that $\tau^t := -\sum \kappa_\nu^t \mathrm{w}_\nu$ is a regular value for the moment map μ for all t. If M_τ is Fano, then one can choose the κ_ν^t so that $\tau^t(A_{d(I)}) > 0$ for all primitive I, and it suffices to check that this condition implies that τ^t is regular. Intuitively, this is so because the condition means that all the edges of the moment polytope Δ have positive length throughout the homotopy.

EXERCISE 11.3.11. In the situation of Example 11.3.7, check that the quantum cohomology of $\mathbb{C}P^2$ is isomorphic to

$$\mathrm{QH}^*(\mathbb{C}P^2; \mathbb{R}[q]) \cong \frac{\mathbb{R}[u_1, u_2, u_3, q]}{\langle u_1 = u_2 = u_3, u_1u_2u_3 = q\rangle}.$$

Moreover, its first Chern class is $u_1 + u_2 + u_3$.

EXAMPLE 11.3.12. The nontrivial 2-sphere bundle $\overline{M} \to S^2$ can be represented in infinitely many ways as a symplectic quotient M_τ of $\mathbb{C}^4$ by a $\mathbb{T}^2$-action. Its standard toric structure is monotone, and identifies it with the one point blow up of $\mathbb{C}P^2$. The corresponding torus action is determined by the vectors

$$\mathrm{w}_1 := (1,0), \qquad \mathrm{w}_2 := (1,1), \qquad \mathrm{w}_3 := \mathrm{w}_4 := (0,1)$$

and $\tau \in \mathfrak{t}^* = \mathbb{R}^2$ is given by

$$\tau := \mathrm{w}_1 + \mathrm{w}_2 + \mathrm{w}_3 + \mathrm{w}_4 = (2,3).$$

Thus the chamber of τ is the open cone $C(\tau) = \{(\tau_1, \tau_2) \in \mathbb{R}^2 \,|\, \tau_2 > \tau_1 > 0\}$ and the effective cone in $H_2(M)$ is the set $\Lambda^{\mathrm{eff}}(\tau) = \{(\lambda_1, \lambda_2) \in \mathbb{Z}^2 \,|\, \lambda_2 \geq 0, \lambda_1 + \lambda_2 \geq 0\}$ (see Figure 1). This cone is spanned by the vectors $v := (1,0)$ and $v' := (-1,1)$. The vectors $d, d' \in \mathbb{Z}^4$ with coordinates $d_\nu := \langle \mathrm{w}_\nu, v\rangle$ and $d'_\nu := \langle \mathrm{w}_\nu, v'\rangle$ are

$$d = (1,1,0,0), \qquad d' = (-1,0,1,1).$$

Hence, by Theorem 11.3.1 and Theorem 11.3.4, we have

$$H^*(M_\tau) \cong \frac{\mathbb{R}[u_1, u_2, u_3, u_4]}{\langle u_3 = u_4 = u_2 - u_1,\ u_1u_2 = 0,\ u_3u_4 = 0\rangle},$$

$$\mathrm{QH}^*(M_\tau, \mathbb{R}[q_1, q_2]) \cong \frac{\mathbb{R}[u_1, u_2, u_3, u_4, q_1, q_2]}{\langle u_3 = u_4 = u_2 - u_1,\ u_1u_2 = q_1,\ u_3u_4 = q_2u_1\rangle}.$$

Here we use $q^\lambda := q_1^{\lambda_1+\lambda_2} q_2^{\lambda_2}$ so that q_1 corresponds to e and q_2 to e'.

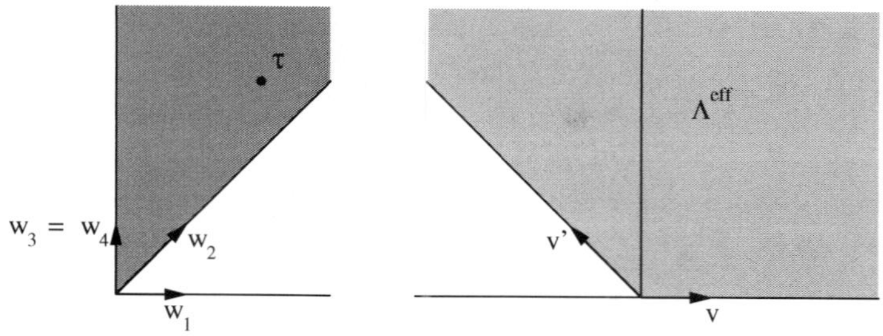

FIGURE 1. A toric variety as a quotient.

Geometrically, the quotient manifold is

$$M_\tau = \frac{\{x \in \mathbb{C}^4 \mid |x_1|^2 + |x_2|^2 = 2,\ |x_2|^2 + |x_3|^2 + |x_4|^2 = 3\}}{[x_1, x_2, x_3, x_4] \equiv [e^{is}x_1, e^{i(s+t)}x_2, e^{it}x_3, e^{it}x_4]}$$

The projection to $S^2 = \mathbb{C}P^1$ is given by $[x_1, x_2, x_3, x_4] \mapsto [x_3 : x_4]$. The cohomology class $\overline{w}_\nu \in H^2(M_\tau)$ is Poincaré dual to the submanifold $\{x_\nu = 0\} \subset M_\tau$. Thus $\overline{w}_3 = \overline{w}_4$ is Poincaré dual to the class $\phi \in H_2(M_\tau)$ of a fiber and $\overline{w}_1$ and $\overline{w}_2$ are Poincaré dual to two section classes $\sigma_1, \sigma_2 \in H_2(M_\tau)$. These classes satisfy

$$\phi = \sigma_2 - \sigma_1, \qquad \sigma_1 \cdot \sigma_1 = -1, \qquad \sigma_2 \cdot \sigma_2 = 1, \qquad \phi \cdot \sigma_1 = \phi \cdot \sigma_2 = 1.$$

The relation $u_1 u_2 = q_1$ implies that the Gromov–Witten invariant of curves in the class of the fiber, passing through the two sections and a generic point, is one. This obviously holds by direct inspection.

The other toric structures on this manifold are nonFano, and, as shown by Spielberg [**384**], Batyrev's formula no longer holds. It is also instructive to compare this approach to calculating the quantum cohomology with that in Ostrover [**314**].

The quantum cohomology ring of a toric manifold can also be described in terms of a Landau–Ginzberg poltential. (We explain this formulation below in the case of Grassmannians.) Even in the monotone case and with appropriate field coefficients, this ring need not be semi-simple: see Ostrover–Tyomkin [**316**]. As shown by Fukaya–Oh–Ohta–Ono [**129, 130**], it is also closely connected with the Lagrangian–Floer homology of the toric fibers of the moment map.

11.3.2. Flag manifolds. In a beautiful early paper [**153**], Givental and Kim computed the quantum cohomology ring of the flag manifold F_{n+1}, and related it to the Toda lattice. To do this they made certain assumptions about the properties of equivariant quantum cohomology that they subsequently established in [**147, 210**]. Another, much more algebraic, proof of this formula was worked out by Ciocan-Fontanine [**70**]. There has been much interest in this example, both because of its very direct connection to integrable systems and because of its relation to mirror symmetry. The ideas have been generalized to partial flag manifolds; a good reference is Buch [**49**]. The extreme case of Grassmannians $G(k, n)$ will be discussed in the next section.

The flag manifold F_{n+1} is the space of all sequences of subspaces

$$E_1 \subset E_2 \subset \cdots \subset E_n$$

of $\mathbb{C}^{n+1}$ with $\dim_{\mathbb{C}} E_j = j$. This manifold has real dimension $n(n+1)$. It is simply connected and carries a natural complex structure. Its topology can be understood from the fact that the forgetful map that takes the flag $E_1 \subset \cdots \subset E_n$ to E_1 is a fibration over $\mathbb{C}P^n$ with fiber F_n. For example, the 6-manifold F_3 embeds into $\mathbb{C}P^2 \times \mathbb{C}P^2$ as the set of pairs (ℓ, P) that satisfy the incidence relation $\ell \subset P$, where the second copy of $\mathbb{C}P^2$ is thought of as the space of complex 2-planes $P \subset \mathbb{C}^3$. Thus

$$F_3 = \big\{\big([z_0 : z_1 : z_2], [w_0 : w_1 : w_2]\big) \mid \sum z_i w_i = 0\big\}.$$

REMARK 11.3.13. Although F_{n+1} supports an effective Hamiltonian action of the n-torus, it is not toric even in the case $n = 2$ because $2n < n(n+1) = \dim F_{n+1}$. On the other hand it can be realized as the homogeneous space $\mathrm{U}(n+1)/\mathbb{T}^{n+1}$, and can also be represented as a symplectic quotient of $\mathbb{C}^N$ for $N := \sum_{k=1}^n k(k+1)$ by a Hamiltonian action of $\mathrm{G} := \mathrm{U}(1) \times \cdots \times \mathrm{U}(n)$. This was explained to us by Shaun Martin.

The cohomology of F_{n+1} is generated by the first Chern classes $u_j \in H^2(M;\mathbb{Z})$ of the canonical line bundles

$$L_j := E_{j+1}/E_j \to F_{n+1}, \qquad j = 0, \ldots, n.$$

Since the Whitney sum $L_0 \oplus \cdots \oplus L_n$ is isomorphic to the trivial bundle $F_{n+1} \times \mathbb{C}^{n+1}$, these classes are related by

$$u_0 + \cdots + u_n = 0.$$

The full cohomology ring of F_{n+1} is the quotient

$$H^*(F_{n+1}) = \frac{\mathbb{Z}[u_0, \ldots, u_n]}{\langle \sigma_1(u), \ldots, \sigma_{n+1}(u)\rangle}$$

where the $\sigma_j(u)$ denote the elementary symmetric functions. To understand this, observe first that the cohomology classes $c_j = \sigma_j(u) \in H^{2j}(F_{n+1})$ are the Chern classes of the trivial bundle $L_0 \oplus \cdots \oplus L_n$ and so must obviously be zero. The above formula asserts that these obvious relations are the only ones and that all the cohomology of F_{n+1} is generated by the classes u_j via the cup product. One way to prove this is to consider the fibration sequence

$$\mathbb{T} \to \mathrm{U} \to \mathrm{U}/\mathbb{T} \to B\mathbb{T} \to B\mathrm{U},$$

where $\mathrm{U} := \mathrm{U}(n+1)$, $\mathbb{T} := \mathbb{T}^{n+1}$ and the quotient $\mathrm{U}/\mathbb{T}$ is identified with F_{n+1}. The induced map $F_{n+1} := \mathrm{U}/\mathbb{T} \to B\mathbb{T}$ classifies $L_0 \oplus \cdots \oplus L_n$ as a $\mathbb{T}$-bundle. An easy spectral sequence argument shows that the relations in $H^*(\mathrm{U}/\mathbb{T})$ are given by the image of $H^*(B\mathrm{U})$ in $H^*(B\mathbb{T})$, that is by the Chern classes when written as functions of the variables u_i.

The cohomology ring of F_{n+1} can also be expressed in terms of the basis

$$p_j = u_j + \cdots + u_n, \qquad j = 1, \ldots, n.$$

In this basis the first Chern class of the tangent bundle TF_{n+1} is given by

$$c_1 = 2(p_1 + \cdots + p_n).$$

The classes p_j can be represented by 2-forms which are Kähler with respect to the obvious complex structure J on F_{n+1}. In fact a cohomology class $a = \sum_j \lambda_j p_j$ can be represented by a Kähler form precisely if the coefficients λ_j are all nonnegative and their sum is positive. In particular, there exists a Kähler form with respect to which the manifold F_{n+1} is monotone. It also follows that $p_j(A) > 0$ for every j and every homology class $A \in K^{\mathrm{eff}}(F_{n+1}) \subset H_2(F_{n+1})$ in the (symplectic) effective cone since J is tamed by the Kähler form in class p_j.

Let $A_1, \dots, A_n$ denote the integer basis of $H_2(F_{n+1})$ that is dual to the basis $p_1, \dots, p_n$ of $H^2(F_{n+1})$ in the sense that $p_j(A_i) = \delta_{ij}$. We work with complex coefficients and choose $\Lambda := \mathbb{C}[q_1, \dots, q_n]$ as our quantum coefficient ring. The homomorphism $\Gamma(F_{n+1}) \to \Lambda(F_{n+1})$ of Definition 11.1.3 is given by

$$A \mapsto e^A := q^d := q_1^{d_1} \cdots q_n^{d_n},$$

where $d := (d_1, \dots, d_n)$ and $d_j = p_j(A) \in \mathbb{Z}$. Since $c_1 = 2(p_1 + \cdots + p_n)$, each q_i has degree $2c_1(A_i) = 4$. We abbreviate

$$\mathrm{QH}^*(F_{n+1}) := \mathrm{QH}^*(F_{n+1}; \mathbb{C}[q_1, \dots, q_n]) = H^*(F_{n+1}) \otimes_{\mathbb{Z}} \mathbb{C}[q_1, \dots, q_n].$$

Since the u_i are multiplicative generators of $H^*(F_{n+1})$, it follows by general principles (as in the proof of Theorem 11.3.14 below) that the quantum cohomology ring is isomorphic to some quotient of the polynomial ring $\mathbb{C}[u_0, \dots, u_n, q_1, \dots, q_n]$.

Givental and Kim proved in [**153**] that there is a natural isomorphism

$$\mathrm{QH}^*(F_{n+1}) \cong \frac{\mathbb{C}[u_0, \dots, u_n, q_1, \dots, q_n]}{\mathcal{I}},$$

where $\mathcal{I} \subset \mathbb{C}[u_0, \dots, u_n, q_1, \dots, q_n]$ denotes the ideal generated by the coefficients of the characteristic polynomial of the matrix

$$A_n = \begin{pmatrix} u_0 & q_1 & 0 & 0 & \cdots & \cdots & 0 \\ -1 & u_1 & q_2 & 0 & & & \vdots \\ 0 & -1 & u_2 & q_3 & \ddots & & \vdots \\ 0 & 0 & -1 & \ddots & \ddots & \ddots & \vdots \\ \vdots & & \ddots & \ddots & \ddots & \ddots & 0 \\ \vdots & & & \ddots & \ddots & u_{n-1} & q_n \\ 0 & \cdots & \cdots & \cdots & 0 & -1 & u_n \end{pmatrix}.$$

More explicitly consider the polynomials

$$c_j = \Sigma_j(u, q)$$

determined by the formula

$$\det(A_n + \lambda \mathbb{1}) = \lambda^{n+1} + c_1 \lambda^n + \cdots + c_n \lambda + c_{n+1}.$$

The ideal $\mathcal{I}$ is generated by these functions Σ_j. In the case $q = 0$ these are the elementary symmetric functions $\sigma_j(u) = \Sigma_j(u, 0)$ and thus the ordinary cohomology ring appears as expected when we specialize to $q = 0$. In other words the classical Chern classes are given by the elementary symmetric functions and the polynomials Σ_j can be regarded as the **quantum deformations of the Chern classes**.

This formula has some very interesting connections with integrable Hamiltonian systems which we now explain. We will see that the quantum Chern classes Σ_j are the Poisson commuting integrals of the Toda lattice (cf. [**292**]).

The Toda lattice is a Hamiltonian differential equation for $n+1$ unit masses on the real line in the positions $x_0, x_1, \dots, x_n$. If

$$u_j = p_j - p_{j+1} = y_j = \dot{x}_j$$

is the momentum of the particle at position x_j and if

$$q_j = e^{x_j - x_{j-1}},$$

then the Hamiltonian function of the Toda lattice is given by

$$H(x,y) = \tfrac{1}{2}\mathrm{trace}(A_n^2) = \tfrac{1}{2}\sum_{j=0}^{n} y_j^2 - \sum_{j=1}^{n} e^{x_j - x_{j-1}}.$$

(The reversal of the usual signs corresponds to considering forces which repel when $x_0 < x_1 < \cdots < x_n$.) In [**292**] Moser discovered that the functions $F_j(x,y) = \mathrm{trace}(A_n^j)$ for $j = 1,\ldots,n+1$ form a complete set of Poisson commuting integrals for this system. As a result the ideal $\mathcal{I}$ generated by the quantum Chern classes Σ_j is invariant under Poisson brackets.

To state this more precisely, observe that the standard symplectic structure $\omega = \sum_{j=0}^n dx_j \wedge dy_j$, when restricted to the set $x_0 + \cdots + x_n = 0$ (zero center of mass) and $y_0 + \cdots + y_n = 0$ (zero momentum) and written in terms of the variables $q_1,\ldots,q_n,p_1,\ldots,p_n$, takes the form

$$\omega = \frac{dq_1}{q_1} \wedge dp_1 + \cdots + \frac{dq_n}{q_n} \wedge dp_n.$$

The Poisson structure is to be understood with respect to this symplectic form.

Geometrically, the symplectic form ω can be interpreted as follows. Consider the complex torus

$$\mathbb{T}_{\mathbb{C}} := H^2(F_{n+1};\mathbb{C}/2\pi i\mathbb{Z}) = \frac{H^2(F_{n+1};\mathbb{C})}{H^2(F_{n+1};2\pi i\mathbb{Z})}$$

which is parametrized by the coordinate functions

$$q_j(a) = e^{\langle a, A_j\rangle}$$

with values in $\mathbb{C}^* = \mathbb{C}\setminus\{0\}$. The cotangent bundle of $\mathbb{T}_{\mathbb{C}}$ can be naturally identified with

$$T^*\mathbb{T}_{\mathbb{C}} = H^2(F_{n+1},\mathbb{C}/2\pi i\mathbb{Z}) \times H_2(F_{n+1},\mathbb{C})$$

where the $p_j : H_2(M,\mathbb{C}) \to \mathbb{C}$ are to be understood as coordinate functions on the cotangent space. These coordinate functions have a geometric meaning in terms of the cone of Kähler forms as explained above. The form ω can be understood as a complex symplectic structure on the cotangent bundle $T^*\mathbb{T}_{\mathbb{C}}$. In this interpretation the above Hamiltonian system with particles $x_0,\ldots,x_n$, when restricted to the set of zero center of mass and zero momentum, lives on the real part of $T^*\mathbb{T}_{\mathbb{C}}$.

This leads to a geometric interpretation of the quantum cohomology itself. The ideal

$$\mathcal{J} \subset \mathbb{C}[p_1,\ldots,p_n,q_1,\ldots,q_n]$$

which corresponds to $\mathcal{I}$ under the identification $u_j := p_j - p_{j+1}$ (with $p_0 = p_{n+1} = 0$) determines an algebraic variety $\mathcal{L} \subset T^*\mathbb{T}_{\mathbb{C}}$. This variety is defined as the common zero set of the polynomials in $\mathcal{J}$. Hence the quantum cohomology ring

$$\mathrm{QH}^*(F_{n+1}) \cong \frac{\mathbb{C}[p_1,\ldots,p_n,q_1,\ldots,q_n]}{\mathcal{J}}$$

with complex coefficients can be interpreted as the space of functions on $\mathcal{L}$. The invariance of the ideal $\mathcal{J}$ under Poisson brackets translates into the condition that the variety $\mathcal{L}$ is Lagrangian. This is the first hint of the structures arising in quantum cohomology (the so called "A-side" of mirror symmetry) that tie it to the "B-side", and was one of the ingredients of Givental's proof [**148**] of the mirror conjecture for flag manifolds.

11.3.3. Grassmannians. The quantum cohomology ring of the Grassmannian $\mathrm{G}(k,n)$ has been studied by Bertram, Daskalopoulos, and Wentworth in [**36**], Witten in [**422**], and Siebert and Tian in [**378**]. In [**422**] Witten also relates the quantum cohomology of $\mathrm{G}(k,n)$ to the Verlinde algebra of representations of $\mathrm{U}(k)$. His explanation for this relation is based on a beautiful, but heuristic, consideration involving path integrals.

Denote by $\mathrm{G}(k,n)$ the Grassmannian of k-planes in $\mathbb{C}^n$. Thus a point in $\mathrm{G}(k,n)$ is a k-dimensional subspace $V \subset \mathbb{C}^n$. A **unitary frame** of V is a matrix $B \in \mathbb{C}^{n\times k}$ such that

$$V = \operatorname{im} B, \qquad B^*B = \mathbb{1}_{k\times k}.$$

Two such frames B and B' represent the same subspace V if there exists a unitary matrix $U \in \mathrm{U}(k)$ such that $B' = BU$. Hence the Grassmannian can be identified with the quotient space

$$\mathrm{G}(k,n) = \mathcal{F}(k,n)/\mathrm{U}(k)$$

where $\mathcal{F}(k,n) \subset \mathbb{C}^{n\times k}$ denotes the set of unitary k-frames. A moment's thought shows that the Grassmannian has real dimension

$$\dim \mathrm{G}(k,n) = 2k(n-k).$$

It can also be interpreted as a symplectic quotient or reduced space. The space $\mathbb{C}^{n\times k}$ is a symplectic manifold and the natural action of the group $\mathrm{U}(k)$ on this space is Hamiltonian with moment map $\mu(B) = B^*B/2i$. To understand this, recall that the moment map μ has image in the dual $\mathfrak{u}(k)^*$ of the Lie algebra of $\mathrm{U}(k)$, and that we can identify $\mathfrak{u}(k)^*$ with $\mathfrak{u}(k)$, the space of all $k\times k$ skew Hermitian matrices via the inner product $\langle \xi,\eta\rangle := \operatorname{trace}(\xi^*\eta)$. Thus the Grassmannian is the quotient $\mathrm{G}(k,n) = \mu^{-1}(\mathbb{1}/2i)/\mathrm{U}(k)$.

There are two natural complex vector bundles, $E \to \mathrm{G}(k,n)$ of rank k and $F \to \mathrm{G}(k,n)$ of rank $n-k$, whose Whitney sum is naturally isomorphic to the trivial bundle $\mathrm{G}(k,n)\times\mathbb{C}^n$. The fiber of E at the point $V \in \mathrm{G}(k,n)$ is just the space V itself and the fiber of F is the quotient $\mathbb{C}^n/V$:

$$E_V = V, \qquad F_V = \mathbb{C}^n/V.$$

Denote the Chern classes of the dual bundles E^* and F^* by

$$x_j = c_j(E^*) \in H^{2j}(\mathrm{G}(k,n)), \qquad y_j = c_j(F^*) \in H^{2j}(\mathrm{G}(k,n)).$$

These classes generate the cohomology of $\mathrm{G}(k,n)$. Since $E\oplus F$ is isomorphic to the trivial bundle there are obvious relations

$$\sum_{i=0}^{j} x_i y_{j-i} = 0$$

for $j = 1,\dots,n$. For $j > n$ this equation is trivially satisfied. For $j = 1,\dots,n-k$ it determines the classes y_j inductively as functions of $x_1,\dots,x_k$ via

$$y_j = -x_1y_{j-1} - \cdots - x_{j-1}y_1 - x_j, \qquad j = 1,\dots,n-k.$$

For $j > n-k$ the classes y_j vanish and this determines relations of the x_j. These are the only relations and hence the cohomology ring of the Grassmannian can be identified with the quotient

$$H^*(\mathrm{G}(k,n),\mathbb{C}) \cong \frac{\mathbb{C}[x_1,\dots,x_k]}{\langle y_{n-k+1},\dots,y_n\rangle}.$$

Moreover, the first Chern class of the tangent bundle is given by $c_1(T\mathrm{G}(k,n)) = nx_1$ and so the minimal Chern number is n. The following theorem was essentially proved by Witten [**422**]. Independently, a rigorous proof with all details was worked out by Siebert and Tian [**378**].

THEOREM 11.3.14 (Siebert–Tian,Witten). *The quantum cohomology ring of the Grassmannian with coefficients in the polynomial ring* $\mathbb{C}[q]$ *is isomorphic to*

$$\mathrm{QH}^*(\mathrm{G}(k,n);\mathbb{C}[q]) \cong \frac{\mathbb{C}[x_1,\dots,x_k,q]}{\langle y_{n-k+1},\dots,y_{n-1},y_n+(-1)^{n-k}q\rangle}.$$

Here x_j *is a generator of degree* $2j$ *and* q *is a generator of degree* $2n$. *The relation* $y_n+(-1)^{n-k}q=0$ *can also be written in the form*

$$x_k y_{n-k} = (-1)^{n-k}q.$$

PROOF. We remark first that, by an easy induction argument, the classes $x_1,\dots,x_k$ still generate the quantum cohomology of $\mathrm{G}(k,n)$. This means that every cohomology class can be expressed as a linear combination of quantum products of the x_i. To prove this one uses induction over the degree and the fact that the difference $x*y-x\smile y$ is a sum of terms of lower degree than $x\smile y$. (See Lemma 2.1 in [**378**] for details.) Now this same argument shows that the original relations in the classical cohomology ring become relations in quantum cohomology by adding certain lower order terms. It follows again by induction over the degree that these new relations generate the ideal of relations in quantum cohomology. (See Theorem 2.2 in [**378**] for details.)

In view of these general remarks we must compute the quantum deformations of the defining relations in the cohomology ring. In this we follow Witten's argument in [**422**]. The quantum cup product of the classes x_i and y_j is a power series of the form

$$x_i*y_j = \sum_d (x_i*y_j)_d q^d$$

where $(x_i*y_j)_d\in H^{2i+2j-2nd}(\mathrm{G}(k,n))$ and $(x_j*y_j)_0 = x_i\smile y_j$. It follows that the quantum cup product x_i*y_j must agree with the ordinary cup product $x_i\smile y_j$ unless $i=k$ and $j=n-k$. Now the relations $y_j=0$ for $j=n-k+1,\dots,n-1$ only involve the products x_iy_j with either $i<k$ or $j<n-k$ and hence they remain valid in quantum cohomology. However the relation $y_n=0$ involves the product x_ky_{n-k} and the only nontrivial contribution to the quantum deformation of this product is the term $(x_k*y_{n-k})_1\in H^0(\mathrm{G}(k,n))$. We claim that

$$(x_k*y_{n-k})_1 = (-1)^{n-k}. \tag{11.3.19}$$

To see this we must examine the moduli space $\mathcal{M}(L;J_0)$ of holomorphic curves $u:\mathbb{C}P^1\to\mathrm{G}(k,n)$ of degree $\deg(u)=c_1(L)=1$, where J_0 denotes the standard complex structure on the Grassmannian. The space of such curves has formal dimension

$$\dim\mathcal{M}(L;J_0) = 2k(n-k)+2n.$$

Every holomorphic curve $u\in\mathcal{M}(L;J_0)$ is of the form

$$u([z_0:z_1]) = \mathrm{span}\,\{z_0v_0+z_1v_1,v_2,\dots,v_k\} \tag{11.3.20}$$

where the vectors $v_0,\dots,v_k\in\mathbb{C}^n$ are linearly independent. Note that these maps form indeed a space of dimension $2n(k+1)-2k^2$ which is in accordance with the dimension formula for $\mathcal{M}(L;J_0)$. Moreover, by Proposition 7.4.3, $J_0\in\mathcal{J}_{\mathrm{reg}}(\mathrm{G}(k,n))$

and so all the curves in $\mathcal{M}(L; J_0)$ are regular. We must prove that

$$\mathrm{GW}^M_{L,3}(x_k, y_{n-k}, a) = (-1)^{n-k}$$

where $a = \mathrm{PD}(pt)$ generates $H^{2k(n-k)}(M)$. Now the Poincaré dual of the top dimensional Chern class of a vector bundle can be represented by the zero set of a generic section. For example fix a vector $w \in \mathbb{C}^n$ and consider the section $\mathrm{G}(k,n) \to E^*$ which assigns to every k-plane $V \subset \mathbb{C}^n$ the linear functional $V \to \mathbb{C} : v \mapsto \langle w, v \rangle$. This section is transverse to the zero section and its zero set is the submanifold

$$X = \{V \in \mathrm{G}(k,n) \,|\, w \perp V\}.$$

This submanifold is a copy of $\mathrm{G}(k, n-1)$ in $\mathrm{G}(k,n)$ and represents the class $\mathrm{PD}(x_k) \in H_{2k}(\mathrm{G}(k,n))$. Now fix a vector $v_0 \in \mathbb{C}^n$ and consider the section $\mathrm{G}(k,n) \to F$ which assigns to every $V \in \mathrm{G}(k,n)$ the equivalence class $[v_0] \in \mathbb{C}^n/V = F_V$. The zero set of this section is the submanifold

$$Y = \{V \in \mathrm{G}(k,n) \,|\, v_0 \in V\}.$$

This submanifold with its natural orientation represents the Poincaré dual of the top Chern class $c_{n-k}(F) = (-1)^{n-k} c_{n-k}(F^*)$. In summary,

$$[X] = \mathrm{PD}(x_k), \qquad [Y] = (-1)^{n-k}\mathrm{PD}(y_{n-k}).$$

Fix any point $V_0 \in \mathrm{G}(k,n) \setminus X$ such that $v_0 \notin V_0$ and choose a basis $v_1, \dots, v_k$ of V_0 such that the vectors $v_2, \dots, v_k$ and $v_0 + v_1$ are perpendicular to w. Then the curve $u([z_0 : z_1]) = \mathrm{span}\{z_0 v_0 + z_1 v_1, v_2, \dots, v_k\}$ satisfies

$$u([0:1]) = V_0, \qquad u([1:1]) \in X, \qquad u([1:0]) \in Y.$$

Moreover, it is easy to see that the intersection number at u is one and that there is no other curve of degree one whose image intersects X, Y, and V_0. This calculates the homology invariant $\mathrm{GW}^M_{L,3}(X, Y, \mathrm{pt}) = 1$, which proves the formula (11.3.19). It follows that in quantum cohomology the deformed relations are $y_j = 0$ for $n-k+1 \le j \le n-1$ and $y_n + (-1)^{n-k} q = 0$. This proves Theorem 11.3.14. □

Theorem 11.3.14 shows once again how the quantum cohomology reduces to the classical cohomology ring if we consider q to be a complex number rather than a variable and specialize to $q = 0$. More explicitly, if we define the Chern polynomials

$$c_t(E^*) = \sum_{i=1}^{k} x_i t^i, \qquad c_t(F^*) = \sum_{j=1}^{n-k} y_j t^j$$

then the classical cohomology ring is determined by the relation $c_t(E^*) c_t(F^*) = 1$ whereas the quantum cohomology ring is determined by

$$c_t(E^*) c_t(F^*) = 1 + (-1)^{n-k} q t^n, \tag{11.3.21}$$

where $q \in \mathbb{C}$.

Landau–Ginzburg formulation. The relations in classical cohomology can be generated by the derivatives of a single function $W_0 = W_0(x_1, \dots, x_k)$. The same is true for the relations in quantum cohomology and the corresponding function $W = W(x_1, \dots, x_k)$ is called the **Landau–Ginzburg potential**. In our discussion of this approach we follow closely the exposition of Witten in [**422**].

Consider the Chern polynomial

$$c_t(E^*) := \sum_{i=1}^{k} x_i t^i$$

where $x_i = c_i(E^*) \in H^{2i}(\mathrm{G}(k,n))$. Define polynomials $y_j = y_j(x_1,\ldots,x_k)$ for $j \geq 0$ by the formula

$$\frac{1}{c_t(E^*)} =: \sum_{j\geq 0} y_j t^j.$$

Then the classical cohomology ring of the Grassmannian is described by the relations $y_j(x) = 0$ for $n-k+1 \leq j \leq n$. These relations imply $y_j(x) = 0$ for $j > n$ and, of course, for $1 \leq j \leq n-k$ the classes y_j are the Chern classes of F^*. In the following we shall not impose the relations $y_j(x) = 0$. Hence the power series $\sum_j y_j t^j$ may no longer be a polynomial.

Consider the holomorphic functions $U_r = U_r(x_1,\ldots,x_k)$ defined by the equation

$$-\log c_t(E^*) =: \sum_{r\geq 0} U_r(x) t^r.$$

Differentiating this expression with respect to x_j we see that

$$-\frac{t^j}{c_t(E^*)} = \sum_{r\geq 0} \frac{\partial U_r}{\partial x_j} t^r.$$

Comparing coefficients we find $\partial U_r/\partial x_j = -y_{r-j}$ for $1 \leq j \leq k$. In particular, when $r-j$ is negative this formula implies the vanishing of the corresponding derivative of U_r. The most interesting case is $r = n+1$. With

$$W_0 := (-1)^{n+1} U_{n+1}$$

we obtain

$$\frac{\partial W_0}{\partial x_j} = (-1)^n y_{n+1-j}$$

for $1 \leq j \leq k$. Hence the defining relations for the classical cohomology of the Grassmannian can be written in the form $dW_0 = 0$.

Now consider the function

$$W := W_0 + (-1)^k q x_1$$

where q is a fixed complex number. The condition $dW = 0$ is equivalent to

$$y_{n-k+1} = 0, \quad \ldots \quad, y_{n-1} = 0, \quad y_n + (-1)^{n-k} q = 0,$$

which are precisely the defining relations for the quantum cohomology ring of $\mathrm{G}(k,n)$. Note that the higher coefficients of the power series $\sum_j y_j t^j$ will no longer vanish. Comparing the coefficients up to order n in the equation $c_t(E^*) \cdot (\sum_{j\geq 0} y_j t^j) = 1$ we obtain

$$\left(\sum_{i=1}^{k} x_i t^i\right) \cdot \left(\sum_{j=1}^{n-k} y_j t^j\right) = 1 + (-1)^{n-k} q t^n$$

and this agrees with (11.3.21). The function W is called the **Landau–Ginzburg potential**. It can be conveniently expressed in terms of the roots $\lambda_1, \dots, \lambda_k$ of the Chern polynomial

$$c_t(E^*) = \sum_{i=1}^{k} x_i t^i = \prod_{i=1}^{k}(1 + \lambda_i t), \tag{11.3.22}$$

namely

$$W(\lambda_1, \dots, \lambda_k) = \sum_{i=1}^{k} \left(\frac{\lambda_i{}^{n+1}}{n+1} + (-1)^k q \lambda_i \right). \tag{11.3.23}$$

Geometrically, the quantum cohomology ring of $\mathrm{G}(k, n)$ can be interpreted as the ring of polynomials in the variables $x_1, \dots, x_k$ restricted to the zero set of dW. Now the function $W : \mathbb{C}^k \to \mathbb{C}$ has only finitely many critical points, and the equivalence class of any polynomial $f \in \mathbb{C}[x_1, \dots, x_k]$ with respect to the ideal

$$\mathcal{J}_q := \langle y_{n-k+1}, \dots, y_{n-1}, y_n + (-1)^{n-k} q \rangle = \left\langle \frac{\partial W}{\partial x_1}, \dots, \frac{\partial W}{\partial x_k} \right\rangle$$

is determined by the values of f at the critical points of W. This gives rise to localization formulas such as

$$I(f) = \frac{(-1)^{k(k-1)/2}}{k!} \sum_{dW(x)=0} \det \left(\frac{\partial^2 W}{\partial x_i \partial x_j} \right)^{-1} f(x) \tag{11.3.24}$$

for every polynomial f of degree $2k(n-k)$. Here the variable x_i is understood to be of degree $2i$ and the functional $I(f)$ denotes the integral of the differential form $\omega_f \in \Omega^{2k(n-k)}(\mathrm{G}(k, n))$ associated to f under the isomorphism

$$H^*(\mathrm{G}(k, n), \mathbb{C}) = \frac{\mathbb{C}[x_1, \dots, x_k]}{\langle dW_0 \rangle}.$$

Note that the class $c_f \in \mathrm{QH}^*(G(k, n))$ that is represented by f is the sum of $[\omega_f]$ with terms involving q.

EXERCISE 11.3.15. Prove that

$$\int_{\mathrm{G}(k,n)} c_k(E^*)^{n-k} = 1$$

by considering intersection points of $n - k$ copies of $\mathrm{G}(k, n - 1)$ in $\mathrm{G}(k, n)$. Now check the formula (11.3.24) by applying it to the polynomial

$$f(x) = x_k{}^{n-k} = \prod_{i=1}^{k} \lambda_i{}^{n-k}$$

which represents the class $c_k(E^*)^{n-k}$. The general case follows from this because $H^{2k(n-k)}(\mathrm{G}(k, n))$ is one dimensional. *Hint:* First use change of variables and the residue calculus to express the sum (11.3.24) as a contour integral of the form

$$I(f) = \frac{(-1)^{k(k-1)/2}}{k!(2\pi i)^k} \int_{|\lambda_j|=R} \frac{f \cdot \prod_{i<j}(\lambda_i - \lambda_j)^2}{\prod_i \partial W / \partial \lambda_i} d\lambda_1 \dots d\lambda_k$$

where R is large. Then note that the term $\partial W/\partial \lambda_i = \lambda_i{}^n + (-1)^k q$ in the denominator can be replaced by $\lambda_i{}^n - \lambda_i{}^{n-k}$ without changing the value of the integral. Now use the residue calculus again to evaluate the new integral. For details see [**422**] and [**378**].

Relation with the Verlinde algebra. There is a remarkable correspondence between the cohomology ring of the Grassmannian $\mathrm{G}(k,n)$ and a suitable quotient of the representation ring of the unitary group $\mathrm{U}(k)$ which we now explain. As a vector space (over the reals) the cohomology of $\mathrm{G}(k,n)$ is generated by the Schubert cycles. A **Schubert cycle** in $\mathrm{G}(k,n)$ is a submanifold associated to a complete flag $\{0\} = \Lambda_0 \subset \Lambda_1 \subset \cdots \subset \Lambda_{n-1} \subset \Lambda_n = \mathbb{C}^n$ with $\dim_{\mathbb{C}} \Lambda_\nu = \nu$ and a finite sequence of nonnegative integers (called a **partition**)

$$n - k \ge a_1 \ge a_2 \ge \cdots \ge a_k \ge 0.$$

The corresponding submanifold

$$\Sigma_a := \Sigma_a(\Lambda) = \Sigma^{k,n}_{a_1,\ldots,a_k}(\Lambda_0,\ldots,\Lambda_n) \subset \mathrm{G}(k,n)$$

consists of all k-dimensional subspaces $V \subset \mathbb{C}^n$ that satisfy the condition

$$\dim(V \cap \Lambda_{n-k+i-a_i}) = i, \qquad \dim(V \cap \Lambda_{n-k+i-a_i-1}) = i-1,$$

for $i = 1,\ldots,k$. In other words, the sequence of dimensions $\dim(V \cap \Lambda_\nu)$ jumps up by one at the points $\nu = n-k+i-a_i$. Each Schubert cycle $\Sigma_a(E)$ is a smooth submanifold of $\mathrm{G}(k,n)$ of complex codimension

$$\mathrm{codim}_{\mathbb{C}}\Sigma_a = a_1 + \cdots + a_k =: |a|.$$

Its closure is given by

$$\overline{\Sigma}_a = \bigcup_{b \preceq a} \Sigma_b,$$

where $b \preceq a$ if and only if $b_i \le a_i$ for all i. Hence Σ_a is a pseudocycle in the sense of Definition 6.5.1 and so represents a homology class $\sigma_a := [\Sigma_a] \in H_{2k(n-k)-2|a|}(\mathrm{G}(k,n))$ via intersection numbers (see Lemma 6.5.7). These homology classes generate $H_*(\mathrm{G}(k,n))$ additively. Thus

$$\dim H_*(\mathrm{G}(k,n)) = \binom{n}{k}.$$

Moreover, for two generic flags Λ and Λ', the Schubert cycles $\Sigma_a(\Lambda)$ and $\Sigma_b(\Lambda')$ intersect transversally and, if they have complementary dimensions, then

$$g_{ab} := \sigma_a \cdot \sigma_b = \begin{cases} 1, & \text{if } a_{k+1-i} + b_i = n-k \text{ for all } i, \\ 0, & \text{otherwise.} \end{cases}$$

Proofs of all these assertions can be found in Griffiths–Harris [**159**] and Milnor–Stasheff [**291**].

REMARK 11.3.16. In terms of the Schubert cycles, the cohomology classes x_i and y_j are given by

$$x_i = c_i(E^*) = \mathrm{PD}(\xi_i), \qquad y_j = c_j(F^*) = (-1)^j \mathrm{PD}(\eta_j),$$

where

$$\xi_i := \sigma_{1,\ldots,1,0,\ldots,0}, \qquad \eta_j := \sigma_{j,0,\ldots,0}$$

(with 1 occurring i times in the first case).

REMARK 11.3.17 (Giambelli's formula). Fix integers $n-k \geq a_1 \geq \cdots \geq a_k \geq 0$ and let $k \geq b_1 \geq \cdots \geq b_{n-k} \geq 0$ be defined by $b_j := \#\{i \,|\, a_i \geq j\}$ for $j = 1, \ldots, n$. Then

$$\sigma_a = \begin{vmatrix} \eta_{a_1} & \eta_{a_1+1} & \eta_{a_1+2} & \cdots & \eta_{a_1+k-1} \\ \eta_{a_2-1} & \eta_{a_2} & \eta_{a_2+1} & \cdots & \eta_{a_2+k-2} \\ \eta_{a_3-2} & \eta_{a_3-1} & \eta_{a_3} & \cdots & \eta_{a_3+k-3} \\ \vdots & \vdots & \vdots & \ddots & \vdots \\ \eta_{a_k-k+1} & \eta_{a_k-k+2} & \eta_{a_k-k+3} & \cdots & \eta_{a_k} \end{vmatrix}$$

$$= \begin{vmatrix} \xi_{b_1} & \xi_{b_1+1} & \xi_{b_1+2} & \cdots & \xi_{b_1+n-k-1} \\ \xi_{b_2-1} & \xi_{b_2} & \xi_{b_2+1} & \cdots & \xi_{b_2+n-k-2} \\ \xi_{b_3-2} & \xi_{b_3-1} & \xi_{b_3} & \cdots & \xi_{b_3+n-k-3} \\ \vdots & \vdots & \vdots & \ddots & \vdots \\ \xi_{b_{n-k}-n+k+1} & \xi_{b_{n-k}-n+k+2} & \xi_{b_{n-k}-n+k+3} & \cdots & \xi_{b_{n-k}} \end{vmatrix}.$$

Here multiplication is to be understood as the intersection product.

The representation ring $\mathcal{R}(k)$ of the group $\mathrm{U}(k)$ is defined as the set of equivalence classes $\rho \ominus \rho'$ of pairs of representations as in K-theory. The **character** of a representation $\rho : \mathrm{U}(k) \to \mathrm{Aut}(V)$ is the function $\theta_\rho : \mathbb{T}^k \to \mathbb{C}$ on the maximal torus $\mathbb{T}^k \subset \mathrm{U}(k)$ of diagonal matrices, defined by

$$\theta_\rho(t) := \mathrm{trace}^c(\rho(t)).$$

Since two diagonal matrices with the same entries (in different order) are conjugate, the character is a symmetric function of the coordinates $t_1, \ldots, t_k$ of t. The character gives rise to a ring isomorphism $\mathcal{R}(k) \to \mathcal{S}(k) : \rho \mapsto \theta_\rho$ from the representation ring (with direct sum and tensor product) to the ring $\mathcal{S}(k)$ of symmetric functions in k variables. Every partition $a_1 \geq \cdots \geq a_k \geq 0$ determines an irreducible representation $\rho_a : \mathrm{U}(k) \to \mathrm{Aut}(V_a)$, unique up to isomorphism, whose character is the function $\theta_{\rho_a} = \theta_a$, defined in Remark 11.3.19 below. Explicitly, the representation V_a is a subspace

$$V_a \subset (\Lambda^1\mathbb{C}^k)^{\otimes m_1} \otimes (\Lambda^2\mathbb{C}^k)^{\otimes m_2} \otimes \cdots \otimes (\Lambda^k\mathbb{C}^k)^{\otimes m_k}$$

where $m_i := a_i - a_{i+1}$; it is defined as the smallest $\mathrm{U}(k)$-invariant subspace which contains the ($\mathbb{T}^k$-invariant) tensor product of the subspaces $(\mathbb{C}e_1 \wedge \cdots \wedge \mathbb{C}e_i)^{\otimes m_i}$.

REMARK 11.3.18. Of particular interest are the special representations

$$V_{1,\ldots,1,0,\ldots,0} = \Lambda^i\mathbb{C}^k, \qquad V_{j,0,\ldots,0} = S^j\mathbb{C}^k,$$

with 1 occurring i times in the first case. In particular, $V_{0,\ldots,0} = \mathbb{C}$ is the trivial representation (the multiplicative unit in the representation ring).

REMARK 11.3.19 (The Jacobi–Trudi identity). Define the **elementary symmetric functions** ϕ_i for $1 \leq i \leq k$ and the **complete symmetric functions** ψ_j for $j \geq 1$, in the variables $t_1, \ldots, t_k$, by

$$\phi_i = \sum_{1 \leq \nu_1 < \cdots < \nu_i \leq k} t_{\nu_1} \cdots t_{\nu_i}, \qquad \psi_j = \sum_{1 \leq \nu_1 \leq \cdots \leq \nu_j \leq k} t_{\nu_1} \cdots t_{\nu_j}.$$

Let a_i and b_j be as in Remark 11.3.17 and define the symmetric function θ_a by

$$\theta_a := \frac{\det\left(({t_i}^{a_j+k-j})_{i,j=1}^k\right)}{\det\left(({t_i}^{k-j})_{i,j=1}^k\right)}.$$

Then

$$\theta_a = \begin{vmatrix} \psi_{a_1} & \psi_{a_1+1} & \cdots & \psi_{a_1+k-1} \\ \psi_{a_2-1} & \psi_{a_2} & \cdots & \psi_{a_2+k-2} \\ \vdots & \vdots & \ddots & \vdots \\ \psi_{a_k-k+1} & \psi_{a_k-k+2} & \cdots & \psi_{a_k} \end{vmatrix} = \begin{vmatrix} \phi_{b_1} & \phi_{b_1+1} & \cdots & \phi_{b_1+n-k-1} \\ \phi_{b_2-1} & \phi_{b_2} & \cdots & \phi_{b_2+n-k-2} \\ \vdots & \vdots & \ddots & \vdots \\ \phi_{b_{n-k}-n+k+1} & \phi_{b_{n-k}-n+k+2} & \cdots & \phi_{b_{n-k}} \end{vmatrix}.$$

The ring $\mathcal{R}(k)$ is generated additively by the irreducible representations ρ_a. Consider the finite dimensional quotient $\mathcal{R}(k,n)$ of $\mathcal{R}(k)$ generated by the irreducible representation ρ_a with $n-k \geq a_1 \geq \cdots \geq a_k \geq 0$. By Remark 11.3.17 and Remark 11.3.19, there is a ring isomorphism

$$H^*(\mathrm{G}(k,n)) \to \mathcal{R}(k,n) : \mathrm{PD}(\sigma_a) \mapsto \rho_a. \tag{11.3.25}$$

The inverse isomorphism can be described as follows. The character $\theta_\rho := \mathrm{trace}^c \circ \rho : \mathrm{U}(k) \to \mathbb{C}$ is a symmetric polynomial in the eigenvalues and hence extends uniquely to a polynomial on $\mathbb{C}^{k\times k}$ which is invariant under conjugation. This extension can then be restricted to the Lie algebra $\mathfrak{u}(k)$. Now fix a $\mathrm{U}(k)$-connection A on the tautological bundle $E \to \mathrm{G}(k,n)$ and denote by $F_A \in \Omega^2(\mathrm{G}(k,n), \mathrm{End}(E))$ its curvature. Then $\theta_\rho(F_A/2\pi i)$ is a closed real valued $2|a|$-form on $\mathrm{G}(k,n)$ which represents a characteristic class of the bundle E. It turns out that

$$[\theta_a(F_a/2\pi i)] = \mathrm{PD}(\sigma_a)$$

and hence the inverse isomorphism is induced by the map

$$\mathcal{R}(k) \to H^*(\mathrm{G}(k,n)) : \rho \mapsto [\theta_\rho(F_A/2\pi i)].$$

These observations are explained in Witten's paper [**422**]. (See also Salamon [**354**] for an exposition.) Note that the isomorphism sends x_i to $\Lambda^i\mathbb{C}^k$ and y_j to $S^j\mathbb{C}^k$ (Remarks 11.3.16 and 11.3.18).

Now the **Verlinde algebra** is a deformation of the ring structure on $\mathcal{R}(k,n)$ just as quantum cohomology is a deformation of the cup product on $H^*(\mathrm{G}(k,n))$. The structure constants of the Verlinde algebra are defined as holomorphic Euler characteristics of certain vector bundles over moduli spaces of flat $\mathrm{U}(k)$-connections associated to an integer d and three representations ρ_a, ρ_b, ρ_c satisfying the same dimension condition as the 3-point Gromov–Witten invariants, namely

$$|a| + |b| + |c| = k(n-k) + nd.$$

Witten's conjecture asserts that the isomorphism (11.3.25) intertwines the two deformed product structures. The conjecture extends to the higher genus Gromov–Witten invariants and the corresponding higher genus Verlinde algebra. One motivation behind it is that the Verlinde gluing rules are the same as those for the Gromov–Witten invariants. Moreover, Witten gives a beautiful heuristic argument, based on path integrals, as to why there should be such a correspondence [**422**]. For the quantum cohomology the conjecture has been confirmed by Agnihotri [**9**]; as in the proof of Theorem 11.3.14, the argument reduces to the computation of a single invariant.

REMARK 11.3.20 (Structure constants). Isomorphisms such as that in Theorem 11.3.14 do not give rise to explicit formulas for the structure constants of quantum cohomology

$$N_d(a,b,c) := \mathrm{GW}^M_{dL,3}(a,b,c).$$

This is already apparent in ordinary cohomology. While $H^*(\mathrm{G}(k,n))$, for example, is easily described as the quotient of a polynomial ring, the structure constants

$$N_0(a,b,c) = \sigma_a \cdot \sigma_b \cdot \sigma_c = \frac{1}{\mathrm{Vol}(\mathrm{G})} \int_{\mathrm{G}} \theta_a(g)\theta_b(g)\theta_c(g) \det(g)^{k-n} dg$$

(integration is understood with respect to a Haar measure on $\mathrm{G} = \mathrm{U}(k)$) are determined by the Littlewood–Richardson rules which involve rather complicated operations on Young tableaus. These formulas have yet to be extended to general d.

11.3.4. Calabi–Yau manifolds. There are some interesting phenomena in the structure of the quantum cohomology ring $\mathrm{QH}^*(M)$ for general semipositive symplectic manifolds which do not appear in the monotone case. For example, the quantum coefficient ring Λ in Definition 11.1.3 can, in general, no longer be chosen as a polynomial ring, but instead we have to work with the Novikov ring Λ_ω or with the universal Novikov ring as in Example 11.1.4 (iv) and (v). An important special case is that of symplectic manifolds with vanishing first Chern class. In this case $\Lambda = \Lambda_0$, i.e. all elements in Λ have degree zero, and $\mathrm{QH}^k(M;\Lambda) = H^k(M) \otimes \Lambda$.

We shall restrict further to the 6-dimensional case. In this situation the dimension $\dim M + 2c_1(A) - 6$ of the moduli space $\mathcal{M}(A;J)/\mathrm{G}$ is zero for every class A and hence, for a generic almost complex structure J, every simple J-holomorphic curve is isolated. It follows that the 3-point invariant $\mathrm{GW}^M_{A,3}(a_1,a_2,a_3)$ is nonzero only when $\deg(a_i) = 2$ for all i. Indeed, condition (7.1.2) shows that the degrees must sum to 6. But we must also have $\deg(a_i) \le 2$ because otherwise the Poincaré dual $\mathrm{PD}(a_i)$ is represented by a cycle of dimension less than four which can be chosen to avoid all the nonconstant J-holomorphic curves since these are isolated. Thus the quantum product equals the usual product except on $H^2(M)$. Moreover, it induces a multiplication

$$H^2(M) \otimes H^2(M) \to H^4(M) \otimes \Lambda.$$

Identifying $H^2(M)$ with $H^4(M)$ by Poincaré duality, we obtain a deformed product structure on $\mathrm{QH}^2(M;\Lambda) = H^2(M) \otimes \Lambda$.

Examples of manifolds satisfying the above conditions are the 6-torus $\mathbb{T}^6$ and the product $\mathbb{T}^2 \times X$ where $X \subset \mathbb{C}P^3$ is a quartic hypersurface (the famous $K3$-surface). In both these cases the Gromov–Witten invariants vanish. However, there is a rich class of examples with nontrivial quantum product, that moreover are simply connected and Kähler. Such manifolds are called **Calabi–Yau 3-folds**. By a theorem of Yau these manifolds admit Kähler metrics with vanishing Ricci tensor, i.e. so-called Kähler–Einstein metrics. Calabi–Yau 3-folds are an essential ingredient of modern field theories, and so have been intensively studied both by physicists and mathematicians: see for example Candelas–Ossa [**57**], Bryan–Pandharipande [**48**], and the references in Cox–Katz [**76**]. They have intricate properties. For example, one might hope that the complex structure J on any Calabi–Yau 3-fold could be deformed (within the class of integrable complex structures) to a complex structure J' for which there are only finitely many J'-holomorphic curves in each given homology class. (A generic almost complex structure on M will have this property.) However, this is not true: an explicit counterexample is given by P.M.H. Wilson in [**419**].

Calabi–Yau 3-folds occur as hypersurfaces in toric 4-folds, and as complete intersections in projective space. Below we study in more detail the special case of hypersurfaces in $\mathbb{C}P^4$. Manifolds (M, J) defined in this way occur in natural families: as one varies the defining equations the complex structure changes. The **mirror symmetry conjecture** in its various formulations expresses the intuition that the structures (such as Gromov–Witten invariants) arising from the symplectic structure of a particular Calabi–Yau manifold (M, ω) (the A-side) should have analogues in the structures arising from deformations of the complex structure J on M (the B-side). One of its consequences is an astonishing prediction due to Candelas–Ossa–Green–Parkes [**58**] for the number of simple curves in a quintic hypersurface in $\mathbb{C}P^4$. The precision of this conjecture aroused great interest, as did the fact that it was based on ideas that were completely new to mathematics at that time. The attempt to put these ideas on firm footing was one of the motivating forces for much of development described in this chapter.

To be more precise, recall from above that the quantum product induces a multiplication on $H^2(M) \otimes \Lambda$. In the Kähler case, we may restrict further to $H^{1,1}(M) \otimes \Lambda$ since the product is trivial on $H^{2,0}$ and $H^{0,2}$. Now the deformation ring of complex structures on M is a ring structure on $H^{1,2}$ which gives rise to a differential equation similar to the WDVV-equation discussed below. One version of the mirror symmetry conjecture states that associated to each Calabi–Yau manifold M there is a mirror manifold M^* such that the rings $H^{1,1}(M^*)$ and $H^{1,2}(M)$ are naturally isomorphic. This conjecture was confirmed by Givental [**150**] in the case of the quintic in $\mathbb{C}P^4$ and more generally for complete intersections in toric manifolds. Givental actually proved much more than this, since there is a much richer structure on each side that is preserved by the correspondence. We will describe some of this additional structure on the A-side in Section 11.5 but the full information includes the gravitational descendants mentioned in Remark 7.5.4. For more information on mirror symmetry, see Cox–Katz [**76**] and Voisin [**409**].

Quintic hypersurfaces in $\mathbf{C}P^4$. Consider the hypersurface of degree k in $\mathbb{C}P^4$

$$Z_k := \left\{ [z_0 : \cdots : z_4] \in \mathbb{C}P^4 \mid \sum_{j=0}^{4} {z_j}^k = 0 \right\}.$$

This manifold is simply connected and has Betti numbers

$$b_2 = b_4 = 1, \qquad b_3 = k^4 - 5k^3 + 10k^2 - 10k + 4.$$

In particular the identity $b_2 = b_4 = 1$ follows from the Lefschetz theorem on hyperplane sections. It follows that $\pi_2(Z_k) \cong H_2(M) \cong \mathbb{Z}$, so that the symplectic form ω does not vanish over $\pi_2(Z)$. Moreover the first Chern class of Z_k is given by

$$c_1 = (5 - k)\iota^* h$$

where $h \in H^2(\mathbb{C}P^4, \mathbb{Z})$ is the standard generator of the cohomology of $\mathbb{C}P^4$ and $\iota : Z_k \to \mathbb{C}P^4$ is the natural embedding of Z_k as a hypersurface in $\mathbb{C}P^4$.

Now let $A \in \pi_2(Z_k)$ be the generator of the homotopy group with $\omega(A) > 0$. An explicit representative of A is given, for example, by the holomorphic curve

$[z_0 : z_1] \mapsto [z_0 : z_1 : -z_0 : -z_1 : 0]$ when k is odd. In all cases A is represented by a line contained in Z_k. Hence evaluating the first Chern class on this generator gives

$$c_1(A) = 5 - k.$$

So for $k \leq 4$ the manifold Z_k is monotone. For $k > 5$ there are no nonconstant J-holomorphic curves for a generic almost complex structure because the only classes on which ω is positive have negative Chern numbers and so are not represented for generic J for dimensional reasons.

Hence the most interesting case is that of the quintic hypersurface $M = Z_5$ with Chern class zero. This is the archetypal example of a Calabi–Yau manifold. In order to compute the quantum product, we need to calculate all nonzero invariants $\mathrm{GW}^M_{dA,3}(a_1, a_2, a_3)$. Since $b_2 = 1$, it suffices to compute the invariant $\mathrm{GW}^M_{dA,3}(h, h, h)$, where h is Poincaré dual to the class $H = [Z_5 \cap \mathbb{C}P^3] \in H_4(Z_5)$ of a hyperplane section. Thus h is the class we previously denoted as $\iota^* h$; it satisfies $h(A) = 1$. Note that for generic (almost complex) J the simple curves will be isolated. Therefore for each ℓ the number N_ℓ of simple J-holomorphic curves in the class ℓA is finite. Since these curves are the basic geometric objects, we would like to compute $\mathrm{GW}^M_{dA,3}(h, h, h)$ in terms of these numbers N_ℓ.

Recall from Remark 7.3.8 that the invariant $\mathrm{GW}^M_{dA,3}(h, h, h)$ is by definition equal to $d^3 \mathrm{GW}^M_{dA,0}$, where $\mathrm{GW}^M_{dA,0}$ is the number of curves in class dA as counted by the Gromov–Witten process. This invariant does not count J-holomorphic curves directly. Rather it is defined by counting graphs which are solutions to a Cauchy–Riemann equation in which J depends on z. Our aim is to understand how the multiple covers of simple J-holomorphic curves contribute to $\mathrm{GW}^M_{dA,0}$.

Observe first that, because Z_5 is a hypersurface in $\mathbb{C}P^4$ of degree 5, we have

$$\mathrm{GW}^M_{0,3}(h, h, h) = H \cdot H \cdot H = [\mathbb{C}P^1] \cdot Z_5 = 5.$$

Moreover, if C is a simple curve in the class dA then $C \cdot H = d$. Thus, if we perturb H to general position, C intersects H transversally in d distinct points. Hence the contribution of the curve C to $\mathrm{GW}^M_{dA,3}(h, h, h)$ is exactly d^3.

Now consider the contribution of the m-fold covers of C to $\mathrm{GW}^M_{dA,3}(h, h, h)$. Choose a holomorphic parametrization $u : \mathbb{C}P^1 \to Z_5$ of C, let $\phi : \mathbb{C}P^1 \to \mathbb{C}P^1$ be a rational map of degree m, and consider the graph $\widetilde{v} : \mathbb{C}P^1 \to \mathbb{C}P^1 \times Z_5$ of $u \circ \phi$:

$$\widetilde{v}(z) = (z, u(\phi(z))).$$

This curve is not regular with respect to the complex structure $i \times J$. It belongs to a family of curves $\mathcal{M}(A_0 + mdA; i \times J)$ of dimension $6 + 4m + 2$ while the dimension in the regular case should only be $8 + 2c_1(mdA) = 8$. (Here $4m + 2$ is the real dimension of the space Rat_m of parametrizations ϕ. One can also verify that $\widetilde{v}$ is nonregular by using Lemma 3.3.1.) Thus, in order to see how C contributes to $\mathrm{GW}^M_{mdA,3}(h, h, h)$, one should (as in Section 7.3) perturb the complex structure $i \times J$ to a generic element $\widetilde{J}$ and count the number of $\widetilde{J}$-holomorphic $(A_0 + mdA)$-curves near C. As shown in Section 6.7 we can assume that $\widetilde{J}$ is given by a z-dependent family $\{J_z\}$ of almost complex structures on M as in (6.7.3). By assumption, there are only finitely many simple J-holomorphic mdA-curves which are all regular, and these are therefore separated from C. Hence under the perturbation $j \times J \to \widetilde{J}$ their graphs move slightly but remain separated from C. Similarly, if $A = m'A'$, the perturbation of any m'-fold cover of an A'-curve is separated from C. Thus one can isolate the contribution of C to $\mathcal{M}(A_0 + mdA; \widetilde{J})$. Note that this contribution can be computed by the methods discussed in Theorem 7.2.3, and reduces to finding the Euler class of the relevant obstruction bundle.

In [**19**] Aspinwall and Morrison proposed a different but very natural calculation of this contribution using methods coming from algebraic geometry. Their answer is that each m-fold cover of C should also contribute d^3 to $\mathrm{GW}^M_{mdA,3}(h,h,h)$. This means that there should be exactly one element of $\mathcal{M}(A_0+mdA; \widetilde{J})$ coming from C. Thus their definition gives the following beautiful formula for $h*h$ when $h \in H^2(M)$ is dual to H. In the notation of Section 11.1 we choose the ring of formal power series in $q := e^A$ as our quantum coefficient ring Λ. Then the formula is

$$\begin{aligned} \int_H h*h &= \sum_{d=0}^{\infty} \int_H (h*h)_{dA}\, q^d \\ &= \sum_{d=0}^{\infty} \mathrm{GW}^M_{dA,3}(h,h,h) q^d \\ &= 5 + \sum_{d=1}^{\infty} \sum_{\deg(C)=d} d^3 (q^d + q^{2d} + q^{3d} + \dots) \\ &= 5 + \sum_{d=1}^{\infty} N_d d^3 \frac{q^d}{1-q^d}. \end{aligned}$$

The second sum in the third line is over all simple holomorphic curves C of degree d. Hence the Aspinwall–Morrison formula is equivalent to the following relation between the numbers N_d and $\mathrm{GW}^M_{dA,0} := \frac{1}{d^3}\mathrm{GW}^M_{dA,3}(h,h,h)$:

$$\sum_{d=1}^{\infty} \mathrm{GW}^M_{dA,0}\, q^d = \sum_{d=1}^{\infty} N_d \frac{q^d}{1-q^d}. \tag{11.3.26}$$

This formula has now been rigorously established by Manin [**285**] and Voisin [**408**]. (See also Voisin [**409**, Section 5.6] for a nice discussion of the issues involved.)

The next question is: how can we calculate the numbers N_d? In distinction to the case $M = \mathbb{C}P^n$, the WDVV equations now give no information. This is clear if one thinks of the underlying geometry; the equations arise by comparing different calculations of the invariants $\mathrm{GW}^{M,I}_{A,k}$ where $|I| \geq 4$, but these must vanish in the present situation. In 1991, using ideas from mirror symmetry Candelas–de la Ossa–Green–Parkes [**58**] predicted values for N_d, the first few being

$$N_1 = 2{,}875, \quad N_2 = 609{,}250, \quad N_3 = 317{,}206{,}375, \quad \dots.$$

Their conjecture was based on a study of the B-side, namely the deformations of mixed Hodge structures on the mirror manifold. It would take us too far afield to describe the ideas behind this: for a thorough discussion see Cox–Katz [**76**] or Voisin [**409**]. Although N_1 was classically known, and N_2 was calculated by Katz [**205**] in 1986, the other N_d were unknown. The conjecture was shown by direct calculation for $d \leq 4$ and is now known to hold for all d thanks to Givental's solution of the mirror conjecture in this case: see [**147**].

We make one further comment. Although the numbers N_d seem to grow very rapidly, Givental's mirror formula implies that the Gromov–Witten potential (11.3.26) is a power series in q with positive radius of convergence. This is a specific example of the convergence problem discussed in Example 11.1.4 (viii). A similar result is now expected in all genera.

11.4. The Seidel representation

In this section we define the Seidel representation for symplectic manifolds that satisfy the strong semipositivity condition (8.5.1). We show how this can be used to understand the structure of the quantum cohomology ring of a general toric manifold. As another application, we calculate the additive cohomology of the total space $\widetilde{M}$ of any Hamiltonian fibration $\widetilde{M} \to S^2$ whose fiber is a semipositive symplectic manifold (M, ω) (Proposition 11.4.5).

In Example 8.6.8 we discussed the Seidel representation in the monotone case. In that case it was possible to define the invariant $\mathcal{S}(\psi)$ of a Hamiltonian loop ψ as an element of the cohomology of M. The definition of the invariant $\mathcal{S}_{\widetilde{A}}(\psi)$ in (8.6.4) carries over verbatim to the semipositive case. However, there may now be infinitely many section classes $\widetilde{A}$ with nonzero invariants $\mathcal{S}_{\widetilde{A}}(\psi)$ and $\mathcal{S}(\psi)$ will be defined as an (invertible) element of quantum cohomology. Here are the details.

Let (M, ω) be a closed symplectic manifold that satisfies (8.5.1). Then every J-holomorphic curve associated to a generic 2-parameter family of almost complex structures on M has nonnegative Chern number. Let $\psi = \{\psi_t\}_{t\in\mathbb{R}/\mathbb{Z}}$ be a loop of Hamiltonian symplectomorphisms of M and $\widetilde{M}_\psi \to S^2$ be the Hamiltonian fibration in (8.2.6) with fibers symplectomorphic to M. Recall from Corollary 9.1.2 that the loop $t \mapsto \psi_t(x)$ is contractible for every $x \in M$. This implies that the fibration $\widetilde{M}_\psi$ has sections. Given a homology class $\widetilde{A} \in H_2(\widetilde{M}_\psi)$ that can be represented by a section, define the cohomology class

$$\mathcal{S}_{\widetilde{A}}(\psi) := \sum_{\nu,\mu} \mathrm{GW}^{\widetilde{M}_\psi,\mathrm{w}}_{\widetilde{A},1}(e_\nu)\, g^{\nu\mu}\, e_\mu \in H^{-2c}(M) \tag{11.4.1}$$

as in (8.6.4). Here $c := c_1^{\mathrm{Vert}}(\widetilde{A}) \le 0$ denotes the vertical Chern number of $\widetilde{A}$, $e_0, \ldots, e_n$ denotes a basis of $H^*(M)$ and $g^{\nu\mu}$ is the inverse of the matrix $g_{\nu\mu}$ defined by

$$g_{\nu\mu} := \int_M e_\nu \smile e_\mu.$$

The Gromov–Witten invariants

$$\mathrm{GW}^{\widetilde{M}_\psi,\mathrm{w}}_{\widetilde{A},1} : H^{2n+2c}(M) \to \mathbb{Z}$$

are understood as in Definition 8.6.6 with one fixed marked point $\mathrm{w} \in S^2$. Thus we identify M with a fiber of $\widetilde{M}_\psi$ over a point $\mathrm{w} \in S^2$ via an embedding $\iota : M \to \widetilde{M}_\psi$ and then $\mathcal{S}_{\widetilde{A}}(\psi)$ is the cohomology class Poincaré dual to the pseudocycle

$$\iota^{-1} \circ \widetilde{\mathrm{ev}}_{\mathrm{w},J,H} : \mathcal{M}(\widetilde{A}; J, H) \to M,$$

defined on the moduli space of J-holomorphic sections of $\widetilde{M}_\psi$ in the class $\widetilde{A}$. Geometrically, its integral over a submanifold $X \subset M$ is the number of J-holomorphic sections of $\widetilde{M}_\psi$ in the class $\widetilde{A}$ passing through $\iota(X)$, counted with appropriate signs. Since the embedding ι is unique up to Hamiltonian isotopy of M, the cohomology class $\mathcal{S}_{\widetilde{A}}(\psi)$ is independent of ι. The cohomology classes $\mathcal{S}_{\widetilde{A}}(\psi)$ give rise to quantum cohomology classes as follows.

Definition 11.4.1. *Let (M,ω) be a closed symplectic manifold that satisfies (8.5.1), $\psi = \{\psi_t\}_{t\in\mathbb{R}/\mathbb{Z}}$ be a loop of Hamiltonian symplectomorphisms, and $\sigma : S^2 \to \widetilde{M}_\psi$ be a section. Fix any quantum coefficient ring Λ as in Definition 11.1.3. Then the* **Seidel element**

$$\mathcal{S}(\psi,\sigma) \in \mathrm{QH}^*(M;\Lambda)$$

is defined by

$$\mathcal{S}(\psi,\sigma) := \sum_{A\in H_2(M)} \mathcal{S}_{[\sigma]+\iota_* A}(\psi)\otimes e^A \tag{11.4.2}$$

As an example let us consider the ring $\Lambda := \Lambda^{\mathrm{univ}}[q,q^{-1}]$ of polynomials in q and q^{-1} with coefficients in the universal Novikov ring (see Example 11.1.4 (v)). The variable q has degree two. The dependence of the Seidel element on the section σ can then be removed via the **coupling class** $u_\psi \in H^2(\widetilde{M}_\psi)$ of Guillemin–Lerman–Sternberg [**163**]. The latter is defined as the cohomology class of a closed 2-form $\widetilde{\omega}\in\Omega^2(\widetilde{M}_\psi)$ that restricts to the given symplectic form on each fiber and such that the fiber integral of $\widetilde{\omega}^{n+1}$ is zero. Its cohomology class is independent of $\widetilde{\omega}$ (see Proposition 8.2.2).

Definition 11.4.2. *Let (M,ω) and ψ be as in Definition 11.4.1 and denote*

$$\Lambda := \Lambda^{\mathrm{univ}}[q,q^{-1}].$$

The **Seidel element** *$\mathcal{S}(\psi)\in \mathrm{QH}^*(M;\Lambda)$ is defined by*

$$\mathcal{S}(\psi) := \sum_{\widetilde{A}} \mathcal{S}_{\widetilde{A}}(\psi)\otimes q^{c(\widetilde{A})}\, t^{\widetilde{\omega}(\widetilde{A})}. \tag{11.4.3}$$

Here $\widetilde{\omega}\in\Omega^2(\widetilde{M}_\psi)$ denotes the the coupling form of a Hamiltonian connection on $\widetilde{M}_\psi$ and $c := c_1^{\mathrm{Vert}} \in H^2(\widetilde{M}_\psi)$ denotes the vertical first Chern class. The sum runs over all homology classes $\widetilde{A}\in H_2(\widetilde{M}_\psi)$ that can be represented by sections of $\widetilde{M}_\psi$ and satisfy $-\dim M \le 2c(\widetilde{A}) \le 0$.

The formulation of the Seidel representation in Definition 11.4.2 is a variation of the one in Lalonde–McDuff–Polterovich [**228**] who replaced $\Lambda^{\mathrm{univ}}[q,q^{-1}]$ by the Novikov ring Λ_ω over the quotient of the group $H_2^S(M)$ of spherical homology classes under the equivalence relation $A\sim B$ if and only if $\omega(A)=\omega(B)$ and $c_1(A)=c_1(B)$. This Novikov ring is a subring of $\Lambda^{\mathrm{univ}}[q,q^{-1}]$ under the inclusion $e^A \mapsto q^{c_1(A)}t^{\omega(A)}$. In other words, dividing the homology group $H_2(M)$ by the kernel of the homomorphism $A\mapsto(\omega(A),c_1(A))$ is equivalent to retaining only the information about $\omega(A)$ and $c_1(A)$ from a homology class A.

Theorem 11.4.3 (Seidel). *Let (M,ω) be a closed symplectic manifold that satisfies the semipositivity condition (8.5.1) and let $\Lambda := \Lambda^{\mathrm{univ}}[q,q^{-1}]$. Then*

$$\mathcal{S}(\phi\psi) = \mathcal{S}(\phi) * \mathcal{S}(\psi), \qquad \mathcal{S}(\mathrm{id}) = 1,$$

for all $\phi,\psi\in\pi_1(\mathrm{Ham}(M,\omega))$. Thus, for each $\psi\in\pi_1(\mathrm{Ham}(M,\omega))$, the Seidel element $\mathcal{S}(\psi)\in \mathrm{QH}^*(M;\Lambda)$ belongs to the group of even multiplicative units in $\mathrm{QH}^*(M;\Lambda)$ and the map $\psi\mapsto\mathcal{S}(\psi)$ is a group homomorphism.

The assertion $\mathcal{S}(\mathrm{id})=1$ is a statement about the Gromov–Witten invariants of the product manifold $\widetilde{M} := S^2\times M$ and holds by Exercise 7.3.4. The identity

$\mathcal{S}(\phi\psi) = \mathcal{S}(\phi)*\mathcal{S}(\psi)$ can be rephrased as an analogue of the *(Splitting)* axiom for the Gromov–Witten invariants and the proof will again be based on the gluing argument of Chapter 10. Although the proof given here applies only in the semipositive case, the theorem does in fact hold for all closed symplectic manifolds: see McDuff [**269**]. Note also that on manifolds for which the quantum multiplication equals the usual cup product, for example if $[\omega]$ vanishes on $\pi_2(M)$, the above identity reduces to the statement $\mathcal{S}(\phi\psi) = \mathcal{S}(\phi) \smile \mathcal{S}(\psi)$ of Theorem 8.6.9.

REMARK 11.4.4. The Seidel element $\mathcal{S}(\phi)$ gives rise to an automorphism (also called $\mathcal{S}(\phi)$) of quantum cohomology via

$$\mathrm{QH}^*(M;\Lambda) \mapsto \mathrm{QH}^*(M;\Lambda) : a \mapsto \mathcal{S}(\phi) * a.$$

We will see in the proof of Theorem 11.4.3 below that

$$\mathcal{S}(\phi) * a = \sum_{\widetilde{A}} \mathrm{GW}^{\widetilde{M}_\phi,\mathbf{w}}_{\widetilde{A},2}(a, e_\nu)\, g^{\nu\mu} e_\mu \otimes q^{c(\widetilde{A})}\, t^{\widetilde{\omega}(\widetilde{A})}, \tag{11.4.4}$$

and so is given in terms of invariants for sections with two fixed marked points $\mathrm{w}_1, \mathrm{w}_2$. Here one should think of the Poincaré dual of the class a as represented by a cycle in the fiber M_1 over the the point $\mathrm{w}_1 \in S^2$ while the Poincaré dual of e_ν is represented by a cycle in the fiber M_2 over $\mathrm{w}_2 \in S^2$. The invariant $\mathrm{GW}^{\widetilde{M},\mathbf{w}}_{\widetilde{A},2}(a, e_\nu)$ counts J-holomorphic sections of $\widetilde{M}$ that pass through these cycles. Since the identification of each fiber with M is well defined up to Hamiltonian symplectomorphisms, $\mathcal{S}(\phi)$ does indeed give rise to a well defined automorphism of $\mathrm{QH}^*(M)$. In terms of homology, given a class $\alpha \in H_*(M)$ represented by a cycle (also called α) in the fiber M_1, the class $\mathcal{S}(\phi)(\alpha)$ is the sum of the terms $\beta_{\widetilde{A}} \otimes q^{-c(\widetilde{A})}\, t^{-\widetilde{\omega}(\widetilde{A})}$ where the homology class $\beta_{\widetilde{A}}$ is represented by the intersection with M_2 of all sections in class $\widetilde{A}$ that go through α. With this interpretation, the Seidel element $\mathcal{S}(\phi) := \mathcal{S}(\phi)(1)$ is the image of the identity element, and Theorem 11.4.3 asserts that $\mathcal{S}(\phi\psi)$ is the composite $\mathcal{S}(\phi)\circ\mathcal{S}(\psi)$ of these automorphisms. Hence the group $\pi_1(\mathrm{Ham}(M,\omega))$ acts on the quantum cohomology of M by right and left $\mathrm{QH}^*(M)$-module isomorphisms, that is

$$\mathcal{S}(\phi) * (a * b) = a * \big(\mathcal{S}(\phi)(b)\big) = \mathcal{S}(a) * b), \qquad a, b \in \mathrm{QH}^*(M;\Lambda).$$

Recently Hutchings [**194**] has discovered an extension of this representation, defining a series of group homomorphisms from $\pi_{2k+1}(\mathrm{Ham}(M,\omega))$ to the additive group $\mathrm{End}_{2k}(\mathrm{QH}^*)$ of $\mathrm{QH}^*(M)$-module endomorphisms of $\mathrm{QH}^*(M)$ of degree $2k$.

The Seidel representation has many interesting applications. We have already seen two: in Section 8.6 we used it to examine the symplectic action of loops of Hamiltonian symplectomorphisms (see Corollary 8.6.10), while in Proposition 9.6.4 it was applied to estimate the Hofer lengths of loops. It is a crucial ingredient in McDuff's work [**273**] on uniruled manifolds because it gives a way to ensure that certain Gromov–Witten invariants do not vanish. As we shall see in Chapter 12 it is the monodromy of the Floer homology bundle over the Hamiltonian group, and hence is an important ingredient in understanding spectral invariants (cf. [**307, 105, 274**]). It also has implications for Hamiltonian fibrations. We give only the easiest consequence here, which is taken from Lalonde–McDuff–Polterovich [**228**]. For further results in this direction see [**228**] and Lalonde–McDuff [**227**].

PROPOSITION 11.4.5. *Let (M,ω) be a closed symplectic manifold satisfying the condition (8.5.1) and let $\pi : \widetilde{M} \to S^2$ be a Hamiltonian fibration whose fibers are symplectomorphic to M. Then the map*

$$H^*(S^2) \otimes H^*(M) \to H^*(\widetilde{M}) : a \otimes b \mapsto \pi^* a \smile \iota_! b$$

is an isomorphism on rational cohomology. Here $\iota : M \to \widetilde{M}$ denotes the inclusion of a fiber and $\iota_!$ is the cohomology pushforward defined in terms of the usual homology pushforward ι_* by $\iota_! := \mathrm{PD}_{\widetilde{M}} \circ i_* \circ \mathrm{PD}_M$.

PROOF. Every Hamiltonian fibration $\widetilde{M} \to S^2$ is defined by gluing together two copies of $\mathbb{C} \times M$ via a Hamiltonian loop ϕ as in (8.2.6). Thus we may identify $\widetilde{M} \to S^2$ with a fibration of the form $\widetilde{M}_\phi \to S^2$. It suffices to show that the inclusion $\iota : M \to \widetilde{M}_\phi$ of a fiber induces an injection on rational homology. But this follows immediately from Theorem 11.4.3. Since the map $a \mapsto \mathcal{S}(\phi) * a$ is an automorphism of quantum cohomology, we have $\mathcal{S}(\phi) * a \neq 0$ if and only if $a \neq 0$. By (11.4.4), $\mathcal{S}(\phi) * a$ is a sum whose coefficients are Gromov–Witten invariants of the form $\mathrm{GW}_{\widetilde{A},2}(a, e_\nu)$ that vanish unless the dual homology class $\mathrm{PD}(a) \in H_*(M)$ has a nonzero image $\iota_*\mathrm{PD}(a)$ in $H_*(\widetilde{M})$. This proves Proposition 11.4.5. □

We now turn to the proof of Theorem 11.4.3. It is based on an analogue of the *(Splitting)* axiom for the Gromov–Witten invariants of Hamiltonian fibrations over the 2-sphere. To formulate it we observe first that, given any two Hamiltonian loops $\phi, \psi \in \pi_1(\mathrm{Ham}(M,\omega))$, the fibration $\widetilde{M}_{\phi\psi}$ determined by their pointwise product $\{\phi_t\psi_t\}$ may be identified with the fiber sum $\widetilde{M}_\phi \# \widetilde{M}_\psi$. This lifts the usual connected sum operation on the base spheres $S^2 \# S^2 = S^2$, and hence is constructed by cutting out a neighbourhood of a fiber F from each of the fibrations $\widetilde{M}_\phi, \widetilde{M}_\psi$ and gluing the remaining two pieces appropriately along their boundary. (Precise formulas are given below. Note also that the ordering chosen for ϕ, ψ is irrelevant since the loops $\{\phi_t\psi_t\}$ and $\{\psi_t\phi_t\}$ are homotopic. Hence $\widetilde{M}_{\phi\psi} = \widetilde{M}_{\psi\phi}$.) Correspondingly there is a connected sum operation on the set $H_2^\sigma(\widetilde{M})$ of homology classes in $\widetilde{M}$ that are represented by sections: represent these classes by sections that go through corresponding points on the deleted fiber and then take the class represented by their connected sum. Denote this map by

$$H_2^\sigma(\widetilde{M}_\phi) \times H_2^\sigma(\widetilde{M}_\psi) \to H_2^\sigma(\widetilde{M}_{\phi\psi}) : (\widetilde{A}, \widetilde{B}) \mapsto \widetilde{A} \# \widetilde{B}.$$

It has the property that

$$c(\widetilde{A}\#\widetilde{B}) = c(\widetilde{A}) + c(\widetilde{B}), \qquad \widetilde{\omega}(\widetilde{A}\#\widetilde{B}) = \widetilde{\omega}(\widetilde{A}) + \widetilde{\omega}(\widetilde{B}). \tag{11.4.5}$$

THEOREM 11.4.6. *Let (M,ω) be a closed symplectic manifold satisfying (8.5.1) and let $\phi, \psi \in \pi_1(\mathrm{Ham}(M,\omega))$. Then, for all $\widetilde{C} \in H_2^\sigma(\widetilde{M}_{\phi\psi})$ and all $a_1, \dots, a_k \in H^*(M)$, and any integer $0 \le \ell \le k$,*

$$\begin{aligned} &\mathrm{GW}^{\widetilde{M}_{\phi\psi}}_{\widetilde{C},k}(a_1, \dots, a_k) \\ &= \sum_{\widetilde{A}\#\widetilde{B}=\widetilde{C}} \sum_{\nu,\mu} \mathrm{GW}^{\widetilde{M}_\phi}_{\widetilde{A},\ell+1}(a_1, \dots, a_\ell, e_\nu) g^{\nu\mu} \mathrm{GW}^{\widetilde{M}_\psi}_{\widetilde{B},k-\ell+1}(e_\mu, a_{\ell+1}, \dots, a_k). \end{aligned}$$

Here the invariants are understood to have fixed marked points; thus

$$\mathrm{GW}^{\widetilde{M}}_{\widetilde{A},j} := \mathrm{GW}^{\widetilde{M},\mathbf{w}}_{\widetilde{A},j}, \qquad \mathbf{w} := \{\mathrm{w}_1, \dots, \mathrm{w}_j\}.$$

This is a very simple case of a general formula (due to Ionel–Parker [**201**] in the symplectic case) that describes the Gromov–Witten invariants for an arbitrary fiber sum.

Proof of Theorem 11.4.3. By Theorem 11.4.6, with $\psi_t \equiv \mathrm{id}$ and $k = 2$, we have

$$\mathrm{GW}^{\widetilde{M}_\phi}_{\widetilde{A},2}(a,b) = \sum_{\widetilde{B}+\iota_* C=\widetilde{A}} \sum_{i,j} \mathrm{GW}^{\widetilde{M}_\phi}_{\widetilde{B},1}(e_i) g^{ij} \mathrm{GW}^{M}_{C,3}(e_j,a,b) \tag{11.4.6}$$

for $\widetilde{A} \in H_2^S(\widetilde{M}_\psi)$ and $a, b \in H^*(M)$. Here we use (8.6.2) and Definition 7.3.7 to identify the invariants $\mathrm{GW}^{\widetilde{M}_\psi}_{\widetilde{C},k}$ with $\mathrm{GW}^{M}_{C,k}$ and also use (11.4.5) to identify $\widetilde{A}$ with $\widetilde{B} + \iota_* C$. Further, from now on the superscript w for GW is understood. This implies

$$\begin{aligned}
\mathcal{S}(\phi) * a &= \sum_{\widetilde{B},C,i,j} \mathrm{GW}^{\widetilde{M}_\phi}_{\widetilde{B},1}(e_i) g^{ij} (e_j * a)_C \otimes q^{c(\widetilde{B}+\iota_* C)}\, t^{\widetilde{\omega}(\widetilde{B}+\iota_* C)} \\
&= \sum_{\widetilde{B},C,i,j,\nu,\mu} \mathrm{GW}^{\widetilde{M}_\phi}_{\widetilde{B},1}(e_i) g^{ij} \mathrm{GW}^{M}_{C,3}(e_j,a,e_\nu) g^{\nu\mu} e_\mu \otimes q^{c(\widetilde{B}+\iota_* C)}\, t^{\widetilde{\omega}(\widetilde{B}+\iota_* C)} \\
&= \sum_{\widetilde{A},\nu,\mu} \mathrm{GW}^{\widetilde{M}_\phi}_{\widetilde{A},2}(a,e_\nu) g^{\nu\mu} e_\mu \otimes q^{c(\widetilde{A})}\, t^{\widetilde{\omega}(\widetilde{A})}.
\end{aligned}$$

Here the last equation follows from (11.4.6). Thus we have proved (11.4.4). It follows that

$$\begin{aligned}
\mathcal{S}(\phi) * \mathcal{S}(\psi) &= \sum_{\widetilde{A},\widetilde{B},\nu,\mu} \mathrm{GW}^{\widetilde{M}_\phi}_{\widetilde{A},2}(\mathcal{S}_{\widetilde{B}}(\psi), e_\nu) g^{\nu\mu} e_\mu \otimes q^{c(\widetilde{A})+c(\widetilde{B})}\, t^{\widetilde{\omega}(\widetilde{A})+\widetilde{\omega}(\widetilde{B})} \\
&= \sum_{\widetilde{A},\widetilde{B},\nu,\mu,i,j} \mathrm{GW}^{\widetilde{M}_\psi}_{\widetilde{B},1}(e_i) g^{ij} \mathrm{GW}^{\widetilde{M}_\phi}_{\widetilde{A},2}(e_j, e_\nu) g^{\nu\mu} e_\mu \otimes q^{c(\widetilde{A}\#\widetilde{B})}\, t^{\widetilde{\omega}(\widetilde{A}\#\widetilde{B})} \\
&= \sum_{\widetilde{C},\nu,\mu} \mathrm{GW}^{\widetilde{M}_{\phi\psi}}_{\widetilde{C},1}(e_\nu) g^{\nu\mu} e_\mu \otimes q^{c(\widetilde{C})}\, t^{\widetilde{\omega}(\widetilde{C})} \\
&= \mathcal{S}(\phi\psi).
\end{aligned}$$

Here the second equality follows uses (11.4.5) and the formula for $\mathcal{S}_{\widetilde{B}}(\psi)$, while the third follows from Theorem 11.4.6 with $k = 1$. Since $\psi\phi = \phi\psi \in \pi_1(\mathrm{Ham}(M))$, the ordering of terms is irrelevant. This proves Theorem 11.4.3. □

Proof of Theorem 11.4.6. Fix two loops

$$\phi^0 = \{\phi^0_t\}_{t\in\mathbb{R}/\mathbb{Z}}, \qquad \phi^\infty = \{\phi^\infty_t\}_{t\in\mathbb{R}/\mathbb{Z}}$$

of Hamiltonian symplectomorphisms of M. The idea of the proof is to extend the gluing theorem of Chapter 10 to the setting of Hamiltonian fibrations. Two J-holomorphic sections of the fibrations

$$\widetilde{M}^0 := \widetilde{M}_{\phi^0}, \qquad \widetilde{M}^\infty := \widetilde{M}_{\phi^\infty}$$

that intersect in a pair of fibers (both to be identified with M) and satisfy a suitable transversality condition can be glued together to give a J-holomorphic section of the fiber sum

$$\widetilde{M}^{\infty 0} := \widetilde{M}_{\phi^\infty \phi^0}.$$

Once the setup and the pregluing construction are understood, the proof from Chapter 10 carries over verbatim to the present situation. Thus we shall content ourselves with describing the main idea.

By definition in (8.2.6), the Hamiltonian fibration $\widetilde{M}^0 \to S^2$ is the quotient

$$\widetilde{M}^0 = \frac{\{\pm 1\} \times \mathbb{C} \times M}{(1, e^{2\pi(s+it)}, x) \ \sim \ (-1, e^{-2\pi(s+it)}, (\phi_t^0)^{-1}(x))}.$$

Throughout we abbreviate

$$z = e^{2\pi(s+it)}$$

so that $2\pi t$ is the argument of z. Note that the base sphere $S^2 = \mathbb{C} \cup \{\infty\}$ is the union of two copies of $\mathbb{C}_\pm$, where $\mathbb{C}_+ \equiv \mathbb{C}$ and $\mathbb{C}_- \equiv (\mathbb{C} \setminus \{0\}) \cup \{\infty\}$. A section of $\widetilde{M}^0$ is a pair of smooth maps $u_\pm^0 : \mathbb{C} \to M$ that satisfy

$$u_+^0(z) = \phi_t^0(u_-^0(z^{-1})),$$

and similarly for $\widetilde{M}^\infty$. To form their fiber sum, we identify the fiber of $\widetilde{M}^0$ over the point $0 \in \mathbb{C}_+$ (the origin) and the fiber of $\widetilde{M}^\infty$ over $0 \in \mathbb{C}_-$ (the point at infinity) with M. Now delete the neighbourhoods $\mathbb{C}_+ \times M$ in $\widetilde{M}^0$ and $\mathbb{C}_- \times M$ in $\widetilde{M}^\infty$ of these fibers, and make the fiber sum by gluing the remaining pieces $\mathbb{C}_- \times M$ in $\widetilde{M}^0$ to $\mathbb{C}_+ \times M$ in $\widetilde{M}^\infty$ by the composite map $\phi_t^\infty \circ \phi_t^0$. This is just $\widetilde{M}^{\infty 0}$; a section is a pair of smooth maps $u_\pm : \mathbb{C} \to M$ such that

$$u_+(z) = \phi_t^\infty \circ \phi_t^0(u_-(z^{-1})).$$

Our strategy is now the following. Starting from two J-holomorphic sections $\widetilde{u}^0$ of $\widetilde{M}^0$ and $\widetilde{u}^\infty$ of $\widetilde{M}^\infty$ that agree in the common fiber we build a family of approximate J^R-holomorphic sections $\widetilde{u}^R$ of $\widetilde{M}^{\infty 0}$ (the pregluing construction). Then the gluing argument of Chapter 10 will produce a nearby family of true J^R-holomorphic sections of $\widetilde{M}^{\infty 0}$. After this is established one can complete the proof by arguing as in Section 10.8. Namely, fix ℓ marked points on $S^2 \setminus \{0\}$ (for $\widetilde{M}^0$) and $k - \ell$ marked points on $S^2 \setminus \{\infty\}$ (for $\widetilde{M}^\infty$), as well as suitable submanifolds in the k fibers that represent our cohomology classes. Then, for R sufficiently large, we must compare the number of J^R-holomorphic sections of $\widetilde{M}^{\infty 0}$ passing at the k fibers through the given submanifolds with the number of connected pairs of J-holomorphic sections of $\widetilde{M}^0$ and $\widetilde{M}^\infty$ satisfying the same constraints at the marked points. As in Section 10.8, this will follow immediately from the gluing theorem. The gluing theorem, in turn, will follow from the pregluing construction by exactly the same arguments as in Sections 10.3 through 10.7 in Chapter 10.

Here are the details of the pregluing construction in the present case. Consider a pair of sections $\widetilde{u}^0 = (u_+^0, u_-^0)$ of $\widetilde{M}^0$ and $\widetilde{u}^\infty = (u_+^\infty, u_-^\infty)$ of $\widetilde{M}^\infty$ satisfying

$$u_+^0(0) = u_-^\infty(0) =: x.$$

For every sufficiently large real number R, we define a section $\widetilde{u}^R = (u_+^R, u_-^R)$ of $\widetilde{M}^{\infty 0}$ such that

$$u_+^R(z) = \begin{cases} u_+^\infty(R^2 z), & \text{if } |z| \le \frac{\delta}{2R}, \\ \phi_t^\infty(x), & \text{if } \frac{\delta}{R} \le |z| \le \frac{1}{\delta R}, \\ \phi_t^\infty(u_+^0(z)), & \text{if } |z| \ge \frac{2}{\delta R}, \end{cases}$$

$$u_-^R(1/z) = \begin{cases} (\phi_t^0)^{-1}(u_-^\infty(1/R^2 z)), & \text{if } |z| \le \frac{\delta}{2R}, \\ (\phi_t^0)^{-1}(x), & \text{if } \frac{\delta}{R} \le |z| \le \frac{1}{\delta R}, \\ u_-^0(1/z), & \text{if } |z| \ge \frac{2}{\delta R}. \end{cases}$$

The reader may check that

$$\phi_t^\infty \circ \phi_t^0(u_-^R(1/z)) = u_+^R(z)$$

for all R and all $z = e^{2\pi(s+it)}$ for which $u_+^R(z)$ and $u_-^R(1/z)$ have been defined. If R is sufficiently large then, as in Section 10.2, we can extend u_+^R to all of $\mathbb{C}$ by interpolating between u_+^∞ and $\phi_t^\infty(x)$ on the annulus $\delta/2R \le |z| \le \delta/R$ using u_-^∞ and between $\phi_t^\infty(x)$ and $\phi_t^\infty(u_+^0)$ on $1/\delta R \le |z| \le 2/\delta R$ using u_+^0. More precisely, since $u_+^0(0) = u_-^\infty(0) = x$, there exist smooth maps $\zeta_+^0, \zeta_-^\infty : \mathbb{C} \to T_xM$ such that

$$u_+^0(z) = \exp_x(\zeta_+^0(z)), \qquad u_-^\infty(z) = \exp_x(\zeta_-^\infty(z)), \qquad \zeta_+^0(0) = \zeta_-^\infty(0) = 0,$$

for $|z|$ sufficiently small. Let $\rho : \mathbb{C} \to [0,1]$ be a smooth cutoff function that vanishes for $|z| \le 1$ and equals one for $|z| \ge 2$, and define

$$u_+^R(z) := \phi_t^\infty\Big(\exp_x\Big(\rho(\frac{\delta}{Rz})\zeta_-^\infty(\frac{1}{R^2 z}) + \rho(\delta R z)\zeta_+^0(z)\Big)\Big),$$
$$u_-^R(1/z) := (\phi_t^0)^{-1}\Big(\exp_x\Big(\rho(\frac{\delta}{Rz})\zeta_-^\infty(\frac{1}{R^2 z}) + \rho(\delta R z)\zeta_+^0(z)\Big)\Big)$$

for $\delta/2R \le |z| \le R/2\delta$. Note that the formula (10.2.2) in Section 10.2 corresponds to the special case $\phi_t^0 = \phi_t^\infty = \mathrm{id}$ for all t.

This pregluing construction defines a family of maps f^R from the moduli space $\mathcal{M}(\widetilde{A}^0, \widetilde{A}^\infty; J^0, J^\infty)$ of pairs of J-holomorphic sections $(\widetilde{u}^0, \widetilde{u}^\infty)$ of the fibrations $\widetilde{M}^0$ and $\widetilde{M}^\infty$ that represent the classes $\widetilde{A}^0$ and $\widetilde{A}^\infty$ and satisfy

$$u_+^0(0) = u_-^\infty(0)$$

into the space of sections of the fibration $\widetilde{M}^{\infty 0}$ that represent the homology class

$$\widetilde{A}^{\infty 0} := \widetilde{A}^\infty \# \widetilde{A}^0.$$

To make the gluing argument work we shall have to assume, as in Chapter 10, that the moduli spaces $\mathcal{M}(\widetilde{A}^0, J^0)$ and $\mathcal{M}(\widetilde{A}^\infty, J^\infty)$ are regular and that the evaluation map

$$\mathcal{M}(\widetilde{A}^0, J^0) \times \mathcal{M}(\widetilde{A}^\infty, J^\infty) \to M \times M : (\widetilde{u}^0, \widetilde{u}^\infty) \mapsto (u_+^0(0), u_-^\infty(0))$$

is transverse to the diagonal. From this point on the proof is word by word the same as in Chapter 10 and will be omitted. This proves Theorem 11.4.6. □

REMARK 11.4.7. (i) Theorem 11.4.3 is a tautology if one defines the Seidel element $\mathcal{S}(\phi)$ as an operator on Floer cohomology. However, the nontrivial part of the story is then to translate that observation back into the present context, i.e. to prove that the operator on Floer cohomology agrees with the Seidel representation defined above. The proof of this correspondence involves the same kind of gluing theorems as the above proof of Theorem 11.4.3. For further discussion see Section 12.5.

(ii) We proved Theorem 11.4.3 above by a calculation based on the splitting rule Theorem 11.4.6. If we think of the Seidel element $\mathcal{S}(\phi)$ as the image of the unit 1 under the automorphism $\mathcal{S}(\phi)$ as in Remark 11.4.4 then we can give the following geometric interpretation of the argument. As always, it is easiest to give the explanation in terms of homology. To this end, denote by $\mathcal{S}_*(\phi)$ the automorphism of $\mathrm{QH}_*(M;\Lambda)$ that takes a homology class α in the fiber over w_1 to the class

$$\mathcal{S}_*(\phi)(\alpha) := \sum_{\widetilde{A}\in H_2^\sigma(\widetilde{M}_\phi)} \mathrm{GW}^{\widetilde{M}_\phi,\mathbf{w}}_{\widetilde{A},2}(\alpha,\varepsilon_i)g^{ij}\varepsilon_j$$

in the fiber over w_2, where $\varepsilon_i := \mathrm{PD}(e_i)$. In other words, we *define* $\mathcal{S}_*(\phi)(\alpha)$ by using (the homology version of) the right hand side of (11.4.4). Thus

$$\mathcal{S}_*(\phi)(\alpha) = \mathrm{PD}\circ\mathcal{S}(\phi)\circ\mathrm{PD}(\alpha),$$

where $\mathrm{PD}:\mathrm{QH}^*(M;\Lambda)\to\mathrm{QH}_*(M;\check{\Lambda})$ is the Poincaré duality isomorphism defined in Remark 11.1.20. Our aim is to prove the composition rule $\mathcal{S}_*(\phi^\infty\phi^0) = \mathcal{S}_*(\phi^\infty)\circ\mathcal{S}_*(\phi^0)$. For $a=0,\infty$, think of $\mathcal{S}_*(\phi^a)$ as an automorphism that takes the quantum homology of the fiber M^a_∞ of $\widetilde{M}^a$ at $0\in\mathbb{C}_-$ (the point at infinity) to that of the fiber M^a_0 of $\widetilde{M}^a$ at $0\in\mathbb{C}_+$ (the origin). Since the image $\mathcal{S}_*(\phi^0)(\alpha)$ of a cycle α in M^0_∞ is represented by the intersection with M^0_0 of sections in $\widetilde{M}^0$, the image of the composite $\mathcal{S}_*(\phi^\infty)\circ\mathcal{S}_*(\phi^0)(\alpha)$ is represented by the intersection with M^∞_0 of pairs of sections that agree in the common fiber. On the other hand, $\mathcal{S}_*(\phi^\infty\phi^0)(\alpha)$ is defined by counting sections of $\widetilde{M}_{\infty 0}$. Therefore the calculation reduces to the gluing statement of Theorem 11.4.6. In this framework, we must also prove that $\mathcal{S}_*(\phi)$ is a module homomorphism, in other words that

$$\mathcal{S}_*(\phi)(\alpha) = \mathcal{S}_*(\phi)(\alpha*1) = \alpha*\mathcal{S}_*(\phi)(1)\ \big(=\mathcal{S}_*(\phi)*\alpha\big).$$

But this follows by arguing as in the first step in the proof of Theorem 11.4.3, taking one of the loops to be trivial. This is the point of view adopted in [**228, 269**].

Example 11.4.8. In most cases where the Seidel element $\mathcal{S}(\psi)\in\mathrm{QH}^*(M)$ has been computed, the loop ψ is represented by a circle action on (M,ω). Let $K:M\to\mathbb{R}$ denote its moment map, normalized so that $\int_M K\omega^n = 0$, and denote the corresponding loop by λ_K. Then the points on which K takes its maximum form a connected symplectic submanifold $F_K\subset M$ that is fixed by the action:

$$K(F_K) = \max_{x\in M} K(x).$$

Suppose in addition this maximal component is simple, in the sense that the circle acts freely in a deleted neighbourhood $U\setminus F_K$ of F_K. Then it is proved in McDuff–Tolman [**279**] that

$$\mathcal{S}(\lambda_K) = \mathrm{PD}(F_K)\otimes q^{-\mathrm{codim}(F_K)/2}t^{\sup K} + \sum_{\omega(\widetilde{A})>\sup K} a_{\widetilde{A}}\otimes q^{c(\widetilde{A})}t^{\omega(\widetilde{A})}. \tag{11.4.7}$$

Thus the leading term in $\mathcal{S}(\lambda_K)$ is determined in a very simple way from the geometry. As an example, consider the rotation of S^2 generated by a height function K. Then $\mathcal{S}(\lambda_K) = \mathrm{PD}([\mathrm{pt}])\otimes q^{-1}t^\mu$ where $\mu = \sup K = \int_{S^2}\omega/2$.

In several interesting cases, there are nonzero lower order terms in (11.4.7) (cf. Seidel [**363**]). However, it is shown in [**279**] that such terms do not occur if (M,ω,λ_K) is Fano, i.e. it has an invariant ω-tame almost complex structure that admits no J-holomorphic spheres of nonpositive Chern number, and if its minimal Chern number N is at least half the codimension of F_K. Thus, for example, these terms do not occur if λ_K is one of the circle actions on a Fano toric variety that fixes one of the facets.

EXAMPLE 11.4.9. We return to the circle action considered in Proposition 9.7.2. This acts on $S^2\times S^2$ by the formula $\psi_t(z,w) = (z, \Phi_{z,t}(w))$, and so rotates the fibers of the projection onto the first sphere, fixing the diagonal and antidiagonal. In Chapter 9 we gave two proofs that when $\lambda > 1$ there is an S^1-invariant symplectic form on $S^2 \times S^2$ that is isotopic to the standard form $\omega_\lambda := \lambda\pi_1^*\sigma + \pi_2^*\sigma$. One can give a third proof by adapting Example 11.3.12 to construct a toric structure on $S^2\times S^2$ that includes this S^1 action: simply take w_2 to be the vector $(2,1)$ instead of $(1,1)$ and choose an appropriate τ. The more interesting part of Proposition 9.7.2 was the claim that it is not possible to homotop the Hamiltonian loop $\{\psi_t\}$ into the symplectomorphism group of $(S^2 \times S^2, \omega_\lambda)$ when $\lambda \le 1$. We now sketch a different proof. By continuity, it suffices to consider the case $\lambda = 1$.

We saw in Example 11.1.13, that in the monotone case the quantum cohomology ring $\mathrm{QH}^*(S^2\times S^2;\Lambda)$ could be constructed to have zero divisors but had none when $\lambda > 1$. This phenomenon is very clear with our current choice of Novikov ring $\Lambda_\lambda := \Lambda^{\mathrm{univ}}[q,q^{-1}]$. Denote

$$A := [S^2\times \mathrm{pt}],\quad B := [\mathrm{pt}\times S^2],\quad a := [\pi_1^*\sigma] = \mathrm{PD}(B),\quad b := [\pi_2^*\sigma] = \mathrm{PD}(A).$$

Then the homomorphism $\phi : \Gamma(M,\omega) \to \Lambda$ is given by

$$\phi(A) := q\,t^\lambda,\quad \phi(B) := q\,t.$$

Thus $a*a = qt$ and $b*b = qt^\lambda$ are equal when $\lambda = 1$. Further

$$(a+b)(a-b) = qt(1-t^{\lambda-1}).$$

Therefore, if $\lambda > 1$, $a+b$ is a unit in $\mathrm{QH}^*(S^2\times S^2;\Lambda_\lambda)$, while if $\lambda = 1$ it is a zero divisor.

The maximal fixed point component F_K of the circle action $\{\psi_t\}$ is the diagonal, which is Poincaré dual to the class $a+b \in H^2(S^2\times S^2)$. Seidel [**363**] showed by direct calculation that there are no lower order terms in $\mathcal{S}(\psi)$; this also holds by the remarks in Example 11.4.8. Hence, when $\lambda > 1$, $\mathcal{S}(\psi)$ has the form $(a+b)\otimes\mu_\lambda$ where $\mu_\lambda \in \Lambda_\lambda$. Now the Seidel element is given by Gromov–Witten invariants in $\widetilde{M}_\psi$ which do not depend on the deformation class of the symplectic form on M. Therefore, if ψ were homotopic to a loop in $\mathrm{Symp}(M,\omega_1)$ its Seidel element $\mathcal{S}(\psi) \in \mathrm{QH}^*(S^2\times S^2;\Lambda_1)$ would still have the form $(a+b)\otimes\mu_1$ for some $\mu_1 \in \Lambda_1$. Since this is not a unit, the homotopy cannot exist.

The small quantum cohomology of smooth toric varieties. We now return to the discussion of the small quantum cohomology ring $\mathrm{QH}^*(M)$ of a toric manifold, started in Section 11.3.1. Our exposition closely follows McDuff–Tolman [**279**]. We assume that (M,ω) is a closed symplectic $2n$-manifold that is equipped with an effective Hamiltonian group action of an n-torus T. Let $\mu : M \to \mathfrak{t}^*$ be a moment map for this action and denote its image by $\Delta := \mu(M)$. Recall that Δ is a convex polytope in $\mathfrak{t}^*$ that can be written in the form

$$\Delta = \left\{\eta \in \mathfrak{t}^* \,|\, \langle \eta, \overline{e}_\nu\rangle \le \kappa_\nu\right\},$$

where $\overline{e}_1,\dots,\overline{e}_N \in \mathfrak{t}$ belong to the integer lattice and $\kappa_1,\dots,\kappa_N \in \mathbb{R}$ (we change here from inward pointing to outward pointing normal vectors of the facets). Recall further the definition of the faces

$$\Delta_I := \left\{\eta \in \Delta \,|\, \nu \in I \Longrightarrow \langle \eta, \overline{e}_\nu\rangle = \kappa_\nu\right\}$$

for $I \subset I_0 := \{1, \ldots, N\}$. We saw in equation (11.3.15) that for every primitive subset $I \subset I_0$ (that satisfies $\Delta_I = \emptyset$ and $\Delta_{I'} \neq \emptyset$ whenever $I' \subsetneq I$) there exists a unique vector $d = (d_1, \ldots, d_N) \in \mathbb{Z}^N$ such that

$$\sum_{\nu=1}^{N} d_\nu \overline{e}_\nu = 0, \qquad d_\nu = 1 \ (\nu \in I), \qquad d_\nu \leq 0 \ (\nu \notin I). \tag{11.4.8}$$

Every such integer vector d corresponds to a homology class $A_d \in H_2(M)$ with first Chern number and symplectic area given by

$$c_1(A_d) = \sum_{\nu=1}^{N} d_\nu, \qquad \omega(A_d) = \sum_{\nu=1}^{N} \kappa_\nu d_\nu. \tag{11.4.9}$$

(See equation (11.3.18) with κ_ν replaced by $-\kappa_\nu$.) We wish to understand the relation in quantum cohomology that corresponds to the relation $\prod_{\nu \in I} u_\nu = 0$ in ordinary cohomology for each primitive index set I.

We give a direct geometric interpretation for the quantum deformations of the Stanley–Reisner relations. Because the dual vector $\overline{e}_\nu$ to the facet Δ_ν is integral, it gives rise to a circle action λ_ν on M, namely the action given by the circle in T that is tangent to $\overline{e}_\nu$. Its moment map is given by $\langle \mu, \overline{e}_\nu \rangle \colon M \to \mathbb{R}$. Hence the relation $\sum_{\nu=1}^{N} d_\nu \overline{e}_\nu = 0$ asserts that the corresponding product in $\pi_1(\mathrm{Ham}(M, \omega))$ is the constant loop:

$$\prod_{\nu=1}^{N} (\lambda_\nu)^{d_\nu} = 1.$$

By Theorem 11.4.3, the image of this product under the Seidel representation must be the identity. In other words

$$\prod_{\nu \in I} \mathcal{S}(\lambda_\nu) = \prod_{\nu \notin I} \mathcal{S}(\lambda_\nu)^{|d_\nu|} \in \mathrm{QH}^*(M; \Lambda). \tag{11.4.10}$$

Here the products are understood in quantum cohomology. As pointed out in Example 11.4.8 above, in the Fano case there are no lower order terms in the expression for $\mathcal{S}(\lambda_\nu)$. Moreover (in all cases) the submanifold $X_\nu \subset M$ on which the moment map of the νth circle action attains its maximum is simply the preimage of the νth facet under the moment map, and so is Poincaré dual to $\overline{\mathrm{w}}_\nu \in H^2(M)$:

$$\kappa_\nu = \sup_{x \in M} \langle \mu, \overline{e}_\nu \rangle = \langle \mu(X_\nu), \overline{e}_\nu \rangle, \qquad X_\nu := \mu^{-1}(\Delta_\nu).$$

Hence it follows from Example 11.4.8 that

$$\mathcal{S}(\lambda_\nu) = \overline{\mathrm{w}}_\nu \otimes q^{-1}\, t^{\kappa_\nu}, \qquad \overline{\mathrm{w}}_\nu := \mathrm{PD}([X_\nu]) \in H^2(M).$$

This gives rise to a proof of Theorem 11.3.4 (in the version of equation (11.3.17)) for the quantum coefficient ring

$$\Lambda := \Lambda^{\mathrm{univ}}[q, q^{-1}].$$

This is not quite the full result because it only retains the information about the Chern number and area of a homology class $A \in H_2(M)$. However, the discussion in the next remark explains how one might prove the full result.

REMARK 11.4.10. Because M is Fano we may choose the polynomial ring $\Lambda_0 := \mathbb{Q}[q_1, \ldots, q_k]$ as coefficients for its quantum homology. For each circle action λ_ν on M there is a canonical section σ_ν in $\widetilde{M}_{\lambda_\nu}$ consisting of the points $\{\pm 1\} \times \mathbb{C} \times \{x\}$

where $x \in X_\nu$. Similarly, for any $d \in \mathbb{Z}$, given the circle action λ_ν^d, we define $(\sigma_\nu)^d$ to be the section of $\widetilde{M}_{\lambda_\nu^d}$ on which the moment map dK_ν takes its maximum. The calculation in [**279**] shows that the Seidel element $\mathcal{S}(\lambda_\nu, \sigma_\nu) \in \mathrm{QH}^*(M; \Lambda_0)$ of Definition 11.4.1 equals $\overline{\mathrm{w}}_\nu$. Consider a relation $\sum d_\nu \overline{e}_\nu = 0$ as above. Then it follows from Theorem 11.4.6 that

$$\prod_{\nu=1}^N \mathcal{S}\big(\lambda_\nu^{d_\nu}, (\sigma_\nu)^{d_\nu}\big) = \mathcal{S}(0, \sigma_d) = 1 \otimes e^A,$$

where σ_d is the section of the trivial bundle formed by concatenating the sections $(\sigma_\nu)^{d_\nu}$ and lies in the class $[S^2 \times \mathrm{pt}] + A \in H_2(S^2 \times M)$. This shows that there is a relation in $\mathrm{QH}^*(M; \Lambda_0)$ of the form

$$\prod_{\nu \in I} \overline{\mathrm{w}}_\nu = e^A \prod_{\nu \notin I} (\overline{\mathrm{w}}_\nu)^{|d_\nu|} \in \mathrm{QH}^*(M; \Lambda_0).$$

Therefore, it remains to identify A with the element previously called A_d. But these two classes have the same image in $\Lambda^{\mathrm{univ}}[q, q^{-1}]$ for each choice of $[\omega]$. Since $[\omega]$ can vary in an open subset of $H^2(M)$ it follows that $A = A_d$.

The virtue of this approach is that the relation (11.4.10) holds for every toric manifold (M, ω), whether or not it is Fano. Namely, by Example 11.4.8, we can always express the Seidel element $\mathcal{S}(\lambda_\nu)$ in the form

$$\mathcal{S}(\lambda_\nu) = y_\nu \otimes q^{-1}\, t^{c_\nu}, \qquad y_\nu = \overline{\mathrm{w}}_\nu + \sum_k \sum_{\varepsilon > 0} a_{k,\varepsilon} q^k t^\varepsilon.$$

As explained in Section 11.3.1, there is a surjective ring homomorphism

$$\frac{\Lambda[u_1, \dots, u_N]}{P(\Delta)} \to \mathrm{QH}^*(M; \Lambda) : u_\nu \mapsto \overline{\mathrm{w}}_\nu, \qquad \Lambda := \Lambda^{\mathrm{univ}}[q, q^{-1}].$$

Hence the element $y_\nu \in \mathrm{QH}^*(M; \Lambda)$ lifts to some element $Y_\nu \in \Lambda[u_1, \dots, u_N]$ of the form $Y_\nu = u_\nu + h.o.t.$ With these polynomials one can define the quantum Stanley–Reisner ideal $SR_Y(\Delta) \subset \Lambda[u_1, \dots, u_N]$ by

$$SR_Y(\Delta) := \left\langle \prod_{\nu \in I} Y_\nu = q^{c_1(A_d)}\, t^{\omega(A_d)} \prod_{\nu \notin I} Y_\nu^{|d_\nu|} \,\middle|\, I \text{ is primitive} \right\rangle. \tag{11.4.11}$$

where $d_\nu = d_\nu(I)$ is as in (11.4.8) and $c_1(A_d)$ and $\omega(A_d)$ are given by (11.4.9). Note that $SR_Y(\Delta)$ depends on the Y_ν, which cannot in general be described without prior knowledge of the ring structure in $\mathrm{QH}^*(M; \Lambda)$. It follows that the small quantum cohomology of M has the form

$$\mathrm{QH}^*(M; \Lambda) \cong \frac{\Lambda[u_1, \dots, u_N]}{P(\Delta) + SR_Y(\Delta)},$$

where $\Lambda := \Lambda^{\mathrm{univ}}[q, q^{-1}]$.

EXAMPLE 11.4.11 (NEF toric varieties). Assume that M is NEF, i.e. that $c_1(B) \geq 0$ for every class $B \in H_2(M)$ with a holomorphic representative. In this case there may be higher order terms in the Seidel elements y_ν. However, it is shown in [**279**] that in this case the cohomology classes $a_{k,\varepsilon}$ occurring in the formula for $\mathcal{S}(\lambda_\nu)$ all have degree either zero or two. Hence they have natural lifts to $\Lambda[u_1, \dots, u_n]$ that can be obtained without prior knowledge of the quantum multiplication. Thus the quantum Stanley–Reisner ring of (11.4.11) can be written down once one has calculated the Seidel elements $\mathcal{S}(\lambda_\nu)$. This result (in particular the substitution of the Y_ν for the u_ν in the Stanley–Reisner ring SR_Y) is related to Givental's change of variable formulas as discussed in [**76**, 11.2.5.2]. Givental's mirror theorem gives a way of calculating the Seidel elements Y_i since it relates two functions $\widetilde{I}$ and $\widetilde{J}$, the first of which can be calculated from combinatorial data and the second of which determines the relations in quantum cohomology: see Gonzalez–Iritani [**155**].

11.5. Frobenius manifolds

We return to the discussion in Section 11.2, assuming as usual that (M, ω) is a semipositive manifold. For simplicity, we will restrict to the even part of the cohomology of M so that we can work with manifolds rather than supermanifolds. Hence we now assume that $e_0 := 1, e_1, \dots, e_N$ is a basis for $H^{ev}(M)$ consisting of elements of pure degree and such that $e_1, \dots, e_m$ span $H^2(M)$. For convenience we also assume that $e_N = \mathrm{PD}(\mathrm{pt})$ is the positive generator of $H^{2n}(M)$. This restriction to even classes suffices for many applications such as flag manifolds and Grassmannians.

Because we are only attempting to give the flavor of the subject, we will make another simplifying assumption, namely that for all $a, b, c \in H^*(M)$ the series

$$\sum_{A \in K^{\mathrm{eff}}} \mathrm{GW}^M_{A,3}(a, b, c) q^{\omega(A)} \tag{11.5.1}$$

converges for sufficiently small $|q| \in \mathbb{C}$. In this case we may choose the quantum coefficient ring $\Lambda := \mathbb{C}$ as in Example 11.1.4(viii). Denote by

$$\mathcal{H} = \mathrm{QH}^{\mathrm{ev}}(M; \mathbb{C}) = \bigoplus_{i=0}^{n} \mathrm{QH}^{2i}(M; \mathbb{C})$$

the even part of the quantum cohomology of M. The above assumptions imply that the quantum cup product is well defined on $\mathcal{H}$. We now show that $\mathcal{H}$ is a Frobenius algebra in the following sense.

DEFINITION 11.5.1. *A* **Frobenius algebra** $\mathcal{A}$ *over a field* $F = \mathbb{R}$ *or* $\mathbb{C}$ *is a commutative associative algebra over* F *with unit* 1, *that is equipped with an* F*-linear function* $\alpha : \mathcal{A} \to F$ *such that the pairing* $\langle a, b \rangle := \alpha(ab)$ *is nondegenerate, that is* $\langle a, b \rangle = 0$ *for all* b *implies* $a = 0$.

Another way of formulating this definition is to require that there be a nondegenerate bilinear pairing $(a, b) \mapsto \langle a, b \rangle \in F$ such that

$$\langle ab, c \rangle = \langle a, bc \rangle, \qquad a, b, c \in \mathcal{A}$$

and then set $\alpha(a) := \langle 1, a \rangle$.

In the case of $\mathcal{H}$, we define the product as the quantum cup product, and we define α to be integration over the fundamental homology class of M. Thus for $a_t = \sum_i t_i e_i$ we have

$$\alpha(a_t) := \int_M a_t = t_N \in \mathbb{C}.$$

The corresponding pairing does not agree with that in Proposition 11.1.11 (iii) since the latter takes values in the ground ring R and is not linear over $\Lambda := \mathbb{C}$. However, the proof of Proposition 11.1.11 applies to show that the above definitions do make

$\mathcal{H}$ into a finite dimensional Frobenius algebra over $\mathbb{C}$. If we must use the Novikov ring as coefficients rather than a field, the above concepts should be appropriately modified: see Abrams [**7**].

REMARK 11.5.2. The Frobenius algebra structure of $QH^*(M)$ can be helpful in understanding its properties; cf. Abrams [**7**] and the discussion in the Appendix to [**273**].

Following Dubrovin [**93**] we now interpret the quantum cup product

$$\mathcal{H} \times \mathcal{H} \to \mathcal{H} : (a, b) \mapsto a * b$$

as a connection on the tangent bundle of $\mathcal{H}$ given by

$$\nabla_Y X(a) = dX(a)Y(a) + X(a) * Y(a) \tag{11.5.2}$$

for $a \in \mathcal{H}$ and two vector fields $X, Y : \mathcal{H} \to \mathcal{H}$. Here we identify all the tangent spaces of $\mathcal{H}$ with $\mathcal{H}$ in the obvious way, so that a vector field X is a map $a \mapsto X(a) \in T_a\mathcal{H} \equiv \mathcal{H}$ and its derivative $dX(a) : T_a\mathcal{H} \to T_{X(a)}\mathcal{H}$ at a is a linear map of $\mathcal{H}$ to itself. The next lemma shows that commutativity of the quantum product can be interpreted as the vanishing of the torsion and associativity as vanishing of the curvature.

LEMMA 11.5.3. (i) *The connection (11.5.2) is torsion free, i.e.*

$$[X, Y] = \nabla_Y X - \nabla_X Y.$$

(ii) *The connection (11.5.2) is flat, i.e.*

$$\nabla_X \nabla_Y Z - \nabla_Y \nabla_X Z + \nabla_{[X,Y]} Z = 0.$$

(iii) *The connection (11.5.2) satisfies*

$$\nabla_X \mathbb{1} = X$$

where $\mathbb{1}$ *denotes the constant vector field defined by the unit element* $1 \in \mathcal{H}$.

PROOF. Our sign convention for the Lie bracket is

$$[X, Y] = dX \cdot Y - dY \cdot X$$

where $dX(a)$ denotes the differential of the map $X : \mathcal{H} \to \mathcal{H}$ and should be thought of as a linear transformation of $\mathcal{H}$ which takes Y to $dX \cdot Y$. The first statement is now obvious. To prove flatness note that

$$\begin{aligned}
\nabla_X \nabla_Y Z &= \nabla_X (dZ \cdot Y + Z * Y) \\
&= d(dZ \cdot Y + Z * Y) \cdot X + (dZ \cdot Y + Z * Y) * X \\
&= d^2 Z(Y, X) + dZ \cdot dY \cdot X + Z * (dY \cdot X) \\
&\quad + (dZ \cdot X) * Y + (dZ \cdot Y) * X + (Z * Y) * X \\
&= \nabla_{dY \cdot X} Z + d^2 Z(X, Y) + Z * (Y * X) \\
&\quad + (dZ \cdot X) * Y + (dZ \cdot Y) * X.
\end{aligned}$$

The third statement is obvious. □

The above connection is rather special because the 1-form

$$A : T\mathcal{H} \to \mathrm{End}(\mathcal{H})$$

given by

$$A_a(x)y = x * y$$

is constant and does not depend on the base point a. A general connection 1-form $A \in \Omega^1(\mathcal{H}, \mathrm{End}(\mathcal{H}))$ can be interpreted as a family of products

$$\mathcal{H} \times \mathcal{H} \to \mathcal{H} : (x, y) \mapsto x *_a y, \tag{11.5.3}$$

parametrized by the elements of $\mathcal{H}$ itself, via the formula

$$A_a(x)y = x *_a y \tag{11.5.4}$$

for $a \in \mathcal{H}$ and $x, y \in T_a\mathcal{H} = \mathcal{H}$. The corresponding connection, when regarded as a differential operator $C^\infty(\mathcal{H}, \mathcal{H}) \to \Omega^1(\mathcal{H}, \mathcal{H})$ is given by

$$\nabla = d + A$$

or, more explicitly, by

$$\nabla_Y X(a) = dX(a)Y(a) + X(a) *_a Y(a) \tag{11.5.5}$$

for two vector fields $X, Y : \mathcal{H} \to \mathcal{H}$. The properties of the **Dubrovin connection** (11.5.5) are related to the products (11.5.3) as follows. We assume here that the map $\mathcal{H} \times \mathcal{H} \times \mathcal{H} \to \mathcal{H} : (a, x, y) \mapsto x *_a y$ is holomorphic and complex bilinear in x and y. We shall consider the case where these products, together with the constant inner product $\langle \cdot, \cdot \rangle$ (which is assumed bilinear rather than Hermitian), determine a family of Frobenius algebra structures, one on each tangent space $T_a\mathcal{H} = \mathcal{H}$.

LEMMA 11.5.4. *Assume that each product $(x, y) \mapsto x *_a y$ satisfies the Frobenius condition. Then the following holds.*

(i) *The connection (11.5.5) is torsion free if and only if the products (11.5.3) are commutative, i.e. for all $a, x, y \in \mathcal{H}$, we have*

$$x *_a y = y *_a x.$$

(ii) *Assume the connection (11.5.5) is torsion free. Then the 1-form (11.5.4) is closed if and only if there exists a holomorphic function $\mathcal{S} : \mathcal{H} \to \mathbb{C}$ such that*

$$\langle x *_a y, z \rangle = \partial^3 \mathcal{S}_a(x, y, z) \tag{11.5.6}$$

(iii) *Assume the connection (11.5.5) is torsion free. Then the 1-form (11.5.4) satisfies $A \wedge A = 0$ if and only if the products (11.5.3) are all associative.*

(iv) *The connection (11.5.5) satisfies $\nabla_X \mathbb{1} = X$ if and only if $\mathbb{1}$ is a unit for all the products (11.5.3).*

PROOF. Statements (i), (iii), and (iv) follow immediately from Lemma 11.5.3. To prove (ii) note that the function $\phi_a(x, y, z) = \langle x *_a y, z \rangle$ is symmetric in x, y, z. Moreover, a simple calculation shows that its derivative

$$\psi_a(w, x, y, z) = \frac{d}{dt}\Big|_{t=0} \phi_{a+tw}(x, y, z)$$

is symmetric in w, x, y, z if and only if the connection 1-form A given by (11.5.4) is closed. Now the symmetry of ψ_a is equivalent to the existence of a holomorphic function $\mathcal{S} : \mathcal{H} \to \mathbb{C}$ such that

$$\phi_a(x, y, z) = \partial^3 \mathcal{S}_a(x, y, z).$$

In fact, an explicit formula for $\mathcal{S}$ is given by

$$\mathcal{S}(a) = \tfrac{1}{2} \int_0^1 (1 - t)^2 \phi_{ta}(a, a, a)\, dt.$$

This proves the lemma. □

REMARK 11.5.5. The proof of statement (ii) in the previous lemma can be formulated more explicitly in terms of a complex basis $e_0, \dots, e_n$ of the cohomology $\mathcal{H} = H^{\mathrm{ev}}(M, \mathbb{C})$. Then $\mathcal{H}$ can be identified with $\mathbb{C}^{n+1}$ via the isomorphism $\mathbb{C}^{n+1} \to \mathcal{H} : x \mapsto \sum_i x^i e_i$. Define the functions $A_{ijk} : \mathbb{C}^{n+1} \to \mathbb{C}$ by

$$A_{ijk}(x) = \langle e_i *_a e_j, e_k \rangle, \qquad a = \sum_{i=0}^{n} x^i e_i.$$

These functions are holomorphic by assumption. They are symmetric under permutations of i, j, k if and only if the products (11.5.3) are symmetric and satisfy the Frobenius condition. If this holds then there exists a holomorphic function $\mathcal{S} : \mathbb{C}^{n+1} \to \mathbb{C}$ such that

$$\frac{\partial^3 \mathcal{S}}{\partial x^i \partial x^j \partial x^k} = A_{ijk}$$

if and only if the derivatives $\partial_\ell A_{ijk}$ are symmetric under permutations of i, j, k, ℓ. An explicit formula for $\mathcal{S}$ is given by

$$\mathcal{S}(x) = \tfrac{1}{2} \int_0^1 (1-t)^2 \sum_{i,j,k} A_{ijk}(tx) x^i x^j x^k \, dt.$$

A holomorphic function

$$\mathcal{S} : \mathcal{H} \to \mathbb{C}$$

is called a **potential function** for the Dubrovin connection (11.5.5) if its third derivatives satisfy the equation (11.5.6). In view of Lemma 11.5.4 such a function exists if and only if the connection satisfies the Frobenius condition, is torsion free, and the connection 1-form is closed. Conversely, if the potential function $\mathcal{S} : \mathcal{H} \to \mathbb{C}$ is given and and the products (11.5.3) are defined by (11.5.6) then these products are automatically commutative and satisfy the Frobenius condition. Moreover, $\mathbb{1}$ is a unit if and only if the function $\mathcal{S}$ satisfies

$$\partial^3 \mathcal{S}_a(\mathbb{1}, x, y) = \langle x, y \rangle. \tag{11.5.7}$$

Associativity translates into the **WDVV equation**

$$\sum_{i,j} \partial^3 \mathcal{S}_a(w, x, e_i) g^{ij} \partial^3 \mathcal{S}_a(e_j, y, z) = \sum_{i,j} \partial^3 \mathcal{S}_a(w, z, e_i) g^{ij} \partial^3 \mathcal{S}_a(e_j, x, y) \tag{11.5.8}$$

where e_i denotes a basis of $\mathcal{H}$, $g_{ij} := \langle e_i, e_j \rangle$, and g^{ij} denotes the inverse matrix.

REMARK 11.5.6. Choose a basis of $\mathcal{H}$ as in Remark 11.5.5 such that $e_0 = 1$. Then condition (11.5.7) takes the form

$$\frac{\partial^3 \mathcal{S}}{\partial x^0 \partial x^i \partial x^j} = g_{ij}. \tag{11.5.9}$$

and the WDVV-equation can be written as

$$\sum_{\nu,\mu} \frac{\partial^3 \mathcal{S}}{\partial x^i \partial x^j \partial x^\nu} g^{\nu\mu} \frac{\partial^3 \mathcal{S}}{\partial x^\mu \partial x^k \partial x^\ell} = \sum_{\nu,\mu} \frac{\partial^3 \mathcal{S}}{\partial x^i \partial x^\ell \partial x^\nu} g^{\nu\mu} \frac{\partial^3 \mathcal{S}}{\partial x^\mu \partial x^j \partial x^k}$$

for all i, j, k, ℓ.

Manifolds with this structure are called **Frobenius manifolds**. More precisely, a (complex) Frobenius manifold is a triple $(\mathcal{H}, g, \nabla)$ consisting of smooth complex manifold, a flat complex bilinear metric $g(x, y) := \langle x, y \rangle$ and a second torsion free complex linear connection ∇ (the Dubrovin connection) that is related to the Levi–Civita connection d of the metric by the equation $\nabla = d + A$ where A is closed and $A \wedge A = 0$. It follows that the tangent spaces of $\mathcal{H}$ are Frobenius algebras and that the connection has a potential function $\mathcal{S}$. The next lemma spells out different equivalent ways of expressing these conditions.

LEMMA 11.5.7. *Let $\mathcal{S} : \mathcal{H} \to \mathbb{C}$ be a holomorphic function that satisfies (11.5.7) and let the quantum multiplication $x *_a y$ be defined by (11.5.3) for $a \in \mathcal{H}$. Then the following are equivalent.*

(i) *The products $x *_a y$ are associative.*

(ii) *The Dubrovin connection (11.5.5) is flat.*

(iii) *The potential function $\mathcal{S}$ satisfies the WDVV-equation (11.5.8).*

PROOF. Statements (i) and (ii) are equivalent because the connection 1-form A of (11.5.4) is closed by part (ii) of Lemma 11.5.4, and, by part (iii) of that lemma, the condition $A \wedge A = 0$ is equivalent to associativity. Since the curvature of ∇ is the endomorphism valued 2-form

$$F_A = dA + A \wedge A$$

we obtain in fact that associativity is equivalent to the condition that all the connections

$$\nabla_\lambda = d + \lambda A$$

with $\lambda \in \mathbb{R}$ are flat.

To prove the equivalence of (i) and (iii) note that $x = \sum_i \langle x, e_i \rangle g^{ij} e_j$ and hence

$$\langle x, y \rangle = \sum_{i,j} \langle x, e_i \rangle g^{ij} \langle e_j, y \rangle$$

for all $x, y \in \mathcal{H}$. This implies

$$\begin{aligned}
\sum_{i,j} \partial^3 \mathcal{S}_a(w, x, e_i) g^{ij} \partial^3 \mathcal{S}_a(e_j, y, z) &= \sum_{i,j} \langle w *_a x, e_i \rangle g^{ij} \langle g_j, y *_a z \rangle \\
&= \langle w *_a x, y *_a z \rangle \\
&= \langle w, x *_a (y *_a z) \rangle
\end{aligned}$$

and so associativity is equivalent to the condition that the left hand side of (11.5.8) is symmetric under permutations of x, y, and z. This proves the lemma. □

If we consider the quantum deformation of the cup product which is independent of a (namely the small quantum product) then the corresponding potential function $\mathcal{S}$ is a cubic polynomial. By definition of the quantum cup product this cubic polynomial is given by

$$\mathcal{S}_3(a) = \frac{1}{3!} \sum_{A \in K^{\mathrm{eff}}} \mathrm{GW}^M_{A,3}(a, a, a).$$

(For this to make sense we are assuming that the series in Example 11.1.4(viii) converge.)

The potential function of the **big quantum cohomology** is the full Gromov–Witten potential

$$\Phi(q, t) := \sum_\alpha \sum_d \frac{\varepsilon(\alpha)}{\alpha!} \mathrm{GW}^M_{A_d, |\alpha|}(e^\alpha) q^d t^\alpha.$$

We saw in Theorem 11.2.4 that Φ is a solution of the WDVV-equation (11.5.8) in the ring of formal power series in the variables $(q,t) \in \mathbb{C}^m \times \mathcal{H} = \mathbb{C}^{m+N}$. If $\Phi(q,t)$ converges on some open neighbourhood of $(0,0) \in \mathbb{C}^m \times \mathcal{H}$ then with appropriate choice of $q \in \mathbb{C}^m$ it defines a holomorphic function $\mathcal{S}_\Phi$ in a some neighbourhood $\mathcal{U} \subset \mathcal{H}$ of zero. Moreover the string equation coincides with (11.5.9). Hence $\mathcal{U}$ becomes a Frobenius manifold. The corresponding family of products $x *_a y, a \in \mathcal{U}$, on $\mathcal{H}$ is called the **big quantum cohomology**. Explicitly, $x *_a y$ is defined by:

$$\langle x *_a y, z \rangle := \sum_p \frac{1}{p!} \sum_A \mathrm{GW}^M_{A,p+3}(x, y, z, a, \ldots, a).$$

If this convergence assumption does not hold then one must think of $\mathcal{U}$ as a formal neighbourhood of the origin in $\mathcal{H}$: see Manin [**286**, §5.2.1].

In general the big quantum cohomology contains considerably more information than the small quantum cohomology. However, Kontsevich and Manin showed in [**216**] that if the cohomology ring $H^*(M)$ of M is generated by $H^2(M)$ then the big quantum product contains no new information.

THEOREM 11.5.8 (Kontsevich–Manin). *Let (M, ω) be a semipositive symplectic manifold such that $H^*(M)$ is generated as a ring by $H^2(M)$. Then the structure constants for $\mathrm{QH}^*(M; \Lambda)$ determine the Gromov–Witten potential.*

For general manifolds, all one can say is that the restriction of the big quantum cohomology to $H^2(M;\mathbb{C})$ is determined by the small one. In other words the family of products $*_a$ for $a \in H^2(M;\mathbb{C})$ on $\mathcal{H}$ are determined by the small product $*$. This is equivalent to saying that the Gromov–Witten invariants

$$\mathrm{GW}^M_{A,p+3}(x, y, z, a, \ldots, a), \qquad a \in H^2(M;\mathbb{C}),$$

are determined by the invariants $\mathrm{GW}^M_{A,3}(x,y,z)$, which is an immediate consequence of the divisor equation.

EXERCISE 11.5.9. Find an explicit formula for the family of products $h *_t h$ on $\mathcal{H} := \mathrm{QH}^2(Z_5;\mathbb{C})$, where Z_5 is the quintic hypersurface in $\mathbb{C}P^4$ discussed in Section 11.3.4 and t is the coordinate on the one dimensional space $\mathcal{H}$.

Frobenius manifolds have a rich and interesting structure. They arise in the study of versal deformations of an isolated singularity, in the study of integrable systems, and, in general, in the study of moduli spaces that have a natural flat connection. We cannot describe the general theory here. All we shall try to do is point out some of the special features of the particular model now under consideration.

We first describe the extra structure that corresponds to the divisor equation. This implies that $\Phi(q,t)$ is the sum of a function of the variables t_0, $q_i e^{t_i}$, $i = 1, \ldots, m$, and t_ν, $\nu > m$ with a quadratic polynomial the variables t_ν, $0 \le \nu \le N$. Thus adding integer multiples of $2\pi i$ to the variables $t_1, \ldots, t_m$ that parametrize $H^2(M, \mathbb{C})$ does not change $\partial^3 \mathcal{S}_\Phi$ and hence does not change the connection. Therefore the Frobenius manifold structure on $\mathcal{U} \subset \mathcal{H}$ descends to a subset of the quotient

$$\overline{\mathcal{H}} := \frac{\mathrm{QH}^{\mathrm{ev}}(M,\mathbb{C})}{H^2(M; 2\pi i\mathbb{Z})} = \mathbb{C}e_0 \oplus H^2(M; \mathbb{C}/2\pi i\mathbb{Z}) \oplus \bigoplus_{i>1} \mathrm{QH}^{2i}(M,\mathbb{C}).$$

Another consequence is that the Frobenius manifold structure on $\mathcal{U}$ does not depend in any essential way on the values $q_i \in \mathbb{C}$ that were chosen for the variables q_i since we can compensate for a change in these values by rescaling the $t_i, 1 \le i \le m$.

Conversely, we can compensate for a change in the values of the t_i by altering $q_i \in \mathbb{C}$, or equivalently by altering the class of the symplectic form.[3] Thus the family of Frobenius algebras $T_a\mathcal{H}$, $a \in H^2(M;\mathbb{C})$, can be identified with the family of Frobenius algebras given by the small quantum cohomology of (M, ω_q) for an appropriate family of symplectic (or Kähler) forms ω_q: cf. Exercise 11.5.9.

Let us apply this remark in the case of Calabi–Yau 3-folds. We showed in Section 11.3.4 that there is an induced quantum product structure on $H^2(M;\mathbb{C})$ that depends on the class $[\omega_q]$ of the symplectic form. Using the divisor equation as above, one can show that this structure descends to the complexified Kähler cone $\mathcal{K} := H^2(M;\mathbb{C}/2\pi i\mathbb{Z})$ of M, defining an analogue of the Dubrovin connection on the trivial bundle $\mathcal{K} \times \mathcal{H} \to \mathcal{K}$. This connection records the variation of the small quantum product as the class $[\omega_q]$ varies in $\mathcal{K}$ and is known as the A-model connection. Though it is flat, it has monodromy around the singular point $q = 0$. This connection turns out to be mirror to the Gauss–Manin connection for deformations of Hodge structures: see Cox–Katz [**76**].

Another important fact about the Gromov–Witten potential Φ is that it is homogeneous of degree $6 - \dim M$ when $\deg(t_\nu) := 2 - \deg(e_\nu)$. Moreover we showed in Exercise 11.2.6 that there is a differential operator $\mathcal{E}$ such that $\mathcal{E}\Phi = (3-n)\Phi$. The formula given for $\mathcal{E}$ involves the terms $\sum c^i q_i \partial/\partial q_i$ (where $\sum c^i e_i$ is the first Chern class) so that $\mathcal{E}$ is not a vector field on $\mathcal{H}$. However if we replace these terms by $\sum_i c^i \partial/\partial t_i$ then we do get a vector field on $\mathcal{H}$, namely

$$E(t) = \sum_{i=1}^m c^i e_i + \sum_{\nu=0}^N (1 - \frac{\deg(e_\nu)}{2}) t_\nu e_\nu.$$

(As before we think of vector fields as maps $\mathcal{H} \to \mathcal{H}$; thus e_ν corresponds to the differentiation $\partial/\partial e_\nu$.) The divisor equation implies that $E\Phi - (3-n)\Phi$ is a quadratic polynomial. It follows that E acts conformally on the Frobenius structure: see Exercise 11.5.10 below. In cases when E satisfies an extra condition called semisimplicity, one can introduce new canonical coordinates on the Frobenius manifold $\mathcal{H}$. This is an important part of the structure of the Frobenius manifolds that arise in quantum cohomology. For more information see Manin [**286**].

EXERCISE 11.5.10. Let E be as defined above. Check that the following identities hold for all vector fields X, Y on $\mathcal{H}$:

$$E(g(X,Y)) + g([E,X],Y) + g(X,[E,Y]) = Dg(X,Y),$$

where $D := 2 - \dim M/2$, and

$$[E, X * Y] + [E,X] * Y + X * [E,Y] = X * Y.$$

The first shows that E is **conformal**, that is $\mathcal{L}_E(g) = Dg$. Here $\mathcal{L}_E$ denotes the Lie derivative, defined using our sign conventions on vector fields by $\mathcal{L}_E(X) := -[X,Y]$. The second identity shows that E also acts conformally on the product structure, namely $\mathcal{L}_E(*) = d_0(*)$, where $d_0 = 1$. Vector fields satisfying these two identities are called Euler fields.

[3] A word of warning about notation: though our variables are called q, t they have a different meaning from the variables q, t in the coefficient ring $\Lambda^{\mathrm{univ}}[q, q^{-1}]$. In particular, as is evident from the formula (11.5.1) the value $q \in \mathbb{C}$ now carries the information about the symplectic form.

We end with a few remarks, pointing in directions that are developed much further in Givental's work **[150]** on the mirror conjecture for toric complete intersections and his development with Coates **[71, 72, 152]** of the Lagrangian cone picture of Frobenius structures.

First of all, recall that in the proof of Lemma 11.5.7 above we observed that a potential function $\mathcal{S}$ satisfying the WDVV equation defines a family of flat Dubrovin connections $\nabla_\lambda := d+\lambda A$ where $\lambda \in \mathbb{C}$. One can think of λ as an element of the ring $\mathbb{C}[\lambda, \lambda^{-1}]$ which can then be interpreted as the ring of functions on $\mathbb{C} \setminus \{0\}$ (Manin) or as the localized cohomology ring of BS^1 (Givental). In some sense this is the algebraic equivalent of replacing J-holomorphic curves $u : S^2 \to M$ by their graphs $\widetilde{u} : S^2 \to S^2 \times M$, and then working with a theory that is equivariant with respect to the circle action on $S^2 \times M$ given by rotating the sphere. However, in **[146]** Givental thinks in terms of an equivariant Floer theory on the loop space on M. Although this version of Floer theory has not yet been set up, the underlying ideas have proved very powerful, inspiring all Givental's subsequent work in this area.

Secondly, Givental discovered that the sections S of $T\mathcal{H}$ that are ∇_λ-flat for all λ play an important role in the theory. He assembles appropriate components of these flat sections to form the J-function, which is an essential ingredient of his work on the mirror conjecture: see Cox–Katz **[76]**. These two aspects of the theory are related: Givental found a formula for the flat sections in terms of k-pointed gravitational descendants which he interprets in **[147]** via localization from the S^1-equivariant theory on $S^2 \times M$. (The gravitational descendants arise from the Euler class of the normal bundle to the fixed point set.)

Therefore it turns out that one gets a much more satisfactory theory by looking at the descendant potential $\mathcal{F}$ (the generating function of the genus zero gravitational descendants) rather than at the Gromov–Witten potential Φ considered here. After a suitable linear change of variables (the dilaton shift), $\mathcal{F}$ can be considered as a (formal) function on the space $\mathcal{H}[z]$ of polynomial functions with coefficients in $\mathcal{H} := H^*(M; \mathbb{C})$. (Here z should be interpreted as $1/\lambda$ where λ parametrizes the family of Dubrovin connections.) It satisfies versions of the string, divisor and WDVV equations. In his work with Coates, Givental **[71]** interprets these fundamental equations entirely in terms of the geometry of the Lagrangian submanifold $\mathcal{L}$ of the cotangent bundle $T^*(\mathcal{H}[z])$ formed by the graph of $d\mathcal{F}$. This gives a very powerful framework in which to understand the structure of the genus zero Gromov–Witten invariants. In particular, it turns out that $\mathcal{L}$ is a cone since $\mathcal{F}$ is homogeneous of degree 2 in the chosen coordinates, while the J-function is simply the intersection of $\mathcal{L}$ with the affine subspace $T^*(\mathcal{H}[z])|_{-z+\mathcal{H}}$ formed by restricting the cotangent bundle to the finite dimensional slice $-z + \mathcal{H}$. Moreover, one can understand the transformations needed to prove the mirror theorem for toric complete intersections in terms of symplectic transformations of the ambient linear space $T^*(\mathcal{H}[z])$ and their effect on $\mathcal{L}$. The corresponding picture for higher genus invariants is not yet understood.

CHAPTER 12

Floer Homology

There is a completely different approach to quantum cohomology arising from Floer's proof of the Arnold conjecture and the resulting notion of Floer homology (cf. [**113**] and [**116**]). In this chapter we give a general outline of the construction of symplectic Floer homology without providing the additional analytical details needed to carry out this program. Floer originally developed his homology theory in the context of monotone symplectic manifolds. This was later extended by Hofer and Salamon in [**180**] and Ono in [**313**] to the semipositive case and it is this extension which we explain in Sections 12.1, 12.2, and 12.3 below. We also explain the Piunikhin–Salamon–Schwarz [**323**] construction of an isomorphism from quantum cohomology to Floer cohomology that intertwines the quantum cup product with the pair-of-pants product on Floer cohomology.

The main missing technical points in our exposition are, first, the Fredholm theory for the solutions of the Floer equations on the cylinder and the Conley–Zehnder index, second, the transversality theory for Floer connecting orbits and, third, the Floer gluing theorem. In addition to Floer's original papers, the details of the proofs can be found in Floer–Hofer–Salamon [**119**], Floer–Hofer [**118**], Schwarz [**358**], Salamon–Zehnder [**356**], Robbin–Salamon [**335, 336**], Salamon [**353**], and the book Audin–Damian [**24**]. As usual, we shall not discuss the extension to general compact symplectic manifolds by Fukaya–Ono [**127**] and Liu–Tian [**249**].

After introducing the basic setup we explain the relevance of Floer homology to the study of the group $\mathrm{Ham}(M,\omega)$ of Hamiltonian symplectomorphisms. In Section 12.4 we describe the construction of the spectral invariants of the elements of the universal cover $\widetilde{\mathrm{Ham}}(M,\omega)$. These invariants are based on the filtration of the Floer chain complex by the symplectic action functional. They give rise to an interesting variant of the Hofer metric due to Schwarz and Oh. Our exposition mostly follows the work of Schwarz [**360**] and Oh [**306, 307**]. As well as the usual semipositivity hypothesis, we shall make the simplifying assumption that $[\omega]$ is integral. In Section 12.5 we return to the topic of the Seidel representation, formulating it in terms of Floer homology and explaining its effect on the spectral invariants.

There is another version of symplectic Floer homology (that we shall not discuss) for intersections of Lagrangian submanifolds. This theory was first developed by Floer [**113, 114**] under the hypotheses that the two Lagrangian submanifolds are Hamiltonian isotopic and $\pi_2(M,L)=0$. An extension to the monotone case is due to Oh [**303**], and later Fukaya–Oh–Ohta–Ono [**128**] developed an obstruction theory for Lagrangian Floer homology that applies to all compact symplectic manifolds.

Donaldson observed that the Floer homology groups can be viewed as the morphisms of a category whose objects are the Lagrangian submanifolds (respectively the symplectomorphisms) of a given symplectic manifold (M,ω). The composition rule is defined by counting holomorphic triangles, an analogue of the pair-of-pants product. This was taken further by Fukaya [**125**] who discovered the underlying A^∞ structure of Floer homology on the chain level.

In Section 12.6 we outline the construction of Donaldson's category for symplectomorphisms. The resulting product structures were used by Seidel [**362**] in an ingenious way to show that there are symplectomorphisms that are smoothly but not symplectically isotopic. He discovered an exact sequence in symplectic Floer homology and was able to use it, among many other applications, to compute the Floer homology groups of certain symplectomorphisms (thereby distinguishing their symplectic isotopy classes). The Seidel exact sequence, especially in connection with the Fukaya category, has many different and remarkable consequences, concerning, for example, mirror symmetry for affine Calabi–Yau manifolds or the topology of exact Lagrangian submanifolds of the cotangent bundles of spheres.

There are yet more versions of Floer homology as invariants of 3-manifolds [**117**] that play an important role as recipients for the Donaldson invariants of smooth 4-manifolds with boundary. In this theory anti-self-dual instantons on 4-manifolds play the role that J-holomorphic curves do in symplectic Floer theory. We shall not discuss this theory in this chapter, although it does have interesting relations to symplectic Floer homology (see Dostoglou–Salamon [**90, 91, 92**]). An excellent reference for this circle of ideas is the recent book by Donaldson [**87**].

The quantum field theory picture that is so apparent in Floer–Donaldson theory in three and four dimensions also appears in symplectic topology in the recent work by Eliashberg–Givental–Hofer [**101**]. They develop an analogue of Floer homology for contact manifolds, called (in its simplest version) contact homology. These homology groups are the recipients for the Gromov–Witten invariants of symplectic manifolds with contact boundaries. Gluing along contact boundaries then gives the Gromov–Witten invariants of closed symplectic manifolds. Again there is a Lagrangian version of this theory that involves Legendrian submanifolds of the contact boundaries. These new developments have many interesting applications. However, describing them in any detail goes far beyond the scope of this book.

We close this chapter with a brief discussion of the symplectic vortex equations (Section 12.7). These are a new set of equations for maps into a symplectic manifold M that is equipped with a Hamiltonian G-action. They were discovered, independently, by Salamon [**64**] and Mundet [**294, 295**]; in the physics literature they are known as gauged sigma models. They generalize the usual vortex equations for sections of a complex line bundle over a Riemann surface. In the trivial case $\mathrm{G} = \{\mathbb{1}\}$ they simply reduce to J-holomorphic curves. They also specialize to the anti-self-dual instantons, or to the Seiberg–Witten equations, on product 4-manifolds $\Sigma \times S$ for suitably chosen (infinite dimensional) ambient manifolds M. In the finite dimensional case they arise naturally in the study of J-holomorphic curves in symplectic quotients. As already mentioned in Section 11.3, the symplectic vortex invariants can be used to calculate the quantum cohomology rings of many interesting symplectic manifolds.

12.1. Floer's cochain complex

Let (M,ω) be a semipositive monotone symplectic manifold and let

$$H_t = H_{t+1} : M \to \mathbb{R}$$

be a smooth 1-periodic family of Hamiltonian functions. Denote by $X_t : M \to TM$ the Hamiltonian vector field defined by

$$\iota(X_t)\omega = dH_t$$

and consider the time dependent Hamiltonian differential equation

$$\dot{x}(t) = X_t(x(t)). \tag{12.1.1}$$

The solutions of (12.1.1) generate a family of symplectomorphisms $\psi_t : M \to M$ via

$$\frac{d}{dt}\psi_t = X_t \circ \psi_t, \qquad \psi_0 = 0.$$

By definition, the fixed points of the time-1-map $\psi = \psi_1$ are in one-to-one correspondence with the 1-periodic solutions of (12.1.1). We now explain how to interpret the contractible 1-periodic solutions as the critical points of the symplectic action functional on the universal cover of the space $\mathcal{L}M$ of contractible loops in M.

For every contractible loop $x : \mathbb{R}/\mathbb{Z} \to M$ there is a smooth map $u : B \to M$, defined on the unit disc $B = \{z \in \mathbb{C} \,|\, |z| \le 1\}$, which satisfies $u(e^{2\pi i t}) = x(t)$. Two such maps u_1 and u_2 are called **equivalent** if their boundary sum $u_1\#(-u_2)$ is homologous to zero (in the space $H_2(M)$ of integral homology divided by torsion). We use the notation

$$[x, u_1] \sim [x, u_2]$$

for equivalent pairs and denote by $\widetilde{\mathcal{L}M}$ the space of equivalence classes. The elements of $\widetilde{\mathcal{L}M}$ will also be denoted by $\widetilde{x}$. The space $\widetilde{\mathcal{L}M}$ is the unique covering space of $\mathcal{L}M$ whose group of deck transformations is the image $H_2^S(M) \subset H_2(M)$ of the Hurewicz homomorphism $\pi_2(M) \to H_2(M)$. We denote by

$$H_2^S(M) \times \widetilde{\mathcal{L}M} \to \widetilde{\mathcal{L}M} : (A, \widetilde{x}) \mapsto A\#\widetilde{x}$$

the obvious action of $H_2^S(M)$ on $\widetilde{\mathcal{L}M}$.

The symplectic action functional $\mathcal{A}_H : \widetilde{\mathcal{L}M} \to \mathbb{R}$ is defined by

$$\mathcal{A}_H([x, u]) = -\int_B u^*\omega - \int_0^1 H_t(x(t))\, dt$$

and satisfies

$$\mathcal{A}_H(A\#\widetilde{x}) = \mathcal{A}_H(\widetilde{x}) - \omega(A). \tag{12.1.2}$$

This function can therefore be interpreted as a closed 1-form on the loop space $\mathcal{L}M$ rather than a function on the covering space $\widetilde{\mathcal{L}M}$, which is exactly the situation considered by Novikov (see Remark 11.1.6).

In Section 9.1 we have seen that the critical points of $\mathcal{A}_H$ are precisely the equivalence classes $[x, u]$ where $x(t) = x(t+1)$ is a contractible periodic solution of (12.1.1) (Lemma 9.1.8). We shall denote by $\widetilde{\mathcal{P}}(H) \subset \widetilde{\mathcal{L}M}$ the set of critical points and by $\mathcal{P}_0(H) \subset \mathcal{L}M$ the corresponding set of periodic solutions. Floer homology is essentially an infinite dimensional version of Morse–Novikov theory for the symplectic action functional. We assume throughout that all the contractible periodic solutions of (12.1.1) are nondegenerate (i.e. the eigenvalues of the differential $d\psi_1(x(0))$ are all different from one for every $x \in \mathcal{P}_0(H)$). This implies that $\mathcal{A}_H$ is a Morse function on $\widetilde{\mathcal{L}M}$, and also that $\mathcal{P}_0(H)$ is a finite set.

Consider the (downwards) gradient flow lines of $\mathcal{A}_H$ with respect to an L^2-metric on $\mathcal{L}M$ which is induced by an almost complex structure on M. These are solutions $\mathbb{R}^2 \to M : (s,t) \mapsto u(s,t)$ of the partial differential equation

$$\partial_s u + J(u)\left(\partial_t u - X_t(u)\right) = 0 \tag{12.1.3}$$

with periodicity condition $u(s,t+1) = u(s,t)$ and limit condition

$$\lim_{s\to\pm\infty} u(s,t) = x^{\pm}(t) \tag{12.1.4}$$

where $x^{\pm} \in \mathcal{P}_0(H)$. We denote by $\mathcal{M}(\widetilde{x}^-, \widetilde{x}^+) = \mathcal{M}(\widetilde{x}^-, \widetilde{x}^+; H, J)$ the space of all solutions of (12.1.3) and (12.1.4) with $\widetilde{x}^- \# u = \widetilde{x}^+$. This space is invariant under the action $u(s,t) \mapsto u(s+s_0, t)$ of the time shift $s_0 \in \mathbb{R}$; equivalence classes of solutions are often called **Floer connecting orbits** or **Floer trajectories.** Each element of $\mathcal{M}(\widetilde{x}^-, \widetilde{x}^+)$ has finite energy

$$\begin{aligned} E(u) &:= \tfrac{1}{2}\int_{-\infty}^{\infty}\int_0^1 \left(|\partial_s u|^2 + |\partial_t u - X_t(u)|^2\right) dtds \\ &= \mathcal{A}_H(\widetilde{x}^-) - \mathcal{A}_H(\widetilde{x}^+). \end{aligned} \tag{12.1.5}$$

Moreover, for a generic Hamiltonian function $H : M \times \mathbb{R}/\mathbb{Z} \to \mathbb{R}$, the space $\mathcal{M}(\widetilde{x}^-, \widetilde{x}^+)$ is a finite dimensional manifold of dimension

$$\dim \mathcal{M}(\widetilde{x}^-, \widetilde{x}^+) = \mu_{\mathrm{CZ}}(\widetilde{x}^-) - \mu_{\mathrm{CZ}}(\widetilde{x}^+).$$

Here the function $\mu_{\mathrm{CZ}} : \widetilde{\mathcal{P}}(H) \to \mathbb{Z}$ is a version of the Maslov index due to Conley and Zehnder. The Conley–Zehnder index $\mu_{\mathrm{CZ}}([x,u])$ is defined by trivializing the tangent bundle over the disc $u(B)$ and considering the path of symplectic matrices generated by the linearized Hamiltonian flow along $x(t)$. We refer to Salamon–Zehnder [**356**], Dostoglou–Salamon [**91**], Robbin–Salamon [**335**], and Salamon [**353**] for more details. Here we only point out that

$$\mu_{\mathrm{CZ}}(A\#\widetilde{x}) = \mu_{\mathrm{CZ}}(\widetilde{x}) - 2c_1(A) \tag{12.1.6}$$

for $\widetilde{x} \in \widetilde{\mathcal{P}}(H)$ and $A \in H_2^S(M)$. Moreover, the Conley–Zehnder index can be normalized so that

$$\mu_{\mathrm{CZ}}(\widetilde{x}) = n - \mathrm{ind}_H(x) = \mathrm{ind}_{-H}(x) - n \tag{12.1.7}$$

whenever $H_t \equiv H$ is a C^2-small Morse function and $\widetilde{x} = [x,u]$, where $x(t) \equiv x$ is a critical point of H of index $\mathrm{ind}_H(x)$ and $u(z) \equiv x$ is the constant disc.

REMARK 12.1.1. There are various choices of normalization in common use. Here we have chosen to emphasize the fact that Floer cohomology can be considered as the "middle dimensional semi-infinite cohomology" of the loop space (cf. Cohen–James–Segal [**73**], Givental [**146**]) and so have centered the grading around zero. Hence the natural isomorphisms from Floer cohomology to quantum cohomology increase degree by n.

In the case $\mu_{\mathrm{CZ}}(\widetilde{x}) - \mu_{\mathrm{CZ}}(\widetilde{y}) = 1$, the space $\mathcal{M}(\widetilde{x}, \widetilde{y})$ is a 1-dimensional manifold, and so each point in the quotient $\mathcal{M}(\widetilde{x}, \widetilde{y})/\mathbb{R}$ (with $\mathbb{R}$ acting by time shift) is isolated. Moreover, for a generic Hamiltonian H we have the following finiteness result.

PROPOSITION 12.1.2. *Assume (M,ω) is semipositive. Then, for a generic almost complex structure J and generic Hamiltonian H, we have*

$$\sum_{\substack{\omega(A)\le c\\ c_1(A)=0}} \#\frac{\mathcal{M}(\widetilde{x}, A\#\widetilde{y})}{\mathbb{R}} < \infty$$

for all $\widetilde{x},\widetilde{y}\in\widetilde{\mathcal{P}}(H)$ with $\mu_{\mathrm{CZ}}(\widetilde{x})-\mu_{\mathrm{CZ}}(\widetilde{y})=1$ and every constant c.

Here the condition on J is that all simple J-holomorphic spheres $v: S^2\to M$ are regular in the sense that the linearized operator D_v is onto. The condition on H is that the linearized operators for the Floer connecting orbits are onto (i.e. the symplectic action functional $\mathcal{A}_H$ is Morse–Smale) and, in addition, the evaluation map

$$\mathcal{M}(\widetilde{x},\widetilde{y};J,H)\times S^1\times\mathcal{M}(A;J)\times_{\mathrm{G}} S^2\to M\times M:(u,t,[v,z])\mapsto(u(0,t),v(z))$$

is transverse to the diagonal for all $\widetilde{x},\widetilde{y}\in\widetilde{\mathcal{P}}(H)$ and all $A\in H_2^S(M)$. To prove Proposition 12.1.2, one has to show that the relevant moduli spaces are compact. This will be the case if no bubbling occurs. The key observation is that, for a generic almost complex structure J, the set of points lying on a J-holomorphic sphere of Chern number zero forms a set in M of codimension four and so, for a generic H, no such sphere will intersect an isolated connecting orbit. Thus it follows from Floer–Gromov compactness that they cannot bubble off. Moreover, J-holomorphic spheres of negative Chern number do not exist by the semipositivity assumption. J-holomorphic spheres of Chern number at least one cannot bubble off because otherwise in the limit there would be a connecting orbit with negative index difference but such orbits do not exist generically. This is the essence of the proof of Proposition 12.1.2. Details are carried out in Hofer–Salamon [**180**].

Whenever $\mu_{\mathrm{CZ}}(\widetilde{x})-\mu_{\mathrm{CZ}}(\widetilde{y})=1$ we denote

$$n(\widetilde{x},\widetilde{y}) := \#\frac{\mathcal{M}(\widetilde{x},\widetilde{y})}{\mathbb{R}},$$

where the connecting orbits are to be counted with appropriate signs determined by a system of coherent orientations of the moduli spaces of connecting orbits as in Floer–Hofer [**118**]. These numbers determine a Floer cochain complex which we now explain. As in quantum cohomology there are many possibilities for coefficient rings. For simplicity, we restrict attention to coefficients in the Novikov ring Λ_ω. It is convenient to slightly modify the definition of Λ_ω (compared to Example 11.1.4 (iii)) and only allow for spherical homology classes. We fix a commutative ring with unit R and denote by Λ_ω the set of all functions

$$\lambda: H_2^S(M)\to R$$

that satisfy the finiteness condition

$$\#\left\{A\in H_2^S(M)\,|\,\lambda(A)\ne 0,\ \omega(A)\le c\right\}<\infty$$

for every $c\in\mathbb{R}$. This is a $2\mathbb{Z}$-graded commutative ring and R-module as explained in Example 11.1.4. The grading is given by $\deg(A) := 2c_1(A)$.

Define the **Floer cochain group** $\mathrm{CF}^* := \mathrm{CF}^*(H)$ as the set of functions

$$\xi:\widetilde{\mathcal{P}}(H)\to R$$

that satisfy the finiteness condition

$$\#\left\{\widetilde{x}\in\widetilde{\mathcal{P}}(H)\,|\,\xi(\widetilde{x})\neq 0,\ \mathcal{A}_H(\widetilde{x})\leq c\right\}<\infty \tag{12.1.8}$$

for every $c\in\mathbb{R}$. This complex CF^* is a module over the Novikov ring Λ_ω with action given by the formula

$$(\lambda*\xi)(\widetilde{x}):=\sum_A\lambda(A)\xi(A\#\widetilde{x}).$$

The grading is given by $\deg(\widetilde{x}):=\mu_{\mathrm{CZ}}(\widetilde{x})$; the degree-$k$ part CF^k consists of all $\xi\in\mathrm{CF}^*$ that are nonzero only on elements $\widetilde{x}\in\widetilde{\mathcal{P}}(H)$ with $\mu_{\mathrm{CZ}}(\widetilde{x})=k$. Thus the action $\xi\mapsto\lambda*\xi$ changes the degree unless $\lambda(A)$ is nonzero only when $c_1(A)=0$. Note that the dimension of $\mathrm{CF}^*(H)$ as a module over Λ_ω is precisely the number $\#\mathcal{P}_0(H)$ of contractible periodic solutions of the Hamiltonian system (12.1.1). In fact, this complex is isomorphic to the cochain complex whose elements are Λ-module homomorphisms $\xi:\widetilde{\mathcal{P}}(H)\to\Lambda$: see Remark 12.5.5.

The above numbers $n(\widetilde{x},\widetilde{y})$ determine a coboundary operator

$$\delta:\mathrm{CF}^*(H)\to\mathrm{CF}^*(H)$$

defined by

$$(\delta\xi)(\widetilde{x})=\sum_{\mu_{\mathrm{CZ}}(\widetilde{y})=\mu_{\mathrm{CZ}}(\widetilde{x})-1}n(\widetilde{x},\widetilde{y})\xi(\widetilde{y}). \tag{12.1.9}$$

Proposition 12.1.2 and the formula (12.1.2) guarantee that the sum on the right is finite for every $\widetilde{x}$ and that $\delta\xi\in\mathrm{CF}^*$ for every $\xi\in\mathrm{CF}^*$. Moreover, if $\xi\in\mathrm{CF}^k$ then $\delta\xi\in\mathrm{CF}^{k+1}$.

Floer [**116**] proved that the square of this operator is zero in the monotone case (see also McDuff [**255**] and Salamon–Zehnder [**356**]). He observed that the coefficients of the operator $\delta\circ\delta$ are given by counting the number of pairs of connecting orbits from $\widetilde{x}$ to $\widetilde{y}$ where $\widetilde{y}$ has index two less than $\widetilde{x}$. In this case the moduli space $\mathcal{M}(\widetilde{x},\widetilde{y})$ of trajectories from $\widetilde{x}$ to $\widetilde{y}$ has dimension two, and hence, after one divides out by the time shift, forms a one-dimensional oriented manifold whose boundary consists of the pairs counted by $\delta\circ\delta$. Conversely, the gluing construction shows that any such pair is the boundary of a component of $\mathcal{M}(\widetilde{x},\widetilde{y})$. Thus the objects counted by $\delta\circ\delta$ cancel out in pairs, so that $\delta\circ\delta=0$. This argument works in the monotone case because any bubbling would lead to Floer connecting orbits of negative index, which do not exist by transversality.

In order to extend this argument to semipositive manifolds, one must check that bubbling does not interfere, for example by causing the one-manifold $\mathcal{M}(\widetilde{x},\widetilde{y})/\mathbb{R}$ to have boundary components other than pairs of connecting orbits. Here the key observation is that 1-parameter families of connecting orbits with index difference two will still avoid the J-holomorphic spheres of Chern number zero because the orbits form a 3-dimensional set in M while the J-holomorphic spheres form a set of codimension four. Similarly, holomorphic spheres of Chern number one can only bubble off if they intersect a periodic solution, and this does not happen for a generic H because the points on these spheres form a set in M of codimension two while the periodic orbits form 1-dimensional sets. If a J-holomorphic sphere with Chern number at least two bubbled off then it would have to be attached to a connecting orbit of index at most zero, and these do not exist for generic pairs (H,J) because the connecting orbits are invariant under the 1-dimensional time shift. Hence no bubbling occurs for connecting orbits with index difference two and hence such orbits can only degenerate by splitting into a pair of orbits each with index difference one. As in the standard theory outlined above this shows that $\delta \circ \delta = 0$.

Hence the solutions of (12.1.3) determine a cochain complex (CF^*, δ), and its homology groups

$$\mathrm{HF}^*(H, J) := \mathrm{HF}^*(M, \omega, H, J) := \frac{\ker \delta}{\operatorname{im} \delta}$$

are called the **Floer cohomology groups** of the pair (H, J). Because the coboundary map is linear over Λ_ω it follows that the Floer cohomology groups form a module over Λ_ω.

The next theorem asserts that the Floer cohomology groups are independent of the almost complex structure J and the Hamiltonian H used to define them.

THEOREM 12.1.3. *Assume (M, ω) is semipositive. Then, for two pairs (H^α, J^α) and (H^β, J^β) that satisfy the regularity requirements for the definition of Floer cohomology, there is a natural isomorphism*

$$\Phi^{\alpha\beta} : \mathrm{HF}^*(M, \omega, H^\beta, J^\beta) \to \mathrm{HF}^*(M, \omega, H^\alpha, J^\alpha).$$

If (H^γ, J^γ) is another such pair then

$$\Phi^{\alpha\gamma} = \Phi^{\alpha\beta} \circ \Phi^{\beta\gamma}, \qquad \Phi^{\alpha\alpha} = \mathrm{id}.$$

These isomorphisms are linear over Λ_ω.

The isomorphisms $\Phi^{\alpha\beta}$ are called **Floer continuation maps**. To prove the theorem, choose a homotopy (H_s, J_s) from (H^α, J^α) to (H^β, J^β). This means that $(H_s, J_s) = (H^\alpha, J^\alpha)$ for s near $-\infty$ and $(H_s, J_s) = (H^\beta, J^\beta)$ for s near ∞. Next consider the finite energy solutions of the following time dependent version of equation (12.1.3)

$$\partial_s u + J_s(u)\left(\partial_t u - X_{s,t}(u)\right) = 0, \tag{12.1.10}$$

where $X_{s.t} \in \mathrm{Vect}(M)$ is defined by $\iota(X_{s,t})\omega = dH_{s,t}$. Here u satisfies the usual periodicity condition $u(s, t+1) = u(s, t)$. Any such solution will have limits

$$\lim_{s \to -\infty} u(s, t) = x^\alpha(t), \qquad \lim_{s \to +\infty} u(s, t) = x^\beta(t)$$

where $x^\alpha \in \mathcal{P}_0(H^\alpha)$ and $x^\beta \in \mathcal{P}_0(H^\beta)$. The crucial point is the energy identity

$$\begin{aligned} E(u) &:= \int_{-\infty}^{\infty} \int_0^1 |\partial_s u|^2 \, dt ds \\ &= \mathcal{A}_{H^\alpha}(x^\alpha) - \mathcal{A}_{H^\beta}(x^\beta) - \int_{-\infty}^{\infty} \int_0^1 (\partial_s H_{s,t})(u) \, dt ds. \end{aligned} \tag{12.1.11}$$

This shows, by the usual compactness and transversality arguments, that the solutions of (12.1.10) determine a chain map $\mathrm{CF}^*(H^\beta) \to \mathrm{CF}^*(H^\alpha)$ which is of degree zero and, choosing a homotopy of homotopies, one can see that the induced map on Floer cohomology is independent of the choice of the homotopy. These arguments are again precisely the same as in Floer's original proof in [**116**] for the monotone case, and for the present case they are carried out in Hofer–Salamon [**180**]. (See also Salamon–Zehnder [**356**] for the case $c_1 = [\omega] = 0$ on $H_2^S(M)$.)

Computing Floer cohomology. Following Floer's ideas one can attempt to compute the Floer cohomology groups by specializing to a (C^2-small) time independent Hamiltonian function and prove that the the relevant solutions of the Floer equations are also independent of t. It then follows that the Floer cohomology groups are naturally isomorphic to the usual cohomology of the underlying manifold M with coefficients in the Novikov ring Λ_ω. But these are precisely the quantum cohomology groups of M. In order not to be concerned with torsion, we take the ground ring R to be either $\mathbb{Q}$, $\mathbb{R}$, or $\mathbb{C}$. Although most of the following results continue to hold for every ground ring, we restrict to coefficients in a field of characteristic zero to make the present discussion consistent with the approach to quantum cohomology in Section 11.1 where we divided the cohomology by the torsion submodule.

The following theorem was proved by Floer [**116**] in the monotone case and by Hofer and Salamon [**180**] in the case where either $c_1(A) = 0$ for all $A \in \pi_2(M)$ or the minimal Chern number is $N \geq n$. The general case is treated with different methods in Piunikhin–Salamon–Schwarz [**323**] and Liu–Tian [**250**].

THEOREM 12.1.4. *Assume (M, ω) is semipositive. Then, for every regular pair (H^α, J^α), there is an isomorphism*

$$\Phi^\alpha : \mathrm{HF}^*(M, \omega, H^\alpha, J^\alpha) \to \mathrm{QH}^*(M; \Lambda_\omega)$$

(of degree n). These maps are natural in the sense that

$$\Phi^\alpha \circ \Phi^{\alpha\beta} = \Phi^\beta$$

and they are linear over Λ_ω.

For reasons that will be clear later, we shall call the isomorphisms Φ^α constructed in the proof of this theorem **PSS maps**. For clarity, we sometimes denote them by Φ^*_{PSS}.

Floer's original proof of Theorem 12.1.4 for the monotone case, as well as the proof in Hofer–Salamon [**180**] for the semipositive case, is based on the following idea. Let $f : M \to \mathbb{R}$ be a Morse function such that the negative gradient flow of f with respect to the metric $\omega(\cdot, J\cdot)$ is Morse–Smale and consider the time independent (or *autonomous*) Hamiltonian

$$H_t := -\varepsilon f, \qquad t \in \mathbb{R}.$$

If ε is sufficiently small then the 1-periodic solutions of (12.1.1) are precisely the critical points of f (i.e. $\mathcal{P}_0(H) = \mathrm{Crit}(f)$) and, by (12.1.7), the Conley–Zehnder index is related to the Morse index by

$$\mu_{\mathrm{CZ}}([x, u_x]) = \mathrm{ind}_f(x) - n$$

where $u_x : B \to M$ denotes the constant map $u_x(z) \equiv x$. The gradient flow lines $u : \mathbb{R} \to M$ of f are solutions of the ordinary differential equation

$$\dot{u}(s) + \nabla f(u(s)) = 0 \tag{12.1.12}$$

and they form special solutions of the partial differential equation (12.1.3), namely those which are independent of t. These solutions determine the Morse–Witten coboundary operator

$$\delta_{\mathrm{MW}} : \mathrm{CM}^*(f; \Lambda_\omega) \to \mathrm{CM}^*(f; \Lambda_\omega).$$

This coboundary operator is defined on the same cochain complex as the Floer coboundary δ and, by (12.1.7), the cochain complex has the same grading for both theories up to a shift by n.

More precisely, the cochain complex $\mathrm{CM}^*(f;\Lambda_\omega)$ can be identified with the graded Λ_ω-module of all functions $\xi : \mathrm{Crit}(f) \times H_2^S(M) \to R$ that satisfy the finiteness condition

$$\#\{(x,A) \mid \xi(x,A) \neq 0,\, \omega(A) \geq c\} < \infty \tag{12.1.13}$$

for every $c \in \mathbb{R}$.[1] The Λ_ω-module structure is given by

$$(\lambda * \xi)(x,A) := \sum_B \lambda(B)\xi(x, A+B);$$

the grading is

$$\deg(x,A) := \mathrm{ind}_f(x) - 2c_1(A)$$

and the coboundary operator has the form

$$(\delta_{\mathrm{MW}}\xi)(x,A) := \sum_y n_f(x,y)\xi(y,A),$$

where $n_f(x,y)$ is the number of connecting orbits from x to y, i.e. of shift equivalence classes of solutions of (12.1.12) satisfying

$$\lim_{s\to-\infty} u(s) = x, \qquad \lim_{s\to\infty} u(s) = y,$$

counted with appropriate signs. Here we assume that the gradient flow of f is Morse–Smale and so the number of connecting orbits is finite whenever $\mathrm{ind}_f(x) - \mathrm{ind}_f(y) = 1$. With these conventions δ_{MW} is a Λ_ω-module homomorphism of degree one and satisfies $\delta_{\mathrm{MW}} \circ \delta_{\mathrm{MW}} = 0$. It is called the **Morse–Witten coboundary operator**. Its homology is naturally isomorphic to the quantum cohomology of M with coefficients in Λ_ω:

$$\mathrm{QH}^*(M;\Lambda_\omega) \cong \mathrm{HM}^*(f;\Lambda_\omega) := \frac{\ker \delta_{\mathrm{MW}}}{\operatorname{im} \delta_{\mathrm{MW}}}.$$

(See Witten [**420**], Floer [**115**], Salamon [**351**], or Schwarz's book [**357**].)

Since the Morse complex $\mathrm{CM}^*(f;\Lambda_\omega)$ and the Floer complex $\mathrm{CF}^*(-\varepsilon f)$ are isomorphic as groups when $\varepsilon > 0$ is sufficiently small, Theorem 12.1.4 will follow from Theorem 12.1.3 if we can show that the boundary operators coincide. For this we need to see that all the solutions of (12.1.3) and (12.1.4) with index

$$\mu(u) := \mu_{\mathrm{CZ}}(\widetilde{x}^-) - \mu_{\mathrm{CZ}}(\widetilde{x}^+) \leq 1$$

are independent of t, provided that $H_t \equiv -\varepsilon f$ is a Morse function which is independent of t and is sufficiently C^2-small. To prove this for generic J without any bound on the energy of the solution requires one to assume that either M is monotone, or $c_1(A) = 0$ for all $A \in \pi_2(M)$, or the minimal Chern number is $N \geq n$. In the other cases of semipositivity (with minimal Chern number $N = n-1$ or $N = n-2$) it has so far only been possible to prove t-independence for the solutions of (12.1.3) with a given bound on the energy, with the required smallness of H depending on

---

[1] In the case $H_t = -f$ the set $\widetilde{\mathcal{P}}(H)$ can be identified with $\mathrm{Crit}(f) \times H_2^S(M)$. Under this correspondence the symplectic action is the function $\mathcal{A}_f : \mathrm{Crit}(f) \times H_2^S(M) \to \mathbb{R}$ given by

$$\mathcal{A}_f(x,A) := f(x) - \omega(A).$$

Hence the finiteness condition (12.1.13) is equivalent to the finiteness condition in the definition of $\mathrm{CF}^*(H)$ with $H_t = -f$.

this bound. To use such a result for the proof of Theorem 12.1.4 one needs an alternative definition of Floer cohomology. (First truncate the chain complex and then take inverse and direct limits.) These ideas are due to Ono and in [**313**] he proved Theorem 12.1.4 with this modified definition of Floer cohomology.

One of the main motivations for Floer's work was the attempt to solve the Arnold conjecture concerning the number of fixed points of a Hamiltonian symplectomorphism. As mentioned in Section 9.1 there are various different forms of this conjecture. The relevant one here concerns the number of fixed points of the time-1 map of a nondegenerate Hamiltonian H. Theorem 12.1.4, even in Ono's modified form, immediately implies that Arnold's conjecture has to hold for such Hamiltonians. For if the homology of the Floer complex $\mathrm{CF}^*(H)$ of H is isomorphic to $H^*(M) \otimes \Lambda_\omega$ then the rank of $\mathrm{CF}^*(H)$ as a module over Λ_ω must be at least equal to the dimension of $H^*(M)$ and, as we pointed out earlier, this rank is precisely the number of elements in $\mathcal{P}_0(H)$. This leads to the following conclusion.

COROLLARY 12.1.5 (Arnold conjecture). [**116, 180, 313**] *Let (M, ω) be a compact semipositive symplectic manifold and $\psi : M \to M$ be a Hamiltonian symplectomorphism with only nondegenerate fixed points. Then*

$$\#\mathrm{Fix}(\psi) \geq \sum_{j=0}^{2n} b_j(M)$$

where $b_j(M) = \dim H^j(M)$ denotes the jth Betti number of M.

REMARK 12.1.6. In the above proof of Corollary 12.1.5 it is crucial that the almost complex structure J is chosen independent of t. If (M, ω) is any semipositive symplectic manifold then the Floer cohomology groups can still be defined with the same techniques if J depends on t. However, in the proof that the Floer cohomology groups are independent of the choice of J_t we need to know that J-holomorphic spheres with negative Chern numbers do not appear in generic 3-parameter families of almost complex structures. Hence we must assume that (M, ω) satisfies the strong semipositivity hypothesis (8.5.1).

The new techniques of the virtual fundamental cycle have been developed precisely to deal with the difficulties that arise from J-holomorphic spheres with negative Chern numbers. With these techniques in place (see Fukaya–Ono [**127**] or Liu–Tian [**249**]) one can define the Floer cohomology groups in full generality and drop the semipositivity hypothesis altogether. In particular, they show that if f is any Morse function then, for sufficiently small $\varepsilon > 0$ the virtual moduli cycle constructed from the Floer complex $\mathrm{CF}^*(-\varepsilon f)$ may be identified with the Morse complex $\mathrm{CM}^*(f; \Lambda_\omega)$. However, with the techniques developed in this book we are restricted to symplectic manifolds where J-holomorphic curves with negative Chern numbers can be avoided.

The PSS isomorphism. To prove Theorem 12.1.4 with the original definition of the Floer groups requires a different approach which was found by Piunikhin–Salamon–Schwarz [**323**]. The idea is to consider perturbed J-holomorphic planes $u : \mathbb{C} \to M$ such that $u(re^{2\pi it})$ converges to a periodic solution $y^\alpha(t)$ of the (time dependent) Hamiltonian system H^α as $r \to \infty$. Note that any such u determines a lift $\widetilde{y} \in \widetilde{\mathcal{P}}(H)$. Let us denote the resulting moduli space by $\mathcal{M}(\widetilde{y}; H^\alpha, J^\alpha)$. With our conventions for the Conley–Zehnder index this space has dimension $n - \mu_{\mathrm{CZ}}(\widetilde{y})$. Now fix a Morse function $f : M \to \mathbb{R}$ such that the (downward) gradient flow (12.1.12) is Morse–Smale and denote by $W^u(x, f)$ the unstable manifold of a critical point $x \in \mathrm{Crit}(H)$ with respect to (12.1.12). It has dimension $\mathrm{ind}_f(x)$ and codimension

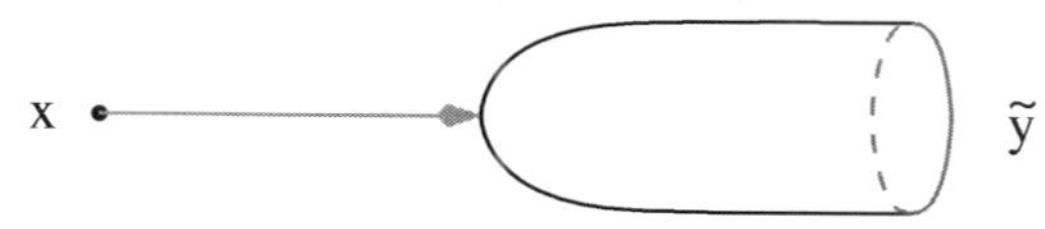

FIGURE 1. The PSS isomorphism $\Phi^\alpha : \mathrm{HF}^* \to \mathrm{QH}^*$.

$2n - \mathrm{ind}_f(x)$. Assuming transversality, the submanifold of all $u \in \mathcal{M}(\widetilde{y}; H^\alpha, J^\alpha)$ with $u(0) \in W^u(x)$ has dimension $\mathrm{ind}_f(x) - \mu_{\mathrm{CZ}}(\widetilde{y}) - n$. One can think of these as J-holomorphic **spiked discs**, where the spike is the gradient flowline from x to $u(0)$ (see Figure 1). In the case

$$\mathrm{ind}_f(x) - n = \mu_{\mathrm{CZ}}(\widetilde{y})$$

the moduli space of spiked discs is 0-dimensional and hence the numbers $n(x, \widetilde{y})$ of its elements can be used to construct a chain map, called for short a **PSS map**,

$$\Phi^\alpha : \mathrm{CF}^*(H^\alpha) \to \mathrm{CM}^*(f; \Lambda_\omega).$$

It is given by

$$(\Phi^\alpha \xi)(x, A) := \sum_{\mu_{\mathrm{CZ}}(\widetilde{y}) = \mathrm{ind}_f(x) - n} n(x, \widetilde{y})\xi(A\#\widetilde{y})$$

for $\xi \in \mathrm{CF}^*(H^\alpha)$. To see that the number of pairs (x, A) with $(\Phi^\alpha\xi)(x, A) \neq 0$ and $\omega(A) \geq -c$ is finite, note that, for every such pair (x, A), there is an element $\widetilde{y} \in \widetilde{\mathcal{P}}(H^\alpha)$ such that $n(x, \widetilde{y}) \neq 0$ and $\xi(A\#\widetilde{y}) \neq 0$. The latter inequality implies that there is a constant $c_\xi > 0$ such that $\mathcal{A}_{H^\alpha}(A\#\widetilde{y}) \geq -c_\xi$, while the former implies that the moduli space $\mathcal{M}(\widetilde{y}; H^\alpha, J^\alpha)$ is nonempty. Hence, by the energy identity (12.1.11), there is a constant $c_E > 0$ (depending only on the choice of (H^α, J^α)) such that

$$E(u) \leq c_E - \mathcal{A}_{H^\alpha}(\widetilde{y}) = c_E - \mathcal{A}_{H^\alpha}(A\#\widetilde{y}) - \omega(A) \leq c_E + c_\xi + c. \tag{12.1.14}$$

This universal energy bound proves the required finiteness condition. The map $\Phi^\alpha : \mathrm{CF}^*(H^\alpha) \to \mathrm{CM}^*(f; \Lambda_\omega)$ is a Λ_ω-module homomorphism and raises the degree by n. The standard arguments in Floer homology (as in the proof that $\delta \circ \delta = 0$) show that it is a chain map and hence induces a homomorphism on cohomology, still denoted by Φ^α. In Piunikhin–Salamon–Schwarz [**323**] it is shown that the induced homomorphism on cohomology is an isomorphism. The proof is based on the construction of an analogous chain map

$$\Psi^\alpha : \mathrm{CM}^*(f; \Lambda_\omega) \to \mathrm{CF}^*(H^\alpha),$$

given by

$$(\Psi^\alpha \xi_0)(\widetilde{x}) := \sum_{\mathrm{ind}_f(y) - 2c_1(A) = \mu_{\mathrm{CZ}}(\widetilde{x}) + n} n((-A)\#\widetilde{x}, y)\xi_0(y, A)$$

for $\xi_0 \in \mathrm{CM}^*(f; \Lambda_\omega)$. One then has to prove that both compositions $\Phi^\alpha \circ \Psi^\alpha$ and $\Psi^\alpha \circ \Phi^\alpha$ are chain homotopic to the identity. The argument is illustrated in Figures 2 and 3. The key technical parts of the argument are the Floer gluing theorem and the gluing theorem for J-holomorphic curves proved in Chapter 10: as indicated in Remark 12.1.7 below the rest of the argument is standard. Note

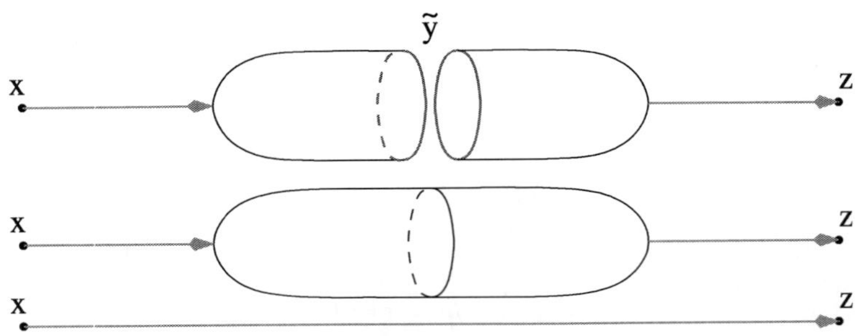

FIGURE 2. $\Phi^\alpha \circ \Psi^\alpha \sim \mathrm{id}$.

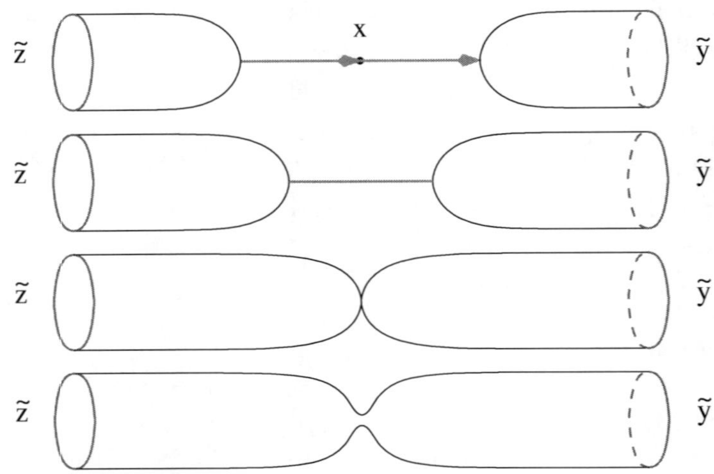

FIGURE 3. $\Psi^\alpha \circ \Phi^\alpha \sim \mathrm{id}$.

also that if $H = -f$ is independent of time, and we are in a situation when Floer's original argument works, then one can show that for sufficiently small $\varepsilon > 0$ the only spiked discs that contribute to the PSS map are constant maps at a critical point of H. Therefore the PSS map $\mathrm{CF}^*(-\varepsilon f) \to \mathrm{CM}^*(f; \Lambda_\omega)$ is the obvious isomorphism. It follows that the PSS isomorphisms Ψ^α are the same as the maps constructed by Floer's approach, whenever the latter maps are defined.

REMARK 12.1.7. One may think of the PSS map

$$\Phi^*_{PSS} : \mathrm{CF}^*(H^\alpha) \to \mathrm{CM}^*(f; \Lambda_\omega)$$

as representing "half" of the Seidel homomorphism $\mathcal{S}(\phi)$. As described in Remark 11.4.4 the Seidel homomorphism is given by counting sections of a bundle over S^2 that intersect cycles lying in fibers over two fixed points $\mathrm{w}_1, \mathrm{w}_2$. If we represent the cycles in the fiber over w_1 by the unstable manifolds of some Morse flow, and those in the fiber over w_2 by the stable manifolds then we are counting spheres with two spikes, one a gradient trajectory from x and the other a gradient trajectory to z; cf. Figure 2. In the present situation we are counting sections over a disc with Floer boundary conditions. Gluing two oppositely oriented such discs along their boundary gives one of the spheres counted by Seidel. Thus the statement that $\Phi^*_{PSS} \circ \Psi^*_{PSS}$ is the identity map on quantum cohomology is equivalent to saying that the Seidel element $\mathcal{S}(\mathrm{id}, \sigma_0)$ of Definition 11.4.1 is the unit in $\mathrm{QH}^0(M; \Lambda)$. (Here id is the identity element of the group $\pi_1(\mathrm{Ham}(M))$ and σ_0 denotes the trivial section class $[S^2 \times \{\mathrm{pt}\}]$ in $S^2 \times M$.) In the present context (and that of Piunikhin–Salamon–Schwarz [**323**]) the gluing is topologically trivial and gives rise to a section of the trivial bundle. Seidel [**363**] noted that one can adapt these ideas to nontrivial bundles: see Section 12.5. This point of view is developed further by Lalonde [**222**].

12.2. Ring structure

There is a natural ring structure on Floer cohomology which is defined by counting perturbed J-holomorphic curves on the 3-punctured 2-sphere. The punctures are understood as cylindrical ends and the perturbations near the punctures are Hamiltonian as in (12.1.3). More precisely, let Σ be a noncompact Riemann surface of genus zero with three cylindrical ends, a so-called pair of pants. We assume that the complex structure on Σ is the standard product structure on the cylindrical ends. We fix parametrizations

$$\phi^\alpha : (-\infty, 0) \times \mathbb{R}/\mathbb{Z} \to \Sigma, \qquad \phi^\beta, \phi^\gamma : (0, \infty) \times \mathbb{R}/\mathbb{Z} \to \Sigma$$

of the cylindrical ends with disjoint images $U^\alpha := \mathrm{im}\,\phi^\alpha$, $U^\beta := \mathrm{im}\,\phi^\beta$, $U^\gamma := \mathrm{im}\phi^\gamma$. Assume that the complement

$$\Sigma' = \Sigma \setminus \left(U^\alpha \cup U^\beta \cup U^\gamma\right)$$

is diffeomorphic to a 2-sphere with three open discs removed. In particular Σ' is compact. Choose a complex structure on Σ which on the three cylindrical ends pulls back to the standard structure $s + it$. In order to define the appropriate moduli spaces choose three Hamiltonian functions $H^\alpha, H^\beta, H^\gamma : \mathbb{R} \times \mathbb{R}/\mathbb{Z} \times M \to \mathbb{R}$ which vanish for $|s| \le 1/2$ and are independent of s for $|s| \ge 1$. Then consider the space of smooth maps

$$u : \Sigma \to M$$

such that the restriction of u to Σ' is a J-holomorphic curve and, in the cylindrical end U^α, the composition $u^\alpha(s,t) := u \circ \phi^\alpha(s,t)$ satisfies

$$\partial_s u^\alpha + J(u^\alpha)\left(\partial_t u^\alpha - X^\alpha_{s,t}(u^\alpha)\right) = 0,$$

where $\iota(X^\alpha_{s,t})\omega = dH^\alpha_{s,t}$ for all s and t, and similarly for $u^\beta := u \circ \phi^\beta$ and $u^\gamma := u \circ \phi^\gamma$. Any finite energy solution of these equations satisfies the limit conditions

$$\lim_{s \to -\infty} u^\alpha(s,t) = x^\alpha(t), \qquad \lim_{s \to \infty} u^\beta(s,t) = x^\beta(t), \qquad \lim_{s \to \infty} u^\gamma(s,t) = x^\gamma(t)$$

where $x^\alpha \in \mathcal{P}(H^\alpha)$, $x^\beta \in \mathcal{P}(H^\beta)$, and $x^\gamma \in \mathcal{P}(H^\gamma)$. Here we slightly abuse notation and denote by $\mathcal{P}(H^\alpha)$ the periodic solutions of the Hamiltonian $H^\alpha_{-\infty,t}$ and similarly for β and γ. The space of such solutions u in the correct homology class will be denoted by $\mathcal{M}(\widetilde{x}^\alpha; \widetilde{x}^\beta, \widetilde{x}^\gamma)$ (see Figure 4). It has dimension

$$\dim \mathcal{M}(\widetilde{x}^\alpha; \widetilde{x}^\beta, \widetilde{x}^\gamma) = \mu_{\mathrm{CZ}}(\widetilde{x}^\alpha, H^\alpha) - \mu_{\mathrm{CZ}}(\widetilde{x}^\beta, H^\beta) - \mu_{\mathrm{CZ}}(\widetilde{x}^\gamma, H^\gamma) - n.$$

In the zero dimensional case we get finitely many solutions by the same argument as above. The signed number of solutions will be denoted by $n(\widetilde{x}^\alpha; \widetilde{x}^\beta, \widetilde{x}^\gamma)$. These numbers give rise to a Λ_ω-module homomorphism

$$\mathrm{CF}^k(H^\beta) \otimes \mathrm{CF}^\ell(H^\gamma) \to \mathrm{CF}^{k+\ell+n}(H^\alpha) : \xi^\beta \otimes \xi^\gamma \mapsto \xi^\beta * \xi^\gamma.$$

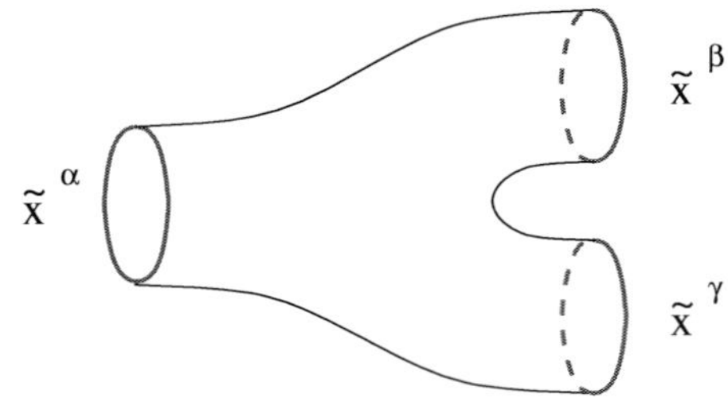

FIGURE 4. The pair-of-pants product.

It is defined by

$$(\xi^\beta * \xi^\gamma)(\widetilde{x}^\alpha) := \sum_{\widetilde{x}^\beta, \widetilde{x}^\gamma} n(\widetilde{x}^\alpha; \widetilde{x}^\beta, \widetilde{x}^\gamma)\, \xi^\beta(\widetilde{x}^\beta)\, \xi^\gamma(\widetilde{x}^\gamma).$$

It follows by the usual gluing techniques in Floer homology that this map is a cochain homomorphism and therefore induces a homomorphism of Floer cohomologies

$$\mathrm{HF}^k(H^\beta) \otimes \mathrm{HF}^\ell(H^\gamma) \to \mathrm{HF}^{k+\ell+n}(H^\alpha).$$

In view of its construction this map is called the **pair-of-pants** product, or briefly the PP product. It follows from the usual deformation arguments in Floer homology that the pair-of-pants product is independent of the complex structure on the Riemann surface Σ used to define it. It is also independent of the Hamiltonian functions $H^\alpha, H^\beta, H^\gamma$, as long as they are not changed at $\pm\infty$. If they are changed at $\pm\infty$, one again uses Floer's gluing techniques to prove that the product is natural with respect to the isomorphisms of Theorem 12.1.3. To prove that the product is skew commutative with the usual sign conventions, choose an orientation preserving diffeomorphism $f : \Sigma \to \Sigma$ which interchanges the two cylindrical ends on the right. The diffeomorphism changes the complex structure of Σ, but this does not effect the resulting map on Floer cohomology. The proof of associativity requires the gluing of two surfaces $\Sigma_{\alpha;\beta\gamma}$ and $\Sigma_{\gamma;\delta\varepsilon}$ with three cylindrical ends over a long cylinder (see Figure 5). This procedure cancels the two ends labelled by γ and results in a Riemann surface $\Sigma_{\alpha;\beta\delta\varepsilon}$ with four cylindrical ends. If the neck is sufficiently long then the resulting triple product corresponds to the composition $\xi^\beta * (\xi^\delta * \xi^\varepsilon)$. Now vary the complex structure on the surface $\Sigma_{\alpha;\beta\delta\varepsilon}$ and decompose it in a different way to obtain the identity $\xi^\beta * (\xi^\delta * \xi^\varepsilon) = (\xi^\beta * \xi^\delta) * \xi^\varepsilon$. This variation of the complex structure corresponds to changing the cross ratio of four fixed marked points in the proof of the associativity of quantum cohomology in Section 11.1.

A comparison theorem. According to Theorem 12.1.4 the Floer cohomology groups $\mathrm{HF}^*(H, J)$ are naturally isomorphic to the quantum cohomology groups $\mathrm{QH}^*(M)$ and one would expect the pair-of-pants product to correspond to the quantum deformation of the cup product under this isomorphism. This question is already nontrivial in the case $\pi_2(M) = 0$ where the quantum deformation agrees with the ordinary cup product. This was proved by Schwarz [**358**] in his thesis for the case $\pi_2(M) = 0$ and by Piunikhin–Salamon–Schwarz [**323**] for the general case. Independently, other proofs were given by Liu–Tian [**250**] and Ruan–Tian [**347**]. A physicist's approach to this problem may be found in Sadov [**350**].

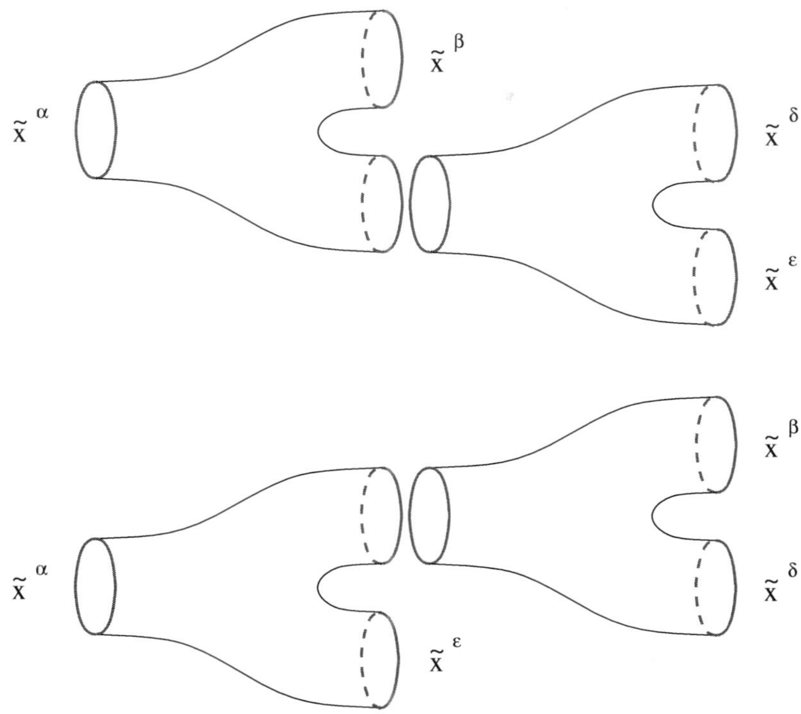

FIGURE 5. Associativity.

THEOREM 12.2.1. *Assume* (M, ω) *is semipositive. Let*

$$\Phi^\alpha : \mathrm{HF}^*(M, \omega, H^\alpha, J^\alpha) \to \mathrm{QH}^*(M)$$

be the PSS isomorphism of Theorem 12.1.4 and similarly for Φ^β *and* Φ^γ. *Then*

$$\Phi^\alpha(\xi^\beta * \xi^\gamma) = \Phi^\beta(\xi^\beta) * \Phi^\gamma(\xi^\gamma)$$

for $\xi^\beta \in \mathrm{HF}^*(H^\beta)$ *and* $\xi^\gamma \in \mathrm{HF}^*(H^\gamma)$. *Here the product on the left is defined by the pair-of-pants construction while the product on the right is the quantum deformation of the cup product.*

The proof of this theorem goes along the following lines. First use the Morse–Witten complex to represent the quantum deformation of the cup product in a somewhat different way. Given three Morse functions $f^\alpha, f^\beta, f^\gamma : M \to \mathbb{R}$, three critical points $x^\alpha, x^\beta, x^\gamma$, three distinct points $\mathrm{w}^\alpha, \mathrm{w}^\beta, \mathrm{w}^\gamma \in \mathbb{C}P^1$, and a regular family of almost complex structures $J = \{J_z\}_{z \in S^2} \in \mathcal{J}_{\mathrm{reg}}(S^2; M, \omega; \mathbf{w})$ as in Definition 6.7.10 (with J_z close to some fixed almost complex structure J_0), consider the space

$$\mathcal{M}_A(x^\alpha; x^\beta, x^\gamma) = \mathcal{M}_A(x^\alpha; x^\beta, x^\gamma; f, J)$$

of all J-holomorphic A-spheres $u : \mathbb{C}P^1 \to M$ such that

$$u(z^\alpha) \in W^u(x^\alpha, f^\alpha), \qquad u(z^\beta) \in W^s(x^\beta, f^\beta), \qquad u(z^\gamma) \in W^s(x^\gamma, f^\gamma).$$

Here $W^s(x, f)$ and $W^u(x, f)$ denote the stable and unstable manifolds of a critical point x of f with respect to the downward gradient flow of the Morse function f. The gradient of f^i is understood with respect to the metric induced by J_{z^i} for

$i = \alpha, \beta, \gamma$ and we assume that these gradient flows are Morse–Smale. One can think of the elements of $\mathcal{M}_A(x^\alpha, x^\beta, x^\gamma)$ as **spiked J-holomorphic spheres** (see Figure 6). Since

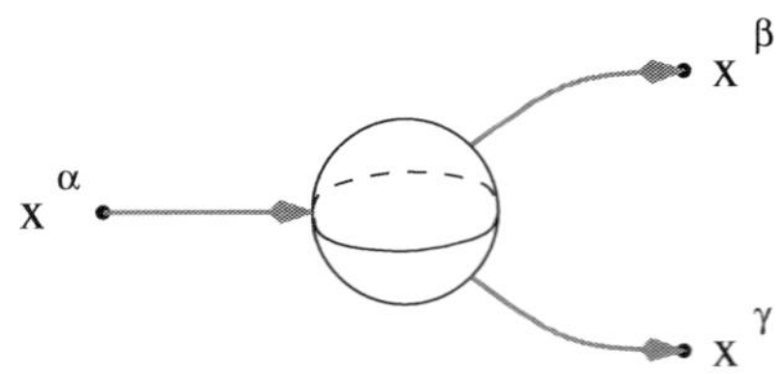

FIGURE 6. The quantum product via Morse theory.

$$\dim W^s(x, f) = 2n - \mathrm{ind}_f(x), \qquad \dim W^u(x, H) = \mathrm{ind}_f(x),$$

we have

$$\dim \mathcal{M}_A(x^\alpha; x^\beta, x^\gamma) = 2c_1(A) + \mathrm{ind}_{f^\alpha}(x^\alpha) - \mathrm{ind}_{f^\beta}(x^\beta) - \mathrm{ind}_{f^\gamma}(x^\gamma)$$

for generic Morse functions $f^\alpha, f^\beta, f^\gamma$. Whenever this dimension is zero, denote by $n_A(x^\alpha; x^\beta, x^\gamma)$ the number of points in $\mathcal{M}_A(x^\alpha; x^\beta, x^\gamma)$, counted with appropriate signs. This gives rise to a chain map

$$\mathrm{CF}^*(f^\beta; \Lambda_\omega) \otimes \mathrm{CF}^*(f^\gamma; \Lambda_\omega) \to C^*(f^\alpha; \Lambda_\omega) : \xi^\beta \otimes \xi^\gamma \mapsto \xi^\beta * \xi^\gamma$$

defined by

$$(\xi^\beta * \xi^\gamma)(x^\alpha, A^\alpha) = \sum_{A^\beta + A^\gamma = A^\alpha} n_A(x^\alpha; x^\beta, x^\gamma)\xi^\beta(x^\beta, A^\beta)\xi^\gamma(x^\gamma, A^\gamma). \tag{12.2.1}$$

This induces a product on Morse–Witten cohomology (with coefficents in Λ_ω and boundary map δ_{MW}) and hence on quantum cohomology. This product agrees with the quantum cup product because the stable and unstable manifolds of critical points of Morse functions represent cohomology classes which generate the cohomology of M.

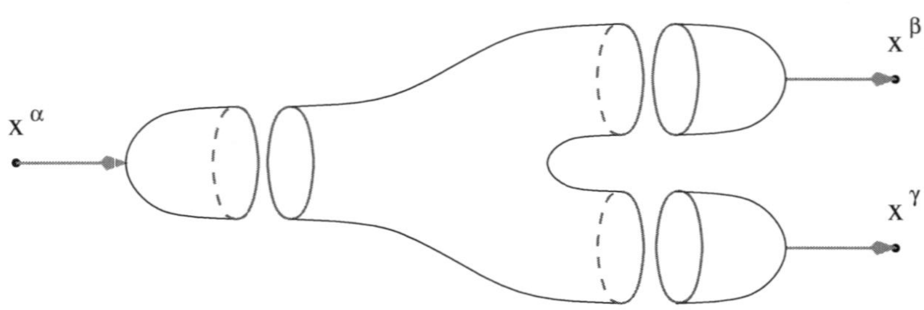

FIGURE 7. Quantum cohomology and the pair-of-pants product.

Now it follows from the standard gluing arguments in Floer homology that the PSS isomorphism $\Phi^\alpha : \mathrm{HF}^*(H^\alpha) \to \mathrm{QH}^*(M; \Lambda_\omega)$ of Theorem 12.1.4, as described in [**323**], intertwines the two product structures. To see this one has to glue three J-holomorphic spiked discs to the boundary of a J-holomorphic pair-of-pants. More precisely, the boundary components are cylindrical ends abutting on periodic solutions and one uses Floer's gluing theorem. As a result one obtains a (perturbed)

J-holomorphic sphere with three spikes as described above (see Figure 7). More details of this argument are given in [**323**]. Note that Theorem 12.2.1 gives rise to an alternative proof of the associativity of the quantum product.

12.3. Poincaré duality

To obtain **Floer homology**, instead of cohomology, one can define the chain complex CF_* as the set of functions $\eta : \widetilde{\mathcal{P}}(H) \to R$ that satisfy the finiteness condition

$$\#\left\{\widetilde{x} \in \widetilde{\mathcal{P}}(H) \mid \eta(\widetilde{x}) \neq 0,\, \mathcal{A}_H(\widetilde{x}) \geq c\right\} < \infty$$

for every $c \in \mathbb{R}$. The action of Λ_ω on CF_* is given by

$$(\lambda * \eta)(\widetilde{x}) := \sum_A \lambda(-A)\xi(A\#\widetilde{x}).$$

There is an obvious pairing $\mathrm{CF}^* \otimes \mathrm{CF}_* \to R$ given by

$$\langle \xi, \eta \rangle := \sum_{\widetilde{x}} \xi(\widetilde{x})\eta(\widetilde{x}). \tag{12.3.1}$$

It is compatible with the Λ_ω-module structures in the sense that

$$\langle \lambda * \xi, \eta \rangle = \langle \xi, \lambda * \eta \rangle.$$

The transpose $\partial : \mathrm{CF}_* \to \mathrm{CF}_*$ of δ with respect to this pairing is given by

$$(\partial\eta)(\widetilde{y}) = \sum_{\mu_{\mathrm{CZ}}(\widetilde{x}) = \mu_{\mathrm{CZ}}(\widetilde{y})+1} \eta(\widetilde{x})n(\widetilde{x}, \widetilde{y}).$$

The grading of CF_* is still given by the Conley–Zehnder index and ∂ lowers the degree by one.

The PSS isomorphisms on Floer homology are induced by chain maps

$$\Phi^\alpha : \mathrm{CM}_*(f; \Lambda_\omega) \to \mathrm{CF}_*(H^\alpha), \qquad \Psi^\alpha : \mathrm{CF}_*(H^\alpha) \to \mathrm{CM}_*(f; \Lambda_\omega).$$

These chain maps are defined as in Section 12.1. Note that Φ^α lowers the degree by n while Ψ^α raises it by n. We have chosen the notation such that the PSS map Φ^α counts the same underlying geometric objects whether we take it in cohomology or homology; we sometimes denote it by Φ^*_{PSS} (on cohomology) and Φ_*^{PSS} (on homology). A similar remark applies to Ψ^α.

Exercise 12.3.1. Let (H, J) be a regular pair as in Proposition 12.1.2 and define $\widehat{H} : \mathbb{R}/\mathbb{Z} \times M \to \mathbb{R}$ by

$$\widehat{H}_t := -H_{-t}.$$

Show that there is a natural isomorphism

$$\mathrm{HF}^k(H, J) \cong \mathrm{HF}_{-k}(\widehat{H}, J).$$

If J depends on t, as in Remark 12.1.6 below, one also has to replace J by $\widehat{J}_t := J_{-t}$. *Hint:* Show that the map $\widetilde{\mathcal{P}}(H) \to \widetilde{\mathcal{P}}(\widehat{H}) : (x, u) \mapsto (\widehat{x}, \widehat{u})$, given by $\widehat{x}(t) := x(-t)$ and $\widehat{u}(z) := u(\bar{z})$, is a bijection that reverses the sign of both the symplectic action and the Conley–Zehnder index. Show that u is a solution of the Floer equation (12.1.3), if and only if $\widehat{u}(s, t) := u(-s, -t)$ is a solution of (12.1.3) with H replaced by $\widehat{H}$.

REMARK 12.3.2. If H generates an element $\widetilde{\phi}$ in the universal cover $\widetilde{\mathrm{Ham}}(M,\omega)$ then $\widehat{H}$ generates $\widetilde{\phi}^{-1}$. More precisely, let $H : \mathbb{R}/\mathbb{Z} \times M \to \mathbb{R}$ be a time dependent Hamiltonian function and $t \mapsto \phi_t$ be the Hamiltonian isotopy generated by H. Then $\phi_{t+1} = \phi_t \circ \phi_1$ for every t. It follows that the paths $t \mapsto \phi_{-t}$ and $t \mapsto \phi_t^{-1}$ in $\mathrm{Ham}(M,\omega)$ on the interval $0 \le t \le 1$ are homotopic with fixed endpoints. An explicit homotopy is given by $\phi_t^\lambda := \phi_{\lambda t}^{-1} \circ \phi_{(\lambda-1)t}$.

The **Poincaré duality** isomorphism

$$\mathrm{PD}^H : \mathrm{HF}^k(H) \cong \mathrm{HF}_{-k}(H)$$

on Floer cohomology can be defined as the composition of the canonical identification $\mathrm{HF}^k(H) \cong \mathrm{HF}_{-k}(\widehat{H})$ of Exercise 12.3.1 with the Floer continuation map from $\mathrm{HF}_{-k}(\widehat{H})$ to $\mathrm{HF}_{-k}(H)$. Composing PD^H on either side with the PSS isomorphisms we obtain the Poincaré duality isomorphism on quantum cohomology:

$$\mathrm{PD} := \Psi_*^{PSS} \circ \mathrm{PD}^H \circ \Psi_{PSS}^* : \ \mathrm{QH}^k(M;\Lambda) \to \mathrm{QH}_{2n-k}(M;\check{\Lambda}),$$

That this composition indeed agrees with the Poincaré duality isomorphism of Remark 11.1.20 follows from Theorem 12.1.4 (together with sign considerations). It follows that

$$\big(a, \mathrm{PD}(b)\big) = \big\langle a * b, 1\big\rangle \in R, \quad a, b \in \mathrm{QH}^*(M;\Lambda),$$

where the pairing on the left is given by (11.1.13) (see Exercise 11.1.22).

REMARK 12.3.3 (Cap product). Fix a cohomology class $a \in H^*(M)$. A result in Piunikhin–Salamon–Schwarz [**323**] asserts that the operator on Floer cohomology, obtained by counting Floer trajectories that intersect a generic cycle Poincaré dual to a, agrees with the pair-of-pants product with the image of a under the PSS-isomorphism. The proof is analogous to the proof that the PSS isomorphism preserves the ring structures: counting spiked Floer cylinders is equivalent to counting pairs of pants with a spiked disc attached to one of the ends. Via Poincaré duality this can be translated into an assertion about operators on Floer homology that can be interpreted as Floer cap products with a. In the case $a = \mathrm{PD}([\mathrm{pt}])$ the relevant assertion can be rephrased as follows.

Assume that (H, J) is a regular pair for Floer homology. Then there is an operator

$$\Psi_0 : \mathrm{HF}_n(H,J) \to \mathrm{HF}_{-n}(H,J)$$

defined by counting the finite energy index zero solutions of the equation

$$\partial_s u + J_t(u)(\partial_t u - X_{H_t}(u)) = 0, \qquad u(s,t) = u(s,t+1), \qquad u(0,0) = x_0,$$

for a generic point $x_0 \in M$. If $n_H(\widetilde{x}, \widetilde{y}; x_0)$ denotes the signed number of solutions of this equation running from $\widetilde{x} \in \widetilde{\mathcal{P}}(H)$ to $\widetilde{y} \in \widetilde{\mathcal{P}}(H)$, then Ψ_0 is given on the chain level by

$$(\Psi_0\eta)(\widetilde{y}) := \sum_{\mu_{\mathrm{CZ}}(\widetilde{x})=n} \eta(\widetilde{x}) n_H(\widetilde{x}, \widetilde{y}; x_0).$$

We claim that

$$\Psi_0 \circ \Phi_*^{PSS}([M]) = \Phi_*^{PSS}([\mathrm{pt}]). \tag{12.3.2}$$

Since $1 = \mathrm{PD}([M])$ this is equivalent in Floer cohomology to the equation

$$\Psi^0 \circ \Phi_{PSS}^*(1) = \Phi_{PSS}^*(\mathrm{PD}([\mathrm{pt}])). \tag{12.3.3}$$

One can see this in two ways. One is to split off a spiked disc from each Floer cylinder contributing to Ψ^0 and deduce that Ψ^0 is the pair-of-pants product with $\Phi^*_{PSS}(\mathrm{PD}([\mathrm{pt}]))$. Then (12.3.3) follows from the fact that Φ^*_{PSS} intertwines the quantum cup product with the pair of pants product. Alternatively, one can glue two spiked discs to the ends of each Floer cylinder contributing to Ψ^0 (as in Remark 12.1.7) and obtain (perturbed) holomorphic spheres with three marked points that, at one of these points, pass through x_0. It then follows that the composition $\Psi^*_{PSS} \circ \Psi^0 \circ \Phi^*_{PSS}$ is given by the quantum cup product with $\mathrm{PD}([\mathrm{pt}])$. Since the quantum product with 1 is the identity on quantum cohomology we obtain

$$\Psi^*_{PSS} \circ \Psi^0 \circ \Phi^*_{PSS}(1) = \mathrm{PD}([\mathrm{pt}])$$

and so equation (12.3.3) follows from the fact that $\Phi^*_{PSS} = (\Psi^*_{PSS})^{-1}$.

12.4. Spectral invariants

The action functional $\mathcal{A}_H$ decreases along Floer connecting orbits and so it is possible to define a filtered version of Floer (co)homology. This gives rise to (Floer) homologically essential critical values of the action functional $\mathcal{A}_H$. These critical values are called **spectral invariants**. In the generality discussed here the spectral invariants were introduced recently by Oh [**306**]. For symplectically aspherical manifolds they were introduced by Schwarz [**360**]. In the context of generating functions similar invariants were introduced earlier by Viterbo [**404, 406**]. We shall work here with the Floer homology groups defined in Section 12.3 since this is the approach taken by Oh [**306**]. But one could equally well use cohomology (see Schwarz [**360**] and Exercise 12.4.7 below). We assume throughout that (M, ω) satisfies the strong semipositivity assumption (8.5.1) and, for simplicity (and with loss of generality), that the integral of ω over every spherical homology class is a rational number:

$$\text{(12.4.1)} \qquad A \in \pi_2(M) \quad \Longrightarrow \quad \omega(A) \in \mathbb{Q}.$$

We denote by $\mathcal{H} = \mathcal{H}(M, \omega)$ the set of all smooth Hamiltonian functions $H : \mathbb{R}/\mathbb{Z} \times M \to \mathbb{R}$ such that $H_t := H(t, \cdot)$ has mean value zero with respect to the volume form ω^n for every t. Given $H \in \mathcal{H}(M, \omega)$ we denote by $\phi^H \in \mathrm{Ham}(M, \omega)$ the time-1 map of the corresponding Hamiltonian differential equation and by $\widetilde{\phi}^H \in \widetilde{\mathrm{Ham}}(M, \omega)$ its lift to the universal cover.

Recall from the discussion preceding Theorem 9.1.6 that the action spectrum $\mathrm{Spec}(H)$ of a Hamiltonian function $H \in \mathcal{H}$ is the set of all critical values of $\mathcal{A}_H$. By Lemma 9.1.9, the action spectrum depends only on the element $\widetilde{\phi} = \widetilde{\phi}^H \in \widetilde{\mathrm{Ham}}(M, \omega)$ determined by H and will also be denoted by $\mathrm{Spec}(\widetilde{\phi})$. Given a real number $\kappa \in \mathbb{R}$ we denote by $\mathcal{H}^\kappa = \mathcal{H}^\kappa(M, \omega)$ the set of all Hamiltonian functions $H \in \mathcal{H}(M, \omega)$ with $\kappa \notin \mathrm{Spec}(H)$ and by $\widetilde{\mathrm{Ham}}^\kappa(M, \omega)$ the set of all $\widetilde{\phi} \in \widetilde{\mathrm{Ham}}(M, \omega)$ with $\kappa \notin \mathrm{Spec}(\widetilde{\phi})$. Abbreviate $\mathcal{J} := \mathcal{J}_\tau(M, \omega)$.

Condition (12.4.1) implies that the spectrum $\mathrm{Spec}(H)$ is a closed and nowhere dense subset of $\mathbb{R}$ for every Hamiltonian H (see Exercise 12.4.1 below). Oh's work also applies to the case where the image of the ω-period map $\pi_2(M) \to \mathbb{R}$ (and hence also the spectrum $\mathrm{Spec}(H)$ of any Hamiltonian H) is dense in $\mathbb{R}$. In this case the proofs of the *(Continuity)* and *(Spectrality)* axioms in Theorem 12.4.4 below are considerably harder.

EXERCISE 12.4.1. Let (M, ω) be a compact symplectic manifold that satisfies (12.4.1). Prove that $\mathrm{Spec}(H)$ is a closed nowhere dense subset of $\mathbb{R}$ for every $H \in \mathcal{H}(M, \omega)$.

EXERCISE 12.4.2. Show that the action spectrum is invariant under conjugation, i.e.

$$\mathrm{Spec}(\psi\widetilde{\phi}\psi^{-1}) = \mathrm{Spec}(\widetilde{\phi})$$

for all $\psi \in \mathrm{Symp}(M, \omega)$ and $\widetilde{\phi} \in \widetilde{\mathrm{Ham}}(M, \omega)$. Deduce that $\mathrm{Spec}(\widetilde{\phi}\widetilde{\psi}) = \mathrm{Spec}(\widetilde{\psi}\widetilde{\phi})$ for all $\widetilde{\phi}, \widetilde{\psi} \in \widetilde{\mathrm{Ham}}(M, \omega)$. *Hint:* To deduce the second identity from the first use the fact that the fundamental group of $\mathrm{Ham}(M, \omega)$ is abelian and so $\widetilde{\psi}\widetilde{\phi}\widetilde{\psi}^{-1} = \psi\widetilde{\phi}\psi^{-1}$. Find an alternative direct proof.

Filtered Floer homology. For $\kappa \in (\mathbb{R} \setminus \mathrm{Spec}(H)) \cup \{\infty\}$ define the R-module

$$\mathrm{CF}^{\kappa}_{*}(H) := \{\eta \in \mathrm{CF}_{*}(H) \,|\, \eta(\widetilde{x}) \neq 0 \Longrightarrow \mathcal{A}_H(\widetilde{x}) \leq \kappa\}.$$

This is not a Λ_ω-module, since it is only invariant under multiplication by those elements $\lambda \in \Lambda_\omega$ that satisfy $\omega(A) > 0$ whenever $\lambda(A) \neq 0$. (See Section 12.3 for the definition of $\lambda * \eta$.) Note that $\mathrm{CF}^{\kappa}_{*}(H)$ is invariant under the Floer boundary operator ∂. Hence we may define

$$\mathrm{HF}^{\kappa}_{*}(H, J) := \frac{\ker \partial|_{\mathrm{CF}^{\kappa}_{*}(H)}}{\mathrm{im}\, \partial|_{\mathrm{CF}^{\kappa}_{*}(H)}}.$$

In particular, if $\kappa = \infty$ we recover the Floer homology groups of Section 12.3.

The invariance of the filtered Floer homology groups $\mathrm{HF}^{\kappa}_{*}(H, J)$ under continuation is more subtle than in the unfiltered case because the filtered Floer homology groups are invariant only along paths (H_s, J_s) for which $\kappa \neq \mathrm{Spec}(H_s)$ for all s. Because the corresponding path spaces may not be connected, the resulting continuation isomorphism may depend on the homotopy class of the path. To be more precise let us assume that (H_0, J_0) and (H_1, J_1) are regular pairs in $\mathcal{H}^{\kappa} \times \mathcal{J}$ (as in Proposition 12.1.2). Let $\mathbb{R} \to \mathcal{H}^{\kappa} \times \mathcal{J} : s \mapsto (H_s, J_s)$ be a smooth homotopy satisfying $(H_s, J_s) = (H_0, J_0)$ for $s \leq 0$ and $(H_s, J_s) = (H_1, J_1)$ for $s \geq 1$. We wish to define a continuation homomorphism

$$\Phi^{\kappa}_{\{H_s, J_s\}} : \mathrm{HF}^{\kappa}_{*}(H_0, J_0) \to \mathrm{HF}^{\kappa}_{*}(H_1, J_1).$$

in terms of the solutions of (12.1.10). The key point is the energy identity (12.1.11). It shows that

$$\mathcal{A}_{H_1}(x_1) \leq \mathcal{A}_{H_0}(x_0) - \int_{-\infty}^{\infty}\int_0^1 \min_M \partial_s H_{s,t}\, dtds$$

and so, because the spectrum is discrete, the Floer continuation map preserves the subcomplexes CF^{κ}_{*} on the chain level when $\int_{-\infty}^{\infty}\int_0^1 \min_M \partial_s H_{s,t}\, dtds \geq -\varepsilon$ for $\varepsilon > 0$ sufficiently small. If this condition is not satisfied we still obtain isomorphisms

$$\Phi^{\kappa}_{\{H_s, J_s\}}(s_1, s_0) : \mathrm{HF}^{\kappa}_{*}(H_{s_0}, J_{s_0}) \to \mathrm{HF}^{\kappa}_{*}(H_{s_1}, J_{s_1})$$

for $|s_1 - s_0|$ sufficiently small. To see this, it suffices to replace (H_s, J_s) by the homotopy $s \mapsto (H_{\beta(s)}, J_{\beta(s)})$ where $\beta(s) := s_0 + \rho(s)(s_1 - s_0)$ and $\rho : \mathbb{R} \to [0, 1]$ is a smooth cutoff function satisfying $\rho(s) = 0$ for $s \leq 0$ and $\rho(s) = 1$ for $s \geq 1$. For a general pair of real numbers s_0, s_1 the isomorphism $\Phi^{\kappa}_{\{H_s, J_s\}}(s_1, s_0)$ can then be defined as a composition of the isomorphisms $\Phi^{\kappa}_{\{H_s, J_s\}}(s_{i+1}, s_i)$ for a suitable partition of the interval $[s_0, s_1]$. It is easy to see that the resulting isomorphism is independent of the partition. Repeating the arguments in the construction of the Floer homology groups we deduce that the continuation isomorphisms on filtered Floer homology have the following properties.

(NATURALITY) If $\mathbb{R} \to \mathcal{H}^\kappa \times \mathcal{J} : s \mapsto (H_s, J_s)$ is a smooth path and (H_{s_0}, J_{s_0}), (H_{s_1}, J_{s_1}), (H_{s_2}, J_{s_2}) are regular pairs then $\Phi^\kappa_{\{H_s,J_s\}}(s_0, s_0) = \mathrm{id}$ and

$$\Phi^\kappa_{\{H_s,J_s\}}(s_2, s_0) = \Phi^\kappa_{\{H_s,J_s\}}(s_2, s_1) \circ \Phi^\kappa_{\{H_s,J_s\}}(s_1, s_0).$$

(HOMOTOPY) The isomorphism $\Phi^\kappa_{\{H_s,J_s\}}(s_1, s_0)$ depends only on the homotopy class (with fixed endpoints) of the path $[s_0, s_1] \to \mathcal{H}^\kappa \times \mathcal{J} : s \mapsto (H_s, J_s)$.

(FILTRATION) The continuation maps commute with the homomorphisms in the long exact sequence

$$\cdots \to \mathrm{HF}^\kappa_*(H, J) \to \mathrm{HF}^{\kappa'}_*(H, J) \to \mathrm{HF}^{[\kappa,\kappa']}_*(H, J) \to \mathrm{HF}^\kappa_{*-1}(H, J) \to \cdots$$

for $\kappa < \kappa'$ and paths $s \mapsto (H_s, J_s) \in (\mathcal{H}^\kappa \cap \mathcal{H}^{\kappa'}) \times \mathcal{J}$. Here $\mathrm{HF}^{[\kappa,\kappa']}_*$ denotes the homology of the quotient complex $\mathrm{CF}^{\kappa'}_*/\mathrm{CF}^\kappa_*$.

(MONOTONICITY) If $\partial_s H_{s,t} \geq 0$ for all s and t then the Floer continuation homomorphism preserves the subcomplexes CF^κ_* and induces a homomorphism $\mathrm{HF}^\kappa_*(H_{s_0}, J_{s_0}) \to \mathrm{HF}^\kappa_*(H_{s_1}, J_{s_1})$ for $s_0 < s_1$. If, in addition, $H_s \in \mathcal{H}^\kappa$ for every $s \in [s_0, s_1]$ then this is an isomorphism and agrees with $\Phi^\kappa_{\{H_s,J_s\}}(s_1, s_0)$.

These axioms show that one can extend the definition of the groups $\mathrm{HF}^\kappa_*(H)$ to an arbitrary, possibly degenerate, Hamiltonian H, provided only that $\kappa \notin \mathrm{Spec}(H)$. The important point is that $\mathrm{Spec}(H)$ is a closed and nowhere dense subset of $\mathbb{R}$ that varies continuously with H. Hence if $\kappa \notin \mathrm{Spec}(H)$ there is a contractible open neighbourhood $\mathcal{U}(H) \subset \mathcal{H}^\kappa$ of H consisting of functions F that do not have κ in their spectrum. (Here we use the C^∞-topology on $\mathcal{H}^\kappa$.) One then defines $\mathrm{HF}^\kappa_*(H)$ to be the group $\mathrm{HF}^\kappa_*(F, J)$ for any such nondegenerate F. The *(Homotopy)* axiom provides a canonical isomorphism between any two such choices. In the language of Conley the Floer homology group $\mathrm{HF}^\kappa_*(H)$ is a **connected simple system**, i.e. a small category in which every morphism is an isomorphism and there is precisely one isomorphism between any two objects. In the case at hand the objects $\mathrm{HF}^\kappa_*(F, J)$ are labelled by the regular pairs $(F, J) \in \mathcal{U}(H) \times \mathcal{J}$ and the morphisms are the continuation isomorphisms.

Note that $\mathrm{HF}^\infty_*(H, J) = \mathrm{HF}_*(H, J)$ agrees with the usual Floer homology group defined in Section 12.3. Moreover, in this case the filtered continuation isomorphisms agree with the usual ones. Hence the *(Filtration)* axiom asserts that there is a well defined homomorphism $\iota^\kappa(H) : \mathrm{HF}^\kappa_*(H) \to \mathrm{HF}_*(H)$ for $H \in \mathcal{H}^\kappa$, induced by the inclusion of chain complexes in the regular case. The *(Filtration)* axiom also shows that every path $[0, 1] \to \mathcal{H}^\kappa : s \mapsto H_s$ determines a commutative diagram

$$\begin{array}{ccc} \mathrm{HF}^\kappa_*(H_0) & \xrightarrow{\Phi^\kappa_{\{H_s,J_s\}}} & \mathrm{HF}^\kappa_*(H_1) \\ \downarrow{\scriptstyle \iota^\kappa(H_0)} & & \downarrow{\scriptstyle \iota^\kappa(H_1)} \\ \mathrm{HF}_*(H_0) & \xrightarrow{\Phi^{10}} & \mathrm{HF}_*(H_1) \end{array}, \tag{12.4.2}$$

where $\Phi^\kappa_{\{H_s,J_s\}}$ is the continuation isomorphism of filtered Floer homology and Φ^{10} is the canonical Floer continuation isomorphism.

Spectral invariants. The filtered Floer homology groups give rise to a function

$$\rho : \widetilde{\mathrm{Ham}}(M,\omega) \times \mathrm{QH}^*(M;\Lambda_\omega) \to \mathbb{R} \cup \{-\infty\}$$

defined by

$$\rho(\widetilde{\phi}^H; a) := \inf\bigl\{\kappa \in \mathbb{R} \setminus \mathrm{Spec}(H) \,|\, \Phi_*^{PSS}(\mathrm{PD}(a)) \in \mathrm{im}\,\iota^\kappa(H)\bigr\}. \tag{12.4.3}$$

Here

$$\Phi_*^{PSS} : \mathrm{QH}_*(M;\check{\Lambda}) \to \mathrm{HF}_*(H)$$

denotes the PSS isomorphism for the Hamiltonian H and

$$\mathrm{PD} : \mathrm{QH}^*(M;\Lambda_\omega) \to \mathrm{QH}_*(M;\check{\Lambda}_\omega)$$

denotes the Poincaré duality isomorphism defined in Remark 11.1.20. The commutative diagram (12.4.2) shows that the number $\rho(\widetilde{\phi}; a)$ is independent of the Hamiltonian function H with $\widetilde{\phi}^H = \widetilde{\phi}$ used to define it. It is called the **spectral invariant** of $\widetilde{\phi} \in \widetilde{\mathrm{Ham}}(M,\omega)$ and $a \in \mathrm{QH}^*(M;\Lambda_\omega)$.

To understand the number $\rho(\widetilde{\phi}; a)$ geometrically, let us assume that there is a regular pair (H,J) such that $\widetilde{\phi} = \widetilde{\phi}^H$. Moreover, let

$$a = \sum_A a_A \otimes e^A \in \mathrm{QH}^{2n-k}(M;\Lambda_\omega)$$

with $a_A \in H^{2n-k-2c_1(A)}(M)$, and assume that each a_A is Poincaré dual to the fundamental cycle of an oriented submanifold $X_A \subset M$ (of dimension $k + 2c_1(A)$). Then the Poincaré dual of a is the quantum homology class

$$\mathrm{PD}(a) = \sum_A [X_A] \otimes e^{-A} \in \mathrm{QH}_k(M;\check{\Lambda}_\omega).$$

Its image under the PSS isomorphism $\Phi_*^{PSS} : \mathrm{QH}_k(M;\check{\Lambda}_\omega) \to \mathrm{HF}_{k-n}(H,J)$ is the Floer homology class of the chain $\eta : \widetilde{\mathcal{P}}(H) \to \mathbb{Z}$ given by

$$\eta(\widetilde{y}) := \sum_A n\bigl(X_A, (-A)\#\widetilde{y}\bigr). \tag{12.4.4}$$

Here $n(X,\widetilde{x})$ denotes the signed number of spiked discs $u \in \mathcal{M}(\widetilde{x}; H, J)$ as in Figure 1 that satisfy $u(0) \in X$. Note that the number $n\bigl(X_A, (-A)\#\widetilde{y}\bigr)$ can only be nonzero if

$$\begin{aligned} n - \mu_{\mathrm{CZ}}(\widetilde{y}) - 2c_1(A) &= \dim \mathcal{M}((-A)\#\widetilde{y}; H, J) \\ &= \mathrm{codim} X_A \\ &= 2n - k - 2c_1(A) \end{aligned}$$

or equivalently $\mu_{\mathrm{CZ}}(\widetilde{y}) = k - n$. Thus the Floer homology class of η belongs to $\mathrm{HF}_{k-n}(H,J)$ as expected. The condition $\kappa \geq \rho(\widetilde{\phi}^H; a)$ is equivalent to the existence of Floer chains $\eta' \in \mathrm{CF}^\kappa_{k-n}(H)$ and $\zeta \in \mathrm{CF}_{k-n+1}(H)$ such that $\eta = \eta' + \partial\zeta$.

Exercise 12.4.3. Show that in diagram (12.4.2) the image of $\iota^\kappa(H)$ is contained in the image of $\iota^{\kappa'}(H)$ whenever $\kappa < \kappa'$. Hence characterize $\rho(\widetilde{\phi}; a)$ as the unique element of $\mathrm{Spec}(\widetilde{\phi})$ such that $\Phi_*^{PSS}(\mathrm{PD}(a)) \in \mathrm{im}\,\iota^\kappa(H)$ for $\kappa > \rho(\widetilde{\phi}; a)$ and $\Phi_*^{PSS}(\mathrm{PD}(a)) \notin \mathrm{im}\,\iota^\kappa(H)$ for $\kappa < \rho(\widetilde{\phi}; a)$.

THEOREM 12.4.4 (Oh). *Let (M, ω) be a closed symplectic manifold that satisfies (8.5.1) and (12.4.1). Then the following holds.*

(SPECTRALITY) *If $a \neq 0$ then $\rho(\widetilde{\phi}; a) \in \mathrm{Spec}(\widetilde{\phi})$ for every $\widetilde{\phi} \in \widetilde{\mathrm{Ham}}(M, \omega)$.*

(SYMPLECTIC INVARIANCE) *For all $\psi \in \mathrm{Symp}(M, \omega)$, $\widetilde{\phi} \in \widetilde{\mathrm{Ham}}(M, \omega)$, and $a \in \mathrm{QH}^*(M; \Lambda)$ we have that*

$$\rho(\psi\widetilde{\phi}\psi^{-1}; a) = \rho(\widetilde{\phi}; a).$$

(CONTINUITY) *If $a \neq 0$ then the map $\widetilde{\mathrm{Ham}}(M, \omega) \to \mathbb{R} : \widetilde{\phi} \mapsto \rho(\widetilde{\phi}; a)$ is continuous with respect to the Hofer metric on $\widetilde{\mathrm{Ham}}(M, \omega)$. More precisely, if $a \neq 0$ and $H_0, H_1 \in \mathcal{H}$ are Hamiltonian functions generating the elements $\widetilde{\phi}_0, \widetilde{\phi}_1 \in \widetilde{\mathrm{Ham}}(M, \omega)$ then*

$$\int_0^1 \min_M (H_{1,t} - H_{0,t})\, dt \leq \rho(\widetilde{\phi}_0; a) - \rho(\widetilde{\phi}_1; a) \leq \int_0^1 \max_M (H_{1,t} - H_{0,t})\, dt. \tag{12.4.5}$$

(ZERO) *If $0 \neq a \in H^*(M)$ and $A \in H_2^S(M)$ then*

$$\rho(\mathrm{id}; a \otimes e^A) = \omega(A).$$

(PRODUCT) *For all $\widetilde{\phi}, \widetilde{\psi} \in \widetilde{\mathrm{Ham}}(M, \omega)$ and all $a, b \in \mathrm{QH}^*(M; \Lambda_\omega)$ we have that*

$$\rho(\widetilde{\phi}\widetilde{\psi}; a * b) \leq \rho(\widetilde{\phi}; a) + \rho(\widetilde{\psi}; b).$$

(NONDEGENERACY) *For every $\phi \in \mathrm{Ham}(M, \omega)$ with $\phi \neq \mathrm{id}$ there is a constant $\delta > 0$ such that, for every $H \in \mathcal{H}(M, \omega)$, we have that*

$$\phi^H = \phi \qquad \Longrightarrow \qquad \rho(\widetilde{\phi}^H; 1) + \rho((\widetilde{\phi}^H)^{-1}; 1) \geq \delta.$$

PROOF. We omit the proof of the *(Product)* axiom which is essentially the same as that given in Schwarz [**360**], and sketch the proofs of the other axioms under the assumption (12.4.1). For the general case the reader is referred to Usher [**398**].

By definition, every Floer homology class belongs to the image of the map $\iota^\kappa(H) : \mathrm{HF}_*^\kappa(H, J) \to \mathrm{HF}_*(H, J)$ for some κ. Hence $\rho(\widetilde{\phi}; a) < \infty$ for every a. If $a \neq 0$ then $\Phi_*^{PSS}(\mathrm{PD}(a))$ is nonzero and so cannot belong to the image of $\iota^\kappa(H)$ for every κ. (The intersection of these images is zero by Exercise 12.4.6 below.) This shows that $\rho(\widetilde{\phi}; a)$ is a real number whenever $a \neq 0$. That this number belongs to the spectrum of $\widetilde{\phi}$ follows from the fact that the image of the homomorphism $\iota^\kappa(H)$ can only change when κ passes through an element of $\mathrm{Spec}(H)$. This proves the *(Spectrality)* axiom. The *(Symplectic invariance)* axiom is obvious from the definitions.

The proof of the *(Continuity)* axiom has two steps. Continuity with respect to the C^1-norm on $\widetilde{\mathrm{Ham}}(M, \omega)$ follows from the *(Homotopy)* axiom for filtered Floer homology: if $H \in \mathcal{H}^\kappa$ and H' is sufficiently close to H in the C^2-norm then $H' \in \mathcal{H}^\kappa$ and, indeed, H' is homotopic to H by a path in $\mathcal{H}^\kappa$. Hence it follows from the commutative diagram (12.4.2) that the Floer continuation homomorphism $\mathrm{HF}_*(H) \to \mathrm{HF}_*(H')$ induces an isomorphism from the image of $\iota^\kappa(H)$ to the image of $\iota^\kappa(H')$. This proves continuity with respect to the C^1-norm.

To prove (12.4.5) assume that there are almost complex structures $J_0, J_1 \in \mathcal{J}$ such that (H_0, J_0) and (H_1, J_1) are regular pairs for Floer homology. Moreover, let

$\mathbb{R} \to \mathcal{J} : s \mapsto J_s$ be a smooth path that agrees with J_0 for $s \le 0$ and with J_1 for $s \ge 1$. Let $\beta : \mathbb{R} \to [0,1]$ be a smooth cutoff function such that

$$\beta(s) = \begin{cases} 0, & \text{for } s \le 0, \\ 1, & \text{for } s \ge 1, \end{cases} \qquad \partial_s \beta \ge 0,$$

and define $H_{s,t} : M \to \mathbb{R}$ by

$$H_{s,t} := \beta(s) H_{1,t} + (1 - \beta(s)) H_{0,t}.$$

Assume first that $\{H_s, J_s\}$ is a regular homotopy for Floer homology. Now let

$$\kappa > \rho(\widetilde{\phi}_0; a).$$

Then there is a Floer homology cycle $\eta_0 \in \mathrm{CF}_*(H_0)$ representing the Floer homology class $\alpha_0 := \Phi_*^{\mathrm{PSS}}(\mathrm{PD}(a)) \in \mathrm{HF}_*(H_0, J_0)$ such that

$$\eta_0(\widetilde{x}_0) \neq 0 \qquad \Longrightarrow \qquad \mathcal{A}_{H_0}(\widetilde{x}_0) \le \kappa$$

for every $\widetilde{x}_0 \in \widetilde{\mathcal{P}}(H_0)$. For $\widetilde{x}_0 \in \widetilde{\mathcal{P}}(H_0)$ and $\widetilde{x}_1 \in \widetilde{\mathcal{P}}(H_1)$ (with the same Conley–Zehnder index) denote by $n(\widetilde{x}_0, \widetilde{x}_1)$ the signed number of solutions of (12.1.10) connecting $\widetilde{x}_0$ to $\widetilde{x}_1$. Define $\eta_1 \in \mathrm{CF}_*(H_1)$ by

$$\eta_1(\widetilde{x}_1) := \sum_{\widetilde{x}_0 \in \widetilde{\mathcal{P}}(H_0)} \eta_0(\widetilde{x}_0) n(\widetilde{x}_0, \widetilde{x}_1).$$

Then η_1 is a Floer chain representing the image of $[\eta_0]$ under the Floer continuation isomorphism, i.e. the Floer homology class $\alpha_1 := \Phi_*^{\mathrm{PSS}}(\mathrm{PD}(a)) \in \mathrm{HF}_*(H_1, J_1)$. Now the energy identity (12.1.11) for the solutions of (12.1.10) shows that $n(\widetilde{x}_0, \widetilde{x}_1)$ can only be nonzero if

$$\mathcal{A}_{H_0}(\widetilde{x}_0) - \mathcal{A}_{H_1}(\widetilde{x}_1) \ge \int_{-\infty}^{\infty} \int_0^1 \min_M \partial_s H_{s,t} \, dt ds = \int_0^1 \min_M (H_{1,t} - H_{0,t}) \, dt.$$

The last identity follows from the special form of our homotopy $H_{s,t}$. Thus we have proved that $\eta_1(\widetilde{x}_1)$ can only be nonzero if there is an element $\widetilde{x}_0 \in \widetilde{\mathcal{P}}(H_0)$ such that $\eta_0(\widetilde{x}_0) \neq 0$ and $n(\widetilde{x}_0, \widetilde{x}_1) \neq 0$ and hence

$$\mathcal{A}_{H_1}(\widetilde{x}_1) \le \mathcal{A}_{H_0}(\widetilde{x}_0) - \int_0^1 \min_M (H_{1,t} - H_{0,t}) \, dt \le \kappa - \int_0^1 \min_M (H_{1,t} - H_{0,t}) \, dt.$$

Since η_1 represents the Floer homology class $\Phi_*^{\mathrm{PSS}}(\mathrm{PD}(a)) \in \mathrm{HF}_*(H_1, J_1)$ this implies

$$\rho(\widetilde{\phi}_1; a) \le \kappa - \int_0^1 \min_M (H_{1,t} - H_{0,t}) \, dt.$$

Taking the infimum over all $\kappa > \rho(\widetilde{\phi}_0; a)$ we obtain

$$\rho(\widetilde{\phi}_1; a) \le \rho(\widetilde{\phi}_0; a) - \int_0^1 \min_M (H_{1,t} - H_{0,t}) \, dt.$$

This proves the first inequality in (12.4.5) under the assumption that $\{H_s, J_s\}$ is a regular homotopy. To prove the inequality in general approximate (H_s, J_s) by a regular homotopy (between regular endpoints) and use the fact that continuity with respect to the C^1-norm on $\widetilde{\mathrm{Ham}}(M, \omega)$ (respectively with respect to the C^2-norm on the space of Hamiltonian functions) has already been established. The second inequality in (12.4.5) follows from the first by interchanging H_0 and H_1.

To prove the *(Zero)* axiom one can consider a sequence of regular Hamiltonians $H_i \in \mathcal{H}(M,\omega)$ converging to zero in the C^∞ topology. Then there is a sequence $\varepsilon_i > 0$ such that ε_i converges to zero and

$$\operatorname{im}\iota^{\kappa-\varepsilon_i}(f) \subset \Psi^{f,H_i}(\operatorname{im}\iota^{\kappa}(H)) \subset \operatorname{im}\iota^{\kappa+\varepsilon_i}(f)$$

in the notation of Exercise 12.4.6 below. In other words one can construct these PSS maps so that the constant c_E in the energy identity (12.1.14) is arbitrarily small. This can be used to show that $\rho(\widetilde{\phi}^{H_i};a)$ converges to zero for every nonzero cohomology class $a \in H^*(M)$. Hence, in the case $A = 0$, the *(Zero)* axiom follows from the *(Continuity)* axiom. To prove the result for $A \neq 0$ one can use the module structure of Floer homology over the Novikov ring.

We sketch a proof that, for every $\phi \in \operatorname{Ham}(M,\omega)$ with $\phi \neq \mathrm{id}$, there is a constant $\delta > 0$ such that, for every $H \in \mathcal{H}(M,\omega)$,

$$\phi^H = \phi \qquad \Longrightarrow \qquad \rho(\widetilde{\phi}^H;1) - \rho(\widetilde{\phi}^H;\mathrm{PD}([\mathrm{pt}])) \geq \delta. \tag{12.4.6}$$

The *(Nondegeneracy)* axiom follows immediately from (12.4.6) and the *(Product)* axiom. Namely, if $\phi^H = \phi$ then

$$\rho(\widetilde{\phi}^H;1) + \rho((\widetilde{\phi}^H)^{-1};1) \geq \rho(\widetilde{\phi}^H;1) - \rho(\widetilde{\phi}^H;\mathrm{PD}([\mathrm{pt}])) + \rho(\mathrm{id};\mathrm{PD}([\mathrm{pt}])) \geq \delta.$$

Here the first inequality follows from the *(Product)* axiom and the last from the *(Zero)* axiom and (12.4.6).

To prove the estimate (12.4.6) we define the constant δ as in the proof of Corollary 9.1.14: choose any smooth family $\mathbb{R} \to \mathcal{J} : t \mapsto \widetilde{J}_t$ of almost complex structures satisfying $\widetilde{J}_{t+1} = \phi^*\widetilde{J}_t$, let $x_0 \in M \setminus \operatorname{Fix}(\phi)$, consider the equation

$$\partial_s\widetilde{u} + \widetilde{J}_t(\widetilde{u})\partial_t\widetilde{u} = 0, \qquad \widetilde{u}(s,t) = \phi(\widetilde{u}(s,t+1)), \qquad \widetilde{u}(0,0) = x_0, \tag{12.4.7}$$

for smooth functions $\widetilde{u} : \mathbb{R}^2 \to M$, and define

$$\delta := \min\{\min_t \hbar(\widetilde{J}_t), \delta_0\}, \qquad \delta_0 := \inf\big\{E(\widetilde{u}) \,|\, \widetilde{u} \text{ satisfies (12.4.7)}, \, E(\widetilde{u}) < \infty\big\}.$$

Here $\hbar(\widetilde{J}_t)$ denotes the smallest area of a nonconstant $\widetilde{J}_t$-holomorpic sphere. That the number $\min_t \hbar(\widetilde{J}_t)$ is positive follows from Proposition 4.1.4 and the fact that $\hbar(\widetilde{J}_{t+1}) = \hbar(\widetilde{J}_t)$. That the number δ_0 is positive follows from the obvious compactness argument: a sequence $\widetilde{u}_i$ of solutions of (12.4.7) with $E(\widetilde{u}_i) \to 0$ has a subsequence that converges, uniformly with all derivatives on compact sets, to a solution $\widetilde{u}$ of (12.4.7) with $E(\widetilde{u}) = 0$; any such solution would have the form $\widetilde{u}(s,t) \equiv x_0 = \phi(x_0)$, which is impossible because x_0 is not a fixed point of ϕ. Thus we have proved that $\delta > 0$. A consequence of the following proof is that $\delta < \infty$, which is not obvious from the definition. (Another proof is given in Theorem 9.1.13.)

Let $H \in \mathcal{H}$ be such that $\phi^H = \phi$, denote by ϕ_t the Hamiltonian isotopy determined by H, and define $J_t = J_{t+1} \in \mathcal{J}$ by the formula $\phi_t^* J_t := \widetilde{J}_t$. We first prove the estimate (12.4.6) under the assumption that (H,J) is a regular pair for Floer homology and the above chosen point x_0. Under this assumption there is an operator $\Psi_0 : \mathrm{HF}_n(H,J) \to \mathrm{HF}_{-n}(H,J)$ defined by

$$(\Psi_0\eta)(\widetilde{y}) := \sum_{\mu_{\mathrm{CZ}}(\widetilde{x})=n} \eta(\widetilde{x})n_H(\widetilde{x},\widetilde{y};x_0),$$

where $n_H(\widetilde{x}, \widetilde{y}; x_0)$ denotes the signed number of Floer connecting orbits from $\widetilde{x}$ to $\widetilde{y}$ passing through x_0. Thus $n_H(\widetilde{x}, \widetilde{y}; x_0)$ counts solutions of the equation

$$\partial_s u + J_t(u)(\partial_t u - X_{H_t}(u)) = 0, \quad u(s,t) = u(s,t+1), \quad u(0,0) = x_0. \tag{12.4.8}$$

As noted in Remark 12.3.3, we have

$$\Psi_0 \circ \Phi_*^{PSS}([M]) = \Phi_*^{PSS}([\mathrm{pt}]). \tag{12.4.9}$$

Now let κ be any real number such that $\kappa \notin \mathrm{Spec}(\widetilde{\phi}^H)$ and

$$\rho(\widetilde{\phi}^H; 1) < \kappa.$$

Then there is a Floer homology cycle $\eta_1 \in \mathrm{CF}_n(H)$ such that $[\eta_1] = \Phi_*^{PSS}([M])$ and

$$\mathcal{A}_H(\widetilde{x}) > \kappa \qquad \Longrightarrow \qquad \eta_1(\widetilde{x}) = 0 \tag{12.4.10}$$

for every $\widetilde{x} \in \widetilde{\mathcal{P}}(H)$. Define $\eta_0 \in \mathrm{CF}_{-n}(H)$ by

$$\eta_0(\widetilde{y}) := \sum_{\mu_{\mathrm{CZ}}(\widetilde{x})=n} \eta_1(\widetilde{x}) n_H(\widetilde{x}, \widetilde{y}; x_0).$$

Then η_0 is a Floer homology cycle and $[\eta_0] = \Phi_*^{PSS}([\mathrm{pt}])$ by (12.4.9). Moreover,

$$\mathcal{A}_H(\widetilde{y}) > \kappa - \delta \qquad \Longrightarrow \qquad \eta_0(\widetilde{y}) = 0.$$

To see this note that every solution u of (12.4.8) is equivalent to a solution

$$\widetilde{u}(s,t) := \phi_t^{-1}(u(s,t))$$

of (12.4.7) and hence has energy $E(u) \geq \delta$. This implies that the coefficient $n_H(\widetilde{x}, \widetilde{y}; x_0)$ in the definition of $\Psi_0(\eta_1)$ can only be nonzero if $\mathcal{A}_H(\widetilde{x}) - \mathcal{A}_H(\widetilde{y}) \geq \delta$. Hence, by (12.4.10), $\eta_0(\widetilde{y})$ can only be nonzero if $\mathcal{A}_H(\widetilde{y}) \leq \mathcal{A}_H(\widetilde{x}) - \delta \leq \kappa - \delta$ for some $\widetilde{x} \in \widetilde{\mathcal{P}}(H)$ with $\mu_{\mathrm{CZ}}(\widetilde{x}) = n$ and $\eta_1(\widetilde{x}) \neq 0$. Thus we have proved that the Floer homology class $\Phi_*^{PSS}([\mathrm{pt}]) = [\eta_0]$ belongs to the image of the homomorphism $\iota^{\kappa-\delta}(H) : \mathrm{HF}_{-n}^{\kappa-\delta}(H, J) \to \mathrm{HF}_{-n}(H, J)$ and hence

$$\rho(\widetilde{\phi}^H; \mathrm{PD}([\mathrm{pt}])) \leq \kappa - \delta.$$

This proves the inequality (12.4.6) in the case where (H, J) is a regular pair.

To prove the result in general choose a sequence of regular pairs (H_i, J_i) converging to (H, J) and denote by δ_i the same constant as above with $(\phi, \widetilde{J}_t)$ replaced by $(\phi_{i1}, \phi_{it}^* J_{it})$ where ϕ_{it} denotes the Hamiltonian isotopy generated by H_i. We claim that

$$\liminf_{i \to \infty} \delta_i \geq \delta. \tag{12.4.11}$$

To see this fix a positive real number

$$\varepsilon < \delta.$$

We must prove that $\varepsilon < \delta_i$ for i sufficiently large. Assume, by contradiction, that this is not the case. Passing to a subsequence, if necessary, we may then assume that $\delta_i \leq \varepsilon$ for every i. If $\min_t \hbar(J_{i_\nu t}) \leq \varepsilon$ for any sequence $i_\nu \to \infty$ then, since $\phi_{i_\nu t}^* J_{i_\nu t}$ converges to $\widetilde{J}_t$, it follows from Gromov compactness that $\min_t \hbar(\widetilde{J}_t) \leq \varepsilon < \delta$. This contradicts the definition of δ. Thus we must have that $\min_t \hbar(J_{it}) > \varepsilon$ for i sufficiently large. Since $\delta_i \leq \varepsilon$ it follows that there is a sequence $\widetilde{u}_i$ of solutions of (12.4.7) with ϕ replaced by ϕ_{i1} and $\widetilde{J}_t$ replaced by $\phi_{it}^* J_{it}$ such that $E(\widetilde{u}_i) \leq \varepsilon$. Since $\varepsilon < \hbar(\widetilde{J}_t)$ for every t by assumption, bubbling cannot occur and so $\widetilde{u}_i$ has a subsequence converging to a solution $\widetilde{u}$ of (12.4.7). It follows that $E(\widetilde{u}) \le \varepsilon < \delta$, again contradicting the definition of δ. This proves (12.4.11).

By the special case, we have

$$\rho(\widetilde{\phi}^{H_i}; 1) - \rho(\widetilde{\phi}^{H_i}; \mathrm{PD}([\mathrm{pt}])) \ge \delta_i$$

for every i. Hence the estimate (12.4.6) follows by taking the limit $i \to \infty$ and using (12.4.11). □

Note that, in general, the Hofer metric may not induce a metric on $\widetilde{\mathrm{Ham}}(M,\omega)$: it induces a metric if and only if every nontrivial homotopy class of Hamiltonian loops has positive minimal Hofer length. No closed manifolds are yet known that support noncontractible loops with minimum length zero. If such loops do exist, then the *(Continuity)* axiom shows that the spectral invariants descend to a smaller covering space of $\mathrm{Ham}(M,\omega)$.

The *(Product)* axiom for ρ can be interpreted as a compatibility between the spectral invariants and quantum multiplication. It was first noted by Schwarz [**360**] in the symplectically aspherical case. This property is the basis for all the most interesting applications, such as their role in defining analogues of the Hofer norm (as in Viterbo [**404**], Schwarz [**360**] and Oh [**307**]) and Entov–Polterovich's construction of a Calabi quasimorphism in [**105**]. We do not attempt to explain these applications here. However, in Lemma 12.5.3 below we describe the action of an extension of the group $\pi_1(\mathrm{Ham}(M,\omega))$ on spectral invariants.

REMARK 12.4.5. The *(Symplectic invariance)* axiom in Theorem 12.4.4 can be restated in the form $\rho(\widetilde{\phi}\widetilde{\psi}; a) = \rho(\widetilde{\psi}\widetilde{\phi}; a)$ for all $\widetilde{\phi}, \widetilde{\psi} \in \widetilde{\mathrm{Ham}}(M,\omega)$. To see this recall from Exercise 12.4.2 that $\psi\widetilde{\phi}\psi^{-1} = \widetilde{\psi}\widetilde{\phi}\widetilde{\psi}^{-1}$ in $\widetilde{\mathrm{Ham}}(M,\omega)$ whenever $\widetilde{\psi}$ is a lift of ψ.

EXERCISE 12.4.6. Prove that

$$\bigcap_{\kappa} \mathrm{im}\, \iota^{\kappa}(H) = \{0\}.$$

Deduce that $\rho(\widetilde{\phi}; a) > -\infty$ for all $(\widetilde{\phi}, a)$ with $a \ne 0$. *Hint:* Define analogous filtered Morse homology groups with coefficients in the Novikov ring and consider the homomorphisms $\iota^{\kappa}(f) : \mathrm{HM}_*^{\kappa}(f; \Lambda_\omega) \to \mathrm{HM}_*(f; \Lambda_\omega)$. Prove that

$$\bigcap_{\kappa} \mathrm{im}\, \iota^{\kappa}(f) = \{0\}.$$

Now let $\Psi^{f,H} : \mathrm{HF}_*(H,J) \to \mathrm{HM}_*(f; \Lambda_\omega)$ be the PSS isomorphism. Use the energy identity (12.1.14) to prove that there is a constant $c > 0$ such that

$$\Psi^{f,H}(\mathrm{im}\, \iota^{\kappa}(H)) \subset \mathrm{im}\, \iota^{\kappa+c}(f)$$

for every $\kappa \in \mathbb{R}$. Then use the fact that $\Psi^{f,H}$ is the composition of the isomorphism $\mathrm{QH}_*(M; \check{\Lambda}) \to \mathrm{HM}_*(f; \Lambda_\omega)$ with $(\Phi_*^{PSS})^{-1}$.

EXERCISE 12.4.7. For $\kappa \in (\mathbb{R} \setminus \mathrm{Spec}(H)) \cup \{-\infty\}$ define the filtered cochain group

$$\mathrm{CF}^*_{\kappa}(H) := \bigl\{\xi \in \mathrm{CF}^*(H) \,|\, \xi(\widetilde{x}) \ne 0 \Longrightarrow \mathcal{A}_H(\widetilde{x}) \ge \kappa\bigr\}.$$

Check that the Floer coboundary operator δ maps $\mathrm{CF}^*_{\kappa}(H)$ into itself and hence define the filtered Floer cohomology groups $\mathrm{HF}^*_{\kappa}(H)$ for $H \in \mathcal{H}^{\kappa}$. Denote by

$$\jmath_{\kappa}(H) : \mathrm{HF}^*_{\kappa}(H) \to \mathrm{QH}^*(M; \Lambda_\omega)$$

the composition of the homomorphism $\mathrm{HF}^*_\kappa(H,J) \to \mathrm{HF}^*(H,J)$ with the PSS isomorphism $\mathrm{HF}^*(H,J) \to \mathrm{QH}^*(M;\Lambda_\omega)$. Prove that

$$\rho((\widetilde{\phi}^H)^{-1};a) = -\sup\{\kappa \in \mathbb{R}\setminus \mathrm{Spec}(H)\,|\, a \in \mathrm{im}\, \jmath_\kappa(H)\} \tag{12.4.12}$$

for every $a \in \mathrm{QH}^*(M;\Lambda_\omega)$.

EXERCISE 12.4.8. Use Theorem 12.4.4 to prove that

$$d(\phi,\psi) := \inf_{\phi^H=\phi\psi^{-1}} \Big(\rho(\widetilde{\phi}^H;1) + \rho((\widetilde{\phi}^H)^{-1};1)\Big)$$

is a metric on $\mathrm{Ham}(M,\omega)$.

12.5. The Seidel representation

Let (M,ω) be a closed symplectic manifold that satisfies the strong semipositivity condition (8.5.1) and $\psi = \{\psi_t\}_{t\in\mathbb{R}/\mathbb{Z}}$ be a loop of Hamiltonian symplectomorphisms of M, generated by a Hamiltonian function $\mathbb{R}/\mathbb{Z}\times M \to \mathbb{R} : (t,x) \mapsto K_t(x)$ via

$$\partial_t\psi_t = Y_t \circ \psi_t, \qquad \iota(Y_t)\omega = dK_t. \tag{12.5.1}$$

We will see below that such a loop acts naturally on the Floer moduli spaces. However, in order to lift this to an action on the Floer cochain complex, one needs to choose a normalization, namely a homotopy class of sections τ of the Hamiltonian fibration $\widetilde{M}_\psi \to S^2$. Recall from (8.2.6) and the proof of Theorem 11.4.6 that a section of $\widetilde{M}_\psi$ can be represented by a pair of smooth maps $u_\pm : \mathbb{C} \to M$ that satisfy $u_+(z) = \psi_t(u_-(1/z))$ for all $z =: e^{2\pi(s+it)} \in \mathbb{C}\setminus\{0\}$. Up to homotopy we may choose the section such that $u_+(z) \equiv x_0$ for $|z| \le 1$. Then the restriction of u_- to the unit disc in $\mathbb{C}$ is a smooth function $\tau : B = \{z \in \mathbb{C}\,|\,|z|\le 1\} \to M$ that satisfies the condition

$$\tau(e^{-2\pi it}) = \psi_t^{-1}(x_0) \tag{12.5.2}$$

for $t \in \mathbb{R}$. We now explain how such a pair (ψ,τ) determines an isomorphism on Floer cohomology.

That there is such an isomorphism follows from the naturality of the Floer equations (12.1.3). Let us assume that (H,J) is a regular pair as in Proposition 12.1.2, except that now $J = J_t$ is allowed to depend on t. Let $u : \mathbb{R}^2 \to M$ be a solution of the Floer equation

$$\partial_s u + J_t(u)\,(\partial_t u - X_t(u)) = 0,$$

where $\iota(X_t)\omega = dH_t$ and define

$$v(s,t) := \psi_t^{-1}(u(s,t)).$$

Then v is a solution of the Floer equation for the pair (ψ^*H, ψ^*J) given by

$$(\psi^*H)_t := (H_t - K_t)\circ\psi_t, \qquad (\psi^*J)_t := \psi_t^*J_t. \tag{12.5.3}$$

Moreover, the limits of v as $s \to \pm\infty$ are (contractible) periodic orbits for ψ^*H. It is obvious from the construction that the pair (ψ^*H,ψ^*J) satisfies the same regularity assumptions as the pair (H,J). Thus pullback under ψ gives an obvious one-to-one correspondence between the Floer connecting orbits of the pairs (H,J)

and (ψ^*H, ψ^*J), and hence between the corresponding sets of 1-periodic orbits. The latter is the map $\mathcal{P}_0(H) \to \mathcal{P}_0(\psi^*H) : x \mapsto \psi^*x$, given by

$$\psi^*x(t) := \psi_t^{-1}(x(t))$$

for $t \in \mathbb{R}$. To obtain an isomorphism of Floer cohomologies we must choose a bijection $\widetilde{\mathcal{P}}(H) \to \widetilde{\mathcal{P}}(\psi^*H)$ that descends to the bijection $\mathcal{P}_0(H) \to \mathcal{P}_0(\psi^*H)$ and is equivariant under the action of $H_2^S(M)$. There are many such bijections and each one of them gives rise to an isomorphism of Floer cohomologies. The section τ provides us with a canonical choice. To see this note that every element $\widetilde{x} \in \widetilde{\mathcal{P}}(H)$ can be represented by a pair (x, u) that satisfies $u(z) = x_0$ for $|z| \le 1/2$. Define the element $(\psi, \tau)^*\widetilde{x} \in \widetilde{\mathcal{P}}(H^\alpha)$ by $(\psi, \tau)^*\widetilde{x} := (\psi^*x, (\psi, \tau)^*u)$, where

$$((\psi, \tau)^*u)(re^{2\pi it}) := \begin{cases} \psi_t^{-1}(u(re^{2\pi it})), & \text{if } 1/2 \le r \le 1, \\ \tau(2re^{-2\pi it}), & \text{if } 0 \le r \le 1/2. \end{cases}$$

Note that the map $\widetilde{\mathcal{P}}(H) \to \widetilde{\mathcal{P}}(\psi^*H) : \widetilde{x} \mapsto (\psi, \tau)^*\widetilde{x}$ is equivariant under the action of $H_2^S(M)$.

EXERCISE 12.5.1. Prove that

$$(\psi, \tau)^*(A\#\widetilde{x}) = (\psi, \tau - \iota_*A)^*\widetilde{x}$$

for $A \in H_2^S(M)$ and

$$\mu_{\mathrm{CZ}}((\psi, \tau)^*\widetilde{x}) = \mu_{\mathrm{CZ}}(\widetilde{x}) + 2c(\tau), \qquad \mathcal{A}_{\psi^*H}((\psi, \tau)^*\widetilde{x}) = \mathcal{A}_H(\widetilde{x}) + \widetilde{\omega}_\psi(\tau).$$

Here $c := c_1^{\mathrm{Vert}} \in H^2(\widetilde{M}_\psi)$ denotes the vertical first Chern class and $\widetilde{\omega}_\psi \in \Omega^2(\widetilde{M}_\psi)$ is the coupling form of a Hamiltonian connection.

There is an isomorphism $\mathcal{S}(\psi, \tau; H) : \mathrm{CF}^*(H) \to \mathrm{CF}^*(\psi^*H)$ of Floer cochain complexes given by

$$\big(\mathcal{S}(\psi, \tau; H)\xi\big)\big((\psi, \tau)^*\widetilde{x}\big) := \xi(\widetilde{x})$$

for $\xi \in \mathrm{CF}^*(H)$ and $\widetilde{x} \in \widetilde{\mathcal{P}}(H)$. It induces an isomorphism

$$\mathcal{S}(\psi, \tau; H, J) : \mathrm{HF}^*(H, J) \to \mathrm{HF}^*(\psi^*H, \psi^*J)$$

of Floer cohomologies whenever (H, J) is a regular pair as in Proposition 12.1.2. The required automorphism

$$\mathcal{S}^{H,J}(\psi, \tau) : \mathrm{HF}^*(H, J) \to \mathrm{HF}^*(H, J)$$

is the composition of the isomorphism $\mathcal{S}(\psi, \tau; H, J)$ (which we can think of as a "twisting") with the continuation isomorphism $\mathrm{HF}^*(\psi^*H, \psi^*J) \to \mathrm{HF}^*(H, J)$ of Theorem 12.1.3.

Following the notation in Section 12.1 we abbreviate $\mathcal{S}^\alpha := \mathcal{S}^{H^\alpha, J^\alpha}$. It is obvious from the construction that the Seidel homomorphisms are natural with respect to the continuation isomorphisms of Theorem 12.1.3 in the sense that

$$\mathcal{S}^\beta(\psi, \tau) \circ \Phi^{\beta\alpha} = \Phi^{\beta\alpha} \circ \mathcal{S}^\alpha(\psi, \tau)$$

for all α and β, and that this is indeed a representation, i.e.

$$\mathcal{S}^\alpha(\phi, \sigma) \circ \mathcal{S}^\alpha(\psi, \tau) = \mathcal{S}^\alpha(\phi\psi, \sigma\#\tau)$$

for any two Hamiltonian loops ϕ and ψ with corresponding sections σ and τ. We leave it to the reader to figure out the precise meaning of the connected sum $\sigma\#\tau$ of two sections. With this operation, the homotopy classes of pairs (ϕ, σ) form a group $\widehat{\mathrm{G}}$ that is an extension of $\mathrm{G} := \pi_1(\mathrm{Ham}(M, \omega))$ by the group $H_2^S(M)$ of deck transformations of the loop space $\widetilde{\mathcal{L}M}$.[2] The assignment $(\phi,\sigma)\mapsto \mathcal{S}^{H,J}(\phi,\sigma)$ defines a representation of this group in the space of isomorphisms of Floer cohomology as a Λ-module. Seidel showed that these isomorphisms $\mathcal{S}(\phi,\sigma)$ respect the structure of $\mathrm{HF}^*(H,J)$ as a right $\mathrm{HF}^*(H,J)$-module, that is

$$\mathcal{S}^{H,J}(\phi,\sigma)(a*b) = (\mathcal{S}^{H,J}(\phi,\sigma)a)*b,$$

for all $a,b\in\mathrm{HF}^*(H,J)$. The difficult part, from this point of view, is to show that the Seidel representation on Floer cohomology agrees with the one on quantum cohomology under the isomorphisms of Theorem 12.1.4. This is the content of the following theorem.

THEOREM 12.5.2 (Seidel). *Assume (M,ω) satisfies (8.5.1). Then, for every regular pair (H^α,J^α), every Hamiltonian loop ψ, and every corresponding section τ, we have*

$$\mathcal{S}(\psi,\tau)*(\Phi^\alpha\xi^\alpha) = \Phi^\alpha(\mathcal{S}^\alpha(\psi,\tau)\xi^\alpha)$$

for every $\xi^\alpha\in\mathrm{HF}^*(H^\alpha,J^\alpha)$, where $\Phi^\alpha:\mathrm{HF}^*(H^\alpha,J^\alpha)\to\mathrm{QH}^*(M;\Lambda_\omega)$ is the PSS isomorphism of Theorem 12.1.4 and $\mathcal{S}(\psi,\tau)\in\mathrm{QH}^*(M;\Lambda_\omega)$ is the Seidel element defined in Section 11.4.

The proof of Theorem 12.5.2 uses the ideas described in Remark 12.1.7. The Seidel element $\mathcal{S}^\alpha(\psi,\tau)$ counts solutions of the s-dependent Floer equations, while Φ^α counts spiked discs. The relation of Theorem 12.5.2 is obtained by gluing two spiked discs to the ends of a Floer cylinder, and thus constructing a J-holomorphic section of the Hamiltonian fibration $\widetilde{M}_\psi\to S^2$. For details see Seidel [**363**].

LEMMA 12.5.3. *Let $\mathbb{R}/\mathbb{Z}\to\mathrm{Ham}(M,\omega):t\mapsto\psi_t$ be a loop of Hamiltonian symplectomorphisms with $\psi_0=\mathrm{id}$ and τ be a homotopy class of sections of $\widetilde{M}_\psi$. Let $\widetilde{\phi}\in\widetilde{\mathrm{Ham}}(M,\omega)$ be the equivalence class of a path $[0,1]\to\mathrm{Ham}(M,\omega):t\mapsto\phi_t$ with $\phi_0=\mathrm{id}$. Then, for every $a\in\mathrm{QH}^*(M;\Lambda_\omega)$, we have*

$$\rho(\psi^*\widetilde{\phi};a) = \rho(\widetilde{\phi};\mathcal{S}(\psi,\tau)a)+\widetilde{\omega}_\psi(\tau),$$

where $\psi^*\widetilde{\phi}$ denotes the equivalence class of the path $t\mapsto\psi_t^{-1}\circ\phi_t$.

PROOF. We use the cohomological description of the spectral invariants in Exercise 12.4.7. Fix a Hamiltonian function $H:\mathbb{R}/\mathbb{Z}\times M\to\mathbb{R}$ that generates $\widetilde{\phi}$. By construction, the homomorphism $\mathcal{S}(\psi,\tau;H,J):\mathrm{HF}^*(H,J)\to\mathrm{HF}^*(\psi^*H,\psi^*J)$ constructed above sends $\mathrm{HF}^*_\kappa(H,J)$ isomorphically to $\mathrm{HF}^*_{\kappa+\widetilde{\omega}_\psi(\tau)}(\psi^*H,\psi^*J)$. Hence there is a commutative diagram

$$\begin{array}{ccc}\mathrm{QH}^*(M;\Lambda_\omega) & \stackrel{\mathcal{S}(\psi,\tau)}{\longrightarrow} & \mathrm{QH}^*(M;\Lambda_\omega)\\ \Phi^H\uparrow & & \uparrow\Phi^{\psi^*H}\\ \mathrm{HF}^*(H,J) & \stackrel{\mathcal{S}(\psi,\tau;H,J)}{\longrightarrow} & \mathrm{HF}^*(\psi^*H,\psi^*J),\end{array}$$

where the top horizontal arrow takes the image of $\jmath_\kappa(H)$ isomorphically onto the image of $\jmath_{\kappa+\widetilde{\omega}_\psi(\tau)}(\psi^*H)$.

[2]Here we should interpret $H_2^S(M)$ as the image of the Hurewicz homomorphism in the torsion free group $H_2(M;\mathbb{Z})/\mathrm{tor}$: see Remark 12.5.4.

Now let $\kappa \in \mathbb{R} \setminus \mathrm{Spec}(H)$. Then $\kappa + \widetilde{\omega}_\psi(\tau) \notin \mathrm{Spec}(\psi^* H)$, by Exercise 12.5.1. Moreover, by Exercise 12.4.7, we have

$$\begin{aligned} \kappa < -\rho(\widetilde{\phi}^{-1}; a) &\iff a \in \mathrm{im}\, \jmath_\kappa(H) \\ &\iff \mathcal{S}(\psi,\tau)a \in \mathrm{im}\, \jmath_{\kappa+\widetilde{\omega}_\psi(\tau)}(\psi^* H) \\ &\iff \kappa + \widetilde{\omega}_\psi(\tau) < -\rho\big((\psi^*\widetilde{\phi})^{-1}; \mathcal{S}(\psi,\tau)a\big) \end{aligned}$$

Here the second equivalence follows from the above commuting diagram and the last condition follows again from Exercise 12.4.7. It follows that

$$\rho(\widetilde{\phi}^{-1}; a) = \rho\big((\psi^*\widetilde{\phi})^{-1}; \mathcal{S}(\psi,\tau)a\big) + \widetilde{\omega}_\psi(\tau).$$

The assertion follows by replacing the path $t \mapsto \phi_t$ with $t \mapsto \phi_t^{-1} \circ \psi_t$. □

Oh [**306**] calls the shift of the spectral invariants under the Seidel action the **monodromy shift**. When working with coefficients in $\Lambda := \Lambda^{\mathrm{univ}}[q, q^{-1}]$, one can eliminate the monodromy shift by standardizing the choice of τ so that $\widetilde{\omega}_\psi(\tau) = 0$ (see Lalonde–McDuff–Polterovich [**228**]). Even so, for manifolds on which the Seidel action on quantum cohomology is nontrivial, only the invariant part of the information contained in the spectral invariants descends to $\mathrm{Ham}(M)$. This is more fully explained in Remark 12.5.5 below. The next remark describes a global framework for the monodromy shift.

REMARK 12.5.4 (Spectral invariants: global perspective). We shall begin by constructing a space $P\mathrm{Ham}(M,\omega)$ that fibers over $\mathrm{Ham}(M,\omega)$ and whose fundamental group is the extension $\widehat{\mathrm{G}}$ of $\mathrm{G} := \pi_1(\mathrm{Ham}(M,\omega))$ considered above. For simplicity, we assume in the following discussion that M is simply connected and that $\pi_2(M)$ is torsion free, so that $\widehat{G}$ is an extension of G by the full group $\pi_1(\mathcal{L}M) \cong \pi_2(M)$ rather than by its quotient $H_2^S(M)$.

Fix a base point $x_0 \in M$ and let $P_0 M$ denote the space of smooth paths $\gamma : [0,1] \to M$ starting at $\gamma(0) = x_0$. Denote by

$$P\mathrm{Ham}(M,\omega) \subset \mathrm{Ham}(M,\omega) \times P_0 M$$

the space of pairs (ϕ,γ) such that $\gamma(1) = \phi^{-1}(x_0)$. There is a fibration

$$\Omega M \to P\mathrm{Ham}(M,\omega) \to \mathrm{Ham}(M,\omega)$$

whose fiber is the based loop space ΩM of M. The boundary map

$$\pi_2(\mathrm{Ham}(M,\omega)) \to \pi_1(\Omega M) \cong \pi_2(M)$$

in the homotopy exact sequence is given by evaluating at the base point. It is well known that its image consists of torsion classes (see [**227**] for example). Therefore our assumptions on M imply that it vanishes. Hence $\pi_1(P\mathrm{Ham}(M,\omega))$ is an extension of G by $\pi_2(M)$ that is isomorphic to $\widehat{\mathrm{G}}$ by (12.5.2). Explicitly, each pair (ψ,τ) determines a loop $\mathbb{R}/\mathbb{Z} \to P\mathrm{Ham}(M,\omega) : t \mapsto (\psi_t, \gamma_t)$ via $\gamma_t(s) := \tau(se^{-2\pi i t})$ and this induces the required isomorphism from $\widehat{\mathrm{G}}$ to $\pi_1(P\mathrm{Ham}(M,\omega))$.

We define $\widetilde{P\mathrm{Ham}}(M,\omega)$ to be the universal cover of $P\mathrm{Ham}(M,\omega)$. Its elements are equivalence classes $[\widetilde{\phi}, \widetilde{\gamma}]$ of paths $(\phi_t,\gamma_t)_{t\in[0,1]}$ in $P\mathrm{Ham}(M,\omega)$ where $\phi_0 = \mathrm{id}$ and γ_0 is the constant path. The group $\widehat{\mathrm{G}}$ acts on $\widetilde{P\mathrm{Ham}}(M,\omega)$ by deck transformations: if (ψ,τ) represents an element of $\widehat{\mathrm{G}}$ and $[\widetilde{\phi}, \widetilde{\gamma}]$ is an equivalence class in $\widetilde{P\mathrm{Ham}}(M,\omega)$ then $(\psi,\tau)\cdot[\widetilde{\phi}, \widetilde{\gamma}]$ denotes the equivalence class of the path in

$P\mathrm{Ham}(M,\omega)$ formed by first going round the loop in $P\mathrm{Ham}(M,\omega)$ determined by ψ and τ and then along the path (ϕ_t,γ_t).

Consider the trivial line bundle

$$\widetilde{P\mathrm{Ham}}(M,\omega)\times \mathrm{QH}^*(M;\Lambda)\times\mathbb{R}\to\widetilde{P\mathrm{Ham}}(M,\omega)\times\mathrm{QH}^*(M;\Lambda),$$

This supports an action of the group $\widehat{\mathrm{G}}$ by deck transformations on $\widetilde{P\mathrm{Ham}}(M,\omega)$, by the Seidel representation on quantum cohomology, and by the monodromy shift on $\mathbb{R}$. More precisely

$$(\psi,\tau)\cdot\Big([\widetilde{\phi},\widetilde{\gamma}];a;\kappa\Big)=\Big((\psi,\tau)\cdot[\widetilde{\phi},\widetilde{\gamma}];\mathcal{S}(\psi,\tau)a;\kappa+\widetilde{\omega}_\psi(\tau)\Big).$$

Lemma 12.5.3 asserts that the spectral invariants define a section

$$\rho:\widetilde{P\mathrm{Ham}}(M,\omega)\times\mathrm{QH}^*(M;\Lambda)\to\widetilde{P\mathrm{Ham}}(M,\omega)\times\mathrm{QH}^*(M;\Lambda)\times\mathbb{R}$$

of this bundle that is $\widehat{\mathrm{G}}$-equivariant. Passing to the quotients with respect to $\widehat{\mathrm{G}}$ we obtain a section of the corresponding line bundle $L\to X$ over the total space X of the fibration

$$\mathrm{QH}^*(M;\Lambda)\to X:=\widetilde{P\mathrm{Ham}}(M,\omega)\times_{\widehat{\mathrm{G}}}\mathrm{QH}^*(M;\Lambda)\to P\mathrm{Ham}(M).$$

There is a similar picture for the spectral values. It follows from Exercise 12.5.1 that the $\widehat{\mathrm{G}}$-action on $\widetilde{P\mathrm{Ham}}(M,\omega)\times\mathbb{R}$ preserves the subset

$$\mathfrak{spec}:=\{([\widetilde{\phi},\widetilde{\gamma}],c)\in\widetilde{P\mathrm{Ham}}(M,\omega)\times\mathbb{R}\,|\,c\in\mathrm{Spec}(\widetilde{\phi})\}$$

of spectral values. Hence the quotient $\mathfrak{spec}/\widehat{\mathrm{G}}$ is a subset of the real line bundle $L:=\widetilde{P\mathrm{Ham}}(M,\omega)\times_{\widehat{\mathrm{G}}}\mathbb{R}\to P\mathrm{Ham}(M,\omega)$.

REMARK 12.5.5. There are many possible variations on the definition of Floer (co)homology. As an introduction to the stripped down version $\mathrm{HF}^*(\phi)$ defined in the next section for an arbitrary symplectomorphism ϕ, we briefly mention two of them here.

(i) First we consider Floer cohomology with coefficients in the ring

$$\Lambda:=\Lambda^{\mathrm{univ}}[q,q^{-1}]$$

of Example 11.1.4 (vi) (over a ground ring R). In this case, for every regular Hamiltonian H, the Floer cochain complex $\mathrm{CF}^*(H;\Lambda)$ is the set of all functions $\xi:\widetilde{\mathcal{P}}(H)\to\Lambda$ that are equivariant in the sense that

$$\xi(A\#\widetilde{x})=\xi(\widetilde{x})t^{\omega(A)}q^{c_1(A)}\tag{12.5.4}$$

for every $\widetilde{x}\in\widetilde{\mathcal{P}}(H)$ and every $A\in H_2(M)$. This is a finitely generated free Λ-module. An element $\xi\in\mathrm{CF}^*(H;\Lambda)$ has degree k if and only if $\xi(\widetilde{x})$ has pure degree and

$$\mu_{\mathrm{CZ}}(\widetilde{x})+\deg(\xi(\widetilde{x}))=k$$

for every $\widetilde{x}\in\widetilde{\mathcal{P}}(H)$. The Floer coboundary operator is still defined by (12.1.9) and it is an easy exercise to show that $\delta\xi$ satisfies (12.5.4) whenever ξ does. The reader may verify that $\delta\xi$ is indeed a function on $\widetilde{\mathcal{P}}(H)$ with values in Λ. The entire discussion of Sections 12.1 and 12.2 carries over to this definition of Floer cohomology.

Now let $t\mapsto\psi_t=\psi_{t+1}$ be a loop of Hamiltonian symplectomorphisms generated by a family of Hamiltonian functions $K_t=K_{t+1}:M\to\mathbb{R}$ via (12.5.1).

Fix a regular pair (H,J) and let (ψ^*H, ψ^*J) be defined by (12.5.3). Then the Hamiltonian loop induces a **twisting operator**

$$\mathcal{S}(\psi; H, J) : \mathrm{HF}^*(H,J) \to \mathrm{HF}^*(\psi^*H, \psi^*J).$$

On the cochain level it is defined by

$$\big(\mathcal{S}(\psi; H, J)\xi\big)\big((\psi,\tau)^*\widetilde{x}\big) := \xi(\widetilde{x})t^{-\widetilde{\omega}_\psi(\tau)}q^{-c(\tau)}$$

for $\xi \in \mathrm{CF}(H;\Lambda)$ and $\widetilde{x} \in \widetilde{\mathcal{P}}(H)$ where τ is some section of $\widetilde{M_\psi}$. (The right hand side is independent of the choice of τ by Exercise 12.5.1 and (12.5.4).) As before, $\mathcal{S}(\psi; H, J)$ descends to a well defined isomorphism on Floer cohomology and hence induces an automorphism $\mathcal{S}^{H,J}(\psi) : \mathrm{HF}^*(H,J) \to \mathrm{HF}^*(H,J)$ via continuation. The upshot is that we obtain a representation of $\mathrm{G} = \pi_1(\mathrm{Ham}(M,\omega))$ on Floer cohomology.

(ii) If one is prepared to work with a mod-2 grading then one can set $q = 1$ in the above discussion, thereby obtaining Floer cohomology groups with coefficients in the universal Novikov ring Λ^{univ} and mod-2 grading. The Floer cochain complex is then the set of all functions $\xi : \widetilde{\mathcal{P}}(H) \to \Lambda^{\mathrm{univ}}$ that satisfy $\xi(A\#\widetilde{x}) = \xi(\widetilde{x})t^{\omega(A)}$. This means that the map $\widetilde{x} \mapsto \xi(\widetilde{x})t^{\mathcal{A}_H(\widetilde{x})}$ is invariant under the action of $H_2^S(M)$ and hence descends to a map from $\mathcal{P}_0(H)$ to Λ^{univ}. Thus the Floer chain complex is the Λ^{univ}-module

$$\mathrm{CF}^*(H) := \mathrm{CF}^*(H;\Lambda^{\mathrm{univ}}) = \mathrm{Map}(\mathcal{P}_0(H), \Lambda^{\mathrm{univ}})$$

and the Floer coboundary is given by

$$(\delta\xi)(x) := \sum_y \sum_\varepsilon n_\varepsilon(x,y)\xi(y)t^\varepsilon.$$

Here $n_\varepsilon(x,y)$ denotes the signed number of index one Floer connecting orbits u from x to y with energy $E(u) = \varepsilon$. The Seidel representation is now given by

$$(\mathcal{S}(\psi; H)\xi)(\psi^*x) := \xi(x)$$

for $x \in \mathcal{P}_0(H)$ and $\xi \in \mathrm{CF}^*(H)$.

It turns out that this complex retains no essential information from the generating Hamiltonian H and may be defined entirely in terms of the time-1 map $\phi \in \mathrm{Ham}(M,\omega)$. However, if we do this the Floer continuation map from $\mathrm{HF}^*(\phi_0)$ to $\mathrm{HF}^*(\phi_1)$ depends in general on the homotopy class of the path ϕ_s used to define it. (This dependence is precisely measured by the Seidel representation.) Moreover, the definition works even if ϕ is not Hamiltonian. Thus we are led to the much more broadly based Floer theory discussed in the next section.

12.6. Donaldson's quantum category

Symplectomorphisms. The Floer homology approach to quantum cohomology can be extended in two directions. Assume for simplicity that the symplectic manifold (M,ω) is compact, simply connected, and monotone. In this case it follows from equations (12.1.2) and (12.1.6) that the energy of a Floer connecting orbit is bounded by its index (compare this with Example 8.6.8). Hence one need not use the Novikov ring; instead one can define the Floer cohomology groups with integer coefficients. The price for doing this is that the groups now only have a grading modulo $2N$ where N is the minimal Chern number. In fact it is useful to reduce the grading mod 2 so that no additional choices need to be made to normalize the grading. The upshot is that, for every symplectomorphism ϕ, we can define a Floer cohomology group $\mathrm{HF}^*(\phi)$ that is graded modulo 2 and whose Euler characteristic

$$\chi(\mathrm{HF}^*(\phi)) = L(\phi)$$

is the Lefschetz number of ϕ. The critical points that generate the Floer complex are now the fixed points of ϕ, which we assume to be all nondegenerate, and the connecting orbits are J-holomorphic maps $u : \mathbb{R} \times [0,1] \to M$ which satisfy $\phi(u(s,1)) = u(s,0)$ for $s \in \mathbb{R}$. We may think of these as J-holomorphic sections of a symplectic fiber bundle $P \to \mathbb{R} \times S^1$ with fiber M and holonomy ϕ around S^1. If there are degenerate fixed points then we must choose a Hamiltonian perturbation as in Section 12.1. The resulting Floer cohomology groups are independent of the Hamiltonian perturbation and the almost complex structure used to define them. (See Floer [**116**], Floer–Hofer–Salamon [**119**], and Dostoglou–Salamon [**90**] for more details.) Since M is simply connected, every symplectic isotopy $\{\phi_s\}_{0\le s\le 1}$ determines an isomorphism $\mathrm{HF}^*(\phi_0) \to \mathrm{HF}^*(\phi_1)$. As noted in Remark 12.5.5, this isomorphism will in general depend on the homotopy class of the symplectic isotopy. Now for any two symplectomorphisms ϕ and ψ there is a natural isomorphism

$$\mathrm{HF}^*(\phi) \to \mathrm{HF}^*(\psi \circ \phi \circ \psi^{-1}).$$

Moreover, according to Donaldson, there is an analogue of the deformed cup product, namely a skew-symmetric and associative pairing

$$\mathrm{HF}^*(\phi) \otimes \mathrm{HF}^*(\psi) \to \mathrm{HF}^*(\psi \circ \phi).$$

This can be defined in terms of J-holomorphic sections of a symplectic fiber bundle $P \to S$ with fiber M where S is a 2-sphere with three punctures and the holonomies around these punctures are conjugate to ϕ, ψ, and $\psi\circ\phi$, respectively. This product structure can be interpreted as a **category** in which the objects are the symplectomorphisms of M and and the morphisms from ϕ to ψ are the elements of the Floer cohomology group $\mathrm{HF}^*(\psi \circ \phi^{-1})$. Composition of two morphisms is given by the quantum product.

Now in the case $\psi = \mathrm{id}$ it follows from the discussion of Section 12.1 that, in the monotone case, the Floer cohomology groups are isomorphic to the ordinary cohomology groups $\mathrm{HF}^*(\mathrm{id}) = \mathrm{QH}^*(M) = H^*(M)$, where the grading is made periodic with period 2. Thus the cohomology of M acts on the Floer cohomology of ϕ:

$$H^*(M) \otimes \mathrm{HF}^*(\phi) \to \mathrm{HF}^*(\phi).$$

In this case there is yet another way to think of the quantum product structure. Namely, one can represent a k-dimensional cohomology class by a codimension-k submanifold (or pseudocycle) $V \subset M$ and then intersect the spaces of connecting orbits in the construction of the Floer cohomology groups of ϕ with this submanifold. More generally, one can intersect these connecting orbit spaces with finite dimensional submanifolds of the path space $\Omega_\phi = \{\gamma : [0,1] \to M \,|\, \phi(\gamma(1)) = \gamma(0)\}$ and obtain an action of the (low dimensional) cohomology $H^*(\Omega_\phi)$ on $\mathrm{HF}^*(\phi)$.

Specializing further to the case $\phi = \mathrm{id}$, we see that the above product construction agrees with the *pair-of-pants product* of Section 12.2, and so in this case the Donaldson category reduces to the quantum cohomology of M.

REMARK 12.6.1. So far the Floer homology groups of symplectomorphisms have been calculated for only a few examples. Even in dimension two this problem is rather nontrivial. In [**361**] Seidel computed the Floer cohomology of a Dehn twist of a Riemann surface Σ, and in [**369**] he proved that the identity map can be distinguished, as an $H^*(\Sigma)$-module, from all other mapping classes: if ϕ is not (symplectically) isotopic to the identity then $H^2(\Sigma)$ acts trivially on $\mathrm{HF}^*(\phi)$. In [**137, 138**] Ralf Gautschi succeeded in computing the Floer cohomology groups of symplectomorphisms $\phi : \Sigma \to \Sigma$ that are of *finite type*, i.e. that have no pseudo-Anosov components in the Thurston decomposition. In [**362, 370**] Seidel discovered an exact sequence in symplectic Floer homology, which is an analogue of Floer's exact sequence for the instanton Floer homology of 3-manifolds. He used the exact sequence to compute the Floer cohomology groups of generalized Dehn twists $\tau : X \to X$ (on symplectic 4-manifolds X) and, as a result, he was able to prove that in many cases the square $\tau \circ \tau$ of such a Dehn twist is not symplectically isotopic to the identity. This is a remarkable result, because $\tau \circ \tau$ is always smoothly isotopic to the identity. (The generalized Dehn twist τ itself acts nontrivially on the cohomology of X.) In this work the product structures on Floer cohomology play a crucial role.

In [**125**] Fukaya studied the product structures on the chain level and discovered the underlying A^∞ structure. The Fukaya A^∞ categories were taken up by Seidel in his recent work [**366, 367**]; they seem to be fundamental for understanding homological versions of mirror symmetry.

Mapping tori. The Donaldson product structures are particularly interesting when the symplectic manifold M is the moduli space M_Σ of flat connections on the nontrivial principle bundle $P \to \Sigma$ with structure group SO(3) over a Riemann surface Σ (of large genus). In this case the mapping class group of Σ acts on M_Σ by symplectomorphisms $\phi_f : M_\Sigma \to M_\Sigma$ (where $f : \Sigma \to \Sigma$ is an orientation preserving diffeomorphism) and one can examine the Floer cohomology groups $\mathrm{HF}^*(\phi_f)$ generated by the mapping class group. Now there is an alternative construction, based on Floer cohomology groups for 3-manifolds and Yang-Mills instantons on 4-dimensional cobordisms. Every diffeomorphism $f : \Sigma \to \Sigma$ induces a mapping torus $Y_f = \Sigma \times \mathbb{R}/\sim$ where $\Sigma \times \{t\}$ is identified with $\Sigma \times \{t+1\}$ via f. These 3-manifolds determine Floer homology groups $\mathrm{HF}^*(Y_f)$ constructed from flat SO(3)-connections on Y_f and anti-self-dual instantons on $Y_f \times \mathbb{R}$. It was conjectured by Atiyah and Floer, and proved in Dostoglou–Salamon [**92**], that there is a natural isomorphism

$$\mathrm{HF}^*(\phi_f) \cong \mathrm{HF}^*(Y_f).$$

Now in Floer–Donaldson theory there is a pairing

$$\mathrm{HF}^*(Y_f) \otimes \mathrm{HF}^*(Y_g) \to \mathrm{HF}^*(Y_{gf})$$

determined by anti-self-dual instantons over the 4-manifold X which is fibered over the 3-punctured sphere S with fiber Σ and holonomy f, g, and gf. That the two products are preserved by the above isomorphisms was proved in Salamon [**352**].

An interesting special case arises when $g = \mathrm{id}$. In this case the symplectic Floer cohomology of ϕ_g and hence the instanton Floer cohomology of Y_g agrees with the ordinary cohomology of the moduli space M_Σ made periodic with period 4. So the quantum cohomology $\mathrm{QH}^*(M_\Sigma)$ acts on the Floer cohomology of Y_f:

$$\mathrm{QH}^*(M_\Sigma) \otimes \mathrm{HF}^*(Y_f) \to \mathrm{HF}^*(Y_f).$$

Now the cohomology of M_Σ (as well as the quantum cohomology) is well understood and is closely related to the homology of the Riemann surface Σ itself; cf.

Example 11.1.15. For example there is a universal construction of a homomorphism $\mu : H_1(\Sigma) \to H^3(\mathcal{B}^*_\Sigma)$ where $\mathcal{B}^*_\Sigma = \mathcal{A}^*_\Sigma/\mathcal{G}_\Sigma$ denotes the infinite dimensional configuration space of irreducible connections on the bundle $P \to \Sigma$ modulo gauge equivalence. This is Donaldson's μ-map. It can be roughly described as the slant product

$$\mu(\gamma) = -\tfrac{1}{4}p_1(\mathcal{P})/\gamma,$$

where $p_1(\mathcal{P}) \in H^4(\mathcal{B}^*_\Sigma \times \Sigma)$ denotes the first Pontryagin class of the universal SO(3)-bundle $\mathcal{P} \to \mathcal{B}^*_\Sigma \times \Sigma$. In more explicit terms, the induced cohomology class in $H^3(M_\Sigma)$ is represented by the codimension-3-submanifold $V_\gamma \subset M_\Sigma$ of those flat connections which have trivial holonomy around γ. In summary, we have the following diagrams

$$\begin{CD}
\mathrm{HF}^*(Y_f) @>{\mu(\gamma)}>> \mathrm{HF}^*(Y_f)\\
@VVV @VVV\\
\mathrm{HF}^*(\phi_f) @>{V_\gamma}>> \mathrm{HF}^*(\phi_f)
\end{CD}
\qquad
\begin{CD}
\mathrm{HF}^*(Y_f) @>{X_\gamma}>> \mathrm{HF}^*(Y_f)\\
@VVV @VVV\\
\mathrm{HF}^*(\phi_f) @>>> \mathrm{HF}^*(\phi_f)
\end{CD}$$

In each diagram the vertical arrows are the natural isomorphisms of [**92**]. In the diagram on the left the horizontal maps are defined by cutting down the moduli spaces of connecting orbits by intersecting them with suitable submanifolds. In the diagram on the right the horizontal maps are given by the product structures which are defined in terms of cobordisms. For example, the class $\gamma \in H_1(\Sigma)$ determines a natural cobordism X_γ with boundary $\partial X_\gamma = (-Y_f) \cup Y_f$. Of course, all four definitions of the product should agree under the natural isomorphisms. The relations between these product structures play an important role in the work of Michael Callaghan about symplectic isotopy problems for symplectomorphisms of M_Σ (induced by separating Dehn twists). Unfortunately, his work never appeared.

These product structures also play an important role in the Floer-Fukaya construction of cohomology groups $\mathrm{HFF}^*(Y,\gamma)$ associated to pairs (Y,γ) where Y is a 3-manifold and $\gamma \in H_1(Y)$. In another direction, the homomorphisms in Floer's exact sequence can be interpreted in terms of these product structures.

Lagrangian intersections. There are similar structures in Floer cohomology for Lagrangian intersections. These form in fact the original context of Donaldson's quantum category construction. With M as above (compact, simply connected, and monotone) there are Floer cohomology groups $\mathrm{HF}^*(L_0, L_1)$ for every pair of Lagrangian submanifolds L_0 and L_1 with $H^1(L_i;\mathbb{R}) = 0$. In this case the critical points are the intersection points $L_0 \cap L_1$ and the connecting orbits are J-holomorphic curves $u : \mathbb{R} \times [0,1] \to M$ with $u(s,0) \in L_0$ and $u(s,1) \in L_1$. This is the context of Floer's original work in [**113**]. The Euler characteristic of Floer cohomology is now the intersection number

$$\chi(\mathrm{HF}^*(L_0, L_1)) = L_0 \cdot L_1.$$

Again there is a pairing

$$\mathrm{HF}^*(L_0, L_1) \otimes \mathrm{HF}^*(L_1, L_2) \to \mathrm{HF}^*(L_0, L_2)$$

defined by holomorphic triangles. In [**113**] Floer proved that $\mathrm{HF}^*(L_0, L_0)$ is isomorphic to the ordinary cohomology of L_0 provided that $\pi_2(M, L_0) = 0$. Under our assumptions, where M is simply connected, this condition is never satisfied and the corresponding assertion is an open question. In [**44**], Biran–Cornea define Lagrangians with $\mathrm{HF}^*(L_0, L_0) \cong H^*(L_0)$ to be **wide** while those with vanishing $\mathrm{HF}^*(L_0, L_0)$ are **narrow**. At the time of writing, this dichotomy is still not well understood.

This multiplicative structure can be interpreted as a quantum category where the objects are the Lagrangian submanifolds $L \subset M$ (with $H^1(L;\mathbb{R}) = 0$) and the morphisms from L_0 to L_1 are the elements of the Floer cohomology group $\mathrm{HF}^*(L_0, L_1)$. The above structure with symplectomorphisms is a special case of this with $M = N \times N$, $L_0 = \Delta$, $L_1 = \mathrm{graph}\,(\phi)$, and $L_2 = \mathrm{graph}\,(\psi)$.

The interpretation of Lagrangian submanifolds as objects of a category and the Floer cohomology groups as the spaces of morphisms is due to Donaldson. Later Fukaya discovered the underlying A^∞ structure on the chain level. It is called the **Fukaya category** and denoted by $\mathcal{F}(M)$. The rigorous construction of the Fukaya category involves quite a lot of subtle analysis, even in simple cases where no bubbling can occur. For example, Seidel [**371**] has developed techniques that compute certain derived categories $\mathcal{D}^b\mathcal{F}(M)$ for affine Calabi–Yau manifolds. His construction involves an induction argument over Lefschetz pencils, vanishing cycles, and the Seidel exact sequence. The paper Fukaya–Seidel–Smith [**131**] uses the Fukaya category of a cotangent bundl to great effect. Here again there is no bubbling, but one has to deal with issues caused by noncompactness. For the general case, see Fukaya–Oh–Ohta–Ono [**128**]. In a recent series of papers [**414, 415, 416, 417**] Wehrheim and Woodward proved functoriality of the Donaldson category under Lagrangian correspondences in the monotone case.

Heegaard splittings. The Floer theory for Lagrangian intersections is related to 3-manifolds as follows. If Y is a homology-3-sphere choose a Heegaard splitting

$$Y = Y_0 \cup Y_1$$

over a Riemann surface Σ. Then each handlebody Y_j determines a (singular) Lagrangian submanifold $L_j = L_{Y_j} \subset M_\Sigma$ in the (singular) moduli space M_Σ of flat SU(2)-connections over Σ. According to Atiyah [**20**] and Floer, there should be a natural isomorphism

$$\mathrm{HF}^*(Y_0 \cup Y_1) \cong \mathrm{HF}^*(L_0, L_1).$$

As before, there is a product structure

$$\mathrm{HF}^*(Y_0 \cup Y_1) \otimes \mathrm{HF}^*(Y_1 \cup Y_2) \to \mathrm{HF}^*(Y_0 \cup Y_2)$$

defined directly in terms of Yang-Mills instantons on a suitable 4-dimensional cobordism and the two product structures should be related by the above (conjectural) isomorphisms. Important progress on this conjecture was made by Katrin Wehrheim [**411, 412, 413**] in her study of anti-self-dual instantons with Lagrangian boundary conditions. This resulted, as an intermediate step, in the definition of Floer homology groups for 3-manifolds with boundary in Salamon–Wehrheim [**355**]. (For another setting, see the earlier work by Fukaya [**126**].)

A Heegaard splitting along a Riemann surface Σ of genus g also determines two Lagrangian tori in the g-fold symmetric product $\mathrm{Sym}^g\Sigma$. In a long series of papers starting with [**318**], Ozsváth–Szabó developed from their Lagrangian Floer theory a powerful new series of invariants for low dimensional manifolds and also for knots called **Heegaard–Floer theory**. Working together with several coauthors, they are well on their way to constructing a four dimensional TQFT, cf. Lipshitz–Ozsváth–Thurston [**247**].

12.7. The symplectic vortex equations

In this section we discuss another recent development that is closely related to the quantum cohomology of symplectic quotients. The starting point is an extension of the notion of J-holomorphic curves to symplectic manifolds M with Hamiltonian G-actions. The new invariants are constructed from the solutions of a version of the vortex equations over a Riemann surface Σ for a pair consisting of a connection on a principal G-bundle $P \to \Sigma$ and a section of the associated bundle with fibers M. The symplectic vortex equations were discovered, independently, by Salamon [**64**] and Ignasi Mundet i Riera [**294, 295**]. In the case where the symplectic manifold in question is $\mathbb{C}^n$ and G is a subgroup of $\mathrm{U}(n)$, they were known in the physics literature as **gauged sigma models**. In the following we describe the basic setup and explain some of the results obtained so far in this subject.

Let (M, ω) be a symplectic manifold equipped with a Hamiltonian action by a compact Lie group G. The action is generated by an equivariant moment map

$$\mu : M \to \mathfrak{g}^*$$

with values in the dual of the Lie algebra $\mathfrak{g} := \mathrm{Lie}(\mathrm{G})$. This means that the infinitesimal action $X_\xi \in \mathrm{Vect}(M)$ of an element $\xi \in \mathfrak{g}$ is given by

$$\iota(X_\xi)\omega = dH_\xi, \qquad H_\xi := \langle \mu, \xi \rangle.$$

Let $(\Sigma, j_\Sigma, \mathrm{dvol}_\Sigma)$ be closed Riemann surface and $P \to \Sigma$ be a principal G-bundle. The **symplectic vortex equations** for a pair (u, A), consisting of an equivariant map $u : P \to M$ and a connection $A \in \mathcal{A}(P)$, depend on a parameter τ and have the form

$$\bar{\partial}_{J,A}(u) = 0, \qquad *F_A + \mu(u) = \tau. \tag{12.7.1}$$

Here $J = \{J_z\}_{z\in\Sigma}$ is a smooth family of G-invariant and ω-compatible almost complex structures on M. We think of the connection as a Lie algebra valued 1-form $A \in \Omega^1(P, \mathfrak{g})$ that is equivariant (i.e. $A_{pg}(vg) = g^{-1}A_p(v)g$ for $v \in T_pP$ and $g \in \mathrm{G}$) and identifies the vertical tangent space with $\mathfrak{g}$ (i.e. $A_p(p\xi) = \xi$ for $p \in P$ and $\xi \in \mathfrak{g}$). The covariant derivative of u with respect to A is the 1-form $d_Au \in \Omega^1(P, u^*TM)$ given by

$$d_Au := du + X_A(u).$$

This 1-form is equivariant and horizontal, and hence descends to a 1-form on Σ with values in the quotient bundle $u^*TM/\mathrm{G} \to \Sigma$. The almost complex structures on M induce a complex structure on this vector bundle and $\bar{\partial}_{J,A}$ denotes the complex anti-linear part of d_Au with respect to this complex structure. The second equation in (12.7.1) couples the curvature

$$F_A := dA + \tfrac{1}{2}[A \wedge A] \in \Omega^2_{\mathrm{ad}}(P, \mathfrak{g}) = \Omega^2(\Sigma, \mathfrak{g}_P)$$

with the moment map. Here $* : \Omega^2(\Sigma, \mathfrak{g}_p) \to \Omega^0(\Sigma, \mathfrak{g}_P^*)$ is the composition of the Hodge $*$-operator with the isomorphism $\mathfrak{g} \to \mathfrak{g}^*$ induced by an invariant inner product on the Lie algebra. We assume that $\tau \in Z(\mathfrak{g}^*)$ belongs to the center and is a regular value of the moment map μ. We emphasize that the volume form on Σ enters into the second equation in (12.7.1) and so the solution space is not invariant under the group of complex automorphisms of Σ (when $J_z = J$ is independent of z).

There is a natural action of the gauge group $\mathcal{G} = \mathcal{G}(P)$ (of smooth maps $g : P \to \mathrm{G}$ that are equivariant under the conjugate action of G on itself) on $C^\infty_{\mathrm{G}}(P, M) \times \mathcal{A}(P)$ via

$$g^*(u, A) := (g^{-1}u, g^{-1}dg + g^{-1}Ag)$$

and this action preserves the space of solutions of (12.7.1).

The solutions of (12.7.1) are the absolute minima of the energy functional

$$E(u, A) := \tfrac{1}{2}\int_\Sigma \Big(|d_A u|^2 + |F_A|^2 + |\mu(u) - \tau|^2\Big)\,\mathrm{dvol}_\Sigma.$$

For solutions of (12.7.1) the energy is given by the topological invariant

$$E(u, A) = \int_\Sigma \Big((d_A u)^*\omega - \langle \mu(u) - \tau, F_A\rangle\Big) = \int_\Sigma \Big(u^*\omega - d\langle \mu(u) - \tau, A\rangle\Big).$$

Here the integrands are equal and descend to a 2-form on Σ. The integral can be interpreted as the evaluation of an equivariant cohomology class $[\omega - \mu + \tau] \in H^2_{\mathrm{G}}(M)$ on the equivariant homology class $[u] \in H_2^{\mathrm{G}}(M)$. (If $\theta : P \to \mathrm{EG}$ is an equivariant classifying map then $[u]$ is the image of the fundamental class $[\Sigma] \in H_2(\Sigma)$ under the map $\Sigma \to M_{\mathrm{G}} = M \times_{\mathrm{G}} \mathrm{EG}$ induced by $u \times \theta : P \to M \times \mathrm{EG}$.)

REMARK 12.7.1. (i) In local holomorphic coordinates $s + it \in U$ on Σ and a local trivialization of P the connection has the form

$$A = \Phi\, ds + \Psi\, dt,$$

where $\Phi, \Psi : U \to \mathfrak{g}$ and u is a smooth map from U to M. The vortex equations then have the form

$$\begin{aligned} \partial_s u + X_\Phi(u) + J_{s,t}(u)\,(\partial_t u + X_\Psi(u)) &= 0, \\ \lambda^{-2} *_{\mathfrak{g}} (\partial_s\Psi - \partial_t\Phi + [\Phi, \Psi]) + \mu(u) &= \tau, \end{aligned} \tag{12.7.2}$$

where $*_{\mathfrak{g}} : \mathfrak{g} \to \mathfrak{g}^*$ denotes the isomorphism induced by the invariant inner product on $\mathfrak{g}$ and $\mathrm{dvol}_\Sigma = \lambda^2\, ds \wedge dt$ is the volume form in local coordinates.

(ii) When $\mathrm{G} = \{\mathbb{1}\}$ the solutions of (12.7.1) are J-holomorphic curves in M.

(iii) The usual vortex equations appear as the special case $M = \mathbb{C}$ and $\mathrm{G} = S^1$. In this case the first equation in (12.7.1) says that u is a holomorphic section of a complex line bundle $L \to \Sigma$ (of degree d). The moduli space of (gauge equivalence classes of) solutions of (12.7.1) in this case can be naturally identified with the d-fold symmetric product of Σ (see Garcia-Prada [**136**]).

(iv) In the case $M = \mathbb{C}^2$ and $\mathrm{G} = \mathrm{U}(2)$ the moduli space of solutions of (12.7.1) is the space of **Bradlow pairs**, i.e. stable pairs consisting of a holomorphic rank-2 bundle over Σ and a holomorphic section [**47**]. The parameter τ then determines the stability condition.

(v) If M is the infinite dimensional manifold of SU(2)-connections on a Riemann surface S and G is the space of SU(2)-gauge transformations over S, then the solutions of (12.7.1) are anti-self-dual instantons over the product $\Sigma \times S$.

(vi) The Seiberg–Witten equations over $\Sigma \times S$ also appear as a special case of (12.7.1) if one chooses M to be the space of pairs $(\Theta, A) \in \Omega^0(S, E) \times \mathcal{A}(E)$ that satisfy $\bar\partial_A \Theta = 0$ and $\mathrm{G} = C^\infty(S, S^1)$, where $E \to S$ is a complex line bundle.

To obtain invariants from solutions of (12.7.1) we must impose the following conditions. In the case $\mathrm{G} = \{\mathbb{1}\}$ these conditions are equivalent to convexity as in Definition 9.2.6.

(H1) The moment map $\mu : M \to \mathfrak{g}$ is proper and (M, ω, μ) is convex at infinity in the following sense. There is a G-invariant ω-compatible almost complex structure $J \in \mathcal{J}_{\mathrm{G}}(M, \omega)$, a G-invariant proper smooth function

$$f : M \to [0, \infty),$$

and, for every $\tau \in Z(\mathfrak{g}^*)$, a compact set $K_\tau \subset M$ such that

$$\omega_f(v, J(x)v) \geq 0, \qquad df(x)J(x)X_{\mu(x)-\tau} \geq 0$$

for every $x \in M \setminus K_\tau$ and every $v \in T_xM$. Here

$$\omega_f := -d(df \circ J).$$

Hypothesis (H1) is needed to obtain any kind of compactness for the solutions of (12.7.1) and without it one cannot expect any meaningful results. The simplest, and nevertheless remarkably useful, class of examples is where, in addition

(H2) (M, ω) is symplectically aspherical.

Conditions (H1) and (H2) together force the ambient manifold M to be noncompact. However, they imply that the moduli space $\mathcal{M}_{B,\Sigma}(\tau)$ of based gauge equivalence classes of solutions of (12.7.1) that represent the equivariant homology class $[u] = B$ is compact for every $B \in H_2^{\mathrm{G}}(M)$ (see [**64, 67**]). Both hypotheses are satisfied for every linear action on $\mathbb{C}^n$ that is generated by a proper moment map. Moreover the moduli space is regular, in the sense that the gauge group acts with finite isotropy, whenever $\tau \in Z(\mathfrak{g}^*)$ is a central regular value of the moment map and Σ has sufficiently large volume. There is a G-equivariant evaluation map

$$\mathrm{ev} : \mathcal{M}_{B,\Sigma}(\tau) \to M$$

and the invariants at the parameter τ take the form of a homomorphism

$$\mathrm{GGW}_{B,\Sigma}^{M,\mu-\tau} : H_{\mathrm{G}}^*(M; \mathbb{Q}) \to \mathbb{Q}.$$

Roughly speaking, this homomorphism is defined by "integrating" the pullback of an equivariant cohomology class $a \in H_{\mathrm{G}}^*(M; \mathbb{Q})$ over the quotient $\mathcal{M}_{B,\Sigma}(\tau)/\mathrm{G}$:

$$\mathrm{GGW}_{B,\Sigma}^{M,\mu-\tau}(a) := \int_{\mathcal{M}_{B,\Sigma}(\tau)/\mathrm{G}} \mathrm{ev}^* a. \tag{12.7.3}$$

These are the **gauged Gromov–Witten invariants**. The precise definition of the integral does not require transversality. It can be understood in terms of the equivariant Euler class of an associated G-moduli problem (see Cieliebak–Mundet–Salamon [**66**]). The details are carried out in Cieliebak–Gaio–Mundet–Salamon [**67**]. There are refinements of this construction, involving varying marked points or pulling back equivariant cohomology classes from the space of connections, which we shall not discuss here.

As in Section 9.1, one can use these invariants to prove the existence of **relative periodic orbits** of time dependent G-equivariant Hamiltonian differential equations, i.e. solutions that close up modulo the G-action (see Cieliebak–Gaio–Mundet–Salamon [**67**]). Another application is the correspondence between the Seiberg–Witten and the gauged Gromov–Witten invariants for a suitable manifold

M (see Cieliebak–Gaio–Mundet–Salamon [**67**] and Okonek–Teleman [**309**]). This recovers the computation of the Seiberg–Witten invariants of ruled surfaces. In fact, because the moduli spaces are compact under assumptions (H1-2), the vortex invariants lend themselves to computations. An interesting feature is the dependence on the parameter τ. In [**69**] Cieliebak and Salamon obtained a wall crossing formula (in the case where G is a torus) which describes how the invariant changes as τ crosses from one chamber into another, and they were able to use this to prove the Batyrev formula for monotone toric manifolds. A crucial ingredient in their proof is the correspondence between the gauged Gromov–Witten invariants and the Gromov–Witten invariants of symplectic quotients which we explain next.

Let us assume for simplicity that $\tau = 0$. To obtain a smooth quotient we must impose a third condition:

(H3) The group G acts freely on $\mu^{-1}(0)$.

Under this assumption the quotient

$$\overline{M} := M /\!\!/ \mathrm{G} := \mu^{-1}(0)/\mathrm{G}$$

is a compact symplectic manifold. A J-holomorphic curve $\overline{u} : \Sigma \to \overline{M}$ can be represented as a pair $(u, A) \in C^{\infty}_{\mathrm{G}}(P, M) \times \mathcal{A}(P)$ that satisfies

$$\bar{\partial}_{J,A}(u) = 0, \qquad \mu(u) = 0. \tag{12.7.4}$$

The idea for establishing a correspondence between the solutions of (12.7.1) and those of (12.7.4) is to study the small ε limit of the equation

$$\bar{\partial}_{J,A}(u) = 0, \qquad *F_A + \varepsilon^{-2}\mu(u) = 0. \tag{12.7.5}$$

This program was carried out by Gaio and Salamon [**134, 135**] in the case where the quotient is monotone and they obtained the following result.

THEOREM 12.7.2 (Gaio–Salamon). *Assume (H1-3) and suppose that $c_1(T\overline{M})$ is a positive multiple of $[\overline{\omega}]$. Let $\overline{B} \in H_2(\overline{M}; \mathbb{Z})$ and $a_i \in H^*_{\mathrm{G}}(M)$ such that*

$$\deg(a_i) < 2N,$$

where N is the minimal Chern number of $\overline{M}$. Denote by $B \in H_2^{\mathrm{G}}(M)$ the image of $\overline{B}$ under the homomorphism $H_2(\overline{M}) \to H_2^{\mathrm{G}}(M)$ and by

$$\overline{a}_i \in H^*(\overline{M})$$

the image of a_i under the Kirwan homomorphism $\kappa : H^*_{\mathrm{G}}(M) \to H^*(\overline{M})$. Then

$$\mathrm{GW}^{\overline{M}}_{\overline{B},\Sigma}(\overline{a}_1, \ldots, \overline{a}_k) = \mathrm{GGW}^{M,\mu}_{B,\Sigma}(a_1 \smile \cdots \smile a_k).$$

Here the Gromov–Witten invariants are understood with Hamiltonian perturbations, fixed marked points, and a fixed complex structure on Σ (as in Chapter 8). The proof of Theorem 12.7.2 relies on the adiabatic limit analysis for equation (12.7.5) as $\varepsilon \to 0$. This involves rather subtle estimates and the proof is reminiscent of the proof of the Atiyah–Floer conjecture in Dostoglou–Salamon [**92**]. Combining Theorem 12.7.2 with the interpretation of the Gromov–Witten invariants as a homomorphism from the quantum cohomology of $\overline{M}$ to $\mathbb{Z}$ in Exercise 11.1.18, one obtains the following corollary.

COROLLARY 12.7.3 (Gaio–Salamon). *Let (M, ω, μ) be as in Theorem 12.7.2 and suppose that Λ is a quantum coefficient ring for $\overline{M}$ as in Definition 11.1.3. Suppose further that the equivariant cohomology of M is generated by classes of degrees less than $2N$. Then there exists a unique ring homomorphism*

$$Q\kappa : H^*_{\mathrm{G}}(M; \Lambda) \to \mathrm{QH}^*(\overline{M}; \Lambda) \tag{12.7.6}$$

that agrees with the Kirwan homomorphism $\kappa : H^*_{\mathrm{G}}(M; \Lambda) \to H^*(\overline{M}; \Lambda)$ for classes of degrees less than $2N$. Moreover, $Q\kappa$ is surjective and

$$\mathrm{GW}^{\overline{M}}_{\overline{B},\Sigma}(Q\kappa(a)) = \mathrm{GGW}^{M,\mu}_{B,\Sigma}(a)$$

for every Σ, every $\overline{B} \in H_2(\overline{M}; \mathbb{Z})$, and every $a \in H^*_{\mathrm{G}}(M)$.

By Remark 11.3.13, the quantum cohomology rings of monotone toric manifolds, flag manifolds, and Grassmannians, discussed in Sections 11.3.1, 11.3.2 and 11.3.3 all provide examples of the ring homomorphism of Corollary 12.7.3 with $M = \mathbb{C}^n$ and $H^*_{\mathrm{G}}(M; \Lambda) \cong H^*(\mathrm{BG}; \Lambda)$.

REMARK 12.7.4 (Quantum Kirwan Homomorphism). (i) If one weakens hypothesis (H3) and assumes only that zero is a regular value of the moment map, then the quotient $\overline{M}$ is a symplectic orbifold. The study of the Gromov–Witten invariants of orbifolds is a new subject (see Chen–Ruan [**62**]). One would expect Theorem 12.7.2, and hence Corollary 12.7.3, to extend in a fairly straightforward manner to the case where the quotient is a monotone symplectic orbifold.

(ii) As conjectured by Gaio–Salamon [**135**], Corollary 12.7.3 should also extend to the nonmonotone case, but the proof requires a refinement of the analysis in [**135**] and involves gluing theorems for the solutions of the symplectic vortex equations. One of the ingredients is a direct geometric definition of the homomorphism $Q\kappa$ via finite energy solutions of the symplectic vortex equations (12.7.2) (with $\lambda \equiv 1$) over the complex plane $\Sigma = \mathbb{C}$ (cf. Ziltener [**430**]).

(iii) It is another matter to drop hypothesis (H2), which asserts that the ambient manifold is symplectically aspherical. This would require the introduction of quantum deformation terms in the equivariant cohomology of M. In recent years a great deal of progress has been made on these conjectures by several researchers, cf. Ziltener [**431, 432**], Gonzalez–Woodward [**156, 157**], Ott [**319**], Nguyen–Woodward–Ziltener [**299**]. In these references the homomorphism $Q\kappa$ in (12.7.6) and its conjectural generalizations are called the **Quantum Kirwan Homomorphism**.

REMARK 12.7.5 (Moment Floer Homology). There is a version of Floer homology that is related to the symplectic vortex equations just as the standard version of symplectic Floer homology is related to J-holomorphic curves. In this theory the action functional is defined on the space of loops $(x, \eta) : \mathbb{R}/\mathbb{Z} \to M \times \mathfrak{g}$ in the product of M with the Lie algebra. It has the form

$$\mathcal{A}_{H,\mu}(x, \eta) = -\int_B u^*\omega + \int_0^1 \Big(\langle \mu(x(t)), \eta(t) \rangle - H_t(x(t)) \Big)\, dt,$$

where $H_t = H_{t+1} : M \to \mathbb{R}$ is a smooth 1-periodic family of G-invariant Hamiltonian functions and, as in Section 12.1, $u : B \to M$ is a smooth map satisfying $u(e^{2\pi i t}) = x(t)$. The (gauge equivalence classes of) critical points of this functional can be identified with the equivalence classes of relative periodic orbits of the Hamiltonian system in $\mu^{-1}(0)$. The Lagrangian version of this Floer theory was used by Frauenfelder [**121, 122**] to prove the Arnold–Givental conjecture for symplectic quotients of symplectically aspherical manifolds with a G-action. This conjecture gives a lower bound for the number of intersections of a Lagrangian submanifold with one of its Hamiltonian deformations, whenever the Lagrangian submanifold in question is the fixed point set of an anti-symplectic involution. An earlier proof, in the monotone case but without the quotient assumption, was given by Fukaya–Oh–Ohta–Ono [**128**].

APPENDIX A

Fredholm Theory

This appendix reviews the necessary functional analytic background for the proof that moduli spaces form smooth finite dimensional manifolds. The first section gives an introduction to Fredholm operators and their stability properties. Section A.2 discusses the determinant line bundle over the space of Fredholm operators between two Banach spaces. This is needed for establishing the orientability of moduli spaces. Section A.3 contains proofs of the inverse and implicit function theorems in a Banach space setting. Section A.4 explains a technique of finite dimensional reduction, sometimes called the *Kuranishi model.* This technique reduces the local analysis of the zero set of a Fredholm map near a singular point to a finite dimensional model. In Section A.5 we give a proof of the Sard–Smale theorem.

A.1. Fredholm theory

In this section we discuss abstract Fredholm operators and their basic properties. A bounded linear operator $D : X \to Y$ between Banach spaces is called a **Fredholm operator** if it has finite dimensional kernel, a closed image, and a finite dimensional cokernel $Y/\mathrm{im}\, D$. The **index** of a Fredholm operator D is defined by

$$\mathrm{index}\, D := \dim \ker D - \dim \mathrm{coker}\, D.$$

Here the kernel and cokernel are to be understood as real vector spaces. If D is a complex linear Fredholm operator between complex Banach spaces then it is important to distinguish between the real and the complex Fredholm index. Obviously, the real Fredholm index is twice the complex Fredholm index. The following lemma plays an important role in establishing the Fredholm property for a given linear operator D.

LEMMA A.1.1. *Let X, Y, Z be Banach spaces. Assume that $D : X \to Y$ is a bounded linear operator and $K : X \to Z$ is a compact operator. Assume that there is a constant $c > 0$ such that*

$$\|x\|_X \le c\,(\|Dx\|_Y + \|Kx\|_Z) \tag{A.1.1}$$

for $x \in X$. Then D has a closed image and finite dimensional kernel.

PROOF. To prove that the kernel of D is finite dimensional it suffices to show that the unit ball in $\ker D$ is compact. To see this, choose a sequence $x_\nu \in X$ such that

$$\|x_\nu\| \le 1, \qquad Dx_\nu = 0.$$

Since x_ν is bounded there exists a subsequence (still denoted by x_ν) such that Kx_ν converges. Since $x_\nu \in \ker D$ it follows from (A.1.1) that x_ν is a Cauchy sequence. Since X is complete x_ν converges.

Now assume without loss of generality that D is injective. Otherwise replace X by a complement of $\ker D$. Such a complement exists by the Hahn-Banach theorem: Pick a basis $x_1, \dots, x_N$ of $\ker D$ and choose $x_k^* \in X^*$ such that $\langle x_k^*, x_j \rangle = \delta_{jk}$. Then $X_0 := \{x \in X \mid \langle x_1^*, x \rangle = \cdots = \langle x_N^*, x \rangle = 0\}$ is the required complement.

Let $y \in \mathrm{cl}(\mathrm{im}\, D)$. Then there exists a sequence $x_\nu \in X$ such that

$$y = \lim_{\nu \to \infty} D x_\nu.$$

We prove first that x_ν is bounded. Suppose, by contradiction, that x_ν is unbounded. Then, passing to a subsequence if necessary, we may assume that $\|x_\nu\|$ diverges to infinity. Consider the sequence

$$\xi_\nu := \|x_\nu\|^{-1} x_\nu.$$

Each term has norm one and $D\xi_\nu$ converges to zero. Passing to a further subsequence we may assume that $K\xi_\nu$ converges. Hence it follows from (A.1.1) that ξ_ν is a Cauchy sequence. The limit $\xi := \lim_{\nu\to\infty} \xi_\nu$ is of norm one and satisfies $D\xi = 0$, a contradiction. Hence the sequence x_ν is bounded, as claimed. Using again the compactness of K, the estimate (A.1.1), and the completeness of X, we deduce that x_ν has a convergent subsequence. Let x be its limit. Then $y = Dx$. □

COROLLARY A.1.2. *Let X and Y be Banach spaces and $D : X \to Y$ be a bounded linear operator with a closed image and finite dimensional kernel.*

(i) *For every compact operator $K : X \to Y$ the operator $D + K$ also has a closed image and finite dimensional kernel.*

(ii) *There exists an $\varepsilon > 0$ such that if $P : X \to Y$ is a bounded linear operator with $\|P\| < \varepsilon$ then $D + P$ has a closed image and finite dimensional kernel.*

PROOF. Suppose $\dim \ker D = n$ and choose a bounded linear operator $K_0 : X \to \mathbb{R}^n$ such that the restriction of K_0 to $\ker D$ is a vector space isomorphism. Then the operator $X \to Y \oplus \mathbb{R}^n : x \mapsto (Dx, K_0 x)$ is injective and has closed image. Hence, by the open mapping theorem, there exists a constant $c > 0$ such that

$$\|x\|_X \le c\left(\|Dx\|_Y + \|K_0 x\|_{\mathbb{R}^n}\right).$$

Hence for any compact operator $K : X \to Y$ we have

$$\|x\|_X \le c\left(\|(D+K)x\|_Y + \|Kx\|_Y + \|K_0 x\|_{\mathbb{R}^n}\right).$$

Similarly, if $\|P\| < 1/c$, then

$$\|x\|_X \le \frac{c}{1 - c\|P\|}\left(\|(D+P)x\|_Y + \|K_0 x\|_{\mathbb{R}^n}\right).$$

Hence the assertions follow from Lemma A.1.1. □

COROLLARY A.1.3. *Let X and Y be Banach spaces and $D : X \to Y$ be a bounded linear operator with a closed image and finite dimensional cokernel.*

(i) *For every compact operator $K : X \to Y$ the operator $D + K$ also has a closed image and finite dimensional cokernel.*

(ii) *There exists an $\varepsilon > 0$ such that if $P : X \to Y$ is a bounded linear operator with $\|P\| < \varepsilon$ then $D + P$ has a closed image and finite dimensional cokernel.*

PROOF. D has closed image if and only if D^* has closed image and it has finite dimensional cokernel if and only if D^* has finite dimensional kernel. Hence the result follows from Corollary A.1.2. □

EXERCISE A.1.4. (i) Prove that a bounded linear operator $D : X \to Y$ is Fredholm if and only if there exists a bounded linear operator $T : Y \to X$ such that both $DT - 1\!\!1$ and $TD - 1\!\!1$ are compact operators. *Hint:* Use Lemma A.1.1 for the "if" part.

(ii) If both $D : X \to Y$ and $T : Y \to Z$ are Fredholm operators prove that $TD : X \to Z$ is a Fredholm operator and

$$\operatorname{index} TD = \operatorname{index} D + \operatorname{index} T. \tag{A.1.2}$$

(iii) Prove that a bounded linear operator $D : X \to Y$ is Fredholm if and only if its dual operator $D^* : Y^* \to X^*$ is and that their indices are related by

$$\operatorname{index} D^* = -\operatorname{index} D.$$

The most important properties of Fredholm operators are related to their stability under perturbations. This is the content of the following theorem. In particular, Theorem A.1.5 asserts that the set of Fredholm operators is open with respect to the norm topology and the index is constant on each connected component.

THEOREM A.1.5. *Let $D : X \to Y$ be a Fredholm operator*

(i) *If $K : X \to Y$ is a compact operator then $D + K$ is a Fredholm operator and* $\operatorname{index}(D + K) = \operatorname{index} D$.

(ii) *There exists an $\varepsilon > 0$ such that if $P : X \to Y$ is a bounded linear operator with $\|P\| < \varepsilon$ then $D + P$ is a Fredholm operator and* $\operatorname{index}(D + P) = \operatorname{index} D$.

PROOF. The assertions about the Fredholm property follow immediately from Corollaries A.1.2 and A.1.3. The assertions about the index are exercises with hints. Choose decompositions $X = X_0 \oplus X_1$ and $Y = Y_0 \oplus Y_1$ such that $X_0 := \ker D$ and $Y_1 := \operatorname{im} D$. Prove that the index of $D + P$ agrees with the index of the finite dimensional operator

$$P_{00} - P_{01}(D_{11} + P_{11})^{-1}P_{10} : X_0 \to Y_0,$$

where $P_{ji} : X_i \to Y_j$ denotes the restriction of P to X_i followed by the projection onto Y_j, and similarly for D. □

A.2. Determinant line bundles

Let X and Y be Banach spaces and denote by $\mathcal{F}(X, Y)$ the space of all Fredholm operators $D : X \to Y$. The **determinant** of a Fredholm operator $D \in \mathcal{F}(X, Y)$ is defined as the 1-dimensional real vector space

$$\det(D) := \Lambda^{\max}(\ker D) \otimes \Lambda^{\max}(\ker D^*).$$

We now show that as D varies the vector spaces $\det(D)$ fit together to form a locally trivial line bundle over $\mathcal{F}(X, Y)$. Here it is important to keep track of the isomorphisms which identify different 1-dimensional vector spaces.

Think of the real line $\mathbb{R}$ as a 1-dimensional real vector space. For any two 1-dimensional real vector spaces V and W we shall use the notation $V \cong W$ to mean that the spaces are *naturally* isomorphic. This means that there is an obvious choice of isomorphism between them. For example, if V is 1-dimensional there is a natural isomorphism

$$V \otimes V^* \to \mathbb{R} : v \otimes v^* \mapsto v^*(v).$$

This notion of natural isomorphism can be more precisely expressed in the language of category theory. Denote by $\mathcal{V}$ the category of 1-dimensional real vector spaces and isomorphisms and let $\mathcal{C}$ be any other category. Two functors $\mathcal{F}_0, \mathcal{F}_1 : \mathcal{C} \to \mathcal{V}$ are called **naturally isomorphic** if there exists a **natural transformation** $\mathcal{T} : \mathrm{Ob}(\mathcal{C}) \to \mathrm{Mor}(\mathcal{V})$ which assigns to every object $A \in \mathrm{Ob}(\mathcal{C})$ a vector space isomorphism $\mathcal{T}(A) : \mathcal{F}_0(A) \to \mathcal{F}_1(A)$ such that $\mathcal{T}(B) \circ \mathcal{F}_0(T) = \mathcal{F}_1(T) \circ \mathcal{T}(A)$ for $A, B \in \mathrm{Ob}(\mathcal{C})$ and $T \in \mathrm{Mor}(A, B)$. In many cases the choice of the isomorphism $\mathcal{T}(A)$ is clear from the context, and we shall simply use the notation $\mathcal{F}_0(A) \cong \mathcal{F}_1(A)$. An example of this is the notation $V \otimes V^* \cong \mathbb{R}$ for the above isomorphism.

The **top exterior power** $\Lambda^{\max} V$ of a finite dimensional real vector space V is the space of equivalence classes $v_1 \wedge \cdots \wedge v_n$ of ordered n-tuples in V, where $\dim V = n$. Two such n-tuples $v_1 \wedge \cdots \wedge v_n$ and $w_1 \wedge \cdots \wedge w_n$ are equivalent iff either both form a basis and the induced isomorphism of V has determinant one, or both n-tuples are linearly dependent. Hence a nonzero vector in $\Lambda^{\max} V$ determines an orientation of V. When V has dimension zero we define $\Lambda^{\max} V := \mathbb{R}$.

The **tensor product** $V \otimes W$ of two 1-dimensional real vector spaces V and W is the space of equivalence classes $v \otimes w$ of ordered pairs $(v, w) \in V \times W$ where $\lambda v \otimes w = v \otimes \lambda w$ for all $\lambda \in \mathbb{R}$. There is a natural isomorphism

$$\Lambda^{\max} V \otimes \Lambda^{\max} W \cong \Lambda^{\max}(V \oplus W)$$

for any two finite dimensional real vector spaces V and W. (Prove this!)

For every n-dimensional real vector space V we think of the top exterior power $\Lambda^{\max} V^* := \Lambda^{\max}(V^*)$ as the space of alternating n-forms on V. There is a natural isomorphism

$$\Lambda^{\max} V \otimes \Lambda^{\max} V^* \cong \mathbb{R},$$

given by $(v_1 \wedge \cdots \wedge v_n) \otimes (v_1^* \wedge \cdots \wedge v_n^*) \mapsto \det(v_k^*(v_j))$. It induces an isomorphism $\Lambda^{\max} V^* \cong (\Lambda^{\max} V)^*$. Likewise, every isomorphism $\Lambda^{\max} V \otimes \Lambda^{\max} W \to \mathbb{R}$ induces an isomorphism $\Lambda^{\max} W \to (\Lambda^{\max} V)^* \cong \Lambda^{\max} V^*$.

LEMMA A.2.1. *Let V be a finite dimensional real vector space and $W \subset V$ be a linear subspace. Then there is a natural isomorphism*

$$\Lambda^{\max} V \cong \Lambda^{\max} W \otimes \Lambda^{\max}(V/W).$$

PROOF. Let $N := \dim V$ and $n := \dim W$. Denote by $\mathcal{B}(V, W)$ the set of all bases $v_1, \ldots, v_N$ of V whose first n elements span W. For $v \in V$ denote $[v] = v + W$. The map $\mathcal{B}(V, W) \to \Lambda^{\max} W \otimes \Lambda^{\max}(V/W)$ defined by

$$(v_1, \ldots, v_N) \mapsto (v_1 \wedge \cdots \wedge v_n) \otimes ([v_{n+1}] \wedge \cdots \wedge [v_N])$$

induces the required isomorphism. It is left to the reader to verify that these isomorphisms define a natural transformation. □

After this preparation, consider once again the determinant lines

$$\det(D) := \Lambda^{\max}(\ker D) \otimes \Lambda^{\max}(\ker D^*).$$

of a family of Fredholm operators D.

THEOREM A.2.2. *The space*

$$\det(X, Y) := \{(D, \sigma) \mid D \in \mathcal{F}(X, Y), \sigma \in \det(D)\}$$

is a real line bundle over $\mathcal{F}(X, Y)$.

PROOF. We must prove that the space $\det(X,Y)$ admits a local trivialization in a neighbourhood of every Fredholm operator $D \in \mathcal{F}(X,Y)$. Assume first that D is onto. Then there is an estimate $\|y^*\|_{Y^*} \le c\|D^*y^*\|_{X^*}$. Hence the operator $D+P : X \to Y$ is onto whenever P is sufficiently small. By Theorem A.1.5 both operators D and $D+P$ have the same Fredholm index and hence their kernels are of the same dimension. Now choose a right inverse $T : Y \to X$ of D so that $DT = \mathbb{1}_Y$. Then there is an isomorphism

$$\ker(D+P) \to \ker D : x \mapsto x + TPx$$

whenever P is sufficiently small. Hence the kernels of $D+P$ form a locally trivial vector bundle over $\mathcal{F}(X,Y)$ in a neighbourhood of a surjective Fredholm operator.

We now reduce the general case to the surjective case. First observe that given any Fredholm operator $D_0 : X \to Y$ there exists a positive integer N and an injective linear map $\Phi : \mathbb{R}^N \to Y$ such that the operator $D_0 \oplus \Phi : X \oplus \mathbb{R}^N \to Y$ defined by $D_0 \oplus \Phi(x,\zeta) := D_0x + \Phi\zeta$ is surjective. To see this let $N = \dim \operatorname{coker} D_0$ and choose elements $y_1, \dots, y_N \in Y$ that span a complement of the image of D_0 in Y. Then the linear map $\Phi\zeta = \sum_j \zeta_j y_j$ is as required.

Now let $D : X \to Y$ be a Fredholm operator and $\Phi : \mathbb{R}^N \to Y$ be a linear map such that $D \oplus \Phi$ is onto. We prove that there is a natural isomorphism $\det(D) \cong \det(D \oplus \Phi)$. To see this, consider the exact sequence

$$0 \longrightarrow \ker D \longrightarrow \ker(D \oplus \Phi) \longrightarrow \Phi^{-1}(\operatorname{im} D) \longrightarrow 0$$

where the second map is $\ker D \to \ker(D \oplus \Phi) : x \mapsto (x,0)$ and the third map is $\ker(D \oplus \Phi) \to \Phi^{-1}(\operatorname{im} D) : (x,\zeta) \mapsto \zeta$. By exactness, this map descends to an isomorphism $\ker(D \oplus \Phi)/\ker D \cong \Phi^{-1}(\operatorname{im} D)$. Hence, by Lemma A.2.1, we have

$$\Lambda^{\max} \ker(D \oplus \Phi) \cong \Lambda^{\max}(\ker D) \otimes \Lambda^{\max}(\Phi^{-1}(\operatorname{im} D)). \tag{A.2.1}$$

It follows also from Lemma A.2.1 that $\Lambda^{\max} W \otimes \Lambda^{\max}(\mathbb{R}^N/W) \cong \Lambda^{\max}\mathbb{R}^N \cong \mathbb{R}$ and hence $\Lambda^{\max} W \cong \Lambda^{\max}(\mathbb{R}^N/W)^*$ for every linear subspace $W \subset \mathbb{R}^N$. With $W = \Phi^{-1}(\operatorname{im} D)$ this gives

$$\Lambda^{\max}(\Phi^{-1}(\operatorname{im} D)) \cong \Lambda^{\max}\left(\frac{\mathbb{R}^N}{\Phi^{-1}(\operatorname{im} D)}\right)^*. \tag{A.2.2}$$

Since $Y = \operatorname{im} D + \operatorname{im} \Phi$, there is an isomorphism

$$\frac{\mathbb{R}^N}{\Phi^{-1}(\operatorname{im} D)} \to \frac{Y}{\operatorname{im} D} : [\zeta] \mapsto [\Phi\zeta].$$

Hence

$$\Lambda^{\max}\left(\frac{\mathbb{R}^N}{\Phi^{-1}(\operatorname{im} D))}\right)^* \cong \Lambda^{\max}\left(\frac{Y}{\operatorname{im} D}\right)^* \cong \Lambda^{\max}(\ker D^*). \tag{A.2.3}$$

Here we use the fact that the dual space of a quotient Y/Y_1 is isomorphic to the annihilator of Y_1 in Y^*. In the case $Y_1 := \operatorname{im} D$ this annihilator is equal to the kernel of D^*. By (A.2.2) and (A.2.3) we have $\Lambda^{\max}(\Phi^{-1}(\operatorname{im} D)) \cong \Lambda^{\max}(\ker D^*)$. Combining this with (A.2.1) we obtain

$$\det(D) = \Lambda^{\max}(\ker D) \otimes \Lambda^{\max}(\ker D^*) \cong \Lambda^{\max} \ker(D \oplus \Phi) = \det(D \oplus \Phi).$$

This proves Theorem A.2.2. □

EXERCISE A.2.3. Let $D : X \to Y$ be a Fredholm operator and $\Phi : \mathbb{R}^N \to Y$ be a linear map (not necessarily injective) such that $D \oplus \Phi$ is onto.

(i) Assume $0 < k := \dim\ker D$ and $0 < \ell := \dim\ker D^* < N$. Given a basis $x_1, \dots, x_k$ of $\ker D$ and a basis $y_1^*, \dots, y_\ell^*$ of $\ker D^*$ prove that there exists a basis $\zeta_1, \dots, \zeta_N$ of $\mathbb{R}^N$ and vectors $\xi_1, \dots, \xi_{N-\ell} \in X$ such that

$$\begin{array}{lll} \langle y_i^*, \Phi\zeta_{N-\ell+j}\rangle & = \delta_{ij}, & i,j = 1,\dots,\ell, \\ D\xi_j + \Phi\zeta_j & = 0, & j = 1,\dots,N-\ell, \\ \det(\zeta_1 \cdots \zeta_N) & = 1. & \end{array}$$

Prove that the map $\Lambda^{\max}(\ker D) \otimes \Lambda^{\max}(\ker D^*) \to \Lambda^{\max}\ker(D \oplus \Phi)$ given by

$$\begin{aligned} &(x_1 \wedge \cdots \wedge x_k) \otimes (y_1^* \wedge \cdots \wedge y_\ell^*) \\ &\quad \mapsto (x_1, 0) \wedge \cdots \wedge (x_k, 0) \wedge (\xi_1, \zeta_1) \wedge \cdots \wedge (\xi_{N-\ell}, \zeta_{N-\ell}) \end{aligned}$$

is a well defined linear isomorphism and agrees with the one constructed in the proof of Theorem A.2.2.

(ii) Examine the isomorphism $\det(D) \cong \det(D \oplus \Phi)$ of Theorem A.2.2 in the cases $k = 0$, $\ell = 0$, and k arbitrary, $\ell = N$.

Crossing numbers. To gain a better understanding of the line bundle

$$\det(X, Y) \to \mathcal{F}(X, Y)$$

in the case of Fredholm index zero we shall interpret trivializations of this line bundle along a path $D : [0,1] \to \mathcal{F}(X, Y)$ as a crossing number. Denote

$$\mathcal{F}^0(X, Y) := \{D \in \mathcal{F}(X, Y) \,|\, \operatorname{index} D = 0\}$$

and for each integer $k \geq 0$ consider the subset

$$\mathcal{F}_k^0(X, Y) := \{D \in \mathcal{F}(X, Y) \,|\, \operatorname{index} D = 0,\ \dim\ker D = k\}$$

of Fredholm operators of index 0 with k-dimensional kernel. We claim that this is a submanifold of codimension k^2. Namely, the tangent space at the operator $D \in \mathcal{F}_k^0(X, Y)$ is given by

$$T_D\mathcal{F}_k^0(X, Y) = \{P \in \mathcal{L}(X, Y) \,|\, Px \in \operatorname{im} D \text{ for all } x \in \ker D\}.$$

Thus it has a k^2-dimensional complement consisting of the space of linear operators from the kernel of D to a complement of the image of D. The union

$$\overline{\mathcal{F}}_1^0(X, Y) = \bigcup_{k \geq 1} \mathcal{F}_k^0(X, Y)$$

is a kind of stratified subvariety of codimension 1 whose complement (in the space of Fredholm operators of index zero) is the space of invertible operators.

Now consider a path $[0,1] \to \mathcal{F}^0(X, Y) : t \mapsto D_t$ with invertible endpoints D_0 and D_1. Assume that the path is continuously differentiable (in the strong operator topology) and define the operator $\dot{D}_t : X \to Y$ by

$$\dot{D}_t x := \frac{d}{dt} D_t x$$

for $x \in X$. Call a point $t \in [0,1]$ a **crossing** if $\ker D_t > 0$. A crossing is called **regular** if

$$x \in \ker D_t, \quad \dot{D}_t x \in \operatorname{im} D_t \qquad \Longrightarrow \qquad x = 0.$$

This means that $\dot{D}_t$ maps the kernel of D_t bijectively onto a complement of $\operatorname{im} D_t$. A **simple crossing** is a regular crossing t with $D_t \in \mathcal{F}_1^0(X, Y)$. Note that the operator D_{t+s} is invertible for small nonzero s whenever t is a regular crossing. Hence every regular crossing is isolated. If $t \mapsto D_t$ is a path with only regular crossings we define the **crossing index** to be the number

$$\mu(\{D_t\}) := \sum_t \dim\ker D_t. \tag{A.2.4}$$

We shall prove that the mod 2 reduction of this number is a homotopy invariant and determines the sign of the map $\det(D_0) \to \det(D_1)$ arising from a trivialization of the determinant line bundle $\det(X,Y)$ along the path $t \mapsto D_t$. More precisely, consider the line bundle $L := \{(t,\sigma) \mid t \in [0,1],\ \sigma \in \det(D_t)\}$ over the unit interval. A trivialization of this line bundle gives rise to an isomorphism $\det(D_0) \to \det(D_1)$. Since the 1-dimensional vector space $\det(D_t)$ inherits a norm from X and Y, this isomorphism can be chosen uniquely as an isometry. Since $\det(D_0) = \det(D_1) = \mathbb{R}$ this isomorphism is given by multiplication with a real number of modulus 1 which we denote by $\nu(\{D_t\}) \in \{\pm 1\}$.

PROPOSITION A.2.4. *Let $[0,1] \to \mathcal{F}^0(X,Y) : t \mapsto D_t$ be a continuously differentiable path with invertible endpoints D_0 and D_1 and only regular crossings. Then any trivialization of the determinant line bundle over this path gives rise to an isomorphism $\det(D_0) = \mathbb{R} \to \det(D_1) = \mathbb{R}$ of sign*

$$\nu(\{D_t\}) = (-1)^{\mu(\{D_t\})} = \prod_t (-1)^{\dim\ker D_t}. \tag{A.2.5}$$

Here the product runs over all crossings t. In particular, the crossing index mod 2 is invariant under homotopies with fixed endpoints.

PROOF. Let us consider a path $t \mapsto D_t$ with a single crossing at $t = 0$. By choosing a suitable splitting of X and Y we may assume without loss of generality that $X = Y = X_0 \oplus \mathbb{R}^k$ and

$$D_0 = \begin{pmatrix} \mathbb{1} & 0 \\ 0 & 0 \end{pmatrix}, \qquad \dot{D}_0 = \begin{pmatrix} A & 0 \\ B & \mathbb{1} \end{pmatrix}.$$

Consider the linear map $\Phi : \mathbb{R}^k \to Y = X_0 \oplus \mathbb{R}^k$ given by $\Phi z = (0,z)$. Then

$$D_t \oplus \Phi = \begin{pmatrix} \mathbb{1} + tA & 0 & 0 \\ tB & t\mathbb{1} & \mathbb{1} \end{pmatrix} + O(t^2).$$

Hence a trivialization of the kernels of the operators $D_t \oplus \Phi$ is given by embeddings $\iota_t : \mathbb{R}^k \to X \oplus \mathbb{R}^k = (X_0 \oplus \mathbb{R}^k) \oplus \mathbb{R}^k$ of the form

$$\iota_t = \begin{pmatrix} 0 \\ \mathbb{1} \\ -t\mathbb{1} \end{pmatrix} + O(t^2).$$

Here the two upper blocks represent the X-component of $\ker(D_t \oplus \Phi) \subset X \oplus \mathbb{R}^k$ while the third block represents the $\mathbb{R}^k$-component. For $\varepsilon > 0$ sufficiently small consider the composition

$$\mathbb{R}^k \to \ker(D_{-\varepsilon} \oplus \Phi) \to \ker(D_\varepsilon \oplus \Phi) \to \mathbb{R}^k,$$

where the last isomorphism is induced by the projection onto $\mathbb{R}^k$, the first by its inverse, and the second by the trivialization. Because of the factor $-t\mathbb{1}$ in ι_t, the composition has the form $-\mathbb{1} + O(\varepsilon)$. This map is orientation reversing when k is odd and is orientation preserving when k is even. This proves the formula (A.2.5) in the case of a single regular crossing. The general case is an obvious consequence. □

EXERCISE A.2.5. Let $X = Y = H$ be an infinite dimensional Hilbert space. Construct a loop of Fredholm operators of index zero with crossing number one. Deduce that the determinant bundle over the space of Fredholm operators on H (of index zero) does not admit a trivialization. *Hint:* Choose an orthonormal basis $e_0, e_1, e_2, \ldots$. For $0 \le t \le 1/2$ define $A_t \in \mathcal{L}(H)$ to be a rotation by angle $2\pi t$ in the (e_{2j}, e_{2j+1})-planes for $j \ge 0$ so that $A_0 = \mathrm{id}$ and $A_{1/2} = -\mathrm{id}$. Then for $1/2 \le t \le 1$ define $A_t \in \mathcal{L}(H)$ by $A_t e_0 = -e_0$ and as a rotation by angle $2\pi t$ in the (e_{2j-1}, e_{2j})-planes for $j \ge 1$. Then A_1 acts by minus the identity on $\mathbb{R}e_0$ and by the identity on the orthogonal complement. Now connect A_1 to the identity by a straight line.

EXERCISE A.2.6. Construct a loop of surjective Fredholm operators of index 1 whose kernels form a Moebius band.

EXERCISE A.2.7. Suppose that X and Y are complex Banach spaces and $[0,1] \to \mathcal{F}(X,Y) : t \mapsto D_t$ is a path of Fredholm operators which are all complex linear. Then the one dimensional real vector space $\det(D_t)$ inherits a natural orientation from the complex structures of $\ker D_t$ and $\ker {D_t}^*$. Prove that any trivialization of the real line bundle $\bigcup_t \det(D_t)$ gives rise to an orientation preserving isomorphism

$$\det(D_0) \to \det(D_1).$$

Hint: Consider the complex line bundle

$$\det^c(X,Y) \longrightarrow \mathcal{F}^c(X,Y)$$

whose fiber over a complex linear Fredholm operator $D \in \mathcal{F}^c(X,Y)$ is the 1-dimensional complex vector space

$$\det^c(D) := \Lambda^{\max}_{\mathbb{C}} \ker D \otimes_{\mathbb{C}} \Lambda^{\max}_{\mathbb{C}} \ker D^*.$$

Here $D^* : Y^* \to X^*$ denotes the real adjoint operator and if $J : X \to X$ is multiplication by i then the complex structure on X^* is given by the dual operator $J^* : X^* \to X^*$. Show that the proof of Theorem A.2.2 carries over to the complex category and hence $\det^c(X,Y)$ is a locally trivial complex line bundle over $\mathcal{F}^c(X,Y)$. Now for every finite dimensional complex vector space V there is a natural quadratic map

$$\Lambda^{\max}_{\mathbb{C}} V \to \Lambda^{\max}_{\mathbb{R}} V : \tau \mapsto i^n (-1)^{\frac{n(n-1)}{2}} \tau \wedge \bar{\tau},$$

where $\bar{\tau} \in \Lambda^{\max}_{\mathbb{C}} \bar{V}$. Here $\bar{V}$ denotes the vector space with the opposite complex structure and both $\Lambda^{\max}_{\mathbb{C}} V$ and $\Lambda^{\max}_{\mathbb{C}} \bar{V}$ are natural linear subspaces of $\Lambda^{\mathrm{mid}}_{\mathbb{R}} V \otimes \mathbb{C}$. With these conventions the above map sends $v_1 \wedge \cdots \wedge v_n$ to $v_1 \wedge (Jv_1) \wedge \cdots \wedge v_n \wedge (Jv_n)$.

A.3. The implicit function theorem

Let X and Y be Banach spaces and $f : X \to Y$ be a smooth map. For every $x \in X$ denote by $df(x) : X \to Y$ the differential of f at x. If this operator is bijective then its inverse $df(x)^{-1} : Y \to X$ is a bounded linear operator by the open mapping theorem. The inverse function theorem asserts that f has a local inverse near every point x at which $df(x)$ is invertible. Denote by $B_r(x;X)$ the open ball of radius r centered at x in the Banach space X. If the Banach space is understood from the context, we abbreviate $B_r(x) := B_r(x;X)$ and $B_r := B_r(0;X)$.

THEOREM A.3.1 (Inverse Function Theorem). *Let X and Y be Banach spaces, $U \subset X$ be an open set, and $f : U \to Y$ be continuously differentiable. Let $x_0 \in U$ and suppose that the differential $df(x_0) : X \to Y$ is bijective. Then there exists an open neighbourhood $U_0 \subset U$ of x_0 such that the restriction of f to U_0 is injective, $V_0 := f(U_0)$ is an open subset of Y, $f^{-1} : V_0 \to U_0$ is continuously differentiable, and*

$$df^{-1}(y) = \left(df(f^{-1}(y))\right)^{-1}$$

for $y \in V_0$. Hence, if f is of class C^ℓ for some positive integer ℓ then so is f^{-1}.

The proof is based on the following lemma about maps $\psi : X \to X$ whose derivative is close to the identity.

LEMMA A.3.2. *Let $\gamma < 1$ and R be positive real numbers. Let X be a Banach space, $x_0 \in X$, and $\psi : B_R(x_0) \to X$ be a continuously differentiable map such that*

$$\|\mathbb{1} - d\psi(x)\| \le \gamma$$

for every $x \in B_R(x_0)$. Then ψ is injective and maps $B_R(x_0)$ onto an open set in X such that

$$B_{R(1-\gamma)}(\psi(x_0)) \subset \psi\left(B_R(x_0)\right) \subset B_{R(1+\gamma)}(x_0).$$

Moreover, the map $\psi^{-1} : \psi(B_R(x_0)) \to B_R(x_0)$ is continuously differentiable and

$$d\psi^{-1}(y) = d\psi(\psi^{-1}(y))^{-1}. \tag{A.3.1}$$

Hence, if ψ is of class C^ℓ for some positive integer ℓ then so is ψ^{-1}.

PROOF. Replacing ψ by the map $x \mapsto \psi(x_0 + x) - \psi(x_0)$ we may assume that $x_0 = 0$ and $\psi(x_0) = 0$. Now the proof has four steps.

STEP 1. *ψ is a homeomorphism onto its image and $\psi(B_R) \subset B_{R(1+\gamma)}$.*

Consider the map

$$\phi := \mathrm{id} - \psi : B_R \to X.$$

It satisfies $\|d\phi(x)\| \le \gamma$ for every $x \in B_R$. Hence

$$\|\phi(x_1) - \phi(x_2)\| \le \gamma \|x_1 - x_2\|. \tag{A.3.2}$$

for all $x_1, x_2 \in B_R$. Hence, by the triangle inequality,

$$(1-\gamma)\|x_1 - x_2\| \le \|\psi(x_1) - \psi(x_2)\| \le (1+\gamma)\|x_1 - x_2\| \tag{A.3.3}$$

for $x_1, x_2 \in B_R$. The second inequality shows that $\psi(B_R) \subset B_{R(1+\gamma)}$ and the first inequality shows that ψ is injective and ψ^{-1} is (Lipschitz) continuous.

STEP 2. *$B_{R(1-\gamma)} \subset \psi(B_R)$.*

Let $y \in B_{R(1-\gamma)}$ and define $\varepsilon > 0$ by $\|y\| =: (1-\gamma)(R-\varepsilon)$. Then, by (A.3.2) with $x_2 = 0$, we have $\|\phi(x)\| \le \gamma\|x\|$ for $x \in B_R$ and hence

$$\|x\| \le R - \varepsilon \qquad \Longrightarrow \qquad \|\phi(x) + y\| \le R - \varepsilon.$$

Hence the map $x \mapsto \phi(x) + y$ is a contraction of the closed ball of radius $R - \varepsilon$. By the contraction mapping principle, this map has a unique fixed point x such that $\|x\| \le R - \varepsilon$. The fixed point satisfies $\psi(x) = x - \phi(x) = y$. Hence $y \in \psi(B_R)$.

STEP 3. *$\psi(B_R)$ is open.*

Let $y = \psi(x)$, where $x \in B_R$, and choose $\varepsilon > 0$ such that $B_\varepsilon(x) \subset B_R$. Then, by Step 2, $B_{\varepsilon(1-\gamma)}(\psi(x)) \subset \psi(B_\varepsilon(x)) \subset \psi(B_R)$.

STEP 4. *ψ^{-1} is continuously differentiable and its derivative is given by* (A.3.1). Let $y_0 = \psi(x_0)$, where $x_0 \in B_R$, and denote $D := d\psi(x_0)$. Then $\|1\!\!1 - D\| \le \gamma$ and hence D has an inverse, given by

$$D^{-1} = \sum_{k=0}^{\infty}(1\!\!1 - D)^k, \qquad \|D^{-1}\| \le \frac{1}{1-\gamma}.$$

We prove that ψ^{-1} is differentiable at y_0 and $d\psi^{-1}(y_0) = D^{-1}$. To see this, fix any constant $\varepsilon > 0$. Since ψ is differentiable at x_0, there exists a constant $\delta > 0$ such that, for every $x \in B_R$,

$$\|x - x_0\| < \frac{\delta}{1-\gamma} \qquad \Longrightarrow \qquad \|\psi(x) - \psi(x_0) - D(x - x_0)\| \le \varepsilon(1-\gamma)^2 \|x - x_0\|.$$

Shrinking δ, if necessary, we may assume, by Step 3, that $B_\delta(y_0) \subset \psi(B_R)$. Now suppose $\|y - y_0\| < \delta$ and denote $x := \psi^{-1}(y) \in B_R$. Then, by (A.3.3),

$$\|x - x_0\| \le \frac{1}{1-\gamma}\|y - y_0\| \le \frac{\delta}{1-\gamma}.$$

Hence

$$\begin{aligned}
\left\|\psi^{-1}(y) - \psi^{-1}(y_0) - D^{-1}(y - y_0)\right\| &= \left\|D^{-1}\left(y - y_0 - D(x - x_0)\right)\right\| \\
&\le \frac{1}{1-\gamma}\|\psi(x) - \psi(x_0) - D(x - x_0)\| \\
&\le \varepsilon(1-\gamma)\|x - x_0\| \\
&\le \varepsilon\|y - y_0\|.
\end{aligned}$$

Hence ψ^{-1} is differentiable at y_0 and its derivative is given by (A.3.1). By Step 1 and (A.3.1), the map $y \mapsto d\psi^{-1}(y)$ is continuous with respect to the norm topology on $\mathcal{L}(X)$. This proves Lemma A.3.2. □

PROOF OF THEOREM A.3.1. Assume without loss of generality that $x_0 = 0$ and $f(0) = 0$. Consider the map $\psi : B_\delta(0; X) \to X$ given by

$$\psi(x) := D^{-1}f(x),$$

where $D := df(x_0)$. Its differential satisfies

$$1\!\!1 - d\psi(x) = 1\!\!1 - D^{-1}df(x) = D^{-1}(D - df(x))$$

and hence

$$\|1\!\!1 - d\psi(x)\| \le c\|D - df(x)\| \le \frac{1}{2}$$

for $x \in B_\delta(0; X)$. Hence it follows from Lemma A.3.2 with $R = \delta$ and $\gamma = 1/2$ that ψ has a continuously differentiable inverse on $B_\delta(0; X)$ and that $\psi(B_\delta(0; X))$ an open set containing $B_{\delta/2}(0; X)$. Thus

$$f(B_\delta(0; X)) = D\psi(B_\delta(0; X)) \supset DB_{\delta/2}(0; X) \supset B_{\delta/2c}(0; Y)$$

and the inverse of f is given by

$$f^{-1}(y) = \psi^{-1}(D^{-1}y).$$

It is continuously differentiable and the formula $df^{-1}(y) = df(f^{-1}(y))^{-1}$ follows from the chain rule. This proves Theorem A.3.1. □

A smooth map

$$f : X \to Y$$

between Banach spaces is called **Fredholm** if the differential $df(x) : X \to Y$ is a Fredholm operator for every $x \in X$. Since the Fredholm index is invariant under small perturbations the index of $df(x)$ is independent of the choice of x. It will be denoted by $\mathrm{index}(f)$. For any smooth map $f : X \to Y$, Fredholm or not, a vector $y \in Y$ is called a **regular value** of f if $df(x) : X \to Y$ is onto and has a right inverse for every $x \in f^{-1}(y)$. The implicit function theorem asserts that $f^{-1}(y)$ is a smooth manifold for every regular value of f. Moreover, if f is a Fredholm map then the dimension of $f^{-1}(y)$ is finite and agrees with the Fredholm index of f.

THEOREM A.3.3 (Implicit function theorem). *Let X and Y be Banach spaces, $U \subset X$ be an open set, and ℓ be a positive integer. If $f : U \to Y$ is of class C^ℓ and y is a regular value of f then*

$$\mathcal{M} := f^{-1}(y) \subset X$$

is a C^ℓ Banach manifold and

$$T_x\mathcal{M} = \ker\, df(x)$$

for every $x \in \mathcal{M}$. Hence, if f is a Fredholm map, $\mathcal{M}$ is finite dimensional and

$$\dim\, \mathcal{M} = \mathrm{index}(f).$$

Before giving the proof, let us translate this geometric statement into more analytic language. Suppose without loss of generality that $y = 0$. If $x_0 \in \mathcal{M}$ then, by assumption, the operator

$$D := df(x_0) : X \to Y$$

is surjective and has a right inverse $Q : Y \to X$ such that

$$DQ = \mathrm{id}_Y.$$

The existence of a right inverse is equivalent to the existence of a splitting

$$X = \ker\, D \oplus \mathrm{im}\, Q.$$

Here $\ker\, D$ consists of the solutions of the linearized equation $df(x_0)\xi = 0$, and we expect the space of solutions of the full nonlinear equation $f(x) = 0$ to *look like* the kernel of D locally near x_0. The implicit function theorem makes this precise. Its claim that $\ker D$ is tangent to $\mathcal{M} = f^{-1}(0)$ is equivalent to saying that there is a smooth map

$$\phi : \ker\, D \to Y$$

with $d\phi(0) = 0$ such that, if x is sufficiently close to x_0, then $f(x) = 0$ if and only if x has the form

$$x = x_0 + \xi + Q\phi(\xi), \quad D\xi = 0,$$

for some sufficiently small vector ξ. Thus, if x_1 denotes $x_0 + \xi$, we are looking for a solution of $f(x_1 + \eta) = 0$ for some $\eta \in \mathrm{im}\, Q$. All we know is that there is an approximate solution of the equation $f(x) = 0$ (at the point $x = x_1$); we do not assume that $f(x_0) = 0$. Nevertheless, the next proposition says that if $f(x_1)$ is sufficiently small there is a true solution on the coset $C = x_1 + \mathrm{im}\, Q$.

PROPOSITION A.3.4. *Let X and Y be Banach spaces, $U \subset X$ be an open set, and $f : U \to Y$ be a continuously differentiable map. Let $x_0 \in U$ be such that $D := df(x_0) : X \to Y$ is surjective and has a (bounded linear) right inverse $Q : Y \to X$. Choose positive constants δ and c such that $\|Q\| \le c$, $B_\delta(x_0; X) \subset U$, and*

$$\|x - x_0\| < \delta \qquad \Longrightarrow \qquad \|df(x) - D\| \le \frac{1}{2c}. \tag{A.3.4}$$

Suppose that $x_1 \in X$ satisfies

$$\|f(x_1)\| < \frac{\delta}{4c}, \qquad \|x_1 - x_0\| < \frac{\delta}{8}. \tag{A.3.5}$$

Then there exists a unique $x \in X$ such that

$$f(x) = 0, \qquad x - x_1 \in \operatorname{im} Q, \qquad \|x - x_0\| < \delta.$$

Moreover, $\|x - x_1\| \le 2c\,\|f(x_1)\|$.

PROOF. Define the map $h : U \to Y$ by

$$h(x) := f(x) - D(x - x_1).$$

Then

$$x + Qh(x) = x_1 \qquad \Longleftrightarrow \qquad f(x) = 0,\ \ x - x_1 \in \operatorname{im} Q. \tag{A.3.6}$$

To see this suppose first that $x + Qf(x) - QD(x - x_1) = x_1$. Then $x - x_1 \in \operatorname{im} Q$. Hence $QD(x - x_1) = x - x_1$ and so $Qf(x) = 0$. Since Q is injective, this means $f(x) = 0$. The converse is similar.

Therefore the problem reduces to finding a solution $x \in B_\delta(x_0; X)$ of the equation $x + Qh(x) = x_1$. Let us rewrite this equation in the form $\psi(x) = x_1$, where $\psi : U \to X$ is defined by

$$\psi(x) := x + Qh(x) = x + Q(f(x) - D(x - x_1))$$

We shall see that ψ satisfies the hypotheses of Lemma A.3.2 and that the point x_1 belongs to the image of $B_\delta(x_0; X)$ under ψ.

More precisely, $d\psi(x) - \mathbb{1} = Q(df(x) - D)$, and hence it follows from (A.3.4) that

$$\|x - x_0\| < \delta \qquad \Longrightarrow \qquad \|\mathbb{1} - d\psi(x)\| \le c\|df(x) - D\| \le 1/2.$$

By Lemma A.3.2, ψ maps $B_\delta(x_0; X)$ bijectively onto some open set in X and

$$B_{\delta/2}(\psi(x_0); X) \subset \psi(B_\delta(x_0; X)) \subset \psi(B_{2\delta}(x_0; X)).$$

Now observe that, by (A.3.3) and (A.3.5),

$$\begin{aligned} \|x_1 - \psi(x_0)\| &= \|\psi(x_1) - \psi(x_0) - Qf(x_1)\| \\ &\le 2\|x_1 - x_0\| + c\|f(x_1)\| \\ &< \delta/2. \end{aligned}$$

Hence there is a unique element $x \in B_\delta(x_0; X)$ such that $\psi(x) = x_1$ or, equivalently, so that $x + Qh(x) = x_1$.

Thus, by (A.3.6), we have found a point $x \in x_1 + \operatorname{im} Q$ such that $f(x) = 0$. Moreover, the inequality (A.3.3) in the proof of Lemma A.3.2 shows that

$$\|x - x_1\| \le 2\|\psi(x) - \psi(x_1)\| = 2\|Qf(x_1)\| \le 2c\|f(x_1)\|.$$

This proves Proposition A.3.4. □

REMARK A.3.5. The proof of Proposition A.3.4 can be interpreted in terms of the Newton–Picard iteration. Given x_1 such that $f(x_1)$ is small we are trying to find a solution $\eta \in Y$ of the equation $g(\eta) := f(x_1 + Q\eta) = 0$. The direct Newton–Picard iteration scheme, for a solution of the equation $g(\eta) = 0$ near $\eta = 0$, is to define a sequence η_i recursively by

$$\eta_{i+1} := \eta_i - dg(0)^{-1}g(\eta_i), \qquad \eta_1 := 0.$$

Let $x_i := x_1 + Q\eta_i$. Then, since $dg(0) = DQ = \mathrm{id}_Y$, this formula is just

$$x_{i+1} := x_i - Qf(x_i).$$

Tracing the proof of Proposition A.3.4 back to the contraction mapping principle in Step 2 of the proof of Lemma A.3.2, the argument reduces to finding a fixed point of the map $x \mapsto x_1 - Qh(x)$, where $h(x) := f(x) - D(x - x_1)$ as before. The iteration for this fixed point problem has the form

$$x_{i+1} = x_1 - Qh(x_i) = x_1 + QD(x_i - x_1) - Qf(x_i).$$

But $x_i = x_1 + QD(x_i - x_1)$ since $x_i - x_1 \in \operatorname{im} Q$. Therefore this is precisely the iteration obtained by the Newton-Picard method.

PROOF OF THEOREM A.3.3. Assume without loss of generality that $y = 0$. Let $x_0 \in U$ such that $f(x_0) = 0$ and denote $D := df(x_0)$. By assumption this operator has a right inverse $Q : Y \to X$. Choose constants c and δ as in Proposition A.3.4. Shrinking δ, if necessary, we may also assume that

$$\|\xi\| < \delta/8 \qquad \Longrightarrow \qquad \|f(x_0 + \xi) - f(x_0) - df(x_0)\xi\| < 2\|\xi\|/c$$

Let $\xi \in \ker D$ with $\|\xi\| < \delta/8$. Then $x_1 := x_0 + \xi$ satisfies (A.3.5). Hence, by Proposition A.3.4, there is a unique element $x = x(\xi) \in X$ such that

$$f(x) = 0, \qquad x - x_0 - \xi \in \operatorname{im} Q, \qquad \|x - x_0\| < \delta.$$

Since Q is injective there is a unique element $\phi(\xi) \in Y$ such that

$$x = x_0 + \xi + Q\phi(\xi).$$

It follows also from Proposition A.3.4 that $\|Q\phi(\xi)\| \le 2c\|f(x_0 + \xi)\| < \delta/2$.

We prove that ϕ is of class C^ℓ. Define $\psi : B_\delta(x_0) \to X$ by

$$\psi(x) := Qf(x) + x - x_0 - QD(x - x_0)$$

Then

$$\psi(x_0) = 0, \qquad d\psi(x) - \mathbb{1} = Q(df(x) - D).$$

Hence it follows from (A.3.4) that $\|d\psi(x) - \mathbb{1}\| \le 1/2$ for every $x \in B_\delta(x_0)$. Hence, by Lemma A.3.2, ψ has a C^ℓ inverse on $B_{\delta/2}(0)$. For $\xi \in \ker D$ with $\|\xi\| < \delta/8$ we have $\|\xi + Q\phi(\xi)\| < 5\delta/8 < \delta$ and $\psi(x_0 + \xi + Q\phi(\xi)) = \xi$, and hence

$$x_0 + \xi + Q\phi(\xi) = \psi^{-1}(\xi).$$

Since $DQ = \mathbb{1}$ this implies

$$\phi(\xi) = D\psi^{-1}(\xi) - Dx_0 \qquad \text{for } \xi \in \ker D \text{ with } \|\xi\| < \delta/8.$$

Hence ϕ is of class C^ℓ as claimed and, since $d\psi(x_0) = \mathbb{1}$, we have $d\phi(0) = 0$. Moreover, if $x \in X$ is such that $f(x) = 0$ and $\|x - x_0\| < \delta/8(1 + c\|D\|) < \delta$, then we can write

$$x = x_0 + \xi + Q\eta \qquad \eta := D(x - x_0), \qquad \xi := x - x_0 - Q\eta \in \ker D.$$

Hence we have $\|\eta\| \le \|D\|\|x - x_0\|$ which implies

$$\|\xi\| \le (1 + c\|D\|)\|x - x_0\| < \delta/8.$$

Therefore $\eta = \phi(\xi)$, so that $x = \psi^{-1}(\xi)$. Thus the map

$$\psi : f^{-1}(0) \cap B_{\delta/8(1+c\|D\|)}(x_0) \to \ker D$$

is a coordinate chart on $f^{-1}(0)$. The transition maps are obviously smooth. This proves Theorem A.3.3. □

In the Banach space setting, the existence of a right inverse does not follow from the fact that D is onto. Such a right inverse exists if and only if the kernel of D has a complement in X. (See Proposition A.4.1 below.) In particular, every surjective Fredholm operator has a right inverse. This generalizes to operators of the form $D \oplus L : X \oplus Z \to Y$ defined by

$$D \oplus L(x, z) = Dx + Lz$$

where $D : X \to Y$ is Fredholm.

LEMMA A.3.6. *Assume $D : X \to Y$ is a Fredholm operator and $L : Z \to Y$ is a bounded linear operator such that $D \oplus L : X \oplus Z \to Y$ is onto. Then $D \oplus L$ has a right inverse. Moreover, the projection $\Pi : \ker(D \oplus L) \to Z$ is a Fredholm operator with* $\ker \Pi \cong \ker D$ *and* $\operatorname{coker} \Pi \cong \operatorname{coker} D$ *and hence*

$$\operatorname{index} \Pi = \operatorname{index} D.$$

PROOF. Choose a complement X_1 of $\ker D$ in X and finitely many vectors $z_\nu \in Z$, $\nu = 1, \ldots, N$, such that the vectors Lz_ν span a complement of $\operatorname{im} D$ in Y. Then a right inverse of $D \oplus L$ is the operator

$$Y \to X \oplus Z : y \mapsto \left(x, \sum_{\nu=1}^{N} \lambda_\nu z_\nu \right)$$

where x and $\lambda_1, \ldots, \lambda_N$ are chosen such that

$$x \in X_1, \qquad y = Dx + \sum_{\nu=1}^{N} \lambda_\nu L z_\nu.$$

Now $\ker \Pi = \ker D \oplus \{0\}$ and $\operatorname{im} \Pi = L^{-1}(\operatorname{im} D)$. Hence

$$\frac{Z}{\operatorname{im} \Pi} = \frac{Z}{L^{-1}(\operatorname{im} D)} \cong \frac{\operatorname{im} L}{\operatorname{im} D \cap \operatorname{im} L} \cong \frac{Y}{\operatorname{im} D}.$$

The second isomorphism is induced by L and the last isomorphism exists because $\operatorname{im} D + \operatorname{im} L = Y$. □

EXERCISE A.3.7. Let X be an infinite dimensional Hilbert space. In contrast to the finite dimensional case a C^∞ smooth function $\phi : X \to \mathbb{R}$ which vanishes outside the unit ball need not be bounded (even though every point in X has a neighbourhood in which the function is bounded). Construct an example of an unbounded smooth function with support in the unit ball. *Hint:* There is an infinite sequence of pairwise disjoint balls of radius $1/4$ which are all contained in the unit ball of radius 1.

A.4. Finite dimensional reduction

In this section we explain how to obtain a finite dimensional model for the local description of the zero set of a smooth map $f : X \to Y$ between two Banach spaces. Such a finite dimensional reduction is sometimes called a *Kuranishi model.* Assume that $f(0) = 0$ and denote $D := df(0) : X \to Y$. By assumption, D is a bounded linear operator. A **pseudoinverse** of D is a bounded linear operator $T : Y \to X$ which satisfies

$$TDT = T, \qquad DTD = D.$$

The next proposition gives a necessary and sufficient criterion for the existence of a pseudoinverse. It shows, in particular, that every Fredholm operator admits a pseudoinverse.

PROPOSITION A.4.1. *A bounded linear operator $D : X \to Y$ admits a pseudoinverse if and only if D satisfies the following*

(i) *D has closed image,*

(ii) *The kernel of D has a complement in X.*

(iii) *The image of D has a complement in Y.*

PROOF. Assume first that D satisfies (i), (ii) and (iii), denote

$$X_0 := \ker D, \qquad Y_1 := \operatorname{im} D,$$

and choose complements X_1 and Y_0 so that

$$X = X_0 \oplus X_1, \qquad Y = Y_0 \oplus Y_1.$$

Then X_1 and Y_1 are Banach spaces and the restriction of D to X_1 determines a bijective bounded linear operator $D_1 : X_1 \to Y_1$. The reader may check that the operator $T : Y \to X$ defined by

$$T(y_0 + y_1) := {D_1}^{-1} y_1$$

for $y_0 \in Y_0$ and $y_1 \in Y_1$ is a pseudoinverse. Conversely, if $T : Y \to X$ is a pseudoinverse of D then the required complements are given by

$$X_1 := \operatorname{im} T, \qquad Y_0 := \ker T.$$

To see this just note that

$$P := TD : X \to X, \qquad Q := DT : Y \to Y$$

are projection operators with

$$\operatorname{im} P = \operatorname{im} T, \qquad \ker P = \ker D, \qquad \operatorname{im} Q = \operatorname{im} D, \qquad \ker Q = \ker T.$$

This proves Proposition A.4.1. □

REMARK A.4.2. Let G be a compact Lie group acting on the Banach spaces X and Y by strongly continuous maps

$$\mathrm{G} \to \mathcal{L}(X) : g \mapsto \Phi_g, \qquad \mathrm{G} \to \mathcal{L}(Y) : g \mapsto \Psi_g.$$

Suppose that $D : X \to Y$ is an equivariant bounded linear operator. If D admits a pseudoinverse then it admits a G-equivariant pseudoinverse. To see this note that if T is any pseudoinverse of D then the operator

$$T_g = \Phi_g T {\Psi_g}^{-1}$$

is also a pseudoinverse. Hence the average

$$S := \int_G T_g \, d\mu(g),$$

with respect to the Haar measure $d\mu$ on G with Vol(G) = 1, is G-equivariant and satisfies $DSD = D$. It follows that the operator

$$R := SDS = \int_G \int_G T_g D T_h \, d\mu(g) d\mu(h)$$

is a G-equivariant pseudoinverse of D.

THEOREM A.4.3. *Let X and Y be Banach spaces, $U \subset X$ be an neighbourhood of zero, and ℓ be a positive integer. Let $f : U \to Y$ be a C^ℓ map such that $f(0) = 0$ and suppose that the operator $D := df(0) : X \to Y$ has a pseudoinverse $T : Y \to X$. Then there is an open neighbourhood $W \subset X$ of zero, a local C^ℓ-diffeomorphism $g : W \to g(W)$ onto an open subset $g(W) \subset U$, and a C^ℓ map $f_0 : W \to Y_0 := \ker T$ such that*

$$f \circ g(x) = f_0(x) + Dx$$

for $x \in W$ and

$$g(0) = 0, \quad dg(0) = \mathbb{1}, \quad f_0(0) = 0, \quad df_0(0) = 0.$$

If f is equivariant with respect to strongly continuous actions of some compact Lie group G *on X and Y, then the maps g and f_0 can be chosen* G*-equivariant.*

PROOF. Consider the C^ℓ map $\psi : U \to X$ defined by

$$\psi(x) := x + T(f(x) - Dx).$$

(This is almost the same formula as in the proof of Proposition A.3.4 except that the term TDx_1 has been dropped.) This map satisfies

$$\psi(0) = 0, \qquad d\psi(0) = \mathbb{1}.$$

Hence, by Theorem A.3.1, ψ has a local inverse defined on some open neighbourhood $U_0 \subset U$ of zero. Let $W := \psi(U_0)$ and define $g : W \to U_0$ and $f_0 : W \to \ker T$ by

$$g := \psi^{-1}, \qquad f_0 := (\mathbb{1} - DT) \circ f \circ \psi^{-1}.$$

The formula

$$D\psi(x) = Dx + DT(f(x) - Dx) = DTf(x)$$

shows that

$$D = DT \circ f \circ \psi^{-1}$$

and hence

$$f \circ g = f \circ \psi^{-1} = (\mathbb{1} - DT) \circ f \circ \psi^{-1} + D = f_0 + D.$$

If f is G-equivariant then, by Remark A.4.2, we can choose a G-equivariant pseudoinverse T of D. In this case the above formulas show that g and f_0 are also G-equivariant. □

Let g and f_0 be as in Theorem A.4.3. Then the local zero set of f near $x = 0$ can be identified with the zero set of $f_0 : W \cap \ker D \to Y_0 = \ker T$ via g:

$$f^{-1}(0) \cap g(W) = \{ g(x) \,|\, x \in W \cap \ker D, \ f_0(x) = 0 \}.$$

To see this just note that $f_0(W) \subset Y_0$ where Y_0 is a complement of the image of D and hence

$$f_0(x) + Dx = 0 \qquad \Longleftrightarrow \qquad Dx = 0, \quad f_0(x) = 0.$$

This observation is interesting when D is a Fredholm operator. In this case the kernel and cokernel of D are finite dimensional and hence $f_0 : U_0 \cap \ker D \to Y_0 \cong \operatorname{coker} D$ is a smooth map between finite dimensional vector spaces.

A.5. The Sard–Smale theorem

Smale [**381**] proved the following infinite dimensional version of Sard's theorem.

THEOREM A.5.1 (Sard–Smale). *Let X and Y be separable Banach spaces and $U \subset X$ be an open set. Suppose that $f : U \to Y$ is a Fredholm map of class C^ℓ, where*

$$\ell \geq \max\{1, \operatorname{index}(f) + 1\}.$$

Then the set

$$Y_{\mathrm{reg}}(f) := \{y \in Y \mid x \in U,\ f(x) = y \implies \operatorname{im} df(x) = Y\}$$

of regular values of f is residual in the sense of Baire (i.e. it contains a countable intersection of open and dense sets).

The separability condition is essential. A Banach space X is called **separable** if it admits a dense sequence. Since every metric space is paracompact so is every Banach space. This means that every open cover of X admits a locally finite refinement. In the separable case every locally finite cover is countable.[1] Since the existence of a countable refinement implies the existence of a countable subcover this proves the following.

PROPOSITION A.5.2. *Let X be a separable Banach space. Then every open cover of X admits a countable subcover.*

PROOF OF THEOREM A.5.1. We prove that every point $x_0 \in U$ has an open neighbourhood U_0 such that, for every closed subset $V_0 \subset U_0$, the set of regular values of the restriction $f|_{V_0}$ is open and dense in Y. Assume without loss of generality that $x_0 = 0 \in U$ and let $T : Y \to X$ be a pseudoinverse of $D := df(0)$. Then, by Theorem A.4.3, there is an open neighbourhood $W \subset X$ of zero, a C^ℓ diffeomorphism $g : W \to g(W)$ onto an open subset $g(W) \subset X$, and a C^ℓ map $f_0 : W \to \ker T$ such that

$$f \circ g = f_0 + D$$

We claim that $U_0 := g(W)$ is the required neighbourhood of $x_0 = 0$.

To see this, recall that there are splittings

$$X = X_0 \oplus X_1, \qquad Y = Y_0 \oplus Y_1,$$

where

$$X_0 := \ker D, \quad X_1 := \operatorname{im} T, \qquad Y_0 := \ker T, \quad Y_1 := \operatorname{im} D.$$

Think of $g : W \to X$ as a coordinate chart on X and write the equation $f(g(x)) = y$ in the form

$$y_0 = f_0(x_0, x_1), \qquad y_1 = D_1 x_1,$$

for $x = x_0 + x_1 \in W$, where $x_i \in X_i$. Here $D_1 : X_1 \to Y_1$ denotes the restriction of D to X_1. It follows from this description that $y = y_0 + y_1$ is a regular value of $f|_{g(W)}$ if and only if y_0 is a regular value of the map

$$\{x_0 \in X_0 \mid x_0 + {D_1}^{-1} y_1 \in W\} \to Y_0 : x_0 \mapsto f_0(x_0, {D_1}^{-1} y_1).$$

[1] If the cover $\{U_\alpha\}_\alpha$ is locally finite and $\{x_i\}_i$ is a dense sequence then the set of pairs (α, i) with $x_i \in U_\alpha$ is countable. Since every U_α contains some point x_i the map $(\alpha, i) \mapsto \alpha$ is surjective.

Since $\dim X_0 - \dim Y_0 = \mathrm{index}(f)$ and $\ell \geq \max\{1, \mathrm{index}(f) + 1\}$, it follows from Sard's theorem for C^ℓ maps between finite dimensional vector spaces (see **[289]** for smooth maps and **[2]** for C^ℓ-maps) that the set

$$Y_{\mathrm{reg}}(f; g(W)) := \{y \in Y \mid x \in g(W),\, f(x) = y \implies \mathrm{im}\, df(x) = Y\}$$

of regular values of $f|_{g(W)}$ is dense in Y. Hence $Y_{\mathrm{reg}}(f; g(V)) \supset Y_{\mathrm{reg}}(f; g(W))$ is dense for every subset $V \subset W$.

We prove that $Y_{\mathrm{reg}}(f; g(V))$ is open whenever $V \subset W$ is closed in X. Let

$$y_\nu = y_{\nu,0} + y_{\nu,1} \in Y$$

be a sequence of singular values of $f|_{g(V)}$ which converges to y. Then there exists a sequence $x_\nu \in V$ such that $f(g(x_\nu)) = y_\nu$ and $df(g(x_\nu))$ is not surjective. The sequence

$$x_{\nu,1} := TDx_\nu = Ty_\nu$$

converges and, since V is bounded, the sequence

$$x_{\nu,0} := x_\nu - TDx_\nu \in \ker D$$

is bounded. Passing to a subsequence we may assume that $x_{\nu,0}$ converges as well, and hence, so does $x_\nu = x_{\nu,0} + x_{\nu,1}$. Since V is closed, the limit point $x := \lim_{\nu\to\infty} x_\nu$ lies again in V and $f(g(x)) = y$. Moreover, $df(g(x))$ is the limit of operators with a nontrivial cokernel and hence cannot be surjective. Hence y is a singular value of $f|_{g(V)}$.

Thus we have proved that every point in U has an open neighbourhood U_0 such that the set $Y_{\mathrm{reg}}(f; V_0)$ is open and dense in Y for every closed subset $V_0 \subset U_0$. Since X is separable, it follows from Proposition A.5.2, that the open set U can be covered by countably many such open neighbourhoods U_i. Hence U can be covered by countable many closed sets V_i such that $Y_{\mathrm{reg}}(f; V_i)$ is open and dense in Y. Hence

$$Y_{\mathrm{reg}}(f) = \bigcap_i Y_{\mathrm{reg}}(f; V_i)$$

is a countable intersection of open and dense sets in Y. □

APPENDIX B

Elliptic Regularity

This appendix gives an introduction to Sobolev spaces and the L^p regularity theory of the Laplace operator that is needed in the theory of pseudoholomorphic curves. In the last section we prove the basic regularity and compactness theorems for pseudoholomorphic curves with totally real boundary conditions, only assuming interior regularity for the two dimensional Laplace operator.

The first section gives an exposition of Sobolev spaces. It contains the basic definitions, a proof that smooth functions are dense in a Sobolev space over a Lipschitz domain, as well as proofs of the Poincaré inequality, the Sobolev embedding theorems, and a version of the trace theorem. Section B.2 contains a proof of the Calderon–Zygmund inequality, following the book of Gilbarg and Trudinger [**140**], and Section B.3 deals with interior regularity for the Laplace operator. Section B.4 explains the elliptic bootstrapping technique for pseudoholomorphic curves. The exposition in the first three sections is carried out for functions of n variables even though for our applications in Section B.4 it suffices to treat the case $n = 2$. However, in this greater generality the basic theory is no more difficult than in the two dimensional case.

B.1. Sobolev spaces

Let $\Omega \subset \mathbb{R}^n$ be an open set. Denote by $C^\infty(\overline{\Omega})$ the space of restrictions of smooth functions on $\mathbb{R}^n$ to $\overline{\Omega}$ and by $C_0^\infty(\Omega)$ the space of smooth compactly supported functions on Ω. We begin by mentioning two fundamental inequalities for smooth compactly supported functions $u, v : \mathbb{R}^n \to \mathbb{R}$, namely, **Hölder's inequality**

$$\|uv\|_{L^1} \le \|u\|_{L^p} \|v\|_{L^q}$$

for $1/p + 1/q = 1$ and **Young's inequality**

$$\|u * v\|_{L^p} \le \|u\|_{L^1} \|v\|_{L^p}.$$

Here

$$u * v(x) = \int_{\mathbb{R}^n} u(x - y)v(y)\, dy$$

denotes the convolution and

$$\|u\|_{L^p} = \left(\int_{\mathbb{R}^n} |u|^p \right)^{1/p}$$

denotes the L^p-norm for $1 \le p < \infty$.

Weak derivatives. Let $u:\Omega\to\mathbb{R}$ be locally integrable and fix a multi-index $\nu=(\nu_1,\dots,\nu_n)$. A locally integrable function $u_\nu:\Omega\to\mathbb{R}$ is called the **weak derivative** of u corresponding to ν if

$$\int_\Omega u(x)\partial^\nu\phi(x)\,dx=(-1)^{|\nu|}\int_\Omega u_\nu(x)\phi(x)\,dx$$

for every $\phi\in C_0^\infty(\Omega)$. The weak derivative, if it exists, is almost everywhere uniquely determined by u and we denote it by

$$\partial^\nu u:=u_\nu.$$

The number $|\nu|:=\nu_1+\cdots+\nu_n$ is called the **order** of the derivative. By the divergence theorem, every C^k-function $u:\Omega\to\mathbb{R}$ has weak derivatives up to order k and these agree with the strong derivatives.

Now fix a nonnegative integer k and a number $1\le p\le\infty$. The **Sobolev space** $W^{k,p}(\Omega)$ is defined as the space of all (equivalence classes of) functions $u\in L^p(\Omega)$ such that the weak derivative $\partial^\nu u$ exists and is p-integrable for every ν such that $|\nu|\le k$. For $1\le p<\infty$ we define the $W^{k,p}$-norm of a function $u\in W^{k,p}(\Omega)$ by

$$\|u\|_{k,p}=\left(\int_\Omega\sum_{|\nu|\le k}|\partial^\nu u(x)|^p\,dx\right)^{1/p}.$$

For $p=\infty$ the $W^{k,\infty}$-norm is defined as the maximum of the L^∞-norms of the weak derivatives $\partial^\nu u$ for $|\nu|\le k$. The space $W^{k,p}_{\mathrm{loc}}(\Omega)$ is defined as the space (equivalence classes of) locally p-integrable functions $u:\Omega\to\mathbb{R}$ whose restrictions to all precompact open subsets Q of Ω are in $W^{k,p}(Q)$. The space $W^{k,p}_0(\Omega)$ is defined as the closure of $C_0^\infty(\Omega)$ in $W^{k,p}(\Omega)$. Thus $W^{k,p}_0(\Omega)$ is the completion of $C_0^\infty(\Omega)$ with respect to the $W^{k,p}$-norm.

EXERCISE B.1.1. Prove that $W^{k,p}(\Omega)$ is a Banach space, and is reflexive for $1<p<\infty$. Prove that $W^{k,p}_0(\Omega)$ is separable for $1\le p<\infty$. *Hint:* Think of $W^{k,p}(\Omega)$ as a closed subspace of $L^p(\Omega,\mathbb{R}^N)$ for a suitable integer N. Every closed subspace of a Banach space is complete. Every closed subspace of a reflexive Banach space is reflexive. For separability use the fact that every smooth function $u:\Omega\to\mathbb{R}$ can be approximated by a sequence of polynomials with rational coefficients, where the convergence is uniform for each derivative on every compact set.

EXERCISE B.1.2. Let $\Omega\subset\mathbb{R}^n$ be a bounded open set.

(i) Show that $u\in W^{k+1,p}(\Omega)$ if and only if $u\in W^{1,p}(\Omega)$ and the weak derivatives $\partial_i u=\partial u/\partial x_i$ lie in $W^{k,p}(\Omega)$ for $i=1,\dots,n$.

(ii) If $u\in W^{k,p}(\Omega)$ and $v\in W^{k,\infty}(\Omega)$ show that $uv\in W^{k,p}(\Omega)$ and

$$\|uv\|_{k,p}\le c\,\|u\|_{k,p}\,\|v\|_{k,\infty}$$

where the constant c depends only on k and n. *Hint:* The proof is surprisingly nontrivial. The hard part is to show that, if $u\in W^{1,p}(\Omega)$ and $v\in W^{1,\infty}(\Omega)$, then the weak derivatives of uv are given by the Leibniz rule $\partial_i(uv)=(\partial_i u)v+u(\partial_i v)$ for $i=1,\dots,n$. These functions are obviously in L^p. With this understood the result follows by induction. The proof of the Leibniz rule requires Proposition B.1.4 below. Prove the result first when u is smooth and then approximate u on a compact subset of Ω by a sequence of smooth functions.

Approximation by smooth functions. Our next goal is to prove that for a large class of domains $\Omega \subset \mathbb{R}^n$ the Sobolev space $W^{k,p}(\Omega)$ can be identified with the completion of $C^\infty(\overline{\Omega})$ with respect to the $W^{k,p}$-norm. An open set $\Omega \subset \mathbb{R}^n$ is called a **Lipschitz domain** if the boundary can locally be represented as the graph of a Lipschitz function. Explicitly, this means that for every $x \in \partial\Omega$ there exist a neighbourhood U of x, a nonzero vector $\xi \in \mathbb{R}^n$, a constant $\delta > 0$, and a Lipschitz continuous function $f : \xi^\perp \to \mathbb{R}$ such that $f(0) = 0$, $|f(\eta)| < \delta$ for $|\eta| < \delta$, and

$$\Omega \cap U = \{x + \eta + t\xi \mid \eta \in \xi^\perp,\ |\eta| < \delta,\ f(\eta) < t < \delta\}.$$

The motivation for this definition lies in the following observation.

EXERCISE B.1.3. Assume $1 \le p < \infty$ and let $\Omega \subset \mathbb{R}^n$ be a bounded Lipschitz domain. Show that for every function $u \in W^{k,p}(\Omega)$ there exists a sequence of open sets $\Omega_j \subset \mathbb{R}^n$ and a sequence of functions $u_j \in W^{k,p}(\Omega_j)$ such that

$$\overline{\Omega} \subset \Omega_j, \qquad \lim_{j\to\infty} \|u_j - u\|_{W^{k,p}(\Omega)} = 0.$$

Hint: Suppose first that $u \in W^{k,p}(\Omega)$ is supported in a sufficiently small neighbourhood U of a boundary point. Prove that the functions $u_\varepsilon(x) := u(x + \varepsilon\xi)$ converge to u in the $W^{k,p}$-norm on Ω. To prove the result in general multiply u by a suitable partition of unity and use Exercise B.1.2 (ii).

PROPOSITION B.1.4. *If $\Omega \subset \mathbb{R}^n$ is a bounded Lipschitz domain then $C^\infty(\overline{\Omega})$ is dense in $W^{k,p}(\Omega)$ for $1 \le p < \infty$ and any nonnegative integer k.*

PROOF. Let $u \in W^{k,p}(\Omega)$. Assume first that u has compact support in Ω. Let $\rho : \mathbb{R}^n \to \mathbb{R}$ be a smooth nonnegative function such that

$$\operatorname{supp}\rho \subset B_1, \qquad \int_{\mathbb{R}^n} \rho(x)\,dx = 1,$$

and, for $\delta > 0$, define

$$\rho_\delta(x) = \delta^{-n}\rho(\delta^{-1}x).$$

Then $\rho_\delta * u$ is smooth with compact support. It is a simple consequence of the definition of weak derivatives that the strong derivatives of $\rho_\delta * u$ are given by

$$\partial^\nu(\rho_\delta * u) = \rho_\delta * \partial^\nu u. \tag{B.1.1}$$

Hence, by Young's inequality,

$$\|\rho_\delta * u\|_{W^{k,p}} \le \|u\|_{W^{k,p}}. \tag{B.1.2}$$

Now one checks easily that for every continuous function $f : \Omega \to \mathbb{R}$ with compact support the functions $\rho_\delta * f$ converge to f uniformly as $\delta \to 0$. Since the continuous functions with compact support form a dense subset of $L^p(\Omega)$ for $1 \le p < \infty$, the inequality (B.1.2) with $k = 0$ shows that $\rho_\delta * f$ converges to f in the L^p-norm for every $f \in L^p(\Omega)$. Hence, by (B.1.1), $\rho_\delta * u$ converges to u in the $W^{k,p}$-norm. Thus we have proved that every $W^{k,p}$-function with compact support can be approximated by a sequence of smooth functions with compact support.

Now consider the general case and let $\varepsilon > 0$. Then, by Exercise B.1.3, there exists an open set $\Omega' \subset \mathbb{R}^n$ and a function $u' \in W^{k,p}(\Omega')$ such that $\overline{\Omega} \subset \Omega'$ and $\|u' - u\|_{W^{k,p}(\Omega)} \le \varepsilon/2$. We may assume without loss of generality that u' has compact support in Ω'. Hence, by what we just proved, there exists a smooth function $v : \Omega' \to \mathbb{R}$ with compact support such that $\|v - u'\|_{W^{k,p}(\Omega')} \le \varepsilon/2$. This proves the proposition. □

A sequence of smooth functions ρ_δ as in the proof of Proposition B.1.4 is called an **approximation to the delta function** and the operator $u \mapsto \rho_\delta * u$ is called a **smoothing operator**.

EXERCISE B.1.5. This exercise shows that the Sobolev space $W^{k,p}$ is preserved by composition (on the right) with $C^{k-1,1}$-diffeomorphisms (i.e. C^{k-1}-diffeomorphisms whose $(k-1)$st derivatives are Lipschitz continuous). Let $\Omega, \Omega' \subset \mathbb{R}^n$ be bounded Lipschitz domains and $\psi : \overline{\Omega'} \to \overline{\Omega}$ be a $C^{k-1,1}$-diffeomorphism (that is ψ is the restriction of a $C^{k-1,1}$-diffeomorphism between suitable open neighbourhoods of the closures). Show that if $u \in W^{k,p}(\Omega)$ then $u \circ \psi \in W^{k,p}(\Omega')$ and

$$\|u \circ \psi\|_{W^{k,p}(\Omega')} \leq c \|u\|_{W^{k,p}(\Omega)}$$

where the constant c is independent of u. *Hint:* Use Exercise B.1.2 and Proposition B.1.4. Prove this by induction over k.

Poincaré's inequality. It is somewhat less than obvious that a function $u \in W^{1,p}(\Omega)$ whose derivatives all vanish must be constant on every component of Ω. The proof requires the following fundamental estimate. As always, ∇u denotes the gradient of the function u. Recall that the mean value of a function is its integral over its domain of definition divided by the volume of the domain.

LEMMA B.1.6 (Poincaré's inequality). *Let $1 \leq p < \infty$ and $\Omega \subset \mathbb{R}^n$ be a bounded open domain. Then, for $u \in W_0^{1,p}(\Omega)$,*

$$\|u\|_{L^p(\Omega)} \leq \operatorname{diam}(\Omega) \|\nabla u\|_{L^p(\Omega)} .$$

If $\Omega = Q^n = (0,1)^n$ is the unit square then every $u \in W^{1,p}(\Omega)$ with $\int_{Q^n} u = 0$ satisfies

$$\|u\|_{L^p(Q^n)} \leq n \|\nabla u\|_{L^p(Q^n)} .$$

PROOF. It suffices to prove the first statement for $u \in C_0^\infty(\Omega)$. Suppose without loss of generality that $\Omega \subset \{x_n > 0\}$ and $0 \in \partial\Omega$. Then

$$u(x) = \int_0^{x_n} \partial_n u(x_1, \ldots, x_{n-1}, t)\, dt.$$

Since $|x_n| \leq \operatorname{diam}(\Omega)$ it follows from Hölder's inequality that

$$|u(x)|^p \leq \operatorname{diam}(\Omega)^{p-1} \int_0^\infty |\partial_n u(x_1, \ldots, x_{n-1}, t)|^p\, dt.$$

Now integrate both sides over $\mathbb{R}^{n-1} \times [0, \operatorname{diam}(\Omega)]$ to obtain

$$\int_\Omega |u|^p \leq \operatorname{diam}(\Omega)^p \int_\Omega |\partial_n u|^p.$$

This proves the first assertion. The second assertion is proved by induction over n. For $n = 1$ the estimate is an easy exercise. Hence assume that the estimate is proved for $n \geq 1$ and let $u \in C^\infty(Q^{n+1})$ be of mean value zero. Define

$$v(t) = \int_{Q^n} u(x_1, \ldots, x_n, t)\, dx_1 \cdots dx_n.$$

Since $v \in C^\infty(Q^1)$ has mean value zero

$$\int_0^1 |v(t)|^p\, dt \leq \int_0^1 |\dot{v}(t)|^p\, dt \leq \int_{Q^{n+1}} |\partial_{n+1} u|^p\, dx.$$

The last step follows from Hölder's inequality. By the induction hypothesis, we have

$$\int_{Q^n} |u(x,t) - v(t)|^p \, dx \le n^p \int_{Q^n} |\nabla u(x,t)|^p \, dx$$

for every t. Integrate over t to obtain $\|u - v\|_{L^p} \le n \|\nabla u\|_{L^p}$. Combining this with the previous estimate gives $\|u\|_{L^p(Q^{n+1})} \le (n+1) \|\nabla u\|_{L^p(Q^{n+1})}$. This proves the second statement for smooth functions. In the general case it follows from Proposition B.1.4. □

COROLLARY B.1.7. *Let $\Omega \subset \mathbb{R}^n$ be a bounded open domain and $u \in W^{1,p}(\Omega)$ with weak derivatives $\partial u/\partial x_j \equiv 0$ for $j = 1, \ldots, n$. Then u is constant on each connected component of Ω. If, moreover, $u \in W_0^{1,p}(\Omega)$ then $u \equiv 0$.*

PROOF. First assume that Ω is a square and u has mean value zero. Approximate u in the $W^{1,p}$-norm by a sequence of smooth functions $u_\nu \in C^\infty(\overline{\Omega})$ with mean value zero. Then Poincaré's inequality shows that u_ν converges to zero in the L^p-norm. Hence $u = 0$. This shows u is constant on every square where its first derivatives vanish. Hence a function $u \in W^{1,p}(\Omega)$ with $\nabla u \equiv 0$ is constant on every connected component of Ω. If $u \in W_0^{1,p}(\Omega)$ choose a sequence of smooth functions $u_\nu \in C_0^\infty(\Omega)$ converging to u in the $W^{1,p}$-norm. Then $\|\nabla u_\nu\|_{L^p}$ converges to zero and hence, by Lemma B.1.6, $\|u_\nu\|_{L^p}$ converges to zero. Hence $u \equiv 0$. □

The previous corollary can also be obtained as a consequence of the next exercise which shows that for any open set $\Omega \subset \mathbb{R}^n$ the Sobolev space $W^{1,\infty}_{\mathrm{loc}}(\Omega)$ can be naturally identified with the space $C^{0,1}_{\mathrm{loc}}(\Omega)$ of locally Lipschitz continuous functions on Ω.

EXERCISE B.1.8. (i) If $u \in L^1_{\mathrm{loc}}(\Omega)$ prove that $u_\delta(x) = \rho_\delta * u(x)$ converges to $u(x)$ for almost every $x \in \Omega$ (see the proof of Proposition B.1.4).

(ii) Show that $C^{0,1}_{\mathrm{loc}}(\Omega) \subset W^{1,\infty}_{\mathrm{loc}}(\Omega)$. *Hint:* Let $u : \Omega \to \mathbb{R}$ be locally Lipschitz continuous and fix a vector $\xi \in \mathbb{R}^n$. Prove that the sequence $u_j(x) = j(u(x-\xi/j) - u(x))$ has a subsequence which converges weakly in $L^2(K)$ for every compact subset $K \subset \Omega$. Prove that the limit function $u^\xi : \Omega \to \mathbb{R}$ is the weak derivative of u in the direction ξ, i.e.

$$\int_\Omega u^\xi \phi = - \int_\Omega u \langle \nabla \phi, \xi \rangle$$

for all $\phi \in C_0^\infty(\Omega)$. Prove that u^ξ is locally bounded.

(iii) Show that $W^{1,\infty}_{\mathrm{loc}}(\Omega) \subset C^{0,1}_{\mathrm{loc}}(\Omega)$. *Hint:* If $u \in W^{1,\infty}_{\mathrm{loc}}(\Omega)$ prove that for $\delta > 0$ the functions $u_\delta = \rho_\delta * u$ are locally Lipschitz continuous with the Lipschitz constant $c = \sup_{B_{r+\delta}(x)} |\nabla u|$ over $B_r(x)$. Now use (i) to prove that u is locally Lipschitz continuous (possibly after redefining it on a set of measure zero).

Extension. The next proposition is a useful auxiliary result which asserts that every function in $W^{k,p}(\Omega)$ can be extended to a compactly supported $W^{k,p}$-function on a larger domain, provided that Ω has a $C^{k-1,1}$ boundary. Moreover the norm of the extended function can be controlled by the norm of the original function.

PROPOSITION B.1.9. *Let $\Omega \subset \mathbb{R}^n$ be a bounded open set with $C^{k-1,1}$-boundary and $\Omega' \subset \mathbb{R}^n$ be an open set with $\overline{\Omega} \subset \Omega'$. Then there exists a bounded linear operator $E : W^{k,p}(\Omega) \to W_0^{k,p}(\Omega')$ such that $Eu|_\Omega = u$ for every $u \in W^{k,p}(\Omega)$.*

PROOF. The proof is taken from [**140**]. First consider the case $\Omega = \mathbb{H}^n = \{x_n > 0\}$ and define the extension operator $E_0 : C^{k-1,1}(\mathbb{H}^n) \to C^{k-1,1}(\mathbb{R}^n)$ by

$$(E_0 u)(x_1, \dots, x_{n-1}, x_n) := \sum_{i=1}^{k} c_i u(x_1, \dots, x_{n-1}, -x_n/i)$$

for $x_n \le 0$ where the constants $c_1, \dots, c_k$ are chosen such that

$$\sum_{i=1}^{k} c_i \left(-\frac{1}{i}\right)^m = 1, \qquad m = 0, \dots, k-1.$$

One checks easily that the derivatives up to order $k-1$ match on the boundary, that if $u(x) = 0$ for $|x| \ge R$ then $E_0 u(x) = 0$ for $|x| \ge kR$, and that for compactly supported functions there is an estimate

$$\|E_0 u\|_{W^{k,p}(\mathbb{R}^n)} \le c_0 \|u\|_{W^{k,p}(\mathbb{H}^n)}.$$

Now let Ω be any bounded open set with $C^{k-1,1}$-boundary and denote $B_r := \{x \in \mathbb{R}^n \mid |x| < r\}$. There exist an open cover

$$\overline{\Omega} \subset U_0 \cup \cdots \cup U_N,$$

open sets U_j', and $C^{k-1,1}$-diffeomorphisms $\psi_j : U_j' \to B_k$ for $j = 1, \dots, N$ such that $\overline{U}_0 \subset \Omega$, $\overline{U}_j \subset U_j' \subset \Omega'$, and

$$\psi_j(U_j) = B_1, \qquad \psi_j(U_j' \cap \Omega) = B_k \cap \mathbb{H}^n$$

for $j = 1, \dots, N$. Choose a partition of unity $\beta_j : \mathbb{R}^n \to [0,1]$ such that

$$\operatorname{supp} \beta_j \subset U_j, \qquad \sum_{j=0}^{n} \beta_j(x) = 1 \text{ for } x \in \Omega.$$

Define $E : C^{k-1,1}(\Omega) \to C_0^{k-1,1}(\Omega')$ by

$$Eu = \beta_0 u + \sum_{j=1}^{N} \left(E_0(\beta_j u \circ \psi_j{}^{-1}) \right) \circ \psi_j.$$

It follows from Proposition B.1.4 and Exercise B.1.5 that E extends to a bounded linear operator from $W^{k,p}(\Omega) \to W_0^{k,p}(\Omega')$. □

Sobolev embedding theorems. A function with weak derivatives need not be continuous. Consider for example the function

$$u(x) := |x|^{-\alpha}$$

with $\alpha \in \mathbb{R}$ in the domain $\Omega = B_1 = \{x \in \mathbb{R}^n \mid |x| < 1\}$. Then

$$\partial_j u = -\alpha x_j |x|^{-\alpha-2}.$$

By induction,

$$|\partial^\nu u(x)| \le c_\nu |x|^{-\alpha-|\nu|}.$$

Now the function $x \mapsto |x|^{-\beta}$ is integrable on B_1 if and only if $\beta < n$. Hence the derivatives of u up to order k will be p-integrable whenever

$$\alpha p + kp < n.$$

If $kp < n$ choose $0 < \alpha < n/p - k$ to obtain a function which is in $W^{k,p}(B_1)$ but not continuous at 0. For $kp > n$ this construction fails and, in fact, in this case every $W^{k,p}$-function is continuous.

EXERCISE B.1.10. Let $u : \mathbb{R}^n \setminus \{0\}$ be a smooth function which vanishes on the complement of the unit ball. Suppose that u and the first derivatives $\partial_i u$ all have finite L^p-norm. Prove that $u \in W^{1,p}(\mathbb{R}^n)$. *Hint:* The function u and its first derivatives need not be bounded near zero. You must show that the weak derivatives agree with the strong derivatives. To see this, let $\phi \in C_0^\infty(\mathbb{R}^n)$ be a test function and consider the integral of $(\partial_i\phi)u$ over $\mathbb{R}^n \setminus B_\varepsilon$. Use Theorem B.1.11 or Theorem B.1.12 below to show that the boundary term tends to zero as $\varepsilon \to 0$.

Define the **Hölder norm**

$$\|u\|_{C^{0,\mu}} = \sup_{x,y\in\Omega} \frac{|u(x)-u(y)|}{|x-y|^\mu} + \sup_{x\in\Omega} |u(x)|$$

for $0 < \mu \le 1$ and

$$\|u\|_{C^{k,\mu}} = \sum_{|\nu|\le k} \|\partial^\nu u\|_{C^{0,\mu}}.$$

Denote by $C^{k,\mu}(\Omega)$ the space of all C^k-functions $u : \Omega \to \mathbb{R}$ with finite Hölder norm $\|u\|_{C^{k,\mu}}$.

THEOREM B.1.11. *Let $\Omega \subset \mathbb{R}^n$ be a bounded Lipschitz domain and suppose that $kp > n$ and $0 < \mu := k - n/p < 1$. Then there exists a constant $c = c(k,p,\Omega) > 0$ such that*

$$\|u\|_{C^{0,\mu}} \le c\,\|u\|_{W^{k,p}}$$

for $u \in C^\infty(\overline{\Omega})$. The inclusion $W^{k,p}(\Omega) \hookrightarrow C^0(\Omega)$ is compact.

THEOREM B.1.12. *Let $\Omega \subset \mathbb{R}^n$ be a bounded Lipschitz domain and suppose that $kp < n$. Then there exists a constant $c = c(k,p,\Omega) > 0$ such that*

$$\|u\|_{L^{np/(n-kp)}} \le c\,\|u\|_{W^{k,p}}$$

for $u \in C^\infty(\overline{\Omega})$. If $q < np/(n-kp)$ then the inclusion $W^{k,p}(\Omega) \hookrightarrow L^q(\Omega)$ is compact.

These are the **Sobolev estimates**. The compactness statement in Theorem B.1.12 is known as **Rellich's theorem**. In particular, Theorem B.1.11 shows that if $\Omega \subset \mathbb{R}^n$ is a bounded Lipschitz domain then

$$C^\infty(\overline{\Omega}) = \bigcap_{k=1}^{\infty} W^{k,p}(\Omega)$$

for $1 \le p \le \infty$. The case $kp = n$ is the borderline situation for these estimates. In this case the space $W^{k,p}$ does not embed into the space of continuous functions. However, applying Theorem B.1.12 with $p' < p$, one sees that it embeds into L^q for $1 \le q < \infty$. Of particular interest here is the case where $n = p = 2$ and $k = 1$.

REMARK B.1.13. In Lemma 10.4.1 we have seen that there exists a sequence of functions $u_j \in W^{1,2}(B_1)$ on the unit disc $B_1 = \{(x,y) \in \mathbb{R}^2 \,|\, x^2 + y^2 < 1\}$ such that

$$u_j(0) = 1, \qquad \lim_{j\to\infty} \|u_j\|_{W^{1,2}} = 0.$$

In two dimensions the relation between $W^{1,2}$ and C^0 is rather subtle. For example the function $f(e^{i\theta}) = \theta$, $0 \le \theta < 2\pi$, (with a single jump discontinuity) does not extend to a $W^{1,2}$-function on B_1. On the other hand there exist discontinuous functions on S^1 which do extend to $W^{1,2}$-functions on B_1. To prove the first assertion, express a function $f : S^1 \to \mathbb{R}$ as a Fourier series

$$f(e^{i\theta}) = \sum_{k\in\mathbb{Z}} a_k e^{ik\theta},$$

where $a_{-k} = \bar{a}_k$. Then f extends to a $W^{1,2}$-function $u \in W^{1,2}(B_1)$ if and only if

$$\sum_{k\in\mathbb{Z}} k|a_k|^2 < \infty.$$

The function $u(e^{i\theta}) = \theta$ does not satisfy this condition. To prove the second assertion, use the argument in the proof of Lemma 10.4.1 to construct an unbounded function $u \in W^{1,2}(\mathbb{R}^2)$.

EXERCISE B.1.14. Show that the inclusion $W^{1,2}(\mathbb{R}^n) \hookrightarrow L^2(\mathbb{R}^n)$ is not compact.

EXERCISE B.1.15. This exercise shows that the assumption of a Lipschitz domain in Theorem B.1.12 cannot be removed. Consider the bounded open set $\Omega \subset \mathbb{R}^2$ defined by

$$\begin{aligned}\Omega \quad := \quad & \left\{(x,y) \in \mathbb{R}^2 \,\middle|\, 0 < x < 1,\, 0 < y < \frac{1}{2}\right\} \\ & \cup \bigcup_{m=0}^{\infty} \left\{(x,y) \in \mathbb{R}^2 \,\middle|\, \frac{1}{2^{2m+1}} < x < \frac{1}{2^{2m}},\, \frac{1}{2} \le y < 1\right\}.\end{aligned}$$

Show that the embedding $W^{1,2}(\Omega) \hookrightarrow L^2(\Omega)$ is not compact. Find a smooth function $u \in W^{1,2}(\Omega)$ such that $u \notin L^q(\Omega)$ for any $q > 2$.

The assertions of Theorems B.1.11 and B.1.12 for $k \ge 2$ follow easily from the case $k = 1$. Moreover, in view of Proposition B.1.9, it suffices to prove these results for $W_0^{1,p}(\Omega)$.

LEMMA B.1.16. *Every $u \in C_0^\infty(\mathbb{R}^n)$ satisfies the estimates*

$$\sup_{x\neq y} \frac{|u(x) - u(y)|}{|x - y|^\mu} \le c \|\nabla u\|_{L^p}, \qquad \sup |u| \le c\left(\|u\|_{L^p} + \|\nabla u\|_{L^p}\right)$$

for $p > n$, where $\mu := 1 - n/p$ and $c := 2^{n+1}{\omega_n}^{-1/p}\left((p-1)/(p-n)\right)^{1-1/p}$. Here ω_n denotes the area of the unit sphere $S^{n-1} \subset \mathbb{R}^n$.

PROOF. First suppose that $B \subset \mathbb{R}^n$ is a bounded convex set with nonempty interior and $u : B \to \mathbb{R}$ is a smooth function with mean value zero. Then u satisfies the inequality

$$|u(x)| \le \frac{d^n \omega_n}{nV{\omega_n}^{1/p}} \left(\frac{p-1}{p-n}\right)^{1-1/p} d^{1-n/p} \|\nabla u\|_{L^p(B)}, \tag{B.1.3}$$

where

$$d := \mathrm{diam}(B), \qquad V := \mathrm{Vol}(B).$$

To see this note first that, since $\int_B u = 0$,

$$u(x) = \frac{1}{V} \int_B \int_0^1 \langle \nabla u(x + t(y - x)), x - y \rangle \, dt dy.$$

Now extend the function $\nabla u : B \to \mathbb{R}^n$ to all of $\mathbb{R}^n$ by declaring $\nabla u(x) := 0$ for $x \notin B$. Then

$$\begin{aligned} V|u(x)| &\le \int_{|y|\le d}\int_0^1 |\nabla u(x+ty)|\,|y|\,dtdy \\ &= \int_0^d r^{n-1}\left(\int_{|\eta|=1}\int_0^1 |\nabla u(x+tr\eta)|\,r\,dt\,dS(\eta)\right) dr \\ &= \int_0^d r^{n-1}\left(\int_{|y|\le r} |y|^{1-n}\,|\nabla u(x+y)|\,dy\right) dr \\ &\le \frac{d^n}{n}\int_B |y-x|^{1-n}\,|\nabla u(y)|\,dy \\ &\le \frac{d^n}{n}\left(\int_{|y|\le d} |y|^{q-nq}\,dy\right)^{1/q} \|\nabla u\|_{L^p(B)}\,. \end{aligned}$$

In the second equality $S(\eta)$ denotes the volume form on an $n-1$-sphere of radius η. The third equality is obtained by substituting $dy = |y|^{n-1}r\,dt\,dS(\eta)$. The last step follows from the Hölder inequality with $1/p + 1/q = 1$. The integral can be easily computed and one obtains (B.1.3). Now apply (B.1.3) to the case

$$B = B_r(x_0), \qquad x_0 := \frac{x+y}{2}, \qquad r := \frac{|x-y|}{2}.$$

Then $d = |x-y|$ and $d^n\omega_n/nV = 2^n$. Hence, with

$$u_B = \frac{1}{V}\int_B u,$$

one obtains

$$\begin{aligned} |u(x)-u(y)| &\le |u(x)-u_B| + |u_B - u(y)| \\ &\le \frac{2^{n+1}}{{\omega_n}^{1/p}}\left(\frac{p-1}{p-n}\right)^{1-1/p} |x-y|^{1-n/p}\,\|\nabla u\|_{L^p(B)} \end{aligned}$$

for every smooth function $u : \mathbb{R}^n \to \mathbb{R}$. This proves the first assertion of the lemma. The second inequality follows easily from (B.1.3) with $B = B_1(x)$ and u replaced by $u - u_B$. □

LEMMA B.1.17. *Assume $p < n$. Then every $u \in C_0^\infty(\mathbb{R}^n)$ satisfies the estimate*

$$\|u\|_{L^{np/(n-p)}} \le \frac{p}{\sqrt{n}}\frac{n-1}{n-p}\,\|\nabla u\|_{L^p}\,.$$

PROOF. The following beautiful argument is due to Nirenberg. The identity

$$u(x) = \int_{-\infty}^{x_i} \partial_i u(x_1,\dots,x_{i-1},t,x_{i+1},\dots,x_n)\,dt$$

shows that

$$|u(x)|^{n/(n-1)} \le \prod_{i=1}^n \left(\int_{-\infty}^{\infty} |\partial_i u(x)| dx_i\right)^{1/(n-1)}.$$

Note that although the product on the right has n terms, the ith factor is independent of x_i. Hence when integrating with respect to x_i, only $n-1$ factors matter. Now integrate over the variables $x_1, \dots, x_n$ in turn, at each step using the *generalized Hölder inequality*

$$\|v_1 \cdots v_m\|_{L^1} \le \|v_1\|_{L^m} \cdots \|v_m\|_{L^m}$$

with $m = n-1$. In the kth step this gives

$$\begin{aligned}&\int |u|^{n/(n-1)}\,dx_1 \cdots dx_k \\ &\le \prod_{i=1}^k \left(\int |\partial_i u|\,dx_1 \cdots dx_k\right)^{1/(n-1)} \prod_{i=k+1}^n \left(\int |\partial_i u|\,dx_1 \cdots dx_k dx_i\right)^{1/(n-1)}.\end{aligned}$$

Here the $(k+1)$st factor does not depend on x_{k+1}. Now integrate over x_{k+1} to obtain the same inequality with k replaced by $k+1$. With $k = n$ this shows that

$$\|u\|_{L^{n/(n-1)}} \le \prod_{i=1}^n \left(\int |\partial_i u|\right)^{1/n} \le \frac{1}{n}\sum_{i=1}^n \int |\partial_i u| \le \frac{1}{\sqrt{n}}\int |\nabla u|.$$

(The last inequality holds because $|a| = \sqrt{\sum_i |a_i|^2}$ denotes the Euclidean norm of the vector a.) This proves the lemma for $p = 1$.

To prove it in general, consider the $L^{n/(n-1)}$-norm of the function

$$v := |u|^\alpha, \qquad \alpha := p\frac{n-1}{n-p}.$$

Since

$$|\nabla v| = \alpha |u|^{\alpha-1}|\nabla u|, \qquad \frac{\alpha n}{n-1} = \frac{np}{n-p},$$

one obtains

$$\begin{aligned}\left(\int |u|^{np/(n-p)}\right)^{1-1/n} &\le \frac{\alpha}{\sqrt{n}}\int |u|^{\alpha-1}|\nabla u| \\ &\le \frac{\alpha}{\sqrt{n}}\left(\int |u|^{\alpha q - q}\right)^{1/q}\left(\int |\nabla u|^p\right)^{1/p} \\ &\le \frac{\alpha}{\sqrt{n}}\left(\int |u|^{np/(n-p)}\right)^{1-1/p}\left(\int |\nabla u|^p\right)^{1/p}.\end{aligned}$$

The second estimate is Hölder's inequality with $1/p + 1/q = 1$ and the last estimate uses the identity $\alpha q - q = np/(n-p)$. This proves the lemma in the general case. □

PROOFS OF THEOREMS B.1.11 AND B.1.12. By Lemma B.1.16, there is an inclusion $W_0^{1,p}(\Omega) \hookrightarrow C^{0,1-n/p}(\Omega)$ for $p > n$ and hence, by the Arzela-Ascoli theorem, the inclusion $W_0^{1,p}(\Omega) \hookrightarrow C^0(\overline{\Omega})$ is compact whenever Ω is bounded. Similarly, by Lemma B.1.17, there is an inclusion $W_0^{1,p}(\Omega) \hookrightarrow L^q(\Omega)$ for $p < n$ and $q = np/(n-p)$. That this inclusion is compact for bounded domains Ω and $q < np/(n-p)$ requires a separate argument. The inequality

$$\|u\|_{L^q} \le \|u\|_{L^1}^\lambda \|u\|_{L^{np/(n-p)}}^{1-\lambda}, \qquad \frac{1}{q} = \lambda + \frac{1-\lambda}{np/(n-p)},$$

for $q < np/(n-p)$ shows that it suffices to prove that the inclusion

$$\iota : W_0^{1,p}(\Omega) \to L^1(\Omega)$$

is compact for bounded domains. To see this denote by $S_\delta : L^1(\Omega) \to L^1(\Omega)$ the smoothing operator

$$S_\delta f := \rho_\delta * f.$$

By the Arzela–Ascoli theorem, S_δ is compact. Namely, if u_i is a bounded sequence in $L^1(\Omega)$ then the sequence $S_\delta u_i \in C^0(\overline{\Omega})$ is bounded and equicontinuous and so has a subsequence which converges in $C^0(\overline{\Omega})$ and hence in $L^1(\Omega)$. It follows that the composition $S_\delta \circ \iota : W_0^{1,p}(\Omega) \to L^1(\Omega)$ is compact. Moreover, integrating the inequality

$$\begin{aligned}|u(x) - u_\delta(x)| &= \left| \int_{|y|\le 1} \rho(y) \int_0^\delta \langle \nabla u(x - ty), y \rangle \, dtdy \right| \\ &\le \int_{|y|\le 1} \rho(y) \int_0^\delta |\nabla u(x - ty)| \, dtdy\end{aligned}$$

one finds

$$\|u - S_\delta u\|_{L^1} \le \delta \|\nabla u\|_{L^1} \le \delta \mathrm{Vol}(\Omega)^{1-1/p} \|\nabla u\|_{L^p}$$

for $u \in W_0^{1,p}(\Omega)$. This shows that the operators

$$S_\delta \circ \iota : W_0^{1,p}(\Omega) \to L^1(\Omega)$$

converge to ι in the uniform operator topology as $\delta \to 0$ and hence the limit operator ι is compact. This proves Theorems B.1.11 and B.1.12 with $W^{k,p}(\Omega)$ replaced by $W_0^{1,p}(\Omega)$ (and without any condition on the regularity of the boundary of Ω). To prove the results for $k = 1$ one simply combines the corresponding embedding theorems for $W_0^{1,p}$ with the extension theorem (Proposition B.1.9). The reduction of the general case to $k = 1$ is left as an exercise. □

Interpolation. To gain an intuitive understanding of Sobolev spaces it is often useful to think of a $W^{k,q}$-function as having $k - n/q$ continuous derivatives. Then the Sobolev embedding theorem B.1.12 can be phrased in the form that there is a continuous inclusion $W^{k,q} \hookrightarrow W^{j,p}$ whenever $W^{k,q}$-functions have more derivatives than $W^{j,p}$-functions, i.e. $j \le k$ and $j - n/p \le k - n/q$. Care must be taken in the borderline case $k - n/q = 0$. A proof of the following interpolation inequality can be found, for example, in [**124**].

Proposition B.1.18 (Gagliardo–Nirenberg). *Let $\Omega \subset \mathbb{R}^n$ be a bounded open domain with C^k boundary. Suppose that $j, k \ge 0$ are integers with $j < k$ and $1 \le p, q, r \le \infty$ with $k - n/q + n/r \ge 0$ and*

$$j - \frac{n}{p} = \lambda \left(k - \frac{n}{q} \right) + (1 - \lambda) \left(-\frac{n}{r} \right), \qquad \frac{j}{k} \le \lambda \le 1.$$

If $(k - j)q = n$ assume also that $\lambda \ne 1$. Then there exists a constant $c > 0$ such that

$$\|u\|_{W^{j,p}} \le c \|u\|_{W^{k,q}}^{\lambda} \|u\|_{L^r}^{1-\lambda}$$

for $u \in W^{k,q}(\Omega)$.

Product estimates. The case $kp > n$ should be viewed as the *good case* where everything works, for example composition with a smooth function and products.

PROPOSITION B.1.19. *Let $\Omega \subset \mathbb{R}^n$ be a bounded open domain with C^k boundary. Assume $kp > n$ and $j \in \{1, \ldots, k\}$ and $j - n/q \leq k - n/p$. Then there is a constant $c > 0$ such that*

$$\|uv\|_{W^{j,q}} \leq c \|u\|_{W^{j,q}} \|v\|_{W^{k,p}},$$

$$\|f \circ u\|_{W^{k,p}} \leq c \|f\|_{C^{k-1,1}} \left(\|u\|_{L^\infty}^{k-1} + 1 \right) \left(\|u\|_{W^{k,p}} + 1 \right)$$

for $u, v \in C^\infty(\Omega)$ and $f \in \mathbb{C}^{k-1,1}(\mathbb{R})$.

The proofs of Proposition B.1.19 and of the next result are straightforward exercises. They make use of Hölder's inequality and of the general interpolation inequality of Gagliardo-Nirenberg in Proposition B.1.18.

PROPOSITION B.1.20. *Let $\Omega \subset \mathbb{R}^n$ be a bounded open domain with C^k boundary. Assume $kp > n$ and $f \in C^{\ell-1,1}(\mathbb{R})$ with $\ell \geq k$. Then the map*

$$W^{k,p}(\Omega) \to W^{k,p}(\Omega) : u \mapsto f \circ u$$

is a $C^{\ell-k}$-map of Banach spaces.

Trace theorems. It is somewhat nontrivial to understand the restriction of functions with weak derivatives to lower dimensional submanifolds. For example the obvious fact that the restriction of a C^k-function to a hyperplane is also of class C^k has no analogue in the realm of Sobolev spaces. A $W^{k,p}$-function *loses derivatives* when restricted to the boundary.

PROPOSITION B.1.21. *Let $1 < p < \infty$ and $\Omega \subset \mathbb{R}^n$ be a bounded open set with smooth boundary. Then there is a $c > 0$ such that every $u \in C^\infty(\overline{\Omega})$ satisfies*

$$\|u\|_{L^p(\partial\Omega)} \leq c \|u\|_{L^p(\Omega)}^{1-1/p} \|u\|_{W^{1,p}(\Omega)}^{1/p}.$$

PROOF. For $x \in \partial\Omega$ let $\nu(x)$ denote the outward unit normal vector. Choose a smooth function $f : \Omega \to \mathbb{R}^n$ such that $f(x) = \nu(x)$ for $x \in \partial\Omega$. Then

$$\begin{aligned}
\int_{\partial\Omega} |u|^p &= \int_\Omega \operatorname{div}(f|u|^p) \\
&= \int_\Omega \left((\operatorname{div} f)|u|^p + p\langle f, \nabla u\rangle u|u|^{p-2} \right) \\
&\leq c_1 \int_\Omega |u|^{p-1} \left(|u| + |\nabla u| \right) \\
&\leq c_1 \left(\int_\Omega |u|^p \right)^{(p-1)/p} \left(\int_\Omega (|u| + |\nabla u|)^p \right)^{1/p} \\
&\leq c_2 \|u\|_{L^p}^{p-1} \|u\|_{W^{1,p}}.
\end{aligned}$$

This proves Proposition B.1.21. □

PROPOSITION B.1.22. *Let p, Ω be as in Proposition B.1.21 and $u \in W^{k,p}(\Omega)$. Then $u \in W_0^{k,p}(\Omega)$ if and only if $\partial^\alpha u$ vanishes on $\partial\Omega$ for $|\alpha| \leq k - 1$.*

PROOF. If $u \in W_0^{k,p}(\Omega)$ then Proposition B.1.21, applied to $\partial^\alpha(u - u_i)$ for a sequence in $C_0^\infty(\Omega)$ with limit u, shows that $\partial^\alpha u$ vanishes on $\partial\Omega$ for $|\alpha| \le k-1$. To prove the converse it suffices to consider the case $k = 1$. Assume that $u \in W^{1,p}(\Omega)$ vanishes on $\partial\Omega$ and extend u to all of $\mathbb{R}^n$ by setting $u(x) := 0$ for $x \in \mathbb{R}^n \setminus \Omega$. Then the extended function is in $W^{1,p}(\mathbb{R}^n)$. To see this approximate u on $\overline{\Omega}$ by a sequence of smooth functions $u_j : \overline{\Omega} \to \mathbb{R}$, using Proposition B.1.4. Then it follows from Proposition B.1.21 that $u_j|_{\partial\Omega}$ converges to zero in $L^p(\partial\Omega)$. Hence it follows from the divergence theorem that

$$\int_\Omega \left(u(\partial_i\phi) + (\partial_i u)\phi\right) = \lim_{j\to\infty}\int_\Omega \left(u_j(\partial_i\phi) + (\partial_i u_j)\phi\right) = \lim_{j\to\infty}\int_{\partial\Omega} \nu_i u_j \phi = 0$$

for every test function $\phi \in C^\infty(\overline{\Omega})$ (and not just for $\phi \in C_0^\infty(\Omega)$). This proves that the extended function u belongs to $W^{1,p}(\mathbb{R}^n)$. Since u vanishes outside of Ω, one can now approximate u by a sequence in $W^{1,p}(\Omega)$ which vanishes near $\partial\Omega$ and hence belongs to $W_0^{1,p}(\Omega)$. (Exercise!) This proves Proposition B.1.22. □

Sections of vector bundles.

REMARK B.1.23. Let M be an n-dimensional smooth compact manifold and $\pi : E \to M$ be a smooth vector bundle. A section $s : M \to E$ is said to be **of class** $W^{k,p}$ if all its local coordinate representations are in $W^{k,p}$. This definition is independent of the choice of the coordinates. To see this note that if $\phi \in \mathrm{Diff}(\mathbb{R}^n)$, $\Phi \in C^\infty(\mathbb{R}^n, \mathbb{R}^{N\times N})$ is a matrix-valued function, and $u \in W^{k,p}_{\mathrm{loc}}(\mathbb{R}^n, \mathbb{R}^N)$, then the composite $\Phi(u \circ \phi)$ belongs to $W^{k,p}(\Omega)$ for every bounded open set $\Omega \subset \mathbb{R}^n$ and there is an estimate

$$\|\Phi(u\circ\phi)\|_{W^{k,p}(\Omega)} \le c\,\|u\|_{W^{k,p}(\Omega)}$$

with $c = c(\phi, \Phi, \Omega)$ independent of u (see Exercises B.1.2 and B.1.5). This holds even when $kp \le n$. To define a norm on the space of $W^{k,p}$-sections one can take the sum of the $W^{k,p}$-norms over finitely many charts which cover M.

REMARK B.1.24. For functions u with *values* in a manifold the situation is quite different since one must consider composites of the form $\phi \circ u$, where ϕ is a coordinate change. Proposition B.1.19 shows that for such functions it is required that $kp > n$. More precisely, let X and M be smooth closed manifolds and suppose that $kp > n := \dim X$. Then the space

$$\mathcal{X}^{k,p} := W^{k,p}(X, M)$$

can be defined as the space of continuous functions $u : X \to M$ which are in local coordinate charts represented by $W^{k,p}$-functions. It follows as in Remark B.1.23 that this definition is independent of the choice of the coordinates. Alternatively, choose an embedding $M \hookrightarrow \mathbb{R}^N$ and define $\mathcal{X}^{k,p}$ as the subset of the space $W^{k,p}(X, \mathbb{R}^N)$ which consists of functions with values in M. We leave it to the reader to check that both definitions of $\mathcal{X}^{k,p}$ are equivalent. The second definition also applies to the case $kp \le n$: the values of $u \in W^{k,p}(X, \mathbb{R}^N)$ are required to be in M almost everywhere. However, in this case the space $\mathcal{X}^{k,p}$ depends on the choice of the embedding.

B.2. The Calderon–Zygmund inequality

The Laplace operator. Denote by

$$\Delta = \frac{\partial^2}{\partial x_1{}^2} + \frac{\partial^2}{\partial x_2{}^2} + \cdots + \frac{\partial^2}{\partial x_n{}^2}$$

the Laplace operator on $\mathbb{R}^n$. A C^2-function $u : \Omega \to \mathbb{R}$ on an open set $\Omega \subset \mathbb{R}^n$ is called **harmonic** if $\Delta u = 0$. Harmonic functions are real analytic. (If $n = 2$ then a function is harmonic iff it is locally the real part of a holomorphic function.) Harmonic functions are characterized by the **mean value property** (see John [**203**])

$$u(x) = \frac{n}{\omega_n r^2} \int_{B_r(x)} u(\xi)\, d\xi, \qquad B_r(x) \subset \Omega.$$

Here $\omega_n = 2\pi^{n/2}\Gamma(n/2)^{-1}$ is the volume of the unit sphere in $\mathbb{R}^n$. In particular, $\omega_2 = 2\pi$.

EXERCISE B.2.1. Prove that harmonic functions have the mean value property. (Cf. the proof of Step 1 of Lemma 4.3.3.)

The **fundamental solution** of Laplace's equation is the function

$$K(x) := \begin{cases} (2\pi)^{-1}\log|x|, & n = 2, \\ (2-n)^{-1}\omega_n^{-1}|x|^{2-n}, & n \geq 3. \end{cases} \tag{B.2.1}$$

Its first and second derivatives are given by

$$K_j(x) = \frac{x_j}{\omega_n |x|^n}, \qquad K_{jk}(x) = \frac{-n x_j x_k}{\omega_n |x|^{n+2}}, \qquad K_{jj}(x) = \frac{|x|^2 - n x_j^2}{\omega_n |x|^{n+2}}$$

where $K_j = \partial K/\partial x_j$ and $K_{jk} = \partial^2 K/\partial x_j \partial x_k$. In particular, $\Delta K = 0$. The function K and its first derivatives K_j are integrable on compact sets while the second derivatives are not. Hence $\partial_j(K * f) = K_j * f$ for compactly supported functions $f : \mathbb{R}^n \to \mathbb{R}$ but there is no such formula for the second derivatives. Moreover, since neither K nor its derivatives are integrable on $\mathbb{R}^n$, care must be taken with functions f which do not have compact support.

Every compactly supported C^2-function $u : \mathbb{R}^n \to \mathbb{R}$ satisfies

$$u = K * \Delta u, \qquad \partial_j u = K_j * \Delta u,$$

where $*$ denotes convolution. Conversely,

$$\Delta(K * f) = f, \qquad \Delta(K_j * f) = \partial_j f$$

for $f \in C_0^\infty(\mathbb{R}^n)$ (see [**203**]). This is **Poisson's identity**. In general $K * f$ will not have compact support. Since the second derivatives of K are not integrable on compact sets there exists a continuous function $f : \mathbb{R}^n \to \mathbb{R}$ such that $K * f \notin C^2$. For such a function f there is no classical solution of $\Delta u = f$. (This follows from the next two lemmas.) The situation is however quite different for weak solutions. Let $f \in L^1_{\mathrm{loc}}(\Omega)$. A function $u \in L^1_{\mathrm{loc}}(\Omega)$ is called a **weak solution** of $\Delta u = f$ if

$$\int_\Omega u(x)\Delta\phi(x)\, dx = \int_\Omega f(x)\phi(x)\, dx$$

for $\phi \in C_0^\infty(\Omega)$. Similarly $u \in L^1_{\mathrm{loc}}(\Omega)$ is called a weak solution of $\Delta u = \partial_j f$ with $f \in L^1_{\mathrm{loc}}$ if

$$\int_\Omega u(x)\Delta\phi(x)\,dx = -\int_\Omega f(x)\partial_j\phi(x)\,dx$$

for $\phi \in C_0^\infty(\Omega)$.

LEMMA B.2.2. *Let $u, f \in L^1(\mathbb{R}^n)$ with compact support.*

(i) *u is a weak solution of $\Delta u = f$ if and only if $u = K * f$.*

(ii) *u is a weak solution of $\Delta u = \partial_j f$ if and only if $u = K_j * f$.*

PROOF. If $u = K * f$. Then

$$\int u\Delta\phi = \int (K * f)\Delta\phi = \int f(K * \Delta\phi) = \int f\phi$$

for $\phi \in C_0^\infty(\mathbb{R}^n)$. Conversely, suppose that u is a weak solution of $\Delta u = f$. Choose $\rho_\delta : \mathbb{R}^n \to \mathbb{R}$ as in the proof of Proposition B.1.4. Then

$$\int (\Delta\rho_\delta * u)\phi = \int u(\Delta\rho_\delta * \phi) = \int f(\rho_\delta * \phi) = \int (\rho_\delta * f)\phi.$$

for every $\phi \in C_0^\infty(\mathbb{R}^n)$ and hence $\Delta(\rho_\delta * u) = (\Delta\rho_\delta) * u = \rho_\delta * f$. This implies that $\rho_\delta * u - K * \rho_\delta * f$ is a bounded harmonic function converging to zero at ∞ and hence $\rho_\delta * u = K * \rho_\delta * f$. Take the limit $\delta \to 0$ to obtain $u = K * f$. This proves (i). The proof of (ii) is similar and is left to the reader. □

LEMMA B.2.3 (Weyl's lemma). *Every weak solution $u \in L^1_{\mathrm{loc}}(\Omega)$ of $\Delta u = 0$ is harmonic.*

PROOF. Let ρ_δ be as in the proof of Lemma B.2.2. The function $u_\delta = \rho_\delta * u$ is harmonic in $\Omega_\delta = \{x \in \Omega \mid B_\delta(x) \subset \Omega\}$. Hence u_δ satisfies the mean value property. Since u_δ converges to u in the L^1-norm on every compact subset of Ω it follows that u has the mean value property. Hence u is harmonic (cf. [**203**]). □

The above results about weak solutions can be restated in the language of distributions as follows. A **distribution** E on Ω is an element of the dual to the space $C_0^\infty(\Omega)$ of test functions. Thus a function u in $L^1_{\mathrm{loc}}(\Omega)$ defines a distribution E_u via the formula

$$E_u(\phi) := \int_\Omega u\phi$$

for $\phi \in C_0^\infty(\Omega)$. Every linear differential operator $L : C_0^\infty(\Omega) \to C_0^\infty(\Omega)$ has a formal adjoint L^* that acts on distributions in the obvious way: $L^*E(\phi) := E(L\phi)$. The identity

$$\int_\Omega u\,(\partial_j\phi) = -\int_\Omega (\partial_j u)\,\phi$$

for $u \in C^\infty(\Omega)$ and $\phi \in C_0^\infty(\Omega)$ shows that the restriction of ∂_j^* to $C_0^\infty(\Omega)$ is equal to $-\partial_j$. It follows that the restriction of L^* to $C_0^\infty(\Omega)$ is again a differential operator and its dual L^{**} agrees with L, when restricted to $C_0^\infty(\Omega)$. Thus a differential operator L on $C_0^\infty(\Omega)$ extends naturally to the space of distributions, and this extension will still be denoted by L. Note that the Laplace operator is symmetric, i.e. the restriction of Δ^* to $C_0^\infty(\Omega)$ agrees with Δ.

Now consider the case $\Omega = \mathbb{R}^n$. The **convolution** of a distribution E on $\mathbb{R}^n$ with a test function $\phi \in C_0^\infty(\mathbb{R}^n)$ is defined by

$$(E * \phi)(x) := E(\phi(x - \cdot)).$$

Note that $E * \phi$ is a smooth function but need not have compact support. The delta function δ acts as the identity operator. Now for any differential operator L with constant coefficients and any two test functions $\phi, \psi \in C_0^\infty(\mathbb{R}^n)$ we have

$$L(\phi * \psi) = (L\phi) * \psi = \phi * (L\psi),$$

Combining this with the identities $\langle E * \psi, \bar{\phi}\rangle = E(\psi * \phi)$ and $L^*\bar{\phi} = \overline{L\phi}$, where $\bar{\phi}(x) := \phi(-x)$, we find

$$L(E * \psi) = (LE) * \psi = E * (L\psi) \tag{B.2.2}$$

for every distribution E and every $\psi \in C_0^\infty(\mathbb{R}^n)$.

In this language, a fundamental solution $K \in L^1_{\mathrm{loc}}(\mathbb{R}^n)$ of the equation $Lu = 0$ is a solution of the distributional equation

$$LK = \delta.$$

Thus the identities $\phi = K * \Delta\phi$ and $\Delta(K * \phi) = \phi$ for the symmetric operator $L = \Delta$ hold for test functions by equation (B.2.2). Moreover, a weak solution $u \in L^1_{\mathrm{loc}}(\Omega)$ of $Lu = 0$ is simply a function whose corresponding distribution E_u solves the equation $LE_u = 0$.

An interpolation lemma. For any measurable function $f : \mathbb{R}^n \to \mathbb{R}$ define

$$\mu(t, f) = |\{x \in \mathbb{R}^n \,|\, |f(x)| > t\}|$$

for $t > 0$ where $|A|$ denotes the Lebesgue measure of the set A.

LEMMA B.2.4. *For $1 \le p < \infty$ and $f \in L^p(\mathbb{R}^n)$*

$$t^p \mu(t, f) \le \int |f(x)|^p \, dx = p \int_0^\infty s^{p-1} \mu(s, f) \, ds.$$

Moreover,

$$\mu(t, f + g) \le \mu(t/2, f) + \mu(t/2, g).$$

PROOF. Integrate the function $F : \mathbb{R}^{n+1} \to \mathbb{R}$ defined by $F(x, t) := pt^{p-1}$ for $0 \le t \le |f(x)|$ and $F(x, t) := 0$ otherwise. □

LEMMA B.2.5 (Marcinkiewicz). *Let $T : L^2(\mathbb{R}^n) \to L^2(\mathbb{R}^n)$ be a bounded linear operator and suppose that there exists a constant $C > 0$ such that*

$$\mu(t, Tf) \le \frac{C \, \|f\|_{L^1}}{t}$$

for every $t > 0$ and every $f \in L^2(\mathbb{R}^n) \cap L^1(\mathbb{R}^n)$. Then, for every real number p such that $1 < p < 2$, there exists a constant $c = c(\|T\|, C, p) > 0$ such that

$$\|Tf\|_{L^p} \le c \, \|f\|_{L^p}$$

for every $f \in L^2(\mathbb{R}^n) \cap L^1(\mathbb{R}^n)$. In particular, the restriction of T to $L^2(\mathbb{R}^n) \cap L^1(\mathbb{R}^n)$ extends (uniquely) to a bounded linear operator from $L^p(\mathbb{R}^n)$ to itself.

PROOF. Denote by $\|T\|$ the norm of T as an operator on $L^2(\mathbb{R}^n)$. Let $f \in L^2(\mathbb{R}^n) \cap L^1(\mathbb{R}^n)$ and consider the functions $f_t, g_t : \mathbb{R}^n \to \mathbb{R}$ defined by

$$f_t(x) := \begin{cases} f(x), & \text{if } |f(x)| > t, \\ 0, & \text{if } |f(x)| \le t, \end{cases} \qquad g_t(x) := \begin{cases} 0, & \text{if } |f(x)| > t, \\ f(x), & \text{if } |f(x)| \le t. \end{cases}$$

Then $f = f_t + g_t$ for every $t \ge 0$ and hence, by Lemma B.2.4,

$$\mu(t, Tf) \le \mu(t/2, Tf_t) + \mu(t/2, Tg_t) \le \frac{2C}{t} \|f_t\|_{L^1} + \frac{4\|T\|^2}{t^2} \|g_t\|_{L^2}^2 .$$

Hence, again by Lemma B.2.4,

$$\begin{aligned} \int_{\mathbb{R}^n} |Tf(x)|^p \, dx &= p \int_0^\infty t^{p-1} \mu(t, Tf) \, dt \\ &\le 2pC \int_0^\infty t^{p-2} \|f_t\|_{L^1} \, dt + 4p \|T\|^2 \int_0^\infty t^{p-3} \|g_t\|_{L^2}^2 \, dt \\ &= \left(\frac{2pC}{p-1} + \frac{4p\|T\|^2}{2-p} \right) \int_{\mathbb{R}^n} |f(x)|^p \, dx. \end{aligned}$$

The last equality follows from Fubini's theorem. This proves Lemma B.2.5. □

EXERCISE B.2.6. Prove the inequality

$$\|f\|_{L^p} \le \|f\|_{L^1}^{2/p-1} \|f\|_{L^2}^{2-2/p}$$

for $f \in C_0^\infty(\mathbb{R}^n)$ and $1 < p < 2$. Deduce that $L^2(\mathbb{R}^n) \cap L^1(\mathbb{R}^n) \subset L^p(\mathbb{R}^n)$. *Hint:* Use Hölder's inequality.

The Calderon–Zygmund inequality.

THEOREM B.2.7 (Calderon–Zygmund inequality). *Let K be the fundamental solution of Laplace's equation on $\mathbb{R}^n$, given by (B.2.1), and suppose that $1 < p < \infty$. Then there exists a constant $c = c(n,p) > 0$ such that*

$$\|\nabla(K_j * f)\|_{L^p} \le c \|f\|_{L^p} \tag{B.2.3}$$

for $f \in C_0^\infty(\mathbb{R}^n)$ and $j = 1, \ldots, n$.

COROLLARY B.2.8. *For every positive integer n and every $p > 1$ there exists a constant $c = c(n,p) > 0$ such that*

$$\sum_{j,k=1}^n \|\partial_j \partial_k u\|_{L^p} \le c \|\Delta u\|_{L^p}$$

for every $u \in C_0^\infty(\mathbb{R}^n)$.

PROOF. Theorem B.2.7 and Lemma B.2.2. □

Theorem B.2.7 is the fundamental estimate for the L^p-theory of elliptic operators. In the proof we follow [**140**]. The first step is the case $p = 2$.

LEMMA B.2.9. *The estimate (B.2.3) holds for $p = 2$ with $c = 1$.*

PROOF. Let $f : \mathbb{R}^n \to \mathbb{R}$ be a smooth function with compact support and define

$$u := K_j * f.$$

Then u need not have compact support. However, since $\partial_i u_j = K_j * \partial_i f$ and $K_j(x) = x_j/\omega_n|x|^n$, there is a constant $c > 0$ such that

$$|x| \geq c \qquad \Longrightarrow \qquad |u(x)| + |\nabla u(x)| \leq \frac{c}{|x|^{n-1}}.$$

Now consider the equation

$$\int_{B_R} |\nabla u|^2 = -\int_{B_R} u\Delta u + \int_{\partial B_R} u\frac{\partial u}{\partial \nu}.$$

By Poisson's identity, we have $\Delta u = \partial_j f$ and so the first term on the right is independent of R for large R. Moreover, the second term converges to zero as R tends to infinity. Hence $\nabla u \in L^2(\mathbb{R}^n, \mathbb{R}^n)$ and

$$\|\nabla u\|_{L^2}^2 = -\langle u, \Delta u\rangle = -\langle u, \partial_j f\rangle = \langle \partial_j u, f\rangle \leq \|\nabla u\|_{L^2} \|f\|_{L^2}.$$

Here all inner products are L^2-inner products. Now divide both sides of the inequality by $\|\nabla u\|_{L^2}$ to obtain the required estimate. □

Lemma B.2.9 shows that if $f \in C_0^\infty(\mathbb{R}^n)$ then the function $\partial_k(K_j * f)$ is in $L^2(\mathbb{R}^n)$ (it does not necessarily have compact support) and that $f \mapsto \partial_k(K_j * f)$ extends to a bounded linear operator (of norm one) from $L^2(\mathbb{R}^n)$ to itself. The next lemma shows that this operator satisfies the requirements of Lemma B.2.5.

LEMMA B.2.10. *There exists a constant $c = c(n) > 0$ such that every function $f \in L^2(\mathbb{R}^n) \cap L^1(\mathbb{R}^n)$ satisfies the following estimate for $j, k = 1, \ldots, n$:*

$$\mu(t, \partial_k(K_j * f)) \leq \frac{c}{t} \int |f(x)|\, dx.$$

PROOF. The proof has three steps. We abbreviate $Tf := \partial_k(\partial_j K * f)$.

STEP 1. *There exists a constant $c = c(n) > 0$ such that the following holds. Let B be a countable union of closed cubes $Q_i \subset \mathbb{R}^n$ with disjoint interiors. Suppose that $h \in L^1(\mathbb{R}^n)$ has support in B and satisfies*

$$\int_{Q_i} h = 0$$

for every i. Then

$$\mu(t, Th) \leq c\left(\mathrm{Vol}(B) + \frac{1}{t}\|h\|_{L^1}\right).$$

Denote by $h_i \in L^1(\mathbb{R}^n)$ the function which is equal to h on Q_i and equal to zero on $\mathbb{R}^n \setminus Q_i$. Let q_i be the center of Q_i and suppose that Q_i has sidelength $2r_i$. Then the maximal distance of any point in Q_i to q_i is $\sqrt{n}r_i$. Hence, for $x \notin Q_i$, we have

$$\begin{aligned}
|Th_i(x)| &= \left|\int_{Q_i} (\partial_k K_j(x-y) - \partial_k K_j(x-q_i))\, h_i(y)\, dy\right| \\
&\leq \max_{y \in Q_i} |\partial_k K_j(x-y) - \partial_k K_j(x-q_i)|\, \|h\|_{L^1(Q_i)} \\
&\leq \sqrt{n} r_i \max_{y \in Q_i} |\nabla \partial_k K_j(x-y)|\, \|h\|_{L^1(Q_i)} \\
&\leq c_1 r_i \max_{y \in Q_i} \frac{1}{|x-y|^{n+1}} \|h\|_{L^1(Q_i)} \\
&\leq \frac{c_1 r_i}{d(x, Q_i)^{n+1}} \|h\|_{L^1(Q_i)}.
\end{aligned}$$

(We denote by c_1, c_2, c_3 constants which depend only on n.) Let

$$P_i := \{x \in \mathbb{R}^n \,|\, |x - q_i| < 2\sqrt{n}r_i\} \supset Q_i.$$

Then $d(x, Q_i) \geq |x - q_i| - \sqrt{n}r_i$ for $x \in \mathbb{R}^n \setminus P_i$. Hence

$$\begin{aligned}
\int_{\mathbb{R}^n \setminus P_i} |Th_i| \, dx &\leq c_1 r_i \int_{|x| > 2\sqrt{n}r_i} \frac{dx}{(|x| - \sqrt{n}r_i)^{n+1}} \|h\|_{L^1(Q_i)} \\
&= c_1 r_i \int_{2\sqrt{n}r_i}^{\infty} \frac{\omega_n \rho^{n-1} \, d\rho}{(\rho - \sqrt{n}r_i)^{n+1}} \|h\|_{L^1(Q_i)} \\
&\leq c_1 \omega_n 2^{n-1} r_i \int_{\sqrt{n}r_i}^{\infty} \frac{d\rho}{\rho^2} \|h\|_{L^1(Q_i)} \\
&= c_2 \|h\|_{L^1(Q_i)} .
\end{aligned}$$

Hence, with $A := \bigcup_i P_i$ we obtain

$$\int_{\mathbb{R}^n \setminus A} |Th| \, dx \leq \sum_i \int_{\mathbb{R}^n \setminus P_i} |Th_i| \, dx \leq c_2 \sum_i \|h\|_{L^1(Q_i)} = c_2 \|h\|_{L^1} .$$

Since $\mathrm{Vol}(A) \leq \sum_i \mathrm{Vol}(P_i) = c_3 \sum_i \mathrm{Vol}(Q_i) = c_3 \mathrm{Vol}(B)$, it follows that

$$\begin{aligned}
t\mu(t, Th) &\leq t\mathrm{Vol}(A) + t \, |\{x \in \mathbb{R}^n \setminus A \,|\, |Th(x)| > t\}| \\
&\leq t\mathrm{Vol}(A) + \int_{\mathbb{R}^n \setminus A} |Th(x)| \, dx \\
&\leq c_4 \left(t\mathrm{Vol}(B) + \|h\|_{L^1}\right),
\end{aligned}$$

where $c_4 := \max\{c_2, c_3\}$. This proves Step 1.

STEP 2. *Let $f \in L^2(\mathbb{R}^n) \cap L^1(\mathbb{R}^n)$ and $t > 0$. Then there exists a countable collection of closed cubes $Q_i \subset \mathbb{R}^n$ with disjoint interiors satisfying the following.*

(i) *$t\mathrm{Vol}(Q_i) < \|f\|_{L^1(Q_i)} \leq 2^n t\mathrm{Vol}(Q_i)$ for every i.*

(ii) *$|f(x)| \leq t$ for almost every $x \in \mathbb{R}^n \setminus B$, where $B := \bigcup_i Q_i$.*

For $k \in \mathbb{Z}^n$ and $\ell \in \mathbb{Z}$ denote

$$Q(k, \ell) := \{x \in \mathbb{R}^n \,|\, 2^{-\ell} k_i \leq x_i \leq 2^{-\ell}(k_i + 1), \, i = 1, \ldots, n\}.$$

Let

$$\mathcal{Q} := \{Q(k, \ell) \,|\, k \in \mathbb{Z}^n, \, \ell \in \mathbb{Z}\}$$

and $\mathcal{Q}_0 \subset \mathcal{Q}$ be the set of all $Q \in \mathcal{Q}$ satisfying

$$t\mathrm{Vol}(Q) < \|f\|_{L^1(Q)}$$

and

$$Q \subsetneq Q' \in \mathcal{Q} \qquad \Longrightarrow \qquad \|f\|_{L^1(Q')} \leq t\mathrm{Vol}(Q').$$

Then every decreasing sequence of cubes in $\mathcal{Q}$ contains at most one element of $\mathcal{Q}_0$. Hence every $Q \in \mathcal{Q}_0$ satisfies assertion (i) and any two cubes in $\mathcal{Q}_0$ have disjoint interiors. Now let

$$B := \bigcup_{Q \in \mathcal{Q}_0} Q.$$

Then

$$x \in \mathbb{R}^n \setminus B, \quad x \in Q \in \mathcal{Q} \qquad \Longrightarrow \qquad \frac{1}{\mathrm{Vol}(Q)} \|f\|_{L^1(Q)} \leq t.$$

(Otherwise take a maximal cube $Q \in \mathcal{Q}$ that satisfies $t\mathrm{Vol}(Q) < \|f\|_{L^1(Q)}$ and contains x. This cube would belong to $\mathcal{Q}_0$ and so $x \in B$.) Thus we have proved that, for every $x \in \mathbb{R}^n \setminus B$, there is a sequence of decreasing cubes $Q_\ell \in \mathcal{Q}$ containing x such that $\mathrm{Vol}(Q_\ell)^{-1}\|f\|_{L^1(Q_\ell)} \le t$. Hence it follows from Lebesgue's differentiation theorem that $|f(x)| \le t$ for almost every $x \in \mathbb{R}^n \setminus B$. This proves Step 2.

STEP 3. *We prove the lemma.*

Fix a constant $t > 0$, let the Q_i be as in Step 2, and denote $B := \bigcup_i Q_i$. Then, by Step 2, $t\mathrm{Vol}(B) \le \|f\|_{L^1}$. Define $g, h : \mathbb{R}^n \to \mathbb{R}$ by

$$g(x) := \begin{cases} f(x), & \text{for } x \notin B, \\ \mathrm{Vol}(Q_i)^{-1} \int_{Q_i} f, & \text{for } x \in Q_i, \end{cases} \qquad h := f - g.$$

Then $\|g\|_{L^1} \le \|f\|_{L^1}$ and $\|h\|_{L^1} \le 2\|f\|_{L^1}$. Moreover, h vanishes in $\mathbb{R}^n \setminus B$ and has mean value zero in each cube Q_i. Hence h satisfies the requirements of Step 1. Hence there exists a constant c, depending only on n, such that

$$\mu(t, Th) \le c\left(\mathrm{Vol}(B) + \frac{1}{t}\|h\|_{L^1}\right) \le \frac{3c}{t}\|f\|_{L^1}.$$

Moreover, it follows from Step 2 that $|g(x)| \le 2^n t$ for almost every $x \in \mathbb{R}^n$. Hence, by Lemma B.2.4,

$$\mu(t, Tg) \le \frac{\|g\|_{L^2}^2}{t^2} \le \frac{2^n\|g\|_{L^1}}{t} \le \frac{2^n\|f\|_{L^1}}{t}.$$

Combining these inequalities we obtain from Lemma B.2.4 that

$$\mu(2t, Tf) \le \mu(t, Tg) + \mu(t, Th) \le \frac{2^{n+1} + 6c}{2t}\|f\|_{L^1}.$$

This proves Lemma B.2.10. □

PROOF OF THEOREM B.2.7. By Lemma B.2.9 and Lemma B.2.10, the linear operator $f \mapsto \partial_k(K_j * f)$ satisfies the requirements of Lemma B.2.5. This proves the theorem in the case $1 < p < 2$. For $2 < p < \infty$ we use duality. Let $1 < q < 2$ such that $1/p + 1/q = 1$. Then

$$\int g(x)\partial_k(K_j * f)(x)\,dx = \int \partial_k(K_j * g)(x)f(x) \le c\|f\|_{L^p}\|g\|_{L^q}$$

for $f, g \in C_0^\infty(\mathbb{R}^n)$. Hence $\|\partial_k(K_j * f)\|_{L^p} \le c\|f\|_{L^p}$ for every $f \in C_0^\infty(\mathbb{R}^n)$. □

B.3. Regularity for the Laplace operator

We are now in a position to establish the crucial regularity theorems and estimates for the Laplace operator.

THEOREM B.3.1 (Interior regularity). *Let $1 < p < \infty$, $k \ge 0$ be an integer, and $\Omega \subset \mathbb{R}^n$ be an open domain. If $u \in L^1_{\mathrm{loc}}(\Omega)$ is a weak solution of*

$$\Delta u = f, \qquad f \in W^{k,p}_{\mathrm{loc}}(\Omega),$$

then $u \in W^{k+2,p}_{\mathrm{loc}}(\Omega)$. Moreover, for every bounded open set $\Omega' \subset \mathbb{R}^n$ with $\overline{\Omega'} \subset \Omega$ there is a constant $c = c(k, p, n, \Omega', \Omega) > 0$ such that, for every $u \in C^\infty(\overline{\Omega})$,

$$\|u\|_{W^{k+2,p}(\Omega')} \le c\left(\|\Delta u\|_{W^{k,p}(\Omega)} + \|u\|_{L^p(\Omega)}\right).$$

PROOF. Let u, f and Ω' be as in the assertion of the theorem. Choose a bounded open neighbourhood U of the closure of Ω' such that the closure of U is contained in Ω. Let $\beta \in C_0^\infty(\Omega)$ be a smooth cutoff function such that $\beta(x) = 1$ for $x \in U$ and define

$$v := K * \beta f.$$

We prove that $v \in W^{k+2,p}(\Omega)$. To see this, choose a sequence of smooth functions $f_i \in C_0^\infty(\Omega)$ such that

$$\lim_{i\to\infty} \|f_i - \beta f\|_{W^{k,p}(\Omega)} = 0.$$

Then the function

$$v_i := K * f_i$$

is smooth for every i (every convolution of a locally integrable function with a compactly supported smooth function is smooth). Since K is locally integrable and f_i converges to βf in $L^p(\Omega)$, it follows from Young's inequality that

$$\lim_{i\to\infty} \|v_i - v\|_{L^p(\Omega)} = 0.$$

Moreover, it follows from Theorem B.2.7 that v_i is a Cauchy sequence in $W^{k+2,p}(\Omega)$. Hence the limit function v belongs to $W^{k+2,p}(\Omega)$. Now it follows from Lemma B.2.2 that v is a weak solution of the equation $\Delta v = \beta f$. Hence the restriction of $u - v$ to U is a weak solution of the equation $\Delta(u - v) = 0$. By Weyl's lemma, $u - v$ is real analytic in U. Hence $u \in W^{k+2,p}(\Omega')$. This proves the first assertion.

To prove the estimate note that, for every bounded open set $\Omega \subset \mathbb{R}^n$ and every integer $k \geq 0$, there exists a constant $c = c(k, n, p, \Omega)$ such that

$$\|u\|_{W^{k+2,p}(\Omega)} \leq c \|\Delta u\|_{W^{k,p}(\Omega)} \tag{B.3.1}$$

for every $u \in C_0^\infty(\Omega)$. For $k = 0$ this follows from Corollary B.2.8 and Lemma B.1.6. For $k \geq 1$ this follows easily by induction (see Exercise B.1.2). Since $C_0^\infty(\Omega)$ is dense in $W_0^{k+2,p}(\Omega)$, the inequality (B.3.1) continues to hold for every $u \in W_0^{k+2,p}(\Omega)$.

Next we prove that, for every open set Ω' such that $\overline{\Omega'} \subset \Omega$ and every $p > 1$, there exists a constant $c = c(n, p, \Omega', \Omega) > 0$ such that, for every $u \in W^{2,p}_{\mathrm{loc}}(\Omega)$,

$$\|u\|_{W^{1,p}(\Omega')} \leq c \left(\|\Delta u\|_{L^p(\Omega)} + \|u\|_{L^p(\Omega)} \right). \tag{B.3.2}$$

To see this consider the function $v := K * \beta\Delta u$ where $\beta \in C_0^\infty(\Omega)$ is equal to one in an open neighbourhood U of the closure of Ω'. Since $\nabla v = \nabla K * \beta\Delta u$ and ∇K is integrable on compact sets, it follows from Young's inequality that

$$\|v\|_{W^{1,p}(\Omega)} \leq c_1 \|\beta\Delta u\|_{L^p(\Omega)} \leq c_1 \|\Delta u\|_{L^p(\Omega)}.$$

Now the function $v - u$ is harmonic in U. Hence, by the mean value property for (the first derivatives of) harmonic functions, there exists a constant $c_2 > 0$ such that

$$\|v - u\|_{W^{1,p}(\Omega')} \leq c_2 \|v - u\|_{L^p(U)} \leq c_2 \left(\|v\|_{L^p(U)} + \|u\|_{L^p(U)} \right).$$

Take the sum of these inequalities to obtain the estimate (B.3.2).

Now let $u \in W^{k+2,p}_{\mathrm{loc}}(\Omega)$, denote $f := \Delta u \in W^{k,p}_{\mathrm{loc}}(\Omega)$, and suppose that Ω' is an open set whose closure is contained in Ω. Choose bounded open sets $\Omega_0, \dots, \Omega_{k+1}$ such that $\overline{\Omega'} \subset \Omega_{k+1}$, $\overline{\Omega}_{j+1} \subset \Omega_j$, and $\overline{\Omega}_0 \subset \Omega$. For $j = 0, \dots, k$ let $\beta_j \in C_0^\infty(\Omega_j)$ be a smooth cutoff function such that $\beta_j(x) = 1$ for $x \in \Omega_{j+1}$. Then $\beta_j u \in W_0^{k+2,p}(\Omega_j)$ and hence, by (B.3.1), we have

$$\begin{aligned}\|u\|_{W^{j+2,p}(\Omega_{j+1})} &\leq \|\beta_j u\|_{W^{j+2,p}(\Omega_j)} \\ &\leq c_j \|\Delta(\beta_j u)\|_{W^{j,p}(\Omega_j)} \\ &\leq c_j' \left(\|\Delta u\|_{W^{j,p}(\Omega_j)} + \|u\|_{W^{j+1,p}(\Omega_j)}\right)\end{aligned}$$

for $j = 0, \ldots, k$. By induction, we obtain

$$\|u\|_{W^{k+2,p}(\Omega_{k+1})} \leq c\left(\|\Delta u\|_{W^{k,p}(\Omega_0)} + \|u\|_{W^{1,p}(\Omega_0)}\right).$$

Now the required estimate follows from (B.3.2), with Ω' replaced by Ω_0. □

Sometimes it is useful to consider weak solutions of $\Delta u = f$ where f is not a function but a distribution in $W^{-1,p}$. We rephrase this in terms of weak solutions of the equation

$$\Delta u = \operatorname{div} f$$

with $f \in L^p$.

THEOREM B.3.2. *Let $1 < p < \infty$ and $\Omega' \subset \Omega \subset \mathbb{R}^n$ be open sets such that $\overline{\Omega'} \subset \Omega$. Then there exists a constant $c = c(p, n, \Omega', \Omega) > 0$ such that the following holds. Assume that $u \in L^1_{\mathrm{loc}}(\Omega)$ and $f = (f_0, \ldots, f_n) \in L^p_{\mathrm{loc}}(\Omega, \mathbb{R}^{n+1})$ satisfy*

$$\int_\Omega u(x)\Delta\phi(x)\,dx = \int_\Omega f_0(x)\phi(x)\,dx - \sum_{j=1}^n \int_\Omega f_j(x)\partial_j\phi(x)\,dx$$

for every $\phi \in C_0^\infty(\Omega)$. Then $u \in W^{1,p}_{\mathrm{loc}}(\Omega)$ and

$$\|u\|_{W^{1,p}(\Omega')} \leq c\left(\|f\|_{L^p(\Omega)} + \|u\|_{L^p(\Omega)}\right).$$

PROOF. Choose an open neighbourhood U of $\overline{\Omega'}$ such that $\overline{U} \subset \Omega$. Let $\beta \in C_0^\infty(\Omega)$ be a smooth cutoff function such that $\beta(x) = 1$ for $x \in U$. Define

$$v := K * \beta f_0 + \sum_{j=1}^n K_j * \beta f_j.$$

It follows from Theorem B.2.7 (and an approximation argument as in the proof of Theorem B.3.1 which we leave to the reader) that $v \in W^{1,p}_{\mathrm{loc}}(\Omega)$ and there is an estimate

$$\|v\|_{W^{1,p}(U)} \leq c_1 \|\beta f\|_{L^p(\Omega)} \leq c_1 \|f\|_{L^p(\Omega)}.$$

Here we have also used Poincaré's inequality. Now, by Lemma B.2.2, v is a weak solution of

$$\Delta v = \beta f_0 + \sum_{j=1}^n \partial_j(\beta f_j)$$

where $\partial_j(\beta f_j)$ is to be understood as a distribution. Hence the restriction of $u - v$ to U is a weak solution of $\Delta(u - v) = 0$. By Weyl's lemma $u - v$ is harmonic in U. Hence $u \in W^{1,p}(\Omega')$. Moreover, by the mean value property for harmonic functions, there exists a constant $c_2 > 0$ such that

$$\|v - u\|_{W^{1,p}(\Omega')} \leq c_2 \|v - u\|_{L^p(U)} \leq c_2\left(\|v\|_{W^{1,p}(U)} + \|u\|_{L^p(U)}\right).$$

Take the sum of these inequalities to obtain the required estimate. □

EXERCISE B.3.3. Let $\Omega \subset \mathbb{C}$ be an open set. The Cauchy–Riemann equations have the form

$$\partial w/\partial \bar{z} = h$$

where $\partial/\partial\bar{z} := \frac{1}{2}(\partial/\partial s + i\partial/\partial t)$. Here $s+it$ denotes the coordinate of Ω. Formulate and prove an analogue of Theorem B.3.1 for the Cauchy–Riemann operator. *Hint:* Write $w = u + iv$ and $h = (f + ig)/2$. Prove that if w is a (weak) solution of the Cauchy–Riemann equations then u and v are (weak) solutions of the second order equations

$$\Delta u = \partial_x f + \partial_y g, \qquad \Delta v = \partial_x g - \partial_y f.$$

Alternatively, show that the function $N(z) := 1/\pi z$ is the fundamental solution of the Cauchy–Riemann equations in the sense that any two functions $w, h \in C_0^\infty(\Omega, \mathbb{C})$ satisfy $\partial w/\partial\bar{z} = h$ if and only if $w = N * h$. Note that $N = 4\partial K/\partial z$, where $K(z) = (2\pi)^{-1}\log|z|$ is the fundamental solution of Laplace's equation.

B.4. Elliptic bootstrapping

In this section we shall prove the following two theorems about the smoothness of J-holomorphic curves with totally real boundary conditions and about compactness for sequences with uniform bounds on the L^p-norm of the derivatives with $p > 2$. Assume that M is a smooth $2n$-dimensional manifold and $L \subset M$ is a closed n-dimensional submanifold. Let $\mathcal{J}^\ell(M, L)$ denote the set of almost complex structures J on M of class C^ℓ such that L is totally real with respect to J. (This means that $TL \cap J(TL) = \{0\}$.) For $\ell = \infty$ we write $\mathcal{J}(M, L) := \mathcal{J}^\infty(M, L)$. Let Σ be an oriented 2-manifold with boundary and denote by $\mathcal{J}(\Sigma)$ the set of complex structures on Σ.

THEOREM B.4.1 (Regularity). *Fix an integer $\ell \geq 2$ and a number $p > 2$. Let $j \in \mathcal{J}(\Sigma)$, $J \in \mathcal{J}^\ell(M, L)$ and suppose that $u : \Sigma \to M$ is a $W^{1,p}$-function such that $du \circ j = J \circ du$ and $u(\partial\Sigma) \subset L$. Then u is of class $W^{\ell,p}$. If $\ell = \infty$ then u is smooth.*

THEOREM B.4.2 (Compactness). *Fix an integer $\ell \geq 2$ and a number $p > 2$. Let $J_\nu \in \mathcal{J}^\ell(M, L)$ be a sequence of almost complex structures on M converging to $J \in \mathcal{J}^\ell(M, L)$ in the C^ℓ topology and j_ν be a sequence of complex structures on Σ converging to j in the C^∞-topology. Let $U_\nu \subset \Sigma$ be an increasing sequence of open sets whose union is Σ and $u_\nu : U_\nu \to M$ be a sequence of (j_ν, J_ν)-holomorphic curves of class $W^{1,p}$ such that*

$$u_\nu(U_\nu \cap \partial\Sigma) \subset L.$$

Assume that for every compact set $Q \subset \Sigma$ (with smooth boundary) there exists a compact set $K \subset M$ and a constant $c > 0$ such that

$$\|du_\nu\|_{L^p(Q)} \leq c, \qquad u_\nu(Q) \subset K$$

for ν sufficiently large. Then there exists a subsequence of u_ν which converges in the $C^{\ell-1}$ topology on every compact subset of Σ.

REMARK B.4.3. If $\partial\Sigma = \emptyset$ we only need $\ell \geq 1$, obtain $u \in W^{\ell+1,p}_{\mathrm{loc}}$ in Theorem B.4.1, and obtain a C^ℓ convergent subsequence in Theorem B.4.2. Moreover, a slightly more careful discussion shows that it suffices to work with almost complex structures of class $C^{\ell-1,1}$. In particular, for interior regularity, one can work with Lipschitz continuous almost complex structures.

Both these theorems are obvious when J is integrable and Σ is closed because each component of a J-holomorphic curve in holomorphic coordinates is a harmonic function. In particular, the compactness theorem follows from the mean value property of harmonic functions. The general case is considerably harder.

The main idea of the proof of both theorems is to replace the nonlinear equation $\partial_s u + J(u)\partial_t u = 0$ by the linear equation $\partial_s u + J'\partial_t u = 0$, where $J'(z) = J(u(z))$ is a matrix valued function on some open subset $\Omega \subset \mathbb{H}$ of the upper half plane. This requires the choice of suitable coordinates on M in which the boundary conditions also become linear. Note that J' will have the same smoothness as u and this leads to the elliptic bootstrapping argument explained below. More precisely, suppose that $J \in W^{1,p}(\Omega, \mathbb{R}^{2n\times 2n})$ such that $J^2 = -\mathbb{1}$ and

$$J(s,0) = J_0 := \begin{pmatrix} 0 & -\mathbb{1} \\ \mathbb{1} & 0 \end{pmatrix} \tag{B.4.1}$$

is the standard complex structure on $\mathbb{R}^{2n}$. Let $u \in W^{1,p}(\Omega, \mathbb{R}^{2n})$ be a solution of the boundary value problem

$$\partial_s u + J\partial_t u = 0, \qquad u(s,0) \in \mathbb{R}^n \times \{0\}.$$

Then the derivative $v := \partial_s u$ solves the inhomogeneous equation

$$\partial_s v + J\partial_t v = \eta, \qquad v(s,0) \in \mathbb{R}^n \times \{0\}, \qquad \eta := -(\partial_s J)\partial_t u.$$

The best we can expect from our regularity theorems is the statement that if $\eta \in L^r(\Omega)$ for some $r > 1$ then v will belong to $W^{1,r}(\Omega)$. We would like to apply this with $r = p$. However, our assumptions imply that the functions $\partial_s J$ and $\partial_t u$ are in L^p and so their product will in general only belong to $L^{p/2}$. To deal with this we improve the regularity of u while that of $\partial_s J$ is fixed. Thus suppose we have already proved that $\partial_t u$ belongs to L^q for some $q > p/(p-1)$. Then η belongs to L^r where $1/p + 1/q = 1/r$. Since $r > 1$ we then find that $v \in W^{1,r}$ and hence $u \in W^{2,r}$. If $1 < r < 2$ then, by Theorem B.1.12, $u \in W^{1,q'}$, where $q' := 2r/(2-r) > q$. Thus we have gained something. Now repeat this process. Once we get to a value for r that is greater than 2 we obtain from Theorem B.1.11 that $\partial_t u$ is continuous and so η belongs to L^p. This iteration is part of the *elliptic bootstrapping* argument. It requires the following two lemmas.

LEMMA B.4.4. *Given $p > 2$ there is a finite sequence $q_0 < q_1 < \cdots < q_m$ such that, for $j = 0, \ldots, m-1$, we have*

$$\frac{p}{p-1} < q_0 \le p, \qquad q_{m-1} < \frac{2p}{p-2} < q_m,$$

$$q_{j+1} = \frac{2r_j}{2-r_j}, \qquad r_j := \frac{pq_j}{p+q_j}.$$

PROOF. Consider the map $h : (p/(p-1), 2p/(p-2)) \to (2,\infty)$ defined by

$$h(q) := \frac{2pq}{2p+2q-pq} = \frac{2r}{2-r}, \qquad r := \frac{pq}{p+q} < 2.$$

The condition $r < 2$ is equivalent to $q < 2p/(p-2)$. The map h is a monotonically increasing diffeomorphism such that $h(q) > q$. Now choose the sequence q_j such that $q_{j+1} = h(q_j)$. □

LEMMA B.4.5. *Assume $p > 2$ and $1 < r \le p$. Then there exists a constant $c = c(p, r) > 0$ such that the following holds. If $f \in W^{1,p}(\mathbb{R}^2)$ and $g \in W^{1,r}(\mathbb{R}^2)$ then $fg \in W^{1,r}(\mathbb{R}^2)$ and*

$$\|fg\|_{W^{1,r}} \le c \|f\|_{W^{1,p}} \|g\|_{W^{1,r}}.$$

PROOF. Examine the L^r-norm of the expression $d(fg) = (df)g + f(dg)$. The term $f(dg)$ can be estimated by the sup-norm of f and the $W^{1,r}$-norm of g. The term $(df)g$ can be estimated by

$$\|(df)g\|_{L^r} \le \|df\|_{L^p} \|g\|_{L^q}, \qquad 1/p + 1/q = 1/r.$$

If $r < 2$ then, since $p > 2$, we have $q = pr/(p-r) < 2r/(2-r)$ and the L^q norm of g can be estimated by the $W^{1,r}$-norm. The latter is obvious when $r \ge 2$. □

We next prove a local estimate for the solutions of the linear Cauchy-Riemann equation. Let $\mathbb{H} := \{s + it \in \mathbb{C} \,|\, t \ge 0\}$ denote the closed upper half space and $\Omega \subset \mathbb{H}$ be an open set. Note that Ω is not necessarily an open subset of $\mathbb{C}$. We denote by $C_0^\infty(\Omega, \mathbb{R}^{2n})$ the space of smooth functions $\phi : \Omega \to \mathbb{R}^{2n}$ whose support is a compact subset of Ω. Such functions need not vanish on the subset $\Omega \cap \mathbb{R}$ of the boundary. As before we denote by $W^{k,p}_{\mathrm{loc}}(\Omega)$ the space of functions on Ω whose restriction to every compact subset of Ω is of class $W^{k,p}$, however, in the present case such functions are $W^{k,p}$-regular near $\Omega \cap \mathbb{R}$. Let $u : \Omega \to \mathbb{R}^{2n}$ be a solution of the linear boundary value problem

$$\partial_s u + J \partial_t u = \eta, \qquad u(\Omega \cap \mathbb{R}) \subset \mathbb{R}^n \times \{0\}, \tag{B.4.2}$$

where $\eta : \Omega \to \mathbb{R}^{2n}$ and $J : \Omega \to \mathbb{R}^{2n \times 2n}$ satisfies $J^2 = -\mathbb{1}$ and (B.4.1). We shall first prove regularity of u under the weakest possible regularity assumptions on the almost complex structure J and the function η. We shall first assume that the almost complex structure J is of class $W^{1,p}$. If u is a solution of (B.4.2) then, by partial integration,

$$\phi(\Omega \cap \mathbb{R}) \subset \mathbb{R}^n \times \{0\} \quad \Longrightarrow \quad \int_\Omega \langle \partial_s \phi + J^T \partial_t \phi, u \rangle = -\int_\Omega \langle \phi, \eta + (\partial_t J) u \rangle \tag{B.4.3}$$

for every test function $\phi \in C_0^\infty(\Omega, \mathbb{R}^{2n})$, where J^T denotes the transpose of J. The next lemma asserts the converse.

LEMMA B.4.6. *Let $\Omega' \subset \Omega \subset \mathbb{H}$ be open sets such that $\overline{\Omega'} \subset \Omega$ and p, q, r be positive real numbers (including possibly plus infinity) such that*

$$2 < p, \qquad 1 < r < \infty, \qquad 1/p + 1/q = 1/r.$$

Then for every constant $c_0 > 0$ there exists a constant $c > 0$ with the following significance. Assume $J \in W^{1,p}(\Omega, \mathbb{R}^{2n \times 2n})$ satisfies $J^2 = -\mathbb{1}$ and (B.4.1) and

$$\|J\|_{W^{1,p}(\Omega)} \le c_0.$$

Then the following holds.

(i) *If $u \in L^q_{\mathrm{loc}}(\Omega, \mathbb{R}^{2n})$ and $\eta \in L^r_{\mathrm{loc}}(\Omega, \mathbb{R}^{2n})$ are such that (B.4.3) holds then $u \in W^{1,r}_{\mathrm{loc}}(\Omega, \mathbb{R}^{2n})$ and u satisfies (B.4.2) almost everywhere.*

(ii) *If $u \in W^{1,r}_{\mathrm{loc}}(\Omega)$ satisfies the boundary condition $u(\Omega \cap \mathbb{R}) \subset \mathbb{R}^n \times \{0\}$ then*

$$\|u\|_{W^{1,r}(\Omega')} \le c \left(\|\partial_s u + J \partial_t u\|_{L^r(\Omega)} + \|u\|_{L^q(\Omega)} \right).$$

REMARK B.4.7. Note that the cases $q = \infty$ or $p = \infty$ are included under the hypotheses of Lemma B.4.6. In the case $q = \infty$ the result holds with $r = p < \infty$. In the case $p = \infty$ it holds with $r = q < \infty$. In all cases $r \le \min\{p, q\}$.

REMARK B.4.8. Under the assumptions of Lemma B.4.6 we have $q \le 2r/(2-r)$ whenever $r < 2$. Hence there is an inclusion $W^{1,r} \hookrightarrow L^q$.

PROOF OF LEMMA B.4.6. Let $\psi : \Omega \to \mathbb{R}^{2n}$ be a smooth test function with compact support such that

$$\psi(s,0) \in \mathbb{R}^n \times \{0\}, \qquad \partial_t \psi(s,0) \in \{0\} \times \mathbb{R}^n \tag{B.4.4}$$

for every $s \in \mathbb{R}$ such that $(s,0) \in \Omega$. Then the function

$$\phi := \partial_s \psi - J^T \partial_t \psi$$

belongs to $W^{1,p}(\Omega, \mathbb{R}^{2n})$ and satisfies $\phi(\Omega \cap \mathbb{R}) \subset \mathbb{R}^n \times \{0\}$ and

$$\partial_s \phi + J^T \partial_t \phi = \Delta \psi - (\partial_s J)^T \partial_t \psi + (\partial_t J)^T J^T \partial_t \psi.$$

Since (B.4.3) continues to hold (under the hypotheses of the Lemma) for test functions ϕ of class $W^{1,p}$ with compact support in Ω, we obtain

$$\int_\Omega \langle \Delta \psi, u \rangle = -\int_\Omega \langle \partial_s \psi, f \rangle - \int_\Omega \langle \partial_t \psi, g \rangle, \tag{B.4.5}$$

where

$$f := (\partial_t J)u + \eta, \qquad g := -(\partial_s J)u - J\eta. \tag{B.4.6}$$

By assumption and Hölder's inequality, the functions f and g are of class L^r_{loc}. Equation (B.4.5) asserts that u is a weak solution of

$$\Delta u = \partial_s f + \partial_t g,$$

where the right hand side is to be understood in the distributional sense.

Now suppose $\Omega \cap \mathbb{R} = \emptyset$. Then, by Theorem B.3.2, we have $u \in W^{1,r}_{\mathrm{loc}}$. Moreover, for every bounded open set Ω' whose closure is contained in Ω there is an estimate

$$\begin{aligned}
\|u\|_{W^{1,r}(\Omega')} &\le c_1 \left(\|f\|_{L^r(\Omega)} + \|g\|_{L^r(\Omega)} + \|u\|_{L^r(\Omega)} \right) \\
&\le c_1 \Big(\|J\|_{W^{1,p}(\Omega)} \|u\|_{L^q(\Omega)} + \|u\|_{L^r(\Omega)} \\
&\qquad + \|J\|_{L^\infty(\Omega)} \|\eta\|_{L^r(\Omega)} + \|\eta\|_{L^r(\Omega)} \Big) \\
&\le c_2 \left(1 + \|J\|_{W^{1,p}(\Omega)} \right) \left(\|\eta\|_{L^r(\Omega)} + \|u\|_{L^q(\Omega)} \right).
\end{aligned}$$

Here the constants c_1 and c_2 depend on p, q, Ω', and Ω, but not on u and J. Moreover, in these estimates we have assumed without loss of generality that Ω is a bounded Lipschitz domain so that the Sobolev embedding theorems apply. It follows from integration by parts that u satisfies (B.4.2) almost everywhere. This proves the lemma in the case $\Omega \cap \mathbb{R} = \emptyset$.

To prove the result in general consider the open set

$$\Omega_0 := \{ s + it \in \mathbb{C} \,|\, s + i|t| \in \Omega \}$$

and extend u, f, and g to Ω_0 by

$$u(s,-t) := Ru(s,t), \qquad f(s,-t) := Rf(s,t), \qquad g(s,-t) := -Rg(s,t)$$

for $t > 0$, where

$$R := \begin{pmatrix} 1\!\!1 & 0 \\ 0 & -1\!\!1 \end{pmatrix}.$$

These extended functions satisfy (B.4.5), with Ω replaced by Ω_0, for every test function $\psi \in C_0^\infty(\Omega_0)$. To see this write

$$\psi(s,t) = \psi_0(s,t) + \alpha_0(s)\beta_0(t) + \alpha_1(s)\beta_1(t),$$

where ψ_0 satisfies (B.4.4), and

$$\alpha_0(s) \in \{0\} \times \mathbb{R}^n, \qquad \alpha_1(s) \in \mathbb{R}^n \times \{0\},$$

$$\beta_0(t) = \beta_0(-t), \qquad \beta_1(t) = -\beta_1(-t), \qquad \beta_0(0) = \partial_t\beta_1(0) = 1.$$

Note that α_0 and α_1 are determined by

$$\psi(s,0) = \psi_0(s,0) + \alpha_0(s), \qquad \partial_t\psi(s,0) = \partial_t\psi_0(s,0) + \alpha_1(s).$$

By the symmetry of u, f, and g, we have

$$\int_{\Omega_0} \langle \Delta(\alpha_i\beta_i), u\rangle = \int_{\Omega_0} \langle \partial_s(\alpha_i\beta_i), f\rangle = \int_{\Omega_0} \langle \partial_t(\alpha_i\beta_i), g\rangle = 0$$

for $i = 0, 1$. Hence we can replace ψ by ψ_0 on both sides of equation (B.4.5), and for $\psi = \psi_0$ the identity has already been established. Therefore it follows from the result for the case $\Omega \cap \mathbb{R} = \emptyset$ that $u \in W^{1,r}_{\mathrm{loc}}(\Omega_0, \mathbb{R}^{2n})$ and that u satisfies the required estimate.

It remains to prove that $u(s,0) \in \mathbb{R}^n \times \{0\}$ for almost every $s \in \mathbb{R} \cap \Omega$. To see this choose a sequence of smooth functions $v_i : \Omega_0 \to \mathbb{R}$ converging to u in the $W^{1,r}$ norm over every compact subset of Ω_0. Then the sequence

$$u_i(s,t) := \frac{1}{2}(v_i(s,t) + Rv_i(s,-t))$$

also converges to u in the $W^{1,r}$ norm over every compact subset of Ω_0. Moreover, $u_i(s,0) \in \mathbb{R}^n \times \{0\}$ for every $s \in \mathbb{R} \cap \Omega$. By Proposition B.1.21, u_i converges to u in the L^r norm over every compact interval in $\mathbb{R} \cap \Omega$. Since the composition of the map $s \mapsto u_i(s,0)$ with the projection $\mathbb{R}^{2n} \to \{0\} \times \mathbb{R}^n$ is zero for every i and converges in L^r_{loc} to the composition of the map $s \mapsto u(s,t)$ with the same projection, it follows that $u(s,0) \in \mathbb{R}^n \times \{0\}$ for almost every $s \in \mathbb{R} \cap \Omega$. □

PROPOSITION B.4.9. *Let $\Omega' \subset \Omega \subset \mathbb{H}$ be open sets such that $\overline{\Omega'} \subset \Omega$, ℓ be a positive integer, and $p > 2$. Then, for every $c_0 > 0$, there is a $c > 0$ with the following significance. Assume $J \in W^{\ell,p}(\Omega, \mathbb{R}^{2n\times 2n})$ satisfies $J^2 = -1\!\!1$ and (B.4.1) and $\|J\|_{W^{\ell,p}(\Omega)} \le c_0$. Then the following holds for every $k \in \{0, \dots, \ell\}$.*

(i) *If $u \in L^p_{\mathrm{loc}}(\Omega, \mathbb{R}^{2n})$ and $\eta \in W^{k,p}_{\mathrm{loc}}(\Omega, \mathbb{R}^{2n})$ are such that (B.4.3) holds then $u \in W^{k+1,p}_{\mathrm{loc}}(\Omega, \mathbb{R}^{2n})$ and u satisfies (B.4.2) almost everywhere.*

(ii) *If $u \in W^{k+1,p}_{\mathrm{loc}}(\Omega)$ satisfies $u(\Omega \cap \mathbb{R}) \subset \mathbb{R}^n \times \{0\}$ then*

$$\|u\|_{W^{k+1,p}(\Omega')} \le c\left(\|\partial_s u + J\partial_t u\|_{W^{k,p}(\Omega)} + \|u\|_{W^{k,p}(\Omega)}\right).$$

PROOF. The proof has three steps.

STEP 1. *The result holds for $k = 0$ and $\ell = 1$.*

By Lemma B.4.6 and the Sobolev embedding theorem we have

$$u \in L^q_{\mathrm{loc}}, \quad \frac{p}{p-1} < q < \frac{2p}{p-2} \qquad \Longrightarrow \qquad u \in W^{1,r}_{\mathrm{loc}} \subset L^{q'}_{\mathrm{loc}},$$

where

$$r := \frac{pq}{p+q} < 2, \qquad q' := \frac{2r}{2-r} > q.$$

Now choose q_j and r_j as in Lemma B.4.4 for $j = 0, \dots, m$. Since $q_0 \le p$ we have $u \in L^{q_0}_{\mathrm{loc}}$ and it follows by induction that $u \in W^{1,r_j}_{\mathrm{loc}}$ for $j = 0, \dots, m$. With $j = m$ we have $q_m > 2p/(p-2)$ and hence

$$r_m = \frac{pq_m}{p+q_m} > 2,$$

so u is continuous. Using Lemma B.4.6 again with $q = \infty$ and $r = p$ (see Remark B.4.7) we obtain $u \in W^{1,p}_{\mathrm{loc}}$. This proves (i) for $k = 0$ and $\ell = 1$.

To prove (ii) choose bounded open Lipschitz domains $\Omega_0, \dots, \Omega_m \subset \Omega$ with

$$\overline{\Omega'} \subset \mathrm{int}_{\mathbb{H}}(\Omega_m), \qquad \overline{\Omega}_j \subset \mathrm{int}_{\mathbb{H}}(\Omega_{j-1}), \qquad \overline{\Omega}_0 \subset \Omega.$$

Then, for $j = 1, \dots, m$, it follows from Lemma B.4.6 that

$$\begin{aligned}
\|u\|_{W^{1,r_j}(\Omega_j)} &\le c_1 \left(\|\partial_s u + J\partial_t u\|_{L^{r_j}(\Omega_{j-1})} + \|u\|_{L^{q_j}(\Omega_{j-1})} \right) \\
&\le c_2 \left(\|\partial_s u + J\partial_t u\|_{L^{p}(\Omega)} + \|u\|_{W^{1,r_{j-1}}(\Omega_{j-1})} \right) \\
&\le c_3 \left(\|\partial_s u + J\partial_t u\|_{L^{p}(\Omega)} + \|u\|_{W^{1,r_0}(\Omega_0)} \right) \\
&\le c_4 \left(\|\partial_s u + J\partial_t u\|_{L^{p}(\Omega)} + \|u\|_{L^{p}(\Omega)} \right).
\end{aligned}$$

The second inequality follows from the Sobolev embedding theorem and the third by induction. In the last inequality we have used $q_0 \le p$. With $j = m$ we have $r_m > 2$ so this gives an estimate of the L^∞-norm of u, by Theorem B.1.11. Now use the estimate of Lemma B.4.6 again with $q = \infty$ and $r = p$.

STEP 2. *The result holds for $k = \ell = 1$.*

Let $q, r > 1$ such that $1/p + 1/q = 1/r$. Then

$$u \in W^{1,q}_{\mathrm{loc}} \qquad \Longrightarrow \qquad u \in W^{2,r}_{\mathrm{loc}} \tag{B.4.7}$$

and u satisfies the obvious estimate. To see this note that the functions

$$u' := \partial_s u \in L^q_{\mathrm{loc}}, \qquad \eta' := \partial_s \eta - (\partial_s J)\partial_t u \in L^r_{\mathrm{loc}}.$$

satisfy (B.4.3) for every $\phi \in C^\infty_0(\Omega, \mathbb{R}^{2n})$. Hence it follows from Lemma B.4.6 that $u' \in W^{1,r}_{\mathrm{loc}}$. By Lemma B.4.5, $\partial_t u = J(\partial_s u - \eta) \in W^{1,r}_{\mathrm{loc}}$ and hence $u \in W^{2,r}_{\mathrm{loc}}$.

To prove the assertions for $k = 1$ choose sequences q_j and r_j as in Lemma B.4.4 and use (B.4.7) inductively as in the proof of Step 1 to obtain $u \in W^{2,r_j}_{\mathrm{loc}}$ for every j. With $j = m$ it follows that u is continuously differentiable and, by (B.4.7) with $q = \infty$, we have $u \in W^{2,p}_{\mathrm{loc}}$. This proves (i) for $k = \ell = 1$. To prove (ii) one argues exactly as in Step 1. The details in the present case are left to the reader.

STEP 3. *The result holds in general.*

Assume, by induction, the proposition is true for some $k \ge 1$ and that $J \in W^{k+1,p}$. Let $\eta \in W^{k+1,p}_{\mathrm{loc}}$. Apply the induction hypothesis to u' and η' (as in Step 2) to obtain that $\partial_s u = u'$ and $\partial_t u = J(\partial_s u - \eta)$ are of class $W^{k+1,p}_{\mathrm{loc}}$ and satisfy an estimate of the form

$$\begin{aligned}\|u\|_{W^{k+2,p}(\Omega')} &\le c_1\left(\|u\|_{W^{k+1,p}(\Omega')} + \|u'\|_{W^{k+1,p}(\Omega')} + \|\eta\|_{W^{k+1,p}(\Omega')}\right)\\ &\le c_2\left(\|\eta\|_{W^{k+1,p}(\Omega)} + \|u\|_{W^{k,p}(\Omega)} + \|\eta'\|_{W^{k,p}(\Omega)} + \|u'\|_{W^{k,p}(\Omega)}\right)\\ &\le c_3\left(\|\eta\|_{W^{k+1,p}(\Omega)} + \|u\|_{W^{k+1,p}(\Omega)}\right).\end{aligned}$$

The first inequality follows from Proposition B.1.19 (since $\partial_t u = J(u' - \eta)$ and $(k+1)p > 2$), the second inequality follows from the induction hypothesis (applied to both u and u') and the last inequality follows from Proposition B.1.19 (since $\eta' = \partial_s \eta - (\partial_s J)\partial_t u$ and $kp > 2$). This proves Proposition B.4.9. □

PROOF OF THEOREM B.4.1. It suffices to prove the result in local holomorphic coordinates on Σ and in local coordinates on M. If J is a C^ℓ almost complex structure, then there is a C^ℓ local coordinate chart on M that identifies L with $\mathbb{R}^n \times \{0\}$ and identifies J with J_0 along L (see Exercise B.4.10 below). Pushing J forward with a C^ℓ coordinate chart gives a $C^{\ell-1}$ almost complex structure on $\mathbb{R}^{2n}$. Hence we shall assume that $\Omega \subset \mathbb{H}$ is an open set and $u : \Omega \to \mathbb{R}^{2n}$ is a solution of the boundary value problem

$$\partial_s u + J(u)\partial_t u = 0, \qquad u(\Omega \cap \mathbb{R}) \subset \mathbb{R}^n \times \{0\},$$

where $J : \mathbb{R}^{2n} \to \mathbb{R}^{2n\times 2n}$ is an almost complex structure of class $C^{\ell-1}$ such that

$$x \in \mathbb{R}^n \times \{0\} \quad \Longrightarrow \quad J(x) = J_0. \tag{B.4.8}$$

Now assume first that $k = 1$. Then $u \in W^{1,p}_{\mathrm{loc}}(\Omega, \mathbb{R}^{2n})$ satisfies the requirements of Proposition B.4.9 with $k = 1$, $\eta = 0$, and J replaced by $J \circ u \in W^{1,p}_{\mathrm{loc}}(\Omega, \mathbb{R}^{2n\times 2n})$. Hence $u \in W^{2,p}_{\mathrm{loc}}$ and $J \circ u \in W^{2,p}_{\mathrm{loc}}$. Continue by induction to obtain that $u \in W^{k+1,p}_{\mathrm{loc}}$ for $k = 1, \dots, \ell - 1$. Hence $u \in W^{\ell,p}_{\mathrm{loc}}$. This proves Theorem B.4.1. □

PROOF OF THEOREM B.4.2. Since the inclusion $W^{1,p} \hookrightarrow C$ is compact for $p > 2$ we may assume that u_ν converges uniformly to a continuous function $u : \Sigma \to M$. Hence, by Exercise B.4.10 below, we may assume that $u_\nu : \Omega \to \mathbb{R}^{2n}$ is a sequence of $W^{1,p}_{\mathrm{loc}}$ solutions of the sequence of boundary value problems

$$\partial_s u_\nu + J_\nu(u_\nu)\partial_t u_\nu = 0, \qquad u_\nu(\Omega \cap \mathbb{R}) \subset \mathbb{R}^{2n} \times \{0\},$$

where $J_\nu \in C^{\ell-1}(\mathbb{R}^{2n}, \mathbb{R}^{2n\times 2n})$ converges to J in the $C^{\ell-1}$-topology and satisfies (B.4.8). It follows from Proposition B.4.9 by induction that $u_\nu \in W^{\ell,p}_{\mathrm{loc}}$ and

$$\sup_\nu \|u_\nu\|_{W^{\ell,p}(Q)} < \infty$$

for every compact subset $Q \subset \Omega$ with smooth boundary. By Theorem B.1.11, the inclusion $W^{\ell,p}(Q) \hookrightarrow C^{\ell-1}(Q)$ is compact. This proves the existence of a subsequence which converges in the $C^{\ell-1}$-topology on any given compact subset of Σ. The theorem follows by choosing a diagonal subsequence associated to an exhausting sequence of compact subsets of Σ. This proves Theorem B.4.2. □

EXERCISE B.4.10. Let $J \in \mathcal{J}^\ell(M, L)$. Prove that, for every point in L, there exists a neighbourhood U and a C^ℓ coordinate chart $\psi : U \to \mathbb{R}^{2n}$ such that

$$\psi(U \cap L) = \psi(U) \cap (\mathbb{R}^n \times \{0\}), \qquad J_0 d\psi(x) = d\psi(x)J(x)$$

$x \in U \cap L$. If $J_\nu \in \mathcal{J}^\ell(M, L)$ converges to J in the C^ℓ topology, prove that ψ_ν can be chosen to converge to ψ in the C^ℓ topology. *Hint:* Let $\phi : V \to \mathbb{R}^n$ be a coordinate chart on an open set $V \subset L$ (which extends to the closure of V), let

$$U_J := \{\exp_x(J(x)\xi) \,|\, x \in V,\, \xi \in T_xL,\, |\xi| < \delta\},$$

and define $\psi_J : U_J \to \mathbb{R}^{2n}$ by

$$\psi_J(\exp_x(J(x)\xi)) := (\phi(x), d\phi(x)\xi).$$

EXERCISE B.4.11. Let $\Omega \subset \mathbb{H}$ be an open subset of the upper half plane. Suppose $J \in C^\infty_{\mathrm{loc}}(\Omega, \mathbb{R}^{2n\times 2n})$ satisfies $J^2 = -\mathbb{1}$ and (B.4.1). Prove that the hypothesis $p > 2$ in Proposition B.4.9 can be dropped.

EXERCISE B.4.12. Find the simplest proof you can of Theorems B.4.1 and B.4.2 when J is integrable and Σ is closed.

APPENDIX C

The Riemann–Roch Theorem

In this appendix we give an introduction to real Cauchy–Riemann operators on Riemann surfaces with boundary and prove the Riemann–Roch theorem in this case. The first section discusses the basic background material and states the Riemann–Roch theozrem. In section C.2 we prove that Cauchy–Riemann operators are Fredholm. Section C.3 gives an introduction to the boundary Maslov index, and the Riemann–Roch theorem is proved in Section C.4. Section C.5 shows how the Riemann mapping theorem can be derived from the Riemann–Roch theorem via an idea in Earle–Eells [**95**]. The final two sections are less important. The first contains a brief discussion of Cauchy–Riemann operators on nonsmooth bundles, while the second derives formulas for D_u in terms of the various connections that one can put on the tangent bundle of a symplectic manifold. The results here were mentioned in Remark 3.1.3. They are not needed in the rest of the book, and are included for the sake of completeness.

In writing up this appendix we greatly benefitted from discussions with Joel Robbin, and he kindly agreed to contribute his beautiful exposition of the boundary Maslov index in Section C.3 to this book.

C.1. Cauchy–Riemann operators

Let Σ be a compact Riemann surface with boundary and $E \to \Sigma$ be a (smooth) complex vector bundle. We denote by $j : T\Sigma \to T\Sigma$ the complex structure on Σ and by $J : E \to E$ the complex structure on E. Denote by

$\Omega^k(\Sigma)$	the smooth complex valued k-forms on Σ,
$\Omega^{p,q}(\Sigma)$	those of type (p,q),
$\Omega^k(\Sigma, E)$	the smooth E valued k-forms on Σ,
$\Omega^{p,q}(\Sigma, E)$	those of type (p,q),

The spaces $\Omega^k(\Sigma)$, $\Omega^{p,q}(\Sigma)$, $\Omega^k(\Sigma, E)$, $\Omega^{p,q}(\Sigma, E)$ are spaces of sections of the complex vector bundles $\Lambda^k T^*\Sigma \otimes \mathbb{C}$, $\Lambda^{p,q}T^*\Sigma$, $\Lambda^k T^*\Sigma \otimes E$, $\Lambda^{p,q}T^*\Sigma \otimes_{\mathbb{C}} E$, respectively, and are thus complex vector spaces. The complex structure j on Σ determines a $\mathbb{C}^*$ action on $\Omega^k(\Sigma)$ and on $\Omega^k(\Sigma, E)$ via $(a+ib, \alpha) \mapsto a\alpha + bj^*\alpha$; the direct sum decompositions

$$\Omega^k(\Sigma) = \bigoplus_{p+q=k} \Omega^{p,q}(\Sigma), \qquad \Omega^k(\Sigma, E) = \bigoplus_{p+q=k} \Omega^{p,q}(\Sigma, E)$$

are the isotypic components of this action. Replacing j by $-j$ interchanges $\Omega^{p,q}$ and $\Omega^{q,p}$.

Denote by $\bar\partial : \Omega^0(\Sigma) \to \Omega^{0,1}(\Sigma)$ the composition of the exterior derivative $d : \Omega^0(\Sigma) \to \Omega^1(\Sigma)$ with the projection $\Omega^1(\Sigma) = \Omega^{1,0}(\Sigma) \oplus \Omega^{0,1}(\Sigma) \to \Omega^{0,1}(\Sigma)$. A (complex linear, smooth) **Cauchy–Riemann operator** on the bundle $E \to \Sigma$ is a $\mathbb{C}$-linear operator

$$D : \Omega^0(\Sigma, E) \to \Omega^{0,1}(\Sigma, E)$$

which satisfies the Leibnitz rule

$$D(f\xi) = f(D\xi) + (\bar{\partial} f)\xi$$

for $\xi \in \Omega^0(\Sigma, E)$ and $f \in \Omega^0(\Sigma)$. For example, $\bar{\partial}$ (acting coordinate wise) is a Cauchy–Riemann operator on the trivial bundle. A Cauchy–Riemann operator on E extends uniquely to a skew wedge derivation $D : \Omega^{p,q}(\Sigma, E) \to \Omega^{p,q+1}(\Sigma, E)$ (same name, different input) satisfying the same Leibnitz rule.

REMARK C.1.1. A holomorphic structure on the bundle $E \to \Sigma$ is a (maximal) atlas of local trivializations for which the transition maps ϕ are holomorphic. Such a structure determines a unique Cauchy–Riemann operator D which agrees with $\bar{\partial}$ locally since $\bar{\partial}\phi = 0$. It is a deep theorem that every (complex linear, smooth) Cauchy–Riemann operator arises in this way (cf. Kobayashi [**212**]). This gives a bijective correspondence between holomorphic structures on E and Cauchy–Riemann operators on E. A holomorphic structure on the bundle E also determines an integrable complex structure on the total space E: the holomorphic local trivializations give rise to coordinate charts on E with holomorphic transition maps.

REMARK C.1.2. A **Hermitian structure** on E is a real inner product $\langle \cdot, \cdot \rangle$ on E such that the complex structure J is orthogonal. A **Hermitian connection** on E is a $\mathbb{C}$-linear operator $\nabla : \Omega^0(\Sigma, E) \to \Omega^1(\Sigma, E)$ such that

$$\nabla(f\xi) = f\nabla\xi + (df)\xi, \qquad d\langle \xi_1, \xi_2 \rangle = \langle \nabla\xi_1, \xi_2 \rangle + \langle \xi_1, \nabla\xi_2 \rangle$$

for all $f \in \Omega^0(\Sigma)$ and $\xi, \xi_1, \xi_2 \in \Omega^0(\Sigma, E)$. Every Hermitian connection ∇ determines a Cauchy–Riemann operator $\bar{\partial}^\nabla$ on E defined by

$$\bar{\partial}^\nabla \xi := (\nabla\xi)^{0,1} = \tfrac{1}{2}(\nabla\xi + J\nabla\xi \circ j)$$

for $\xi \in \Omega^0(\Sigma, E)$. Conversely, for every Cauchy–Riemann operator D there exists a unique Hermitian connection ∇ on E such that $D = \bar{\partial}^\nabla$. (Prove this!)

REMARK C.1.3. Fix a Hermitian structure on E, a Hermitian connection ∇ and a volume form dvol on Σ which is compatible with the orientation determined by j (i.e. $\mathrm{dvol}(v, jv) > 0$ for $0 \neq v \in T\Sigma$). The volume form and Hermitian structure determine L^2 inner products on the spaces $\Omega^k(\Sigma, E)$. We claim that the formal adjoint operator of $\bar{\partial}^\nabla$ with respect to these inner products is given by

$$(\bar{\partial}^\nabla)^* = - * \partial^\nabla * = - * J\partial^\nabla : \Omega^{0,1}(\Sigma, E) \to \Omega^0(\Sigma, E). \tag{C.1.1}$$

Here $*$ denotes the Hodge $*$-operator, and the second identity follows from the fact that $*\eta = -\eta \circ j = J\eta$ for $\eta \in \Omega^{0,1}(\Sigma, E)$. (If ds, dt is an orthonormal positive basis of $T_z^*\Sigma$ then $*ds = dt = -j^*ds$ and $*dt = -ds = -j^*dt$.) To see this, note that the formal adjoint operator is characterized by the identity

$$\int_\Sigma \langle (\bar{\partial}^\nabla)^* \eta, \xi \rangle \,\mathrm{dvol} = \int_\Sigma \langle \eta, \bar{\partial}^\nabla \xi \rangle \,\mathrm{dvol} \tag{C.1.2}$$

for $\xi \in \Omega^0(\Sigma, E)$ and $\eta \in \Omega^{0,1}(\Sigma, E)$ such that the 1-form $\langle \eta, J\xi \rangle$ vanishes on $T\partial\Sigma$. Hence the above formula for the adjoint follows from the identity

$$-\langle *J\partial^\nabla \eta, \xi \rangle \,\mathrm{dvol} - \langle \eta, \bar{\partial}^\nabla \xi \rangle \,\mathrm{dvol} = d\langle \eta, J\xi \rangle.$$

REMARK C.1.4. Consider the bundle E in a unitary trivialization along local holomorphic coordinates on Σ. We denote the coordinates by $s + it \in \Omega \subset \mathbb{H}$, where $\mathbb{H}$ is the closed upper half plane. In such a trivialization J is identified with multiplication by i and the connection ∇ is given by

$$\nabla_s = \partial_s + \Phi, \qquad \nabla_t = \partial_t + \Psi,$$

where $\Phi, \Psi : \Omega \to \mathfrak{u}(n)$. The Cauchy–Riemann operator $\bar{\partial}^\nabla$ then has the form

$$\bar{\partial}^\nabla \xi = \tfrac{1}{2}\left(\nabla_s\xi + i\nabla_t\xi\right) ds + \tfrac{1}{2}\left(\nabla_t\xi - i\nabla_s\xi\right) dt$$

for $\xi : \Omega \to \mathbb{C}^n$. Elements in $\Omega^{0,1}(\Omega, E)$ have the form $\eta = \zeta ds - J\zeta dt$, and the dual operator is given by

$$\left(\bar{\partial}^\nabla\right)^* (\zeta\, ds - i\zeta\, dt) = \lambda^{-2}\left(-\nabla_s\zeta + i\nabla_t\zeta\right),$$

where the induced volume form on Ω is given by $\mathrm{dvol} = \lambda^2\, ds \wedge dt$. Thus, considered as an operator on the function ζ (in local coordinates on Σ), we see that the adjoint is simply $\zeta \mapsto -2\lambda^{-2}\partial^\nabla\zeta$.

Real linear Cauchy–Riemann operators. For the applications in Chapter 3 we need to consider more general Cauchy–Riemann operators whose zeroth order term is $\mathbb{R}$-linear rather than $\mathbb{C}$-linear. It will also be useful to weaken our smoothness assumptions. Thus we make the following definition. As in Remark B.1.23 we denote by $W^{k,q}_F(\Sigma, E)$ the closure of

$$\Omega^0_F(\Sigma, E) := \left\{\xi \in \Omega^0(\Sigma, E) \,|\, \xi(\partial\Sigma) \subset F\right\}$$

in the Sobolev space $W^{k,q}(\Sigma, E)$ and by $W^{k,q}_F(\Sigma, \Lambda^{0,1}T^*\Sigma \otimes E)$ the closure of

$$\Omega^{0,1}_F(\Sigma, E) := \left\{\eta \in \Omega^{0,1}(\Sigma, E) \,|\, \eta(T\partial\Sigma) \subset F\right\}$$

in $W^{k,q}(\Sigma, \Lambda^{0,1}T^*\Sigma \otimes E)$.

DEFINITION C.1.5. *Let ℓ be a positive integer and $p > 1$ such that $\ell p > 2$. A* **real linear Cauchy–Riemann operator of class $W^{\ell-1,p}$** *on E is an operator of the form $D = D_0 + \alpha$, where $\alpha \in W^{\ell-1,p}(\Sigma, \Lambda^{0,1}T^*\Sigma \otimes \mathrm{End}_\mathbb{R}(E))$ and D_0 is a smooth complex linear Cauchy–Riemann operator on E.*

Real linear Cauchy–Riemann operators satisfy the equation

$$D(f\xi) = f(D\xi) + (\bar{\partial}f)\xi$$

only for real valued functions f. As in the complex case, they arise via the formula

$$\bar{\partial}^\nabla \xi := (\nabla\xi)^{0,1} = \tfrac{1}{2}(\nabla\xi + J\nabla\xi \circ j)$$

from connections on (E, J), but this time ∇ is no longer required to be Hermitian and its connection potentials need only be of class $W^{\ell-1,p}$. Thus ∇ may not preserve either the complex structure or the metric on E. More precisely, let ∇_0 be any smooth Hermitian connection on E. Then we may write

$$\nabla = \nabla_0 + A, \qquad A \in W^{\ell-1,p}(\Sigma, T^*\Sigma \otimes_\mathbb{R} \mathrm{End}_\mathbb{R}(E)),$$

so that $D = \bar{\partial}^\nabla$ has the form

$$D\xi = D_0\xi + (A\xi)^{0,1},$$

where $D_0 = \bar{\partial}^{\nabla_0}$ is the (complex linear) Cauchy–Riemann operator associated to ∇_0.

EXERCISE C.1.6. Show that every real linear Cauchy–Riemann operator of class $W^{\ell-1,p}$ equals $\bar{\partial}^\nabla$ for some connection ∇ of class $W^{\ell-1,p}$.

EXERCISE C.1.7. Show that every real linear Cauchy–Riemann operator splits as a sum of a complex linear Cauchy–Riemann operator and a zeroth order complex antilinear operator.

The next exercise shows that every complex linear Cauchy–Riemann operator on a line bundle is gauge equivalent to a smooth complex linear Cauchy–Riemann operator.

EXERCISE C.1.8. Let $E \to S$ be a complex line bundle over a closed Riemann surface and a complex Cauchy–Riemann operator D of class L^p over S, where $p > 2$. Show that

$$D = D_0 + \alpha^{0,1},$$

where D_0 is smooth and complex linear and $\alpha \in L^p(\Sigma, T^*\Sigma \otimes_{\mathbb{R}} i\mathbb{R})$. By Hodge theory, there is a decomposition $\alpha = \alpha_0 + df + *dg$, where $f, g \in W^{1,p}(S, i\mathbb{R})$ and $\alpha_0 \in \Omega^1(S, i\mathbb{R})$ is harmonic and hence smooth. Show that the function $u \in W^{1,p}(S, \mathbb{C}^*)$, given by $u := e^{-f-ig}$, satisfies $u^{-1}\bar{\partial}u = -(df + *dg)^{0,1}$ and hence

$$u^{-1} \circ D \circ u = D_0 + \alpha_0^{0,1}.$$

The Riemann–Roch theorem. A Riemann–Roch boundary value problem on a vector bundle $E \to \Sigma$ is an operator

$$D_F : W^{\ell,p}_F(\Sigma, E) \to W^{\ell-1,p}(\Sigma, \Lambda^{0,1}T^*\Sigma \otimes E)$$

where $F \subset E|_{\partial\Sigma}$ is a totally real subbundle and D_F is the restriction of a real linear Cauchy–Riemann operator D of class $W^{\ell-1,p}$ to the space

$$W^{\ell,p}_F(\Sigma, E) := \{\xi \in \mathrm{W}^{\ell,p}(\Sigma, E) \,|\, \xi(\partial\Sigma) \subset F\}.$$

As always, we assume $p > 1$ and $\ell p > 2$. A Hermitian inner product $\langle \cdot, \cdot \rangle$ on E is called **compatible with** F if $JF = F^\perp$. The dual operator of D_F with respect to such a Hermitian form on E and a volume form dvol on Σ is the restriction

$$D^*_F : W^{\ell,p}_F(\Sigma, \Lambda^{0,1}T^*\Sigma \otimes E) \to W^{\ell-1,p}(\Sigma, E),$$

of the formal adjoint operator D^* to the subspace

$$W^{\ell,p}_F(\Sigma, \Lambda^{0,1}T^*\Sigma \otimes E) := \{\eta \in W^{\ell,p}(\Sigma, \Lambda^{0,1}T^*\Sigma \otimes E) \,|\, \eta(T\partial\Sigma) \subset F\}.$$

We denote the smooth analogs of these reduced spaces by $\Omega^0_F(\Sigma, E)$ and $\Omega^{0,1}_F(\Sigma, E)$. The following lemma is the key to showing that these boundary conditions are Fredholm. In other words, they guarantee that the corresponding operators D_F and D^*_F are Fredholm (see Step 3 in the proof of Theorem C.2.3).

LEMMA C.1.9. *For every $\xi \in \Omega^0_F(\Sigma, E)$ and every $\eta \in \Omega^{0,1}_F(\Sigma, E)$*

$$\int_\Sigma \langle D^*_F\eta, \xi \rangle \,\mathrm{dvol} = \int_\Sigma \langle \eta, D_F\xi \rangle \,\mathrm{dvol}.$$

PROOF. Since $JF = F^\perp$, the 1-form $\langle \eta, J\xi \rangle$ vanishes on $T\partial\Sigma$ for every $\xi \in \Omega^0_F(\Sigma, E)$ and every $\eta \in \Omega^{0,1}_F(\Sigma, E)$. Hence when D is smooth and complex linear the result follows from Remark C.1.3. The general case is an immediate consequence since the formal adjoint of a zeroth order operator α coincides with its actual functional analytic adjoint (i.e. its transpose α^T, which satisfies the equation $\langle \alpha^T\eta, \xi \rangle = \langle \eta, \alpha\xi \rangle$ pointwise). □

Here is the main theorem of this appendix.

THEOREM C.1.10 (Riemann–Roch). *Let $E \to \Sigma$ be a complex vector bundle over a compact Riemann surface with boundary and $F \subset E|_{\partial\Sigma}$ be a totally real subbundle. Let D be a real linear Cauchy–Riemann operator on E of class $W^{\ell-1,p}$, where ℓ is a positive integer and $p > 1$ such that $\ell p > 2$. Then the following holds for every integer $k \in \{1, \dots, \ell\}$ and every real number $q > 1$ such that $k - 2/q \le \ell - 2/p$.*

(i) *The operators*

$$D_F : W^{k,q}_F(\Sigma, E) \to W^{k-1,q}(\Sigma, \Lambda^{0,1}T^*\Sigma \otimes E),$$

$$D_F^* : W^{k,q}_F(\Sigma, \Lambda^{0,1}T^*\Sigma \otimes E) \to W^{k-1,q}(\Sigma, E)$$

are Fredholm. Moreover, their kernels are independent of k and q, and we have

$$\eta \in \operatorname{im} D_F \iff \int_\Sigma \langle \eta, \eta_0 \rangle \mathrm{dvol} = 0 \quad \forall \eta_0 \in \ker D_F^* \tag{C.1.3}$$

for $\eta \in W^{k-1,q}(\Sigma, \Lambda^{0,1}T^*\Sigma \otimes E)$ and

$$\xi \in \operatorname{im} D_F^* \iff \int_\Sigma \langle \xi, \xi_0 \rangle \mathrm{dvol} = 0 \quad \forall \xi_0 \in \ker D_F \tag{C.1.4}$$

for $\xi \in W^{k-1,q}(\Sigma, E)$.

(ii) *The real Fredholm index of D_F is given by*

$$\operatorname{index}(D_F) = n\,\chi(\Sigma) + \mu(E, F),$$

where $\chi(\Sigma)$ is the Euler characteristic of Σ, $\mu(E,F)$ is the boundary Maslov index (see Section C.3), and n is the complex rank of E. In the case $\partial\Sigma = \emptyset$ the boundary Maslov index is equal to twice the first Chern number, i.e. $\mu(E) = 2\langle c_1(E), [\Sigma]\rangle$.

(iii) *If $n = 1$ then*

$$\mu(E, F) < 0, \quad \Longrightarrow \quad D_F \text{ is injective},$$

$$\mu(E, F) + 2\chi(\Sigma) > 0 \quad \Longrightarrow \quad D_F \text{ is surjective}.$$

Serre duality. In Theorem C.1.10 we have used the L^2 inner products on $\Omega^0(\Sigma, E)$ and $\Omega^{0,1}(\Sigma, E)$ induced by the Riemannian metric on Σ and the Hermitian structure on E to write the dual operator of $D_F : \Omega^0_F(\Sigma, E) \to \Omega^{0,1}(\Sigma, E)$ as an operator from $\Omega^{0,1}_F(\Sigma, E)$ to $\Omega^0(\Sigma, E)$. Here is an alternative description of the dual operator in the complex setting which does not require a volume form on Σ or a Hermitian structure on E. Consider the complex vector bundle

$$E^* := \operatorname{Hom}^{\mathbb{C}}(E, \mathbb{C})$$

and the totally real subbundle

$$F^* := \Big\{(z, \zeta) \,|\, z \in \partial\Sigma,\ \zeta \in \operatorname{Hom}^{\mathbb{C}}(E_z, \mathbb{C}),\ \zeta(F_z) \subset \mathbb{R}\Big\} \subset E^*|_{\partial\Sigma}.$$

Slightly abusing notation we denote in this subsection, and in this subsection only, the complex dual space of $T_z\Sigma$ by $T_z^*\Sigma = \operatorname{Hom}^{\mathbb{C}}(T_z\Sigma, \mathbb{C})$ (instead of $\Lambda^{1,0}T_z^*\Sigma$ as elsewhere in this book). Thus

$$T^*\Sigma \otimes_{\mathbb{C}} E^* := \operatorname{Hom}^{\mathbb{C}}(T\Sigma, E^*), \qquad T^*\partial\Sigma \otimes_{\mathbb{R}} F^* \subset \operatorname{Hom}^{\mathbb{C}}(T\Sigma, E^*)|_{\partial\Sigma}.$$

A section of $T^*\Sigma \otimes_{\mathbb{C}} E^*$ is a complex linear 1-form on Σ with values in E^*, i.e. $\Omega^0(\Sigma, T^*\Sigma \otimes_{\mathbb{C}} E^*) = \Omega^{1,0}(\Sigma, E^*)$. A section of the totally real subbundle $T^*\partial\Sigma \otimes_{\mathbb{R}} F^*$

over the boundary sends tangent vectors of the boundary to F^*. There is a natural complex bilinear pairing

$$\Omega^0(\Sigma, T^*\Sigma \otimes_{\mathbb{C}} E^*) \times \Omega^{0,1}(\Sigma, E) \to \mathbb{C} : (\zeta, \eta) \mapsto \int_\Sigma \zeta \wedge \eta, \tag{C.1.5}$$

where the 2-form $\zeta \wedge \eta \in \Omega^{1,1}(\Sigma)$ is defined via the pairing between E^* and E. By the Riesz representation theorem this pairing identifies the space of L^2-sections of $T^*\Sigma \otimes_{\mathbb{C}} E^*$ with the space of complex linear functionals on $\Omega^{0,1}(\Sigma, E)$ that are continuous with respect to the L^2-topology. Likewise, there is a natural isomorphism

$$\Omega^{0,1}(\Sigma, T^*\Sigma \otimes_{\mathbb{C}} E^*) \cong \Omega^{1,1}(\Sigma, E^*).$$

Explicitly, if $\phi : T_z\Sigma \to \mathrm{Hom}^{\mathbb{C}}(T_z\Sigma, E_z^*)$ is a complex anti-linear homomorphism the corresponding 2-form with values in E_z^* is given by

$$\omega(\widehat{z}_1, \widehat{z}_2) := \phi(\widehat{z}_1)(\widehat{z}_2) - \phi(\widehat{z}_2)(\widehat{z}_1) \in E_z^*, \qquad \widehat{z}_1, \widehat{z}_2 \in T_z\Sigma. \tag{C.1.6}$$

The inverse sends $\omega \in \Lambda^{1,1}T_z^*\Sigma \otimes E_z^*$ to the complex anti-linear homomorphism $T_z\Sigma \to \mathrm{Hom}^{\mathbb{C}}(T_z\Sigma, E_z^*) : \widehat{z} \mapsto \phi(\widehat{z}) := \frac{1}{2}\big(\omega(\widehat{z}, \cdot) - i\omega(\widehat{z}, i\cdot)\big)$. (*Exercise:* Prove this and translate it into local coordinates.) Hence there is a complex bilinear pairing

$$\Omega^{0,1}(\Sigma, T^*\Sigma \otimes_{\mathbb{C}} E^*) \times \Omega^0(\Sigma, E) \to \mathbb{C} : (\omega, \xi) \mapsto \int_\Sigma \omega \wedge \xi \tag{C.1.7}$$

where the 2-form $\omega \wedge \xi \in \Omega^{1,1}(\Sigma)$ is defined via the pairing between E^* and E.

Fix a holomorphic structure on E, i.e. local trivializations with holomorphic transition maps $g_{\beta\alpha} : U_\alpha \cap U_\beta \to \mathrm{GL}(n, \mathbb{C})$ and denote by $\bar\partial : \Omega^0(\Sigma, E) \to \Omega^{0,1}(\Sigma, E)$ be the corresponding Cauchy–Riemann operator. There are corresponding trivializations of E^* with transition maps $g_{\beta\alpha}^T$ and they determine a Cauchy–Riemann operator $\bar\partial : \Omega^0(\Sigma, E^*) \to \Omega^{0,1}(\Sigma, E^*)$. Combining this with the holomorphic structure on $T^*\Sigma$ we obtain a Cauchy–Riemann operator

$$\bar\partial : \Omega^0(\Sigma, T^*\Sigma \otimes_{\mathbb{C}} E^*) \to \Omega^{0,1}(\Sigma, T^*\Sigma \otimes_{\mathbb{C}} E^*).$$

With $\Omega^0(\Sigma, T^*\Sigma \otimes_{\mathbb{C}} E^*) \cong \Omega^{1,0}(\Sigma, E^*)$ and $\Omega^{0,1}(\Sigma, T^*\Sigma \otimes_{\mathbb{C}} E^*) \cong \Omega^{1,1}(\Sigma, E^*)$ this is the extended operator $\bar\partial : \Omega^{1,0}(\Sigma, E^*) \to \Omega^{1,1}(\Sigma, E^*)$. Now let

$$D = \bar\partial + A : \Omega^0(\Sigma, E^*) \to \Omega^{0,1}(\Sigma, E^*)$$

be a general Cauchy–Riemann operator with $A \in \Omega^{0,1}(\Sigma, \mathrm{End}_{\mathbb{R}}(E))$ and denote

$$D^* := -\bar\partial + A^* : \Omega^0(\Sigma, T^*\Sigma \otimes_{\mathbb{C}} E^*) \to \Omega^{0,1}(\Sigma, T^*\Sigma \otimes_{\mathbb{C}} E^*).$$

(This notation should not be confused with the formal adjoint operator in (C.1.1).) Then D and D^* satisfy the identity

$$\zeta \wedge (D\xi) - (D^*\zeta) \wedge \xi = d(\zeta \wedge \xi) \in \Omega^{1,1}(\Sigma, E^*), \tag{C.1.8}$$

for $\xi \in \Omega^0(\Sigma, E)$ and $\zeta \in \Omega^0(\Sigma, T^*\Sigma \otimes_{\mathbb{C}} E^*) \cong \Omega^{1,0}(\Sigma, E^*)$, where the $(1,0)$-form $\zeta \wedge \xi \in \Omega^{1,0}(\Sigma)$ is again understood in terms of the pairing between E^* and E. (*Exercise:* Prove equation (C.1.8), assuming first that $A = 0$.) This continues to hold when A is only of class $W^{\ell,p}$. Integrating equation (C.1.8) we obtain

$$\int_\Sigma \zeta \wedge (D\xi) - \int_\Sigma (D^*\zeta) \wedge \xi = \int_{\partial\Sigma} \zeta \wedge \xi \tag{C.1.9}$$

for $\xi \in \Omega^0(\Sigma, E)$ and $\zeta \in \Omega^0(\Sigma, T^*\Sigma \otimes_{\mathbb{C}} E^*) \cong \Omega^{1,0}(\Sigma, E^*)$. Note that the restriction of the 1-form to the boundary $\partial\Sigma$ is real valued whenever ξ and η satisfy the boundary conditions $\xi(\partial\Sigma) \subset F$ and $\zeta(\partial\Sigma) \subset T^*\partial\Sigma \otimes_{\mathbb{R}} F^*$.

COROLLARY C.1.11 (Serre Duality). *Let $E \to \Sigma$ be a complex vector bundle over a compact Riemann surface with boundary and $F \subset E|_{\partial\Sigma}$ be a totally real subbundle. Let D be a real linear Cauchy–Riemann operator on E of class $W^{\ell-1,p}$, where ℓ is a positive integer and $p > 1$ such that $\ell p > 2$. Let $\zeta \in L^r(\Sigma, T^*\Sigma \otimes_{\mathbb{C}} E^*)$ where $r > 1$. Then the following are equivalent.*

(i) $\int_\Sigma \zeta \wedge D\xi \in \mathbb{R}$ *for every* $\xi \in \Omega^0_F(\Sigma, E)$.

(ii) ζ *is of class* $W^{\ell,p}$, $D^*\zeta = 0$, *and* $\zeta|_{\partial\Sigma}$ *is a section of the subbundle* $T^*\partial\Sigma \otimes_{\mathbb{R}} F^*$.

PROOF. That (ii) implies (i) is obvious from the above discussion. To prove the converse, fix a Riemannian metric on Σ compatible with the complex structure and choose a real valued Hermitian inner product $\langle\cdot,\cdot\rangle_E$ on E as in Remark C.1.2 such that JF is orthogonal to F. Denote by

$$\langle\cdot,\cdot\rangle^{\mathbb{C}}_E := \langle\cdot,\cdot\rangle_E + i\langle J\cdot,\cdot\rangle_E$$

the corresponding complex valued Hermitian form on E that is complex anti-linear in the first argument and is complex linear in the second argument. This Hermitian form induces a complex anti-linear isomorphism

$$\Omega^{0,1}_F(\Sigma, E) \to \Omega^0_{T^*\Sigma\otimes_{\mathbb{R}}F^*}(\Sigma, T^*\Sigma \otimes_{\mathbb{C}} E^*) : \eta \mapsto \zeta = \langle\eta,\cdot\rangle^{\mathbb{C}}_E.$$

If $\eta \in L^r(\Sigma, \Lambda^{0,1}T^*\Sigma \otimes_{\mathbb{C}} E)$ and $\zeta := \langle\eta,\cdot\rangle^{\mathbb{C}}_E$ is the corresponding L^r-section of $T^*\Sigma \otimes_{\mathbb{C}} E^*$ then

$$\operatorname{Im} \int_\Sigma \zeta \wedge D\xi = \int_\Sigma \langle J\eta \wedge D\xi\rangle_E = -\int_\Sigma \langle \eta \wedge JD\xi\rangle_E = -\int_\Sigma \langle\eta, D\xi\rangle_E \,\mathrm{dvol}_\Sigma$$

for every $\xi \in \Omega^0_F(\Sigma, E)$. Here we have used the fact that $JD\xi = -(D\xi)\circ j = *D\xi$, where $*$ denotes the Hodge $*$-operator on 1-forms. Hence, if ζ satisfies (i) and $q > 1$ is chosen such that $1/q + 1/r = 1$, it follows that $\eta \in L^r = (L^q)^*$ annihilates the image of the operator $D_F : W^{1,q}_F(\Sigma, E) \to L^q(\Sigma, \Lambda^{0,1}T^*\Sigma \otimes E)$. By Theorem C.1.10 (i), this image is equal to the annihilator of the kernel of the operator $D_F^* : W^{1,r}_F(\Sigma, \Lambda^{0,1}T^*\Sigma \otimes E) \to L^r(\Sigma, E)$. Since the annihilator of the annihilator of a closed subspace of a reflexive Banach space is equal to the original subspace, it follows that η belongs to the kernel of $D_F^* : W^{1,r}_F(\Sigma, \Lambda^{0,1}T^*\Sigma \otimes E) \to L^r(\Sigma, E)$. Since this kernel is independent of the choice of the Sobolev completion, again by Theorem C.1.10 (i), we have $\eta \in W^{\ell,p}_F(\Sigma, \Lambda^{0,1}T^*\Sigma \otimes E)$. Hence ζ is of class $W^{\ell,p}$ and satisfies $D^*\zeta = 0$ and the boundary condition. This proves Corollary C.1.11. □

REMARK C.1.12. In the case of a smooth complex linear Cauchy–Riemann operator on a closed Riemann surface, Corollary C.1.11 asserts that the kernel of $\bar\partial : \Omega^0(\Sigma, T^*\Sigma \otimes_{\mathbb{C}} E^*) \to \Omega^{0,1}(\Sigma, T^*\Sigma \otimes_{\mathbb{C}} E^*)$ is isomorphic to the cokernel of $\bar\partial : \Omega^0(\Sigma, E) \to \Omega^{0,1}(\Sigma, E)$. This is a special case of Serre duality. In general, the Serre Duality Theorem asserts that for a closed complex manifold M of complex dimension n and a holomorphic vector bundle $E \to M$ there is an isomorphism

$$H^q(M, \Omega^p(E^*)) \cong H^{n-q}(M, \Omega^{n-p}(E))^*$$

of complex vector spaces. These are sheaf cohomology groups and $\Omega^p(E)$ denotes the sheaf of germs of holomorphic p-forms on M with values in E.

C.2. Elliptic estimates

In this section we shall prove the Fredholm property for Cauchy–Riemann operators and the duality statement in assertion (i) of Theorem C.1.10. The reader who is mainly interested in the proof of the index formula may wish to go directly to Sections C.3 and C.4.

The proof of the Fredholm property is based on Lemma A.1.1 which states that an operator D has closed image and finite dimensional kernel provided that a certain estimate hold. The required estimate for smooth Cauchy–Riemann operators is established in Lemma C.2.1 below. Hence in the smooth case the operators D_F and D_F^* have finite dimensional kernel and closed image. The same result holds in general because, as we show in Lemma C.2.2 below, every Cauchy–Riemann operator of class $W^{\ell-1,p}$ is a compact perturbation of a smooth Cauchy–Riemann operator. It remains to prove that the cokernels are finite dimensional as well. This follows from a duality theorem which asserts that the cokernel of D_F can be identified with the kernel of D_F^* and vice versa. A refined version of this duality statement is formulated in Theorem C.2.3 below. The proof is based on elliptic regularity and requires a number of steps under our weak regularity assumptions on the coefficients of the operator D. However, as we explain in Steps 2 and 3, the smooth case is a straightforward consequence of Lemma C.1.9 and the regularity results in Section B.4.

We assume throughout that $(\Sigma, j, \mathrm{dvol})$ is a compact Riemann surface with boundary, that E is a complex vector bundle over Σ, $F \subset E|_{\partial\Sigma}$ is a totally real subbundle, and $\langle\cdot,\cdot\rangle$ is a Hermitian inner product on E such that $JF = F^\perp$.

LEMMA C.2.1. *Let D be a smooth real linear Cauchy–Riemann operator. Then, for every positive integer k and every $q > 1$ there exists a constant $c > 0$ such that*

$$\|\xi\|_{W^{k,q}} \le c\,(\|D_F\xi\|_{W^{k-1,q}} + \|\xi\|_{W^{k-1,q}})$$

for every $\xi \in W^{k,q}_F(\Sigma, E)$ and

$$\|\eta\|_{W^{k,q}} \le c\,(\|D_F^*\eta\|_{W^{k-1,q}} + \|\eta\|_{W^{k-1,q}})$$

for every $\eta \in W^{k,q}_F(\Sigma, \Lambda^{0,1}T^*\Sigma \otimes_{\mathbb{C}} E)$.

PROOF. In a local trivialization of E along a holomorphic coordinate chart on Σ the equation $D_F\xi = \zeta$ has the form

$$\partial_s\xi + \Phi\xi + i(\partial_t\xi + \Psi\xi) = \zeta, \qquad \xi(s,0) \in \mathbb{R}^n. \tag{C.2.1}$$

Here $s + it$ denotes the coordinate on an open subset Ω of the upper half plane

$$\mathbb{H} := \{s + it \in \mathbb{C}\,|\,t \ge 0\}$$

and $\Phi\,ds + \Psi\,dt$ is the connection potential of the corresponding connection ∇ (see Exercise C.1.6). Thus $\Phi, \Psi \in C^\infty(\Omega, \mathrm{End}_{\mathbb{R}}(\mathbb{C}^n))$. The function $\zeta \in W^{k-1,q}(\Omega, \mathbb{C}^n)$ represents the $(0,1)$-form $\frac{1}{2}(\zeta\,ds - i\zeta\,dt)$. By Proposition B.4.9 there exist, for every open set Ω' whose closure is contained in Ω, positive constants c and c' such that

$$\begin{aligned}\|\xi\|_{W^{k,q}(\Omega')} &\le c\left(\|\zeta - \Phi\xi - J\Psi\xi\|_{W^{k-1,q}(\Omega)} + \|\xi\|_{W^{k-1,q}(\Omega)}\right)\\ &\le c'\left(\|\zeta\|_{W^{k-1,q}(\Omega)} + \|\xi\|_{W^{k-1,q}(\Omega)}\right)\end{aligned}$$

for every solution $\xi \in W^{k,q}_{\mathrm{loc}}(\Omega, \mathbb{C}^n)$ of (C.2.1).

To prove the estimate, cover Σ by finitely many open sets each of which is biholomorphic to an open ball in $\mathbb{H}$ and then choose a unitary trivialization of E over each of the open sets in this cover. This proves the first assertion.

Now let E^* denote the bundle E with the complex structure $-J$. Then, by Remark C.1.3, the operator D^* is conjugate to a Cauchy–Riemann operator on the bundle $T\Sigma \otimes_{\mathbb{C}} E^*$ with the totally real subbundle $T\partial\Sigma \otimes_{\mathbb{R}} F$ over the boundary. Hence the second assertion follows from the first. □

It follows from Lemmas A.1.1 and C.2.1 that, in the smooth case, both operators D_F and D_F^* have a finite dimensional kernel and a closed image. That this continues to hold for Cauchy–Riemann operators of class $W^{\ell-1,p}$ is a consequence of Corollary A.1.2 and the next lemma.

LEMMA C.2.2. *Let ℓ be an integer and $p > 1$ such that $\ell p > 2$. Let*

$$A \in W^{\ell-1,p}(\Sigma, T^*\Sigma \otimes_{\mathbb{R}} \mathrm{End}_{\mathbb{R}}(E)).$$

Then the linear operator

$$W^{k,q}(\Sigma, E) \to W^{k-1,q}(\Sigma, \Lambda^{0,1}T^*\Sigma \otimes E) : \xi \mapsto (A\xi)^{0,1}$$

is well defined and compact whenever $1 \le k \le \ell$ and $q > 1$ and $k - 2/q \le \ell - 2/p$.

PROOF. If $k = \ell$ then $q \le p$ and hence $W^{\ell-1,p} \subset W^{k-1,q}$, by Hölder's inequality. If $k \le \ell - 2$ then, by Theorem B.1.11, we obtain

$$W^{\ell-1,p} \subset C^{k-1} \subset W^{k-1,q}$$

for every q. If $k = \ell - 1$, then the inequality $k - 2/q \le \ell - 2/p$ is equivalent to $2/p - 1 \le 2/q$ and hence, in the case $p < 2$, to

$$q \le \frac{2p}{2-p}.$$

Hence, by Theorem B.1.12, we have $W^{\ell-1,p} \subset W^{k-1,q}$. This holds in all cases. Thus we have proved that $A \in W^{k-1,q}$.

Now suppose that $kq > 2$. Then the map $W^{k,q} \to W^{k-1,q} : \xi \mapsto (A\xi)^{0,1}$ factors through the inclusion $W^{k,q} \to C^{k-1}$. By Theorem B.1.11, this inclusion is compact. This proves the lemma in the case $kq > 2$.

If $kq \le 2$ we have $k = 1$ and $q \le 2$. Moreover, by Theorem B.1.12, we have $A \in L^s$ for some $s > 2$. Let

$$r := \frac{sq}{s-q} < \frac{2q}{2-q}.$$

Then, by Theorem B.1.12, the inclusion $W^{1,q} \to L^r$ is compact. Moreover, since

$$\frac{1}{r} + \frac{1}{s} = \frac{1}{q},$$

it follows from Hölder's inequality that multiplication by A defines a bounded linear operator from L^r to L^q. This proves the lemma in the case $kq \le 2$. □

To establish the Fredholm property of Cauchy–Riemann operators we must prove that the cokernel of D_F can be identified with the kernel of D_F^*. This is the content of the following theorem which restates assertion (i) of Theorem C.1.10.

THEOREM C.2.3. *Let D be a real linear Cauchy–Riemann operator of class $W^{\ell-1,p}$, where ℓ is a positive integer and $p>1$ such that $\ell p>2$. Let k be an integer such that $1\le k\le \ell$ and $q,r>1$ such that $2/r-1\le k-2/q\le \ell-2/p$. Then the following holds.*

(i) *The operators*

$$D_F : W^{k,q}_F(\Sigma,E)\to W^{k-1,q}(\Sigma,\Lambda^{0,1}T^*\Sigma\otimes_{\mathbb{C}} E),$$
$$D^*_F : W^{k,q}_F(\Sigma,\Lambda^{0,1}T^*\Sigma\otimes_{\mathbb{C}} E)\to W^{k-1,q}(\Sigma,E).$$

are Fredholm and $\mathrm{index}(D_F)+\mathrm{index}(D^*_F)=0$. *Moreover, the Fredholm index of D_F is independent of the choice of D and of the complex structure on Σ. It is also independent of k and q.*

(ii) *If $\eta\in L^r(\Sigma,\Lambda^{0,1}T^*\Sigma\otimes_{\mathbb{C}} E)$ and $\xi\in W^{k-1,q}(\Sigma,E)$ satisfy*

$$\int_\Sigma\langle\eta,D_F\zeta\rangle\mathrm{dvol}=\int_\Sigma\langle\xi,\zeta\rangle\mathrm{dvol} \tag{C.2.2}$$

for every $\zeta\in W^{k,q}_F(\Sigma,E)$ then $\eta\in W^{k,q}_F(\Sigma,\Lambda^{0,1}T^*\Sigma\otimes_{\mathbb{C}} E)$ and $D^*_F\eta=\xi$.

(iii) *If $\xi\in L^r(\Sigma,E)$ and $\eta\in W^{k-1,q}(\Sigma,\Lambda^{0,1}T^*\Sigma\otimes_{\mathbb{C}} E)$ satisfy*

$$\int_\Sigma\langle D^*_F\zeta,\xi\rangle\mathrm{dvol}=\int_\Sigma\langle\zeta,\eta\rangle\mathrm{dvol} \tag{C.2.3}$$

for every $\zeta\in W^{k,q}_F(\Sigma,\Lambda^{0,1}T^*\Sigma\otimes_{\mathbb{C}} E)$ then $\xi\in W^{k,q}_F(\Sigma,E)$ and $D_F\xi=\eta$.

PROOF. The proof has eight steps. The first step shows that the hypotheses in (ii) and (iii) are meaningful. Steps 2 and 3 prove the result for smooth Cauchy–Riemann operators.

STEP 1. (a) *Let $s>1$ such that $1/s+1/r=1$. Then there are inclusions $W^{k-1,q}\hookrightarrow L^s$ and $L^r\hookrightarrow (W^{k-1,q})^*$.*

(b) *There is an inclusion $W^{k,q}\hookrightarrow (W^{\ell-1,p})^*$.*

(c) *There is an inclusion $W^{k,q}\hookrightarrow (W^{k-1,q})^*$ if and only if $-1/2\le k-2/q$.*

The inequality $-2/s=2/r-2\le k-1-2/q$ shows that if $(k-1)q<2$ then $s\le 2q/(2-(k-1)q)$. Hence the inclusion $W^{k-1,q}\hookrightarrow L^s$ follows from the Sobolev embedding theorem B.1.12. This proves (a).

Assertion (b) is obvious whenever either $kq\ge 2$ or $(\ell-1)p\ge 2$. Hence assume $kq<2$ and $(\ell-1)p<2$. Then, by Theorem B.1.12, there is an inclusion $W^{k,q}\hookrightarrow L^{2q/(2-kq)}$. It follows also from Theorem B.1.12 that there are (dense) inclusions

$$W^{\ell-1,p}\hookrightarrow L^{2p/(2+p-\ell p)},\qquad L^{2p/(p+\ell p-2)}\hookrightarrow \left(W^{\ell-1,p}\right)^*.$$

Since $\ell p>2$ we have $2/q-k\le 1\le 1+\ell-2/p$ and hence $2q/(2-kq)\ge 2p/(p+\ell p-2)$. This implies

$$W^{k,q}\subset L^{2q/(2-kq)}\subset L^{2p/(p+\ell p-2)}\subset (W^{\ell-1,p})^*.$$

This proves (b).

Assertion (c) is obvious when $kq\ge 2$. Hence assume $kq<2$. Then $k=1$ and $1<q<2$. By the Sobolev embedding theorem B.1.12 we have

$$W^{1,q}\hookrightarrow (L^q)^*\cong L^{q/(q-1)}\qquad\iff\qquad \frac{q}{q-1}\le\frac{2q}{2-q}.$$

This holds if and only if $q\ge 4/3$ which is equivalent to $-1/2\le 1-2/q=k-2/q$. This proves (c) and Step 1.

STEP 2. *We prove (ii) and (iii) under the assumption that D is a smooth Cauchy–Riemann operator.*

Consider the weak equation in a local trivialization along local holomorphic coordinates $s+it$ on Σ with volume form $\lambda^2\, ds \wedge dt$. Then (C.2.2) has the form

$$\int_\Omega \langle \eta, \partial_s\zeta + i\partial_t\zeta\rangle\, dsdt = \int_\Omega \langle \lambda^2\xi - \Phi^*\eta + \Psi^* i\eta, \zeta\rangle\, dsdt, \tag{C.2.4}$$

where $\eta \in L^r_{\mathrm{loc}}(\Omega, \mathbb{C}^n)$ represents the $(0,1)$-form $\frac{1}{2}(\eta\, ds - i\eta\, dt)$, $\xi \in W^{k-1,q}_{\mathrm{loc}}(\Omega, \mathbb{C}^n)$, and the test function $\zeta \in C^\infty_0(\Omega, \mathbb{C}^n)$ satisfies the boundary condition $\zeta(s,0) \in \mathbb{R}^n$ for $s \in \mathbb{R}$ such that $(s,0) \in \Omega$. By assumption, the functions $\Phi, \Psi : \Omega \to \mathrm{End}_{\mathbb{R}}(\mathbb{C}^n)$ are smooth. We denote by $\Phi^*, \Psi^* : \Omega \to \mathrm{End}_{\mathbb{R}}(\mathbb{C}^n)$ the (real) adjoint matrices.

Suppose $\eta \in L^r_{\mathrm{loc}}(\Omega, \mathbb{C}^n)$ satisfies (C.2.4) for every $\zeta \in C^\infty_0(\Omega, \mathbb{C}^n)$ such that $\zeta(\Omega \cap \mathbb{R}) \subset \mathbb{R}^n$. We must prove that $\eta \in W^{k,q}_{\mathrm{loc}}(\Omega, \mathbb{C}^n)$. To see this, note that[1]

$$\lambda^2\xi - \Phi^*\eta + \Psi^* i\eta \in L^{r_0}_{\mathrm{loc}}(\Omega, \mathbb{C}^n), \qquad r_0 := \min\{r, q\} > 1,$$

Hence it follows from Lemma B.4.6 (with $p = \infty$) that $\eta \in W^{1,r_0}_{\mathrm{loc}}$ and

$$-\partial_s\eta + i\partial_t\eta = \lambda^2\xi - \Phi^*\eta + \Psi^* i\eta, \qquad \eta(s,0) \in \mathbb{R}^n.$$

If $r_0 \geq 2$ we obtain from Theorem B.1.12 that $\eta \in L^q_{\mathrm{loc}}$. If $r_0 < 2$, we obtain $\eta \in L^{2r_0/(2-r_0)}_{\mathrm{loc}}$ and hence

$$\lambda^2\xi - \Phi^*\eta + \Psi^* i\eta \in L^{r_1}_{\mathrm{loc}}(\Omega, \mathbb{C}^n), \qquad r_1 := \min\{2r_0/(2-r_0), q\}.$$

The number r_1 is either equal to q or bigger than two. In the latter case it follows from Lemma B.4.6 that $\eta \in W^{1,2}_{\mathrm{loc}}$ and hence, by Theorem B.1.12, $\eta \in L^q_{\mathrm{loc}}$. Now it follows from Proposition B.4.9 that, if $\eta \in W^{j-1,q}_{\mathrm{loc}}$ for some $j \in \{1, \dots, k\}$, then $\eta \in W^{j,q}_{\mathrm{loc}}$. Hence $\eta \in W^{k,q}_{\mathrm{loc}}$ as claimed. We emphasize that the condition $p > 2$ in Proposition B.4.9 is redundant since in the present situation the (almost) complex structure is smooth (see Exercise B.4.11).

To complete the proof of Step 2, cover the surface by finitely many holomorphic coordinate charts and apply the local result to unitary trivializations of E along these charts.

STEP 3. *We prove (i) under the assumption that D is a smooth Cauchy–Riemann operator.*

Assume first that $k = 1$ and $1/r + 1/q = 1$. Then $2/r - 1 = 1 - 2/q$ and so r satisfies the hypotheses of the theorem with $k = 1$. With D_F and D_F^* understood as operators from $W^{1,q}_F$ to L^q, we claim that

$$L^q(\Sigma, \Lambda^{0,1}T^*\Sigma \otimes_{\mathbb{C}} E) = \mathrm{im}\, D_F \oplus \ker D_F^*. \tag{C.2.5}$$

We prove that the intersection is zero. Let $\eta_0 \in \mathrm{im}\, D_F \cap \ker D_F^*$. Then $\eta_0 \in W^{1,q}_F(\Sigma, \Lambda^{0,1}T^*\Sigma \otimes_{\mathbb{C}} E)$ and, since $D_F^*\eta_0 = 0$, it follows from Lemma C.1.9, that η_0 belongs to the annihilator of the image of the operator $D_F : W^{1,r}_F \to L^r$. Hence, by Step 2 with the roles of r and q interchanged, η_0 is smooth. (Note that the condition $k - 2/r \leq \ell - 2/p$ might not be satisfied. However, this condition is immaterial

[1]Here one can clearly see the main difficulty of the proof. Suppose $k = \ell = 1$, $q = p$, and $1/r + 1/q = 1$. Then, if D is a Cauchy–Riemann operator of class L^p, we only know that the function $\lambda^2\xi - \Phi^*\eta + \Psi^* i\eta$ is locally integrable. However, there is no L^1 elliptic regularity theory. The Calderon-Zygmund inequality only holds for $p > 1$. This situation occurs in our main application.

for smooth Cauchy–Riemann operators.) Hence, again by Lemma C.1.9, η_0 also annihilates the image of $D_F : W^{1,q}_F \to L^q$. Since $\eta_0 \in \operatorname{im} D_F$, this shows that $\|\eta_0\|_{L^2} = 0$ and so $\eta_0 = 0$.

We prove that the direct sum $\operatorname{im} D_F \oplus \ker D_F^*$ is equal to $L^q(\Sigma, \Lambda^{0,1}T^*\Sigma \otimes_{\mathbb{C}} E)$. Since, by Lemma C.2.1, D_F has a closed image and D_F^* has a finite dimensional kernel, this direct sum is a closed subspace of $L^q(\Sigma, \Lambda^{0,1}T^*\Sigma \otimes_{\mathbb{C}} E)$. We must prove that it is dense. By the Hahn–Banach theorem, it suffices to show that its annihilator is zero. Let $\zeta \in L^r(\Sigma, \Lambda^{0,1}T^*\Sigma \otimes_{\mathbb{C}} E)$, and suppose that

$$\zeta \perp \operatorname{im} D_F \oplus \ker D_F^*.$$

Since $\zeta \perp \operatorname{im} D_F$ it follows from Step 2 that ζ is smooth and $D_F^*\zeta = 0$. Since $\zeta \perp \ker D_F^*$ this implies $\zeta = 0$. This proves (C.2.5). A similar formula holds with D_F and D_F^* interchanged. It follows from these two identities that D_F and D_F^* are Fredholm operators and that $\operatorname{index}(D_F) + \operatorname{index}(D_F^*) = 0$. This proves the assertion in the case $k = 1$.

Now suppose $k \geq 2$. With D_F and D_F^* understood as operators from $W^{k,q}_F$ to $W^{k-1,q}$ we claim that

$$\begin{aligned} W^{k-1,q}(\Sigma, \Lambda^{0,1}T^*\Sigma \otimes_{\mathbb{C}} E) &= \operatorname{im} D_F \oplus \ker D_F^*, \\ W^{k-1,q}(\Sigma, E) &= \operatorname{im} D_F^* \oplus \ker D_F. \end{aligned} \tag{C.2.6}$$

To prove the first equation in (C.2.6) note first that the intersection is zero, since the kernel of D_F^* consists of smooth sections and the image of D_F is contained in the image of the operator $D_F : W^{1,q}_F \to L^q$. Now let $\eta \in W^{k-1,q}(\Sigma, \Lambda^{0,1}T^*\Sigma \otimes_{\mathbb{C}} E)$. Then, by (C.2.5), there exists an $\eta_0 \in W^{1,q}(\Sigma, \Lambda^{0,1}T^*\Sigma \otimes_{\mathbb{C}} E)$ and a $\xi \in W^{1,q}_F(\Sigma, E)$ such that

$$\eta = D_F\xi + \eta_0, \qquad D_F^*\eta_0 = 0.$$

As noted above, η_0 is smooth and so

$$D_F\xi = \eta - \eta_0 \in W^{k-1,q}(\Sigma, \Lambda^{0,1}T^*\Sigma \otimes_{\mathbb{C}} E).$$

Moreover, by the Sobolev embedding theorems B.1.11 and B.1.12, ξ is continuous. Hence it follows from assertion (iii) for smooth Cauchy–Riemann operators (which was proved in Step 2) that $\xi \in W^{k,q}_F(\Sigma, E)$. This proves the first identity in (C.2.6). The second identity follows by interchanging the roles of D_F and D_F^*. It follows from (C.2.6) that D_F and D_F^* are Fredholm operators. That their indices are independent of k and q follows from the fact that the kernels of D_F and D_F^* consist of smooth sections. That the index of D_F is independent of the choice of D follows from the fact that the difference of two Cauchy–Riemann operators is a compact operator from $W^{k,q}_F(\Sigma, E)$ to $W^{k-1,q}(\Sigma, \Lambda^{0,1}T^*\Sigma \otimes E)$ by Rellich's theorem (see Theorem B.1.12 and Theorem A.1.5). To prove that it is also independent of the complex structure on Σ use parallel transport along a smooth family $\lambda \mapsto j_\lambda$ of complex structures on Σ to construct a smooth family of bundle isomorphisms $\Phi_\lambda : T\Sigma \to T\Sigma$ such that $\Phi_0 = \mathrm{id}$ and $\Phi_\lambda j_0 = j_\lambda \Phi_\lambda$ for all λ. Fix a Hermitian connection ∇ on E and, for each λ, let D_λ be the Cauchy–Riemann operator associated to ∇ and j_λ. Then we obtain a continuous family of Fredholm operators

$$W^{k,q}_F(\Sigma, E) \to W^{k-1,q}(\Sigma, \Lambda^{0,1}_{j_0}T^*\Sigma \otimes E) : \xi \mapsto (D_\lambda\xi) \circ \Phi_\lambda.$$

By Theorem A.1.5 the Fredholm index of this operator (and hence that of D_λ) is independent of λ. This proves Step 3.

At this point we have proved the theorem for smooth Cauchy–Riemann operators. The remainder of the proof extends the result to operators of class $W^{\ell-1,p}$. Step 4 establishes the Fredholm property, Steps 5 and 6 deal with the case $-1/2 \le k - 2/q$, and Steps 7 and 8 with the general case.

Standing assumptions. *From now on we assume that D is a Cauchy–Riemann operator on E of class $W^{\ell-1,p}$, where ℓ is a positive integer and $p > 1$ such that $\ell p > 2$. We also assume that $k \in \{1, \dots, \ell\}$ and $q > 1$ such that $k - 2/q < \ell - 2/p$, and we think of D_F and D_F^* as operators from $W_F^{k,q}$ to $W^{k-1,q}$.*

STEP 4. *Assertion (i) holds in general.*

For Cauchy–Riemann operators of class $W^{\ell-1,p}$ assertion (i) follows from Step 3, Lemma C.2.2, and Theorem A.1.5. Namely, Lemma C.2.2 shows that every Cauchy–Riemann operator of class $W^{\ell-1,p}$ is a compact perturbation of a smooth Cauchy–Riemann operator and hence is a Fredholm operator and has the same index.

STEP 5. *Assume $-1/2 \le k - 2/q$. Then D_F and D_F^* satisfy (C.2.6). Moreover, the inclusions* $\ker D_F \hookrightarrow \operatorname{coker} D_F^*$ *and* $\ker D_F^* \hookrightarrow \operatorname{coker} D_F$ *are isomorphisms.*

By Step 4, the operators are Fredholm and $\operatorname{index}(D_F) + \operatorname{index}(D_F^*) = 0$. Hence

$$\dim \ker D_F + \dim \ker D_F^* = \dim \operatorname{coker} D_F + \dim \operatorname{coker} D_F^*. \tag{C.2.7}$$

By Step 1 (c), the condition $-1/2 \le k - 2/q$ guarantees the existence of a natural inclusion of $W^{k,q}$ into the dual space of $W^{k-1,q}$. With this understood, it follows from Lemma C.1.9 that the kernel of D_F, under the inclusion

$$W_F^{k,q}(\Sigma, E) \hookrightarrow (W_F^{k-1,q}(\Sigma, E))^*,$$

is mapped to the cokernel (that is to the annihilator of the image) of D_F^*. Likewise, the kernel of D_F^* is mapped to the cokernel of D_F. Now the dimension formula (C.2.7) shows that these inclusions must be isomorphisms. Thus we have proved that equation (C.2.6) continues to hold for Cauchy–Riemann operators of class $W^{\ell-1,p}$, provided that $k \le \ell$ and $-1/2 \le k - 2/q \le \ell - 2/p$.

STEP 6. *Assume $-1/2 \le k - 2/q$. Then D_F and D_F^* satisfy (ii) and (iii).*

Suppose that $\eta \in L^r(\Sigma, \Lambda^{0,1}T^*\Sigma \otimes_{\mathbb{C}} E)$ and $\xi \in W^{k-1,q}(\Sigma, F)$ satisfy (C.2.2). In the case $kq < 2$ we have $k = 1$ and may assume, without loss of generality, that $1/r + 1/q = 1$. (Thus r is the smallest real number greater than one satisfying $2/r - 1 \le k - 2/q$ with $k = 1$.) By the second identity in (C.2.6), proved in Step 5 under the present assumptions, there exist $\xi_0 \in W_F^{k,q}(\Sigma, E)$ and $\eta_1 \in W_F^{k,q}(\Sigma, \Lambda^{0,1}T^*\Sigma \otimes_{\mathbb{C}} E)$ such that

$$\xi = \xi_0 + D_F^*\eta_1, \qquad D_F\xi_0 = 0. \tag{C.2.8}$$

Now choose s such that $1/r + 1/s = 1$. Then, by Step 1, there is an inclusion $W^{k-1,q} \hookrightarrow L^s$ and the bounded linear functional $\langle \eta, \cdot \rangle \colon L^s \to \mathbb{R}$ satisfies

$$\langle \eta, D_F\zeta \rangle = \langle \xi, \zeta \rangle = \langle \xi_0 + D_F^*\eta_1, \zeta \rangle = \langle \xi_0, \zeta \rangle + \langle \eta_1, D_F\zeta \rangle \tag{C.2.9}$$

for every $\zeta \in W_F^{k,q}(\Sigma, E)$. Here the last identity follows from Lemma C.1.9 and the fact that $W^{k,q}$ embeds into the dual space of $W^{k-1,q}$. With $\zeta = \xi_0$ we obtain

$$\|\xi_0\|_{L^2}^2 = \langle \xi_0, \xi_0 \rangle = \langle \eta - \eta_1, D_F\xi_0 \rangle = 0.$$

Hence $\xi_0 = 0$ and hence $\langle \eta - \eta_1, D_F\zeta\rangle = 0$ for every $\zeta \in W_F^{k,q}(\Sigma, E)$. By Step 5, the inclusion $\ker D_F^* \hookrightarrow \operatorname{coker} D_F$ is an isomorphism, so $\eta - \eta_1 \in \ker D_F^*$. Thus there exists an $\eta_0 \in W_F^{k,q}(\Sigma, \Lambda^{0,1}T^*\Sigma \otimes_{\mathbb{C}} E)$ such that

$$\eta = \eta_0 + \eta_1, \qquad D_F^*\eta_0 = 0.$$

Hence $\eta \in W_F^{k,q}(\Sigma, \Lambda^{0,1}T^*\Sigma \otimes_{\mathbb{C}} E)$ and $D_F\eta = D_F\eta_1 = \xi$. This proves (ii). Assertion (iii) follows by interchanging the roles of D_F and D_F^*.

STEP 7. *For every* $\xi_0 \in W_F^{k,q}(\Sigma, E)$,

$$D_F\xi_0 = 0 \qquad \Longrightarrow \qquad \xi_0 \in W_F^{\ell,p}(\Sigma, E). \tag{C.2.10}$$

Similarly for D_F^*. *Moreover, the inclusions* $\ker \mathrm{D}_F \hookrightarrow \operatorname{coker} D_F^*$ *and* $\ker \mathrm{D}_F^* \hookrightarrow \operatorname{coker} D_F$ *are isomorphisms and (C.2.6) holds.*

Assume first that $kq < 2$. Let $\xi_0 \in W_F^{k,q}(\Sigma, E)$ such that $D_F\xi_0 = 0$. Then, by Theorem B.1.12,

$$\xi_0 \in L^r(\Sigma, E), \qquad r := \frac{2q}{2 - kq}.$$

The proof of Step 1 (b) shows that $2/r - 1 \le \ell - 2/p$ and that there is an inclusion $L^r \hookrightarrow (W^{\ell-1,p})^*$. Hence, by Lemma C.1.9, $\langle \xi_0, D_F^*\zeta\rangle = 0$ for every $\zeta \in W_F^{\ell,p}(\Sigma, \Lambda^{0,1}T^*\Sigma \otimes_{\mathbb{C}} E)$. Now it was proved in Step 6 that D_F and D_F^*, when understood as operators from $W_F^{\ell,p}$ to $W^{\ell-1,p}$, satisfy assertion (iii) of Theorem C.2.3. Applying this in the case $\eta = 0$ we obtain $\xi_0 \in W_F^{\ell,p}(\Sigma, E)$. This proves (C.2.10) in the case $kq < 2$. In the case $kq \ge 2$ the argument is similar.

Since $\ell p > 2$ there is an inclusion $W^{\ell,p}(\Sigma, E) \hookrightarrow (W^{k-1,q}(\Sigma, E))^*$. Hence it follows from (C.2.10) and Lemma C.1.9 that there is a natural inclusion of the kernel of D_F into the cokernel of D_F^*. Similarly for the kernel of D_F^*. Hence it follows from the dimension formula (C.2.7) that these inclusions are isomorphisms. This shows that the identities in (C.2.6) continue to hold when $-1/2 > k - 2/q$.

STEP 8. *Assertions* (ii) *and* (iii) *hold in general.*

In view of Step 7 we know that the kernel of D_F embeds into $W^{\ell,p}$ and hence into the dual space of $W^{k-1,q}$. This is all that is needed to repeat the argument in the proof of Step 6 with slight modifications which we now explain.

Suppose that $\eta \in L^r(\Sigma, \Lambda^{0,1}T^*\Sigma \otimes_{\mathbb{C}} E)$ and $\xi \in W^{k-1,q}(\Sigma, F)$ satisfy (C.2.2). By the second identity in (C.2.6), proved in Step 7 under the present assumptions, we can write ξ in the form (C.2.8), where now $\xi_0 \in W_F^{\ell,p}(\Sigma, E)$ (and $\eta_1 \in W_F^{k,q}(\Sigma, \Lambda^{0,1}T^*\Sigma \otimes_{\mathbb{C}} E)$ as before). Then (C.2.9) holds for every $\zeta \in W_F^{\ell,p}(\Sigma, E)$. Note that we no longer assume $-1/2 \le k - 2/q$ and hence there might not be an inclusion of $W^{k,q}$ into the dual space of $W^{k-1,q}$. However, there is an inclusion of $W^{\ell,p}$ into the dual space of $W^{k-1,q}$ and, by Step 1, an inclusion of $W^{k,q}$ into the dual space of $W^{\ell-1,p}$. The last identity in (C.2.9) follows from Lemma C.1.9 and these observations. As in Step 6 we choose $\zeta = \xi_0$ in (C.2.9) to obtain $\xi_0 = 0$ and hence $\langle \eta - \eta_1, D_F\zeta\rangle = 0$ for $\zeta \in W_F^{\ell,p}(\Sigma, E)$. Now, by Step 7, the inclusion $\ker D_F^* \hookrightarrow \operatorname{coker} D_F$ is an isomorphism and hence $\eta - \eta_1 \in W_F^{\ell,p}(\Sigma, \Lambda^{0,1}T^*\Sigma \otimes_{\mathbb{C}} E)$ and $D_F(\eta - \eta_1) = 0$. Thus $\eta \in W_F^{k,q}(\Sigma, \Lambda^{0,1}T^*\Sigma \otimes_{\mathbb{C}} E)$ and $D_F\eta = D_F\eta_1 = \xi$. This proves (ii). Assertion (iii) follows by interchanging the roles of D_F and D_F^*. □

PROOF OF THEOREM C.1.10 (I). Suppose $\eta \in W^{k-1,q}(\Sigma, \Lambda^{0,1}T^*\Sigma \otimes_{\mathbb{C}} E)$ satisfies

$$\eta_0 \in \ker D_F^* \subset W^{\ell,p}(\Sigma, \Lambda^{0,1}T^*\Sigma \otimes_{\mathbb{C}} E) \quad \Longrightarrow \quad \langle \eta, \eta_0 \rangle = 0$$

By (C.2.6) and Step 7 in the proof of Theorem C.2.3, there exist a $\xi \in W_F^{k,q}(\Sigma, E)$ and an $\eta_1 \in W^{\ell,p}(\Sigma, \Lambda^{0,1}T^*\Sigma \otimes_{\mathbb{C}} E)$ such that

$$\eta = D_F\xi + \eta_1, \qquad D_F^*\eta_1 = 0.$$

With $\eta_0 = \eta_1$, it follows that $\eta_1 = 0$ and hence $\eta \in \operatorname{im} D_F$. □

C.3. The boundary Maslov index (by Joel Robbin)

In this section we give a definition of the boundary Maslov index that appears in the index formula of Theorem C.1.10 and establish its properties. The basic technique is to argue by induction over a decomposition of the surface Σ into pairs of pants. That such decompositions exist is an easy application of Morse theory. This allows one to reduce the relevant results to the case of bundles over a disc.

Pair of pants decomposition. By a **2-manifold** we mean a compact, oriented, connected 2-dimensional manifold Σ, possibly with boundary.

DEFINITION C.3.1. *A* **decomposition** *of a 2-manifold Σ_{02} is a pair of submanifolds Σ_{01}, Σ_{12} of Σ_{02} such that*

$$\Sigma_{02} = \Sigma_{01} \cup \Sigma_{12}, \qquad \Sigma_{01} \cap \Sigma_{12} = \partial\Sigma_{01} \cap \partial\Sigma_{12}.$$

It follows that

$$\partial\Sigma_{ij} = \Gamma_i \cup \Gamma_j, \qquad \Gamma_i \cap \Gamma_j = \emptyset,$$

where Γ_i is a disjoint union of circles in Σ_{02} and $\Gamma_1 = \Sigma_{01} \cap \Sigma_{12}$.

DEFINITION C.3.2. *A* **disc with h holes** *is a 2-manifold with boundary obtained from a closed disc by removing the interiors of h disjoint closed subdiscs. A* **pair of pants** *is a disc with two holes. A* **pair of pants decomposition** *of a connected 2-manifold Σ is a sequence*

$$\Sigma_0 \subset \Sigma_1 \subset \Sigma_2 \subset \cdots \subset \Sigma_n = \Sigma$$

of submanifolds such that Σ_0 is a disc, for $k = 1, 2, \ldots, n$, Σ_k decomposes (in the sense of Definition C.3.1) into Σ_{k-1} and a pair of pants (or possibly a disc for $k = n$).

A pair of pants decomposition can be constructed via Morse theory. In the closed case there is a Morse function $f : \Sigma \to [0,1]$ having precisely one local minimum and one local maximum. Perturb f so that it has only one critical point on each critical level and cut Σ at a sequence of regular levels to obtain a decomposition of Σ that starts and ends with a disc (neighbourhoods of the minimum and maximum) and adds a pair of pants each time one passes a critical level. If Σ has a nonempty boundary choose f such that $f^{-1}(1) = \partial\Sigma$, 1 is a regular value, and proceed as above.

Pair of pants induction. Suppose that we want to prove a theorem about 2-manifolds or about all the 2-submanifolds of a given 2-manifold. Suppose further that we have proved that

(Basis Step): the theorem holds for a disc, and
(Inductive Step): if Σ_{02} decomposes into Σ_{01} and Σ_{12} where the theorem holds for two of the three manifolds Σ_{ij}, then the theorem holds for the third.

Then the theorem holds in general. This may be seen as follows. First, the theorem holds for a disc with holes by induction on the number of holes and thus for a manifold with at least two boundary components by induction on the length n of a pair of pants decomposition. Finally, it holds for all manifolds since a manifold with boundary is obtained by removing a disc from a manifold with one less boundary component.

The Maslov index. We review Arnold's definition of the Maslov index. (For another treatment see [**277**, Chapter 2].) Let

$$R(n) := \mathrm{GL}(n, \mathbb{C})/\mathrm{GL}(n, \mathbb{R})$$

be the manifold of totally real subspaces of $\mathbb{C}^n$. Note that the manifold

$$L(n) := \mathrm{U}(n)/\mathrm{O}(n) = \mathrm{Sp}\,(2n)/\mathrm{GL}(n, \mathbb{R}) \subset R(n)$$

of Lagrangian subspaces in $\mathbb{R}^{2n} = \mathbb{C}^n$ is a deformation retract of $R(n)$. Define $\rho : R(n) \to S^1$ by

$$\rho\,(a \cdot \mathrm{GL}(n, \mathbb{R})) := \frac{\det(a^2)}{\det(a^* a)}$$

where a^* denotes the conjugate transpose. Let Γ be a compact oriented 1-manifold without boundary, i.e. a disjoint union of circles. The **Maslov index** of a map $\Lambda : \Gamma \to R(n)$ is defined by

$$\mu(\Lambda) := \deg(\rho \circ \Lambda). \tag{C.3.1}$$

When Γ is connected we have

THEOREM C.3.3 (Arnold [**17**]). *Two loops in $R(n)$ are homotopic if and only if they have the same Maslov index.*

A complex vector bundle E over the circle S^1 admits a trivialization because $\mathrm{GL}(n, \mathbb{C})$ is connected. When $E = S^1 \times \mathbb{C}^n$, a totally real subbundle $F \subset E$ determines a map $\Lambda : S^1 \to R(n)$ via

$$\Lambda(z) := F_z.$$

It is important to distinguish between Λ and the real vector bundle F. Since $\mathrm{GL}(n, \mathbb{R})$ has two connected components there are two real rank n vector bundles over a circle. Thus two totally real subbundles F_1, F_2 of E are isomorphic as real vector bundles when their Maslov indices have the same parity. Since $E = F_j \oplus iF_j$, a real isomorphism $F_1 \to F_2$ extends to a complex isomorphism; i.e. there is a loop $U : S^1 \to \mathrm{GL}(n, \mathbb{C})$ with

$$\Lambda_2(z) = U(z)\Lambda_1(z).$$

Any such loop U is contractible if and only if $\mu(\Lambda_1) = \mu(\Lambda_2)$.

DEFINITION C.3.4. *A* **bundle pair** (E,F) *over* Σ *consists of a complex vector bundle* $E\to\Sigma$ *and a totally real subbundle* $F\subset E|_{\partial\Sigma}$ *over the boundary. A* **decomposition** *of a bundle pair* (E_{02},F_{02}) *over* Σ_{02} *consists of two bundle pairs,* $(E_{01},F_0\cup F_1)$ *over* Σ_{01} *and* $(E_{12},F_1\cup F_2)$ *over* Σ_{12}*, such that* Σ_{01},Σ_{12} *is a decomposition of* Σ_{02} *as in Definition C.3.1 and* $F_i\subset E_{02}|_{\Gamma_i}$.

The next theorem gives an axiomatic definition of the boundary Maslov index.

THEOREM C.3.5. *There is a unique operation, called the* **boundary Maslov index***, that assigns an integer* $\mu(E,F)$ *to each bundle pair* (E,F) *and satisfies the following axioms:*

(ISOMORPHISM) *If* $\Phi: E_1\to E_2$ *is a vector bundle isomorphism covering an orientation preserving diffeomorphism* $\phi:\Sigma_1\to\Sigma_2$*, then*

$$\mu(E_1,F_1)=\mu(E_2,\Phi(F_1)).$$

(DIRECT SUM)

$$\mu(E_1\oplus E_2,F_1\oplus F_2)=\mu(E_1,F_1)+\mu(E_2,F_2).$$

(COMPOSITION) *For a composition as in Definition C.3.4 we have*

$$\mu(E_{02},F_{02})=\mu(E_{01},F_{01})+\mu(E_{12},F_{12})$$

(NORMALIZATION) *For* $\Sigma=D$ *the unit disc,* $E=D\times\mathbb{C}$ *the trivial bundle, and* $F_z=\mathbb{R}e^{ik\theta/2}$ *for* $z=e^{i\theta}\in\partial D=S^1$ *we have*

$$\mu(D\times\mathbb{C},F)=k.$$

Here are its main properties. Recall that any complex vector bundle over a connected Riemann surface with nonempty boundary admits a trivialization; for a proof see Lemma C.3.8 below.

THEOREM C.3.6. *The boundary Maslov index satisfies the following:*

(TRIVIAL BUNDLE) *If* $\partial\Sigma\neq\emptyset$ *and* $E=\Sigma\times\mathbb{C}^n$*, then*

$$\mu(\Sigma\times\mathbb{C}^n,F)=\mu(\Lambda)\tag{C.3.2}$$

where μ *is the Maslov index defined in (C.3.1) and* Λ *is defined by*

$$\Lambda(z):=F_z,\qquad z\in\partial\Sigma.\tag{C.3.3}$$

(CHERN CLASS) *If* $\partial\Sigma=\emptyset$*, then* $\mu(E,\emptyset)$ *is twice the value of the first Chern class* $c_1(E)\in H^2(\Sigma)$ *on the fundamental class* $[\Sigma]\in H_2(\Sigma)$*:*

$$\mu(E,\emptyset)=2\langle c_1(E),[\Sigma]\rangle,\tag{C.3.4}$$

THEOREM C.3.7. *Two bundle pairs* (E_1,F_1) *and* (E_2,F_2) *over the same manifold* Σ *are isomorphic (over the identity) if and only if* E_1 *and* E_2 *have the same rank,* $\mu(E_1,F_1)=\mu(E_2,F_2)$*, and for each component* C *of the boundary* $\partial\Sigma$ *the real vector bundles* $F_1|_C$ *and* $F_2|_C$ *are isomorphic.*

The proofs of Theorems C.3.5, C.3.6, and C.3.7 use the following auxiliary lemma. Let $F\to\Gamma$ be a vector bundle of real rank n over a compact 1-manifold. A **partial framing** of F is a decomposition $F=L\oplus F'$ into a real line bundle $L\to\Gamma$ and a subbundle $F'\to\Gamma$ together with a framing of F', i.e. an $(n-1)$-tuple of sections $s_2,\dots,s_n:\Gamma\to F'$ that form a basis at every point.

LEMMA C.3.8. *Assume $\partial\Sigma \neq \emptyset$ and let (E, F) be a bundle pair over Σ, equipped with a partial framing $(L, F', s_2, \ldots, s_n)$. Then there exists a trivialization*

$$\Phi : E \to \Sigma \times \mathbb{C}^n$$

and a map $\lambda : \partial\Sigma \to S^1$ such that

$$\Phi(L_z) = \sqrt{\lambda(z)}\mathbb{R}e_1, \qquad \Phi(s_i(z)) = e_i, \qquad i = 2, \ldots, n,$$

for $z \in \partial\Sigma$, where $e_1, \ldots, e_n$ denote the standard basis of $\mathbb{C}^n$. In particular,

$$\Phi(F_z) = \sqrt{\lambda(z)}\mathbb{R} \times \mathbb{R}^{n-1}, \qquad z \in \partial\Sigma.$$

PROOF. Since $\partial\Sigma \neq \emptyset$, we have that Σ is homotopy equivalent to a bouquet of circles. Since $\pi_0(\mathrm{GL}(n, \mathbb{C})) = 0$, any complex vector bundle $E \to \Sigma$ is trivial. Hence we may suppose that $E = \Sigma \times \mathbb{C}^n$. Define Λ as in (C.3.3) and choose λ so $\mu(\Lambda|_C) = \mu(\Lambda'|_C)$ for every component C of $\partial\Sigma$ where $\Lambda'(z) = \sqrt{\lambda(z)}\mathbb{R} \times \mathbb{R}^{n-1}$. By Arnold's theorem the maps Λ and Λ' are homotopic. Choose a smooth map $U : \Gamma \to \mathrm{GL}(n, \mathbb{C})$ such that

$$U(z)L_z = \sqrt{\lambda(z)}\mathbb{R}e_1, \qquad U(z)s_i(z) = e_i, \qquad i = 2, \ldots, n,$$

for every $z \in \partial\Sigma$. Then $U(z)\Lambda(z) = \Lambda'(z)$ for $z \in \partial\Sigma$. Hence it follows from Arnold's theorem that the restriction of U to each component of $\partial\Sigma$ is contractible. Therefore U extends to a map (still denoted by U) from a collar neighbourhood of $\partial\Sigma$ to $\mathrm{GL}(n, \mathbb{C})$ such that U is the identity on the inner boundary of the collar. Now extend U by the identity to the rest of Σ and define $\Phi(z, v) = (z, U(z)v)$. □

COROLLARY C.3.9. *For any rank one bundle pair (E, F) over a disc $\mathbb{D}$, there is a trivialization $\Phi : E \to \mathbb{D} \times \mathbb{C}$ such that $\Phi(F_z) = \mathbb{R}e^{ik\theta/2}$ for $z = e^{i\theta} \in \partial\mathbb{D} = S^1$.*

PROOF. Take $\lambda(e^{i\theta}) = e^{ik\theta}$ in the proof of Lemma C.3.8. □

PROOFS OF THEOREMS C.3.5 AND C.3.6. Uniqueness in Theorem C.3.5 is an immediate consequence of the axioms, Corollary C.3.9, and pair of pants induction. For existence we will define μ as required by Theorem C.3.6 and then prove the axioms. To justify this we must first show that if $\partial\Sigma \neq \emptyset$ and (E, F) is a bundle pair over Σ, then the boundary Maslov index

$$\mu(E, F) = \mu(\Lambda_\Phi), \qquad \Lambda_\Phi(z) = \Phi(F_z), \tag{C.3.5}$$

is independent of the trivialization $\Phi : E \to \Sigma \times \mathbb{C}^n$ used to define it. Another trivialization $\Psi : E \to \Sigma \times \mathbb{C}^n$ determines $U : \Sigma \to \mathrm{GL}(n, \mathbb{C})$ such that

$$\Phi \circ \Psi^{-1}(z, v) = (z, U(z)v), \qquad z \in \Sigma, \ v \in \mathbb{C}^n,$$

and

$$U(z)\Lambda_\Psi(z) = \Lambda_\Phi(z), \qquad z \in \partial\Sigma.$$

If $a \in \mathrm{GL}(n, \mathbb{C})$ satisfies $a(\mathbb{R}^n) = \Lambda_\Psi(z)$, then $U(z)a \in \mathrm{GL}(n, \mathbb{C})$ satisfies

$$U(z)a(\mathbb{R}^n) = \Lambda_\Phi(z).$$

Hence by (C.3.1) we have

$$\rho(\Lambda_\Phi(z)) = g(z)\rho(\Lambda_\Psi(z)), \qquad g(z) = \frac{\det(U(z)^2)}{\det(U(z)^*U(z))}, \tag{C.3.6}$$

for $z \in \partial\Sigma$. But $g : \partial\Sigma \to S^1$ has degree zero as it extends to Σ. Hence $\mu(\Lambda_\Phi) = \deg(\rho \circ \Lambda_\Phi) = \deg(\rho \circ \Lambda_\Psi) = \mu(\Lambda_\Psi)$ as claimed.

We can now prove existence in Theorem C.3.5 and Theorem C.3.6 at the same time by showing that $\mu(E,F)$, defined by (C.3.2) when $\partial\Sigma \neq \emptyset$ and by (C.3.4) when $\partial\Sigma = \emptyset$, satisfies the axioms of Theorem C.3.5. All the axioms except the *(Composition)* axiom are immediate. In defining the degree of $\rho\circ\Lambda$ the orientation of $\partial\Sigma$ which is inherited from the orientation of Σ is understood. Thus for a decomposition as in the *(Composition)* axiom we have

$$\mu(\Lambda_{02}) = \mu(\Lambda_{01}) + \mu(\Lambda_{12}), \qquad \Lambda_{ij}(z) = (F_{ij})_z,$$

because the contributions over the intersection $\Sigma_{01}\cap\Sigma_{12}$ cancel. This proves the *(Composition)* axiom in case $\partial\Sigma_{02}\neq\emptyset$. The case $\partial\Sigma_{02}=\emptyset$ requires more argument, and we reformulate it in the next theorem to simplify the notation. □

THEOREM C.3.10. *Suppose $\partial\Sigma=\emptyset$. Let $\Sigma=\Sigma_0\cup\Sigma_1$ be a decomposition of Σ as in Definition C.3.1 and define*

$$\Gamma := \Sigma_0\cap\Sigma_1 = \partial\Sigma_0 = \partial\Sigma_1.$$

Suppose E is a vector bundle over Σ, $F\subset E|_\Gamma$ a totally real subbundle, and let $E_i = E|_{\Sigma_i}$. Then

$$2\langle c_1(E),[\Sigma]\rangle = \mu(E_0,F)+\mu(E_1,F).$$

PROOF. We conclude from Lemma C.3.8 that E splits as a line bundle and a trivial bundle $\Sigma\times\mathbb{C}^{n-1}$ with $F\cap(\Gamma\times\mathbb{C}^{n-1})=\Gamma\times\mathbb{R}^{n-1}$. Thus we assume without loss of generality that $n=1$. We assume also that E is equipped with a Hermitian form. A unitary trivialization $\Phi_i : E|_{\Sigma_i}\to\Sigma_i\times\mathbb{C}$ has the form $\Phi_i^{-1}(z,v)=vs_i(z)$ for $z\in\Sigma_i$ and $v\in\mathbb{C}$, where $s_i:\Sigma_i\to E_i$ is a section of norm one. Define $u:\Gamma\to S^1$ by

$$s_1(z) =: u(z)s_0(z), \qquad z\in\Gamma.$$

Moreover, define the Lagrangian subspace $\Lambda_i(z)\subset\mathbb{C}$ by $\Lambda_i(z):=\Phi_i(F_z)$ for $z\in\Gamma$. Then $\Lambda_0(z)=u(z)\Lambda_1(z)$ and hence, as in (C.3.6), we have

$$\rho(\Lambda_0)=u^2\rho(\Lambda_1).$$

Now orient Γ as the boundary of Σ_0. Then $\mu(E_0,F)=\deg(\rho\circ\Lambda_0)$ and $\mu(E_1,F)= -\deg(\rho\circ\Lambda_1)$. Hence, taking degrees,

$$\mu(E_0,F)+\mu(E_1,F)=2\deg(u). \tag{C.3.7}$$

Now let ∇ be a Hermitian connection on E and define

$$\alpha_0 := s_0^{-1}\nabla s_0\in\Omega^1(\Sigma_0,i\mathbb{R}), \qquad \alpha_1 := s_1^{-1}\nabla s_1\in\Omega^1(\Sigma_1,i\mathbb{R}).$$

Then $F^\nabla|_{\Sigma_0}=d\alpha_0$, $F^\nabla|_{\Sigma_1}=d\alpha_1$, and $(\alpha_1-\alpha_0)|_\Gamma = u^{-1}du$. Hence

$$\begin{aligned}\langle c_1(E),[\Sigma]\rangle &= \frac{i}{2\pi}\int_\Sigma F^\nabla\\ &= \frac{i}{2\pi}\int_{\Sigma_0}\alpha_0+\frac{i}{2\pi}\int_{\Sigma_1}\alpha_1\\ &= \frac{i}{2\pi}\int_\Gamma(\alpha_0-\alpha_1)\\ &= \frac{1}{2\pi i}\int_\Gamma u^{-1}du\\ &= \deg(u).\end{aligned}$$

So the assertion follows from (C.3.7). □

Proof of Theorem C.3.7. As in the proof of Theorem C.3.10 we may assume that the rank is one. Assume that $\partial\Sigma \neq \emptyset$ and that $E_1 = E_2 = \Sigma \times \mathbb{C}$. By hypothesis we may find $u : \partial\Sigma \to \mathrm{GL}(1, \mathbb{C})$ so that $u(z)F_1(z) = F_2(z)$; we must prove that u extends to Σ. Without loss of generality assume $|u| = 1$. By hypothesis, the degree of u is zero, but the restriction to a boundary component need not have degree zero. We use induction on the number of boundary components. If there is only a single boundary component, the map u is homotopic to a constant (since the degree is zero) and the homotopy may be used to extend to a collar neighbourhood and extend the constant map on the inner boundary as a constant map to the rest of Σ. If $u|_C$ has degree zero for some boundary component C, then we can cap that component with a disc and extend to that disc, thereby reducing the number of boundary components. If the degree of $u|_C$ is nonzero for every boundary component C, then there are two points on distinct boundary components where u takes the same value. We connect these two points with an arc, extend by the common value to the arc, and cut open along the arc, thereby reducing the number of boundary components.

The case where $\partial\Sigma = \emptyset$ is well known (a line bundle is characterized by its first Chern class) but this can also be proved by decomposing Σ into Σ' and a disc D, introducing totally real subbundles $F_i \subset E_i|_{\partial D}$ so that $\mu(E_1|_D, F_1) = \mu(E_2|_D, F_2)$, and applying the case $\partial \neq \emptyset$ to $E_i|_{\Sigma'}$. Use the proof of Theorem C.3.10 to relate $c_1(E_i)$ to $\mu(E_i|_D, F_i)$. □

C.4. Proof of the Riemann–Roch theorem

The first part of Theorem C.1.10 was proved in Theorem C.2.3. Therefore we just have to establish the index formula and prove the injectivity/surjectivity statements in (iii) for line bundles. Note that when proving the index formula it suffices to consider smooth complex linear operators. Namely, by Lemma C.2.2, every Cauchy–Riemann operator of class $W^{\ell-1,p}$ differs from such an operator by a compact operator and so, by Theorem A.1.5, both operators have the same index. Moreover, by Theorem C.1.10 (i), it suffices to prove the index formula for $k = 1$ and $q = 2$. However, the statements in (iii) are more delicate and require one to consider the particular operators given.

Theorem C.4.1. *Theorem C.1.10 holds when $\Sigma = \mathbb{D}$ is the closed unit disc in $\mathbb{C}$ and the operator D is complex linear.*

Proof. By the direct sum axiom for the boundary Maslov index and the obvious corresponding property of the Fredholm index, we may assume that $E \to \mathbb{D}$ is a complex line bundle. By Corollary C.3.9, we may further assume that

$$E = \mathbb{D} \times \mathbb{C}, \qquad F_{e^{i\theta}} = \mathbb{R}e^{ik\theta/2}$$

for every $\theta \in \mathbb{R}$ and some integer k. The proof now consists of four steps. Let

$$X_F := W^{1,2}_F(\mathbb{D}, \mathbb{C}), \qquad Y := L^2(\mathbb{D}, \Lambda^{0,1}T^*\mathbb{D} \otimes \mathbb{C}),$$

and denote by $D_F : X_F \to Y$ the Cauchy–Riemann operator associated to this subbundle and the zero connection:

$$D_F\xi = \tfrac{1}{2}\left(\partial_s\xi + i\partial_t\xi\right)(ds - idt).$$

The first three steps prove the theorem for D_F by direct calculation, and the last extends the result to other operators.

STEP 1: *The orthogonal complement of the image of D_F is the space of all $(0,1)$-forms $\zeta d\bar{z}$ such that $\zeta : \mathbb{D} \to \mathbb{C}$ is a smooth function and*

$$\partial_s \zeta - i\partial_t \zeta = 0, \qquad \zeta(e^{i\theta}) \in ie^{i\theta + ik\theta/2}\mathbb{R}. \tag{C.4.1}$$

That the orthogonal complement consists of smooth $(0,1)$-forms $\zeta d\bar{z}$ that satisfy $\partial_s\zeta - i\partial_t\zeta = 0$ follows from (C.1.3) and the formula for the dual operator in Remark C.1.4. Now let $\xi \in X_F$ and $\zeta : \mathbb{D} \to \mathbb{C}$ be such that $\partial_s\zeta - i\partial_t\zeta = 0$. Then

$$\begin{aligned}\int_{\mathbb{D}} \langle \zeta\, d\bar{z}, D_F\xi\rangle\, dsdt &= \mathrm{Re}\int_{\mathbb{D}} \bar{\zeta}(\partial_s\xi + i\partial_t\xi)\, dsdt + \mathrm{Re}\int_{\mathbb{D}} \overline{(\partial_s\zeta - i\partial_t\zeta)}\xi\, dsdt \\ &= \mathrm{Re}\int_{\mathbb{D}} \big(\partial_s(\bar{\zeta}\xi) + i\partial_t(\bar{\zeta}\xi)\big)\, dsdt \\ &= \mathrm{Re}\int_0^{2\pi} e^{i\theta}\overline{\zeta(e^{i\theta})}\xi(e^{i\theta})\, d\theta.\end{aligned}$$

The right hand side vanishes for all $\xi \in X_F$ if and only if ζ satisfies (C.4.1).

STEP 2: *If $k \le -1$ then D_F is injective. If $k \ge 0$ then* $\dim\ker D_F = 1 + k$.

If $\xi \in X_F$ and $\bar{\partial}\xi = 0$ then ξ can be expressed as a power series

$$\xi(z) = \sum_{n\ge 0} a_n z^n$$

with convergence radius bigger than or equal to 1. The boundary condition takes the form $\xi(e^{i\theta})e^{-ik\theta/2} \in \mathbb{R}$ or, equivalently,

$$\xi(e^{i\theta})e^{-ik\theta} = \overline{\xi(e^{i\theta})}$$

for $\theta \in \mathbb{R}$. The Fourier coefficients are given by

$$a_n = \lim_{r\to 1}\frac{1}{2\pi}\int_0^{2\pi} \xi(re^{i\theta})e^{-in\theta}\, d\theta = \frac{1}{2\pi}\int_0^{2\pi} \xi(e^{i\theta})e^{-in\theta}\, d\theta.$$

Hence

$$a_{k-n} = \frac{1}{2\pi}\int_0^{2\pi} \xi(e^{i\theta})e^{-ik\theta}e^{in\theta}\, d\theta = \frac{1}{2\pi}\int_0^{2\pi} \overline{\xi(e^{i\theta})}e^{in\theta}\, d\theta = \bar{a}_n.$$

This implies $\xi = 0$ whenever $k < 0$ and

$$\xi(z) = a_0 + a_1 z + \cdots + a_k z^k, \qquad a_{k-n} = \bar{a}_n$$

whenever $k \ge 0$. Hence $\dim\ker D_F = 1 + k$ for $k \ge 0$.

STEP 3: *If $k \ge -1$ then D_F is surjective. If $k \le -2$ then*

$$\dim\operatorname{coker} D_F = -k - 1.$$

Let us temporarily write $D_F(k)$ for the operator $D_F = \bar{\partial}$ considered above with boundary conditions

$$F(k) := F_{e^{i\theta}} = \mathbb{R}e^{ik\theta/2}.$$

If ζ satisfies (C.4.1), then $\bar{\zeta}$ is in the kernel of $D_F(-k-2)$. Therefore, Step 1 shows that $D_F(k)$ is surjective precisely when $D_F(-k-2)$ is injective. By Step 2 this happens when $-k-2 \le -1$. This proves the first statement. The second is left to the reader.

STEP 4: *We prove the theorem.*

An arbitrary complex linear Cauchy–Riemann operator on the disc has the form $D_F + \alpha$ where $\alpha \in W^{\ell-1,p}(\mathbb{D}, \Lambda^{0,1}T^*\mathbb{D})$. By Step 3 with $k = 0$, there exists a function $f \in W^{\ell,p}(D, \mathbb{C})$ such that $f(e^{i\theta}) \in \mathbb{R}$ and $\bar{\partial} f = \alpha$. Hence the function $w := e^f : \mathbb{D} \to \mathbb{C}^*$ satisfies

$$wF = F, \qquad w^{-1}\bar{\partial} w = \alpha.$$

Hence $w^{-1} \circ D_F \circ w = D_F + \alpha$. Thus $D_F + \alpha$ is injective or surjective precisely when D_F is. This completes the proof of Theorem C.4.1. □

The proof of the Riemann–Roch theorem, given below, uses the *pair of pants induction* explained in Section C.3. Here is the inductive setup. Let

$$\Sigma_{02} = \Sigma_{01} \cup \Sigma_{12}, \qquad \Sigma_{01} \cap \Sigma_{12} = \partial\Sigma_{01} \cap \partial\Sigma_{12}$$

be a decomposition as in Definition C.3.1. Write

$$\partial\Sigma_{ij} = (-\Gamma_i) \cup \Gamma_j$$

where $\Gamma_i \cap \Gamma_j = \emptyset$ and $\Gamma_1 = \Sigma_{01} \cap \Sigma_{12}$. The sign indicates that we orient Γ_1 as the boundary of Σ_{01}, Γ_2 as the boundary of Σ_{02} or Σ_{12}, and Γ_0 has the opposite orientation as the boundary of Σ_{01} or Σ_{02}. Let (E_{02}, F_{02}) be a bundle pair over Σ_{02} with a decomposition into (E_{01}, F_{01}) and (E_{12}, F_{12}) as in Definition C.3.4, where $F_{ij} = F_i \cup F_j$ and $F_i \subset E_{02}|_{\Gamma_i}$ is a totally real subbundle. Denote

$$X_{ij} := W^{1,2}_{F_{ij}}(\Sigma_{ij}, E_{ij}), \qquad Y_{ij} := L^2(\Sigma_{ij}, \Lambda^{0,1}T^*\Sigma_{ij} \otimes_{\mathbb{C}} E_{ij})$$

and let $D_{ij} : X_{ij} \to Y_{ij}$ be the Cauchy–Riemann operator obtained by restriction of a smooth Cauchy–Riemann operator on E_{02}.

THEOREM C.4.2. *In the above situation the Fredholm indices are related by*

$$\text{index}(D_{02}) = \text{index}(D_{01}) + \text{index}(D_{12}). \tag{C.4.2}$$

PROOF. The basic idea of the proof is quite straightforward. Let $E_1 \to \Gamma_1$ be the restriction of E_{02} to Γ_1 and $\overline{E}_1 \to \Gamma_1$ be the bundle E_1 with the reversed complex structure. Then the totally real subbundle $F_1 \oplus F_1$ of $\overline{E}_1 \oplus E_1$ has the same Maslov index as the diagonal in any trivialization. Hence they are homotopic through totally real subbundles. Each of the totally real subbundles in the homotopy gives rise to a nonlocal elliptic boundary condition for our Cauchy–Riemann operator and the assertion follows from the deformation invariance of the Fredholm index. To make this idea precise, it is convenient to translate the homotopy argument for the boundary conditions into a homotopy argument for lower order perturbations of a fixed operator. For the sake of the exposition we assume that Γ_1 is connected. The general case follows by repeating the construction below for each component of Γ_1. We explain the argument in five steps, the first two of which normalize the given Cauchy–Riemann operator near Γ_1.

STEP 1. *Let $U \subset \Sigma_{02}$ be a closed tubular neighbourhood of Γ_1 and choose a diffeomorphism*

$$\phi : [-1, 1] \times \mathbb{R}/\mathbb{Z} \to U$$

such that $\phi(0 \times \mathbb{R}/\mathbb{Z}) = \Gamma_1$. By Theorem C.2.3 we may assume without loss of generality that ϕ is holomorphic. Thus $i = \phi^*j$ is the standard complex structure on $[-1, 1] \times \mathbb{R}/\mathbb{Z}$ with coordinates $s + it$, where $-1 \le s \le 1$ and $t \in \mathbb{R}/\mathbb{Z}$.

STEP 2. *Choose a complex trivialization $U \times \mathbb{C}^n \to E|_U : (z,\zeta) \mapsto \Phi(z)\zeta$. We may assume without loss of generality that $D \circ \Phi = \Phi \circ \bar{\partial}$.*

Define $A \in \Omega^{0,1}(U, \mathrm{End}_{\mathbb{R}}(\mathbb{C}^n))$ by $A\zeta_0 := \Phi^{-1}D(\Phi\zeta_0)$ for the constant map $\zeta_0 \in \mathbb{C}^n$. Then

$$D(\Phi\zeta) = \Phi\left(\bar{\partial}\zeta + A\zeta\right)$$

for every smooth map $\zeta : U \to \mathbb{C}^n$. Extend $\Phi A \Phi^{-1} \in \Omega^{0,1}(U, \mathrm{End}_{\mathbb{R}}(E))$ to a global endomorphism valued $(0,1)$-form $B \in \Omega^{0,1}(\Sigma, \mathrm{End}_{\mathbb{R}}(E))$ and replace D by the Cauchy–Riemann operator $D - B$ (of the same index). This proves Step 2.

STEP 3. *Define the spaces*

$$X := W^{1,2}(\Sigma_{01}, E_{01}) \oplus W^{1,2}(\Sigma_{12}, E_{12}),$$
$$Y := L^2(\Sigma_{01}, \Lambda^{0,1}T^*\Sigma_{01} \otimes E_{01}) \oplus L^2(\Sigma_{12}, \Lambda^{0,1}T^*\Sigma_{12} \otimes E_{12}),$$

and the subspaces

$$X_0 := \left\{ (\xi_{01}, \xi_{12}) \in X \,\middle|\, \begin{array}{l} \xi_{01}(\Gamma_0) \subset F_0, \xi_{01}(\Gamma_1) \subset F_1, \\ \xi_{12}(\Gamma_1) \subset F_1, \xi_{12}(\Gamma_2) \subset F_2 \end{array} \right\},$$
$$X_1 := \left\{ (\xi_{01}, \xi_{12}) \in X \,\middle|\, \begin{array}{l} \xi_{01}(\Gamma_0) \subset F_0, \xi_{12}(\Gamma_2) \subset F_2, \\ \xi_{01}(z) = \xi_{12}(z) \;\; \textit{for } z \in \Gamma_1 \end{array} \right\}.$$

(See the remark preceding Definition C.1.5.) Then the operators $D_0 : X_0 \to Y$ and $D_1 : X_1 \to Y$ determined by D are Fredholm operators with indices

$$\mathrm{index}(D_0) = \mathrm{index}(D_{01}) + \mathrm{index}(D_{12}), \qquad \mathrm{index}(D_1) = \mathrm{index}(D_{02}).$$

The operator D_0 is the direct sum of D_{01} and D_{12}, and is Fredholm by part (i) of Theorem C.1.10. The operator D_1 is isomorphic to D_{02}. The isomorphism of the domains sends $\xi_{02} \in W^{1,2}_{F_{02}}(\Sigma_{02}, E_{02})$ to the pair $(\xi_{02}|_{\Sigma_{01}}, \xi_{02}|_{\Sigma_{12}}) \in X_1$ and similarly for the target spaces. (*Warning:* We chose $k=1$ since the map $\xi_{0,2} \mapsto (\xi_{02}|_{\Sigma_{01}}, \xi_{02}|_{\Sigma_{12}})$ is not surjective for $W^{k,2}$ with $k \geq 2$.) This proves Step 3.

STEP 4. *For $t \in \mathbb{R}/\mathbb{Z}$ define the totally real subspace $\Lambda(t) \subset \mathbb{C}^n$ by*

$$\Lambda(t) := \Phi(z_t)^{-1}F_{1,z_t}, \quad z_t := \phi(0,t) \in U.$$

Then there exists a smooth map $\Psi : [0,1] \times \mathbb{R}/\mathbb{Z} \to \mathrm{End}_{\mathbb{R}}(\mathbb{C}^n \oplus \mathbb{C}^n)$ satisfying

$$\Psi(s,t)I = I\Psi(s,t), \qquad I := \begin{pmatrix} -i & 0 \\ 0 & i \end{pmatrix}, \tag{C.4.3}$$

for all s and t,

$$\Psi(0,t)^{-1}\Delta = \Lambda(t) \oplus \Lambda(t), \tag{C.4.4}$$

and $\Psi(s,t) = \mathbb{1}$ for $1/2 \leq s \leq 1$.

The loop $\Lambda_0(t) := \Lambda(t) \oplus \Lambda(t)$ of totally real subspaces in $(\mathbb{C}^n \oplus \mathbb{C}^n, I)$ has Maslov index zero and hence is homotopic to the constant loop

$$\Lambda_1(t) := \Delta := \{(\zeta,\zeta) \,|\, \zeta \in \mathbb{C}^n\}.$$

Choose a smooth homotopy $[0,1] \times \mathbb{R}/\mathbb{Z} \to \mathcal{R}(\mathbb{C}^n \oplus \mathbb{C}^n, I) : (s,t) \mapsto \Lambda(s,t)$ such that $\Lambda(0,t) = \Lambda_0(t)$ and $\Lambda(s,t) = \Lambda_1(t)$ for $1/2 \leq s \leq 1$. Next choose a global smooth frame $e_1(s,t), \ldots, e_{2n}(s,t)$ of $\Lambda(s,t)$ such that $e_i(s,t) = e_i(1,t)$ for all i and $1/2 \leq s \leq 1$. Then define $\Psi(s,t) \in \mathrm{End}_{\mathbb{R}}(\mathbb{C}^n \oplus \mathbb{C}^n)$ by $\Psi(s,t)e_i(s,t) := e_i(1,t)$ and $\Psi(s,t)Ie_i(s,t) := Ie_i(1,t)$ for all i, s, t. Then Ψ satisfies (C.4.3) and (C.4.4) and $\Psi(s,t) = \mathbb{1}$ for $1/2 \leq s \leq 1$. This proves Step 4.

STEP 5. *There are Hilbert space isomorphisms $\Psi_X : X_0 \to X_1$ and $\Psi_Y : Y \to Y$ such that the operator $D_1 \circ \Psi_X - \Psi_Y \circ D_0 : X_0 \to Y$ is compact.*

Let $\Psi : [0,1] \times \mathbb{R}/\mathbb{Z} \to \mathrm{End}_{\mathbb{R}}(\mathbb{C}^n \oplus \mathbb{C}^n)$ be as in Step 4 and write

$$\Psi(s,t) =: \begin{pmatrix} A(s,t) & B(s,t) \\ C(s,t) & D(s,t) \end{pmatrix}.$$

By (C.4.3), the endomorphisms $A(s,t)$ and $D(s,t)$ are complex linear and equal $\mathbb{1}$ near $s=1$, while $B(s,t)$ and $C(s,t)$ are complex anti-linear and vanish near $s=1$. Define $\Psi_X : X_0 \to X_1$ by

$$\Psi_X(\xi_{01}, \xi_{12}) := (\widetilde{\xi}_{01}, \widetilde{\xi}_{12})$$

where $\widetilde{\xi}_{ij} = \xi_{ij}$ on $\Sigma_{ij} \setminus U$ and $\widetilde{\xi}_{ij}|_{\Sigma_{ij}\cap U}$ are defined by the equations

$$\widetilde{\xi}_{ij}(\phi(s,t)) := \Phi(\phi(s,t))\widetilde{\zeta}_{ij}(s,t), \quad \xi_{ij}(\phi(s,t)) =: \Phi(\phi(s,t))\zeta_{ij}(s,t), \tag{C.4.5}$$

where

$$\begin{aligned} \widetilde{\zeta}_{01}(s,t) &:= A(-s,t)\zeta_{01}(s,t) + B(-s,t)\zeta_{12}(-s,t), & -1 \le s \le 0, \\ \widetilde{\zeta}_{12}(s,t) &:= C(s,t)\zeta_{01}(-s,t) + D(s,t)\zeta_{12}(s,t), & 0 \le s \le 1. \end{aligned} \tag{C.4.6}$$

If $(\xi_{01}, \xi_{12}) \in X_0$ then $\xi_{01}(z), \xi_{12}(z) \in F_{1,z}$ for every $z \in \Gamma_1$. Hence

$$\big(\zeta_{01}(0,t), \zeta_{12}(0,t)\big) \in \Lambda(t) \oplus \Lambda(t)$$

for every t, hence $\big(\widetilde{\zeta}_{01}(0,t), \widetilde{\zeta}_{12}(0,t)\big) \in \Delta$ for every t, and hence $\widetilde{\xi}_{01}(z) = \widetilde{\xi}_{12}(z)$ for every $z \in \Gamma_1$. This shows that Ψ_X sends X_0 to X_1.

Next define $\Psi_Y : Y \to Y$ by $\Psi_Y(\eta_{01}, \eta_{12}) := (\widetilde{\eta}_{01}, \widetilde{\eta}_{12})$ where $\widetilde{\eta}_{ij} = \eta_{ij}$ on $\Sigma_{ij} \setminus U$ and $\widetilde{\eta}_{ij}|_{\Sigma_{ij}\cap U}$ are defined by the equations

$$\phi^*\widetilde{\eta}_{ij} := (\Phi \circ \phi)\widetilde{\beta}_{ij}, \qquad \phi^*\eta_{ij} =: (\Phi \circ \phi)\beta_{ij}(s,t), \tag{C.4.7}$$

where

$$\begin{aligned} \widetilde{\beta}_{01}(s,t) &:= A(-s,t)\beta_{01}(s,t) - B(-s,t)\beta_{12}(-s,t), & -1 \le s \le 0, \\ \widetilde{\beta}_{12}(s,t) &:= -C(s,t)\beta_{01}(-s,t) + D(s,t)\beta_{12}(s,t), & 0 \le s \le 1. \end{aligned} \tag{C.4.8}$$

Let $(\xi_{01}, \xi_{12}) \in X_0$ and define $\zeta_{ij}(s,t), \widetilde{\zeta}_{ij}(s,t) \in \mathbb{C}^n$ by (C.4.5) and (C.4.6). Then, because B is complex antilinear, we have

$$\begin{aligned} &\partial_s\widetilde{\zeta}_{01}(s,t) + i\partial_t\widetilde{\zeta}_{01}(s,t) \\ &= A(-s,t)\left(\partial_s\zeta_{01} + i\partial_t\zeta_{01}\right)(s,t) - B(-s,t)\left(\partial_s\zeta_{12} + i\partial_t\zeta_{12}\right)(-s,t) \\ &\quad + (-\partial_s A + i\partial_t A)(-s,t)\zeta_{01}(s,t) + (-\partial_s B + i\partial_t B)(-s,t)\zeta_{12}(s,t) \end{aligned}$$

for $-1 \le s \le 0$ and

$$\begin{aligned} &\partial_s\widetilde{\zeta}_{12}(s,t) + i\partial_t\widetilde{\zeta}_{12}(s,t) \\ &= -C(s,t)\left(\partial_s\zeta_{01} + i\partial_t\zeta_{01}\right)(-s,t) + D(s,t)\left(\partial_s\zeta_{12} + i\partial_t\zeta_{12}\right)(s,t) \\ &\quad - (\partial_s C + i\partial_t C)(s,t)\zeta_{01}(-s,t) + (\partial_s D + i\partial_t D)(s,t)\zeta_{12}(s,t) \end{aligned}$$

for $0 \le s \le 1$. But $D_1(\Psi_X\xi) = \Phi(\phi)\bar{\partial}\widetilde{\zeta}$ by Steps 1 and 2. Hence it follows from the definition of Ψ_Y that $D_1\Psi_X - \Psi_Y D_0$ is a compact operator. This proves Step 5.

It follows from Step 5 and Theorem A.1.5 that the operators D_0 and D_1 have the same Fredhom index. Hence the assertion follows from Step 3. This proves Theorem C.4.2. □

Proof of Theorem C.1.10. Assertion (i) was proved in Section C.2. The proof of the index formula in (ii) is based on Theorem C.4.2. It follows from (C.4.2) and the composition axiom for the boundary Maslov index that, if the index formula holds for two of the three surfaces Σ_{ij} in the decomposition then it holds for the third. Hence assertion (ii) of Theorem C.1.10 follows from Theorem C.4.1 and the pair of pants induction in Section C.3.

We prove assertion (iii). The first step is to reduce the injectivity statement for $\mu(E,F)<0$ to the case of closed Riemann surfaces by gluing Σ to its opposite surface. Here are the details. Let Σ be a compact connected Riemann surface with nonempty boundary

$$\partial\Sigma = \Gamma,$$

equipped with the trivial bundle $\Sigma\times\mathbb{C}$, and let $F\subset\Gamma\times\mathbb{C}$ be a totally real subbundle over the boundary. Then there exists a function $\lambda:\Gamma\to S^1/\{\pm1\}$ such that

$$F_z = \lambda(z)\mathbb{R}, \qquad z\in\Gamma.$$

A section $\xi:\Sigma\to\mathbb{C}$ satisfies the boundary condition $\xi(z)\in F_z$ if and only if

$$\overline{\xi(z)} = \lambda(z)^{-2}\xi(z), \qquad z\in\Gamma.$$

Consider the closed Riemann surface

$$S := \Sigma\times\{0,1\}/\sim$$

where the complex structure on $\Sigma\times\{1\}$ is reversed and $(z,0)\sim(z,1)$ for $z\in\Gamma$. Let $\gamma:=\lambda^{-2}:\Gamma\to S^1$ and consider the line bundle

$$E := E_0\times_\gamma E_1 \to S,$$

where $E_0 := (\Sigma\times\{0\})\times\mathbb{C}$ and $E_1 := (\Sigma\times\{1\})\times\mathbb{C}$ are the trivial bundles and $(z,0,\zeta)\sim(z,1,\gamma(z)\zeta)$ for $z\in\Gamma$ and $\zeta\in\mathbb{C}$. A section $\zeta:S\to E$ is given by a pair of functions $\zeta_0,\zeta_1:\Sigma\to\mathbb{C}$ such that

$$\zeta_1(z) = \gamma(z)\zeta_0(z), \qquad z\in\Gamma.$$

By Theorem C.3.10, the Chern number of E is given by

$$2\langle c_1(E),[S]\rangle = \mu(E_0,F)+\mu(E_1,F) = 2\mu(\Sigma\times\mathbb{C},F) < 0.$$

To make this construction precise (in the smooth category), we must choose coordinate charts of S and local trivializations of E across the boundary curve C. Then the smooth sections of E have to satisfy additional gluing conditions on all their higher derivatives along C. However, as long as we only deal with $W^{1,p}$-sections and Cauchy–Riemann operators of class L^p it is not necessary to make such choices.

Now consider a Cauchy–Riemann operator of the form $\bar\partial+\alpha$ on Σ, where

$$\alpha\in L^p(\Sigma,\Lambda^{0,1}T^*\Sigma\otimes\mathrm{End}_{\mathbb{R}}(\mathbb{C}))$$

for some $p>2$. This operator determines a Cauchy–Riemann operator D of class L^p on E given by $\bar\partial+\alpha$ on $\Sigma\times\{0\}$ and by $\partial+\bar\alpha$ on $\Sigma\times\{1\}$. Here

$$\bar\alpha(z,\hat z) := \tau\circ\alpha(z,\hat z)\circ\tau, \qquad \hat z\in T_z\Sigma,$$

where $\tau:\mathbb{C}\to\mathbb{C}$ denotes complex conjugation. A function $\xi\in W^{1,p}(\Sigma,\mathbb{C})$ that satisfies $\bar\partial\xi+\alpha\xi=0$ and $\xi(\partial\Sigma)\subset F$ gives rise to a section $\zeta\in W^{1,p}(S,E)$ in the kernel of D given by $\zeta_0(z):=\xi(z)$ and $\zeta_1(z):=\overline{\xi(z)}$. Hence it suffices to prove assertion (iii) for Cauchy–Riemann operators of class L^p on closed surfaces.

Now let $E \to S$ be a complex line bundle over a closed Riemann surface S and D be a Cauchy–Riemann operator of class L^p over S, where $p > 2$. We aim to show that D is injective when $\mu(E) < 0$. First consider the case when D is complex linear. Then, as we saw in Exercise C.1.8, D is gauge equivalent to a smooth complex linear Cauchy–Riemann operator D_0. By Remark C.1.1 there is a holomorphic structure on E such that D_0 is the usual delbar operator on E. Thus any element $\xi : S \to E$ in the kernel of D_0 can be represented locally as a holomorphic function on an open set. Hence, if ξ is nonzero, it follows that the zeros of ξ are isolated and have positive index. Since the Chern number of E is the sum of the indices of the zeros of a section with only isolated zeros, D_0 must have trivial kernel if $\mu(E) = 2c_1(E) < 0$. Since the kernels of D and D_0 are isomorphic, this proves the first assertion of Theorem C.1.10 (iii) for complex linear Cauchy–Riemann operators.

To deal with the real linear case we use a trick from Hofer–Lizan–Sikorav [**179**]. (The same idea occurs in the proof of Step 2 of Theorem 2.3.5.) Choose a smooth complex linear Cauchy–Riemann operator D_0 on the complex line bundle $E \to S$ and write

$$D = D_0 + a, \qquad a \in L^p(S, \Lambda^{0,1}T^*S \otimes \mathrm{End}_{\mathbb{R}}(E)).$$

Let $\xi \in W^{1,p}_F(S, E)$ such that $D\xi = D_0\xi + a\xi = 0$. Since ξ is continuous and the ratio of two elements in E is a well defined complex number, we may define $b \in L^p(S, \Lambda^{0,1}T^*S)$ by setting

$$b(z, \widehat{z}) := \begin{cases} (a(z, \widehat{z})\xi(z))/\xi(z), & \text{if } \xi(z) \neq 0, \\ 0, & \text{otherwise,} \end{cases} \qquad \widehat{z} \in T_zS.$$

Then $D_0 + b$ is a complex linear Cauchy–Riemann operator and $D_0\xi + b\xi = 0$ since $a\xi = b\xi$ by construction. In other words every element in the kernel of a real linear Cauchy–Riemann operator also belongs to the kernel of a complex linear Cauchy–Riemann operator (of class L^p) on the same line bundle. Hence, by what we have proved, every real linear Cauchy–Riemann operator on a complex line bundle $E \to S$ with negative Chern number is injective.

This shows that the first assertion of Theorem C.1.10 (iii) holds in general. To prove the second assertion, note that, by Corollary C.1.11 of Theorem C.1.10 (i), the cokernel of D is isomorphic to the kernel of a Cauchy–Riemann operator on the pair $(\Lambda^{1,0}T^*\Sigma \otimes_{\mathbb{C}} E^*, T\partial\Sigma \otimes_{\mathbb{R}} (E/F)^*)$ with boundary Maslov index

$$\mu(\Lambda^{1,0}T^*\Sigma \otimes_{\mathbb{C}} E^*, T\partial\Sigma \otimes_{\mathbb{R}} (E/F)^*) = -\mu(E, F) - 2\chi(\Sigma).$$

Hence the cokernel vanishes whenever $\mu(E, F) + 2\chi(\Sigma) > 0$. This proves Theorem C.1.10. □

C.5. The Riemann mapping theorem

In this section we use the Riemann–Roch theorem for real Cauchy–Riemann operators on the disc to prove two different versions of the Riemann mapping theorem for simply connected domains with smooth boundary. Throughout we denote by

$$\mathbb{D} := \{z \in \mathbb{C} \mid |z| \leq 1\}$$

the closed unit disc in $\mathbb{C}$, by $\mathcal{J}(\mathbb{D})$ the space of almost complex structures on $\mathbb{D}$ that induce the standard orientation, and by $\mathrm{Diff}(\mathbb{D})$ the group of orientation preserving diffeomorphisms of $\mathbb{D}$.

The Riemann mapping theorem in its classical form asserts that if $\Omega \subset \mathbb{C}$ is a connected simply connected open set that is nonempty and not equal to $\mathbb{C}$, then there is a holomorphic diffeomorphism $\phi : \mathrm{int}(\mathbb{D}) \to \Omega$. Assuming that there exists some diffeomorphism from $\mathrm{int}(\mathbb{D})$ to Ω, one can pull back the standard complex structure on Ω to $\mathrm{int}(\mathbb{D})$ and rephrase the result in terms of integrable almost complex structures on $\mathrm{int}(\mathbb{D})$. In our version of the theorem we drop the integrability assumption, but assume that the almost complex structure extends up to the boundary. Corollaries of this version are boundary regularity for the Riemann mapping and integrability in dimension two. On the other hand, we do not treat arbitrary open domains.

THEOREM C.5.1 (The Riemann mapping theorem). *For every $j \in \mathcal{J}(\mathbb{D})$ there is a unique diffeomorphism $\psi \in \mathrm{Diff}(\mathbb{D})$ such that*

$$\psi^* i = j, \qquad \psi(0) = 0, \qquad \psi(1) = 1. \tag{C.5.1}$$

The resulting map $\mathcal{J}(\mathbb{D}) \to \mathrm{Diff}(\mathbb{D}) : j \mapsto \psi$ is continuous.

COROLLARY C.5.2. *Let $\Omega \subset \mathbb{C}$ be a connected simply connected bounded open set with smooth boundary. Then the classical Riemann mapping theorem holds for Ω. Moreover, every holomorphic diffeomorphism $\phi : \mathrm{int}(\mathbb{D}) \to \Omega$ extends to a diffeomorphism $\mathbb{D} \to \overline{\Omega}$.*

PROOF. Choose any diffeomorphism $f : \mathbb{D} \to \overline{\Omega}$. Let $\psi : \mathbb{D} \to \mathbb{D}$ be as in Theorem C.5.1 with $j := f^* i$. Then $f \circ \psi^{-1} : \mathbb{D} \to \overline{\Omega}$ is a holomorphic diffeomorphism. Hence its restriction to the interior of $\mathbb{D}$ differs from ϕ by composition with a Möbius transformation. □

Another important corollary of Theorem C.5.1 is the integrability of complex structures in dimension two even on manifolds with boundary.

COROLLARY C.5.3 (Integrability). *Every almost complex structure j on a 2-manifold Σ (with boundary) is integrable.*

PROOF. Let $p_0 \in \Sigma$ and choose a coordinate chart $\phi : U \to \Omega$ on an open neighbourhood $U \subset \Sigma$ of p_0, where $\Omega \subset \mathbb{H}$ is an open set in the closed upper half plane $\mathbb{H} = \{\zeta \in \mathbb{C} \,|\, \mathrm{Im}\,\zeta \geq 0\}$. Compose ϕ with the Möbius transformation

$$\rho : \mathbb{H} \to \mathbb{D}, \qquad \rho(\zeta) := \frac{i - \zeta}{i + \zeta} \tag{C.5.2}$$

to obtain a diffeomorphism $\rho \circ \phi : U \to \rho(\Omega) \subset \mathbb{D}$. Shrinking U, if necessary, we may assume without loss of generality that $\rho_* \phi_* j$ extends to an almost complex structure j' on $\mathbb{D}$. By Theorem C.5.1, there exists a diffeomorphism $\psi : \mathbb{D} \to \mathbb{D}$ such that $\psi^* i = j'$. Hence $\phi^* \rho^* \psi^* i = j$ and hence

$$\rho^{-1} \circ \psi \circ \rho \circ \phi : U \to \rho^{-1} \circ \psi \circ \rho(\Omega) \subset \mathbb{H}$$

is a holomorphic coordinate chart on U with values in an open subset of $\mathbb{H}$. This proves Corollary C.5.3. □

EXERCISE C.5.4. Let (Σ, j) be a Riemann surface with boundary and let $\Gamma \subset \partial\Sigma$ be a connected component of the boundary that is diffeomorphic to the circle. Prove that, for $\varepsilon > 0$ sufficiently small, there is a holomorphic diffeomorphism from a neighbourhood of Γ onto the annulus $\{z \in \mathbb{C} \,|\, 1 - \varepsilon < |z| \leq 1\}$.

COROLLARY C.5.5. *Consider the inclusion*

$$\mathrm{PSL}(2,\mathbb{R}) \to \mathrm{Diff}(\mathbb{D}), \tag{C.5.3}$$

which assigns to every element of $\mathrm{PSL}(2,\mathbb{R})$ *the conjugation of the corresponding Möbius transformation of* $\mathbb{H}$ *with the diffeomorphism* $\rho : \mathbb{H} \cup \{\infty\} \to \mathbb{D}$ *in* (C.5.2). *The map* (C.5.3) *is a homotopy equivalence. In particular,* $\mathrm{Diff}(\mathbb{D})$ *is connected.*

PROOF. Let $\mathcal{J}(\mathbb{D}) \to \mathrm{Diff}(\mathbb{D}) : j \mapsto \psi_j$ be the map that assigns to every almost complex structure j on $\mathbb{D}$ the diffeomorphism $\psi_j := \psi \in \mathrm{Diff}(\mathbb{D})$ of Theorem C.5.1. Then the map $\mathrm{PSL}(2,\mathbb{R}) \times \mathcal{J}(\mathbb{D}) \to \mathrm{Diff}(\mathbb{D}) : (\phi, j) \mapsto \phi \circ \psi_j$ is a homeomorphism with inverse $f \mapsto (f \circ \psi_{f^*i}^{-1}, f^*i)$. Hence the assertion of Corollary C.5.5 follows from the fact that $\mathcal{J}(\mathbb{D})$ is contractible. □

REMARK C.5.6. We give another proof of a local version of Theorem C.5.1 in Appendix E (see Theorem E.3.1). Note that an almost complex structure $j \in \mathcal{J}(\mathbb{D})$ corresponds to a unique conformal class of metrics for which the vectors ξ and $j\xi$ are related by rotation through a quarter turn. (This conformal class consists of all metrics that are compatible with j.) Thus Theorem C.5.1 can also be thought of as the uniformization theorem for discs; cf. Corollary E.3.2.

Our first proof of Theorem C.5.1 is based on an idea of Earle and Eells [**95**]. It applies only when j is integrable: this is needed to establish the Fredholm property of the Cauchy–Riemann operator $\bar{\partial}_j : \Omega^0_{T\partial\mathbb{D}}(\mathbb{D}, T\mathbb{D}) \to \Omega^{0,1}_j(\mathbb{D}, T\mathbb{D})$. In exchange it is geometric in spirit, using very little analysis. Our second proof is based on a continuation argument and does not require integrability.

FIRST PROOF OF THEOREM C.5.1: INTEGRABLE CASE. Following Earle and Eells [**95**] we consider the diagram

$$\mathrm{PSL}(2,\mathbb{R}) \to \mathrm{Diff}(\mathbb{D}) \to \mathcal{J}(\mathbb{D})$$

where the first map is the inclusion (C.5.3) and the second map is given by $\psi \mapsto \psi^*i$.

STEP 1. *The map* $\mathrm{Diff}(\mathbb{D}) \to \mathcal{J}(\mathbb{D}) : \psi \mapsto \psi^*i$ *is a fibration.*

We must prove that the map has the path lifting property. Thus assume that $\mathbb{R} \to \mathcal{J}(\mathbb{D}) : t \mapsto j_t$ is a smooth path of almost complex structures on $\mathbb{D}$. Given a diffeomorphism $\psi_0 \in \mathrm{Diff}(\mathbb{D})$ such that $\psi_0^* j_0 = i$ we must find a smooth path $\mathbb{R} \to \mathrm{Diff}(\mathbb{D}) : t \mapsto \psi_t$ starting at ψ_0 such that

$$\psi_t^* j_t = i. \tag{C.5.4}$$

Following Moser, we look for a smooth family of vector fields $\mathbb{R} \to \mathrm{Vect}(\mathbb{D}) : t \mapsto \xi_t$ such that

$$\frac{d}{dt}\psi_t = \xi_t \circ \psi_t. \tag{C.5.5}$$

The resulting isotopy satisfies (C.5.4) if and only if

$$\mathcal{L}_{\xi_t} j_t + \frac{d}{dt} j_t = 0, \qquad \xi(\partial\mathbb{D}) \subset T\partial\mathbb{D}. \tag{C.5.6}$$

Note that the tangent space $T_j\mathcal{J}(\mathbb{D})$ is the space of smooth maps $\alpha : \mathbb{D} \to \mathbb{R}^{2\times 2}$ that anti-commute with j. Think of this as the space of complex anti-linear 1-forms on $\mathbb{D}$ with values in the tangent bundle $T\mathbb{D} = \mathbb{D} \times \mathbb{R}^2$ with the complex structure j:

$$T_j\mathcal{J}(\mathbb{D}) = \Omega^{0,1}_j(\mathbb{D}, T\mathbb{D}).$$

Think of a vector field ξ on $\mathbb{D}$ as a section of $T\mathbb{D}$. Then the Lie derivative is related to the Cauchy–Riemann operator

$$\bar{\partial}_j : \mathrm{Vect}(T\mathbb{D}) = \Omega^0(\mathbb{D}, T\mathbb{D}) \to \Omega^{0,1}_j(\mathbb{D}, T\mathbb{D}) = T_j\mathcal{J}(\mathbb{D})$$

of the Riemann surface $(\mathbb{D}, j)$ by the formula

$$\begin{aligned}(\mathcal{L}_\xi j)\eta &= [j\eta, \xi] - j[\eta, \xi] \\ &= \nabla_\xi(j\eta) - \nabla_{j\eta}\xi - j\nabla_\xi\eta + j\nabla_\eta\xi \\ &= j\nabla_\eta\xi - \nabla_{j\eta}\xi \\ &= 2j\bar{\partial}_j\xi(\eta).\end{aligned}$$

Here ∇ denotes the Levi–Civita connection on $\mathbb{D}$ with respect to a Riemannian metric that is compatible with j, the third equation follows from the fact that $\nabla j = 0$, and the term $\bar{\partial}_j$ in the last equation denotes the Cauchy–Riemann operator of the (holomorphic) tangent bundle of the 2-manifold $(\mathbb{D}, j)$. Thus $\bar{\partial}_j\xi = \frac{1}{2}(\nabla\xi + j\nabla\xi \circ j)$ as in Remark C.1.2. Since the bundle pair $(T\mathbb{D}, T\partial\mathbb{D})$ has boundary Maslov index 2 it follows from Theorem C.1.10 that equation (C.5.6) has a solution ξ_t for every t. Moreover, if we impose the conditions $\xi_t(0) = \xi_t(1) = 0$ then the solution is unique and depends smoothly on t. This proves the path lifting property.

STEP 2. *If $u : \mathbb{D} \to \mathbb{D}$ is a holomorphic diffeomorphism such that $u(0) = 0$ and $u(1) = 1$ then $u = \mathrm{id}$.*

We use the reflection principle. Define $v : \mathbb{D} \to \mathbb{D}$ by $v(z) := \overline{u(\bar{z})}$. Then v is a holomorphic diffeomorphism that satisfies $v(0) = 0$ and $u(1/z)v(z) = 1$ for $z \in S^1$. Hence u extends to a holomorphic diffeomorphism of the 2-sphere via

$$u(z) := \frac{1}{v(1/z)}, \qquad |z| \geq 1.$$

The derivative of u has the form

$$u'(z) = \frac{v'(1/z)}{z^2 v(1/z)^2}, \qquad |z| \geq 1.$$

Since $zv(1/z)$ converges to $v'(0) \neq 0$ as $z \to \infty$ it follows that u' is a bounded holomorphic function and hence is constant by Liouville's theorem. Since $u(0) = 0$ and $u(1) = 1$ it follows that $u = \mathrm{id}$ as claimed.

Now existence follows from Step 1 and uniqueness from Step 2. The proof of continuity is left as an exercise. This proves Theorem C.5.1 under the assumption that every almost complex structure on $\mathbb{D}$ is integrable. □

Our second proof of Theorem C.5.1 is technically more demanding than the first and has more important consequences. It implies integrability of almost complex structures in dimension two, even near boundary points. To carry out the proof we must address the following fundamental question. Given an almost complex structure j on $\mathbb{C}$ and an (i, j)-holomorphic disc $u : (\mathbb{D}, \partial\mathbb{D}) \to (\mathbb{C}, S^1)$ such that the restriction $u|_{\partial\mathbb{D}} : \partial\mathbb{D} \to S^1$ has degree one, why is u a diffeomorphism of the closed disc? It is fairly easy to prove that u restricts to a diffeomorphism of the open unit disc. However, a more subtle point is to show that u cannot have a critical point on the boundary. A related issue is unique continuation at a boundary point. We give a proof below which is based on the Carleman Similarity Principle in Theorem 2.3.5. The technique extends to higher dimensions.

LEMMA C.5.7. *Let j be a C^2 almost complex structure on $\mathbb{C}$ and $u : \mathbb{D} \to \mathbb{C}$ be a C^1 function satisfying*

$$\partial_s u + j(u)\partial_t u = 0, \qquad u(\partial\mathbb{D}) \subset S^1, \qquad \deg\left(u|_{\partial\mathbb{D}} : \partial\mathbb{D} \to S^1\right) = 1. \tag{C.5.7}$$

Then u is a C^1 diffeomorphism of $\mathbb{D}$.

PROOF. For $\zeta \in \mathbb{C} \setminus S^1$ denote by $\deg(u, \zeta)$ the local degree of u. Since the map $u|_{\partial\mathbb{D}} : \partial\mathbb{D} \to S^1$ has degree one so does the map

$$u_\zeta : \partial\mathbb{D} \to S^1, \qquad u_\zeta(z) := \frac{u(z) - \zeta}{|u(z) - \zeta|},$$

for every $\zeta \in \mathbb{C}$ with $|\zeta| < 1$. Hence $\deg(u, \zeta) = \deg(u_\zeta) = 1$ for $\zeta \in \mathbb{C}$ with $|\zeta| < 1$ and $\deg(u, \zeta) = \deg(u_\zeta) = 0$ for $\zeta \in \mathbb{C}$ with $|\zeta| > 0$. By the Carleman Similarity Principle in Theorem 2.3.5, applied to the function $z \mapsto u(z_0 + z) - \zeta$, every point $z_0 \in u^{-1}(\zeta)$ contributes positively to the local degree and contributes a number greater than one if z_0 is a critical point. Hence $u^{-1}(\zeta) = \emptyset$ for $|\zeta| > 1$, $\#u^{-1}(\zeta) = 1$ for $|\zeta| < 1$, and every point $\zeta \in \mathbb{C}$ with $|\zeta| < 1$ is a regular value of u. It follows also from Theorem 2.3.5 that $u(\mathrm{int}(\mathbb{D}))$ is an open set. Since $u(\mathbb{D}) \subset \mathbb{D}$ this implies that $u(\mathrm{int}(\mathbb{D})) = \mathrm{int}(\mathbb{D})$. Hence u restricts to a C^1 diffeomorphism of $\mathrm{int}(\mathbb{D})$.

It remains to prove that u cannot have a critical point on the boundary. We first reduce to the case when u has a boundary fixed point and $j = i$ along the boundary. It is convenient also to work with $\mathbb{H}$ rather than $\mathbb{D}$. To this end, fix a point $z_0 = e^{i\theta} \in \partial\mathbb{D}$, define $\rho : \mathbb{H} \to \mathbb{D}$ by

$$\rho(\zeta) := z_0 \frac{i - \zeta}{i + \zeta}, \qquad \rho(0) = z_0, \qquad \rho(\mathbb{H}) = \mathbb{D} \setminus \{-z_0\},$$

define the Möbius transformation ψ by

$$\psi(z) := i\frac{1 - \overline{u(z_0)}z}{1 + \overline{u(z_0)}z}, \qquad \psi(u(z_0)) = 0, \qquad \psi(\mathbb{D} \setminus \{-u(z_0)\}) = \mathbb{H},$$

and denote

$$j' := \psi_* j, \qquad u' := \psi \circ u \circ \rho : U' \to \mathbb{C}, \qquad U' := \{\zeta \in \mathbb{H} \,|\, u(\rho(\zeta)) \neq -u(z_0)\}.$$

Then j' is a C^2 almost complex structure on $\mathbb{C} \setminus \{-i\}$, and $u' : U' \to \mathbb{C} \setminus \{-i\}$ is a j'-holomorphic C^1 function such that $u'(U' \cap \mathbb{R}) \subset \mathbb{R}$ and $u'(0) = 0$. Choose a C^2 diffeomorphism $\chi : \Omega'' \to \Omega'$ between two neighbourhoods $\Omega'', \Omega' \subset \mathbb{C}$ of zero such that

$$\chi(\zeta) = \zeta, \qquad (\chi^* j')(\zeta) = i \qquad \forall \zeta \in \Omega'' \cap \mathbb{R}. \tag{C.5.8}$$

An explicit formula for χ is

$$\chi(\zeta) := \mathrm{Re}\,\zeta + (\mathrm{Im}\,\zeta) j'(\zeta) 1.$$

The reader may verify that the restriction of χ to a sufficiently small neighbourhood of the real axis is a C^2 diffeomorphism satisfying (C.5.8). Define

$$u'' := \chi^{-1} \circ u' : U'' := (u')^{-1}(\Omega') \to \Omega'', \qquad j'' = \chi^* j'.$$

Then u'' and j'' satisfy the hypotheses of Lemma C.5.8 below, namely j'' is a C^1 almost complex structure on Ω'', standard on the real axis, and $u'' : U'' \to \Omega''$ is a C^1 function satisfying

$$\partial_s u'' + j''(u'')\partial_t u'' = 0, \qquad u''(U'' \cap \mathbb{R}) \subset \mathbb{R}, \qquad u''(0) = 0.$$

Hence there is a positive integer ν and real numbers $\delta > 0$ and $c \neq 0$ such that

$$u''(\zeta) = c\zeta^\nu + O\left(|\zeta|^{\nu+\delta}\right)$$

for ζ near zero. Since u maps the disc $\mathbb{D}$ to itself, it follows that u'' maps the upper halfplane $\mathbb{H}$ to itself. This implies $\nu = 1$ and $c > 0$. In particular, $du''(0) \neq 0$ and thus $du(z_0) \neq 0$. So u has no critical point on the boundary and therefore is a C^1 diffeomorphism of the closed unit disc $\mathbb{D}$ to itself. This proves Lemma C.5.7. □

LEMMA C.5.8. *Fix a real number $p > 2$ and denote by $\mathbb{H} \subset \mathbb{C}$ the closed upper halfplane. Let j be a $W^{1,p}$ almost complex structure on a connected open neighbourhood $\Omega \subset \mathbb{C}$ of zero, let $W \subset \mathbb{H}$ be a connected open neighbourhood of zero, and let $w : W \to \Omega$ be a $W^{1,p}$ function satisfying*

$$\partial_s w + j(w)\partial_t w = 0, \qquad w(\mathbb{R} \cap W) \subset \mathbb{R}, \qquad w(0) = 0. \tag{C.5.9}$$

Suppose that $j(\zeta) = i$ for $\zeta \in \mathbb{R}$ and that w is nonconstant. Then there is a positive integer ν and a real number $c \neq 0$, such that

$$w(\zeta) = c\zeta^\nu + O\left(|\zeta|^{\nu+1-2/p}\right). \tag{C.5.10}$$

PROOF. The result follows by combining a Schwarz reflection type argument with the Carleman Similarity Principle in Theorem 2.3.5 (which in turn is proved by using the easy Riemann Roch Theorem C.4.1 on the unit disc). Let

$$\tau : \mathbb{C} \to \mathbb{C}$$

denote complex conjugation. Define

$$U \subset \mathbb{C}, \qquad u : U \to \mathbb{C}, \qquad J : U \to \mathrm{End}_{\mathbb{R}}(\mathbb{C})$$

by $U := W \cup \tau(W)$,

$$u(\zeta) := \begin{cases} w(\zeta), & \text{for } \zeta \in W, \\ \overline{w(\bar\zeta)}, & \text{for } \zeta \in \tau(W), \end{cases}$$

and

$$J(\zeta) := \begin{cases} j(w(\zeta)), & \text{for } \zeta \in W, \\ -\tau \circ j(w(\bar\zeta)) \circ \tau, & \text{for } \zeta \in \tau(W). \end{cases}$$

Then $U \subset \mathbb{C}$ is an open neighbourhood of zero. Since $w(\zeta) \in \mathbb{R}$ for $\zeta \in \mathbb{R}$ and $j(w(\zeta)) \in \mathrm{End}_{\mathbb{R}}(\mathbb{C})$ is multiplication by i for $\zeta \in \mathbb{R}$, it follows that

$$u \in W^{1,p}(U, \mathbb{C}), \qquad J \in W^{1,p}(U, \mathrm{End}_{\mathbb{R}}(\mathbb{C})).$$

Moreover, u and J satisfy

$$\partial_s u(\zeta) + J(\zeta)\partial_t u(\zeta) = 0, \qquad u(0) = 0, \qquad J(0) = i. \tag{C.5.11}$$

for $\zeta = s + it \in U$. Hence, by Theorem 2.3.5, there exists a constant $r > 0$ with

$$B_r := \{\zeta \in \mathbb{C} \mid |\zeta| < r\} \subset U,$$

a map $\Phi \in W^{1,p}(B_r, \mathrm{End}_{\mathbb{R}}(\mathbb{C}))$, and a holomorphic function $\sigma : B_r \to \mathbb{C}$ such that $\Phi(\zeta)$ is invertible and

$$u(\zeta) = \Phi(\zeta)\sigma(\zeta), \qquad \sigma(0) = 0, \qquad J(\zeta)\Phi(\zeta) = \Phi(\zeta)i, \tag{C.5.12}$$

for every $\zeta \in B_r$. Since $J(0)$ is multiplication by i and $J\Phi = \Phi i$ it follows that $\Phi(0)$ is complex linear. Hence there exists a nonzero complex number a such that $\Phi(0) : \mathbb{C} \to \mathbb{C}$ is multiplication by a. Since $u = \Phi\sigma$ is nonconstant and Φ is everywhere invertible it follows that σ is nonconstant. Since $\sigma(0) = 0$ there exists a positive integer ν and a nonzero complex number b such that

$$\sigma(\zeta) = b\zeta^\nu + O\left(|\zeta|^{\nu+1}\right).$$

Denote $c := ab \in \mathbb{C} \setminus \{0\}$. Then, by (C.5.12), we have

$$\begin{aligned} w(\zeta) &= u(\zeta) \\ &= ab\zeta^\nu + (\Phi(\zeta) - \Phi(0))\, b\zeta^\nu + \Phi(\zeta)(\sigma(\zeta) - b\zeta^\nu) \\ &= c\zeta^\nu + O\left(|\zeta|^{\nu+1-2/p}\right) \end{aligned}$$

for $\zeta \in B_r \cap \mathbb{H}$. Here the last equation follows from the Sobolev Embedding Theorem B.1.11, which shows that Φ is Hölder continuous with the exponent $1 - 2/p$ and hence $\Phi(\zeta) - \Phi(0) = O(|\zeta|^{1-2/p})$. Since $u(U \cap \mathbb{R}) \subset \mathbb{R}$ the number c must be real. This proves Lemma C.5.8. □

Second Proof of Theorem C.5.1: General Case. Fix an integer $\ell \geq 2$ and a real number $p > 2$. Let $\mathcal{J}^\ell(\mathbb{C})$ be the space of almost complex structures on $\mathbb{C}$ of class C^ℓ that are equal to i outside of the disc of radius two. Then $\mathcal{J}^\ell(\mathbb{C})$ is a Banach manifold. Denote by $\mathrm{Diff}^1(\mathbb{D})$ the group of C^1 diffeomorphisms of $\mathbb{D}$. Define

$$\mathcal{M}^\ell := \left\{(j, u) \in \mathcal{J}^\ell(\mathbb{C}) \times \mathrm{Diff}^1(\mathbb{D}) \,|\, u^*j = i,\, u(0) = 0,\, u(1) = 1\right\}$$

and

$$\mathcal{J}_0^\ell := \left\{j \in \mathcal{J}^\ell(\mathbb{C}) \,|\, \text{there is an element } u \in \mathrm{Diff}^1(\mathbb{D}) \text{ such that } (j, u) \in \mathcal{M}^\ell\right\}.$$

We prove in six steps that $\mathcal{J}_0^\ell = \mathcal{J}^\ell(\mathbb{C})$.

Step 1. *$\mathcal{J}_0^\ell$ is nonempty.*

The pair (i, id) is an element of $\mathcal{M}^\ell$ and hence $i \in \mathcal{J}_0^\ell$.

Step 2. *The set*

$$\mathcal{U}^{1,p} := \left\{ u \in W^{1,p}(\mathbb{D}, \mathbb{C}) \,\middle|\, \begin{array}{l} u(\partial\mathbb{D}) \subset S^1, \\ \deg(u|_{\partial\mathbb{D}} : \partial\mathbb{D} \to S^1) = 1, \\ u(0) = 0,\, u(1) = 1 \end{array} \right\}$$

is a smooth Banach submanifold of $W^{1,p}(\mathbb{D}, \mathbb{C})$ and its tangent space at a point $u \in \mathcal{U}^{1,p}$ is the Banach space

$$T_u\mathcal{U}^{1,p} = \left\{ \widehat{u} \in W^{1,p}(\mathbb{D}, \mathbb{C}) \,\middle|\, \begin{array}{l} \widehat{u}(z) \in T_{u(z)}S^1 \;\forall z \in \partial\mathbb{D}, \\ \widehat{u}(0) = \widehat{u}(1) = 0 \end{array} \right\}.$$

This is standard and the proof is left to the reader.

Step 3. *If $(j, u) \in \mathcal{J}^\ell(\mathbb{C}) \times \mathcal{U}^{1,p}$ and $\partial_s u + j(u)\partial_t u = 0$ then $u \in \mathrm{Diff}^1(\mathbb{D})$.*

Since j is a C^ℓ almost complex structure with $\ell \geq 2$ and u is a $W^{1,p}$ solution of the boundary value problem (C.5.7) with $p > 2$, it follows from Theorem B.4.1 that u is a $W^{\ell,p}$-function. Since $p > 2$ it follows from the Sobolev embedding theorem B.1.11 that u is of class $C^{\ell-1}$. Hence the assertion of Step 3 follows from Lemma C.5.7.

Step 4. *If $(j, u) \in \mathcal{M}^\ell$ then the operator $D_{j,u} : T_u\mathcal{U}^{1,p} \to L^p(\mathbb{D}, \mathbb{C})$, defined by*

$$D_{j,u}\widehat{u} := \partial_s\widehat{u} + j(u)\partial_t\widehat{u} + (dj(u)\widehat{u})\partial_t u,$$

is bijective.

Every element $\widehat{u} \in T_u\mathcal{U}^{1,p}$ has the form $\widehat{u} = du \cdot \xi$, where $\xi \in W^{1,p}(\mathbb{D}, \mathbb{C})$ satisfies

$$\xi(z) \in T_zS^1 \quad \forall\, z \in \partial\mathbb{D}, \qquad \xi(0) = 0, \qquad \xi(1) = 1.$$

Moreover, since $\partial_s u + j(u)\partial_t u = 0$, we have $D_{j,u}\widehat{u} = du \cdot (\partial_s\xi + i\partial_t\xi)$ for $\widehat{u} = du \cdot \xi$, and thus

$$D_{j,u}\widehat{u} = 0 \qquad \Longleftrightarrow \qquad \partial_s\xi + i\partial_t\xi = 0.$$

Hence it follows from Step 2 in the proof of Theorem C.4.1 with $k = 2$ that every element in the kernel of $D_{j,u}$ has the form

$$\widehat{u}(z) = du(z)i(a + bz + \bar{a}z^2)$$

with $a \in \mathbb{C}$ and $b \in \mathbb{R}$. Since $\widehat{u}(0) = 0$ we have $a = 0$ and since $\widehat{u}(1) = 0$ we have $b = 0$. Hence $D_{j,u}$ is injective. It follows also from Theorem C.4.1 that $D_{j,u}$ has Fredholm index zero (because of the conditions $\widehat{u}(0) = \widehat{u}(1) = 0$ for elements in the domain of $D_{j,u}$) and hence $D_{j,u}$ is bijective. This proves Step 4.

STEP 5. *$\mathcal{J}_0^\ell$ is an open subset of $\mathcal{J}^\ell(\mathbb{C})$.*

By Step 2, the set $\mathcal{J}^\ell(\mathbb{C}) \times \mathcal{U}^{1,p}$ is a smooth Banach manifold. Define the map

$$\mathcal{F} : \mathcal{J}^\ell(\mathbb{C}) \times \mathcal{U}^{1,p} \to L^p(\mathbb{D}, \mathbb{C})$$

by

$$\mathcal{F}(j, u) := \partial_s u + j(u)\partial_t u.$$

This map is ℓ times continuously differentiable, and its derivative at a pair (j, u) is the linear operator $d\mathcal{F}(j, u) : T_j\mathcal{J}^\ell(\mathbb{C}) \times T_u\mathcal{U}^{1,p} \to L^p(\mathbb{D}, \mathbb{C})$ given by

$$d\mathcal{F}(j, u)(\widehat{\jmath}, \widehat{u}) = D_{j,u}\widehat{u} + \widehat{\jmath}(u)\partial_t u.$$

By Step 3, the zero set of $\mathcal{F}$ is $\mathcal{M}^\ell$. If $(j, u) \in \mathcal{M}^\ell$ then the operator

$$D_{j,u} : T_u\mathcal{U}^{1,p} \to L^p(\mathbb{D}, \mathbb{C})$$

is bijective by Step 4 and hence, by Lemma A.3.6, the operator $d\mathcal{F}(j, u)$ is surjective and has a right inverse. Hence it follows from the Implicit Function Theorem A.3.3 that $\mathcal{M}^\ell$ is a C^ℓ Banach submanifold of $\mathcal{J}^\ell(\mathbb{C}) \times \mathcal{U}^{1,p}$. Its tangent space at (j, u) is the Banach space

$$T_{(j,u)}\mathcal{M}^\ell = \left\{(\widehat{\jmath}, \widehat{u}) \in T_j\mathcal{J}^\ell(\mathbb{C}) \times T_u\mathcal{U}^{1,p} \,|\, D_{j,u}\widehat{u} + \widehat{\jmath}(u)\partial_t u = 0\right\}.$$

Since $D_{j,u}$ is bijective, so is the projection

$$T_{(j,u)}\mathcal{M}^\ell \to T_j\mathcal{J}^\ell(\mathbb{C}) : (\widehat{\jmath}, \widehat{u}) \mapsto \widehat{\jmath}.$$

This implies that the projection $\mathcal{M}^\ell \to \mathcal{J}^\ell(\mathbb{C}) : (j, u) \mapsto j$ is a local diffeomorphism between C^ℓ Banach manifolds. Hence it follows from the Inverse Function Theorem A.3.1 that its image $\mathcal{J}_0^\ell$ is an open subset of $\mathcal{J}^\ell(\mathbb{C})$. This proves Step 5.

STEP 6. *$\mathcal{J}_0^\ell$ is a closed subset of $\mathcal{J}^\ell(\mathbb{C})$.*

Let j^ν be a sequence in $\mathcal{J}_0^\ell$ converging in the C^ℓ topology to an element $j \in \mathcal{J}^\ell(\mathbb{C})$. Choose a sequence $u^\nu \in \mathrm{Diff}^1(\mathbb{D})$ such that $(j^\nu, u^\nu) \in \mathcal{M}^\ell$ for every ν. Then it follows from Theorem B.4.1 that $u^\nu \in W^{\ell,p}(\mathbb{D}, \mathbb{C})$ for every ν. We claim that the first derivatives of u_ν are uniformly bounded. Once this is established, it follows from Theorem B.4.2 that u^ν is uniformly bounded in the C^ℓ topology and hence a subsequence of u^ν converges in the $C^{\ell-1}$ topology to a map $u : \mathbb{D} \to \mathbb{C}$. The limit is an element of $\mathcal{U}^{1,p}$ and satisfies $\partial_s u + j(u)\partial_t u = 0$. Hence $(j, u) \in \mathcal{M}^\ell$ by Step 3 and hence $j \in \mathcal{J}_0^\ell$.

To prove the required bound on the first derivatives we choose a sequence of Riemannian metrics on $\mathbb{C}$ determined by the almost complex structures j_ν and the standard volume form $\omega := ds \wedge dt$ via $\langle\cdot,\cdot\rangle_{j^\nu} := \omega(\cdot, j_\nu\cdot)$. The energy of u^ν with respect to this metric is given by

$$E(u_\nu) = \int_{\mathbb{D}} |\partial_s u^\nu|_{j^\nu}^2 \, dsdt = \int_{\mathbb{D}} (u^\nu)^*\omega = \pi.$$

Here the last equation follows from the fact that the degree of $u^\nu|_{\partial\mathbb{D}} : \partial\mathbb{D} \to S^1$ is equal to one. Moreover, the energy of every j-holomorphic disc with boundary in S^1 is an integer multiple of π.

Now suppose, by contradiction, that the sequence $\sup_{\mathbb{D}} |du^\nu|$ is unbounded. Then Theorem 4.6.1 asserts that, after passing to a subsequence, u^ν has uniformly bounded first derivatives on every compact subset of the complement of a nonempty finite set $Z \subset \mathbb{D}$ such that, for each $\zeta \in Z$, there is a sequence $z^\nu \in \mathbb{D}$ satisfying $z^\nu \to \zeta$ and $c^\nu := |du^\nu(z^\nu)| \to \infty$. Here $\zeta \notin \mathrm{int}(\mathbb{D})$; otherwise a further subsequence of the rescaled sequence $v_\nu(z) := u_\nu(z_\nu + z/c_\nu)$ would converge, by Theorem B.4.2 and Theorem 4.1.2, to a nonconstant j-holomorphic sphere in $\mathbb{C}$, which does not exist. Hence $Z \subset \partial\mathbb{D}$ and, by Theorem 4.6.1, we have that

$$\lim_{\varepsilon\to 0}\lim_{\nu\to\infty} E(u^\nu, B_\varepsilon(\zeta)\cap\mathbb{D}) \geq \pi \qquad \forall \zeta \in Z.$$

Since the first derivatives of u^ν are uniformly bounded on every compact subset of $\mathbb{D}\backslash Z$, Theorem B.4.2 asserts that a further subsequence of u^ν converges (in the $C^{\ell-1}$ topology on every compact subset of $\mathbb{D}\setminus Z$) to a j-holomorphic map $u : \mathbb{D}\setminus Z \to \mathbb{C}$. Since the energy $E(u^\nu) = \pi$ is used up at the bubbling points (of which there can only be one) it follows that the limit u is constant. But this is impossible because $u(0) = 0$ and $u(\partial\mathbb{D}\setminus Z) \subset S^1$. This contradiction shows that the first derivatives of u^ν must have been uniformly bounded as claimed. This proves Step 6.

Since $\mathcal{J}^\ell(\mathbb{C})$ is connected, it follows from Steps 1, 5, and 6 that $\mathcal{J}_0^\ell = \mathcal{J}^\ell(\mathbb{C})$. Now let j be a smooth almost complex structure on $\mathbb{D}$. Extend j to a smooth almost complex structure on $\mathbb{C}$, still denoted by j, that agrees with i outside the disc of radius two (See Exercise C.5.9 below). Then $j \in \mathcal{J}^\ell(\mathbb{C})$ and hence $j \in \mathcal{J}_0^\ell$ by what we have just proved. Hence there is a C^1 diffomorphism $u : \mathbb{D} \to \mathbb{D}$ such that $(j, u) \in \mathcal{M}^\ell$. Since j is smooth it follows from Theorem B.4.1 that u is smooth. Hence $\psi := u^{-1} : \mathbb{D} \to \mathbb{D}$ is a diffeomorphism satisfying (C.5.1). This proves existence. Uniqueness follows from Step 2 in the first proof of Theorem C.5.1. To prove continuity one can use the fact that the projection $\mathcal{M}^\ell \to \mathcal{J}^\ell(\mathbb{C})$ is a C^ℓ diffeomorphism of C^ℓ Banach manifolds for every ℓ. This completes the proof of the general case of Theorem C.5.1. □

EXERCISE C.5.9. Let $f : (-\infty, 0] \to \mathbb{R}$ be a smooth function. This exercise shows how to extend f to a smooth function on all of $\mathbb{R}$ that vanishes for $t \geq 1$. Choose a smooth cutoff function $\beta : [0,\infty) \to [0,1]$ such that

$$\beta(t) = \begin{cases} 1, & \text{for } 0 \leq t \leq 1/2, \\ 0, & \text{for } t \geq 1. \end{cases}$$

For $k = 0, 1, 2, 3, \ldots$ choose $a_k \in \mathbb{R}$ and $\varepsilon_k > 0$ such that

$$a_k := \frac{f^{(k)}(0)}{k!}, \qquad \varepsilon_k \leq \frac{1}{|a_k|\, k^k}, \qquad \varepsilon_k \leq \frac{1}{2}.$$

For $t \ge 0$ define

$$\text{(C.5.13)} \qquad f(t) := \sum_{k=0}^{\infty} f_k(t), \qquad f_k(t) := a_k \left(\beta\left(\frac{t}{\varepsilon_k}\right) t \right)^k.$$

Prove that $f : \mathbb{R} \to \mathbb{R}$ is smooth. *Hint:* For every $\ell \in \mathbb{N}$ there exists a constant $c_\ell > 0$ such that, for every $t \ge 0$ and every $k \in \mathbb{N}$, we have

$$\text{(C.5.14)} \qquad \left| f_k^{(\ell)} \right| \le c_\ell \left| a_k \right| k^\ell \varepsilon_k^{k-\ell}.$$

Prove that every smooth almost complex structure $j \in \mathcal{J}(\mathbb{D})$ extends to a smooth almost complex structure on $\mathbb{C}$ that is equal to i outside the disc of radius two.

C.6. Nonsmooth bundles

For some of our applications it is important to consider bundles with weaker smoothness properties. For example, in Chapter 3 we must deal with bundles of class $W^{\ell,p}$ and connections of class $W^{\ell-1,p}$ for some integer ℓ and a number $p > 1$ such that $\ell p > 2$. More precisely, let (E, F) be a bundle pair over a compact Riemann surface (Σ, j) with boundary as in Definition C.3.4 and suppose that $\langle \cdot, \cdot \rangle$ is a Hermitian inner product on E (i.e. an inner product satisfying $J^* = -J$) such that $JF = F^\perp$. We call the triple

$$(E, F, \langle \cdot, \cdot \rangle)$$

a **Hermitian bundle pair**. A Hermitian bundle pair $(E, F, \langle \cdot, \cdot \rangle)$ admits a system of unitary local trivializations

$$\phi_\alpha : \pi^{-1}(U_\alpha) \to U_\alpha \times \mathbb{C}^n,$$

for a suitable open cover $\{U_\alpha\}_\alpha$ of Σ, such that

$$\phi_\alpha(\pi^{-1}(U_\alpha) \cap F) = (U_\alpha \cap \partial\Sigma) \times \mathbb{R}^n.$$

The transition functions

$$g_{\beta\alpha} : U_\alpha \cap U_\beta \to \mathrm{U}(n)$$

are defined by

$$(z, g_{\beta\alpha}(z)v) := \phi_\beta \circ \phi_\alpha^{-1}(z, v)$$

and they satisfy

$$\text{(C.6.1)} \qquad g_{\gamma\beta} g_{\beta\alpha} = g_{\gamma\alpha}, \qquad g_{\alpha\alpha} = 1\!\!1, \qquad g_{\beta\alpha}(U_\alpha \cap U_\beta \cap \partial\Sigma) \subset \mathrm{O}(n).$$

Conversely, every system of transition functions $g_{\beta\alpha} : U_\alpha \cap U_\beta \to \mathrm{U}(n)$ satisfying (C.6.1) determines a Hermitian bundle pair $(E, F, \langle \cdot, \cdot \rangle)$. (Prove this!) In the local trivializations a Hermitian connection ∇ has the form

$$(\nabla\xi)_\alpha = d\xi_\alpha + A_\alpha \xi_\alpha \in \Omega^1(U_\alpha, \mathbb{C}^n),$$

where $\xi_\alpha : U_\alpha \to \mathbb{C}^n$ and $(\nabla\xi)_\alpha \in \Omega^1(U_\alpha, \mathbb{C}^n)$ are defined by

$$\phi_\alpha \circ \xi(z) =: (z, \xi_\alpha(z)), \qquad \phi_\alpha^*(\nabla\xi)_\alpha =: \nabla\xi|_{U_\alpha}.$$

The 1-forms $A_\alpha \in \Omega^1(U_\alpha, \mathfrak{u}(n))$ are called the **connection potentials**. Note that the ξ_α and A_α satisfy

$$g_{\beta\alpha}\xi_\alpha = \xi_\beta, \qquad A_\alpha = g_{\beta\alpha}^{-1} dg_{\beta\alpha} + g_{\beta\alpha}^{-1} A_\beta g_{\beta\alpha}.$$

DEFINITION C.6.1. *Let ℓ be an integer and $p > 1$ such that $\ell p > 2$. A (continuous) Hermitian bundle pair (E, F) is said to be* **of class** $W^{\ell,p}$ *if it admits trivializations as above such that the transition functions $g_{\alpha\beta}$ are of class $W^{\ell,p}$. A connection ∇ is said to be* **of class** $W^{\ell-1,p}$ *if the connection potentials A_α are of class $W^{\ell-1,p}$. A* **Cauchy–Riemann operator of class** $W^{\ell-1,p}$ *on a Hermitian bundle pair of class $W^{\ell,p}$ is an operator of the form $D = \bar{\partial}^\nabla$, where ∇ is a Hermitian connection of class $W^{\ell-1,p}$.*

The following exercise shows that every Hermitian bundle pair (E, F) of class $W^{\ell,p}$ is (isometrically) $W^{\ell,p}$-isomorphic to a smooth Hermitian bundle pair (E_0, F_0).

EXERCISE C.6.2 (Bundles of class $W^{\ell,p}$). Let (E, F) be a Hermitian bundle pair of class $W^{\ell,p}$ with local trivializations $\phi_\alpha : \pi^{-1}(U_\alpha) \to U_\alpha \times \mathbb{C}^n$.

(i) Choose a collection of smooth cutoff functions $\sigma_\alpha : \Sigma \to [0, 1]$ such that $\operatorname{supp} \sigma_\alpha \subset U_\alpha$ and the functions $\rho_\alpha := \sigma_\alpha^2$ form a partition of unity. Define an embedding $E \to \Sigma \times \mathbb{C}^N : (z, e) \mapsto (z, \Phi(z)e)$ by

$$\Phi(z)e := \{\sigma_\alpha(z)\Phi_\alpha(z)e\}_\alpha,$$

where $\Phi_\alpha(z) : E_z \to \mathbb{C}^n$ is given by $\phi_\alpha(z, e) = (z, \Phi_\alpha(z)e)$. Prove that $\Phi(z)$ is a linear isometric embedding and $\Phi(z)F_z = \operatorname{im} \Phi(z) \cap \mathbb{R}^N$ for $z \in \partial\Sigma$.

(ii) Let Φ be as in (i) and define $\Pi : \Sigma \to \mathbb{C}^{N\times N}$ by

$$\Pi(z) := \Phi(z)(\Phi(z)^*\Phi(z))^{-1}\Phi(z)^*.$$

Prove that $\Pi(z)$ is the orthogonal projection of $\mathbb{C}^N$ onto the image of $\Phi(z)$. Prove that $\Pi \in W^{\ell,p}(\Sigma, \mathbb{C}^{N\times N})$ and $\Pi(\partial\Sigma) \subset \mathbb{R}^{N\times N}$.

(iii) Let $\Pi_0 : \Sigma \to \mathbb{C}^{N\times N}$ be a smooth function such that

$$\Pi_0(z)^2 = \Pi_0(z) = \Pi_0(z)^*, \qquad \Pi_0(\partial\Sigma) \subset \mathbb{R}^{N\times N}, \qquad \max_{z\in\Sigma} |\Pi(z) - \Pi_0(z)| < 1.$$

Prove that $\operatorname{rank} \Pi_0(z) = \operatorname{rank} \Pi(z) = n$ for every $z \in \Sigma$.

(iv) Define the smooth Hermitian bundle pair (E_0, F_0) over Σ by $E_{0z} := \operatorname{im} \Pi_0(z)$ for $z \in \Sigma$ and $F_{0z} := \operatorname{im} \Pi_0(z) \cap \mathbb{R}^N$ for $z \in \partial\Sigma$. Prove that the map $\Psi_0 : (E_0, F_0) \to (E, F)$, given by $\Psi_0 := \Phi^{-1}\Pi$, is a $W^{\ell,p}$ bundle isomorphism. Prove that $\Phi_0 := \Psi_0(\Psi_0^*\Psi_0)^{-1/2}$ is a Hermitian bundle isomorphism of class $W^{\ell,p}$.

COROLLARY C.6.3. *Let ℓ be a positive integer and $p > 1$ such that $\ell p > 2$. Let (E, F) be a Hermitian bundle pair of class $W^{\ell,p}$ and D be a Cauchy–Riemann operator on E of class $W^{\ell-1,p}$. Then the assertions of Theorem C.1.10 continue to hold for $k \in \{1, \dots, \ell\}$ and $q > 1$ such that $k - 2/q \le \ell - 2/p$.*

PROOF. By Exercise C.6.2, we may assume that the Hermitian bundle pair (E, F) is smooth. Hence the result follows from Theorem C.1.10. □

C.7. Almost complex structures

As we remarked in Section 2.1, a general almost complex manifold (M, J) does not have connections that are both torsion free and compatible with J in the sense that parallel translation commutes with J. However, it does have natural connections $\widehat{\nabla}$ that are compatible with J and whose torsion T is a multiple of the Nijenhuis tensor. In this section we derive a formula for $\widehat{\nabla}$, and show that when J is ω-compatible it reduces to the connection

$$\widetilde{\nabla}_X Y := \nabla_X Y - \tfrac{1}{2} J(\nabla_X J)Y$$

considered in Section 3.1. We shall also establish the explicit formulas for D_u in terms of $\widehat{\nabla}$ that are stated in Remark 3.1.3.

Let (M, J) be an almost complex manifold. Suppose that M is equipped with a Riemannian metric that is preserved by J and let ∇ denote the Levi–Civita connection of this metric. The next lemma shows that $\nabla J = 0$ if and only M is Kähler, i.e. J is integrable and the 2-form $\omega := \langle J\cdot, \cdot\rangle$ is closed.

LEMMA C.7.1. *Let (M, J) be an almost complex manifold equipped with a Riemannian metric that is preserved by J and let $\omega := \langle J\cdot, \cdot\rangle \in \Omega^2(M)$. Then*

$$(\nabla_X J)J + J(\nabla_X J) = 0, \qquad \langle(\nabla_X J)Y, Z\rangle + \langle Y, (\nabla_X J)Z\rangle = 0, \tag{C.7.1}$$

$$d\omega(X, Y, Z) = \langle(\nabla_X J)Y, Z\rangle + \langle(\nabla_Y J)Z, X\rangle + \langle(\nabla_Z J)X, Y\rangle, \tag{C.7.2}$$

$$N(X, Y) = (J\nabla_Y J - \nabla_{JY} J)X - (J\nabla_X J - \nabla_{JX} J)Y \tag{C.7.3}$$

for $X, Y, Z \in \mathrm{Vect}(M)$. If ω is closed then

$$\langle X, N(Y, Z)\rangle = 2\langle J(\nabla_X J)Y, Z\rangle, \tag{C.7.4}$$

$$J(\nabla_{JX} J) = \nabla_X J, \qquad N(X, Y) = 2J(\nabla_Y J)X - 2J(\nabla_X J)Y, \tag{C.7.5}$$

$$\langle X, N(Y, Z)\rangle + \langle Y, N(Z, X)\rangle + \langle Z, N(X, Y)\rangle = 0 \tag{C.7.6}$$

for $X, Y, Z \in \mathrm{Vect}(M)$.

PROOF. To prove (C.7.1) differentiate the identities $\langle Y, JZ\rangle + \langle JY, Z\rangle = 0$ and $J^2 = -1\!\!1$. To prove (C.7.2) we choose vector fields X, Y, and Z such that all six covariant derivatives of the form $\nabla_X Y$ vanish at a given point $x \in M$. Then the three Lie brackets $[X, Y]$, $[Y, Z]$, and $[Z, X]$ vanish at x and $\nabla_X(\omega(Y, Z)) = \langle(\nabla_X J)Y, Z\rangle$ at x. Hence, at the point $x \in M$, we have

$$\begin{aligned} d\omega(X, Y, Z) &= \nabla_X(\omega(Y, Z)) + \nabla_Y(\omega(Z, X)) + \nabla_Z(\omega(X, Y)) \\ &\quad + \omega([X, Y], Z) + \omega([Y, Z], X) + \omega([Z, X], Y) \\ &= \langle(\nabla_X J)Y, Z\rangle + \langle(\nabla_Y J)Z, X\rangle + \langle(\nabla_Z J)X, Y\rangle. \end{aligned}$$

This proves (C.7.2). The formula (C.7.3) for the Nijenhuis tensor follows directly from the definition of N by

$$N(X, Y) := [X, Y] + J[JX, Y] + J[X, JY] - [JX, JY]$$

and the formula $[X, Y] = \nabla_Y X - \nabla_X Y$. (Our sign conventions are explained in Section 2.1.) Using (C.7.3) and (C.7.2) with $d\omega = 0$ we obtain

$$\begin{aligned} \langle X, N(Y, Z)\rangle &= \langle X, J(\nabla_Z J)Y\rangle - \langle X, J(\nabla_Y J)Z\rangle \\ &\quad - \langle X, (\nabla_{JZ} J)Y\rangle + \langle X, (\nabla_{JY} J)Z\rangle \\ &= \langle X, J(\nabla_Z J)Y\rangle - \langle X, J(\nabla_Y J)Z\rangle \\ &\quad + \langle JZ, (\nabla_Y J)X\rangle + \langle Y, (\nabla_X J)JZ\rangle \\ &\quad - \langle JY, (\nabla_Z J)X\rangle - \langle Z, (\nabla_X J)JY\rangle \\ &= 2\langle J(\nabla_X J)Y, Z\rangle. \end{aligned}$$

Here the last equality follows from (C.7.1). Thus we have proved (C.7.4). Now

$$\begin{aligned} 2\langle J(\nabla_{JX}J)Y, Z\rangle &= \langle JX, N(Y,Z)\rangle \\ &= -\langle X, JN(Y,Z)\rangle \\ &= \langle X, N(Y,JZ)\rangle \\ &= 2\langle J(\nabla_X J)Y, JZ\rangle \\ &= 2\langle (\nabla_X J)Y, Z\rangle. \end{aligned}$$

Here the first and fourth equalities follow from (C.7.4) and the third follows from the fact that the Nijenhuis tensor is complex anti-linear in both variables. This proves the formula $J\nabla_{JX}J = \nabla_X J$ and so (C.7.5) follows from (C.7.3). To prove (C.7.6) note that

$$\begin{aligned} &\langle X, N(Y,Z)\rangle + \langle Y, N(Z,X)\rangle + \langle Z, N(X,Y)\rangle \\ &= 2\langle J(\nabla_X J)Y, Z\rangle + 2\langle J(\nabla_Y J)Z, X\rangle + 2\langle J(\nabla_Z J)X, Y\rangle \\ &= -2\langle (\nabla_{JX} J)Y, Z\rangle - 2\langle (\nabla_Y J)Z, JX\rangle - 2\langle (\nabla_Z J)JX, Y\rangle \\ &= 0. \end{aligned}$$

Here the first equation follows from (C.7.4), the second from (C.7.5), and the last from (C.7.2) with $d\omega = 0$. □

Let us now assume that J is tamed by a symplectic form ω and that the Riemannian metric is given by

$$\langle v, w\rangle := \tfrac{1}{2}\Big(\omega(v, Jw) + \omega(w, Jv)\Big) \tag{C.7.7}$$

as in (2.1.1). We emphasize that the 2-form $\langle J\cdot, \cdot\rangle$ does not agree with ω unless J is ω-compatible. Let ∇ be the Levi–Civita connection of (C.7.7). Then the formula

$$\widetilde{\nabla}_X Y := \nabla_X Y - \tfrac{1}{2} J(\nabla_X J)Y$$

defines a connection on TM which preserves J. The formula (C.7.1) shows that $\widetilde{\nabla}$ also preserves the metric. However, this connection will not, in general, be torsion free. If J is ω-compatible then the formula (C.7.5) shows that the torsion of $\widetilde{\nabla}$ is minus a quarter of the Nijenhuis tensor:

$$\begin{aligned} T^{\widetilde{\nabla}}(X,Y) &:= \widetilde{\nabla}_Y X - \widetilde{\nabla}_X Y - [X,Y] \\ &= \tfrac{1}{2} J(\nabla_X J)Y - \tfrac{1}{2} J(\nabla_Y J)X \\ &= -\tfrac{1}{4} N(X,Y). \end{aligned}$$

If J is only ω-tame it is convenient to introduce a connection $\widehat{\nabla}$ on TM via

$$\widehat{\nabla}_X Y := \widetilde{\nabla}_X Y - \tfrac{1}{4}\Big(\nabla_{JY} J + J\nabla_Y J\Big)X = \nabla_X Y - Q(X,Y), \tag{C.7.8}$$

where

$$4Q(X,Y) := (\nabla_{JY} J)X + J(\nabla_Y J)X + 2J(\nabla_X J)Y. \tag{C.7.9}$$

The connection $\widehat{\nabla}$ will not, in general, preserve the metric. However, the next lemma shows that it preserves J and that its torsion is always equal to minus a quarter times the Nijenhuis tensor.

LEMMA C.7.2. *Let (M,ω) be a symplectic manifold and $J \in \mathcal{J}_\tau(M,\omega)$. Denote by ∇ the Levi-Civita connection of the metric (C.7.7) and define $\widehat{\nabla}$ by (C.7.8) Then $\widehat{\nabla}$ preserves the almost complex structure J and its torsion is*

$$T^{\widehat{\nabla}}(X,Y) := \widehat{\nabla}_Y X - \widehat{\nabla}_X Y - [X,Y] = -\tfrac{1}{4}N(X,Y).$$

If J is ω-compatible then $\widehat{\nabla} = \widetilde{\nabla}$ and so $\widehat{\nabla}$ preserves the metric.

PROOF. That $\widehat{\nabla}$ preserves J follows by direct calculation. To compute its torsion note that

$$\begin{aligned} 4(Q(X,Y) - Q(Y,X)) &= (\nabla_{JY}J)X - J(\nabla_Y J)X - (\nabla_{JX}J)Y + J(\nabla_X J)Y \\ &= -N(X,Y). \end{aligned}$$

Here the last equation follows from (C.7.5). Now the formula for the torsion of $\widehat{\nabla}$ follows from the fact that ∇ is torsion free. In the ω-compatible case it follows from (C.7.5) that $\widehat{\nabla} = \widetilde{\nabla}$. This proves the lemma. □

We now establish the formulas for D_u given in Remark 3.1.3.

LEMMA C.7.3. (i) *If J is ω-tame and $u : \Sigma \to M$ is a smooth map then*

$$D_u\xi = (\widehat{\nabla}\xi)^{0,1} + \tfrac{1}{4}N_J(\xi, \partial_J(u)) + \tfrac{1}{4}\big(J(\nabla_{\bar\partial_J(u)}J) + \nabla_{J\bar\partial_J(u)}J\big)\xi. \tag{C.7.10}$$

In particular,

$$D_u\xi = (\widetilde{\nabla}\xi)^{0,1} + \tfrac{1}{4}N_J(\xi, \partial_J(u)). \tag{C.7.11}$$

whenever J is ω-compatible.

PROOF. Note that

$$\nabla\xi = \widehat{\nabla}\xi + \tfrac{1}{2}J(\nabla_{du}J)\xi + \tfrac{1}{4}J(\nabla_\xi J)du + \tfrac{1}{4}(\nabla_{J\xi}J)du.$$

and, by (C.7.3),

$$N_J(\xi, \partial_J(u)) = J(\nabla_{\partial_J(u)}J)\xi - (\nabla_{J\partial_J(u)}J)\xi - J(\nabla_\xi J)\partial_J(u) + (\nabla_{J\xi}J)\partial_J(u).$$

Hence

$$\begin{aligned} D_u\xi &= \tfrac{1}{2}\big(\nabla\xi + J\nabla\xi\circ j - J(\nabla_\xi J)\partial_J(u)\big) \\ &= \tfrac{1}{2}\big(\widehat{\nabla}\xi + J(u)\widehat{\nabla}\xi\circ j\big) - \tfrac{1}{2}J(\nabla_\xi J)\partial_J(u) \\ &\quad + \tfrac{1}{4}J(\nabla_{du}J)\xi + \tfrac{1}{8}J(\nabla_\xi J)du + \tfrac{1}{8}(\nabla_{J\xi}J)du \\ &\quad - \tfrac{1}{4}(\nabla_{du\circ j}J)\xi - \tfrac{1}{8}(\nabla_\xi J)du\circ j + \tfrac{1}{8}J(\nabla_{J\xi}J)du\circ j \\ &= (\widehat{\nabla}\xi)^{0,1} + \tfrac{1}{4}J(\nabla_{du}J)\xi - \tfrac{1}{4}(\nabla_{du\circ j}J)\xi - \tfrac{1}{4}J(\nabla_\xi J)\partial_J(u) + \tfrac{1}{4}(\nabla_{J\xi}J)\partial_J(u) \\ &= (\widehat{\nabla}\xi)^{0,1} + \tfrac{1}{4}N_J(\xi,\partial_J(u)) + \tfrac{1}{4}J(\nabla_{\bar\partial_J(u)}J)\xi - \tfrac{1}{4}(\nabla_{\bar\partial_J(u)\circ j}J)\xi. \end{aligned}$$

This proves (C.7.10). Equation (C.7.11) follows from (C.7.10) and (C.7.5). □

APPENDIX D

Stable Curves of Genus Zero

In this appendix we give a self-contained proof of the fact that the Grothendieck–Knudsen compactification of the moduli space of marked genus zero curves is a complex algebraic manifold.

Section D.1 begins with a preliminary discussion of cross ratios and fractional linear transformations of the Riemann sphere. In Section D.2 we consider stable trees T in which the vertices are labelled by integers between 1 and n. Each integer must appear exactly once and a vertex may carry several labels or no label, but at each vertex the total number of labels and edges must be at least three. Cutting an edge gives rise to a splitting of the index set $\{1, \dots, n\}$ into two subsets, and we show how the (isomorphism class of the) tree can be recovered from the associated set S of splittings of the index set.

Section D.3 consists mainly of definitions. It introduces the moduli space $\overline{\mathcal{M}}_{0,n}$ of stable Riemann surfaces of genus zero with n marked points. Section D.4 shows how to use certain functions $w_{ijk\ell} : \overline{\mathcal{M}}_{0,n} \to S^2$, defined in terms of cross ratios, to construct an embedding of $\overline{\mathcal{M}}_{0,n}$ into a product of 2-spheres. In Section D.4 we also prove that the image $\overline{M}_{0,n}$ of this map is a smooth submanifold of $(S^2)^N$ and that each labelled tree determines a smooth submanifold (Knudsen). Keel proved that these submanifolds generate the cohomology of $\overline{\mathcal{M}}_{0,n}$ and computed the ring structure. Section D.5 examines the topology of $\overline{\mathcal{M}}_{0,n}$, Section D.6 discusses its universal properties, and Section D.7 explains the relation to the Mumford quotient.

Our approach to the Grothendieck–Knudsen compactification via cross ratios is apparently new. The mathematical techniques are elementary but subtle. The embedding of $\overline{\mathcal{M}}_{0,n}$ into a product of 2-spheres is reminiscent of a construction by Fulton and MacPherson for higher dimensional varieties (cf. [**132**]). The extended cross ratios $w_{ijk\ell} : \overline{\mathcal{M}}_{0,n} \to S^2$ also appear in the work of Keel [**208**].

D.1. Möbius transformations and cross ratios

Throughout we identify the unit sphere $S^2 \subset \mathbb{R}^3$ with the Riemann sphere $\mathbb{C} \cup \{\infty\}$ via stereographic projection from the north pole

$$S^2 \to \mathbb{C} \cup \{\infty\} : x \mapsto z = \frac{x_1 + ix_2}{1 - x_3}.$$

This diffeomorphism identifies the two standard complex structures (namely $\xi \mapsto x \times \xi$ on S^2 and i on $\mathbb{C}$) and it identifies one fourth of the standard metric on S^2 with the Fubini-Study metric on $\mathbb{C} \cup \{\infty\}$ (see Exercise 4.2.1). The corresponding volume form is given by

$$\mathrm{dvol} = \frac{dx \wedge dy}{(1 - x^2 + y^2)^2},$$

where $z = x + iy$. With this metric the 2-sphere has area π.

Möbius transformations. A **Möbius transformation** is a holomorphic diffeomorphism of the 2-sphere. Such a diffeomorphism has the form

$$\phi(z) = \frac{az+b}{cz+d}, \qquad ad - bc = 1,$$

and is also called a **fractional linear transformation**. Throughout we shall denote by $\mathrm{G} = \mathrm{PSL}(2, \mathbb{C})$ the group of Möbius transformations. **Isometries** are characterized by the condition $d = \bar{a}$, $c = -\bar{b}$ and they form the subgroup

$$\mathrm{SO}(3) = \mathrm{SU}(2)/\{\pm 1\!\!1\} \subset \mathrm{G}.$$

This is a maximal compact subgroup of G.

Cross ratios. For any three distinct points $z_0, z_1, z_2 \in S^2 = \mathbb{C} \cup \{\infty\}$ there is a unique Möbius transformation $\phi \in \mathrm{G}$ which maps z_0 to 0, z_1 to 1, and z_2 to ∞. It is given by the formula $z \mapsto w(z_0, z_1, z_2, z)$, where

$$w(z_0, z_1, z_2, z_3) := \frac{(z_1 - z_2)(z_3 - z_0)}{(z_0 - z_1)(z_2 - z_3)}.$$

This number is called the **cross ratio** of z_0, z_1, z_2, and z_3. It is well defined for all quadruples of distinct points in S^2 and, on this domain, takes values in $S^2 \setminus \{0, 1, \infty\}$. Note that $w(0, 1, \infty, z) = z$. Moreover, the function w extends continuously to the set of all quadruples among which no more than two points are equal. No triple collisions are allowed. Thus $w : (S^2)^4 \setminus \Delta_3 \to S^2$ where

$$\Delta_3 := \{(z_0, z_1, z_2, z_3) \,|\, \exists\, i < j < k \;\; s.t. \;\; z_i = z_j = z_k\}.$$

This extension has the following properties.

(Fractional linearity) For any three distinct points among z_0, z_1, z_2, z_3, the cross ratio is a fractional linear transformation of the fourth.

(Invariance) For $\phi \in \mathrm{G}$ and $(z_0, z_1, z_2, z_3) \in (S^2)^4 \setminus \Delta_3$,

$$w(\phi(z_0), \phi(z_1), \phi(z_2), \phi(z_3)) = w(z_0, z_1, z_2, z_3).$$

(Symmetry) For any permutation $\{i, j, k, \ell\} = \{0, 1, 2, 3\}$ abbreviate

$$w_{ijk\ell} := w(z_i, z_j, z_k, z_\ell).$$

Then

$$w_{jik\ell} = w_{ij\ell k} = 1 - w_{ijk\ell}, \qquad w_{ikj\ell} = \frac{w_{ijk\ell}}{w_{ijk\ell} - 1}. \tag{D.1.1}$$

(Normalization) For $(z_0, z_1, z_2, z_3) \in (S^2)^4 \setminus \Delta_3$,

$$w_{0123} = \begin{cases} \infty, & \text{if } z_0 = z_1 \text{ or } z_2 = z_3, \\ 1, & \text{if } z_0 = z_2 \text{ or } z_3 = z_1, \\ 0, & \text{if } z_0 = z_3 \text{ or } z_1 = z_2. \end{cases} \tag{D.1.2}$$

(Recursion) If $(z_0, z_1, z_2, z_3), (z_0, z_1, z_2, z_4), (1, \infty, w_{0123}, w_{0124}) \in (S^2)^4 \setminus \Delta_3$ then

$$w_{1234} = \frac{w_{0124} - 1}{w_{0124} - w_{0123}}. \tag{D.1.3}$$

EXERCISE D.1.1. Prove the *(Recursion)* property. *Hint:* If z_0, z_1, z_2 are distinct there is a unique Möbius transformation that takes the points z_0, z_1, z_2, z_3, z_4 to $(0, 1, \infty, u, v)$, where $u := w_{0123}$ and $v := w_{0124}$. If z_1, z_2, z_3 are distinct then so are $1, \infty, u$ and the unique element $\phi \in \mathrm{G}$ taking $(1, \infty, u)$ to $(0, 1, \infty)$ has the form $\phi(z) = (z-1)/(z-u)$. Hence $w_{1234} = w(1, \infty, u, v) = \phi(v) = (v-1)/(v-u)$.

Bubbling. The group G is noncompact. By Exercise 4.2.3, every sequence $\phi^\nu \in \mathrm{G}$ with uniformly bounded derivative has a subsequence which converges uniformly with all derivatives on all of S^2. Observe that the sequence $\phi^\nu(z) := z\nu + \nu^2$ has no such subsequence, although it converges to ∞ for each $z \in S^2$. The convergence is uniform in every bounded set $\{|z| \le c\}$ but not in any neighbourhood of $z = \infty$ since $\phi^\nu(-\nu) = 0$. Next we shall characterize sequences in G that do not have convergent subsequences.

LEMMA D.1.2. *Let $\phi^\nu : S^2 \to S^2$ be a sequence of Möbius transformations which does not have a uniformly convergent subsequence. Then there exist points $x, y \in S^2$ and a subsequence ϕ^{ν_i} which converges to y uniformly in compact subsets of $S^2 \setminus \{x\}$.*

PROOF. Choose a sequence $\rho^\nu \in \mathrm{SO}(3)$ such that $\rho^\nu \circ (\phi^\nu)^{-1}(\infty) = \infty$. Passing to a subsequence, we may assume that ρ^ν converges to ρ, uniformly on all of S^2. The sequence $\psi^\nu := \phi^\nu \circ (\rho^\nu)^{-1}$ satisfies $\psi^\nu(\infty) = \infty$ and hence has the form

$$\psi^\nu(z) = a^\nu z + b^\nu, \qquad a^\nu \neq 0.$$

Replacing ψ^ν by $(\psi^\nu)^{-1}$, if necessary, we may assume that a^ν is bounded, and, passing to a subsequence, that a^ν converges to $a \in \mathbb{C}$. Passing to a further subsequence, we may assume that either b^ν converges to $b \in \mathbb{C}$ or $|b^\nu| \to \infty$. If $|b^\nu| \to \infty$ then ψ^ν converges to ∞, uniformly on compact subsets of $\mathbb{C} = S^2 \setminus \{\infty\}$. If $b^\nu \to b$ then $a^\nu \to 0$ and so ψ^ν converges to b, uniformly on compact subsets of $\mathbb{C}$. In either case, the sequence ψ^ν satisfies the assertion of the lemma, and hence so does $\phi^\nu = \psi^\nu \circ \rho^\nu$. This proves Lemma D.1.2. □

We will say that the sequence ϕ^ν **converges to** y **u.c.s. on** $S^2 \setminus \{x\}$ if it converges to y uniformly on compact subsets of $S^2 \setminus \{x\}$. Note that nothing is asserted about the behaviour of the sequence $\phi^\nu(x)$.

EXERCISE D.1.3. Let $\phi^\nu \in \mathrm{G}$ be a sequence that converges to $y \in S^2$ u.c.s. on $S^2 \setminus \{x\}$. For each $w \in S^2$ give an example where $\phi^\nu(x)$ converges to w. Prove also that in all cases $(\phi^\nu)^{-1}$ converges u.c.s. to x on $S^2 \setminus \{y\}$.

The following technical lemma will be needed in Section D.5 and Chapter 5.

LEMMA D.1.4. *Let $\phi^\nu : S^2 \to S^2$ be a sequence of Möbius transformations. Suppose that $x_0, y_0 \in S^2$ and $x_1^\nu, x_2^\nu, y^\nu \in S^2$ are convergent sequences such that*

$$x_0 \neq \lim_{\nu\to\infty} x_1^\nu \neq \lim_{\nu\to\infty} x_2^\nu \neq x_0, \qquad y_0 \neq \lim_{\nu\to\infty} y^\nu,$$

and

$$\lim_{\nu\to\infty} \phi^\nu(x_1^\nu) = \lim_{\nu\to\infty} \phi^\nu(x_2^\nu) = y_0, \qquad \lim_{\nu\to\infty} (\phi^\nu)^{-1}(y^\nu) = x_0.$$

Then ϕ^ν converges to y_0, uniformly on compact subsets of $S^2 \setminus \{x_0\}$.

PROOF. Denote $y = \lim_{\nu\to\infty} y^\nu$ and $x_i = \lim_{\nu\to\infty} x_i^\nu$ for $i = 1, 2$. Choose sequences of isometries $\sigma^\nu, \rho^\nu \in \mathrm{SO}(3) \subset \mathrm{G}$, converging uniformly to the identity, such that

$$\rho^\nu(y^\nu) = y, \qquad \sigma^\nu \circ (\phi^\nu)^{-1}(y^\nu) = x_0.$$

Denote $\xi_1^\nu = \sigma^\nu(x_1^\nu)$ and $\xi_2^\nu = \sigma^\nu(x_2^\nu)$. Then the sequence

$$\psi^\nu := \rho^\nu \circ \phi^\nu \circ (\sigma^\nu)^{-1}$$

satisfies

$$\psi^\nu(x_0) = y, \qquad \lim_{\nu\to\infty} \psi^\nu(\xi_1^\nu) = \lim_{\nu\to\infty} \psi^\nu(\xi_2^\nu) = y_0.$$

Suppose, without loss of generality, that $y_0 = 0$ and $x_0 = y = \infty$. Then we have $\psi^\nu(\infty) = \infty$, and hence

$$\psi^\nu(z) = a^\nu z + b^\nu$$

for some $a^\nu, b^\nu \in \mathbb{C}$. These sequences satisfy

$$\lim_{\nu\to\infty} (a^\nu \xi_1^\nu + b^\nu) = \lim_{\nu\to\infty} (a^\nu \xi_2^\nu + b^\nu) = 0,$$

where

$$\lim_{\nu\to\infty} \xi_1^\nu = x_1 \neq x_2 = \lim_{\nu\to\infty} \xi_2^\nu.$$

Hence both a^ν and b^ν converge to zero. This means that ψ^ν converges to $y_0 = 0$, uniformly on compact subsets of $\mathbb{C} = S^2 \setminus \{\infty\}$. Since σ^ν and ρ^ν both converge to the identity, uniformly on all of S^2, it follows that ϕ^ν also converges to zero, uniformly on compact subsets of $\mathbb{C}$. This proves Lemma D.1.4. □

D.2. Trees, labels, and splittings

A **tree** is a connected graph without cycles. The vertices form a finite set T and the edges a relation $E \subset T \times T$. (The vertices α and β are connected by an edge whenever $\alpha E \beta$). The *"connected graph without cycles"* condition translates into the following axioms.

(SYMMETRY) If $\alpha E \beta$ then $\beta E \alpha$.

(ANTIREFLEXIVITY) If $\alpha E \beta$ then $\alpha \neq \beta$.

(CONNECTEDNESS) For all $\alpha, \beta \in T$ with $\alpha \neq \beta$ there exist $\gamma_0, \dots, \gamma_m \in T$ with $\gamma_0 = \alpha$ and $\gamma_m = \beta$ such that $\gamma_i E \gamma_{i+1}$ for all i.

(NO CYCLES) If $\gamma_0, \dots, \gamma_m \in T$ with $\gamma_i E \gamma_{i+1}$ and $\gamma_i \neq \gamma_{i+2}$ for all i then $\gamma_0 \neq \gamma_m$.

A map $f : (T, E) \to (\tilde{T}, \tilde{E})$ is called a **tree homomorphism** if $f^{-1}(\tilde{\alpha})$ is a tree for all $\tilde{\alpha} \in \tilde{T}$ and $\alpha E \beta$ implies that either $f(\alpha) = f(\beta)$ or $f(\alpha) \tilde{E} f(\beta)$. It is called a **tree isomorphism** if it is bijective and both f and f^{-1} are tree homomorphisms. Geometrically, one can think of a tree homomorphism as a map that collapses edges.

Let (T, E) be a tree. Then for every pair $\alpha, \beta \in T$ there exists a unique ordered set of vertices $\gamma_0, \dots, \gamma_m \in T$ with $m \geq 0$ such that

$$\gamma_i E \gamma_{i+1}, \qquad \gamma_i \neq \gamma_{i+2}, \qquad \gamma_0 = \alpha, \qquad \gamma_m = \beta.$$

We call this the **chain (of edges) running from α to β** and denote the set of vertices belonging to this chain by

$$[\alpha, \beta] := [\beta, \alpha] := \{\gamma_i \,|\, i = 0, \dots, m\}.$$

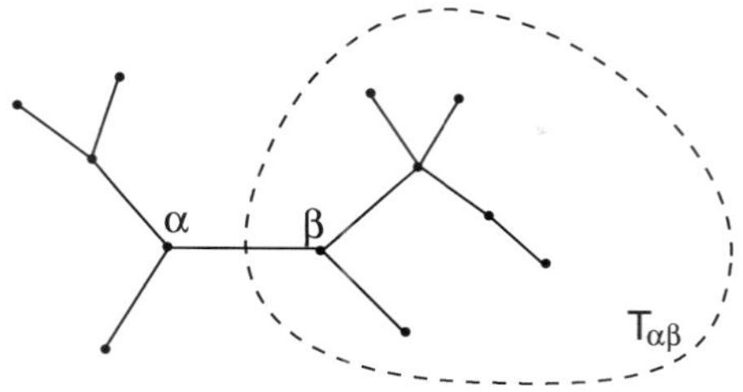

FIGURE 1. Trees

In particular $[\alpha, \alpha] = \{\alpha\}$. Cutting (i.e. removing) any edge $\alpha E \beta$ decomposes the tree T into two components. The component containing β will be denoted by $T_{\alpha\beta}$ (see Figure 1). This is the set of all vertices that can be reached from α by a chain of edges through β. In other words

$$T_{\alpha\beta} := \{\gamma \in T \,|\, \beta \in [\alpha, \gamma]\}.$$

This set is called a **branch** of the tree T. For every edge $\alpha E \beta$ the tree T decomposes as a disjoint union of the branches $T_{\alpha\beta}$ and $T_{\beta\alpha}$.

We emphasize that we think of the edges of a tree T as *unoriented.* Thus we can identify the set of edges with the set of two element subsets $\{\alpha, \beta\} \subset T$ such that $\alpha E \beta$. The number of edges will be denoted by

$$e(T) := \#\text{edges} = \#\text{vertices} - 1.$$

In contrast, an ordered pair $(\alpha, \beta) \in E$ is an oriented edge and the number of oriented edges is $2e(T) = \#E$.

EXERCISE D.2.1. Let (T, E) and $(\tilde{T}, \tilde{E})$ be trees. Prove that $f : T \to \tilde{T}$ is a tree homomorphism if and only if $f([\alpha, \beta]) = [f(\alpha), f(\beta)]$ for all $\alpha, \beta \in T$.

EXERCISE D.2.2. Let $f : T \to \tilde{T}$ be a surjective tree homomorphism. Prove that for every edge $\tilde{\alpha}\tilde{E}\tilde{\beta}$ in $\tilde{T}$ there exists a unique edge $\alpha E \beta$ in T such that $f(\alpha) = \tilde{\alpha}$ and $f(\beta) = \tilde{\beta}$.

Labels. Fix an integer $n \geq 3$. An **n-labelled tree** is a tuple (T, E, Λ), where (T, E) is a tree and $\Lambda = \{\Lambda_\alpha\}_{\alpha \in T}$ is a decomposition of the index set $\{1, \ldots, n\}$ into a disjoint union

$$\{1, \ldots, n\} = \bigcup_{\alpha \in T} \Lambda_\alpha.$$

A labelling is said to be **stable** if

$$n_\alpha = \#\Lambda_\alpha + \#\{\beta \in T \,|\, \alpha E \beta\} \geq 3 \tag{D.2.1}$$

for every $\alpha \in T$. A tree with stable n-labelling is called a **stable tree**. Two n-labelled trees (T, E, Λ) and $(\tilde{T}, \tilde{E}, \tilde{\Lambda})$ are called **isomorphic** if there exists a tree isomorphism $f : T \to \tilde{T}$ such that $\tilde{\Lambda}_{f(\alpha)} = \Lambda_\alpha$ for every $\alpha \in T$. The set of isomorphism classes of stable n-labelled n trees will be denoted by $\mathcal{L}_n$.[1]

[1]Here we abuse notation. Strictly speaking, n-labelled trees form a category and not a set, so that the isomorphism classes of n-labelled trees do not form a set. However, we shall simply ignore this point and pretend that these isomorphism classes form a set. Theorem D.2.6 below shows that, in fact, there are only finitely such isomorphism classes. To make this terminology

LEMMA D.2.3. *Let (T, E, Λ) be an n-labelled tree. Then*

$$\sum_{\alpha \in T} (n_\alpha - 3) + e(T) = n - 3.$$

PROOF. Denote $E_\alpha = \{\beta \in T \,|\, \alpha E \beta\}$. Then $n_\alpha = \#\Lambda_\alpha + \#E_\alpha$ for $\alpha \in T$ and $\sum_{\alpha \in T} \#E_\alpha = \#E = 2e(T)$. Hence

$$\sum_{\alpha \in T} (n_\alpha - 3) = \sum_{\alpha \in T} \#\Lambda_\alpha + \sum_{\alpha \in T} \#E_\alpha - 3(e(T) + 1) = n - 3 - e(T).$$

This proves the lemma. □

An n-label of a tree (T, E) can be expressed as a function

$$\{1, \ldots, n\} \to T : i \mapsto \alpha_i$$

defined by $\alpha_i = \alpha$ iff $i \in \Lambda_\alpha$. Given any three distinct indices $i, j, k \in \{1, \ldots, n\}$ we denote by $\alpha(i, j, k) \in T$ the unique vertex such that

$$\{\alpha(i,j,k)\} = [\alpha_i, \alpha_j] \cap [\alpha_j, \alpha_k] \cap [\alpha_k, \alpha_i]. \tag{D.2.2}$$

(See Exercise D.2.4 and Figure 2.) If α_i, α_j, and α_k are pairwise distinct then $\alpha = \alpha(i, j, k)$ is characterized by the fact that α_i, α_j, and α_k lie in distinct sets of the decomposition

$$T = \{\alpha\} \cup \bigcup_{\beta \in T,\, \alpha E \beta} T_{\alpha\beta}.$$

On the other hand, if two or more of the vertices α_i, α_j, and α_k are equal to β then $\alpha(i, j, k) = \beta$.

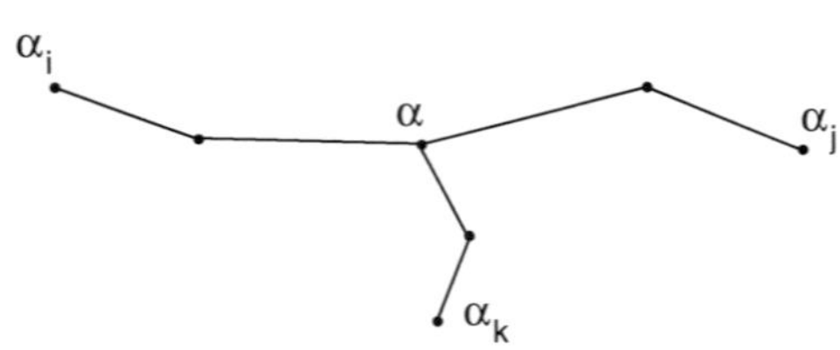

FIGURE 2. A tree triangle

EXERCISE D.2.4. Let (T, E) be a tree. Prove that for every three vertices α, β, γ the intersection $[\alpha, \beta] \cap [\beta, \gamma] \cap [\gamma, \alpha]$ consists of precisely one point.

EXERCISE D.2.5. The stability condition (D.2.1) is equivalent to the assertion that every $\alpha \in T$ has the form $\alpha = \alpha(i, j, k)$ for some distinct indices i, j, k.

Manin introduces the following terminology in [**286**, Ch III §2.1]. He defines a labelled tree as a tuple (F, V, δ, j), where F and V are finite sets, $\delta : F \to V$ is a map, and $j : F \to F$ is an involution. The elements of F are called *flags*, those of V are called vertices and δ is called the boundary map. The fixed points of j correspond to the marked points while the period 2 orbits are the edges. Thus the set of flags includes all ordered pairs (α, β) of vertices such that $\alpha E \beta$ as well as n other flags that indicate the positions of the marked points. Geometrically a flag $f \in F$ can be thought of as a half-open interval that is attached to the vertex $\delta(f)$ at its endpoint.

completely rigorous, one can restrict attention to trees whose vertices belong to some given set, but we shall not do this here.

Splittings. In the following we shall give an alternative description of n-labelled trees in terms of certain partially ordered sets (**posets**), whose elements are splittings of the index set $\{1, \dots, n\}$ into two disjoint subsets which arise from cutting an edge. We shall prove that the isomorphism class of the stable tree can be recovered from the associated system of splittings.

We describe a splitting in terms of the subset I that does not contain the last index n. More precisely, a subset $I \subset \{1, \dots, n\}$ is called an **n-splitting** if

$$2 \le \# I \le n-2, \qquad n \notin I.$$

Note that if I is a splitting both sets I and $\{1, \dots, n\} \setminus I$ contain at least two elements. As will become clear, this condition corresponds to stability. The set of n-splittings will be denoted by $\mathcal{I}_n$. A **network of n-splittings** is a collection $S \subset \mathcal{I}_n$ of n-splittings which satisfy the compatibility condition

$$I, J \in S,\ I \cap J \ne \emptyset \qquad \Longrightarrow \qquad I \subset J \text{ or } J \subset I. \tag{D.2.3}$$

This condition asserts that under the ordering given by inclusion S is a poset in which every element has at most one neighbour above but possibly several neighbours below. The set of all networks of n-splittings will be denoted by $\mathcal{S}_n$. Observe that the empty network $S = \emptyset$ is an element of $\mathcal{S}_n$.

Now let (T, E, Λ) be a stable n-labelled tree. A **(T, E, Λ)-splitting** is an n-splitting $I \in \mathcal{I}_n$ such that

$$i, i' \in I,\ j, j' \notin I \qquad \Longrightarrow \qquad [\alpha_i, \alpha_{i'}] \cap [\alpha_j, \alpha_{j'}] = \emptyset. \tag{D.2.4}$$

The next theorem shows that one can think of such a splitting as a decomposition of T into two connected subtrees, obtained by removing an edge. It also shows that the collection of all such splittings is a network of n-splittings and establishes a one-to-one correspondence between stable n-labelled trees and networks of n-splittings.

THEOREM D.2.6. (i) *For every stable n-labelled tree (T, E, Λ) the set*

$$S(T, E, \Lambda) := \{ I \in \mathcal{I}_n \mid \text{(D.2.4)} \}$$

of (T, E, Λ)-splittings is a network of n-splittings.

(ii) *Let (T, E, Λ) be a stable n-labelled tree and $I \subset \{1, \dots, n\}$. Then $I \in S(T, E, \Lambda)$ if and only if there exists an edge $\alpha E \beta$ such that $\alpha_n \in T_{\alpha\beta}$ and*

$$I = I_{\beta\alpha}(T, E, \Lambda) := \{ i \in \{1, \dots, n\} \mid \alpha_i \in T_{\beta\alpha} \}.$$

(iii) *The map $\mathcal{L}_n \to \mathcal{S}_n : [T, E, \Lambda] \mapsto S(T, E, \Lambda)$ is bijective.*

(iv) *Let (T, E, Λ) and $(\tilde{T}, \tilde{E}, \tilde{\Lambda})$ be stable n-labelled trees. For $i = 1, \dots, n$ define $\alpha_i \in T$ and $\tilde{\alpha}_i \in \tilde{T}$ by $i \in \Lambda_{\alpha_i} \cap \tilde{\Lambda}_{\tilde{\alpha}_i}$. Then*

$$S(\tilde{T}, \tilde{E}, \tilde{\Lambda}) \subset S(T, E, \Lambda)$$

if and only if there is a tree homomorphism $f : T \to \tilde{T}$ such that $f(\alpha_i) = \tilde{\alpha}_i$ for all i. Moreover, if this holds then the tree homomorphism f is necessarily surjective and satisfies $f(\alpha(i, j, k)) = \tilde{\alpha}(i, j, k)$ for all i, j, k.

PROOF. We prove (i). Let (T, E, Λ) be a stable n-labelled tree and suppose, by contradiction, that $S(T, E, \Lambda)$ is not a network of n-splittings. Then there exist $I, J \in S(T, E, \Lambda)$ such that $I \cap J \ne \emptyset$, $I \not\subset J$ and $J \not\subset I$. Choose $i \in I \setminus J$, $j \in J \setminus I$, $\nu \in I \cap J$. Since $\nu, i \in I$ and $j, n \notin I$ we have $[\alpha_\nu, \alpha_i] \cap [\alpha_j, \alpha_n] = \emptyset$ and since $\nu, j \in J$ and $i, n \notin J$, we have $[\alpha_\nu, \alpha_j] \cap [\alpha_i, \alpha_n] = \emptyset$. But these two conditions cannot both hold at the same time. This proves (i).

We prove (ii). Suppose first that $I = I_{\beta\alpha}(T, E, \Lambda)$ for some edge $\alpha E \beta$ such that $\alpha_n \in T_{\alpha\beta}$. Then obviously $n \notin I$. By (D.2.1) each endpoint of the tree carries at least two labels. Since every branch $T_{\alpha\beta}$ contains at least one endpoint it follows that $2 \le \#I \le n-2$. That I satisfies (D.2.4) follows from the fact that $[\alpha_i, \alpha_{i'}] \subset T_{\beta\alpha}$ whenever $i, i' \in I$ and $[\alpha_j, \alpha_{j'}] \subset T_{\alpha\beta}$ whenever $j, j' \notin I$. Hence $I \in S(T, E, \Lambda)$.

Conversely, suppose that $I \in S(T, E, \Lambda)$ and define

$$T_I = \{\alpha(i,j,k) \,|\, \#(\{i,j,k\} \cap I) \ge 2\} \subset \bigcup_{i,i' \in I} [\alpha_i, \alpha_{i'}],$$

$$T_J = \{\alpha(i,j,k) \,|\, \#(\{i,j,k\} \cap I) \le 1\} \subset \bigcup_{j,j' \notin I} [\alpha_j, \alpha_{j'}].$$

Then $T_I \cup T_J = T$ and, by (D.2.4), $T_I \cap T_J = \emptyset$. Hence both inclusions are equalities. With this established one checks easily that T_I and T_J are trees. Namely, if $\alpha, \alpha' \in T$ then there must be indices $i, i' \in I$ such that

$$[\alpha_i, \alpha] \cup [\alpha, \alpha'] \cup [\alpha', \alpha_{i'}] = [\alpha_i, \alpha_{i'}] \subset T_I.$$

Since T_I and T_J are trees, there exists a unique edge $\alpha E \beta$ with $\alpha \in T_I$ and $\beta \in T_J$. Hence $T_I = T_{\beta\alpha}$ and $T_J = T_{\alpha\beta}$. Since $\alpha_i \in T_I$ for $i \in I$ and $\alpha_j \in T_J$ for $j \notin I$, we obtain $I = I_{\beta\alpha}(T, E, \Lambda)$. This proves (ii).

To prove (iii), we construct an inverse map $\mathcal{S}_n \to \mathcal{L}_n$. Given a network of n-splittings $S \in \mathcal{S}_n$, define

$$T := S \cup \{\{1, \dots, n\}\}. \tag{D.2.5}$$

Thus the vertices $\alpha \in T$ are the elements $I \in S$ together with the full index set $\{1, \dots, n\}$. The edge relation $E \subset T \times T$ is given by

$$\alpha E \beta \quad \Longleftrightarrow \quad \begin{cases} \text{either } \alpha \subsetneq \beta & \text{and } \alpha \subsetneq \gamma \subset \beta \Longrightarrow \gamma = \beta, \\ \text{or } \beta \subsetneq \alpha & \text{and } \beta \subsetneq \gamma \subset \alpha \Longrightarrow \gamma = \alpha. \end{cases} \tag{D.2.6}$$

The n-label Λ is given by

$$\Lambda_\alpha := \alpha \setminus \bigcup_{\beta \subsetneq \alpha} \beta. \tag{D.2.7}$$

We prove that (T, E, Λ) is a labelled tree. The relation E obviously satisfies the *(Symmetry)*, *(Antireflexivity)* and *(Connectedness)* axioms. To establish the *(No cycles)* axiom, suppose that $\gamma_0, \dots, \gamma_\ell \in T$ are given with $\gamma_i E \gamma_{i+1}$ and $\gamma_i \ne \gamma_{i+2}$ for all i. Then it follows from the definition of E that there exists a $k \in \{0, \dots, \ell\}$ such that

$$\gamma_0 \subsetneq \cdots \subsetneq \gamma_{k-1} \subsetneq \gamma_k \supsetneq \gamma_{k+1} \supsetneq \cdots \supsetneq \gamma_\ell, \qquad \gamma_{k-1} \cap \gamma_{k+1} = \emptyset.$$

This implies $\gamma_0 \ne \gamma_\ell$. Hence (T, E) is a tree.

We prove that the sets Λ_α are pairwise disjoint. If $\alpha \cap \beta = \emptyset$ then, since $\Lambda_\alpha \subset \alpha$ and $\Lambda_\beta \subset \beta$ we have $\Lambda_\alpha \cap \Lambda_\beta = \emptyset$. If $\alpha \subsetneq \beta$ then

$$\Lambda_\beta = \beta \setminus \bigcup_{\gamma \subsetneq \beta} \gamma \subset \beta \setminus \alpha \subset \beta \setminus \Lambda_\alpha$$

and hence again $\Lambda_\alpha \cap \Lambda_\beta = \emptyset$. But if $\alpha \ne \beta$ then, by (D.2.3), we have either $\alpha \cap \beta = \emptyset$ or $\alpha \subsetneq \beta$ or $\beta \subsetneq \alpha$.

We prove that

$$\bigcup_{\alpha \in T} \Lambda_\alpha = \{1, \ldots, n\}.$$

Let $j \in \{1, \ldots, n\}$ and consider the set

$$A_j := \{\alpha \in T \,|\, j \in \alpha\}.$$

Then $\alpha \cap \beta \neq \emptyset$ for all $\alpha, \beta \in A_j$ and hence it follows from the compatibility condition (D.2.3) that A_j is totally ordered. Hence there is a smallest element $\alpha_j \in A_j$. If $\beta \subsetneq \alpha_j$ then $\beta \notin A_j$ and hence $j \notin \beta$. Therefore

$$j \in \alpha_j \setminus \bigcup_{\beta \subsetneq \alpha_j} \beta = \Lambda_{\alpha_j}$$

as claimed.

We prove that

$$n_\alpha = \#\Lambda_\alpha + \#\{\beta \in T \,|\, \alpha E \beta\} \geq 3$$

for every $\alpha \in T$. If there is no edge, then $\Lambda_\alpha = \{1, \ldots, n\}$ and hence $n_\alpha = n \geq 3$. Next suppose that there is precisely one vertex β with $\alpha E \beta$. If $\alpha \subsetneq \beta$ then $\Lambda_\alpha = \alpha$. If $\beta \subsetneq \alpha$ then $\alpha = \{1, \ldots, n\}$ and $\Lambda_\alpha = \{1, \ldots, n\} \setminus \beta$. In either case we have $\#\Lambda_\alpha \geq 2$ and so $n_\alpha = 1 + \#\Lambda_\alpha \geq 3$. Finally, suppose that there are precisely two distinct vertices β_1 and β_2 with $\alpha E \beta_1$ and $\alpha E \beta_2$. If $\beta_1 \subsetneq \alpha$ and $\beta_2 \subsetneq \alpha$ then $\alpha = \{1, \ldots, n\}$ and so $n \in \Lambda_\alpha$. If $\beta_1 \subsetneq \alpha \subsetneq \beta_2$ then $\Lambda_\alpha = \alpha \setminus \beta_1$ contains at least one element. In either case we have $\#\Lambda_\alpha \geq 1$ and so $n_\alpha = 2 + \#\Lambda_\alpha \geq 3$. This proves that (T, E, Λ) is a stable n-labelled tree.

We prove that the composition $\mathcal{S}_n \to \mathcal{L}_n \to \mathcal{S}_n$ is the identity. Let S be a network of n-splittings and suppose that (T, E, Λ) is given by (D.2.5), (D.2.6) and (D.2.7). Let $\alpha, \beta \in T$ such that $\alpha E \beta$ and $\alpha_n = \{1, \ldots, n\} \in T_{\alpha\beta}$. We must prove that

$$\alpha = I_{\beta\alpha}(T, E, \Lambda) = \{i \in \{1, \ldots, n\} \,|\, \alpha_i \in T_{\beta\alpha}\}.$$

To see this note first that $\alpha \subsetneq \beta$ since otherwise $\alpha_n \notin T_{\alpha\beta}$. Hence, by (D.2.6), we obtain

$$\gamma \in T_{\beta\alpha} \qquad \Longleftrightarrow \qquad \gamma \subset \alpha.$$

This shows immediately that $I_{\beta\alpha}(T, E, \Lambda) \subset \alpha$. Conversely, suppose that $i \in \alpha$ and choose $\gamma \in T$ such that $i \in \Lambda_\gamma$. Then $\alpha \cap \gamma \neq \emptyset$ and so either $\gamma \subset \alpha$ or $\alpha \subsetneq \gamma$. If $\alpha \subsetneq \gamma$ then $\Lambda_\gamma \subset \gamma \setminus \alpha$ and so $i \notin \Lambda_\gamma$, a contradiction. Hence $\gamma \subset \alpha$, hence $\gamma \in T_{\beta\alpha}$, and hence $i \in I_{\beta\alpha}(T, E, \Lambda)$. Thus we have proved that $\alpha = I_{\beta\alpha}(T, E, \Lambda)$ as claimed.

We prove that the composition $\mathcal{L}_n \to \mathcal{S}_n \to \mathcal{L}_n$ is the identity. Let (T, E, Λ) be an n-labelled tree and suppose that $S := S(T, E, \Lambda)$ is the corresponding network of n-splittings. There is a natural bijection $f : T \to S \cup \{\{1, \ldots, n\}\}$ defined by

$$f(\alpha) := \begin{cases} \{1, \ldots, n\}, & \text{if } \alpha = \alpha_n, \\ I_{\beta\alpha}(T, E, \Lambda), & \text{if } \alpha \neq \alpha_n, \ \alpha E \beta, \text{ and } \alpha_n \in T_{\alpha\beta}. \end{cases}.$$

This map obviously identifies the edge relation E with the one given by (D.2.6) and satisfies

$$\Lambda_\alpha = f(\alpha) \setminus \bigcup_{f(\beta) \subsetneq f(\alpha)} f(\beta) = \Lambda_{f(\alpha)}.$$

This proves (iii).

We prove (iv). Abbreviate

$$S := S(T, E, \Lambda), \qquad \tilde{S} := S(\tilde{T}, \tilde{E}, \tilde{\Lambda}).$$

Assume first that $f : T \to \tilde{T}$ is a tree homomorphism which satisfies $f(\alpha_i) = \tilde{\alpha}_i$ for all i. Then, by Exercise D.2.1,

$$f([\alpha_i, \alpha_j]) = [\tilde{\alpha}_i, \tilde{\alpha}_j] \tag{D.2.8}$$

for all i and j. Hence it follows from (D.2.2) that

$$f(\alpha(i,j,k)) = \tilde{\alpha}(i,j,k) \qquad \forall\, i,j,k.$$

In particular, f is surjective. Now let $I \in \tilde{S}$. Then $[\tilde{\alpha}_i, \tilde{\alpha}_{i'}] \cap [\tilde{\alpha}_j, \tilde{\alpha}_{j'}] = \emptyset$ for $i, i' \in I$ and $j, j' \notin I$. Hence, by (D.2.8), we have $[\alpha_i, \alpha_{i'}] \cap [\alpha_j, \alpha_{j'}] = \emptyset$ for $i, i' \in I$ and $j, j' \notin I$. Hence $I \in S$. This shows that $\tilde{S} \subset S$.

To prove the converse assume $\tilde{S} \subset S$. If $i, j, k, \ell \in \{1, \dots, n\}$ are pairwise distinct, then $\alpha(i,j,k) = \alpha(i,j,\ell)$ if and only if an edge which separates i, j cannot separate k, ℓ or, equivalently, $\#(\{i,j\} \cap I) = 1$ implies $\#(\{k,\ell\} \cap I) \neq 1$ for every $I \in S$. Since $\tilde{S} \subset S$,

$$\alpha(i,j,k) = \alpha(i,j,\ell) \qquad \Longrightarrow \qquad \tilde{\alpha}(i,j,k) = \tilde{\alpha}(i,j,\ell). \tag{D.2.9}$$

If $i, j, k \in \{1, \dots, n\}$ are pairwise distinct, then $\alpha_i = \alpha(i,j,k)$ if and only if $\alpha_i \in [\alpha_j, \alpha_k]$, or equivalently $j, k \in I$ implies $i \in I$ for every $I \in S$. Since $\tilde{S} \subset S$,

$$\alpha_i = \alpha(i,j,k) \qquad \Longrightarrow \qquad \tilde{\alpha}_i = \tilde{\alpha}(i,j,k). \tag{D.2.10}$$

By (D.2.9) and (D.2.10), there is a well-defined function $f : T \to \tilde{T}$ which satisfies

$$f(\alpha_i) = \tilde{\alpha}_i, \qquad f(\alpha(i,j,k)) = \tilde{\alpha}(i,j,k).$$

for all i, j, k.

It remains to show that f is a tree homomorphism. To see this, let $\alpha, \beta \in T$ be distinct vertices and denote $\tilde{\alpha} := f(\alpha)$ and $\tilde{\beta} := f(\beta)$. Choose four distinct indices $i, j, k, \ell \in \{1, \dots, n\}$ such that the chains of edges $[\alpha_i, \alpha_j]$ and $[\alpha_k, \alpha_\ell]$ are disjoint, $\alpha \in [\alpha_i, \alpha_j]$, $\beta \in [\alpha_k, \alpha_\ell]$, and the chain of edges $[\alpha, \beta]$ intersects $[\alpha_i, \alpha_j]$ only in α and intersects $[\alpha_k, \alpha_\ell]$ only in β. This can be expressed in the form

$$\alpha = \alpha(i,j,k) = \alpha(i,j,\ell), \qquad \beta = \alpha(i,k,\ell) = \alpha(j,k,\ell).$$

Then $\gamma \in [\alpha, \beta]$ if and only if there exists an $m \in \{1, \dots, n\} \setminus \{i,j,k,\ell\}$ such that

$$\gamma = \alpha(i,m,k) = \alpha(j,m,k) = \alpha(i,m,\ell) = \alpha(j,m,\ell).$$

Similarly, $\tilde{\gamma} \in [\tilde{\alpha}, \tilde{\beta}]$ if and only if there exists an $m \in \{1, \dots, n\} \setminus \{i,j,k,\ell\}$ such that

$$\tilde{\gamma} = \tilde{\alpha}(i,m,k) = \tilde{\alpha}(j,m,k) = \tilde{\alpha}(i,m,\ell) = \tilde{\alpha}(j,m,\ell).$$

This shows that $f([\alpha, \beta]) = [f(\alpha), f(\beta)]$. Hence it follows from Exercise D.2.1 that f is a tree homomorphism. This proves (iv) and Theorem D.2.6. □

D.3. Stable curves

DEFINITION D.3.1 (**Stable Curves**). *Let $T = (T, E, \Lambda)$ be a stable n-labelled tree, $z_{\alpha\beta} \in S^2$ be a finite collection of points, indexed by ordered pairs $(\alpha, \beta) \in T \times T$ with $\alpha E \beta$, and $z_1, \ldots, z_n \in S^2$. The tuple*

$$\mathbf{z} = (\{z_{\alpha\beta}\}_{\alpha E\beta}, \{\alpha_i, z_i\}_{1\le i\le n})$$

is called a **stable curve of genus zero with n marked points** *if for each $\alpha \in T$ the points $z_{\alpha\beta}$ (for $\alpha E\beta$) and z_i (for $\alpha_i = \alpha$) are pairwise distinct. These points form the set*

$$Y_\alpha := Y_\alpha(\mathbf{z}) := \{z_{\alpha\beta} \,|\, \beta \in T,\ \alpha E\beta\} \cup \{z_i \,|\, 1 \le i \le n,\ \alpha_i = \alpha\}$$

of **special points** *on the α-sphere.*

Here the stability condition is hidden in the assumption that the labelled tree is stable. It says that each Y_α contains at least three points. Hence the only automorphism of a stable curve is the identity (see Definition D.3.4 below).

REMARK D.3.2. It is useful to introduce the notation

$$z_{\alpha i} := \begin{cases} z_i, & \text{if } \alpha_i = \alpha, \\ z_{\alpha\beta}, & \text{if } \alpha_i \in T_{\alpha\beta}, \end{cases} \tag{D.3.1}$$

for $\alpha \in T$ and $i \in \{1, \ldots, n\}$. If $\alpha \ne \alpha_i$, then one can think of $z_{\alpha i}$ as the unique nodal point on the α-sphere, through which it is connected to the sphere on which z_i lies, namely, the α_i-sphere (see Figure 4).

REMARK D.3.3. Given distinct indices $i, j, k \in \{1, \ldots, n\}$, the vertex $\alpha(i, j, k)$ defined by (D.2.2) can be characterized as the unique vertex $\alpha \in T$, for which the points $z_{\alpha i}$, $z_{\alpha j}$, and $z_{\alpha k}$ are pairwise distinct. This holds because, when $i \ne j$,

$$\alpha \in [\alpha_i, \alpha_j] \qquad \Longleftrightarrow \qquad z_{\alpha i} \ne z_{\alpha j}.$$

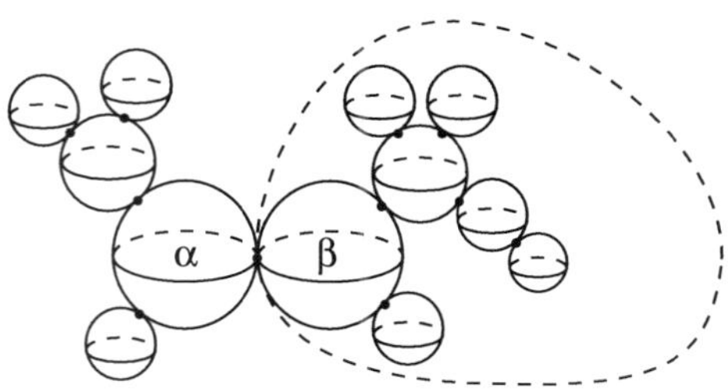

FIGURE 3. Stable curves.

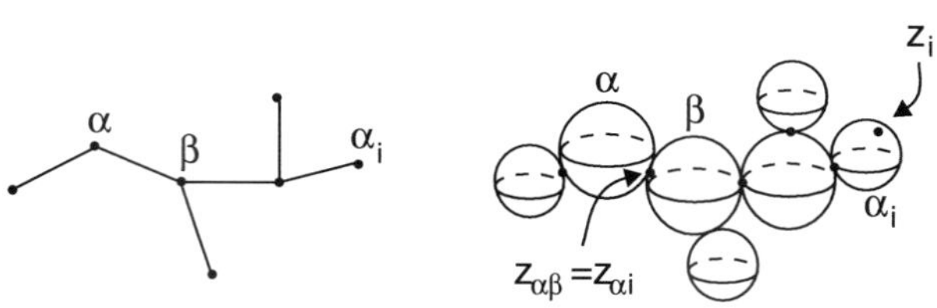

FIGURE 4. Marked points.

DEFINITION D.3.4 (**Equivalence**). *Two stable curves*

$$\mathbf{z} = (\{z_{\alpha\beta}\}_{\alpha E\beta}, \{\alpha_i, z_i\}_{1\le i\le n}), \qquad \tilde{\mathbf{z}} = (\{\tilde z_{\alpha\beta}\}_{\alpha\tilde E\beta}, \{\tilde\alpha_i, \tilde z_i\}_{1\le i\le n})$$

of genus zero with n marked points are called **equivalent** *if there exists a tree isomorphism $f: T\to \tilde T$ and a collection of Möbius transformations $\phi = \{\phi_\alpha\}_{\alpha\in T}$ such that*

$$\tilde z_{f(\alpha)f(\beta)} = \phi_\alpha(z_{\alpha\beta}), \qquad \tilde\alpha_i = f(\alpha_i), \qquad \tilde z_i = \phi_{\alpha_i}(z_i)$$

for $i = 1,\dots,n$ and $\alpha,\beta\in T$ with $\alpha E\beta$.

For every equivalent pair $\mathbf{z}\sim\tilde{\mathbf{z}}$ there exists a unique tree isomorphism $f: T\to \tilde T$ and a unique collection of Möbius transformations $\phi = \{\phi_\alpha\}_{\alpha\in T}$ which satisfy the requirements of Definition D.3.4. The uniqueness of f follows from the fact that $f(\alpha(i,j,k;\mathbf{z})) = \alpha(i,j,k;\tilde{\mathbf{z}})$ for all i,j,k (see Remark D.3.3 and Exercise D.2.5). The uniqueness of ϕ_α follows from the fact that $\phi_\alpha(z_{\alpha i}) = \tilde z_{f(\alpha)i}$, where $\tilde z_{\tilde\alpha i}$ is determined by (D.3.1) with T, α, z_i replaced by $\tilde T$, $\tilde\alpha$, $\tilde z_i$.

The moduli space. Let us denote by $\mathcal{SC}_{0,n}$ the category of all stable curves of genus zero with n marked points. The Grothendieck–Knudsen moduli space $\overline{\mathcal{M}}_{0,n}$ is defined as the quotient

$$\overline{\mathcal{M}}_{0,n} := \mathcal{SC}_{0,n}/\sim$$

under the equivalence relation of Definition D.3.4. The elements of $\overline{\mathcal{M}}_{0,n}$ will be denoted by $\zeta = [\mathbf{z}]$ where $\mathbf{z}$ is as in Definition D.3.1. The space $\overline{\mathcal{M}}_{0,n}$ is sometimes also called the Deligne–Mumford space, however, this name should perhaps more accurately be reserved for the higher genus case.

One can think of a stable curve $\mathbf{z}$ of genus zero as a finite collection of 2-spheres $\{S_\alpha\}_{\alpha\in T}$, glued together at the points $z_{\alpha\beta}\in S_\alpha$ and equipped with the marked points $z_i\in S_{\alpha_i}$. More formally, associated to every $\mathbf{z}$ as in Definition D.3.1, there is a **nodal Riemann surface**

$$\Sigma(\mathbf{z}) := T\times S^2/\sim, \tag{D.3.2}$$

where the equivalence relation is defined by $(\alpha, z)\sim(\beta,w)$ iff either $\alpha=\beta$ and $z = w$, or $\alpha E\beta$, $z = z_{\alpha\beta}$, and $w = z_{\beta\alpha}$. Let us denote by $[\alpha, z]$ the equivalence class of the pair $(\alpha,z)\in T\times S^2$. The points $[\alpha, z_{\alpha\beta}] = [\beta, z_{\beta\alpha}]$ are called the **nodal points** of $\Sigma(\mathbf{z})$. The **marked points** $[\alpha_i, z_i]\in\Sigma(\mathbf{z})$ are required to be distinct and to be different from the nodal points. Moreover, each component S_α contains at least three **special points**, i.e. either nodal or marked points. For every equivalent pair $\mathbf{z},\tilde{\mathbf{z}}\in\mathcal{SC}_{0,n}$, there is a unique tuple $(f,\{\phi_\alpha\}_{\alpha\in T})$, that satisfies the requirements of Definition D.3.4. This tuple determines a holomorphic diffeomorphism $\psi_{\tilde{\mathbf{z}}\mathbf{z}}:\Sigma(\mathbf{z})\to\Sigma(\tilde{\mathbf{z}})$ defined by

$$\psi_{\tilde{\mathbf{z}}\mathbf{z}}([\alpha,z]) := [f(\alpha),\phi_\alpha(z)], \qquad \alpha\in T,\ z\in S^2.$$

Hence, associated to each equivalence class $\zeta\in\overline{\mathcal{M}}_{0,n}$, there is a space

$$\Sigma_\zeta := \{(\mathbf{z},x)\,|\,\mathbf{z}\in\zeta,\ x\in\Sigma(\mathbf{z})\}/\equiv, \tag{D.3.3}$$

where $(\mathbf{z}_0,x_0)\equiv(\mathbf{z}_1,x_1)$ iff $x_1 = \psi_{\mathbf{z}_1\mathbf{z}_0}(x_0)$. Thus the moduli space $\overline{\mathcal{M}}_{0,n}$ can be understood as a collection of nodal Riemann surfaces Σ_ζ, each equipped with n marked points. In Section D.7 we give explicit descriptions of $\overline{\mathcal{M}}_{0,n}$ for $n = 3,4,5$.

D.4. The Grothendieck–Knudsen manifold

Construction of an embedding. This section constructs a natural injection $\overline{\mathcal{M}}_{0,n} \to (S^2)^N$ with image denoted $\overline{M}_{0,n}$. The main tool is the cross ratio

$$w(z_0, z_1, z_2, z_3) = \frac{(z_1 - z_2)(z_3 - z_0)}{(z_0 - z_1)(z_2 - z_3)}$$

of four points in the Riemann sphere $S^2 = \mathbb{C} \cup \{\infty\}$. It is defined in the complement of the set $\Delta_3 \subset (S^2)^4$ where at least three of the four points are equal. The relevant properties of the cross ratio are listed in Section D.1. This definition extends to any set $\{i, j, k, \ell\}$ of four distinct marked points on a stable curve: simply compose the map $\overline{\mathcal{M}}_{0,n} \to \overline{\mathcal{M}}_{0,4}$ that forgets all points except z_i, z_j, z_k, z_ℓ with the cross ratio $\overline{\mathcal{M}}_{0,4} \to S^2$. The next lemma will allow us to give an explicit formula for this cross ratio.

LEMMA D.4.1. *Let* $\mathbf{z} = (\{z_{\alpha\beta}\}_{\alpha E \beta}, \{\alpha_i, z_i\}_{1 \le i \le n}) \in \mathcal{SC}_{0,n}$ *and suppose that the integers* $i, j, k, \ell \in \{1, \dots, n\}$ *are pairwise distinct. Then the following holds.*

(i) *There exists an* $\alpha \in T$ *such that* $(z_{\alpha i}, z_{\alpha j}, z_{\alpha k}, z_{\alpha\ell}) \notin \Delta_3$.

(ii) *If* $\alpha, \beta \in T$ *such that* $(z_{\alpha i}, z_{\alpha j}, z_{\alpha k}, z_{\alpha\ell}), (z_{\beta i}, z_{\beta j}, z_{\beta k}, z_{\beta\ell}) \notin \Delta_3$, *then*

$$w(z_{\alpha i}, z_{\alpha j}, z_{\alpha k}, z_{\alpha\ell}) = w(z_{\beta i}, z_{\beta j}, z_{\beta k}, z_{\beta\ell}).$$

PROOF. Assertion (i) follows from Remark D.3.3: simply choose $\alpha = \alpha(i, j, k)$ such that (D.2.2) holds and note that $z_{\alpha i}$, $z_{\alpha j}$, $z_{\alpha k}$ are pairwise distinct. To prove (ii), suppose first that the points $z_{\alpha i}$, $z_{\alpha j}$, $z_{\alpha k}$, $z_{\alpha\ell}$ are pairwise distinct. We then claim that α is the only vertex that satisfies (i). To see this let $\beta \ne \alpha$. Then there exists a vertex $\gamma \in T$ such that $\alpha E \gamma$ and $\beta \in T_{\alpha\gamma}$. Now at least three of the vertices α_i, α_j, α_k, α_ℓ lie in $T_{\gamma\alpha}$ and hence are reached from β through the same edge. Hence the corresponding three points among $z_{\beta i}$, $z_{\beta j}$, $z_{\beta k}$, and $z_{\beta\ell}$ coincide. Thus we have proved that $(z_{\beta i}, z_{\beta j}, z_{\beta k}, z_{\beta\ell}) \in \Delta_3$ whenever $\beta \ne \alpha$. This shows that (ii) holds whenever the four points $z_{\alpha i}$, $z_{\alpha j}$, $z_{\alpha k}$, and $z_{\alpha\ell}$ are pairwise distinct.

If the four points $z_{\alpha i}$, $z_{\alpha j}$, $z_{\alpha k}$, $z_{\alpha\ell}$ are not pairwise distinct, then by assumption at most two coincide, say $z_{\alpha\ell}$ and $z_{\alpha k}$. It follows that

$$[\alpha_i, \alpha_j] \cap [\alpha_k, \alpha_\ell] = \emptyset, \qquad w(z_{\alpha i}, z_{\alpha j}, z_{\alpha k}, z_{\alpha\ell}) = \infty.$$

Suppose that $\beta \in T$ is any other vertex with $(z_{\beta i}, z_{\beta j}, z_{\beta k}, z_{\beta\ell}) \notin \Delta_3$. If $\beta \notin [\alpha_i, \alpha_j]$ then $z_{\beta i} = z_{\beta j}$. If $\beta \notin [\alpha_k, \alpha_\ell]$ then $z_{\beta k} = z_{\beta\ell}$. One of these two conditions is satisfied, and hence $w(z_{\beta i}, z_{\beta j}, z_{\beta k}, z_{\beta\ell}) = \infty$. □

The previous lemma gives rise to a collection of maps $w_{ijk\ell} : \overline{\mathcal{M}}_{0,n} \to S^2$, one for any four pairwise distinct integers $i, j, k, \ell \in \{1, \dots, n\}$, defined as follows. Given $\mathbf{z} \in \mathcal{C}_{0,n}$ and four distinct integers $i, j, k, \ell \in \{1, \dots, n\}$ choose a vertex $\alpha \in T$ such that $(z_{\alpha i}, z_{\alpha j}, z_{\alpha k}, z_{\alpha\ell}) \notin \Delta_3$ and define

$$w_{ijk\ell}(\mathbf{z}) := w(z_{\alpha i}, z_{\alpha j}, z_{\alpha k}, z_{\alpha\ell}) = \frac{(z_{\alpha j} - z_{\alpha k})(z_{\alpha\ell} - z_{\alpha i})}{(z_{\alpha i} - z_{\alpha j})(z_{\alpha k} - z_{\alpha\ell})}. \tag{D.4.1}$$

Lemma D.4.1 asserts that this map is well defined. Moreover, the *(Invariance)* property of the cross ratio shows that the right hand side in (D.4.1) is invariant under Möbius transformations. Hence, by Definition D.3.4, $w_{ijk\ell}(\mathbf{z}) = w_{ijk\ell}(\tilde{\mathbf{z}})$ whenever $\mathbf{z}$ is equivalent to $\tilde{\mathbf{z}}$.

Note that the effect of permuting the indices i,j,k,ℓ is governed by the *(Symmetry)* property of the cross ratio:

$$w_{jik\ell} = w_{ij\ell k} = 1 - w_{ijk\ell}, \qquad w_{ikj\ell} = \frac{w_{ijk\ell}}{w_{ijk\ell} - 1} \tag{D.4.2}$$

(see Section D.1). The *(Recursion)* property (D.1.3) of the cross ratio gives rise to the condition

$$(1, \infty, w_{ijk\ell}, w_{ijkm}) \notin \Delta_3 \quad \Longrightarrow \quad w_{jk\ell m} = \frac{w_{ijkm} - 1}{w_{ijkm} - w_{ijk\ell}} \tag{D.4.3}$$

for any five pairwise distinct integers $i,j,k,\ell,m \in \{1,\dots,n\}$. This gives rise to the following definition. Let $N := n(n-1)(n-2)(n-3)$ and write an element of $(S^2)^N$ as a tuple $w = \{w_{ijk\ell}\}_{1\le i,j,k,\ell\le n}$, indexed by quadruples (i,j,k,ℓ) of pairwise distinct integers between 1 and n. Denote the set of all tuples $w \in (S^2)^N$ that satisfy the symmetry condition (D.4.2) and the recursive condition (D.4.3) by

$$\overline{M}_{0,n} := \left\{ w = \{w_{ijk\ell}\}_{1\le i,j,k,\ell\le n} \in (S^2)^N \,\middle|\, \begin{array}{l} w \text{ satisfies} \\ \text{(D.4.2) and (D.4.3)} \end{array} \right\}. \tag{D.4.4}$$

The next theorem is the main result of this appendix.

THEOREM D.4.2. (i) *If $\mathbf{z} \in \mathcal{SC}_{0,n}$ then the tuple $w_{ijk\ell} = w_{ijk\ell}(\mathbf{z})$ defined by* (D.4.1) *satisfies* (D.4.2) *and* (D.4.3) *and hence is an element of $\overline{M}_{0,n}$.*

(ii) *The map $\overline{\mathcal{M}}_{0,n} \to \overline{M}_{0,n}$ defined by* (D.4.1) *is bijective.*

(iii) *$\overline{M}_{0,n}$ is a complex submanifold of $(S^2)^{n(n-1)(n-2)(n-3)}$.*

PROOF OF THEOREM D.4.2 (I). Let $\alpha = \alpha(i,j,k) \in T$ be the unique vertex for which the points $z_{\alpha i}$, $z_{\alpha j}$, and z_{α_k} are pairwise distinct (see Remark D.3.3). We may assume without loss of generality that $z_{\alpha i} = 0$, $z_{\alpha j} = 1$, $z_{\alpha k} = \infty$, $z_{\alpha \ell} = w_{ijk\ell}$, and $z_{\alpha m} = w_{ijkm}$. Then, by assumption, $(z_{\alpha j}, z_{\alpha k}, z_{\alpha \ell}, z_{\alpha m}) \notin \Delta_3$ and hence

$$w_{jk\ell m} = \frac{(z_{\alpha k} - z_{\alpha \ell})(z_{\alpha m} - z_{\alpha j})}{(z_{\alpha j} - z_{\alpha k})(z_{\alpha \ell} - z_{\alpha m})} = \frac{w_{ijkm} - 1}{w_{ijkm} - w_{ijk\ell}},$$

as claimed. This proves Theorem D.4.2 (i). □

Construction of the inverse. The main tool for the proof of assertion (ii) in Theorem D.4.2 is the notion of a w-splitting. This allows us to recover the tree structure corresponding to a point $w \in \overline{M}_{0,n}$ in the form of a network of splittings of the index set $\{1,\dots,n\}$. Fix a point $w \in \overline{M}_{0,n}$. An n-splitting $I \in \mathcal{I}_n$ is called a **w-splitting** (or a **(w,n)-splitting**) if $i,i' \in I,\ j,j' \notin I \implies w_{ii'jj'} = \infty$. The set of w-splittings will be denoted by $S(w)$.

LEMMA D.4.3. *If $w \in \overline{M}_{0,n}$ then $S(w)$ is a network of n-splittings. Moroever, there is a commuting diagram*

$$\begin{array}{ccc} \overline{\mathcal{M}}_{0,n} & \longrightarrow & \mathcal{L}_n \\ \downarrow & & \downarrow \\ \overline{M}_{0,n} & \longrightarrow & \mathcal{S}_n \end{array} \tag{D.4.5}$$

where the map $\overline{\mathcal{M}}_{0,n} \to \overline{M}_{0,n}$ is defined by (D.4.1), the bijection $\mathcal{L}_n \to \mathcal{S}_n$ is given by Theorem D.2.6, the map $\overline{\mathcal{M}}_{0,n} \to \mathcal{L}_n$ assigns to every stable curve $(T,E,\mathbf{z})$ the labelled tree $(T,E,\Lambda(\mathbf{z}))$ with $\Lambda_\alpha(\mathbf{z}) = \{i \,|\, \alpha_i = \alpha\}$, and the map $\overline{M}_{0,n} \to \mathcal{S}_n$ is given by $w \mapsto S(w)$.

PROOF. Let $w \in \overline{M}_{0,n}$ and suppose, by contradiction, that $S(w)$ is not a network of n-splittings. Then there exist $I, J \in S(w)$ such that $I \cap J \neq \emptyset$, $I \not\subset J$ and $J \not\subset I$. Choose $i \in I \setminus J$, $j \in J \setminus I$, $\nu \in I \cap J$. Then $w_{\nu ijn} = w_{\nu jin} = \infty$, in contradiction to (D.4.2). Thus there is a well defined map $M_{0,n} \to \mathcal{S}_n$. (Recall that it is permissible for $S(w)$ to be empty.)

Let $\mathbf{z}$ be a stable curve of genus zero with corresponding n-labelled tree (T, E, Λ) and denote $w = w(\mathbf{z}) \in \overline{M}_{0,n}$. We must prove that $S(w) = S(T, E, \Lambda)$. Let $I \in \mathcal{I}_n$ be an n-splitting and suppose that $i, i' \in I$ and $j, j' \notin I$ such that $i \neq i'$ and $j \neq j'$. Then

$$w_{ii'jj'} = \infty \qquad \iff \qquad [\alpha_i, \alpha_{i'}] \cap [\alpha_j, \alpha_{j'}] = \emptyset.$$

Hence $I \in S(w)$ if and only if $I \in S(T, E, \Lambda)$. This proves Lemma D.4.3. □

The next lemma describes some important consequences of condition (D.4.3).

LEMMA D.4.4. *Let $w \in \overline{M}_{0,n}$.*

(i) *Let $I \in S(w)$ be a w-splitting and let $i, i', i'' \in I$ be pairwise distinct. Then $w_{ii'i''j} = w_{ii'i''j'}$ for all $j, j' \notin I$.*

(ii) *If $w_{ijk\ell} = \infty$ then there exists a w-splitting $I \in S(w)$ such that either $i, j \in I$ and $k, \ell \notin I$ or $i, j \notin I$ and $k, \ell \in I$.*

PROOF. We prove (i). Suppose, by contradiction, that there exist $j, j' \notin I$ such that $w_{ii'i''j} \neq w_{ii'i''j'}$. Then $(1, \infty, w_{ii'i''j}, w_{ii'i''j'}) \notin \Delta_3$ and it follows from (D.4.3) that

$$w_{i'i''jj'} = \frac{w_{ii'i''j'} - 1}{w_{ii'i''j'} - w_{ii'i''j}} \neq \infty.$$

This contradicts the assumption that I be a w-splitting.

To prove (ii), observe first that, by the (*Symmetry*) property of the cross ratio, we have

$$w_{ijk\ell} = \infty \qquad \Longrightarrow \qquad w_{ij\ell k} = w_{k\ell ij} = w_{k\ell ji} = \infty.$$

Hence we can reorder the indices so that $\ell > i, j, k$. With that ordering, $I := \{i, j\}$ is a splitting of the index set $\{i, j, k, \ell\}$ whenever $w_{ijk\ell} = \infty$. Now we can use the following claim inductively to construct a (w, n)-splitting which separates the pair $\{i, j\}$ from the pair $\{k, \ell\}$.

CLAIM. *If $I \in \mathcal{I}_{n-1}$ is a $(w, n-1)$-splitting then either I or $\{1, \ldots, n-1\} \setminus I$ is a (w, n)-splitting.*

To prove this, let $I \in \mathcal{I}_{n-1}$ be a $(w, n-1)$-splitting. We shall prove that either

$$i, i' \in I, \; j \notin I \qquad \Longrightarrow \qquad w_{ii'jn} = \infty,$$

in which case I is a (w, n)-splitting, or

$$i \in I, \; j, j' \notin I \qquad \Longrightarrow \qquad w_{injj'} = \infty,$$

in which case $\{1, \ldots, n-1\} \setminus I$ is a (w, n)-splitting.

Let us assume that

$$w_{i_0 i_1 j_0 n} \neq \infty$$

for some $i_0, i_1 \in I$ and some $j_0 \notin I$. Since $w_{i_0 i_1 j_0 j} = \infty$ for all $j \notin I$ we have $(1, \infty, w_{i_0 i_1 j_0 j}, w_{i_0 i_1 j_0 n}) \notin \Delta_3$ and hence, by (D.4.3),

$$w_{i_1 j_0 j n} = \frac{w_{i_0 i_1 j_0 n} - 1}{w_{i_0 i_1 j_0 n} - w_{i_0 i_1 j_0 j}} = 0.$$

This shows that $(1,\infty,w_{i_1j_0ji},w_{i_1j_0jn})\notin\Delta_3$ for $i\in I$ and $j\notin I$. Hence, by (D.4.3),

$$w_{j_0jin}=\frac{w_{i_1j_0jn}-1}{w_{i_1j_0jn}-w_{i_1j_0ji}}=\infty.$$

Here we have used the fact that $w_{ii_1j_0j}=\infty$ and hence $w_{i_1j_0ji}=0$. Since $w_{j_0jin}=\infty$ for $j\notin I$ we have $w_{j_0inj}=w_{j_0inj'}=0$ for $j,j'\notin I$. Hence $(1,\infty,w_{j_0inj},w_{j_0inj'})\notin\Delta_3$ and, by (D.4.3),

$$w_{injj'}=\frac{w_{j_0inj'}-1}{w_{j_0inj'}-w_{j_0inj}}=\infty$$

for $i\in I$ and $j,j'\notin I$. This proves (ii) and Lemma D.4.4. □

PROOF OF THEOREM D.4.2 (II). We prove that the map $\overline{\mathcal{M}}_{0,n}\to\overline{M}_{0,n}$ is injective. Let $\mathbf{z},\tilde{\mathbf{z}}\in\mathcal{SC}_{0,n}$ and suppose that $w_{ijk\ell}=w_{ijk\ell}(\mathbf{z})=w_{ijk\ell}(\tilde{\mathbf{z}})$ for all quadruples of pairwise distinct integers $i,j,k,\ell\in\{1,\dots,n\}$. By Theorem D.2.6 and Lemma D.4.3, $\mathbf{z}$ and $\tilde{\mathbf{z}}$ are modelled over isomorphic n-labelled trees. Hence we may assume, without loss of generality, that

$$T=\tilde{T},\qquad E=\tilde{E},\qquad \alpha_i=\tilde{\alpha}_i$$

for all $i\in\{1,\dots,n\}$.

Let $\alpha\in T$ be any vertex. For each edge $\alpha E\beta$ choose an index $i_{\alpha\beta}\in\{1,\dots,n\}$ with $\alpha_{i_{\alpha\beta}}\in T_{\alpha\beta}$. Denote

$$I_\alpha:=\{i\,|\,\alpha_i=\alpha\}\cup\{i_{\alpha\beta}\,|\,\alpha E\beta\}.$$

Then $\#I_\alpha\ge 3$ and

$$w(z_{\alpha i},z_{\alpha j},z_{\alpha k},z_{\alpha\ell})=w(\tilde{z}_{\alpha i},\tilde{z}_{\alpha j},\tilde{z}_{\alpha k},\tilde{z}_{\alpha\ell})\notin\{0,1,\infty\}$$

for any four pairwise distinct indices $i,j,k,\ell\in I_\alpha$. Hence there is a unique Möbius transformation ϕ_α such that $\phi_\alpha(z_{\alpha i})=\tilde{z}_{\alpha i}$ for all $i\in I_\alpha$. This implies $\phi_\alpha(z_i)=\tilde{z}_i$ whenever $\alpha_i=\alpha$, and $\phi_\alpha(z_{\alpha\beta})=\tilde{z}_{\alpha\beta}$ whenever $\alpha E\beta$. Hence $\mathbf{z}$ is equivalent to $\tilde{\mathbf{z}}$. This proves that the map $\overline{\mathcal{M}}_{0,n}\to\overline{M}_{0,n}$ is injective.

We prove that the map $\overline{\mathcal{M}}_{0,n}\to\overline{M}_{0,n}$ is surjective. Let $w\in\overline{M}_{0,n}$ be given. By Theorem D.2.6 and Lemma D.4.3, there exists an n-labelled tree (T,E,Λ) such that $S(T,E,\Lambda)=S(w)$. We prove the existence of $\mathbf{z}$ with $w(\mathbf{z})=w$ by induction over the number of edges.

Suppose first that $T=\{\mathrm{pt}\}$ so that there is no w-splitting. Then it follows from Lemma D.4.4 (ii) that $w_{ijk\ell}\notin\{0,1,\infty\}$ for all i,j,k,ℓ. Under this assumption we claim that the assertion holds with $\mathbf{z}=(z_1,\dots,z_n)\in(S^2)^n$ given by

$$z_1=0,\qquad z_2=1,\qquad z_3=\infty,\qquad z_\ell=w_{123\ell},$$

for $\ell=4,\dots,n$. We prove that these points are pairwise distinct and satisfy

$$w_{ijk\ell}=w(z_i,z_j,z_k,z_\ell)$$

for all i,j,k,ℓ. The latter holds by assumption whenever $\{i,j,k\}=\{1,2,3\}$. Hence

$$w_{12k\ell}=\frac{w_{312\ell}-1}{w_{312\ell}-w_{312k}}=w(z_1,z_2,z_k,z_\ell).$$

for all $k,\ell\in\{4,\dots,n\}$. Here the first equation follows from (D.4.3) and the second from (D.1.3). Since $w_{12k\ell}\neq\infty$, this shows in particular that $z_k\neq z_\ell$ for $k\neq\ell$. A similar argument can be used to replace 2 by j and 1 by i, so that $w_{ijk\ell}=w(z_i,z_j,z_k,z_\ell)$ for all i,j,k,ℓ. This proves the assertion in the case $T=\{\mathrm{pt}\}$.

Now consider the general case and let $\alpha \in T$ be any vertex. As in the proof of injectivity, denote

$$I_\alpha := \{i \,|\, \alpha_i = \alpha\} \cup \{i_{\alpha\beta} \,|\, \alpha E \beta\},$$

where $i_{\alpha\beta} \in \{1, \ldots, n\}$ has been chosen such that $\alpha_{i_{\alpha\beta}} \in T_{\alpha\beta}$. Then it follows from Lemma D.4.4 (ii) that $w_{ijk\ell} \notin \{0, 1, \infty\}$ for $i, j, k, \ell \in I_\alpha$, and from Lemma D.4.4 (i) that these numbers are independent of the choices of the indices $i_{\alpha\beta}$. Now the first step of the proof shows that there exists a set of pairwise distinct points ζ_i for $i \in I_\alpha$, such that $w_{ijk\ell} = w(\zeta_i, \zeta_j, \zeta_k, \zeta_\ell)$ for all $i, j, k, \ell \in I_\alpha$. Define $z_i := \zeta_i$ whenever $\alpha_i = \alpha$, and $z_{\alpha\beta} := \zeta_{i_{\alpha\beta}}$ whenever $\alpha E \beta$, to obtain the set

$$Y_\alpha = \{z_i \,|\, \alpha_i = \alpha\} \cup \{z_{\alpha\beta} \,|\, \alpha E \beta\}.$$

This defines a stable curve $\mathbf{z} \in \mathcal{SC}_{0,n}$ and, since the numbers $w_{ijk\ell}$ for $i, j, k, \ell \in I_\alpha$ are independent of the choices of the $i_{\alpha\beta}$, it follows that $w(\mathbf{z}) = w$. This proves Theorem D.4.2 (ii). □

Construction of coordinate charts. The proof of assertion (iii) in Theorem D.4.2 is based on an explicit construction of coordinate charts. Given a point $w^0 = w(\mathbf{z}^0) \in \overline{M}_{0,n}$ with corresponding labelled tree (T, E, Λ) we choose $n-3$ cross ratios as coordinates from which the others can be reconstructed as smooth functions in a neighbourhood of w^0. The choice of the $n-3$ coordinates is geometric. Firstly, for each vertex α we choose $n_\alpha - 3$ cross ratios which, up to a complex automorphism, determine the position of the n_α special points. Note that these cross ratios take values in $S^2 \setminus \{0, 1, \infty\}$ since special points are distinct. Secondly, for each edge $\alpha E \beta$ we choose a *"normal"* coordinate $w_{ijk\ell}$ that equals ∞ at w_0 and is such that $w_{ijk\ell}(w) = \infty$ if and only if w is modelled on a tree with an edge corresponding to $\alpha E \beta$. That this gives $n-3$ coordinates follows from the formula

$$\sum_{\alpha \in T} (n_\alpha - 3) + e(T) = n - 3$$

of Lemma D.2.3.

Proof of Theorem D.4.2 (iii). Fix a point $w^0 = w(\mathbf{z}^0) \in \overline{M}_{0,n}$ with corresponding labelled tree (T, E, Λ) and write

$$I_{\alpha\beta} := \{i \in \{1, \ldots, n\} \,|\, \alpha_i \in T_{\alpha\beta}\}.$$

As in the proof of Theorem D.4.2 (ii), let $E \to \{1, \ldots, n\} : (\alpha, \beta) \mapsto i_{\alpha\beta}$ be a function which satisfies $i_{\alpha\beta} \in I_{\alpha\beta}$. We prove that this function can be chosen such that

$$\text{(D.4.6)} \qquad i_{\alpha\beta} \in I_\beta, \qquad I_\beta := \Lambda_\beta \cup \{i_{\beta\gamma} \,|\, \gamma \in T,\, \beta E \gamma\}$$

for $(\alpha, \beta) \in E$. The function will be constructed inductively on an increasing sequence of subsets of E. The first step is to define

$$E^0 := \{(\alpha, \beta) \in E \,|\, \Lambda_\beta \neq \emptyset\}.$$

Choose any function $\lambda^0 : E^0 \to \{1, \ldots, n\}$ which satisfies $\lambda^0(\alpha, \beta) \in \Lambda_\beta$. We shall then define, recursively, a sequence of subsets

$$\Lambda_\beta =: \Lambda^0(\alpha, \beta) \subset \Lambda^1(\alpha, \beta) \subset \cdots \subset \Lambda^k(\alpha, \beta)$$

for $\alpha E \beta$ and denote

$$E^k := \big\{(\alpha, \beta) \in E \,|\, \Lambda^k(\alpha, \beta) \neq \emptyset\big\}.$$

Let $k \geq 0$ and suppose, by induction, that a function $\lambda^k : E^k \to \{1, \ldots, n\}$ has been constructed such that $\lambda^k(\alpha, \beta) \in \Lambda^k(\alpha, \beta)$. Define

$$\Lambda^{k+1}(\alpha, \beta) := \Lambda_\beta \cup \{\lambda^k(\beta, \gamma) \,|\, (\beta, \gamma) \in E^k, \gamma \neq \alpha\} \tag{D.4.7}$$

and choose $\lambda^{k+1} : E^{k+1} \to \{1, \ldots, n\}$ to be equal to λ^k on E^k and arbitrarily with $\lambda^{k+1}(\alpha, \beta) \in \Lambda^{k+1}(\alpha, \beta)$ elsewhere. Continue by induction. A moment's thought shows that E^k is the set of all edges $(\alpha, \beta) \in E$ which form the beginning of a chain of at most $k+1$ edges ending at a vertex γ with $\Lambda_\gamma \neq \emptyset$. Hence the process terminates after finitely many steps when $\Lambda^k(\alpha, \beta) \neq \emptyset$ for all edges $\alpha E \beta$ and hence $E^k = E$. It follows from (D.4.7) by induction that $\Lambda^k(\alpha, \beta) \subset I_{\alpha\beta}$ for all k and all edges $\alpha E \beta$. When $E^k = E$ we define $i_{\alpha\beta} := \lambda^k(\alpha, \beta)$ and note that, for every edge $\alpha E \beta$, we have $\Lambda^{k+1}(\alpha, \beta) = I_\beta \setminus \{i_{\beta\alpha}\}$.

Now we choose $n-3$ coordinates near a point $w^0 \in \overline{M}_{0,n}$ with tree structure (T, E, Λ). Suppose $\alpha \in T$ is a vertex with $n_\alpha \geq 4$ special points and write the set I_α in the form

$$I_\alpha =: \{i_1(\alpha), \ldots, i_{n_\alpha}(\alpha)\}.$$

Then the points $z^0_{\alpha i_\nu(\alpha)}$ for $\nu = 1, \ldots, n_\alpha$ are pairwise distinct and we select the components

$$x_{\alpha\nu}(w) := w_{i_1(\alpha) i_2(\alpha) i_3(\alpha) i_\nu(\alpha)} \tag{D.4.8}$$

for $\nu = 4, \ldots, n_\alpha$. If $\alpha E \beta$ is an edge which satisfies the normalization condition $\alpha_n \in T_{\beta\alpha}$, we choose four distinct indices

$$i, j \in I_\alpha \setminus \{i_{\alpha\beta}\}, \qquad k, \ell \in I_\beta \setminus \{i_{\beta\alpha}\}, \qquad i = i_{\beta\alpha}, \qquad k = i_{\alpha\beta},$$

(see Figure 5) and select the component

$$y_{\alpha\beta}(w) := w_{ijk\ell}. \tag{D.4.9}$$

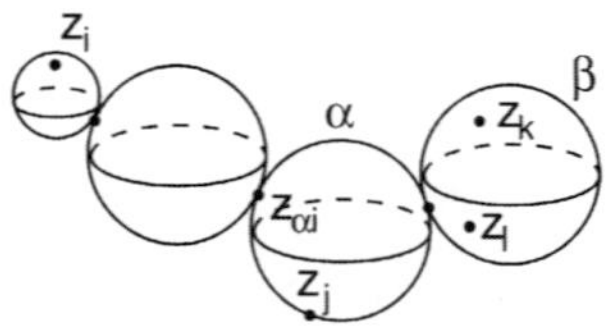

FIGURE 5. The coordinate $w_{ijk\ell}$ for an edge $\alpha E \beta$

We must prove that, in a neighbourhood of the given point $w^0 \in \overline{M}_{0,n}$, all the components of w can be obtained as smooth functions of the $n-3$ coordinates given by (D.4.8) and (D.4.9). We prove this by induction over the number of edges. Assume first that $T = \{\text{pt}\}$. Then the $n-3$ chosen coordinates are

$$w_{1234}, w_{1235}, \ldots, w_{123n}.$$

It follows from the recursive formula (D.1.3) in Section D.1 that each coordinate $w_{ijk\ell}$ can be expressed as a smooth function of the w_{123m}, $m = 4, \ldots, n$, in the open set

$$U := \{w \in \overline{M}_{0,n} \,|\, w_{ijk\ell}(\mathbf{z}) \notin \{0, 1, \infty\} \; \forall \, i, j, k, \ell\}.$$

For $i, j, k, \ell \in \{4, \ldots, n\}$ an explicit formula is given by

$$w_{ijk\ell} = \frac{(w_{123j} - w_{123k})(w_{123\ell} - w_{123i})}{(w_{123i} - w_{123j})(w_{123k} - w_{123\ell})}.$$

This formula extends to all i, j, k, ℓ if we use the rule (as in (D.1.2) in Section D.1)

$$w_{ijk\ell} = \begin{cases} \infty, & \text{if } i = j \text{ or } k = \ell, \\ 1, & \text{if } i = k \text{ or } j = \ell, \\ 0, & \text{if } i = \ell \text{ or } j = k. \end{cases}$$

Now suppose that $e(T) = \#\text{edges} \geq 1$ and that the result has been proved for all trees with strictly fewer edges. Fix any edge $\alpha E \beta$. By reordering the indices, if necessary, we may assume, without loss of generality, that

$$I_{\beta\alpha} = \{1, \dots, m\}, \qquad I_{\alpha\beta} = \{m+1, \dots, n\}, \qquad i_{\beta\alpha} = m, \qquad i_{\alpha\beta} = m+1,$$

and that the coordinate associated to the edge $\alpha E \beta$ is given by

$$y_{\alpha\beta}(w) = w_{1,m,m+1,n}.$$

Consider the projections $w \mapsto w_I$ and $w \mapsto w_J$ where

$$w_I := \{w_{ijk\ell}\}_{i,j,k,\ell \leq m+1}, \qquad w_J := \{w_{ijk\ell}\}_{i,j,k,\ell \geq m}.$$

Note that $w_I^0 \in M_{m+1}$ is modelled over the tree $T_{\beta\alpha}$, where the $(m+1)$-st marked point corresponds to the edge $\alpha E \beta$ in w^0. By assumption, the $m-2$ coordinates associated to the vertices and edges in $T_{\beta\alpha}$ do not involve any indices bigger than $m+1$ and so belong to w_I. Hence the induction hypothesis asserts that all components $w_{ijk\ell}$ with $i, j, k, \ell \in \{1, \dots, m+1\}$ can, in a neighbourhood of w^0, be smoothly reconstructed from these $m-2$ coordinates associated to $T_{\beta\alpha}$. Likewise, all components $w_{ijk\ell}$ with $i, j, k, \ell \in \{m, \dots, n\}$ can, in a neighbourhood of w^0, be smoothly reconstructed from the $n-m-2$ coordinates associated to the vertices and edges in $T_{\alpha\beta}$. Hence it remains to express all coordinates of w (in some neighbourhood of w^0) as smooth functions of $w_{1,m,m+1,n}$ and all those $w_{ijk\ell}$ which satisfy either $i, j, k, \ell \leq m$ or $i, j, k, \ell \geq m+1$.

To see this, note first that $w_{1,m+1,n,m}(\mathbf{z}^0) = 0$ and hence, by (D.4.3),

$$w_{m+1,n,i,m} = \frac{w_{1,m+1,n,m} - 1}{w_{1,m+1,n,m} - w_{1,m+1,n,i}}$$

for all i (and all w in some neighbourhood of w^0). If we choose $i \leq m-1$ then $w_{m,m+1,n,i}(\mathbf{z}^0) = 0$ and hence, by (D.4.3),

$$w_{m+1,n,i,j} = \frac{w_{m,m+1,n,j} - 1}{w_{m,m+1,n,j} - w_{m,m+1,n,i}}$$

for all j. If we choose $i, j \leq m$ then $w_{m+1,i,j,n}(\mathbf{z}^0) = 0$ and hence, by (D.4.3),

$$w_{i,j,k,n} = \frac{w_{m+1,i,j,n} - 1}{w_{m+1,i,j,n} - w_{m+1,i,j,k}}$$

for all k. Finally, $w_{nij\ell}(\mathbf{z}^0) = 0$ whenever $\ell \geq m+1$. Hence, again by (D.4.3),

$$w_{ijk\ell} = \frac{w_{nij\ell} - 1}{w_{nij\ell} - w_{nijk}}$$

for all $\ell \geq m+1$, all $i, j \leq m$ and all k. Thus we have expressed $w_{ijk\ell}$ in the required form (for all w in some neighbourhood of w^0) whenever $i, j \leq m$ and $k, \ell \geq m+1$ or $i, j, k \leq m$ and $\ell \geq m+1$. The case $i \leq m$ and $j, k, \ell \geq m+1$ follows from the latter by duality. This completes the proof of Theorem D.4.2. □

EXERCISE D.4.5. Above we constructed explicit coordinate charts near each element $\mathbf{z} \in \overline{\mathcal{M}}_{0,n}$. Interpret these in the language of gluing and gluing parameters that is used in Chapter 10.

Stratification. The group G acts on the space $(S^2)^n \setminus \Delta_n$ of n-tuples of distinct points on S^2. The quotient

$$\mathcal{M}_{0,n} := \frac{(S^2)^n \setminus \Delta_n}{\mathrm{G}} \tag{D.4.10}$$

is the moduli space of complex structures on the 2-sphere with n punctures. This quotient is not compact and $\overline{\mathcal{M}}_{0,n}$, with the topology induced by the embedding $\overline{\mathcal{M}}_{0,n} \to (S^2)^N$, is its **Grothendieck–Knudsen compactification**. Under the identification $\overline{\mathcal{M}}_{0,n} \to \overline{M}_{0,n}$ the subset $\mathcal{M}_{0,n}$ corresponds to

$$M_{0,n} := \left\{ w \in \overline{M}_{0,n} \,|\, w_{ijk\ell} \notin \{0, 1, \infty\} \;\forall\, i, j, k, \ell \right\}. \tag{D.4.11}$$

The next proposition shows that the complement of the top stratum $M_{0,n}$ in $\overline{M}_{0,n}$ is a finite union of smooth submanifolds of real codimension 2. Here is some relevant notation. Given an n-splitting $I \in \mathcal{I}_n$ denote the set of all points $w \in \overline{M}_{0,n}$ that admit I as a w-splitting by

$$V_I := \left\{ w \in \overline{M}_{0,n} \,|\, I \in S(w) \right\}. \tag{D.4.12}$$

Thus a stable curve $\mathbf{z} \in \mathcal{SC}_{0,n}$ has $w(\mathbf{z}) \in V_I$ if and only if there exists an edge which separates the marked points in I from those not in I. More generally, for any network of n-splittings $S \in \mathcal{S}_n$ define

$$V_S := \left\{ w \in \overline{M}_{0,n} \,|\, S \subset S(w) \right\} = \bigcap_{I \in S} V_I. \tag{D.4.13}$$

Note that $w(\mathbf{z}) \in V_S$ if and only if the tree structure of $\mathbf{z}$ is a refinement of the one given by S. Note also that $V_\emptyset = \overline{M}_{0,n}$.

EXERCISE D.4.6. Let $n = 8$ and $S = \{\{1,2\}, \{3,4\}, \{5,6\}, \{1,2,3,4,5,6\}\}$. Describe the elements in V_S and verify part (iv) of the next theorem.

THEOREM D.4.7 (Knudsen). (i) *For every n-splitting $I \in \mathcal{I}_n$ the subset*

$$V_I \subset \overline{M}_{0,n}$$

is a compact complex submanifold of complex codimension one.

(ii) *Let $S \subset \mathcal{I}_n$. Then*

$$V_S \neq \emptyset \qquad \Longleftrightarrow \qquad S \in \mathcal{S}_n.$$

(iii) *For every network of n-splittings $S = \{I_1, \dots, I_m\} \in \mathcal{S}_n$ the submanifolds $V_{I_1}, \dots, V_{I_m}$ intersect transversally and hence*

$$V_S = \bigcap_{j=1}^m V_{I_j} \subset \overline{M}_{0,n}$$

is a complex submanifold of complex codimension $\operatorname{codim} V_S = \#S$.

(iv) *If $S \in \mathcal{S}_n$ is a network of n-splittings with corresponding labelled tree (T, E, Λ) and n_α is defined by (D.2.1) for $\alpha \in T$ then there is a diffeomorphism*

$$V_S \cong \prod_{\alpha \in T} \overline{M}_{0,n_\alpha}.$$

PROOF. Assertions (i) and (iii) follow directly from the construction of the coordinate charts in the proof of Theorem D.4.2 (iii). To see this fix a point

$$w^0 = w(\mathbf{z}^0) \in \overline{M}_{0,n},$$

let $I \in S(w^0)$ be a w^0-splitting with corresponding edge $\alpha E \beta$, and suppose that the indices i, j, k, ℓ satisfy

$$\text{(D.4.14)} \qquad i, j \in I, \qquad k, \ell \notin I, \qquad z^0_{\alpha i} \neq z^0_{\alpha j}, \qquad z^0_{\beta k} \neq z^0_{\beta \ell}.$$

We shall use three observations. Firstly, condition (D.4.14) is satisfied if and only if I is the only $w(\mathbf{z}^0)$-splitting which separates the pair $\{i, j\}$ from $\{k, \ell\}$. Secondly, if w is sufficiently close to w^0 then

$$S(w) \subset S(w^0).$$

Thirdly, if $w_{ijk\ell} = \infty$ then, by Lemma D.4.4 (ii), there exists a w-splitting which separates $\{i, j\}$ from $\{k, \ell\}$. The previous two observations show that this can only be I. Thus we have proved that, for w near w^0, we have

$$\text{(D.4.15)} \qquad I \in S(w) \qquad \Longleftrightarrow \qquad w_{ijk\ell} = \infty.$$

Now the proof of Theorem D.4.2 (iii) shows that near w^0 there is a coordinate chart on $\overline{M}_{0,n}$ in which each w^0-splitting $I \in S(w^0)$ is represented by one coordinate which satisfies (D.4.14). This proves (i) and (iii).

To prove (ii) suppose that $S \subset \mathcal{I}_n$ is a collection of n-splittings which does not satisfy (D.2.3). Then there exist $I, J \in S$ such that

$$I \cap J \neq \emptyset, \qquad I \setminus J \neq \emptyset, \qquad J \setminus I \neq \emptyset.$$

In this case Lemma D.4.3 shows that $V_I \cap V_J = \emptyset$.

To prove (iv) choose, for each $\alpha \in T$, an index set $I_\alpha \subset \{1, \ldots, n\}$ which contains Λ_α and precisely one element of the set $I_{\alpha\beta}(T, E, \Lambda)$ for each edge $\alpha E \beta$ (as in the proof of Theorem D.4.2). Then consider the projection

$$\pi_\alpha : \overline{M}_{0,n} \to \overline{M}_{0,n_\alpha}$$

which only remembers the indices in I_α. These projections determine a map

$$\pi_S : V_S \to \prod_{\alpha \in T} \overline{M}_{0,n_\alpha}$$

which is obviously injective. To see that it is surjective choose any tuple $\{w_\alpha\}_{\alpha \in T}$ with $w_\alpha \in \overline{M}_{0,n_\alpha}$ and reconstruct all the other components $w_{ijk\ell}$ by using the tree structure. Namely, if $i, j \in I$ and $k, \ell \notin I$ for some $I \in S$ define $w_{ijk\ell} = \infty$, and similarly for other splittings into two pairs. If there is no splitting into pairs then there is a vertex α with four distinct indices $i', j', k', \ell' \in I_\alpha$ such that either $i = i'$ or $i, i' \in T_{\alpha\beta}$ for some edge $\alpha E \beta$, and similarly for j', k', ℓ'. In this case define $w_{ijk\ell} := w_{i'j'k'\ell'}$. One checks easily that the tuple w, thus defined, satisfies (D.4.2) and (D.4.3) and hence lies in $\overline{M}_{0,n}$. This shows that the map π_S is surjective and that its inverse is smooth. This proves Theorem D.4.7. □

D.5. The Gromov topology

In this section we give an intrinsic description of the topology on the moduli space $\overline{\mathcal{M}}_{0,n}$ and prove that it is equivalent to the extrinsic topology, i.e. the one defined by the embedding into the product of 2-spheres. The following definition is the special case of Gromov convergence for stable maps in which the target manifold is a point (see Definition 5.5.1).

DEFINITION D.5.1. *A sequence* $\mathbf{z}^\nu = (\{z^\nu_{\alpha\beta}\}_{\alpha E^\nu \beta}, \{\alpha^\nu_i, z^\nu_i\}_{1\le i\le n}) \in \mathcal{SC}_{0,n}$ *is said to* **Gromov converge** *to* $\mathbf{z} = (\{z_{\alpha\beta}\}_{\alpha E\beta}, \{\alpha_i, z_i\}_{1\le i\le n})$ *if, for* ν *sufficiently large, there exists a surjective tree homomorphism* $f^\nu : T \to T^\nu$ *and a collection of Möbius transformations* $\{\phi^\nu_\alpha\}^{\nu\in\mathbb{N}}_{\alpha\in T}$ *such that the following holds.*

(RESCALING) *If* $\alpha, \beta \in T$ *such that* $\alpha E\beta$ *and* ν_j *is a subsequence such that* $f^{\nu_j}(\alpha) = f^{\nu_j}(\beta)$ *then the sequence* $\phi^{\nu_j}_{\alpha\beta} := (\phi^{\nu_j}_\alpha)^{-1} \circ \phi^{\nu_j}_\beta$ *converges to* $z_{\alpha\beta}$ *u.s.c. on* $S^2 \setminus \{z_{\beta\alpha}\}$.

(NODAL POINT) *If* $\alpha, \beta \in T$ *such that* $\alpha E\beta$ *and* ν_j *is a subsequence such that* $f^{\nu_j}(\alpha) \neq f^{\nu_j}(\beta)$ *then* $z_{\alpha\beta} = \lim_{j\to\infty} (\phi^{\nu_j}_\alpha)^{-1}(z^{\nu_j}_{f^{\nu_j}(\alpha) f^{\nu_j}(\beta)})$.

(MARKED POINT) $\alpha^\nu_i = f^\nu(\alpha_i)$ *and* $z_i = \lim_{\nu\to\infty} (\phi^\nu_{\alpha_i})^{-1}(z^\nu_i)$ *for all* i.

Because each tree T has only finitely many isomorphism classes of surjective images, any Gromov convergent sequence is the finite union of subsequences for which the trees T^ν and maps f^ν are all isomorphic. These are the subsequences relevant to the *(Rescaling)* and *(Nodal point)* axioms above.

Here is the main result of this section.

THEOREM D.5.2. *There is a unique topology on* $\overline{\mathcal{M}}_{0,n}$ *whose convergent sequences are precisely the Gromov convergent sequences. This coincides with the topology induced on* $\overline{\mathcal{M}}_{0,n}$ *by the embedding of Theorem D.4.2.*

With this topology $\overline{\mathcal{M}}_{0,n}$ is homeomorphic to $\overline{M}_{0,n}$ and thus admits the structure of a compact metric space, and indeed of a compact smooth algebraic variety.

PROPOSITION D.5.3. *Let* $\mathbf{z}^\nu, \mathbf{z} \in \mathcal{SC}_{0,n}$. *The following are equivalent.*

(i) $\mathbf{z}^\nu$ *Gromov converges to* $\mathbf{z}$.

(ii) *For* ν *sufficiently large there exists a surjective tree homomorphism* $f^\nu : T \to T^\nu$ *and Möbius transformations* $\phi^\nu_\alpha \in \mathrm{G}$ *such that* $f^\nu(\alpha_i) = \alpha^\nu_i$ *and*

$$z_{\alpha i} = \lim_{\nu\to\infty} (\phi^\nu_\alpha)^{-1}(z^\nu_{f^\nu(\alpha) i}) \tag{D.5.1}$$

for $i \in \{1, \ldots, n\}$ *and* $\alpha \in T$.

(iii) *For any four distinct integers* $i, j, k, \ell \in \{1, \ldots, n\}$

$$w_{ijk\ell}(\mathbf{z}) = \lim_{\nu\to\infty} w_{ijk\ell}(\mathbf{z}^\nu). \tag{D.5.2}$$

PROPOSITION D.5.3 IMPLIES THEOREM D.5.2. By Lemma 5.6.4, to see that there is a unique topology on $\overline{\mathcal{M}}_{0,n}$ whose convergent sequences are precisely the Gromov convergent sequences, we just need to check that the five axioms *(Constant)*, *(Subsequence)*, *(Subsubsequence)*, *(Diagonal)*, and *(Uniqueness)* hold. But the equivalence of (i) and (iii) in Proposition D.5.3 shows that a sequence Gromov converges exactly if its image in $\overline{M}_{0,n}$ converges. Hence these axioms hold. The resulting topology must then coincide with the pullback topology since they have the same convergent sequences. □

PROOF OF PROPOSITION D.5.3. We prove that (i) implies (ii). Hence assume that $\mathbf{z}^\nu$ Gromov converges to $\mathbf{z}$ and suppose, by contradiction, that there is an $\alpha \in T$ and an $i \in \{1, \dots, n\}$ such that (D.5.1) does not hold. Passing to a subsequence, we may assume that

$$\lim_{\nu \to \infty} (\phi_\alpha^\nu)^{-1}(z_{f^\nu(\alpha)i}^\nu) = z_0 \neq z_{\alpha i}. \tag{D.5.3}$$

Passing to a further subsequence, we may assume that $T^\nu = T'$ and the tree homomorphism $f^\nu = f : T \to T'$ are independent of ν. We write the map f in the form $\alpha \mapsto \alpha' := f(\alpha)$. This is consistent with our notation for the labels since $f(\alpha_i) = \alpha_i'$, where $\alpha_i \in T$, respectively $\alpha_i' \in T'$, is the vertex that carries the ith marked point.

By (D.5.3) and the *(Marked point)* axiom, we have $\alpha \neq \alpha_i$. Choose a chain

$$\gamma_{-\ell}, \dots, \gamma_{-1}, \gamma_0, \dots, \gamma_k$$

running from $\gamma_{-\ell} = \alpha_i$ to $\gamma_k = \alpha$. The labelling is chosen such that $\gamma_j' = \alpha'$ for $j \geq 0$ and $\gamma_j' \neq \alpha'$ for $j < 0$. We first claim that

$$z_{\gamma_0 i} = \lim_{\nu \to \infty} (\phi_{\gamma_0}^\nu)^{-1}(z_{\gamma_0' i}^\nu). \tag{D.5.4}$$

If $\alpha' = \alpha_i'$ then $\gamma_0 = \alpha_i$ and hence $z_{\gamma_0 i} = z_i$ and $z_{\gamma_0' i}^\nu = z_i^\nu$. In this case (D.5.4) follows from the *(Marked point)* axiom in Definition D.5.1. If $\alpha' \neq \alpha_i'$ then $z_{\gamma_0 i} = z_{\gamma_0 \gamma_{-1}}$ and hence $z_{\gamma_0' i}^\nu = z_{\gamma_0' \gamma_{-1}'}^\nu$. In this case (D.5.4) follows from the *(Nodal point)* axiom in Definition D.5.1.

With (D.5.4) established, we prove by induction that

$$z_{\gamma_j i} = \lim_{\nu \to \infty} (\phi_{\gamma_j}^\nu)^{-1}(z_{\gamma_j' i}^\nu) \tag{D.5.5}$$

for $j = 1, \dots, k$. Suppose that (D.5.5) holds for $j \leq k-1$. Note that

$$z_{\gamma_j i} = z_{\gamma_j \gamma_{j-1}} \neq z_{\gamma_j \gamma_{j+1}}.$$

By the *(Rescaling)* axiom, $(\phi_{\gamma_{j+1}}^\nu)^{-1} \circ \phi_{\gamma_j}^\nu$ converges to $z_{\gamma_{j+1}\gamma_j} = z_{\gamma_{j+1} i}$, uniformly on compact subsets of $S^2 \setminus \{z_{\gamma_j \gamma_{j+1}}\}$. Since $z_{\gamma_j i} \neq z_{\gamma_j \gamma_{j+1}}$, this implies (D.5.5) with j replaced by $j+1$. This completes the induction. Hence (D.5.5) holds for $j = k$, and this contradicts (D.5.3). This contradiction shows that our assumption that (D.5.1) is false must have been wrong. This proves that (i) implies (ii).

We prove that (ii) implies (iii). Again it suffices to consider the case when $T^\nu = T'$ and $f^\nu = f : T \to T'$ for every ν. For any four distinct integers $i, j, k, \ell \in \{1, \dots, n\}$ choose $\alpha \in T$ such that $(z_{\alpha i}, z_{\alpha j}, z_{\alpha k}, z_{\alpha \ell}) \notin \Delta_3$. By assumption, we have

$$z_{\alpha i} = \lim_{\nu \to \infty} (\phi_\alpha^\nu)^{-1}(z_{\alpha' i}^\nu),$$

where $\alpha' = f(\alpha)$, and similarly for j, k, ℓ. Hence $(z_{\alpha' i}^\nu, z_{\alpha' j}^\nu, z_{\alpha' k}^\nu, z_{\alpha' \ell}^\nu) \notin \Delta_3$ for ν sufficiently large and

$$\begin{aligned} w_{ijk\ell}(\mathbf{z}) &= w(z_{\alpha i}, z_{\alpha j}, z_{\alpha k}, z_{\alpha \ell}) \\ &= \lim_{\nu \to \infty} w\left((\phi_\alpha^\nu)^{-1}(z_{\alpha' i}^\nu), \dots, (\phi_\alpha^\nu)^{-1}(z_{\alpha' \ell}^\nu)\right) \\ &= \lim_{\nu \to \infty} w(z_{\alpha' i}^\nu, z_{\alpha' j}^\nu, z_{\alpha' k}^\nu, z_{\alpha' \ell}^\nu) \\ &= \lim_{\nu \to \infty} w_{ijk\ell}(\mathbf{z}^\nu). \end{aligned}$$

This proves (D.5.2).

We prove that (iii) implies (ii). If (iii) holds then, for ν sufficiently large, we have

$$w_{ijk\ell}(\mathbf{z}) \neq \infty \qquad \Longrightarrow \qquad w_{ijk\ell}(\mathbf{z}^\nu) \neq \infty,$$

and hence $S(\mathbf{z}^\nu) \subset S(\mathbf{z})$. For such ν there exists, by Theorem D.2.6 (iv), a unique surjective tree homomorphism $f^\nu : T \to T^\nu$ which satisfies

$$f^\nu(\alpha_i) = \alpha_i^\nu$$

for all i. In the following we denote

$$\alpha^\nu := f^\nu(\alpha) \in T^\nu$$

for every $\alpha \in T$ and every ν. Now for each $\alpha \in T$ choose indices $i, j, k \in \{1, \dots, n\}$ such that $\alpha = \alpha(i, j, k; T)$. Then, by Theorem D.2.6 (iv), $\alpha^\nu = \alpha(i, j, k; T^\nu)$. By Remark D.3.3, this means that the points $z_{\alpha i}$, $z_{\alpha j}$, $z_{\alpha k}$ are pairwise distinct, and so are the points $z^\nu_{\alpha^\nu i}$, $z^\nu_{\alpha^\nu j}$, $z^\nu_{\alpha^\nu k}$ for large ν. Hence there exists a unique Möbius transformation $\phi^\nu_\alpha \in \mathrm{G}$ such that

$$\phi^\nu_\alpha(z_{\alpha i}) = z^\nu_{\alpha^\nu i}, \qquad \phi^\nu_\alpha(z_{\alpha j}) = z^\nu_{\alpha^\nu j}, \qquad \phi^\nu_\alpha(z_{\alpha k}) = z^\nu_{\alpha^\nu k}.$$

We claim that this sequence satisfies (D.5.1). To see this note that

$$\begin{aligned} w_{ijk\ell}(\mathbf{z}^\nu) &= w(z^\nu_{\alpha^\nu i}, z^\nu_{\alpha^\nu j}, z^\nu_{\alpha^\nu k}, z^\nu_{\alpha^\nu \ell}) \\ &= \frac{(z_{\alpha j} - z_{\alpha k})(\zeta^\nu_{\alpha\ell} - z_{\alpha i})}{(z_{\alpha i} - z_{\alpha j})(z_{\alpha i} - \zeta^\nu_{\alpha\ell})}, \end{aligned}$$

where

$$\zeta^\nu_{\alpha\ell} := (\phi^\nu_\alpha)^{-1}(z^\nu_{\alpha^\nu \ell})$$

for $\alpha \in T$ and all ν. Now use the convergence of $w_{ijk\ell}(\mathbf{z}^\nu)$ to $w_{ijk\ell}(\mathbf{z})$ to obtain

$$z_{\alpha\ell} = \lim_{\nu\to\infty} \zeta^\nu_{\alpha\ell}$$

for all $\alpha \in T$ and all $\ell \in \{1, \dots, n\}$. This proves (ii).

We prove that (ii) implies (i). If (ii) holds then the *(Marked point)* axiom in in Definition D.5.1 is obviously satisfied (choose $\alpha = \alpha_i$ in (D.5.1)). To verify the *(Nodal point)* axiom choose i such that $z_{\alpha\beta} = z_{\alpha i}$, and hence $z^\nu_{\alpha^\nu\beta^\nu} = z^\nu_{\alpha^\nu i}$. Then apply (D.5.1). To verify the *(Rescaling)* axiom suppose that $\alpha E \beta$ and

$$f^\nu(\alpha) = f^\nu(\beta) = \alpha^\nu$$

for all ν. Choose indices j and j' such that the three points $z_{\beta\alpha}$, $z_{\beta j}$, $z_{\beta j'}$ are pairwise distinct and hence $z_{\alpha j} = z_{\alpha j'} = z_{\alpha\beta}$. By (D.5.1), we have

$$\lim_{\nu\to\infty} (\phi^\nu_\beta)^{-1}(z^\nu_{\alpha^\nu j}) = z_{\beta j} \neq z_{\beta j'} = \lim_{\nu\to\infty} (\phi^\nu_\beta)^{-1}(z^\nu_{\alpha^\nu j'}),$$

$$\lim_{\nu\to\infty} (\phi^\nu_\alpha)^{-1}(z^\nu_{\alpha^\nu j}) = z_{\alpha\beta} = \lim_{\nu\to\infty} (\phi^\nu_\alpha)^{-1}(z^\nu_{\alpha^\nu j'}).$$

Interchanging the roles of α and β, we find an index i such that

$$\lim_{\nu\to\infty} (\phi^\nu_\alpha)^{-1}(z^\nu_{\alpha^\nu i}) = z_{\alpha i} \neq z_{\alpha\beta}, \qquad \lim_{\nu\to\infty} (\phi^\nu_\beta)^{-1}(z^\nu_{\alpha^\nu i}) = z_{\beta\alpha}.$$

These facts taken together show, by Lemma D.1.4, that $(\phi^\nu_\alpha)^{-1} \circ \phi^\nu_\beta$ converges to $z_{\alpha\beta}$, u.s.c. on $S^2 \setminus \{z_{\beta\alpha}\}$. □

D.6. Cohomology

The Grothendieck–Knudsen manifolds $\overline{\mathcal{M}}_{0,n}$ have many remarkable properties. We have seen in Theorem D.4.7 that every network of n-splittings determines a complex submanifold. It turns out that these submanifolds generate the cohomology of $\overline{\mathcal{M}}_{0,n}$ and give useful information about its ring structure. In the following discussion, it is helpful to think of $\overline{\mathcal{M}}_{0,n}$ at some times as a set of decorated trees and at others as the manifold $\overline{M}_{0,n}$ with coordinates $w \in (S^2)^N$.

The universal curve. There is a natural projection

$$\pi : \overline{M}_{0,n+1} \to \overline{M}_{0,n} \tag{D.6.1}$$

onto those coordinates $w_{ijk\ell}$ that do not contain the index $n+1$. This map is obviously holomorphic and takes values in $\overline{M}_{0,n}$ (see equation (D.4.4)). Under the bijection $\overline{\mathcal{M}}_{0,n+1} \cong \overline{M}_{0,n+1}$ of Theorem D.4.2, it corresponds to deleting the $(n+1)$st marked point. In geometric terms the map can be described as follows. If the vertex α carrying the $(n+1)$st marked point has at least four special points the curve obtained by deleting this marked point is still stable. Otherwise the vertex carries exactly three marked points, at least one of them nodal. If it carries two nodal points, replace it by a new pair of nodal points on the two vertices connected to α by the two edges. If it carries one nodal point and two marked points indexed by m and $n+1$, replace it by the mth marked point, to be placed on the unique vertex connected to α by an edge.

For $i = 1, \ldots, n$ let

$$\mathcal{Z}_i \subset \overline{\mathcal{M}}_{0,n+1}$$

be the subset associated to the $(n+1)$-splitting $I_i := \{1, \ldots, n\} \setminus \{i\}$. This set consists of all equivalence classes of stable curves of genus zero with $n+1$ marked points, such that the marked points with indices i and $n+1$ belong to the same vertex that, moreover, has precisely three special points. The bijection $\overline{\mathcal{M}}_{0,n+1} \to \overline{M}_{0,n+1}$ of Theorem D.4.2 maps $\mathcal{Z}_i$ onto the set

$$Z_i := \left\{ w \in \overline{M}_{0,n+1} \,|\, w_{jj'in+1} = \infty \; \forall j, j' \in \{1, \ldots, n\} \setminus \{i\} \right\}. \tag{D.6.2}$$

By Theorem D.4.7 this is a complex submanifold of $\overline{M}_{0,n+1}$ and the restriction of the projection (D.6.1) to this submanifold is a diffeomorphism $\pi : Z_i \to \overline{M}_{0,n}$. Its inverse is a section of π denoted by

$$\sigma_i : \overline{M}_{0,n} \to \overline{M}_{0,n+1}, \quad \pi \circ \sigma_i = \mathrm{id}, \quad \sigma_i(\overline{M}_{0,n}) = Z_i, \quad i = 1, \ldots, n. \tag{D.6.3}$$

Explicitly, σ_i sends the tuple $w \in \overline{M}_{0,n}$ to the tuple $\sigma_i(w) \in \overline{M}_{0,n+1}$ given by

$$(\sigma_i(w))_{jk\ell m} := \begin{cases} w_{jk\ell m}, & \text{if } j,k,\ell,m \le n, \\ w_{jk\ell i}, & \text{if } m = n+1 \text{ and } j,k,\ell \ne i, \\ w_{jkim}, & \text{if } \ell = n+1 \text{ and } j,k,m \ne i, \\ w_{ji\ell m}, & \text{if } k = n+1 \text{ and } j,\ell,m \ne i, \\ w_{ik\ell m}, & \text{if } j = n+1 \text{ and } k,\ell,m \ne i, \\ 0, & \text{if } \{j,m\} = \{i,n+1\} \text{ or } \{k,\ell\} = \{i,n+1\}, \\ 1, & \text{if } \{j,\ell\} = \{i,n+1\} \text{ or } \{k,m\} = \{i,n+1\}, \\ \infty, & \text{if } \{j,k\} = \{i,n+1\} \text{ or } \{\ell,m\} = \{i,n+1\}, \end{cases}$$

for every quadruple of pairwise distinct indices $j, k, \ell, m \in \{1, \ldots, n+1\}$. Geometrically, the section σ_i doubles the ith marked point. Thus it introduces a new vertex in the tree which is connected to α_i by an edge and carries the marked points with indices i and $n+1$.

EXERCISE D.6.1. Find explicit formulas for the projection (D.6.1) and the section (D.6.3) in terms of the tuples in $\mathcal{SC}_{0,n}$ modelled over labelled trees.

EXERCISE D.6.2. Prove that the map

$$\pi : \overline{M}_{0,n+1} \to \overline{M}_{0,n}$$

is a nodal family in the sense that each critical point of π is nodal (see [**337**, Definition 4.2]). Prove that $M_{0,n+1}$ is the set of regular points and $M_{0,n}$ is the set of regular values of π (see equation (D.4.11)).

It follows from Exercise D.6.2 that each fiber of $\pi : \overline{M}_{0,n+1} \to \overline{M}_{0,n}$ is a nodal Riemann surface with n marked points associated to the sections $Z_1, \dots, Z_n$. In fact for every stable curve $\mathbf{z} \in \mathcal{SC}_{0,n}$ the fiber over the point $w(\mathbf{z}) \in \overline{M}_{0,n}$ defined by (D.4.1) is isomorphic to the space $\Sigma(\mathbf{z})$ itself as defined in (D.3.2):

$$\pi^{-1}(w(\mathbf{z})) \cong \Sigma(\mathbf{z}), \qquad \pi^{-1}(w(\mathbf{z})) \cap Z_i \cong [\alpha_i, z_i]. \tag{D.6.4}$$

To see this, note that if $[\alpha, z]$ is not a nodal point on $\Sigma(\mathbf{z})$ and is not equal to any of the n marked points, then we can include it as the $(n+1)$st marked point to obtain an element of the fiber. If it is equal to either a nodal point or one of the marked points, then we can introduce an additional sphere to obtain an element in the fiber.

A nodal family $(\pi_B : Q \to B, Z_1, \dots, Z_n)$ with n sections is called a **universal family (of nodal Riemann surfaces of genus zero with n marked points)** if it satisfies the following conditions (see [**337**, Definitions 5.1 and 6.2]).

(I) For every other nodal family $(\pi_A : P \to A, R_1, \dots, R_n)$ and every fiber isomorphism $f : P_a \to Q_b$ with $f(P_a \cap R_i) = Q_b \cap Z_i$ for $i = 1, \dots, n$, there is a unique extension of f to a morphism of nodal families on a neighbourhood of P_a in P.

(II) Every fiber of Q is a stable curve of genus zero with n marked points and, conversely, every stable curve of genus zero with n marked points is isomorphic to precisely one of the fibers.

In [**337**, Theorem 5.4] it is shown that a nodal family

$$(\pi_B : Q \to B, Z_1, \dots, Z_n)$$

with n sections satisfies (I) if and only if the linearized Cauchy–Riemann operator associated to each fiber is bijective. The domain of this operator is the space of vector fields along the fiber that project to a constant tangent vector of the base and are tangent to the sections Z_i. The target space is the space of $(0,1)$-forms on the fiber with values in the vertical tangent bundle. The nodal family

$$(\pi : \overline{M}_{0,n+1} \to \overline{M}_{0,n}, Z_1, \dots, Z_n)$$

defined by (D.6.1) and (D.6.2) satisfies (II) by definition. That it also satisfies (I) is the content of the next exercise. It is often called the **universal curve** over $\overline{M}_{0,n}$.

EXERCISE D.6.3. Prove that $(\pi : \overline{M}_{0,n+1} \to \overline{M}_{0,n}, Z_1, \dots, Z_n)$ is a universal family of nodal Riemann surfaces of genus zero with n marked points.

Generators of the cohomology. In [**208**] Keel proved that the complex submanifolds V_S associated to networks of n-splittings $S \in \mathcal{S}_n$ generate (additively) the cohomology of $\overline{M}_{0,n}$.

THEOREM D.6.4 (Keel). *$\overline{M}_{0,n}$ is simply connected and $H^{\mathrm{odd}}(\overline{M}_{0,n};\mathbb{Z}) = \{0\}$. Moreover, if $a \in H^{2k}(\overline{M}_{0,n};\mathbb{Z})$ satisfies $\langle a, [V_S]\rangle = 0$ for every network of n-splittings $S \in \mathcal{S}_n$ with $\#S = n-3-k$ then $a = 0$.*

PROOF OF SIMPLE CONNECTIVITY. The space $\overline{\mathcal{M}}_{0,3} = \{\mathrm{pt}\}$ is obviously simply connected. Now let $n \geq 4$ and suppose, by induction, that $\overline{\mathcal{M}}_{0,n-1}$ is simply connected. Let $\gamma : \mathbb{R}/\mathbb{Z} \to \overline{\mathcal{M}}_{0,n}$ be a smooth loop. By standard transversality theory we may assume that γ is tranverse to V_I for every n-splitting $I \in \mathcal{I}_n$, and hence $\gamma(t)$ lies in the top stratum $\mathcal{M}_{0,n}$ for all t. Now choose a homotopy of loops

$$\mathbb{R}/\mathbb{Z} \times [0,1] \to \overline{\mathcal{M}}_{0,n} : (t,\lambda) \mapsto \gamma_\lambda(t)$$

such that $w_{ijk\ell}(\gamma_\lambda(t)) = w_{ijk\ell}(\gamma(t))$ for $i,j,k,\ell \leq n-1$ and

$$w_{123n}(\gamma_\lambda(t)) = (1-\lambda)w_{123n}(\gamma(t)) + \lambda w_{123n-1}(\gamma(t)).$$

These conditions determine γ_λ uniquely, and ensure that $\gamma_0(t) = \gamma(t)$ and that $\gamma_1(t)$ lies in the image of a smooth section of the projection $\pi : \overline{\mathcal{M}}_{0,n} \to \overline{\mathcal{M}}_{0,n-1}$. Geometrically, γ_0 lies in a subset of $\overline{\mathcal{M}}_{0,n}$ which can be identified with a complex line bundle over $\mathcal{M}_{0,n-1}$ (namely the subset where $w_{ijk\ell} \notin \{0,1,\infty\}$ for $i,j,k,\ell \leq n-1$ and $w_{123n} \neq \infty$) and is therefore homotopic to a loop in the zero section of this line bundle. Now the induction hypothesis asserts that $\overline{\mathcal{M}}_{0,n-1}$ is simply connected and hence γ_1 is contractible. This shows that $\overline{\mathcal{M}}_{0,n}$ is simply connected. □

The full proof of Keel's theorem involves the blow up construction and we shall not discuss this argument here. It would be interesting to find an explicit formula for a perfect Morse function on the moduli space $\overline{\mathcal{M}}_{0,n}$. We are not aware of any such construction in the existing literature. That a perfect Morse function exists follows from the discussion in Section D.7 for $n = 3,4,5$, and for $n \geq 6$ by the standard handle canceling procedures used to prove the h-cobordism theorem.

Gravitational descendants. For $i = 1,\dots,n$ there is a complex line bundle

$$L_i \to \overline{M}_{0,n}$$

defined as the pullback of the conormal bundle of $Z_i \subset \overline{M}_{0,n+1}$ under the section $\sigma_i : \overline{M}_{0,n} \to Z_i$. Geometrically, one can think of the fiber of L_i over $w(\mathbf{z})$ as the tangent space of $\Sigma(\mathbf{z})$ at the ith marked point. Denote the first Chern class of this line bundle by

$$\psi_i := c_1(L_i) \in H^2(\overline{M}_{0,n};\mathbb{Z}). \tag{D.6.5}$$

These cohomology classes are the simplest special cases of the so-called **gravitational descendents**. The following result is proved in Givental [**149**].

THEOREM D.6.5. *Let $d_1,\dots,d_n$ be nonnegative integers such that*

$$d_1 + d_2 + \cdots + d_n = n-3.$$

Then

$$\int_{\overline{M}_{0,n}} \psi_1^{d_1} \smile \psi_2^{d_2} \smile \cdots \smile \psi_n^{d_n} = \frac{(n-3)!}{d_1!\cdots d_n!}. \tag{D.6.6}$$

The identities (D.6.6) can be expressed in the form of a single equation

$$\int_{\overline{M}_{0,n}} \prod_{i=1}^{n} \frac{1}{1-t_i\psi_i} = (t_1+\cdots+t_n)^{n-3}. \tag{D.6.7}$$

Both sides of this equation are formal power series in the variables $t_1, \dots, t_n$. Hence it follows by comparing coefficients that equation (D.6.6) is equivalent to (D.6.7).

The proof of Theorem D.6.5 is based on a recursion formula for the classes ψ_i. For $n \geq 3$ and $i = 1, \dots, n$ denote by

$$\pi_{n+1} : \overline{M}_{0,n+1} \to \overline{M}_{0,n}$$

the projection (D.6.1), by

$$\sigma_{i,n} : \overline{M}_{0,n} \to \overline{M}_{0,n+1}$$

the section (D.6.3), by $Z_{i,n+1} \subset \overline{M}_{0,n+1}$ its image, by

$$L_{i,n} \to \overline{M}_{0,n}$$

the pullback of the conormal bundle of $Z_{i,n+1}$ under $\sigma_{i,n}$, by $\psi_{i,n} \in H^2(\overline{M}_{0,n}; \mathbb{Z})$ its first Chern class (D.6.5), and by

$$e_{i,n+1} := \mathrm{PD}\left([Z_{i,n+1}]\right) \in H^2(\overline{M}_{0,n+1}; \mathbb{Z})$$

the Poincaré dual of the section of $Z_{i,n+1} \subset \overline{M}_{0,n+1}$. Thus $\psi_{i,n} = -\sigma_{i,n}^* e_{i,n+1}$.

PROOF OF THEOREM D.6.5. The proof is taken from [**149**]. It has three steps.

STEP 1 *The cohomology class $\psi_{i,n+1}$ and $\psi_{i,n}$ are related by the recursion formula*

$$\psi_{i,n+1} = \pi_{n+1}^* \psi_{i,n} + e_{i,n+1} \tag{D.6.8}$$

for $i = 1, \dots, n$ and $n \geq 3$.

We sketch the argument. The bundles $L_{i,n+1}$ and $\pi_{n+1}^* L_{i,n}$ are isomorphic over the complement $\overline{M}_{0,n+1} \setminus Z_{i,n+1}$. Hence their first Chern classes differ by an integer multiple of $e_{i,n+1}$. Thus there is an integer k such that $\psi_{i,n+1} - \pi_{n+1}^*\psi_{i,n} = ke_{i,n+1}$. We must prove that $k = 1$. To see this, note that the restriction of $L_{i,n+1}$ to $Z_{i,n+1}$ admits a trivialization and the restriction of $\pi_{n+1}^* L_{i,n}$ to $Z_{i,n+1}$ is the conormal bundle. Hence the restriction of $\psi_{i,n+1} - \pi_{n+1}^*\psi_{i,n}$ to $Z_{i,n+1}$ is the Euler class of the normal bundle and therefore agrees with the restriction of $e_{i,n+1}$ to $Z_{i,n+1}$. Since this class is nonzero, it follows that $k = 1$. This proves Step 1.

STEP 2. *If $d_i \geq 0$ and $d_1 + \cdots + d_n = n - 2$ then*

$$\int_{\overline{M}_{0,n+1}} \prod_{i=1}^{n} \psi_{i,n+1}^{d_i} = \sum_{d_i>0} \int_{\overline{M}_{0,n}} \psi_{i,n}^{d_i-1} \prod_{j\neq i} \psi_{j,n}^{d_j}. \tag{D.6.9}$$

Since the bundle $L_{i,n+1} \to \overline{M}_{0,n+1}$ admits a trivialization over $Z_{i,n+1}$ its first Chern class $\psi_{i,n+1}$ is Poincaré dual to a submanifold of $\overline{M}_{0,n+1}$ that is disjoint from $Z_{i,n+1}$. Hence the cup product of $\psi_{i,n+1}$ and $e_{i,n+1}$ vanishes and hence, by (D.6.8), we have

$$\pi_{n+1}^*\psi_{i,n}^d = \psi_{i,n+1}^d + (-e_{i,n+1})^d, \qquad d > 0.$$

This implies

$$\begin{aligned}
0 &= \int_{\overline{M}_{0,n+1}} \prod_{i=1}^{n} \pi_{n+1}^* \psi_{i,n}^{d_i} \\
&= \int_{\overline{M}_{0,n+1}} \prod_{d_i>0} \left(\psi_{i,n+1}^{d_i} + (-e_{i,n+1})^{d_i}\right) \\
&= \int_{\overline{M}_{0,n+1}} \prod_{i=1}^{n} \psi_{i,n+1}^{d_i} + \sum_{d_i>0} \int_{\overline{M}_{0,n+1}} (-e_{i,n+1})^{d_i} \prod_{j\neq i} \pi_{n+1}^* \psi_{j,n}^{d_j} \\
&= \int_{\overline{M}_{0,n+1}} \prod_{i=1}^{n} \psi_{i,n+1}^{d_i} - \sum_{d_i>0} \int_{Z_{i,n+1}} (-e_{i,n+1})^{d_i-1} \prod_{j\neq i} \pi_{n+1}^* \psi_{j,n}^{d_j} \\
&= \int_{\overline{M}_{0,n+1}} \prod_{i=1}^{n} \psi_{i,n+1}^{d_i} - \sum_{d_i>0} \int_{\overline{M}_{0,n}} \psi_{i,n}^{d_i-1} \prod_{j\neq i} \psi_{j,n}^{d_j}.
\end{aligned}$$

Here the fourth equation follows from the fact that $e_{i,n+1}$ is the Thom class of the submanifold $Z_{i,n+1}$ and the last equation follows from the degree theorem for the integral over $Z_{i,n+1}$ by pulling the integrand back to $\overline{M}_{0,n}$ with the holomorphic diffeomorphism $\sigma_{i,n} : \overline{M}_{0,n} \to Z_{i,n+1}$, This proves (D.6.9).

Equation (D.6.9) can be written in the form

$$\text{(D.6.10)} \qquad \int_{\overline{M}_{0,n+1}} \prod_{i=1}^{n} \frac{1}{1-t_i\psi_i} = (t_1 + \cdots + t_n) \int_{\overline{M}_{0,n}} \prod_{i=1}^{n} \frac{1}{1-t_i\psi_i}$$

This is a special case of the **string equation.** We have reverted to the notation ψ_i instead of $\psi_{i,n}$ or $\psi_{i,n+1}$ as the second index is clear from the context.

STEP 3. *We prove equation* (D.6.6).

The proof is by induction on n. For $n = 3$ the manifold $\overline{M}_{0,3}$ is a point and (D.6.6) is obvious. Now assume, by induction, that (D.6.6) has been established for some integer $n \geq 3$. Then (D.6.7) holds for n. Hence it follows from the string equation (D.6.10) that

$$\int_{\overline{M}_{0,n+1}} \prod_{i=1}^{n} \frac{1}{1-t_i\psi_i} = (t_1 + \cdots + t_n)^{n-2}.$$

This implies

$$\text{(D.6.11)} \qquad \int_{\overline{M}_{0,n+1}} \psi_1^{d_1} \smile \psi_2^{d_2} \smile \cdots \smile \psi_n^{d_n} = \frac{(n-2)!}{d_1! \cdots d_n!}.$$

for every n-tuple of nonnegative integers $d_1, \ldots, d_n$ such that $d_1 + \cdots + d_n = n-2$. Now let $d_1, \ldots, d_{n+1}$ be nonnegative integers such that

$$d_1 + \cdots + d_{n+1} = n - 2.$$

Then one of the integers d_i must vanish. Hence it follows from (D.6.11) by permuting the indices that

$$\text{(D.6.12)} \qquad \int_{\overline{M}_{0,n+1}} \psi_1^{d_1} \smile \psi_2^{d_2} \smile \cdots \smile \psi_{n+1}^{d_{n+1}} = \frac{(n-2)!}{d_1! \cdots d_{n+1}!}.$$

This proves Step 3 and Theorem D.6.5. □

D.7. Examples

The space $\overline{\mathcal{M}}_{0,3}$ consists of a single point since the three marked points z_1, z_2, z_3 can be moved to $0, 1, \infty$ by a Möbius transformation. In the case of $\overline{\mathcal{M}}_{0,4}$ we can remove the action of the group on the top component $\mathcal{M}_{0,4}$ by fixing three points at 0, 1, and ∞. Since the fourth marked point z may be any other point, $\mathcal{M}_{0,4}$ can be identified with $S^2 \setminus \{0, 1, \infty\}$. It is easy to check that there are precisely three other elements τ_0, τ_1, and τ_∞ in $\overline{\mathcal{M}}_{0,4}$. These are modelled on the tree T_2 with two vertices, and differ only in the placing of the marked points. By stability, there must be two marked points at each vertex of T_2 and so what distinguishes these three stable curves are the decompositions of the index set: In τ_0 the 4th marked point $z_4 = z$ is paired with $z_1 = 0$, while τ_1 pairs it with $z_2 = 1$ and τ_∞ with $z_3 = \infty$. Then the stable curve τ_z converges to τ_i with respect to the GK-topology on $\overline{\mathcal{M}}_{0,4}$ if and only if z converges to i for $i = 0, 1, \infty$. In other words, as two of the marked points come together a bubble forms that separates these two marked points from the other two. This discussion shows that

$$\overline{\mathcal{M}}_{0,4} \cong \mathbb{C}P^1.$$

Now, for any four pairwise distinct indices $i, j, k, \ell \in \{1, \ldots, n\}$, there is a natural projection

$$\pi_{ijk\ell} : \overline{\mathcal{M}}_{0,n} \to \overline{\mathcal{M}}_{0,4}$$

given by forgetting the other marked points. If we identify $\overline{\mathcal{M}}_{0,4}$ with the Riemann sphere $S^2 = \mathbb{C} \cup \{\infty\}$ as above then these are precisely our cross ratio maps $w_{ijk\ell} : \overline{\mathcal{M}}_{0,n} \to S^2$. These projections also play a crucial role in the work of Keel [**208**].

Next consider the space $\overline{\mathcal{M}}_{0,5}$. Forgetting the last point defines a map

$$\overline{\mathcal{M}}_{0,5} \to \overline{\mathcal{M}}_{0,4} = \mathbb{C}P^1.$$

This is a singular fibration. In other words after deleting the inverse images of a finite number of points (the singular fibers) one obtains a locally trivial fiber bundle whose fiber is called the "generic fiber". In the case at hand, this generic fiber is $\mathbb{C}P^1$, and each of the three exceptional fibers consist of a pair of intersecting copies of $\mathbb{C}P^1$. The projection $\overline{\mathcal{M}}_{0,5} \to \overline{\mathcal{M}}_{0,4}$ can be described explicitly by choosing two generic sections of a line bundle $L \to \mathbb{C}P^2$ of degree 2, and blowing up the four intersection points of the two quadrics given by their zero sections. Now $\mathbb{C}P^2$ with the four points blown up is swept out by a family of quadrics, parametrized by $\mathbb{C}P^1$, and this family contains precisely three nodal curves, each consisting of a pair of lines. Hence

$$\overline{\mathcal{M}}_{0,5} \cong \mathbb{C}P^2 \# 4\overline{\mathbb{C}P}^2.$$

One can visualize this by drawing 4 points in the plane, in general position, and a family of ellipses or hyperbolas, passing through these points. The nodal curves are given by the three pairs of lines with their three intersection points (see Figure 6).

Here is an alternative description of the moduli space $\overline{\mathcal{M}}_{0,5}$. Fix three points at $z_0 = 0$, $z_1 = 1$, $z_2 = \infty$. Then $\overline{\mathcal{M}}_{0,5}$ consists of pairs (z_3, z_4) which are not equal to the three pairs $(0,0)$, $(1,1)$, (∞, ∞), together with three 2-spheres representing the possible configurations which arise from a collision of both z_3 and z_4 with z_i for $i = 0, 1, 2$. Examining neighbourhoods of these 2-spheres one obtains the product $S^2 \times S^2$ with three points on the diagonal blown up: see [**267**] for example. This

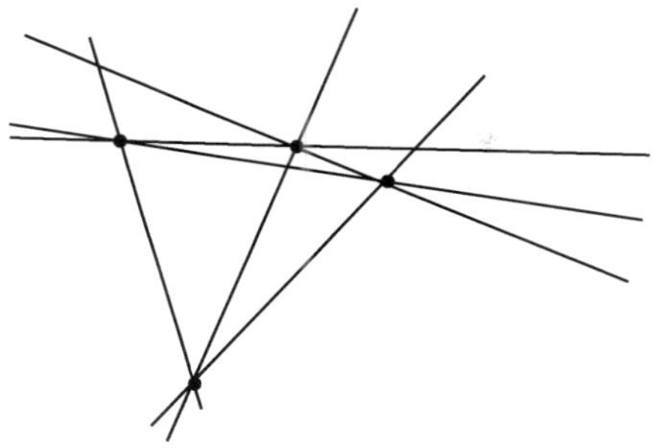

FIGURE 6. Three pairs of lines determined by four generic points

shows again that

$$\overline{\mathcal{M}}_{0,5} \cong (S^2 \times S^2)\#3\overline{\mathbb{C}P}^2 \cong (\mathbb{C}P^2)\#4\overline{\mathbb{C}P}^2.$$

A similar construction can be used to study moduli spaces of marked points on surfaces of higher genus with a fixed complex structure. More precisely, fix a Riemann surface Σ of genus $g \geq 1$ and denote by j_Σ the complex structure on Σ. Then the compactified moduli space $\overline{\mathcal{M}}_{g,3}(j_\Sigma)$ of three points on Σ with a fixed complex structure j_Σ is diffeomorphic to a fiber bundle over Σ where the fiber over $z \in \Sigma$ is the product $\Sigma \times \Sigma$ with the point (z, z) blown up. Thus there is a fibration

$$(\Sigma \times \Sigma)\#\overline{\mathbb{C}P}^2 \hookrightarrow \overline{\mathcal{M}}_{g,3}(j_\Sigma) \longrightarrow \Sigma.$$

Equivalently, the space $\overline{\mathcal{M}}_{g,3}(j_\Sigma)$ is naturally diffeomorphic to the triple product $\Sigma \times \Sigma \times \Sigma$ with the diagonal $\Delta = \{(z, z, z) \mid z \in \Sigma\}$ blown up. Now one can use the constructions of this appendix to embed the moduli space $\overline{\mathcal{M}}_{g,n}(j_\Sigma)$ as a submanifold into the product $(\overline{\mathcal{M}}_{g,3}(j_\Sigma))^N$ with $N = \binom{n}{3}$. The embedding

$$\overline{\mathcal{M}}_{g,n}(j_\Sigma) \hookrightarrow \left(\overline{\mathcal{M}}_{g,3}(j_\Sigma)\right)^N$$

is a special case of a construction by Fulton and MacPherson in [**132**] for higher dimensional varieties. For any complex manifold X they define X_3 as the product $X \times X \times X$ with the diagonal $\Delta = \{(x, x, x) \mid x \in X\}$ blown up. Then they consider the obvious embedding of the configuration space of n ordered distinct points in X into $(X_3)^N$ and prove that the closure of the image is a smooth submanifold.

The Mumford quotient. It is interesting to examine the relation between $\overline{\mathcal{M}}_{0,n}$ and the **Marsden-Weinstein quotient** $X/\!\!/\mathrm{G}$ of $X = (S^2)^n$ by the diagonal action of $\mathrm{G} = \mathrm{SO}(3)$. This action is Hamiltonian. For each factor the moment map is the inclusion of S^2 into $\mathbb{R}^3 \cong \mathfrak{so}(3)$ as the unit sphere [**293**]. Hence the SO(3)-action on X is generated by the moment map $\mu : X \to \mathbb{R}^3$ given by

$$\mu(x_1, \ldots, x_n) = x_1 + \cdots + x_n.$$

Thus $\mu^{-1}(0)$ is the set of n-tuples $(x_1, \ldots, x_n) \in (S^2)^n$ with mean value zero. The critical points of the moment map are the n-tuples $x = (x_1, \ldots, x_n)$ such that the vectors $x_i \in \mathbb{R}^3$ are all pairwise linearly dependent, and hence lie on a single line through 0. Hence zero is a regular value of μ if and only if n is odd. Moreover, in order for $\mathrm{G} = \mathrm{SO}(3)$ to act freely on $\mu^{-1}(0)$ we must have that, for every $(x_1, \ldots, x_n) \in \mu^{-1}(0)$, at least three of the x_i are distinct. When n is even both these statements fail; the set $\mu^{-1}(0)$ has singularities at points $(x_1, \ldots, x_n)$ where exactly half the x_i equal a and the other half equal $-a$; and the action of G is not free at such points. On the other hand, if n is odd then the Marsden-Weinstein quotient

$$X/\!\!/\mathrm{G} := \mu^{-1}(0)/\mathrm{G}$$

is a smooth compact symplectic manifold (also called the **reduced space**).

Kirwan showed in [**293**] that in a good case like this the symplectic quotient $X/\!\!/\mathrm{G}$ can be identified with the so-called **Mumford quotient** of X by the action of the complexified group G^c that is defined in the framework of geometric invariant theory. As shown in Exercise D.7.1 below, the usual quotient X/G^c can be badly behaved. Mumford discovered that there is a subset X^{ss} of X consisting of so-called semistable elements whose quotient X^{ss}/G^c is, in good cases, a compact smooth manifold. In general, the construction of the Mumford quotient in complex geometry is quite subtle and we shall not discuss it here. In the symplectic setting a point is called **semistable** if the closure of its complexified group orbit intersects $\mu^{-1}(0)$, it is called **polystable** if its complexified group orbit intersects $\mu^{-1}(0)$, and it is called **stable** if it is polystable and its isotropy subgroup is finite. The semistable and the stable points form open subsets of X, and the complex quotient X^{ps}/G^c is naturally isomorphic to the symplectic quotient $X/\!\!/\mathrm{G}$. The good case is when zero is a regular value of the moment map and G acts freely on $\mu^{-1}(0)$. In this case the symplectic quotient is a smooth manifold (compact when the moment map is proper), every semistable point is stable, and Kirwan's theorem provides a natural identification between the symplectic and complex quotients:

$$X/\!\!/\mathrm{G} \cong X^s/\mathrm{G}^c.$$

In the case at hand the complexified group $\mathrm{G}^c = \mathrm{PSL}(2,\mathbb{C})$ acts diagonally by Möbius transformations on $X = (S^2)^n$. A tuple $z = (z_1, \dots, z_n) \in X$ is semistable if and only if no more than $n/2$ of the z_i are equal, i.e.

$$X^{ss} = \left\{ z \in (S^2)^n \,|\, \text{if } z_{i_1} = \cdots = z_{i_m} \text{ and } i_1 < \cdots < i_m \text{ then } 2m \le n \right\}.$$

It is stable if and only if less than $n/2$ of the z_i are equal, i.e.

$$X^{s} = \left\{ z \in (S^2)^n \,|\, \text{if } z_{i_1} = \cdots = z_{i_m} \text{ and } i_1 < \cdots < i_m \text{ then } 2m < n \right\}.$$

It is polystable if and only if it is either stable or there are precisely two points, each with weight $n/2$. If n is even the quotient X^{ss}/G^c is not even a Hausdorff space, while X^s/G^c is a smooth manifold but not compact and not diffeomorphic to $X/\!\!/\mathrm{G}$. On the other hand, if $n \ge 3$ is odd, then $X^{ss} = X^{ps} = X^s$ and the quotient X^s/G^c is a compact smooth manifold of dimension $2n - 6$ and is diffeomorphic to $X/\!\!/\mathrm{G}$.

Now observe that the open stratum $\mathcal{M}_{0,n} = ((S^2)^n \setminus \Delta_n)/\mathrm{G}^c$ in the moduli space $\overline{\mathcal{M}}_{0,n}$ can be embedded into X^s/G^c. When n is odd this inclusion extends to a natural projection

$$\overline{\mathcal{M}}_{0,n} \to X^s/\mathrm{G}^c.$$

Here is an explicit description of this map as explained to us by Mukai. Let

$$\mathbf{z} = (\{z_{\alpha\beta}\}_{\alpha E \beta}, \{\alpha_i, z_i\}_{1 \le i \le n})$$

be a stable curve modelled on a tree T. Call α a **Mumford component** of T if $(z_{\alpha 1}, \dots, z_{\alpha n}) \in X^s$. If n is odd then every stable curve $\mathbf{z}$ has a unique Mumford component $\alpha_{\mathbf{z}}$. (Prove this!) The resulting map $[\mathbf{z}] \mapsto [z_{\alpha_{\mathbf{z}} 1}, \dots, z_{\alpha_{\mathbf{z}} n}]$ is smooth. This map is a diffeomorphism for $n = 3$ and $n = 5$ but not for $n > 5$. Thus the Grothendieck–Knudsen compactification $\overline{\mathcal{M}}_{0,n}$ of $\mathcal{M}_{0,n}$ is a refinement of the Mumford quotient $(S^2)^n /\!\!/ \mathrm{SO}(3)$ when n is odd.

EXERCISE D.7.1. Let $X = (\mathbb{C}P^1)^n$, X^{ss}, and $\mathrm{G}^c = \mathrm{PSL}(2,\mathbb{C})$ be as above. Give a direct proof of the fact that X^{ss}/G^c is compact for every n, even though X^{ss} is noncompact. Show that the quotient topology on X/G^c is nonHausdorff. *Hint:* Consider the orbit of a point $z = (z_1, \ldots, z_n)$, where $z_1 = \cdots = z_n$. Show that the only neighbourhood of the equivalence class of z in X/G^c is the whole space.

EXERCISE D.7.2. Show that the map $\mathcal{M}_{0,6} \to X^{ss}/\mathrm{G}^c$ does not extend continuously to $\overline{\mathcal{M}}_{0,6}$. Show that X^{ss}/G^c is nonHausdorff when $n > 4$ is even. Show that the map $\overline{\mathcal{M}}_{0,n} \to X^{ss}/\mathrm{G}^c$ is well defined when n is odd. Show that it is a diffeomorphism for $n = 5$ but not for $n = 7$.

Comments on the literature. The spaces $\overline{\mathcal{M}}_{0,n}$ were first discovered by Grothendieck and Knudsen. The corresponding moduli spaces $\overline{\mathcal{M}}_{g,n}$ of curves of higher genus were constructed by Deligne and Mumford. In [**211**] Knudsen proved that $\overline{\mathcal{M}}_{0,n}$ is a smooth projective variety. In [**206**] Kapranov gives a description of the space $\overline{\mathcal{M}}_{0,n}$ as a *Chow quotient* of the complex Grassmannian $\mathrm{G}(2,n)$ by the maximal torus $H \subset \mathrm{GL}(n,\mathbb{C})$. He also represents this moduli space as a blow up. In [**208**] Keel computes the cohomology ring of $\overline{\mathcal{M}}_{0,n}$ and proves that it is isomorphic to the Chow ring.

Constructions similar to those in this appendix can be found in the work of Naraki on cubic surfaces in $\mathbb{C}P^3$ (cf. [**297**]). Every such cubic surface $S \subset \mathbb{C}P^3$ contains 27 lines, and for every line $L \subset S$ there are precisely five planes $H \subset \mathbb{C}P^3$ which contain L as well as two further lines in S. Now any four of these five planes give rise to a cross ratio. Namely choose any line L' which is transverse to all four planes. Then the cross ratio of the four intersection points with L' is independent of L'. This gives rise to 135 cross ratios of which 45 are independent. Naraki proves that this construction determines an embedding of the moduli space of cubic surfaces in $\mathbb{C}P^3$ together with an ordering of the 27 lines into the product of 45 copies of $\mathbb{C}P^1$.

APPENDIX E

Singularities and Intersections (written with Laurent Lazzarini)

The purpose of this appendix is to explain two related phenomena for J-holomorphic curves. The first concerns the precise analysis of the local structure near the singular points. (We assume throughout that our curves have no boundary.) The main result in this part asserts that a singularity of a pseudo-holomorphic curve is conjugate to a singularity of a holomorphic curve by a C^1-diffeomorphism [**287**]. It follows that every simple J-holomorphic curve has only finitely many self-intersections and that two distinct J-holomorphic curves can only intersect in finitely many points. The second phenomenon is specific to dimension 4. In this case every intersection of two distinct J-holomorphic curves contributes positively to the intersection number and every singularity contributes positively to the self-intersection number. This leads to the adjunction inequality

$$2g - 2 \leq C \cdot C - \langle c_1(TM), C \rangle$$

for every simple J-holomorphic curve $C \subset M^4$ with equality if and only if C is embedded. Another consequence of the local analysis is that every homology class $A \in H_2(M;\mathbb{Z})$ that can be represented by a J-holomorphic curve can also be represented by an immersed J'-holomorphic curve for some almost complex structure J' which is C^1-close to J. This holds in every dimension.

All these results are well known and the goal of this appendix is to give a comprehensive exposition. Finiteness and positivity of intersections were already noted by Gromov in [**160**]. A first proof appeared in McDuff [**256**] (with corrections in [**264**]); this paper also was the first to state and prove the adjunction inequality. Micallef and White [**287**] took another approach to these questions. Their work is more general as they deal with arbitrary minimal surfaces and then apply those results to the special case of J-holomorphic curves. A third exposition of this circle of ideas was given by Sikorav [**379**]. Using the Carleman similarity principle he was able to obtain sharp regularity results for the diffeomorphism that conjugates a pseudoholomorphic singularity to a holomorphic one. (It can be chosen of class $C^{1+\mu}$ with $\mu < 1$ but not necessarily of class C^2.) His paper appeared while work was being completed on the proofs given below.

This appendix is organized as follows. We first state the Micallef–White theorem and the main results on finiteness and positivity of intersections. In Section E.2 we show how these results follow from the Micallef–White theorem. The proof of the Micallef–White theorem is based, in turn, upon the Hartman–Wintner theorem that describes the lowest order term of an approximately harmonic function. To prove this in the proper generality requires the use of classical results on the integrability of almost complex structures in dimension two (the Newlander–Nirenberg theorem) and the existence of isothermal coordinates. After establishing these in

Section E.3, we prove the Hartman–Wintner theorem in Section E.4. All this is fairly standard. The originality of the present approach lies in the way we apply the Hartman–Wintner theorem to understand the lowest order term of the difference $u - v$ between two intersecting J-holomorphic curves. This step occupies Sections E.5 and E.6. The final section E.7 explains the Micallef–White argument for conjugating a J-holomorphic curve locally to a complex polynomial.

E.1. The main results

In this section we state the main results proved in this appendix. Throughout (M, J) is a smooth (not necessarily compact) almost complex manifold of dimension $2n$. A **J-holomorphic curve** in M is a smooth map $u : \Sigma \to M$, defined on a (not necessarily compact) complex 2-manifold (Σ, j), such that

$$du \circ j = J \circ du.$$

It is called **simple** if there are no disjoint nonempty open subsets $U_0, U_1 \subset \Sigma$ such that $u(U_0) = u(U_1)$. If Σ is closed this means that the restriction of u to each component of Σ is not multiply covered, and different components of Σ have different images under u. Given a J-holomorphic curve $u : \Sigma \to M$, a point $z \in \Sigma$ is called a **critical point** of u if $du(z) = 0$ and is called an **injective point** of u if it is not critical and $u^{-1}(u(z)) = \{z\}$.

Recall from Lemma 2.4.1 that every J-holomorphic curve $u : \Sigma \to M$ defined on a closed Riemann surface Σ has a finite number of critical points and is finite to one, i.e. $u^{-1}(x)$ is a finite set for every $x \in M$. Moreover, Proposition 2.5.1 asserts that the set of noninjective points of a simple J-holomorphic curve u is at most countable and can only accumulate at the finite set of critical points of u. This is all that is needed to develop the transversality theory of Chapters 3 and 6. However, many interesting applications of J-holomorphic curves in dimension four are based on positivity of intersections and the adjunction formula (see Chapter 9). The proofs of these results require the refined local analysis explained in this appendix.

Let J be an almost complex structure on $\mathbb{C}^n$ such that $J(0) = J_0$, where $J_0 := i$ denotes the standard complex structure on $\mathbb{C}^n$, denote the closed unit disc in $\mathbb{C}$ by

$$\mathbb{D} := \{z \in \mathbb{C} \,|\, |z| \leq 1\},$$

and let $u : (\mathbb{D}, 0) \to (\mathbb{C}^n, 0)$ be a nonconstant J-holomorphic curve. Examining the Taylor expansion of u at the origin (see Lemma 2.4.1) one observes that u has the form

$$u(z) = az^k + O(|z|^{k+1}) \tag{E.1.1}$$

for some integer $k \geq 1$ and some nonzero vector $a \in \mathbb{C}^n$. This implies that u has a tangent space at $z = 0$ namely the line $\mathbb{C}a$ (see Remark E.5.11). The results of this appendix are based on a much more precise local normal form for u that is due to Micallef–White [**287**]. Their theorem asserts that, by a C^2-coordinate change in the source and a C^1-coordinate change in the target, u is conjugate to a holomorphic map of the form

$$v(z) = z^k(a + p(z))$$

where $a \in \mathbb{C}^n$ is a nonzero vector and $p : \mathbb{C} \to \mathbb{C}^n$ is a polynomial that vanishes at $z = 0$ and takes values in the hyperplane orthogonal to a. Moreover, this normal form can be achieved for a finite collection of curves by a simultaneous coordinate change in the target. We shall see that positivity of intersections and the adjunction formula can be derived easily from this normal form.

THEOREM E.1.1 (Micallef–White). *Let $u : \Sigma \to M$ be a J-holomorphic curve, $\Omega \subset \Sigma$ be an open set such that $\overline{\Omega} \subset \Sigma$ is compact, $x_0 \in u(\Omega) \setminus u(\partial\Omega)$, and*

$$\{z_1, \ldots, z_N\} = u^{-1}(x_0) \cap \Omega.$$

Then there are C^2-coordinate charts $\varphi_i : (U_i, z_i) \to (\mathbb{C}, 0)$ defined on disjoint open neighbourhoods $U_1, \ldots, U_N \subset \Omega$ of the points $z_i \in U_i$, unitary matrices $L_i \in \mathbb{C}^{n\times n}$, and a C^1-coordinate chart $\Psi : (W, x_0) \to (\mathbb{C}^n, 0)$ defined on an open neighbourhood $W \subset M$ of x_0, such that the following holds.

(i) $\Psi_* J(x_0) = J_0$, *that is* $d\Psi(x_0)J(x_0) = J_0 d\Psi(x_0)$.

(ii) $u_i := \Psi \circ u \circ \varphi_i^{-1} : \varphi_i(U_i) \to \mathbb{C}^n$ *is a polynomial in the variable z.*

(iii) $L_i u_i(z) = (z^{k_i}, z^{k_i} p_i(z))$, *where k_i is a positive integer and $p_i : \mathbb{C} \to \mathbb{C}^{n-1}$ is a polynomial in z such that $p_i(0) = 0$. Moreover, the matrix L_i depends only on the tangent space of u_i at the origin, i.e.*

$$L_i^{-1}(\mathbb{C} \times \{0\}) = L_j^{-1}(\mathbb{C} \times \{0\}) \quad \Longrightarrow \quad L_i = L_j.$$

The proof has three main steps. Since the result is local we may identify (W, x_0) with $(\mathbb{C}^n, 0)$, where $\mathbb{C}^n$ is equipped with an almost complex structure that is standard at the origin. Then the functions $u \circ \varphi_i^{-1} : (\mathbb{D}, 0) \to (\mathbb{C}^n, 0)$ can be expanded in powers of z and $\overline{z}$. To control these expansions we consider in Section E.5 a decreasing sequence $\mathcal{O}_\ell \supset \mathcal{O}_{\ell+1}$ of function spaces, where $\mathcal{O}_\ell$ denotes the space of smooth functions $f : \mathbb{D} \setminus \{0\} \to \mathbb{C}^n$ on the punctured disc that are bounded by a constant times $|z|^\ell$ near the origin (the precise definition also includes growth conditions on all the derivatives of f). The first step uses a refinement of (E.1.1) to find C^2-charts φ_i and complex matrices L_i so that the composites $u_i := L_i \circ u \circ \varphi_i^{-1}$ have the form $z \mapsto (z^{k_i}, F_i(z))$ where $F_i \in \mathcal{O}_{k_i+1}$. This is relatively straightforward (see Proposition E.5.10).

The second step, which is much deeper, is to show that for any $i \neq j$ the lowest order term in the difference $F_i - F_j$ is a multiple of a power z^ℓ of z for some $\ell > \min(k_i, k_j)$ (see Proposition E.6.1). The argument also applies when $F_j = 0$, and hence implies that the lowest order term in F_i is holomorphic. This step uses the Hartman–Wintner theorem of Section E.4. This is a refinement of Aronszajn's theorem in [**18**] for the 2-dimensional case. It shows that the lowest order term in $F_i - F_j$ is a homogeneous polynomial in the variables s, t (where $z = s + it$); that this polynomial is holomorphic follows from a local analysis of the Cauchy–Riemann equation (see Lemma E.5.9).

The final step is a beautiful argument due to Micallef–White which uses the information gained in step 2 to construct a suitable C^1-coordinate chart on the target space M. The idea is to "bring the range into focus" so that functions such as $F_i - F_j$ that were approximately equal to powers of z become precisely equal to these powers. Note that although the first two steps are straightforward in the integrable case the last step is not. Namely, if $u(z) = (z^k, z^{k+1}F(z))$, where F is an arbitrary holomorphic function, it is not obvious how to find a C^1-diffeomorphism $\Psi : (\mathbb{C}^n, 0) \to (\mathbb{C}^n, 0)$ such that $\Psi \circ u$ is polynomial. This step is carried out in Section E.7 and can be read independently of the rest of the appendix.

A first application of Theorem E.1.1 is the following result about the finiteness of the number of noninjective points for simple J-holomorphic curves.

THEOREM E.1.2. *Let (M, J) be an almost complex manifold, (Σ, j) be a closed (not necessarily connected) Riemann surface, and $u : \Sigma \to M$ be a simple J-holomorphic curve. Then the set $Z \subset \Sigma$ of noninjective points of u is finite.*

PROOF. We shall prove the following two assertions.

(i) There is an open neighbourhood $U \subset \Sigma \times \Sigma$ of the diagonal such that

$$u(z_0) = u(z_1), \quad (z_0, z_1) \in U \qquad \Longrightarrow \qquad z_0 = z_1.$$

(ii) The set of self-intersections

$$\mathcal{S}(u) := \{(z_0, z_1) \in \Sigma \times \Sigma \,|\, u(z_0) = u(z_1),\, z_0 \neq z_1\}$$

is discrete.

The theorem follows immediately: by (i), $\mathcal{S}(u)$ is a (closed) subset of the compact space $\Sigma \times \Sigma \backslash U$ and hence, by (ii), it is a finite set. Now the set Z of noninjective points is the union of the projection of $\mathcal{S}(u)$ onto the first factor with the set of critical points of u, which is finite by Lemma 2.4.1. Hence Z is a finite set.

To prove (i) it suffices to show that every point $z_0 \in \Sigma$ has a neighbourhood U_0 such that the restriction of u to U_0 is injective. Once this is established, cover the diagonal $\Delta \subset \Sigma \times \Sigma$ by finitely many such product neighbourhoods $U_j \times U_j$, $j = 1, \ldots, N$. Then the union $U := \bigcup_j (U_j \times U_j)$ satisfies (i).

By the implicit function theorem, the local injectivity of u near z_0 is obvious whenever $du(z_0) \neq 0$. Hence assume $du(z_0) = 0$. Then, by Theorem E.1.1, we may assume without loss of generality that $z_0 = 0$, $M = \mathbb{C}^n$, and $u : \mathbb{D} \to \mathbb{C}^n$ has the form

$$u(z) = (z^k, z^k p(z)),$$

where k is a positive integer and $p : \mathbb{C} \to \mathbb{C}^{n-1}$ is a polynomial that vanishes at the origin. Assume, by contradiction, that u is not locally injective near zero. Then there exist nonzero sequences $z_\nu \to 0$ and $w_\nu \to 0$ such that $u(z_\nu) = u(w_\nu)$ and $z_\nu \neq w_\nu$ for every ν, and hence

$$z_\nu^k = w_\nu^k, \qquad p(z_\nu) = p(w_\nu).$$

Passing to a subsequence, if necessary, we may assume that there exists a kth root of unity $\zeta \neq 1$ such that $w_\nu = \zeta z_\nu$ for all ν. It then follows that the polynomial $z \mapsto p(z) - p(\zeta z)$ vanishes at the infinite number of points z_ν and hence vanishes identically. This implies that $u(\zeta z) = u(z)$ for every z, so that u is not simple, a contradiction. This proves (i).

We prove (ii). Let $(z_0, z_1) \in \mathcal{S}(u)$ and denote $x := u(z_0) = u(z_1)$. By Theorem E.1.1 we may choose C^1-coordinate charts $\varphi_i : (U_i, z_i) \to (\mathbb{C}, 0)$ and $\Psi : (W, x) \to (\mathbb{C}^n, 0)$ so that the maps $u_i := \Psi \circ u \circ \phi_i^{-1}$ have the form

$$u_0(z) = z^k(a + p(z)), \qquad u_1(z) = z^\ell(b + q(z)), \qquad p(0) = q(0) = 0.$$

where $a, b \in \mathbb{C}^n$ are nonzero vectors, k, ℓ are positive integers and $p, q : \mathbb{C} \to \mathbb{C}^n$ are polynomials. There are two cases to consider. Assume first that a and b are linearly independent. Then the curves are not tangent and so the intersection is isolated. More precisely, there is a constant $\delta > 0$ such that

$$|\lambda a + \mu b| \geq \delta\,(|\lambda| + |\mu|)$$

for all $\lambda, \mu \in \mathbb{C}$. Moreover, there exists a constant $c > 0$ such that $|p(z)| \le c|z|$ and $|q(z)| \le c|z|$ for all $z \in B$. Hence, for all $z, w \in \mathbb{D}$, we have

$$\begin{aligned} |u_0(z) - u_1(w)| &\ge \left|z^k a - w^\ell b\right| - \left|z^k p(z) - w^\ell q(w)\right| \\ &\ge |z|^k (\delta - c|z|) + |w|^\ell (\delta - c|w|). \end{aligned}$$

The right hand side is positive whenever $|z| < \delta/c$ and $|w| < \delta/c$. This shows that (z_0, z_1) is an isolated self-intersection whenever a and b are linearly independent.

Now assume that a and b are linearly dependent. Then, by Theorem E.1.1 (iii), we may assume that $a = b = (1, 0) \in \mathbb{C} \times \mathbb{C}^{n-1}$ and that $p(z)$ and $q(z)$ take values in $\{0\} \times \mathbb{C}^{n-1}$. Assume, by contradiction, that (z_0, z_1) is not an isolated self-intersection of u. Then there exist nonzero sequences $z_\nu \to 0$ and $w_\nu \to 0$ such that $u_0(z_\nu^\ell) = u_1(w_\nu^k)$ for every ν and hence $z_\nu^m = w_\nu^m, p(z_\nu^\ell) = q(w_\nu^k)$, and $m := k\ell$. Passing to a subsequence, if necessary, we may assume that there is an mth root of unity ζ such that $w_\nu = \zeta z_\nu$ for every ν. Hence the polynomial $z \mapsto p(z^\ell) - q(\zeta^k z^k)$ has infinitely many zeros at $z = z_\nu$ and so vanishes identically. It follows that $u_0(z^\ell) = u_1(\zeta^k z^k)$ for every $z \in \mathbb{D}$ which contradicts our assumption that u is simple. This proves (ii) and the theorem. □

REMARK E.1.3. The proof of Theorem E.1.2 shows that every critical point $z_0 \in \Sigma$ of a simple J-holomorphic curve $u : \Sigma \to M$ has an open neighbourhood U such that the restriction of u to U is injective and its restriction to $U \setminus \{z_0\}$ is an embedding. The hypothesis that Σ be closed is not required.

The most important consequences of the Micallef–White theorem occur in dimension 4. We prove in Section E.2 that in this dimension it implies positivity of intersections for J-holomorphic curves. Although the relevant results are stated in Section 2.6, they are restated here for the convenience of the reader.

Let Σ, Σ_0, Σ_1 be closed Riemann surfaces, not necessarily connected, and $u : \Sigma \to M$, $u_0 : \Sigma_0 \to M$, $u_1 : \Sigma_1 \to M$ be J-holomorphic curves. Denote the number of intersections of u_0 and u_1 by

$$\delta(u_0, u_1) := \#\{(z_0, z_1) \in \Sigma_0 \times \Sigma_1 \,|\, u(z_0) = u(z_1)\},$$

and the number of self-intersections of u by

$$\delta(u) := \frac{1}{2}\#\{(z_0, z_1) \in \Sigma \times \Sigma \,|\, u(z_0) = u(z_1),\ z_0 \ne z_1\}.$$

By Theorem E.1.2 the number $\delta(u)$ is finite whenever u is simple, and the number $\delta(u_0, u_1)$ is finite whenever the disjoint union of u_0 and u_1 is simple.

EXERCISE E.1.4. Suppose that Σ_0 and Σ_1 are two closed connected Riemann surfaces and let $u_0 : \Sigma_0 \to M$ and $u_1 : \Sigma_1 \to M$ be two J-holomorphic curves such that $u_0(\Sigma_0) \ne u_1(\Sigma_1)$. Prove that $\delta(u_0, u_1) < \infty$.

THEOREM E.1.5 (Positivity of intersections). *Let (M, J) be an almost complex 4-manifold and $A_0, A_1 \in H_2(M; \mathbb{Z})$ be homology classes that are represented by simple J-holomorphic curves $u_0 : \Sigma_0 \to M$ and $u_1 : \Sigma_1 \to M$, respectively. Suppose that $u_0(U_0) \ne u_1(U_1)$ for any two nonempty open subsets $U_0 \subset \Sigma_0$ and $U_1 \subset \Sigma_1$. Then*

$$\delta(u_0, u_1) \le A_0 \cdot A_1,$$

with equality if and only if all intersections are transverse (i.e. if $u_0(z_0) = u_1(z_1) =: x$ then $\operatorname{im} du_0(z_0) + \operatorname{im} du_1(z_1) = T_xM$).

THEOREM E.1.6 (Adjunction Formula). *Let (M, J) be an almost complex 4-manifold, (Σ, j) be a closed Riemann surface, not necessarily connected, and $u : \Sigma \to M$ be a simple J-holomorphic curve. Denote by $A \in H_2(M; \mathbb{Z})$ the homology class represented by u. Then*

$$2\delta(u) \leq A \cdot A - c_1(A) + \chi(\Sigma), \tag{E.1.2}$$

with equality if and only if u is an immersion and all self-intersections are transverse (i.e. if $u(z_0) = u(z_1) =: x$ and $z_0 \neq z_1$ then $\operatorname{im} du(z_0) + \operatorname{im} du(z_1) = T_xM$).

COROLLARY E.1.7 (McDuff). *Let M, Σ, u and A be as in Theorem E.1.6. Then*

$$A \cdot A - c_1(A) + \chi(\Sigma) \geq 0, \tag{E.1.3}$$

with equality if and only if u is an embedding.

PROOF. The inequality (E.1.3) follows from (E.1.2). Equality in (E.1.3) is equivalent to equality in (E.1.2) with $\delta(u) = 0$. □

REMARK E.1.8. For a further discussion of the defect $A \cdot A - c_1(A) + \chi(\Sigma) - 2\delta(u)$ in (E.1.2) see Milnor [**290**].

E.2. Positivity of intersections

We now deduce Theorems E.1.5 and E.1.6 from the Micallef–White theorem. Our starting point is the definition of the local intersection number.

DEFINITION E.2.1. *Let M be an oriented 4-manifold, Σ_0 and Σ_1 be two oriented 2-manifolds, and $u_0 : \Sigma_0 \to M$, $u_1 : \Sigma_1 \to M$ be smooth maps. A pair $(z_0, z_1) \in \Sigma_0 \times \Sigma_1$ is called an* **isolated intersection** *of u_0 and u_1 if $u_0(z_0) = u_1(z_1)$ and there exist open neighbourhoods $U_i \subset \Sigma_i$ of z_i with compact closures such that*

$$u_0(w_0) = u_1(w_1) \qquad \Longrightarrow \qquad w_0 = z_0,\ w_1 = z_1$$

for $w_i \in \bar{U}_i$. The **local intersection number** *of u_0 and u_1 at an isolated intersection (z_0, z_1) is defined by*

$$\iota(u_0, u_1; z_0, z_1) := v_0 \cdot v_1,$$

where $W \subset M$ is a compact contractible neighbourhood of $u_0(z_0) = u_1(z_1)$ such that

$$W \cap u_0(\partial U_0) = W \cap u_1(\partial U_1) = \emptyset,$$

and $v_i : U_i \to M$ for $i = 0, 1$ are transversally intersecting smooth maps such that v_i agrees with u_i on $U_i \setminus u_i^{-1}(W)$ and satisfies $v_i(U_i \cap u_i^{-1}(W)) \subset W$.

Because W is contractible any two choices of v_0 are homotopic by a homotopy with support in $U_0 \cap u_0^{-1}(W)$ and with values in W. Hence the local intersection number is independent of the choice of the perturbations v_0 and v_1 used to define it. It follows that it is also independent of the choice of the neighbourhoods U_0, U_1 and W. Note that Definition E.2.1 extends verbatim to self-intersections $(z_0, z_1) \in \Sigma$ of a smooth map $u : \Sigma \to M$ with $u(z_0) = u(z_1)$ and $z_0 \neq z_1$. (Replace Σ_0 and Σ_1 by disjoint neighbourhoods of z_0 and z_1, respectively.)

PROPOSITION E.2.2. *Let (M, J) be an almost complex 4-manifold, (Σ, j) be a Riemann surface, not necessarily closed or connected, and $u : \Sigma \to M$ be a simple J-holomorphic curve. Then every self-intersection $(z_0, z_1) \in \Sigma \times \Sigma \setminus \Delta$ of u is isolated and satisfies $\iota(u, u; z_0, z_1) \geq 1$, with equality if and only if the intersection is transverse.*

To prove this proposition we construct suitable perturbations of u. A polynomial $u : \mathbb{C} \to \mathbb{C} \times \mathbb{C}^{n-1}$ of the form

$$u(z) = z^k(1, p(z))$$

with $p(0) = 0$ is called **simple** if its restriction to a neighbourhood of zero is injective. Equivalently, $p(z) \not\equiv p(\zeta z)$ for every kth root of unity ζ (see the proof of Theorem E.1.2). Another equivalent statement is that u cannot be expressed in the form $u(z) = v(z^m)$ where v is a polynomial and $m > 1$ is an integer that divides k.

LEMMA E.2.3. *Let $u : \mathbb{D} \to \mathbb{C} \times \mathbb{C}^{n-1}$ be a simple polynomial of the form $u(z) = z^k(1, p(z))$ where k is a positive integer and $p(0) = 0$. Let $U \subset \mathbb{C}$ be an open neighbourhood of zero and $V \subset \mathbb{C}$, $W \subset \mathbb{C}^n$ be closed neighbourhoods of zero such that*

$$V \subset U, \qquad u(V) \subset \operatorname{int}(W), \qquad W \cap u(\partial U) = \emptyset.$$

Then, for every $\varepsilon > 0$, there is an immersion $v : U \to \mathbb{C}^n$ with the following properties.

(i) *v agrees with u on $U \setminus u^{-1}(W)$, and satisfies $v(U \cap u^{-1}(W)) \subset W$;*
(ii) *v is holomorphic in V and $\|u - v\|_{C^1} < \varepsilon$.*
(iii) *$\#\{v^{-1}(0)\} = k$.*

PROOF. Define

$$u_\varepsilon(z) := f_\varepsilon(z)(1, p(z)), \qquad f_\varepsilon(z) := \prod_{j=0}^{k-1}(z + j\varepsilon). \tag{E.2.1}$$

Let $U_1 \subset \mathbb{C}$ be an open neighbourhood of zero such that

$$z \in U_1 \setminus \{0\},\ \lambda^k = 1,\ \lambda \neq 1 \qquad \Longrightarrow \qquad p'(z) \neq 0,\ p(z) \neq p(\lambda z),$$

Since $f'_\varepsilon(0) \neq 0$, f_ε and f'_ε do not have any common zeros, and

$$u'_\varepsilon = (f'_\varepsilon, f'_\varepsilon p + f_\varepsilon p'),$$

it follows that the restriction of u_ε to U_1 is an immersion. Moreover, for every $\varepsilon \neq 0$ we have $u_\varepsilon^{-1}(0) = \{j\varepsilon \,|\, j = 0, \dots, k-1\}$.

We now form v by patching u_ε together with u. Let $V' \subset U$ be an open set such that $V \subset V'$ and $\bar{V}' \subset \operatorname{int}(u^{-1}(W))$. Choose a smooth cutoff function $\beta : U \to [0, 1]$ such that

$$\beta(z) = \begin{cases} 1, & \text{for } z \in V, \\ 0, & \text{for } z \in U \setminus V'. \end{cases}$$

Define

$$v_\varepsilon(z) := u(z) + \beta(z)(u_\varepsilon(z) - u(z)). \tag{E.2.2}$$

Then, for $\varepsilon > 0$ sufficiently small, the map $v := v_\varepsilon : U \to \mathbb{C}^2$ is an immersion that agrees with u on $U_1 \setminus u^{-1}(W)$ and satisfies $v_\varepsilon(U_1 \cap u^{-1}(W)) \subset W$. Moreover it satisfies conditions (ii) and (iii) by construction: v is J-holomorphic except possibly in the support of β, and at points in this support v is arbitrarily C^1-close to u. □

PROOF OF PROPOSITION E.2.2. For $i = 0, 1$ denote by $u_i : \Sigma_i \to M$ the restriction of u to a neighbourhood of z_i. By Theorem E.1.2, (z_0, z_1) is an isolated intersection of u_0 and u_1. By Theorem E.1.1 we may assume without loss of generality that $M = \mathbb{C}^2$, $\Sigma_0 = \Sigma_1 = \mathbb{C}$, $z_0 = z_1 = 0$, and that there are matrices L_i such that $L_i u_i$ is a polynomial of the form

$$L_i u_i(z) = z^{k_i}(1, p_i(z))$$

where k_i is a positive integer and $p_i(0) = 0$.

STEP 1. *The assertion holds when* $k_0 = k_1 = 1$.

If the two curves intersect transversally then the result is obvious. If not, then by Theorem E.1.1 we may assume that $L_0 = L_1 = \mathbb{1}$. Replacing u_i by $\Psi \circ u_i$ where $\Psi(z, w) = (z, w - zp_0(z))$ we may assume in addition that $u_0(z) = (z, 0)$. Thus

$$u_0(z) = (z, 0), \qquad u_1(z) = (z, z^k q(z))$$

where $k \geq 2$ and $q : \mathbb{C} \to \mathbb{C}$ is a polynomial with $q(0) \neq 0$. The second component $z^k q$ of u_1 is a polynomial with a k-fold zero at $z = 0$. Hence any perturbation with k distinct zeroes in the disc $|z| \leq \varepsilon$ will give a map v_1 that intersects u_0 k times. Here are the details. Choose an open neighbourhood $U_1 \subset \mathbb{C}$ of zero such that $q(z) \neq 0$ for $z \in \bar{U}_1$, let $\beta : \mathbb{C} \to [0, 1]$ be a smooth cutoff function with support in U_1 such that $\beta(z) = 1$ near zero, let f_ε be as in (E.2.1), and define

$$v_1(z) := (z, (z^k + \beta(z)(f_\varepsilon(z) - z^k))q(z)).$$

Then, for $\varepsilon > 0$ sufficiently small, v_1 is an immersion and $\iota(u_0, u_1; 0, 0) = u_0 \cdot v_1 = k \geq 2$ as required.

STEP 2. *The assertion holds when* $k_0 k_1 > 1$.

By assumption, there is an open disc $U \subset \mathbb{C}$ centered at zero such that $u_0(w_0) \neq u_1(w_1)$ for $w_0, w_1 \in U \setminus \{0\}$. Choose a closed ball $W \subset \mathbb{C}^2$ centered at zero such that $W \cap u_0(\partial U) = \emptyset$ and $W \cap u_1(\partial U) = \emptyset$. Now let $V_0, V_1 \subset U$ be closed discs such that $u_0(V_0) \subset \operatorname{int}(W)$ and $u_1(V_1) \subset \operatorname{int}(W)$. Choose $\varepsilon > 0$ such that

$$\inf_{w_0 \in U, w_1 \in U \setminus V_1} |u_0(w_0) - u_1(w_1)| > 3\varepsilon, \qquad \inf_{w_0 \in U \setminus V_0, w_1 \in U} |u_0(w_0) - u_1(w_1)| > 3\varepsilon.$$

Let $v_0, v_1 : U \to \mathbb{C}^2$ be the immersions constructed in Lemma E.2.3. Then $v_0(w_0) \neq v_1(w_1)$ unless $w_0 \in V_0$ and $w_1 \in V_1$. Hence v_0 and v_1 intersect only in the set where they are both holomorphic. Hence by Step 1 each intersection point contributes at least 1 to the intersection number $v_0 \cdot v_1$. Moreover, by construction v_i has precisely k_i branches through the origin. Hence

$$\iota(u_0, u_1; 0, 0) = v_0 \cdot v_1 \geq k_0 k_1 > 1,$$

as required. This proves the proposition. □

This completes the discussion of local intersection numbers. To prove the adjunction formula we need a global perturbation result.

PROPOSITION E.2.4. *Let (M, J) be an almost complex 4-manifold, (Σ, j) be a closed Riemann surface, and $u : \Sigma \to M$ be a simple J-holomorphic curve. Let $\langle \cdot, \cdot \rangle$ be a Riemannian metric on M with respect to which J is skew-adjoint and denote*

$$\omega := \langle J\cdot, \cdot \rangle \in \Omega^2(M).$$

(This form is not required to be closed.) Then, for every $\varepsilon > 0$ and every neighbourhood $U \subset \Sigma$ of the set of noninjective points, there exists an immersion $v : \Sigma \to M$ satisfying the following conditions.

(i) *$v^*\omega$ is a positive area form on Σ.*

(ii) *$\|u - v\|_{C^1} < \varepsilon$.*

(iii) *$v(z) = u(z)$ for $z \in \Sigma \setminus U$.*

(iv) *v has only transverse self-intersections.*

(v) *v satisfies the inequality*

$$\sum_{\substack{z_0 \neq z_1 \\ v(z_0)=v(z_1)}} \iota(v, v; z_0, z_1) \geq \sum_{\substack{z_0 \neq z_1 \\ u(z_0)=u(z_1)}} \iota(u, u; z_0, z_1), \tag{E.2.3}$$

with equality if and only if u is an immersion.

PROOF. Let $z_1, \ldots, z_N$ be the singular points of u. By Theorem E.1.1, there exist open neighbourhoods $W_i \subset M$ of $u(z_i)$ and $U_i \subset \Sigma$ of z_i, C^1-coordinate charts $\Psi_i : W_i \to \mathbb{C}^2$, C^2-coordinate charts $\varphi_i : U_i \to \mathbb{C}$, integers $k_i > 1$, and polynomials $p_i : \mathbb{C} \to \mathbb{C}$ such that $\varphi_i(z_i) = 0$, $\Psi_i(u_i(z_i)) = 0$,

$$U_i \subset U, \qquad u_i(U_i) \subset W_i, \qquad d\Psi_i(u(z_i))J(u(z_i)) = J_0 d\Psi_i(u(z_i)), \qquad p_i(0) = 0,$$

and

$$u_i(z) := \Psi_i \circ u \circ \varphi_i^{-1}(z) = (z^{k_i}, z^{k_i} p_i(z)), \qquad z \in \varphi_i(U_i).$$

Shrinking U_i and W_i, if necessary, we may assume that the C^1-almost complex structure $\Psi_i^* J_0$ on W_i is tamed by the (not necessarily closed) 2-form ω. For every i fix a smooth cutoff function $\beta_i : \varphi_i(U_i) \to [0, 1]$ which is equal to zero near the boundary and equal to one near the origin. For $\varepsilon > 0$ define $u_\varepsilon : \Sigma \to M$ by

$$u_\varepsilon(z) := \begin{cases} \Psi_i^{-1} \circ u_{i,\varepsilon} \circ \varphi_i(z), & \text{for } z \in U_i, \\ u(z), & \text{for } z \in \Sigma \setminus \bigcup_i U_i, \end{cases}$$

where

$$u_{i,\varepsilon}(z) := u_i(z) + \beta_i(z) \left(\prod_{j=0}^{k_i - 1} (z + j\varepsilon) - z^{k_i} \right) (1, p_i(z))$$

as in the proof of Lemma E.2.3. Let

$$V_i := \{ z \in U_i \,|\, \beta_i(\varphi_i(z)) = 1 \}, \qquad V := \bigcup_i V_i.$$

Then the restriction of u_ε to V_i is a $\Psi_i^* J_0$-holomorphic immersion. Since $\Psi_i^* J_0$ is tamed by ω it follows that the restriction of $u_\varepsilon^* \omega$ to V is a positive area form. Moreover, $u^*\omega$ is a positive area form on $\Sigma \setminus V$ and u_ε converges to u in the C^1 topology. Hence $u_\varepsilon^* \omega$ is a positive area form for $\varepsilon > 0$ sufficiently small. Thus we have proved that u_ε is a C^1-immersion which satisfies (i-iii) for small positive ε. Since conditions (i) and (ii) are open and the restriction of u to $\Sigma \setminus U$ is an embedding, we can perturb u_ε in U to obtain a smooth immersion $v : \Sigma \to M$ with only transverse double points. Thus v satisfies (i-iv). To prove (v) note first that

$$\sum_{\substack{z_0 \neq z_1 \\ v(z_0)=v(z_1)}} \iota(v, v; z_0, z_1) = v \cdot v = u_\varepsilon \cdot u_\varepsilon = \sum_{\substack{z_0 \neq z_1 \\ u_\varepsilon(z_0)=u_\varepsilon(z_1)}} \iota(u_\varepsilon, u_\varepsilon; z_0, z_1).$$

Now there are two kinds of self-intersections of u_ε. Namely those pairs (z_0, z_1) where z_0 and z_1 do not belong to the same U_i (which correspond to self-intersection points of u) and those where $z_0, z_1 \in U_i$ for some i (corresponding to its critical points). Denote

$$\mathcal{S}_{\varepsilon 0} := \bigcup_i \{(z_0, z_1) \in U_i \times U_i) \,|\, u_\varepsilon(z_0) = u_\varepsilon(z_1),\ z_0 \neq z_1\},$$

$$\mathcal{S}_{\varepsilon 1} := \{(z_0, z_1) \in \Sigma \times \Sigma \,|\, u_\varepsilon(z_0) = u_\varepsilon(z_1),\ z_0 \neq z_1\} \setminus \mathcal{S}_{\varepsilon 0}.$$

Then, by Definition E.2.1, we have

$$\sum_{(z_0,z_1)\in\mathcal{S}_{\varepsilon 1}} \iota(u_\varepsilon, u_\varepsilon; z_0, z_1) = \sum_{\substack{z_0\neq z_1\\ u(z_0)=u(z_1)}} \iota(u, u; z_0, z_1).$$

Further, by construction of u_ε,

$$\sum_{(z_0,z_1)\in\mathcal{S}_{\varepsilon 0}} \iota(u_\varepsilon, u_\varepsilon; z_0, z_1) \geq 0$$

with equality if and only if $\mathcal{S}_{\varepsilon 0} = \emptyset$, i.e. u is an immersion. To see this, note that if $(z_0, z_1) \in \mathcal{S}_{\varepsilon 0}$ is a pair of points in U_i then the contribution to the above sum from the pairs in U_i is at least $k_i(k_i - 1)/2$. This proves the proposition. □

PROOF OF THEOREM E.1.5. Since the union of the curves u_0 and u_1 is simple, it follows from Theorem E.1.2 that all the intersections of u_0 and u_1 are isolated. Hence

$$A_0 \cdot A_1 = \sum_{u_0(z_0)=u_1(z_1)} \iota(u_0, u_1; z_0, z_1).$$

By Proposition E.2.2, applied to the union of the curves u_0 and u_1, the sum on the right is greater than or equal to $\delta(u_0, u_1)$, with equality if and only if all intersections are transverse. This proves the theorem. □

PROOF OF THEOREM E.1.6. Let $u : \Sigma \to M$ be a J-holomorphic curve representing the homology class $A \in H_2(M; \mathbb{Z})$. We shall prove that

$$A \cdot A - c_1(A) + \chi(\Sigma) \geq \sum_{\substack{z_0\neq z_1\\ u(z_0)=u(z_1)}} \iota(u, u; z_0, z_1) \tag{E.2.4}$$

with equality if and only if u is immersed. By Proposition E.2.2, the sum on the right is bigger than or equal to $2\delta(u)$ with equality if and only if all self-intersections are transverse. Hence the result follows from (E.2.4).

To prove (E.2.4) choose any Riemannian metric $\langle \cdot, \cdot \rangle$ on M such that J is skew-adjoint and denote $\omega := \langle J\cdot, \cdot \rangle \in \Omega^2(M)$. Let $v : \Sigma \to M$ be an immersion satisfying the conditions of Proposition E.2.4. Then $[v] = A$ because v is a perturbation of u. Moreover, $v^*\omega$ is an area form on Σ and hence there is a splitting

$$v^*TM \simeq T\Sigma \oplus \nu_v,$$

where $\nu_v \to \Sigma$ denotes the symplectic vector bundle whose fiber over $z \in \Sigma$ is the symplectic complement of the image of $dv(z)$ in $T_{v(z)}M$. Moreover,

$$A \cdot A = \langle c_1(\nu_v), [\Sigma] \rangle + \sum_{\substack{z_0\neq z_1\\ v(z_0)=v(z_1)}} \iota(v, v; z_0, z_1).$$

Hence the identity $c_1(v^*TM) = c_1(T\Sigma) + c_1(\nu_v)$ shows that

$$A \cdot A - c_1(A) + \chi(\Sigma) = \sum_{\substack{z_0 \neq z_1 \\ v(z_0)=v(z_1)}} \iota(v, v; z_0, z_1).$$

Therefore (E.2.4) follows from Proposition E.2.4 (iv-v). □

REMARK E.2.5. Suppose that $v : \Sigma \to M$ is a perturbation of the curve u as constructed in Proposition E.2.4. Since v is homotopic to u the sum

$$v \cdot v = \sum_{\substack{z_0 \neq z_1 \\ v(z_0)=v(z_1)}} \iota(v, v; z_0, z_1)$$

is the self-intersection number $A \cdot A$ of the homology class $A := [u_*(\Sigma)]$. It is possible to define a local intersection number $\sigma(u; z_0)$ of u at a critical point z_0 that counts the contribution of z_0 to the total self-intersection number of A. Thus $A \cdot A$ decomposes into a sum of contributions, one from each noninjective point of u. For details see McDuff [**256**].

REMARK E.2.6. The intersection number of two holomorphic curves can be interpreted as the multiplicity of a zero of a suitable holomorphic function. (This is the approach taken in Micallef–White [**287**].) More precisely, let $u : \mathbb{C} \to \mathbb{C}^2$ be a polynomial of the form

$$u(z) = (z^k, p(z))$$

where k is a positive integer and p is divisible by z^{k+1}. Define $f_u \in \mathbb{C}[x, y]$ by

$$f_u(x, y) := \prod_{\lambda^k = x} (y - p(\lambda)).$$

Then $f_u(x, y) = 0$ if and only if $(x, y) \in u(\mathbb{C})$. Now define $m := m(u) \in \mathbb{N}$ by

$$m(u) := |\Gamma(u)|, \qquad \Gamma(u) := \left\{\lambda \in S^1 \,|\, u(\lambda z) \equiv u(z)\right\}.$$

Then m divides k. Moreover, $m(u) = 1$ if and only if u is injective near zero, if and only if f_u is irreducible. (We omit the proof.) It follows that there is a polynomial $v : \mathbb{C} \to \mathbb{C}^2$ of the form $v(z) = (z^\ell, q(z))$ with $\ell := k/m$ such that

$$u(z) = v(z^m), \qquad f_u = f_v^m,$$

v is injective near the origin, and f_v is irreducible. Now one can prove the following general result about local intersection numbers.

If $u_0, u_1 : (\mathbb{D}, 0) \to (\mathbb{C}^2, 0)$ and $f_0 : (\mathbb{D} \times \mathbb{D}, 0) \to (\mathbb{C}, 0)$ are holomorphic maps, f_0 is irreducible (in the ring of holomorphic power series in two variables), u_0 is injective, $f_0 \circ u_1 \not\equiv 0$, and

$$f_0(x, y) = 0 \qquad \Longleftrightarrow \qquad (x, y) \in u_0(\mathbb{D})$$

for all $x, y \in \mathbb{D}$, then $(0, 0)$ is an isolated intersection of u_0 and u_1 and the local intersection number

$$\iota(u_0, u_1; 0, 0) = m(f_0 \circ u_1)$$

is the multiplicity of $z = 0$ as a zero of the holomorphic function $f_0 \circ u_1$.

Again we omit the proof (which uses the Nullstellenatz and unique factorization). These observations give rise to an alternative proof of Proposition E.2.2.

E.3. Integrability

The Newlander–Nirenberg theorem asserts that every almost complex structure whose Nijenhuis tensor vanishes is integrable. In dimension two the Nijenhuis tensor vanishes always and so every almost complex structure is integrable. This result is equivalent to the statement that all metrics on Riemannn surfaces are conformally flat. We now prove these classical results for almost complex structures and metrics of class $C^{1+\alpha}$.

Denote by $\mathscr{J} \subset \mathbb{R}^{2\times 2}$ the set of matrices $j \in \mathbb{R}^{2\times 2}$ such that $j^2 = -\mathbb{1}$ and $\{v, jv\}$ is a positively oriented basis of $\mathbb{R}^2$ for every nonzero vector $v \in \mathbb{R}^2$. Denote by $i \in \mathscr{J}$ the standard complex structure on $\mathbb{R}^2 \cong \mathbb{C}$. The coordinates on $\mathbb{D} \subset \mathbb{C}$ are $s + it$.

THEOREM E.3.1 (Newlander–Nirenberg). *Assume $0 < \alpha < 1$. Let $j : \mathbb{D} \to \mathscr{J}$ be a $C^{1+\alpha}$-almost complex structure. Then there is a local $C^{2+\alpha}$-diffeomorphism $\psi : (\mathbb{D}, 0) \to (\mathbb{D}, 0)$ such that $\psi^* i = j$. Moreover, if j is smooth then so is ψ. If $j(0) = i$ then ψ can be chosen such that $d\psi(0) = \mathbb{1}$.*

The equation $\psi^* i = j$ for ψ is equivalent to the equation $(\psi^{-1})^* j = i$ for its inverse. Thus we are looking for a local j-holomorphic curve $\psi^{-1} : (\mathbb{C}, i) \to (\mathbb{D}, j)$. In Appendix C we developed techniques for solving Cauchy–Riemann equations. The proof of Theorem E.3.1 consists in manipulating the equation $\psi^* i = j$ so that these results can be used. Before carrying out the details of the proof, we restate the theorem in the form that will be needed in Section E.4.

A Riemannian metric on $\mathbb{D}$ has the form

$$g = Eds^2 + 2Fdsdt + Gdt^2$$

where $E, F, G : \mathbb{D} \to \mathbb{R}$ are (smooth or $C^{1+\alpha}$ as the case may be) functions satisfying

$$E > 0, \qquad G > 0, \qquad EG - F^2 > 0.$$

A metric g induces a volume form $\mathrm{dvol} = \mathrm{dvol}_g \in \Omega^2(\mathbb{D})$ and an almost complex structure $j = j_g : \mathbb{D} \to \mathscr{J}$ given by

$$\mathrm{dvol}_g := \sqrt{EG - F^2}\, ds \wedge dt, \qquad j_g := \frac{1}{\sqrt{EG - F^2}} \begin{pmatrix} -F & -G \\ E & F \end{pmatrix}.$$

Conversely, the metric can be recovered from the volume form and almost complex structure (both inducing the standard orientation on $\mathbb{D}$) via the usual formula $g = \mathrm{dvol}(\cdot, j\cdot)$. Note that $j = i$ is the standard complex structure if and only if $E = G =: \lambda$ and $F = 0$, i.e. the metric $g = \lambda(ds^2 + dt^2)$ is conformally flat. Hence Theorem E.3.1 can be rephrased as follows.

COROLLARY E.3.2 (Isothermal coordinates). *Assume $0 < \alpha < 1$. Let*

$$g = Eds^2 + 2Fdsdt + Gdt^2$$

be a $C^{1+\alpha}$-metric on $\mathbb{D}$. Then there is an orientation preserving local $C^{2+\alpha}$-diffeomorphism $\psi : (\mathbb{D}, 0) \to (\mathbb{D}, 0)$ such that

$$\psi^*(ds^2 + dt^2) = \lambda\left(Eds^2 + 2Fdsdt + Gdt^2\right), \qquad \lambda := \frac{\det(d\psi)}{\sqrt{EG - F^2}}$$

or, equivalently,

$$|\partial_s\psi|^2 = \lambda E, \qquad \langle \partial_s\psi, \partial_t\psi \rangle = \lambda F, \qquad |\partial_t\psi|^2 = \lambda G.$$

Moreover, if g is smooth then so is ψ. If $E(0) = G(0)$ and $F(0) = 0$ then ψ can be chosen such that $d\psi(0) = 1\!\!1$.

Denote by Δ_g the Laplace–Beltrami operator of the metric g:

$$\begin{aligned}\Delta_g &:= -d^*d \\ &= \frac{1}{EG-F^2}\left(G\frac{\partial^2}{\partial s^2} - 2F\frac{\partial^2}{\partial s\partial t} + E\frac{\partial^2}{\partial t^2}\right) \\ &\quad + \frac{1}{\sqrt{EG-F^2}}\left(\frac{\partial}{\partial s}\frac{G}{\sqrt{EG-F^2}} - \frac{\partial}{\partial t}\frac{F}{\sqrt{EG-F^2}}\right)\frac{\partial}{\partial s} \\ &\quad + \frac{1}{\sqrt{EG-F^2}}\left(\frac{\partial}{\partial t}\frac{E}{\sqrt{EG-F^2}} - \frac{\partial}{\partial s}\frac{F}{\sqrt{EG-F^2}}\right)\frac{\partial}{\partial t}.\end{aligned}$$

If $g = ds^2 + dt^2$ is the standard metric then $\Delta_g = \Delta = \partial_s^2 + \partial_t^2$ is the standard Laplacian and, if $\psi : \mathbb{D} \to \mathbb{D}$ is a diffeomorphism, then

$$\begin{aligned}\Delta_{\psi^*(ds^2+dt^2)}f &= (\Delta(f\circ\psi^{-1}))\circ\psi \\ &= \frac{1}{\det(d\psi)^2}\left(|\partial_t\psi|^2\frac{\partial^2 f}{\partial s^2} - 2\langle\partial_s\psi,\partial_t\psi\rangle\frac{\partial^2 f}{\partial s\partial t} + |\partial_s\psi|^2\frac{\partial^2 f}{\partial t^2}\right) \\ &\quad + \frac{1}{\det(d\psi)}\left(\frac{\partial}{\partial s}\frac{|\partial_t\psi|^2}{\det(d\psi)} - \frac{\partial}{\partial t}\frac{\langle\partial_s\psi,\partial_t\psi\rangle}{\det(d\psi)}\right)\frac{\partial f}{\partial s} \\ &\quad + \frac{1}{\det(d\psi)}\left(\frac{\partial}{\partial t}\frac{|\partial_s\psi|^2}{\det(d\psi)} - \frac{\partial}{\partial s}\frac{\langle\partial_s\psi,\partial_t\psi\rangle}{\det(d\psi)}\right)\frac{\partial f}{\partial t}.\end{aligned}$$

The second order part of this operator will be denoted by

$$(\psi^*\Delta) := \frac{|\partial_t\psi|^2\partial_s^2 - 2\langle\partial_s\psi,\partial_t\psi\rangle\partial_s\partial_t + |\partial_s\psi|^2\partial_t^2}{\det(d\psi)^2}. \tag{E.3.1}$$

With this notation Corollary E.3.2 can be rephrased as follows.

Corollary E.3.3. *Assume $0 < \alpha < 1$. Let $a, b, c : \mathbb{D} \to \mathbb{R}$ be $C^{1+\alpha}$-functions satisfying*

$$a > 0, \qquad c > 0, \qquad ac - b^2 > 0.$$

Then there is a local $C^{2+\alpha}$-diffeomorphism $\psi : (\mathbb{D}, 0) \to (\mathbb{D}, 0)$ such that

$$\psi^*\Delta = \lambda\left(a\partial_s^2 + 2b\partial_s\partial_t + c\partial_t^2\right), \qquad \lambda := \frac{1}{\det(d\psi)\sqrt{ac-b^2}}.$$

Moreover, if a, b, c are smooth then so is ψ. If $a(0) = c(0)$ and $b(0) = 0$ then ψ can be chosen such that $d\psi(0) = 1\!\!1$.

Proof of Theorem E.3.1. Throughout the proof we denote

$$\partial f := \frac{1}{2}(\partial_s f - i\partial_t f), \qquad \bar\partial f := \frac{1}{2}(\partial_s f + i\partial_t f) \tag{E.3.2}$$

for any C^1-map $f : \mathbb{D} \to \mathbb{C}$. The proof has five steps. We first reduce the problem to finding a solution of the Cauchy–Beltrami equation and then solve this equation.

Step 1. *Let $j : \mathbb{D} \to \mathcal{J}$ be a C^1-almost complex structure of the form*

$$j = \begin{pmatrix} b & -a \\ c & -b \end{pmatrix}, \qquad a > 0, \qquad c > 0, \qquad ac - b^2 = 1,$$

define $\mu : \mathbb{D} \to \mathbb{C}$ *by*

$$\mu := \frac{c - a - 2ib}{a + c + 2},$$

and let $\psi : \mathbb{D} \to \mathbb{C}$ *be a* C^2*-function. Then the following are equivalent.*

(i) *The restriction of* ψ *to a neighbourhood of zero is a diffeomorphism onto an open neighbourhood of zero such that* $\psi^*i = j$, $\psi(0) = 0$, *and* $\partial_s\psi(0) = 1$.

(ii) ψ *satisfies the Cauchy–Beltrami equation*

$$\bar{\partial}\psi = \mu\partial\psi, \qquad \psi(0) = 0, \qquad \partial_s\psi(0) = 1, \tag{E.3.3}$$

in a neighbourhood of zero.

Note that

$$d\psi \circ j = (b\partial_s\psi + c\partial_t\psi)\, ds - (a\partial_s\psi + b\partial_t\psi)\, dt.$$

Hence $id\psi = d\psi \circ j$ if and only if

$$(b - i)\partial_s\psi + c\partial_t\psi = 0, \qquad a\partial_s\psi + (b + i)\partial_t\psi = 0.$$

Since $ac - b^2 = 1$ these two equations are equivalent. That they are also equivalent to the Cauchy–Beltrami equation $\bar{\partial}\psi = \mu\partial\psi$ follows from the identity

$$\bar{\partial}\psi - \mu\partial\psi = \frac{a + 1 + ib}{a + c + 2}\left(\partial_s\psi + \frac{b + i}{a}\partial_t\psi\right).$$

If $\bar{\partial}\psi = \mu\partial\psi$ and $\partial_s\psi(0) = 1$ then $\partial_s\psi(0)$ and $\partial_t\psi(0)$ are linearly independent over the reals and so ψ is a local diffeomorphism near zero by the inverse function theorem. This proves Step 1.

Now let $j : \mathbb{D} \to \mathcal{J}$ be a $C^{1+\alpha}$-almost complex structure and let $\mu : \mathbb{D} \to \mathbb{C}$ be the $C^{1+\alpha}$-function defined in Step 1. Then $|\mu| < 1$ everywhere. We prove in the next four steps that there is a local $C^{2+\alpha}$-solution $\psi : (\mathbb{D}, 0) \to \mathbb{C}$ of (E.3.3) and that the solution is smooth whenever μ is smooth. We deal with the smoothness question first.

STEP 2. *Let* $\psi : \mathbb{D} \to \mathbb{C}$ *be a* $C^{1+\alpha}$*-solution of (E.3.3) and* $W \subset \mathbb{D}$ *be an open neighbourhood of zero such that the restriction of* ψ *to* W *is a diffeomorphism onto an open set* $U \subset \mathbb{D}$. *If* μ *is of class* $C^{k+\alpha}$ *for some integer* $k > 0$ *then* ψ *is of class* $C^{k+1+\alpha}$ *in* W.

Let $j : \mathbb{D} \to \mathcal{J}$ be the almost complex structure associated to μ and $u : U \to W$ be the inverse of ψ. Then j is of class $C^{k+\alpha}$ and, by Step 1, we have $u^*j = i$ or equivalently $\bar{\partial}_j(u) = 0$. In particular, $j \in W^{k,p}$ for every $p > 2$ and hence, by Theorem B.4.1 and Remark B.4.3, we have $u \in W^{k+1,p}$ for every $p > 2$. By Theorem B.1.11, this implies $u \in C^{k+\alpha}$. Moreover, u satisfies the equation

$$\Delta u = (\partial_t(j \circ u))\partial_s u - (\partial_s(j \circ u))\partial_t u.$$

Since u and j are (locally) of class $C^{k+\alpha}$, it follows that Δu is of class $C^{k-1+\alpha}$ and hence, by Schauder regularity (see Gilbarg–Trudinger [**140**]), u is of class $C^{k+1+\alpha}$. Hence ψ is of class $C^{k+1+\alpha}$ as claimed.

STEP 3. *If* $\mu(0) = 0$ *then equation (E.3.3) has a local* $C^{2+\alpha}$*-solution* ψ *near zero.*

For $\beta > 0$ with $\beta \notin \mathbb{Z}$ denote by $C^\beta(\mathbb{D}) = C^\beta(\mathbb{D}, \mathbb{C})$ the corresponding Hölder space of complex valued functions on $\mathbb{D}$. By Theorem C.4.1 and Schauder regularity (see [**140**]), the Cauchy-Riemann operator $\bar{\partial} : C^{1+\alpha}(\mathbb{D}) \to C^\alpha(\mathbb{D})$ restricts to surjective Fredholm operator of (real) index four from the Banach space

$$C_0 := \{\xi \in C^{1+\alpha}(\mathbb{D}) \,|\, \xi(e^{i\theta}) \in \mathbb{R}e^{3i\theta/2}\}$$

to $C^\alpha(\mathbb{D})$. The kernel of this operator consists of all polynomials of the form $\xi(z) = a_0 + a_1 z + \bar{a}_1 z^2 + \bar{a}_0 z^3$. Hence the linear map

$$C_0 \to C^\alpha(\mathbb{D}) \times \mathbb{C} \times \mathbb{C} : \xi \mapsto (\bar{\partial}\xi, \xi(0), \partial_s \xi(0))$$

is a Banach space isomorphism. Let $T : C^\alpha(\mathbb{D}) \times \mathbb{C} \times \mathbb{C} \to C^{1+\alpha}(\mathbb{D})$ denote its inverse. Then $T(0,0,1)(z) = z + z^2$ and, if $f \in C^\alpha(\mathbb{D})$, then the function $\psi := T(f, 0, 1)$ is such that

$$\bar{\partial}\psi = f, \qquad \psi(0) = 0, \qquad \partial_s \psi(0) = 1.$$

In particular ψ satisfies the normalization conditions in (E.3.3). (In fact we constructed the above operator to have index four precisely so that we can accommodate these conditions.) Now define the smooth nonlinear map $\Phi : C^\alpha(\mathbb{D}) \times C^\alpha(\mathbb{D}) \to C^\alpha(\mathbb{D}) \times C^\alpha(\mathbb{D})$ by

$$\Phi(f, \lambda) := (f - \lambda\partial(T(f, 0, 1)), \lambda).$$

Note that, if $\Phi(f, \lambda) = (0, \lambda)$ then the function $\psi := T(f, 0, 1)$ satisfies $\bar{\partial}\psi = f = \lambda\partial\psi$ as well as $\psi(0) = 0$ and $\partial_s\psi(0) = 1$. These are the three conditions in (E.3.3) with $\mu = \lambda$. Thus it remains to show that the equation $\Phi(f, \lambda) = (0, \lambda)$ has local solutions when $\lambda(0) = 0$.

To this end, note first that

$$(d\Phi(0,0)(\hat{f}, \hat{\lambda}))(z) = (\hat{f} - \hat{\lambda}(1 + 2z), \hat{\lambda}).$$

This operator is invertible. Hence, by the inverse function theorem, Φ has a smooth inverse defined on some open neighbourhood $\mathcal{U}^\alpha \subset C^\alpha(\mathbb{D}) \times C^\alpha(\mathbb{D})$ of zero, mapping zero to itself. Now choose a smooth cutoff function $\chi : \mathbb{D} \to [0, 1]$ such that $\chi(z) = 1$ for z near 0 and $\chi(z) = 0$ for $|z| \geq 1/2$. For $0 < \varepsilon \leq 1$ define $\mu_\varepsilon : \mathbb{D} \to \mathbb{C}$

$$\mu_\varepsilon(z) := \mu(z)\chi(z/\varepsilon).$$

Since $\mu(0) = 0$ and μ is continuously differentiable, the functions μ_ε converge to zero in the C^α-norm as $\varepsilon \to 0$. Hence $(0, \mu_\varepsilon) \in \mathcal{U}^\alpha$ for ε sufficiently small. It follows that the function

$$\psi := T(f, 0, 1), \qquad (f, \mu_\varepsilon) := \Phi^{-1}(0, \mu_\varepsilon),$$

is a local $C^{1+\alpha}$-solution of equation (E.3.3). By Step 2, ψ is of class $C^{2+\alpha}$.

STEP 4. *If μ is constant then equation (E.3.3) has a smooth solution on $\mathbb{D}$.*

An explicit formula is

$$\psi(s + it) := s + i\frac{1 - \mu}{1 + \mu}t.$$

STEP 5. *We prove the theorem.*

Let $\mu : \mathbb{D} \to \mathbb{C}$ be a $C^{1+\alpha}$-function such that $|\mu| < 1$ everywhere. Let $\varphi : \mathbb{D} \to \mathbb{C}$ be the solution of the equation

$$\bar{\partial}\varphi = \mu(0)\partial\varphi, \qquad \varphi(0) = 0, \qquad \partial_s\varphi(0) = 1,$$

constructed in Step 4. Set

$$\tilde{\mu} := \left(\frac{\partial\varphi}{\bar{\partial}\bar{\varphi}}\frac{\mu-\mu(0)}{1-\bar{\mu}(0)\mu}\right)\circ\varphi^{-1}.$$

Then $\tilde{\mu}(0)=0$ and so, by Step 3, there is a local $C^{2+\alpha}$-solution $\tilde{\psi}:(\mathbb{D},0)\to(\mathbb{D},0)$ of (E.3.3) with μ replaced by $\tilde{\mu}$. Define

$$\psi := \tilde{\psi}\circ\varphi.$$

Then ψ is a $C^{2+\alpha}$-solution of (E.3.3). This follows from the identity

$$\mu_{\psi\circ\varphi^{-1}} = \left(\frac{\partial\varphi}{\bar{\partial}\bar{\varphi}}\frac{\mu_\psi-\mu_\varphi}{1-\bar{\mu}_\varphi\mu_\psi}\right)\circ\varphi^{-1}, \qquad \mu_f := \frac{\bar{\partial}f}{\partial f}.$$

Moreover, if μ is smooth then $\tilde{\mu}$ is smooth, hence $\tilde{\psi}$ is smooth by Step 2, and hence ψ is smooth. Thus ψ satisfies the requirements of the theorem by Step 1. □

E.4. The Hartman–Wintner theorem

The Hartman–Wintner theorem describes the lowest order term of an approximately harmonic real valued function. Throughout we abbreviate $z := s + it$ and write $u(z)$ instead of $u(s,t)$. In particular, the real homogeneous polynomial function $h(z)$ mentioned in the theorem should be considered as a polynomial in the variables s and t (or z and $\bar{z}$); it is not holomorphic.

THEOREM E.4.1 (Hartman–Wintner). *Assume* $0<\alpha<1$. *Let* $a,b,c:\mathbb{D}\to\mathbb{R}$ *be three* $C^{1+\alpha}$*-functions such that*

$$a>0, \qquad c>0, \qquad ac-b^2>0. \tag{E.4.1}$$

Let $u:\mathbb{D}\to\mathbb{R}^N$ *be a* C^2*-function satisfying* $u(0)=0$ *and the estimate*

$$\left|a\partial_s^2u+2b\partial_s\partial_tu+c\partial_t^2u\right|\le C\left(|u|+|du|\right). \tag{E.4.2}$$

Assume that u *does not vanish identically. Then there is an integer* $m\ge1$ *and a nonzero homogeneous polynomial* $h:\mathbb{C}\to\mathbb{R}^N$ *of degree* m *such that*

$$u(z)=h(z)+o(z^m), \qquad du(z)=dh(z)+o(z^{m-1}).$$

REMARK E.4.2. Note that this theorem does not follow from Aronszajn's theorem since this concludes only that u is not flat at the origin. The statement that the lowest order terms in u form a polynomial is quite strong; even if u were C^2 such a polynomial need not exist (consider for instance the function $(s,t)\mapsto|t|^3$). Moreover, the assertion of Aronszajn's theorem does not give any information about the derivative du. Hartman and Wintner's theorem also cannot be derived from the Carleman similarity principle, although they both imply a strong unique continuation theorem. Good references for the proof of Theorem E.4.1 are the original paper [**169**] and Jost [**204**, Lemma 8.2.6]. In [**204**] the hypothesis is slightly stronger, but the proof goes through for all functions satisfying (E.4.2) with $a=c=1$ and $b=0$. We show below that the general case reduces to this special case by using Corollary E.3.3.

The next proposition is a reformulation of the Hartman–Wintner theorem for the case $a=c=1$ and $b=0$. It concerns a real valued map $v:\mathbb{D}\to\mathbb{R}^N\subset\mathbb{C}^N$. Note that the value $dv(z)\zeta$ of the derivative $dv(z)$ at the tangent vector $\zeta\in T_z\mathbb{D}=\mathbb{C}$

can be written as $2\Re e(\zeta \partial v(z))$ where $\partial v(z) \in \mathbb{C}^N$ is as in (E.3.2) and ζ acts by multiplication.

PROPOSITION E.4.3. *Let $v : \mathbb{D} \to \mathbb{R}^N \subset \mathbb{C}^N$ be a C^2-function satisfying*

$$|\Delta v| \le C(|v| + |dv|). \tag{E.4.3}$$

If n is a positive integer and

$$v(z) = o(z^n) \tag{E.4.4}$$

then there is a vector $c \in \mathbb{C}^N$ such that

$$dv(z)\zeta = \Re e(\zeta z^n c) + o(z^n)\zeta$$

for every $\zeta \in \mathbb{C}$. If (E.4.4) holds for every n then v vanishes identically in a neighbourhood of zero.

PROOF. The first statement is equivalent to saying that $z^{-n}\partial v(z)$ has a limit as z tends to zero. We prove this by induction over n in four steps. We need two basic tools, a Green's formula and a Cauchy formula, which are proved in Step 1. From these we derive a pointwise estimate in Step 2, which implies by the bootstrapping argument of Step 3 that $z^{-n}\partial v(z)$ is bounded. Step 4 combines this with the Green's formula to get the first statement. The second statement follows from the explicit estimate in Step 3.

Let $k \in \{0, \dots, n\}$. We shall prove that if

$$\partial v(z) = o(z^{k-1}) \tag{E.4.5}$$

then the function $z^{-k}\partial v$ has a limit as z tends to zero. Throughout we denote

$$\mathbb{D}_R := \{\zeta \in \mathbb{C} \,|\, |\zeta| \le R\}$$

and

$$w(z) := |v(z)| + |\partial v(z)|.$$

Since v takes values in $\mathbb{R}^N$ and $\partial v = (\partial_s v - i\partial_t v)/2$ we have $|\partial v| = |\bar\partial v| = |dv|/2$.

STEP 1. *If $0 \le k \le n$ and v satisfies (E.4.3-E.4.5) then*

$$z^{-k}v(z) = \frac{1}{2\pi i}\int_{\partial\mathbb{D}_R} \frac{\zeta^{-k}v(\zeta)}{\zeta - z}\,d\zeta - \frac{1}{\pi}\int_{\mathbb{D}_R} \frac{\zeta^{-k}\bar\partial v(\zeta)}{\zeta - z}\frac{d\bar\zeta \wedge d\zeta}{2i}, \tag{E.4.6}$$

$$z^{-k}\partial v(z) = \frac{1}{2\pi i}\int_{\partial\mathbb{D}_R} \frac{\zeta^{-k}\partial v(\zeta)}{\zeta - z}\,d\zeta - \frac{1}{4\pi}\int_{\mathbb{D}_R} \frac{\zeta^{-k}\Delta v(\zeta)}{\zeta - z}\frac{d\bar\zeta \wedge d\zeta}{2i} \tag{E.4.7}$$

for $0 < R \le 1$ and $0 < |z| < R$.

Stokes' formula can be expressed in the form

$$\int_{\partial\Omega} f(\zeta)\,d\zeta = \int_{\Omega} \bar\partial f(\zeta)\,d\bar\zeta \wedge d\zeta$$

for every bounded open set $\Omega \subset \mathbb{C}$ with smooth boundary and every smooth function $f : \bar\Omega \to \mathbb{R}$. Fix a complex number z such that $0 < |z| < R$ and a real number $\varepsilon > 0$ such that $2\varepsilon < |z|$ and $|z| + \varepsilon < R$. Applying Stokes' formula to the domain

$$\Omega_\varepsilon := \{\zeta \in \mathbb{C} \,|\, \varepsilon < |\zeta| < R, |\zeta - z| > \varepsilon\}.$$

and the function $f(\zeta) := \zeta^{-k}(\zeta - z)^{-1}v(\zeta)$ we obtain

$$\int_{\partial\Omega_\varepsilon} \frac{\zeta^{-k}v(\zeta)}{\zeta - z}\,d\zeta = \int_{\Omega_\varepsilon} \frac{\zeta^{-k}\bar\partial v(\zeta)}{\zeta - z}\,d\bar\zeta \wedge d\zeta. \tag{E.4.8}$$

By (E.4.5), $|\bar\partial v(\zeta)| = |\partial v(\zeta)| = o(|\zeta|^{k-1})$. Hence the function

$$\zeta \mapsto \zeta^{-k}(\zeta - z)^{-1}\bar\partial v(\zeta)$$

is integrable on the unit disc for each fixed $z \neq 0$, and so the right hand side of (E.4.8) converges to the integral over $\mathbb{D}_R$ as ε tends to zero. Moreover,

$$\lim_{\varepsilon\to 0}\int_{|\zeta|=\varepsilon}\frac{\zeta^{-k}v(\zeta)}{\zeta - z}\,d\zeta = 0, \qquad \lim_{\varepsilon\to 0}\int_{|\zeta-z|=\varepsilon}\frac{\zeta^{-k}v(\zeta)}{\zeta - z}\,d\zeta = 2\pi i z^{-k}v(z). \tag{E.4.9}$$

Here the second identity holds always while the first identity follows from the fact that $|v(\zeta)| = o(|\zeta|^{k-1})$. Now (E.4.6) follows from (E.4.8) and (E.4.9) by taking the limit $\varepsilon \to 0$. To prove (E.4.7) replace v by ∂v and $\bar\partial v$ by $\bar\partial\partial v = \Delta v/4$ in (E.4.6). The same argument works because $\partial v(\zeta) = o(\zeta^{k-1})$, by (E.4.5), and the function $\zeta \mapsto \zeta^{-k}(\zeta - z)^{-1}\Delta v(\zeta)$ is integrable, by (E.4.3-E.4.5).

STEP 2. *If $0 \le k \le n$ and v satisfies (E.4.3-E.4.5) then*

$$|z^{-k}w(z)| \le 2\max_{R/2\le|\zeta|\le R}|\zeta^{-k}w(\zeta)| + \frac{C_1}{4\pi}\left\|\frac{\zeta^{-k}w(\zeta)}{\zeta - z}\right\|_{L^1(\mathbb{D}_R)} \tag{E.4.10}$$

for $0 < R \le 1$ and $0 < |z| \le R$. Here $C_1 := C + 4$, where C is the constant in (E.4.3).

For $0 < |z| < R$ we have

$$\left|\int_{\partial\mathbb{D}_R}\frac{d\zeta}{\zeta - z}\right| \le \frac{2\pi R}{R - |z|}.$$

Hence it follows from (E.4.3), (E.4.6), and (E.4.7) that

$$\begin{aligned}
|z^{-k}w(z)| &\le \left(\frac{R}{R-|z|}\right)\max_{\zeta\in\partial\mathbb{D}_R}|\zeta^{-k}w(\zeta)| \\
&\quad + \frac{1}{4\pi}\int_{\mathbb{D}_R}\frac{|\zeta^{-k}|}{|\zeta - z|}\Big(Cw(\zeta) + 4|\bar\partial v(\zeta)|\Big)\frac{d\bar\zeta\wedge d\zeta}{2i} \\
&\le 2\max_{\zeta\in\partial\mathbb{D}_R}|\zeta^{-k}w(\zeta)| + \frac{C+4}{4\pi}\int_{\mathbb{D}_R}\frac{|\zeta^{-k}|w(\zeta)}{|\zeta - z|}\frac{d\bar\zeta\wedge d\zeta}{2i}.
\end{aligned}$$

The last inequality holds for $0 < |z| \le R/2$. This proves (E.4.10) for $0 < |z| \le R/2$. For $R/2 \le |z| \le R$ the estimate is obvious.

STEP 3. *If $0 \le k \le n$ and v satisfies (E.4.3-E.4.5) then*

$$0 < |z| \le R \quad\Longrightarrow\quad |z^{-k}w(z)| \le \frac{2 - 2C_1R}{1 - 2C_1R}\max_{R/2\le|\zeta|\le R}|\zeta^{-k}w(\zeta)|. \tag{E.4.11}$$

for $0 < R < 1/2C_1$.

Fix a positive real number $R < 1/2C_1$ and a point $z_0 \in \mathbb{D}_R \setminus \{0\}$. Multiply both sides of the estimate (E.4.10) by $(z - z_0)^{-1}$ and integrate over $z \in \mathbb{D}_R$. Combining this with the inequalities

$$\left|\frac{1}{z - z_0}\frac{1}{\zeta - z}\right| \le \frac{1}{|\zeta - z_0|}\left(\frac{1}{|z - z_0|} + \frac{1}{|\zeta - z|}\right),$$

and

$$0 < |\zeta| \le R \quad\Longrightarrow\quad \int_{\mathbb{D}_R}\frac{d\bar z\wedge dz}{2i|\zeta - z|} \le \int_{\mathbb{D}_{2R}}\frac{d\bar z\wedge dz}{2i|z|} = 4\pi R$$

we obtain

$$\begin{aligned}\int_{\mathbb{D}_R} \frac{\left|z^{-k}\right| w(z)\, d\bar{z}\wedge dz}{2i|z-z_0|} &\leq 2 \max_{R/2\leq|\zeta|\leq R} |\zeta^{-k}w(\zeta)| \int_{\mathbb{D}_R} \frac{d\bar{z}\wedge dz}{2i|z-z_0|} \\ &\quad + \frac{C_1}{4\pi}\int_{\mathbb{D}_R}\left(\int_{\mathbb{D}_R}\frac{d\bar{z}\wedge dz}{2i|z-z_0|}\right)\frac{\left|\zeta^{-k}\right| w(\zeta)\, d\bar{\zeta}\wedge d\zeta}{2i|\zeta-z_0|} \\ &\quad + \frac{C_1}{4\pi}\int_{\mathbb{D}_R}\left(\int_{\mathbb{D}_R}\frac{d\bar{z}\wedge dz}{2i|\zeta-z|}\right)\frac{\left|\zeta^{-k}\right| w(\zeta)\, d\bar{\zeta}\wedge d\zeta}{2i|\zeta-z_0|} \\ &\leq 8\pi R \max_{R/2\leq|\zeta|\leq R} |\zeta^{-k}w(\zeta)| \\ &\quad + 2C_1R\int_{\mathbb{D}_R}\frac{\left|\zeta^{-k}\right| w(\zeta)\, d\bar{\zeta}\wedge d\zeta}{2i|\zeta-z_0|}.\end{aligned}$$

Since $0 < R < 1/2C_1$ we obtain

$$\left\|\frac{\zeta^{-k}w(\zeta)}{\zeta - z}\right\|_{L^1(\mathbb{D}_R)} \leq \frac{8\pi R}{1-2C_1R} \max_{R/2\leq|\zeta|\leq R} |\zeta^{-k}w(\zeta)|$$

for $0 < |z| \leq R$. Hence (E.4.11) follows from (E.4.10).

STEP 4. *The limit* $c := 2\lim_{z\to 0} z^{-n}\partial v(z)$ *exists.*

We prove by induction on k that

$$\lim_{z\to 0} z^{-k}\partial v(z) = 0, \qquad k = 0,\dots,n-1.$$

For $k = 0$ this is obvious because v is a C^2-function. Let $k \in \{1,\dots,n-1\}$ and assume, by induction, that $z^{1-k}\partial v(z)$ converges to zero as z tends to zero. Then v satisfies (E.4.5). Hence it follows from Step 3 with, say, $R := 1/4C_1$ that that the function $z \mapsto z^{-k}w(z)$ is bounded. Hence the function $z \mapsto z^{-k}\Delta v(z)$ is bounded by (E.4.3). Now the functions $\zeta \mapsto (\zeta - z)^{-1}$ converge to the function $\zeta \mapsto \zeta^{-1}$ in $L^1(\mathbb{D})$ as z tends to zero. Hence the right hand side of (E.4.7) converges as z tends to zero. This shows that the limit $\lim_{z\to 0} z^{-k}\partial v(z)$ exists. Denote

$$c := \lim_{z\to 0}\frac{2\partial v(z)}{z^k} \in \mathbb{C}^N.$$

Then

$$\begin{aligned} v(z) &= 2\Re e\left(\int_0^1 \partial v(\tau z) z\, d\tau\right) \\ &= \Re e\left(\int_0^1 c(\tau z)^k z\, d\tau\right) + o(z^{k+1}) \\ &= \Re e\left(\frac{c}{k+1}z^{k+1}\right) + o(z^{k+1}).\end{aligned}$$

Since $k \leq n-1$ and $v(z) = o(z^n)$ this implies $c = 0$. This completes the induction. Repeat the first part of the induction step for $k = n$ to deduce that $z^{-n}\partial v(z)$ converges as z tends to zero. This proves Step 4.

Step 4 shows that $\partial v(z) = \frac{1}{2}z^n c + o(z^n)$ and hence

$$dv(z)\zeta = 2\Re e(\partial v(z)\zeta) = \Re e(\zeta z^n c) + o(z^n).$$

This proves the first assertion of the proposition.

We now assume that (E.4.4) holds for all n.

Recall that the constant $C_1 = C + 4$ in Step 3 is independent of k and n. Fix a positive real number $R < 1/2C_1$ and let $z \in \mathbb{C}$ with $|z| < R/2$. Then, by (E.4.11), we have

$$|v(z)| \le |z^n| \frac{2 - 2C_1 R}{1 - 2C_1 R} \max_{R/2 \le |\zeta| \le R} \left|\zeta^{-n} w(\zeta)\right| \le \frac{2 - 2C_1 R}{1 - 2C_1 R} \max_{\zeta \in \mathbb{D}_R} w(\zeta) \left(\frac{2|z|}{R}\right)^n$$

for every integer $n > 0$. Hence $v(z) = 0$. □

PROOF OF THEOREM E.4.1. Let $a, b, c : \mathbb{D} \to \mathbb{R}$ be three $C^{1+\alpha}$-functions satisfying (E.4.1) and $u : \mathbb{D} \to \mathbb{R}^N$ be a C^2-function satisfying (E.4.2).

CASE 1. *Assume that u does not vanish identically in a neighbourhood of zero.*

Let $\psi : (\mathbb{D}, 0) \to (\mathbb{D}, 0)$ be a local $C^{2+\alpha}$-diffeomorphism that satisfies the requirements of Corollary E.3.3. Then the C^2-function $v := u \circ \psi^{-1}$ satisfies (E.4.3). Hence, by Proposition E.4.3, there is an integer $n > 0$ such that $v(z) = o(z^n)$ and $v(z) \ne o(z^{n+1})$. Proposition E.4.3 also shows that there is a vector $c \in \mathbb{C}^N$ such that

$$dv(z)\zeta = \Re e(\zeta z^n c) + o(z^n)\zeta.$$

Moreover, $\psi(z) = Az + o(z) \in \mathbb{R}^2 = \mathbb{C}$ where $A := d\psi(0) \in \mathrm{GL}(2, \mathbb{R})$. Define the polynomial $h : \mathbb{C} \to \mathbb{R}^N$ by

$$h(z) := \Re e \left(\frac{(Az)^{n+1}}{n+1} c \right).$$

Then

$$\begin{aligned} du(z)\zeta &= dv(\psi(z)) d\psi(z)\zeta \\ &= dv(Az)A\zeta + o(z^n)\zeta \\ &= \Re e(A\zeta(Az)^n c) + o(z^n)\zeta \\ &= dh(z)\zeta + o(z^n)\zeta \end{aligned}$$

and hence

$$u(z) = \int_0^1 du(\tau z) z \, d\tau = \int_0^1 dh(\tau z) z \, d\tau + o(z^{n+1}) = h(z) + o(z^{n+1}).$$

Since $v(z) \ne o(z^{n+1})$ we have $u(z) \ne o(z^{n+1})$ and hence h does not vanish. This proves the first part of the Hartman–Wintner theorem.

CASE 2. *Assume that u vanishes identically in a neighbourhood of zero.*

Denote by R_{max} the maximal radius $R \le 1$ such that u vanishes identically on the disc $\mathbb{D}_R$. If $R_{max} < 1$, then there is a point $z_0 \in \partial\mathbb{D}_{R_{max}}$ such that u does not vanish identically in any neighbourhood of z_0. Therefore, according to the first part of the proof, there is a nonzero homogeneous polynomial h of some degree $m \ge 1$ such that

$$u(z) = h(z - z_0) + o((z - z_0)^m).$$

But such a function cannot vanish on the interior of $\mathbb{D}_R$. This contradiction shows that $R_{max} = 1$ and so u is identically zero. □

REMARK E.4.4. Let J be a smooth almost complex structure on $\mathbb{R}^{2n}$ and $u : \mathbb{D} \to \mathbb{R}^{2n}$ be a nonconstant J-holomorphic curve (with respect to the standard complex structure i on $\mathbb{D}$). Applying the operator $\partial_s - J\partial_t$ to $\bar{\partial}_J(u)$ we find that

$$\Delta u + \partial_s(J \circ u)\partial_t u - \partial_t(J \circ u)\partial_s u = 0.$$

Hence u satisfies (E.4.3) and so it follows from Proposition E.4.3 that u is not flat at the origin, i.e. the partial derivatives of u cannot all vanish at the origin. As pointed out in Section 2.3, a similar argument proves the Unique Continuation Theorem 2.3.2.

E.5. Local behaviour

This section carries out the first step in the proof of Theorem E.1.1. The goal is to find coordinates on M and Σ in which the J-holomorphic curve can be written as

$$u(z) = (z^k, F(z))$$

where the error term $F = O(|z|^{k+1})$ predicted by (E.1.1) takes values in $0 \times \mathbb{C}^{n-1}$. To achieve this we must deal with smooth (but not necessarily holomorphic) functions $f : \mathbb{C} \setminus \{0\} \to \mathbb{C}$ and control their behaviour near zero. To this end we introduce the following notation.

Let $U \subset \mathbb{C}$ be an open neighbourhood of zero and $\ell \in \mathbb{Z}$. A smooth function $f : U \setminus \{0\} \to \mathbb{C}$ is said to be **of order** ℓ **(near zero)** if, for every pair of integers $j, k \geq 0$, there exists a constant $c_{jk} > 0$ such that

$$\left|\partial_s^j \partial_t^k f(z)\right| \leq c_{jk} |z|^{\ell-j-k} \tag{E.5.1}$$

for every $z \in U \setminus \{0\}$. Denote by $\mathcal{O}_\ell(U)$ the set of smooth functions $f : U \setminus \{0\} \to \mathbb{C}$ of order ℓ and by

$$\mathcal{O}_\ell := \bigcup_U \mathcal{O}_\ell(U)$$

the union over all open neighbourhoods of zero. Whenever convenient we extend the notation $\mathcal{O}_\ell$ to functions with values in a complex vector space.

REMARK E.5.1. Let $U \subset \mathbb{C}$ be an open neighbourhood of zero, $f : U \to \mathbb{C}$ be a smooth function, and ℓ be a nonnegative integer. Then $f \in \mathcal{O}_\ell(U)$ if and only if all partial derivatives of f up to order $\ell - 1$ vanish at $z = 0$ or, equivalently, $f(z) = O(|z|^\ell)$.

EXERCISE E.5.2. Let ℓ be a nonnegative integer and $p \in [1, 2)$. Prove that

$$\mathcal{O}_\ell(U) \subset W^{\ell,\infty}(U) \cap W^{\ell+1,p}(U).$$

The next lemma lists the properties of the function spaces $\mathcal{O}_\ell$.

LEMMA E.5.3. (MONOTONE) *$\mathcal{O}_\ell \subset \mathcal{O}_{\ell-1}$ for every $\ell \in \mathbb{Z}$.*

(DERIVATIVE) *If $f \in \mathcal{O}_\ell$ then $\partial_s f, \partial_t f \in \mathcal{O}_{\ell-1}$.*

(POLE–ZERO) *The map $z \mapsto z^m$ belongs to $\mathcal{O}_m$ for every $m \in \mathbb{Z}$.*

(LIPSCHITZ) *If $f \in \mathcal{O}_1$ then f is Lipschitz continuous near zero.*

(SUM) *If $f, g \in \mathcal{O}_\ell$ then $f + g \in \mathcal{O}_\ell$.*

(PRODUCT) *If $f \in \mathcal{O}_k$ and $g \in \mathcal{O}_\ell$ then $fg \in \mathcal{O}_{k+\ell}$.*

(COMPOSITION) *Let ℓ and m be positive integers and $U \subset \mathbb{C}$ be an open neighbourhood of zero. If $h : U \to \mathbb{C}$ is a smooth function in $\mathcal{O}_m$ and $f \in \mathcal{O}_\ell$ then $h \circ f \in \mathcal{O}_{m\ell}$.*

(INVERSE) *If $\phi : (\mathbb{C}, 0) \to (\mathbb{C}, 0)$ is a local C^1-diffeomorphism and $\phi \in \mathcal{O}_1$ then $\phi^{-1} \in \mathcal{O}_1$.*

PROOF. All the assertions apart from the *(Composition)* and *(Inverse)* axioms follow directly from the definitions. We prove the *(Composition)* axiom. Let $\mathbb{N}$ denote the set of nonegative integers and, for $\alpha = (j,k) \in \mathbb{N}^2$, abbreviate $\partial^\alpha := \partial_s^j \partial_t^k$ and $|\alpha| := j + k$. Then the derivatives of $h \circ f$ can be expressed in the form

$$\partial^\alpha (h \circ f)(z) = \sum_{N=1}^{|\alpha|} \sum_\beta m_\beta d^N h(f(z))(\partial^{\beta_1} f(z), \ldots, \partial^{\beta_N} f(z))$$

for $\alpha \in \mathbb{N}^2$ where the second sum runs over all tuples $\beta = (\beta_1, \ldots, \beta_N) \in (\mathbb{N}^2)^N$ with $|\beta_1| + \cdots + |\beta_N| = |\alpha|$ and the coefficients $m_\beta \geq 0$ are integers. Since

$$|d^N h(f(z))| \leq c_N |z|^{\ell(m-N)}$$

all terms in the sum are of order $m\ell - |\alpha|$.

We prove the *(Inverse)* axiom. Let $\psi := \phi^{-1}$. One can prove by induction that

$$\partial^\alpha \psi(\phi) = \frac{1}{\det(\phi)^{2|\alpha|-1}} \sum_{\beta,n} m_{\beta,n} \prod_{i=1}^{|\alpha|} (\partial^{\beta_i} \phi^1)^{n_i} \prod_{j=1}^{|\alpha|} (\partial^{\beta_{\ell+j}} \phi^2)^{n_{\ell+j}}$$

for $\alpha \in \mathbb{N}^2$ where ϕ^1, ϕ^2 are the components of ϕ, $m_{\beta,n} \in \mathbb{Z}^2$, and the sum runs over all tuples $\beta = (\beta_1, \ldots, \beta_{2|\alpha|}) \in (\mathbb{N}^2)^{2|\alpha|}$ and $n = (n_1, \ldots, n_{2|\alpha|}) \in \mathbb{N}^{2|\alpha|}$ such that

$$\sum_{j=1}^{2|\alpha|} |\beta_j| n_j = 4|\alpha| - 3, \qquad \sum_{j=1}^{2|\alpha|} n_j = 3|\alpha| - 2.$$

Since $\phi \in \mathcal{O}_1$ this implies that $|\partial^\alpha \psi(z)| \leq c_\alpha |z|^{1-|\alpha|}$ for $|z|$ sufficiently small and some constant $c_\alpha > 0$. □

REMARK E.5.4. The proof shows that the *(Composition)* axiom extends to the following situation.

Let ℓ be a positive integer and m be an integer. If $f \in \mathcal{O}_\ell$ satisfies

$$z \neq 0 \qquad \Longrightarrow \qquad f(z) \neq 0$$

for z near zero and $h \in \mathcal{O}_m$ then $h \circ f \in \mathcal{O}_{m\ell}$.

The condition $\ell > 0$ guarantees that f takes values in an arbitrarily small neighbourhood of zero. Since $f(z) \neq 0$ for $z \neq 0$ it follows that the composition $h \circ f$ is a smooth function defined on a punctured neighbourhood of zero. With this understood the proof of the (*Composition*) axiom carries over to the present case.

REMARK E.5.5. Both versions of the *(Composition)* axiom extend to functions with values in complex vector spaces. We shall need the following version.

If E, F are complex vector spaces, $h : E \to F$ is a smooth map satisfying $h(0) = 0$, and $f : U \to E$ belongs to $\mathcal{O}_\ell$ with $\ell \geq 1$ then $h \circ f : U \to F$ belongs to $\mathcal{O}_\ell$.

We now examine the regularity property of the Cauchy–Riemann operator with respect to the sets $\mathcal{O}_\ell$. The following lemma deals with the same point as Exercise B.1.10.

LEMMA E.5.6. *Assume $p > 1$. Let $h : \mathbb{D} \to \mathbb{C}$ be a function which belongs to $W^{1,p}(\mathbb{D})$, is smooth away from the origin, and satisfies $\bar{\partial} h = 0$ on $\mathbb{D} \setminus \{0\}$. Then h extends to a holomorphic function on $\mathbb{D}$.*

PROOF. We must prove that $\bar{\partial}h = 0$ as a distribution. Then h is a weak solution of the Cauchy–Riemann equations and the result follows from Lemma B.4.6 (i). Let $\varphi : \mathbb{D} \to \mathbb{R}$ be a smooth function with compact support in $\mathbb{D}$. Set

$$\mathbb{D}_\epsilon := \{z \in \mathbb{C} \,|\, |z| \le \epsilon\}; \qquad C_\epsilon := \{z \in \mathbb{C} \,|\, |z| = \epsilon\}.$$

Then

$$\int_{\mathbb{D}\setminus\mathbb{D}_\varepsilon} \bar{\partial}h \cdot \varphi \, d\bar{z} \wedge dz + \int_{\mathbb{D}\setminus\mathbb{D}_\varepsilon} h \cdot \bar{\partial}\varphi \, d\bar{z} \wedge dz = -\int_{C_\varepsilon} h\varphi \, dz.$$

Hence

$$\int_{\mathbb{D}} h \cdot \bar{\partial}\varphi = -\lim_{\varepsilon \to 0} \frac{1}{2i} \int_{C_\varepsilon} h\varphi \, dz$$

and it suffices to show this limit is zero. By Hölder's inequality we have

$$\|h\varphi\|_{L^1(C_\varepsilon)} \le \|h\|_{L^p(C_\varepsilon)} \|\varphi\|_{L^q(C_\varepsilon)}, \qquad \frac{1}{p} + \frac{1}{q} = 1.$$

The trace theorem (Proposition B.1.21) shows that there exists a constant $C > 0$ such that $\|h\|_{L^p(C_\varepsilon)} \le C\|h\|_{W^{1,p}(\mathbb{D}_\varepsilon)}$. Therefore

$$\|h\varphi\|_{L^1(C_\varepsilon)} \;\le\; C(2\pi\varepsilon)^{1/q}\|h\|_{W^{1,p}(\mathbb{D})} \max_{z\in\mathbb{D}} |\varphi(z)| \;\to\; 0,$$

as required. □

LEMMA E.5.7. *For every integer ℓ the Cauchy–Riemann operator $\bar{\partial}$ maps $\mathcal{O}_{\ell+1}$ into $\mathcal{O}_\ell$ and its image contains $\mathcal{O}_{\ell+1}$.*

PROOF. That $\bar{\partial} := \frac{1}{2}(\partial_s + i\partial_t)$ maps $\mathcal{O}_{\ell+1}$ into $\mathcal{O}_\ell$ follows from the *(Derivative)* axiom. We prove that its image contains $\mathcal{O}_{\ell+1}$. Multiplying all the functions by $z^{-(\ell+1)}$ we only need to consider the case $\ell = -1$.

Let $U \subset \mathbb{C}$ be an open disc centered at zero and $f \in \mathcal{O}_0(U)$. Fix a real number q such that $1 < q < 2$. Then $f \in W^{1,q}(U)$ (see Exercise E.5.2). By Theorem C.1.10 there is a function $g \in W^{2,q}(U)$ such that $g(0) = 0$ and $\bar{\partial}g = f$. By elliptic regularity (Proposition B.4.9), g is smooth on $U \setminus \{0\}$. Now let $\partial := \frac{1}{2}(\partial_s - i\partial_t)$ and n be a nonnegative integer. Then by the *(Pole–Zero)* and *(Product)* axioms

$$\bar{\partial}(z^n\partial^n g) = z^n\partial^n\bar{\partial}g = z^n\partial^n f \in \mathcal{O}_0(U) \subset W^{1,q}(U)$$

and hence $z^n\partial^n g \in W^{2,q}_{\mathrm{loc}}(U)$ by elliptic regularity (Proposition B.4.9). Since $W^{2,q} \subset W^{1,p} \subset C^{1-2/p}$ for $p := 2q/(2-q) > 2$ (see Theorems B.1.11 and B.1.12) it follows that $z^n\partial^n g$ is continuous at the origin. Since this holds for every integer $n \ge 0$ we deduce that $g \in \mathcal{O}_0$. □

REMARK E.5.8. Define $g(z) := z\sqrt{-\log|z|}$. An easy calculation shows that g and $\bar{\partial}g$ belong to $\mathcal{O}_0$. Assume that there is $h \in \mathcal{O}_1$ such that $\bar{\partial}h = \bar{\partial}g$. Then $g - h \in W^{1,q}(\mathbb{D})$ for some $q > 1$ and so Lemma E.5.6 implies that $h - g$ is smooth near the origin. In particular g should already belong to $\mathcal{O}_1$, which is impossible because of the singularity of $\sqrt{-\log|z|}$. This proves that none of the operators $\bar{\partial} : \mathcal{O}_{\ell+1} \to \mathcal{O}_\ell$ is onto.

LEMMA E.5.9. *Let k be a positive integer and $f \in \mathcal{O}_k$. If $u : \mathbb{D}\setminus\{0\} \to \mathbb{C}$ is a smooth function such that $\bar{\partial}u = f$ then the following holds.*

(i) *If u extends to a smooth function on $\mathbb{D}$ then there is a complex polynomial $p(z) = a_0 + a_1 z + \cdots + a_k z^k$ such that $u - p$ belongs to $\mathcal{O}_{k+1}$.*

(ii) *If u is bounded near the origin then there is a complex polynomial $p(z) = a_0 + a_1 z + \cdots + a_{k-1} z^{k-1}$ such that $u - p$ belongs to $\mathcal{O}_k$.*

PROOF. We prove (i). Denote by p the Taylor polynomial of u (in the variables s, t or $z, \bar{z}$) up to order k. Then

$$u(z) = p(z) + R(z)$$

where $R \in \mathcal{O}_{k+1}$. Moreover,

$$\bar{\partial} p(z) = \bar{\partial} u - \bar{\partial} R = f - \bar{\partial} R \in \mathcal{O}_k.$$

Hence by Remark E.5.1 all the partial derivatives of $\bar{\partial} p$ up to order $k-1$ vanish at the origin. Since $\bar{\partial} p$ is a polynomial of degree at most $k-1$ it follows that $\bar{\partial} p = 0$ and so p is a complex polynomial.

We prove (ii). Since $f = \bar{\partial} u$ belongs to $\mathcal{O}_k$, Lemma E.5.7 shows that $f = \bar{\partial} v$ with $v \in \mathcal{O}_k$. Hence $u - v$ is holomorphic away from the origin and bounded. By the Cauchy integral formula, there is a constant $c > 0$ such that $|du(z) - dv(z)| \le c|z|^{-1}$. Hence $u - v \in W^{1,p}$ for $1 < p < 2$ and therefore, by Lemma E.5.6, $u - v$ extends to a holomorphic function on $\mathbb{D}$. It follows that u has the form

$$u(z) = a_0 + a_1 z + \cdots + a_{k-1} z^{k-1} + z^k w(z) + v(z),$$

where $w : \mathbb{D} \to \mathbb{C}$ is holomorphic. □

We are finally in a position to prove the local structure theorem we have been aiming for. We can be more precise than in Chapter 2 (cf. the proof of Lemma 2.4.1) because our discussion of the function spaces $\mathcal{O}_\ell$ allows us to make sharp statements about the properties of the coordinate change φ.

PROPOSITION E.5.10. *Let J be a smooth almost complex structure on $\mathbb{C}^n$ standard at the origin and $u : \mathbb{D} \to \mathbb{C}^n$ be a nonconstant solution of the Cauchy–Riemann equations*

$$\partial_s u + J(u) \partial_t u = 0, \qquad u(0) = 0. \tag{E.5.2}$$

Then there is a positive integer k and a nonzero vector $a_0 \in \mathbb{C}^n$ such that the following holds.

(i) *There are vectors $a_1, \ldots, a_{k-1} \in \mathbb{C}^n$ such that*

$$u(z) = z^k a_0 + z^{k+1} a_1 + \cdots + z^{2k-1} a_{k-1} + R(z)$$

where $R \in \mathcal{O}_{2k}$.

(ii) *If $\Pi \subset \mathbb{C}^n$ is a complex hyperplane transverse to the line $\mathbb{C} a_0$ then there is a local $C^{k,1}$-diffeomorphism $\varphi : (\mathbb{D}, 0) \to (\mathbb{D}, 0)$ such that*

$$u \circ \varphi(z) = z^k a_0 + F(z), \qquad \varphi(z) = z + \varphi_1(z),$$

where $F : \mathbb{D} \to \Pi$ belongs to $\mathcal{O}_{k+1}$ and $\varphi_1 \in \mathcal{O}_2$. If $k = 1$ then φ can be chosen smooth.

PROOF. Since u is smooth and nonconstant, its derivatives at the origin cannot all vanish (see Remark E.4.4 or Theorem 2.3.2). Hence there is an integer $k \ge 1$ such that $d^k u(0) \ne 0$ and $d^j u(0) = 0$ for all $j < k$. This implies that u belongs to $\mathcal{O}_k$. Since u satisfies (E.5.2), we have

$$\bar{\partial} u(z) := \frac{1}{2}\Big(\partial_s u(z) + J_0 \partial_t u(z) \Big) = \frac{1}{2}\Big(J_0 - J(u(z)) \Big) \partial_t u(z).$$

By the *(Composition)* axiom, $(J_0 - J) \circ u$ belongs to $\mathcal{O}_k$, and the *(Product)* and *(Derivative)* axioms show that $\bar{\partial} u$ belongs to $\mathcal{O}_{2k-1}$. Hence, by Lemma E.5.9 (i), there is a complex polynomial P of degree at most $2k-1$ such that $u - P \in \mathcal{O}_{2k}$. By definition of k and by uniqueness of the Taylor expansion, it follows that P has the form

$$P(z) = z^k p(z), \qquad p(z) = a_0 + za_1 + \cdots + z^{k-1}a_{k-1}, \qquad a_0 \neq 0.$$

This proves (i). To prove (ii) we choose linear complex coordinates on $\mathbb{C}^n$ so that

$$a_0 = (1, 0, \ldots, 0), \qquad \Pi = \{0\} \times \mathbb{C}^{n-1} \subset \mathbb{C}^n.$$

Define $E : \mathbb{D} \to \mathbb{C}^n$ by

$$E(z) := z^{-k} R(z), \qquad R := u - P,$$

for $z \neq 0$ and $E(0) := 0$. Then, by the *(Pole–Zero)* and *(Product)* axioms in Lemma E.5.3, $E \in \mathcal{O}_k$. Note that E is not necessarily smooth at the origin. Let $\pi : \mathbb{C}^n \to \mathbb{C}$ denote the projection onto the first coordinate. Then $\pi \circ p(0) = 1$ and $\pi \circ E(0) = 0$. Let $U \subset \mathbb{D}$ be an open disc centered at zero such that $\pi(p(z))$ and $\pi(p(z) + E(z))$ have positive real part for every $z \in U$. Define $\psi : U \to \mathbb{C}$ and $\psi_0 : U \to \mathbb{C}$ by

$$\psi(z) := z \cdot \Big(\pi \circ p(z) + \pi \circ E(z)\Big)^{1/k}, \qquad \psi_0(z) := z \cdot \big(\pi \circ p(z)\big)^{1/k}.$$

Then ψ_0 is a local C^∞-diffeomorphism and ψ is smooth in $U \setminus \{0\}$. Moreover,

$$\psi(z) = \psi_0(z)(1 + f(z))^{1/k}, \qquad f := (\pi \circ p)^{-1} \pi \circ E.$$

By the *(Product)* axiom $f \in \mathcal{O}_{k+1}$ and hence, by the *(Composition)* axiom with $h(z) := (1+z)^{1/k} - 1$, we have $(1+f)^{1/k} - 1 \in \mathcal{O}_{k+1}$. Hence it follows from the *(Product)* axiom that

$$\psi - \psi_0 \in \mathcal{O}_{k+1}.$$

In particular, ψ is a local $C^{k,1}$-diffeomorphism (Exercise E.5.2). Hence $\varphi := \psi^{-1}$ is a local $C^{k,1}$-diffeomorphism and $\varphi \in \mathcal{O}_1$ by the *(Inverse)* axiom. By definition of ψ we have

$$\psi(z)^k = \pi(z^k p(z) + z^k E(z)) = \pi(P(z) + R(z)) = \pi(u(z))$$

and hence

$$\pi \circ u(\varphi(z)) = z^k.$$

Moreover, the function

$$F(z) := u \circ \varphi(z) - z^k = (\mathrm{id} - \pi) \circ u \circ \varphi(z)$$

takes values in $\Pi = \{0\} \times \mathbb{C}^{n-1}$. Since $(\mathrm{id} - \pi)(u(z)) = (\mathrm{id} - \pi)(u(z) - z^k a_0)$ we deduce that $(\mathrm{id} - \pi) \circ u$ is a smooth function in $\mathcal{O}_{k+1}$. Hence it follows from the *(Composition)* axiom that $F \in \mathcal{O}_{k+1}$.

It remains to check that $\varphi_1(z) := \varphi(z) - z$ belongs to $\mathcal{O}_2$. Since $\varphi(z) \neq 0$ for $z \neq 0$ and $\phi \in \mathcal{O}_1$ it follows from Remark E.5.4 that

$$f \in \mathcal{O}_m \qquad \Longrightarrow \qquad f \circ \varphi \in \mathcal{O}_m.$$

Define

$$\psi_1(z) := \psi(z) - z.$$

Then ψ_1 belongs to $\mathcal{O}_2$ and $(\mathbb{1} + d\psi_1(\varphi(z))) \cdot d\varphi(z) = \mathbb{1}$ by definition of φ. Hence

$$d\varphi_1(z) = d\phi(z) - \mathbb{1} = -d\psi_1(\varphi(z)) \cdot d\varphi(z).$$

Since $d\psi_1 \in \mathcal{O}_1$ we have $d\psi_1 \circ \varphi \in \mathcal{O}_1$ and so the (*Product*) axiom shows that $d\varphi_1 \in \mathcal{O}_1$. Moreover, φ_1 is continuous at the origin and $\varphi_1(0) = 0$. Hence φ_1 belongs to $\mathcal{O}_2$. That φ is smooth in the case $k = 1$ is obvious from the construction. □

REMARK E.5.11. Given a continuous map $u : \mathbb{D} \to \mathbb{R}^N$ such that $u(0) = 0$, the tangent cone to u at the origin is the set

$$\mathcal{C}_0(u) := \left\{ \lambda\xi \,\middle|\, \lambda \geq 0 \text{ and } \exists z_\nu \to 0 \text{ such that } \frac{u(z_\nu)}{|u(z_\nu)|} \to \xi \right\}.$$

If $\psi : \mathbb{R}^N \to \mathbb{R}^N$ is a local diffeomorphism such that $\psi(0) = 0$ then

$$\mathcal{C}_0(\psi \circ u) = d\psi(0)(\mathcal{C}_0(u)).$$

This means that every continuous map $u : \mathbb{D} \to M$ has an intrinsic tangent cone $\mathcal{C}_z(u) \subset T_{u(z)}M$ at every point $z \in \mathbb{D}$. When u is a nonconstant J-holomorphic curve, Proposition E.5.10 shows that this cone is a 1-dimensional complex subspace of TM. We call it **the tangent space of** u **at** z and denote it by $\mathcal{T}_z(u)$.

The **order** of a nonconstant J-holomorphic curve $u : \Sigma \to M$ at a point $z_0 \in \Sigma$ is the unique integer k such that $\psi \circ u \circ \phi^{-1} \in \mathcal{O}_k \backslash \mathcal{O}_{k+1}$, where $\phi : (\Sigma, z_0) \to (\mathbb{C}, 0)$ is a coordinate chart on Σ and $\psi : (M, u(z_0)) \to (\mathbb{C}^n, 0)$ is a coordinate chart on M. By Proposition E.5.10, there is such an integer k and, by the *(Composition)* axiom in Remarks E.5.4 and E.5.5, the order is independent of the local coordinate charts used to define it. Note that z_0 is a regular point of u (i.e. $du(z_0) \neq 0$) if and only if the order of u at z_0 is $k = 1$, and is a singular point if and only if $k \geq 2$.

Now assume that $u : \mathbb{D} \to M$ is a J-holomorphic curve with a singularity of order $k \geq 2$ at $z = 0$. Identify M locally with $\mathcal{T}_0(u) \times \mathbb{C}^{n-1}$ and $\mathcal{T}_0(u)$ with $\mathbb{C}$. Then Proposition E.5.10 asserts that there is a local C^2-reparametrization $\varphi : \mathbb{D} \to \mathbb{D}$ such that

$$\varphi(0) = 0, \qquad u \circ \varphi(z) = (z^k, F(z))$$

with $F(z) = o(z^k)$. Thus the image of a J-holomorphic curve u can be seen locally as the graph of a multivalued map over its tangent space at the origin, mapping ζ to the set $\{F(z) \,|\, z^k = \zeta\}$. The price one has to pay for this nice picture is that F is not holomorphic but only of class C^2. However, this is enough since we only study J-holomorphic singularities up to C^1-conjugation.

E.6. Contact between branches

This section carries out the second step in the proof of Theorem E.1.1. It contains a precise analysis of the distance between two distinct J-holomorphic curves near an intersection. The only nontrivial case is when both curves have the same tangent space and order of singularity at their intersection point. Since the statement is local we will identify M with $\mathbb{C}^n$. We shall assume that the curves have the form provided by Proposition E.5.10.

PROPOSITION E.6.1. *Let J be an almost complex structure on $\mathbb{C}^n$ that is standard at 0 and k be a positive integer. For $j = 1, 2$ let $U_j \subset \mathbb{C}$ be an open neighbourhood of zero, $u_j : \bar{U}_j \to \mathbb{C}^n$ be a J-holomorphic curve, and $\varphi_j : \mathbb{D} \to \bar{U}_j$ be a C^2-diffeomorphism such that $\varphi_j - \mathrm{id} \in \mathcal{O}_2$, φ_j is smooth in the case $k = 1$, and*

$$u_j \circ \varphi_j(z) = (z^k, F_j(z)), \qquad j = 1, 2,$$

where $F_j : \mathbb{D} \to \mathbb{C}^{n-1}$ *belongs to* $\mathcal{O}_{k+1}$. *If* $F_2 - F_1$ *does not vanish identically in any neighbourhood of zero then there is an integer* $m > k$ *and a nonzero vector* $a \in \mathbb{C}^{n-1}$ *such that*

$$F_2(z) - F_1(z) = z^m a + R(z)$$

where $R \in \mathcal{O}_{m+1}$.

The idea of the proof is as follows. If the reparametrizations φ_j were holomorphic, then one could apply the Hartman–Wintner theorem to the difference $F_1 - F_2$ to conclude that for some integer $m > k$ there is a nonzero homogeneous polynomial h of degree m such that $F_1 - F_2 - h(z) \in \mathcal{O}_{m+1}$. It would then follow from Lemma E.5.9 that h is holomorphic and hence of the form $z^m a$. As it is, we know only that the composite maps $u_j \circ \varphi_j$ have complex tangent spaces at all points, and therefore must work quite hard to show that $F_1 - F_2$ satisfies the hypotheses of the Hartman–Wintner theorem. First we must choose suitable coordinates near 0 in $\mathbb{C}^n$, that are called normal coordinates, and then we must investigate the properties of the maps F_j in these coordinates.

Normal coordinates. The next proposition constructs an embedded J-holomorphic disc tangent to a given one-dimensional complex subspace $\Pi_0 \subset T_{x_0}M$ which extends to an embedded family of J-holomorphic discs of codimension one. We shall apply this proposition to the case where x_0 is a singular point on a J-holomorphic curve and Π_0 is the tangent space to this curve. As a result, one obtains a useful coordinate chart on M near the singularity.

PROPOSITION E.6.2. *Let* (M, J) *be a smooth almost complex* $2n$*-manifold. Let* $x_0 \in M$ *and* $\Pi_0 \subset T_{x_0}M$ *be a complex line. Then there exists a smooth coordinate chart* $\Psi : (M, x_0) \to (\mathbb{C}^n, 0)$ *such that*

$$\Psi(x_0) = 0, \qquad d\Psi(x_0)\Pi_0 = \mathbb{C} \times \{0\}, \qquad \Psi_* J = \begin{pmatrix} i & J_3 \\ 0 & J_4 \end{pmatrix} \in \mathrm{GL}_{\mathbb{R}}(\mathbb{C} \times \mathbb{C}^{n-1}),$$

where $J_3(0) = 0$ *and* $J_4(0)$ *is the standard complex structure* i *on* $\mathbb{C}^{n-1}$.

PROOF. Assume without loss of generality that $M = \mathbb{C}^n$, $x_0 = 0$, and J is an almost complex structure on $\mathbb{C}^n$ such that $J(0) = i$. Fix a number $p > 2$ and define

$$\mathcal{X} := \left\{ u \in W^{2,p}(\mathbb{D}, \mathbb{C}^n) \,|\, u(e^{i\theta}) \in e^{3i\theta/2}\mathbb{R}^n \right\}, \qquad \mathcal{Y} := W^{1,p}(\mathbb{D}, \mathbb{C}^n) \times \mathbb{C}^n \times \mathbb{C}^n.$$

Define $\mathcal{F} : \mathcal{X} \to \mathcal{Y}$ by

$$\mathcal{F}(u) := \left(\frac{1}{2}(\partial_s u + J(u)\partial_t u), u(0), \partial_s u(0) \right).$$

Then

$$d\mathcal{F}(0)\xi = (\bar{\partial}\xi, \xi(0), \partial_s \xi(0))$$

for $\xi \in \mathcal{X}$. The same argument as in Step 3 in the proof of Theorem E.3.1 shows that $d\mathcal{F}(0)$ is a Banach space isomorphism. Hence, by the inverse function theorem, $\mathcal{F}$ has a smooth inverse defined on some open neighbourhood $\mathcal{U} \subset \mathcal{Y}$ of zero, mapping zero to itself.

Choose a complex complement Π of Π_0 in $\mathbb{C}^n$ and a point $\xi_0 \in \Pi_0 \backslash \{0\}$ and an open neighbourhood $W \subset \Pi$ of zero such that $(0, \xi_0, x) \in \mathcal{U}$ for every $x \in W$. Define the map $\Phi : \mathbb{D} \times W \to \mathbb{C}^n$ by

$$\Phi(z, x) := \mathcal{F}^{-1}(0, x, \xi_0)(z), \qquad z \in \mathbb{D}, \ x \in W.$$

For $x \in W$ define the map $u_x : \mathbb{D} \to \mathbb{C}^n$ by

$$u_x(z) := \Phi(z, x)$$

Then $u_0(z) = z\xi_0 + z^2\bar{\xi}_0$ and, for $x \in W$, u_x is the unique J-holomorphic disc near u_0 that satisfies $u_x(0) = x$ and $\partial_s u_x(0) = \xi_0$ and $u_x(e^{i\theta}) \in e^{3i\theta/2}\mathbb{R}^n$. By elliptic regularity, u_x depends smoothly on x and so Φ is smooth. Moreover,

$$d\Phi(0)(\zeta, \xi) = \zeta\xi_0 + \xi$$

and so $\Phi(0) = 0$ and $d\Phi(0) \in \mathrm{GL}(n, \mathbb{C})$. By the inverse function theorem, Φ restricts to a diffeomorphism from a sufficiently small open neighbourhood of zero in $\mathbb{D} \times \Pi$ to an open neighbourhood of zero in $\mathbb{C}^n$. Identify Π with $\mathbb{C}^{n-1}$. Then $\Psi := \Phi^{-1}$ satisfies the requirements of the proposition. □

Maps with complex tangent spaces. We now turn to the proof of Proposition E.6.1. The first goal is to show that, although the reparametrization

$$u \circ \varphi(z) = (z^k, F(z))$$

is not required to be J-holomorphic, the map F satisfies an elliptic equation in normal coordinates. The proofs are straightforward, but involve considerable algebraic manipulation. We denote by

$$\{\Phi, F\} := \partial_s\Phi \cdot \partial_t F - \partial_t\Phi \cdot \partial_s F$$

the Poisson bracket of two functions $\Phi : \mathbb{D} \to \mathrm{Hom}_{\mathbb{R}}(\mathbb{C}^k, \mathbb{C}^\ell)$ and $F : \mathbb{D} \to \mathbb{C}^k$.

LEMMA E.6.3. *Let*

$$J := \begin{pmatrix} i & J_3 \\ 0 & J_4 \end{pmatrix} \in \mathrm{End}_{\mathbb{R}}(\mathbb{C} \oplus \mathbb{C}^{n-1}) \tag{E.6.1}$$

be a C^1-almost complex structure on $\mathbb{C}^n$ and $u : \mathbb{D} \to \mathbb{C}^n = \mathbb{C} \times \mathbb{C}^{n-1}$ be a C^2-function of the form

$$u(z) =: (f(z), F(z)).$$

Assume that $f : \mathbb{D} \to \mathbb{C}$ is a nonconstant holomorphic function, denote $Y := \{z \in \mathbb{D} \,|\, f'(z) = 0\}$, and assume that u has a one-dimensional J-complex tangent space at every point $z \in \mathbb{D}\backslash Y$. Define the functions $A, B, C : \mathbb{D}\backslash Y \to \mathbb{C}$ by

$$A := \frac{1}{f'}J_3(u)\partial_s F, \qquad B := \frac{1}{f'}J_3(u)\partial_t F, \qquad C := \left\{\frac{1}{f'}J_3(u), F\right\}. \tag{E.6.2}$$

Then F satisfies the equations

$$\partial_s F + J_4(u) \cdot \partial_t F = \Re e(B)\partial_s F + \Im m(B)\partial_t F, \tag{E.6.3}$$

$$J_4(u) \cdot \partial_s F - \partial_t F = \Re e(A)\partial_s F + \Im m(A)\partial_t F, \tag{E.6.4}$$

and

$$\begin{aligned} &\Delta F - \Re e(B)\partial_s^2 F + \Im m(A)\partial_t^2 F + \Re e(A + iB)\partial_s\partial_t F \\ &= \Re e(C)\partial_s F + \Im m(C)\partial_t F - \{J_4(u), F\} \end{aligned} \tag{E.6.5}$$

on $\mathbb{D}\backslash Y$.

PROOF. Since u has J-complex tangencies, there exist functions $a, b, c, d : \mathbb{D}\backslash Y \to \mathbb{R}$ such that

$$J(u) \cdot (\partial_s f, \partial_s F) = a \cdot (\partial_s f, \partial_s F) + b \cdot (\partial_t f, \partial_t F), \tag{E.6.6}$$

$$J(u) \cdot (\partial_t f, \partial_t F) = c \cdot (\partial_s f, \partial_s F) + d \cdot (\partial_t f, \partial_t F). \tag{E.6.7}$$

Since $\partial_s f = f'$, $\partial_t f = if'$, the first components of equations $(E.6.6)$ and $(E.6.7)$, imply the two following relations.

$$\begin{aligned} if'(z) + J_3\big(f(z), F(z)\big)\partial_s F &= af'(z) + ibf'(z), \\ -f'(z) + J_3\big(f(z), F(z)\big)\partial_t F &= cf'(z) + idf'(z). \end{aligned}$$

Divide these equations by f' and use (E.6.2) to obtain

$$\begin{aligned} a &= \Re e(A), & b &= 1 + \Im m(A), \\ c &= -1 + \Re e(B), & d &= \Im m(B). \end{aligned}$$

Hence (E.6.3) and (E.6.4) follow from the second components of (E.6.6) and (E.6.7). Equation (E.6.5) follows from (E.6.3) and (E.6.4) by computing $\partial_s(\partial_s F + J_4 \partial_t F) + \partial_t(\partial_t F - J_4 \partial_s F)$ and using the fact that $C = \partial_s B - \partial_t A$. □

Next we consider two maps $u_j = (f_j, F_j)$ as in Lemma E.6.3 with $f_1(z) = f_2(z) = z^k$ and examine the difference $H := F_1 - F_2$. The main point is equation (E.6.9) below which shows that H satisfies the hypotheses of the Hartman–Wintner theorem for suitable functions $a, b, c : \mathbb{D} \to \mathbb{R}$.

LEMMA E.6.4. *Let J be as in Lemma E.6.3 and assume $J(0) = i$. Let k be a positive integer and $F_1, F_2 : \mathbb{D} \to \mathbb{C}^{n-1}$ be two maps which belong to $\mathcal{O}_{k+1}$. Denote*

$$u_j(z) := (z^k, F_j(z)), \qquad H := F_2 - F_1,$$

and assume that each u_j has J-complex tangent spaces at all points $z \in \mathbb{D}\backslash\{0\}$. Then the following holds.

(i) *The maps $J_3(u_2) - J_3(u_1)$ and $J_4(u_2) - J_4(u_1)$ have the form*

$$J_3(u_2)(z) - J_3(u_1)(z) = \alpha(z) \cdot H(z), \qquad J_4(u_2)(z) - J_4(u_1)(z) = \beta(z) \cdot H(z),$$

where

$$\alpha : \mathbb{D} \to \mathrm{Hom}_{\mathbb{R}}\big(\mathbb{C}^{n-1}, \mathrm{Hom}_{\mathbb{R}}(\mathbb{C}^{n-1}, \mathbb{C})\big), \quad \beta : \mathbb{D} \to \mathrm{Hom}_{\mathbb{R}}\big(\mathbb{C}^{n-1}, \mathrm{End}_{\mathbb{R}}(\mathbb{C}^{n-1})\big),$$

both belong to $\mathcal{O}_0$.

(ii) *The map $\bar\partial H := \frac{1}{2}(\partial_s H + J_0 \partial_t H)$ has the form*

$$\bar\partial H(z) = \varepsilon_1(z) \cdot H(z) + \varepsilon_2(z) \cdot \partial_s H(z) + \varepsilon_3(z) \cdot \partial_t H(z), \tag{E.6.8}$$

where $\varepsilon_1, \varepsilon_2, \varepsilon_3 : \mathbb{D} \to \mathrm{End}_{\mathbb{R}}(\mathbb{C}^{n-1})$ belong to $\mathcal{O}_k$.

(iii) *There is a constant $C > 0$ and functions $a, b, c : \mathbb{D} \to \mathbb{R}$ in $\mathcal{O}_{k+1}$ satisfying $a(0) = c(0) = 1$, $b(0) = 0$, and*

$$\left| a\partial_s^2 H + 2b\partial_s\partial_t H + c\partial_t^2 H \right| \le C(|H| + |\partial_s H| + |\partial_t H|). \tag{E.6.9}$$

in a neighbourhood of zero.

PROOF. For $j = 1, 2$ let $A_j, B_j, C_j : \mathbb{D} \setminus \{0\} \to \mathbb{C}$ be defined by (E.6.2) with $f(z) := z^k$ and (u, F) replaced by (u_j, F_j). Then A_j and B_j belong to $\mathcal{O}_{k+1}$ and C_j belongs to $\mathcal{O}_k$. To see this, note that $1/f' \in \mathcal{O}_{-k+1}$, $\partial_s F_j \in \mathcal{O}_k$, and $J_3(0) = 0$. Hence the *(Composition)* axiom in Lemma E.5.3 and Remark E.5.5 shows that $J_3(f, F_j) \in \mathcal{O}_k$. Therefore the *(Product)* axiom shows that $A_j = (1/f')J_3(f, F_j)\partial_s F_j$ belongs to $\mathcal{O}_{k+1}$. Similarly, $B_j \in \mathcal{O}_{k+1}$ and hence $C_j = \partial_s B_j - \partial_t A_j \in \mathcal{O}_k$.

We prove (i). Define the map $\alpha : \mathbb{D} \to \mathrm{Hom}_{\mathbb{R}}\big(\mathbb{C}^{n-1}, \mathrm{Hom}_{\mathbb{R}}(\mathbb{C}^{n-1}, \mathbb{C})\big)$ by

$$\alpha(z)\xi := \left(\int_0^1 dJ_3\Big(u_1(z) + \tau\big(0, H(z)\big)\Big) d\tau \right) \cdot (0, \xi)$$

for $z \in \mathbb{D}$ and $\xi \in \mathbb{C}^{n-1}$. Since $(0, H(z)) = u_2(z) - u_1(z)$ we have

$$J_3(u_2)(z) - J_3(u_1)(z) = \alpha(z) \cdot H(z).$$

(Compare this with the formula for J in the proof of Theorem 2.3.2.) We must prove that α belongs to $\mathcal{O}_0$. To see this, define $\alpha_\tau(z)$ and $\widehat{\alpha}$ by

$$\alpha_\tau(z)\xi := dJ_3\Big(u_1(z) + \tau\big(0, H(z)\big)\Big) \cdot (0, \xi), \qquad \widehat{\alpha}\xi := dJ_3(0) \cdot (0, \xi)$$

By the *(Composition)* axiom of Remark E.5.5, we have $\alpha_\tau - \widehat{\alpha} \in \mathcal{O}_k$ for every τ. Moreover, the constants in the definition of $\mathcal{O}_k$ can be chosen independent of τ. Hence $\alpha - \widehat{\alpha} = \int_0^\tau (\alpha_\tau - \widehat{\alpha})\, d\tau$ belongs to $\mathcal{O}_k$ and so $\alpha \in \mathcal{O}_0$. This proves the first assertion in (i). The proof of the second assertion is similar.

We prove (ii). By definition of α we have

$$B_2 - B_1 = \frac{1}{f'}\Big((\alpha \cdot H) \cdot \partial_t F_1 + J_3(u_2) \cdot \partial_t H\Big).$$

Moreover, $1/f' \in \mathcal{O}_{-k+1}$, $\partial_t F_1 \in \mathcal{O}_k$, and $\alpha \in \mathcal{O}_0$ by (ii). Hence it follows from the *(Product)* axiom in Lemma E.5.3 that $B_2 - B_1$ has the form

$$(B_2 - B_1)(z) = \beta_1(z) \cdot H(z) + \beta_2(z) \cdot \partial_t H(z),$$

where $\beta_1, \beta_2 : \mathbb{D} \to \mathrm{Hom}_{\mathbb{R}}(\mathbb{C}^{n-1}, \mathbb{C})$ belong to $\mathcal{O}_1$. By Lemma E.6.3, F_1 and F_2 satisfy (E.6.3), i.e.

$$\partial_s F_j + J_4(u_j) \cdot \partial_t F_j = \Re e(B_j)\partial_s F_j + \Im m(B_j)\partial_t F_j$$

for $j = 1, 2$. Taking the difference of these equations we find

$$\begin{aligned} &2\bar{\partial} H + (J_4(u_2) - J_4(u_1))\, \partial_t F_1 + \big(J_4(u_2) - i\big)\, \partial_t H \\ &= \Re e\big(B_2 - B_1\big)\partial_s F_1 + \Re e(B_2)\partial_s H \\ &\quad + \Im m\big(B_2 - B_1\big)\partial_t F_1 + \Im m(B_2)\partial_t H \end{aligned}$$

and hence

$$\begin{aligned} &2\bar{\partial} H + (\beta \cdot H)\, \partial_t F_1 + \big(J_4(u_2) - i\big)\, \partial_t H \\ &= \Re e\big(\beta_1 \cdot H + \beta_2 \cdot \partial_t H\big)\partial_s F_1 + \Re e(B_2)\partial_s H \\ &\quad + \Im m\big(\beta_1 \cdot H + \beta_2 \cdot \partial_t H\big)\partial_t F_1 + \Im m(B_2)\partial_t H. \end{aligned}$$

This proves (ii) because $J_4(u_2) - i$ belongs to $\mathcal{O}_k$.

We prove (iii). Similar calculations show that $A_1 - A_2$ and $C_1 - C_2$ have the form

$$\begin{aligned} A_1(z) - A_2(z) &= \alpha_1(z) \cdot H(z) + \alpha_2(z) \cdot \partial_s H(z), \\ C_1(z) - C_2(z) &= \gamma_1(z) \cdot H(z) + \gamma_2(z) \cdot \partial_s H(z) + \gamma_3(z) \cdot \partial_t H(z), \end{aligned}$$

where $\alpha_1, \alpha_2 \in \mathcal{O}_1$ and $\gamma_1, \gamma_2, \gamma_3 \in \mathcal{O}_0$ take their values in $\mathrm{Hom}_{\mathbb{R}}(\mathbb{C}^{n-1}, \mathbb{C})$. By Lemma E.6.3, F_1 and F_2 satisfy (E.6.5), i.e.

$$\begin{aligned} &\Delta F_j - \Re e(B_j)\partial_s^2 F_j + \Im m(A_j)\partial_t^2 F_j + \Re e(A_j + iB_j)\partial_s\partial_t F_j \\ &= \Re e(C_j)\partial_s F_j + \Im m(C_j)\partial_t F_j - \{J_4(u_j), F_j\} \end{aligned}$$

for $j = 1, 2$. Taking the difference of these equations we find

$$\begin{aligned}
&\Delta H - \Re e(B_2 - B_1)\partial_s^2 F_1 - \Re e(B_2)\partial_s^2 H \\
&\qquad + \Im m(A_2 - A_1)\partial_t^2 F_1 + \Im m(A_2)\partial_t^2 H \\
&\qquad + \Re e\big(A_2 - A_1 + i(B_2 - B_1)\big)\partial_s\partial_t F_1 + \Re e(A_2 + iB_2)\partial_s\partial_t H \\
&= \Re e(C_2 - C_1)\partial_s F_1 + \Re e(C_2)\partial_s H \\
&\qquad + \Im m(C_2 - C_1)\partial_t F_1 + \Im m(C_2)\partial_t H \\
&\qquad - \{J_4(u_2) - J_4(u_1), F_1\} - \{J_4(u_2), H\}.
\end{aligned}$$

Therefore, if we set $a := 1 - \Re e(B_2)$, $2b := \Re e(A_2 + iB_2)$ and $c := 1 + \Im m(A_2)$, we find that

$$a\partial_s^2 H + 2b\partial_s\partial_t H + c\partial_t^2 H = \lambda_1 \cdot H + \lambda_2 \cdot \partial_s H + \lambda_3 \cdot \partial_t H,$$

where $\lambda_1 \in \mathcal{O}_k$ and $\lambda_2, \lambda_3 \in \mathcal{O}_{k-1}$ take their values in $\mathrm{End}_{\mathbb{R}}(\mathbb{C}^{n-1})$. In particular, the maps λ_i are bounded. This proves the lemma. □

PROOF OF PROPOSITION E.6.1. Let J be an almost complex structure on $\mathbb{C}^n$ that is standard at 0 and k be a positive integer. For $j = 1, 2$ let $U_j \subset \mathbb{C}$ be an open neighbourhood of zero, $u_j : \bar{U}_j \to \mathbb{C}^n$ be a J-holomorphic curve, and $\varphi_j : \mathbb{D} \to \bar{U}_j$ be a C^2-diffeomorphism such that $\varphi_j - \mathrm{id} \in \mathcal{O}_2$, φ_j is smooth in the case $k = 1$, and

$$u_j \circ \varphi_j(z) = (z^k, F_j(z)), \qquad j = 1, 2,$$

where $F_j : \mathbb{D} \to \mathbb{C}^{n-1}$ belongs to $\mathcal{O}_{k+1}$. Note that F_j is smooth in the case $k = 1$. Suppose that $F_2 - F_1$ does not vanish identically in any neighbourhood of zero. We must show that $H := F_1 - F_2$ may be written in the form $z^m a + R(z)$ where $R \in \mathcal{O}_{m+1}$.

Assume first that J satisfies the normal form condition (E.6.1). Then, by Lemma E.6.4 (iii), the C^2-function $H := F_2 - F_1$ satisfies the hypotheses of the Hartman–Wintner theorem in a neighbourhood of zero for suitable functions $a, b, c : \mathbb{D} \to \mathbb{R}$ (that have Lipschitz continuous first derivatives and satisfy (E.4.1). Since H does not vanish in any neighbourhood of zero it follows from Theorem E.4.1 that there is a positive integer m and a nonzero homogeneous polynomial $h : \mathbb{C} \to \mathbb{C}^{n-1}$ of degree m such that

$$H(z) = h(z) + o(z^m).$$

Since H belongs to $\mathcal{O}_{k+1}$ and h is nonzero it follows that $m > k$. We need to see that $h(z)$ is holomorphic and that $H - h \in \mathcal{O}_{m+1}$.

Assume that H belongs to $\mathcal{O}_\ell$ where $\ell \le m$. Then (E.6.8) shows that $\bar{\partial} H$ belongs to $\mathcal{O}_{\ell+1}$ if $k > 1$, and to $\mathcal{O}_\ell$ if $k = 1$. In both cases it follows from Lemma E.5.9 that there is a complex polynomial p of degree at most ℓ such that

$$H - p \in \mathcal{O}_{\ell+1}.$$

Since $\ell \le m$ this implies $h(z) - p(z) = (h(z) - H(z)) + (H(z) - p(z)) = o(z^\ell)$. If $\ell < m$ we deduce that $p(z) = o(z^\ell)$ and therefore $p = 0$ and $H \in \mathcal{O}_{\ell+1}$. So by induction $h \in \mathcal{O}_m$. Now use the same argument again with $\ell = m$. Then the condition $h(z) - p(z) = o(z^m)$ implies that

$$h(z) = p(z) = az^m$$

for some nonzero vector $a \in \mathbb{C}^{n-1}$ and we have $H(z) - az^m \in \mathcal{O}_{m+1}$. This proves the proposition in the case where J satisfies condition (E.6.1).

Now consider the general case. By Proposition E.6.2 with $M = \mathbb{C}^n$, $x_0 = 0$, $\Pi_0 = \mathbb{C} \times \{0\}$, there is a smooth local diffeomorphism $\Phi : (\mathbb{C}^n, 0) \to (\mathbb{C}^n, 0)$ such that

$$d\Phi(x_0)(\mathbb{C} \times \{0\}) = \mathbb{C} \times \{0\}, \qquad \widetilde{J} := \Phi^* J = \begin{pmatrix} i & \widetilde{J}_3 \\ 0 & \widetilde{J}_4 \end{pmatrix}$$

where $\widetilde{J}_3(0) = 0$ and $\widetilde{J}_4(0) = i$. Hence $d\Phi(0)$ has the form (E.6.10). For $j = 1, 2$ the function

$$\widetilde{u}_j := \Phi^{-1} \circ u_j : U_j \to \mathbb{C}^n$$

is a $\widetilde{J}$-holomorphic curve. By assumption, the complex tangent space of the curve $\widetilde{u}_j$ at $z = 0$ is the subspace $\mathbb{C} \times \{0\}$ and $\widetilde{u}_j$ has a singularity of order k at $z = 0$ (see Remark E.5.11). Hence it follows from Proposition E.5.10 that, for $j = 1, 2$, there is a local $C^{k,1}$-diffeomorphism $\widetilde{\varphi}_j : (\mathbb{D}, 0) \to (U_j, 0)$ such that

$$\widetilde{\varphi}_j - \mathrm{id} \in \mathcal{O}_2,$$

$\widetilde{\varphi}_j$ is smooth in the case $k_j = 1$, and

$$\widetilde{u}_j \circ \widetilde{\varphi}_j(z) = (z^k, \widetilde{F}_j(z)),$$

where $\widetilde{F}_j : \mathbb{D} \to \mathbb{C}^{n-1}$ belongs to $\mathcal{O}_{k+1}$. Since $\widetilde{J}$ satisfies (E.6.1) it follows from the first part of the proof that there is an integer $m > k$ and a nonzero vector $\widetilde{a} \in \mathbb{C}^{n-1}$ such that

$$\widetilde{H} - \widetilde{p} \in \mathcal{O}_{m+1}, \qquad \widetilde{H} := \widetilde{F}_2 - \widetilde{F}_1, \qquad \widetilde{p}(z) := \widetilde{a} z^m.$$

Hence it follows from Lemma E.6.5 below (with $v_j := \widetilde{u}_j \circ \widetilde{\varphi}_j$, $G_j := \widetilde{F}_j$, and $\psi_j := \widetilde{\varphi}_j^{-1} \circ \varphi_j$) that

$$H - p \in \mathcal{O}_{m+1}, \qquad H := F_2 - F_1, \qquad p(z) := C\widetilde{a} z^m.$$

This proves the proposition. □

Given an integer k and two functions $f : \mathbb{D} \to \mathbb{C}^m$ and $g : \mathbb{D} \to \mathbb{C}^\ell$ we write

$$f = \mathcal{O}_k \cdot g$$

whenever there is a function $\alpha : \mathbb{D} \backslash \{0\} \to \mathrm{Hom}_{\mathbb{R}}(\mathbb{C}^\ell, \mathbb{C}^m)$ which belongs to $\mathcal{O}_k$ and satisfies $f(z) = \alpha(z) \cdot g(z)$ for z sufficiently small.

LEMMA E.6.5. *Let $k \geq 1$ be an integer and $\Phi : (\mathbb{C}^n, 0) \to (\mathbb{C}^n, 0)$ be a local diffeomorphism such that*

$$d\Phi(0) =: \begin{pmatrix} A & B \\ 0 & C \end{pmatrix} \tag{E.6.10}$$

where $A \in \mathbb{C} \backslash \{0\}$, $B \in \mathrm{Hom}_{\mathbb{C}}(\mathbb{C}^{n-1}, \mathbb{C})$, and $C \in \mathrm{GL}(n-1, \mathbb{C})$. For $j = 1, 2$ let $v_j : (\mathbb{C}, 0) \to (\mathbb{C}^n, 0)$ and $\psi_j : (\mathbb{C}, 0) \to (\mathbb{C}, 0)$ be such that $\psi_j - \mathrm{id} \in \mathcal{O}_2$ and

$$v_j(z) = (z^k, G_j(z)), \qquad \Phi \circ v_j \circ \psi_j(z) = (z^k, F_j(z))$$

where $F_j, G_j \in \mathcal{O}_{k+1}$. Then, for every $b \in \mathbb{C}^{n-1}$ and every $m \in \mathbb{Z}$,

$$G_2(z) - G_1(z) - bz^m \in \mathcal{O}_{m+1} \qquad \Longleftrightarrow \qquad F_2(z) - F_1(z) - Cbz^m \in \mathcal{O}_{m+1}.$$

PROOF. Denote by Φ_2 the last $n-1$ coordinates of Φ and by Ψ_1 the first coordinate of $\Psi := \Phi^{-1}$. Then $\Phi_2(\psi_j(z)^k, G_j(\psi_j(z))) = F_j(z)$ and hence

$$H := F_2 - F_1 = \Phi_2(\psi_2^k, G_2 \circ \psi_2) - \Phi_2(\psi_1^k, G_1 \circ \psi_1) = H_1 + H_2,$$

where

$$\begin{aligned} H_1 &:= \Phi_2(\psi_2^k, G_1 \circ \psi_2) - \Phi_2(\psi_1^k, G_1 \circ \psi_1), \\ H_2 &:= \Phi_2(\psi_2^k, G_2 \circ \psi_2) - \Phi_2(\psi_2^k, G_1 \circ \psi_2). \end{aligned}$$

Denote by d_1 the derivative with respect to the coordinates $z_1 = s_1 + it_1$ of the first component in $\mathbb{C}^n := \mathbb{C} \times \mathbb{C}^{n-1}$ and by d_2 the derivative with respect to the other components. Then the maps $z \mapsto \partial_1 \Phi_2(z^k, G_j(z))$ and $z \mapsto d_2\Phi_2(z^k, G_j(z)) - C$ belong to $\mathcal{O}_k$ by Remark E.5.5. Hence

$$H_2 = \left(\int_0^1 d_2\Phi_2\Big(\psi_2^k, G_1 \circ \psi_2 + \tau K \circ \psi_2\Big)\, d\tau\right) \cdot (K \circ \psi_2) = (C + \mathcal{O}_k) \cdot (K \circ \psi_2),$$

where $K := G_2 - G_1$. Now denote

$$\Gamma(z) := \Phi_2(z^k, G_1(z)).$$

Then $H_1 = \Gamma \circ \psi_2 - \Gamma \circ \psi_1$ and

$$d\Gamma(z) = kz^{k-1} d_1\Phi_2(z^k, G_1(z)) + d_2\Phi_2(z^k, G_1(z))\, dG_1(z),$$

which belongs to $\mathcal{O}_k$ by the *(Product)* axiom. Hence

$$H_1 = \left(\int_0^1 d\Gamma\Big(\psi_1 + \tau(\psi_2 - \psi_1)\Big)\, d\tau\right) \cdot (\psi_2 - \varphi_1) = \mathcal{O}_k \cdot (\psi_2 - \psi_1).$$

On the one hand, we have $\psi_j(z)^k = \Psi_1(z^k, F_j(z))$ and hence

$$\psi_2(z)^k - \psi_1(z)^k = \left(\int_0^1 d_2\Psi_1(z^k, F_1(z) + \tau H(z))\, d\tau\right) \cdot H(z) = \mathcal{O}_0 \cdot H(z).$$

On the other hand, $\psi_j - \mathrm{id} \in \mathcal{O}_2$ and hence

$$\begin{aligned} \psi_2(z)^k - \psi_1(z)^k &= (kz^{k-1} + \mathcal{O}_k)(\psi_2(z) - \psi_1(z)) \\ &= kz^{k-1}(1 + \mathcal{O}_1)(\psi_2(z) - \psi_1(z)). \end{aligned}$$

These two equations imply $\psi_2 - \psi_1 = \mathcal{O}_{1-k} \cdot (\psi_2^k - \psi_1^k) = \mathcal{O}_{1-k} \cdot H$ and hence

$$H_1 = \mathcal{O}_1 \cdot H.$$

Since $H = H_1 + H_2 = \mathcal{O}_1 \cdot H + (C + \mathcal{O}_k) \cdot K$ we deduce

$$H = (\mathbb{1} + \mathcal{O}_1)^{-1} \cdot (C + \mathcal{O}_k) \cdot K = (C + \mathcal{O}_1) \cdot K.$$

Thus, for every vector $b \in \mathbb{C}^{n-1}$ and every integer m, we have

$$K(z) = z^m b + \mathcal{O}_{m+1} \qquad \Longrightarrow \qquad H(z) = z^m Cb + \mathcal{O}_{m+1}.$$

The converse follows by replacing Φ with Φ^{-1} and v_j by $\Phi \circ v_j \circ \psi_j$. This proves the lemma. $\square$

E.7. Singularities of J-holomorphic curves

We now complete the proof of the Micallef–White theorem which asserts that J-holomorphic singularities cannot be worse than singularities in the algebraic case: the local branches of a J-curve passing through a given point are simultaneously C^1-conjugate to germs of polynomials in one complex variable.

This is not obvious even when $M = \mathbb{C}^n$ and when the curve is holomorphic with exactly one branch. Indeed assume that u has the form

$$u(z) = (z^k, f(z))$$

where $f(z) = a_{k+1}z^{k+1} + \cdots$ is analytic (with infinitely many nonzero coefficients). The task at hand is to find a local diffeomorphism $\Psi : (\mathbb{C}^n, 0) \to (\mathbb{C}^n, 0)$ such that

$$\Psi \circ u(z) = (z^k, a_{k+1}z^{k+1} + \cdots + a_m z^m)$$

where $m > k$ is a sufficiently large integer. We shall see below that in the case when u is injective this is possible as soon as $z \mapsto (z^k, a_{k+1}z^{k+1} + \cdots + a_m z^m)$ is also injective. We shall look for a map Ψ of the form

$$\Psi(x, y) = (x, y - \eta(x, y) f_m(x^{1/k}))$$

where $\eta : \mathbb{C}^n \to [0, 1]$ is a cutoff function with $\eta = 1$ on the image of u and where

$$f_m(z) = a_{m+1}z^{m+1} + a_{m+2}z^{m+2} + \cdots . \tag{E.7.1}$$

Such a map Ψ is not well defined unless we explain how we choose $x^{1/k}$, which is a priori multivalued.

Though it is important that Ψ extends to a global C^1-diffeomorphism its defining condition only involves the points on $\operatorname{im} u$. This set can be considered as the graph of the multivalued function $x \mapsto f(x^{1/k})$. We show below that the different branches of this function are sufficiently far apart for it to be possible to specify Ψ by different formulas near each branch. To explain this more precisely we need some preparation. Throughout this section we fix an integer $n \geq 2$. For $\ell \in \mathbb{Z}$ we use the notation $\mathcal{O}_\ell$ for the space of pairs (U, f) where $U \subset \mathbb{C}$ is an open neighbourhood of zero and $f : U \setminus \{0\} \to \mathbb{C}^{n-1}$ is a smooth map that satisfy (E.5.1) (see Section E.5). We think of the elements of $\mathcal{O}_\ell$ as maps, rather than germs, and write $f \in \mathcal{O}_\ell$. Thus the notation $f = g$ for $f, g \in \mathcal{O}_\ell$ means that the maps have the same domain and agree at each point in this domain (rather than just in a neighbourhood of zero). For every positive integer k and every integer ℓ the group $\mathrm{G}(k) := \{\zeta \in \mathbb{C} \mid \zeta^k = 1\}$ of kth roots of unity acts on $\mathcal{O}_\ell$ by

$$\mathrm{G}(k) \times \mathcal{O}_\ell \to \mathcal{O}_\ell : (\zeta, f) \mapsto f \circ \zeta,$$

where $f \circ \zeta(z) := f(\zeta z)$.

DEFINITION E.7.1. *Let k be a positive integer and ℓ, m be integers. A subset $\mathcal{S} \subset \mathcal{O}_\ell$ is called k-**invariant** if it is invariant under the action of $\mathrm{G}(k)$. If $\mathcal{S} \subset \mathcal{O}_\ell$ is k-invariant then a map $\Gamma : \mathcal{S} \to \mathcal{O}_m$ is called k-**equivariant** if*

$$\Gamma(f \circ \zeta) = \Gamma(f) \circ \zeta$$

*for every $\zeta \in \mathrm{G}(k)$. A subset $\mathcal{S} \subset \mathcal{O}_\ell$ is called m-**separated** if the following holds for any two elements $f, g \in \mathcal{S}$: if $f \neq g$ then there is a constant $\varepsilon > 0$ such that*

$$|f(z) - g(z)| \geq \varepsilon |z|^m$$

for z sufficiently small.

The next lemma deals with the case of N curves $u_i := (z^k, f_i(z))$. The set $\mathcal{S}_m$ should be thought of as the $\mathrm{G}(k)$-orbit of the corresponding set of leading order terms.

LEMMA E.7.2. *Let $m \geq k \geq 1$ be two integers and*

$$f_1, \dots, f_N \in \mathcal{O}_{k+1}.$$

Let $\mathcal{S}_m \subset \mathcal{O}_{k+1}$ be a k-invariant m-separated set and $\widetilde{f}_1, \dots, \widetilde{f}_N \in \mathcal{S}_m$ such that

$$\widetilde{f}_j - f_j \in \mathcal{O}_{m+1}, \qquad j = 1, \dots, N.$$

Let $\Gamma : \mathcal{S}_m \to \mathcal{O}_{m+1}$ be a k-equivariant map and $\eta > 0$. Then there exists a local C^1-map

$$E_\Gamma : (\mathbb{C}^n, 0) \to (\mathbb{C}^{n-1}, 0)$$

such that $dE_\Gamma(0) = 0$,

$$|y| > \eta|x| \qquad \Longrightarrow \qquad E_\gamma(x, y) = 0$$

for $x \in \mathbb{C}$ and $y \in \mathbb{C}^{n-1}$ sufficiently small, and

$$E_\Gamma(z^k, f_j(z)) = \Gamma(\widetilde{f}_j)(z) \tag{E.7.2}$$

for $j = 1, \dots, N$ and z sufficiently small.

With this lemma in place we can now return to our search for a suitable diffeomorphism Ψ. We just need Lemma E.7.2 for $N = 1$ and a single map $f \in \mathcal{O}_{k+1}$. If this map is holomorphic we may choose

$$\mathcal{S}_m := \{(f - f_m) \circ \zeta^j \,|\, 1 \leq j \leq k\}, \qquad \Gamma(f - f_m) := f_m,$$

where f_m denotes the higher order terms of f (see equation (E.7.1)). Here we assume that the integer m is chosen so large that $f \circ \zeta \not\equiv f$ implies $(f - f_m) \circ \zeta \not\equiv f - f_m$ for every $\zeta \in \mathrm{G}(k)$. Let $E_\Gamma : (\mathbb{C}^n, 0) \to (\mathbb{C}^{n-1}, 0)$ be the C^1-map provided by Lemma E.7.2 and define

$$\Psi(x, y) := (x, y - E_\Gamma(x, y)).$$

This map is a local C^1-diffeomorphism near zero and satisfies

$$\Psi(z^k, f(z)) = (z^k, f(z) - E_\Gamma(z^k, f(z))) = (z^k, f(z) - f_m(z))$$

by (E.7.2) with $\widetilde{f} = f - f_m$. Hence Ψ is the required diffeomorphism which removes the higher order terms from u.

In the general case, when f is not holomorphic, it is not clear whether such a truncated complex polynomial $f - f_m$ exists. Here comes the second main idea in Micallef and White's proof. For every $m > k$ they construct an element

$$A_m^f \in \mathcal{O}_{k+1}$$

as an average of appropriate maps in the orbit of f with the property that the set

$$\mathcal{S}_m := \{A_m^g \,|\, g \in \mathrm{G}(k) \cdot f\}$$

is k-invariant and m-separating and furthermore $f - A_m^f \in \mathcal{O}_{m+1}$. This is the content of the following lemma. Note that even in the holomorphic case this lemma will allow us to find a more refined normal form for u known as the Puiseau expansion: see Exercise E.7.4 below and the appendix to Chapter 1 in the book [**96**] by Eisenbud and Neumann.

LEMMA E.7.3. *Let k be a positive integer and let $\mathcal{S} \subset \mathcal{O}_{k+1}$ be a finite k-invariant set of maps with the same domain. Assume that for every pair $f, g \in \mathcal{S}$ with $f \neq g$ there exists an integer $m(f,g) > k$ and a nonzero vector $a_{f,g} \in \mathbb{C}^{n-1}$ such that*

$$f(z) - g(z) = z^{m(f,g)} a_{f,g} + \mathcal{O}_{m(f,g)+1}.$$

We set $m(f,f) := \infty$ and $a_{f,f} := 0$. For $f \in \mathcal{S}$ and $m \geq k$ let A_m^f denote the average among $\mathcal{S}$ of the maps which approximate f up to order $m+1$:

$$A_m^f := \frac{1}{|\mathcal{S}_m^f|} \sum_{g \in \mathcal{S}_m^f} g, \qquad \mathcal{S}_m^f := \{g \in \mathcal{S} \,|\, g - f \in \mathcal{O}_{m+1}\}.$$

Then the set

$$\mathcal{S}_m := \{A_m^f \,|\, f \in \mathcal{S}\}$$

is k-invariant and m-separated and the map $\mathcal{S} \to \mathcal{S}_m : f \mapsto A_m^f$ is k-equivariant. Moreover, the following holds for all $f, g \in \mathcal{S}$ with $f \neq g$ and every integer $m > k$.

(i) *$f - A_m^f \in \mathcal{O}_{m+1}$.*
(ii) *If $m \geq \max\{m(f,g) \,|\, g \in \mathcal{S} \backslash \{f\}\}$ then $A_m^f = f$.*
(iii) *If $m \geq m(f,g)$ then $A_m^f(z) - A_m^g(z) = z^{m(f,g)} a_{f,g} + \mathcal{O}_{m(f,g)+1}$.*
(iv) *$A_m^f = A_m^g$ if and only if $m < m(f,g)$.*
(v) *There is a vector $a_m^f \in \mathbb{C}^{n-1}$ such that $A_m^f(z) = A_{m-1}^f(z) + z^m a_m^f + \mathcal{O}_{m+1}$.*

With this lemma in place one can proceed as follows in the case where f is not holomorphic. Let $f \in \mathcal{O}_{k+1}$ and suppose the set $\mathcal{S} := \mathrm{G}(k) \cdot f$ satisfies the hypothesis of Lemma E.7.3. Define $\Gamma_m : \mathcal{S}_m \to \mathcal{O}_{m+1}$ by

$$\Gamma_m(A_m^f)(z) := A_m^f(z) - A_{m-1}^f(z) - a_m^f z^m$$

for $m > k$ and by $\Gamma_k(A_k^f) := A_k^f$. By Lemma E.7.3, Γ_m satisfies the hypotheses of Lemma E.7.2 for every $m \geq k$. The crucial point is that $A_m^f = f$ and so $a_m^f = 0$ for m sufficiently large, say $m \geq N$. Define

$$\Psi(x,y) := \left(x, y - \sum_{m=k}^{N} E_{\Gamma_m}(x,y) \right).$$

Then, as we explain in more detail below, the formula

$$f(z) = A_k^f(z) + \sum_{m=k+1}^{N} \left(A_m^f(z) - A_{m-1}^f(z) - a_m^f z^m \right) + \sum_{m=k+1}^{N} a_m^f z^m$$

implies that

$$\Psi(z^k, f(z)) = \left(z^k, \sum_{m=k+1}^{N} a_m^f z^m \right).$$

Again, the effect of each E_{Γ_m} is to remove the corresponding set of higher order terms $A_m^f(z) - A_{m-1}^f(z) - a_m^f z^m$ in this expression for f. We shall now give the proofs of the two lemmas and show how they imply Theorem E.1.1 in its general form.

EXERCISE E.7.4. Let $f : \mathbb{C} \to \mathbb{C}^2$ be the function $f(z) = (z^4, z^6 + z^8 + z^9 e^z)$. Apply the above process to find Ψ such that $\Psi \circ f(z) = (z^4, z^6 + z^9)$.

PROOF OF LEMMA E.7.2. Fix a constant $\eta > 0$. Choose a smooth cutoff function $\chi : [0, \infty) \to [0, 1]$ such that

$$\chi(r) = \begin{cases} 1, & \text{if } r \leq 1/4, \\ 0, & \text{if } r \geq 1/3. \end{cases}$$

Since $\mathcal{S}_m$ is a finite subset of $\mathcal{O}_{k+1}$ we may assume, shrinking the domains if necessary, that each function $f \in \mathcal{S}_m$ is defined on the same open set $U \subset \mathbb{C}$ and that U is an open disc centered at zero. Moreover, for every $\alpha \in \mathbb{N}^2$, there is a constant $C_\alpha > 0$ such that

$$|\partial^\alpha f(z)| \leq C_\alpha |z|^{m+1-|\alpha|}$$

for every $f \in \mathcal{S}_m$ and every $z \in U$. Now recall that $\mathcal{S}_m$ is m-separated (see Definition E.7.1). Hence, shrinking U if necessary, we may assume that there is a constant $\varepsilon > 0$ such that every pair $f, g \in \mathcal{S}_m$ with $f \neq g$ satisfies

$$|f(z) - g(z)| \geq \varepsilon |z|^m \tag{E.7.3}$$

for $z \in U$. This constant ε can be chosen smaller than η.

It is very important to control the support of the perturbation $E_\Gamma(x, y)$. To this end, note that because $m \geq k$ and $\varepsilon < \eta$ there is a constant $\delta > 0$ such that

$$C_0 \delta^{(m+1-k)/k} + \frac{\varepsilon}{2} \delta^{(m-k)/k} < \eta.$$

This constant has the following significance: if $f \in \mathcal{S}_m$, $z \in U$, $x \in \mathbb{C}$, and $y \in \mathbb{C}^{n-1}$ satisfy

$$z^k = x, \qquad |y - f(z)| \leq \frac{\varepsilon}{2} |z|^m, \qquad |x| < \delta, \tag{E.7.4}$$

then $|y| \leq \eta |x|$; indeed

$$|y| \leq |f(z)| + \frac{\varepsilon}{2}|z|^m \leq \left(C_0 |x|^{(m+1-k)/k} + \frac{\varepsilon}{2} |x|^{(m-k)/k} \right) |x| \leq \eta |x|. \tag{E.7.5}$$

Given $(x, y) \in \mathbb{C} \times \mathbb{C}^{n-1}$ with $0 < |x| < \delta$ we define $E_\Gamma(x, y)$ by

$$E_\Gamma(x, y) := \chi\left(\frac{|y - f(z)|}{\varepsilon |z|^m} \right) \Gamma(f)(z)$$

if there is a pair $(f, z) \in \mathcal{S}_m \times U$ that satisfies (E.7.4). If no such pair exists (or if $x = 0$ and $y = 0$) we set $E_\Gamma(x, y) := 0$. It follows from (E.7.3) that, for every $z \in U$ with $z^2 = x$ and $|x| < \delta$ and every $y \in \mathbb{C}^{n-1}$, there is at most one $f \in \mathcal{S}_m$ that satisfies (E.7.4). Moreover, the pair (f, z) satisfies (E.7.4) if and only if $(f \circ \zeta^{-1}, \zeta(z))$ does as well. Therefore the equivalence class $[f, z] \in \mathcal{S}_m \times_{\mathrm{G}(k)} U$ of pairs satisfying (E.7.4) depends only on (x, y). It follows that $f(z)$ and $\Gamma(f)(z)$ are independent of the choice of f and z and so the map E_Γ is well defined.

That $E_\Gamma(x, y)$ vanishes for $|y| \geq \eta |x|$ and $|x| < \delta$ is obvious: in this case it follows from (E.7.5) that there is no pair (f, z) that satisfies (E.7.4). That E_Γ satisfies (E.7.2) is also obvious from the construction: since $f_j - \widetilde{f}_j \in \mathcal{O}_{m+1}$, we have

$$|f_j(z) - \widetilde{f}_j(z)| \leq \frac{\varepsilon}{4} |z|^m$$

for $z \in U$ sufficiently small; hence the pair $(\widetilde{f}_j, z)$ satisfies (E.7.4) with $x := z^k$ and $y := f_j(z)$, and so

$$E_\Gamma(z^k, f_j(z)) = \Gamma(\widetilde{f}_j)(z).$$

It remains to prove that E_Γ is continuously differentiable and $dE_\Gamma(0) = 0$. Define the open set $\Omega \subset \mathbb{C} \times \mathbb{C}^{n-1}$ by

$$\Omega := \left\{ (x, y) \,\middle|\, |x| < \delta,\ |y| > \eta|x| \text{ or } \min_{f \in \mathcal{S}_m, z^k = x} \frac{|y - f(z)|}{\varepsilon |z|^m} > \frac{1}{3} \right\}$$

Since $\chi(r) = 0$ for $r \geq 1/3$ it follows that E_Γ vanishes on Ω. So it suffices to examine E_Γ near a point in the complement of Ω. Let $(x_0, y_0) \in \mathbb{C} \times \mathbb{C}^n$ and $z_0 \in U$ such that

$$0 < |x_0| < \delta, \qquad |y_0| \leq \eta |x_0|, \qquad |y_0 - f(z_0)| \leq \frac{\varepsilon}{3} |z_0|^m, \qquad z_0^k = x_0.$$

Then there exists a smooth map $x \mapsto x^{1/k}$ for x close to x_0 that takes the value z_0 when $x = x_0$ and satisfies

$$|y - f(x^{1/k})| < \frac{\varepsilon}{2} |x|^{m/k}$$

in a neighbourhood of (x_0, y_0). This shows that the map E_Γ is locally given by

$$E_\Gamma(x, y) = \chi \left(\frac{|y - f(x^{1/k})|}{\varepsilon |x|^{m/k}} \right) \Gamma(f)(x^{1/k}) \tag{E.7.6}$$

and hence is smooth away from the origin. That E_Γ is differentiable at the origin and $dE_\Gamma(0) = 0$ follows from the inequality

$$|E_\Gamma(x, y)| \leq c |x|^{(m+1)/k} \tag{E.7.7}$$

for x sufficiently small. The constant $c > 0$ is chosen such that

$$|\Gamma(f)(z)| \leq c |z|^{m+1}$$

for $f \in \mathcal{S}_m$ and z sufficiently small. (It exists because $\Gamma(\mathcal{S}_m) \subset \mathcal{O}_{m+1}$.) The inequality (E.7.7) then follows from the fact that either $E_\Gamma(x, y) = 0$ or $E_\Gamma(x, y) = \chi(r)\Gamma(f)(z)$ for some $r \geq 0$, some $f \in \mathcal{S}_m$, and some $z \in U$ with $z^k = x$.

We prove that dE_Γ is continuous at the origin. This follows from the inequality

$$|dE_\Gamma(x, y)| \leq C \left(|x|^{(m+1-k)/k} + |x|^{1/k} \right) \tag{E.7.8}$$

for x sufficiently small and a suitable constant $C > 0$. To prove (E.7.8) choose a local kth root $x \mapsto x^{1/k}$ near a point $x_0 \neq 0$, note that the derivative of this map has the form $x \mapsto x^{(1-k)/k}/k$, and differentiate the right hand side of (E.7.6). The term $\Gamma(f)'(x^{1/k}) x^{(1-k)/k}$ can be estimated by $|x|^{(m+1-k)/k}$. The derivative of the map $y \mapsto |x|^{-m/k} |y - f(x^{1/k})|$ is bounded by $|x|^{-m/k}$ and the derivative of the map $x \mapsto |x|^{-m/k} |y - f(x^{1/k})|$ can be estimated by $|x|^{-1}$. Since both terms get multiplied by $|\Gamma(f)(x^{1/k})| \leq c|x|^{(m+1)/k}$ this proves (E.7.8) and the lemma. □

Proof of Lemma E.7.3. We prove (i): A_m^f is the average of finitely many functions $g \in \mathcal{O}_{k+1}$ such that $f - g \in \mathcal{O}_{m+1}$; hence $f - A_m^f \in \mathcal{O}_{m+1}$.

We prove (ii). Suppose that $m \geq m(f, g)$ for every $g \in \mathcal{S} \setminus \{f\}$. Then $\mathcal{S}_m^f = \{f\}$. Namely, if $g \in \mathcal{S}$ and $g \neq f$ then $g - f \notin \mathcal{O}_{m(f,g)+1}$ by assumption. This implies $g - f \notin \mathcal{O}_{m+1}$ and hence $g \notin \mathcal{S}_m^f$. Therefore $A_m^f = f$.

We prove (iii). If $m \geq m(f, g)$ then it follows from (i) that

$$A_m^f - A_m^g - (f - g) \in \mathcal{O}_{m+1} \subset \mathcal{O}_{m(f,g)+1}.$$

By assumption $f - g - z^{m(f,g)} a_{f,g} \in \mathcal{O}_{m(f,g)+1}$ and hence

$$A_m^f - A_m^g - z^{m(f,g)} a_{f,g} \in \mathcal{O}_{m(f,g)+1},$$

as claimed.

We prove (iv). If $m < m(f,g)$ then $f - g \in \mathcal{S}_{m(f,g)+1} \subset \mathcal{S}_{m+1}$ and so g belongs to $\mathcal{S}^f_m$. Hence $\mathcal{S}^f_m = \mathcal{S}^g_m$, and hence $A^f_m = A^g_m$. If $m \geq m(f,g)$ (and $f \neq g$) then it follows from (iii) that $A^f_m - A^g_m \notin \mathcal{O}_{m(f,g)+1}$ and so $A^f_m \neq A^g_m$.

We prove (v). First note that

$$g \in \mathcal{S}^f_{m-1} \backslash \mathcal{S}^f_m \qquad \Longleftrightarrow \qquad m(f,g) = m \tag{E.7.9}$$

for every $g \in \mathcal{S}$. For $g = f$ this is obvious. For $g \neq f$ we have $f - g \in \mathcal{O}_{m(f,g)} \backslash \mathcal{O}_{m(f,g)+1}$ by assumption. Hence $m = m(f,g)$ if and only if $f - g \in \mathcal{O}_m \backslash \mathcal{O}_{m+1}$ if and only if $g \in \mathcal{S}^f_{m-1} \backslash \mathcal{S}^f_m$. Define

$$a^f_m := \frac{1}{|\mathcal{S}^f_{m-1}|} \sum_{m(f,g)=m} a_{f,g}$$

where the sum runs over all $g \in \mathcal{S}$ with $m(f,g) = m$. Now observe that

$$A^f_m - A^f_{m-1} = f - A^f_{m-1} + \mathcal{O}_{m+1}$$

by (i). Hence

$$\begin{aligned} A^f_m - A^f_{m-1} &= \frac{1}{|\mathcal{S}^f_{m-1}|} \sum_{g \in \mathcal{S}^f_{m-1}} (f - g) + \mathcal{O}_{m+1} \\ &= z^m a^f_m + \mathcal{O}_{m+1}. \end{aligned}$$

To prove the last equation note that $m(f,g) = m$ for every $g \in \mathcal{S}^f_{m-1} \backslash \mathcal{S}^f_m$ by (E.7.9), and $m > m(f,g)$ for every $g \in \mathcal{S}^f_m$, also by (E.7.9).

This we have proved assertions (i-v). We prove that the set $\mathcal{S}_m$ is m-separated. If $f, g \in \mathcal{S}$ and $A^f_m \neq A^g_m$ then $m \geq m(f,g)$ by (iv). Hence it follows from (iii) that there is a constant $\varepsilon > 0$ such that

$$A^f_m(z) - A^f_g(z)| \geq \varepsilon |z|^{m(f,g)} \geq \varepsilon |z|^m$$

for z sufficiently small. That the map $\mathcal{S} \to \mathcal{O}_{k+1} : f \mapsto A^f_m$ is k-equivariant is obvious from the definition. Hence the set $\mathcal{S}_m$ is k-invariant and this proves the lemma. □

PROOF OF THEOREM E.1.1. Let $u : \Sigma \to M$ be a J-holomorphic curve, $\Omega \subset \Sigma$ be an open set such that $\overline{\Omega} \subset \Sigma$ is compact, $x_0 \in u(\Omega) \setminus u(\partial\Omega)$, and

$$\{z_1, \ldots, z_N\} = u^{-1}(x_0) \cap \Omega.$$

Assume without loss of generality that $M = \mathbb{C}^n$, $x_0 = 0$, and J is an almost complex structure on $\mathbb{C}^n$ that is standard at the origin. For $j = 1, \ldots, N$ let

$$T_j = \mathbb{C} c_j \subset \mathbb{C}^n$$

be the tangent space of u_j at $z = 0$ and suppose that u_j has order $k_j \geq 1$ at $z = 0$ (see Remark E.5.11). Thus $du_j(0) = 0$ if and only if $k_j \geq 2$. Denote by

$$\Pi_j := T_j^\perp$$

the orthogonal complement of T_j. By Proposition E.5.10, there exist disjoint neighbourhoods $U_j \subset \Omega$ and C^2-coordinate charts $\varphi_j : (U_j, z_j) \to (\mathbb{D}, 0)$ such that $\varphi_j - \mathrm{id} \in \mathcal{O}_2$, φ_j is smooth in the case $k_j = 1$, and the composition

$$\widetilde{u}_j := u \circ \varphi_j^{-1}$$

has the form

$$\widetilde{u}_j(z) = z^{k_j} c_{\nu(j)} + f_j(z),$$

where $f_j : \mathbb{D} \to \Pi_j$ belongs to $\mathcal{O}_{k_j+1}$. We first consider the case where all the tangent spaces are equal and all the curves have the same order of zero.

STEP 1. *Fix a positive integer k. Assume*

$$T_j = \mathbb{C} \times \{0\}, \qquad k_j = k$$

for every j.

Modifying Ψ by a further linear coordinate change of $\mathbb{C}^n$ and φ_j by a rescaling in the target, if necessary, we may assume that

$$\widetilde{u}_j(z) = (z^k, f_j(z))$$

where $f_j : \mathbb{D} \to \mathbb{C}^{n-1}$ belongs to $\mathcal{O}_{k+1}$. Define the set $\mathcal{S} \subset \mathcal{O}_{k+1}$ by

$$\mathcal{S} := \{f_j \circ \zeta^\ell \,|\, 1 \le j \le N,\, 1 \le \ell \le k\},$$

where $\zeta(z) := \exp(2\pi i/k)z$. Note that the maps $f_j \circ \zeta^\ell$ are related to u_j by the same formula as f_j: just replace φ_j with $\zeta^{-\ell} \circ \varphi_j$. Hence it follows from Proposition E.6.1 that there is an open disc $U \subset \mathbb{D}$ centered at zero such that the following holds. if $f, g \in \mathcal{S}$ and $f - g$ does not vanish on U then there is an integer $m(f,g) > k$ and a vector $a_{f,g} \in \mathbb{C}^{n-1}$ such that

$$f(z) - g(z) - z^{m(f,g)} a_{f,g} \in \mathcal{O}_{m(f,g)+1}.$$

From now on we denote by $f \in \mathcal{S}$ the restriction of the function to U. Then the set $\mathcal{S}$ satisfies the hypothesis of Lemma E.7.3.

For every $f \in \mathcal{S}$ and every integer $m \ge k$ let $A^f_m \in \mathcal{O}_{k+1}$ and $a^f_m \in \mathbb{C}^{n-1}$ be defined as in Lemma E.7.3, denote

$$\mathcal{S}_m := \left\{A^f_m \,|\, f \in \mathcal{S}\right\},$$

and define the map $\Gamma_m : \mathcal{S}_m \to \mathcal{O}_{m+1}$ by $\Gamma_k(A^f_k) := A^f_k$ and

$$\Gamma_m(A^f_m)(z) := A^f_m(z) - A^f_{m-1}(z) - a^f_m z^m, \qquad m > k.$$

We prove that this map is well defined. If $A^f_m = A^g_m$ then $m < m(f,g)$ and hence $A^f_{m-1} = A^g_{m-1}$ by Lemma E.7.3 (iv). Hence it follows from Lemma E.7.3 (v) that the polynomial $z \mapsto z^m(a^f_m - a^g_m)$ belongs to $\mathcal{O}_{m+1}$ and so $a^f_m = a^g_m$. Thus Γ_m is well defined. Lemma E.7.3 also asserts that $\mathcal{S}_m$ is k-invariant and m-separated. That the map Γ_m is k-equivariant follows from the fact that it is well defined and that the map $f \mapsto A^f_m$ is k-equivariant. Thus we have proved that the map $\Gamma_m : \mathcal{S}_m \to \mathcal{O}_{m+1}$ satisfies the requirements of Lemma E.7.2 with $\widetilde{f}_j := A^{f_j}_m$.

For $m \ge k$ let $E_{\Gamma_m} : (\mathbb{C}^n, 0) \to \mathbb{C}^{n-1}, 0)$ be the local C^1-map in the assertion of Lemma E.7.2. Let

$$m_0 := \max\{m(f,g) \,|\, f, g \in \mathcal{S},\, f \ne g\}$$

and define the local C^1-diffeomorphism $\Psi : (\mathbb{C}^n, 0) \to (\mathbb{C}^n, 0)$ by

$$\Psi(x,y) := \left(x, y - \sum_{m=k}^{m_0} E_{\Gamma_m}(x,y)\right)$$

for $x \in \mathbb{C}$ and $y \in \mathbb{C}^{n-1}$ sufficiently small. We claim that

$$\Psi(z^k, f(z)) = \left(z^k, \sum_{m=k+1}^{m_0} a_m^f z^m \right) \tag{E.7.10}$$

for every $f \in \mathcal{S}$. To see this, note that

$$E_{\Gamma_k}(z^k, f(z)) = A_k^f, \qquad E_{\Gamma_m}(z^k, f(z)) = A_m^f(z) - A_{m-1}^f(z) - a_m^f z^m$$

for $m > k$ by Lemma E.7.2. Moreover, by Lemma E.7.3,

$$f(z) = A_k^f(z) + \sum_{m=k+1}^{m_0} \left(A_m^f(z) - A_{m-1}^f(z) - a_m^f z^m \right) + \sum_{m=k+1}^{m_0} a_m^f z^m.$$

Here we have used the fact that $A_{m_0}^f = f$ for every $f \in \mathcal{S}$. Inserting the last two equations in the definition of Ψ we obtain

$$\Psi(z^k, f(z)) = \left(z^k, f(z) - \sum_{m=k+1}^{m_0} E_{\Gamma_m}(z^k, f(z)) \right) = \left(z^k, \sum_{m=k+1}^{m_0} a_m^f z^m \right)$$

as claimed. Since $f_j \in \mathcal{S}$ for $j = 1, \dots, N$ it follows from (E.7.10) that Ψ and φ_j satisfy the requirements of Theorem E.1.1 under the assumptions of Step 1.

STEP 2. *Assume $T_j = \mathbb{C} \times \{0\}$ for every j.*

Let k be the least common multiple of the k_j. Applying Step 1 to the functions $v_j(z) := \widetilde{u}_j(z^{k/k_j})$ we obtain a C^1-diffeomorpism $\Psi : \mathbb{C}^n \to \mathbb{C}^n$ such that

$$\Psi \circ v_j(z) = \Psi(z^k, f_j(z^{k/k_j})) = (z^k, P_j(z)), \qquad j = 1, \dots, N,$$

where P_j is a complex polynomial of the form

$$P_j(z) = a_{j,k+1} z^{k+1} + \cdots + a_{j,m_0} z^{m_0}.$$

By definition of v_j we have $P_j \circ \zeta^{k_j} = P_j$. This shows that $a_{j,\ell} = 0$ for $\ell \notin (k/k_j)\mathbb{Z}$. It follows that

$$P_j(z) = Q_j(z^{k/k_j})$$

where Q_j is a complex polynomial. Hence

$$\Psi(z^{k_j}, f_j(z)) = (z^{k_j}, Q_j(z))$$

for $j = 1, \dots, N$.

STEP 3. *Assume the general case.*

Lemma E.7.2 shows that the support of Ψ (the closure of the set of points that are not fixed under Ψ) can be chosen arbitrarily close to $\mathbb{C} \times \{0\}$. More precisely

$$\operatorname{supp}(\Psi) \subset \{(x, y) \in \mathbb{C} \times \mathbb{C}^{n-1} \,|\, |y| \le \eta |x|\},$$

where $\eta > 0$ can be chosen arbitrary small. Choose a finite collection of pairwise linearly independent vectors $a_1, \dots, a_r$ such that each subspace T_j is equal to $\mathbb{C} a_\rho$ for some $\rho \in \{1, \dots, r\}$. Then choose a constant $\eta > 0$ so that the punctured cones

$$V_\rho := \{\lambda a_\rho + v \,|\, \lambda \in \mathbb{C} \backslash \{0\},\, v \perp a_\rho,\, 0 \le |v| \le \eta |\lambda a_\rho|\}, \qquad \rho = 1, \dots, r,$$

are pairwise disjoint. For each ρ let $\Lambda_\rho : \mathbb{C}^n \to \mathbb{C}^n$ be a unitary matrix such that

$$\Lambda_\rho(a_\rho \mathbb{C}) = \mathbb{C} \oplus \{0\}.$$

By Step 2 there is a collection of C^1-diffeomorphisms Ψ_ρ with support in $\Lambda_\rho(\bar{V}_\rho)$ such that

$$c_j \in \mathbb{C}a_\rho \qquad \Longrightarrow \qquad \Psi_\rho \circ \Lambda_\rho \circ u_j \circ \varphi_j^{-1}(z). = (z^{k_j}, z^{k_j} p_j(z))$$

Here each p_j is a complex polynomial that vanishes at zero. Now set

$$\Psi := \prod_\rho \Lambda_\rho^{-1} \circ \Psi_\rho \circ \Lambda_\rho$$

and define

$$L_j := \Lambda_\rho, \qquad c_j \in \mathbb{C}a_\rho.$$

Since the punctured cones V_ρ are pairwise disjoint these maps satisfy all the conditions of Theorem E.1.1. □

Bibliography

[1] M. Abouzaid, Nearby Lagrangians with vanishing Maslov class are homotopy equivalent, arXiv:1005.0358.

[2] R. Abraham and J.W. Robbin *Transversal Mappings and Flows*, Benjamin, New York–Amsterdam, 1970.

[3] M. Abreu, Topology of symplectomorphism groups of $S^2 \times S^2$, *Inv. Math.*, **131** (1998), 1–23.

[4] M. Abreu, G. Granja, and N. Kitchloo, Compatible complex structures on symplectic rational ruled surfaces, *Duke Math.J.* **148** (2009), no. 3, 539–600.

[5] M. Abreu and D. McDuff, Topology of symplectomorphism groups of rational ruled surfaces, SG/9910057, *J. Amer. Math. Soc.* **13**, (2000) 971–1009.

[6] M. Abreu and L. Macarini, Remarks on Lagrangian intersections in toric manifolds, arXiv:1105.0640.

[7] L. Abrams, The quantum Euler class and the quantum cohomology of the Grassmannian, *Israel Math.J.* **117** (2000), 335–352.

[8] B. Aebischer, M. Borer, Ch. Leuenberger, M. Kälin, and H.M. Reimann, *Symplectic geometry, an introduction based on the Seminar at Berne, 1992*, Birkhäuser, Basel, 1994.

[9] S. Agnihotri, *Quantum Cohomology and the Verlinde Algebra*, PhD thesis, Oxford University, September 1995.

[10] A. Akhmedov and B.D. Park, Exotic smooth structures on small 4-manifolds with odd signatures, *Invent. Math.* **181** (2010), 577–603.

[11] M. Akveld and D.A. Salamon, Loops of Lagrangian submanifolds and pseudoholomorphic disks, *Geom. Funct. Anal.* **11** (2001), 609–650.

[12] P. Albers and H. Hofer, On the Weinstein conjecture in higher dimensions, *Comm. Math. Helv.* **84** (2009), no 2, 429–436.

[13] S. Anjos, The homotopy type of symplectomorphism groups of $S^2 \times S^2$, *Geom. and Top.* **6** (2002), 195–218.

[14] S. Anjos and G. Granja, Homotopy decomposition of a group of symplectomorphisms of $S^2 \times S^2$, *Topology* **43** (2004) (3), 599–618.

[15] V.I. Arnold, *Mathematical methods in classical mechanics,* Springer, Berlin, 1978.

[16] V.I. Arnold, Sur une proprieté topologique des applications globalement canoniques et à mécanique classique, *C.R. Acad. Sci. Paris* **261** (1965), 3719–3722.

[17] V.I. Arnold, On a characteristic class entering into conditions of quantization, *Functional analysis* **1** (1967), 1-8.

[18] N. Aronszajn, A unique continuation theorem for elliptic differential equations or inequalities of the second order, *J. Math. Pures Appl.* **36** (1957), 235–249.

[19] P.S. Aspinwall and D.R. Morrison, Topological Field Theory and rational curves, hep-th/9110048, *Comm. Math. Physics* **151** (1993), 245–262.

[20] M.F. Atiyah, New invariants for three and four dimensional manifolds, *Proc. Symp. Pure Math.* **48** (1988), 285–299.

[21] M.F. Atiyah and I.G. MacDonald, *Introduction to Commutative Algebra*, Addison Wesley, 1969.

[22] M. Audin, Fibrés normaux d'immersions en dimension moitié, points doubles d'immersions lagrangiennes et plongements totalement réels, *Comment. Math. Helv.* **63** (1988), 593-623.

[23] M. Audin, *The Topology of Torus Actions on Symplectic Manifolds*, Progress in Mathematics **93**, Birkhäuser, 1991.

[24] M. Audin and M. Damian, Théorie de Morse et homologie de Floer, Savoirs Actuels (les Ulis), EDP Sciences, Les Ulis, CNRS Éditions, Paris 2010.

[25] M. Audin and F. Lafontaine, ed., *Holomorphic Curves in Symplectic Geometry,* Progress in Mathematics **117**, Birkhäuser 1994.

[26] D. Auroux, Symplectic 4-manifolds as branched coverings of $\mathbb{C}P^2$, *Invent. Math.* **139** (2000), 551–602.

[27] D. Auroux, Monodromy invariants in symplectic topology, SG/0304113, notes of lectures given at IPAM, Los Angeles.

[28] D. Auroux and L. Katzarkov, Branched coverings of $\mathbb{C}P^2$ and invariants of symplectic 4-manifolds, *Invent. Math.* **142** (2000), 631–673.

[29] D. Auroux, V. Muñoz, and F. Presas, Lagrangian submanifolds and Lefschetz pencils, *J. Symplectic Geom.* **3** (2005), no 2, 171–219.

[30] S. Baldridge and P. Kirk, A symplectic manifold homeomorphic but not diffeomorphic to $\mathbb{C}P^2\#3\overline{\mathbb{C}P}^2$, *Geom. Topol.* **12** (2008), 919–940.

[31] V. Batyrev, Quantum cohomology rings of toric manifolds, *Astérisque* **218** (1993), 9–34.

[32] S. Bauer, Almost complex 4-manifolds with vanishing first Chern class, *J. Differential Geom.* **79** (2008), (1), 25–32.

[33] A. Beauville, *Surfaces Algébriques Complexes, Astérisque* **54**, 1978.

[34] K. Behrend and Yu. Manin, Stacks of stable maps and Gromov-Witten invariants, *Duke Math. J.* **85** (1996), 1–60.

[35] R. Berndt, *An Introduction to Symplectic Geometry,* Graduate Studies in Math. vol 26, A.M.S. (2001).

[36] A. Bertram, G. Daskalopoulos, and R. Wentworth, Gromov invariant for holomorphic maps from Riemann surfaces to Grassmannians, *J. Amer. Math. Soc.* **9** (1996), 529–571.

[37] M. Bialy and L. Polterovich, Geodesics of Hofer's metric on the group of Hamiltonian diffeomorphisms, *Duke Math. J.* **76** (1994), 273–292.

[38] P. Biran, Symplectic packing in dimension 4, *Geometric and Functional Analysis* **7** (1997), 420–437.

[39] P. Biran, A stability property of symplectic packing, *Invent. Math.* **136** (1999), 123–155.

[40] P. Biran, From symplectic packing to algebraic geometry and back, *European Congress of Mathematics, Barcelone 2000*, 507–524, in Progress in Math. **202**, Birkhäuser, Basel, 2001.

[41] P. Biran and K. Cieliebak, Symplectic topology on subcritial manifolds, *Comm. Math. Helv* **76** (2001), 712–753.

[42] P. Biran and K. Cieliebak, Lagrangian embeddings into subcritical Stein manifolds, *Israel J. Math.*, **127** (2002), 221-244.

[43] P. Biran and O. Cornea, Rigidity and uniruling for Lagrangian submanifolds, *Geom. Topol.* **13** (2009), no. 5, 2881–2989.

[44] P. Biran and O. Cornea, A Lagrangian Quantum homology, *New perspectives and challenges in symplectic field theory,* 1–44, *CRM Proc. Lecture Notes* **49**, Amer. Math. Soc., Providence, RI, 2009.

[45] R. Bott and L. Tu, *Differential Forms in Algebraic Topology,* Springer, New York, Heidelberg, Berlin 1982.

[46] F. Bourgeois, Y. Eliashberg, H. Hofer, K. Wysocki and E. Zehnder, Compactness results in Symplectic Field theory, *Geom. and Top.* **7** (2003), 799-833.

[47] S. Bradlow, Special metrics and stability for holomorphic bundles with global sections, *J. Diff. Geom.* **33** (1991), 169–213.

[48] J. Bryan and R. Pandharipande, Curves in Calabi–Yau 3-folds and topological quantum field theories, *Duke Math. J.* **126** (2005), no. 2, 369–396.

[49] A. Buch, Quantum cohomology of partial flag manifolds, *Trans. Amer. Math. Soc.* **357** (2005), no. 2, 443-458.

[50] L. Buhovsky, The Maslov class of Lagrangian tori and quantum products in Floer cohomology, *J. Topol. Anal.* **2** (2010), 57-75.

[51] L. Buhovsky and Y. Ostrover, On the uniqueness of Hofer's Geometry, arXiv:1008.06661

[52] L. Buhovsky, M. Entov and L. Polterovich, Poisson brackets and symplectic invariants, arXiv:1102.3198.

[53] L. Buhovsky and S. Seyfaddini, Uniqueness of generating Hamiltonian for continuous Hamiltonian flows, arXiv:1003.2612.

[54] O. Buse, Relative family Gromov–Witten invariants and symplectomorphisms, *Pacific J. Math.* **218** (2005), no. 2, 315–341.

[55] O. Buse, Negative inflation and stability in symplectomorphism groups of ruled surfaces, *J. Symplectic Geom.* **9** (2011), no. 2, 147–160.

[56] O. Buse and R. Hind, Symplectic embeddings of ellipsoids in dimension greater than four, arXiv:1112.1452.

[57] P. Candelas and X.C. de la Ossa, Moduli space of Calabi-Yau manifolds, *Nuclear Physics* **B355** (1991), 455–481.

[58] P. Candelas, X.C. de la Ossa, P.S. Green, and L. Parkes, A pair of Calabi-Yau manifolds as an exactly soluble superconformal field theory, *Nuclear Phys* **B 359** (1991), 21–74.

[59] A. Cannas da Silva, *Lectures on Symplectic Geometry*, Lecture Notes in Mathematics vol 1764, Springer, Berlin, 2001.

[60] P. Cascini and D. Panov, Symplectic generic complex structures on 4-manifolds with $b^+ = 1$, arXiv:1012.3644.

[61] Yu. Chekanov, Invariant Finsler metrics on the space of Lagrangian embeddings, *Math. Zeitschrift* **234** (2000), 605–619.

[62] W. Chen and Y. Ruan, Orbifold Gromov–Witten theory, AG/0103156, in: *Orbifolds in Mathematics and Physics (Madison 2001), Contemp. Math.* **310** Amer. Math. Soc. (2002), 25–85.

[63] K. Cieliebak and Ya. Eliashberg, *From Stein to Weinstein and back: Symplectic geometry of affine complex manifolds*, Amer. Math. Soc., Providence, RI, 2012.

[64] K. Cieliebak, A.R. Gaio, D.A. Salamon, J-holomorphic curves, moment maps, and invariants of Hamiltonian group actions, *Int. Math. Res. Notes* **10** (2000), 831–882.

[65] K. Cieliebak, H. Hofer, J. Latschev and F. Schlenk, Quantitative symplectic geometry, *Dynamics, Ergodic Theory, Geometry MSRI*, **54** (2007), 1–44.

[66] K. Cieliebak, I. Mundet i Riera and D.A. Salamon, Equivariant moduli problems and the Euler class, *Topology* **42** (2003), 641–700.

[67] K. Cieliebak, R.A. Gaio, I. Mundet i Riera and D.A. Salamon, The symplectic vortex equations and invariants of Hamiltonian group actions, *J. Sympl. Geom.* **1** (2002), 543–645.

[68] K. Cieliebak and K. Mohnke, Symplectic hypersurfaces and transversality in Gromov–Witten theory, *J. Symplectic Geom.* **3** (2005), no. 4, 589–654.

[69] K. Cieliebak and D.A. Salamon, Wall crossing for symplectic vortices and quantum cohomology, *Math. Ann.* **335** (2006), no. 1, 133–192.

[70] I. Ciocan-Fontanine, Quantum cohomology of flag varieties, *Int. Math. Res. Notes*, **2** (1995), 263–277.

[71] T. Coates and A.B. Givental, Quantum Riemann–Roch, Lefschetz and Serre, *Ann. of Math.* (2) **165** (2007), no. 1, 15–53.

[72] T. Coates, Riemann-Roch theorems in Gromov–Witten theory, Ph.D. thesis, Berkeley (2003).

[73] R. Cohen, J.D.S. Jones and G. Segal, Floer's infinite dimensional Morse theory and homotopy theory, in [**181**], 297–325.

[74] V. Colin, P. Ghiggini, and Ko Honda, Equivalence of Heegaard Floer homology and embedded contact homology via open book decompositions, *Proc. Natl. Acad. Sci. USA* **108** (2011), no, 20, 8100–8105.

[75] C. Conley and E. Zehnder, The Birkhoff–Lewis fixed point theorem and a conjecture of V.I. Arnold, *Invent. Math.*, **73** (1983), 33–49.

[76] D. Cox and S. Katz, *Mirror Symmetry and Algebraic Geometry*, Math. Surveys and Monographs vol 68, AMS, Providence 1999.

[77] M. Damian, Floer homology on the universal cover, a proof of Audin's conjecture and other constraints on Lagrangian submanifolds. Preprint 2010, to appear in *Comment. Math. Helv.*.

[78] M. W. Davis and T. Januszkiewicz, Convex polytopes, Coxeter orbifolds and torus actions, *Duke Math Journ.* **62** (1991), 417–451.

[79] T. Delzant, Hamiltoniens périodiques et images convexes de l'application moment, *Bull. Soc. Math. France* **116** (1988), 315–339.

[80] R. Dijkgraaf, C. Faber, G. van der Geer, eds., *The moduli space of curves*, Progress in Mathematics **129** Birkhäuser, 1995.

[81] S.K. Donaldson, The orientation of Yang-Mills moduli spaces and 4-manifold topology, *J. Diff. Geom.* **26** (1987), 397–428.

[82] S.K. Donaldson, Gluing techniques in the cohomology of moduli spaces, in: *Topological methods in modern mathematics (Stony Brook 1991)*, 137–170, Publish or Perish, TX, 1993.

[83] S.K. Donaldson, Floer homology and algebraic geometry, in *Vector Bundles in Algebraic Geometry (Durham 1993)*, 119–138, London Math. Soc. Lecture Notes **208**, Cambridge Univ. Press, Cambridge 1995.

[84] S.K. Donaldson, Symplectic submanifolds and almost-complex geometry, *J. Diff. Geom.* **44** (1996), 666–705.

[85] S.K. Donaldson, Lefschetz pencils on symplectic manifolds, *J. Diff. Geom.* **53** (1999), 205–236.

[86] S.K. Donaldson, Moment maps and diffeomorphisms, *Asian J. Math.* **3** (1999), 1–15.

[87] S.K. Donaldson, *Floer homology groups in Yang–Mills Theory*, Cambridge University Press, Cambridge, 2002.

[88] S.K. Donaldson and P. Kronheimer, *The Geometry of Four-Manifolds*, Oxford University Press, Oxford, 1990.

[89] S.K. Donaldson and I. Smith, Lefschetz pencils and the canonical class for symplectic 4-manifolds, *Topology* **42** (2003), 743–785.

[90] S. Dostoglou and D.A. Salamon, Instanton homology and symplectic fixed points, in *Symplectic Geometry*, edited by D. Salamon, 57–93, LMS Lecture Notes Series **192**, Cambridge University Press, 1993.

[91] S. Dostoglou and D.A. Salamon, Cauchy–Riemann operators, self-duality, and the spectral flow, in *First European Congress of Mathematics, Volume I, Invited Lectures (Part 1)*, edited by Joseph et al., 511–545, Birkhäuser, Progress in Mathematics, **Vol. 119** (1994).

[92] S. Dostoglou and D.A. Salamon, Self-dual instantons and holomorphic curves, *Annals of Math.* **139** (1994), 581–640.

[93] B. Dubrovin, Integrable systems in topological field theory, *Nucl. Phys. B* **379** (1992), 627–689.

[94] B. Dubrovin, Geometry of $2D$ topological field theories, in *Springer Lecture Notes in Mathematics* **1620** (1996), 120–348.

[95] C. Earle and J. Eells: A fibre bundle approach to Teichmüller theory, *J. Diff. Geo.* **3** (1969), 19–43.

[96] D. Eisenbud and W. Neumann, *Three Dimensional Link Theory and Invariants of Plane Curve Singularities*, Annals of Mathematics Studies # 110, Princeton Univ. Press, 1968.

[97] I. Ekeland and H. Hofer, Symplectic topology and Hamiltonian dynamics, *Math. Zeit.* **200** (1989), 355–378.

[98] Y. Eliashberg, A theorem on the structure of wave fronts and its applications in symplectic topology, *Functional Analysis and Applications* **21** (1987), 65–72.

[99] Y. Eliashberg, D. Fuchs, T. Ratiu and A Weinstein, eds., *Northern California Symplectic Geometry Seminar, Amer. Math. Soc. Transl.* (2) **196** (1999).

[100] Y. Eliashberg and M. Gromov, Convex symplectic manifolds, in *Several Complex Variables and Complex Geometry*, Part 2 (Santa Cruz CA 1989), Proc. Sympos. Pure Math. **52**, Part 2, Amer. Math. Soc., Providence, RI (1991), 135–162.

[101] Y. Eliashberg, A.B. Givental, and H. Hofer, Introduction to symplectic field theory, *Geom. Funct. Anal.* **10** (2000), 560–673.

[102] Y. Eliashberg and N. Mishachev, *An introduction to the h-principle*, Graduate Texts # 48, Amer. Math. Soc., Providence, RI, 2002.

[103] Y. Eliashberg and L. Traynor, eds., *Symplectic Geometry and Topology*, IAS/Park City Mathematics Series, Vol. **7**, Amer. Math. Soc., Providence, RI, 1999.

[104] M. Entov, K-area, Hofer metric and geometry of conjugacy classes in Lie groups, *Invent. Math.* **146** (2001), 93–141.

[105] M. Entov and L. Polterovich, Calabi quasimorphism and quantum homology, *Int. Math. Res. Notes* (2003), no. 30, 1635–1676.

[106] M. Entov and L. Polterovich, Rigid subsets of symplectic manifolds, *Compos. Math.* **145** (2009), no. 3, 773–826.

[107] M. Entov, L. Polterovich and F. Zapolsky, An "anit-Gleason" phenomenon and simultaneous measurement in Classical Mechanics, *Foundations of Physics,* **37** (8), (1007), 1306–1316.

[108] J.D. Evans, Symplectic mapping class groups of some Stein and rational surfaces, *J. Symplectic Geom.* **9** (2011), no.1, 45–82.

[109] J. Fine and D. Panov, The diversity of Calabi–Yau six-manifolds, arXiv:1108.5944.

[110] R. Fintushel and R. Stern, Knots, links and 4-manifolds, *Invent. Math.* **134** (1998), 363–400.

[111] R. Fintushel and R. Stern, Pinwheels and nullhomologous surgery on 4-manifolds with $b^+ = 1$, *Algeb. Geom. Topol.* **11** (2011), 1649–1699.

[112] A. Floer, Proof of the Arnold conjecture for surfaces and generalizations to certain Kähler manifolds, *Duke Math. J.* **53** (1986), 1–32.

[113] A. Floer, Morse theory for Lagrangian intersections, *J. Diff. Geom.* **28** (1988), 513–547.

[114] A. Floer, The unregularized gradient flow of the symplectic action, *Comm. Pure Appl. Math.* **41** (1988), 775–813.

[115] A. Floer, Wittens complex and infinite dimensional Morse theory, *J. Diff. Geom.* **30** (1989), 207–221.

[116] A. Floer, Symplectic fixed points and holomorphic spheres, *Comm. Math. Phys.* **120** (1989), 575–611.

[117] A. Floer, An instanton invariant of 3-manifolds, *Comm. Math. Phys.* **118** (1989), 215–240.

[118] A. Floer and H. Hofer, Coherent orientations for periodic orbit problems in symplectic geometry, *Math. Zeit.* **212** (1993), 13–38.

[119] A. Floer, H. Hofer, and D.A. Salamon, Transversality in elliptic Morse theory for the symplectic action, *Duke Math. J.* **80** (1995), 251–292.

[120] U. Frauenfelder, Gromov convergence of pseudoholomorphic discs, Diploma thesis, ETH Zürich, February 2000.

[121] U. Frauenfelder, Floer homology of symplectic quotients and the Arnold–Givental conjecture, PhD thesis, ETH-Zürich, January 2003.

[122] U. Frauenfelder, The Arnold-Givental conjecture and moment Floer homology, *IMRN* (2004), 2179–2269.

[123] U. Frauenfelder and F. Schlenk, Hamiltonian dynamics on convex symplectic manifolds, *Israel J. Math.* **159** (2007), 1–56.

[124] A. Friedman, *Partial Differential Equations*, Holt, Rinehart, and Winston, 1969.

[125] K. Fukaya, Floer homology, A^∞-categories, and topological field theory, in *Proceedings of GARC Workshop on Geometry and Topology '93 (Seoul)*, ed H. Kim, Lecture Notes Ser **18**, Seoul Nat. Univ, Seoul 1993, 1–102.

[126] K. Fukaya, Floer homology for 3-manifolds with boundary, *Topology, geometry and field theory*, 1–21, World Sci. Publ., River Edge, NJ, 1994. with update 1997 at http://www.kusm.kyoto-u.ac.jp/ fukaya/fukaya.html.

[127] K. Fukaya and K. Ono, Arnold conjecture and Gromov–Witten invariants for general symplectic manifolds, *Topology* **38** (1999), 933–1048.

[128] K. Fukaya, Y.-G. Oh, H. Ohta, K. Ono, *Lagrangian Intersection Theory, Anomaly and Obstruction, Parts I and II*, AMS/IP Studies in Advanced Mathematics, Amer. Math. Soc. (Providence RI) and Internat. Press (Somerville MA), 2009.

[129] K. Fukaya, Y.-G. Oh, H. Ohta, K. Ono, Lagrangian FLoer theory on compact toric manifolds, I, *Duke Math. J.* **151** (2010), no 1, 23-174.

[130] K. Fukaya, Y.-G. Oh, H. Ohta, K. Ono, Lagrangian FLoer theory on compact toric manifolds, II: bulk deformations, *Selecta Math.* (N.S.) **17** (2011), no.3, 609–711.

[131] K. Fukaya, P. Seidel and I. Smith, The symplectic geometry of cotangent bundles from a categorical viewpoint, in *Homological mirror symmetry*, 1–26, Lecture Notes in Physics, vol 757, Springer, Berlin, 2009.

[132] W. Fulton and R. MacPherson, A compactification of configuration space, *Annals of Math.* **139** (1994), 183–225.

[133] W. Fulton and R. Pandharipande, *Notes on Stable Maps and Quantum Cohomology*, in *Algebraic Geometry – Santa Cruz 1995*, eds. Kollar, Lazarsfeld and Morrison, *Proc. Symp. Pure. Math.* **62** part 2 (1997), 45–96.

[134] A.R. Gaio, J-holomorphic curves and moment maps, PhD-thesis, University of Warwick, April 2000.

[135] A.R. Gaio and D.A. Salamon, Gromov–Witten invariants of symplectic quotients and adiabatic limits, *J. Symplectic Geom.* **3** (2005), no 1, 55–159. Erratum: *J. Symplectic Geom.* **7** (2009), no 3, 377–379.

[136] O. Garcia-Prada, A direct existence proof for the vortex equations over a compact Riemann surface, *Bull. London Math. Soc.* **26** (1994), 88–96.

[137] R. Gautschi, Floer homology and surface diffeomorphisms, PhD thesis, ETH-Zürich, April 2002.

[138] R. Gautschi, Floer homology of algebraically finite mapping classes, *J. Symplectic Geom.* **1** (2003), no. 4, 715–765.

[139] E. Getzler, Operads and moduli spaces of genus zero Riemann surfaces, in [**80**], 199–230.

[140] D. Gilbarg and N.S. Trudinger, *Elliptic Partial Differential Equations of the Second Order*, Springer, Berlin, 1983.

[141] V.L. Ginzburg, An embedding $S^{2n-1} \to \mathbb{R}^{2n}$, $2n-1 \geq 7$, whose Hamiltonian flow has no periodic trajectories. *Int. Math. Res. Notes* **2** (1995), 83–98.

[142] V.L. Ginzburg, A smooth counterexample to the Hamiltonian Seifert conjecture in $\mathbb{R}^6$, *Int. Math. Res. Notes* (1997), 641–650.

[143] V.L. Ginzburg, The Conley conjecture, *Ann. of Math.* (2) **172** (2010), 1127–1180.

[144] A.B. Givental, Lagrangian embeddings of surfaces and unfolded Whitney umbrella, *Funkt. Anal. and Appl.* **20** (1986), 35–41.

[145] A.B. Givental, Nonlinear generalization of the Maslov index, *Theory of singularities and its applications*, 71-103, *Adv. Soviet Math.* **1**, Amer. Math. Soc., Providence, RI, 1990.

[146] A.B. Givental, Homological geometry I, Projective Hypersurfaces, *Selecta Math.* **1** (1995), 325–345.

[147] A.B. Givental, Equivariant Gromov–Witten invariants, *Int. Math. Res. Notes* **3** (1996), 613–663.

[148] A.B. Givental, Stationary phase integrals, quantum Toda lattices, flag manifolds, and the mirror conjecture, *Amer. Math. Soc. Transl. (2)*, **180** (1997), 103–215.

[149] A.B. Givental, *Topics in Enumerative Algebraic Geometry* Lecture Notes (taken by Y.-P. Lee), August 1997.

[150] A.B. Givental, A mirror theorem for toric complete intersections, AG/9701016, in: *Topological Field Theory, Primitive Forms and Related Topics*, Kashiwara et al. eds, 141–175, Birkhäuser 1998.

[151] A.B. Givental, A tutorial on quantum cohomology, in [**103**], 231–264.

[152] A.B. Givental, Symplectic geometry of Frobenius structures, *Frobenius manifolds*, 91–112, *Aspects Math.*, **E36**, Vieweg, Weisbaden, 2004.

[153] A. Givental and B. Kim, Quantum cohomology of flag manifolds and Toda lattices, *Comm. Math. Physics* **168** (1995), 609–641.

[154] R. Gompf, A new construction for symplectic 4-manifolds, *Annals of Mathematics*, **142** (1995), 527–595.

[155] E. Gonzalez and H. Iritani, Seidel elements and mirror transformations, arXiv:1103.4171.

[156] E. Gonzalez and C. Woodward. Deformations of symplectic vortices, *Ann. Global Anal. Geom.* **39** (2011), 45-?82.

[157] E. Gonzalez and C. Woodward, Gauged Gromov-Witten theory for small spheres, arXiv:0907.3869, to appear in *Math. Zeit.*.

[158] T. Graber and R. Pandharipande, Localization of virtual classes, *Invent. Math.* **135** (1999), 487–518.

[159] P. Griffiths and J. Harris, *Principles of Algebraic Geometry*, Wiley 1978.

[160] M. Gromov, Pseudo holomorphic curves in symplectic manifolds, *Invent. Math.* **82** (1985), 307–347.

[161] M. Gromov, *Partial differential relations*, Springer, Berlin, 1986.

[162] A. Grothendieck, Sur la classification des fibrés holomorphes sur la sphère de Riemann, *Amer. J. Math.* **76** (1957), 121–138.

[163] V. Guillemin, E. Lerman, and S. Sternberg, *Symplectic Fibrations and Multiplicity Diagrams*, Cambridge University Press 1996.

[164] V. Guillemin, and S. Sternberg, *Supersymmetry and Equivariant deRham Theory*, Springer, Berlin, 1999.

[165] V. Guillemin and A. Pollack, *Differential Topology*, Prentice Hall 1974.

[166] S. Guillermou, M. Kashiwara, and P. Shapira, Sheaf quantization of Hamiltonian isotopies and applications to non displaceability problems, arXiv:1005.1517, to appear in *Duke Math Journ.*

[167] L. Guth, Symplectic embeddings of polydiscs, arXiv:math/0709.1957, *Invent. Math.* **172** (2008), 477–489.

[168] J. Harris and I. Morrison, *Moduli of curves*, Graduate Texts in Mathematics # 187, Springer, Berlin, 1998.

[169] P. Hartman and A. Wintner, On the local behavior of solutions of non-parabolic partial differential equations, *Amer. J. Math.* **75** (1953), 449–476.

[170] R. Hartshorne, Ample vector bundles, *Publications mathématiques de l'I.H.É.S.* **29** (1966), 63–94.

[171] R. Hartshorne, *Algebraic Geometry*, Graduate Texts in Mathematics # 52, Springer, Berlin, 1977.

[172] M. Herman, Exemple de flots hamiltoniens dont aucune perturbation en topologie C^∞ n'a d'orbites périodiques sur un ouvert de surfaces d'energie, *Comptes Rendus A.S.* **312** ser. 1 (1991), 989–994.

[173] D. Hill and M. Taylor, Integrability of rough almost complex structures, *J. Geom. Anal.* **13** (2003), 163–172.

[174] R. Hind, Lagrangian spheres in $S^2 \times S^2$, *Geom. Funct. Anal,* **14** (2004), no 2, 303–318.

[175] R. Hind and E. Kerman, New obstructions to symplectic embeddings, arXiv:0906.4296.

[176] N. Hingston, Subharmonic solutions of Hamiltonian equations on tori, to appear in *Ann. of Math.* (2), **170** (2009), no. 2, 529–560.

[177] H. Hofer, On the topological properties of symplectic maps, *Proc. Royal Soc. Edin.* **115** (1990), 25–38.

[178] H. Hofer, Pseudoholomorphic curves in symplectizations with applications to the Weinstein conjecture in dimension three, *Invent. Math.* **114** (1993), 515–563.

[179] H. Hofer, V. Lizan, and J.C. Sikorav, On genericity of holomorphic curves in 4-dimensional almost complex manifolds, *J. Geom. Anal.* **7** (1997), 149–159.

[180] H. Hofer and D. Salamon, Floer homology and Novikov rings, in [**181**], 483–524.

[181] H. Hofer, C.H. Taubes, A. Weinstein, and E. Zehnder, eds., *The Floer Memorial Volume*, Birkhäuser 1995.

[182] H. Hofer and C. Viterbo, The Weinstein conjecture in the presence of holomorphic spheres, *Comm. on Pure and Applied Math.* **XLV** (1992), 583–622.

[183] H. Hofer, K. Wysocki, and E. Zehnder, Finite energy foliations of tight three-spheres and Hamiltonian dynamics, *Ann. of Math.* (2) **157** (2003), 125–255.

[184] H. Hofer, K. Wysocki, and E. Zehnder, A general Fredholm theory I, A splicing based differential geometry, *J. Eur. Math. Soc. (JEMS)* **9** (2007), no 4, 841–876.

[185] H. Hofer, K. Wysocki, and E. Zehnder, A general Fredholm theory II, Implicit Function theorems, *Geom. Funct. Anal.* **19** (2009), no. 1, 206–293.

[186] H. Hofer, K. Wysocki, and E. Zehnder, A general Fredholm theory III, Fredholm functors and polyfolds, *Geom. Topol.* **13** (2009), no. 4, 2279–2387.

[187] H. Hofer, K. Wysocki, and E. Zehnder, Integration theory on the zero sets of polyfold Fredholm sections, *Math. Ann.* **346** (2010), no.1, 139–198.

[188] H. Hofer, K. Wysocki, and E. Zehnder, sc-smoothness, retractions and new models for smooth spaces, *Discrete Contin. Dyn. Syst.* **28** (2010), no. 2, 665–788.

[189] H. Hofer, K. Wysocki, and E. Zehnder, Applications of Polyfold theory I: the Polyfolds of Gromov–Witten theory, arXiv:1107.2097.

[190] H. Hofer and E. Zehnder, *Symplectic capacities and Hamiltonian Dynamics*, Birkhäuser, Basel 1994.

[191] Jianxun Hu, Gromov–Witten invariants of blow-ups along surfaces, *Compos. Math.* **125** (2001), 345–352.

[192] J. Hu, T.-J. Li and Yongbin Ruan, Birational cobordism invariance of uniruled symplectic manifolds, *Invent. Math.* **172** 2008, no. 2, 231–275.

[193] C. Hummel, *Gromov's Compactness Theorem for Pseudoholomorphic curves*, Progress in Mathematics **151**, Birkhäuser 1997.

[194] M. Hutchings, Floer homology of families I, *Algebr. Geom. Topol.* **8** (2008), 435–492.

[195] M. Hutchings and M. Sullivan, Rounding corners of polygons and the embedded contact homology of T^3, *Geom. Topol.* (2006), 169–266.

[196] E. Ionel, Genus 1 enumerative invariants in $\mathbb{C}P^n$ with fixed j invariant, *Duke Math. J.* **94** (1998), 279–324.

[197] E. Ionel, GW invariants relative normal crossings divisors, arXiv:1103.3977

[198] E. Ionel and T. Parker, The Gromov invariants of Ruan–Tian and Taubes, *Math. Res. Lett.* **4** (1997), 521–532.

[199] E. Ionel and T. Parker, Gromov invariants and symplectic maps, *Math. Annalen* **314** (1999), 127–158.

[200] E. Ionel and T. Parker, Relative Gromov–Witten invariants, *Ann. of Math.* (2) **157** (2003), 45–96.

[201] E. Ionel and T. Parker, The symplectic sum formula for Gromov–Witten invariants, *Ann. of Math.* (2) **159** (2004), 935–1025.

[202] S. Ivashkovich and V. Shevchishin, Structure of the moduli space in a neighborhood of a cusp-curve and meromorphic hulls, *Invent. Math.* **136** (1999), 571–602.

[203] F. John, *Partial Differential Equations*, Springer, Berlin, 1984.

[204] J. Jost, *Riemannian Geometry and Geometric Analysis*, Springer, Berlin, 1998.

[205] Sheldon Katz, On the finiteness of rational curves on quintic three folds, *Compos. Math.* **60** (1986), no. 2, 151–162.

[206] M.M. Kapranov, Chow quotients of Grassmannians I, *Adv. Soviet Math.* **16** (1993), 29–110.

[207] R. Kaufmann, The intersection form in $H^*(\overline{M}_{0,n})$ and the explicit Künneth formula in quantum cohomology, *Internat. Math. Res. Notices* **19** (1996), 929–952.

[208] S. Keel, Intersection theory of moduli spaces of n-pointed curves of genus zero. *Trans. Amer. Math. Soc.* **330** (1992), 545–574.

[209] J. L. Kelley, *General Topology*, Van Norstrand, 1955.

[210] B. Kim, Quantum cohomology of flag manifolds, G/B and quantum Toda lattices, *Ann. of Math.* **149** (1999), 129–148.

[211] F.F. Knudsen, The projectivity of moduli spaces of stable curves II, *Funct. Anal. Appl.* **19** (1983), 161–199.

[212] S. Kobayashi, *Differential Geometry of Complex Vector Bundles*, Publications of the Mathematical Society of Japan **15**, Princeton University Press 1987.

[213] S. Kobayashi and K. Nomizu, *Foundations of Differential Geometry*, Wiley 1963.

[214] J. Kollár, Low degree polynomial equations: arithmetic, geometry, topology, *European Congress of Mathematics*, Vol 1, (Budapest 1996), 255–288. Progress in Math **168**, Birhauser, Basel, 1998.

[215] M. Kontsevich, Enumeration of rational curves via torus actions, in [**80**], 335–368.

[216] M. Kontsevich and Yu. Manin, Gromov-Witten classes, quantum cohomology, and enumerative geometry, *Comm. Math. Phys.* **164** (1994), 525–562.

[217] Parametrized ring spectra and the nearby Lagrangian conjecture, arXiv:1107.4674.

[218] A. Kriegl and P. Michor, *The convenient setting of global analysis*, Mathematical Surveys and Monographs **53**, Amer. Math. Soc. Providence 1997.

[219] C. Kutluhan, Yi-Jen Lee and C. Taubes, $HF = HM1$: Heegaard Floer homology and Seiberg–Witten–Floer homology, I, arXiv:1007.1979.

[220] D. Kwon and Y. G.Oh, Structure of the image of (pseudo)-holomorphic discs with totally real boundary condition. With an appendix by Jean-Pierre Rosay. *Comm. Anal. Geom.* **8** (1) (2000), 31– 82.

[221] F. Lalonde, Isotopy of symplectic balls, Gromov's radius, and structure of ruled symplectic manifolds, *Mathematische Annalen* **300** (1994), 273–296.

[222] F. Lalonde, A field theory for symplectic fibrations over surfaces, *Geom. Topol.* **8** (2004), 1189–1226.

[223] F. Lalonde and D. McDuff, The Geometry of Symplectic Energy, *Annals of Math.* **141** (1995), 349–371.

[224] F. Lalonde and D. McDuff, Hofer's L^∞ geometry: geodesics and stability, I, II. *Invent. Math.* **122** (1995), 1–33, 35–69.

[225] F. Lalonde and D. McDuff, The classification of ruled symplectic 4-manifolds, *Math. Res. Lett.* **3** (1996), 769–778.

[226] F. Lalonde and D. McDuff, J-curves and the classification of rational and ruled symplectic 4-manifolds, in: *Symplectic and Contact Geometry* ed. C. Thomas, Camb. Univ. Press 1996, pp 3–42.

[227] F. Lalonde and D. McDuff, Symplectic structures on fiber bundles, SG/0010275, *Topology* **42** (2003), 309–347.

[228] F. Lalonde, D. McDuff and L. Polterovich, Topological rigidity of Hamiltonian loops and quantum homology, *Invent. Math* **135** (1999), 369–385.

[229] F. Lalonde and M. Pinsonnault, The topology of the space of symplectic balls in rational 4-manifolds, *Duke Math. J.* **122** (2004), no.2, 347–397.

[230] F. Lalonde and J.-P. Sikorav, Sous-Variétés lagrangiennes et lagrangiennes exactes des fibrers cotangents, *Comm. Math. Helv.* **66** (1991), 18–33.

[231] J. Latschev, D. McDuff, and F. Schlenk, The Gromov width of 4-dimensional tori, arXiv:1111.6566

[232] B. Lawson, *Minimal Varieties in Real and Complex Geometry*, Sem. Math. Sup. Vol **57**, Presses Université de Montréal, 1974.

[233] R. Lazersfeld *Positivity in Algebraic Geometry I; Classical Settings: Linear Bundles and Linear Series*, Ergebnisse der Mathematik **48**, Springer, Berlin, 2004.

[234] L. Lazzarini, Existence of a somewhere injective pseudoholomorphic disc, *Geom. Funct. Anal.* **10** (2000), 829–862.

[235] L. Lazzarini, Decomposition of a J-holomorphic curve, preprint. Can be downloaded at http://www.math.jussieu.fr/ lazzarin/articles.html

[236] Hong-Van Le and Kaoru Ono, Parameterized Gromov–Witten invariants and topology of symplectomorphism groups, *Advanced Studies in Pure Math.* **52** (2008), 51–75.

[237] E. Lerman, Symplectic Cuts, *Math. Res. Lett.* **2** (1995), 247–258.

[238] A.M. Li and Y. Ruan, Symplectic surgery and Gromov–Witten invariants of Calabi–Yau 3-folds, *Invent. Math.* **145** (2001), 151–218.

[239] Jun Li and G. Tian, Virtual moduli cycles and GW invariants of general symplectic manifolds, in [**385**], 47–83.

[240] T.J. Li and A. Liu, Symplectic structure on ruled surfaces and generalized adjunction formula, *Math. Res. Lett.* **2** (1995), 453–471.

[241] T.J. Li, Smoothly embedded spheres in symplectic 4-manifolds, *Proc. Amer. Math. Soc.* **127** (1999), 609–13.

[242] Bang-He Li and T.-J. Li, Symplectic genus, minimal genus and diffeomorphisms, *Asian J. Math.* **6** (2002), 123-144.

[243] T.J. Li, The Kodaira dimension of symplectic 4-manifolds, *Floer homology, gauge theory and low-dimensional topology*, 249–261, *Clay Math. Proc. 5* Amer. Math. Soc., Providence, RI, 2006.

[244] T.J. Li, Quaternionic bundles and Betti numbers of symplectic 4-manifolds with Kodaira dimension zero, *Int. Math. Res. Not.* **2006** Art ID 37385, 28pp.

[245] T.J. Li, The space of symplectic structures on closed 4-manifolds, *Third International Conference of Chinese mathematicians*, AMS/IP Stud. Adv. Math., **42**, part I,II, 259–277, Amer. Math. Soc., Providence, RI, 2008.

[246] T.J. Li and Weiwei Wu, Lagrangian spheres, symplectic surfaces and the symplectic mapping class group, arXiv:1012.4146.

[247] R. Lipshitz, P. Ozsváth and D. Thurston, A tour of bordered Floer homology, *Proc. Nat. Acad. Sci.* **108**, (2011), 8085–8092.

[248] G. Liu, Associativity of quantum multiplication, *Comm. Math. Phys.* **191** (1998), 265–282.

[249] G. Liu and G. Tian, Floer homology and Arnold conjecture, *J. Diff. Geom.*, **49** (1998), 1–74.

[250] G. Liu and G. Tian, On the equivalence of multiplicative structures in Floer homology and quantum homology, *Acta Math. Sinica* **15** (1999), 53–80.

[251] G. Liu and G. Tian, Weinstein conjecture and GW invariants, *Commun. Contemp. Math.* **2** (2000), 405–459.

[252] V. Lizan, About the bubbling off phenomenon in the limit of a sequence of J-curves, in: *Geometry, Topology and Dynamics*, Lalonde, ed., 87–100, CRM Proceedings and Lecture Notes, Vol **15**, Amer. Math. Soc. 1998.

[253] GuangCun Lu, The Weinstein conjecture in the uniruled manifolds, *Math. Res. Lett.* **7** (4) (2000), 383–387.

[254] D. McDuff, Examples of symplectic structures, *Invent. Math.* **89** (1987), 13–36.

[255] D. McDuff, Elliptic methods in symplectic geometry, *Bull. A.M.S.* **23** (1990), 311–358.

[256] D. McDuff, The local behaviour of holomorphic curves in almost complex 4-manifolds, *J. Diff. Geom.* **34** (1991), 143–164.

[257] D. McDuff, Blowing up and symplectic embeddings in dimension 4, *Topology*, **30**, (1991), 409–421.

[258] D. McDuff, The structure of rational and ruled symplectic 4-manifolds, *J. Amer. Math. Soc.* **3** (1990), 679–712; Erratum: *J. Amer. Math. Soc* **5** (1992), 987–988.

[259] D. McDuff, Symplectic manifolds with contact-type boundaries, *Invent. Math* **103** (1991), 651–671.

[260] D. McDuff, Immersed spheres in symplectic 4-manifolds, *Annal. de l'Inst Fourier* **42** (1991), 369–392.

[261] D. McDuff, Singularities of J-holomorphic curves in almost complex 4-manifolds, *J. Geom. Anal.* **3** (1992), 249–266.

[262] D. McDuff, Notes on ruled symplectic 4-manifolds, *Trans. Amer. Math. Soc.* **345** (1994), 623–639.

[263] D. McDuff and L. Polterovich, Symplectic packings and algebraic geometry, *Invent. Math.*, **115**, (1994) 405–429.

[264] D. McDuff, Singularities and positivity of intersections of J-holomorphic curves, with Appendix by G. Liu, in [**25**], 191–216.

[265] D. McDuff, Lectures on Gromov invariants, in *Gauge Theory and Symplectic Geometry*, ed. Hurtubise and Lalonde, NATO ASI Series, Kluwer, Dordrecht/Boston 1997.

[266] D. McDuff, From symplectic deformation to isotopy, in [**385**], 85–100.

[267] D. McDuff, The virtual moduli cycle, in [**99**], 73 – 102.

[268] D. McDuff, Almost complex structures on $S^2 \times S^2$, *Duke Math. J.* **101** (2000), 135–177.

[269] D. McDuff, Quantum homology of fibrations over S^2, *Internat. J. of Math.* **11** (2000), 665–721.

[270] D. McDuff, Symplectomorphism Groups and Almost Complex Structures, SG/0010274, *Enseignement Math.* **38** (2001), 1–30.

[271] D. McDuff, Geometric variants of the Hofer norm, SG/0103089, *J. Sympl. Geom.* **1** (2002), 197–252.

[272] D. McDuff, Lectures on groups of symplectomorphisms, in Proceedings of Winter School (Srni), *Rendiconti di Circolo di Mat, Palermo* Serie II Suppl. **72** (2004),43–78.

[273] D. McDuff, Hamiltonian S^1-manifolds are uniruled, *Duke Math. Journ.* **146** (2009), 449–507.

[274] D. McDuff, Monodromy in Hamiltonian Floer theory, *Comment. Math. Helv.* **85** (2010), 95–133.

[275] D. McDuff, Displacing Lagrangian toric fibers via probes, *Low-dimensional and Symplectic Topology*, 131–160, *Proc. Sympos. Pure Math.* **82** Amer. Math. Soc., Providence, RI, 2011.

[276] D. McDuff and D.A. Salamon, *J-holomorphic curves and quantum cohomology*, AMS Lecture Series **6**, Amer. Math. Soc., Providence, RI, 1994.

[277] D. McDuff and D.A. Salamon, *Introduction to Symplectic Topology*, 2nd edition, Oxford University Press, 1998.

[278] D. McDuff and F. Schlenk, The embedding capacity of 4-dimensional symplectic ellipsoids, to appear in the *Annals of Math.* (2012).

[279] D. McDuff and S. Tolman, Topological properties of Hamiltonian circle actions, *International Mathematical Research Papers* Volume 2006 (2006). article ID 72826, pp 77.

[280] D. McDuff and S. Tolman, Polytopes with mass linear functions, Part I, arxiv:math/0807.0900 *IMRN* (2009) doi: 10.1093/imrn/rnp179

[281] D. McDuff and K. Wehrheim, Smooth Kuranishi structures with trivial isotropy, Preprint 2012.

[282] B. McKay, Dual curves and pseudoholomorphic curves, *Selecta Math.* **9** (2003), 251–311.

[283] M. McLean, Lefschetz fibrations and symplectic homology, *Geom. Topol.* **13** (2009), no.4, 1877–1944.

[284] C.T. McMullen and C.H. Taubes, 4-manifolds with inequivalent symplectic forms and 3-manifolds with inequivalent fibrations, *Math. Res. Lett.* **6** (1999), 681–696.

[285] Yu. Manin, Generating functions in algebraic geometry and sums over trees, in [**80**], 401–418.

[286] Yu. Manin, *Frobenius Manifolds, Quantum Cohomology and Moduli Spaces*, A.M.S. Colloq. Publ. Vol. **47**, 1999.

[287] M. Micallef and B. White, The structure of branch points in area minimizing surfaces and in pseudo-holomorphic curves, *Annals of Math.* **141** (1995), 35–85.

[288] J.W. Milnor, On spaces having the homotopy type of a CW-complex, *Trans. A.M.S.* **90** (1959), 272–280.

[289] J.W. Milnor, *Topology from the Differential Viewpoint*, Univ. Press of Virginia, 1965.

[290] J.W. Milnor, *Singular Points of Complex Hypersurfaces*, Annals of Mathematics Studies # 61, Princeton Univ. Press, 1968.

[291] J.W. Milnor and J.D. Stasheff, *Characteristic Classes*, Princeton Univ. Press, 1974.

[292] J. Moser, Finitely many mass points on the line under the influence of an exponential potential – an integrable system, in *Dynamical Systems, Theory and Applications*, 467–497 Springer, Berlin, 1975.

[293] D. Mumford, J. Fogarty, and F. Kirwan, *Geometric Invariant Theory*, Springer, Berlin, 1994.

[294] I. Mundet i Riera, Yang–Mills theory for symplectic fibrations, PhD thesis, Madrid, April 1999.

[295] I. Mundet i Riera, Hamiltonian Gromov–Witten invariants, *Topology* **42** (2003), 525–553.

[296] V. Munoz, On the Gromov–Witten invariants of the moduli of bundles on a surface, AG/9910105, *Forum Math.* **14** (2002), 759–774.

[297] I. Naraki, Cross ratio variety as a moduli space of cubic surfaces, *Proc. London Math. Soc.* **45** (1982), 1–30.

[298] S. Nemirovski, Lagrangian Klein bottles in $\mathbb{R}^{2n}$, *Grom. Funct. Anal.* **19** (2009), no. 3, 902–909.

[299] K.L. Nguyen, C. Woodward, F. Ziltener, Morphisms of CohFT algebras and quantization of the Kirwan map, Preprint 2009, revised 2011. arXiv:0903.4459

[300] A. Nijenhuis and W. Woolf, Some integration problems in almost-complex and complex manifolds, *Annals of Math.* **77** (1963), 424–489.

[301] S.P. Novikov, Multivalued functions and functionals – an analogue of the Morse theory, *Soviet Math. Dokl.* **24** (1981), 222–225.

[302] Y.-G. Oh, Removal of boundary singularities of pseudo-holomorphic curves with Lagrangian boundary conditions, *Comm. Pure Appl. Math.* **45** (1992), 121–139.

[303] Y.-G. Oh, Floer cohomology of Lagrangian intersections discs and pseudoholomorphic discs I, II. *Comm. Pure Appl. Math.*, **46** (1993), 949–94, 995–1012. III. Arnold–Givental conjecture, in [**181**], 555–574.

[304] Y.-G. Oh, Floer cohomology, spectral sequences and the Maslov class of Lagrangian embeddings, *Int. Math. Res. Notes* **7** (1996), 305–346.

[305] Y.-G. Oh, Chain level Floer theory and Hofer's geometry on the Hamiltonian diffeomorphism group, *Asian J. Math.* **6** (2003), 579–624.

[306] Y.-G. Oh, Spectral invariants, analysis of the Floer moduli space and geometry of the Hamiltonian diffeomorphism group, *Duke Math. J.* **130** (2005), no. 2, 199–295.

[307] Y.-G. Oh, Spectral invariants and length minimizing property of Hamiltonian paths, *Asian J. Math.* **9** (2005), no. 1, 1–18.

[308] Y.-G. Oh, The group of Hamiltonian homeomorphisms and C^0 symplectic topology, *J. Symplectic Geom.* **5** (2007), no.2, 167–219.

[309] C. Okonek and A. Teleman, Gauge theoretic equivariant Gromov–Witten invariants and the full Seiberg–Witten invariants of ruled surfaces, *Commun. Math. Phys.* **227** (2002), 551–585.

[310] C. Okonek and A. Teleman, Gauge theoretical Gromov–Witten invariants and virtual fundamental classes, *The Fano conference*, 591–623, *Univ. Torino, Turin*, (2004).

[311] A. Okounkov and R. Pandharipande, Gromov–Witten theory, Hurwitz numbers, and completed cycles, *Ann. of Math.* (2) **163** (2006), no. 2, 517–560.

[312] A. Okounkov and R. Pandharipande, The equivariant Gromov–Witten theory of $\mathbb{P}^1$, *Ann. of Math.* (2) **163** (2006), no. 2, 561–605.

[313] K. Ono, The Arnold conjecture for weakly monotone symplectic manifolds, *Invent. Math.* **119** (1995), 519–537.

[314] Y. Ostrover, Calabi quasi-morphisms for some non-monotone symplectic manifolds, *Alg. Geom. Top.* **6** (2006), 405–434.

[315] Y. Ostrover, A comparison of Hofer's metrics on Hamiltonian diffeomorphisms and Lagrangian submanifolds, *Commun. Contemp. Math* **5** (2003), no. 5, 803–811.

[316] Y. Ostrover and I. Tyomkin, On the quantum homology algebra of toric Fano manifolds, *Selecta Math.* (N.S.) **15** (2009), no. 1, 131–149.

[317] Y. Ostrover and R. Wagner, On the extremality of Hofer's metric on the group of Hamiltonian diffeomorphisms, *Int. Math. Res. Not.* **2005**, no 35, 2123–2141.

[318] P. Ozsváth and Z. Szabó, Holomorphic discs and three-manifold invariants for closed 3-manifolds, *Ann. Math.* **159** (2004), 1027–1158.

[319] A. Ott. Removal of singularities and Gromov compactness for symplectic vortices, arXiv:0912.2500, to appear in *J. Symplectic Geom.*

[320] P. Pansu, Pseudo-holomorphic curves in symplectic manifolds, in [**25**], 233–250.

[321] T.H. Parker and J.G. Wolfson, Pseudoholomorphic maps and bubble trees, *J. Geom. Anal.* **3** (1993), 63–98.

[322] M. Pinsonnault, Symplectomorphism groups and embeddings of balls into rational ruled 4-manifolds, *Compos. Math.* **144** (2008), no. 3, 787–810.

[323] S. Piunikhin, D. Salamon, M. Schwartz Symplectic Floer-Donaldson theory and quantum cohomology, in *Contact and Symplectic Geometry*, edited by C.B. Thomas, Publications of the Newton Institute, Cambridge University Press 1996, 171–200.

[324] L. Polterovich, Monotone Lagrangian submanifolds of linear spaces and the Maslov class in cotangent bundles, *Math. Zeit.* **207** (1991), 217–222.

[325] L. Polterovich, Gromov's K-area and symplectic rigidity, *Geom. Funct. Anal.* **6** (1996), 726–739.

[326] L. Polterovich, Symplectic aspects of the first eigenvalue, *J. Reine Angew. Math.* **502** (1998), 1–17.

[327] L. Polterovich, Precise measurements in symplectic topology, in *European Congress of Mathematics*, Vol. II (Budapest, 1996), 159–166, *Progr.Math.* **169**, Birkhäuser, Basel, 1998.

[328] L. Polterovich, Geometry on the group of Hamiltonian diffeomorphisms, Proceedings of the International Congress of Mathematicians, (Berlin 1998), *Doc. Math.*, Extra Vol. II, 401–410.

[329] L. Polterovich, Growth of maps, distortion in groups and symplectic geometry, *Invent. Math.* **150** (2002), 655–686.

[330] L. Polterovich, *The Geometry of the Group of Symplectic Diffeomorphisms*, Lectures in Math, ETH, Birkhauser, 2001.

[331] L. Polterovich, Quantum unsharpness and symplectic rigidity, arXiv:1110.5247

[332] P. Py, Quasimorphisms and Calabi invariant, *Ann. Sci. École Norm. Sup.* (4) **39** (2006), 177–194.

[333] J.W. Robbin, Y. Ruan, and D.A. Salamon, The moduli space of regular stable maps *Math. Zeit.* **259** (2008), 525–574.

[334] J.W. Robbin and D.A. Salamon, The Maslov index for paths, *Topology* **32** (1993), 827–844.

[335] J.W. Robbin and D.A. Salamon, The spectral flow and the Maslov index, *Bull. L.M.S.* **27** (1995), 1–33.

[336] J.W. Robbin and D.A. Salamon, Asymptotic behaviour of holomorphic strips, *Annales de l'Institute Henri Poincaré – Analyse Nonlineaire* **18** (2001), 573–612.

[337] J.W. Robbin and D.A. Salamon, A construction of the Deligne-Mumford orbifold, *J. Europ. Math. Soc.* **8** (2006), 611–699.

[338] H.L. Royden, *Real Analysis*, 3rd Edition, Macmillan, New York, 1988.

[339] Y. Ruan, Symplectic topology and extremal rays, *Geom. Funct. Anal.* **3** (1993), 395–430.

[340] Y. Ruan, Symplectic Topology on Algebraic 3-folds, *J. Diff. Geom.* **39** (1994), 215–227.

[341] Y. Ruan, Symplectic topology and complex surfaces, in: *Geometry and Analysis on Complex manifolds*, ed Mabuchi, Noguchi, Ochiai, World Scientific (1994), 171–197.

[342] Y. Ruan, Topological sigma model and Donaldson type invariants in Gromov theory, *Duke Math J.* **83** (1996), 461–500.

[343] Y. Ruan, Virtual neighborhoods and pseudoholomorphic curves, in [**385**], 101–116.

[344] Y. Ruan, Surgery, quantum cohomology, and birational geometry, in [**99**], 183–199.

[345] Y. Ruan and G. Tian, A mathematical theory of quantum cohomology, *Math. Res. Letters* **1** (1994), 269–278.

[346] Y. Ruan and G. Tian, A mathematical theory of quantum cohomology, *J. Diff. Geom.* **42** (1995), 259–367.

[347] Y. Ruan and G. Tian, Bott–type symplectic Floer cohomology and its multiplicative structures, *Math. Res. Letters*, **2** (1995), 203–219.

[348] Y. Ruan and G. Tian, Higher genus symplectic invariants and sigma model coupled with gravity, *Invent. Math.* **130** (1997), 455–516.

[349] J. Sacks and K. Uhlenbeck, The existence of minimal immersions of 2-spheres, *Annals of Math.* **113** (1981), 1–24.

[350] V. Sadov, On equivalence of Floer's and quantum cohomology, hep-th/9310153, *Commun. Math. Phys.* **173** (1995), 77–100.

[351] D.A. Salamon, Morse theory, the Conley index and Floer homology, *Bull. London.M.S.* **22** (1990), 113–140.

[352] D.A. Salamon, Quantum products for mapping tori and the Atiyah-Floer conjecture, *Amer. Math. Soc. Transl.* **196** (1999), 199–235.
[353] D.A. Salamon, Lectures on Floer homology, in [**103**], 145–229.
[354] D.A. Salamon, A survey of symmetric functions, Grassmannians, and representations of the unitary group. Unpublished Notes, University of Warwick, 1996. http://www.math.ethz.ch/ salamon/publications.html
[355] D.A. Salamon and K. Wehrheim, Instanton Floer homology with Lagrangian boundary conditions. *Geometry & Topology* **12** (2008), 747–918.
[356] D.A. Salamon and E. Zehnder, Morse theory for periodic solutions of Hamiltonian systems and the Maslov index, *Comm. Pure Appl. Math.* **45** (1992), 1303–1360.
[357] M. Schwarz, *Morse Homology*, Birkhäuser, 1993.
[358] M. Schwarz, Cohomology operations from S^1-cobordisms in Floer theory, PhD thesis, ETH-Zürich, 1995.
[359] M. Schwarz, Equivalences for Morse homology, in *Geometric Topology in Dynamics*, ed. Barge and Kuperberg, *Contemporary Mathematics* **246**, Amer. Math. Soc. 1999.
[360] M. Schwarz, On the action spectrum for closed symplectically aspherical manifolds, *Pacific J. Math* **193** (2000), 419–461.
[361] P. Seidel, Floer homology of a Dehn twist, *Math. Res. Letters* **3** (1997), 829–834.
[362] P. Seidel, *Floer homology and the symplectic isotopy problem*, PhD thesis, Oxford, 1997.
[363] P. Seidel, π_1 of symplectic automorphism groups and invertibles in quantum homology rings, *Geom. Funct. Anal.* **7** (1997), 1046–1095.
[364] P. Seidel, On the length of Hamiltonian loops, unpublished preprint (1997).
[365] P. Seidel, On the group of symplectic automorphisms of $\mathbb{C}P^m \times \mathbb{C}P^n$, in [**99**], 237–250.
[366] P. Seidel, Vanishing cycles and mutations, in: *Third European Congress of Mathematics (Barcelona 2000)*, Progress in Mathematics **202** Birkhäuser (2001), 65–86.
[367] P. Seidel, More on vanishing cycles and mutations, in: *Symplectic Geometry and Mirror Symmetry (Seoul 2000)*, World Scientific (2001), 249–265.
[368] P. Seidel, Symplectic automorphisms of T^*S^2, arXiv:DG/9803084.
[369] P. Seidel, Symplectic Floer homology and the mapping class group, *Pac. J. Math.* **206** (2002), 219–229.
[370] P. Seidel, A long exact sequence for symplectic Floer homology, *Topology* **42** (2003), 1003–1064.
[371] P. Seidel, *Fukaya Categories and Picard–Lefschetz theory*, Zurich Lectures in Advanced Mathematics, European Math. Soc. (EMS), Zurich, 2008.
[372] P. Seidel, Lectures on four-dimensional Dehn twists, *Symplectic four manifolds and algebraic surfaces*, 231–267, Lecture Notes in Math, **1938**, Springer, Berlin, 2008.
[373] P. Seidel, Abstract analogs of flux as symplectic invariants, arXiv:1108.0394.
[374] E. Shelukhin, The action homomorphism, quasimorphisms, and moment maps on the space of almost complex structures, arXiv:1105.5814.
[375] V. V. Shevchishin, Pseudoholomorphic curves and the symplectic isotopy problem, Habilitation thesis, Ruhr-Univ, Bochum, arXiv:0010262.
[376] B. Siebert, Symplectic Gromov–Witten invariants, in: *New Trends in Algebraic Geometry*, ed. Catanese, Peters, Reid, L.M.S. Lect. Notes **264**, Camb. Univ. Press 1999, 375–424.
[377] B. Siebert and G. Tian, Recursive relations for the cohomology ring of moduli spaces of stable bundles, *Turkish Journal of Mathematics* **19** (1995), 131–144.
[378] B. Siebert and G. Tian, On quantum cohomology rings of Fano manifolds and a formula of Vafa and Intriligator, *Asian J. Math.* **1** (1997), 679–695.
[379] J.-C. Sikorav, Singularities of J-holomorphic curves, *Math. Zeitschrift* **226** (1997), 359–373.
[380] J.-C. Sikorav, Dual elliptic planes, *Actes des Journées Math. à la mémoire de Jean Leray*, 185–207, *Sém. Congr., 9*, Soc. Math. France, Paris, 2004.
[381] S. Smale, An infinite dimensional version of Sard's theorem, *Am. J. Math.* **87** (1973), 213–221.
[382] I. Smith, On moduli spaces of symplectic forms, *Math. Res. Letters* **7** (2000), 779–788.
[383] I. Smith, Torus fibrations on symplectic 4-manifolds, *Turkish J. Math.* **25** (1) (2001), 69–95.
[384] H. Spielberg, Counting Generic Genus 0 curves on Hirzebruch surfaces, *Proc. Amer. Math. Soc.* **130** (2001), 1257–1264.
[385] R. Stern, ed., *Topics in Symplectic 4-manifolds (Irvine CA 1996)*, Internat. Press Lectire Series, Vol **1**, Internat. Press, Cambridge, MA, 1998.

[386] J. Streets and G. Tian, Symplectic Curvature Flow, arXiv:1012.2104.

[387] Z. Szabo, Simply connected irreducible 4-manifolds with no symplectic structure, *Invent. Math.* **132** (1998), 457–466.

[388] D. Tamarkin, Microlocal condition for non-displaceability, arXiv:0809.1584

[389] C.H. Taubes, Counting pseudoholomorphic submanifolds in dimension 4, *J. Diff. Geom.* **44** (1996), 818–893. (republished in [**393**])

[390] C.H. Taubes, The Seiberg-Witten invariants and symplectic forms, *Math. Res. Lett.* **1** (1994), 809–822.

[391] C.H. Taubes, More constraints on symplectic forms from Seiberg-Witten invariants, *Math. Res. Lett.* **2** (1995), 9–14.

[392] C.H. Taubes, The Seiberg-Witten and the Gromov invariants, *Math. Res. Lett.* **2** (1995), 221–238.

[393] C.H. Taubes, *Seiberg–Witten and Gromov Invariants for Symplectic* 4- *Manifolds*, First Internat. Press Lecture Series vol 2, Internat. Press. Cambridge MA, 2000.

[394] C.H. Taubes, The Seiberg–Witten equations and the Weinstein conjecture, *Geom. Topol.* **11** (2007), 2117–2202.

[395] R. Thom, Quelques propriétés globales des variétés différentiables, *Comment. Math. Helv.* **29** (1954), 17–85.

[396] Zhiyu Tian, Some examples of symplectic rationally connected 4-folds, arXiv:1112.1369.

[397] V. Tosatti and B. Weinkove, The Calabi–Yau equation on the Kodaira–Thurston manifold, *J. Inst. Math. Jussieu* **10** (2011), 437–447.

[398] M. Usher, Spectral numbers in Floer theories, *Compos. Math.* **144** (2008), 1581–1592.

[399] M. Usher, Deformed Hamiltonian Floer theory, capacity estimates and Calabi quasimorphisms, *Geom. Topol.* **15** (2011), 1313–1417.

[400] M. Usher, Hofer's metrics and boundary depth, arXiv:1107.4599

[401] C. Vafa, Topological mirrors and quantum rings, in: *Essays on Mirror Manifolds*, ed. S.-T. Yau, International Press, Hong Kong 1992.

[402] C. Viterbo, A proof of the Weinstein conjecture in $\mathbb{R}^{2n}$, *Annales de l' Institut Henri Poincaré – Analyse Nonlinéaire* **4** (1987), 337–357.

[403] C. Viterbo, A new obstruction to embedding Lagrangian tori, *Invent. Math.* **100** (1990), 301–320.

[404] C. Viterbo, Symplectic topology as the geometry of generating functions. *Math. Annalen*, **292** (1992), 685–710.

[405] C. Viterbo, The cup-product on the Thom-Smale-Witten complex, and Floer cohomology, in [**181**], 609–626.

[406] C. Viterbo, Functors and computations in Floer homology with applications, part I, *Geom. Funct. Anal.* **9** (1999), 985–1033.

[407] C. Viterbo, On the uniqueness of generating Hamiltonian for continuous limits of Hamiltonian flows, *Int. Math. Res. Not* **2006**, Art ID 34028, erratum Art ID 38784

[408] C. Voisin, A mathematical proof of a formula of Aspinwall and Morrison, *Compositio Math.*, **104**, (1996), 135–151.

[409] C. Voisin, *Mirror Symmetry*, SMF/AMS texts and Monographs, Vol **1**, Amer. Math. Soc., Providence, RI, 1996.

[410] C. Voisin, Rationally connected 3-folds and symplectic geometry, *Astérisque* **322** (2008), 1–21.

[411] K. Wehrheim, Anti-self-dual instantons with Lagrangian boundary conditions, PhD thesis, ETH-Zürich, July 2002.

[412] K. Wehrheim, Banach space valued Cauchy–Riemann equations with totally real boundary conditions, *Commun. Contemp. Math.* **6** (2004), no. 4, 601–635.

[413] K. Wehrheim, Anti-self-dual instantons with Lagrangian boundary conditions: elliptic theory, *Comm. Math. Phys.* **254** (2005), no. 1, 45–89.

[414] K. Wehrheim and C. Woodward, Functoriality for Lagrangian correspondences in Floer theory, *Quantum Topol.* **1** (2010) no. 2, 129–170.

[415] K. Wehrheim and C. Woodward, Floer Cohomology and Geometric Composition of Lagrangian Correspondences, arXiv:0905.1368

[416] K. Wehrheim and C. Woodward, Pseudoholomorphic Quilts, arXiv:0905.1369

[417] K. Wehrheim and C. Woodward, Quilted Floer Cohomology, *Geom. Topol.* **14** (2010), no. 2, 192–210.

[418] C. Wendl, Strongly fillable contact manifolds and J-holomorphic foliations, *Duke Math. J.* **151** (2010), no.3, 337–384.

[419] P.M.H. Wilson, The Kähler cone on Calabi-Yau threefolds, *Invent. Math.* **107** (1992), 561–584.

[420] E. Witten, Supersymmetry and Morse theory, *J. Diff. Geom.* **17** (1982), 661–692.

[421] E. Witten, Two dimensional gravity and intersection theory on moduli space, *Surveys in Diff. Geom.* **1** (1991), 243–310.

[422] E. Witten, The Verlinde algebra and the cohomology of the Grassmannian, hep-th/9312104, in *Geometry, Topology and Physics* Conf. Proc. Lecture Notes Geom. Topology VI, Internat. Press, Cambridge MA, 1995, 357–422.

[423] R. Ye, Gromov's compactness theorem for pseudoholomorphic curves, *Trans. Amer. Math. Soc.* **342** (1994), 671–694.

[424] J.-C. Yoccoz, Travaux de Herman sur les tores invariants, Sem. Bourbaki, 1991-2, no 754, *Astérisque* **206** (1992), 311–344.

[425] K. Zehmisch, The annulus property of simple holomorphic discs, *J. of Symplectic Geom.* (2012).

[426] A. Zinger, Enumerative vs. Symplectic Invariants and Obstruction Bundles, *J. Symplectic Geom.* **2** (2004), no. 4, 445–543.

[427] A. Zinger, Enumeration of genus 2 curves with a fixed complex structure in $\mathbb{P}^2$ and $\mathbb{P}^3$, *J. Differential Geom.* **65** (2003) no. 3, 341–467.

[428] A. Zinger, Counting rational curves of arbitrary shape in projective spaces, *Geom. Topol.* **9** (2005), no. 1, 35–81.

[429] A. Zinger, Pseudocycles and integral homology, *Trans. Amer. Math. Soc.* **360** (2008), no. 5, 2741–2765.

[430] F. Ziltener Symplectic vortices on the complex plane and quantum cohomology, Ph.D. thesis, ETH Zürich, 2006.

[431] F. Ziltener, The Invariant Symplectic Action and Decay for Vortices, *J. Symplectic Geom.* **7** (2009) no. 3, 357–376.

[432] F. Ziltener, A Quantum Kirwan Map: Bubbling and Fredholm Theory, arXiv:0905.4047 and arXiv:1106.1729, to appear in *Mem. Amer. Math. Soc.*

List of Symbols

$A(r, R)$ 106
$A\#\widetilde{x}$ 489
$A \in H_2(M,\mathbb{Z})$ 4
A_d 422
$\mathcal{A}_H : \widetilde{L_0M} \to \mathbb{R}$ 302
$\mathcal{A}_H : \widetilde{\mathcal{L}_0M} \to \mathbb{R}$ 489
$\widetilde{A} := \iota_* A + [\Sigma \times \mathrm{pt}]$ 184
$(a * b)_A$ 14
$[\alpha, \beta]$ 622
$\alpha E \beta$ 622
$|a|$ 427
$a * b$ 14, 425
$a = 0, \infty$ 389
$a = \sum_A a_A \otimes e^A$ 424
$a_t := t_0e_0 + t_1e_1 + \cdots + t_Ne_N$ 436

$B^{2n}(r)$ 8
B_ε 21
$\mathcal{B} := C^\infty(\Sigma, M)$ 19
$\mathcal{B}^*$ 40
$\mathcal{B}^{k,p}_*$ 46
$\beta_{k,I}$ 245

$C = \operatorname{Im} u$ 3
$C = u(\Sigma)$ 26
$C^{k,\mu}(\Omega)$ 555
$\mathbb{C}P^1$ 3
$\mathbb{C}^n$ 2
$C^\infty(\overline{\Omega}), C_0^\infty(\Omega)$ 549
$\operatorname{coker} D$ 5
$c_1(A)$ 2
$c_i(TM)$ 2

$D : \Omega^0(\Sigma, E) \to \Omega^{0,1}(\Sigma, E)$ 580
D_F^* 582
D_u 40
$D_{0,\infty,r}$ 383
$D_{0,\infty}$ 383
$D_{\widetilde{u}}$ 185
$D_{\widetilde{u}} = D_{J,H,\widetilde{u}}$ 275
$\mathbb{D}$ 654
$\Delta = (\partial_s)^2 + (\partial_t)^2$ 22
Δ 562
$\Delta^E \subset M^E$ 159
$\Delta^m \subset M^m$ 62
$(\delta, R) \in \mathcal{A}(\delta_0)$ 371
$\bar\partial^\nabla \xi := (\nabla\xi)^{0,1}$ 580
$\bar\partial_J$ 19
$\bar\partial_{J,H}(u) = d_H(u)^{0,1}$ 258
δ_A 421
$\delta(u), \delta(u_0, u_1)$ 657
$\delta_{\mathrm{MW}} : \mathrm{CM}^*(f;\Lambda_\omega) \to \mathrm{CM}^*(f;\Lambda_\omega)$ 494
$\det(D)$ 53
$\det(D) = \Lambda^{\max}(\ker D) \otimes \Lambda^{\max}(\ker D^*)$ 533
∂_s, ∂_t 20
$|du|_J^2$ 20
$d\mathcal{F}(x)$ 5
$d\pi(u, J)$ 54

$E(u)$ 6, 20
$E(u; B)$ 81
$E(u; B_r)$ 96
$E^{\mathrm{Vert}}(\widetilde{u})$ 268
$E_H(u)$ 262
$E_i(x,\xi) : T_xM \to T_{\exp_x(\xi)}M$ 389
$\mathcal{E}_u^p$ 68
$\mathcal{E}^{k-1,p} \to \mathcal{B}^{k,p} \times \mathcal{J}^\ell$ 50
$\mathcal{E}_u = \Omega^{0,1}(\Sigma, u^*TM)$ 19
$\operatorname{End}(TM, J, \omega)$ 49
$\operatorname{End}(\mathbb{R}^{2n})$ 23
$\operatorname{End}_{\mathbb{R}}(\mathbb{C}^n)$ 24
$\mathrm{ev} : \mathcal{M}^*_{0,k}(A; J) \to M^k$ 177
$\mathrm{ev} : \mathcal{M}_{0,k}(A; J) \to M^k$ 9
$\mathrm{ev} = \mathrm{ev}_J$ 8
$\mathrm{ev} \times \pi : \overline{\mathcal{M}}_{0,k}(A; J) \to M^k \times \overline{\mathcal{M}}_{0,k}$ 197
$\mathrm{ev}^E(\mathbf{u}, \mathbf{z})$ 159
$\mathrm{ev}_{\mathbf{w}} : \mathcal{M}^*(A; J) \to M^k$ 180
$\mathrm{ev}_{\mathbf{w},\mathcal{J}}$ 62
$\widetilde{\mathrm{ev}} : \mathcal{M}_{\Sigma,k}(\widetilde{A}; J, H) \to \widetilde{M}^k$ 280
$e(T)$ 623
$e \cdot f$ 173
e^A 420
e^α 437
e_ν 12, 218

$\mathcal{F} : X \to Y$ 5
$\mathcal{F}_d$ 231
$\mathcal{F}_u$ 41

$|\cdot|_{\mathrm{FS}}$ 82
f^R 387

$\mathrm{G}(T)$ 118
$\mathrm{G} := \mathrm{PSL}(2,\mathbb{C})$ 6, 81
$\mathrm{GW}^M_{A,k}(a_1,\dots,a_k;\beta_{k,I})$ 245
$\mathrm{GW}^M_{A,k}(a_1,\dots,a_k;\beta)$ 239
$\mathrm{GW}^{\widetilde{M},\mathbf{w}}_{\widetilde{A},k}$ 288
$\mathrm{GW}^{\widetilde{M}}_{\widetilde{A},k}$ 285
$\mathrm{GW}^M_{A,k}(a_1,\dots,a_k)$ 11, 203
$\mathrm{GW}^M_{A,k} : H^*(M)^{\otimes k} \to \mathbb{Z}$ 203
$\mathrm{GW}^{M,I}_{A,k}$ 11, 223
$\Gamma(M,\omega)$ 420
$\widehat{\mathrm{G}}$ 515
$\widetilde{\mathrm{GW}}^{M,I}_{A,k}$ 221
$g_J(v,w) = \langle v,w\rangle_J$ 17
$g_{\nu\mu}, g^{\nu\mu}$ 12, 218

$H^*(M)$ 174, 203
$H_*(M)$ 14
$H_2^S(M)$ 84
H_ζ 258
H_t 488
$\mathrm{HF}^*(L_0,L_0)$ 522
$\mathrm{HF}^*(M,\omega,H,J)$ 493
$\mathrm{HF}^*(\phi)$ 520
$\mathbb{H}\subset\mathbb{C}$ 77
$\mathrm{HM}^*(f;\Lambda_\omega)$ 495
$\mathrm{Ham}(M,\omega)$ 297
$\mathcal{H}$ 478
$\mathcal{H}(H_0,H_1)$ 272
$\mathcal{H}_{\mathrm{reg}}(\widetilde{A},\{J_\lambda\}_\lambda;H_0,H_1)$ 273
$\widetilde{\mathrm{Ham}}(M,\omega)$ 302
$\hbar := \hbar(M,\omega,L,J)$ 80

$\mathrm{index}\, D_u$ 44
$\mathcal{I}_n$ 625
$\mathrm{index}\, D$ 5
$\iota_c^R = \iota_c^{\delta,R}$ 388

J 1
J^0, J^∞ 370
J_z^R 371
J_0 20
$\mathcal{J}, \mathcal{J}^\ell$ 44
$\mathcal{J}(J_0,J_1)$ 45
$\mathcal{J}(M,\omega)$ 17
$\mathcal{J}^\ell = \mathcal{J}^\ell(M,\omega)$ 46
$\mathcal{J}_+(M,\omega)$ 169
$\mathcal{J}_+(M,\omega;\kappa)$ 169
$\mathcal{J}_+(\Sigma;M,\omega;\kappa)$ 182
$\mathcal{J}_\tau(M,\omega)$ 2, 17
$\mathcal{J}_\tau(\Sigma;M,\omega)$ 182
$\mathcal{J}_{\mathrm{reg}}(A)$ 5
$\mathcal{J}_{\mathrm{reg}}(A) := \mathcal{J}_{\mathrm{reg}}(A,S^2)$ 45
$\mathcal{J}_{\mathrm{reg}}(A,\Sigma;J_0,J_1)$ 45
$\mathcal{J}_{\mathrm{reg}}(M,\omega)$ 160
$\mathcal{J}_{\mathrm{reg}}(S^2;A^{0,\infty})$ 370
$\mathcal{J}_{\mathrm{reg}}(S^2;M,\omega;\mathbf{w})$ 183
$\mathcal{J}_{\mathrm{reg}}(T,\{A_\alpha\})$ 160
$\mathcal{J}_{\mathrm{reg}}(\Sigma;A)$ 187
$\mathcal{J}_{\mathrm{reg},\mathrm{K}}$ 54
$\mathcal{J}^{\mathrm{Vert}}_{\mathrm{reg}}(\Sigma;A)$ 189
$\widetilde{J}(z,x)$ 183
$\widetilde{J}_H$ 261

$K^{\mathrm{eff}}(M,\omega)$ 418
$\ker D$ 5

(Λ,ϕ,ι) 420
$L\in H_2(\mathbb{C}P^n;\mathbb{Z})$ 208
L^* 57
$L^p_{u,J}$ 382
Λ^{univ} 422
Λ_ω 422
$\widetilde{\mathcal{L}M}$ 489
$\check{\Lambda}^k$ 435
$\ell(\gamma_r)$ 77
$\lambda * \xi$ 492

(M,J) 2
(M,ω) 1
$(\widetilde{M},\pi,\widetilde{\omega})$ 264
M_Σ 521
$\mathcal{M}(A,\Sigma;\{J_z\})$ 184
$\mathcal{M}(A;J)$ 4
$\mathcal{M}(\widetilde{x}^-,\widetilde{x}^+,H,J)$ 490
$\mathcal{M}(\widetilde{x}^\alpha;\widetilde{x}^\beta,\widetilde{x}^\gamma)$ 499
$\mathcal{M}^*(A,\Sigma;\mathcal{J}^\ell)$ 44
$\mathcal{M}^*(A,\Sigma;\mathcal{J}^\ell)\subset\mathcal{B}^{k,p}\times\mathcal{J}^\ell$ 49
$\mathcal{M}^*(A;J)$ 4
$\mathcal{M}^*_{0,T}(\{A_\alpha\};J)$ 155
$\mathcal{M}^*_{0,k}(A;\mathbf{w},J)$ 10
$\mathcal{M}_A(x^\alpha;x^\beta,x^\gamma)$ 501
$\mathcal{M}_{0,T}(M,A;J)$ 119
$\mathcal{M}_{0,k}(A;J)$ 9
$\mathcal{M}^*_{0,k}(A;J)$ 9
$\mathcal{M}_{0,n}(A;J)$ 119
$\widetilde{M} := \Sigma\times M$ 183
$\widetilde{M}_\psi$ 269
$\widetilde{\mathcal{M}}_{0,T}(A;J)$ 119, 120
$\widetilde{\mathcal{M}}_{0,k}(A;J)$ 9
$\overline{\mathcal{M}}_{0,T}$ 120
$\overline{\mathcal{M}}_{0,n} := \mathcal{SC}_{0,n}/\sim$ 630
$\overline{\mathcal{M}}_{0,n}$ 11, 120
$\overline{\mathcal{M}}_{0,n}(M,A;J)$ 119
$\mu(A,k)$ 9
$\mu(E,F)$ 583
$\mu_{\mathrm{CZ}}([x,u])$ 490

N_J 18
$\widehat{\nabla},\widetilde{\nabla}$ 614
$\widetilde{\nabla}_X Y$ 18
$\widetilde{\nabla}$ 616

∇ 18
∇ 580
∇^z 184
$\nu(u,v)$ 173
$n(\widetilde{x},\widetilde{y})$ 491
$n_A(x^\alpha;x^\beta,x^\gamma)$ 502

$\Omega^{0,1}(u^*TM)$ 19
$\Omega^{p,q}(\Sigma,E)$ 579
Ω_ϕ 520
Ω_f 172
$\mathcal{O}_\ell(U)$,$\mathcal{O}_\ell$ 673
$\omega(X_H,\cdot)=dH$ 303
ω_0 1
ω_{FS} 324
$\widetilde{\omega}_H$ 259

$\mathrm{PD}(a)$ 174
$\Phi(t)$ 436
Φ^*_{PSS},Φ^{PSS}_* 503
$\Phi^\alpha:\mathrm{CF}^*(H^\alpha)\to\mathrm{CM}^*(f;\Lambda_\omega)$ 497
$\Phi^{\beta\alpha}$ 493
$\Phi_e(b)$ 174
$\Phi_u(\xi)$ 41
$\mathcal{P}_d$ 231
Ψ^*_{PSS},Ψ^{PSS}_* 503
$\Psi^\alpha:\mathrm{CM}^*(f;\Lambda_\omega)\to\mathrm{CF}^*(H^\alpha)$ 497
$\widetilde{P\mathrm{Ham}}(M,\omega)$ 517
$\pi_{0,k}$ 241
$\pi_{k,I}$ 242
$\widetilde{\mathcal{P}}(H),\mathcal{P}_0(H)$ 489
$\phi\in\mathrm{G}$ 81
ϕ_f 521
$\pi:\mathcal{M}^*(A,\Sigma;\mathcal{J}^\ell)\to\mathcal{J}^\ell$ 54
$\pi:\mathcal{M}_{0,n}(A;J)\to\overline{\mathcal{M}}_{0,n}$ 11
$\pi:\overline{\mathcal{M}}_{0,n+1}\to\overline{\mathcal{M}}_{0,n}$ 643
$\pi:\overline{\mathcal{M}}_{0,n}(A;J)\to\overline{\mathcal{M}}_{0,n}$ 121
$\psi\in\mathrm{G}$ 6

$Q_u=D_u^*(D_uD_u^*)^{-1}$ 68
$Q_{0,\infty,r}$ 383
$Q_{0,\infty}:=Q_{u^0,u^\infty}$ 383
Q_{u^R} 387
$\mathrm{QH}^*(M;\Lambda)$ 424
q^d 422

R 420
$R_H\mathrm{dvol}_\Sigma$ 260
$\mathbb{R}^{2n}$ 1
$\rho(\widetilde{\phi};a)$ 508

$(\Sigma,j_\Sigma,\mathrm{dvol}_\Sigma)$ 17
$S^2=\mathbb{C}\cup\{\infty\}$ 3
$\Sigma(\mathbf{z})$ 630
$\mathcal{S}(I_0,I_1)$ 248
$\mathcal{S}(\psi,\tau;H)$ 515
$\mathcal{S}(\psi;H,J)$ 519
$\mathcal{S}:\mathcal{H}\to\mathbb{C}$ 480
$\mathcal{SC}_{0,n}(M;J)$ 118
$\mathcal{S}_*(\phi)$ 473
$\mathcal{S}_n$ 625
$\mathrm{Symp}_0(M,\omega)$ 346
$\sigma_{A,k}$ 241
Spec(H),Spec($\widetilde{\phi}$) 302
$\widetilde{\sigma}_H$ 259

$T=(T,E,\Lambda)$ 623
TM 1
$T^{\mathrm{Vert}}\widetilde{M}$ 267
$T_\ell(u)$ 26
$T_u\mathcal{B}$ 40
T_xM 2
$T_{\alpha\beta}$ 623
T_{u^R} 384
t^α 437
t_ν,t_μ 436

(u,J) 49
$(\mathbf{u},\mathbf{z})$ 116
$[\mathbf{u},\mathbf{z}]$ 11, 118
$\|u\|_{k,p}$ 550
$\widetilde{u}(z):=(z,u(z))$ 183
$\widetilde{u}^R$ 388
$u:(\Sigma,j)\to(M,J)$ 3
$u^R:=u^{\delta,R}$ 374
$u^{0,r}$,$u^{\infty,r}$ 375
$u_\alpha=u\circ\phi_\alpha^{-1}$ 19

$\langle v,w\rangle$ 6

$W^s(x,f),W^u(x,f)$ 501
$W^{1,p}_u$ 382
$W^{2,2}_{\mathrm{loc}}(\Omega,\mathbb{R}^m)$ 23
$W^{k,p}(M,E)$ 43
$W^{k,p}(\Sigma,M)$ 47
$W^{k,p}_{\mathrm{loc}}(\Omega)$ 550
$W^{k,q}_F(\Sigma,E)$ 581
$W^{k-1,p}(\Sigma,\Lambda^{0,1}\otimes_J u^*TM)$ 43, 50
$\mathcal{W}(A,\Sigma;\{J_\lambda\}_\lambda)$ 45
$\mathcal{W}(\widetilde{A};J,\{H_\lambda\}_\lambda)$ 272
$w_{ijk\ell}(\mathbf{z})$ 631
$\mathbf{w}=\{\mathrm{w}_i\}_{i\in I}$ 191

$X_H(u)^{0,1}$ 258
X_t 489
(ξ,Y) 51
$[x,u_1]\sim[x,u_2]$ 489
$\widetilde{\xi}^R\in\mathrm{im}\,Q_{u^R}$ 388
$\xi\in\mathrm{CF}^*(H)$ 491
$x*_a y$ 480

$Y_\alpha:=Y_\alpha(\mathbf{z})$ 629
Y_f 521

$[\mathbf{z}]$ 12
$\langle\cdot,\cdot\rangle_z$ 184
$z=s+it$ 19
$\mathbf{z}$ 629

Index

action, 302–306
  equivariant, 528
  filters Floer (co)homology, 505
  for paths with Lagrangian boundary, 91
  functional, 302
  local, 91
  monodromy shift, 516
  Morse–Novikov theory for, 489
  on loop space, 302, 489
  relative, 305
  spectrum, 302, 505
    and Hofer norm, 359
    global properties, 518
  vanishes, 291
adiabatic limit, 527
adjunction formula, 36, 658
  in $\mathbb{C}P^2$, 226
  proof, 662
almost complex structure
  κ-semipositive, 169
  ω-compatible, 17
  ω-tame, 1, 17
  adapted to boundary, 315
  adapted to Lagrangian, 88
  compatible with fibration, 267
  condition to be regular, 56
  connections, 42, 614–617
  depending on z, 182
  family
    regular for $\mathbf{w}$, 191
    regular for A, 45
    regular for T, 164
  including perturbation, 261
  integrability, 2, 18
  integrable in 2-dimensions, 604–613, 664–668
  Nijenhuis tensor, 18
  normal coordinates, 679
  recognizing regular, 206
  regular for A, 45
  regular for T, 159
  semipositive, 169
  smooth homotopy, 45
  tame versus compatible, 2, 20, 170
  various on $S^2 \times S^2$, 58
  vertical, 267
  vertically regular, 189
ample line bundle, 419
anti-canonical bundle, 419
approximation
  by smooth functions, 551
  to J-curve, 373
  to delta function, 552
  to right inverse, 382
apriori estimate
  for energy density, 85–90
Arnold conjecture, 297–302, 423
  Lagrangian version, 319, 529
  sketch proof, 496
Arnold–Givental conjecture, 529
Aronszajn's theorem, 22, 668
Aspinwall–Morrison formula, 466
Atiyah–Floer conjecture, 521, 523
Audin conjecture, 311
axioms
  for boundary Maslov index, 595
  for convergence
    of sphere, 122
    of stable curve, 640
    of stable map, 139
  for filtered Floer homology, 507
  for Gromov–Witten invariants, 240
  for spectral invariants, 508
  for tree, 622
  for vertical Levi-Civita connection, 273

balanced area form, 212
Banach space
  $C^{k,\mu}(\Omega)$, 555
  separable, 547
  $W^{k,p}(\Omega)$, 550
Barlow surface, 361
Batyrev
  on toric quantum cohomology, 445
blowing up and down, 323–326
boundary
  of contact type, 316
  of convex type, 315

boundary Maslov index, 593
boundary value problem, 582
bubble
  energy of sequence, 123
  ghost, 116
  ghosts and GW invariant, 292
  in the fiber, 188, 277
bubbling, 75, 80–85
  is finite, 99
  on 2-disc, 82
  on 2-sphere, 82, 621–622
  without energy loss, 105–114
bundle
  of class $W^{\ell,p}$, 613
  pair, 594

$\mathbb{C}P^2$
  lines in, 334, 352
  one point blowup, 431
  symplectomorphism group, 350
$\mathbb{C}P^n$
  counting curves in, 208, 225–239
  Gromov–Witten potential, 252
  Hofer length of circle action, 353, 356
  quantum cohomology, 428
Calabi–Yau 3-fold, 463–467, 484
Calderon–Zygmund inequality, 565–568
Carleman similarity principle, 23, 668
Cauchy–Riemann equation, 2, 5, 19–20
  and Laplacian, 571
  Hamiltonian perturbation, 13, 257–264
  perturbed linear, 573
  regularity, 571
  with J depending on z, 182
Cauchy–Riemann operator
  and connection, 581
  and Hermitian structure, 580
  and holomorphic structure, 580
  complex linear, 580
  dual, 582
  explicit formula, 581
  formal adjoint, 580
  index computation, 602
  is Fredholm, 586
  linearized
    for graphs, 184
    for sections, 275
    for spheres, 40
  real linear, 41, 581
chain of edges, 622
Chern class, 2
  quantum, 453
Chern number, 57
  minimal, 169
  zero, 463
Cieliebak–Mundet–Salamon
  on Euler class, 212
Cieliebak–Salamon
  on Fano toric manifolds, 445
circle action
  minimizes length, 359
  on $S^2 \times S^2$, 362, 475
  Seidel element of, 474
class, *see also* homology class
cohomology
  quantum, 424
comments
  Givental's work on mirror symmetry, 485
  on $\overline{\mathcal{M}}_{0,n}$, 651
  on bubbling, 80
  on distinguishing manifolds, 359
  on fibrations, 293
  on gluing, 375
  on Gromov compactness, 123
  on Gromov–Witten invariants, 197, 239
  on Hofer norm, 358
  on Lagrangian embeddings, 310
  on nonsqueezing, 328
  on periodic orbits, 297
  on presence of holomorphic spheres, 343
  on Seidel representation, 469
  on symplectic 4-manifolds, 330, 343, 345
  on symplectomorphism groups, 345
  on transversality argument, 44
  various approaches to symplectic
    topology, 16
compactness, 6–7, 74–114, pheres(126
  for spheres, 134
  for stable maps, 139–146
condition
  $kp > n$, 47, 560, 561
  stability, 629
cone
  symplectic versus Kähler, 333
conformal
  coordinates, 19
  field theory, 244
  map, 19, 75
  rescaling, 75
conformally flat metric, 664
Conley conjecture, 298
Conley–Zehnder
  index, 490
  on Arnold conjecture, 297
connected simple system, 507
connection
  form, 259
  Hermitian, 18, 580
  Levi–Civita, 18
  with torsion $N(X,Y)$, 616
convergence
  modulo bubbling, 99–104
  u.c.s, 621
  u.c.s., 122
  uniform implies smooth, 104
  with bubbling, 123

convexity, 315–318
in equivariant case, 526
counting
constrained curves, 61
curves in $\mathbb{C}P^1$, 225
curves in $\mathbb{C}P^2$, 229–235
curves in $\mathbb{C}P^3$, 254
curves in $\mathbb{C}P^n$, 236–239
curves in Calabi–Yau manifold, 466
graphs, 220
higher genus curves, 182
sections
with fixed marked points, 287
spheres, 203
tori, 60, 288
coupling class, 468
coupling form, 265
critical point
finite number of, 26
stable and unstable manifolds, 501
cross ratio, 620–621
as forgetful map, 121, 648
on stable curve, 631
recursion property, 620
symmetry, 620, 632
crossing number, 53, 536–538
cup product
comparing PP and quantum, 500
pair-of-pants (PP), 500
quantum, 425
curvature, 266
Hofer norm of, 262
in local coordinates, 262
of Hamiltonian connection, 260
curve, *see also* disc, *see also* spheres
J-holomorphic, 3
adjunction formula, 36, 58
algebraic approach, 3, 119, 230
and divisors, 230
approximate J-holomorphic, 68
as graph, 182–199
as harmonic map, 21, 90
as section, 279–285
as symplectic submanifold, 3
bubble, 81
counting discrete curves, 205
critical point, 26–29
cusp-curve, 9, 63, 119
determined by ∞-jet, 22
energy, 6, 20
has tangent, 678
implicit function theorem for, 68
in dimension 4, 59
injective point, 4, 30
intersections of, 27, 29
irreducible, 120
isolated, 201
J-curve, 19
J-holomorphic, 19
local normal form, 676
local normal form for two curves, 678
minimal energy, 84
minimizes energy, 21
multiply covered, 4, 30
nonregular
example, 58
order of singularity, 678
parametrized, 3
perturbed to immersed, 659, 660
positivity of intersections, 35, 60, 209
preglued, 373
definition, 374
prestable, 117
rational, 120
reduced, 120
reducible, 115, 120
regularity criterion, 56
simple, 4, 30–34
is somewhere injective, 30
singular, 36, 60
somewhere injective, 4, 30–34
stable, 120, 629
universal, 644
zero versus high genus, 59
cusp-curve, 115
cutoff functions, 380–382
cycle
smooth, 172
cylinder
long with small energy, 106

D_u
definition, 40
definition for graphs, 184
formula for, 68, 617
index
independent of smoothness, 48
quadratic estimates for, 72
relation to adjoint D_u^*, 48
right inverse, 68, 382–396
surjectivity of, 68
$D_{\tilde{u}}$
compared with D_u, 185
definition for graphs, 185
definition for sections, 275
Darboux's theorem, 1
decomposition, 595
rule, 12
deformation equivalence, 331
deformed cup product, *see also* cup product
Deligne–Mumford space, 11
determinant bundle, 53, 533
crossing numbers, 536–538
trivialization, 536–538
dimension
of subset in manifold, 9

dimension condition, 437
disc
  as path in loop space, 302
  bubbling, 82, 99
  existence of holomorphic, 310
  general comments, 6
  integrability for, 605
  multiply covered, 34
  Riemann–Roch theorem for, 598
  spiked, 497
  transversality issues, 311
distribution, 563
divisor
  axiom, 205
  equation, 439
    in Frobenius manifold, 483
Donaldson's four-six conjecture, 360
Donaldson's quantum category, 519–524
  for Lagrangians, 522
  for mapping tori, 521
Dostoglou–Salamon
  on Atiyah–Floer conjecture, 521
Dubrovin connection, 479
  and quantum products, 480
  explicit formula, 481
  family of, 485
  flatness, 482
  potential function, 481

edge
  chain of, 622
  diagonal, 159
edge evaluation map, 159
  is submersion, 165
  surjectivity for graphs, 192
  surjectivity for sections, 282
effective cone
  complex, 418
  symplectic, 418
Ekeland–Hofer
  on capacities, 298
Eliashberg
  on Arnold conjecture, 297
elliptic
  bootstrapping, 75, 76, 571–578
  bootstrapping explained, 572
  bounded implies compact, 571
  regularity, 46–48, 549, 571
embedding problems, 329
energy, 20, 76–80
  as topological invariant, 20
  bounds derivative, 85
  conformal invariance, 75
  density, 20
  exponential decay on long cylinder, 106, 405
  identity, 20–21, 75, 76
    for fibration, 262
    vertical, 268
  lower bound, 78
  no loss, 109
  vertical, 268
  with Hamiltonian perturbation, 262
  with Lagrangian boundary conditions, 263
Entov–Polterovich
  on Calabi quasimorphism, 359, 513
evaluation map, 7
  and orientations, 206
  as pseudo-cycle, 10, 177–196
  at edges, 159
  fixed marked points, 10
  image of, 178
    compactification, 178
  transversality, 61–65, 165–168
  universal, 62
  with fixed marked points, 180
examples
  bubbling of discs and spheres, 82
  comparing GW^M and $\mathrm{GW}^{M,I}$, 219
  comparing C^0 with $W^{1,2}$, 555
  computing with WDVV equation, 254
  conics in $\mathbb{C}P^n$, 236
  counting sections of mapping tori, 288
  counting tori, 60
  curves in $\mathbb{C}P^n$, 238
  evaluating $\mathrm{GW}^M_{A,k}$, 207
  GW potential for $\mathbb{C}P^n$, 251
  higher genus curves in $\mathbb{C}P^n$, 292
  lack of control of marked points, 180
  moduli space of bundles, 431, 521
  N_d is positive, 229–235
  nonregular J, 58
  of $\overline{\mathcal{M}}_{0,n}$, 648
  of full GW pseudocycle, 199
  of quantum cohomology, 442–467
  on Möbius group, 82
  quantum cohomology, 428
  rational surface
    as toric manifold, 450
  using obstruction bundle, 215
  $W^{k,p}$-functions, 554
exceptional
  divisor, 324
  sphere, 331

$\mathcal{F}_u$
  definition, 41
  quadratic estimates for, 70
facets, 446
Fano, *see also* symplectic manifold,Kähler manifold
Fano manifold, 419
fat diagonal, 62
fiber sum
  of fibrations, 470

fibration
fiber sum, 470
Hamiltonian, 264, 293
Lefschetz, 293
locally Hamiltonian, 264
over S^2, 269
finite dimensional reduction, 545
finiteness result, 490
first Chern class, 2
fixed point
contractible, 297
flag manifolds, 451–454
Floer
connecting orbit, 490
for symplectomorphism, 520
continuation map, 493
equation, 306
s-dependent, 516
in Hamiltonian context, 263
trajectory, 490
constructed as limit, 308
Floer (co)homology, 488–505
and Morse–Witten theory, 494
and quantum cohomology, 494
cap product, 504
chain complex, 503
cochain complex, 491, 518, 519
Euler characteristic, 522
filtered, 505, 506
J depending on t, 496
mod-2 grading, 519
of (H, J), 493
of symplectomorphism, 519–521
on loop space, 485
pair-of-pants product, 500
Poincaré duality, 503–505
PSS map, 494, 497
PSS maps and products, 500
ring structure, 499–500
small autonomous Hamiltonian, 494
spectral invariants, 505
with various coefficients, 518
Floer theory, *see also* Floer (co)homology, 6, 15, 297, 487–488
for Lagrangians, 319, 522
for vortex equations, 528
Floer–Donaldson theory, 521
Floer–Hofer
on Floer theory, 491
forest, 165
forgetful map, 11, 121, 643
Frauenfelder metric, 86
Fredholm
map, 541
regular value, 541
operator, 5
determinant bundle, 53
index, 5
regular value, 5, 49
stable under perturbation, 533
Fredholm theory, 4, 531–533
Frobenius algebra, 478
Frobenius manifold, 478–485
Fubini–Study metric
on 2-sphere, 82
Fukaya category, 488, 521, 523
Fulton–MacPherson compactification, 649
functor
natural isomorphism, 534
fundamental class axiom, 205
becomes string equation, 439
fundamental solution, 564

Gagliardo–Nirenberg interpolation, 559
Gaio–Salamon theorem
on equivariant and quantum cohomology, 527
gauged Gromov–Witten invariants, 526
Gauss–Manin connection, 484
ghost, *see also* bubble
Giambelli's formula, 460
Givental theorem
on mirror conjecture, 478, 485
Givental–Kim formula for flag manifolds, 451
gluing, 12, 239, 369–405
as converse to convergence, 375
construction of map, 387
cutoff functions, 380
estimate for $Q_{0,\infty,r}$, 392
estimate for T_{u^R}, 382
estimating covariant derivatives, 394–399
example in $\mathbb{C}P^2$, 375
in Hamiltonian fibrations, 471
pregluing, 373
statement of theorem, 371–373
for fixed J, 414
weighted norms, 376
gluing map
construction, 387
is embedding, 388
is surjective, 400–405
Grassmannians, 455–463
gravitational descendants, 245, 485, 645–647
Gromov compactness
for spheres, 126–134
for stable maps, 141
Gromov convergence, 115, 122–126
axioms, 122
of stable maps, 139–143
topology given by, 149
uniqueness of limit, 134–138, 140
Gromov theorem
on compactness, 9
on Lagrangian intersections, 319

on Lagrangians in $\mathbb{C}^n$, 310
on nonsqueezing, 8, 323
on star-shaped domains, 330
on symplectomorphism groups, 344
Gromov topology, 147–151, 640
Gromov–McDuff theorem
on $S^2 \times S^2$, 338
Gromov–Witten invariants, 10–13, 201–205
algebraic reformulations, 244
as Euler number, 211
as integrals, 244
as iterated product, 433
axioms, 239–249
axioms for $\mathrm{GW}^M_{A,k}$, 247
axioms for $\mathrm{GW}^{M,I}_{A,k}$, 247
comparing GW and $\widetilde{\mathrm{GW}}$, 222
comparing GW^M and $\mathrm{GW}^{M,I}$, 219
counting graphs, 220
counting sections, 285
definition of $\mathrm{GW}^{\widetilde{M},\mathbf{w}}_{\widetilde{A},k}$, 288
definition of $\mathrm{GW}^M_{A,k}$, 203, 223
definition of $\mathrm{GW}^{M,I}_{A,k}$, 223
definition of $\widetilde{\mathrm{GW}}^{M,I}_{A,k}$, 220
deformation invariance, 11, 207
for 6-manifolds, 360
for M using sections, 288
for discrete curves, 205
for fiber sum, 471
for Kähler manifolds, 224
for lines in $\mathbb{C}P^n$, 208
full, 11, 197
gauged, 526
gluing rules, 249
gravitational descendants, 245, 485
in class $A = 0$, 204
in homology, 286
iteration formula for $\mathbb{C}P^2$, 249
mixed, 202
mod 2 coefficients, 225
modelled on tree, 217
no marked points, 205
nontrivial for odd classes, 431
on blowup, 208, 328
parametric, 279
signs, 206
splitting axiom, 406, 427, 436
using nonregular J, 210
with < 3 marked points, 224, 243
with $A = 0$ is cup product, 222
with fixed marked points, 11, 218, 223
Gromov–Witten potential, 436–441
and descendant potential, 485
for $\mathbb{C}P^n$, 251
homogeneity, 437, 441, 484
Gromov–Witten pseudocycle, 177
defined by tree, 179
full, 197
of graphs, 183
of sections, 280
with fixed marked points, 180
Grothendieck–Knudsen
compactification, 638
space, *see also* $\overline{\mathcal{M}}_{0,n}$, 11

Hamiltonian
connection, 258
coupling form, 265
explicit formulas, 261
families, 268
dynamics, 14
fibration
cohomology of, 279, 469
explicit formulas, 269
local, 264
over S^2, 269
flow, 489
periodic orbit, 297
loop, 269
perturbation, 257–264
symplectomorphism
has fixed point, 297
vector field, 296
Hamiltonian group
Hofer diameter, 359
Hofer length of loop, 355
loop evaluation map is trivial, 297
loop of infinite order, 362
quasimorphisms on, 359, 430
harmonic function, 562
mean value property, 562
Hartman–Wintner theorem, 668–672
Heegaard splitting, 523
Heegaard–Floer theory, 523
Heinz trick, 87
Hermitian metric
compatible with F, 582
higher genus curves
counting, 60, 182, 258, 292
regularity, 59
Hirzebruch surface, 279, 345, 367
Hofer length
of Hamiltonian loop, 353
of Lagrangian loops, 263, 358
Hofer norm
and symplectic area, 354
distortion in, 358
nondegeneracy, 308
of curvature, 262
of function, 307
on space of Lagrangians, 358
one sided, 354
uniqueness, 358
Hofer–Salamon
on Floer cohomology, 491

Hölder
inequality, 380, 549
generalized, 558
norm, 555
space, 555
homology class
indecomposable, 206
represented by pseudocycle, 172
spherical, 7, 84
Hopf number, 364
Hurewicz homomorphism, 7
Hutchings representation, 469

implicit function theorem, 541–544
for J-holomorphic curves, 69
for moduli spaces, 68–73
index, *see also*
Conley–Zehnder index,Maslov index
calculation, 45, 583, 595–604
of Fredholm operator, 531
index set
w-splitting, 632
network of splittings, 625
splitting, 625
splittings are labellings, 625
integrability theorem, 18, 664
surfaces with boundary, 604–613
integrable systems, 453
interpolation inequality
for L^p, 565
for $W^{k,p}$, 559
intersection
isolated, 658
number of, 657
number of local, 658
transverse, 35, 658
inverse function theorem, 539
isoperimetric inequality, 75, 91–96, 112
for almost complex manifolds, 78
for punctured disc, 96
for symplectic vector spaces, 93
isothermal coordinates, 664
isotopy
framed, 350
smooth versus symplectic, 345

(J, J')-holomorphic, 2
J-holomorphic curve, *see also* curve
Jacobi–Trudi identity, 461

$K3$-surface
counting tori, 60
Kähler manifold
and symplectic deformations, 360
Fano variety, 168, 419
Gromov–Witten invariants, 224
NEF variety, 419
Kodaira vanishing theorem, 57
Kodaira–Serre duality, 57

Kontsevich
compactification, 9
recursion formula, 249, 254
Kontsevich–Manin axioms, 12, 201, 239
comments, 242
Kontsevich–Manin theorem, 483
Kuranishi model, 545

labelling
stable, 623
Lagrangian
Audin conjecture, 311
boundary conditions, 6
embeddings
obstructions to, 13, 310–322
embeddings in $\mathbb{C}^n$, 310
exact, 310, 319
Floer (co)homology of, 522
intersections, 319–322, 522
Klein bottle, 310
nearby Lagrangian conjecture, 319
Lalonde–McDuff
on Hofer length, 353
Lalonde–Sikorav theorem
on Lagrangian intersections, 320
Landau–Ginzberg potential, 451
Landau–Ginzburg potential, 457, 459
Laplace operator, 562–571
fundamental solution, 562
interior regularity, 568
Lazzarini
appendix on singularities, 653
Leibnitz rule, 580
Levi-Civita connection
vertical, 273
line bundle
ample, 419
very ample, 419
Liouville vector field, 315
Lipschitz domain, 551
Liu
on gluing, 369
local intersection number, 658
as multiplicity of zero, 663
loop
of symplectomorphisms, 269
loop space
of manifold, 489
universal cover, 489

$\overline{\mathcal{M}}_{0,n}$, 120, 619–651
as Chow quotient, 651
as Grothendieck–Knudsen
compactification, 638
cohomology, 643, 645
comments on literature, 651
compared with symplectic quotient, 649
definition, 630
embedded in product of spheres, 632

examples, 648
intrinsic topology on, 640
strata in, 638
the cycles $\beta_{k,I}$, 246
$\overline{\mathcal{M}}_{0,n}(A;J)$
examples, 121
forgetful map, 121
Gromov topology, 149
simple strata are regular, 160
strata, 119
manifold
Fréchet versus Banach, 5
Frobenius, 481
mapping tori
and Atiyah-Floer conjecture, 521
counting sections, 288
Marcinkiewicz interpolation, 564
Maslov index, 490
for boundary, 593–598
of Lagrangian loop, 594
Maslov number, 311
McDuff theorem
adjunction formula, 658
on nonisotopic symplectic forms, 363
on rational and ruled 4-manifolds, 330
McDuff–Tolman
on circle actions, 474
mean value inequality, 85
metric
Frauenfelder, 86
Micallef–White
averaging argument, 686
Micallef–White theorem
on singularities, 655
minimal
area and Hofer norm, 354
Chern number, 169
energy spheres, 84
length of loop, 354
surface, 21, 75
monotonicity, 77, 323
singularities of, 37
symplectic 4-manifold, 331
mirror symmetry conjecture, 423, 454, 464, 478
Möbius transformations, 81–84, 619–621
moduli space, 4–6
as pseudocycle, 153–155
cobordism of, 46
complex structure on, 53
main theorems, 39–46
marked points, 9
notation, 120
of flat connections, 521
of graphs, 184
of sections, 271
orientation, 52, 206
regularity for simple stratum, 158
universal, 49
is manifold, 44, 50
Yang-Mills, 53
moment poltyope, 446
monotone, *see also* symplectic manifold
Morse–Novikov theory, 423, 489
Morse–Witten complex, 495
Moser isotopy, 341
Mumford quotient, 649

Nakai–Moishezon criterion, 419
NEF, *see also* symplectic manifold,Kähler manifold
Newlander–Nirenberg integrability theorem, 18
in dimension 2, 604–613, 664–668
Newton–Picard iteration, 542
Nijenhuis tensor, 18, 614–617
Noetherian ring, 231
noncompactness
consequences of, 295, 311
nonsqueezing theorem, 8, 323–329
generalizations, 328
norm
$W^{k,p}$, 550
normal coordinates, 27, 679
notation
for moduli spaces, 120
smoothness convention, 50
Novikov ring, 422, 489
dual, 435
universal, 422
Nullstellensatz, 231
number of curves, *see also* counting

obstruction bundle, 211–216
Oh
on spectral invariants, 508
omega limit set, 172
Ono
on Floer theory, 496
orientation
and crossing numbers, 53
and spectral flow, 53
of gluing map, 399
of moduli space, 53
orientation and crossing numbers, 536–538

pair of pants decomposition, 593
periodic orbit
of Hamiltonian flow, 297
relative, 526
perturbation
Hamiltonian 1-form, 258
symmetry breaking, 221
plurisubharmonic function, 315
Poincaré
dual, 174
duality

for Floer cohomology, 503
for quantum cohomology, 434
inequality, 552
point
critical, 26–29
marked, 630
nodal, 630
singular for sequence, 103
special, 630
Poisson's identity, 562
Polterovich
on group actions, 303
on Hofer length, 354–358
poset, 625
positivity of intersections, 35, 59, 657
easy case, 35
proof, 662
pregluing, 373–376
definition, 374
in Hamiltonian fibration, 472
symmetry, 386
prestable map, 156
product, *see also* cup product, quantum cohomology
pseudocycle, 10, 171–177
bordant, 172
Gromov–Witten, 154
of graphs, 182
of dimension d, 172
of graphs, 182, 199
of sections, 279–285
products, 175
strong transversality, 173
weak representative, 174
weak transversality, 176
pseudoinverse, 545
equivariant, 545

quantum
category, 519, 520
Lagrangian, 523
Chern classes, 453
coefficient ring, 420
cup product
as action on $\mathrm{HF}^*(\phi)$, 520
associativity, 426
homology, 434–435
Kirwan Homomorphism, 528
quantum cohomology, 14–15, 417–435
and Floer cohomology, 494
application to eigenvalues, 433
as Lagrangian variety, 454
big, 418
big versus small, 418, 483
coefficients, 420
convergence problem, 423, 466
cup product, 245, 425–428
as connection, 479
associativity, 12, 503
iterated, 433
same as Floer product, 500
definition, 424
dependence on $[\omega]$, 429, 475
effect of coefficients, 429, 475
examples, 428, 442–467
Frobenius structure of, 428
geometric interpretation, 426
intersection pairing, 424
Künneth formula, 433
mod 2 grading, 423
of $\mathbb{C}P^n$, 15, 428
of flag manifold, 15, 451–454
of Grassmannian, 455–462
of toric manifold, 475–478
Fano case, 442–451
NEF case, 477
Poincaré duality, 434
structure constants, 426, 431, 442, 463, 483
quasimorphism, 359
quintic hypersurface, 464, 483

rational surface
as toric manifold, 431
regular
almost complex structure, 45, 159
family of almost complex structures, 186, 191
homotopy $\{J_\lambda\}$, 45, 164
pair (J, H), 282
value of Fredholm operator, 5, 49, 541
regularity
and transitive group action, 62, 226
dealing with disconnected regular set, 301
different meanings of, 46
effect of adding marked points, 160
for discs, 311
for graphs, 191
for sections, 282
for simple strata, 158
for spheres, 45
of J-holomorphic curve, 56, 58
of J_0 on projective space, 226
of curve in 4-manifold, 58
of genus zero constant maps, 186
of simple strata, 160
outline of main argument, 44
recognizing, 206
with pointwise constraints, 61
Rellich's theorem, 47, 555
removal of singularities, 75, 77, 96–99
for almost complex manifolds, 78
represention ring of $\mathrm{U}(n)$, 460
rescaling
source and target, 123

residual, 45, 547
Riemann mapping theorem, 310, 604–613
Riemann surface
  decomposition, 593
  nodal, 630
Riemann–Hurwitz numbers, 225
Riemann–Roch theorem, 44, 582–583, 598–604
  boundary value problem, 582
  for discs, 598
  proof, 602
right inverse
  existence of, 544
Robbin
  on Maslov index, 593
Ruan theorem
  on nonequivalent 6-manifolds, 360
Ruan–Tian
  on associativity of quantum product, 426
  on gluing, 369

$S^2 \times S^2$
  circle action on, 475
  Gromov–Witten invariants, 210, 279
  Hamiltonian loop in, 362–368
  quantum cohomology, 429
  symplectic forms on, 338
  symplectomorphism group, 345
Sacks–Uhlenbeck bubbling, 75
Sard–Smale theorem, 5, 49, 54, 547
Schubert cycle, 460
Schwarz
  on spectral invariants, 303, 505
Schwarz theorem
  on fixed points, 302
section
  horizontal, 272
Seiberg–Witten invariants, 288, 359, 527
Seidel
  element, 290, 467
    as automorphism, 469
    in Floer cohomology, 516
  exact sequence, 488, 521
  on generalized Dehn twists, 345, 520
Seidel representation, 14, 467–475
  action on spectral invariants, 516
  and PSS maps, 498
  application to Hofer length, 356
  applications, 469
  for circle actions, 474
  for Fano toric manifolds, 474
  in Floer cohomology
    twisting operator, 519
  in Floer theory, 514–519
  in homology, 473
  monotone case, 289–292
self-intersection
  number of, 657
semipositivity, 168–171, 205, 419
  stronger condition, 280
Serre duality, 583–585
Siebert–Tian formula for Grassmannians, 456
sigma model, 202
signs, 406
  convention for Lie bracket, 18, 479
  conventions for Hamiltonian flow, 303
  of intersection, 206
  on supermanifold, 438
Sikorav
  on singularities, 653
simple
  curve, 30–34
  stable map, 155–158
smoothing operator, 552
smoothness
  of class $W^{k,p}$, 561
Sobolev
  embedding theorem, 47, 554
  estimate
    borderline case, 555
    with uniform constants, 69, 377
  interpolation inequality, 559
  norm, 47
  space, 549–561
    trace theorem, 560
Sobolev estimate
  borderline case, 6, 75
special point, 629
spectral invariants, 16, 359, 487, 505–514
  and the Hamiltonian group, 517
  axioms, 508
  monodromy shift, 516
spheres
  in the fiber, 188, 277
  spiked, 502
splitting
  of index set, 625–628
splitting axiom, 12, 241, 243, 249
  becomes WDVV equation, 253, 439
  for 4 points, 416
  proof, 406–413
stable curve, 629
  n marked points, 629
  equivalence, 630
stable map, 9, 115
  distance between, 143
  equivalence, 117
  for graphs, 190
  modelled on tree, 115
  reducing to simple, 156
  reparametrization group, 117
  represents homology class, 118
  restriction to subtree, 140
  simple, 120, 155
  strata, 119

weighted, 157
Stanley–Reisner ideal, 448
quantum, 449, 477
star-shaped, 330
Stein domain, 314
stereographic projection, 82
stratum
nonsimple, 168
simple, 120
top, 119
string equation, 439, 441, 483, 647
submanifold
totally real, 76
subset
of dimension $\leq d$, 171
residual, 45, 547
star-shaped, 330
symplectically convex, 315
supermanifold, 436
sign conventions, 436, 438
symplectic
blowup, 323
capacity, 298
cone, 333
cutting, 325
embedding questions, 329
fiber sum, 470
field theory, 6, 488
compactness results, 75
Floer homology, 288
form
deformation, 207
nonisotopic, 363
on $S^2 \times S^2$, 338, 365–368
inflation, 331, 367
isotopy problem, 37
mapping class group, 345
packing, 367
quotient, 215, 443, 649
rigidity, 323
ruled surface, 330
versus Kähler, 344
vortex equations, 445, 524–529
symplectic 4-manifold
Hirzebruch surface, 279, 345, 367
symplectic action, *see also* action
symplectic manifold, 1–3
4-dimensional, 13, 329–343
comments, 330
aspherical, 302
comments, 360
contact type boundary, 316
convex at infinity, 315
deformation equivalent, 331
Fano, 419
Fano versus monotone, 168, 445
geography problem, 360
group action on, 303
monotone, 14, 168, 344
NEF, 419
non-equivalent, 360
nonFano, 442, 463
rationally conncted, 344
semimonotone, 168
semipositive, 168
Arnold conjecture for, 496
toric, 215, 431, 442–451
uniruled, 344
weakly monotone, 168
symplectomorphism
action spectrum, 302
Floer homology of, 519–522
Lefschetz number, 288, 520
symplectomorphism group, 13
components of, 345
evaluation map, 297
homotopy type of CW complex, 348
of $\mathbb{C}P^2$, 350
of $S^2 \times S^2$, 345
of rational 4-manifold, 344–353

tame versus compatible J, 21
taming condition, 1, 5
Taubes' argument, 44, 54
Taubes–Seiberg–Witten theory, 331, 359
Toda lattice, 225, 451, 453
topology
defined by sequences, 147, 640
Gromov, 147, 640
toric manifold, 215, 442–451
as symplectic quotient, 442
cohomology, 443
moment polytope of, 446
quantum cohomology, 445, 475–478
totally real, 571
boundary conditions, 76
transitive group action
implies regularity, 62, 226
transversality, *see also* regularity
for discs, 311
for Fredholm maps, 49–56
of edge evaluation map, 165
of evaluation map, 61–65
tree, 622
n-labelled, 623
as ordered set, 117
branch, 623
edges unoriented, 623
forest, 165
homomorphism, 622
isomorphism, 622
labellings are splittings, 625
stable, 623
weighted, 117

U(n) representation ring, 460
u.c.s. convergence, 122, 621

unique continuation, 21–25
unique factorization domain, 231
universal
  curve, 643–644
  evaluation map
    is submersion, 62
  graph moduli space
    is manifold, 192
  moduli space
    is a manifold, 50
    with Hamiltonian perturbations, 276
  section moduli space
    is manifold, 282

vanish to infinite order, 22
vector field
  Hamiltonian, 296
  Liouville, 315
Verlinde algebra, 460–463
vertical differential, 41
very ample line bundle, 419
virtual fundamental cycle, 155
virtual moduli cycle, 197, 245, 257, 496
Viterbo
  on filtered complexes, 513
  on Weinstein conjecture, 298
vortex equations, *see also* symplectic vortex equations

$W^{k,p}$ spaces
  bundles, 613
  extending domain, 553
  loss of derivatives, 560
  on manifolds, 561
  product estimates, 560, 573
  properties of, 550
WDVV equation, 426, 439, 481, 482
  gives no information, 466
  in $\mathbb{C}P^n$, 253–255
weak derivative, 550
weak solution, 22, 562, 564
  regularity of, 570
Weierstrass $\wp$-function, 84
weighted
  prestable map, 156
  tree, 117
weighted norm, 376–379
  and conformal rescaling, 376
  and uniform Sobolev estimates, 378
  symmetry, 376
Weinstein conjecture, 298
Weyl's lemma, 563
Witten
  formula for Grassmannians, 456
  on Verlinde algebra, 455

Young's inequality, 549

Published Titles in This Series

52 **Dusa McDuff and Dietmar Salamon,** J-holomorphic Curves and Symplectic Topology, Second Edition, 2012

51 **Alexander Beilinson and Vladimir Drinfeld,** Chiral Algebras, 2004

50 **E. B. Dynkin,** Diffusions, Superdiffusions and Partial Differential Equations, 2002

49 **Vladimir V. Chepyzhov and Mark I. Vishik,** Attractors for Equations of Mathematical Physics, 2002

48 **Yoav Benyamini and Joram Lindenstrauss,** Geometric Nonlinear Functional Analysis, 2000

47 **Yuri I. Manin,** Frobenius Manifolds, Quantum Cohomology, and Moduli Spaces, 1999

46 **J. Bourgain,** Global Solutions of Nonlinear Schrödinger Equations, 1999

45 **Nicholas M. Katz and Peter Sarnak,** Random Matrices, Frobenius Eigenvalues, and Monodromy, 1999

44 **Max-Albert Knus, Alexander Merkurjev, Markus Rost, and Jean-Pierre Tignol,** The Book of Involutions, 1998

43 **Luis A. Caffarelli and Xavier Cabré,** Fully Nonlinear Elliptic Equations, 1995

42 **Victor W. Guillemin and Shlomo Sternberg,** Variations on a Theme by Kepler, 1991

41 **Alfred Tarski and Steven Givant,** A Formalization of Set Theory without Variables, 1987

40 **R. H. Bing,** The Geometric Topology of 3-Manifolds, 1983

39 **Nathan Jacobson,** Structure and Representations of Jordan Algebras, 1968

38 **O. Ore,** Theory of Graphs, 1962

37 **N. Jacobson,** Structure of Rings, 1956

36 **Walter Helbig Gottschalk and Gustav Arnold Hedlund,** Topological Dynamics, 1955

35 **A. C. Schaeffer and D. C. Spencer,** Coefficient Regions for Schlicht Functions, 1950

34 **J. L. Walsh,** The Location of Critical Points of Analytic and Harmonic Functions, 1950

33 **Joseph Fels Ritt,** Differential Algebra, 1950

32 **R. L. Wilder,** Topology of Manifolds, 1949

31 **E. Hille and R. S. Phillips,** Functional Analysis and Semi-groups, 1996

30 **Tibor Radó,** Length and Area, 1948

29 **A. Weil,** Foundations of Algebraic Geometry, 1946

28 **G. T. Whyburn,** Analytic Topology, 1942

27 **S. Lefschetz,** Algebraic Topology, 1942

26 **N. Levinson,** Gap and Density Theorems, 1940

25 **Garrett Birkhoff,** Lattice Theory, 1940

24 **A. A. Albert,** Structure of Algebras, 1939

23 **G. Szego,** Orthogonal Polynomials, 1939

22 **Charles N. Moore,** Summable Series and Convergence Factors, 1938

21 **Joseph Miller Thomas,** Differential Systems, 1937

20 **J. L. Walsh,** Interpolation and Approximation by Rational Functions in the Complex Domain, 1935

19 **R. C. Paley and N. Wiener,** Fourier Transforms in the Complex Domain, 1934

18 **M. Morse,** The Calculus of Variations in the Large, 1934

17 **J. H. M. Wedderburn,** Lectures on Matrices, 1934

16 **Gilbert Ames Bliss,** Algebraic Functions, 1933

15 **M. H. Stone,** Linear Transformations in Hilbert Space and Their Applications to Analysis, 1932

14 **Joseph Fels Ritt,** Differential Equations from the Algebraic Standpoint, 1932

13 **R. L. Moore,** Foundations of Point Set Theory, 1932

PUBLISHED TITLES IN THIS SERIES

12 **Solomon Lefschetz,** Topology, 1930

11 **D. Jackson,** The Theory of Approximation, 1930

10 **Arthur B. Coble,** Algebraic Geometry and Theta Functions, 1929

9 **George D. Birkhoff,** Dynamical Systems, 1927

8 **L. P. Eisenhart,** Non-Riemannian Geometry, 1927

7 **Eric T. Bell,** Algebraic Arithmetic, 1927

6 **Griffith Conrad Evans,** The Logarithmic Potential: Discontinuous Dirichlet and Neumann Problems, 1927

5.2 **O. Veblen,** Analysis Situs, 1922

5.1 **G. C. Evans,** Functionals and Their Applications. Selected Topics, Including Integral Equations, 1918

5 **Griffith Conrad Evans and Oswald Veblen,** The Cambridge Colloquium, 1918

4 **Leonard Eugene Dickson and William Fogg Osgood,** The Madison Colloquium, 1914

3.2 **E. Kasner,** Differential-Geometric Aspects of Dynamics, 1913

3.1 **G. A. Bliss,** Fundamental Existence Theorems, 1913

3 **Gilbert Ames Bliss and Edward Kasner,** The Princeton Colloquium, 1913

2 **Eliakim Hastings Moore, Ernest Julius Wilczynski, and Max Mason,** The New Haven Colloquium, 1910

1 **Henry Seely White, Frederick Shenstone Woods, and Edward Burr Van Vleck,** The Boston Colloquium, 1905